JOAN CARR

P9-CAG-971

INTERMEDIATE ALGEBRA

INTERMEDIATE ALGEBRA

Ronald Hatton
Sacramento City College

Gene R. Sellers
Sacramento City College

Harcourt Brace Jovanovich, Publishers
and its subsidiary, Academic Press
San Diego New York Chicago Austin Washington, D.C.
London Sydney Tokyo Toronto

To my sons Curt and Scott
for their support and patience
—Ronald Hatton

To my wife Linda
for her love and encouragement
—Gene R. Sellers

Copyright © 1991 by Harcourt Brace Jovanovich, Inc.

All rights reserved. No part of this publication may be reproduced or transmitted in any form or by any means, electronic or mechanical, including photocopy, recording, or any information storage and retrieval system, without permission in writing from the publisher.

Requests for permission to make copies of any part of the work should be mailed to: Permissions Department, Harcourt Brace Jovanovich, Inc., 8th Floor, Orlando, Florida 32887.

ISBN: 0-15-541379-1 (Student Edition)
0-15-541380-5 (Instructor's Edition)

Printed in the United States of America

Preface

The mathematical concepts and principles presented in *Intermediate Algebra* can be covered in a one-semester or two-quarter course in intermediate algebra. Extent and depth of coverage are among the strongest features of this book. Students who successfully complete a course using this textbook will be well prepared to continue their mathematical studies in trigonometry, finite mathematics, statistics, nonengineering physics, and other applied mathematics courses.

WRITING STYLE AND USE OF GRAPHICS

Throughout this textbook the aim has been to write mathematical principles as concisely as possible while maintaining clarity. A new principle is frequently introduced with a specific example to help students grasp the evolution of the principle from specific to general. The use of graphics, including lines, arrows, arcs, and marginal comments, promotes visual awareness. The book is greatly enhanced by the use of four colors, with each color having a well-defined pedagogical function. The result is a textbook that is easy to read, easy to follow, and appealing to the eye.

1. Each section begins with a list of **key topics** identifying the skills to be learned within the section. The key topics are then used to separate the section into subsections.
2. Most sections begin with an **introductory statement** that links current topics to previously learned topics or to topics to be learned in future sections.
3. **Definitions** and **important concepts** are clearly identified by color boxes. Guided solutions are also printed in color. Students have found these displays to be a useful way of identifying important items in a clear, concise, and organized manner.
4. Each section contains numerous worked examples. The solutions to these examples contain several features that enhance their effectiveness as a teaching tool: a **discussion** that sets forth a plan

of attack, **marginal comments** that help explain the solution, and a **step-by-step approach** that leaves no room for confusion.

RELATIONS AND FUNCTIONS

Most intermediate algebra textbooks cover relations and functions at the end of the course. Since linear and quadratic functions are studied in introductory algebra courses, there is a review of these functions in Chapter 1. The absolute value function is also studied at this time. Graphs of these and other functions are then used throughout the textbook to help students understand why certain solutions of equations and inequalities in one variable are obtained. While students are required to solve equations and inequalities by the traditional algebraic methods found in most intermediate algebra textbooks, the geometric models give visual evidence of the solutions obtained. For example, a graph of an absolute value function can be used to demonstrate why some absolute value equations in x have two solutions, while others have one, and some have none. Similarly, a graph of a quadratic function can be used to illustrate the solutions of quadratic inequalities in x. The graphical illustrations are integrated only as enhancement features, but *do not* replace the algebraic methods that are traditionally used to solve equations and inequalities in one variable.

SIX STEPS TO SUCCESS

One of the basic assumptions underlying the writing of this textbook is that the mastery of algebra is best achieved by working problems. To that end, *Intermediate Algebra* offers a wealth of problems, with each section containing four different kinds of exercise sets. The exercise sets themselves are described in some detail below; here we offer six problem-solving steps which, if followed consistently, will maximize a student's chances for success.

1. Read each section carefully, following the problem-solving strategy in the examples.
2. Work the odd-numbered **Practice Exercises** and check the answers at the back of the book.
3. Go to the text-specific tutorial software **(Hattonware)** as needed for additional practice.
4. Work all 10 **Review Problems.**
5. Work all **Summary Exercises.** (A student who successfully completes all 8 for a given section will have mastered that section.)
6. Work any **Supplementary Exercises** the instructor may assign.

EXERCISES

We are aware that students in a typical intermediate algebra class have different educational goals as well as different mathematical abilities. Some students want to learn only enough mathematics to get a passing grade, while others want to achieve a deeper understanding of the material covered. Many students have personal objectives that place them somewhere between these two extremes. To meet such diverse needs, this textbook has four different kinds of exercise sets in each section. An instructor can select exercises from these sets to design a course that meets the specific needs of every student in his or her class. In short, this textbook can be used to teach students who want to achieve mathematical competence at a minimum level, moderate level, or maximum level. Answers to all exercises can be found at the back of the Instructor's Edition.

1. **Practice Exercises** are keyed to the examples in the text and progress from easy to difficult. Each Practice Exercise set appears in a parallel odd–even format. That is, every odd-numbered exercise has a corresponding even-numbered problem that has the same level of difficulty, covers the same principle, and has a similar type of solution. Answers to the odd-numbered exercises can be found at the back of the Student Edition.
2. Beginning with Chapter 2, each section has 10 **Review Exercises,** which are intended to serve two primary purposes: they provide a continuous review of material learned in previous sections and they link related topics from different sections. For example, several review sets have each type of equation and inequality previously studied. This requires students to recognize the type of equation and then apply the correct technique to solve it. Answers to all of these exercises can be found at the back of the Student Edition.
3. **Supplementary Exercises** offer additional problems that are more difficult than those in the Practice Exercises, and they are not keyed to examples. These exercises frequently have guided solutions that enable students to discover new mathematical concepts. Other exercises will challenge students to extend the concepts they learned in a given section to related problems of a more difficult nature. Occasionally an illustrative example is given to show students how to work these exercises, but other times only written instructions are given. Many of these exercises can be used by instructors who vary classroom instruction by using small study groups. Some of the exercises can be assigned for extra-credit homework problems. The variety in the types and levels of difficulty provide a challenge to instructors to think of ways in which they can be used to improve the quality of their program. Answers to the odd-numbered exercises can be found at the back of the Student Edition.

4. **Summary Exercises** covering all the concepts studied in a particular section appear, by section, at the end of each chapter. They are easily located by the color edges of the pages. There are typically 8 Summary Exercises for each section, with space between exercises for students to show their work. Furthermore, each exercise set has an answer column and a heading for name, date, and score. These exercises can be easily removed and used for homework assignments or as in-class quizzes since there are no answers in the Student Edition. Complete solutions to these exercises can be found at the back of the Instructor's Edition.
5. An extensive set of **Review Exercises** can be found at the end of each chapter. These exercises include two of each type of problem found in each Practice Exercise set in each chapter. Answers to the odd-numbered exercises can be found at the back of the Student Edition.

WORD APPLICATIONS

In any mathematics course, applied problems are a source of frustration to most students and an area of concern for all instructors. In this textbook, we have improved the chances for success in this area in three ways. First, exposure to applied problems is increased by including them in more sections. Second, many applied problems in Supplementary Exercise sets have guided solutions that lead a student step-by-step to the answer. Third, the discussion component of examples gives almost every example the look of an applied problem.

EXTENDED COVERAGE

At the end of the textbook are seven appendixes. Appendix A is a complete study of sequences and series. Appendix B is a complete study of the binomial theorem. Appendix C is a study of linear interpolation. Appendixes A, B, and C contain exercises with complete answers at the back of the book. Appendix D discusses significant digits. Appendix E is a table of common logarithms. Appendix F is a table of values for e^x and e^{-x}. Appendix G provides students with blank graphs for use with problems requiring graphing.

SUPPLEMENTS FOR INSTRUCTORS

- *Instructor's Edition* provides answers to all the exercises and problems in the textbook.
- *Videotapes* provide 15 hours of lessons to complement classroom lectures.

- *Instructor's Manual with Solutions* provides detailed solutions for all exercises and problems in the textbook.
- *Computerized Test Bank (Micro-Pac Genie)* provides a test-generating and test-writing system, with graphics, that can print five types of questions and multiple versions of a given test. Also available in a hardcopy format, it accommodates 10,000 questions and creates almost any graph.

SUPPLEMENT FOR STUDENTS

A *Student Study Guide with Solutions* contains three sections for every chapter in the textbook. Section A is a summary of the topics covered in the chapter, with some additional information and all definitions and procedures. This part of the manual could be used in place of the notes students frequently take in class but seldom use. Section B contains detailed solutions for one-half of the odd-numbered Practice and Supplementary Exercise sets. The solutions are similar to those that accompany examples in the textbook. The structure of the exercise sets enables students to study a solution and then apply the technique to similar exercises. Section C is a test readiness section containing sample tests that will allow a student to determine whether he or she has the speed and skills needed to succeed in testing on a particular chapter.

COMPUTER SUPPLEMENTS FOR INSTRUCTORS AND STUDENTS

Hattonware provides a computer-assisted instructional program that gives students an opportunity to practice on an unlimited number of exercises. This software package was developed by Ronald Hatton to be used exclusively with this textbook. As a consequence, it identifies topics by section and page number. The disks are not copy-protected, and instructors are encouraged to make copies for students.

Other software tutorials available are EXPERT ALGEBRA TUTOR, MATH PAC, ALGEBRA Student Tutorial by TRUE BASIC, and BEGINNING ALGEBRA by MATH LAB.

ACKNOWLEDGMENTS

The authors would like to acknowledge several individuals who made significant contributions to this project. First, we would like to thank all the reviewers of the manuscript for suggestions that resulted in many improvements, but would especially like to thank Allan Bluman, Community College of Allegheny County; Joan Dykes, Edison Community College; and Richard Semmler, Northern Virginia Community College.

We would also like to express our deepest appreciation to the staff at Harcourt Brace Jovanovich for their efforts in the development of this project. We especially want to thank Chris Nelson, Judi McClellan, Don Fujimoto, Florence Kawahara, Lynne Bush, and Paulette Russo. We also want to express our appreciation to Michael Johnson and Nancy Evans, mathematics editors, and Ted Barnett, marketing manager, who worked diligently to get the support materials needed for the project and who checked and rechecked to make sure the textbook covered all of the topics taught in mainstream intermediate algebra courses. Without the support and encouragement of these team members, the project would not have achieved the goals we envisioned.

RONALD HATTON
GENE R. SELLERS

Contents

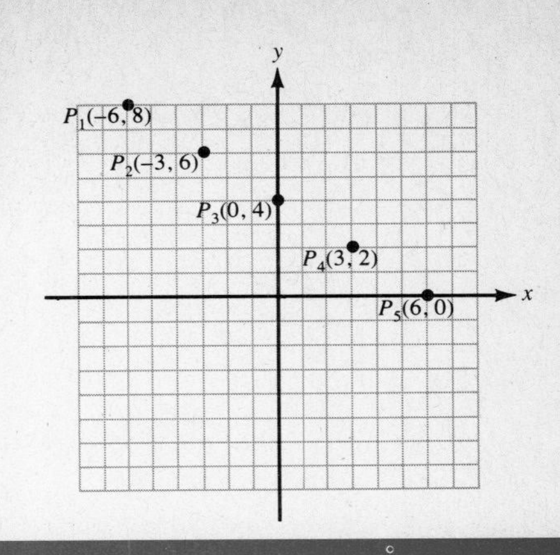

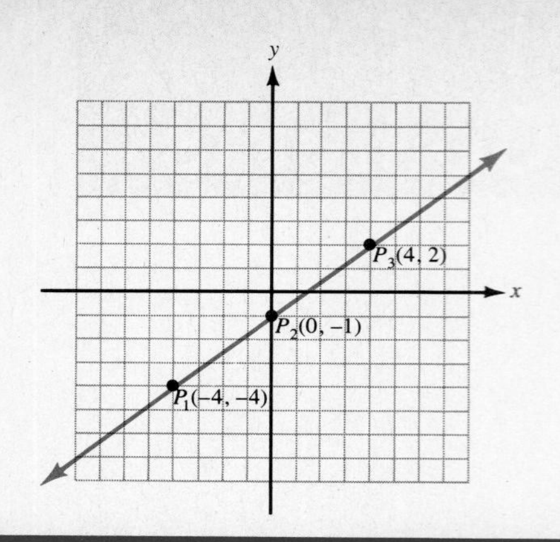

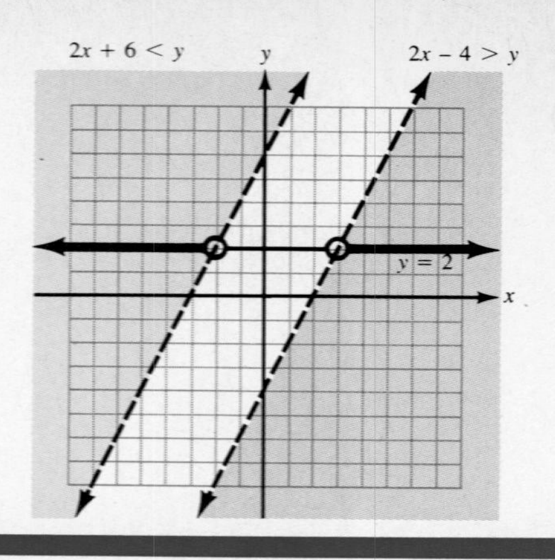

CHAPTER

4

CHAPTER

5

CHAPTER

6

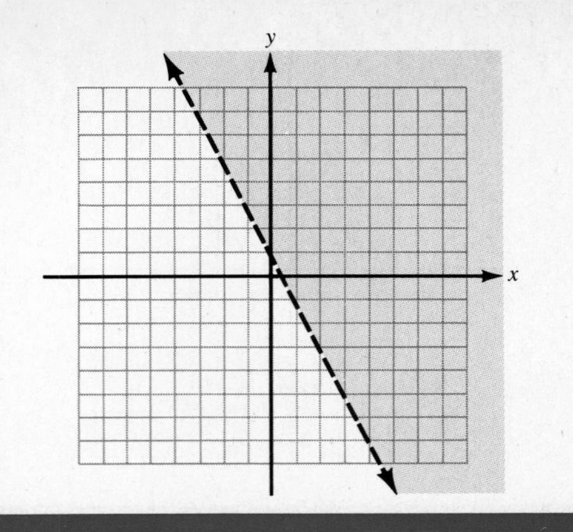

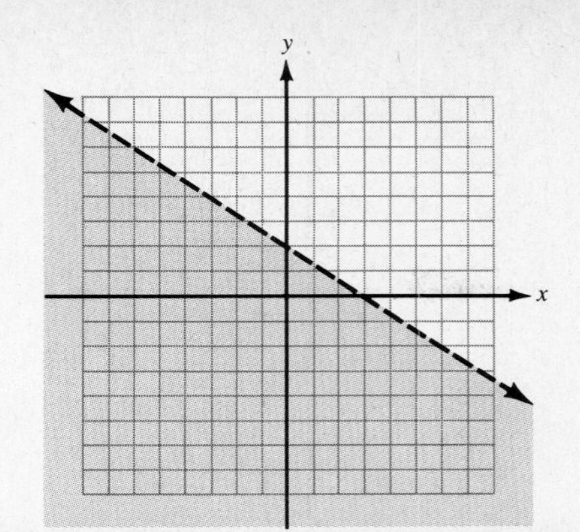

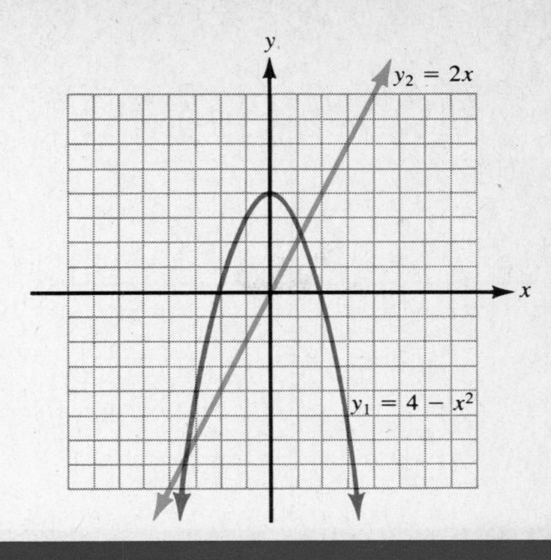

CHAPTER

7

First-Degree Equations in Two Variables **463**

CHAPTER

8

Graphs of Other Functions **524**

CHAPTER

9

Systems of Linear Equations and Inequalities **575**

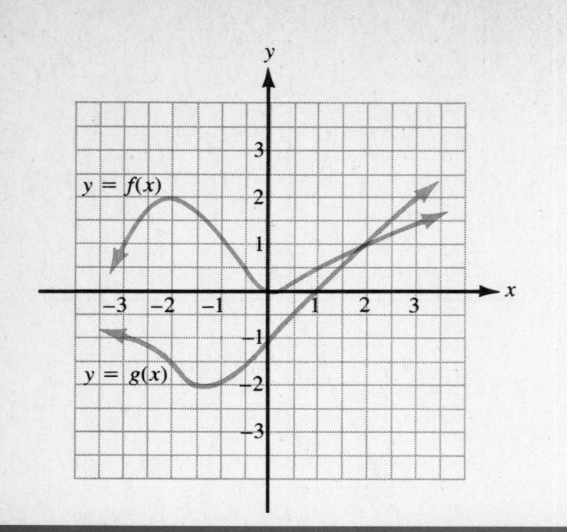

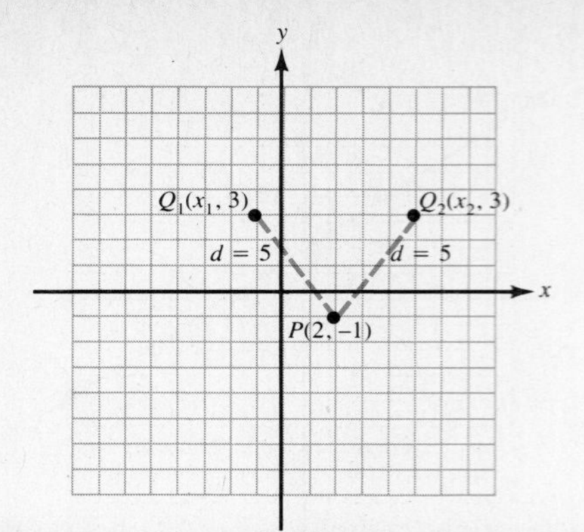

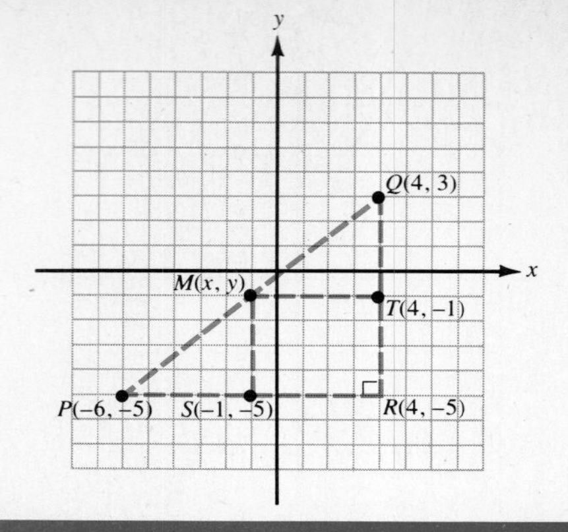

CHAPTER

10

CHAPTER

11

Introduction

Ridgewood Community College is located near a medium-sized city in the Midwest. The focus of area employment opportunites has been changing from industrial manufacturing to work involving electronic components and computer software. The college has been an important avenue for retraining residents so that their abilities match the requisite skills of the more technical jobs. The academic areas that have felt the impact most are mathematics and English.

Ms. Sharon Glaston has been a member of Ridgewood's mathematics department for 12 years. She recently has been studying the content of the intermediate-level algebra courses with three other department members to determine whether these courses meet the needs of the changing student body. The committee's preliminary report suggests that these courses should include more "problem-solving techniques" and problem sets that offer a greater variety and a greater challenge. (Many of these changes are found in this textbook.)

All chapters in this textbook begin with a short look into Ms. Glaston's Intermediate Algebra classroom. These stories give an overview of the material to be covered in each chapter. The questions and answers given by class members remind us that successful learning in mathematics requires active participation on the part of the student. The exchange of ideas among teacher and students provides a learning environment that will benefit all participants.

CHAPTER

1

Essential Topics from Elementary Algebra

As the students entered Ms. Glaston's Intermediate Algebra class, they were greeted by the following display on the overhead projector screen:

Please write an answer to each question.

1. Identify the numbers in set A that are:

a. Integers
b. Irrational

$$A = \left\{-7, -3.25, -\sqrt{2}, 0, \frac{1}{2}, 0.\overline{85}, \pi, 10\right\}$$

2. a. To simplify the following expression, what operation is performed first?

$$-13 + 7 \cdot 5 - 48 \div (-6) + 5^2 \cdot 3 - 2^3$$

b. What is the value of the expression?

3. What property justifies writing $13t^2$ for $5t^2 + 8t^2$?

4. Is $(-5, 7)$ a solution of $3x + 4y = 13$? Justify your answer.

5. For a line defined by $y = \frac{3}{2}x - 5$:

a. Identify the slope.
b. Identify the y-intercept.
c. Graph the function.

6. Simplify $3(7t^2 - 4t + 1) - 5(3t^2 - 2 - 4t) + 2(7 - 3t^2 - 3t)$.

7. Evaluate $\dfrac{3a^2 + 2ab - b^2}{a^2 - 5ab - 2b^2}$ for $a = -3$ and $b = 5$.

"Good morning," Ms. Glaston said as she entered the room. "I'm sure you have all been studying the material on the screen. Is anyone prepared to share their answers at this time?"

After a moment's silence, Mike Daniels raised his hand and said, "Ms. Glaston, I was one of your students last term in Beginning Algebra, and I got a good grade, too. I recognize most of what the questions ask for, but I don't remember enough to answer all seven questions. For example, to graph the line in number 5, I would first have to review the chapter on that topic from Beginning Algebra."

"First of all, Mike," Mrs. Glaston replied, "let me assure you—and everyone else in the class—that I do not expect anyone to be able to answer all parts of every question right now. I'm satisfied that you can recognize most of what is contained in these questions. However, I'm quite confident that, with perhaps a little prompting from me, we could pool our knowledge and collectively answer all of these questions as a class project."

Cathy Lankenau then asked, "Are we going to review enough to individually answer questions like this? As you know, math tests are usually taken alone, and not together as a class."

A chorus of agreement followed Cathy's comment, and Ms. Glaston smiled as she answered, "Yes, Cathy, we will spend some time reviewing. But you must be aware that the review will be rapid. In some cases, one section will cover a topic that took a chapter in Beginning Algebra. The purpose of the review will be to sharpen the skills we need for the current course, Intermediate Algebra. At the completion of this first chapter, I expect everyone in the class to answer correctly all of the questions on the screen. In other words, these questions could very well be part of a test at the end of this chapter."

SECTION 1-1. A Review of the Real-Number System

1. The subsets of the real numbers

2. Multiple and factor defined

3. The number line

4. Equality and inequality

5. Properties of equality and inequality

6. Graphing intervals of numbers

For many reasons, mathematicians find it useful to separate numbers into sets.* The numbers in a set or subset are identified by specific characteristics shared by each number in that set or subset. In this section, we will examine the characteristics that identify the subsets of the real numbers.

The Subsets of the Real Numbers

The ten symbols used to write numbers are called **digits.** If D stands for the set of digits, then:

$$D = \{0, 1, 2, 3, 4, 5, 6, 7, 8, 9\}$$

The three sets of numbers usually studied first in mathematics are natural numbers (N), whole numbers (W), and integers (I):

$N = \{1, 2, 3, 4, 5, \ldots\}$ — The smallest natural number is 1.

$W = \{0, 1, 2, 3, 4, 5, \ldots\}$ — The whole numbers include 0 and the set of natural numbers.

$I = \{\ldots, -3, -2, -1, 0, 1, 2, 3, \ldots\}$ — The **integers** include all whole numbers, and the **negatives** (or **opposites**) of the natural numbers.

* See Appendix C for a review of those terms from set theory used in this text.

The **rational numbers** comprise one of the two major subsets of the real numbers. The numbers in this set can all be written in a specific form.

> **Definition 1.1. The set of rational numbers**
>
> Every rational number can be written in the form $\frac{a}{b}$, where a and b are integers, and $b \neq 0$.

Example 1. Verify that each number is rational.

a. -13 **b.** $3\frac{2}{5}$ **c.** 0.37 **d.** $0.181818\ldots$

Solution. **a.** Notice that -13 is an integer, and can also be written as $\frac{-13}{1}$, $\frac{-26}{2}$, $\frac{-39}{3}$, and so on. Since these are ratios of integers, the number is rational.

> All integers are rational numbers.

b. Notice that $3\frac{2}{5}$ is a **mixed number,** and can also be written as $\frac{17}{5}$. Since 17 and 5 are integers, the number is **rational.**

> All mixed numbers are rational numbers.

c. Notice that 0.37 is a **terminating decimal,** and can also be written as $\frac{37}{100}$. Since 37 and 100 are integers, the number is **rational.**

> All terminating decimals are rational numbers.

d. Notice that $0.181818\ldots$ is a **periodic decimal,** also called a **repeating decimal.** For this number, the 18 repeats forever. When $\frac{2}{11}$ is changed to a decimal form, the periodic decimal $0.181818\ldots$ is obtained. In the Supplementary Exercises of this section, there is a procedure demonstrating how to convert any periodic decimal into the ratio of two integers (see p. 13).

> All periodic decimals are rational numbers.

The **irrational numbers** comprise the other major subset of the real numbers. The decimal representations of the numbers in this set neither terminate (as do some rational numbers), nor are periodic (as are the rest of the rational numbers.) In other words, *irrational numbers cannot be written in the* $\frac{a}{b}$ *form of a rational number, where* a *and* b *are integers and* $b \neq 0$.

Examples **a–e** illustrate five irrational numbers.

a. $0.12112111211112\ldots$ **b.** $-1.57557555755557\ldots$

c. $\sqrt{2} \approx 1.414213562\ldots$ **d.** $-\sqrt{3} \approx -1.7320508\ldots$

e. $\pi \approx 3.14159265\ldots$

> **Rational and irrational numbers**
>
> **1.** The decimal representation of any rational number either terminates or is periodic.
>
> **2.** The decimal representation of any irrational number neither terminates nor is periodic.

If a number is rational, then it cannot be irrational; if a number is irrational, then it cannot be rational. The **union*** of the sets of rational and irrational numbers is the set of **real numbers.** In other words, any real number is either a rational number or an irrational number. Figure 1-1 shows the subsets of numbers that comprise the set of real numbers.

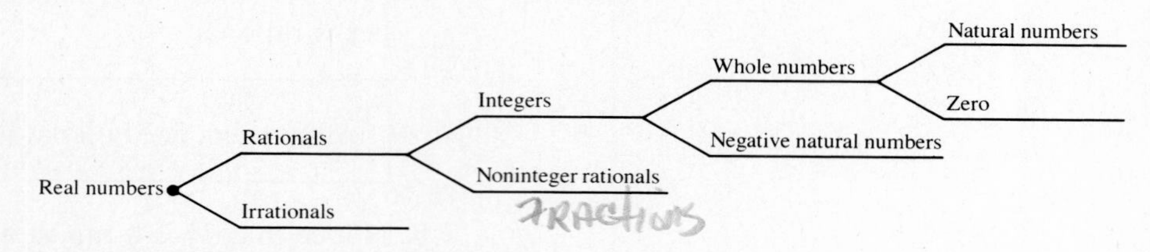

Figure 1-1. The subsets of the real numbers.

Example 2. Identify all of the sets to which each number belongs.

a. -37 **b.** $\frac{72}{9}$ **c.** $\sqrt{12}$

Solution.

a. The number -37 is an integer. Therefore, it is also a rational and real number.

b. The ratio $\frac{72}{9}$ is equal to 8, a natural number. Therefore, it is also a whole number, integer, rational, and real number.

c. The number 12 is not a perfect square, such as 4, 9, 16, and so on. Therefore, $\sqrt{12}$ is an irrational number. It is also a real number.

Multiple and Factor Defined

Consider the number 12. If 12 is multiplied by the natural numbers, then the products are called multiples of 12. For example:

$$1 \cdot 12 = 12, \quad 2 \cdot 12 = 24, \quad 3 \cdot 12 = 36, \quad 4 \cdot 12 = 48, \quad 5 \cdot 12 = 60, \quad 6 \cdot 12 = 72$$

Thus, 12, 24, 36, 48, 60, and 72 are multiples of 12.

Now consider pairs of natural numbers whose product is 12:

$$1 \cdot 12 = 12, \qquad 2 \cdot 6 = 12, \qquad 3 \cdot 4 = 12$$

Therefore, 1, 2, 3, 4, 6, and 12 are called factors of 12.

* See Appendix C for a discussion on the union of sets.

Definition 1.2. Multiple and factor
Let n be a natural number.

1. $1 \cdot n, 2 \cdot n, 3 \cdot n, 4 \cdot n, \ldots$ are **multiples** of n.

2. If x and y are natural numbers such that $x \cdot y = n$, then x and y are **factors** of n.

Example 3. **a.** List the first five multiples of 48.

b. List the natural number factors of 48.

Solution. **a.** $1 \cdot 48 = 48$, $2 \cdot 48 = 96$, $3 \cdot 48 = 144$, $4 \cdot 48 = 192$, $5 \cdot 48 = 240$

The first five multiples of 48 are 48, 96, 144, 192, and 240.

b. $1 \cdot 48 = 48$, $2 \cdot 24 = 48$, $3 \cdot 16 = 48$, $4 \cdot 12 = 48$, $6 \cdot 8 = 48$

The natural number factors of 48 are 1, 2, 3, 4, 6, 8, 12, 16, 24, and 48.

The Number Line

A geometric line can be used as a visual model of the set of real numbers. To each point of the line is assigned a number, called the **coordinate** of the point. For each real number there is a point, called the **graph** of the number. The set of points of a line and the set of real numbers form a **one-to-one correspondence;** that is, for each point there is a number coordinate and for each number there is a graph.

Usually a few integers are shown on a number line to indicate the **unit length** of the line, the distance between any two consecutive integers. An arrow on the end of a number line shows the direction of the increasing values of numbers on the line. A number line is shown in Figure 1-2.

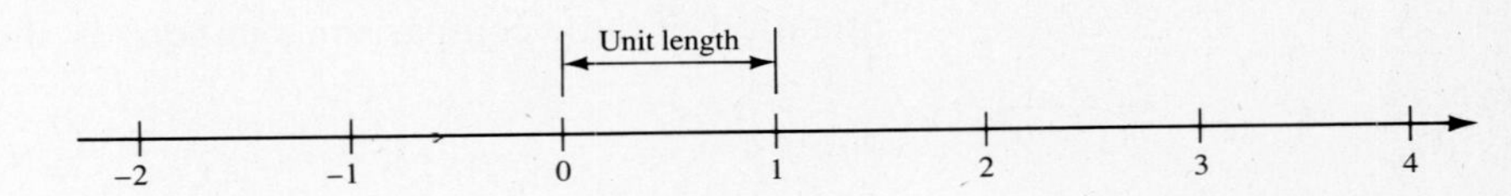

Figure 1-2. A number line.

Example 4. Graph each number on a number line.

a. 4 **b.** $\sqrt{2}$ **c.** $-\pi$ **d.** $4\frac{2}{3}$ **e.** $-\sqrt{31}$

Solution. **a.** The number 4 is four units to the right of 0. Therefore, point A in Figure 1-3 is the graph of 4.

b. Using a calculator, $\sqrt{2} \approx 1.414213\ldots$. Therefore, the graph of $\sqrt{2}$ is between 1 and 2. Point B in Figure 1-3 is the approximate location of the graph of $\sqrt{2}$.

c. Using a calculator, $-\pi \approx -3.141592\ldots$. Therefore, the graph of $-\pi$ is between -3 and -4. Point C in Figure 1-3 is the approximate location of the graph of $-\pi$.

d. The mixed number $4\frac{2}{3}$ is between the points with coordinates 4 and 5. Point D in Figure 1-3 is the approximate location of the graph of $4\frac{2}{3}$.

e. Using a calculator, $-\sqrt{31} \approx -5.56776436\ldots$. Therefore, the graph of $-\sqrt{31}$ is between -5 and -6. Point E in Figure 1-3 is the approximate location of the graph of $-\sqrt{31}$.

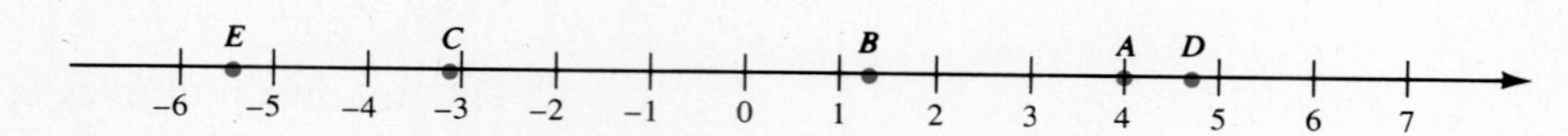

Figure 1-3. Graphs of 4, $\sqrt{2}$, $-\pi$, $4\frac{2}{3}$, and $-\sqrt{31}$.

Equality and Inequality

If x and y are any two real numbers, then there is exactly one of only three ways that x and y can be compared.

> If x and y are real numbers, then exactly one of the following statements is true:
>
> **1.** $x < y$, which is read "x is less than y"
>
> **2.** $x = y$, which is read "x is equal to y"
>
> **3.** $x > y$, which is read "x is greater than y"

The graphs of x and y on a number line can be used to give a visual interpretation of these comparison symbols, as shown in Figure 1-4.

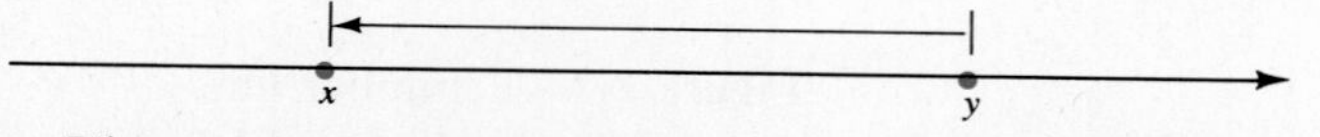

a. The graph of x is *to the left of* the graph of y when $x < y$.

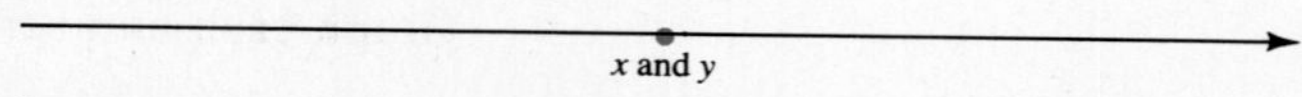

b. The graphs of x and y are *the same point* when $x = y$.

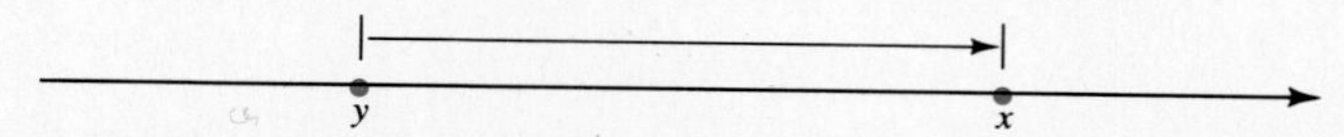

c. The graph of x is *to the right of* the graph of y when $x > y$.

Figure 1-4. Graphical interpretations of $<$, $=$, and $>$.

Example 5. Place $<$, $=$, or $>$ between each pair of numbers to make a true statement.

a. $-1.5 \quad -15$ **b.** $\sqrt{10} \quad \sqrt{13}$

c. $\frac{42}{7} \quad 6$

Solution.

a. A graph of -1.5 is to the right of a graph of -15. Therefore, $-1.5 > -15$.

b. Using a calculator, $\sqrt{10} \approx 3.16227\ldots$ and $\sqrt{13} \approx 3.60555\ldots$. Since a graph of $3.16227\ldots$ is to the left of a graph of $3.60555\ldots$, $\sqrt{10} < \sqrt{13}$.

c. A graph of $\frac{42}{7}$ and 6 is the same point, so $\frac{42}{7} = 6$.

Properties of Equality and Inequality

The relationships *less than, equals,* and *greater than* have some properties that are useful in solving equations and inequalities.

Properties of equality (=) and inequality (< or >)
Let x, y, and z represent real numbers.

Property		Example
1. Reflexive Property of Equality	$x = x$	$-8 = -8$
2. Symmetric Property of Equality	If $x = y$, then $y = x$.	If $15 = 8 + 7$, then $8 + 7 = 15$.
3. Transitive Property of Equality	If $x = y$ and $y = z$, then $x = z$.	If $2 \cdot 9 = 9 + 9$, and $9 + 9 = 18$, then $2 \cdot 9 = 18$.
4. Antisymmetric Property of Inequality	If $x < y$, then $y > x$. If $x > y$, then $y < x$.	If $7 < 10$, then $10 > 7$. If $-2 > -6$, then $-6 < -2$.
5. Transitive Property of Inequality	If $x < y$ and $y < z$, then $x < z$. If $x > y$ and $y > z$, then $x > z$.	If $3 < 5$ and $5 < 7$, then $3 < 7$. If $8 > 6$ and $6 > 4$, then $8 > 4$.

Example 6. Identify each property illustrated.

a. If $t = 5 + 9$ and $5 + 9 = 14$, then $t = 14$.

b. If $3 > u$, then $u < 3$.

Solution.

a. The three equalities illustrate the transitive property of equality.

b. The two inequalities illustrate the antisymmetric property of inequality.

Graphing Intervals of Numbers

Much of the study of algebra is devoted to *solving equations and inequalities.* In Chapter 2, we will solve several types of equations and inequalities. As we will see, the answer to an equation or inequality may be a set of numbers, called the

solution set. A solution set may be written using **set-builder notation,*** or by a graph of the numbers in the set using a number line. Examples **1–5** below illustrate five possible solution sets using set-builder notation and a corresponding graph. In each example, x represents a real-number variable and a and b are real numbers with $a < b$.

Set of numbers	Corresponding graph	Meaning in words
1. $\{x \mid x > a\}$	a	x is greater than a
2. $\{x \mid x < b\}$	b	x is less than b
3. $\{x \mid x \geq b\}$	b	x is greater than or equal to b
4. $\{x \mid a < x < b\}$	a b	x is between a and b
5. $\{x \mid a \leq x \leq b\}$	a b	x is between a and b, inclusive

Example 7. Graph the numbers in each set.

a. $\{x \mid x \leq 2\}$ **b.** $\{x \mid -4 < x < 3\}$

Solution. **a.** The set $\{x \mid x \leq 2\}$ contains all numbers less than 2, but including 2. A graph of the set is shown in Figure 1-5.

Figure 1-5. A graph of $x \leq 2$.

b. The set $\{x \mid -4 < x < 3\}$ contains all numbers less than 3 and greater than -4. In other words, x is between -4 and 3. A graph of the set is shown in Figure 1-6.

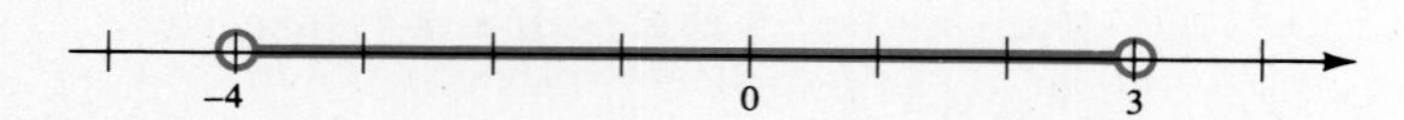

Figure 1-6. A graph of $-4 < x < 3$.

Example 8. Use x and set-builder notation to identify the numbers in each graph.

a.

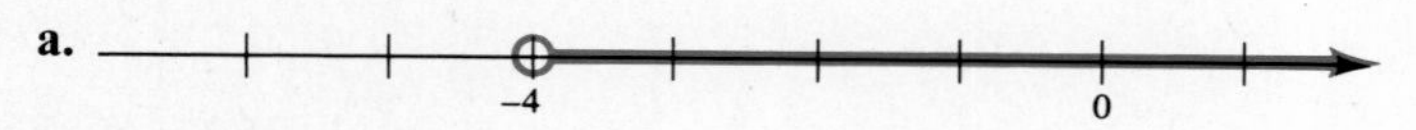

Figure 1-7.

b.

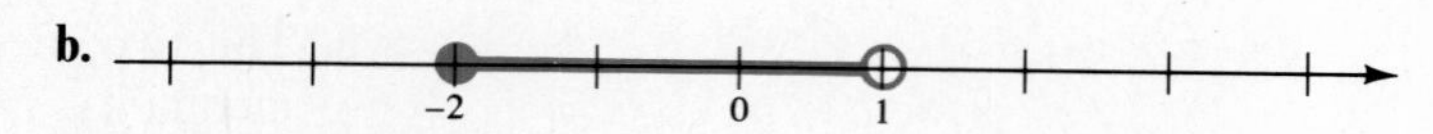

Figure 1-8.

* See Appendix C for a discussion on set-builder notation.

Solution.

a. Figure 1-7 is a graph of all numbers greater than -4. In set-builder notation, $\{x \mid x > -4\}$.

b. Figure 1-8 is a graph of all numbers between -2 and 1, including -2 but excluding 1. In set-builder notation, $\{x \mid -2 \leq x < 1\}$.

SECTION 1-1. Practice Exercises

In exercises **1–16**, verify that each number is rational.

[Example 1]

1. 39 

2. 17

3. $4\frac{3}{5}$

4. $5\frac{2}{3}$

5. 0.25

6. 0.6

7. -8

8. -15

9. $-6\frac{1}{2}$

10. $-10\frac{3}{4}$

11. -3.75

12. -7.8

13. 0

14. -1

15. $\frac{1.5}{8}$

16. $\frac{0.03}{2}$

In exercises **17–36**, by each number place an X in the column that identifies the set to which that number belongs. A given number may belong to more than one set.

[Example 2]

	Whole numbers	Integers	Rational numbers	Irrational numbers	Real numbers
17. 89					
18. -25					
19. $\frac{-3}{7}$					
20. $\frac{5}{8}$					
21. 0					
22. 1					
23. 0.21836 . . .					

	Whole numbers	Integers	Rational numbers	Irrational numbers	Real numbers
24. $-2.05719\ldots$					
25. $10\frac{2}{3}$					
26. $-9\frac{1}{2}$					
27. $-0.999\ldots$					
28. $8.252525\ldots$					
29. $\sqrt{5}$					
30. $-\sqrt{7}$					
31. 0.3					
32. -9.25					
33. $\frac{20}{4}$					
34. $\frac{-6}{3}$					
35. $\sqrt{9}$					
36. $\sqrt{12}$					

In exercises **37–46**, list the first five multiples of each number.

[Example 3] **37.** 5 

38. 9

39. 13

40. 10

41. 19

42. 21

43. 25

44. 33

45. 39

46. 37

In exercises **47–56**, list the natural number factors of each number.

47. 12

48. 18

49. 21

50. 15

51. 35

52. 42

53. 43

54. 53

55. 78

56. 96

In exercises **57–70**, graph each number on a number line.

[Example 4]

57. -6

58. -4

59. $\frac{7}{8}$

60. $\frac{5}{6}$

61. $-\frac{1}{3}$

62. $-\frac{13}{2}$

63. $4\frac{1}{2}$

64. $5\frac{1}{4}$

65. $\sqrt{5}$

66. $\sqrt{8}$

67. $-\frac{7}{4}$

68. $-\frac{9}{5}$

69. $-\sqrt{13}$

70. $-\sqrt{12}$

In exercises **71–80**, place $<$, $=$, or $>$ between each pair of numbers to make a true statement.

[Example 5]

71. $-9 \quad -12$

72. $-19 \quad -16$

73. $-3.7 \quad -3.4$

74. $-7.2 \quad -7.5$

75. $\sqrt{21} \quad \sqrt{26}$

76. $\sqrt{45} \quad \sqrt{41}$

77. $-\sqrt{17} \quad -\sqrt{15}$

78. $-\sqrt{27} \quad -\sqrt{22}$

79. $2\frac{7}{8} \quad 2\frac{9}{16}$

80. $3\frac{5}{6} \quad 3\frac{7}{12}$

In exercises **81–90**, identify each property illustrated.

[Example 6]

81. If $10 = x + 3$, then $x + 3 = 10$.

82. If $3x + 5 = y$, then $y = 3x + 5$.

83. If $x = 8 + 9$ and $8 + 9 = 17$, then $x = 17$.

84. If $x = 10 \cdot 6$ and $10 \cdot 6 = 60$, then $x = 60$.

85. If $-5 > x$, then $x < -5$.

86. If $y > 3$, then $3 < y$.

87. $3x - 2y = 3x - 2y$

88. $4x + 7 = 4x + 7$

89. If $y < x + 2$ and $x + 2 < 6$, then $y < 6$.

90. If $x > 4 - y$ and $4 - y > 9$, then $x > 9$.

In exercises **91–102**, graph the numbers in each set.

[Example 7]

91. $\{x | x \leq 10\}$

92. $\{x | x \geq 3\}$

93. $\{y | -5 < y < 5\}$

94. $\{y | -10 \leq y \leq 10\}$

95. $\{y | y \leq -1\}$

96. $\{y | y \geq 9\}$

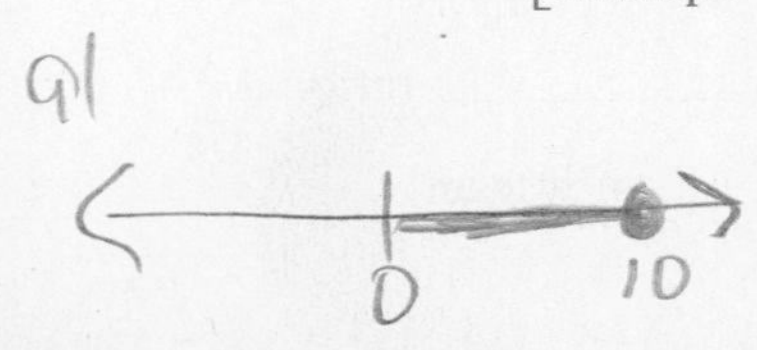

97. $\{x \mid -12 \leq x \leq -8\}$

98. $\{x \mid 5 < x < 9\}$

99. $\{y \mid 0 \leq y < 10\}$

100. $\{y \mid -3 \leq y < 0\}$

101. $\{x \mid -4 < x \leq 1\}$

102. $\{x \mid -2 \leq x < 9\}$

In exercises **103–112**, use x and set-builder notation to identify the numbers in each graph.

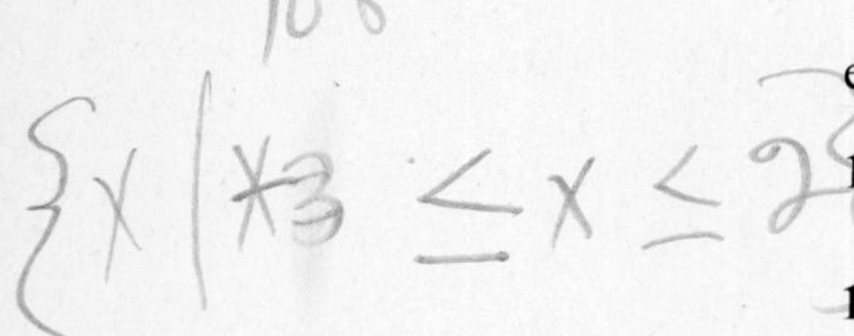

103.

104.

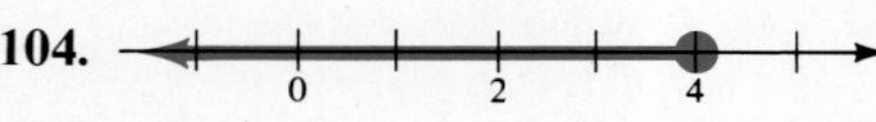

105. 0 2 4 6

106.

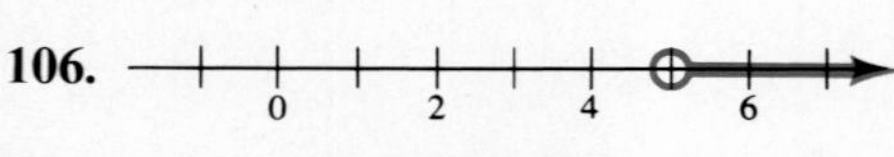

107. 0 2 4 6 8

108. −4 −2 0 2

109. −6 −4 −2 0

110. −8 −4 0 4 8

111. −20 −10 0 10

112.

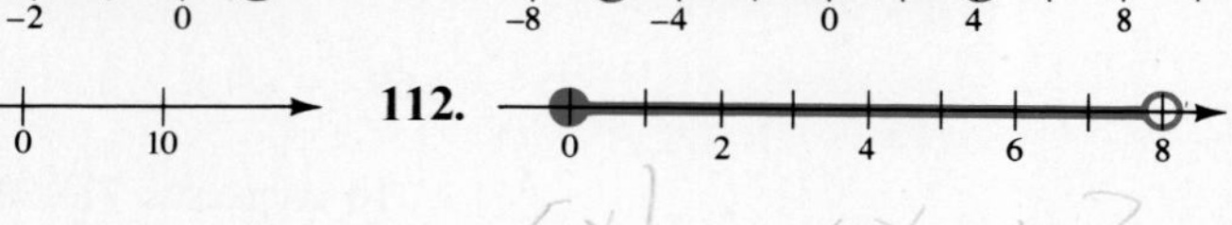

SECTION 1-1. Supplementary Exercises

In Exercises **1–20**, fill in the blank with *always, sometimes,* or *never* to make a true statement.

1. A rational number is ________________ an irrational number.
2. An integer is ________________ a whole number.
3. An integer is ________________ a rational number.
4. Zero is ________________ a real number.
5. A whole number is ________________ an irrational number.
6. A real number is ________________ an irrational number.
7. A rational number is ________________ an integer.
8. A negative integer is ________________ a whole number.
9. An irrational number is ________________ an integer.
10. A rational number is ________________ a real number.
11. An integer is ________________ negative.
12. A rational number is ________________ positive.
13. A whole number is ________________ negative.
14. An irrational number is ________________ zero.
15. A square root of a positive number is ________________ irrational.
16. A ratio of any two whole numbers is ________________ rational.
17. The width of a rectangle is ________________ an integer.

18. The height of a triangle is ________________ irrational.

19. The radius of a circle is ________________ rational.

20. The perimeter of a trapezoid is ________________ a whole number.

A **prime number** is a natural number that has exactly two factors—namely, 1 and the number itself. Below is a list of the first 24 prime numbers:

2, 3, 5, 7, 11, 13, 17, 19, 23, 29, 31, 37,

41, 43, 47, 53, 59, 61, 67, 71, 73, 79, 83, 89

In exercises **21–32**, write each number as a product of only prime numbers. (For example, $15 = 3 \cdot 5$ and $36 = 2 \cdot 2 \cdot 3 \cdot 3$.)

21. 12 **22.** 18 **23.** 28

24. 24 **25.** 51 **26.** 57

27. 60 **28.** 78 **29.** 132

30. 156 **31.** 348 **32.** 372

Example. Write each number as a ratio of integers.

a. 0.4444 . . . **b.** 0.261261261 . . .

Solution. **a.** Let $N = 0.4444\ldots$ (1)

$10N = 4.4444\ldots$ (2) There is *one digit* in the period.

$$\begin{array}{lrl} (2) & 10N &= 4.4444\ldots \\ (1) & N &= 0.4444\ldots \\ \hline & 9N &= 4.0000\ldots \end{array}$$

Subtract equation (1) from (2).

$$N = \frac{4}{9}$$

Divide both sides by 9.

Thus, $0.4444\ldots = \frac{4}{9}$.

b. Let $N = 0.261261261\ldots$ (1)

$1{,}000N = 261.261261261\ldots$ (2) There are *three digits* in the period.

$$\begin{array}{lrl} (2) & 1{,}000N &= 261.261261261\ldots \\ (1) & N &= 0.261261261\ldots \\ \hline & 999N &= 261.000000000\ldots \end{array}$$

Subtract (1) from (2).

$$N = \frac{261}{999} = \frac{29}{111}$$

Divide by 999 and reduce.

In exercises **33–38**, write each number as a ratio of integers.

33. 0.7777 . . . **34.** 0.5555 . . .

35. 0.121212 . . . **36.** 0.505050 . . .

37. 0.081081 . . . **38.** 0.027027 . . .

Common multiple

1. If m and n are natural numbers and k is a multiple of both m and n, then k is a **common multiple** of m and n.
2. If m and n are natural numbers and k is the smallest common multiple of m and n, then k is called the **least common multiple,** written LCM.

In exercises **39–46**, find the LCM of each set of numbers.

39. 6 and 15
40. 8 and 12
41. 12 and 15
42. 10 and 14
43. 10, 15, and 25
44. 12, 18, and 30
45. 14, 21, and 35
46. 16, 24, and 40

In exercises **47–60**, use x to write an inequality for each word statement.

47. x is less than -12
48. x is greater than 20
49. x is less than or equal to 9
50. x is greater than or equal to -2
51. x is between -10 and -3, excluding -10 and -3
52. x is between 0 and 8, including 0 and 8
53. x is between -6 and 5, including -6 and excluding 5
54. x is between -7 and 4, excluding -7 and including 4
55. x is a positive real number
56. x is a negative real number
57. x is a nonpositive real number
58. x is a nonnegative real number
59. x is a positive number or 0
60. x is a negative number or 0

SECTION 1-2. Simplifying Numerical Expressions

1. The absolute value of a number
2. Adding real numbers
3. Subtracting real numbers
4. Multiplying real numbers
5. Dividing real numbers
6. Raising a number to a power

7. A rule for the order of operations

8. Grouping symbols

In this section, we will review the rules for adding, subtracting, multiplying, dividing, and raising to powers real numbers. We will also study the rule that governs the order of operations.

The Absolute Value of a Number

In arithmetic, we learn number facts that help us compute the sums of positive numbers. However, real numbers can be positive, negative, or 0. We therefore need a procedure that will:

1. specify how to find a sum of real numbers.

2. specify whether the sum is positive, negative, or 0.

The absolute value of a number is useful for writing such a procedure.

Definition 1.3. The absolute value of a number
If b is a real number, then the **absolute value** of b:

1. $|b| = b$, if $b \geq 0$ $\quad |8| = 8$ and $|0| = 0$

2. $|b| = -b$, if $b < 0$ $\quad |-6| = -(-6) = 6$

Example 1. Simplify each expression.

a. $|-(-4)|$ **b.** $-|-9|$

Solution. **a.** Inside the absolute value bars:

$-(-4) = 4$ The opposite of -4 is 4.

$|-(-4)| = |4|$ Replace $-(-4)$ by 4.

$= 4$ The absolute value of a positive number is the number itself.

b. $|-9| = -(-9)$ The absolute value of a negative number is the opposite of the number.

$= 9$ The opposite of -9 is 9.

$-|-9| = -9$ Replace $|-9|$ by 9.

The minus sign *outside* the absolute value bars makes the number negative.

Adding Real Numbers

The results of adding real numbers is called the **sum.** The procedure for adding real numbers depends on whether the numbers being added have the same sign (both positive or both negative), or opposite signs (one positive and the other negative).

To add real numbers x and y:

1. If x and y have the **same sign,** then add their absolute values. Give the sum the common sign.
2. If x and y have **opposite signs,** then subtract their absolute values, the smaller from the larger. Give the sum the sign of the number with the greater absolute value.

Example 2. Simplify each expression.

a. $18 + (-39)$ **b.** $-32 + (-49)$

Solution. **a.** $|18| = 18$ and $|-39| = 39$ — The numbers have opposite signs.

$39 - 18 = 21$ — Subtract the absolute values.

$18 + (-39) = -21$ — The negative number has the greater absolute value.

b. $|-32| = 32$ and $|-49| = 49$ — The numbers are both negative.

$32 + 49 = 81$ — Add the absolute values.

$-32 + (-49) = -81$ — The common sign is negative.

Subtracting Real Numbers

Subtraction is the **inverse operation** of addition. That is:

$$x - y = z \text{ if, and only if, } x = y + z$$

For example, to compute $26 - 18$, we mentally look for the number to add to 18 to get 26:

$$26 - 18 = 8, \text{ because } 26 = 18 + 8$$

The relationship between subtraction and addition permits us to rewrite a subtraction in terms of an addition. Once this change has been made, the procedure for adding real numbers can be applied.

Definition 1.4. Subtraction in terms of addition
If x and y are real numbers, then:

$$x - y = x + (-y)$$

Indicates subtraction (the $-$ in $x - y$) **Indicates the opposite of y** (the $-$ in $-y$)

Example 3. Simplify each expression.

a. $-19 - 46$ **b.** $36 - (-54)$

Solution. **Discussion.** To apply Definition 1-4, there are two tasks to perform:

Task 1. Change the subtraction to addition.

Task 2. Change the number being subtracted to its opposite.

a.	$-19 - 46$	The given subtraction
	$= -19 + (-46)$	The opposite of 46 is -46.
	$= -65$	Both numbers are negative.
b.	$36 - (-54)$	The given subtraction
	$= 36 + 54$	The opposite of -54 is 54.
	$= 90$	Both numbers are positive.

Multiplying Real Numbers

The result of multiplying real numbers is called the **product.** The absolute value of a product of any two real numbers is the same, regardless of whether the numbers are positive, negative, or 0. That is:

$$|x| \cdot |y| = |x \cdot y|, \text{ for all } x \text{ and } y$$

However, the product of x and y will be positive or negative depending on the signs of x and y.

To multiply real numbers x and y:

Step 1. Multiply the absolute values of x and y.

Step 2. Make the product positive if x and y are both positive or both negative. Make the product negative if either x or y is positive and the other is negative.

Step 2 can be remembered as follows:

$x \cdot y > 0$ if x and y have the *same sign*

$x \cdot y < 0$ if x and y have *opposite signs*

Example 4. Simplify each expression.

a. $-40(15)$ **b.** $\frac{-7}{12} \cdot \frac{4}{15} \cdot \frac{-9}{10}$

Solution.

a.	$-40(15) = -600$	The numbers have opposite signs.
b.	$\frac{-7}{12} \cdot \frac{4}{15} \cdot \frac{-9}{10}$	Two negatives and one positive
	$= \frac{(-7)(4)(-9)}{12 \cdot 15 \cdot 10}$	Multiply the numerators. / Multiply the denominators.
	$= \frac{252}{1800}$	The products are positive.
	$= \frac{7}{50}$	Reduce the fraction.

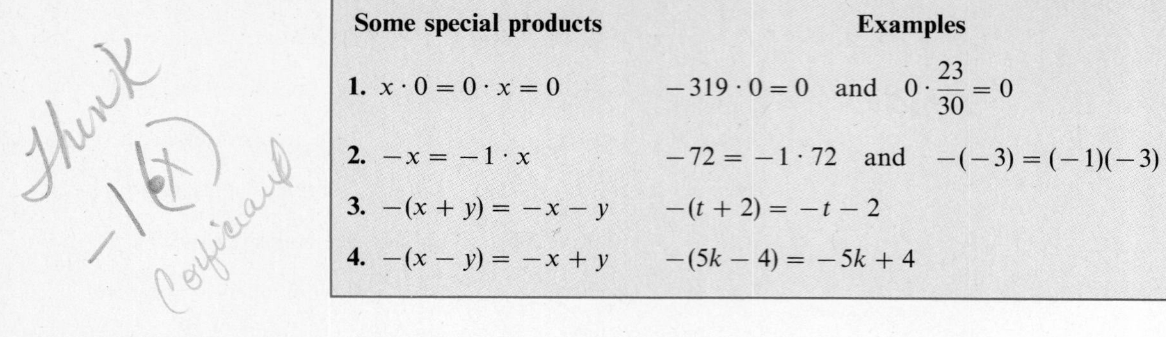

Some special products	Examples
1. $x \cdot 0 = 0 \cdot x = 0$	$-319 \cdot 0 = 0$ and $0 \cdot \frac{23}{30} = 0$
2. $-x = -1 \cdot x$	$-72 = -1 \cdot 72$ and $-(-3) = (-1)(-3)$
3. $-(x + y) = -x - y$	$-(t + 2) = -t - 2$
4. $-(x - y) = -x + y$	$-(5k - 4) = -5k + 4$

Dividing Real Numbers

Division is the inverse operation of multiplication. That is, if $y \neq 0$:

$$x \div y = z \text{ if, and only if, } x = y \cdot z$$

For example, to compute $54 \div 9$, we mentally look for the number to multiply by 9 to get 54:

$$54 \div 9 = 6, \text{ because } 54 = 9 \cdot 6$$

The relationship between division and multiplication permits us to rewrite a division in terms of a multiplication. Once this change has been made, the rule for the sign of a product can be applied to the quotient.

Definition 1.5. Division in terms of multiplication
If x and y are real numbers and $y \neq 0$, then:

$$x \div y = x \cdot \frac{1}{y}$$

Indicates division (points to $\div$) — **Indicates the reciprocal of y** (points to $\frac{1}{y}$)

As in multiplication:

$$x \div y > 0 \quad \text{if} \quad x \text{ and } y \text{ have the } \textit{same sign}$$

$$x \div y < 0 \quad \text{if} \quad x \text{ and } y \text{ have } \textit{opposite signs}$$

Example 5. Simplify each expression.

a. $\frac{5}{7} \div \frac{-11}{14}$ **b.** $(-0.063) \div (-0.03)$

Solution. **a.**

$$\frac{5}{7} \div \frac{-11}{14}$$ The numbers have opposite signs.

$$= \frac{5}{7} \cdot \frac{-14}{11}$$ The reciprocal of $\frac{-11}{14}$ is $\frac{-14}{11}$.

$$= \frac{-70}{77}$$ $\frac{\text{Multiply the numerators.}}{\text{Multiply the denominators.}}$

$$= \frac{-10}{11}$$ Reduce the fraction.

b. $(-0.063) \div (-0.03)$ — The numbers have the same sign.

$= (-6.3) \div (-3)$ — Multiply both numbers by 100.

$= 2.1$ — The quotient is positive.

Some special quotients	Examples
1. If $x \neq 0$, then $0 \div x = 0$.	$\frac{0}{15} = 0$ and $\frac{0}{-8} = 0$
2. If $x \neq 0$, then $x \div 0$ is undefined.	$\frac{15}{0}$ and $\frac{-8}{0}$ are undefined
3. If $x \neq 0$, then $1 \div \frac{1}{x} = x$.	$1 \div \frac{1}{5} = 5$ and $1 \div \frac{-1}{9} = -9$
4. If $x \neq 0$ and $y \neq 0$, then $1 \div \frac{x}{y} = \frac{y}{x}$.	$1 \div \frac{4}{9} = \frac{9}{4}$ and $1 \div \frac{-8}{3} = \frac{-3}{8}$

Raising a Number to a Power

Consider the following indicated products:

a. $7 \cdot 7 \cdot 7 \cdot 7$ **b.** $(-3)(-3)(-3)(-3)(-3)(-3)$

Both expressions can be written using a symbol called an **exponent.**

> If b is a real number and n is a positive integer, then:
>
> b is called the **base** ⟶ $b^n = \underbrace{b \cdot b \cdot b \cdots b}_{n\text{-factors}}$ (← **exponent**)
>
> If $n = 1$, then b^1 is written b.

Using an exponent:

a. $7 \cdot 7 \cdot 7 \cdot 7$ can be written 7^4

b. $(-3)(-3)(-3)(-3)(-3)(-3)$ can be written $(-3)^6$

To illustrate how an exponent is read:

c. $b \cdot b = b^2$ is read "b-squared" or "b to the second power"

d. $b \cdot b \cdot b = b^3$ is read "b-cubed" or "b to the third power"

e. $b \cdot b \cdot b \cdot b = b^4$ is read "b to the fourth power"

f. $b \cdot b \cdot b \cdot b \cdot b = b^5$ is read "b to the fifth power"

Example 6. Simplify each expression:

a. 3^4 **b.** $(-2)^6$

Solution. **a.** $3^4 = 3 \cdot 3 \cdot 3 \cdot 3 = 81$

b. $(-2)^6 = (-2)(-2)(-2)(-2)(-2)(-2) = 64$

The sign of b^n	Example
1. If $b > 0$, then $b^n > 0$ for all n.	$6^5 = 7776$ and $\left(\frac{5}{3}\right)^4 = \frac{625}{81}$
2. If $b = 0$, then $b^n = 0$ for all n.	$0^8 = 0$ and $0^{11} = 0$
3. If $b < 0$ and n is even, then $b^n > 0$.	$(-3)^4 = 81$ and $\left(\frac{-1}{2}\right)^8 = \frac{1}{256}$
4. If $b < 0$ and n is odd, then $b^n < 0$.	$(-7)^3 = -343$ and $\left(\frac{-3}{2}\right)^5 = \frac{-243}{32}$

A Rule for the Order of Operations

When two or more operations are used in the same expression, that expression can be simplified by observing the rule that governs the order of operations.

Rule for the order of operations
Unless grouping symbols indicate otherwise:

Step 1. Do any indicated powers.

Step 2. Do the multiplication and division operations in order from left to right.

Step 3. Do the addition and subtraction operations in order from left to right.

Example 7. Simplify $3^2 \cdot 5 - 2^5 + 75 \div 15$.

Solution. **Step 1.** Do any powers first.

$9 \cdot 5 - 32 + 75 \div 15$ — $3^2 = 9$ and $2^5 = 32$

Step 2. Do multiplication and division operations in order from left to right.

$= 45 - 32 + 5$ — $9 \cdot 5 = 45$ and $75 \div 15 = 5$

Step 3. Do addition and subtraction operations in order from left to right.

$= 13 + 5$ — $45 - 32 = 13$

$= 18$ — $13 + 5 = 18$

Grouping Symbols

The order in which operations should be performed can be changed by using one or more **grouping symbols.** Four symbols are commonly used to group terms in an expression.

Name	Symbol	Examples
1. Parentheses	( )	$27 - (5 - 12)$
2. Brackets	[ ]	$9 + [3 - (8 + 5)]$
3. Bar (or vinculum)	—	$\dfrac{12 + 35}{4 + 7}$
4. Braces	{ }	$z\{1 - [6 + (3 - 8)]\}$

When an expression contains grouping symbols, the operations within the grouping symbols must be performed first, always following the order-of-operations rule. Furthermore, if one set of grouping symbols occurs within a second set, the innermost set must be simplified first. In other words, work from the inside out.

Example 8. Simplify $7(5 \cdot 2 - 3^2) + 5[6^2 + (-2)^5]$.

Solution. **Discussion.** First simplify the two parenthetical expressions, using the rule for order of operations.

$$7(5 \cdot 2 - 3^2) + 5[6^2 + (-2)^5]$$

$= 7(5 \cdot 2 - 9) + 5[36 + (-32)]$ — $3^2 = 9$, $6^2 = 36$, and $(-2)^5 = -32$

$= 7(10 - 9) + 5(4)$ — $5 \cdot 2 = 10$ and $36 + (-32) = 4$

$= 7(1) + 5(4)$ — $10 - 9 = 1$

$= 7 + 20$ — Perform the multiplications.

$= 27$ — The expression is 27.

Example 9. Simplify $\dfrac{-2(15 - 3 \cdot 2^2)}{5^2 - 3^2} + \dfrac{11 + 12 \div 4 \cdot 9}{2(13 - 5)}$.

Solution. **Discussion.** The fraction bars are grouping symbols. Therefore, the numerators and denominators are simplified first.

$$\frac{-2(15 - 3 \cdot 2^2)}{5^2 - 3^2} + \frac{11 + 12 \div 4 \cdot 9}{2(13 - 5)}$$

$$= \frac{-2(15 - 3 \cdot 4)}{25 - 9} + \frac{11 + 3 \cdot 9}{2(8)}$$

$$= \frac{-2(15 - 12)}{16} + \frac{11 + 27}{16}$$

$$= \frac{-2(3)}{16} + \frac{38}{16}$$

$$= \frac{-6}{16} + \frac{38}{16}$$

$= \dfrac{32}{16} = 2$ — The value of the expression is 2.

SECTION 1-2. Practice Exercises

In exercises **1–126**, simplify each expression.

[Example 1]

1. $|23|$

2. $|52|$

3. $|-39|$

4. $|-17|$

5. $|-(-44)|$

6. $|-(-63)|$

7. $-|-12|$

8. $-|-18|$

9. $|x|$, if $x < 0$

10. $|x|$, if $x > 0$

11. $|a - b|$, if $a > b$

12. $|a - b|$, if $a < b$

13. $|3a|$, if $a < 0$

14. $|5a|$, if $a > 0$

[Example 2]

15. $21 + 15$

16. $10 + 43$

17. $-18 + (-12)$

18. $-3 + (-37)$

19. $32 + (-19)$

20. $57 + (-20)$

21. $0 + (-16)$

22. $0 + (-10)$

23. $-80 + 35$

24. $-45 + 41$

25. $-63 + 70$

26. $-17 + 29$

27. $-13 + (-17)$

28. $-19 + (-11)$

29. $75 + (-35)$

30. $-75 + 35$

31. $12 + 46$

32. $15 + 23$

33. $(-61) + (-42)$

34. $(-39) + (-53)$

[Example 3]

35. $12 - 18$

36. $9 - 20$

37. $8 - (-9)$

38. $13 - (-7)$

39. $-10 - 6$

40. $-17 - 19$

41. $-5 - 12$

42. $-11 - 15$

43. $-4 - (-14)$

44. $-21 - (-3)$

45. $11 - 7$

46. $19 - 4$

47. $-11 - (-7)$

48. $-8 - (-12)$

49. $15 - (-25)$

50. $17 - (-13)$

51. $44 - 14$

52. $35 - 15$

53. $-8 - 10$

54. $-6 - 13$

[Example 4]

55. $-3 \cdot 7$

56. $-6 \cdot 4$

57. $(-9)(-2)$

58. $(-10)(-8)$

59. $-5(12)$

60. $-8(13)$

61. $(-4)(5)(-1)$

62. $(-3)(-6)(5)$

63. $-1(7)$

64. $-1(4)$

65. $(-2)(5)(-3)(-6)$

66. $(-4)(2)(10)(6)$

67. $(-8)(-6)(0)$

68. $(-5)(-9)(0)$

69. $\left(\frac{-1}{2}\right)\left(\frac{4}{9}\right)$

70. $\left(\frac{5}{8}\right)\left(\frac{-6}{7}\right)$ **71.** $\left(\frac{-7}{10}\right)\left(\frac{-15}{14}\right)$ **72.** $\left(\frac{-3}{5}\right)\left(\frac{-10}{21}\right)$

73. $\left(\frac{1}{3}\right)\left(\frac{-6}{7}\right)\left(\frac{-1}{4}\right)$ **74.** $\left(\frac{-3}{4}\right)\left(\frac{8}{9}\right)\left(\frac{-1}{2}\right)$ **75.** $\left(-1\frac{2}{3}\right)\left(-2\frac{1}{2}\right)$

76. $\left(3\frac{1}{7}\right)\left(-1\frac{3}{11}\right)$ **77.** $\left(\frac{-3}{8}\right)\left(1\frac{1}{7}\right)\left(\frac{-1}{3}\right)$ **78.** $\left(1\frac{1}{2}\right)\left(\frac{-2}{3}\right)\left(\frac{100}{101}\right)$

[Example 5] **79.** $(-48) \div (-8)$ **80.** $(-28) \div (-7)$ **81.** $\frac{-60}{6}$

82. $\frac{-39}{3}$ **83.** $\frac{-52}{-13}$ **84.** $\frac{-63}{-21}$

85. $180 \div (-60)$ **86.** $70 \div (-14)$ **87.** $\frac{-2}{5} \div \frac{-1}{10}$

88. $\frac{-5}{7} \div \frac{-5}{6}$ **89.** $\frac{3}{8} \div \frac{-15}{4}$ **90.** $\frac{-8}{9} \div \frac{4}{21}$

91. $\frac{-10}{3} \div (-5)$ **92.** $\frac{21}{8} \div 7$ **93.** $16 \div \frac{4}{5}$

94. $-26 \div \frac{13}{7}$ **95.** $\left(-7\frac{1}{7}\right) \div \frac{5}{14}$ **96.** $\left(-5\frac{1}{4}\right) \div \left(-2\frac{4}{5}\right)$

[Example 6] **97.** 5^3 **98.** 9^2 **99.** $(-10)^3$

100. $(-7)^3$ **101.** -4^2 **102.** -6^2

103. $(-4)^2$ **104.** $(-6)^2$ **105.** $3^3(-8)^2$

106. $4^3(-10)^4$

[Example 7] **107.** $15 - 3 \cdot 2$ **108.** $21 + 5 \cdot 3$

109. $20 - 24 \div 2^3 - 3 \cdot 7$ **110.** $-9 - 9 \div 3 + 2 \cdot 5^2$

111. $30 + 100 \div 5 \cdot 2 - 8 \cdot 5$ **112.** $15 - 90 \div 3 \cdot 2 + 5 \cdot 7$

113. $40 \div 2 \cdot 10 - 35 \div 7 - 2$ **114.** $48 \div 6 \cdot 2 - 12 \cdot 8 \div 4$

[Example 8] **115.** $5(3 \cdot 7 - 4^2) + 3(10 - 8 \cdot 2)$ **116.** $4(9 - 2 \cdot 8) - (20 + 3^3)$

117. $-2[10 - 3(4 - 2^3)]$ **118.** $-3[(-2)^4 - 5(10 - 8)]$

119. $3 + 2[4(12 - 7) + (2^3 - 3^2)^2]$ **120.** $4 + 6[(9 - 3)^2 + (2^2 - 7)^3]$

[Example 9] **121.** $\frac{5^3}{3^2 - 2^2} + \frac{5^2 - 3}{2^4 - 5}$ **122.** $\frac{7^2 - 3^2}{3^2 + 1} + \frac{11 + 3^3}{5^2 - 6}$

123. $\frac{10 - 2(2^4 - 3^2)}{2^3 - 10} + \frac{3(7 - 5)^2 - 2}{5(10 - 3^2)}$ **124.** $\frac{18 - 3(3^3 - 5^2)}{2(1 - 3)} + \frac{3^3 - 3(12 - 3^2)}{20 - 2 \cdot 7}$

125. $\frac{5^3 - (2^3 - 2)^2 + 1}{10 + 4 \cdot 5} - \frac{10^2 - (2 \cdot 5 - 1)^2}{6^2 - (10 \cdot 2 - 3)}$

126. $\frac{13^2 - (5 + 2)^2}{(6 + 2^2) - 2} - \frac{11^2 - (2^2 + 7)}{5^2 - (2 + 3)7}$

SECTION 1-2. Supplementary Exercises

The operations of addition and multiplication have several properties over the set of real numbers.

Properties of the real numbers

Let x, y, and z be real numbers.

Name of property	Addition	Multiplication
Closure	$x + y$ is a real number "A sum of real numbers is a real number."	$x \cdot y$ is a real number "A product of real numbers is a real number."
Commutative	$x + y = y + x$ "Numbers can be added in reverse order and not change the sum."	$x \cdot y = y \cdot x$ "Numbers can be multiplied in reverse order and not change the product."
Associative	$x + (y + z) = (x + y) + z$ "Numbers added can be regrouped and not change the sum."	$x \cdot (y \cdot z) = (x \cdot y) \cdot z$ "Numbers multiplied can be regrouped and not change the product."
Identity elements	$x + 0 = 0 + x = x$ "The sum of any number x and 0 is x."	$x \cdot 1 = 1 \cdot x = x$ "The product of any number x and 1 is x."
Inverse elements	$x + (-x) = -x + x = 0$ "The sum of x and the *opposite* of x is 0."	$x \cdot \frac{1}{x} = \frac{1}{x} \cdot x = 1, x \neq 0$ "The product of x and the *reciprocal* of x is 1."
Distributive property of multiplication over addition	$x(y + z)$ "Add first, then multiply."	$= \quad x \cdot y + x \cdot z$ "Multiply first, then add."

In exercises **1–20**, identify each property illustrated. Assume that all variables represent real numbers.

1. $10x$ is a real number

2. $x + 10$ is a real number

3. $5y + 8 = 8 + 5y$

4. $y9 = 9y$

5. $25(4 \cdot 13) = (25 \cdot 4)13$

6. $(43 + 62) + 38 = 43 + (62 + 38)$

7. $8 + 0 = 8$

8. $-37 \cdot 1 = -37$

9. $-15\left(\frac{-1}{15}\right) = 1$

10. $-15 + 15 = 0$

11. $6(x + 3) = 6x + 6 \cdot 3$

12. $2(5 + y) = 2 \cdot 5 + 2y$

13. $(x + 3) + 9 = x + (3 + 9)$

14. $3(9x) = (3 \cdot 9)x$

15. $20 + (-10) = -10 + 20$

16. $20 + (-10)$ is a real number

17. $20(-10)$ is a real number

18. $20(-10) = -10(20)$

19. $8x + (-8x) = 0$

20. $\frac{1}{7} \cdot 7x = 1 \cdot x$

In exercises **21–28**, given each number:

a. Find the opposite.

b. Find the reciprocal.

21. $\frac{3}{8}$

22. $\frac{5}{7}$

23. $\frac{-2}{9}$

24. $\frac{-3}{5}$

25. -7

26. -11

27. $2\frac{2}{3}$

28. $5\frac{2}{9}$

Example. Use the distributive property to simplify $35\left(\frac{1}{7} + \frac{1}{5}\right)$.

Solution. $35\left(\frac{1}{7} + \frac{1}{5}\right)$

$= 35 \cdot \frac{1}{7} + 35 \cdot \frac{1}{5}$ $a(b + c) = a \cdot b + a \cdot c$

$= 5 + 7$ Multiply first.

$= 12$ Add second.

In exercises **29–34**, use the distributive property to simplify each expression.

29. $12\left(\frac{1}{3} + \frac{1}{4}\right)$

30. $15\left(\frac{3}{5} + \frac{2}{3}\right)$

31. $21\left(\frac{5}{7} - \frac{1}{3}\right)$

32. $10\left(\frac{1}{2} - \frac{2}{5}\right)$

33. $6\left(\frac{1}{2} + \frac{1}{3} + \frac{1}{6}\right)$

34. $30\left(\frac{2}{5} + \frac{1}{2} + \frac{2}{3}\right)$

Example. Use the distributive property to simplify $\frac{5}{8} \cdot 31 + \frac{5}{8} \cdot 17$.

Solution. $\frac{5}{8} \cdot 31 + \frac{5}{8} \cdot 17$

$= \frac{5}{8}(31 + 17)$ $a \cdot b + a \cdot c = a(b + c)$

$= \frac{5}{8}(48)$ Add first.

$= 30$ Multiply second.

In exercises **35–40**, use the distributive property to simplify each expression.

35. $\frac{3}{4} \cdot 17 + \frac{3}{4} \cdot 11$

36. $\frac{5}{6} \cdot 19 + \frac{5}{6} \cdot 11$

37. $\frac{3}{8} \cdot 53 - \frac{3}{8} \cdot 13$

38. $\frac{5}{9} \cdot 31 - \frac{5}{9} \cdot 4$

39. $\frac{2}{7} \cdot 4 + \frac{2}{7} \cdot 9 + \frac{2}{7} \cdot 8$

40. $\frac{7}{10} \cdot 14 + \frac{7}{10} \cdot 11 + \frac{7}{10} \cdot 5$

In exercises **41–46**, use the expression:

$$15 - 7 \cdot 4 + 16 \div 4 \cdot 2$$

41. Find the value of the expression as written.

42. Insert one grouping symbol to change the value of the expression to 40.

43. Insert one grouping symbol to change the value of the expression to -11.

44. Insert one grouping symbol to change the value of the expression to -55.

45. Insert two grouping symbols to change the value of the expression to 34.

46. Insert two grouping symbols to change the value of the expression to 80.

SECTION 1-3. Simplifying Algebraic Expressions

1. Parts of algebraic expressions
2. Combining terms in expressions
3. Multiplying and dividing monomials
4. Simplifying expressions
5. Evaluating expressions

In the last section, we studied **numerical expressions.** These expressions consist of real numbers and certain operations on those numbers. Numerical expressions can be simplified, frequently to a single number, using the rule for the order of operations. In this section, we will make a similar study of **algebraic expressions.**

Parts of Algebraic Expressions

An algebraic expression differs from a numerical expression in that an algebraic expression can contain symbols called **variables.** A variable is a symbol that represents any one of the numbers in a set of numbers. For our purposes, variables represent numbers from the set of real numbers, and are called **real-number variables.** Letters are the most common symbol used to represent variables.

An **algebraic expression** can contain some, or all, of the following:

1. **Numbers,** usually from the set of real numbers
2. **Variables,** usually letters that represent numbers
3. **Operations,** such as addition, multiplication, and so on
4. **Grouping symbols,** such as parentheses, brackets, and so on

There are two commonly used ways of identifying an expression.

Method 1. By the variable (or variables) in the expression

Method 2. By the number of terms in the expression:

Monomial, if one term

Binomial, if two terms

Trinomial, if three terms

In an algebraic expression, terms are separated by + and − signs.

Examples **a–d** illustrate four expressions.

a. $\frac{-5t^2}{8}$ A monomial in t

b. $16x^2 - 25y^2$ A binomial in x and y

c. $2a^2 - 7ab + 10b^2$ A trinomial in a and b

d. $u^3 - \frac{3}{2}u^2 + \frac{2}{3}u - 10$ An expression of four terms in u

There are two important items of information related to a term.

Item 1. Numerical coefficient and literal coefficient
The numerical coefficient is the number factor in the term. The literal coefficient is the variable (or variables) factor in the term.

Item 2. The conventional form for a term
A term is written in conventional form when the numerical coefficient is first and the variables follow in alphabetical order.

Example 1. **(i)** Write each term in conventional form.

(ii) Identify the numerical coefficient.

(iii) Identify the literal coefficient.

a. $\frac{-nm^2}{5}$ **b.** $-k^2i7j^3$

Solution. **a.** **(i)** $\frac{-nm^2}{5}$ can be written in the form $\frac{-1}{5}\,m^2n$ or $\frac{-m^2n}{5}$

(ii) The numerical coefficient is $\frac{-1}{5}$.

(iii) The literal coefficient is m^2n.

b. **(i)** $-k^2i7j^3$ can be written in the form $-7ij^3k^2$

(ii) The numerical coefficient is -7.

(iii) The literal coefficient is ij^3k^2.

Combining Terms in Expressions

Consider the expressions in examples **e** and **f**:

e. $7t + 9t - 10t$ A trinomial in t

f. $2x^2 - 8x + 5x^2 - 3x$ An expression of four terms in x

Based on the rule for the order of operations, the multiplication operations in example **e** and the powers and multiplication operations in example **f** have priority over the addition and subtraction operations. However, the distributive property of multiplication over addition and subtraction can change the order by removing common literal factors from some of the terms. *The result is a reduction in the number of terms without a change in the value that the expression represents.*

e. $7t + 9t - 10t$ The common literal coefficient is t.

$= (7 + 9 - 10)t$ Use the distributive property.

$= 6t$ Simplify within the parentheses.

Thus, for any real number replacement for t, the given trinomial and $6t$ will have the same value.

f. $2x^2 - 8x + 5x^2 - 3x$ The common coefficients are x^2 and x.

$= 2x^2 + 5x^2 - 8x - 3x$ Use the commutative property.

$= (2 + 5)x^2 + (-8 - 3)x$ Use the distributive property.

$= 7x^2 + (-11)x$ Simplify within the parentheses.

or $7x^2 - 11x$

Notice that $7x^2 - 11x$ cannot be combined, because the literal coefficients are not the same. As a consequence, the distributive property cannot remove the literal coefficients, and the rule for the order of operations mandates the priority of powers and multiplication over subtraction. Hence, the terms cannot be combined.

> Terms in an expression that have identical literal coefficients can be **combined**—that is, added or subtracted—by adding or subtracting as indicated the numerical coefficients. Such terms are frequently referred to as **like terms.**

Example 2. Simplify each expression.

a. $-2cd + 13cd + 9cd - 16cd$

b. $9y^2 + y - 7 - y + 16y^2 - 2$

Solution. **a.** $-2cd + 13cd + 9cd - 16cd$ Four cd-terms

$= (-2 + 13 + 9 - 16)cd$ Use the distributive property.

$= 4cd$ Simplify within parentheses.

b. $9y^2 + y - 7 - y + 16y^2 - 2$ The given expression.

$= 9y^2 + 16y^2 + y - y - 7 - 2$ Rearrange the terms.

$= 25y^2 - 9$ $y - y = 0 \cdot y = 0$

Multiplying and Dividing Monomials

Consider the expressions in examples **g** and **h**.

g. $(-5t^2u)(4t^3u)$ A multiplication of two monomials

h. $\dfrac{75x^3y^2}{15xy^2}$ A division of two monomials

The monomials in examples **g** and **h** do not have the same literal coefficients. Therefore, they are not alike and cannot be added or subtracted. However, they can be multiplied and divided in the sense that the numbers of symbols in the product and quotient are less. The commutative property of multiplication and the definition of division can be used to rewrite these expressions as follows:

g. $(-5t^2u)(4t^3u)$ can be written $(-5 \cdot 4)(t^2 \cdot t^3)(u \cdot u)$

h. $\dfrac{75x^3y^2}{15xy^2}$ can be written $\dfrac{75}{15} \cdot \dfrac{x^3}{x} \cdot \dfrac{y^2}{y^2}$

The variable factors with the same bases can be simplified by using the following properties of exponents:

Some properties of exponents

Let a and b be nonzero numbers and m and n positive integers.

Property 1. $a^m \cdot a^n = a^{m+n}$ $\qquad k^4 \cdot k^2 = k^6$ and $2^5 \cdot 2^3 = 2^8$

Property 2. $\dfrac{a^m}{a^n} = a^{m-n}$, and $m > n > 0$ $\qquad \dfrac{k^4}{k} = k^{4-1} = k^3$ and $\dfrac{2^7}{2^4} = 2^3$

Property 3. $\dfrac{a^m}{a^m} = 1$ $\qquad \dfrac{2^6}{2^6} = 1$ and $\dfrac{(-5)^3}{(-5)^3} = 1$

g. $(-5t^2u)(4t^3u) = -20t^5u^2$

$-5 \cdot 4 = -20$

$t^2 \cdot t^3 = t^5$

$u \cdot u = u^2$

h. $\dfrac{75x^3y^2}{15xy^2} = 5x^2$

$\dfrac{75}{15} = 5$

$\dfrac{x^3}{x} = x^2$

$\dfrac{y^2}{y^2} = 1$

Example 3. Simplify $-6p^2q^3(5p^2 + pq - 3q^2)$.

Solution. **Discussion.** The distributive property is used to distribute $(-6p^2q^3)$ to each term of the trinomial.

$$\begin{aligned}&-6p^2q^3(5p^2 + pq - 3q^2)\\ &= (-6p^2q^3)(5p^2) + (-6p^2q^3)(pq) - (-6p^2q^3)(3q^2)\\ &= -30p^4q^3 + (-6p^3q^4) - (-18p^2q^5)\\ &= -30p^4q^3 - 6p^3q^4 + 18p^2q^5\end{aligned}$$

Simplifying Expressions

In Examples 4 and 5, the given algebraic expressions are simplified by following the order-of-operations rule and combining any like terms.

Example 4. Simplify $\frac{1}{2}(8t^2 + 2t - 10) - \frac{1}{3}(9t + 3) + \frac{3}{5}(20t^2 + 10)$.

Solution.

$$\frac{1}{2}(8t^2 + 2t - 10) - \frac{1}{3}(9t + 3) + \frac{3}{5}(20t^2 + 10)$$

$$= 4t^2 + t - 5 - 3t - 1 + 12t^2 + 6$$

$$= 16t^2 - 2t$$

$4t^2 + 12t^2 = 16t^2$

$t - 3t = -2t$

$-5 - 1 + 6 = 0$

Example 5. Simplify $2x(x^2 + 3x - 9) - 6x^2(x - 2) - 5(2x^2 - 7)$.

Solution.

$$2x(x^2 + 3x - 9) - 6x^2(x - 2) - 5(2x^2 - 7)$$

$$= 2x^3 + 6x^2 - 18x - 6x^3 + 12x^2 - 10x^2 + 35$$

$$= -4x^3 + 8x^2 - 18x + 35$$

Evaluating Expressions

If the variable factors in an algebraic expression are replaced by numbers, then the expression becomes numerical and can be further simplified. The value of an expression usually changes when the variables are replaced by different numbers. Keep in mind that the order-of-operations rule must be followed when the resulting numerical expressions are simplified.

Example 6. Evaluate $t^3 - 6t^2 + 12t - 8$ for $t = -3$.

Solution.

$(-3)^3 - 6(-3)^2 + 12(-3) - 8$	Replace t by -3.
$= -27 - 6(9) + 12(-3) - 8$	First find the powers.
$= -27 - 54 + (-36) - 8$	Then compute the products.
$= -81 - 36 - 8$	$-27 - 54 = -81$
$= -117 - 8$	$-81 - 36 = -117$
$= -125$	The value of the expression is -125.

SECTION 1-3. Practice Exercises

In exercises **1–8**:

a. Write each term in conventional form.

b. Identify the numerical coefficient.

c. Identify the literal coefficient.

[Example 1] **1.** $c^3 5ba^2$

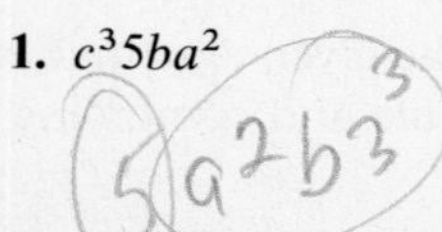

2. $b8c^4a^3$

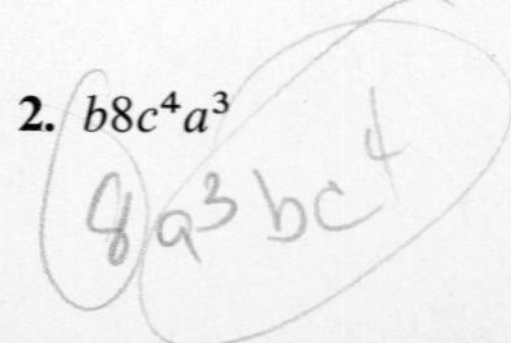

3. $-y^2x\frac{3}{4}$

4. $-x^4\frac{2}{3}y$

5. $n^7\frac{m^4}{3}l^3$

6. $ml^5\frac{n^2}{6}$

7. $r^5p(-q^2)$

8. $q^3(-r^4)p$

In exercises **9–40**, simplify each expression.

[Example 2] 9. $-8x + 21x - 5x$

10. $-4x + 19x - 7x$

11. $5yz - 12yz + 17yz - 9yz$

12. $12yz + 8yz - 6yz - 11yz$

13. $7p^2 + 6p - 13p^2 + p$

14. $9p^2 + p - 11p^2 + 5p$

15. $m^2 - 4m + 21 + 4m^2 + 7m - 35$

16. $3m^2 + 9m - 13 + m^2 - 2m + 8$

[Example 3] 17. $(6x^2y)(-2xy^4)$

18. $(-7xy^3)(3x^2y^3)$

19. $\frac{80p^2q^8}{5pq^6}$

20. $\frac{65p^3q^{10}}{13pq^8}$

21. $3a^2b(8a^2b - 2ab^3)$

22. $5ab^3(11a^2b^2 - 3ab^3)$

23. $-7s^5t^9(9s^3t^4 - 2s^2t + 6st^4)$

24. $-9s^4t^7(6s^3t^2 + 5s^2t^3 - 4st^5)$

[Example 4] 25. $\frac{1}{2}(10x^2 - 6x) + \frac{1}{5}(15x^2 + 5x)$

26. $\frac{1}{3}(9x^2 + 12x) + \frac{1}{4}(20x^2 - 4x)$

27. $\frac{2}{7}(28t^2 - 21t + 49) - \frac{3}{4}(36t^2 + 12)$

28. $\frac{3}{5}(35t^2 - 20t) - \frac{4}{9}(18t^2 - 36t + 72)$

29. $\frac{1}{2}(2p^2 - 12p - 10) - \frac{1}{3}(18p^2 - 9) + \frac{2}{5}(15p^2 - 30p)$

30. $\frac{1}{4}(16p^2 + 24p - 4) - \frac{1}{2}(22p^2 - 12) + \frac{3}{7}(42p^2 - 35p)$

[Example 5] 31. $5(u^3 - 2u^2 + u - 1) + 3(3u^2 - u^3 + 2 - 2u)$

32. $-2(5u - u^3 + 7 - u^2) + 9(u^3 + 2 - u^2 + u)$

33. $x(x^2 - 5x + 4) - x^2(2x - 7)$

34. $x^2(5x - 3) - x(x^2 + 8x - 3)$

35. $7y(2y^2 - 5) + 4y(3y - 8) - 2(-32y + 7y^3 + 6y^2)$

36. $-5y(7y + 6) + 9y(y^2 - 3y) - 3(3y^3 - 10y - 20y^2)$

37. $3(2t^2 + 4t - 1) + 2t(13t - 3) - 5t(t^2 + 9t - 8)$

38. $7(5t^2 + 2t + 3) - 3t(t + 7) + 4t(3t^2 - 6t + 1)$

39. $4(m^2 - 3mn - 2n^2) - 3m(m - 3n) + 5n(2n + m)$

40. $6n(5m - n) + m(10m - 3n) - 4(3m^2 + 7mn - 2n^2)$

In exercises **41–50**, evaluate each expression for the given values.

[Example 6] 41. $m^2 + 3m - 5$
a. $m = -2$ **b.** $m = 5$

42. $m^2 - 4m + 3$
a. $m = -3$ **b.** $m = 7$

43. $16p^3 - 10p - 13$

a. $p = 2$ **b.** $p = \frac{-1}{2}$

44. $27p^3 + 6p - 19$

a. $p = 3$ **b.** $p = \frac{-1}{3}$

45. $7u^2 - 3uv - 2v^2$, for $u = -2$ and $v = 5$

46. $4u^2 + uv - 3v^2$, for $u = 6$ and $v = -7$

47. $\frac{x^2}{2} + \frac{3xy}{10} - \frac{3y^2}{5}$, for $x = 6$ and $y = -15$

48. $\frac{3x^2}{4} - \frac{2xy}{28} + \frac{9y^2}{7}$, for $x = -8$ and $y = -14$

49. $\frac{8m^2 - 48mn + 9n^2}{4m^2 + 42mn + 18n^2}$, for $m = \frac{1}{2}$ and $n = \frac{1}{3}$

50. $\frac{48m^2 + 24mn - 54n^2}{16m^2 - 48mn + 72n^2}$, for $m = \frac{-3}{2}$ and $n = \frac{-2}{3}$

SECTION 1-3. Supplementary Exercises

In exercises **1–4**, find the distance d given the rate r and the time t, and $d = rt$.

1. A rate of 43 mph and a time of 3 hours

2. A rate of 67 mph and a time of 5 hours

3. A time of 15 minutes and a rate of 600 mph

4. A time of 45 minutes and a rate of 440 mph

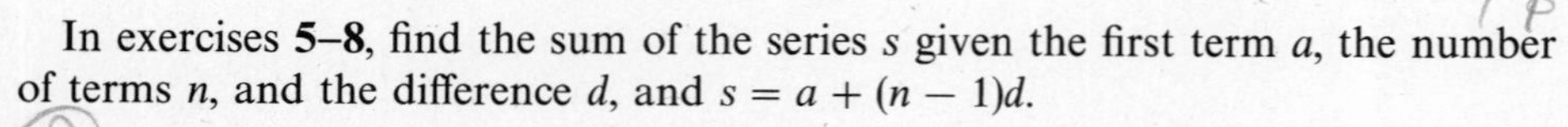

In exercises **5–8**, find the sum of the series s given the first term a, the number of terms n, and the difference d, and $s = a + (n - 1)d$.

5. The first term is 1, the number of terms is 50, and the difference is 2.

6. The first term is 4, the number of terms is 20, and the difference is 5.

7. The difference is $\frac{1}{2}$, the first term is 45, and the number of terms is 11.

8. The difference is $\frac{2}{3}$, the first term is 12, and the number of terms is 22.

In exercises **9–12**, find the total resistance (measured in ohms) in a parallel circuit R_T given the first resistance R_1 and the second resistance R_2, and $\frac{1}{R_1} + \frac{1}{R_2} = \frac{1}{R_T}$.

9. $R_1 = 125$ ohms, and $R_2 = 250$ ohms

10. $R_1 = 330$ ohms, and $R_2 = 660$ ohms

11. The first resistance is $\frac{3}{2}$ ohms, and the second resistance is $\frac{9}{7}$ ohms.

12. The first resistance is $1\frac{3}{4}$ ohms, and the second resistance is $2\frac{1}{8}$ ohms.

In exercises **13–16**, find the discriminant D given the values of a, b, and c, and $D = b^2 - 4ac$.

13. $a = 7$, $b = 4$, and $c = -9$

14. $a = 5$, $b = 2$, and $c = -8$

15. $a = 10$, $b = 0$, and $c = 13$

16. $a = 3$, $b = -7$, and $c = 15$

The DK Manufacturing Company measures one type of production in number of tons produced. Over the next planning cycle process A will produce $-t^2 + 16t - 54$ tons, where t is a particular week. Process B will produce $-t^2 + 2t + 4$ tons and process C will produce $30t^2$ tons.

In exercises **17–20**, find an algebraic expression for:

17. The output of processes A and B

18. The output of processes A, B, and C

19. The output of process A and two outputs of process B

20. The output of three A processes and two B processes

In exercises **21–26**, find an algebraic expression for the perimeter of each figure.

21.

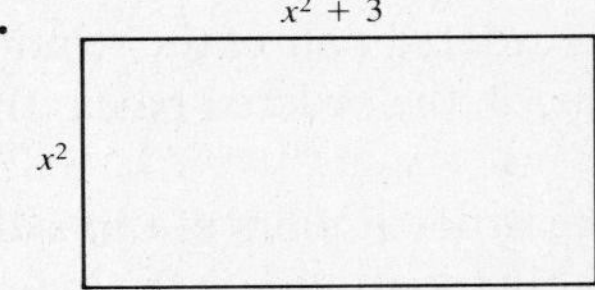

22.

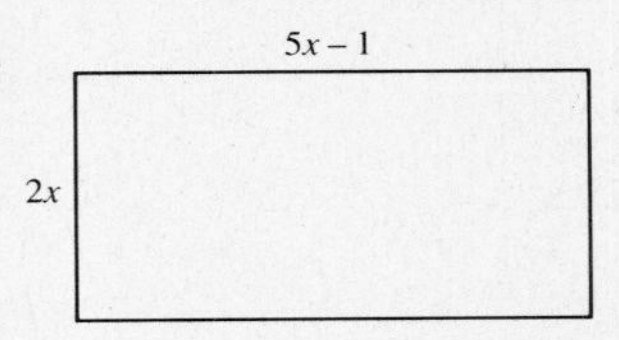

23.

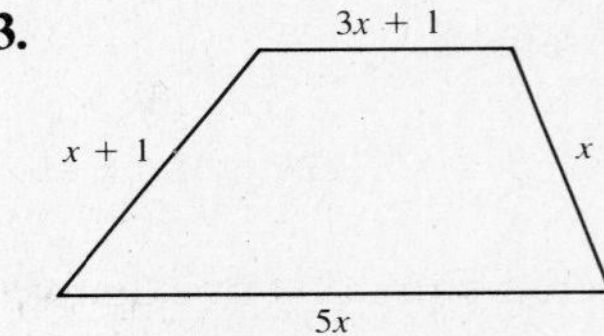

24.

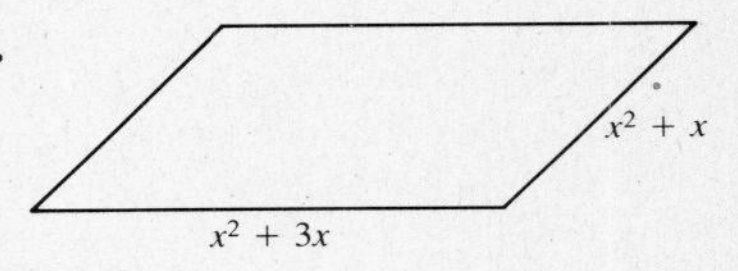

25.

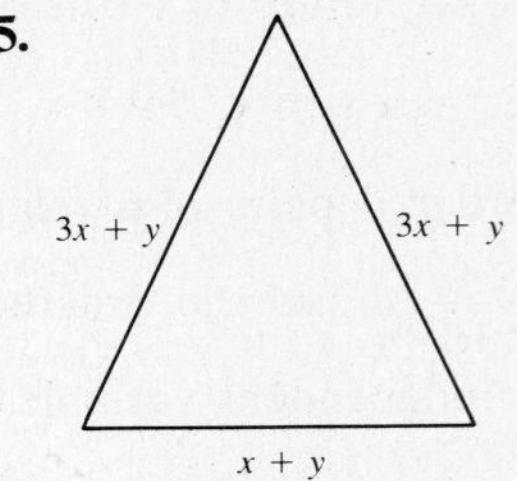

26.

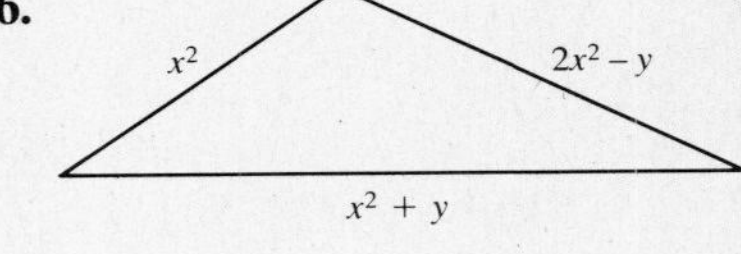

SECTION 1-4. A Review of Linear Relations and Functions

1. A definition of a relation
2. A definition of a function
3. The rectangular coordinate system
4. The linear function in x
5. A definition of the slope of a line
6. A definition of the y-intercept

In this section, we will first review some of the general concepts related to relations and functions. We will then review the linear function in detail.

A Definition of a Relation

Definition 1.6 is one of the ways in which a mathematical relation can be defined.

> **Definition 1.6. A mathematical relation**
> A **relation** is a set of ordered pairs of real numbers.

Examples **a** and **b** are two relations represented by r and s.

a. $r = \{(-6, -14), (-4, -11), (-2, -8), (0, -5), (2, -2), (4, 1), (6, 4)\}$

b. $s = \{(-3, 6), (-2, 0), (-1, -4), (0, -6), (1, -6), (2, -4), (3, 0)\}$

Many relations are specified by equations or inequalities in two variables. The first number in an ordered pair of the relation is a value of the **independent variable.** The second number in the ordered pair is the corresponding value of the **dependent variable.**

To illustrate, the ordered pairs of r in example **a** can be obtained by using equation (1) below. In this equation, x is the independent variable and y is the dependent variable. If x is replaced by -6, -4, -2, 0, 2, 4, and 6, then the corresponding values of y are obtained:

(1) $y = \frac{3}{2}x - 5$ For example, if $x = -6$, then $y = \frac{3}{2}(-6) - 5 = -14$.

Similarly, the ordered pairs of s in example b can be obtained by using equation (2) below. If x (the independent variable) is replaced by -3, -2, -1, 0, 1, 2, and 3, then the corresponding values of y (the dependent variable) are obtained:

(2) $y = x^2 - x - 6$ If $x = 3$, then $y = 3^2 - 3 - 6 = 0$.

Example 1. The ordered pairs of a relation are determined by the equation: $y = \frac{2x}{x+3}$. Find the ordered pairs of the relation for each value of the independent variable.

a. $x = -4$ **b.** $x = 3$

Solution. **a.** If $x = -4$, then $y = \frac{2(-4)}{-4+3} = \frac{-8}{-1} = 8$. Thus, $(-4, 8)$ is an ordered pair of the relation.

b. If $x = 3$, then $y = \frac{2(3)}{3+3} = \frac{6}{6} = 1$. Thus, $(3, 1)$ is an ordered pair of the relation.

The first and second components of the ordered pairs of a relation are separated into two sets of numbers.

> **Definition 1.7. The domain and range of a relation**
> The **domain** is the set of all first components of a relation. The **range** is the set of all second components of a relation.

Example 2. Given: $r = \{(-6, -14), (-4, -11), (-2, -8), (0, -5), (2, -2)\}$

a. List the numbers in the domain.

b. List the numbers in the range.

Solution. **a.** The domain is the set of first components:

Domain $= \{-6, -4, -2, 0, 2\}$

b. The range is the set of second components:

Range $= \{-14, -11, -8, -5, -2\}$

A Definition of a Function

The following definition identifies a special type of relation:

Definition 1.8. Function
A **function** is a relation in which each domain element is paired with exactly one range element.

Example 3. Determine whether each relation is a function.

a. $r = \{(1, 16), (2, 10), (3, 8), (4, 8)\}$

b. $s = \{(3, -2), (2, 1), (2, -5), (1, 2), (1, -6)\}$

Solution. **a.** Each domain element (1, 2, 3, and 4) is paired with exactly one range element. Thus, r is a function.
b. The domain element 2 is paired with 1 and -5. Also, the domain element 1 is paired with 2 and -6. Therefore, s is not a function.

The Rectangular Coordinate System

Figure 1-9 on page 36 shows a coordinate system on which relations and functions can be graphed. The coordinate system is usually called **rectangular,** but it is also called **Cartesian,** in honor of the French mathematician René Descartes (1596–1650). As shown, the number lines that form the system are labeled x and y, because many relations and functions defined by equations are stated in terms of x and y. Some of the more important features of the coordinate system are:

1. A **horizontal number line,** labeled the ***x*-axis**
2. A **vertical number line,** labeled the ***y*-axis**
3. The **origin,** the point of intersection of the axes
4. Four **quadrants,** labeled I, II, III, and IV, as shown
5. Every point in the plane of the axes has an ordered pair of numbers that is the **coordinate of the point.**
6. The first component of the coordinate of a point, called the **abscissa,** is the coordinate of the point on the x-axis that is directly above or below the point. For example, $P_1(3, 4)$ in the figure is *above* the point with coordinate 3 on the x-axis.

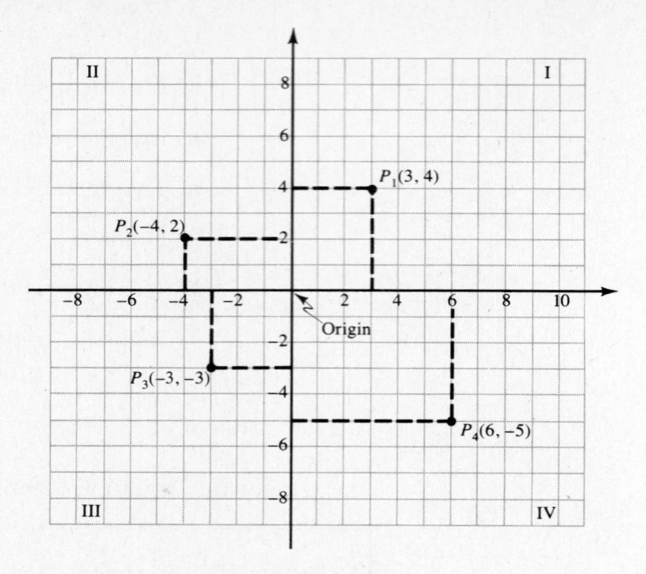

Figure 1-9. A rectangular coordinate system.

7. The second component of the coordinate of a point, called the **ordinate,** is the coordinate of the point on the y-axis that is directly to the right or left of the point. For example, $P_2(-4, 2)$ in the figure is *to the left* of the point with coordinate 2 on the y-axis.

8. Four points are plotted in Figure 1-9.
   a. $P_1(3, 4)$ is 3 units to the right and 4 units up from the origin.
   b. $P_2(-4, 2)$ is 4 units to the left and 2 units up from the origin.
   c. $P_3(-3, -3)$ is 3 units to the left and 3 units down from the origin.
   d. $P_4(6, -5)$ is 6 units to the right and 5 units down from the origin.

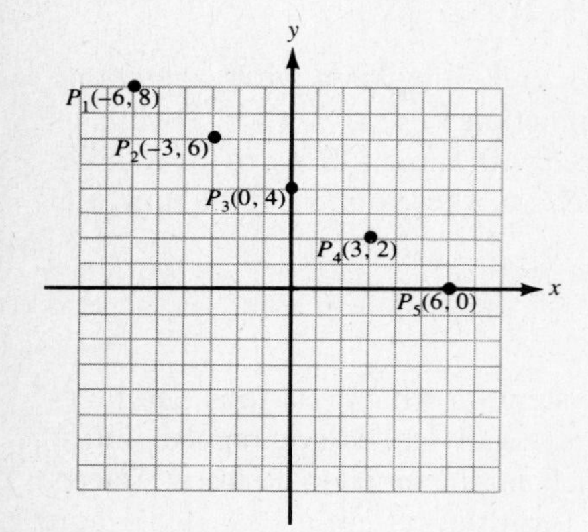

Figure 1-10. A graph of f.

Example 4. Graph $f = \{(-6, 8), (-3, 6), (0, 4), (3, 2), (6, 0)\}$.

Solution. **Discussion.** The f stands for "function", which in this example consists of five ordered pairs. To graph f means to locate the points in a coordinate system with these ordered pairs as coordinates. In Figure 1-10, the points P_1 through P_5 are a graph of f.

The Linear Function in x

One of the most important functions is the linear function.

Definition 1.9. A linear function
A **linear function** in x can be defined by an equation written in the form

$$y = mx + b$$

where m and b are real numbers. The **domain** is the set of real numbers, which can be written "all x".

Example 5. Graph $y = \frac{3}{4}x - 1$.

Solution. **Discussion.** The given equation defines a linear function in which $m = \frac{3}{4}$ and $b = -1$. A graph of any linear function is a line. To locate a line we use three points, two to determine the line and a third to check it. Using three arbitrarily selected values of x, we compute the corresponding values of y. These **ordered-pair solutions** of the equation are then plotted and a line is drawn through them.

x	$y = \frac{3}{4}x - 1$
-4	$y = \frac{3}{4}(-4) - 1 = -4$
0	$y = \frac{3}{4}(0) - 1 = -1$
4	$y = \frac{3}{4}(4) - 1 = 2$

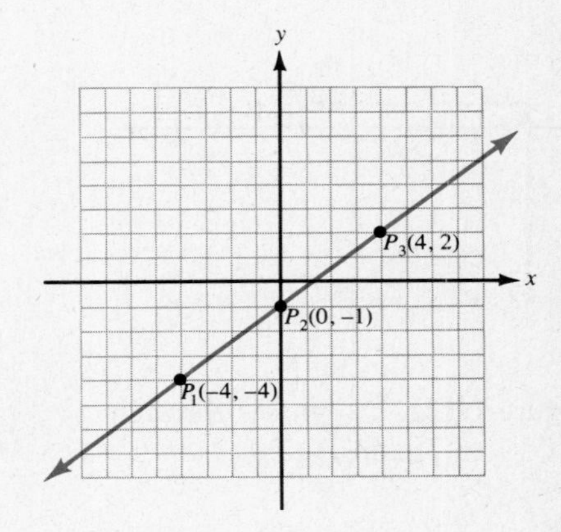

Figure 1-11. A graph of $y = \frac{3}{4}x - 1$.

The points $P_1(-4, -4)$, $P_2(0, -1)$, and $P_3(4, 2)$ are plotted in Figure 1-11. The line through these points is a graph of the equation. This means that *the coordinates of every point on the line are a solution of the equation that defines the function.* Using set notation and f for the function, we can write:

$$f = \left\{(x, y) \middle| y = \frac{3}{4}x - 1\right\}$$

This is read as "f is the function consisting of the set of ordered pairs of the form (x, y), such that $y = \frac{3}{4}x - 1$".

A Definition of the Slope of a Line

If $\mathscr{L}$ is not a vertical line, then $\mathscr{L}$ inclines upward, or downward, or is horizontal as the line is traced from left to right. In Figure 1-12 on page 38, $\mathscr{L}_1$, $\mathscr{L}_2$, and $\mathscr{L}_3$ are, respectively, three such lines. The amount and direction of inclination is measured by the slope of the line.

Definition 1.10. The slope of a line
If $\mathscr{L}$ is a line with $P_1(x_1, y_1)$ and $P_2(x_2, y_2)$ on $\mathscr{L}$, then the slope of $\mathscr{L}$ is m, where

$$m = \frac{y_2 - y_1}{x_2 - x_1}, \text{ that is, } \frac{\text{difference in } y\text{-coordinates}}{\text{difference in } x\text{-coordinates}}, \text{ or } \frac{\text{rise}}{\text{run}}$$

provided $x_1 \neq x_2$.

For the three lines in Figure 1-12:

Line	$y_2 - y_1$ = rise	$x_2 - x_1$ = run	$m = \dfrac{y_2 - y_1}{x_2 - x_1}$
$\mathscr{L}_1$	$1 - (-7) = 8$	$2 - (-7) = 9$	$m = \dfrac{8}{9}$
$\mathscr{L}_2$	$-3 - 6 = -9$	$4 - (-2) = 6$	$m = \dfrac{-9}{6} = \dfrac{-3}{2}$
$\mathscr{L}_3$	$3 - 3 = 0$	$4 - (-6) = 10$	$m = \dfrac{0}{10} = 0$

Figure 1-12a. $\mathscr{L}_1$ passes through $P_1(-7, -7)$ and $P_2(2, 1)$.

Example 6. Find the slope of the line that passes through each pair of points.

a. $P_1(-4, 5)$ and $P_2(5, -4)$

b. $P_1(-8, -3)$ and $P_2(1, 3)$

c. $P_1(6, 9)$ and $P_2(6, -8)$

Solution.

a. $m = \dfrac{-4 - 5}{5 - (-4)}$ $\qquad m = \dfrac{y_2 - y_1}{x_2 - x_1}$

$= \dfrac{-9}{9} = -1$ $\qquad$ The line drops one unit for each one-unit run to the right.

b. $m = \dfrac{3 - (-3)}{1 - (-8)}$ $\qquad m = \dfrac{\text{rise}}{\text{run}}$

$= \dfrac{6}{9} = \dfrac{2}{3}$ $\qquad$ The line rises two units for each three-unit run to the right.

c. $m = \dfrac{-8 - 9}{6 - 6}$ $\qquad$ The x-coordinates are the same.

$= \dfrac{-17}{0}$ $\qquad$ Division by 0 is undefined.

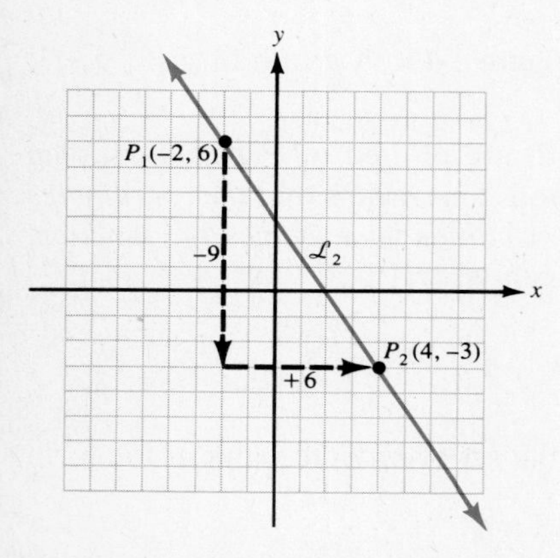

Figure 1-12b. $\mathscr{L}_2$ passes through $P_1(-2, 6)$ and $P_2(4, -3)$.

In example **c**, the two points lie on a vertical line. The attempt to compute m yielded a ratio with a zero denominator. Division by 0 is undefined. As a consequence, we say that *slope is undefined for vertical lines.*

The slope of a line is valuable because it provides directions on how to get from one point to another point on a line. To illustrate, consider again the slopes of $\mathscr{L}_1$, $\mathscr{L}_2$, and $\mathscr{L}_3$ in Figure 1-12:

Figure 1-12c. $\mathscr{L}_3$ passes through $P_1(-6, 3)$ and $P_2(4, 3)$.

From any point on	$m = \dfrac{\text{vertical rise}}{\text{horizontal run}}$	To arrive at a different point on the line:
$\mathscr{L}_1$	$m = \dfrac{8}{9}$	First move 8 units up (the 8 is positive), then move 9 units to the right (the 9 is positive).
$\mathscr{L}_2$	$m = \dfrac{-3}{2}$	First move 3 units down (the 3 is negative), then move 2 units to the right (the 2 is positive).
$\mathscr{L}_3$	$m = 0$	Do not move vertically upward or downward, but move only to the left or right.

Example 7. Graph the line that passes through $P(-3, -5)$ and has slope $\frac{7}{5}$.

Solution. **Discussion.** Plot $P(-3, -5)$, then use the slope to find a second point on the line. Draw the line through these points.

In Figure 1-13, $P(-3, -5)$ is plotted. For the given line, $m = \frac{7}{5}$. Begin at P and count up 7 units. Now turn right and count 5 units, ending at $Q(2, 2)$. Since P and Q are two different points on the line, a line through them is the desired graph.

Figure 1-13.

A Definition of the y-Intercept

A graph of every linear function intersects the y-axis at some point. The y-coordinate of this point is called the y-intercept.

> **Definition 1.11. The y-intercept of a linear function**
> If $P(0, b)$ is a point on a graph of $y = mx + b$, then b is the **y-intercept.**

When a linear function is written in the $y = mx + b$ form, then:

1. m, the coefficient of x, is the slope of a graph of the function.
2. b, the constant in the equation, is the y-intercept.

Example 8. Given: $2x + 5y = 25$

a. Write the equation in the $y = mx + b$ form.

b. Identify the slope.

c. Identify the y-intercept.

d. Graph the function.

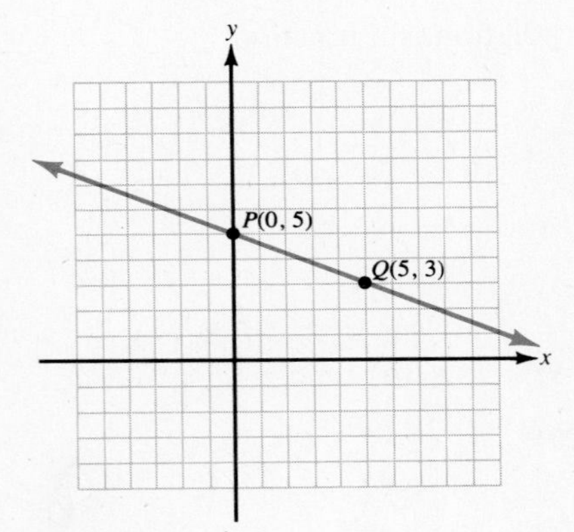

Figure 1-14. A graph of $2x + 5y = 25$.

Solution. a. $2x + 5y = 25$ — The given equation

$5y = -2x + 25$ — Subtract $2x$ from both sides.

$y = \frac{-2}{5}x + 5$ — Divide each term by 5.

b. The slope is $\frac{-2}{5}$. — $m = \frac{-2}{5}$

c. The y-intercept is 5. — $b = 5$

d. In Figure 1-14, plot $P(0, 5)$. From P count down 2 units, then to the right 5 units to $Q(5, 3)$. The line through P and Q is a graph of the function.

SECTION 1-4. Practice Exercises

In exercises **1–8**, find the ordered pairs of the relation for each value of the independent variable.

[Example 1]

1. $y = 6x - 2$
   a. $x = 5$ b. $x = -8$

2. $y = -3x + 7$
   a. $x = 4$ b. $x = -9$

3. $y = x^2 - 5x$
 a. $x = 7$ b. $x = -3$

4. $y = x^2 + 6x$
 a. $x = -6$ b. $x = 5$

5. $y = \frac{x}{2x + 3}$
 a. $x = 0$ b. $x = -3$

6. $y = \frac{5x}{x - 7}$
 a. $x = 6$ b. $x = 0$

7. $y = \frac{x^2 + 4}{x^2 - 3x}$
 a. $x = 4$ b. $x = 2$

8. $y = \frac{x^2 + 5x}{x^2 - 10}$
 a. $x = -5$ b. $x = 3$

In exercises **9–16**, for each relation:

a. List the numbers in the domain.

b. List the numbers in the range.

[Example 2] **9.** $r = \{(1, -5), (-3, 2), (7, 0), (-11, 3)\}$

10. $r = \{(4, 2), (9, 5), (-6, -3), (-12, 4)\}$

11. $r = \{(-3, 0), (7, -3), (5, 0), (21, -3)\}$

12. $r = \{(8, -4), (9, 5), (-6, -4), (0, 5)\}$

13. $r = \{(0, 3), (0, 7), (8, 3), (2, 7), (2, 5)\}$

14. $r = \{(11, 5), (-9, -4), (11, -4), (-9, 5), (2, 5)\}$

15. $r = \{(-7, 3), (-7, -8), (-7, 12), (-7, -1), (-7, 0)\}$

16. $r = \{(8, 1), (8, 0), (8, -12), (8, -5), (8, 11)\}$

In exercises **17–24**, determine whether each relation is a function.

[Example 3] **17.** $r = \{(6, 3), (9, 2), (4, -1), (-10, 63)\}$

18. $r = \{(-4, -2), (5, 13), (-18, 4), (-2, 1)\}$

19. $s = \{(3, 2), (-6, 2), (4, 2)\}$

20. $s = \{(-9, 7), (-11, 7), (8, 7), (3, 7)\}$

21. $t = \{(21, 12), (-36, 19), (-18, -5), (-36, 9)\}$

22. $t = \{(-42, 19), (25, -33), (16, 41), (-42, 29)\}$

23. $r = \{(24, 6), (-8, -15), (17, -12), (-8, 0)\}$

24. $r = \{(35, 1), (29, 33), (52, 21), (19, 37), (35, 6)\}$

In exercises **25–32**, graph each function.

[Example 4] **25.** $f = \{(-7, -6), (-5, -2), (-3, 2), (-2, 4), (0, 8)\}$

26. $f = \{(-2, -6), (-1, -3), (0, 0), (1, 3), (2, 6)\}$

27. $g = \{(-7, -4), (-5, -1), (-3, 2), (-1, 5), (1, 8)\}$

28. $g = \{(-6, -5), (-4, -2), (-2, 1), (0, 4), (2, 7)\}$

29. $h = \{(-7, 4), (-4, 3), (-1, 2), (2, 1), (5, 0), (8, -1)\}$

30. $h = \{(-8, 5), (-5, 2), (-2, -1), (1, -4), (4, -7)\}$

31. $f = \{(-6, 4), (-3, 4), (0, 4), (3, 4), (7, 4)\}$

32. $f = \{(-8, -2), (-5, -2), (1, -2), (4, -2), (7, -2)\}$

In exercises **33–40**, graph each function.

[Example 5] **33.** $y = -4x$ **34.** $y = 3x$

35. $y = 2x - 3$ **36.** $y = -3x + 4$

37. $y = \frac{-2}{3}x - 1$ **38.** $y = \frac{3}{5}x + 2$

39. $y = \frac{1}{2}x + 5$ **40.** $y = \frac{-1}{4}x - 3$

In exercises **41–50**, find the slope of the line that passes through each pair of points. If P_1 and P_2 lie on a vertical line, write "undefined".

[Example 6] **41.** $P_1(2, -3), P_2(4, 7)$ **42.** $P_1(-4, 2), P_2(-1, 8)$

43. $P_1(4, -1), P_2(-2, 8)$ **44.** $P_1(8, -4), P_2(4, 6)$

45. $P_1(-5, 2), P_2(-5, 6)$ **46.** $P_1(4, 8), P_2(4, -6)$

47. $P_1(-7, 7), P_2(-5, 5)$ **48.** $P_1(9, -9), P_2(-6, 6)$

49. $P_1(6, 2), P_2(-3, 2)$ **50.** $P_1(-8, -1), P_2(2, -1)$

In exercises **51–58**, graph the line that passes through each point and has the given slope.

[Example 7] **51.** slope $\frac{1}{2}$, $P_1(-5, -1)$ **52.** slope $\frac{1}{4}$, $P_1(-6, -4)$

53. slope $\frac{-2}{3}$, $P_1(-4, 3)$ **54.** slope $\frac{-4}{5}$, $P_1(-7, 4)$

55. slope 0, $P_1(1, -3)$ **56.** slope 0, $P_1(4, 2)$

57. slope -3, $P_1(4, -1)$ **58.** slope -1, $P_1(6, -3)$

In exercises **59–66**:

a. Write each equation in the $y = mx + b$ form.

b. Identify the slope.

c. Identify the y-intercept.

d. Graph the function.

[Example 8] **59.** $3x + y = 6$ **60.** $2x + y = -5$

61. $x - 2y = 8$ **62.** $x - 3y = 18$

63. $4x + 3y = 3$ **64.** $6x + 5y = 10$

65. $7x - 9y = 27$ **66.** $4x - 5y = 20$

SECTION 1-4. Supplementary Exercises

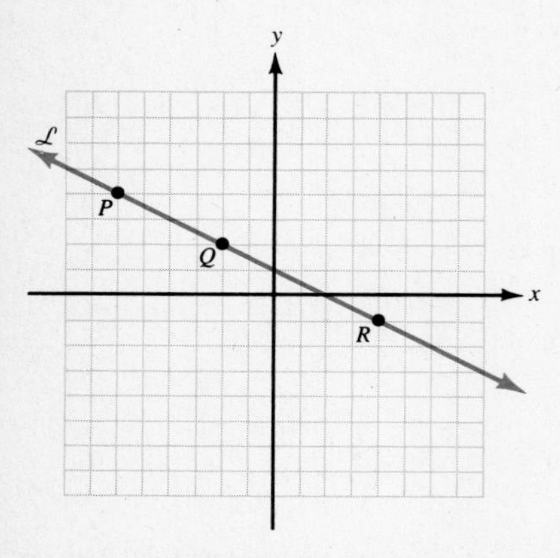

Figure 1-15. Points P, Q, and R on line $\mathscr{L}$.

In exercises **1–22**, refer to line $\mathscr{L}$ in Figure 1-15.

1. Identify the coordinates of points P, Q, and R.

2. Find the slope of $\mathscr{L}$ using the coordinates of P and Q.

3. Find the slope of $\mathscr{L}$ using the coordinates of P and R.

4. Comparing the answers obtained to exercises **2** and **3**, what conclusions can you make regarding the slope of any nonvertical line?

5. Find the x-intercept.

6. Find the y-intercept.

7. Find the value of y when $x = 6$.

8. Find the value of y when $x = -4$.

9. Find the value of y when $x = 1$.

10. Find the value of y when $x = -5$.

11. Find the value of x when $y = 3$.

12. Find the value of x when $y = -1$.

13. For what values of x is y positive?

14. For what values of x is y negative?

15. For what values of y is x positive?

16. For what values of y is x negative?

17. For what value(s) of x is y zero?

18. For what value(s) of y is x zero?

19. For what values of x is $2 < y < 4$?

20. For what values of x is $-3 < y < 1$?

21. For what values of x is $y < 3$?

22. For what values of x is $y > 2$?

SECTION 1-5. Other Functions

KEY TOPICS IN THIS SECTION

1. The definition of a quadratic function in x
2. The graph of a quadratic function in x
3. A procedure for graphing a quadratic function in x
4. The definition of an absolute value function in x
5. The graph of an absolute value function in x

6. A procedure for graphing an absolute value function in x

7. Graphs of functions defined by $y = ax^3$

In this section we will review the forms of equations that define some other frequently studied functions. We will also look at the shapes of the graphs of these functions.

The Definition of a Quadratic Function in x

If a, b, and c are real numbers and $a \neq 0$, then a quadratic equation in x can be written in the form:

$$ax^2 + bx + c = 0$$

If 0 is replaced by y, then the equation defines a quadratic function in x.

Definition 1.12. The standard form of a quadratic function in x
If a, b, and c are real numbers ($a \neq 0$), then an equation that can be written in the form

$$y = ax^2 + bx + c$$

defines a **quadratic function.** The domain is the set of real numbers, written "all x".

The Graph of a Quadratic Function in x

Figure 1-16 and Figure 1-17 contain graphs of $y = 2x^2 - 6$ and $y = 4 - 2x - x^2$, respectively. For $y = 2x^2 - 6$: $a = 2$, $b = 0$, and $c = -6$. For $y = 4 - 2x - x^2$: $a = -1$, $b = -2$, and $c = 4$.

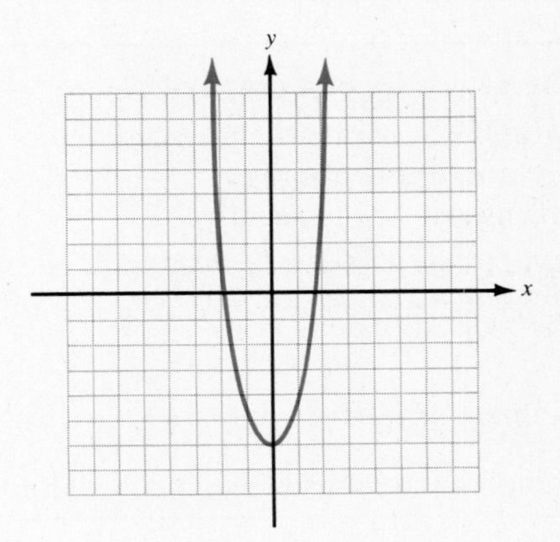

Figure 1-16. A graph of $y = 2x^2 - 6$.

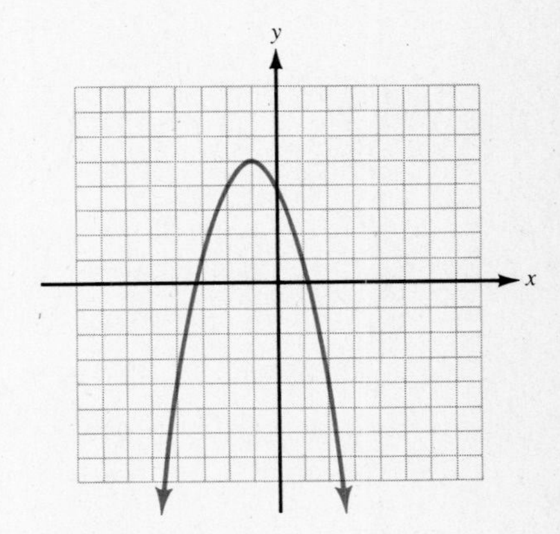

Figure 1-17. A graph of $y = 4 - 2x - x^2$.

Notice that the graphs of both functions are cup-shaped curves. In Figure 1-16, the cup is turned upright because the a in the equation is positive ($2 > 0$). In Figure 1-17, the cup is turned downward because the a in the equation is negative ($-1 < 0$). These smooth curves are called **parabolas.**

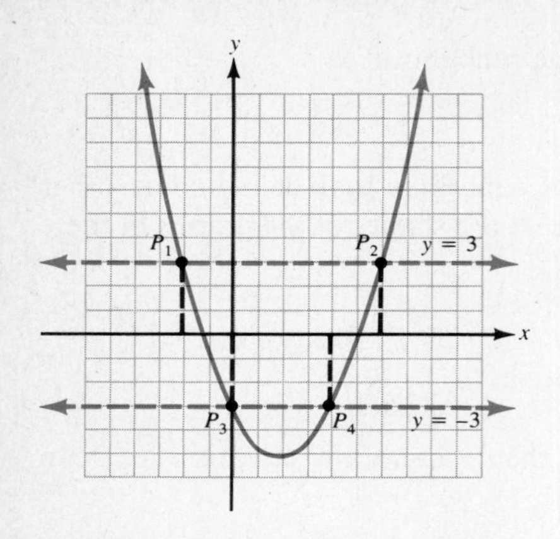

Figure 1-18. A graph of $y = \frac{1}{2}x^2 - 2x - 3$.

Example 1. Figure 1-18 contains a graph of $y = \frac{1}{2}x^2 - 2x - 3$. Use the graph to find the values of x when:

a. $y = 3$ **b.** $y = -3$

Solution. **Discussion.** Graphs of $y = 3$ and $y = -3$ are horizontal lines, shown as dotted lines in Figure 1-18. Each of these lines intersects the parabola at two points. The x-coordinates of these points are the values of x for the given values of y.

a. The line defined by $y = 3$ intersects the parabola at P_1 and P_2. The x-coordinates of P_1 and P_2 are -2 and 6, respectively. Thus, $x = -2$ and 6 when $y = 3$.

Checking $x = -2$: $3 = \frac{1}{2}(-2)^2 - 2(-2) - 3$

$3 = 2 - (-4) - 3$

$3 = 6 - 3$, *true*

Checking $x = 6$: $3 = \frac{1}{2}(6^2) - 2(6) - 3$

$3 = 18 - 12 - 3$

$3 = 6 - 3$, *true*

b. The line defined by $y = -3$ intersects the parabola at P_3 and P_4. The x-coordinates of P_3 and P_4 are 0 and 4, respectively. Thus, $x = 0$ and 4 when $y = -3$.

A Procedure for Graphing a Quadratic Function in x

A graph of a quadratic function has three characteristics that can be utilized to simplify the task of graphing such a function.

Three characteristics of a graph of $y = ax^2 + bx + c$

1. The graph is a **parabola** that opens:
   a. Upward, if a is positive
   b. Downward, if a is negative
2. The graph has a **vertex** (a turning point of the graph):
   a. The x-coordinate is $\frac{-b}{2a}$.
   b. Evaluate $ax^2 + bx + c$ for $x = \frac{-b}{2a}$ to get the y-coordinate.
3. The graph has an **axis of symmetry** that is a vertical line through the vertex.

Example 2. Graph $y = x^2 - 6x + 4$.

Solution. For the given quadratic function: $a = 1$, $b = -6$, and $c = 4$.

Thus, $\frac{-b}{2a}$ becomes $\frac{-(-6)}{2(1)} = 3$ The x-coordinate of the vertex

$3^2 - 6(3) + 4 = -5$ The y-coordinate of the vertex

The following table of values of x and y provides the points for the graph in Figure 1-19:

x	$y = x^2 - 6x + 4$
0	4
1	-1
2	-4
3	-5 (*The vertex*)
4	-4
5	-1
6	4

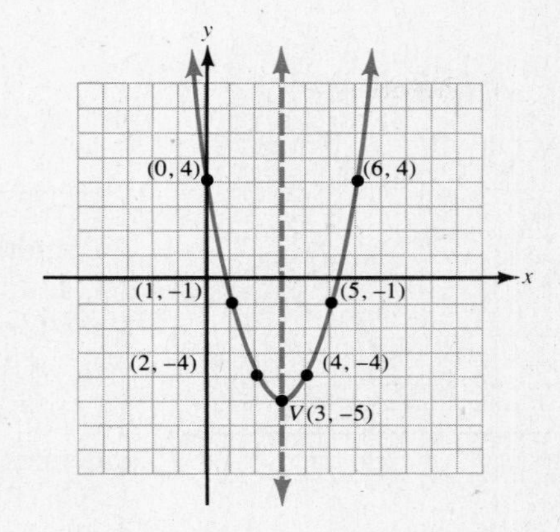

Figure 1-19. A graph of $y = x^2 - 6x + 4$.

Notice that the parabola opens upward, which is expected since $a = 1$, a positive number. If the page on which the parabola is drawn is folded along the vertical line through $V(3, -5)$, then the left and right branches of the parabola would coincide.

Example 3. Graph $y = 8x - 2x^2$.

Solution. For the given quadratic function: $a = -2$, $b = 8$, and $c = 0$.

Thus, $\dfrac{-b}{2a}$ becomes $\dfrac{-8}{2(-2)} = 2$. The x-coordinate of the vertex

$8(2) - 2(2)^2 = 8$ The y-coordinate of the vertex

The following table of values for x and y provides the points for the graph in Figure 1-20:

x	$y = 8x - 2x^2$
0	0
1	6
2	8 (*The vertex*)
3	6
4	0

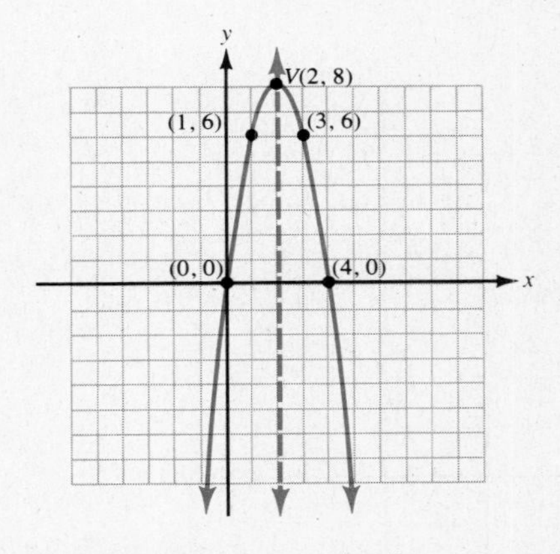

Figure 1-20. A graph of $y = 8x - 2x^2$.

Since $a = -2$, a negative number, the parabola opens downward.

The Definition of an Absolute Value Function in x

If a, b, and c are real numbers ($a \neq 0$), then an absolute value equation in x can be written in the form:

$$a|x - b| = c$$

If c is replaced by y, then the equation defines an absolute value function in x.

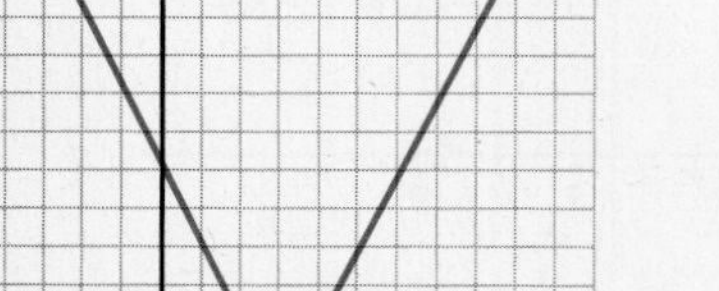

Figure 1-21. A graph of $y = 2|x - 3|$.

> **Definition 1.13. An absolute value function in x**
> If a and b are real numbers ($a \neq 0$), then an equation that can be written in the form
>
> $$y = a|x - b|$$
>
> defines an **absolute value function.** The domain is the set of real numbers, written "all x".

The Graph of an Absolute Value Function in x

Figure 1-21 and Figure 1-22 are graphs of $y = 2|x - 3|$ and $y = -|x + 1|$, respectively. For $y = 2|x - 3|$, $a = 2$ and $b = 3$. For $y = -|x + 1|$, $a = -1$ and $b = -1$, because $-|x + 1|$ can be written $-1 \cdot |x - (-1)|$.

Notice that the graphs of both functions are V-shaped. In Figure 1-21, the V is turned upright because the a in the equation is positive ($2 > 0$). In Figure 1-22, the V is turned downward because the a in the equation is negative ($-1 < 0$).

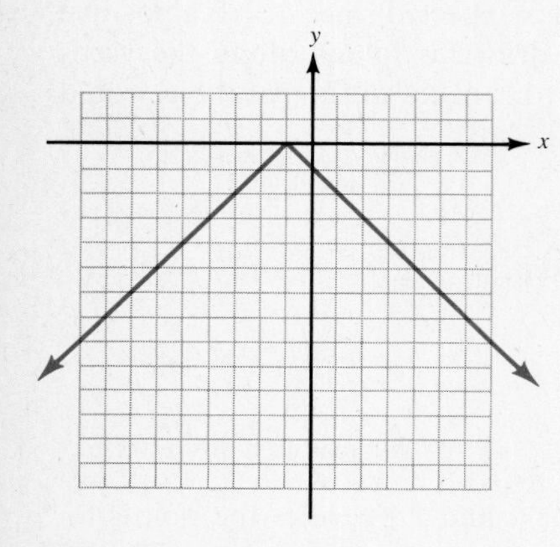

Figure 1-22. A graph of $y = -|x + 1|$.

Example 4. Figure 1-23 is a graph of $y = \frac{3}{2}|x + 2|$. Use the graph to find the values of x when:

a. $y = 9$ **b.** $y = 3$

Solution. **Discussion.** Graphs of $y = 9$ and $y = 3$ are horizontal lines, shown as dotted lines in Figure 1-23. Each of these lines intersects the V at two points. The x-coordinates of these points are the values of x for the given values of y.

a. The line defined by $y = 9$ intersects the V at P_1 and P_2. The x-coordinates of P_1 and P_2 are -8 and 4, respectively.

Checking $x = -8$: $9 = \frac{3}{2}|-8 + 2|$

$9 = \frac{3}{2}(6)$, *true*

Checking $x = 4$: $9 = \frac{3}{2}|4 + 2|$

$9 = \frac{3}{2}(6)$, *true*

Figure 1-23. A graph of $y = \frac{3}{2}|x + 2|$.

b. The line defined by $y = 3$ intersects the V at P_3 and P_4. The x-coordinates of P_3 and P_4 are -4 and 0, respectively. Thus, $x = -4$ and 0 when $y = 3$.

A Procedure for Graphing an Absolute Value Function in x

A graph of an absolute value function has three characteristics that can be utilized to simplify the task of graphing such a function.

Three characteristics of a graph of $y = a|x - b|$

1. The graph is a V that opens:

a. Upward, if a is positive
b. Downward, if a is negative

2. The graph has a **vertex** (the point of the V):

a. The x-coordinate is b.
b. The y-coordinate is 0.
} $V(b, 0)$ is the vertex.

3. The **slopes of the sides** of the V are a and $-a$.

Example 5. Graph $y = 2|x - 3|$.

Solution. For the given absolute value function, $a = 2$ and $b = 3$. The following table of values for x and y provides the points for the graph in Figure 1-24:

x	$y = 2\|x - 3\|$
-1	8
1	4
3	0 (*The vertex*)
5	4
7	8

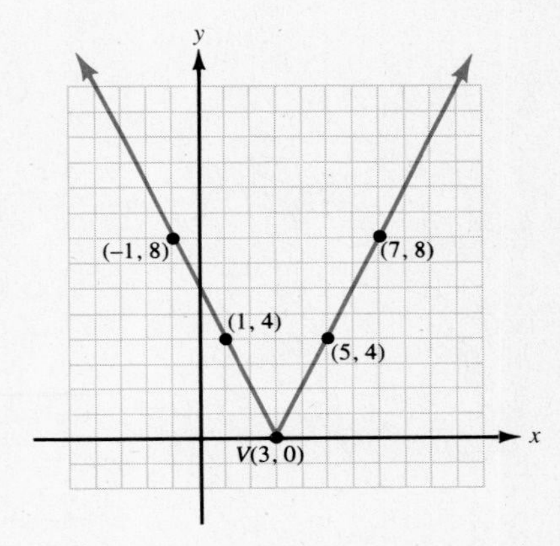

Figure 1-24. A graph of $y = 2|x - 3|$.

The vertex is the point $V(3, 0)$. The slopes of the sides of the V are 2 and -2. Also, since a is positive, the V opens upward.

Example 6. Graph $y = \frac{-2}{3}|x + 2|$.

Solution. For the given absolute value function, $a = \frac{-2}{3}$ and $b = -2$. [*Note:* $x + 2$ can be written $x - (-2)$]. The following table of values for

x and y provides the points for the graph in Figure 1-25:

x	$y = \frac{-2}{3}\lvert x + 2\rvert$
-8	-4
-5	-2
-2	0 (*The vertex*)
1	-2
4	-4

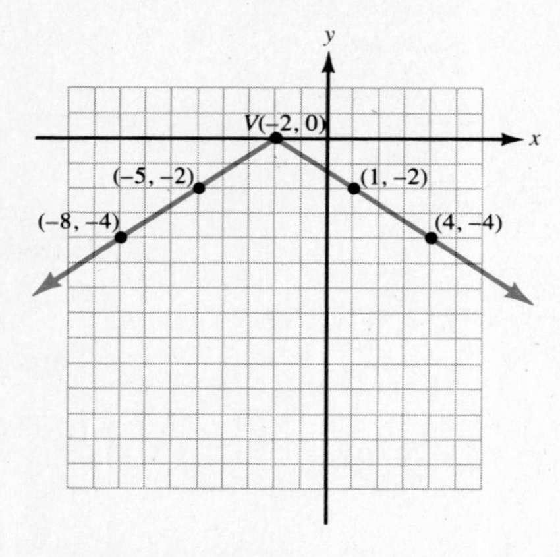

Figure 1-25. A graph of $y = \frac{-2}{3}\lvert x + 2\rvert$.

The vertex is the point $V(-2, 0)$. The slope of the sides of the V are $\frac{2}{3}$ and $\frac{-2}{3}$. Also, since a is negative, the V opens downward.

Graphs of Functions Defined by $y = ax^3$

Figure 1-26 and Figure 1-27 are graphs of $y = \frac{1}{2}x^3$ and $y = \frac{-3}{2}x^3$, respectively. In $y = \frac{1}{2}x^3$, the coefficient of x^3 is positive ($\frac{1}{2} > 0$). Therefore, the graph "rises" as the curve is traced from left to right. In $y = \frac{-3}{2}x^3$, the coefficient of x^3 is negative ($\frac{-3}{2} < 0$). Therefore, the graph "falls" as the curve is traced from left to right.

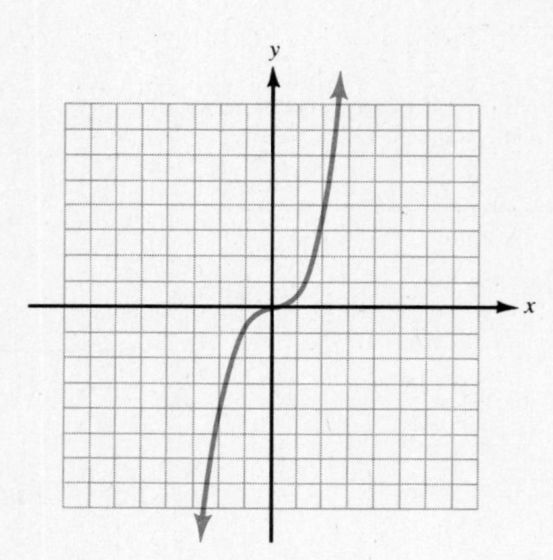

Figure 1-26. A graph of $y = \frac{1}{2}x^3$.

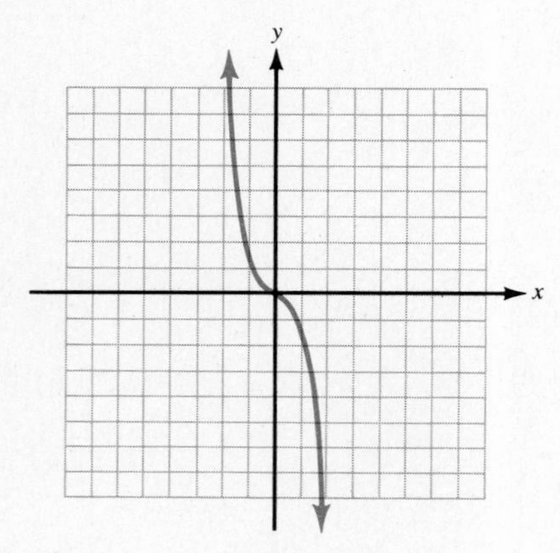

Figure 1-27. A graph of $y = \frac{-3}{2}x^3$.

Example 7. Figure 1-28 is a graph of $y = \frac{3}{4}x^3$. Use the graph to find the value of x when:

a. $y = 6$ **b.** $y = -6$

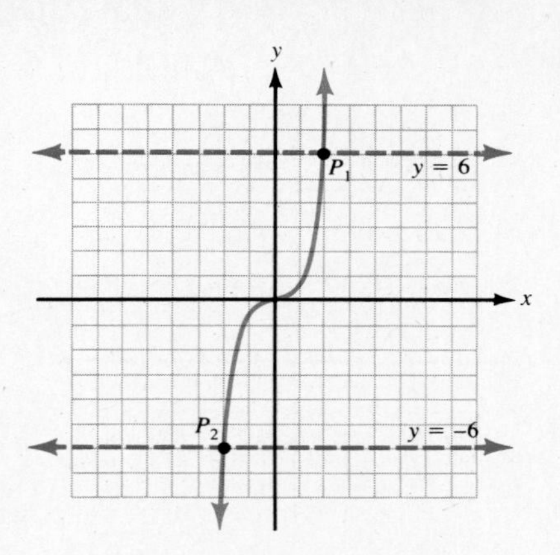

Figure 1-28. A graph of $y = \frac{3}{4}x^3$.

Solution.

Discussion. Graphs of $y = 6$ and $y = -6$ are horizontal lines, shown as dotted lines in Figure 1-28. Each of these lines intersects the curve at one point. The x-coordinates of these points are the values of x for the given values of y.

a. The line defined by $y = 6$ intersects the curve at P_1. The x-coordinate of P_1 is 2.

Checking $x = 2$: $6 = \frac{3}{4}(2)^3$

$6 = \frac{3}{4}(8)$, *true*

b. The line defined by $y = -6$ intersects the curve at P_2. The x-coordinate of P_2 is -2. Thus, $x = -2$ when $y = -6$.

SECTION 1-5. Practice Exercises

In exercises **1–4**, use the graph in Figure 1-29.

[Example 1] **1.** Find the value(s) of x when $y = 4$.

2. Find the value(s) of x when $y = 1$.

3. Find the value(s) of y when $x = 2$.

4. Find the value(s) of y when $x = 0$.

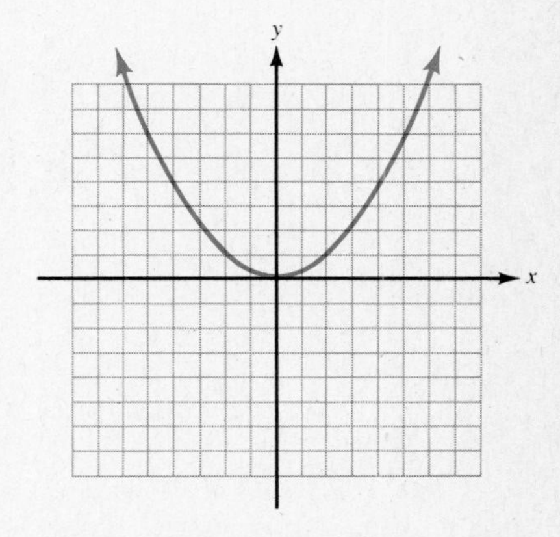

Figure 1-29. A graph of $y = \frac{1}{4}x^2$.

In exercises **5–8**, use the graph in Figure 1-30.

5. Find the value(s) of x when $y = -3$.

6. Find the value(s) of x when $y = 5$.

7. Find the value(s) of y when $x = 3$.

8. Find the value(s) of y when $x = 1$.

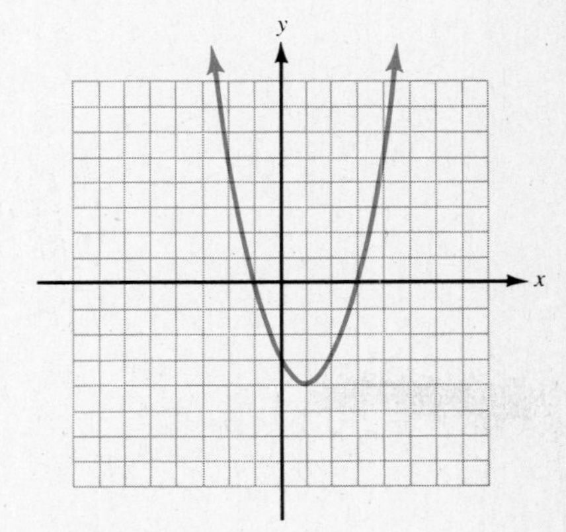

Figure 1-30. A graph of $y = x^2 - 2x - 3$.

In exercises **9–12**, use the graph in Figure 1-31.

9. Find the value(s) of x when $y = 2$.

10. Find the value(s) of x when $y = -3$.

11. Find the value(s) of y when $x = -3$.

12. Find the value(s) of y when $x = -1$.

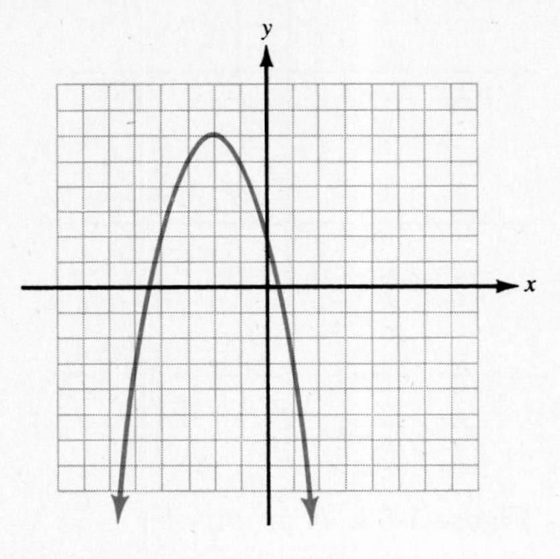

Figure 1-31. A graph of $y = -x^2 - 4x + 2$.

In exercises **13–28**, graph each function.

[Examples 2 and 3]

13. $y = x^2 - 2x + 4$

14. $y = x^2 - 8x + 4$

15. $y = x^2 + 10x + 25$

16. $y = x^2 + 6x + 5$

17. $y = 4 - x^2$

18. $y = -5 - x^2$

19. $y = -x^2 + 4x - 1$

20. $y = -x^2 - 2x + 1$

21. $y = 2x^2 + 4x$

22. $y = 3x^2 + 12x$

23. $y = -3x^2 - 12x - 6$

24. $y = -2x^2 + 4x + 3$

25. $y = \frac{1}{3}x^2 - 2x - 1$

26. $y = \frac{1}{2}x^2 + 4x + 2$

27. $y = 8x - 4x^2$

28. $y = 10x - 5x^2$

In exercises **29–32**, use the graph in Figure 1-32.

[Example 4]

29. Find the value(s) of x when $y = 5$.

30. Find the value(s) of x when $y = 1$.

31. Find the value(s) of y when $x = -1$.

32. Find the value(s) of y when $x = 6$.

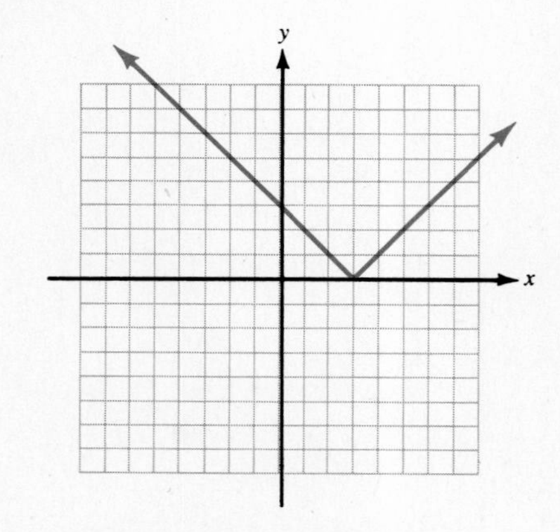

Figure 1-32. A graph of $y = |x - 3|$.

In exercises **33–36**, use the graph in Figure 1-33.

33. Find the value(s) of x when $y = -2$.

34. Find the value(s) of x when $y = -5$.

35. Find the value(s) of y when $x = -2$.

36. Find the value(s) of y when $x = 3$.

Figure 1-33. A graph of $y = -|x + 2|$.

Figure 1-34. A graph of $y = \frac{-2}{3}|x - 1|$.

In exercises **37–40**, use the graph in Figure 1-34.

37. Find the value(s) of x when $y = -4$.

38. Find the value(s) of x when $y = 0$.

39. Find the value(s) of y when $x = -8$.

40. Find the value(s) of y when $x = 7$.

In exercises **41–54**, graph each function.

[Examples 5 and 6]

41. $y = |x + 3|$
42. $y = |x + 2|$
43. $y = 3|x - 5|$
44. $y = 2|x - 1|$
45. $y = -|x + 1|$
46. $y = -|x + 4|$
47. $y = \frac{1}{2}|x - 2|$
48. $y = \frac{1}{3}|x - 3|$
49. $y = \frac{-1}{5}|x + 1|$
50. $y = \frac{-1}{4}|x + 2|$
51. $y = \frac{-3}{4}|x + 7|$
52. $y = \frac{-3}{5}|x + 6|$
53. $y = \frac{5}{2}|x - 1|$
54. $y = \frac{3}{2}|x - 4|$

In exercises **55–56**, use the graph in Figure 1-35.

55. Find the value(s) of x when $y = 2$.

56. Find the value(s) of x when $y = -2$.

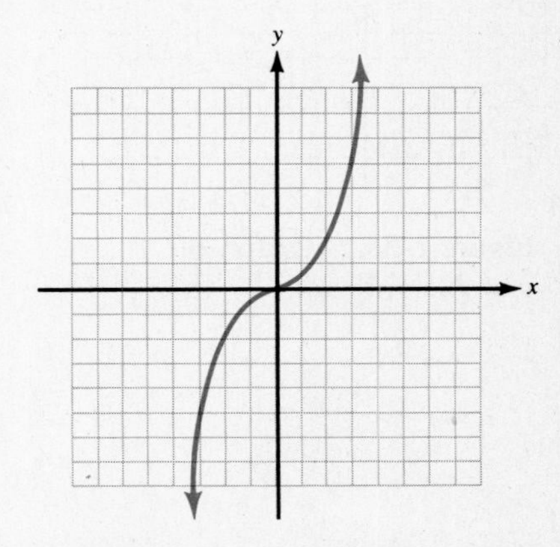

Figure 1-35. A graph of $y = \frac{1}{4}x^3$.

In exercises **57–58**, use the graph in Figure 1-36.

57. Find the value(s) of x when $y = -3$.

58. Find the value(s) of x when $y = 3$.

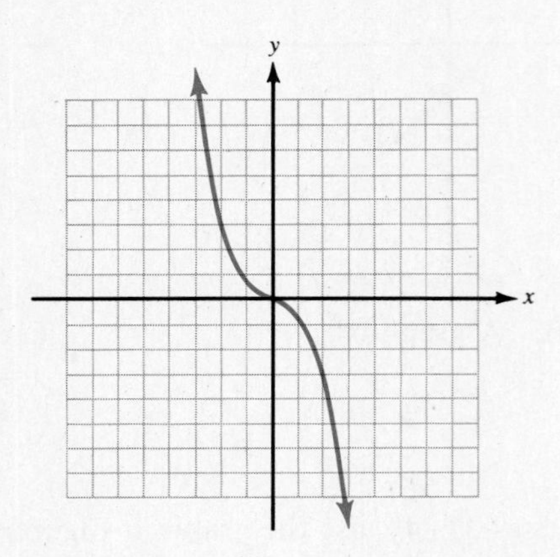

Figure 1-36. A graph of $y = \frac{-3}{8}x^3$.

Figure 1-37. A graph of $y = \frac{-5}{4}x^3$.

In exercises **59–60**, use the graph in Figure 1-37.

59. Find the value(s) of x when $y = 10$.

60. Find the value(s) of x when $y = -10$.

SECTION 1-5. Supplementary Exercises

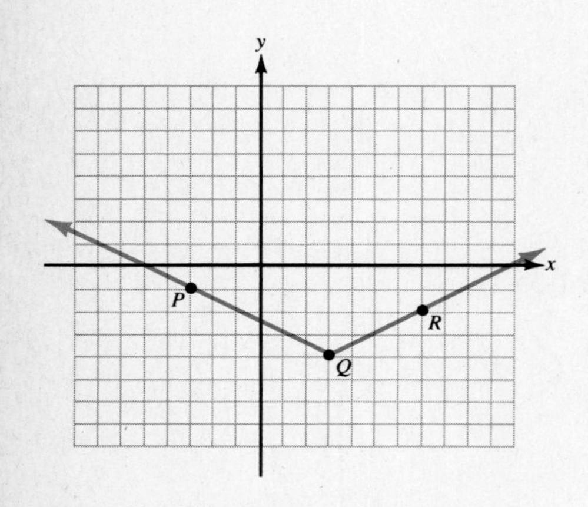

Figure 1-38. A graph of $y = \frac{1}{2}|x + 3| - 4$.

For exercises **1–20**, use the graph in Figure 1-38.

1. Find the value(s) of y when $x = 9$.

2. Find the value(s) of y when $x = -7$.

3. Find the value(s) of y when $x = 1$.

4. Find the value(s) of y when $x = 5$.

5. Find the value(s) of x when $y = -3$.

6. Find the value(s) of x when $y = -2$.

7. For what values of x is y positive?

8. For what values of x is y not positive?

9. For what values of y is x positive?

10. For what values of y is x negative?

11. For what value(s) of x is y zero?

12. For what value(s) of y is x zero?

13. For what values of x is $-4 < y < -2$?

14. For what values of x is $-3 < y < 0$?

15. For what values of x is $y < -1$?

16. For what values of x is $y > -1$?

17. Find the slope of the line that passes through points P and Q.

18. Find the slope of the line that passes through points Q and R.

19. Find the x-intercept(s).

20. Find the y-intercept(s).

In exercises **21–26**, use the graph in Figure 1-39.

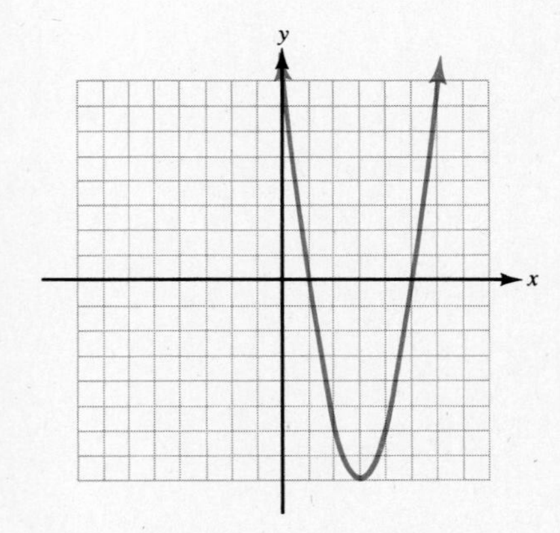

Figure 1-39. A graph of $y = 2x^2 - 12x + 10$.

21. For what values of x is y negative?

22. For what values of x is y positive?

23. For what values of x is y equal to 0?

24. For what values of x is $-8 < y < -6$?

25. For what values of x is $-6 < y < 0$?

26. For what values of x is $y > -10$?

Name

Date

Score

SECTION 1-5. Summary Exercises

Answer

1. Use the graph in Figure 1-A.

   a. Find the value(s) of x when $y = -9$.

   b. Find the value(s) of y when $x = 2$.

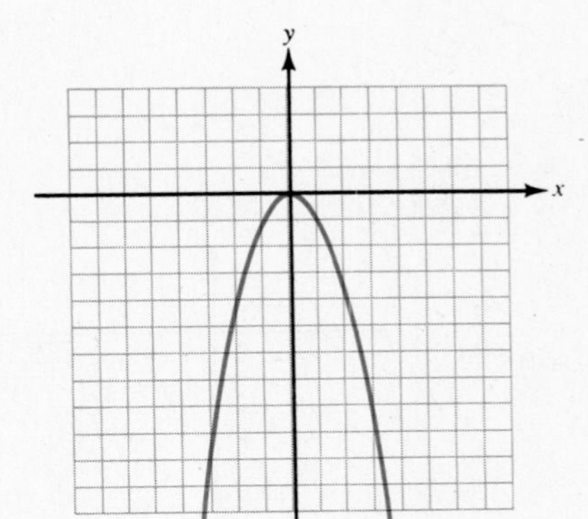

Figure 1-A. A graph of $y = -x^2$.

1. a. ____________

   b. ____________

2. Use the graph in Figure 1-B.

   a. Find the value(s) of x when $y = 3$.

   b. Find the value(s) of y when $x = 8$.

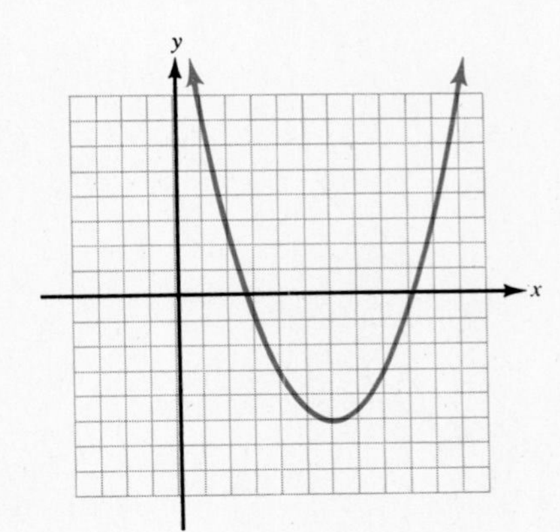

Figure 1-B. A graph of $y = \frac{1}{2}x^2 - 6x + 13$.

2. a. ____________

   b. ____________

3. Graph $y = x^2 + 4x + 3$.

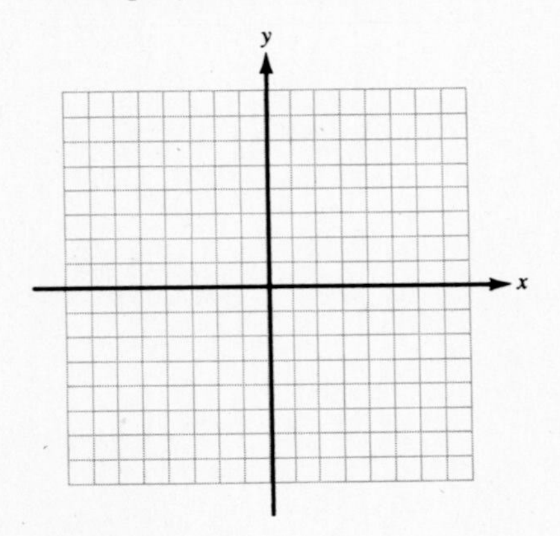

4. Graph $y = 6x - 3x^2$.

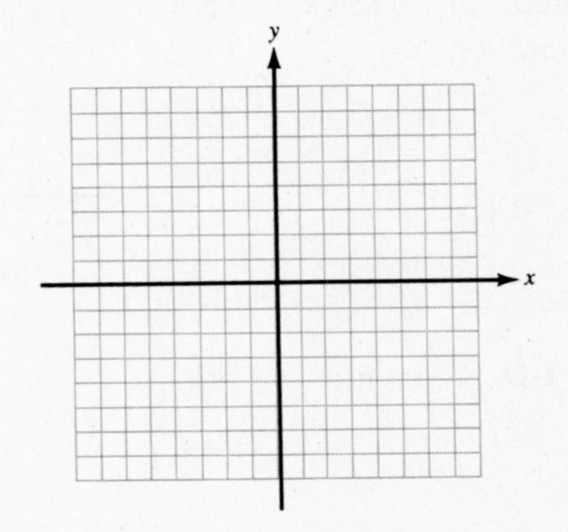

5. Graph $y = 6 - x^2$.

5.

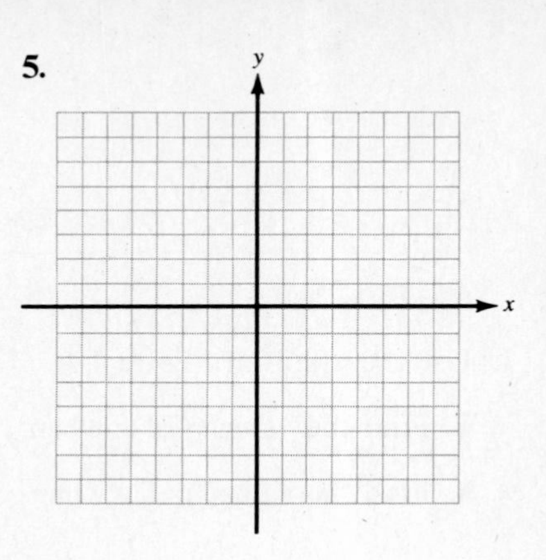

6. Use the graph in Figure 1-C.

a. Find the value(s) of x when $y = -4$.

b. Find the value(s) of y when $x = 0$.

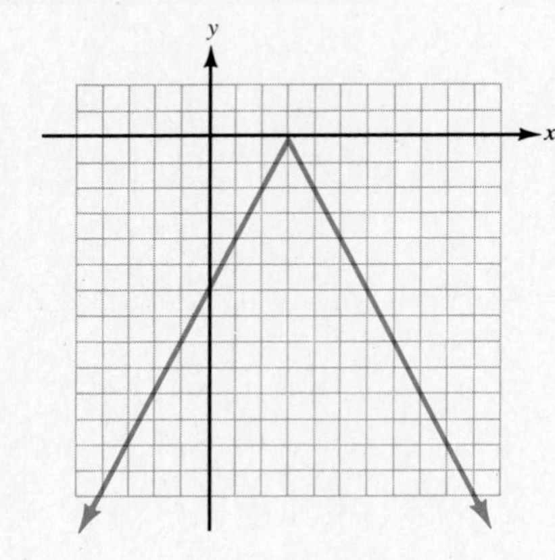

Figure 1-C. A graph of $y = -2|x - 3|$.

6. a. ________________

b. ________________

7. Graph $y = \frac{-1}{2}|x + 4|$.

7.

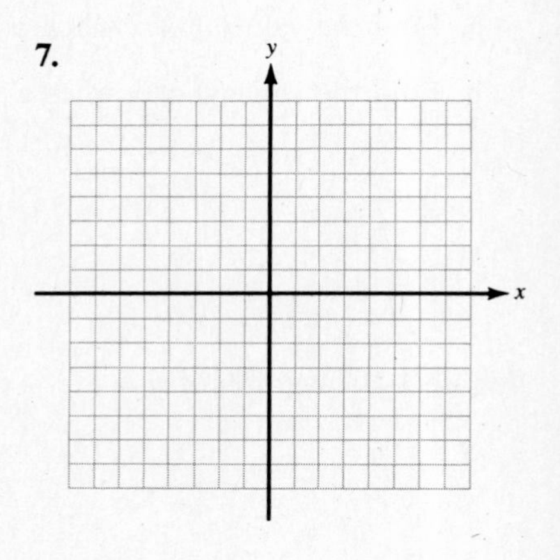

8. Use the graph in Figure 1-D.

a. Find the value(s) of x when $y = 8$.

b. Find the value(s) of y when $x = -2$.

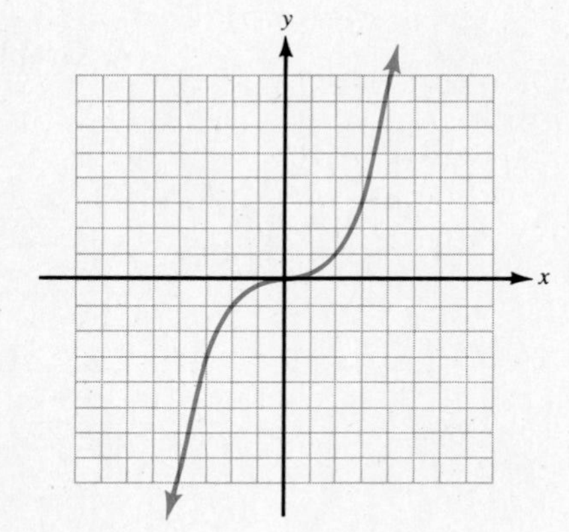

Figure 1-D. A graph of $y = \frac{1}{8}x^3$.

8. a. ________________

b. ________________

CHAPTER 1. Review Exercises

In exercises **1** and **2**, verify that each number is rational.

1. $4\frac{2}{5}$

2. -0.75

In exercises **3–8**, by each number place an X in the column that identifies the set to which that number belongs. A given number may belong to more than one set.

	Whole numbers	Integers	Rational numbers	Irrational numbers	Real numbers
3. $\frac{-7}{12}$					
4. 3.22					
5. 21					
6. $\sqrt{7}$					
7. $-0.585858\ldots$					
8. $0.595595559\ldots$					

In exercises **9** and **10**, find the LCM of each set of numbers.

9. 10 and 18

10. 60, 90, and 150

In exercises **11** and **12**, list the natural number factors of each number.

11. 30

12. 85

In exercises **13–18**, graph each number on a number line.

13. 4

14. -2.7

15. $\frac{8}{5}$

16. $-\frac{14}{3}$

17. $-4\frac{1}{2}$

18. $\sqrt{5}$

In exercises **19–22**, place <, =, or > between each pair of numbers to make a true statement.

19. $-4.3 \quad -4.8$

20. $\frac{2}{3} \quad \frac{6}{9}$

21. $\frac{-6}{5} \quad 0.01$

22. $-\sqrt{7} \quad -\sqrt{10}$

In exercises **23–26**, identify each property illustrated.

23. If $9 \cdot 2 = 18$, then $18 = 9 \cdot 2$.

24. If $8 + 10 = x$ and $x = 2y$, then $8 + 10 = 2y$.

25. If $p > -3$ and $-3 > w$, then $p > w$.

26. $3x + 2 = 3x + 2$

In exercises **27–30**, graph the numbers in each set.

27. $\{x \mid -2 \le x \le 3\}$

28. $\{x \mid x > 1\}$

29. $\{x \mid -6 < x < -5\}$

30. $\{x \mid x < 4\}$

In exercises **31** and **32**, use x and set-builder notation to identify the numbers in each graph.

31. 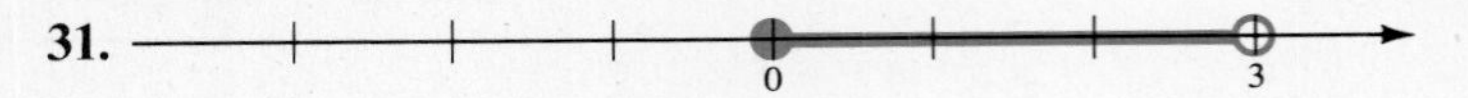

32.

−3 −2 −1 0 1 2 3

In exercises **33–48**, simplify each expression.

33. $|-72|$

34. $-\left|\dfrac{-6}{5}\right|$

35. $41 + (-53)$

36. $-16 - (-23)$

37. $\dfrac{-5}{12} - \dfrac{1}{6}$

38. $0.91 + (-2.00)$

39. $(-5)(-8)(2)$

40. $\left(\dfrac{3}{7}\right)\left(\dfrac{-14}{15}\right)$

41. $108 \div (-4)$

42. $-63 \div (-0.7)$

43. 7^3

44. $-3^2 \cdot 2^4$

45. $-20 \div 4 + 7 - 9 \cdot 2 + 3^2$

46. $-3[5 - 2(2^3 - 5) + 4^2 - (-2)(5)]$

47. $\dfrac{3 \cdot 2^2 - 6 \cdot 3}{3 \cdot 4 - 7 \cdot 2 - 6}$

48. $\dfrac{2 \cdot 5^2 + 5^2 - 3}{3(10) - 6} \cdot \dfrac{5^3 \cdot 3 - 15 \cdot 10}{5^3 \cdot 2 - 25}$

In exercises **49** and **50**:

a. Write each term in conventional form.

b. Identify the numerical coefficient.

c. Identify the literal coefficient.

49. $y^3x(-3)z^5$

50. $-m^3 3n^2\left(\dfrac{1}{7}\right)$

In exercises **51–56**, simplify each expression.

51. $5x - 2y + 4 - 3x + 2y - 4$

52. $7z^2 - 5z + z^2 + z - 3z^2$

53. $3a^2b(7a^3b + ab^2 - 9b^3)$

54. $\frac{2}{7}(28t^2 - 49t) - \frac{2}{5}(15t - 45t^2)$

55. $6m(m^2 - 2m + 6) - m(2m^2 + 9m - 1) + 5(m^2 - 2)$

56. $\frac{-84p^6q^7r}{12p^5qr}$

In exercises **57** and **58**, evaluate each expression for the indicated values.

57. $-3b^2 + 10b - 2$, for $b = -4$

58. $\frac{m^2n - mn^2 + mn}{m^2n + mn^2 - mn}$, for $m = 5$ and $n = -2$

In exercises **59** and **60**, find the ordered pairs of the relation for each value of the independent variable.

59. $y = x^3 - 9$
a. $x = -1$
b. $x = 9$

60. $y = \frac{x + 5}{x}$
a. $x = -5$
b. $x = 1$

In exercises **61–64**, let:

$$r = \{(0, 3), (8, 6), (0, -2), (5, -1)\}$$

$$s = \{(15, 3), (-9, 7), (-8, -1), (2, 14)\}$$

61. List the numbers in the range of r.

62. List the numbers in the domain of s.

63. Determine whether r is a function.

64. Determine whether s is a function.

In exercises **65** and **66**, graph each function.

65. $y = \frac{-1}{2}x$

66. $y = 4x - 5$

In exercises **67** and **68**, find the slope of the line that passes through P_1 and P_2.

67. $P_1(4, -3)$ and $P_2(-2, 3)$

68. $P_1(6, -5)$ and $P_2(6, 8)$

In exercises **69** and **70**, graph the line that passes through P_1 with the given slope.

69. $P_1(-4, 1)$ and slope $\frac{-1}{3}$

70. $P_1(-6, -3)$ and slope $\frac{2}{5}$

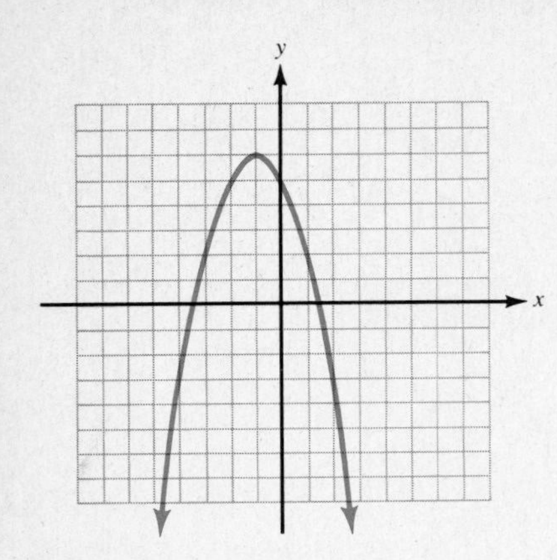

Figure 1-40. A graph of $y = -x^2 - 2x + 5$.

In exercises **71** and **72**, for each equation:

a. Identify the slope.

b. Identify the y-intercept.

c. Graph the function.

71. $x - 2y = 6$

72. $4x + 3y = -3$

In exercises **73** and **74**, use the graph in Figure 1-40.

73. Find the value(s) of x when $y = 2$.

74. Find the value(s) of y when $x = -4$.

In exercises **75** and **76**, graph each function.

75. $y = x^2 + 10x + 21$

76. $y = \frac{-1}{4}x^2 - 2$

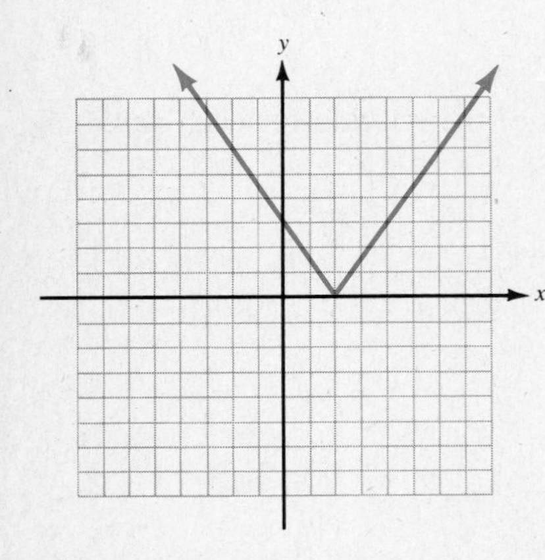

Figure 1-41. A graph of $y = \frac{3}{2}|x - 2|$.

In exercises **77** and **78**, use the graph in Figure 1-41.

77. Find the value(s) for x when $y = 6$.

78. Find the value(s) for y when $x = 4$.

In exercises **79** and **80**, graph each function.

79. $y = -|x - 5|$

80. $y = \frac{-4}{3}|x + 1|$

In exercises **81** and **82**, use the graph in Figure 1-42.

81. Find the value(s) of x when $y = 36$.

82. Find the value(s) of y when $x = -3$.

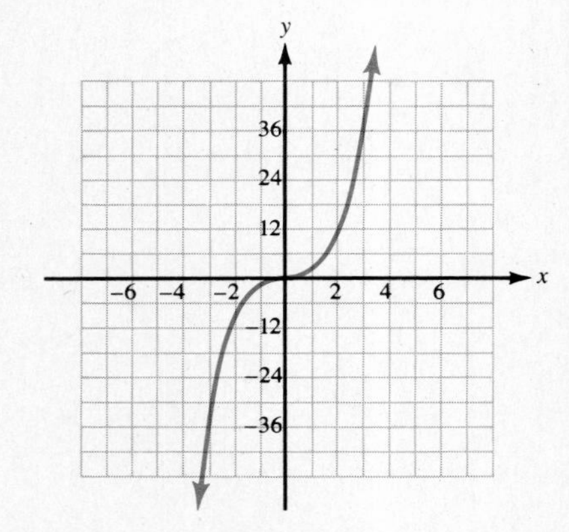

Figure 1-42. A graph of $y = \frac{4}{3}x^3$.

CHAPTER

2

Equations and Inequalities in One Variable

Ms. Glaston was preparing some material before class when a student from last term's Beginning Algebra class came in. "Good morning, Ted," she said, "I hope you're here to continue the study of algebra we began last term. As I recall, you did quite well."

Ted Barnett smiled and said, "Yeah, I did okay. But I know I could have gotten a higher grade if I had put in a little more time studying. That's why I stopped by to see you before class. This term I have a busy schedule. So I wanted to get some advice on how much time I should plan on studying math this term. It seemed like we learned a ton of stuff in Beginning Algebra last term. Is this course basically a repeat of all that, only with bigger numbers?"

"Well, Ted," Ms. Glaston replied, "much of the material will be a review of Beginning Algebra. But much is also new. Let's take equations in one variable as an example. Let me list some of the kinds of equations that we will learn to solve this term."

She then went to the board and listed the equations in examples **a–h**.

a. $3(x + 1) + 2x = 9 - 7(2 - x)$ **b.** $3t^2 = 10 - 13t$

c. $\dfrac{2y + 15}{3y} = \dfrac{1}{6} + \dfrac{5 - y}{3y}$ **d.** $\sqrt{3z - 2} = 2 + \sqrt{z}$

e. $2^{k+1} = 12^{1-k}$ **f.** $\log(4u - 2) - \log(x + 3) = \log 2$

g. $|5u - 3| = 2$ **h.** $\sqrt[3]{8v^3 + 4v^2} = 2v + 1$

"Well, Ted," Ms. Glaston said as she turned from the board, "how many of these equations do you recognize?"

"Gee, Ms. Glaston," Ted replied, "a couple of them look familiar, but some of them I'm sure I've never seen before. And when you put them all in a group, I'm not even sure how to solve the ones that look familiar."

"These equations," Ms. Glaston added, "illustrate the general intent of this algebra course. First, we want to review much of the content learned in Beginning Algebra. Second, we want to build on the skills learned in that course by introducing new material.

"Each of the equations on the board is known by a different name, because there is something unique about its structure. Although it may not be important to know the name of each one, it is vital to recognize its structure. As you will learn, Ted, the procedure for solving each equation depends on what kind of equation it is. Well, does this answer your thoughts on study time and this present course?"

"Thanks, Ms. Glaston," Ted replied, "but you probably raised more than one question in what you said. One thing is for certain—I'm going to have to allow for more study time than I originally planned on."

SECTION 2-1. Linear Equations in One Variable

1. A definition of a linear equation in x
2. Basic terminology associated with equations
3. A procedure for solving linear equations in one variable
4. Equations with no solution
5. Equations with all real numbers as solutions

In this section, we will review the procedure for solving linear equations in one variable. We will also study special equations that have no solution and those that have all real numbers as solutions.

A Definition of a Linear Equation in x

The following definition identifies the general form of a linear equation in x:

Definition 2.1. A definition of a linear equation in x
A **linear equation in x** is one that can be written in the form

$$ax + b = c$$

where a, b, and c are real numbers and $a \neq 0$.

Examples **a** and **b** are linear equations in t and z, respectively.

a. $2t + 4(2t + 3) = 14 - 6(t - 1)$ A linear equation in t

b. $5z - 6(z + 1) = 3(2 - z) - 9$ A linear equation in z

Basic Terminology Associated with Equations

The task usually associated with an equation is to determine what real-number replacements for the variable (if, in fact, there are any) will make the equation true.

Definition 2.2. Solution, or root, of an equation
A **solution,** or **root,** of an equation is a number replacement for the variable that makes the equation *true*. The **solution set** is the set that contains all the solutions of the equation.

Example 1. **a.** Verify that -8 is a root of $10 - 2(3y + 17) = 3(y + 5) + 33$.

b. Given that -8 is the only solution of the equation, state the solution set.

Solution. **Discussion.** To verify that -8 is a root of the equation, replace y by -8 and simplify the number expressions on both sides of the equation. If both expressions simplify to the same number, then -8 is a solution.

a. $10 - 2(3(-8) + 17) = 3(-8 + 5) + 33$ — Replace y by -8.

$10 - 2(-7) = 3(-3) + 33$ — Simplify within parentheses.

$10 - (-14) = -9 + 33$ — Multiply first.

$24 = 24$, *true* — Thus, -8 is a solution.

b. Given that -8 is the only solution of the equation, the solution set is $\{-8\}$.

The process of finding any solutions of an equation usually involves changing the given equation to equivalent equations until any root, or roots, can be easily identified.

Definition 2.3. Equivalent equations
If two or more equations have the same solution set, then they are **equivalent equations.**

Example 2. Verify that equations (1) and (2) are equivalent, given that (1) has only one solution.

(1) $19 - 5(4k + 1) = 12k + 6$

(2) $k = \frac{1}{4}$

Solution. **Discussion.** The root of (2) is $\frac{1}{4}$. Now (1) and (2) are equivalent if $\frac{1}{4}$ is also a root of (1), and (1) has no other roots.

(1) $19 - 5\left[4\left(\frac{1}{4}\right) + 1\right] = 12\left(\frac{1}{4}\right) + 6$ — Replace k by $\frac{1}{4}$.

$19 - 5(2) = 3 + 6$

$19 - 10 = 9$

$9 = 9$, *true* — $\frac{1}{4}$ is a root of (1)

Given that (1) has only one root, (1) and (2) are equivalent.

To solve a linear equation in x means to change it, if possible, to an equivalent equation in the form

$$x = c \qquad \text{or} \qquad c = x$$

where c is a real number.

There are two properties of equality that can be used to change an equation to an equivalent one. That is, *these properties will change the form of an equation but will not change the solution set.*

Addition and multiplication properties of equality

Addition property	**Multiplication property**
If a, b, and c are numbers with $a = b$, then:	If a, b, and c are numbers ($c \neq 0$) with $a = b$, then:
$a + c = b + c$	$a \cdot c = b \cdot c$
Add the same number to both sides.	Multiply both sides by the same nonzero number.

A Procedure for Solving Linear Equations in One Variable

The following six steps are a recommended procedure for solving a linear equation in one variable:

Step 1. Remove any grouping symbols in the equation.

Step 2. Combine any like terms on the left side of the equation; do the same on the right side.

Step 3. Use the addition property of equality to separate the variable and constant terms to opposite sides of the equal sign.

Step 4. If necessary, use the multiplication property of equality to change the coefficient of the variable term to 1.

Step 5. Check the solution by substituting it into the given equation.

Step 6. State the solution set.

Example 3. Solve $3x + 3 + 2x = 9 + 7x - 14$.

Solution. **Step 1.** Remove any grouping symbols.
There are no grouping symbols to remove.

Step 2. Combine like terms on both sides.

$5x + 3 = -5 + 7x$ $\quad$ $3x + 2x = 5x$ and $9 - 14 = -5$

Step 3. Combine x-terms on the right side and constants on the left side.

$3 = -5 + 2x$ $\quad$ Add $-5x$ to both sides.

$8 = 2x$ $\quad$ Add 5 to both sides.

Step 4. Change the coefficient of x from 2 to 1.

$4 = x$ $\quad$ Multiply both sides by $\frac{1}{2}$.

Step 5. Check 4 in the given equation.

$$3(4) + 3 + 2(4) = 9 + 7(4) - 14 \qquad \text{Replace } x \text{ by 4.}$$

$$12 + 3 + 8 = 9 + 28 - 14$$

$$15 + 8 = 37 - 14$$

$$23 = 23, \textit{ true} \qquad \text{The root is 4.}$$

Step 6. State the solution set.
The solution set is $\{4\}$.

Example 4. Solve $4(2y + 3) + 2y = 14 - 6(y - 1)$.

Solution.

Step 1. $8y + 12 + 2y = 14 - 6y + 6$ — Remove parentheses.

Step 2. $10y + 12 = 20 - 6y$ — Combine like terms.

Step 3. $16y + 12 = 20$ — Add $6y$ to both sides.

$16y = 8$ — Add -12 to both sides.

Step 4. $y = \frac{8}{16}$ — Multiply both sides by $\frac{1}{16}$.

$= \frac{1}{2}$ — Reduce the fraction.

Step 5. The check is omitted.

Step 6. The solution set is $\left\{\frac{1}{2}\right\}$.

Equations with No Solution

When solving an equation, the variable terms may combine to yield 0, and the resulting number equation is false (that is, a **contradiction**). For such an equation there is no solution, so the solution set is $\varnothing$.

Example 5. Solve $3(b + 1) + 4b = 18 - 7(2 - b)$.

Solution.

$3b + 3 + 4b = 18 - 14 + 7b$ — Remove parentheses.

$7b + 3 = 4 + 7b$ — Combine like terms.

$7b + 3 + (-7b) = 4 + 7b + (-7b)$ — Add $-7b$ to each side.

$3 = 4$ — Combine like terms.

Since $3 \neq 4$ (a contradiction) and the variable terms combined to yield 0, there is no solution for this equation. The solution set is $\varnothing$.

Equations with All Real Numbers as Solutions

When solving an equation, the variable terms may combine to yield 0, and the resulting number equation is true (that is, an **identity**). For such an equation all real numbers are solutions, and the solution set is the set of real numbers.

Example 6. Solve $5z - 8(z + 1) = 3(2 - z) - 14$.

Solution.

$5z - 8z - 8 = 6 - 3z - 14$	Remove parentheses.
$-3z - 8 = -8 - 3z$	Combine like terms.
$-3z - 8 + 3z = -8 - 3z + 3z$	Add $3z$ to each side.
$-8 = -8$	Combine like terms.

Since $-8 = -8$ (an identity) and the variable terms combined to yield 0, any real number is a solution. The solution set is the set of real numbers, written R.

If in solving a linear equation in one variable the variable terms combine to 0, then:

1. The solution set is $\varnothing$ if the resulting number statement is *false*.
2. The solution set is R if the resulting number statement is *true*.

SECTION 2-1. Practice Exercises

In exercises **1–8**, determine whether the given value for the variable is a root of the equation.

[Example 1]

1. $6a - 2 = 3(a + 1) + 1;\ a = 2$

2. $5a + 4 = 4(a + 2) + 3;\ a = 7$

3. $9s - 3(3 - s) = 4s - 17;\ s = -1$

4. $11s - 16 = 15 + 5(3s + 1);\ s = -9$

5. $4(2y + 1) - 3 = 2(4y - 1);\ y = -5$

6. $5y - 5(1 - y) = 10(y - 2);\ y = 3$

7. $2 - 9(x - 2) = 9 + 11(x + 1);\ x = 0$

8. $39 - 2(3 - x) = 38 - 5(1 - 4x);\ x = 0$

In exercises **9–16**, determine whether equations (1) and (2) are equivalent, given that (1) has only one solution.

[Example 2]

9. (1) $3y + 2 + 7y = 6y - 18$
(2) $y = -5$

10. (1) $4 + 7y - 30 = 40 + 13y$
(2) $y = -11$

11. (1) $4(5b - 6) - 2(4b + 2) = 30$
(2) $b = 5$

12. (1) $6(b - 5) - 2(3 - b) = 12$
(2) $b = 6$

13. (1) $7p - 2(p - 4) = 29 - 3p$
(2) $p = 3$

14. (1) $8p - 10 = 3(p + 5) + 28$
(2) $p = 5$

15. (1) $10q - 11 = 3 + 2q - 10$
(2) $q = \frac{1}{2}$

16. (1) $7 - 4q + 5 = 39 - 40q$
(2) $q = \frac{3}{4}$

In exercises **17–44**, solve each equation.

[Examples 3 and 4]

17. $6x - 11 = 21 - 2x$

18. $46 + 8x = 13x - 14$

19. $b + 15 - 21b = 15 - 8b + 8$

20. $9 - 4b + 1 = 6 + 26b + 10$

21. $29 - 2z = 7z - 2(z - 4)$

22. $3(z + 5) + 25 = 8z - 10$

23. $2 - 12y = 6(3y + 2) - 20$

24. $10(2y + 1) - 1 = 11 + 6y + 19$

25. $5(4x - 1) - 5 = 2(4x - 3) - 19$

26. $5(2x + 5) = 18 + 3(10x + 3)$

27. $2 - 9(a - 2) = 9 + 11(a + 1)$

28. $39 - 2(3 - a) = 38 - 5(1 - 4a)$

29. $\frac{3}{5}(10p - 35) = 15 - \frac{3}{4}(12 - 4p)$

30. $23 - \frac{5}{9}(27 - 18p) = \frac{3}{5}(25p + 5)$

31. $3[2(q + 5) - (3q - 2)] = 48$

32. $5[3(2q - 9) - 4(q - 5)] = -45$

33. $2[3 + 4(t - 7)] = 6$

34. $3[2(4t + 5) - 3t] = 30$

[Examples 5 and 6]

35. $12 + 11x - 2 = 5x - 7 + 6x$

36. $x - 5 + 20x = 12x - 3 + 9x$

37. $7y + 6 - 2y = 8 + y - 2 + 4y$

38. $8 - 5y + 7 - y = 2y + 15 - 8y$

39. $4(2p + 1) - 3 = 2(4p - 1)$

40. $5p - 5(1 - p) = 10(p - 2)$

41. $6 - 4(z - 3) = 2(1 - 2z) + 16$

42. $5z + 2(z - 5) = 7(z - 1) - 3$

43. $4(2q + 3) - 3(q - 3) = 5(q + 3)$

44. $3(2q + 3) - 4(q - 1) = 2(q + 3)$

SECTION 2-1. Ten Review Exercises

1. Evaluate $2(3k - 1) - (7k - 9)$ for $k = -3$.
2. Simplify $2(3k - 1) - (7k - 9)$.
3. Evaluate the expression obtained in exercise **2** for $k = -3$ and compare with the answer obtained to exercise **1**.
4. Solve $2(3k - 1) = 7k - 9$.
5. Evaluate $8(m - 1) + 7(2m - 9) - (6m + 5)$ for $m = 5$.
6. Simplify $8(m - 1) + 7(2m - 9) - (6m + 5)$.
7. Without evaluating, state the value of the expression obtained in exercise **6** for $m = 5$ based on the answer to exercise **5**.
8. Solve $8(m - 1) + 7(2m - 9) = 6m + 5$.

In exercises **9** and **10**, simplify each expression.

9. $\dfrac{5 \cdot 2 - (2^5 - 2 \cdot 3^2)}{1 - 5}$

10. $-3[(19 - 5^2)^2 + (2^2 - 7)^3]$

SECTION 2-1. Supplementary Exercises

In exercises **1–8**, solve each equation.

1. $0.3z - 1.5 = 1.5 - 1.2z$

2. $3.1z + 2.6 = 1.1 - 1.4z$

3. $0.1x - 1.2 = 0.1(0.5 - 3x)$

4. $0.2(3x - 0.7) = 0.4x + 1.1$

5. $2.4 - (1 - a) = 0.7(2 - a) - 0.3a$

6. $0.5(0.2 - 3a) = 1.2a + 0.5(1 - 1.4a)$

7. $3.2(4 - b) = 2(6.4 - b) - 1.2b$

8. $8.1(6 + b) = 4(2b + 9.5) + 0.1(b + 106)$

In exercises **9–16**, find the value of b so that the given number is a solution of each equation.

9. $7t - 5 = 3t + b; t = 6$

10. $4t + 3 = 2t - b; t = -2$

11. $5 + 2p = b - p; p = 0$

12. $8 + 5p = b - 2p; p = 0$

13. $3x - 1 = 5x - b; x = \frac{3}{2}$

14. $2x - 1 = 4x + b; x = \frac{3}{4}$

15. $4s - 2(b - s) = 12; s = 3$

16. $5s - 2(b - s) = 6; s = 2$

In exercises **17–22**, find the value of the unknown variable in each equation given the other values.

17. $ax - b = c$; $x = 3$, $a = 9$, and $c = 20$

18. $ax - b = c$; $x = 7$, $a = 8$, and $c = 47$

19. $ax - b = cx + d$; $x = 9$, $a = 3$, $c = 2$, and $d = 5$

20. $ax + b = c + dx$; $x = 2$, $a = 6$, $c = 3$, and $d = 8$

21. $ax - b = b - cx$; $x = 1$, $a = 7$, and $b = 8$

22. $ax - b = b - cx$; $x = 1$, $a = 5$, and $b = 7$

In exercises **23–30**, solve each equation for x.

23. $5x + 6 = a$

24. $7x - 3 = a$

25. $ax - 9 = 7$, and $a \neq 0$

26. $ax + 5 = 12$, and $a \neq 0$

27. $ax - 5x = 7$, and $a \neq 5$

28. $9x - ax = 12$, and $a \neq 9$

29. $3(x + a) = 5(x - 2)$

30. $4(x - a) = 3(x + 3)$

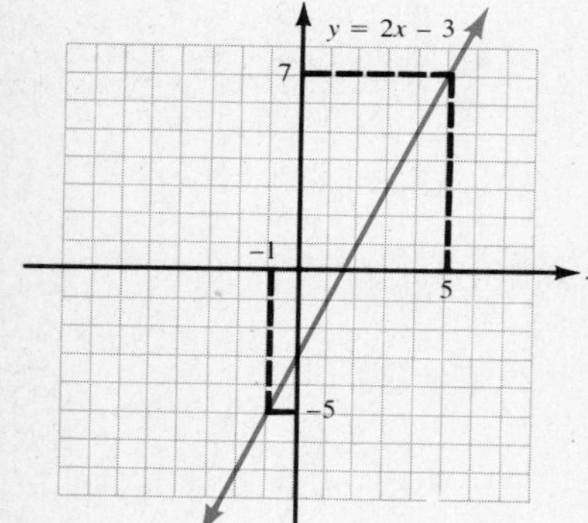

Figure 2-1. A graph of $y = 2x - 3$.

Example. The line in Figure 2-1 is a graph of $y = 2x - 3$. Use the graph to find the values of x for the given values of y.

a. $y = 7$ **b.** $y = -5$

Solution. **Discussion.** To find the corresponding values of x, locate the given values of y on the vertical axis. From this point move horizontally to the line, then vertically up or down as needed to the horizontal axis. The x-coordinate of the point is the corresponding value. Notice that *the x-coordinate is the root of the linear equation in* x *for the given value of* y.

a. Using the graph, $x = 5$ when $y = 7$.

b. Using the graph, $x = -1$ when $y = -5$.

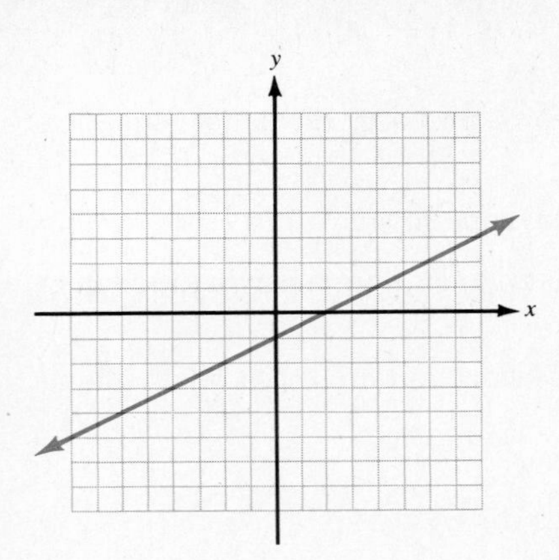

Figure 2-2. A graph of $y = \frac{1}{2}x - 1$.

In exercises **31–34**, use the graph in each figure to find the values of x for the given values of y.

31. Figure 2-2 is a graph of $y = \dfrac{1}{2}x - 1$.

a. $y = 3$ **b.** $y = -4$

32. Figure 2-3 is a graph of $y = \dfrac{5}{3}x - 4$.

a. $y = 6$ **b.** $y = -9$

33. Figure 2-4 is a graph of $y = \dfrac{-3}{2}x + 4$.

a. $y = 7$ **b.** $y = -5$

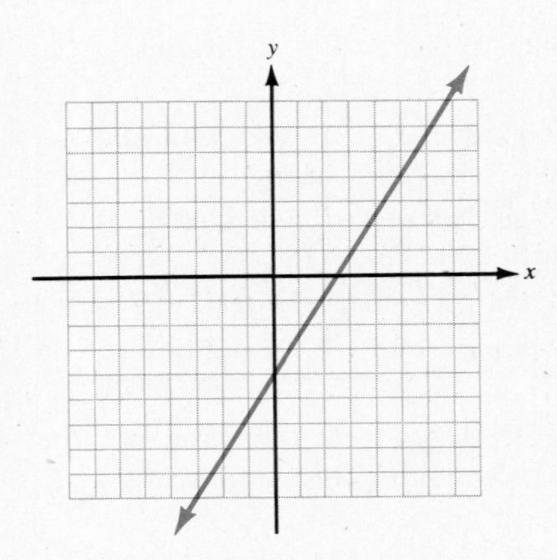

Figure 2-3. A graph of $y = \frac{5}{3}x - 4$.

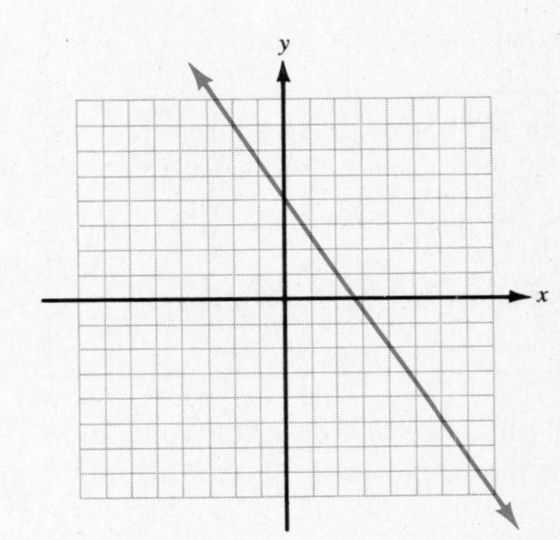

Figure 2-4. A graph of $y = \frac{-3}{2}x + 4$.

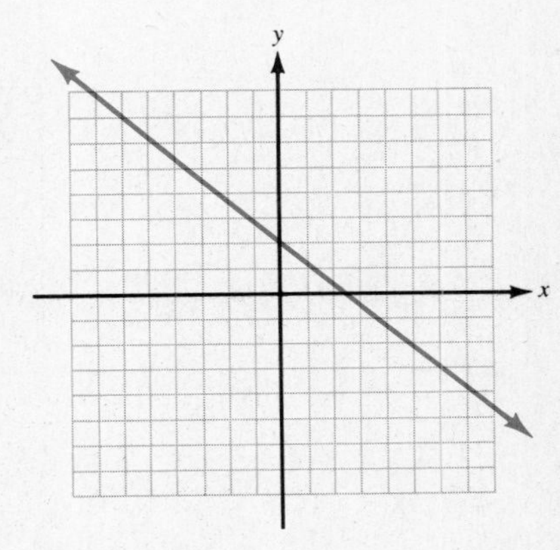

Figure 2-5. A graph of $y = \frac{-5}{6}x + 2$.

34. Figure 2-5 is a graph of $y = \dfrac{-5}{6}x + 2$.

a. $y = 7$ **b.** $y = -3$

SECTION 2-2. Applied Problems

KEY TOPICS IN THIS SECTION

1. Changing word phrases into mathematical symbols
2. A procedure for solving applied problems
3. Number problems
4. Geometry problems

One of the goals in any mathematics course is to have students apply the skills learned in that course to some type of word problem. To achieve this goal, students must develop a skill to do the following tasks:

1. Accurately translate the problem given in words into a mathematical equation using symbols.
2. Determine any solutions to the mathematical equation.
3. Write in words a correct interpretation of what the solutions mean in the context of the given problem.

These three tasks are used to solve word problems that are traditionally found in Intermediate Algebra courses.

Changing Word Phrases into Mathematical Symbols

Certain word phrases are frequently used to describe addition, subtraction, multiplication, division, and raising to powers. The following table contains some of the more commonly used phrases and corresponding symbolic expressions:

Addition: $t + 5$	**Subtraction: $t - 5$**	**Multiplication: $5t$**	**Division: $\frac{t}{5}$**
The sum of t and 5	The difference between t and 5	The product of 5 and t	The quotient of t and 5
5 more than t	5 less than t	The product of t and 5	t divided by 5
t increased by 5	t decreased by 5	5 times t	t over 5
t combined with 5	t minus 5	t multiplied by 5	One-fifth of t

Word phrases that describe raising to powers depend on the base and the exponent. Examples **a–h** illustrate some of the more commonly used phrases.

a. The square of t $\quad t^2$

b. The cube of t $\quad t^3$

c. t to the fourth power $\quad t^4$

d. The square of 5 times t $\quad (5t)^2$

e. The square of the quantity t plus 5 $\quad (t + 5)^2$

f. The cube of t over 5 $\quad \left(\frac{t}{5}\right)^3$

g. The cube of the quantity t minus 5 $\quad (t - 5)^3$

h. The product of 5 and t-squared $\quad 5t^2$

Example 1. Write each word phrase in mathematical symbols.

a. The square of x minus 10 is divided by the product of 10 and x.

b. The difference in the squares of y and z is increased by 25.

Solution. **a.** "The square of x minus 10" is written $(x - 10)^2$.
"The product of 10 and x" is written $10x$.
"The square . . . is divided by the product . . ." is written $\frac{(x - 10)^2}{10x}$.

b. "The difference . . . y and z" is written $y^2 - z^2$.
"The difference . . . is increased by 25" is written $(y^2 - z^2) + 25$.

A Procedure for Solving Applied Problems

The following six steps give a procedure for solving applied problems:

To solve an applied problem:

Step 1. Read the problem slowly and thoroughly. If necessary, read it several times until the content of the problem is understood.

Step 2. Assign a variable to an unknown quantity in the problem. If possible, be descriptive in the choice. For example, let d represent *distance*, ℓ represent *length*, and so on.

Step 3. Write an equation using the variable and expressing its relationship to other quantities in the problem. If the equation is difficult to write, try the **guess and test method** until an equation can be written.*

Step 4. Solve the equation of Step 3.

Step 5. Check the solution of Step 4 in the context of the problem.

Step 6. Answer the problem.

Number Problems

A number problem provides some information about one or more numbers. The information should first be changed to two or more mathematical expressions. An equation can then be written using these expressions that accurately describes the problem. The solution of the equation is either the unknown number, or can be used to determine the number(s).

Example 2. If 8 is added to the product of 3 times the sum of a number and 2, the result is the same as when 5 times the number is decreased by 10. Find the number.

Solution. **Discussion.** A key to writing a correct equation for the given information is locating the words for the equal sign. In this problem, the phrase "the result is the same as" identifies the equal sign.

Step 1. Notice the following:

a. If "8 is added [+] to the product [·] of 3 times the sum [use parentheses around the sum] of a number [variable] and 2 . . ."

b. ". . . 5 times [·] the number [variable] is decreased [−] by 10."

Step 2. Let n represent the number.

* See this section's Supplementary Exercises for a discussion on the guess and test method (p. 83–84).

Step 3. **a.** "If 8 is added to the product of 3 times the sum of n and 2" is written $8 + 3(n + 2)$.
b. "5 times n is decreased by 10" is written $5n - 10$.
c. An equation of the problem is:

$$8 + 3(n + 2) = 5n - 10$$

Step 4. $8 + 3n + 6 = 5n - 10$ Remove parentheses.

$3n + 14 = 5n - 10$ Combine like terms.

$14 = 2n - 10$ Add $-3n$ to both sides.

$24 = 2n$ Add 10 to both sides.

$12 = n$ Multiply both sides by $\frac{1}{2}$.

Step 5. If the number is 12, then:
a. 8 added to 3 times 12 plus 2 becomes $8 + 3(14) = 50$
b. 5 times 12 is decreased by 10 becomes $5(12) - 10 = 50$

Step 6. Since $50 = 50$, the number is 12.

Geometry Problems

In a typical geometry problem, facts are given about the dimensions of some figure. To solve the problem, a formula related to the figure may be needed. Some of the more commonly used geometric figures and related formulas are listed on the inside cover of this text.

Example 3. Hector Rodriguez needs a deck in his back yard. The shape of the deck is a trapezoid whose nonparallel sides are equal in length. The shorter of the parallel sides is 10 feet more than twice the length of a nonparallel side. The longer of the parallel sides is 3 times of sum of the length of a nonparallel side and 5 feet. If the perimeter of the deck is 200 feet, find the lengths of the four sides.

Solution. **Discussion.** It is quite helpful to sketch the figure. Then the given information can be used to label the measurements of certain parts in the figure. A sketch of the trapezoid is shown in Figure 2-6.

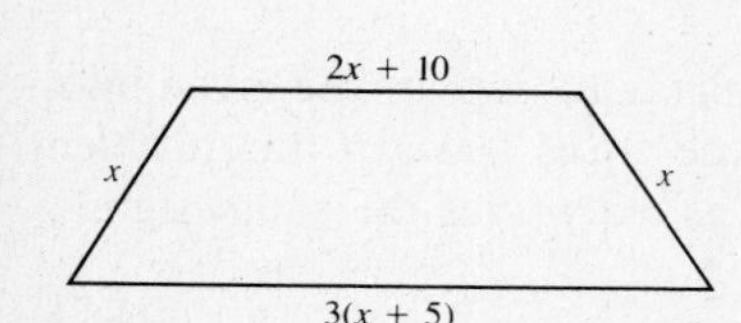

Figure 2-6. A trapezoid with nonparallel sides equal in length.

Step 1. Notice the following:
a. ". . . nonparallel sides are equal [the same] in length."
b. "The shorter of the parallel sides is 10 feet more [+] than twice [2 times] the length of a nonparallel side."
c. "The longer of the parallel sides is 3 times the sum [use parentheses to show the sum] of the length of a nonparallel side [+] and 5 feet."

Step 2. Let x represent the length of each of the nonparallel sides.
a. $2x + 10$ represents the shorter base
b. $3(x + 5)$ represents the longer base

Step 3. The perimeter is the sum of the lengths of the four sides:

$x + 3(x + 5) + x + 2x + 10 = 200$ The perimeter is 200 feet.

Step 4. $x + 3x + 15 + x + 2x + 10 = 200$ Remove parentheses.

$7x + 25 = 200$ Combine like terms.

$7x = 175$ Add -25 to both sides.

$x = 25$ Multiply both sides by $\frac{1}{7}$.

Step 5. The nonparallel sides are both 25 feet long.
The shorter base is $2x + 10$ or 60 feet long.
The longer base is $3(x + 5)$ or 90 feet long.

$2(25) + 60 + 90 = 200$, *check*

Step 6. The lengths of the sides are 25 feet, 90 feet, 25 feet, and 60 feet.

SECTION 2-2. Practice Exercises

In exercises **1–10**, write each word phrase in mathematical symbols.

[Example 1] **1.** The product of x and y is decreased by 3 times the value of x.

2. The quotient of x and y is increased by twice the value of y.

3. The cube of z is divided by the sum of x and y.

4. The square of z is divided by the difference between x and y.

5. The square of the sum of twice x and 5 is decreased by one-fourth of x.

6. The cube of the difference between 7 and twice x is increased by one-half of x.

7. Twelve more than y is multiplied by z.

8. Nine less than y is divided by z.

9. The product of a number and 11 is added to the quotient of 3 and the number.

10. The difference between a number and 3 is multiplied by the sum of 7 and the number.

In exercises **11–26**, solve each word problem.

[Example 2] **11.** When 7 times a number is decreased by 2, the result is the same as 5 times the sum of the number and 10. Find the number.

12. The product of 8 times the difference between a number and 3 is the same as 6 more than 5 times the number. Find the number.

13. When 5 is added to 10 times the sum of a number and 2, the result is the same as when 2 times the number is increased by 1. Find the number.

14. When 3 is subtracted from 8 times a number, the result is the same as 5 times the difference between the number and 6. Find the number.

15. When 4 times a number is decreased by 2, and this difference is added to the sum of 6 times the number and 7, the result is 35. Find the number.

16. The difference between 9 times a number increased by 1 and 3 times the number increased by 2 is -1. Find the number.

17. If twice the sum of a number and 1 is then decreased by one-half the difference of the number and 1, the result is 23 more than the number. Find the number.

18. If 5 times the sum of a number and 1 is increased by the sum of the number and 10, the result is 7 times the number. Find the number.

[Example 3] 19. A slab of concrete is in the shape of a rectangle. The length is 5 feet less than twice the width. If the perimeter of the slab is 110 feet, find the dimensions of the slab.

20. The label on a bed sheet gives the dimensions of the sheet in inches. The distance around the edge of the sheet is 360 inches. Three times the width is the same as twice the length. Find the width and length of the sheet.

21. The gable end of a house is in the shape of a triangle. The sum of the lengths of the three sides is 76 feet. The length of the base of the triangle is the width of the house. If the base is 4 feet less than twice the length of one of the equal sides, find the width of the house.

22. A flower garden is in the shape of a triangle. The middle-length side is 1 foot less than twice the length of the shortest side. The longest side is 4 feet less than 3 times the shortest side. The perimeter of the garden is 43 feet. Find the lengths of the three sides of the garden.

23. One of the supports for a railroad bridge has the shape of a trapezoid. The nonparallel sides are equal in length. The base of the support is the longer of the two parallel sides. The perimeter of the support is 80 feet. Each of the nonparallel sides is 5 feet less than the length of the top support. The base is 10 feet less than twice the length of the top support. Find the lengths of the four sides of the support.

24. The lengths of the four sides of a trapezoid are all different. The top base is the side with the shortest length, and the bottom base has the longest length. The shorter of the nonparallel sides is 3 feet longer than the top base, and the other nonparallel side is twice the length of the top base. The bottom base is 4 feet less than 3 times the length of the top base. The sum of the lengths of the nonparallel sides is 1 foot less than the sum of the lengths of the parallel bases. Find the lengths of the four sides of the trapezoid.

25. Exactly 9 yards of fringe are to be sewn around the four sides of a rectangular-shaped tablecloth. If the length of the tablecloth is 18 inches longer than it is wide, find the dimensions of the tablecloth.

26. A farmer has 550 meters of fencing to use to enclose three sides of a rectangular field (one length and two widths). The fourth side has a natural barrier that does not need fencing. If the length is 25 meters longer than the width, find the length and width of the field.

SECTION 2-2. Ten Review Exercises

In exercises **1–5**, identify each property illustrated.

1. $20 + 5y = 20 + 5y$

2. If $z < -20$, then $-20 > z$.

3. If $-20 = t$, then $t = -20$.

4. If $2x = 5y$ and $5y = 7z$, then $2x = 7z$.

5. If $x > y$ and $y > 3$, then $x > 3$.

In exercises **6–10**, simplify each expression.

6. $\frac{-2}{3} + \frac{1}{6} + \frac{3}{4} - \frac{-1}{12}$

7. $\left(\frac{-5}{6}\right)\left(\frac{1}{10}\right)\left(\frac{-3}{4}\right)\left(\frac{-1}{2}\right)$

8. $\left(-4\frac{1}{6}\right) \div 25$

9. $-36 \div \left(-2\frac{2}{5}\right)$

10. $\left(7\frac{5}{12}\right) \div \frac{89}{12}$

SECTION 2-2. Supplementary Exercises

The most difficult part of solving a word problem is often writing an equation that accurately describes the relationships in the problem. A problem-solving technique that is helpful in writing equations is the so-called guess and test method. The method suggests that a **guess** be made as to the answer to the question. The arbitrarily selected number is then **tested** in the problem. (The number will, in most cases, *not* be the answer to the problem.) However, the process of checking the number will generate an equation that may be the one needed to solve the problem.

In exercises **1–4**:

a. Fill in each table based on the guesses provided for each problem.

b. Write the equation.

1. If 5 times a number is decreased by 10 the result is 40. Find the number.

Guess	5 (Number)	5 (Number) − 10	Is	Equal to 40?
3	5(3)	5(3) − 10	5	No
6	5(6)	5(6) − 10	20	No
8				
15				
12				
x			40	

2. If 3 times the sum of a number and 2 is increased by the number, the result is 70. Find the number.

Guess	Number + 2	3 (Number + 2)	Increased by number	Is	Equal to 70?
5	5 + 2	3(5 + 2)	3(5 + 2) + 5	26	No
7	7 + 2	3(7 + 2)	3(7 + 2) + 7	34	No
9					
11					
15					
x				70	

3. A rancher wants to enclose a rectangular area of land with exactly 110 feet of fencing. The length of the area is to be 9 feet longer than the width. Find the length and the width of the rectangular area.

Width	Length	2 (Width)	2 (Length)	Perimeter	Is	Equal to 82?
5	$5 + 9$	$2(5)$	$2(5 + 9)$	$2(5) + 2(5 + 9)$	38	No
7						
10						
15						
20						
x					110	

4. A piece of steel wire 57 inches long is to be bent into the shape of a triangle. Two sides are equal in length, and the third side is 6 inches longer than each of the equal sides. Find the length of each side.

First side	Second side	Third side	Sum of sides	Is	Equal to 57?
5	5	$5 + 6$	$5 + 5 + (5 + 6)$	21	No
6					
8					
12					
15					
x				57	

In exercises **5–8**, solve each word problem, if possible. Some problems provide more information than is needed, while other problems do not provide enough information to be solved.

5. A new house under construction at 8883 Sandburg Lane has four windows in the main room. Each window has the shape of a square surmounted by an isosceles triangle (a triangle having two equal sides), as shown in Figure 2-7. Each of the two equal sides of the triangle is 6 inches longer than a side of the square. Each of the windows is evenly spaced along a wall that is 40 feet long. The perimeter of the window is 182 inches. Find the length of each side of the triangle.

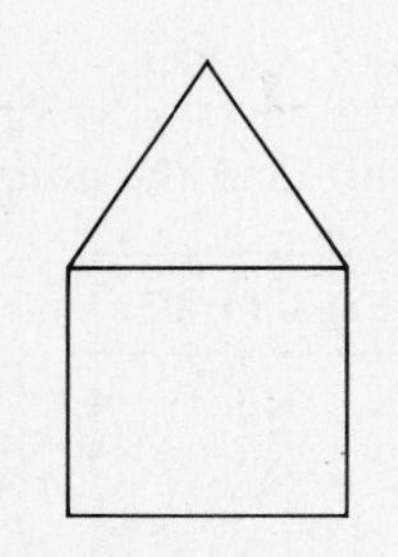

Figure 2-7. A window.

6. An 80 by 100 centimeter rectangular picture is to be framed for the Accident and Industrial Clinic, established in 1959. The picture is to have a frame of uniform width. The perimeter of the framed picture is 400 centimeters. The framing shop charges 59¢ per inch for this type of frame. Find the width of the frame.

7. A number can be increased by 5 more than itself. This value can then be divided by twice the number. What is the number?

8. The Certified Chimney Sweep Service, state license number #479319, uses a long pole and brush to clean the inside of a standard size chimney. The pole is 19 times as long as the brush. The handle on the pole is the same size as the brush. How long is the pole?

In exercises **9–12**:

a. Put each equation into words. (In each equation, let x represent a number.)

b. Find the number.

Example. Given: $3x - 8 = x + 2$

Solution. **a.** The difference between 3 times a number and 8 is the same as the sum of the number and 2.

b. $2x - 8 = 2$ Add $-x$ to both sides.

$2x = 10$ Add 8 to both sides.

$x = 5$ The number is 5.

9. $2x + 6 = x - 12$

10. $x - 30 = 4x$

11. $6(x + 1) - (4 + x) = 102$

12. $(x - 3) + 4x = x + 5$

SECTION 2-3. More Applied Problems

KEY TOPICS IN THIS SECTION

1. Mixture problems
2. Motion problems
3. Age problems

Section 2-2 demonstrated a procedure for solving applied or word problems. In this section, we will use the procedure to solve what are classified as mixture, motion, and age problems.

Mixture Problems

In a typical mixture problem, facts are given about two or more different items being combined. Usually the facts include data on the quantity of each item being mixed. There are also data on the value of the items being mixed. In some problems, the values may actually be amounts of money. In other problems, the values may be percents, such as the percent of copper in an ore or the percent of alcohol in a liquid.

Example 1. Janet Lazdowski has \$20,000 to invest. There is a high-risk investment A that pays 12.5% returns each year. There is a safer investment B that pays 8.6% returns each year. If Janet earned \$2,051.50 on the two investments in one year, then how much did she invest in A and B, respectively?

Solution. **Step 1.** The \$20,000 to be invested is "mixed" into the two investments A and B. The returns from the investments depend in part on the *quantity of dollars* invested in each one. A table is one way to organize the information.

Step 2. Let d represent the number of dollars invested in A.

Items	Quantities mixed	Values of quantities
Investment A	$\$d$ in A	12.5% of A; $0.125d$
Investment B	$\$(20{,}000 - d)$ in B	8.6% of B; $0.086\,(20{,}000 - d)$

Step 3. $\begin{pmatrix}\text{Income}\\ \text{from A}\end{pmatrix} + \begin{pmatrix}\text{Income}\\ \text{from B}\end{pmatrix} = \begin{pmatrix}\text{Total income}\\ \text{earned}\end{pmatrix}$

$$0.125d + 0.086(20{,}000 - d) = 2{,}051.50$$

Step 4.

$$0.125d + 1{,}720 - 0.086d = 2{,}051.50$$

$$0.039d + 1{,}720 = 2{,}051.50$$

$$0.039d = 331.50$$

$$d = 8{,}500$$

$$20{,}000 - d = 11{,}500$$

Step 5. $\$8{,}500 \cdot 0.125 = \$1{,}062.50$ return on A
$\$11{,}500 \cdot 0.086 = \$\ \ 989.00$ return on B
$\$2{,}051.50$ total return, *check*

Step 6. Janet invested $8,500 in A and $11,500 in B.

Example 2. Scott Hatton is a student assistant for the chemistry department at High Tech University. He was instructed to prepare 5 liters of an 8% sulfuric acid (H_2SO_4) solution. In the supply room, flask A has a solution that is 5% H_2SO_4, and flask B has a solution that is 10% H_2SO_4. How many liters from flasks A and B, respectively, should Scott mix to get the required concentration?

Solution.

Step 1. In this problem, the percents of H_2SO_4 in the two flasks are mixed. Thus, the values of the mixtures are not amounts of money, but amounts of acid.

Step 2. Let x represent the amount of solution from flask A.

Solution	Quantity	Percent acid	Amount of acid
Flask A	x liters	$5\% = 0.05$	$0.05x$ liters
Flask B	$(5 - x)$ liters	$10\% = 0.10$	$0.10(5 - x)$ liters
Final solution	5 liters	$8\% = 0.08$	$0.08(5)$ liters

Step 3. $\begin{pmatrix}\text{Acid from}\\ \text{flask A}\end{pmatrix} + \begin{pmatrix}\text{Acid from}\\ \text{flask B}\end{pmatrix} = \begin{pmatrix}\text{Amount of acid in}\\ \text{final solution}\end{pmatrix}$

$$0.05x + 0.10(5 - x) = 0.08(5)$$

Step 4. $0.05x + 0.5 - 0.10x = 0.4$

$$-0.05x = -0.1$$

$$x = 2$$

$$5 - x = 3$$

Step 5. $0.05(2) = 0.10$ liters of acid from flask A.
$0.10(3) = 0.30$ liters of acid from flask B.
0.40 liters of acid in final solution, *check*

Step 6. Scott should mix 2 liters from flask A and 3 liters from flask B.

Motion Problems

In a typical motion problem, some details are given about one or more moving objects. The information may include data on the distance an object moves, the time of the motion, or the rate at which the object is moving. The following formula relates the three items:

$$d = r \cdot t \begin{cases} d \text{ represents } \textit{distance,} \text{ such as miles, kilometers, etc.} \\ r \text{ represents } \textit{rate,} \text{ such as mph, mps, etc.} \\ t \text{ represents } \textit{time,} \text{ such as hours, seconds, etc.} \end{cases}$$

Example 3. Global Flight 242 left Kansas City International, heading west. One hour and 30 minutes later, Luwanna Johnson left the same airport in her Bonanza aircraft, heading east at a rate that was 180 mph slower than Flight 242. Four hours after Global Flight 242 left Kansas City, the two aircraft were 3,190 miles apart. Find the rates at which the two aircraft were flying.

Solution.

Step 1. This problem is an example of *motion in opposite directions.* Thus, the distances the two aircraft covered during the times they were flying are *added* to find the total distance between them.

Step 2. Let r represent the rate of the Global aircraft.

Aircraft	Rate	Time	Distance
Flight 242	r mph	4 hours	$4r$ miles
Bonanza	$(r - 180)$ mph	2.5 hours	$2.5(r - 180)$ miles

Step 3. $\begin{pmatrix}\text{Distance}\\ \text{covered by}\\ \text{Flight 242}\end{pmatrix} + \begin{pmatrix}\text{Distance covered}\\ \text{by Bonanza}\end{pmatrix} = \begin{pmatrix}\text{Total}\\ \text{distance}\end{pmatrix}$

$$4r \quad + \quad 2.5(r - 180) \quad = 3{,}190$$

Step 4. $4r + 2.5r - 450 = 3{,}190$

$$6.5r = 3{,}640$$

$$r = 560 \text{ mph}$$

$$r - 180 = 380 \text{ mph}$$

Step 5. $560 \text{ mph} \cdot 4 \text{ hours} = 2{,}240 \text{ miles (Flight 242)}$
$380 \text{ mph} \cdot 2.5 \text{ hours} = 950 \text{ miles (Bonanza)}$
3,190 miles, *check*

Step 6. Flight 242 was flying at 560 mph, and Luwanna Johnson was flying at 380 mph.

Example 4. Jessie Andrews left Salt Lake City and headed west in her late model Mustang at a speed of 72 mph. Six hours later, Dwayne Pepper left Salt Lake City in his Piper airplane and headed west, following the interstate highway that Jessie was driving. If Dwayne was flying at 360 mph, how far from Salt Lake City were Jessie and Dwayne when the plane overtook the car?

Solution. **Step 1.** This problem is an example of *motion in the same direction,* also referred to as *pursuit-overtake.* That is, Jessie has a head start, but Dwayne overtakes the car because his rate is faster.

Step 2. Let t represent the time that Jessie was driving.

Vehicles	Rate	Time	Distance
Jessie's car	72 mph	t hours	$72t$
Dwayne's plane	360 mph	$(t - 6)$ hours	$360(t - 6)$

Step 3. $\begin{pmatrix}\text{Distance covered} \\ \text{by car}\end{pmatrix} = \begin{pmatrix}\text{Distance covered} \\ \text{by plane}\end{pmatrix}$

$$72t = 360(t - 6)$$

Step 4.

$$72t = 360t - 2{,}160$$

$$2{,}160 = 288t$$

$$7.5 = t \qquad \text{Jessie's time}$$

$$1.5 = t - 6 \qquad \text{Dwayne's time}$$

Step 5. $(7.5 \text{ hours})(72 \text{ mph}) = 540 \text{ miles}$ (distance covered by car)

$(1.5 \text{ hours})(360 \text{ mph}) = 540 \text{ miles}$ (distance covered by plane)

The distances are the same, so the answer checks.

Step 6. Jessie and Dwayne were 540 miles west of Salt Lake City when Dwayne overtook Jessie.

Age Problems

In a typical age problem, the ages of two or more items are compared. Usually a statement is made that compares the current ages of the items. An additional comparison is then made of their ages at some different point in time.

Example 5. The Petersons have three children, Alice, Beth, and Candy. Alice is 2 years older than Beth, and Candy is 3 years younger than Beth. Three times Alice's age in 5 years will be 7 less than 2 times the sum of the ages of Beth and Candy at that time. How old is each child now?

Solution. **Step 1.** Beth's age is the key to this problem. Once her age is determined, the ages of Alice and Candy can then be calculated.

Step 2. Let a represent Beth's current age.

Individuals	Current age	Age in 5 years
Alice	$(a + 2)$ years	$(a + 2) + 5 = (a + 7)$ years
Beth	a years	$a + 5$ years
Candy	$(a - 3)$ years	$(a - 3) + 5 = (a + 2)$ years

Step 3. "Three times Alice's age in 5 years" is written $3(a + 7)$. "Twice the sum of the ages of Beth and Candy in 5 years" is written $2[(a + 5) + (a + 2)]$. An equation of the problem is $3(a + 7) + 7 = 2[(a + 5) + (a + 2)]$.

Step 4. $3a + 21 + 7 = 2[2a + 7]$

$$3a + 28 = 4a + 14$$

$$14 = a$$

Hence, Beth is 14, Alice is $14 + 2 = 16$, and Candy is 11.

Step 5. In 5 years, Alice's age will be 21, Beth's age will be 19, and Candy will be 16.

$$3(21) + 7 = 2(19 + 16)$$

$$63 + 7 = 2(35), \textit{ check}$$

Step 6. Beth is 14 years old, Alice is 16 years old, and Candy is 11 years old.

SECTION 2-3. Practice Exercises

In exercises **1–20**, solve each word problem.

[Examples 1 and 2]

1. For last Saturday's football game in Ashland, general admission tickets cost $6.50 each. Student tickets cost $3.00 each. The number of student tickets sold for the game was 700 less than 3 times the number of general admission tickets sold. If the total receipts for the game were $18,050, how many of each kind of ticket was sold?

2. The Foxy Lady Dress Shoppe got a shipment of ladies' skirts, sweaters, and belts. The wholesale prices were $32 each for the skirts, $21 each for the sweaters, and $9 each for the belts. The number of skirts in the shipment was 5 more than twice the number of belts. The number of sweaters was 3 times the number of belts decreased by 5. If the total value of the merchandise was $5,223, how many skirts, sweaters, and belts, respectively, were in the shipment?

3. Frieda King owns some municipal bonds that pay 8% per year in interest. She also owns some stock in a utility company that pays 13% each year in dividends. The amount she has in bonds is $1,400 less than twice the amount in utility company stocks. Last year Frieda earned $2,440 on her investments. How much does she have invested in bonds and how much in stocks?

4. Art Levinski has been saving nickels, dimes, and quarters for a short time. He currently has $13.85 in his savings. The number of nickels is 3 more than 3 times the number of quarters. The number of dimes is 1 less than twice the number of quarters. How many of each kind of coin does he currently have in his collection?

5. In a hospital stock room, flask A contains a mixture with 40% alcohol. Flask B contains a similar mixture, but with 15% alcohol. Sue Ellen needs 20 liters of the mixture with 25% alcohol. How much mixture should she take from flask A and how much from flask B?

6. Pete Jablonski needs 400 pounds of concrete mix that contains 15% cement to set some fence posts. In the construction yard, pile A contains concrete mix with 20% cement. Pile B contains concrete mix with 12.5% cement. How much mix, to the nearest pound, should he take from each pile?

[Examples 3 and 4]

7. A train left Philadelphia and headed north. Three hours later a bus left Philadelphia and headed south, traveling 8 mph less than the train's rate. Six hours after the train left the city, the two vehicles were 516 miles apart. Find the rates of the train and bus.

8. Nancy left City Park on a bicycle and headed west. Thirty minutes later, Fred left the park heading east, jogging at a rate of 8 mph less than Nancy's rate. Forty-five minutes after Fred began jogging, the two were 18 miles apart. Find the rates of Nancy and Fred.

9. Al and Betty go to work in opposite directions when they leave their home in the morning. To get to work by 8:00, Al must leave home at 7:00, since he must use cross streets that limit his average speed to 25 mph. Betty leaves home at 7:30 to get to work by 8:00, since she uses the expressway, on which she averages 55 mph. How far apart are Al and Betty when they are at work?

10. After dinner at the River Front Restaurant, the Kubota family motored their boat upriver at an average rate of 15 mph. The Andrews motored their boat downriver at 25 mph. Both families left the restaurant at the same time.
a. How long would it take for them to be 24 miles apart?
b. How far did the Kubota family travel in this time?

11. Jacob left Atlanta, Georgia, in his car and traveled west on an interstate highway at an average speed of 56 mph. Exactly 30 minutes later, Claudine, a highway patrol officer, left Atlanta to patrol the same interstate highway. She also headed west at an average speed of 70 mph.
a. How long did it take Claudine to overtake Jacob?
b. How far west of Atlanta were they at that time?

12. A cargo ship left Boston harbor at 5:00 A.M, heading east at 24 knots (nautical mph). At 11:00 A.M., a Coast Guard amphibious helicopter left the harbor, heading east to overtake the ship at a speed of 216 knots.
a. How many minutes did it take the helicopter to overtake the ship?
b. How many miles east of the harbor were they at that time?

13. A Trans Global Airway flight left O'Hare International airport, heading south at an average speed of 520 mph. A second Trans Global Airway flight left O'Hare 30 minutes later, heading south at an average speed of 460 mph.
a. In how many hours were the two planes 500 miles apart?
b. How many miles south of O'Hare was the first plane at that time?
c. How many miles south of O'Hare was the second plane at that time?

14. Mr. Bercut's car was removed from a "No Parking" zone by International Towing Service. The truck headed to the storage yard at an average rate of 5 mph through town. Mr. Bercut was able to hail a taxi and head for the storage yard half an hour after the tow truck left the zone. If the taxi can average 10 mph, how long will it take Mr. Bercut to overtake the tow truck?

[Example 5]

15. Gladys and Harry Porter recently celebrated their 25th wedding anniversary. Harry is 5 years older than Gladys. Four times Harry's age when they were married is 4 years less than 2 times Harry's age now. How old are Gladys and Harry?

16. The Post Office building in Terrytown was 10 years old when the Court House building was built. Three times the age of the Post Office 15 years ago is the same as two times the age of the Court House in 20 years. How old is each building?

17. Becky, Carol, and Darlene are sisters. Becky is 6 years older than Carol, and Darlene is 8 years younger than Carol. Three times the sum of the ages of Carol and Darlene 5 years ago is the same as 3 times Becky's age now. How old are Becky, Carol, and Darlene?

18. Kevin, Lance, and Mike are brothers. Kevin is 4 years older than Lance, and Mike is 7 years younger than Lance. Nine times Mike's age in 2 years is the same as 2 times the sum of Kevin's and Lance's ages in 3 years. How old are Kevin, Lance, and Mike?

19. Valerie inherited a ring and pin that belonged to her grandmother. The ring is 25 years older than the pin. The product of 3 times the difference between the current age of the pin and 5 is 10 less than 2 times the current age of the ring. How old is each piece of jewelry?

20. A table, a desk, and a lamp are three items in an antique shop. The age of the desk is 20 years more than the table, and the age of the lamp is 15 years less than the table. Three times the sum of the ages of the table and lamp in 12 years is 3 years less than 5 times the age of the desk at that time. How old is each item?

SECTION 2-3. Ten Review Exercises

In exercises **1** and **2**, find the LCM of each set of numbers.

1. 14, 21, and 35

2. 10, 52, and 65

In exercises **3** and **4**, list the whole number factors of each number.

3. 42

4. 110

In exercises **5** and **6**, simplify each expression.

5. $5^2 - [-3(2 - 7) + 10(1 - 2^2)]$

6. $\dfrac{3 \cdot 6 - 5^2}{1 + 3^2} - \dfrac{2(5 - 3^2)}{2^2 + 3 \cdot 2}$

In exercises **7** and **8**, solve each equation.

7. $-3[5 - 4(1 - t) + 7] = 2(1 - 5t) - 8$

8. $3(u + 5) + 2 = 10 + 2(3 + u) + 1$

In exercises **9** and **10**, solve each word problem.

9. If 7 is added to the product of 10 and some number, the result is the same as when 4 times the number is subtracted from 12. Find the number.

10. The product of 3 times Tim's age 8 years ago is the same as 2 times Tim's age in 1 year. How old is Tim?

SECTION 2-3. Supplementary Exercises

In exercises **1** and **2**, solve each word problem.

1. An amount of a 60% solution is to be mixed with an amount of 20% solution to produce a 40% mixture. How much of the 60% solution will be needed if the amount of 20% solution used is:
a. 1 gallon?
b. 2 gallons?
c. 4 gallons?

2. An amount of a 60% solution is to be mixed with an amount of 20% solution to produce a 50% mixture. How much of the 60% solution will be needed if the amount of 20% solution used is:
a. 1 gallon?
b. 2 gallons?
c. 4 gallons?

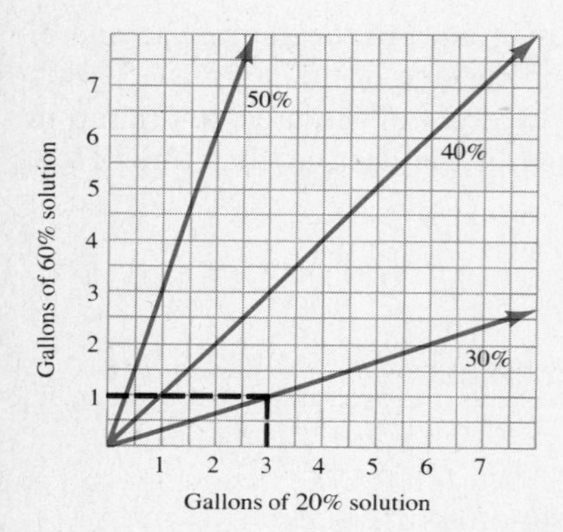

Figure 2-8.

The graphs in Figure 2-8 can be used to find the amount of 60% solution needed to mix with a 20% solution to produce either a 30%, 40% or 50% mixture.

Example. How many gallons of 60% solution are needed to mix with 3 gallons of 20% solution to produce a 30% mixture?

Solution. Locate 3 gallons on the 20% axis. Move up to the 30% line. Now move to the left to the 60% axis and read 1 gallon. Thus, 1 gallon of 60% solution mixed with 3 gallons of 20% solution will produce a 30% mixture.

In exercises **3–6**, use the graphs in Figure 2-8 to determine the number of gallons of 60% solution needed to mix with the given amounts of 20% solution to obtain the indicated percent solution.

3. A 30% solution, given:
a. 1.5 gallons of 20% solution
b. 4.5 gallons of 20% solution

4. A 40% solution, given:
a. 1 gallon of 20% solution
b. 3 gallons of 20% solution

5. A 50% solution, given:
a. 1 gallon of 20% solution
b. 2.5 gallons of 20% solution

6. A 40% solution, given:
a. 2.5 gallons of 20% solution
b. 5.5 gallons of 20% solution

In exercises **7–10**, use the graphs in Figure 2-8 to determine the number of gallons of 20% solution needed to mix with the given amounts of 60% solution to obtain the indicated percent solution.

7. A 30% solution, given:
a. 2 gallons of 60% solution
b. 2.5 gallons of 60% solution

8. A 40% solution, given:
a. 4.5 gallons of 60% solution
b. 3.5 gallons of 60% solution

9. A 50% solution, given:
a. 3 gallons of 60% solution
b. 7.5 gallons of 60% solution

10. A 50% solution, given:
a. 4.5 gallons of 60% solution
b. 6.0 gallons of 60% solution

We can use the graph in Figure 2-9 to approximate the number of gallons of 20% solution needed to mix with 20 gallons of 60% solution to obtain the required percent solution.

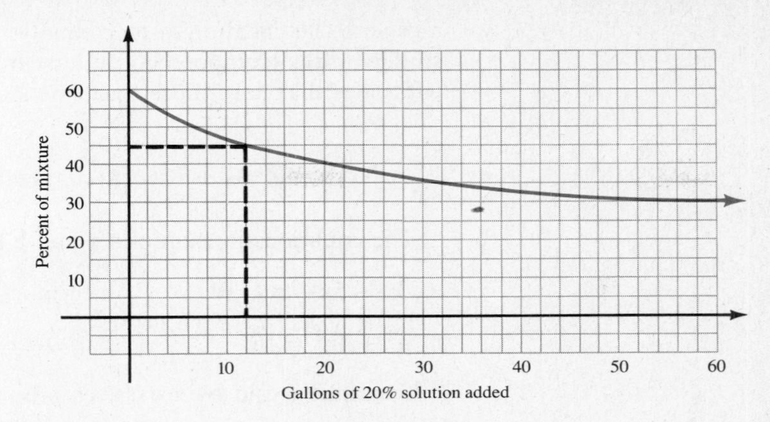

Figure 2-9.

Example. If 12 gallons of a 20% solution are mixed with a 20 gallons of a 60% solution, what will be the percent solution of the mixture?

Solution. Locate 12 gallons on the Gallons of 20% Solution Added axis and move up to the graph. Now move left to the Percent of Mixture axis and read 45%. Thus, the mixture will be a 45% solution.

In exercises **11–14**, use the graph in Figure 2-9 to estimate the percent of the solution for the given number of gallons of 20% solution added to the 20 gallons of 60% solution.

11. 6 gallons

12. 10 gallons

13. 30 gallons

14. 50 gallons

In exercises **15–18**, use the graph in Figure 2-9 to answer each question.

15. If 0 gallons of 20% is added, then what is the percent of the mixture?

16. What happens to the percent of the mixture as the quantity of 20% solution is increased?

17. **a.** Theoretically, what is the smallest possible percent of the solution?
b. How much 20% solution must be added to reach the theoretical limit?

18. **a.** Theoretically, what is the largest possible percent of the solution?
b. How much 20% solution must be added to reach the theoretical limit?

SECTION 2-4. Literal Equations

1. The meaning of a literal equation
2. Solving a literal equation for one variable
3. Literal equations that are called formulas
4. Evaluating a formula

The equations we studied in Section 2-1 are examples of equations in one variable. Equations can also be written with more than one variable. Although many of these types of equations are put into categories based on specific characteristics, a general classification of such equations is **literal.** The fact that a literature course studies *words* (composed of letters) may help you to associate literal equations with ones that have more than one *letter* (that is, variables).

The Meaning of a Literal Equation

The equations in examples **a** and **b** have more than one variable.

a. $3x + 5y - z = 10$ An equation in x, y, and z

b. $u^2 + 4v^2 = 16$ An equation in u and v

Examples **a** and **b** illustrate equations that can be called literal.

> A **literal equation** is one that has more than one variable.

A solution of a linear equation in one variable is a real number that makes the equation true. A solution of a literal equation is not a single real number, but a *group of numbers,* one for each variable in the equation. For example, (1, 2, 3) is a solution for $3x + 5y - z = 10$ when it is understood that 1 is the replacement for x, 2 is the replacement for y, and 3 is the replacement for z:

$$3(1) + 5(2) - 3 = 10$$

$$3 + 10 - 3 = 10, \textit{ true}$$

In this section we will not be primarily concerned with determining solutions of literal equations, but rather with manipulating the equations using the addition and multiplication properties of equality to change their forms. To illustrate, examples **c–e** are three different, but **equivalent equations,** of the one in example **a.** They are equivalent equations in that any solution of one of these equations is also a solution of the others. We can verify that (1, 2, 3) is a solution of examples **c–e** by replacing x with 1, y by 2, and z by 3.

c. $x = \dfrac{10 - 5y + z}{3}$ Solved for x in terms of y and z

d. $y = \dfrac{10 - 3x + z}{5}$ Solved for y in terms of x and z

e. $z = 3x + 5y - 10$ Solved for z in terms of x and y

Solving a Literal Equation for One Variable

Examples **c–e** illustrate that it may be possible to write a literal equation in an equivalent form with one of the variables isolated on one side of the equation. This form of the equation is said to be solved for that variable in terms of the remaining variables in the equation.

> **To solve a literal equation for one of the variables in the equation:**
>
> **Step 1.** Write the term (or terms) containing the specified variable on one side of the equation and the other terms on the opposite side.
>
> **Step 2.** If necessary, change the coefficient of the specified variable to one.

Example 1. Solve $4u - v - 2w = 9$ for w.

Solution. **Discussion.** There is only one w-term in the equation. To change the operation of subtraction on this term to addition, isolate the term on the right side.

$$4u - v - 2w = 9 \quad \text{The given equation}$$

$$4u - v = 9 + 2w \quad \text{Add } 2w \text{ to both sides.}$$

$$4u - v - 9 = 2w \quad \text{Subtract 9 from both sides.}$$

$$\frac{4u - v - 9}{2} = w \quad \text{Divide both sides by 2.}$$

$$w = \frac{4u - v - 9}{2} \quad \text{The preferred form}$$

Example 2. Solve $3t + a = bt + 2$ for t.

Solution. **Discussion.** There are two t-terms in this equation. To change the two t-terms to one, factor the t using the distributive property (See Section 1-3).

$$3t + a = bt + 2 \quad \text{The given equation}$$

$$3t - bt = 2 - a \quad \text{Subtract } bt \text{ and } a \text{ from both sides.}$$

$$(3 - b)t = 2 - a \quad \text{Factor a } t \text{ on the left side.}$$

$$t = \frac{2 - a}{3 - b} \quad \text{Divide both sides by } 3 - b.$$

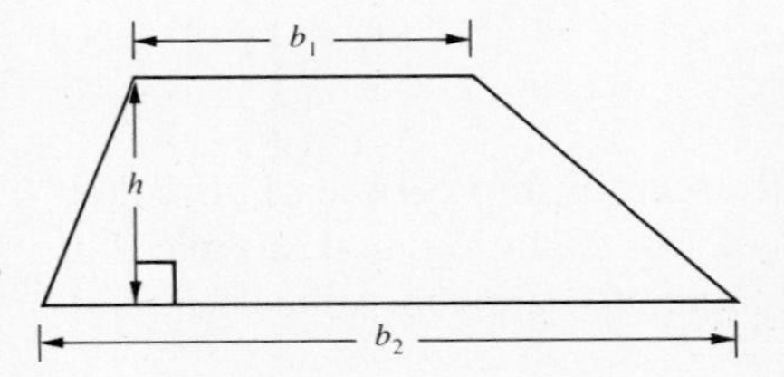

Figure 2-10. The area of a trapezoid is $A = \frac{1}{2}h(b_1 + b_2)$.

Literal Equations That Are Called Formulas

Many literal equations are related to specific quantities such as area, volume, interest, and distance. For example, $A = \frac{1}{2}h(b_1 + b_2)$ is an equation for computing the area A of a trapezoid with altitude h and parallel bases b_1 and b_2, as shown in Figure 2-10. Because this equation always represents a true relationship between the bases, altitude, and area of any trapezoid, it is called a **formula.**

Example 3. Solve $A = \frac{1}{2}h(b_1 + b_2)$ for b_1.

Solution. **Discussion.** Whenever a formula has a fraction in the equation, it is a good idea to eliminate it.

$$A = \frac{1}{2}h(b_1 + b_2) \quad \text{The given equation}$$

$$2A = h(b_1 + b_2) \quad \text{Multiply both sides by 2.}$$

$$2A = hb_1 + hb_2 \quad x(y + z) = xy + xz$$

$$2A - hb_2 = hb_1 \quad \text{Subtract } hb_2 \text{ from both sides.}$$

$$\frac{2A - hb_2}{h} = b_1 \quad \text{Divide both sides by } h.$$

$$b_1 = \frac{2A - hb_2}{h} \quad \text{The formula solved for } b_1$$

The equation $5F - 9C = 160$ is a formula that shows a relationship between the Fahrenheit and Celsius scales for measuring temperature. In the Fahrenheit scale, water freezes at 32° and boils at 212°. In the Celsius scale, water freezes at 0° and boils at 100°.

Example 4. Solve $5F - 9C = 160$ for C.

Solution. **Discussion.** To change the indicated subtraction to addition, add the $9C$-term to both sides. In this way, we can ultimately divide both sides of the equation by 9, and not -9. The preferred form of any expression is one which has a positive number for a denominator.

$5F - 9C = 160$ The given equation

$5F = 160 + 9C$ Add $9C$ to both sides.

$5F - 160 = 9C$ Subtract 160 from both sides.

$\frac{5F - 160}{9} = C$ Divide both sides by 9.

Another form of this equation frequently seen is obtained by factoring a 5 in the numerator and forming the fraction $\frac{5}{9}$:

$$C = \frac{5}{9}(F - 32)$$

Evaluating a Formula

The equation $5F - 9C = 160$ shows an *implied* relationship between Fahrenheit and Celsius temperature readings. The form of this equation in Example 4 is *explicit*, in that a value of C can readily be obtained for a given value of F.

Value of F	$C = \frac{5}{9}(F - 32)$	Corresponding value of C
$-4°$	$C = \frac{5}{9}(-4 - 32) = -20$	$-20°$
32°	$C = \frac{5}{9}(32 - 32) = 0$	0°
98.6°	$C = \frac{5}{9}(98.6 - 32) = 37$	37°
212°	$C = \frac{5}{9}(212 - 32) = 100$	100°

Example 5. Given: $A = P + Prt$

a. Solve for P.

b. Find P if $A = 18{,}875$, $r = 8.5\%$, and $t = 6$.

Solution. **Discussion.** This formula shows the relationship between:

A—the *amount* in an account

P—the *principal* or lump-sum deposit into the account

r—the annual *rate* of interest earned by the principal

t—the *time,* in years, the principal draws interest

a. $A = P + Prt$ — The given equation

$A = P(1 + rt)$ — Factor a P.

$\frac{A}{1 + rt} = P$ — Divide both sides by $(1 + rt)$.

$P = \frac{A}{1 + rt}$ — If $x = y$, then $y = x$.

b. $P = \frac{18{,}875}{1 + (0.085)(6)}$ — Replace A by 18,875, r by 0.085, and t by 6.

$= \frac{18{,}875}{1.510}$ — Simplify the denominator.

$= 12{,}500$

Thus, a lump-sum deposit of \$12,500 into an account with an annual rate of interest of 8.5% for 6 years will grow to \$18,875.

SECTION 2-4. Practice Exercises

In exercises **1–12**, solve each equation for the indicated variable.

[Example 1] Given: $2a + 3b = 7$

1. Solve for a.

2. Solve for b.

Given: $P = abc$

3. Solve for b.

4. Solve for c.

Given: $13 - 5x = -4y + z$

5. Solve for x.

6. Solve for y.

Given: $y = mx + b$

7. Solve for m.

8. Solve for x.

Given: $D = b^2 - 4ac$

9. Solve for a.

10. Solve for c.

Given: $m = \frac{a - b}{2}$

11. Solve for b.

12. Solve for a.

In exercises **13–36**, solve each equation for the indicated variable.

[Example 2] **13.** $ax - bx = c$, for x

14. $py - qy = 7$, for y

15. $S = 5 + nd - d$, for d

16. $p = 9 + s - st$, for s

17. $5m + 7 = am - b$, for m

18. $pm - 3 = am + k$, for m

19. $2b + 4y = b^2y + 5$, for y

20. $10y - b^2 = 3by + 2$, for y

[Examples 3 and 4] **21.** $V = \frac{1}{3}\pi r^2 h$, for h

22. $V = s^2h$, for h

23. $P = 2l + 2w$, for w

24. $P = 2l + 2w$, for l

25. $P = 2a + b + c$, for a

26. $E = 4l + 4w + 4h$, for w

27. $Fr = mv^2$, for m

28. $Fr^2 = GMN$, for G

29. $T = I + Id$, for I

30. $a = S - Sr$, for S

31. $xy + x = y$, for y

32. $2xy + x = y - 1$, for y

33. $a^2 + b^2 = c^2$, for b^2

34. $D = b^2 - 4ac$, for b^2

35. $Fr^2 = GMN$ for r^2

36. $Fr = mv^2$, for v^2

In exercises **37–44**, answer parts **a–c**.

[Example 5] **37. a.** Solve $x + 2y = 8$ for y.
b. Find y when $x = -6$.
c. Find y when $x = 10$.

38. a. Solve $4x - 3y = 12$ for y.
b. Find y when $x = -9$.
c. Find y when $x = 6$.

39. a. Solve $2E = mv^2$ for m.
b. Find m for $E = 48$ and $v^2 = 16$.
c. Find m for $E = 294$ and $v^2 = 49$.

40. a. Solve $gT^2 = 2\pi\ell$ for g.
b. Find g for $\ell = 64$ and $T^2 = 40$. Leave π in your answer.
c. Find g for $\ell = 12$ and $T^2 = 8$. Leave π in your answer.

41. a. Solve $S - Sr = a$ for r.
b. Find r for $S = 20$ and $a = 10$.
c. Find r for $S = -18$ and $a = -6$.

42. a. Solve $at = v - s$ for v.
b. Find v for $a = 8$, $t = 10$, and $s = 30$.
c. Find v for $a = -5$, $t = 15$, and $s = 160$.

43. a. Solve $y - k = a(x - h)^2$ for a.
b. Find a for $x = 5$, $y = 50$, $h = 1$, and $k = 2$.
c. Find a for $x = -6$, $y = -11$, $h = -4$, and $k = -3$.

44. **a.** Solve $x^2 + y^2 - 2ax = 0$ for a.
b. Find a when $x = 8$ and $y = 4$.
c. Find a when $x = 1$ and $y = -3$.

SECTION 2-4. Ten Review Exercises

In exercises **1–4**, simplify each expression.

1. $12 - (-8) + 15 - 13$

2. $\dfrac{-3}{8} - \dfrac{-5}{12} + \dfrac{7}{24}$

3. $\left(5\dfrac{1}{2}\right) \div \left(-7\dfrac{1}{3}\right)$

4. $\dfrac{-420}{-35}$

5. Evaluate $3(1 - 4k) + 7 - 8k + 4(k + 2) + 14$ for $k = 2$.

6. Simplify $3(1 - 4k) + 7 - 8k + 4(k + 2) + 14$.

7. Evaluate the expression obtained in Exercise **6** for $k = 2$.

8. Solve $3(1 - 4k) + 7 = 8k + 4(k + 2) + 14$.

9. Solve $y = ax^2 + c$ for x^2.

10. Solve $y = ax^2 + bx + c$ for b.

SECTION 2-4. Supplementary Exercises

In exercises **1–10**, solve each equation for the indicated variable.

1. $P = \dfrac{x}{n}$, for n

2. $z = \dfrac{w}{s}$, for s

3. $z = \dfrac{x - m}{p}$, for p

4. $w = \dfrac{x - 3}{y}$, for y

5. $F = \dfrac{mv^2}{r}$, for r

6. $F = \dfrac{GMN}{r^2}$, for r^2

7. $a^2x = 3x + a - 2$, for x

8. $1 - b + b^2y = 4y$, for y

9. $ax - ay + az = 128$, for a

10. $sp + sq - sr = -30$, for s

In exercises **11–16**, use the graph in each figure to find the value of the missing variable.

11. Use the graph in Figure 2-11.
a. If $x = 0$, then $y = ?$
b. If $x = 5$, then $y = ?$
c. If $y = 8$, then $x = ?$
d. If $y = -3$, then $x = ?$

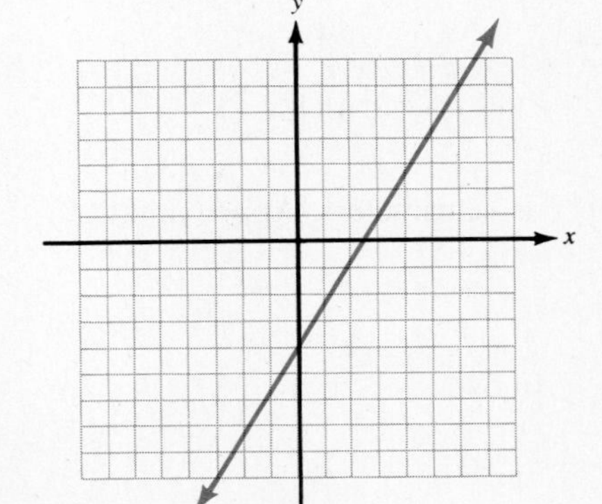

Figure 2-11. A graph of $y = mx + b$.

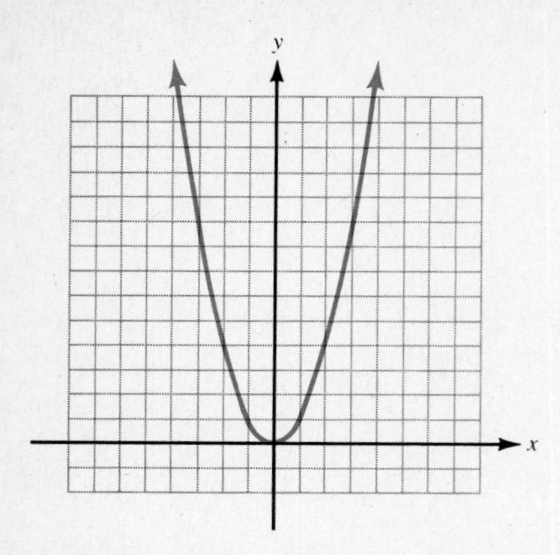

Figure 2-12. A graph of $y = ax^2$.

12. Use the graph in Figure 2-12.
   a. If $x = -3$, then $y = ?$
   b. If $x = -2$, then $y = ?$
   c. If $y = 0$, then $x = ?$
   d. If $y = 1$, then $x = ?$ or ?

13. Use the graph in Figure 2-13.
   a. If $x = 1$, then $y = ?$
   b. If $y = 1$, then $x = ?$
   c. If $x = 5$, then $y = ?$
   d. If $y = 0$, then $x = ?$

14. Use the graph in Figure 2-14.
   a. If $x = 1$, then $y = ?$
   b. If $x = 0$, then $y = ?$
   c. If $y = -1$, then $x = ?$
   d. If $y = 8$, then $x = ?$

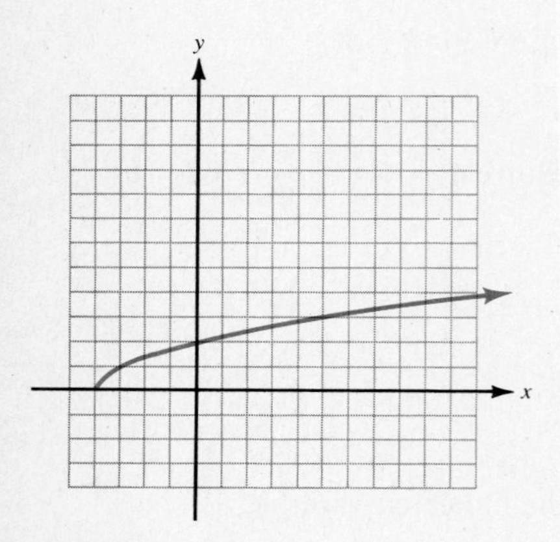

Figure 2-13. A graph of $y^2 = x + h$.

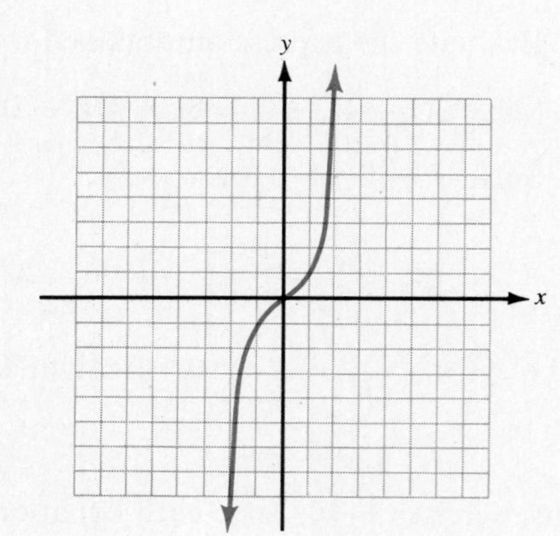

Figure 2-14. A graph of $y = ax^3$.

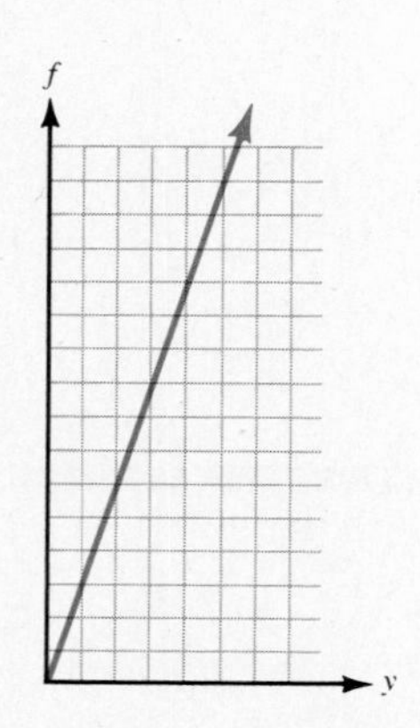

Figure 2-15. A graph of $f = 3y$.

15. In the graph in Figure 2-15, y is yards and f is feet.
   a. If $y = 3$, then $f = ?$
   b. If $y = 4.5$, then $f = ?$
   c. If $f = 1$, then $y = ?$
   d. If $f = 12$, then $y = ?$

16. In the graph in Figure 2-16, i is inches and c is centimeters. Approximate to the nearest whole number.
   a. If $i = 3$, then $c \approx ?$
   b. If $i = 5$, then $c \approx ?$
   c. If $c = 10$, then $i \approx ?$
   d. If $c = 15$, then $i \approx ?$

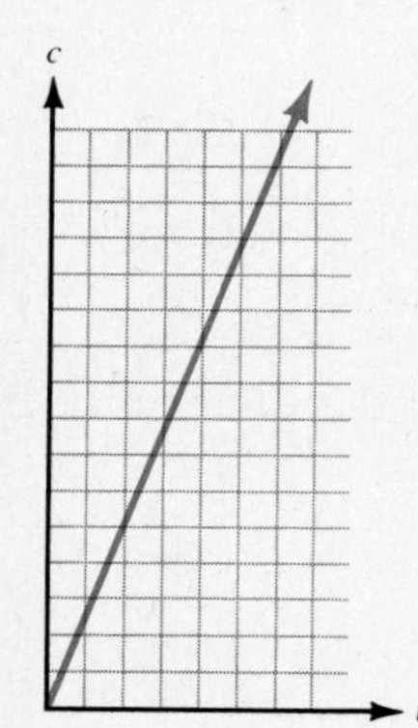

Figure 2-16. A graph of $c = 2.54i$.

In exercises **17–22**, write a formula for the perimeter of each figure using the fewest symbols.

17.

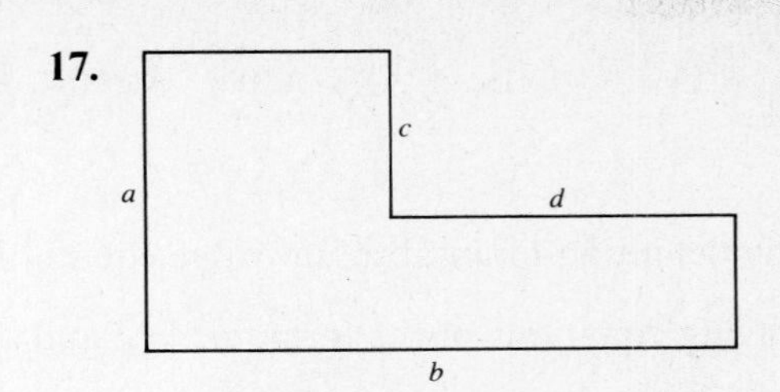

18.

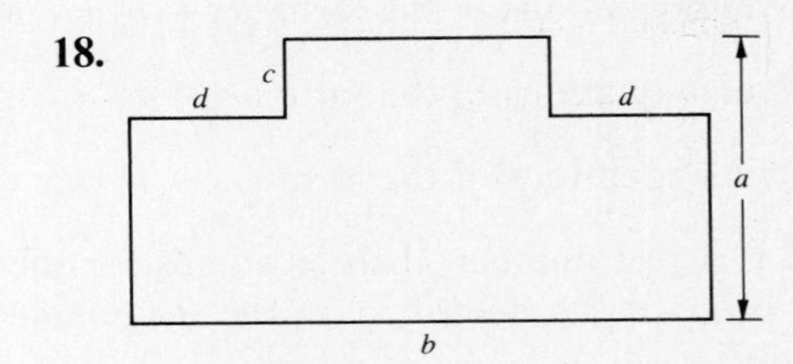

19.

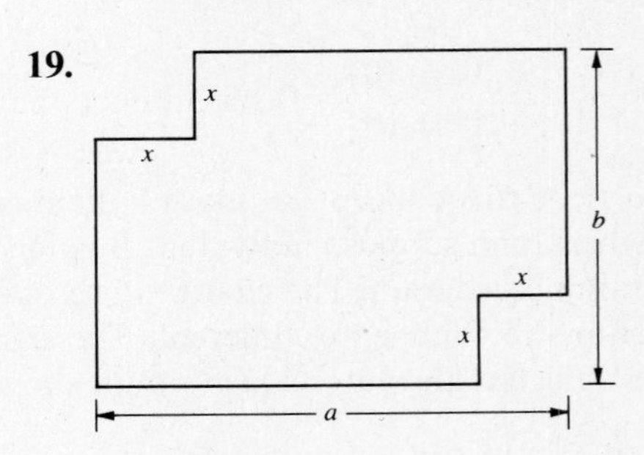

20.

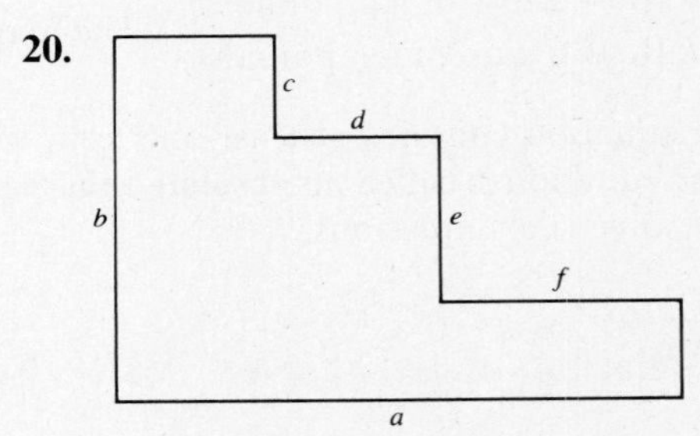

21.

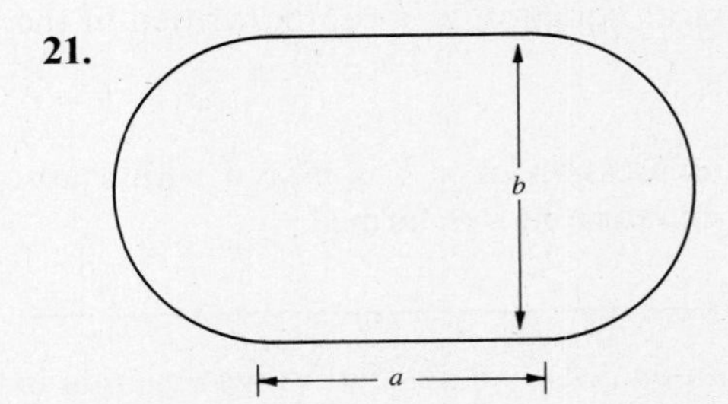

22.

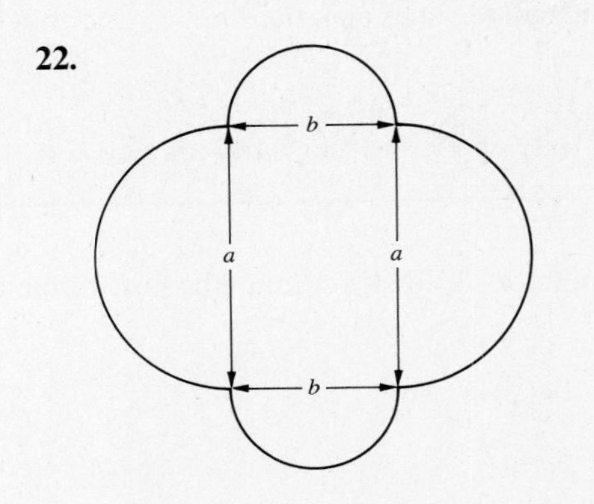

SECTION 2-5. Absolute Value Equations

1. The definition of an absolute value equation
2. Solving equations of the form $|ax| = c$ and $c \geq 0$
3. Solving equations of the form $|ax + b| = c$ and $c \geq 0$
4. Solving equations of the form $|ax + b| = c$ and $c < 0$
5. Solving equations of the form $|ax + b| = |cx + d|$

If k is a real number, then $|k|$ stands for the **absolute value** of k. The absolute value of k can be described as the **magnitude** of k, but not the sign of k. For example:

$$\left.\begin{aligned} |+10| &= 10 \\ |-10| &= 10 \end{aligned}\right\} \quad \text{Both numbers have a magnitude of 10.}$$

To illustrate this concept, suppose individual A lost 10 pounds as a result of a diet. Furthermore, suppose individual B gained 10 pounds as a result of a special weight training program. The change in the weights of A and B is 10 pounds, but the directions of change are different. The signs of the numbers would show the *directions*, but the absolute values would show the *amounts*.

$$\left.\begin{aligned} -10 &\rightarrow \text{a loss of ten pounds} \\ +10 &\rightarrow \text{a gain of ten pounds} \end{aligned}\right\} \quad \text{The magnitude of change is the same.}$$

If an equation contains at least one term with a variable inside absolute value bars, the equation is called an **absolute value equation.** In this section, we will study how to solve such equations.

The Definition of an Absolute Value Equation

A linear equation in x can be written in the form:

$$ax + b = c$$

If the expression $ax + b$ is placed within absolute value bars, then an absolute value equation in x is formed.

Definition 2.4. An absolute value equation in x
An **absolute value equation** in x is one that can be written in the form

$$|ax + b| = c$$

where a, b, and c are real numbers and $a \neq 0$.

Examples **a–d** illustrate four absolute value equations.

a. $|2x| = \dfrac{3}{4}$ An absolute value equation in x

b. $|3y - 5| = 7$ An absolute value equation in y

c. $10 + |6 - t| = 3$ An absolute value equation in t

d. $|2u - 1| = |5 - u|$ An absolute value equation in u

Solving Equations of the Form $|ax| = c$ and $c \geq 0$

Consider the following absolute value equation in which $a = 1$, $b = 0$, and $c = 15$:

$$|x| = 15$$

The solutions of this equation are numbers whose absolute values are 15. Since $|15| = 15$ and $|-15| = 15$, the solution set of the equation is $\{-15, 15\}$.

Example 1. Solve $|2t| = 3$.

Solution. **Discussion.** The roots of this equation are real numbers t, such that the absolute value of $2t$ is 3. Since the absolute values of -3 and 3 are both 3, we remove the absolute value bars and write two equations.

(1) $2t = 3$ and (2) $2t = -3$

$t = \frac{3}{2}$ $t = \frac{-3}{2}$

The solution set is $\left\{\frac{-3}{2}, \frac{3}{2}\right\}$.

Solving Equations of the Form $|ax + b| = c$ and $c \geq 0$

If the expression inside the absolute value bars is a binomial, then additional steps are needed to find the solutions after writing equations (1) and (2).

> If $|ax + b| = c$, where a, b, and c are real numbers with $a \neq 0$ and $c \geq 0$, then remove the absolute value bars and write equations (1) and (2):
>
> (1) $ax + b = c$ and (2) $ax + b = -c$
>
> Now solve (1) and (2) for x. If $c = 0$, then equations (1) and (2) are the same.

Example 2. Solve $|3y - 5| = 7$.

Solution. **Discussion.** Since $7 > 0$, we can apply the recommended procedure for solving.

$|3y - 5| = 7$ The given equation

(1) $3y - 5 = 7$ or (2) $3y - 5 = -7$

$3y = 12$ $3y = -2$

$y = 4$ $y = \frac{-2}{3}$

The solution set is $\left\{\frac{-2}{3}, 4\right\}$.

Example 3. Solve $6 = 19 - \left|15 - \frac{1}{3}x\right|$

Solution. **Discussion.** Before removing the absolute value bars and writing equations (1) and (2), we must first write the equation with the absolute value expression by itself on one side with a positive coefficient.

$$6 = 19 - \left|15 - \frac{1}{3}x\right| \qquad \text{The given equation}$$

$$6 + \left|15 - \frac{1}{3}x\right| = 19 \qquad \text{Add } \left|15 - \frac{1}{3}x\right| \text{ to both sides.}$$

$$\left|15 - \frac{1}{3}x\right| = 13 \qquad \text{Subtract 6 from both sides.}$$

$$(1)\quad 15 - \frac{1}{3}x = 13 \qquad \text{or} \qquad (2)\quad 15 - \frac{1}{3}x = -13$$

$$\frac{-1}{3}x = -2 \qquad\qquad \frac{-1}{3}x = -28$$

$$x = 6 \qquad\qquad x = 84$$

The solution set is $\{6, 84\}$.

Solving Equations of the Form $|ax + b| = c$ and $c < 0$

Consider the following equation:

$$|5u| = -8.$$

On the left side of the equation is an absolute value expression. For any real number u, $|5u|$ will be greater than or equal to 0. On the right side of the equation is a negative number. Since no value of u can yield a negative number on the left side, the solution set is empty, that is, $\varnothing$.

> If $|ax + b| = c$ and $c < 0$, then the solution set is $\varnothing$.

Example 4. Solve $10 + |6 - t| = 3$.

Solution.

$$10 + |6 - t| = 3 \qquad \text{The given equation}$$

$$|6 - t| = -7 \qquad \text{Subtract 10 from both sides.}$$

Since $-7 < 0$, the solution set is $\varnothing$.

Solving Equations of the Form $|ax + b| = |cx + d|$

Consider the following equation:

$$|x| = |c|, \text{ where } c \text{ is a real number}$$

To remove both absolute value bars, we need to write four equations:

$$(1)\quad x = c \qquad \text{or} \qquad (2)\quad x = -c \qquad \text{or}$$

$$(3)\quad -x = c \qquad \text{or} \qquad (4)\quad -x = -c$$

By multiplying both sides of equations (3) and (4) by -1, they can be changed to the forms of equations (2) and (1), respectively. Therefore, the solutions of $|x| = |c|$ can be found by solving only equations (1) and (2).

Example 5. Solve $|2u - 1| = |5 - u|$.

Solution. **Discussion.** Remove the absolute value bars and write equations (1) and (2). Write $5 - u$ as a grouped expression preceded by a minus bar:

(1) $2u - 1 = 5 - u$	or	(2) $2u - 1 = -(5 - u)$
$3u = 6$		$2u - 1 = -5 + u$
$u = 2$		$u = -4$

The solution set is $\{-4, 2\}$.

SECTION 2-5. Practice Exercises

In exercises **1–46**, solve each equation. If an equation has no solution, write $\varnothing$.

[Examples 1–4]

1. $|x| = 6$

2. $|x| = 8$

3. $|y| = \sqrt{3}$

4. $|y| = \pi$

5. $|z| = -10$

6. $|z| = -20$

7. $|3t| = 0$

8. $|4t| = 4$

9. $|5k| = 1$

10. $|6k| = 2$

11. $\left|\frac{u}{6}\right| = 12$

12. $\left|\frac{u}{4}\right| = 8$

13. $|x + 1| = 13$

14. $|x + 7| = 3$

15. $|y - 6| = 2$

16. $|y - 10| = 20$

17. $|2t - 1| = 7$

18. $|2t - 3| = 13$

19. $|4u + 3| = 3$

20. $|7u + 2| = 2$

21. $\left|\frac{2y - 1}{3}\right| = 5$

22. $\left|\frac{3y + 2}{5}\right| = 2$

23. $\left|\frac{z - 6}{2}\right| = 0$

24. $\left|\frac{z + 5}{3}\right| = 0$

25. $|2 + 5k| - 7 = 10$

26. $|1 + 3k| - 8 = 2$

27. $5 + \left|\frac{1}{2}x + 7\right| = 5$

28. $\left|\frac{1}{3}x - 2\right| + 3 = 3$

29. $9 + \left|\frac{3}{2}y - 1\right| = 4$

30. $7 + \left|\frac{5}{4}y + 3\right| = 2$

31. $12 - \left|\frac{2}{3}t + 6\right| = 8$

32. $18 - \left|\frac{8}{3}t + 4\right| = 16$

33. $3|3u + 1| + 7 = 9$

34. $2|2u - 1| + 5 = 6$

35. $13 + 5|4 - v| = 38$

36. $6 + 3|7 - 2v| = 39$

[Example 5] **37.** $|x| = |3x - 8|$

38. $|2x| = |5 - x|$

39. $|x + 2| = |2x - 3|$

40. $|x - 5| = |3x + 1|$

41. $|5x - 3| = |x + 7|$

42. $|3x - 4| = |2x + 1|$

43. $|x - 2| = |x + 10|$

44. $|2x + 5| = |2x - 7|$

45. $\left|\frac{x + 1}{2}\right| = \left|\frac{2x - 1}{3}\right|$

46. $\left|\frac{2x - 5}{2}\right| = \left|\frac{10 - x}{5}\right|$

SECTION 2-5. Ten Review Exercises

1. Find the LCM of 9, 15, and 10.

2. List all the positive integer factors of 66.

In exercises **3–6**, simplify each expression.

3. $\frac{-7}{3} \div \frac{7}{2}$

4. $-3^2 + (-4) - 7 + 5 \cdot 2 - (-6)$

5. $\left(\frac{-3}{10}\right)\left(\frac{-5}{6}\right) \div \left(\frac{1}{2}\right)\left(\frac{-7}{4}\right)$

6. $\frac{-9}{4} + \frac{7}{6} - \frac{7}{12} - \left(\frac{-1}{3}\right)$

In exercises **7–10**, solve each equation.

7. $\frac{2}{3}(9t - 3) + \frac{3}{4}(20 - 8t) = \frac{-3}{5}(10t - 15) - 14$

8. $\frac{6u + 1}{5} - \frac{4 - 3u}{4} = \frac{5}{2} - \frac{9u - 5}{10}$

9. $\frac{1}{2}|5 - 3x| = 8$

10. $9 - |2y + 3| = 15$

SECTION 2-5. Supplementary Exercises

In exercises **1–18**, solve each equation. If the equation has no solution, write $\varnothing$.

1. $|3 - x| = 12$

2. $|x - 3| = 12$

3. $5|y + 4| = 35$

4. $|7y - 3| = \frac{1}{2}$

5. $\left|\dfrac{1-2z}{4}\right| = 3$

6. $15 - |6 - z| = 11$

7. $|0.1t + 3| = 2.1$

8. $-|0.1t + 3| = 2.1$

9. $|u - 3| = |3 - u|$

10. $|4v - 5| = |5 - 4v|$

11. $\left|\dfrac{6k-5}{6}\right| = \left|k - \dfrac{1}{3}\right|$

12. $|3a - 5| = |3a + 2|$

13. $|6 - 7b| = |7b - 9|$

14. $|x| = |x + 1|$

15. $3|t| - 5 = |t| + 7$

16. $10 + 2|t + 3| = 5|t + 3| - 8$

17. $5|2k - 3| + 9 = 3(7 + |2k - 3|) - 4$

18. $23 - 9|3k + 2| = 2(1 - 3|3k + 2|) - 3$

In exercises **19–24**, solve each equation for x. Assume a, b and d are nonzero numbers and $c > 0$.

19. $|ax| = c$

20. $|x + b| = c$

21. $|ax + b| = c$

22. $\left|\dfrac{ax+b}{d}\right| = c$

23. $|b - x| = c$

24. $|b - ax| = c$

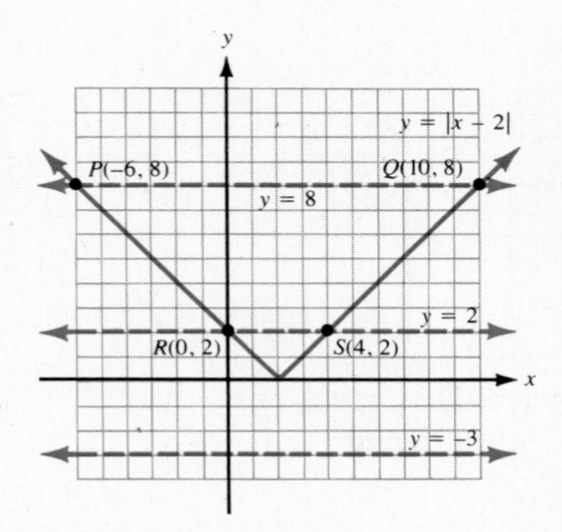

Figure 2-17. Graphs of $y = |x - 2|$, $y = 8$, $y = 2$, and $y = -3$.

In Section 1-4, we reviewed the graphs of $y = |ax + b|$. Graphs of these functions can be used to solve some absolute value equations.

Example. Use the graphs in Figure 2-17 to solve each equation.

a. $|x - 2| = 8$

b. $|x - 2| = 2$

c. $|x - 2| = -3$

Solution. **Discussion.** Consider the constants in the three equations—namely, 8, 2, and -3—as defining the horizontal lines (shown as dotted lines) in the figure. *The x-coordinates of the points where the lines intersect the graph of* $y = |x - 2|$ *are the solutions to the equations.*

a. The line $y = 8$ intersects $y = |x - 2|$ at $P(-6, 8)$ and $Q(10, 8)$. The solution set of $|x - 2| = 8$ is $\{-6, 10\}$.

b. The line $y = 2$ intersects $y = |x - 2|$ at $R(0, 2)$ and $S(4, 2)$. The solution set of $|x - 2| = 2$ is $\{0, 4\}$.

c. The line $y = -3$ does not intersect $y = |x - 2|$. The solution set of $|x - 2| = -3$ is $\varnothing$.

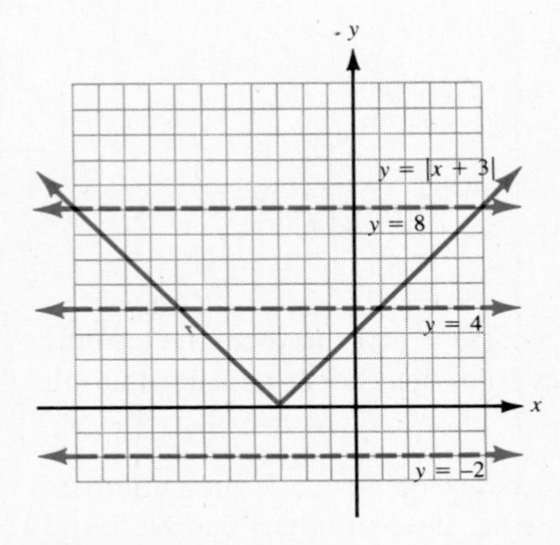

Figure 2-18. Graphs of $y = |x + 3|$, $y = 8$, $y = 4$, and $y = -2$.

In exercises **25–28**, use the graphs in the figure to solve each equation.

25. Use the graphs in Figure 2-18.

a. $|x + 3| = 8$

b. $|x + 3| = 4$

c. $|x + 3| = -2$

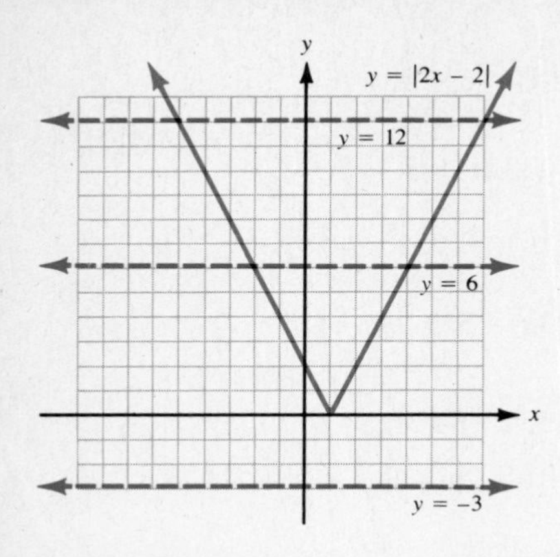

Figure 2-19. Graphs of $y = |2x - 2|$, $y = 12$, $y = 6$, and $y = -3$.

26. Use the graphs in Figure 2-19.

a. $|2x - 2| = 12$

b. $|2x - 2| = 6$

c. $|2x - 2| = -3$

27. Use the graphs in Figure 2-20.

a. $\left|\frac{3}{2}x + 3\right| = 12$

b. $\left|\frac{3}{2}x + 3\right| = 6$

c. $\left|\frac{3}{2}x + 3\right| = 0$

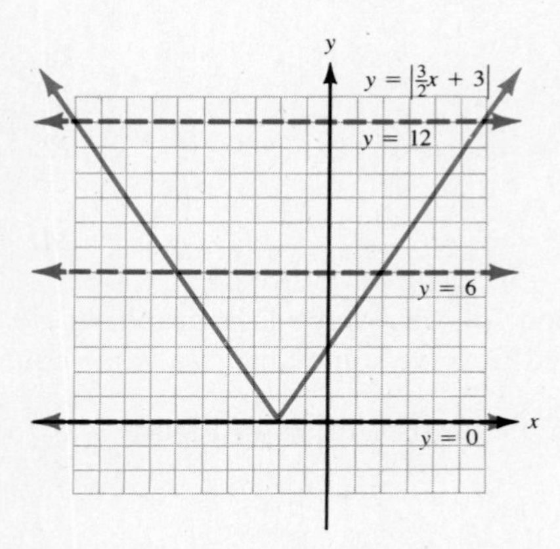

Figure 2-20. Graphs of $y = |\frac{3}{2}x + 3|$, $y = 12$, $y = 6$, and $y = 0$.

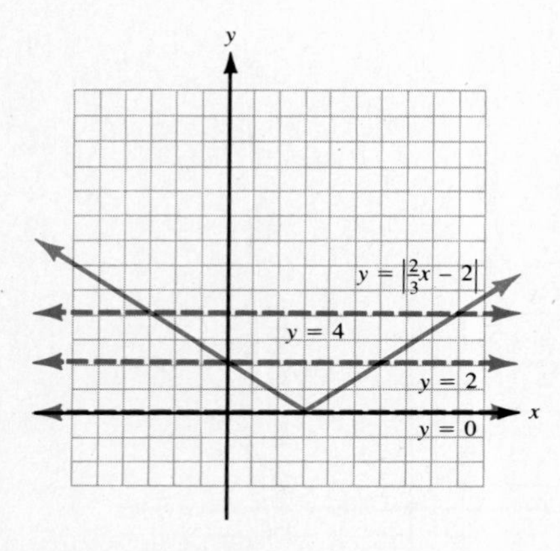

Figure 2-21. Graphs of $y = |\frac{2}{3}x - 2|$, $y = 4$, $y = 2$, and $y = 0$.

28. Use the graphs in Figure 2-21.

a. $\left|\frac{2}{3}x - 2\right| = 4$

b. $\left|\frac{2}{3}x - 2\right| = 2$

c. $\left|\frac{2}{3}x - 2\right| = 0$

In exercises **29–36**, the variables a, b, and c represent real numbers. Determine whether each equation is *always true*, *sometimes true*, or *never true* using the following procedure:

1. Try to find one or more numbers for which the equation is true. If such numbers exist, then the equation is at least sometimes true. If no numbers can be found (try positive and negative numbers, and 0), then assert that the equation is never true.

2. If you can show that the equation is sometimes true, now try to find numbers for which the equation is not true. Again, try positive and negative numbers, and 0. If no such numbers can be found, then assert that the equation is always true.

3. Compare your answers with those in the back of the text.

29. $|a \cdot b| = |a| \cdot |b|$

30. $\left|\frac{a}{b}\right| = \frac{|a|}{|b|}$, and $b \neq 0$

31. $|c \cdot a| = c|a|$, and $c \neq 0$

32. $|c(a + b)| = |ca| + |cb|$

33. $|a + b| = |a| + |b|$

34. $|a + b| = a + b$

35. $|a - b| = |a| - |b|$

36. $|a - b| = a - b$

SECTION 2-6. Linear Inequalities in One Variable

1. The definition of a linear inequality in x
2. Three properties of inequality
3. A procedure for solving a linear inequality in x
4. Using a number line to graph a solution set
5. Applied problems

Consider the following problem: "Find a number such that the sum of twice the number and 2 times the difference between the number and 5 is always less than 43."

To write in mathematical symbols a statement for this problem, we must consider the phrase, "must always be less than". This phrase cannot be represented by = (the equal sign). This symbol is used for phrases such as "is the same as". This problem can, however, be written as an **inequality,** in which $<$ is used to represent "less than".

Thus, if n represents the unknown number, the word problem above can be written as follows:

$$2n + 2(n - 5) < 43$$

In this section, we will learn a procedure for solving such inequalities.

The Definition of a Linear Inequality in x

As Definition 2.5 shows, there are four different forms for a linear inequality in x.

> **Definition 2.5. A linear inequality in x**
> If a, b, and c are real numbers and $a \neq 0$, then a **linear inequality** in x can be written in one of the following forms:
>
> (1) $ax + b < c$ or (2) $ax + b > c$ or
>
> (3) $ax + b \leq c$ or (4) $ax + b \geq c$

Inequalities (3) and (4) are **compound,** in that they include the equality relation in the statement. For example:

$$(3) \quad ax + b \leq c \quad \text{means} \quad ax + b < c \quad \text{or} \quad ax + b = c$$

Thus, the solution set of (3) and (4) contains those values of x that are solutions of either the inequality or the equality.

Examples **a–d** illustrate inequalities that can be written in one of the forms of Definition 2.5.

a. $5x + 12 < 3(x + 7) - 1$ — A linear inequality in x

b. $2(5y - 4) > 6 - 8(1 - 2y)$ — A linear inequality in y

c. $\dfrac{2t + 1}{3} - \dfrac{4t - 3}{2} \leq \dfrac{-1}{6}$ — A linear inequality in t

d. $5 - 3[u - (3u - 4)] \geq u + 8$ — A linear inequality in u

Three Properties of Inequality

A linear inequality in x is considered solved when it has been changed to one of the following forms, in which k is a real number:

(1*) $x < k$ or (2*) $x > k$ or (3*) $x \leq k$ or (4*) $x \geq k$

There are three properties of inequality that can be used to change a given inequality to one of the forms 1*–4* and not change the solution set.

In the following properties, a, b, and c are real numbers.

Addition property of inequality

If $a < b$, then $a + c < b + c$.

Multiplication property of inequality

Part 1. If $a < b$ and $c > 0$, then $a \cdot c < b \cdot c$.

Part 2. If $a < b$ and $c < 0$, then $a \cdot c > b \cdot c$.

Antisymmetric property of inequality

If $a < b$, then $b > a$.

These three properties of inequality are also valid for the other three order relations—namely, $>$, $\leq$, and $\geq$.

Examples **e–i** illustrate these properties of inequality.

e. $3 < 9$ and $c = 5$ — $3 + 5 < 9 + 5$ — Addition of inequality

$8 < 14$

f. $3 < 9$ and $c = -5$ — $3 + (-5) < 9 + (-5)$ — Addition of inequality

$-2 < 4$

g. $3 < 9$ and $c = 5$ — $3 \cdot 5 < 9 \cdot 5$ — Multiply by a positive number.

$15 < 45$

h. $3 < 9$ and $c = -5$ $\quad 3(-5) > 9(-5)$ $\quad$ Multiply by a negative number.

$$-15 > -45$$

i. $-9 < -3$ and $c = -5$ $\quad (-9)(-5) > (-3)(-5)$ $\quad$ Multiply by a negative number.

$$45 > 15$$

j. $3 < 9$, therefore $9 > 3$ $\quad$ Antisymmetric property

A Procedure for Solving a Linear Inequality in x

The following six steps can be used to solve a linear inequality in one variable:

To solve a linear inequality in x:

Step 1. Remove any grouping symbols.

Step 2. Combine any like terms on both sides of the inequality.

Step 3. Use the addition property of inequality to combine the variable terms on one side and the constant terms on the opposite side.

Step 4. If necessary, use the multiplication property of inequality to change the coefficient of the variable term to 1.

Step 5. Check a possible solution from the solution set in the given inequality.

Step 6. State the solution set and graph the solutions on a number line.

Example 1. Solve $5x + 12 < 3(x + 7) - 1$.

Solution.

Step 1. $5x + 12 < 3x + 21 - 1$ $\quad$ Remove the parentheses.

Step 2. $5x + 12 < 3x + 20$ $\quad$ Combine like terms.

Step 3. $2x + 12 < 20$ $\quad$ Add $-3x$ to both sides.

$2x < 8$ $\quad$ Add -12 to both sides.

Step 4. $x < 4$ $\quad$ Multiply both sides by $\frac{1}{2}$.

(*Note:* Since $\frac{1}{2} > 0$, the order of the inequality is not reversed.)

Step 5. Check the solution set by replacing x with any number less than 4—for example, 3.

$$5(3) + 12 < 3(3 + 7) - 1$$

$$27 < 29, \textit{true}$$

Step 6. The solution set is any real number x, where $x < 4$. Using set notation: $\{x \mid x < 4\}$, which is read "the set of x's and $x < 4$".

Using a Number Line to Graph a Solution Set

The solution set of a linear inequality in one variable frequently contains an infinite number of solutions. Therefore, a number line is used to display the solutions. Such a display is called a **graph of the solution set.** A graph of the solution set in Example 1 is shown in Figure 2-22. The **hollow dot** at 4 indicates that 4 is not a solution of the inequality.

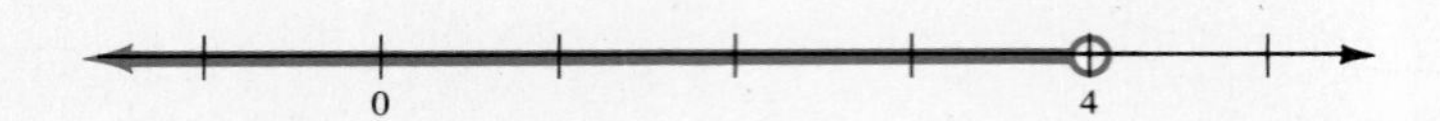

Figure 2-22. A graph of $\{x|x < 4\}$.

Example 2. Solve and graph $2(5y - 4) > 6 - 8(1 - 2y)$.

Solution.

Step 1. $10y - 8 > 6 - 8 + 16y$ Remove parentheses.

Step 2. $10y - 8 > 16y - 2$ Combine like terms.

Step 3. $-6y - 8 > -2$ Add $-6y$ to both sides.

$-6y > 6$ Add 8 to both sides.

Step 4. $y < -1$ Multiply both sides by $\frac{-1}{6}$.

(Since $\frac{-1}{6} < 0$, the order of the inequality is reversed.)

Step 5. Arbitrarily selecting -2:

$2(5(-2) - 4) > 6 - 8(1 - 2(-2))$

$-28 > -34$, *true*

Step 6. The solution set is $\{y|y < -1\}$. A graph of the solution set is shown in Figure 2-23.

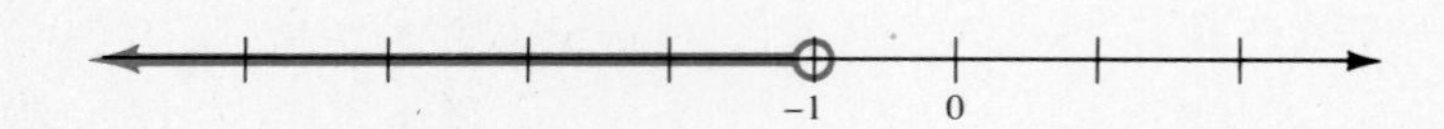

Figure 2-23. A graph of $\{y|y < -1\}$.

Example 3. Solve and graph $2(2t + 1) - 3(4t - 3) \leq -1$.

Solution.

Step 1. $4t + 2 - 12t + 9 \leq -1$ Remove parentheses.

Step 2. $11 - 8t \leq -1$ Combine like terms.

Step 3. $-8t \leq -12$ Add -11 to both sides.

Step 4. $t \geq \frac{12}{8}$ Multiply both sides by $\frac{-1}{8}$.

$t \geq \frac{3}{2}$ Reduce the fraction.

Step 5. The check is omitted.

Step 6. The solution set is $\{t \mid t \geq \frac{3}{2}\}$. A graph of the solution set is shown in Figure 2-24. A solid dot is used at $\frac{3}{2}$ because $\frac{3}{2}$ is included in the solution set.

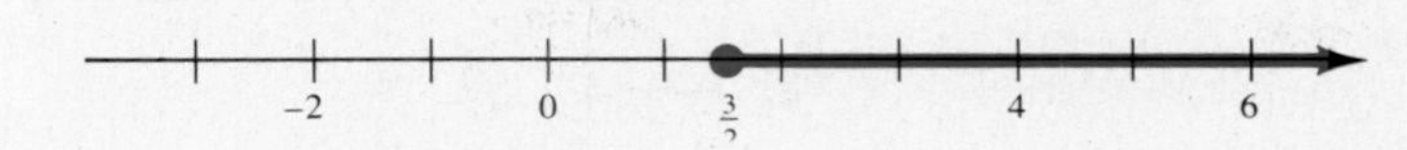

Figure 2-24. A graph of $\{t \mid t \geq \frac{3}{2}\}$.

Applied Problems

Examples 4 and 5 illustrate two types of applied problems that can be solved using a linear inequality in one variable.

Example 4. The difference between 5 times an integer and 3 times the next consecutive integer is always less than 7. Find the possible values for the smaller of the two integers.

Solution. **Discussion.** If n is the smaller of two consecutive integers, then $(n + 1)$ is the next consecutive integer.

$$5n - 3(n + 1) < 7$$

$$5n - 3n - 3 < 7$$

$$2n < 10$$

$$n < 5$$

The possible values for the smaller integer are 4, 3, 2, ...

Example 5. The length of a rectangle is 5 centimeters more than the width. If the perimeter is less than 46 centimeters, find the possible values for the width.

Solution. **Discussion.** If w represents the width, then $w + 5$ represents the length. The formula for the perimeter of a rectangle is:

$$2(\text{width}) + 2(\text{length}) = \text{perimeter}$$

Based on the given data:

$$2w + 2(w + 5) < 46$$

$$2w + 2w + 10 < 46$$

$$4w < 36$$

$$w < 9$$

Since w represents the width of a rectangle, the smallest possible value is 0. Thus, the possible values for w are any numbers between 0 and 9, which can be written:

$$0 < w < 9$$

The width can be between 0 and 9 centimeters.

SECTION 2-6. Practice Exercises

In exercises **1–30**, solve and graph each inequality.

[Examples 1–3]

1. $7y - 3 \geq 39$

2. $8y + 9 \leq 81$

3. $11 - 4p < 59$

4. $10 - 6p > 40$

5. $5x + 3 < 2x + 9$

6. $6x - 5 \leq 35 - 2x$

7. $4 - 3a > a + 4$

8. $10 - a \geq 10 + 3a$

9. $6 + 8b + 3 \leq b + 10 + 5b$

10. $9 + 12b - 4 > 1 - 9b - 10$

11. $5 - 2(y - 3) > y + 8$

12. $4y + 18 \geq 1 - 3(2y + 1)$

13. $16 - 5z < 4(z - 1) - 7$

14. $10 - (7 - z) \geq 3z - 9$

15. $21 - (6t - 1) < 2(9t + 7)$

16. $3(5 - 2t) > 18 - 2(4t - 1)$

17. $2(1 - 3u) \leq 32 + 5(u - 6)$

18. $3(1 - 4u) - 35 \geq 7(2u + 1)$

19. $3(2z - 1) - 2(2z + 5) < -10$

20. $4(2z + 3) - (6z - 5) > 10$

21. $3[2(y + 1) - 5] > 4y - 7$

22. $-2[4(y - 2) + 3] \geq y + 55$

23. $3(x + 2) - [2(2x - 1) + 6] < 0$

24. $2(2x - 3) + [7(x - 6) - (2x - 12)] < 0$

25. $5(3 - q) - 2(2q + 3) \geq 21$

26. $4 + q - 2(3 - 2q) \geq -2$

27. $5(2p + 3) - (2p + 9) \leq 2(15 - 2p)$

28. $2(6p + 5) - (3p + 1) \leq 6(4 - p)$

29. $2(m + 2) - (3m - 2) \leq 2(3m + 4) - 3(m - 2)$

30. $6(m - 3) - (m - 8) \geq 5 - 3(m - 3)$

In exercises **31–44**, solve each word problem.

[Example 4]

31. The sum of 2 times an integer and 9 is less than 15. Find the possible values for the integer.

32. The difference between 3 times an integer and 7 is greater than 20. Find the possible values for the integer.

33. The product of 2 and an integer decreased by 5 is greater than or equal to the integer increased by 8. Find the possible values for the integer.

34. The product of 3 and an integer increased by 4 is less than or equal to the integer decreased by 6. Find the possible values for the integer.

35. Jerry currently weighs 240 pounds. He is on a diet that will allow him to lose up to 3 pounds per week. How long will it take Jerry to reach a weight of 198 pounds?

36. Sandy is doing odd jobs after school to earn money. She can earn up to $12 per week. How long will it be before she can purchase a $150 stereo system if she has already saved $90?

[Example 5]

37. The length of a rectangle is 5 times the width. If the perimeter of the rectangle is less than 60 inches, find the possible values for the width.

38. The length of a rectangle is 3 times the width. If the perimeter of the rectangle is less than or equal to 72 centimeters, find the possible values for the width.

39. The length of a rectangle is 1 centimeter more than twice the width. The perimeter is not more than 62 centimeters. Find the possible values for the length.

40. The length of a rectangle is 1 inch less than 3 times the width. The perimeter is not more than 46 inches. Find the possible values for the length.

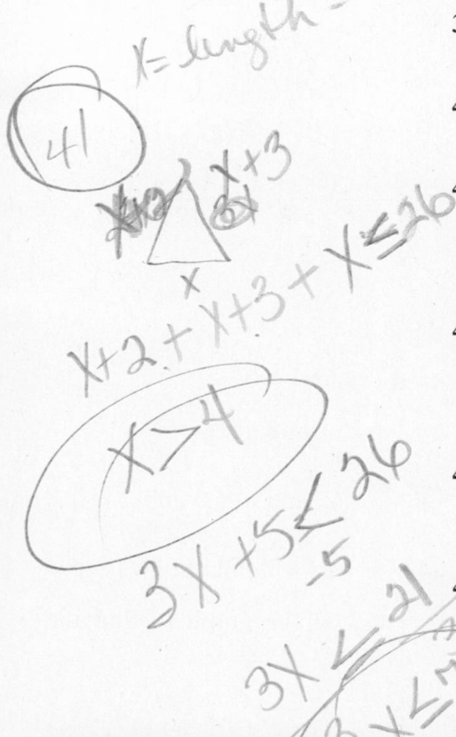

41. One side of a triangle is 2 centimeters longer than the smallest side. The largest side is 3 centimeters longer than the smallest. The perimeter of the triangle is not more than 26 centimeters. If the smallest side must be larger than 4 centimeters, find the possible values for the smallest side.

42. The second side of a triangle is 1 foot longer than the first side. The third side is 10 feet longer than the first. The perimeter of the triangle is not more than 47 feet. If the first side must be at least 10 feet long, find the possible values for the first side.

43. The sides of a triangle are three consecutive odd integers. The smallest side must be more than 3 units long and the perimeter can be no more than 33 units long. Find the possible integer values for the smallest side.

44. The sides of a triangle are three consecutive even integers. The smallest side must be at least 4 units long and the perimeter can be no more than 30 units long. Find the possible integer values for the smallest side.

SECTION 2-6. Ten Review Exercises

In exercises **1–5**, simplify each expression.

1. $3 \cdot 5 - (5^2 - 7 \cdot 3) - 2^3$

2. $3^2 - (7 \cdot 2 - 2^4) - 5 \cdot 2$

3. $3[(2 - 3) \cdot 5 + 2^3] - 7$

4. $5[-17 + 5 \cdot 2^2 - 3^2)] + 19$

5. $3(t - 4) - 9t - 12 + 6(1 - 3t)$.

6. Evaluate $3(t - 4) - 9t - 12 + 6(1 - 3t)$ for $t = -2$.

7. Solve $3(t - 4) - 9t = 12 - 6(1 - 3t)$.

8. Solve and graph $3(t - 4) - 9t > 12 - 6(1 - 3t)$.

In exercises **9** and **10**, solve each equation.

9. $|4t - 2| = 1$.

10. $|4t - 2| = |t + 6|$.

SECTION 2-6. Supplementary Exercises

If the variable terms combine to 0 in the process of solving, then:

a. The solution set is $\varnothing$ if the resulting number statement is *false*.

b. The solution set is R, the set of real numbers, if the resulting number statement is *true*.

In exercises **1–8**, determine whether the solution set of each inequality is $\varnothing$ or R.

1. $5x < 5x + 2$

2. $3x < 3x - 5$

3. $3(y + 5) \geq 3y + 16$

4. $4(y - 3) \geq 4y - 15$

5. $2(3p + 2) < 3(2p + 5) - 10$

6. $10 - (7 - 3p) \geq 5(2p - 3) - 7p$

7. $\dfrac{2w + 1}{2} - \dfrac{3w - 1}{3} \geq 1$

8. $\dfrac{6w + 5}{6} - \dfrac{6w + 1}{12} \geq \dfrac{4 + w}{2}$

In exercises **9–18**, solve each inequality for x.

9. $x + b < c$

10. $x - b \geq c$

11. $ax < c$, and $a > 0$

12. $ax \leq c$, and $a < 0$

13. $ax + b > c$, and $a > 0$

14. $ax + b \leq c$, and $a < 0$

15. $\dfrac{ax + b}{d} < c$, and $a > 0$ and $d < 0$

16. $\dfrac{ax + b}{d} < c$, and $a < 0$ and $d < 0$

17. $ax - b > cx - d$, and $(a - c) < 0$

18. $ax + b < cx + d$, and $(a - c) > 0$

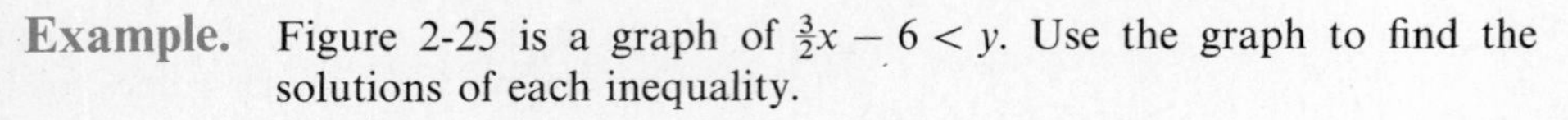

Example. Figure 2-25 is a graph of $\frac{3}{2}x - 6 < y$. Use the graph to find the solutions of each inequality.

a. $\dfrac{3}{2}x - 6 < 3$ **b.** $\dfrac{3}{2}x - 6 < -3$

Solution. **a.** The horizontal line defined by $y = 3$ intersects the edge of the half-plane at $P(6, 3)$. Thus, the solution set for the given inequality is $x < 6$.

b. The horizontal line defined by $y = -3$ intersects the edge of the half-plane at $Q(2, -3)$. Thus, the solution set for the given inequality is $x < 2$.

Figure 2-25. A graph of $\frac{3}{2}x - 6 < y$.

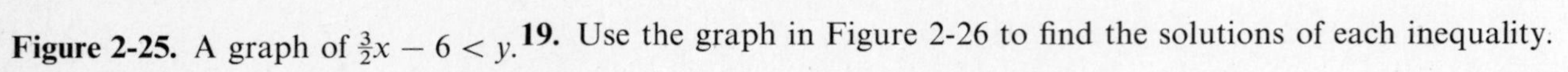

19. Use the graph in Figure 2-26 to find the solutions of each inequality.

a. $\dfrac{2}{3}x - 4 < 0$ **b.** $\dfrac{2}{3}x - 4 < -6$

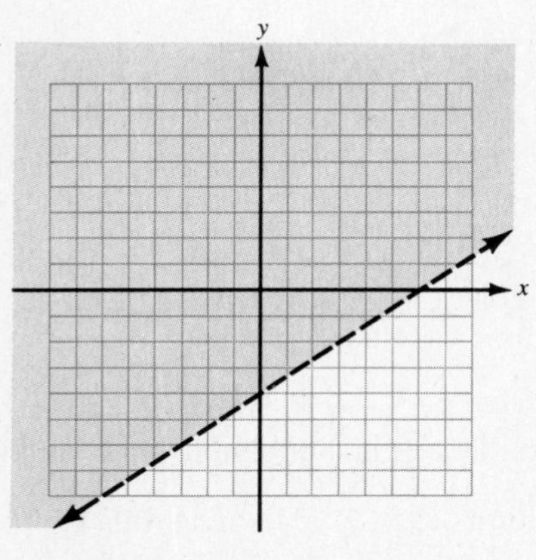

Figure 2-26. A graph of $\frac{2}{3}x - 4 < y$.

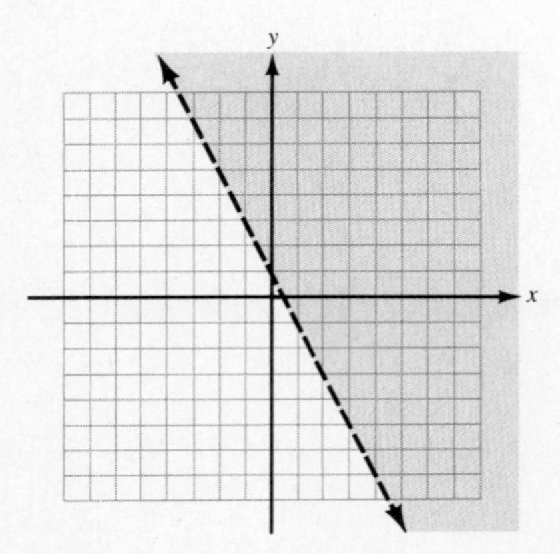

Figure 2-27. A graph of $1 - 2x < y$.

20. Use the graph in Figure 2-27 to find the solutions of each inequality.

a. $1 - 2x < 5$ **b.** $1 - 2x < -1$

21. Use the graph in Figure 2-28 to find the solutions of each inequality.

a. $2 - \frac{2}{3}x \geq 4$ **b.** $\frac{2}{3}x - 2 \leq 0$

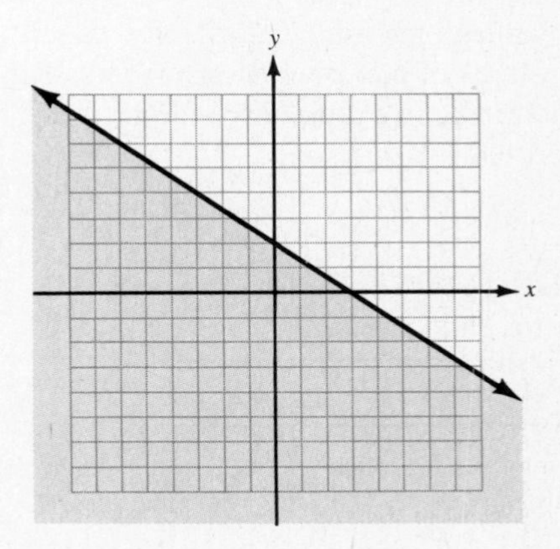

Figure 2-28. A graph of $2 - \frac{2}{3}x \geq y$.

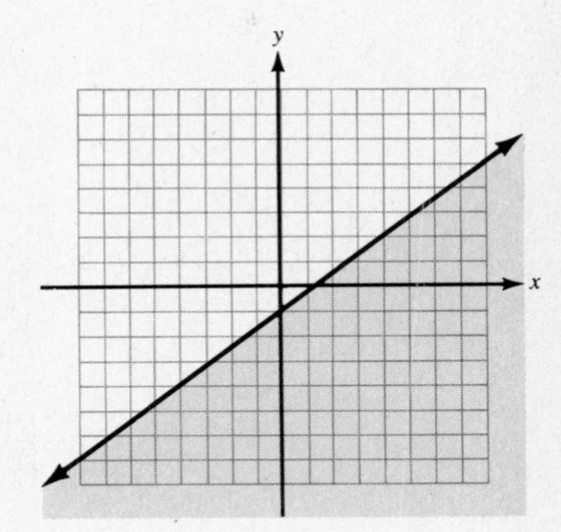

Figure 2-29. A graph of $\frac{3}{4}x - 1 \geq y$.

22. Use the graph in Figure 2-29 to find the solutions of each inequality.

a. $\frac{3}{4}x - 1 \geq 5$ **b.** $1 - \frac{3}{4}x \leq 7$

SECTION 2-7. Compound Relations and Inequalities

1. Truth values for compound sentences joined with OR and AND
2. Solving compound relations joined with OR
3. Solving compound relations joined with AND
4. Compound relations with no solution or all real numbers as solutions

In the English language, we use sentences that are simple, and sentences that are compound. Similarly, in mathematics we write equations and inequalities that are simple and compound. In this section, we will use the procedure for solving simple inequalities to find solutions of compound inequalities.

Truth Values for Compound Sentences Joined with OR and AND

In logic, a **simple statement** is one that is either true or false, but not both at the same time. The symbols p, q, r, and so on are frequently used to represent simple statements. Examples **a** and **b** are simple statements.

a. p is the statement "Today is Monday."

b. q is the statement "The sun is shining."

Simple statements can be used to form **compound statements.** Two types of compound statements are formed using the connectives OR and AND. Examples **c** and **d** illustrate the use of connectives with the p and q of examples **a** and **b**.

c. p or q: "Today is Monday or the sun is shining."

d. p and q: "Today is Monday and the sun is shining."

Based on rules of logic, compound statements are true or false depending on the truth values of p and q. A display called a **truth table** is often used to show the truth values of compound statements. Table 2-1 is a truth table for the compound statements in examples **c** and **d**. In this table, T represents a *true* statement and F represents a *false* one.

TABLE 2-1. Truth table for p, q, (p or q), and (p and q).

p	q	(p or q)	(p and q)
T	T	T	T
T	F	T	F
F	T	T	F
F	F	F	F

(p or q) is true (T) if p is true, or q is true, or both are true

(p and q) is true (T) only if both p and q are true

Using the simple statements of **a** and **b** to illustrate:

"Today is Monday or the sun is shining" is true if
- Today is Monday.
- The sun is shining.
- Today is Monday and the sun is shining.

"Today is Monday and the sun is shining" is true if, and only if, today is in fact Monday and the sun is shining.

Example 1. Given the following simple statements:

p: x is an even number

q: x is a number greater than 10

For each value of x, determine whether or not:
(i) (p or q) is true
(ii) (p and q) is true

a. $x = 16$

b. $x = 15$

Solution. **Discussion.** First determine whether p is true or false, and whether q is true or false, for the given values of x. Then use the truth values in Table 2-1 to state whether (p or q) and (p and q) are true or false.

a. If $x = 16$, then p is *true* and q is *true*.
(i) (p or q) becomes T or $T \rightarrow$ True
(ii) (p and q) becomes T and $T \rightarrow$ True

b. If $x = 15$, then p is *false* and q is *true*.
(i) (p or q) becomes F or $T \rightarrow$ True
(ii) (p and q) becomes F and $T \rightarrow$ False

Solving Compound Relations Joined with OR

A compound relation in mathematics can be formed by joining two equations, two inequalities, or an equation and an inequality with OR. The solutions, if any, of such a compound relation are determined by the same set of truth values for (p or q) in Table 2-1. Examples **e–g** illustrate three compound relations joined by OR, including ways in which each statement should be read and a corresponding graph of the solution set.

Compound relation	Word statement	Graph of solution set
e. $x = -2$ or $x = 3$	"x is -2 or 3"	
f. $x < 0$ or $x = 0$	"x is less than or equal to 0"	
g. $x < -1$ or $x > 2$	"x is less than -1 or greater than 2"	

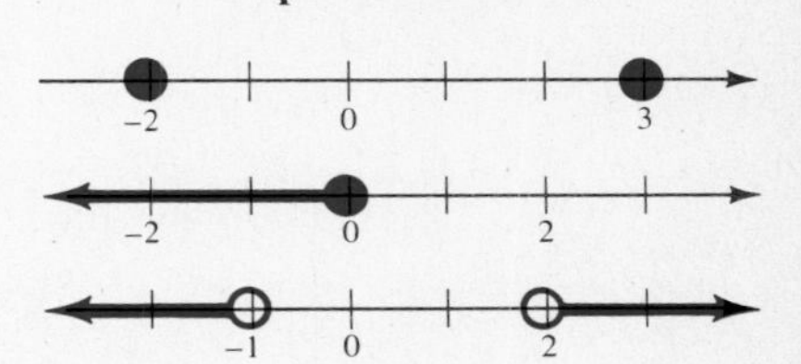

> **To solve a compound relation joined by OR:**
>
> **Step 1.** Solve each equation or inequality.
>
> **Step 2.** A solution of the compound statement is a solution of either of the equations or inequalities. That is, the solution set is the **union** of the solution sets of the equations or inequalities in the compound statement.

Example 2. Solve and graph: $9 - 2t = 6 - 3t$ or $3 < 4t - 5$

Solution. **Step 1.** Solve the equation and the inequality.

$9 - 2t = 6 - 3t$	$3 < 4t - 5$
$9 + t = 6$	$8 < 4t$
$t = -3$	$2 < t$
The solution set is $\{-3\}$.	The solution set is $\{t \mid t > 2\}$

Step 2. The union of the solution sets is the solution set of the compound statement.

$\{-3\} \cup \{t \mid t > 2\} = \{t \mid t = -3 \text{ or } t > 2\}$

A graph of the solution set is shown in Figure 2-30.

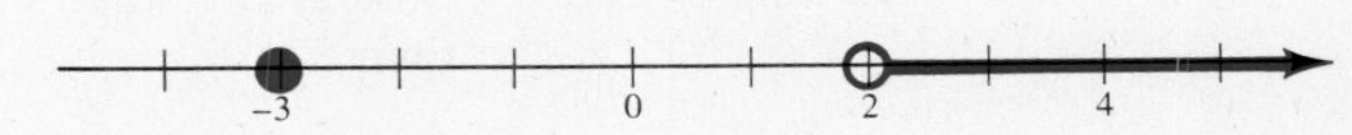

Figure 2-30. A graph of $\{t \mid t = -3 \text{ or } t > 2\}$.

Example 3. Solve and graph each compound relation.

a. $x \leq -7$ or $x > -1$

b. $3(y + 5) < 3 - y$ or $7y + 3 \geq 2(y + 4)$

Solution. **a.** Graphs of the solution sets of the given inequalities are shown in Figure 2-31. By combining these graphs we get a graph of the compound inequality, as shown in Figure 2-32.

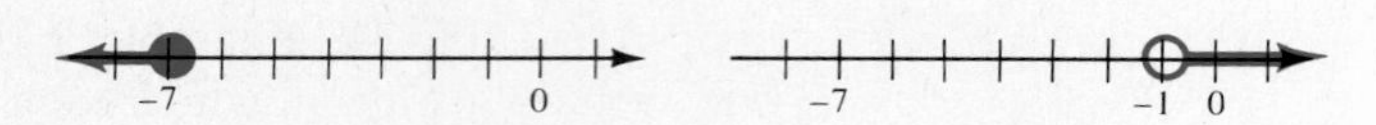

Figure 2-31. Graphs of $x \leq -7$ and $x > -1$.

Figure 2-32. A graph of $\{x | x \leq -7$ or $x > -1\}$.

b. Step 1. $3y + 15 < 3 - y$ or $7y + 3 \geq 2y + 8$

$4y < -12$ or $5y \geq 5$

$y < -3$ or $y \geq 1$

Step 2. The solution set of the compound statement is $\{y | y < -3$ or $y \geq 1\}$. A graph of the solution set is shown in Figure 2-33.

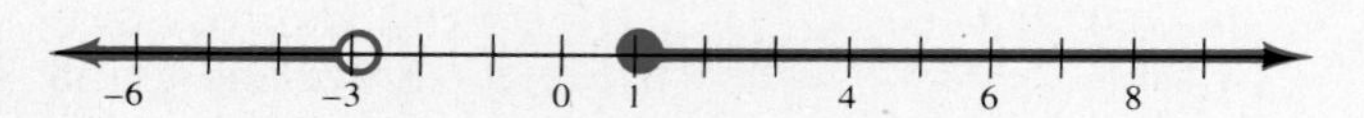

Figure 2-33. A graph of $\{y | y < -3$ or $y \geq 1\}$.

Solving Compound Relations Joined with AND

A compound relation in mathematics can also be formed by joining two equations, two inequalities, or an equation and an inequality with AND. The solutions, if any, of such a compound relation are determined by the same set of truth values for (p and q) in Table 2-1. Examples **h–j** illustrate two compound relations joined by AND.

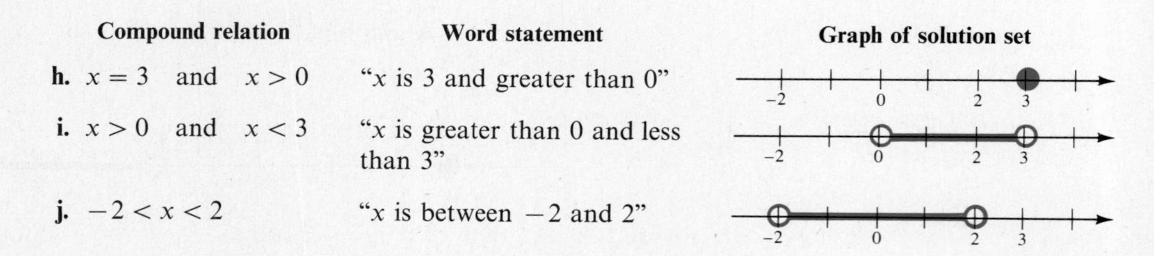

Compound relation	Word statement	Graph of solution set
h. $x = 3$ and $x > 0$	"x is 3 and greater than 0"	
i. $x > 0$ and $x < 3$	"x is greater than 0 and less than 3"	
j. $-2 < x < 2$	"x is between -2 and 2"	

> **To solve a compound relation joined by AND:**
>
> **Step 1.** Solve each equation or inequality.
>
> **Step 2.** A solution of the compound statement is a solution of both of the equations or inequalities. That is, the solution set is the **intersection** of the solution sets of the equations or inequalities in the compound statement.

Example 4. Solve and graph each compound relation.

a. $p > 2$ and $p \le 8$

b. $3(q + 2) > 1 - 2q$ and $q + 3 > 6(q - 2)$

Solution. **a.** Graphs of the solution sets of the given inequalities are shown in Figure 2-34. By taking only those points that are in both graphs, we get a graph of the compound statement, as shown in Figure 2-35.

Figure 2-34. Graphs of $p > 2$ and $p \le 8$.

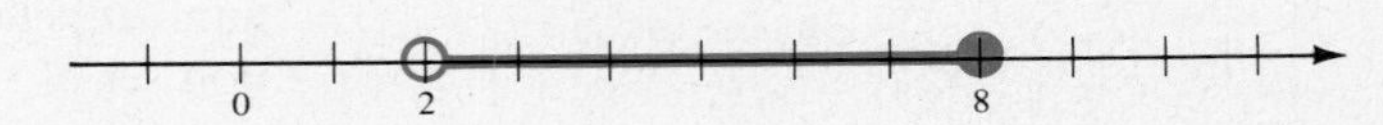

Figure 2-35. A graph of $\{p|2 < p \le 8\}$.

b. Step 1.

$3q + 6 > 1 - 2q$	and	$q + 3 > 6q - 12$
$5q > -5$	and	$15 > 5q$
$q > -1$	and	$3 > q$
$-1 < q$	and	$q < 3$

Thus, $-1 < q < 3$. A graph of the solution set is shown in Figure 2-36.

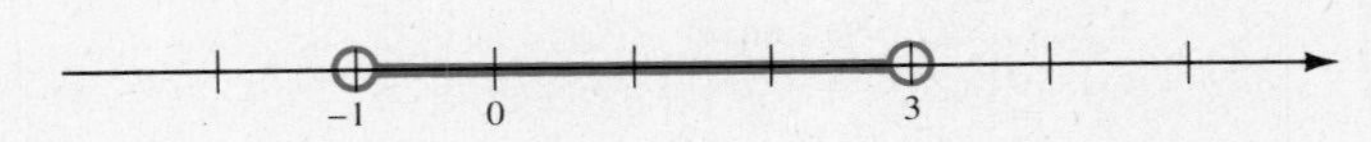

Figure 2-36. A graph of $\{q| -1 < q < 3\}$.

Example 5. Solve and graph $7 < 3z + 4 < 19$.

Solution. **Discussion.** The given inequality is a shorthand version of $7 < 3z + 4$ and $3z + 4 < 19$. To solve the equality, we need to operate on the expression $3z + 4$ to change it to z.

$7 < 3z + 4 < 19$	The given inequalities
$3 < 3z < 15$	Add -4 to each part.
$1 < z < 5$	Divide each part by 3.

The solution set is $\{z \mid 1 < z < 5\}$. A graph of the solution set is shown in Figure 2-37.

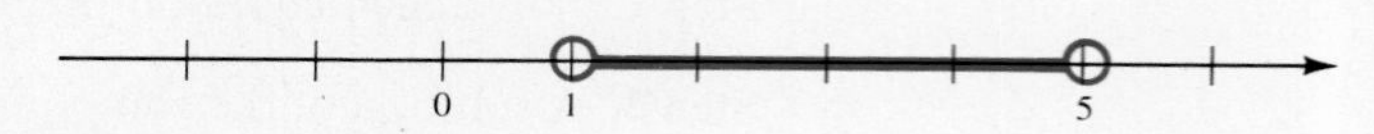

Figure 2-37. A graph of $\{z \mid 1 < z < 5\}$.

Compound Relations with No Solutions or All Real Numbers as Solutions

The solution set of a compound relation may contain no solutions, or it may contain all real numbers.

Example 6. Solve and graph: $2k + 5(6 - k) > 0$ and $2(k + 6) + 3 < 9(k - 10)$

Solution. **Step 1.** $2k + 30 - 5k > 0$ and $2k + 12 + 3 < 9k - 90$

$-3k > -30$ and $-7k < -105$

$k < 10$ and $k > 15$

Step 2. Graphs of the solution sets of both inequalities are shown in Figure 2-38. From the graphs it is apparent that no real number is less than 10 and greater than 15. Thus, the solution set of the compound inequality is $\varnothing$.

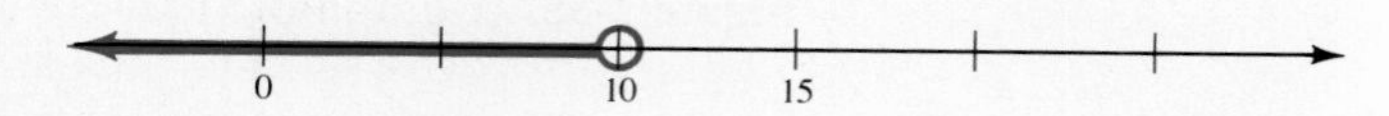

Figure 2-38a. A graph of $k < 10$.

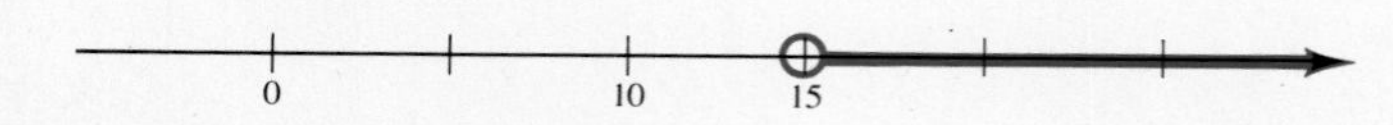

Figure 2-38b. A graph of $k > 15$.

Example 7. Solve and graph: $5 - 3r > 11$ or $2r + 3 > -9$

Solution. **Step 1.** $-3r > 6$ or $2r > -12$

$r < -2$ or $r > -6$

Step 2. Graphs of the solution sets of both inequalities are shown in Figure 2-39. From the graphs it is apparent that every real number is either less than -2 or greater than -6. Thus, the solution set of the compound inequality is R.

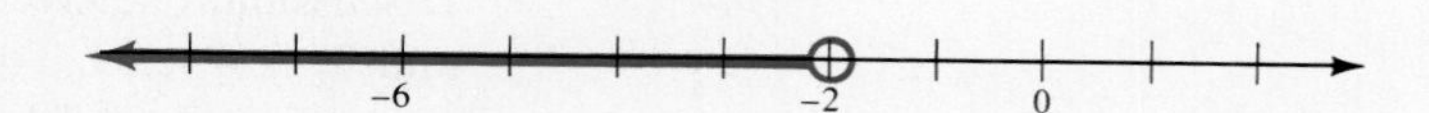

Figure 2-39a. A graph of $r < -2$.

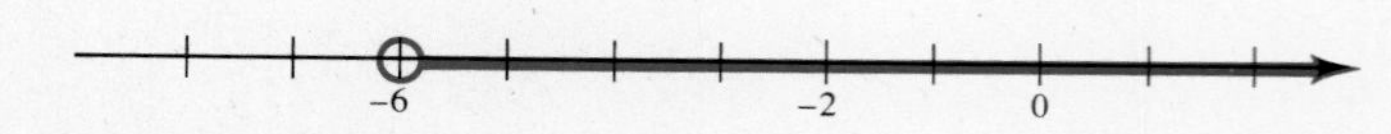

Figure 2-39b. A graph of $r > -6$.

SECTION 2-7. Practice Exercises

In exercises **1–10**, given the statements p and q, for each value of x determine the truth value of:

(i) p or q

(ii) p and q

[Example 1]

1. p: x is an odd integer
q: x is positive

a. $x = 5$
b. $x = 8$

2. p: x is an even integer
q: x is less than 9

a. $x = 4$
b. $x = 7$

3. p: x is an odd integer
q: x is an even integer

a. $x = 11$
b. $x = 0.5$

4. p: x is a positive integer
q: x is a negative integer

a. $x = -4$
b. $x = -1.25$

5. p: x is greater than 1
q: x is less than 8

a. $x = 0$
b. $x = 5$

6. p: x is less than 20
q: x is greater than 7

a. $x = 34$
b. $x = 13$

7. p: x is greater than 17
q: x is less than 10

a. $x = 12$
b. $x = 6$

8. p: x is less than 14
q: x is greater than 19

a. $x = 17$
b. $x = 22$

9. p: x is 4
q: x is greater than 1

a. $x = -4$
b. $x = 4$

10. p: x is less than 35
q: x is 11

a. $x = 2$
b. $x = 11$

In exercises **11–56**, solve and graph each compound relation. For any compound inequality with an empty solution set, write $\varnothing$. For any compound inequality with solution set all real numbers, write R.

[Examples 2 and 3]

11. $8x - 11 = 21$ or $5x = x - 20$

12. $46 = 5x - 14$ or $4 = 25 + 7x$

13. $2 + 10y = 6y - 18$ or $5y + 3 > 9 - y$

14. $7y - 26 = 40 + 13y$ or $10y + 27 > 3 + 2y$

15. $z < 1$ or $z > 5$

16. $z < 0$ or $z > 10$

17. $t \leq -5$ or $t \geq -2$

18. $t \leq -1$ or $t \geq 6$

19. $u > -6$ or $u < 6$

20. $u > -3$ or $u < 3$

21. $w \geq 5$ or $w \geq 7$

22. $w \leq 8$ or $w \leq 11$

23. $p \leq -6$ or $p = 1$

24. $p \geq 5$ or $p = -2$

25. $\frac{3q}{2} \leq -3$ or $\frac{2q}{3} > 4$

26. $\frac{q}{2} < -2$ or $\frac{q}{5} \geq 1$

27. $2x + 3 \geq 9$ or $4x + 5 = 1$

28. $3x - 5 < -8$ or $5x - 2 = 3$

29. $3y + 21 < 1 - y$ or $3y + 8 > 3 - 2y$

30. $5z - 7 < 3(z - 9)$ or $3(z + 5) > 15$

[Example 4] **31.** $t > 1$ and $t < 4$

32. $t > 0$ and $t < 5$

33. $u \leq -2$ and $u \geq -10$

34. $u \leq 3$ and $u \geq -3$

35. $v < 5$ and $v > -5$

36. $v < 4$ and $v > -8$

37. $p < 11$ and $p = 5$

38. $p < 17$ and $p = 10$

39. $2q + 3 \leq 9$ and $q - 5 > -6$

40. $3q - 5 \leq -11$ and $2q + 3 \geq -9$

41. $4x + 9 > 11 - 2(x - 2)$ and $7x - 5 < 2x + 10$

42. $4(3x - 1) < 2(x + 3)$ and $2x + 5 > 5 - x$

[Examples 5 and 6] **43.** $-3 < 2y + 3 < 1$

44. $7 \leq 5 + 2y \leq 13$

45. $1 \leq 7 - 3z \leq 22$

46. $30 < 6 - 4z < 46$

47. $2 < 4t + 5 \leq 6$

48. $3 \leq 6t + 7 < 11$

49. $u < -3$ or $u = 0$ or $u > 3$

50. $u < 1$ or $u = 2$ or $u > 7$

51. $2k > 8$ or $\frac{k}{3} < 2$

52. $\frac{k}{2} < 5$ or $3k > 6$

53. $m - 2 > -5$ or $m + 2 < 5$

54. $m + 3 > 2$ or $m - 1 < 3$

55. $3(x + 4) < 16 - x$ and $3(x + 2) > 2(10 - x) + 6$

56. $2(x + 4) < 8 - 2(8 + x)$ and $8(x + 2) > 4(x + 4) - 4x$

SECTION 2-7. Ten Review Exercises

1. Simplify $5(2u - 1) - 4 + 7(1 - u)$.

2. Evaluate $5(2u - 1) - 4 + 7(1 - u)$ for $u = -5$.

3. Solve $5(2u - 1) = 4 - 7(1 - u)$.

4. Solve and graph $5(2u - 1) > 4 - 7(1 - u)$.

5. Solve and graph: $5(2u - 1) > 15$ or $7(1 - u) > 14$

6. Solve $|2u - 1| = 13$.

In exercises **7–10**, simplify each expression.

7. $-5^2 + 3 \cdot 2 - 3 + (-2)^3$

8. $2^3 - 10^2 + 91 \div (-13) + (-5)(-3)$

9. $-6^2 + (-3) \cdot 2^2 - 3^3$

10. $\dfrac{7 - 6^2}{-2 + 7 \cdot 2} + \dfrac{-2 + 5^2}{-6 + 3 \cdot 6}$

SECTION 2-7. Supplementary Exercises

In exercises **1–6**, solve and graph each compound relation.

1. $-8 \leq 2x \leq 10$ or $x > 12$

2. $6 < x + 5 < 9$ or $10 \leq x + 5 < 13$

3. $y < 5$ or $y > 10$ or $y = 6$

4. $y > -2$ or $y < -5$ or $y = -3$

5. $2z < 8$ and $z + 3 < 10$ and $3z - 1 \geq 2$

6. $\frac{z}{3} \geq -1$ and $z + 9 > 4$ and $5z < 25$

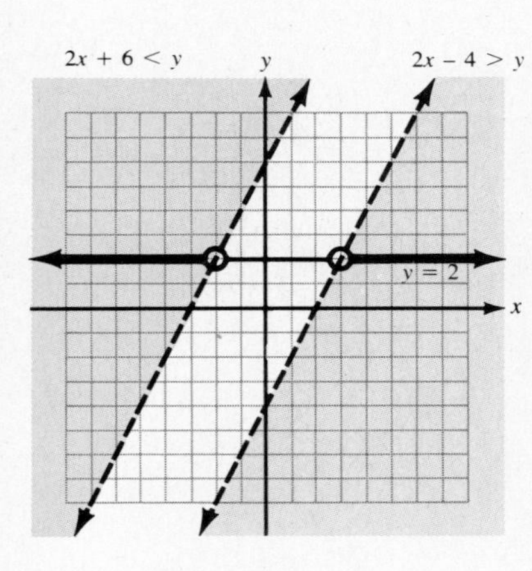

Figure 2-40. Graphs of $2x - 4 > y$ and $2x + 6 < y$.

Example. Figure 2-40 contains graphs of $2x - 4 > y$ and $2x + 6 < y$. Use the graphs to solve: $2x - 4 > 2$ or $2x + 6 < 2$.

Solution. The horizontal line defined by $y = 2$ intersects the graph in the regions where $x < -2$ or $x > 3$. Thus, the solution set of the compound statement is: $\{x | x < -2 \text{ or } x > 3\}$.

In exercises **7–10**, use the graphs in Figure 2-40 to solve each compound relation.

7. $2x - 4 > 4$ or $2x + 6 < 4$

8. $2x - 4 > 0$ or $2x + 6 < 0$

9. $2x - 4 > -2$ or $2x + 6 < -2$

10. $2x - 4 > -6$ or $2x + 6 < -6$

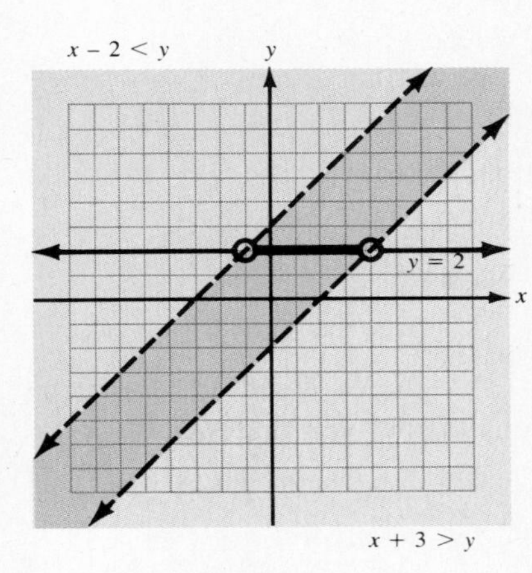

Figure 2-41. Graphs of $x - 2 < y$ and $x + 3 > y$.

Example. Figure 2-41 contains graphs of $x - 2 < y$ and $x + 3 > y$. Use the graph to solve: $x - 2 < 2$ and $x + 3 > 2$.

Solution. The horizontal line defined by $y = 2$ intersects the regions of the graph where $x < 4$ and $x > -1$. Thus, the solution set of the compound statement is $\{x | -1 < x < 4\}$.

In exercises **11–14**, use the graphs in Figure 2-41 to solve each compound relation.

11. $x - 2 < 4$ and $x + 3 > 4$

12. $x - 2 < 0$ and $x + 3 > 0$

13. $x - 2 < -3$ and $x + 3 > -3$

14. $x - 2 < 6$ and $x + 3 > 6$

In exercises **15** and **16**, answer parts **a–e**.

15. Consider the inequality $\frac{8}{t} > 4$, where $t \neq 0$.

- **a.** Assume $t > 0$ and multiply both sides of the inequality by t.
- **b.** Solve the inequality obtained in part **a**.
- **c.** Now assume $t < 0$ and multiply both sides of the inequality by t.
- **d.** Solve the inequality obtained in part **c**.
- **e.** Compare the solution sets obtained in parts **b** and **d**.

16. Consider the inequality $\frac{7}{t - 1} < 2$, where $(t - 1) \neq 0$.

- **a.** Assume $(t - 1) > 0$ and multiply both sides of the inequality by $t - 1$.
- **b.** Solve the inequality obtained in part **a**.
- **c.** Now assume $(t - 1) < 0$ and multiply both sides of the inequality by $t - 1$.
- **d.** Solve the inequality obtained in part **c**.
- **e.** Compare the solution sets obtained in parts **b** and **d**.

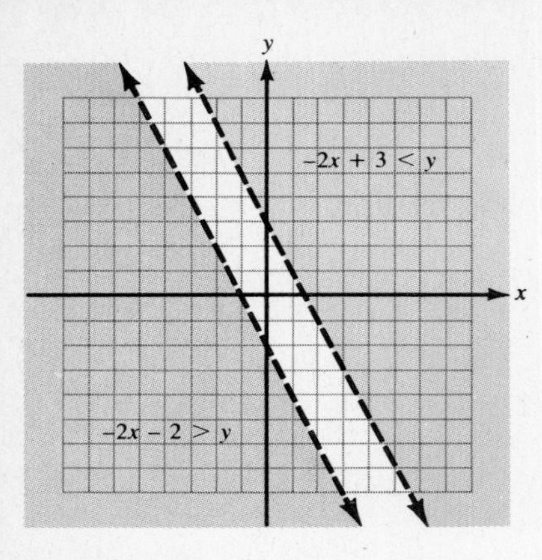

Figure 2-42. Graphs of $-2x + 3 < y$ and $-2x - 2 > y$.

In exercises **17–22**, use the graphs in Figure 2-42 to solve. (*Hint:* one line is needed for *each* inequality).

17. $-2x + 3 < 1$ or $-2x - 2 > -2$

18. $-2x + 3 < 1$ or $-2x - 2 > 2$

19. $-2x + 3 < -1$ or $-2x - 2 > 0$

20. $-2x + 3 < -1$ or $-2x - 2 > 4$

21. $-2x + 3 < 3$ and $-2x - 2 > 2$

22. $-2x + 3 < 5$ and $-2x - 2 > 6$

SECTION 2-8. Absolute Value Inequalities

1. The definition of an absolute value inequality

2. Solving inequalities of the form $|ax + b| < c$, where $c > 0$

3. Solving inequalities of the form $|ax + b| > c$, where $c > 0$

4. Writing absolute value inequalities for intervals

In Section 2-5, we solved absolute value equations. In this section, we will study procedures for solving absolute value inequalities.

The Definition of an Absolute Value Inequality

Examples **a** and **b** are absolute value equations of the type that we solved in Section 2-5.

a. $|2t + 3| = 9$ **b.** $|2 - 3z| = 5$

If the equals signs are replaced by $<$, $\leq$, $>$, or $\geq$, then an **absolute value inequality** is formed. Examples **a*** and **b*** illustrate two such inequalities.

a*. $|2t + 3| < 9$ A "less than" absolute value inequality

b*. $|2 - 3z| \geq 5$ A "greater than or equal to" absolute value inequality

Definition 2.6. An absolute value inequality in x
If a, b, and c are real numbers and $a \neq 0$, then an **absolute value inequality** in x can be written in one of the following forms:

$$(1)\ |ax + b| < c \quad \text{or} \quad (2)\ |ax + b| > c \quad \text{or}$$
$$(3)\ |ax + b| \leq c \quad \text{or} \quad (4)\ |ax + b| \geq c$$

Solving Inequalities of the Form $|ax + b| < c$, Where $c > 0$

Consider inequalities **c** and **d**.

c. $-5 < x < 5$ — A compound inequality whose solution set contains all numbers between -5 and 5

d. $|x| < 5$ — A "less than" absolute value inequality in which $a = 1$, $b = 0$, and $c = 5$. The solution set contains all real numbers whose absolute values are less than 5.

In Figure 2-43, suppose the point P represents every point between P_1 and P_2. Furthermore, suppose k is the coordinate of P.

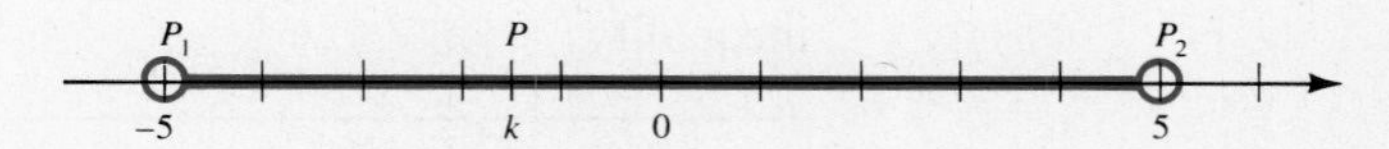

Figure 2-43. A graph of the numbers between -5 and 5.

Note 1. Since P is between P_1 and P_2, we may conclude that k is between -5 and 5; that is, the interval in the figure is a graph of the solution set of $-5 < x < 5$.

Note 2. Since P is within 5 units of 0, the absolute value of k must be less than 5; that is, $|k| < 5$. Thus, the interval in the figure is also a graph of the solution set of $|x| < 5$.

The fact that $-5 < x < 5$ and $|x| < 5$ have the same solution sets suggests a procedure for solving absolute value inequalities with $<$ or $\leq$ relations. The procedure requires that the absolute value inequality be replaced by an equivalent compound inequality with the form of one that we solved in Section 2-7.

To solve $|ax + b| < c$ or $|ax + b| \leq c$, where $c > 0$:

Step 1. Remove the absolute value bars and write:

$$-c < ax + b < c \qquad \text{or} \qquad -c \leq ax + b \leq c$$

Step 2. Solve the compound inequality.

Example 1. Solve and graph each inequality.

a. $|t| \leq 3$ **b.** $|2t + 3| < 9$

Solution. **a. Step 1.** $-3 \leq t \leq 3$ — If $|t| \leq 3$, then $-3 \leq t \leq 3$.

Step 2. A graph of the solution set is shown in Figure 2-44.

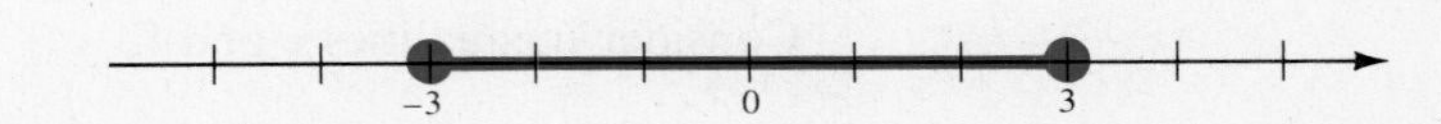

Figure 2-44. A graph of $|t| \leq 3$.

b. Step 1. $|2t + 3| < 9$ — If $|ax + b| < c$,

$-9 < 2t + 3 < 9$ — then $-c < ax + b < c$.

Step 2. $-12 < 2t < 6$ Add -3 to each part.

$-6 < t < 3$ Multiply each part by $\frac{1}{2}$.

A graph of the solution set is shown in Figure 2-45.

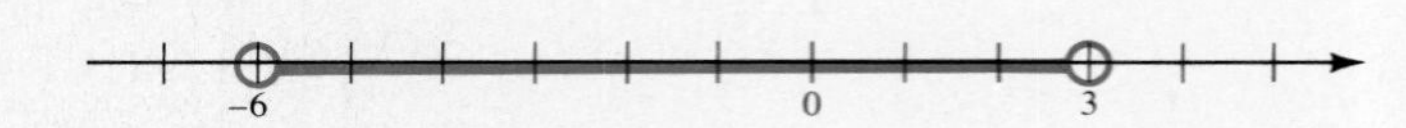

Figure 2-45. A graph of $|2t + 3| < 9$.

The antisymmetric property of inequality can be extended to a compound inequality.

> If a, b, and c are real numbers such that:
>
> (1) $a > b > c$ or (2) $a \geq b \geq c$
>
> then the inequalities can also be written as:
>
> (1*) $c < b < a$ or (2*) $c \leq b \leq a$

Example 2. Solve and graph $|4 - 3u| \leq 7$.

Solution. **Step 1.** $|4 - 3u| \leq 7$ If $|ax + b| \leq c$,

$-7 \leq 4 - 3u \leq 7$ then $-c \leq ax + b \leq c$.

$-11 \leq -3u \leq 3$ Add -4 to each part.

$\frac{11}{3} \geq u \geq -1$ Multiply each part by $\frac{-1}{3}$.

$-1 \leq u \leq \frac{11}{3}$ Antisymmetric property

A graph of the solution set is shown in Figure 2-46.

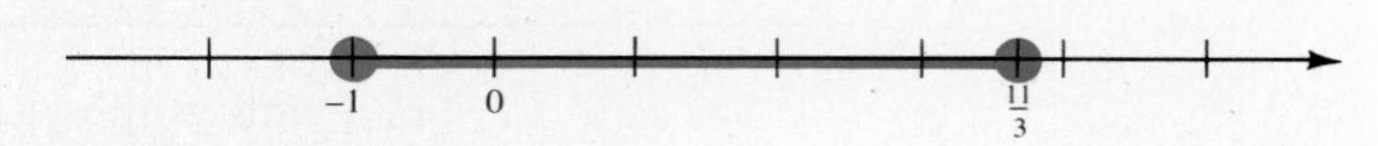

Figure 2-46. A graph of $|4 - 3u| \leq 7$.

Solving Inequalities of the Form $|ax + b| > c$, Where $c > 0$

Consider inequalities **e** and **f**.

e. $x < -3$ or $x > 3$ A compound inequality whose solution set contains all real numbers less than -3 or greater than 3

f. $|x| > 3$ An absolute value inequality in which $a = 1$, $b = 0$, and $c = 3$. The solution set contains all real numbers whose absolute values are greater than 3.

In Figure 2-47, suppose P represents every point to the left of P_1 or to the right of P_2. Furthermore, suppose k is the coordinate of P.

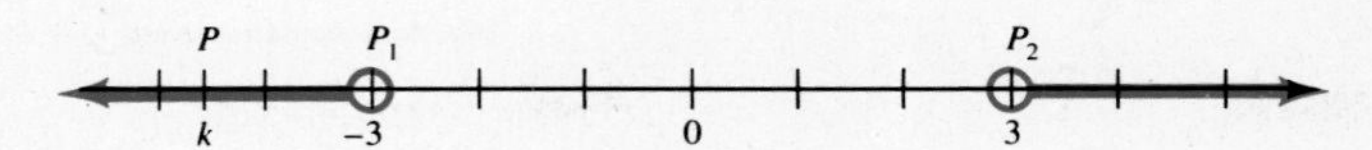

Figure 2-47. A graph of the numbers less than -3 or greater than 3.

Note 1. Since P is to the left of P_1 or to the right of P_2, it follows that $k < -3$ or $k > 3$. Thus, the intervals in Figure 2-47 are a graph of the solution set of $x < -3$ or $x > 3$.

Note 2. If P is to the left of P_1 or to the right of P_2, then P is more than 3 units from 0. Thus, in absolute value the coordinate of P must be greater than 3; that is, $|k| > 3$. The intervals in Figure 2-47 are also a graph of the solution set of $|x| > 3$.

The similarity in the solution sets of the inequalities in examples **e** and **f** suggests a procedure for solving an absolute value inequality formed by $>$ or $\geq$.

To solve $|ax + b| > c$ or $|ax + b| \geq c$, where $c > 0$:

Step 1. Remove the absolute value bars and write:

$ax + b < -c$ or $ax + b > c$ For $|ax + b| > c$

$ax + b \leq -c$ or $ax + b \geq c$ For $|ax + b| \geq c$

Step 2. Solve the compound inequality.

Example 3. Solve and graph each inequality.

a. $|t| \geq 7$ **b.** $|4y + 1| > 9$

Solution.

a. Step 1. $t \leq -7$ or $t \geq 7$ If $|t| \geq c$, then $t \leq -c$ or $t \geq c$.

Step 2. A graph of the solution set is shown in Figure 2-48.

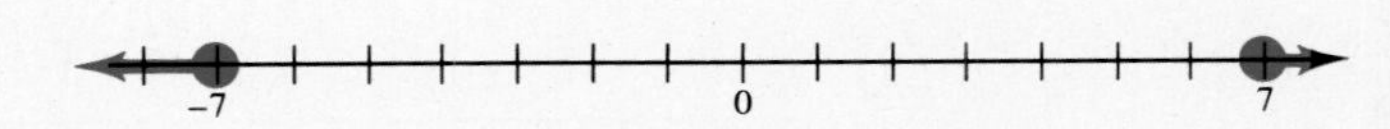

Figure 2-48. A graph of $|t| \geq 7$.

b. Step 1. $|4y + 1| > 9$ If $|ax + b| > c$, then

$4y + 1 < -9$ or $4y + 1 > 9$ $ax + b < -c$ or $ax + b > c$.

Step 2. $4y < -10$ $4y > 8$ Add -1 to each part.

$y < \dfrac{-10}{4}$ $y > 2$ Multiply each part by $\dfrac{1}{4}$.

$y < \dfrac{-5}{2}$ Reduce the fraction.

A graph of the solution set is shown in Figure 2-49.

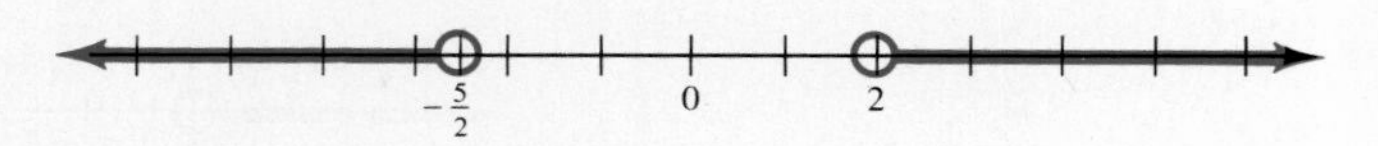

Figure 2-49. A graph of $|4y + 1| > 9$.

Writing Absolute Value Inequalities for Intervals

In Figure 2-50, a and b, two intervals are graphed. Suppose the task is to write absolute value inequalities in x that define these intervals.

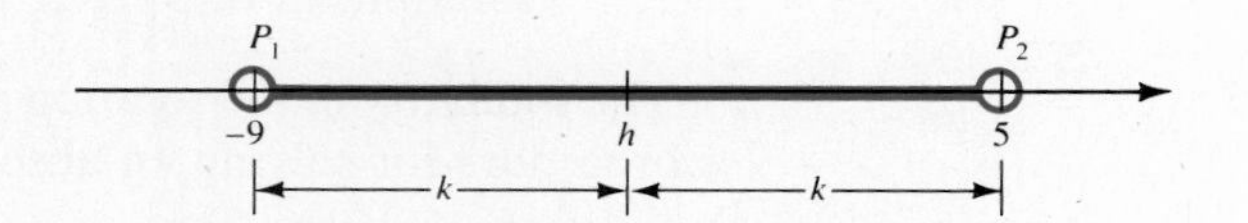

Figure 2-50a. An interval between P_1 and P_2.

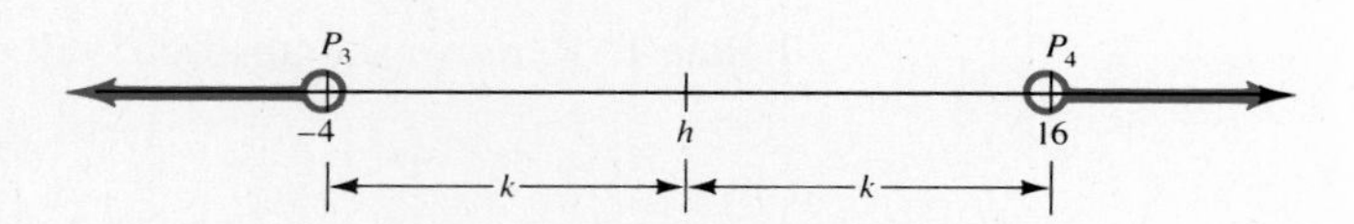

Figure 2-50b. Intervals to the left of P_3 and to the right of P_4.

In Figure 2-50a, h is the coordinate of the point midway between P_1 and P_2. The distance from this midpoint to P_1 and also P_2 is k, where $k > 0$. Any point between P_1 and P_2 will have a coordinate x, such that:

$$|x - h| < k$$

In Figure 2-50b, h is the coordinate of the point midway between P_3 and P_4. The distance from this midpoint to P_3 and P_4 is also k, where $k > 0$. Any point to the left of P_3 or to the right of P_4 will have a coordinate x, such that:

$$|x - h| > k$$

Let P_1 and P_2 be two points on a number line with coordinates x_1 and x_2, respectively.

1. The coordinate of the point midway between P_1 and P_2 is h, and

$$h = \frac{x_1 + x_2}{2}.$$ The coordinate of the midpoint is the mean of the coordinates of P_1 and P_2.

2. The distance between P_1 and the midpoint, or P_2 and the midpoint is k, and

$$k = \frac{1}{2}|x_2 - x_1|.$$ Use absolute values, because $k > 0$.

Example 4. **a.** Write an absolute value inequality in x that defines the interval in Figure 2-50a.

b. Write an absolute value inequality in x that defines the intervals in Figure 2-50b.

Solution. **a.** The coordinates of the endpoints of the interval are -9 and 5.

$$h = \frac{-9 + 5}{2} = -2$$ The midpoint of the interval has coordinate -2.

$$k = \frac{1}{2}|5 - (-9)|$$

$$= 7$$ The distance from the midpoint to the endpoints is 7.

Therefore, an inequality that defines the interval is

$$|x - (-2)| < 7 \quad \text{or} \quad |x + 2| < 7.$$

Notice that the interval is *between the endpoints.*

b. The coordinates of the endpoints of the interval are -4 and 16.

$$h = \frac{-4 + 16}{2} = 6$$ The midpoint of the interval has coordinate 6.

$$k = \frac{1}{2}|16 - (-4)|$$

$$= 10$$ The distance from the midpoint to the endpoints is 10.

Therefore, an inequality that defines the intervals is $|x - 6| > 10$. Notice that the intervals are *beyond the endpoints.*

SECTION 2-8. Practice Exercises

In exercises **1–40**, solve and graph each inequality.

[Examples 1 and 2]

1. $|x| < 3$

2. $|x| < 7$

3. $|y| \le 0.75$

4. $|y| \le 0.5$

5. $|z - 1| \le 3$

6. $|z - 4| \le 7$

7. $|t + 4| < 2$

8. $|t + 2| < 4$

9. $|-5 - u| < 6$

10. $|-2 - u| < 10$

11. $|2v - 3| \le 5$

12. $|2v - 5| \le 3$

13. $|5 - 8k| < 13$

14. $|8 - 5k| < 2$

15. $|3p + 2| \le 2$

16. $|3p + 1| \le 1$

17. $2|q - 2| \le 5$

18. $2|q - 5| \le 3$

19. $|2x - 5| < \sqrt{2}$ **20.** $|3x - 2| < \sqrt{5}$

[Example 3] **21.** $|y| \geq 6$ **22.** $|y| \geq 9$

23. $|z - 3| > 10$ **24.** $|z - 5| > 7$

25. $|t + 2| \geq 2$ **26.** $|t + 6| \geq 6$

27. $|-10 - u| > 7$ **28.** $|-3 - u| > 9$

29. $|2v - 3| > 5$ **30.** $|2v - 1| > 7$

31. $2|4 - w| > 10$ **32.** $6|8 - w| > 30$

33. $|3 - 5x| \geq 2$ **34.** $|13 - 2x| \leq 11$

35. $5|2y - 5| > 1$ **36.** $3|2y - 3| > 2$

37. $|z - 5| > \pi$ **38.** $|z + 4| > \pi$

39. $|3t + 1| - 5 > 8$ **40.** $|2 + 5t| + 8 > 25$

In exercises **41–50**, write an absolute value inequality in x that defines the interval in each figure.

[Example 4] **41.**

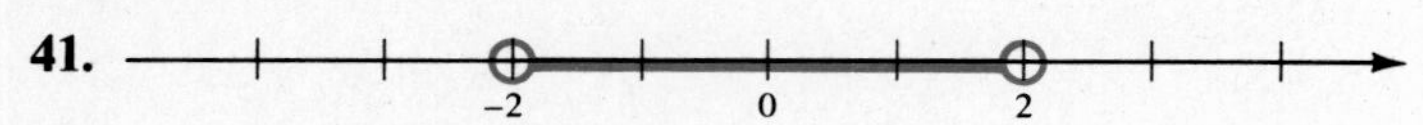

42.

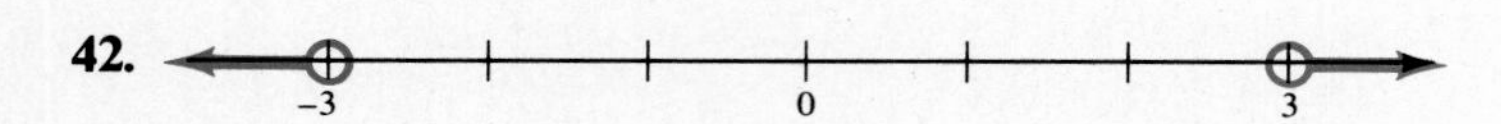

43. −1 0 1

44. −4 0 4

45. −3 0 1

46. −4 0 2

47. 0 4

48. −2 0 4

49. −10 −4 0

50. 0 3 13

SECTION 2-8. Ten Review Exercises

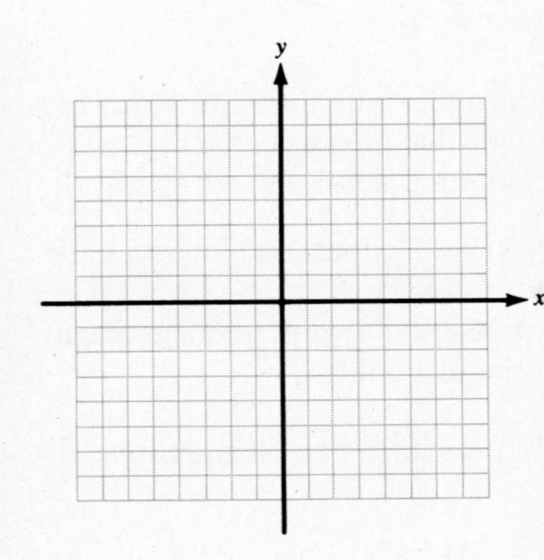

1. On the grid, plot $P(-6, 6)$ and $Q(6, -2)$.

2. Draw a line $\mathscr{L}$ through P and Q.

3. Determine the slope of $\mathscr{L}$ using the graph.

4. Compute the slope of $\mathscr{L}$ using the formula.

5. Identify the y-intercept of the line.

6. Identify the x-intercept of the line.

7. Write an equation for $\mathscr{L}$ in the $y = mx + b$ form.

In exercises **8** and **9**, solve each equation.

8. $2x - 3 = 9$

9. $|2x - 3| = 9$

10. Solve $|2x - 3| < 9$.

SECTION 2-8. Supplementary Exercises

In exercises **1–10**, solve and graph each inequality.

1. $\left|\frac{3}{2}x + 1\right| < 5$

2. $\left|\frac{1}{3}x - 10\right| < 4$

3. $|4y - 5| > \frac{1}{3}$

4. $|4 - 3y| > \frac{1}{4}$

5. $\left|\frac{1 + 2z}{4}\right| \leq 3$

6. $\left|\frac{z - 1}{2}\right| \leq 5$

7. $\left|\frac{3 - t}{2}\right| \geq 5$

8. $\left|\frac{1 - 2t}{3}\right| \geq 7$

9. $\left|\frac{2u - 1}{3}\right| < 6$

10. $\left|\frac{2u - 8}{5}\right| > 4$

A graph of $|ax + b| = y$ is V-shaped. A graph of:

(1) $|ax + b| < y$ is the *interior region* of the V

(2) $|ax + b| > y$ is the *exterior region* to the V

A graph of $y = k$ is a horizontal line. The portion of a horizontal line that lies in the regions of relations defined by (1) or (2) corresponds to the solution set of a linear inequality in one variable.

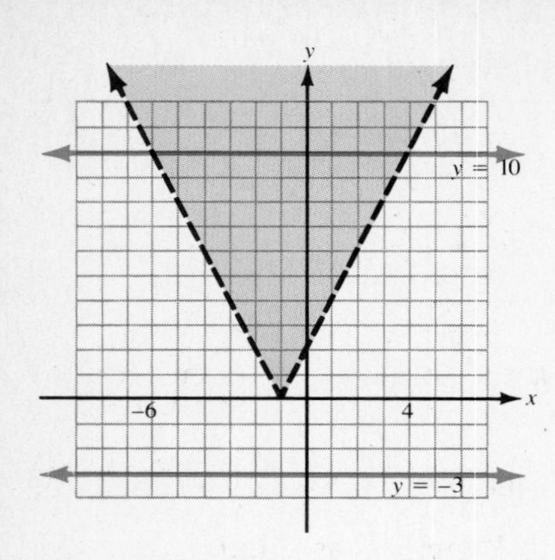

Figure 2-51. Graphs of $|2x + 2| < y$, $y = 10$, and $y = -3$.

Example. Figure 2-51 contains graphs of $|2x + 2| < y$, $y = 10$, and $y = -3$. Use these graphs to solve:

a. $|2x + 2| < 10$ **b.** $|2x + 2| < -3$

Solution. **a. Discussion.** The solution set contains all values of x for which the line intersects the shaded region. The line intersects the region for x between -6 and 4. Thus, the solution set is $\{x \mid -6 < x < 4\}$.

b. Discussion. The line defined by $y = -3$ does not intersect the shaded region. Thus, the solution set of $|2x + 2| < -3$ is $\varnothing$. Use the fact that an absolute value expression is always nonnegative to justify the empty set solution to the inequality.

In exercises **11–14**, use the graphs in Figure 2-52 to solve each inequality.

11. $\left|\frac{3}{4}x - \frac{3}{2}\right| < 6$

12. $\left|\frac{3}{4}x - \frac{3}{2}\right| < 3$

13. $\left|\frac{3}{4}x - \frac{3}{2}\right| \leq 0$

14. $\left|\frac{3}{4}x - \frac{3}{2}\right| \leq -2$

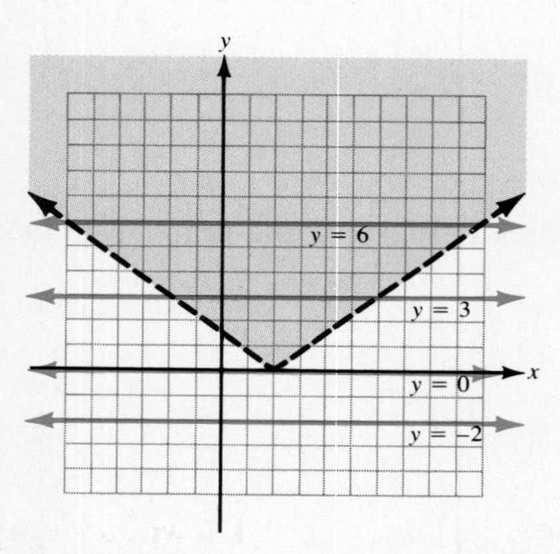

Figure 2-52. Graphs of $|\frac{3}{4}x - \frac{3}{2}| < y$, $y = 6$, $y = 3$, $y = 0$, and $y = -2$.

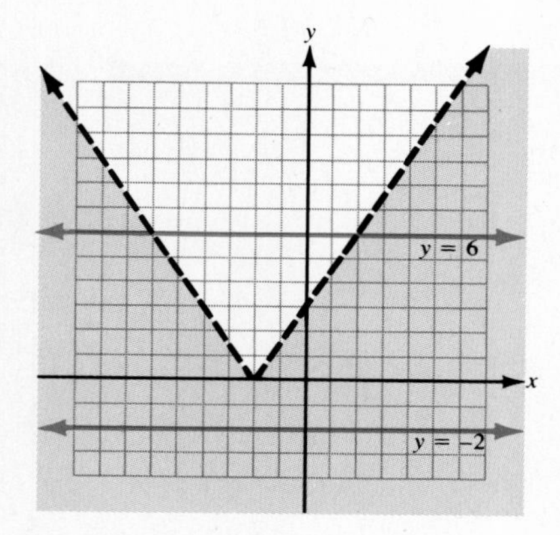

Figure 2-53. Graphs of $|\frac{3}{2}x + 3| > y$, $y = 6$, and $y = -2$.

Example. Figure 2-53 contains graphs of $|\frac{3}{2}x + 3| > y$, $y = 6$, and $y = -2$. Use these graphs to solve:

a. $\left|\frac{3}{2}x + 3\right| > 6$ **b.** $\left|\frac{3}{2}x + 3\right| > -2$

Solution. **a. Discussion.** The solution set contains all values of x for which the line intersects the shaded region. The line intersects the region for $x < -6$ or $x > 2$. The solution set is $\{x \mid x < -6 \text{ or } x > 2\}$.

b. Discussion. The line defined by $y = -2$ intersects the shaded region at every point. Thus, the solution set of $|\frac{3}{2}x + 3| > -2$ is R.

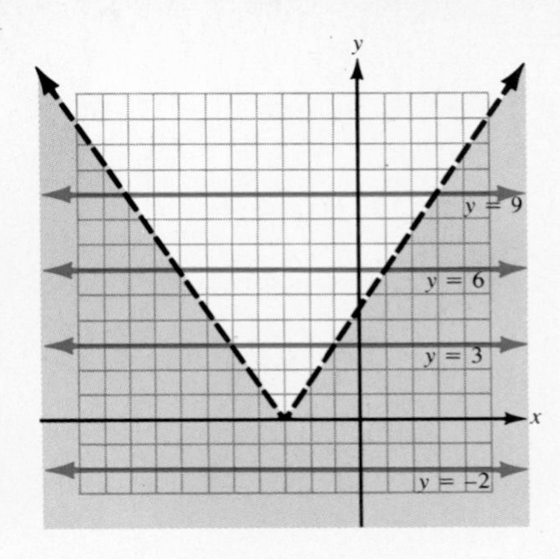

Figure 2-54. Graphs of $\left|\frac{3}{2}x + \frac{9}{2}\right| > y$, $y = 9$, $y = 6$, $y = 3$, and $y = -2$.

In exercises **15–18**, use the graphs in Figure 2-54 to solve each inequality.

15. $\left|\frac{3}{2}x + \frac{9}{2}\right| > 9$

16. $\left|\frac{3}{2}x + \frac{9}{2}\right| > 6$

17. $\left|\frac{3}{2}x + \frac{9}{2}\right| > 3$

18. $\left|\frac{3}{2}x + \frac{9}{2}\right| > -2$

If a, b, and c are real numbers and $a \neq 0$, the following theorems can be established:

Theorem A. $|ax - b| < c$ and $c < 0$ has no solution

Theorem B. $|ax - b| > c$ and $c < 0$ has all real numbers as solutions

Theorem C. $|ax - b| \leq 0$ has one solution, namely, $\frac{b}{a}$

Theorem D. $|ax - b| \geq 0$ has all real numbers as solutions

Theorem E. $|ax - b| < 0$ has no solution

Theorem F. $|ax - b| > 0$ has all real numbers except $\frac{b}{a}$ as solutions

In exercises **19–30**, use the appropriate theorem to find the solutions (if any exist) of each inequality.

19. $|2x - 5| < 0$

20. $|8x + 4| < 0$

21. $|9x + 7| > -2$

22. $|x - 2| > -2$

23. $|x - 10| > 0$

24. $|2x + 4| > 0$

25. $|3x + 8| < -1$

26. $|5x - 2| < -2$

27. $|3x| \geq 0$

28. $|10x| \geq 0$

29. $|2x - 6| \leq 0$

30. $|5x + 10| \leq 0$

The following inequality is known as the **triangle inequality theorem:**

If a and b are any real numbers, then $|a + b| \leq |a| + |b|$.

In exercises **31–34**, verify the triangle inequality theorem for the given values of a and b.

31. $a = 5$ and $b = 8$

32. $a = 12$ and $b = -4$

33. $a = -17$ and $b = 9$

34. $a = -13$ and $b = -11$

Name

Date

Score

SECTION 2-3. Summary Exercises

Answer

In exercises **1–4**, solve each word problem.

1. At 7:00 A.M. a truck leaves a warehouse and travels due east. At 9:00 A.M. a car leaves the warehouse and travels due west. If the truck has a rate of 45 mph and the car has a rate of 65 mph, how long will it take for the two vehicles to be 420 miles apart?

1. ____________

2. A local train left a station and traveled west at a rate of 24 mph. Two hours later, an express train left the same station and traveled due west at a rate of 56 mph until it overtook the local train.

a. Find the distance traveled.

2. a. ____________

b. Find the time it took for the express train to overtake the local train.

b. ____________

3. How many kilograms of select coffee worth \$12 a kilogram should be mixed with 60 kilograms of an ordinary coffee worth \$8 a kilogram to obtain a mixture worth \$9 a kilogram?

3. \$12 coffee: ____________

\$8 coffee: ____________

4. Currently the age of a mother is 5 years less than 3 times her daughter's age. Four years ago, the mother's age was 5 years less than 4 times the daughter's age at that time. Find the present ages of mother and daughter.

4. Mother's age: ____________

Daughter's age: ____________

Name

Date

Score

SECTION 2-4. Summary Exercises

Answer

In exercises **1–7**, solve each equation for the indicated variable.

1. $I = Prt$, for r

1. ______

2. $12a - 5b = 3$, for b

2. ______

3. $P = 2a + b + c$, for b

3. ______

4. $S = a + (n - 1)d$, for n

4. ______

5. $5x + 7 = ax + t$, for x

5. ____________

6. $2A = h(a + b)$, for a

6. ____________

7. $S = at^2 + vt + h$, for v

7. ____________

8. a. Solve $Fr^2 = kmn$ for n.

8. a. ____________

b. Find n for $F = 133$, $k = 7$, $m = 90$, and $r^2 = 360$.

b. ____________

Name

Date

Score

SECTION 2-7. Summary Exercises

Answer

1. Given the following statements:

   p: x is less than 100
   q: x is odd

   For each value of x, determine the truth value of:

   (i) p or q
   (ii) p and q

   a. $x = -5$

   b. $x = 68$

   c. $x = 124$

   d. $x = 111$

1. a. (i) ______

(ii) ______

b. (i) ______

(ii) ______

c. (i) ______

(ii) ______

d. (i) ______

(ii) ______

In exercises **2–8**, solve and graph each compound relation.

2. $x < 7$ or $x \geq 8$

2. ______

3. $x - 5 \geq -1$ and $x + 3 \leq 10$

3. ______

4. $5 - 3x > 11$ or $2x + 5 = 3$

4. ______

5. $-5 \leq 2x - 7 \leq 5$

5. ______________________

6. $2(x + 2) \geq -12$ and $3(x + 2) \leq 6$

6. ______________________

7. $6 + 2(2 - x) \geq 10 + 2x$ or $3x - 10 > 6 - x$

7. ______________________

8. $5 - 3x < 11$ and $2x + 3 < -9$

8. ______________________

Name

Date

Score

SECTION 2-8. Summary Exercises

Answer

In exercises **1–6**, solve and graph each inequality.

1. $|x| < 5$

1. ______________________

2. $|y| \geq 3$

2. ______________________

3. $|z - 7| \leq 1$

3. ______________________

4. $|p + 4| > 7$

4. ______________________

5. $|3q - 4| \geq 4$

5. ______________________

6. $\left|\frac{10 - 3t}{2}\right| < 8$

6. ______________________

In exercises **7** and **8**, write an absolute value inequality in x that defines the interval in each figure.

7. 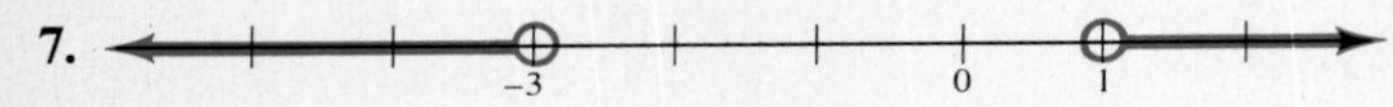

7. ______________________

8. 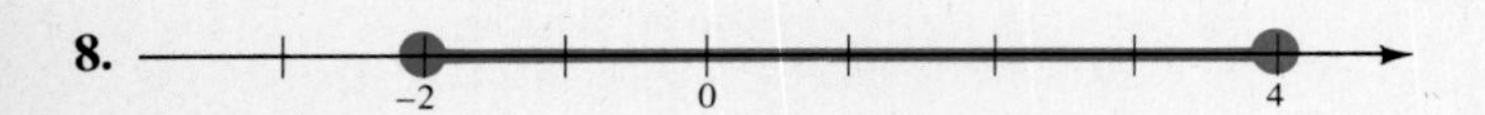

8. ______________________

CHAPTER 2. Review Exercises

In exercises **1–6**, solve each equation.

1. $7x + 5 - 3x = 2x - 1$

2. $10 - 6x - 3 = 12 - 15x + 1$

3. $a - 2(5 - 3a) = 6 - (a + 4) - 16$

4. $5b - 3(b + 7) = 20 - 2(7 - b)$

5. $1 - 5(2k + 7) = 2(3 - 5k) - 40$

6. $y(y + 3) - y(1 - y) = 2(y^2 + 5)$

In exercises **7–13**, solve each word problem.

7. When the product of 5 times a number is subtracted from 10, the result is 2 less than the product of 9 times the sum of the number and 20. Find the number.

8. The sum of the interior angles of any pentagon (a plane figure with five sides) is 540°. In pentagon $ABCDE$, the measures of angles A and B are equal. The measure of angle C is 15° more than angle A. The measure of angle D is three-fourths the measure of angle A. The measure of angle E is 15° more than one-half the measure of angle A. Find the measure of each angle.

9. Nancy Bain has $4,000 invested that is earning 9% interest each year. She has an opportunity to invest some additional money at 12% interest per year. If she wants to earn $1,260 in interest a year on these two investments, how much must she invest at the 12% rate?

10. A chemist wants to produce a 40% solution of alcohol. She has a container with 8 gallons of a 25% solution. She also has a large quantity of a 60% alcohol solution she can use. How much of the 60% solution should be mixed with the 8 gallons of the 25% mixture to produce a 40% solution?

11. A ship leaves New York harbor at 7:00 A.M. traveling east at 20 mph. At 11:00 A.M., a helicopter leaves the same port at 120 mph to overtake the ship. In approximately how many minutes will the helicopter reach the ship?

12. The cities of Pittsburgh and Philadelphia are about 300 miles apart. A passenger train leaves Pittsburgh at an average speed of 60 mph and heads for Philadelphia. Thirty minutes later a freight train leaves Philadelphia at 40 mph and heads for Pittsburgh. How long does the freight train travel before the trains meet?

13. The age of a mother is currently 5 years more than 4 times her daughter's age. Four years ago, the mother's age was 1 year less than 13 times the daughter's age at that time. Find the present ages of mother and daughter.

In exercises **14–18**, solve each equation for the indicated variable.

14. $6x - 5y = 25$ for y

15. $s = at^2 + vt$ for a

16. $V = \frac{4}{3}\pi r^3$ for r^3

17. $k + 2y = my + 4$ for y

18. a. Solve $x^2 + y^2 - 2by = 10$ for b.

b. Find b when $x = 6$ and $y = 4$.

c. Find b when $x = 10$ and $y = 3$.

In exercises **19–24**, solve each equation.

19. $|x| = 20$

20. $|y| = -5$

21. $|a - 8| = 3$

22. $|2x + 5| = 19$

23. $\left|\frac{2t}{5} + 7\right| = 7$

24. $|5x + 3| = |3x - 7|$

In exercises **25–28**, solve and graph each inequality.

25. $2y + 5 - 7y \leq 35$

26. $3 - 2(a + 4) < 3(a - 2) - 4$

27. $10(b + 4) - 4(2b - 1) \geq 5b + 20$

28. $k(2k + 5) - 10 > 15 + 2k^2$

In exercises **29** and **30**, solve each word problem.

29. Twice the first and twice the third of 3 consecutive integers is less than 44. Find the possible values for the smallest integer.

30. A rectangular air vent has a length that is 1 inch more than 3 times as large as the width. If the perimeter cannot be larger than 26 inches, find the possible values for the width.

31. Given the following statements:

p: x is greater than 100
q: x is a multiple of 5

If $x = 139$, determine the truth value of:

a. p or q
b. p and q

32. Given the following statements:

p: x is odd
q: x is a prime number

If $x = 53$, determine the truth value of:

a. p or q
b. p and q

In exercises **33–36**, solve and graph each compound relation.

33. $x + 1 > 5$ or $4x - 1 < 7$

34. $7x + 5 \leq 40$ and $2x \geq x + 1$

35. $x - 2 > 5$ or $x + 2 = 12$

36. $0 \leq 6 - 9x \leq 24$

In exercises **37–42**, solve and graph each inequality.

37. $|x + 3| < 13$

38. $|4x - 3| \geq 9$

39. $|8 - 5x| \leq 13$

40. $2|x + 5| > 3$

41. $|10x + 9| < -2$

42. $|x - 6| > 0$

In exercises **43** and **44**, write an absolute value inequality in x that defines the interval in each figure.

43. 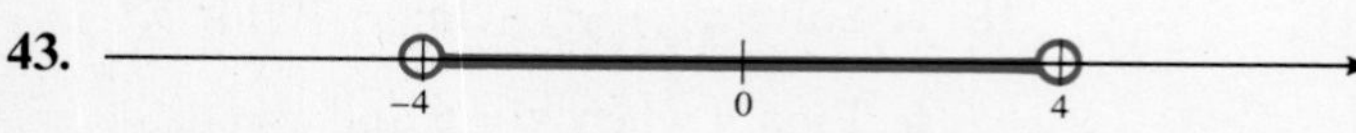

44. 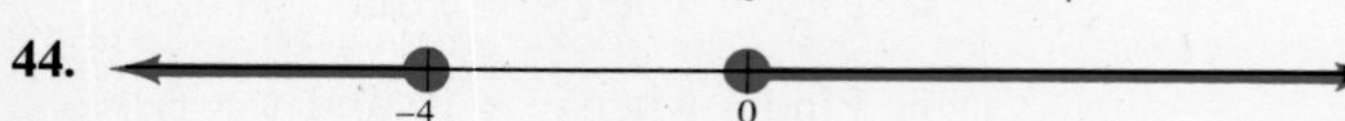

CHAPTER

3

Polynomial Expressions and Equations

The following display was on the board as the class entered the room:

a. $11t + 3(t - 3) + 1 = 5(t + 2) - (t + 3)$

b. $11t + 3(t - 3) + 1 - 5(t + 2) + (t + 3)$

"Okay, class," Ms. Glaston said as she arrived, "what is the major difference between what you see in **a** and **b**?"

Virginia Colburn was the first with her hand up. "There's an equal sign in **a**, but not in **b**," she said. "That makes **a** an equation, but not **b**."

"That's right, Virginia," Ms. Glaston replied. "So what instructions would be given for **a**, and what instructions would be given for **b**?"

"Since **a** is an equation," Marshall McKissick added, "wouldn't we be told to solve it? That is, find the value for t that would result in a true statement."

"I agree," Janet Peters said. "The equation in **a** is just like the linear equations we worked on in the last chapter. But **b** isn't an equation, so what would we do with it?"

"Well, Janet," Ms. Glaston replied, "the collection of symbols in **b** is called a **mathematical expression.** If t is replaced by a number and the indicated operations carried out, the result is a number. The number obtained is called **the value of the expression** for the given value of t. The instructions that might be given for such an expression would be "simplify." That is, reduce the number of symbols needed to represent the expression. For example, I simplified this expression and reduced it to $10t - 15$. Therefore, when t is replaced by any number in either expression, the same number is obtained."

"In mathematics," Ms. Glaston continued, "it is convenient to put expressions in categories, just like we put equations and inequalities in categories. A geometric diagram can be used to represent the set of all expressions, and some of the principal types that are studied separately."

She then turned to the board and drew the diagram shown in Figure 3-1.

"As you can see," Ms. Glaston continued, "I have identified six specific categories of mathematical expressions. In this chapter, we will study what are called polynomial expressions. Since expressions represent numbers, we operate on them as we operate on numbers. Specifically, we will add, subtract, multiply, divide, and factor polynomials. However, since polynomial expressions are not equations, we will not be instructed to solve them. At this time are there any questions? If not, let's begin our study by reviewing some properties of exponents."

Figure 3-1. A geometric display of the set of all mathematical expressions.

SECTION 3-1. Basic Laws of Exponents

1. A definition of a positive integer exponent
2. Five properties of exponents
3. A definition of b^0, where $b \neq 0$
4. Using the properties to simplify expressions

An indicated sum of the same term can be written using fewer symbols as a multiplication.

An indicated sum		**An indicated product**
$\underbrace{b + b + b \ldots b}_{n \text{ terms of } b}$	can be written as	$b \cdot n$

An indicated product of the same factor can be written using fewer symbols as a power:

An indicated product		**An indicated power**
$\underbrace{b \cdot b \cdot b \ldots b}_{n \text{ factors of } b}$	can be written as	b^n

In this section, we will review five basic properties of exponents. A definition will also be given for 0 as an exponent.

A Definition of a Positive Integer Exponent

Definition 3.1 identifies the meaning of a positive integer exponent.

> **Definition 3.1.** If b is a real number and n is a positive integer, that is, 1, 2, 3, . . . , then:
>
> $$b^n = \underbrace{b \cdot b \cdot b \ldots b}_{n \text{ factors of } b}$$
>
> If $n = 1$, then b^1 is written b.

Examples **a–f** evaluate several exponential expressions.

a. $6^4 = 6 \cdot 6 \cdot 6 \cdot 6$

$= 1{,}296$

b. $(-3)^6 = (-3)(-3)(-3)(-3)(-3)(-3)$

$= 729$

c. $\left(\frac{2}{5}\right)^3 = \frac{2}{5} \cdot \frac{2}{5} \cdot \frac{2}{5}$

$= \frac{8}{125}$

d. $(-0.7)^3 = (-0.7)(-0.7)(-0.7)$

$= -0.343$

e. $(-10)^2 = (-10)(-10) = 100$

f. $-10^2 = -1 \cdot 10^2 = -1 \cdot 10 \cdot 10 = -100$

Five Properties of Exponents

The definition of a positive integer exponent can be used to verify five basic properties of exponents. These properties are stated below, and the first property is verified. In this section's Supplementary Exercises, guided solutions are given to verify the remaining properties (see p. 163).

Five properties of exponents

Let a and b be nonzero numbers, with m and n positive integers.

Property 1. $a^m \cdot a^n = a^{m+n}$ — When multiplying like bases, add exponents.

Property 2. $(a^m)^n = a^{mn}$ — When raising a power to a power, multiply exponents.

Property 3. $\frac{a^m}{a^n} = \begin{cases} a^{m-n}, & \text{if } m > n \\ 1, & \text{if } m = n \\ \frac{1}{a^{n-m}}, & \text{if } m < n \end{cases}$ — When dividing like bases, subtract exponents, the smaller from the larger. If the exponents are equal, write 1 for the quotient.

Property 4. $(ab)^m = a^m \cdot b^m$ — When raising a product to a power, raise each factor to the power.

Property 5. $\left(\frac{a}{b}\right)^m = \frac{a^m}{b^m}$ — When raising a quotient to a power, raise each factor to the power.

Verification of Property 1

Write a^m as $a \cdot a \ldots a \rightarrow m$ factors of a.

Write a^n as $a \cdot a \ldots a \rightarrow n$ factors of a

$$a^m \cdot a^n = \underbrace{(a \cdot a \ldots a)}_{m \text{ factors}} \cdot \underbrace{(a \cdot a \ldots a)}_{n \text{ factors}}$$

$$= \underbrace{(a \cdot a \ldots a) \cdot (a \cdot a \ldots a)}_{m + n \text{ factors}}$$

$$= a^{m+n}$$

Example 1. Simplify each expression.

a. $t^5 \cdot t^3 \cdot t$ **b.** $(r^5s^3)(rs^4)$

Solution. **a.** $t^5 \cdot t^3 \cdot t = t^{5+3+1}$ t can be written t^1

$= t^9$ Add exponents.

b. $(r^5s^3)(rs^4) = (r^5 \cdot r)(s^3 \cdot s^4)$ Group like factors.

$= r^6s^7$ Add exponents.

Example 2. Simplify each expression.

a. $(z^3)^5$ **b.** $(t^2)^3(t^5)^2$

Solution. **a.** $(z^3)^5 = z^{3 \cdot 5}$ Multiply exponents.

$= z^{15}$ Simplify the product.

b. $(t^2)^3(t^5)^2 = t^6 \cdot t^{10}$ Multiply exponents.

$= t^{16}$ Now add the exponents.

Example 3. Simplify each expression.

a. $\frac{u^9}{u^5}; u \neq 0$ **b.** $\frac{(k^4)^3}{(k^7)^2}; k \neq 0$

Solution. **a.** $\frac{u^9}{u^5} = u^{9-5}$ Subtract exponents.

$= u^4$ Simplify the difference.

b. $\frac{(k^4)^3}{(k^7)^2} = \frac{k^{12}}{k^{14}}$ Multiply exponents.

$= \frac{1}{k^{14-12}}$ Subtract smaller from larger.

$= \frac{1}{k^2}$ Simplify the difference.

Example 4. Simplify each expression.

a. $(-3y)^4$ **b.** $(2uv^2)^5$

Solution. **a.** $(-3y)^4 = (-3)^4 \cdot y^4$ Raise both factors to the power 4.

$= 81y^4$ Simplify $(-3)^4$.

b. $(2uv^2)^5 = 2^5 \cdot u^5 \cdot (v^2)^5$ Raise each factor to the power 5.

$= 32u^5v^{10}$ $(v^2)^5 = v^{2 \cdot 5} = v^{10}$

Example 5. Simplify each expression.

a. $\left(\frac{p}{3}\right)^3$ **b.** $\left(\frac{rs^2}{2}\right)^4$

Solution. **a.** $\left(\frac{p}{3}\right)^3 = \frac{p^3}{3^3}$ Raise both factors to the power 3.

$= \frac{p^3}{27}$ Simply the indicated power.

b. $\left(\frac{rs^2}{2}\right)^4 = \frac{r^4 \cdot (s^2)^4}{2^4}$ Raise each factor to the power 4.

$= \frac{r^4 s^8}{16}$ $(s^2)^4 = s^{2 \cdot 4} = s^8$

A Definition of b^0, Where $b \neq 0$

Definition 3.1 applies only to exponents that are positive integers. In mathematics and applied areas, the use of exponents has been extended beyond this initial intent. However, each time numbers other than positive integers are used as exponents, precise definitions are given to specify what exactly the exponent means.

Definition 3.2. Zero as an exponent
If b is a real number and $b \neq 0$, then $b^0 = 1$.

Example 6. Simplify each expression.

a. 19^0 **b.** $(-10)^0$ **c.** -10^0

Solution. **a.** $19^0 = 1$ $b^0 = 1$, provided $b \neq 0$

b. $(-10)^0 = 1$ A negative number to the 0 power is 1.

c. $-10^0 = -1$ The base is 10, not -10.

Part **c** of Example 6 is a negative 1 because the minus bar is not part of the base. We should read -10^0 as "The opposite of 10^0". Thus:

$$-10^0 = -1 \cdot 10^0 = -1 \cdot 1 = -1$$

Using the Properties to Simplify Expressions

In Examples 7 and 8, two or more properties of exponents are used to simplify each expression. The rule for the order of operations is followed in doing the simplification.

Example 7. Simplify $(-2u^2v)^3(-v^2)^4(3u)^2$.

Solution. **Discussion.** Raising to powers has priority over multiplication. Therefore, any indicated powers are simplified before any indicated products.

$(-2u^2v)^3(-v^2)^4(3u)^2$ — The given expression

$= (-2)^3(u^2)^3v^3(-1)^4(v^2)^4 3^2u^2$ — $(-v^2)^4 = (-1 \cdot v^2)^4$

$= -8u^6v^3(1)v^8 9u^2$ — Simplify the indicated powers.

$= -72u^8v^{11}$ — Add exponents on like bases.

Example 8. Simplify $\left(\frac{6p^3}{q^2}\right)^3\left(\frac{q^3}{3p^4}\right)^2$, where $p \neq 0$ and $q \neq 0$.

Solution.

$\left(\frac{6p^3}{q^2}\right)^3\left(\frac{q^3}{3p^4}\right)^2$ — The given expression

$= \frac{(6p^3)^3}{(q^2)^3} \cdot \frac{(q^3)^2}{(3p^4)^2}$ — $\left(\frac{a}{b}\right)^m = \frac{a^m}{b^m}$

$= \frac{6^3(p^3)^3}{q^6} \cdot \frac{q^6}{3^2(p^4)^2}$ — $(ab)^m = a^m \cdot b^m$, and $(a^m)^n = a^{mn}$

$= \frac{216p^9q^6}{9p^8q^6}$ — Write using one fraction bar.

$= 24p$ — $q^{6-6} = q^0 = 1$

Suppose you need to evaluate the given expression in Example 8 for $p = 5$ and $q = -4$. The answer is 120, because $24(5) = 120$. That is, *the simplified form of the expression has the same value as the given expression for any nonzero values of* p *and* q. Clearly, the simplified form is easier to evaluate!

SECTION 3-1. Practice Exercises

In exercises **1–64**, simplify each expression. Assume all variables are not 0.

[Example 1]

1. $x^4 \cdot x^2$ **2.** $x^{10} \cdot x^3$

3. $y^5 \cdot y^3 \cdot y^2$ **4.** $y^7 \cdot y^4 \cdot y^3$

5. $3^8 \cdot 3 \cdot 3^3$ **6.** $3^4 \cdot 3^6 \cdot 3$

7. $(a^3b)(a^5b^2)$ **8.** $(a^6b^2)(a^4b)$

[Example 2]

9. $(x^2)^3$ **10.** $(x^5)^2$

11. $(5^6)^3$ **12.** $(5^4)^7$

13. $(y^3)^3(z^2)^4$ **14.** $(y^6)^6(z^3)^2$

15. $(a^3)^2(a^2)^4a$ **16.** $(a^4)^2(a^2)^3a$

[Example 3]

17. $\dfrac{x^{12}}{x^4}$

18. $\dfrac{x^9}{x^3}$

19. $\dfrac{y^5}{y^{10}}$

20. $\dfrac{y^4}{y^{16}}$

21. $\dfrac{a^4b^3}{ab^3}$

22. $\dfrac{a^7b^5}{a^7b}$

23. $\dfrac{(z^6)^2}{z^3}$

24. $\dfrac{(z^4)^3}{z^2}$

25. $\dfrac{10^8(x^2)^3}{10^2(x^4)^2}$

26. $\dfrac{10^8(x^3)^3}{10^4(x^2)^5}$

[Example 4]

27. $(2t)^5$

28. $(5t)^3$

29. $(-3u)^7$

30. $(-10u)^5$

31. $(2v^2w^3)^2$

32. $(5v^4w^2)^2$

33. $(-4ab^2)^4$

34. $(-3a^2b)^2$

[Example 5]

35. $\left(\dfrac{x}{10}\right)^6$

36. $\left(\dfrac{x}{2}\right)^8$

37. $\left(\dfrac{y^2}{3}\right)^3$

38. $\left(\dfrac{y^3}{5}\right)^4$

39. $\left(\dfrac{a^2b}{4}\right)^2$

40. $\left(\dfrac{ab^3}{6}\right)^3$

41. $\left(\dfrac{z}{2}\right)^3\left(\dfrac{z^2}{2}\right)^2$

42. $\left(\dfrac{z^2}{3}\right)^2\left(\dfrac{z}{3}\right)$

[Example 6]

43. 18^0

44. 15^0

45. $(-31)^0$

46. $(-25)^0$

47. $-\left(\dfrac{2x}{3}\right)^0$

48. $-\left(\dfrac{5x}{4}\right)^0$

49. -11^0

50. -19^0

[Example 7]

51. $(-3xy)^3(xy)^2$

52. $(-2xy)^4(-xy)^5$

53. $(-5a^2)(2a)^4(-a)^3$

54. $(8a)^2(a^2)^3(-a)$

55. $(2pq^2)^2(-3p^3q)^3$

56. $(-p^2q)^3(2pq)^5$

[Example 8]

57. $\left(\dfrac{2a}{b^3}\right)^3\left(\dfrac{b}{a^2}\right)\left(\dfrac{b^2}{4}\right)^3$

58. $\left(\dfrac{9b}{a^2}\right)\left(\dfrac{2a}{b^2}\right)\left(\dfrac{a^3b}{3}\right)^3$

59. $\left(\dfrac{t^2}{2}\right)^3\left(\dfrac{t}{3}\right)^2\left(\dfrac{6}{t^3}\right)$

60. $\left(\dfrac{t}{5}\right)^2\left(\dfrac{-t^3}{2}\right)\left(\dfrac{10}{t^4}\right)$

61. $(-4u^2)^4\left(\dfrac{u}{2}\right)^2$

62. $(-6u^3)^2\left(\dfrac{u}{2}\right)^3$

63. $(a^2b)^2(-ab)^3\left(\dfrac{-1}{a^2b^2}\right)$

64. $(-a^3b^2)^3(a^2b)^2\left(\dfrac{-1}{a^4b^3}\right)$

SECTION 3-1. Ten Review Exercises

1. Evaluate $3t - 2 - 4t + 9$ for $t = 7$. **2.** Solve $3t - 2 = 4t - 9$.

3. Solve and graph $3t - 2 < 4t - 9$.

4. Solve and graph: $2t - 5 < 3$ or $3t - 2 < 4t - 9$

5. Solve $|5u + 1| = 26$. **6.** Solve and graph $|5u + 1| < 26$.

7. Evaluate $|2p + 7| - |4 - p|$ for $p = -3$.

8. Solve $|2p + 7| = |4 - p|$.

In exercises **9** and **10**, use the distributive property to simplify each expression.

9. $\frac{1}{3}(15x^2 - 9x + 3)$ **10.** $\frac{3}{7} \cdot 11 + \frac{3}{7} \cdot 18 - \frac{3}{7}$

SECTION 3-1. Supplementary Exercises

In exercises **1–10**, simplify each expression. Assume a and b are restricted so that each exponent is a positive integer.

1. $(x^{a+b})(x^{a-b})$ **2.** $(x^{2a-b})(x^{3a+b})$

3. $(x^a)^2(x^{2a})^3$ **4.** $(x^{a+b})^2(x^{a-2b})$

5. $\frac{x^{a+b}}{x^{a-b}}$ **6.** $\frac{x^{3a+2b}}{x^{3a+b}}$

7. $\frac{(x^a)^3}{(x^2)^a}$ **8.** $\frac{(x^{a+b})^3}{(x^{a-b})^2}$

9. $\frac{(x^{2a})^b}{(x^b)^a}$ **10.** $\frac{(x^{a+b})^a}{(x^{a+b})^b}$

In exercises **11–14**, simplify each expression.

11. **a.** $5^2 \cdot 5^3$
b. $5^2 + 5^3$
c. $2^5 \cdot 5^3$

12. **a.** $4^2 \cdot (-4)^2$
b. $4^2 + (-4)^2$
c. $4^2 - 4^2$

13. **a.** $4z^2 + 5z^2$
b. $(4z)^2 + (5z)^2$
c. $(4z)^2 \cdot (5z)^2$

14. Assume $p \neq 0$.
a. $6p^2 \div 3p^2$
b. $(6p^2 \div 3p)^2$
c. $(6p^2) \div (3p)^2$

In exercises **15** and **16**, answer parts **a–f**.

15. **a.** Evaluate $12x^3$ for $x = 2$.
b. Evaluate $3x$ for $x = 2$.
c. Find the quotient of the values found in parts **a** and **b**.

d. Simplify $\dfrac{12x^3}{3x}$

e. Evaluate the result of part **d** for $x = 2$.
f. Compare the results of parts **c** and **e**.

16. **a.** Evaluate $(3x^2)^2$ for $x = 2$.
b. Evaluate $(5x^3)^2$ for $x = 2$.
c. Find the product of parts **a** and **b**.
d. Simplify $(3x^2)^2(5x^3)^2$.
e. Evaluate the result of part **d** for $x = 2$.
f. Compare the results of parts **c** and **e**.

17. Verify Property 2 of exponents.
a. Write $(a^m)^n$ as n factors of a^m.
b. Use Property 1 to write the n factors of a^m using one base.
c. Use the definition of exponents to simplify.

18. Verify Property 3 of exponents.
a. Write a^m as m factors of a above the fraction bar, and write a^n as n factors of a below the fraction bar.
b. Supposing $m > n$, change n factors of a to 1 and simplify.
c. Supposing $m = n$, change all the factors of a to 1 and simplify.
d. Supposing $m < n$, change m factors of a to 1 and simplify.

19. Verify Property 4 of exponents.
a. Write $(ab)^m$ as m factors of ab.
b. Use the associative property of multiplication to group the a factors on the left and the b factors on the right.
c. Use the definition of an exponent to simplify.

20. Verify Property 5 of exponents.
a. Write $\left(\dfrac{a}{b}\right)^m$ as m factors of $\dfrac{a}{b}$.
b. Write the indicated product as m factors of a over m factors of b.
c. Use the definition of an exponent to simplify.

The exponent can yield deceptively large numbers. Exercises **21** and **22** are two such examples.

21. A piece of ordinary paper is about 0.003 inches thick. Suppose the paper is torn in two pieces and stacked one on top of the other. The two pieces together measure 0.006 inches thick. Now suppose *these* pieces are torn into two pieces and stacked together. The stack now measures 0.012 inches thick. Suppose the process continues for 50 trials. The stack of paper now measures $0.003 \cdot 2^{50}$ inches.
a. Do not calculate, but guess the height of the paper. Express your answer in inches.
b. Use a calculator with a $\boxed{y^x}$ key to compute the height in inches.
c. Divide the number in b by 63,360 inches per mile to convert the height to miles.

22. Suppose an individual contracts to do a particular job that will last exactly 30 days. There are two possible payment plans for the job:

Plan A. A fixed fee of $10,000

Plan B. 1¢ on day one, double (or 2¢) on day two, double again (or 4¢) on day three, and so on:

Work day	Amount of wages on that day
1	1¢
2	$2 \cdot 1¢ = 2¢$
3	$4 \cdot 1¢ = 4¢$
4	$8 \cdot 1¢ = 8¢$
5	$16 \cdot 1¢ = 16¢$
⋮	⋮
30	$2^{29} \cdot 1¢ = ?$

a. Do not calculate, but guess which plan yields the higher pay.
b. Use a calculator to determine the wages on day 30 using Plan B.

SECTION 3-2. Polynomials Defined, Added, and Subtracted

1. The definition of a polynomial in x
2. Terminology related to a polynomial
3. Polynomials in more than one variable
4. Adding polynomials
5. Subtracting polynomials
6. Combined operations

In the introduction to this chapter, Ms. Glaston separated mathematical expressions into several categories. One of the categories she identified is studied extensively in Beginning and Intermediate Algebra courses. Specifically, the expressions are called **polynomial,** and they are identified in this section.

The Definition of a Polynomial in x

A polynomial is frequently identified by the variable, or variables, in the polynomial. Definition 3.3 identifies the form of each term of a polynomial in x.

Definition 3.3. A polynomial in x
If $a_n, a_{n-1}, \ldots, a_0$ are real numbers ($a_n \neq 0$) and n is a positive integer, then a **polynomial** in x is an expression that can be written in the form:

$$a_nx^n + a_{n-1}x^{n-1} + \ldots + a_1x + a_0$$

The common characteristics of every term of a polynomial are the numbers represented by the a's and n's. Any factor a must be a real number. (If $a = 1$, then it is usually not written.) The exponent n in any term must be a **positive integer.** (If $n = 1$, then it is usually not written.)

Terminology Related to a Polynomial

Items **a–f** identify some of the terminology associated with a polynomial written in the form:

$$a_n x^n + a_{n-1} x^{n-1} + \ldots + a_1 x + a_0$$

a. The number a_n is called the **leading coefficient.**

b. The exponents $n, n - 1, \ldots$ identify the **degree** of each term.

c. The **degree of the polynomial** is n, the largest exponent on x in the expression.

d. The number a_0 is called the **constant term.** If $a_0 \neq 0$, then the degree is 0. If $a_0 = 0$, then it has no degree.

e. With the highest-degree term written on the left and each successive term decreasing in power, the polynomial is written in **descending powers.**

f. If every term from $n, n - 1, \ldots, 1$ is present together with a_0, then the polynomial is called **complete.**

Example 1. Given: $7x - x^3 + 10x^5 - 2$

a. Write the polynomial in descending powers.

b. Identify the degree.

c. Identify the leading coefficient.

d. Identify the constant term.

e. State whether it is complete.

Solution. **Discussion.** For the given form of the polynomial, the degree of each term from left to right is 1, 3, 5, and 0. To write it in descending powers, the terms are rearranged so that the degree of each term from left to right is 5, 3, 1, and 0.

a. $10x^5 - x^3 + 7x - 2$ — In descending powers

b. The degree is 5. — The largest exponent on x

c. The leading coefficient is 10. — The coefficient of x^5

d. The constant term is -2. — Write $7x - 2$ as $7x + (-2)$.

e. The polynomial is not complete. — There is no x^4- or x^2-term.

A polynomial is frequently referred to by the number of terms in it.

1. A **monomial** is a polynomial of one term.
2. A **binomial** is a polynomial of two terms.
3. A **trinomial** is a polynomial of three terms.
4. A **polynomial of n terms** is a polynomial with n terms, where n is a positive integer greater than 3.

Polynomials in More Than One Variable

Examples **g** and **h** are expressions that qualify as polynomials in that the numerical coefficients are real numbers, and exponents on all variables are positive integers. Polynomials such as these are identified by the variables in the expression.

g. $2x^3 - x^2y + 5xy^2 - 10y^3$ — A polynomial in x and y

h. $\frac{1}{2}t^5 + \frac{4}{3}t^3u - \frac{9}{10}tu^2v + u^3v^3$ — A polynomial in t, u, and v

If a polynomial has more than one variable, then the following adjustments are made to the terminology associated with it.

b*. The **degree of each term** is the sum of the exponents on the variables in that term.

c*. The **degree of the polynomial** is the largest sum obtained in **b***.

e*. The polynomial can be written in descending powers of any of the variables in the expression by treating the other variables as constants.

Example 2. Given: $5yz^2 + 2x^2z^3 + x^4y^3z - 8 - 3xy^5$

a. State the degree of the polynomial.

b. Write the polynomial in descending powers of x.

c. Write the polynomial in descending powers of y.

d. Write the polynomial in descending powers of z.

Solution. **a.** The degree of $5yz^2$ is $1 + 2 = 3$.
The degree of $2x^2z^3$ is $2 + 3 = 5$.
The degree of x^4y^3z is $4 + 3 + 1 = 8$.
The degree of -8 is 0.
The degree of $-3xy^5$ is $1 + 5 = 6$.

The largest degree term is 8; therefore, the degree of the polynomial is 8.

b. Discussion. To write the expression in descending powers of x, we treat y and z as constants. Any term not containing an x factor can be written last. Thus, the terms $5yz^2$ and -8 are written on the right. These two terms can be listed as $5yz^2 - 8$, or $-8 + 5yz^2$.
In descending powers of x: $x^4y^3z + 2x^2z^3 - 3xy^5 + 5yz^2 - 8$

c. In descending powers of y: $-3xy^5 + x^4y^3z + 5yz^2 + 2x^2z^3 - 8$

d. In descending powers of z: $2x^2z^3 + 5yz^2 + x^4y^3z - 3xy^5 - 8$

Adding Polynomials

Consider the polynomial expressions in examples **i** and **j**.

i. $(2t^2 + 5t - 1) + (3 - 4t + t^2) + (3t^2 + 5 - t)$

j. $6t^2 + 7$

Example **i** is an indicated sum of three trinomials in t. Example **j** is a binomial in t. Notice the values obtained when these expressions are evaluated for $t = 5$:

i. $(2(5)^2 + 5(5) - 1) + (3 - 4(5) + 5^2) + (3(5)^2 + 5 - 5)$

$= (50 + 25 - 1) + (3 - 20 + 25) + (75 + 5 - 5)$

$= 74 + 8 + 75$

$= 157$

j. $6(5)^2 + 7 = 150 + 7 = 157$

The evaluations of both expressions yielded 157. Furthermore, for every real number replacement of t, the expressions would similarly yield the same number. The binomial in **j** is called the **sum** of the three trinomials in **i**. The sum is obtained by combining (adding or subtracting as indicated) the numerical coefficients of the t-terms with the same degree, as well as the constant terms. Arcs can be drawn to link the terms that can be combined:

$$(2t^2 + 5t - 1) + (3 - 4t + t^2) + (3t^2 + 5 - t)$$

$$= 6t^2 + 7 \quad \begin{cases} 2t^2 + t^2 + 3t^2 = 6t^2 \\ 5t - 4t - t = 0t = 0 \\ -1 + 3 + 5 = 7 \end{cases}$$

Evaluating $6t^2 + 7$ was much easier than evaluating the three trinomials in example **i**. As a consequence, the instruction for adding or subtracting polynomials is frequently "simplify". These instructions require writing the given expression as an equivalent expression with the minimum number of symbols.

Example 3. Simplify each expression.

a. $(8u^3 - 1) + (9 - 4u^2) + (4u^2 - 12u + 9) + (10u - 20)$

b. $(9x^2y^2 + 1 - xy) + (3 - 5x^2y^2 + 2xy) + (3xy - x^2y^2 - 4)$

Solution. **Discussion.** First write the expressions without grouping symbols. Then arcs may be drawn to indicate the terms that may be combined.

a. $8u^3 - 1 + 9 - 4u^2 + 4u^2 - 12u + 9 + 10u - 20$

$= 8u^3 + 0u^2 - 2u - 3$

$= 8u^3 - 2u - 3$

b. $9x^2y^2 + 1 - xy + 3 - 5x^2y^2 + 2xy + 3xy - x^2y^2 - 4$

$= 3x^2y^2 + 4xy + 0$

$= 3x^2y^2 + 4xy$

Subtracting Polynomials

In example **k**, the right polynomial is to be subtracted from the left one.

k. $(2t^3 + 3t - 5 - t^2) - (5t + 2t^2 - 3 + t^3)$

As in addition, subtraction of polynomials is performed by operating on the coefficients of the terms with the same degree. To remove the parentheses on the right polynomial, we can multiply each term of the polynomial by -1:

$(2t^3 + 3t - 5 - t^2) - 1 \cdot (5t + 2t^2 - 3 + t^3)$	The given expression
$= 2t^3 + 3t - 5 - t^2 - 5t - 2t^2 + 3 - t^3$	Distribute the -1.
$= t^3 - 3t^2 - 2t - 2$	Combine like terms.

Keep in mind that the given expression and the simplified expression have the same value for any real number replacement for t.

Example 4. Simplify each expression.

a. $(4p^2 - q^2 + 3pq) - (q^2 - pq + 2p^2) - (2pq - p^2 - 2q^2)$

b. $(5z^3 - z + 7) - (4 - z - 3z^2) + (2z^2 - 3z^3 - 3)$

Solution. **a.**

$$(4p^2 - q^2 + 3pq) - 1 \cdot (q^2 - pq + 2p^2) - 1 \cdot (2pq - p^2 - 2q^2)$$

$$= 4p^2 - q^2 + 3pq - q^2 + pq - 2p^2 - 2pq + p^2 + 2q^2$$

$$= 3p^2 + 2pq$$

$4p^2 - 2p^2 + p^2 = 3p^2$

$3pq + pq - 2pq = 2pq$

$-q^2 - q^2 + 2q^2 = 0q^2$

b.

$$(5z^3 - z + 7) - 1 \cdot (4 - z - 3z^2) + (2z^2 - 3z^3 - 3)$$

$$= 5z^3 - z + 7 - 4 + z + 3z^2 + 2z^2 - 3z^3 - 3$$

$$= 2z^3 + 5z^2$$

$5z^3 - 3z^3 = 2z^3$

$3z^2 + 2z^2 = 5z^2$

$-z + z = 0z$

$7 - 4 - 3 = 0$

Combined Operations

In Example 5, the distributive property is used to remove parentheses. The expression is then simplified by combining like terms.

Example 5. Simplify $2(4m^2 - 5m + 2) - 3(3m^2 - m - 2) + (m^2 + 6m - 3)$.

Solution.

$$2(4m^2 - 5m + 2) - 3(3m^2 - m - 2) + (m^2 + 6m - 3)$$

$$= 8m^2 - 10m + 4 - 9m^2 + 3m + 6 + m^2 + 6m - 3$$

$$= -m + 7$$

$8m^2 - 9m^2 + m^2 = 0m^2$

$-10m + 3m + 6m = -1 \cdot m = -m$

$4 + 6 - 3 = 10 - 3 = 7$

SECTION 3-2. Practice Exercises

In exercises **1–10**, for each polynomial in x:

a. Write it in descending powers.

b. Identify the degree.

c. Identify the leading coefficient.

d. Identify the constant term.

e. Identify the polynomial by the number of terms in it.

[Example 1] **1.** $5x - x^2 + 10$ **2.** $1 + 6x^4 - 2x^2$

3. $100x^8$ **4.** $-x^5$

5. $12x^3 - 3 + 7x - x^2$ **6.** $15x - 4x^3 - 5x^5 + 1000$

7. $36 - 25x^8$ **8.** $16x^{10} - x^8$

9. $ax^3 + 5 + bx^2 + abx$ **10.** $kx^3 + 7 + kx + x^2$

In exercises **11–18**, for each polynomial:

a. State the degree of each term.

b. Write the polynomial in descending powers of x.

c. Write the polynomial in descending powers of y.

d. Write the polynomial in descending powers of z.

[Example 2] **11.** $3xz^2 + x^2y - 2y^2z$ **12.** $y^2z - 10x^2y - xz^2$

13. $xyz + 2x^3y^2 - x^4z^3 + 5y^3z^2$ **14.** $xy^2z - 7x^3y^3 + 9x^4z^4 - yz^2$

15. $4x^3z - y^2 + x^5y - z^3$ **16.** $2xy^2z - 7x^3y^3 + 9x^4z^4 - yz^2$

17. $xyz - y^2z^2 + x^2y^2 + 2 + x^2z^2$ **18.** $xy - 5 + x^2y^2z^2 - yz + xz$

In exercises **19–60**, simplify each expression. Write the answers in descending powers of one of the variables.

[Example 3] **19.** $(x^2 - 5x - 6) + (2x^2 - x - 3)$ **20.** $(4x^2 + 4x + 1) + (9x^2 - 6x + 1)$

21. $(x - 3x^2 + 2) + (x^2 - 10x + 25)$ **22.** $(36x^2 - 12x + 1) + (5x - 6x^2 + 4)$

23. $(c^3 - 5c) + (4c^2 - 3) + (c^2 + 3c) + (2c - 1)$

24. $(4 - 9c^2) + (2c - 1) + (2c^3 - 3) + (4c^2 - 3c)$

25. $(16x^2 - 1) + (x^3 + 8) + (x^2 - 4x + 4) + (4x - 12)$

26. $(8x^3 - 1) + (9 - 4x^2) + (4x^2 - 12x + 9) + (10x - 20)$

[Example 4] **27.** $(x^2 - 11xy - 12y^2) + (y^2 - 5xy + 6x^2)$

28. $(3x^2 + 7xy + 2y^2) + (4y^2 - 4xy + x^2)$

29. $(8x^2 - 3y^2 - 5xy) + (5xy + 3x^2 + 2y^2)$

30. $(2y^2 - 7xy + 5x^2) + (2x^2 - 2y^2 - 3xy)$

31. $(a^3 - a^2b + 2b^3 + 3ab^2) + (b^3 - ab^2 - 2a^3 + 3a^2b)$

32. $(2a^3 + 3a^2b - ab^2 + b^3) + (a^3 + ab^2 - 3a^2b + 2b^3)$

33. $(x^3y^3 - 5xy + 7) + (2x^2y^2 + xy - 5) + (3x^3y^3 - x^2y^2 - 2)$

34. $(9x^2y^2 - xy + 1) + (5x^3y^3 - 6x^2y^2 - 4) + (3xy - 4x^3y^3 + 3)$

[Example 5] **35.** $(3x^2 + x - 1) - (x^2 - x + 1)$ **36.** $(4x^2 - 4x + 1) - (9x^2 - 6x + 1)$

37. $(2y^2 - 3y - 5) - (3y^2 + y - 2)$ **38.** $(6y^2 + 5y + 1) - (4y^2 + 12y + 9)$

39. $(3x^3 - 3x + 2) - (2x^2 + x + 1) - (x^3 + 1 - x^2 - 4x)$

40. $(2x^2 + x - 3) - (3x^3 - x^2 + 1) - (2x - 5x^3 - 3x^2 - 7)$

41. $(2z^4 + z^2 - 7) - (z^3 - z - 5) - (2z + z^4 - z^3 + 2)$

42. $(4z^3 - z^2 + 10) - (z^4 + z + 12) - (3 - z^4 + 3z^3 - z)$

43. $(2ab - 3b^2 + a^2) - (b^2 - ab - a^2) + (3a^2 + 4b^2 - ab)$

44. $(5b^2 - a^2 - 3ab) - (2ab - 4a^2 + b^2) + (a^2 + 5ab - 3b^2)$

45. $5(x^2 + 2x + 5) + (3x^2 - x - 2) - 4(2x^2 - x - 1)$

46. $(2x^2 - x - 3) + 4(x^2 - 6x + 9) - 3(4x^2 - 5x + 1)$

47. $2(4y^2 - 5y + 2) + (y^2 + 6y - 3) - 5(3y^2 - y - 2)$

48. $6(10y^2 + 6y + 7) + (3y^2 - 7y - 4) - 3(12y^2 + y - 3)$

49. $(5a - 1) - 4(3a^2 + 2) - (-4a^2 - 3) + 3(a^2 - 3a)$

50. $(2b - 9) - 3(b^2 - 5) + 2(b^2 - 1) - (4b - 7)$

51. $(c^3 - 2c - 3) - 3(2c^2 - c + 5) + 5(2c^3 + 5c + 10)$

52. $4(c^2 - 10c + 25) - 2(4c^3 - 6c + 13) + (6c^3 - c^2 + 4c - 12)$

53. $5(t^2 - 4u^2 + 2tu) - 3(3tu - 5u^2 + 2t^2) + 2(2u^2 - tu + t^2)$

54. $-3(tu - 2t^2 + 3u^2) + 2(4u^2 + 2tu - t^2) - (4t^2 - tu - u^2)$

55. $\frac{1}{2}(8z^3 + 4z - 10) - \frac{2}{3}(9 - 6z^2 + 12z^3) + \frac{3}{5}(10z^2 - 5z^3 + 30 - 5z)$

56. $\frac{-5}{12}(96z^2 + 108 - 36z) + \frac{1}{6}(18 + 108z^2 - 90z) - \frac{5}{8}(-144 + 8z - 32z^2)$

57. $12\left(\frac{5}{6}a^2 - \frac{3}{4}b^2 + \frac{5}{3}ab - \frac{7}{2}\right) - 15\left(\frac{4}{3}ab - \frac{9}{5} + \frac{8}{15}a^2 - b^2\right)$

58. $8\left(\frac{3}{2}ab - \frac{3}{4}b^2 + \frac{9}{8} + \frac{1}{4}a^2\right) - 24\left(\frac{1}{12}a^2 + \frac{3}{8} - \frac{1}{4}b^2 + \frac{3}{2}ab\right)$

59. $0.2(1.5k^2 + 4.1 - 0.6k) + 1.3(0.8k - 0.2k^2 - 0.6)$

60. $1.5(3k - 0.6 - 2k^2) - 0.9(5k - 3.3k^2 - 1)$

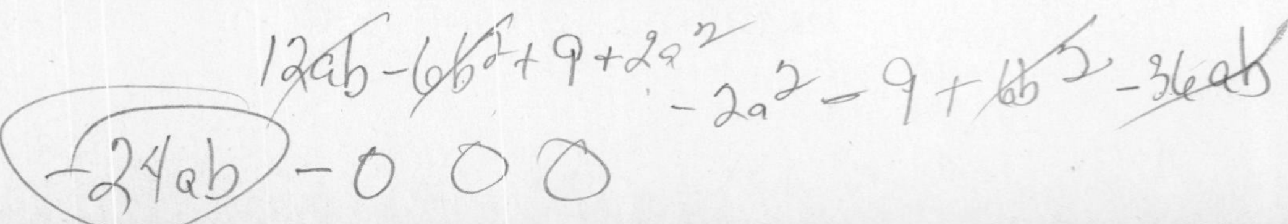

SECTION 3-2. Ten Review Exercises

1. Use the distributive property to simplify $30\left(\frac{1}{2}+\frac{7}{15}-\frac{3}{5}+\frac{1}{10}\right)$.

In exercises **2** and **3**, simplify each expression.

2. $\left(-3\frac{3}{5}\right)\left(\frac{7}{9}\right)\left(\frac{-15}{28}\right)$

3. $-5(3-7)+10(4-8)-6(-3)$

In exercises **4** and **5**, solve each equation.

4. $10k-2(5k+2)=3(15k-4)-7(5k-2)$

5. $|3x+2|=7$

6. Solve $3y-8=5x+ky$ for y.

In exercises **7** and **8**, solve and graph each inequality.

7. $5(3-a)<a+45$

8. $|4-x|\geq 5$

In exercises **9** and **10**, simplify each expression. Assume all variables are not 0.

9. $\left(\frac{-3x^2y}{5}\right)^2\left(\frac{-10xy^2}{3}\right)$

10. $\frac{(2x^3y)^2}{6x^5y^2}$

SECTION 3-2. Supplementary Exercises

A polynomial may use letters such as $a, b, c, \ldots$ to represent some, or all, of the numerical coefficients.

Example. Simplify $(ax+5)+(3x+b)$.

Solution. **Discussion.** The distributive property of multiplication over addition can be used to write the two x-terms as one x-term:

$$ax+5+3x+b$$

$$=ax+3x+5+b \quad \text{Group the like terms.}$$

$$=(a+3)x+5+b \quad \text{Use the distributive property.}$$

In exercises **1–6**, simplify each expression in x.

1. $(ax^2+5x+6)+(9x^2+bx+4)$

2. $(ax^3+bx^2)+(cx^3+2x^2+x)$

3. $(9x^2+4x+a)-(2x^2-bx-c)$

4. $(ax+11)+(4x+b)-(cx+d)$

5. $(ax^2+bx+c)+(dx^2-ex+f)+(gx^2+hx-i)$

6. $(ax^3-1)+(bx^2-4)+(cx^2+dx+9)$

A polynomial is frequently referred to by a form of the function notation. For example, the trinomial $2x^2-x+7$ can be referred to by $P(x)$, which is read

"P of x", where:

$$P(x) = 2x^2 - x + 7$$

Then, for example:

$$P(3) = 2(3)^2 - 3 + 7 = 18 - 3 + 7 = 22$$

In exercises **7–14**, use the polynomials $P(x)$ and $Q(x)$, where:

$$P(x) = 2x^2 + 5x - 1 \quad \text{and} \quad Q(x) = 3x^2 - 4x + 2$$

7. **a.** Find $P(x) + Q(x)$.
b. Evaluate the polynomial of part **a** for $x = 5$.

8. **a.** Compute $P(5)$.
b. Compute $Q(5)$.
c. Compute $P(5) + Q(5)$.
d. Compare the answers obtained for **7b** and **8c**.

9. **a.** Find $P(x) - Q(x)$.
b. Evaluate the polynomial of **a** for $x = -3$.

10. **a.** Compute $P(-3)$.
b. Compute $Q(-3)$.
c. Compute $P(-3) - Q(-3)$.
d. Compare the answers obtained for **9b** and **10c**.

11. Compare $2P(x) + 3Q(x)$.

12. Compute $4P(x) + 5Q(x)$.

13. Compute $3P(x) - 2Q(x)$.

14. Compute $5P(x) - 6Q(x)$.

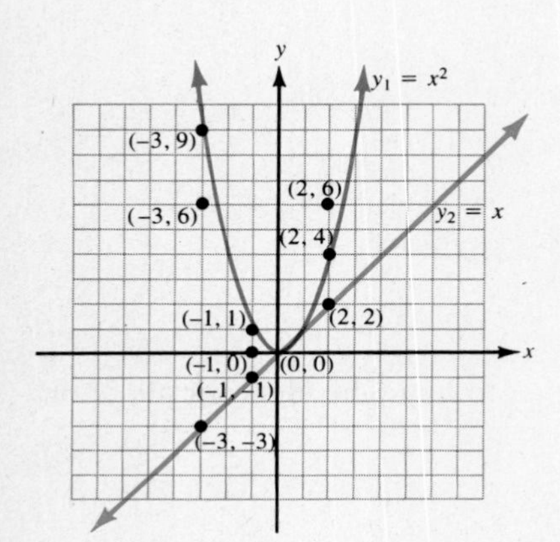

Figure 3-2. Graphs of (1) $y_1 = x^2$ and (2) $y_2 = x$.

Example. Figure 3-2 contains graphs of (1) $y_1 = x^2$ and (2) $y_2 = x$. A graph of

$$(3) \quad y_1 + y_2 = x^2 + x$$

is a curve representing the "sum" of the graphs of (1) and (2). Use the graphs in the figure to find the coordinates of (3) for each value of x.

a. $x = -3$ **b.** $x = -1$ **c.** $x = 2$

Solution. **a.** For $x = -3$:

(1) $(-3, 9)$ is on the graph
(2) $(-3, -3)$ is on the graph $\Big\}$ $(-3, 9 + (-3)) = (-3, 6)$ is on (3)

b. For $x = -1$:

(1) $(-1, 1)$ is on the graph
(2) $(-1, -1)$ is on the graph $\Big\}$ $(-1, 1 + (-1)) = (-1, 0)$ is on (3)

c. For $x = 2$:

(1) $(2, 4)$ is on the graph
(2) $(2, 2)$ is on the graph $\Big\}$ $(2, 4 + 2) = (2, 6)$ is on (3)

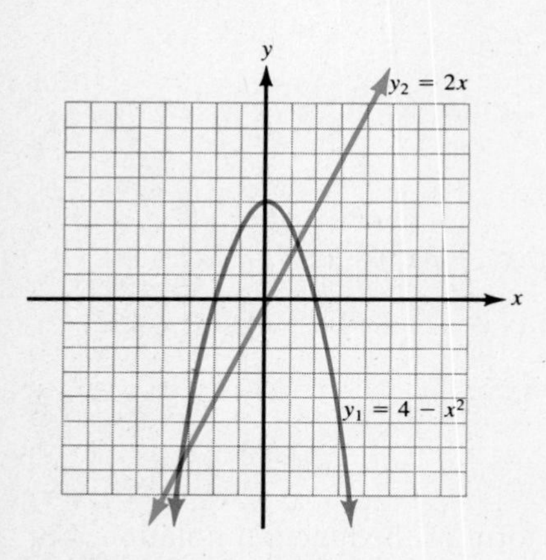

Figure 3-3. Graphs of (1) $y_1 = 4 - x^2$ and (2) $y_2 = 2x$.

Figure 3-3 contains graphs of (1) $y_1 = 4 - x^2$ and (2) $y_2 = 2x$. In exercises **15–18**, use the graphs in the figure to find the coordinates of a graph of

$$(3) \quad y_1 + y_2 = 4 + 2x - x^2$$

for each value of x.

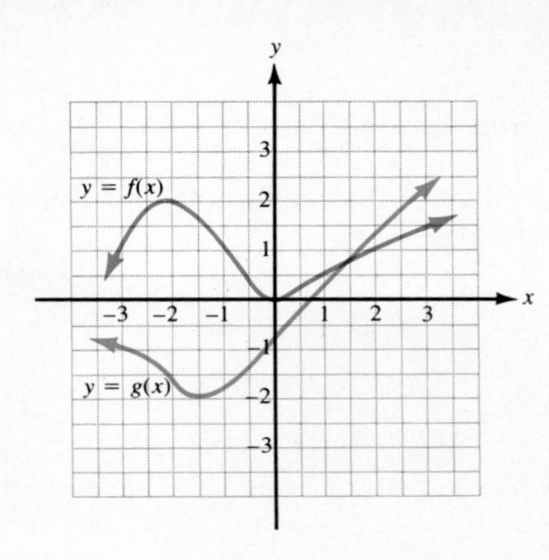

Figure 3-4. Graphs of (1) $y_1 = f(x)$ and (2) $y_2 = g(x)$.

15. $x = -3$ **16.** $x = -1$

17. $x = 0$ **18.** $x = 2$

Figure 3-4 contains graphs of (1) $y_1 = f(x)$ and (2) $y_2 = g(x)$. In exercises **19–22**, use the graphs in the figure to find the coordinates of a graph of

$$(3) \quad y_1 + y_2 = f(x) + g(x)$$

for each value of x.

19. $x = 2$ **20.** $x = -3$

21. $x = -1$ **22.** $x = 0$

SECTION 3-3. Multiplication of Polynomials

1. Multiplying monomials
2. Multiplying polynomials with two or more terms
3. Multiplying binomials using the FOIL method
4. Special products
5. Combined operations

In this section, we will review the techniques for multiplying polynomials. Because multiplication and addition are commutative and associative, we can rearrange factors and terms to simplify the products.

Multiplying Monomials

Consider the indicated products of monomials in examples **a** and **b**.

a. $(7t^3)(5t^4)$ **b.** $(2u^2v)(-5uv)(10v^3)$

The commutative and associative properties of multiplication can be used to rewrite these expressions as follows:

a. $(7 \cdot 5)(t^3 \cdot t^4)$ **b.** $(2(-5)10)(u^2 \cdot u)(v \cdot v \cdot v^3)$

In these forms, the products can be simplified.

a. $35t^7$ **b.** $-100u^3v^5$

These examples suggest a two-step procedure for multiplying monomials.

> **To multiply monomials:**
>
> **Step 1.** Compute the product of the numerical coefficients.
>
> **Step 2.** Add the exponents on the like bases.

Example 1. Find the indicated products.

a. $(-4xy^2)(-x^2y^3)$ **b.** $(-5tu)(-t^2v^3)(3u^4v)$

Solution. **a. Discussion.** To compute the product of the coefficients, think of $-x^2y^3$ as having the form $-1 \cdot x^2y^3$.

$(-4xy^2)(-1 \cdot x^2y^3)$ The given expression

$= (-4)(-1)(x \cdot x^2)(y^2 \cdot y^3)$ Group like factors.

$= 4x^3y^5$ Add exponents on like bases.

b. $(-5tu)(-1 \cdot t^2v^3)(3u^4v)$

$= (-5)(-1)(3)(t \cdot t^2)(u \cdot u^4)(v^3 \cdot v)$

$= 15t^3u^5v^4$

Multiplying Polynomials with Two or More Terms

In Examples 2 and 3, the distributive property is used to multiply polynomials.

Example 2. Multiply $2t(5t^3 + 3t - 9)$.

Solution.

$2t(5t^3 + 3t - 9)$ The given expression

$= (2t)(5t^3) + (2t)(3t) - (2t)(9)$ Distribute $2t$ to each term.

$= 10t^4 + 6t^2 - 18t$ Simplify the indicated products.

Example 3. Multiply $(3u + 5)(2u^2 + 7u - 4)$.

Solution. **Discussion.** The distributive property is first used to multiply each term of the trinomial by $(3u + 5)$. It is then used again to distribute each term of the trinomial to $(3u + 5)$.

$(3u + 5)(2u^2 + 7u - 4)$

$= (3u + 5)(2u^2) + (3u + 5)(7u) - (3u + 5)(4)$

$= (3u)(2u^2) + 5(2u^2) + (3u)(7u) + 5(7u) - (3u)(4) - (5)(4)$

$= 6u^3 + 10u^2 + 21u^2 + 35u - 12u - 20$

$= 6u^3 + 31u^2 + 23u - 20$ The simplified product

The procedure used to simplify the products in Example 3 can be extended to polynomials of several terms. However, using a **vertical format** is frequently preferred when multiplying polynomials in which the number of terms in both is at least two, and the number of terms in one is at least three.

Example 4. Multiply $(u^2 + 3 - 3u + 2u^3)(2u^2 - 1 - u)$ with a vertical format.

Solution. **Discussion.** To use a vertical format, first rewrite both polynomials in descending powers. Then write on one line the polynomial with the greatest number of terms. Write on a second line below it the other polynomial. Now multiply each term of the first polynomial by each term of the second one and place the products in columns of like terms. As a final step combine the like terms in each column.

$$\begin{array}{rrrrrrl}
 & 2u^3 & + u^2 & - 3u & + 3 & & \text{Write both polynomials in descending powers of } u. \\
 & & 2u^2 & - u & - 1 & & \\
\hline
4u^5 & + 2u^4 & - 6u^3 & + 6u^2 & & & \longrightarrow 2u^2(2u^3 + u^2 - 3u + 3) \\
 & - 2u^4 & - u^3 & + 3u^2 & - 3u & & \longrightarrow -u(2u^3 + u^2 - 3u + 3) \\
 & & - 2u^3 & - u^2 & + 3u & - 3 & \longrightarrow -1(2u^3 + u^2 - 3u + 3) \\
\hline
4u^5 & & - 9u^3 & + 8u^2 & & - 3 & \text{The } u^4\text{- and } u\text{-terms combine to 0.}
\end{array}$$

Thus, $(u^2 + 3 - 3u + 2u^3)(2u^2 - 1 - u) = 4u^5 - 9u^3 + 8u^2 - 3$.

Multiplying Binomials Using the FOIL Method

A task frequently encountered in mathematics requires multiplying two binomials, such as $(3t^2 + 4)(2t - 5)$. The acronym FOIL can be used as a reminder that four products are needed to write the simplified product.

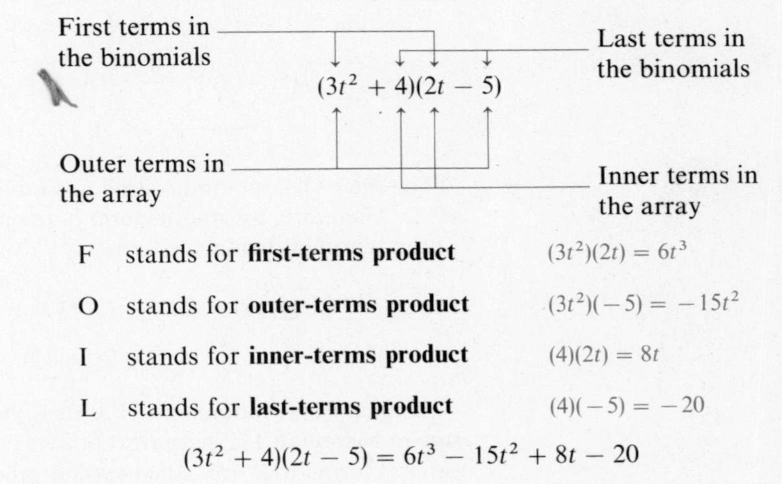

F stands for **first-terms product** $(3t^2)(2t) = 6t^3$

O stands for **outer-terms product** $(3t^2)(-5) = -15t^2$

I stands for **inner-terms product** $(4)(2t) = 8t$

L stands for **last-terms product** $(4)(-5) = -20$

$$(3t^2 + 4)(2t - 5) = 6t^3 - 15t^2 + 8t - 20$$

Example 5. Find the indicated products.

a. $(2p^2 - 3q)(q + 7p)$ **b.** $(5mn - 3)(7mn - 10)$

Solution. **a. Discussion.** It is recommended that both binomials be written in descending powers of the same variable.

$(2p^2 - 3q)(q + 7p)$ The given problem

$= (2p^2 - 3q)(7p + q)$ In descending powers of p

$= 14p^3 + 2p^2q - 21pq - 3q^2$ FOIL method

No terms are alike; therefore, the final expression cannot be simplified.

b. $(5mn - 3)(7mn - 10)$ The given problem

$= 35m^2n^2 - 50mn - 21mn + 30$ FOIL method

$= 35m^2n^2 - 71mn + 30$ Combine the mn − terms.

Special Products

Consider examples **c** and **d**.

c. $(5t + 12)(5t - 12)$ **d.** $(5t + 12)^2$

The binomials in example **c** have the form called "the sum and difference of terms".

$5t + 12 \longleftarrow$ *The sum of 5t and 12*

$5t - 12 \longleftarrow$ *The difference between 5t and 12*

When the FOIL method is used to simplify the product, the O- and I-terms combine to 0.

e. $(5t + 12)(5t - 12) = (5t)^2 - 60t + 60t - 12^2$

$= (5t)^2 - 12^2$ A difference of two squares

A product of a sum and difference of two terms will always yield a difference in the squares of the two terms.

The binomials in example **d** have the form called "the square of a binomial".

$(5t + 12)^2 \longleftarrow$ The exponent 2 can be read "square".

$5t + 12$ is a binomial.

When the FOIL method is used to simplify the product, the O- and I-terms are the same. Therefore, the middle term of the product is 2 times the product of the terms of the binomial being squared.

f. $(5t + 12)(5t + 12) = (5t)^2 + (5t)(12) + (5t)(12) + 12^2$

$= (5t)^2 + 2(5t)(12) + 12^2$ A perfect square trinomial

A square of a binomial will always yield an expression that is called a **perfect-square trinomial.** The patterns of the products illustrated in examples **c** and **d** have general forms that are called **special products.**

General forms of special products

Sum and difference of terms		Difference of two squares
$(x + y)(x - y)$	$=$	$x^2 - y^2$
Square of a sum or difference		**Perfect-square trinomial**
$(x + y)^2$	$=$	$x^2 + 2xy + y^2$
$(x - y)^2$	$=$	$x^2 - 2xy + y^2$

Example 6. Multiply $(7p + 9q)(7p - 9q)$.

Solution. $(7p + 9q)(7p - 9q)$ The indicated product

$= (7p)^2 - (9q)^2$ $(x + y)(x - y) = x^2 - y^2$

$= 49p^2 - 81q^2$ Simplify.

Example 7. Multiply $(7r - s)^2$

Solution.

$$(7r - s)^2 \qquad (x - y)^2$$
$$= (7r)^2 - 2(7r)(s) + s^2 \qquad = x^2 - 2xy + y^2$$
$$= 49r^2 - 14rs + s^2 \qquad \text{The simplified product}$$

Combined Operations

In the expression in Example 8, more than one operation is indicated. The order of operations is followed in simplifying this expression.

Example 8. Simplify $5 + (2z + 1)^2 - 3z(z + 3)$.

Solution.

$$5 + (2z + 1)^2 - 3z(z + 3) \qquad \text{The given expression}$$
$$= 5 + 4z^2 + 4z + 1 - 3z(z + 3) \qquad \text{Raise to powers first.}$$
$$= 5 + 4z^2 + 4z + 1 - 3z^2 - 9z \qquad \text{Now do the multiplication.}$$
$$= z^2 - 5z + 6 \qquad \text{Combine like terms.}$$

SECTION 3-3. Practice Exercises

In exercises **1–60**, find the indicated products.

[Example 1]

1. $(3x^2)(2x)$ **2.** $(-2x^2)(4x^2)$

3. $(-3m^3n^3)(7m^2n^4)$ **4.** $(6m^2)(4mn^5)$

5. $(-5xz^4)(2yz^2)(x^2y^3)$ **6.** $(2y^3z)(-10z^2)(4y^4z^3)$

7. $(-10p^2q)(-5p^3q^4)(2p^4q^5)$ **8.** $(9pq)(-3p^2q^2)(-4p^3q^3)$

9. $(3a^2b^2)(-bc^3)(-8a^4c^2)(-3a^2b^3c^4)$ **10.** $(7ab^2)(-5a^3b)(11b^4c^3)(-2b^5c^2)$

[Examples 2–4]

11. $(x + 2)(x^2 + 3)$ **12.** $(2x^2 - 3)(x + 1)$

13. $(3m^2 + 2m + 1)(3m - 2)$ **14.** $(m^2 - 5m + 7)(3m + 1)$

15. $(3r^2 - 5r + 7)(2r + 5)$ **16.** $(5r^2 + r - 3)(2r - 3)$

17. $(y^3 + 3y^2 - y - 2)(y^2 + 1)$ **18.** $(8y^3 - 2y^2 - 4y + 1)(2y^2 + 1)$

19. $(2z^2 - z + 1)(2z^2 + z - 3)$ **20.** $(z^2 - 5z + 3)(z^2 + 5z + 3)$

21. $(a^3 - a^2b + 2ab^2 + 3b^3)(a^2 + ab - b^2)$

22. $(2a^3 + a^2b - 3ab^2 + 3b^3)(2a^2 - ab - b^2)$

23. $(t^6 + 7t^3 + 49)(t^3 - 7)$ **24.** $(t^4 - t^2 + 1)(t^2 + 1)$

25. $(x^2 + x + 5)(x^3 - 2x^2 + 1)$ **26.** $(x^3 + 3x + 1)(x^3 - x^2 + 5)$

27. $(5y^3 + 2y - 6)(y^2 + 3y + 9)$ **28.** $(5y^4 + 7y^2 + 9)(5y^3 - 7y^2 + 8)$

29. $(u^2 - uv + v^2 + u + v)(u - v)$

30. $(u^2 + v^2 + w^2 + uv - uw - vw)(u - v)$

[Example 5] 31. $(x^2 + 2)(x - 5)$

32. $(2x - 3)(x + 2)$

33. $(x + 2y)(3x - y)$

34. $(x^2 - 2y)(x + y)$

35. $(2a - 3)(a^2 - 5)$

36. $(a - 2b)(3a + 7b)$

37. $(3m^2 + n^2)(m^2 - 2n^2)$

38. $(m^2n + 4)(mn^2 - 2)$

39. $(t^3 - 6)(t^2 + 10)$

40. $(3t^2 - 2)(7t^2 + 2)$

41. $(2s^2 + t)(s^2 + 3t)$

42. $(2s^2 - 3t)(3s + 7t)$

[Examples 6 and 7] 43. $(x - 3)(x + 3)$

44. $(x + 7)(x - 7)$

45. $(y - 2)^2$

46. $(y + 4)^2$

47. $(2r - s)(2r + s)$

48. $(3r + 5s)(3r - 5s)$

49. $(2a^2 + 3)^2$

50. $(5a^3 - 1)^2$

51. $(t^2 + 4)(t^2 - 4)$

52. $(9 - 2t^2)(9 + 2t^2)$

53. $(st - 9)^2$

54. $(s^2 + 4t)^2$

55. $(u^2v + 12)(u^2v - 12)$

56. $(u^3v - 20)(20 + u^3v)$

57. $(m^4n^2 + 3)^2$

58. $(2m^2n - 5)^2$

59. $(4w^5 + 9z^2)(4w^5 - 9z^2)$

60. $(10w^3 + 3z)(10w^3 - 3z)$

In exercises **61–70**, do the indicated operations and simplify.

[Example 8] 61. $5a(a^2 + 1) + (a + 1)^2$

62. $10(a + 3)(a - 3) - 2a(5a - 7)$

63. $(3x^2 - 1)^2 + (x + 1)(x - 1)$

64. $(x^2 - 1)(2x - 1) - (x^2 + 1)(2x + 1)$

65. $(5m + n)^2 + (5m - n)^2$

66. $(2m - n)(2m + n) + 5n^2$

67. $(8t - 3)(9t + 2) - (8t + 3)(9t - 2)$

68. $(t + 2)^2 - (t + 1)^2 + (2t + 1)(2t - 1)$

69. $(u^4 + u^2 + 1)(u^4 - u^2 + 2) - (u^4 - u^2 + 1)(u^4 + u^2 + 2)$

70. $\frac{1}{4}(10w + 8)(4w - 8) - \frac{1}{3}(5w + 7)(6w - 9)$

SECTION 3-3. Ten Review Exercises

1. Evaluate $(2k + 1)(2k - 1) - (4k + 1)(k - 3)$ for $k = 5$.
2. Simplify $(2k + 1)(2k - 1) - (4k + 1)(k - 3)$.
3. Evaluate the simplified expression of exercise **2** for $k = 5$.
4. Solve $(2k + 1)(2k - 1) = (4k + 1)(k - 3)$.
5. Identify the subsets of the real numbers to which $5\frac{1}{3}$ belongs.

In exercises **6–10**, given: $5x + x^4 - 2 + 3x^6$

6. Write the polynomial in descending powers.

7. Identify the degree of the polynomial.

8. Identify the leading coefficient.

9. Identify the constant term.

10. Evaluate the polynomial for $x = -2$.

SECTION 3-3. Supplementary Exercises

In exercises **1–10**, find the indicated products.

1. $\left(\frac{1}{2}m^2n\right)\left(\frac{2}{3}mn^3\right)\left(\frac{9}{10}m^2n^2\right)\left(\frac{-5}{3}m^4n^5\right)$

2. $(3a^2 + 2b)(a^2 - 2b)$

3. $(6m^6 + 2m^4 - 3m^2 - 1)(m^2 - 2)$

4. $(x^2y^2 + 10)(x^2y^2 - 10)$

5. $(9m^2 - n)^2$

6. $(11a^4 - 3b^2)(a^3 - b)$

7. $(3x^2 + 2)(9x^2 - 5)$

8. $(3x^3 - x^2 + x - 4)(3x^3 - x^2 - x + 4)$

9. $(3x^2 - y)(3x + y) + (2x - 3y)^2$

10. $(y^5 + y^4 + y^3 + y^2 + y + 1)(y - 1)$

Example. Compute (324)(21) as a product of two polynomials.

Solution. $324 = 3(100) + 2(10) + 4$

$21 = 2(10) + 1$

Let $x = 10$ and $x^2 = 100$.
(324)(21) takes on the form $(3x^2 + 2x + 4)(2x + 1)$

$$\begin{array}{r} 3x^2 + 2x + 4 \\ 2x + 1 \\ \hline 6x^3 + 4x^2 + 8x \quad\quad \\ 3x^2 + 2x + 4 \\ \hline 6x^3 + 7x^2 + 10x + 4 \end{array}$$

Replacing x by 10, x^2 by 100, and x^3 by 1,000:

$6{,}000 + 700 + 100 + 4 = 6{,}804$

In exercises **11–16**, compute each product using the method in the example.

11. (32)(21)

12. (30)(43)

13. (121)(23)

14. (231)(12)

15. (304)(21)

16. (200)(402)

In exercises **17–20**:

a. Evaluate each polynomial for the given value of x.

b. Compute the product based on the values obtained in **a**.

c. Multiply the polynomials.

d. Evaluate the product polynomial for the given value of x.

e. Compare the values obtained in **b** and **d**.

17. $(2x + 3)$ and $(x^2 - 1)$, for $x = 3$

18. $(x^2 + 3)$ and $(3x - 1)$, for $x = -2$

19. $(x + 2)$ and $(x^2 - 2x + 4)$, for $x = 5$

20. $(x - 3)$ and $(x^2 + 3x + 9)$, for $x = -4$

In exercises **21–24**, compute the squares using the square of a binomial multiplication form.

21. $(9 + 5)^2$

22. $(4 + 11)^2$

23. $(20 + 5)^2$

24. $(20 + 10)^2$

In exercises **25–28**, compute each product using the sum and difference of terms.

25. (18)(22) [*Hint:* write as $(20 - 2)(20 + 2)$.]

26. (24)(16) [*Hint:* write as $(20 + 4)(20 - 4)$.]

27. (39)(41) [*Hint:* write as $(40 - 1)(40 + 1)$.]

28. (45)(55) [*Hint:* write as $(50 - 5)(50 + 5)$.]

In exercises **29–36**, find the indicated products. Assume a and b are positive integers.

29. $(2x^a + 5)(x^a + 3)$

30. $(4x^a - 9)(2x^a - 3)$

31. $(x^a + 3y^b)(2x^a - 7y^b)$

32. $(5x^a - 4y^b)(x^a + 6y^b)$

33. $(3x^a - 2y^b)^2$

34. $(4x^{2a} + y^{2b})^2$

35. $(x^{2a} + 2x^a - 3)(2x^a + 1)$

36. $(x^{2a} - 3x^a + 2)(x^a - 3)$

SECTION 3-4. Division of Polynomials

1. Dividing monomials
2. Dividing a polynomial by a monomial
3. Dividing a polynomial by a polynomial

In Sections 3-2 and 3-3, we added, subtracted, and multiplied polynomials. In this section, we will review the procedure for dividing polynomials.

Dividing Monomials

Division by 0 is not defined. Therefore, *in this section all variables have been suitably restricted so that no divisor is 0.* Consider, then, the indicated quotients of monomials in examples **a** and **b**.

a. $\dfrac{-78t^5}{13t^3}$ **b.** $\dfrac{196u^2v^4w}{28uvw}$

These expressions can be rewritten as follows:

a. $\dfrac{-78}{13} \cdot \dfrac{t^5}{t^3}$ **b.** $\dfrac{196}{28} \cdot \dfrac{u^2}{u} \cdot \dfrac{v^4}{v} \cdot \dfrac{w}{w}$

In this form, the numerical coefficients can be reduced, and the exponents on the like bases can be subtracted. Thus, the simplified quotients can be written as:

a. $-6t^2$ **b.** $7uv^3$

These examples suggest a two-step procedure for dividing monomials.

To divide monomials:

Step 1. Compute the quotient of the numerical coefficients.

Step 2. Subtract the exponents on the like bases.

Example 1. Divide $\dfrac{96x^8y^3}{8x^5y}$.

Solution. **Step 1.** Divide the numerical coefficients:

$$\frac{96}{8} = 12$$

Step 2. Subtract the exponents on the like bases:

$$\frac{x^8}{x^5} = x^{8-5} = x^3 \quad \text{and} \quad \frac{y^3}{y} = y^{3-1} = y^2$$

Thus, $\dfrac{96x^8y^3}{8x^5y} = 12x^3y^2$.

Dividing a Polynomial by a Monomial

In arithmetic, we learn that the numerators of two or more fractions with the same denominator can be combined over the common denominator:

$$\text{If } z \neq 0, \text{ then } \frac{x}{z} + \frac{y}{z} \text{ can be written as } \frac{x+y}{z}.$$

This property can also be used in reverse:

$$\text{If } z \neq 0, \text{ then } \frac{x+y}{z} \text{ can be written as } \frac{x}{z} + \frac{y}{z}.$$

Example 2. Divide $\dfrac{147t^5 - 35t^4 + 21t^3 - 7t^2}{7t^2}$.

Solution. **Discussion.** By first rewriting each term on the top over the monomial divisor, the quotient can be simplified using the two-step procedure of Example 1.

$$\frac{147t^5 - 35t^4 + 21t^3 - 7t^2}{7t^2} \qquad \text{The given expression}$$

$$= \frac{147t^5}{7t^2} - \frac{35t^4}{7t^2} + \frac{21t^3}{7t^2} - \frac{7t^2}{7t^2} \qquad \text{Write each term over } 7t^2.$$

$$= 21t^3 - 5t^2 + 3t - 1 \qquad \text{Simplify each quotient.}$$

Keep in mind that the division carried out does not change the value of the given expression; it merely *simplifies* it. That is, *for all nonzero replacements of* t, *the given expression and the simplified one have the same value.* For example, if t is replaced by 2:

$$\frac{147(2^5) - 35(2^4) + 21(2^3) - 7(2^2)}{7(2^2)} = \frac{4{,}284}{28} = 153$$

and

$$21(2^3) - 5(2^2) + 3(2) - 1 = 168 - 20 + 6 - 1 = 153$$

Thus, both expressions are 153 for t replaced by 2.

Dividing a Polynomial by a Polynomial

Using the FOIL method, it can be shown that $(t + 5)(t - 8) = t^2 - 3t - 40$. It therefore follows that $(t^2 - 3t - 40) \div (t + 5) = t - 8$. A **long-division format** can be used to carry out such divisions.

Example 3. Divide $(t^2 - 3t - 40)$ by $(t + 5)$.

Solution. **Step 1.** Divide the first term of the dividend by the first term of the divisor.

$$\frac{t^2}{t} = t \qquad t + 5\overline{)\,t^2 - 3t - 40} \quad \text{(quotient: } t\text{)}$$

Step 2. Multiply the divisor by the monomial of Step 1 and subtract the product from the dividend.

$$t(t + 5) = t^2 + 5t$$

$$\begin{array}{r} t \\ t + 5\,\overline{)\,t^2 - 3t - 40} \\ (-)\,t^2 + 5t \\ \hline -8t - 40 \end{array}$$

Step 3. Repeat Steps 1 and 2 until a difference is obtained that has degree less than the degree of the divisor.

$$\frac{-8t}{t} = -8$$

$$-8(t + 5) = -8t - 40$$

$$\begin{array}{r} t - 8 \\ t + 5\,\overline{)\,t^2 - 3t - 40} \\ (-)\,t^2 + 5t \\ \hline -8t - 40 \\ (-)\,-8t - 40 \\ \hline 0 \end{array}$$

Example 4. Divide $(8x^4 - 22x^3 - x^2 + 11x - 3)$ by $(4x + 3)$.

Solution.

Step 1. Divide $8x^4$ by $4x$.

$$\begin{array}{r|l} & 2x^3 \\ \hline 4x + 3 & 8x^4 - 22x^3 - x^2 + 11x - 3 \end{array}$$

Step 2. Multiply $(4x + 3)(2x^3)$ and subtract the product from the dividend.

$$2x^3(4x + 3) = 8x^4 + 6x^3$$

$$\begin{array}{rr} & 2x^3 \qquad\qquad\qquad\qquad \\ \hline 4x + 3 \,) & 8x^4 - 22x^3 - x^2 + 11x - 3 \\ (-) & 8x^4 + 6x^3 \qquad\qquad\qquad \\ \hline & -28x^3 - x^2 + 11x - 3 \end{array}$$

Step 3. Repeat Steps 1 and 2, and continue the process until a difference is obtained that has degree less than the divisor.

$$\begin{array}{rr} & 2x^3 - 7x^2 + 5x - 1 \quad \\ \hline 4x + 3 \,) & 8x^4 - 22x^3 - x^2 + 11x - 3 \\ (-) & 8x^4 + 6x^3 \qquad\qquad\qquad \\ \hline & -28x^3 - x^2 + 11x - 3 \\ (-) & -28x^3 - 21x^2 \qquad\qquad \\ \hline & 20x^2 + 11x - 3 \\ (-) & 20x^2 + 15x \qquad \\ \hline & -4x - 3 \\ (-) & -4x - 3 \\ \hline & 0 \end{array}$$

In each of the division problems solved with a long-division format, the remainder has been 0. Thus, the dividends have been **evenly divisible** by the divisors. If a nonzero remainder is obtained, a term written as a fraction with the divisor as denominator must be added to the quotient.

Example 5. Divide $\dfrac{2x^5 - 13x^3 + 8x - 5}{x^2 + 2x - 1}$.

Solution. **Discussion.** The dividend is missing x^4- and x^2-terms. As an aid in finding the quotient, $0x^4$ and $0x^2$ are included in the dividend polynomial.

$$\begin{array}{rrl} & 2x^3 - 4x^2 - 3x + 2 \qquad\qquad\qquad & \\ \hline x^2 + 2x - 1 \,) & 2x^5 + 0x^4 - 13x^3 + 0x^2 + 8x - 5 & \quad 2x^5 \div x^2 = 2x^3 \\ (-) & 2x^5 + 4x^4 - 2x^3 \qquad\qquad\qquad\quad & \\ \hline & -4x^4 - 11x^3 + 0x^2 + 8x - 5 & \quad -4x^4 \div x^2 = -4x^2 \\ (-) & -4x^4 - 8x^3 + 4x^2 \qquad\qquad\quad & \\ \hline & -3x^3 - 4x^2 + 8x - 5 & \quad -3x^3 \div x^2 = -3x \\ (-) & -3x^3 - 6x^2 + 3x \qquad & \\ \hline & 2x^2 + 5x - 5 & \quad 2x^2 \div x^2 = 2 \\ (-) & 2x^2 + 4x - 2 & \\ \hline & x - 3 & \end{array}$$

Since the degree of $x - 3$ is less than the degree of $x^2 + 2x - 1$ (the divisor), the division stops and this remainder is written as the numerator of a fraction that is added to the quotient:

$$\frac{2x^5 - 13x^3 + 8x - 5}{x^2 + 2x - 1} = 2x^3 - 4x^2 - 3x + 2 + \frac{x - 3}{x^2 + 2x - 1}$$

SECTION 3-4. Practice Exercises

In exercises **1–20**, find the indicated quotients.

[Example 1] **1.** $\dfrac{72x^3}{9x}$ **2.** $\dfrac{64x^7}{4x^3}$

3. $\dfrac{13t^9u^7}{t^2u^5}$ **4.** $\dfrac{-39t^8u^3}{13t^3u^3}$

5. $\dfrac{27a^3b^2c^3}{3b^2c}$ **6.** $\dfrac{-8a^7b^5c^2}{4a^6b^2c}$

7. $\dfrac{-16u^6v^4}{8u^3v^3}$ **8.** $\dfrac{-96u^5v^3}{-12u^2v^2}$

9. $\dfrac{28p^4q^3r^2}{7p^2q^2r^2}$ **10.** $\dfrac{-33p^4q^4r^4}{11p^4r^2}$

[Example 2] **11.** $\dfrac{6x^4 - 4x^3 + 2x^2}{2x^2}$ **12.** $\dfrac{125x^3y^3 - 5x^2y^2 + 35xy}{5xy}$

13. $\dfrac{45m^4n - 9m^3n^2 + 6m^2n^3 - 3mn}{3mn}$ **14.** $\dfrac{96m^3n - 64m^2n - 16mn + 8n}{8n}$

15. $\dfrac{33s^4t^2 - 42s^3t^3 + 18s^2t^4}{3s^2t^2}$ **16.** $\dfrac{18s^5t^2 + 24s^4t^3 - 30s^3t^4}{6s^3t^2}$

17. $\dfrac{147a^5b^2 - 35a^4b^3 + 21a^3b^4 - 28a^2b^5}{7a^2b^2}$

18. $\dfrac{-24a^2b^2 + 8a^3b^3 - 26a^4b^4 + 2a^5b^5}{-2a^2b^2}$

19. $\dfrac{30x^8y^7z^6 - 3x^6y^8z^7 + 9x^5y^9z^8}{3x^5y^7z^6}$ **20.** $\dfrac{12x^3y^4z^5 - 4x^4y^3z^4 + 16x^5y^2z}{4x^3y^2z}$

In exercises **21–40**, find the indicated quotients. Write any nonzero remainder as a fraction with the divisor as denominator.

[Examples 3–5] **21.** $(x^3 + 4x^2 + x - 6) \div (x + 2)$ **22.** $(6x^3 - 11x^2 + 10x - 8) \div (3x - 4)$

23. $(6y^3 - y^2 - 5y + 2) \div (2y - 1)$ **24.** $(8y^3 - 6y^2 - 5y + 3) \div (4y + 3)$

25. $(2a^3 - 3a^2b + 4ab^2 + 3b^3) \div (2a + b)$

26. $(12a^3 - 5a^2b - 11ab^2 + 6b^3) \div (3a - 2b)$

27. $(6k^4 + 7k^3 + 2k^2 + 16k - 5) \div (2k^2 - k + 3)$

28. $(10k^4 + 3k^3 - 3k + 3) \div (2k^2 + k + 1)$

29. $(8x^4 + 3x^2 - 5x + 3) \div (x^2 + x + 1)$

30. $(10x^4 + x^3 - 2x^2 + 5x - 2) \div (2x^2 - x + 1)$

31. $\frac{y^5 + 32}{y + 2}$

32. $\frac{y^5 - 3125}{y - 5}$

33. $\frac{64a^6 - 729}{2a - 3}$

34. $\frac{729a^6 - 64}{3a + 2}$

35. $\frac{p^4 - 6p^2 + 9}{p^2 - 3}$

36. $\frac{p^4 + 16p^2 + 64}{p^2 + 8}$

37. $\frac{64z^6 - 16z^3 + 8}{2z^2 + 1}$

38. $\frac{64z^6 + 16z^3 - 1}{4z^2 - 1}$

39. $\frac{b^6 - 16b^3 + 30}{b^2 + b + 2}$

40. $\frac{b^6 + 54b^3 + 729}{b^2 - 3b + 9}$

SECTION 3-4. Ten Review Exercises

1. Evaluate each expression for $t = 5$.
   a. $2t^2 + t - 5$
   b. $3t^2 - 4t + 1$
2. Simplify $(2t^2 + t - 5) + (3t^2 - 4t + 1)$.
3. Evaluate the sum in exercise **2** for $t = 5$, and compare the value with the sum of the values obtained in exercise **1**.
4. Simplify $(2t^2 + t - 5) - (3t^2 - 4t + 1)$.
5. Evaluate the difference in exercise **4** for $t = 5$, and compare the value with the difference in the values obtained in exercise **1**.
6. Multiply $(2t^2 + t - 5)(3t^2 - 4t + 1)$.
7. Evaluate the product in exercise **6** for $t = 5$, and compare the value with the product of the values obtained in exercise **1**.

In exercises **8** and **9**, solve each equation.

8. $5(6k + 1) - 4(12k - 5) = 10$
9. $\left|\frac{1 - 2u}{3}\right| = 5$
10. Solve and graph $5(2y + 3) > 2y + 9 - 2(2y - 15)$.

SECTION 3-4. Supplementary Exercises

Consider the divisions in examples **a** and **b**.

a. $\frac{x^3 + 5x^2 + 8x + 4}{x + 2}$ **b.** $\frac{1{,}584}{12}$

If x is replaced by 10 in example **a**, the division in example **b** is obtained. Compare the two divisions using similar long-division formats.

a.

$$\begin{array}{r}
x^2 + 3x + 2 \\
x + 2\overline{)\,x^3 + 5x^2 + 8x + 4} \\
(-)\,x^3 + 2x^2 \qquad\qquad \\
\hline
3x^2 + 8x + 4 \\
(-)\,3x^2 + 6x \qquad \\
\hline
2x + 4 \\
(-)\,2x + 4 \\
\hline
0
\end{array}$$

b.

$$\begin{array}{r}
13 \\
12\overline{)\,1584} \\
12 \\
\hline
384 \\
360 \\
\hline
24 \\
24 \\
\hline
0
\end{array}$$

In exercises **1–4**:

a. Evaluate the dividend and divisor for $x = 10$.

b. Find the quotient of the polynomials.

c. Find the quotient of the numbers obtained in part **a**.

d. Compare the divisions line by line in parts **b** and **c**.

1. $x^2 + 5x + 6 \quad \div \quad x + 3$

2. $6x^2 + 7x + 2 \quad \div \quad 2x + 1$

3. $x^4 + 2x^3 + 5x^2 + 4x + 9 \quad \div \quad x^2 + 2x + 3$

4. $12x^4 + 4x^3 + 8x^2 \quad \div \quad 4x^2$

Synthetic division is a process whereby a polynomial dividend can be divided by a linear divisor of the form $x - c$ by using only the coefficients of the dividend. The six-step method given below is a recommended procedure for using synthetic division. The steps are illustrated by finding the indicated quotient.

Given: $(x^4 + 2x^3 - 5x - 22) \div (x - 2)$

Step 1. Array the coefficients of the dividend. Insert a 0 for any missing power of x.

The array of the coefficients of $x^4 + 2x^3 - 5x - 22$ provide the following array:

$$\begin{array}{ccccc} 1 & 2 & 0 & -5 & -22 \end{array}$$

↑ for the missing x^2 term (under the 0)

Step 2. Put the value of c in a box to the left of the array.

$$\boxed{2}\quad \begin{array}{ccccc} 1 & 2 & 0 & -5 & -22 \end{array}$$

↑ c is 2 since the divisor is $x - 2$

Step 3. Skip a space, then draw a line and write the first number in the array of coefficients below this line.

$$\begin{array}{c|ccccc}
\boxed{2} & 1 & 2 & 0 & -5 & -22 \\
 & \multicolumn{5}{c}{\text{(skipping a space)}} \\
\hline
 & 1 & & & &
\end{array}$$

Step 4. Multiply the number brought down by c, then add the product to the second number in the array of coefficients.

$$\begin{array}{c|ccccc}
\boxed{2} & 1 & 2 & 0 & -5 & -22 \\
 & \downarrow & 2 & (=1 \cdot 2) & & \\
\hline
 & 1 & 4 & & &
\end{array}$$

Step 5. Continue the process stated in Step 4, using the remaining numbers in the array of coefficients.

$$\begin{array}{c|ccccc}
\boxed{2} & 1 & 2 & 0 & -5 & -22 \\
 & & 2 & 8 & 16 & 22 \\
\hline
 & 1 & 4 & 8 & 11 & 0
\end{array}$$

Step 6. The numbers below the line are the coefficients of the quotient polynomial, except the last number, which is the numerator of the remainder.

$1x^3 + 4x^2 + 8x + 11 + \dfrac{0}{x-2}$ is the quotient

Notice that the quotient is a polynomial of degree 3, since the dividend has degree 4 and the divisor has degree 1. That is, *the degree of the quotient is one less than the degree of the dividend.*

Example. Divide $(x^4 + 10x^3 + 32x^2 + 37x + 10)$ by $(x + 5)$ using synthetic division.

Solution. Since $x + 5$ can be written $x - (-5)$, the value of c is -5.

$$\begin{array}{r|rrrrr} \boxed{-5} & 1 & 10 & 32 & 37 & 10 \\ & & -5 & -25 & -35 & -10 \\ \hline & 1 & 5 & 7 & 2 & 0 \end{array}$$

← The 0 indicates that there is no remainder.

The quotient is $x^3 + 5x^2 + 7x + 2$.

Example. Divide $(2x^4 - 19x^2 + 10)$ by $(x - 3)$ using synthetic division.

Solution. Zeros are inserted in the array for the missing x^3- and x-terms.

$$\begin{array}{r|rrrrr} \boxed{3} & 2 & 0 & -19 & 0 & 10 \\ & & 6 & 18 & -3 & -9 \\ \hline & 2 & 6 & -1 & -3 & 1 \end{array}$$

← The 1 is the remainder.

The quotient is $2x^3 + 6x^2 - x - 3 + \dfrac{1}{x-3}$.

The next example illustrates that c can be a rational number that is not an integer.

Example. Divide $(4x^3 + 3x - 2)$ by $\left(x - \frac{1}{2}\right)$ using synthetic division.

Solution. Notice that $c = \frac{1}{2}$.

$$\begin{array}{r|rrrr} \boxed{\frac{1}{2}} & 4 & 0 & 3 & -2 \\ & & 2 & 1 & 2 \\ \hline & 4 & 2 & 4 & 0 \end{array}$$

The quotient is $4x^2 + 2x + 4$.

In exercises **5–20**, find the indicated quotients using synthetic division.

5. $(x^3 - 2x^2 - x + 2) \div (x - 2)$

6. $(x^3 - 6x^2 + 25) \div (x - 5)$

7. $(x^3 + x^2 - 4x - 4) \div (x + 1)$

8. $(x^3 + 3x^2 - 2x - 6) \div (x + 3)$

9. $(x^4 - x^3 - 5x^2 - x - 6) \div (x - 3)$

10. $(x^4 - 10x^2 + 9) \div (x + 3)$

11. $(3x^4 + 14x^3 + 18x - 35) \div (x + 5)$

12. $(2x^4 - 7x^3 - 50x^2 + 49) \div (x - 7)$

13. $(6x^5 - 48x^4 - x^3 + 11x^2 - 26x + 16) \div (x - 8)$

14. $(5x^5 + 32x^4 + 12x^3 - 3x^2 - 26x - 48) \div (x + 6)$

15. $(4x^5 + 16x^4 - 9x^3 - 36x^2 + 7x + 28) \div (x + 4)$

16. $(7x^5 - 40x^4 - 12x^3 - 10x + 60) \div (x - 6)$

17. $(2x^4 - 7x^3 - 47x^2 - 17x + 21) \div \left(x - \frac{1}{2}\right)$

18. $(6x^3 - 5y^2 + 4) \div \left(x + \frac{2}{3}\right)$

19. $(4x^4 - 37x^2 + 10) \div \left(x + \frac{1}{2}\right)$

20. $(4x^5 - 69x^3 - 172x^2 - 135x - 17) \div \left(x + \frac{3}{2}\right)$

SECTION 3-5. Factoring a Polynomial: Basic Concepts and Special Factoring Forms

1. A definition of a prime polynomial
2. Factoring a common monomial factor
3. Factoring a simple trinomial
4. Factoring a difference of squares
5. Factoring a perfect-square trinomial
6. Factoring a sum or difference of cubes

Consider the equalities in examples **a** and **b**.

a. $5 \cdot 13 = 65$ **b.** $2 \cdot 3 \cdot 7 = 42$

The numbers on the right sides are called the **products** of the numbers on the left sides. That is, 65 and 42 are products. The numbers on the left sides are called the **factors** of the products. For example, 5 and 13 are factors of 65 in that both numbers divide 65 with a remainder of 0.

These basic concepts can also be applied to polynomials. To illustrate, consider the equations in examples **a*** and **b***. The left sides are called the **factored forms** and the right sides are called the **multiplied forms** of the polynomials.

Factored forms

a*. $(t + 2)(5t - 2) = 5t^2 + 8t - 4$

b*. $u(u + 1)(4u - 1) = 4u^3 + 3u^2 - u$

Multiplied forms

The similarities between examples **a** and **a*** can be seen by replacing t by 3 in **a*** and getting the numbers in example **a**. Similarly, when u is replaced by 2 in **b*** the numbers in example **b** are obtained.

A Definition of a Prime Polynomial

Suppose we are told to write 24 as an indicated product of two or more numbers. With no specific directions as to what numbers we are permitted to use, we could find several ways of accomplishing this task. Examples **c–j** suggest eight of the many possible choices.

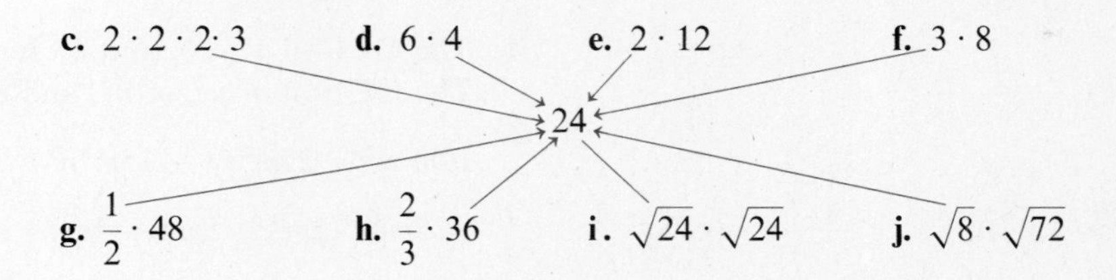

If we are told to use only **integers** as factors, then examples **g**, **h**, **i**, and **j** would be eliminated, because some of the factors are not integers. If we are further told to use only **prime integers,** then examples **d**, **e**, and **f** would be eliminated, because 6, 4, 12, and 8 are not prime numbers. Thus, if instructed to use only prime numbers, then the only factored form for 24 is $2 \cdot 2 \cdot 2 \cdot 3$.

Similarly, if we are told to **factor a polynomial,** then we need to be guided in what numbers can be used as coefficients.

> **Definition 3.4. A prime polynomial *P***
> If P is a polynomial with **integer coefficients,** then P is **prime** if and only if the degree of P is greater than 0, and P can be written as a product of polynomial factors with integer coefficients in exactly two ways:
>
> $$(1)\ \ 1 \cdot P \qquad \text{or} \qquad (2)\ \ (-1)(-P)$$

Based on Definition 3.4, binomials such as $2x + 5$ and $x^2 - 3$ are classified as prime polynomials that cannot be factored. However, using the techniques for multiplying polynomials, it can be shown that $2x + 5$ can be written as $2(x + \frac{5}{2})$, and $x^2 - 3$ can be written as $(x + \sqrt{3})(x - \sqrt{3})$. The factored forms of these binomials contain numbers that are not integers—namely, $\frac{5}{2}$ and $\sqrt{3}$. As a consequence, these binomials will be *prime* for our purposes in factoring polynomials.

Factoring a Common Monomial Factor

If each term of a polynomial P contains a **common factor,** then P can be written in factored form by removing the common factor. The common factor can be an *integer*, a *power of one or more variables in* P, or a *product of an integer and variables.*

Example 1. Factor each expression.

a. $12t^5 + 42t^4$ **b.** $10m^6n^2 - 25m^4n^3 - 15m^3n^5$

Solution. **Discussion.** In removing a common factor, determine the **greatest common factor,** written GCF.

a. The GCF of 12 and 42 is 6.
The GCF of t^5 and t^4 is t^4.

$12t^5 + 42t^4$

$= 6t^4(2t) + 6t^4(7)$

$= 6t^4(2t + 7)$

b. The GCF of 10, 25, and 15 is 5.
The GCF of m^6n^2, m^4n^3, and m^3n^5 is m^3n^2.

$10m^6n^2 - 25m^4n^3 - 15m^3n^5$

$= 5m^3n^2(2m^3 - 5mn - 3n^3)$

Factoring a Simple Trinomial

Consider the trinomials in examples **k** and **l**.

k. $t^2 - 8t + 12$ **l.** $u^2 - 10u - 75$

Trinomials such as these have leading coefficients that are one; that is, $1 \cdot t^2$ and $1 \cdot u^2$. *Trinomials with leading coefficients of one can be called simple.* The general form of a simple trinomial in x is $x^2 + bx + c$, where b and c are nonzero integers.

> To factor $x^2 + bx + c$, find (if they exist), integers m and n such that:
>
> (1) $m \cdot n = c$ and (2) $m + n = b$

Example 2. Factor each expression.

a. $t^2 - 8t + 12$ **b.** $u^2 - 10u - 75$

Solution. **a. Discussion.** For the given trinomial:

(1) $m \cdot n = 12$ and (2) $m + n = -8$

Since the product of m and n is positive and the sum is negative, we may conclude that both are negative numbers. By trial and error:

(1) $(-6)(-2) = 12$ and (2) $(-6) + (-2) = -8$

Thus, $t^2 - 8t + 12 = (t - 6)(t - 2)$.

b. Discussion. For the given trinomial:

(1) $m \cdot n = -75$ and (2) $m + n = -10$

Since the product of m and n is negative, we may conclude that m and n are *opposite in signs.* Since the sum is negative, we know the number with the greater absolute value is negative. By trial and error:

(1) $(-15)(5) = -75$ and (2) $(-15) + 5 = -10$

Thus, $u^2 - 10u - 75 = (u - 15)(u + 5)$.

Example 3. Factor $30p^2q^2 - 84pq^3 + 6p^3q$.

Solution. **Discussion.** Before attempting to determine m and n for the given trinomial, there are two tasks that need to be carried out first.

Task 1. Remove the common factor $6pq$.

Task 2. Write the resulting trinomial in descending powers of p.

$$30p^2q^2 - 84pq^3 + 6p^3q$$
$$= 6pq(5pq - 14q^2 + p^2)$$
$$= 6pq(p^2 + 5pq - 14q^2)$$

(1) $m \cdot n = -14$ and (2) $m + n = 5$

By trial and error:

(1) $(-2)(7) = -14$ and (2) $(-2) + 7 = 5$

The final factored form is $6pq(p - 2q)(p + 7q)$.

Factoring a Difference of Squares

Example **m** is an indicated product of *a sum and difference of the terms* $5k$ *and* 13. The multiplied form is therefore *the difference in the squares of these terms.*

m. $(5k + 13)(5k - 13) = (5k)^2 - (13)^2$

$$= 25k^2 - 169$$

If a binomial can be written as a difference of two squares, then the factored form is an indicated product of the sum and difference of terms.

Difference of squares factoring form

Difference of squares		**Sum and difference of terms**
$x^2 - y^2$	$=$	$(x + y)(x - y)$

Example 4. Factor $4z^2 - 49$.

Solution. **Discussion.** First write both terms as squares of monomials.

$4z^2 - 49$	The given binomial
$(2z)^2 - (7)^2$	In the form $x^2 - y^2$
$= (2z + 7)(2z - 7)$	In factored form

Factoring a Perfect-Square Trinomial

Examples **n** and **o** are indicated **squares of binomials.** The multiplied forms are therefore **perfect-square trinomials.**

n. $(5k + 3)^2$
$= 25k^2 + 30k + 9$

o. $(3u - 4v)^2$
$= 9u^2 - 24uv + 16v^2$

If a trinomial can be written in the form of a perfect-square trinomial, then the factored form is the square of a binomial.

Perfect-square trinomial factoring forms

Perfect-square trinomial		Square of a sum or difference
$x^2 + 2xy + y^2$	=	$(x + y)^2$
$x^2 - 2xy + y^2$	=	$(x - y)^2$

To determine whether a given trinomial qualifies as a perfect square, follow these four steps:

Step 1. Write the first term in the form x^2.

Step 2. Write the last term in the form y^2.

Step 3. Write the middle term in the form $2xy$.

Step 4. Check the signs to make sure they are either (+ +) or (− +).

If any of Steps 1–4 cannot be carried out, then the given trinomial is not a perfect square and cannot be written as a square of a binomial.

Example 5. Factor each expression.

a. $16z^2 + 56z + 49$ **b.** $32t + 2t^3 - 16t^2$

Solution. **a.** **Step 1.** $16z^2$ can be written as $(4z)^2$ $x = 4z$

Step 2. 49 can be written as $(7)^2$ $y = 7$

Step 3. $56z$ can be written as $2(4z)(7)$ $2xy = 56z$

Step 4. The signs are (+ +).
In factored form: $16z^2 + 56z + 49 = (4z + 7)^2$

b. **Discussion.** Remove the common factor $2t$ and write the trinomial in descending powers of t.

$32t + 2t^3 - 16t^2$

$= 2t(t^2 - 8t + 16)$

Step 1. t^2 can be written as $(t)^2$

Step 2. 16 can be written as $(4)^2$

Step 3. $8t$ can be written as $2(t)(4)$

Step 4. The signs are (− +).
In factored form: $32t + 2t^3 - 16t^2 = 2t(t - 4)^2$

Factoring a Sum or Difference of Cubes

Examples **p** and **q** are indicated products that yield a sum of cubes and a difference of cubes, respectively.

p. $(3t + 2)(9t^2 - 6t + 4)$
$= 27t^3 + 8$

q. $(5u - 1)(25u^2 + 5u + 1)$
$= 125u^3 - 1$

If a binomial can be written in the form of a sum or difference of cubes, then the factored form is the indicated product of a binomial and corresponding trinomial.

Sum and difference of cubes factoring forms

Sum or difference of cubes		Binomial multiplied by corresponding trinomial
$x^3 + y^3$	$=$	$(x + y)(x^2 - xy + y^2)$
$x^3 - y^3$	$=$	$(x - y)(x^2 + xy + y^2)$

Example 6. Factor $375y^4 - 24y$.

Solution. **Discussion.** First remove the common factor $3y$.

$375y^4 - 24y$ — The given binomial

$= 3y(125y^3 - 8)$ — Common factor $3y$

$= 3y[(5y)^3 - (2)^3]$ — In the form $x^3 - y^3$

$= 3y(5y - 2)(25y^2 + 10y + 4)$ — $= (x - y)(x^2 + xy + y^2)$

SECTION 3-5. Practice Exercises

In exercises **1–60**, factor each expression.

[Example 1]

1. $x^4y^2 - 2x^3y^3$ **2.** $5x^5y^2 + 4x^4y^3$

3. $2k^5 + 14k^4 - 6k^3$ **4.** $3k^6 - 15k^4 - 30k^2$

5. $6x^6y^2 - 21x^4y^3 + 3x^2y^4$ **6.** $12x^8y^4 + 4x^6y^5 - 32x^4y^6$

7. $10p^2q^2r^3 - 5pq^3r^3 + 25pq^2r^4 - 15pq^2r^3$

8. $35p^3q^2r^2 + 14p^2q^3r^2 - 7p^2q^2r^3 + 28p^2q^2r^2$

[Example 2]

9. $x^2 + 16x - 36$ **10.** $x^2 + x - 56$

11. $y^2 - 7y + 10$ **12.** $y^2 - y - 6$

13. $p^2 - 13p + 42$ **14.** $p^2 - 13p + 40$

15. $w^2 - 12w - 28$ **16.** $w^2 - 9w - 36$

17. $z^2 + 21z + 90$ **18.** $z^2 + 20z + 96$

[Example 3]

19. $5p^3q + 45p^2q^2 + 100pq^3$ **20.** $6p^4q^2 - 42p^3q^3 + 36p^2q^4$

21. $2x^6 + 6x^5 + 4x^4$ **22.** $3x^9 + 6x^8 - 45x^7$

23. $5t^3 - 50t^2 + 45t$ **24.** $2t^5 - 6t^4 - 8t^3$

25. $12y^4 - 24y^3 - 36y^2$

26. $4y^5 - 32y^4 + 48y^3$

[Example 4] **27.** $x^2 - 81$

28. $4x^2 - 121$

29. $16 - 9y^2$

30. $361 - 25y^2$

31. $a^2 - 49b^2$

32. $a^2 - 100b^2$

33. $7p^2q^2 - 7\pi^2$

34. $72p^2q^2 - 2$

35. $3r^3s - 192rs^3$

36. $5r^5 - 125r$

[Example 5] **37.** $x^2 + 20x + 100$

38. $25x^2 + 10x + 1$

39. $81a^2 - 18ab + b^2$

40. $9a^2 - 42ab + 49b^2$

41. $9k^4 - 60k^2 + 100$

42. $121k^6 + 110k^3 + 25$

43. $128r^4 + 32\pi r^2 + 2\pi^2$

44. $48 - 72r^2s^2 + 27r^4s^4$

45. $3p^3q + 18p^2q^2 + 27pq^3$

46. $32mn - 16m^3n^3 + 2m^5n^5$

[Example 6] **47.** $y^3 - 729$

48. $27y^3 - 1$

49. $343t^3 + u^3$

50. $t^3u^3 - 729$

51. $27x^3 + 125$

52. $125x^3 + 8$

53. $a^3 + 1000b^3$

54. $a^3b^6 + 216c^9$

55. $625x^4y - 5xy^4$

56. $16x^5y^2 - 54x^2y^5$

57. $40k^5 + 5k^2$

58. $10k^4 + 1250k$

59. $2a^9b^6c^3 - 16$

60. $48a^6b^6 - 6c^6$

SECTION 3-5. Ten Review Exercises

In exercises **1–4**, given: $5x + x^4 - 2 + 3x^6$

1. Write the polynomial in descending powers.

2. Identify the degree of the polynomial.

3. Identify the leading coefficient.

4. Identify the constant term.

In exercises **5–7**, given: $2(t^2 - 5t + 1) - 3(2 - t^2 - 3t) + 5(t - t^2 - 2)$

5. Evaluate for $t = 4$.

6. Simplify the expression.

7. Evaluate the expression of exercise **6** for $t = 4$, and compare with the result obtained in exercise **5**.

8. Evaluate $2k - 5(1 - k) - 10(k - 2)$ for $k = 5$.

9. Solve $2k - 5(1 - k) = 10(k - 2)$.

10. Simplify $2k - 5(1 - k) - 10(k - 2)$.

SECTION 3-5. Supplementary Exercises

Example. Factor $x^4 + 5x^2 - 36$.

Solution. **Discussion.** The degree of this trinomial is 4. Because the leading coefficient is 1, we may consider it to be simple. As an aid to factoring it, we can change it to a trinomial of degree 2.

Let $k = x^2$ and $k^2 = (x^2)^2 = x^4$.

$x^4 + 5x^2 - 36$ can be written:	The given trinomial
$k^2 + 5k - 36$	Second degree in k
$= (k + 9)(k - 4)$	Factored in terms of k
$= (x^2 + 9)(x^2 - 4)$	Replace k by x^2.
$= (x^2 + 9)(x + 2)(x - 2)$	$x^2 + 9$ is a prime binomial

In exercises **1–14**, factor each expression

1. $k^4 - 16k^2 + 28$

2. $x^4 - 16x^2 - 36$

3. $2y^6 + 6y^4 + 4y^2$

4. $3m^9 + 6m^6 - 45m^3$

5. $16a^4 - 1$

6. $x^4 - 8x^2 + 16$

7. $b^6 + 64$

8. $p^4 - 81$

9. $x^4 - 50x^2 + 625$

10. $64s^3t^2 + 8t^5$

11. $1 - 64y^6$

12. $8p^5 - 40p^3 + 32p$

13. $8x^4 - 162y^2$

14. $t^4 - 2t^2 - 63$

Example. Factor $(t - 1)^2 - (u + 2)^2$.

Solution. **Discussion.** This expression is in the general form of the difference of two squares, $x^2 - y^2$.

Let $x = t - 1$ and $y = u + 2$.

$(t - 1)^2 - (u + 2)^2$ can be written:	The given expression
$x^2 - y^2$	Difference of squares
$= (x + y)(x - y)$	Factored in x and y
$= [(t - 1) + (u + 2)][(t - 1) - (u + 2)]$	Replace original expressions.
$= [t + u + 1][t - u - 3]$	Simplify each factor.

Thus, $(t - 1)^2 - (u + 2)^2 = (t + u + 1)(t - u - 3)$.

In exercises **15–22**, factor each expression.

15. $49 - (p + q)^2$

16. $(m - n)^3 + 125$

17. $(x + a)^2 + 2(x + a) + 1$

18. $(2t + u)^2 - 10(2t + u) + 25$

19. $(m - n)^2 - 7(m - n) + 12$

20. $10(2a + b)^2 - 30(2a + b)$

21. $(x + y)^2 - (x - y)^2$

22. $(k^2 + 4k)^2 + 14(k^2 + 4k) + 49$

In exercises **23–34**, write each expression in factored form. Assume a and b represent positive integers.

23. $x^{2a} - 7x^a + 10$

24. $x^{2a} - 13x^a - 30$

25. $x^{2a} + 7x^a y^b - 8y^{2b}$

26. $x^{2a} + 2x^a y^b - 15y^{2b}$

27. $x^{2a+2b} - 10x^{a+b} + 21$

28. $x^{6a} - 4x^{3a}y^{2b} + 3y^{4b}$

29. $64p^{3a} - 1$

30. $p^{3a} - 27q^{3b}$

31. $25m^{4a} - 10m^{2a} + 1$

32. $m^{4a} + 14m^{2a}n^{2b} + 49n^{4b}$

33. $(x^{2a} - 5x^a)^2 - 36$

34. $(x^{2a} - 3x^a)^2 - 16$

SECTION 3-6. Factoring by Grouping

1. Factoring by grouping in pairs
2. Factoring by grouping a trinomial
3. Factoring a general trinomial by grouping

In Section 3-5, we studied a few special techniques for factoring polynomials. In this section, we will look at a more general technique for factoring, called "grouping". In this section's Supplementary Exercises, this technique is also applied to the methods we have previously studied (see pp. 202–203).

Factoring by Grouping in Pairs

In example **a**, the multiplied form of two binomials is obtained using the FOIL method.

Factored form	**Multiplied form**
a. $(t^2 + 2)(3t + 5)$	$= 3t^3 + 5t^2 + 6t + 10$

Suppose now that the multiplied form is given, and we need to find the factored form. (That is, pretend the two binomials of the factored form are not known.) A **grouping technique** suggests that we separate the four-term polynomial into groups of two terms each. Each pair of terms is then examined for a common monomial factor. After the common factors have been removed, the expected outcome is a common binomial factor.

$3t^3 + 5t^2 + 6t + 10$	The four-term polynomial
$= (3t^3 + 5t^2) + (6t + 10)$	Two groups of two-terms each
$= t^2(3t + 5) + 2(3t + 5)$	Common factors t^2 and 2
Common binomial factor	
$= (3t + 5)(t^2 + 2)$	Common factor $(3t + 5)$

This example suggests the following four-step procedure for factoring a polynomial by grouping pairs of terms:

To factor a polynomial by grouping in pairs:

Step 1. Group the terms into pairs in which the terms of each binomial have a common factor.

Step 2. Remove the common factor in each binomial.

Step 3. Check the resulting binomials to see whether they are the same.

Step 4. If the binomials are alike, then factor them using the distributive property. If the binomials are not alike, then regroup the given polynomial into different pairs and repeat Steps 1–3.

Example 1. Factor $15k^3 + 21k^2 + 5k + 7$.

Solution. **Discussion.** As written, the first pair of terms has a common factor $3k^2$, and the second pair has a common factor 1.

Step 1. $(15k^3 + 21k^2) + (5k + 7)$ Group into two pairs.

Step 2. $3k^2(5k + 7) + 1(5k + 7)$ Factor a $3k^2$ and 1.

Same binomial

Step 3. $(5k + 7)(3k^2 + 1)$ Factor the binomial.

In factored form: $15k^3 + 21k^2 + 5k + 7 = (5k + 7)(3k^2 + 1)$

Example 2. Factor each expression.

a. $6xy - 9y - 4x + 6$ **b.** $2p^3 + 15 - 5p^2 - 6p$

Solution. **a. Discussion.** In the given order, the first two terms have a common factor $3y$. The last two terms have a common factor of either 2 or -2. We will factor the number that yields the same binomial obtained when the first pair is factored.

Step 1. $(6xy - 9y) + (-4x + 6)$ Group into pairs.

Step 2. $3y(2x - 3) - 2(2x - 3)$ Factor a (-2) to get $(2x - 3)$.

Step 3. $(2x - 3)(3y - 2)$ Factor the $(2x - 3)$.

b. Discussion. In the given order, the first two terms have no common factor. Writing the polynomial in descending powers of p is therefore recommended.

Step 1. In descending powers: $(2p^3 - 5p^2) + (-6p + 15)$

Step 2. $p^2(2p - 5) - 3(2p - 5)$ Factor a (-3) to get $(2p - 5)$.

Step 3. $(2p - 5)(p^2 - 3)$ Factor the $(2p - 5)$.

This technique can sometimes be used to factor polynomials with more than four terms.

Example 3. Factor $t^2u - 3t^2 + 2tu - 6t - 5u + 15$.

Solution. **Discussion.** In the given order, the first, middle, and last pairs of terms all have common factors. Therefore, the given order will be checked before another order is tried.

Step 1. $(t^2u - 3t^2) + (2tu - 6t) + (-5u + 15)$

Step 2. $t^2(u - 3) + 2t(u - 3) - 5(u - 3)$

Step 3. $(u - 3)(t^2 + 2t - 5)$

The trinomial $t^2 + 2t - 5$ cannot be factored. Thus, the factored form of the given six-term polynomial is $(u - 3)(t^2 + 2t - 5)$.

Factoring by Grouping a Trinomial

Example **b** is an indicated product of two trinomials. The multiplied form can more easily be obtained by using two special product forms. To do this, we temporarily group the first two terms of both trinomials. The expression now takes on the form $(x + y)(x - y)$, where x is actually a binomial. The first term of the product is therefore the square of the binomial x.

b. $(t + 2u + 5)(t + 2u - 5)$ — The given trinomials

$= [(t + 2u) + 5][(t + 2u) - 5]$ — Group the first two terms.

$= (t + 2u)^2 - 5^2$ — $(x + y)(x - y) = x^2 - y^2$

$= t^2 + 4tu + 4u^2 - 25$ — $(x + y)^2 = x^2 + 2xy + y^2$

Suppose now that the four-term polynomial is given, and the task is to write it in factored form. We would need to use a grouping-of-terms technique. However, instead of grouping pairs of terms, we need to recognize the three terms that have the perfect-square trinomial form. That is, *we separate the polynomial into a trinomial and a monomial.*

$t^2 + 4tu + 4u^2 - 25$ — The given polynomial

$= (t^2 + 4tu + 4u^2) - 25$ — Group the perfect-square trinomial.

$= (t + 2u)^2 - 5^2$ — Write as a difference of squares.

$\quad x^2 - y^2$

$= [(t + 2u) + 5][(t + 2u) - 5]$ — $x^2 - y^2 = (x + y)(x - y)$

$\quad (x + y)(x - y)$

$= (t + 2u + 5)(t + 2u - 5)$ — Remove inside parentheses.

Example 4. Factor $9m^2 - 6m + 1 - 4n^2$.

Solution. **Discussion.** Recognize the perfect-square trinomial $9m^2 - 6m + 1$.

$(9m^2 - 6m + 1) - 4n^2$	Group the perfect-square trinomial.
$= (3m - 1)^2 - (2n)^2$	$(3m)^2 - 2(3m)(1) + 1^2 = (3m - 1)^2$
$= [(3m - 1) + 2n][(3m - 1) - 2n]$	Factor as a difference of squares.
$= (3m - 1 + 2n)(3m - 1 - 2n)$	Remove inside parentheses.

Example 5. Factor $16 - 9p^2 + 30pq - 25q^2$.

Solution. **Discussion.** We must recognize that the last three terms have the possibility of forming a perfect-square trinomial:

$$9p^2 \square\ 30pq \square\ 25q^2 \text{ can be written } (3p)^2 \square\ 2(3p)(5q) \square\ (5q)^2$$

However, the operational signs in the empty boxes are not appropriate for either form of the perfect-square trinomial. By "factoring a negative one", we can get the correct signs.

$16 - 9p^2 + 30pq - 25q^2$	The given polynomial
$= 16 - 1 \cdot (9p^2 - 30pq + 25q^2)$	Factor a (-1).
$= 4^2 - (3p - 5q)^2$	Write as difference of squares
$= [4 + (3p - 5q)][4 - (3p - 5q)]$	Factor as $x^2 - y^2$.
$= (4 + 3p - 5q)(4 - 3p + 5q)$	Remove inside parentheses.

Factoring a General Trinomial by Grouping

The grouping-in-pairs method can be used to factor a **general trinomial.** The procedure requires writing the trinomial as a four-term polynomial. Steps 1–3 in the following procedure tell us how to find the appropriate terms to make the change from three to four terms.

To factor $ax^2 + bx + c$, where $a \neq 0$ and $a \neq 1$, using the grouping technique:

Step 1. Form the product $a \cdot c$.

Step 2. List the pairs of integers m and n, such that:

$$m \cdot n = a \cdot c$$

Step 3. If one exists, find the m and n from the list in Step 2, such that:

$$m + n = b$$

Step 4. Replace bx in the trinomial by $mx + nx$ selected in Step 3.

Step 5. Factor the four-term polynomial formed in Step 4, using the grouping-in-pairs technique.

Example 6. Factor each expression.

a. $7t^2 + 37t + 10$ **b.** $12x^2 + 5xy - 3y^2$

Solution. **a. Discussion.** The signs of the trinomial indicate that both m and n are *positive numbers.*

Step 1. With $a = 7$ and $c = 10$, $a \cdot c = 7 \cdot 10 = 70$.

Step 2. List pairs of positive integers m and n whose product is 70:

(1)(70) (2)(35) (5)(14) (7)(10)

Step 3. Since $b = 37$, we look for a pair of integers in the list with sum 37.

$2 + 35 = 37$

Step 4. Replace $37t$ by $(2t + 35t)$ and factor by grouping.

Step 5. $7t^2 + 2t + 35t + 10$ Write $37t$ as $2t + 35t$.

$= t(7t + 2) + 5(7t + 2)$ Factor a t and 5.

$= (7t + 2)(t + 5)$ Factor the $(7t + 2)$.

b. Discussion. The signs of the trinomial indicate that m and n are *opposite in signs:* that is, one is positive and the other is negative.

Step 1. With $a = 12$ and $c = -3$, $a \cdot c = 12(-3) = -36$.

Step 2. List pairs of integers whose product is -36:

(1)(−36) (−1)(36) (2)(−18) (−2)(18)

(3)(−12) (−3)(12) (4)(−9) (−4)(9) (6)(−6)

Step 3. Since $b = 5$, we look for a pair of integers in the list with sum 5.

$-4 + 9 = 5$

Step 4. Replace $5xy$ by $(-4xy + 9xy)$ and factor by grouping.

Step 5. $12x^2 + 5xy - 3y^2$

$= 12x^2 - 4xy + 9xy - 3y^2$ Write $5xy$ as $-4xy + 9xy$.

$= 4x(3x - y) + 3y(3x - y)$ Factor a $4x$ and $3y$.

$= (3x - y)(4x + 3y)$ Factor the $(3x - y)$.

SECTION 3-6. Practice Exercises

In exercises **1–50**, factor each expression.

[Example 1]

1. $2t^3 + t^2 + 6t + 3$

2. $10t^3 + 15t^2 + 2t + 3$

3. $6u^3 + 4u^2 + 27u + 18$

4. $35u^3 + 28u^2 + 15u + 12$

5. $3m^3 + mn + 4n^2 + 12m^2n$

6. $50m^2n + 3mn + 10m^3 + 15n^2$

7. $21xy^2 + 20x + 28y + 15x^2y$

8. $36x^2y + 5y + 45xy^2 + 4x$

[Example 2] **9.** $2x^3 + 3x^2 + 2x + 3$

10. $3x^3 - x^2 + 6x - 2$

11. $a^3 + 3a^2 - 5a - 15$

12. $2a^3 - 3a^2 - 12a + 18$

13. $5xy - 25y + x - 5$

14. $6xy - 9y + 4x - 6$

15. $3 - 10a - 15b + 50ab$

16. $10 + 12ab - 40b - 3a$

[Example 3] **17.** $2x^2y + 4x^2 + xy + 2x + 3y + 6$

18. $2y^2x + y^2 - 2xy - y - 2x - 1$

19. $3z^5 + 2z^4 - 3z^3 - 2z^2 - 3z - 2$

20. $4z^5 - 10z^4 + 6z^3 - 15z^2 - 2z + 5$

21. $3a^5 - a^3b^2 - 9a^4b - 3a^2b^3 + ab^4 - 3b^5$

22. $3a^4b + 20a^3b^2 + 10a^5 + 6a^2b^3 - 50ab^4 - 15b^5$

23. $m^4 + m^3n - 2m^2n - 2mn^2 - m^2n^2 - 2n^3$

24. $4m^3 + 6m^2n^2 - 3mn^3 - 2m^2n + 6mn^2 + 9n^4$

[Examples 4 and 5] **25.** $25a^2 - 10a + 1 - b^2$

26. $a^2 + 16a + 64 - b^2$

27. $p^2 - 2pq + q^2 - 9$

28. $16p^2 - 24pq + 9q^2 - 4$

29. $100 - x^2 + 18xy - 81y^2$

30. $225 - 4x^2 - 4xy - y^2$

31. $1 - 4m^4 - 8m^2n - 4n^2$

32. $121 - 9m^2 + 36mn - 36n^2$

[Example 6] **33.** $8p^2 + 14p + 5$

34. $12p^2 + 16p + 5$

35. $2m^2 - 5m - 12$

36. $3m^2 - 17m - 6$

37. $4a^2 - 13ab + 3b^2$

38. $6a^2 - 11ab - 2b^2$

39. $10z^2 + 13z - 3$

40. $9z^2 - 21z + 10$

41. $5k^2 + 13k - 6$

42. $5k^2 - 17k + 6$

43. $15t^2 - 7tu - 2u^2$

44. $15t^2 - 89tu - 6u^2$

45. $6r^2 - 13r - 15$

46. $6r^2 + 11r - 2$

47. $4x^2 - 9y^2 + 16xy$

48. $15xy + 25x^2 - 4y^2$

49. $9ab - b^2 + 36a^2$

50. $64a^2 - b^2 - 12ab$

SECTION 3-6. Ten Review Exercises

1. Evaluate $3t^2 - 13t - 10$ for $t = 5$.

2. Is 5 a root of $3t^2 - 13t - 10 = 0$?

3. Evaluate $3t^2 - 13t - 10$ for $t = \frac{-2}{3}$.

4. Is $\frac{-2}{3}$ a root of $3t^2 - 13t - 10 = 0$?

In exercises **5** and **6**, factor each expression.

5. $3t^2 - 13t - 10$.

6. $6t^4 - 20t^2 - 26t^3$

In exercises **7–10**, simplify each expression.

7. $(5ab^2)(-7a^3)$

8. $(-x^2y)^2(-2xy^3)$

9. $\left(\frac{-ij^2k^3}{3}\right)^3\left(\frac{i^2j^2}{3}\right)$

10. $\frac{-42r^4s^3}{6r^4s}$

SECTION 3-6. Supplementary Exercises

In exercises **1–10**, factor each expression.

1. $4n - 9m + 12mn - 3$

2. $a^2 - b^2 + 16b - 64$

3. $12s^2 - 28s - 5$

4. $c^2 + 2c + 1 - d^2 - 4d - 4$

5. $2x^3 + 2x^2y + 3x^2y + 3xy^2 + xy^2 + y^3$

6. $15m^2 + 88m - 12$

7. $5pq^3 - 4 - pq + 20q^2$

8. $y^4 + 20y^2 + 100 - x^4$

9. $225p^2 - 30pq + q^2 - r^2 + 2rs - s^2$

10. $2t^2u + 3t^2 - 10tu - 15t + 12u + 18$

To factor the difference of squares by grouping:

Given: $x^2 - y^2$

Write: $x^2 + xy - xy - y^2$

Factor by grouping: $x(x + y) - y(x + y) = (x + y)(x - y)$

In exercises **11–14**, for each expression:

a. Write each binomial as a four-term polynomial.

b. Factor by grouping.

11. $x^2 - 25$

12. $4y^2 - 49$

13. $100p^2 - 81q^2$

14. $64m^4 - 9$

To factor a perfect-square trinomial by grouping:

Given: $x^2 + 2xy + y^2$

Write: $x^2 + xy + xy + y^2$

Factor by grouping: $x(x + y) + y(x + y) = (x + y)(x + y)$

In exercises **15–18**, for each expression:

a. Write each trinomial as a four term polynomial.

b. Factor by grouping.

15. $x^2 + 10x + 25$

16. $4y^2 + 12y + 9$

17. $49p^2 - 84pq + 36q^2$

18. $m^4 - 6m^2n + 9n^2$

To factor a sum of cubes by grouping:

Given: $x^3 + y^3$

Write: $x^3 + x^2y - x^2y - xy^2 + xy^2 + y^3$

Factor by grouping: $(x^3 + x^2y) + (-x^2y - xy^2) + (xy^2 + y^3)$

$$= x^2(x + y) - xy(x + y) + y^2(x + y)$$

$$= (x + y)(x^2 - xy + y^2)$$

In exercises **19–22**, for each expression:

a. Write each binomial as a polynomial of six terms.

b. Factor by grouping.

19. $a^3 + b^3$

20. $p^3 + 8$

21. $27m^3 + 1$

22. $125s^3 + t^3$

SECTION 3-7. Solving Quadratic Equations by Factoring

1. The definition of a quadratic equation in one variable
2. The zero-product property
3. Solving quadratic equations by factoring
4. Applied problems

The equations in **a** and **b** are examples of **quadratic equations in one variable.**

a. $t^2 - t - 6 = 0$ **b.** $5 - 13u = 6u^2$

Equations such as these require a technique different from the one used to solve linear equations in one variable. For quadratic equations in this section, we can factor the quadratic expression and then use the zero-product property to solve. We will learn other techniques for solving such equations in Chapter 6.

The Definition of a Quadratic Equation in One Variable

As indicated by Definition 3.5, every quadratic equation in one variable must have a term of degree 2. Furthermore, it has no term of degree more than 2.

Definition 3.5. The standard form of a quadratic equation in *x*
A quadratic equation in x is one that can be written in the form:

$$ax^2 + bx + c = 0 \qquad \text{or} \qquad 0 = ax^2 + bx + c$$

where a, b, and c are real numbers and $a > 0$.

The equation in example **a** is written in the standard form with $a = 1$, $b = -1$, and $c = -6$. The equation in example **b** can be changed to $0 = 6u^2 + 13u - 5$. For this equation, $a = 6$, $b = 13$, and $c = -5$.

The Zero-Product Property

The multiplication property of 0 states that 0 times any number is 0.

Multiplication property of 0
For any a, $a \cdot 0 = 0$ and $0 \cdot a = 0$

The zero-product property states that a product can only be 0 if at least one of the numbers being multiplied is 0.

Zero-product property
If a and b are numbers and $a \cdot b = 0$, then $a = 0$, $b = 0$, or a and b are both 0.

Example 1. Solve each equation.

a. $7(2p + 9) = 0$ **b.** $(3z - 1)(z + 8) = 0$

Solution. **a. Discussion.** The left side of the equation indicates a product of 7 and $2p + 9$ equal to 0. A solution of this equation must be a solution of one of the following equations:

$$7 = 0 \qquad \text{or} \qquad 2p + 9 = 0$$

$$2p = -9$$

$$p = \frac{-9}{2}$$

Since $7 \neq 0$, the solution must come from the other factor. If $p = \frac{-9}{2}$, then $2p + 9 = 0$, and $7 \cdot 0 = 0$ is true. Thus, the solution set is $\{\frac{-9}{2}\}$.

b. Discussion. The left side of the equation indicates a product of two factors equal to 0. A solution of this equation must be a solution of one of the following equations:

$$3z - 1 = 0 \quad \text{or} \quad z + 8 = 0$$

$$z = \frac{1}{3} \qquad\qquad z = -8.$$

If $z = \frac{1}{3}$, then the first factor is 0.
If $z = -8$, then the second factor is 0.
} For each value of z, the product is 0.

The solution set is $\left\{\frac{1}{3}, -8\right\}$.

Solving Quadratic Equations by Factoring

If the nonzero expression of a quadratic equation in standard form can be factored, then the zero-product property can be used to solve the equation.

Example 2. Solve each equation.

a. $t^2 - t - 6 = 0$

b. $5 - 13u = 6u^2$

Solution. **a.** $t^2 - t - 6 = 0$ — The given equation

$(t - 3)(t + 2) = 0$ — Factor the simple trinomial.

$t - 3 = 0 \quad \text{or} \quad t + 2 = 0$ — Use the zero-product property.

$t = 3 \qquad\qquad t = -2$ — Solve the linear equations.

The check is omitted. The solution set is $\{-2, 3\}$.

b. $5 - 13u = 6u^2$ — The given equation

$0 = 6u^2 + 13u - 5$ — Add -5 and $13u$ to both sides.

$0 = (3u - 1)(2u + 5)$ — Factor the general trinomial.

$3u - 1 = 0 \quad \text{or} \quad 2u + 5 = 0$ — Use the zero-product property.

$u = \frac{1}{3} \qquad u = \frac{-5}{2}$ — Solve the linear equations.

The check is omitted. The solution set is $\left\{\frac{-5}{2}, \frac{1}{3}\right\}$.

Applied Problems

Examples 3 and 4 are applied problems that can be solved using quadratic equations. The required equations can be solved by factoring and the zero-product property.

Example 3. If the sum of two numbers is 10 and their product is -39, then what are the numbers?

Solution. **Discussion.** Use the fact that the sum of the numbers is 10 to write expressions for both numbers using only one variable. Use the fact that their product is -39 to write an equation.

Let n be one of the numbers.

$10 - n$ is the other number. — Their sum is 10.

$n(10 - n) = -39$ — Their product is -39.

$10n - n^2 = -39$ — Remove parentheses.

$0 = n^2 - 10n - 39$ — In standard form

$0 = (n - 13)(n + 3)$ — Factor the trinomial.

$n - 13 = 0 \quad \text{or} \quad n + 3 = 0$ — The zero-product property

$n = 13 \qquad n = -3$ — Solve for n.

$10 - n = -3 \qquad 10 - n = 13$ — Solve for $10 - n$.

Thus, the numbers are -3 and 13.
Notice that $-3 + 13 = 10$ and $(-3)(13) = -39$.

Example 4. Wendy Eledge has a triangular-shaped area for a vegetable garden. The altitude to the base of the triangle is 6 feet less than the length of the base. If the area of the garden is 80 square feet, find the lengths of the base and altitude.

Solution. **Discussion.** Use the relationship between the lengths of the base and altitude to write expressions for both using only one variable. Use the formula for the area of a triangle to write an equation.

Let h represent the height and $h + 6$ the base (see Figure 3-5). For any triangle:

Figure 3-5.

$$A = \frac{1}{2}(\text{base})(\text{height})$$

$$80 = \frac{1}{2}(h + 6)(h)$$

$160 = h^2 + 6h$ — Multiply both sides by 2.

$0 = h^2 + 6h - 160$ — Add -160 to both sides.

$0 = (h - 10)(h + 16)$ — Factor the trinomial.

$h - 10 = 0 \quad \text{or} \quad h + 16 = 0$ — The zero-product property

$h = 10 \qquad h = -16$ — Solve for h.

$h + 6 = 16$ — Reject the -16.

Thus, the altitude (height) of the garden is 10 feet and the base is 16 feet.

SECTION 3-7. Practice Exercises

In exercises **1–30**, solve each equation.

[Example 1] **1.** $12(x - 7) = 0$ **2.** $17(x - 5) = 0$

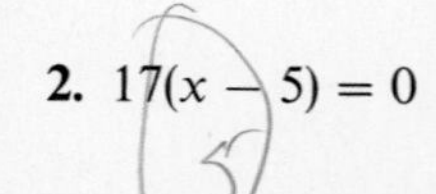

3. $(y + 2)(y - 3) = 0$

4. $(y + 8)(y - 1) = 0$

5. $(2z - 1)(z + 5) = 0$

6. $(5z + 3)(z - 4) = 0$

7. $m(3m - 7) = 0$

8. $m(2m + 11) = 0$

9. $t(4t - 1)(3t + 1) = 0$

10. $t(5t + 2)(3t - 2) = 0$

[Example 2] **11.** $u^2 - 10u - 24 = 0$

12. $u^2 + u - 56 = 0$

13. $0 = 4t^2 - 3t$

14. $0 = 2t^2 + 5t$

15. $3v^2 + 45v - 300 = 0$

16. $2v^2 + 40v + 200 = 0$

17. $x^2 = 26x - 25$

18. $4x^2 = 7x + 2$

19. $25z^2 = 4$

20. $169 = 4z^2$

21. $u(4u - 5) = 6$

22. $u(2u - 5) = 3$

23. $5v(v + 3) = 2(v - 3)$

24. $6v(v + 3) = 5v - 6$

25. $(z - 2)(z - 3) = 6$

26. $(z + 1)(z + 2) = 2$

27. $(t + 4)^2 = 49$

28. $(2t - 1)^2 = 4$

29. $\dfrac{(5u + 3)(5u - 3)}{5} = -8u - 5$

30. $\dfrac{(2u + 1)(2u - 1)}{2} = 6u - 5$

In exercises **31–42**, solve each word problem.

[Example 3] **31.** The sum of two numbers is 1 and their product is -12. Find both numbers.

32. The sum of two numbers is -2 and their product is -35. Find both numbers.

33. The difference between the larger and the smaller of two numbers is 8, and their product is 48. Find both numbers.

34. The difference between the larger and the smaller of two numbers is 12, and their product is 45. Find both numbers.

35. The larger of two numbers is 3 less than twice the smaller. If their product is 54, find both numbers.

36. The larger of two numbers is 5 more than 3 times the smaller. If their product is 42, find both numbers.

[Example 4] **37.** The altitude of a triangle is 3 inches more than the base. The area of the triangle is 35 square inches. Find the lengths of the base and the altitude.

38. The base of a triangle is 8 inches more than the altitude. The area of the triangle is 42 square inches. Find the lengths of the base and the altitude.

39. Mr. S. K. Graff has a triangular card made for his investment firm. The base is 1 centimeter more than twice the length of the altitude. The area is 18 square centimeters. Find the length of the base of the card.

40. Ms. Sedlar has a triangular sign made for her real estate firm. The base is 3 inches more than twice the length of the altitude. The area is 45 square inches. Find the length of the base of the sign.

41. The lengths of the base and altitude of a triangle are two consecutive even integers. The area of the triangle is 220 square feet. Find the lengths of the base and altitude.

42. The lengths of the base and altitude of a triangle are two consecutive even integers. The area of the triangle is 84 square feet. Find the lengths of the base and altitude.

SECTION 3-7. Ten Review Exercises

1. Multiply $(t - 3)(2t^2 + t - 5)$.

2. Evaluate $2t^3 - 5t^2 - 8t + 15$ for $t = 3$.

3. Divide $(2t^3 - 5t^2 - 8t + 15)$ by $(t - 3)$.

4. Evaluate the quotient of exercise **3** for $t = -2$.

In exercises **5–10**, factor each expression.

5. $3m^3 + 12m^2 - 9m$

6. $3n^3 - 12n$

7. $u^2 - 40 - 3u$

8. $x^3 + 125$

9. $16y^2 + 9 - 24y$

10. $2z^2 - 15z + 18$

SECTION 3-7. Supplementary Exercises

Example. Solve $x^2 + 5xy + 4y^2 = 0$ for y.

Solution. **Discussion.** The trinomial in the equation can be factored and the zero-product property can be used. The solutions are found by solving the two equations for the indicated variable.

$x^2 + 5xy + 4y^2 = 0$ The given equation

$(x + y)(x + 4y) = 0$ Factor the trinomial.

$x + y = 0 \quad \text{or} \quad x + 4y = 0$ The zero-product property

$y = -x \qquad y = \dfrac{-x}{4}$ Solve for y.

If $x^2 + 5xy + 4y^2 = 0$, then $y = -x$ or $y = \frac{-x}{4}$.

In exercises **1–8**, solve each equation for the indicated variable.

1. $x^2 - 2xy - 3y^2 = 0$, for y

2. $p^2 - 26pq + 25q^2 = 0$, for p

3. $4s^2 - 7st - 2t^2 = 0$, for t

4. $y^2 = 5yz + 14z^2$, for y

5. $36x^2 = y^2$, for y

6. $5m^2 - 22mn = 15n^2$, for m

7. $a^2b + 2abc + 3a + 6c = 0$, for a

8. $3x + x^2z + 6y + 2xyz = 0$, for x

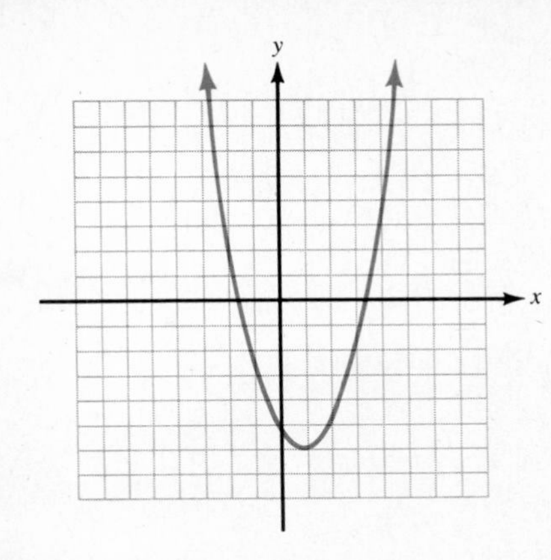

Figure 3-6. A graph of $y = x^2 - 2x - 5$.

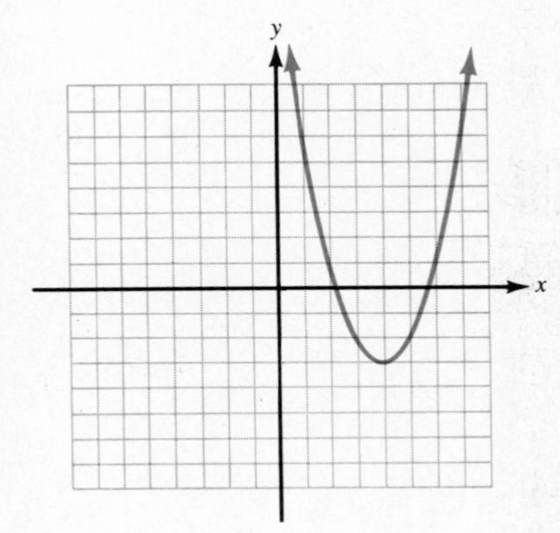

Figure 3-7. A graph of $y = x^2 - 8x + 13$.

In exercises **9–14**, solve each equation. (*Hint:* Write each trinomial as a product of *four linear factors.* Then use the zero-product property.)

9. $x^4 - 5x^2 + 4 = 0$

10. $p^4 - 29p^2 + 100 = 0$

11. $y^4 + 144 = 25y^2$

12. $z^4 = 50z^2 - 49$

13. $4 = 13t^2 - 9t^4$

14. $29m^2 = 25 + 4m^4$

15. Figure 3-6 is a graph of $y = x^2 - 2x - 5$.
a. From the graph find the values of x for $y = 3$.
b. Check the answers to part **a** by solving $3 = x^2 - 2x - 5$.
c. From the graph find the values of x for $y = -2$.
d. Check the answers to part **c** by solving $-2 = x^2 - 2x - 5$.

16. Figure 3-7 is a graph of $y = x^2 - 8x + 13$.
a. From the graph find the values of x for $y = 6$.
b. Check the answers to part **a** by solving $6 = x^2 - 8x + 13$.
c. From the graph find the values of x for $y = -2$.
d. Check the answers to part **c** by solving $-2 = x^2 - 8x + 13$.

17. Figure 3-8 is a graph of $y = -x^2 + 2x + 3$.
a. From the graph, find the values of x for $y = 0$.
b. Check the answers to part **a** by solving $0 = -x^2 + 2x + 3$.
c. From the graph find the values of x for $y = 4$.
d. Check the answer to part **c** by solving $4 = -x^2 + 2x + 3$.
e. Based on the graph, will $y = -x^2 + 2x + 3$ have any real-number solutions when y is replaced by numbers greater than 4?

18. Figure 3-9 is a graph of $y = x^3 - 8$.
a. Write the equation with $x^3 - 8$ in factored form.
b. Based on the zero-product property, one of the factors in **a** must be 0 for y to equal 0. From the graph, what is the value of x for which $y = 0$?
c. In how many places does the graph cross the x-axis?
d. Based on the answer to part **c**, what conclusion can you make about the possibility that the other factor of $x^3 - 8$ has real-number solutions when set equal to 0?

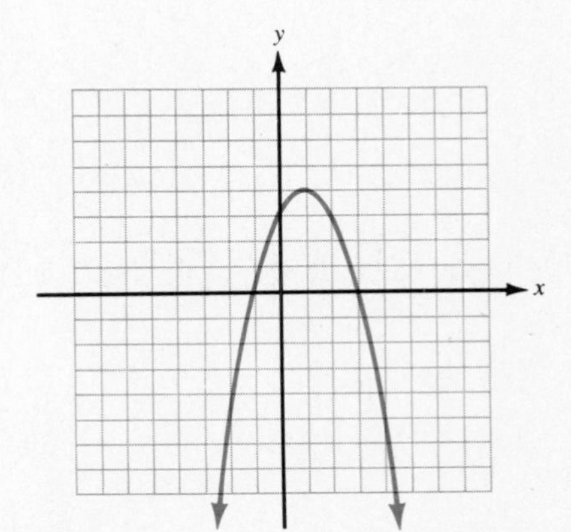

Figure 3-8. A graph of $y = -x^2 + 2x + 3$.

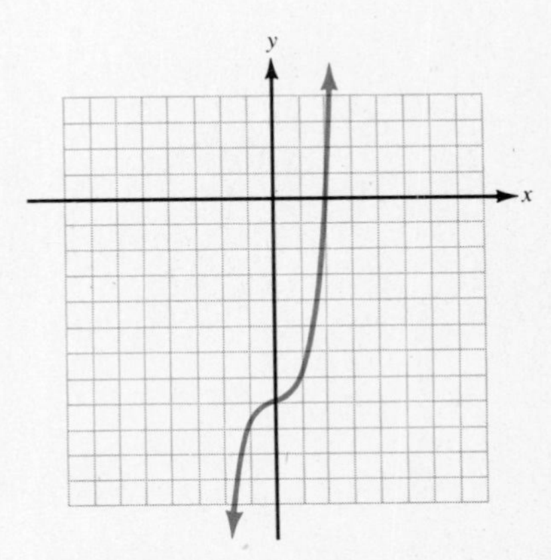

Figure 3-9. A graph of $y = x^3 - 8$.

Name

Date

Score

SECTION 3-1. Summary Exercises

Answer

In exercises **1–8**, simplify each expression. Assume all variables are not 0.

1. $y^6 \cdot y^8 \cdot y$

1. ____________________

2. $(x^2)^4(-x)^3$

2. ____________________

3. $\dfrac{2^6 p^{15} q^3}{2^8 p^5 q^3}$

3. ____________________

4. $(-3s^2t^3u)^4$

4. ____________________

5. 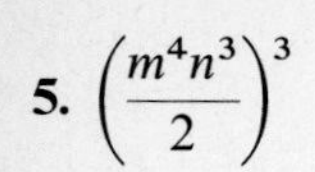 $\left(\frac{m^4n^3}{2}\right)^3$

5. ____________________

6. $(17a^2)^0$

6. ____________________

7. $(5x^2y)^2(-x^3y^4)^3$

7. ____________________

8. $\left(\frac{6s}{t}\right)^2\left(\frac{3s^2}{t}\right)\left(\frac{t^2}{9}\right)^2$

8. ____________________

Name

Date

Score

SECTION 3-6. Summary Exercises

Answer

In exercises **1–8**, factor each expression.

1. $35m + 7mn - 20 - 4n$

1. ____________________

2. $p^3q - 12 + 2p^2 - 6pq$

2. ____________________

3. $s^3 + 2s^2t + 5s^2t + 10st^2 - st^2 - 2t^3$

3. ____________________

4. $4x^2 + 4xy + y^2 - 25$

4. ____________________

5. $49 - m^2 + 6mn - 9n^2$

5. ____________________

6. $2y^2 - 5y - 3$

6. ____________________

7. $6a^2 - 7a - 5$

7. ____________________

8. $15t^2 + 14t - 8$

8. ____________________

CHAPTER 3. Review Exercises

In exercises **1–10**, do the indicated operations.

1. $(xy^2z^3)(-x^2z)(y^3z^2)$

2. $(-2a^3b)^3$

3. $\dfrac{-27m^5n^3}{3m^3n}$

4. $\dfrac{(3p^2q)^2}{p^3q}$

5. $\left(\dfrac{3s^2}{t^4}\right)^2$

6. $\left(\dfrac{-rs^2}{2}\right)\left(\dfrac{r^2s}{3}\right)^2$

7. $(-2t^3)^2(-t^4)(3t^2)^3$

8. $\dfrac{(u^2v)^4(-5uv^2)^2}{(-2u^3v)^3}$

9. $(-0.2)^0$

10. $\left(\dfrac{10y}{3}\right)^0$; $y \neq 0$

11. Given: $3x^2 - 10 + 8x^5 + 2x - x^3 + 5x^4$
- **a.** Write the polynomial in descending powers.
- **b.** Identify the degree.
- **c.** Identify the leading coefficient.
- **d.** Identify the constant term.
- **e.** Identify the polynomial by the number of terms it has.

12. Given: $a^3b - 3b^3c^2 + 5ab^2c - 10a^2c^3$
- **a.** State the degree of each term of the polynomial.
- **b.** Write the polynomial in descending powers of a.
- **c.** Write the polynomial in descending powers of b.
- **d.** Write the polynomial in descending powers of c.

In exercises **13–34**, do the indicated operations.

13. $(3x^3 + 5 - 2x - x^2) + (3x - 1 + x^3 + 2x^2) + (x^2 - 3x^3 + 2x - 7)$

14. $(2a^3 - ab^2 + 5a^2b - 10b^3) - (a^3 - 8b^3 + ab^2 - a^2b)$

15. $(mn - n^2 + 3m^2) - (2m^2 - n^2 - 4mn) + (3n^2 + m^2 - 3mn)$

16. $(p^3 - 4p^2 + 3) + (5p - 2 - 2p^3) - (4 - p^3 - 3p - 7p^2)$

17. $4(y^2 + 2y - 3) - 2(y^2 - 6) + 5(8y + 1)$

18. $\dfrac{1}{2}(10t^3 - 8t + 16) - \dfrac{2}{3}(6t^2 + 21)$

19. $(-3x^3yz^2)(-5y^3z)(6x^5y^2)(2x^3y^4)$

20. $\left(\dfrac{3}{4}a^3b^2c\right)\left(\dfrac{-1}{6}ab^4c\right)\left(\dfrac{4}{5}ac^2\right)\left(\dfrac{-10}{13}a^3bc^5\right)$

21. $(x^2 + xy + y^2)(x - y + 1)$

22. $(k^3 - k^2 + k - 1)(2k^2 + 3k + 1)$

23. $(10x^2 + 3)(10x^2 - 3)$

24. $\left(\dfrac{y}{2} + \dfrac{3}{4}\right)\left(\dfrac{y}{2} - \dfrac{3}{4}\right)$

25. $(4k^3 - 1)^2$

26. $(3a + 7b^2)^2$

27. $(2x - 3y)^2 - (x - 3y)(x + 4y)$

28. $2[(3a - 1)(3a + 1) - 5a] - [(2a + 1)^2 - 4a(a + 1)]$

29. $\dfrac{-36x^4y^2}{9x^2y}$

30. $\dfrac{12a^5b^3c}{-4abc}$

31. $\dfrac{14m^3n^2 - 35m^2n^3 + 7mn^4}{7mn^2}$

32. $\dfrac{24a^4b^2c^2 - 12a^3b^3c^4 + 18a^2b^3c^3 - 6a^2b^4c^2}{6a^2b^2c^2}$

33. $(6x^4 - 5x^3 - 17x^2 + 15x - 3) \div (3x - 1)$

34. $(y^6 + y^5 - 9y^4 - 17y^3 - 8y^2 + 72y + 72) \div (y^2 + y - 6)$

In exercises **35–60**, factor each expression.

35. $2x^2y - 2xy^2 + 8xy$

36. $12a^4b^2 + 6a^3b^3 - 3a^2b^4 + 15a^2b^2$

37. $p^2 - 10pq + 24q^2$

38. $3p^5 - 3p^3 - 168p$

39. $4m^2 + 5m - 6$

40. $6m^2 - 19m - 7$

41. $45 - 3k^4 - 6k^2$

42. $4m^3n - 8mn^3 + 14m^2n^2$

43. $100x^4 - 9y^4$

44. $6a^3b^3 - 150ab$

45. $1 - 20mn + 100m^2n^2$

46. $72c^3d - 24c^2d^2 + 2cd^3$

47. $125c^3 - d^3$

48. $c^9d^6 - \pi^3$

49. $1{,}000k^3 + 1$

50. $64k^6 + 27$

51. $6ac - 9a - 15bc + 15b$

52. $5abc - 3c - 18d + 30abd$

53. $x^2 + 6xy + 9y^2 - 169$

54. $x^4 - x^2 + 12x - 36$

55. $r^4 - 9s^2 - 16r^2 + 64$

56. $324r^6 + 1 - 36r^3 - 25s^4$

57. $18a^2 - 9a - 2$

58. $20a^2 - 23a + 6$

59. $12x^2 + 11x - 15$

60. $6x^2 - 35x + 36$

In exercises **61–66**, solve each equation.

61. $(3p + 2)(p) = 0$

62. $(a + 5)(2 - 3a) = 0$

63. $x^2 - x - 30 = 0$

64. $z^2 = 144$

65. $14t^2 = 5t + 1$

66. $7u = 21u^2$

In exercises **67** and **68**, solve each word problem.

67. The product of the second and third positive consecutive even integers is 48. Find all three integers.

68. The area of a triangle is 35 square centimeters. The height is 3 centimeters more than the base. Find the length of the base of the triangle.

CHAPTER

4

Rational Expressions and Equations

"By now we should all be aware of the fact that the set of real numbers has subsets with specific definitions," Ms. Glaston announced as she began class. "For example, what name do we give to set I?" She turned to the board and wrote:

$$I = \{\ldots, -5, -4, -3, -2, -1, 0, 1, 2, 3, 4, 5, \ldots\}$$

Joyce Pierscinski raised her hand and said, "I think the I stands for integers. If I remember correctly, the integers include all the whole numbers and the negatives of the whole numbers."

"That's right, Joyce," Ms. Glaston replied. "Suppose now that we write an integer over a nonzero integer. What kind of number is formed? For example, let a and b be integers with b not 0. What name is given to the number?" She then turned to the board and wrote under set I:

If a and b are integers and $b \neq 0$, then $\frac{a}{b}$ is a __?__ number.

After a brief period of silence, Scott Cooney raised his hand and said, "That number looks like a fraction, but I don't recall fractions as being one of the subsets of the real numbers."

Ms. Glaston smiled and said, "You're right, Scott. The a over b does have the form we frequently call a common fraction in arithmetic. However, in algebra we use the more precise term rational number. That is, a rational number is a ratio of integers, provided the denominator is not 0. We might think of the rational numbers as being an extension of the set of integers." She then turned to the board and wrote:

If P and Q are polynomials, then $\frac{P}{Q}$ is a __?__ expression,
provided __?__.

"In Chapter 3," Ms. Glaston continued, "we studied expressions called polynomials. We operated on such expressions by adding, subtracting, multiplying, dividing, and factoring them. We also solved some polynomial equations of degree 2. I suggest that we now extend this study by forming a ratio of polynomials, as I have indicated on the board. First, what name would be appropriate for such expressions, and second, what provision must be included?"

Karen Shepherd raised her hand and said, "Well, if we call a ratio of integers a rational number, then a ratio of polynomials could be called a rational expression. And, since the denominator of a rational number cannot be 0, I suppose the denominator of a rational expression cannot be 0 either."

"Right on both counts, Karen," Ms. Glaston replied. "A ratio of two polynomials is called a rational expression. And, as you indicated, Q is restricted from taking on the value of 0. In Chapter 4 we will manipulate—that is, reduce, add, subtract, multiply, and divide—rational expressions. Equations can be written that have rational expressions as terms. We will therefore learn a special technique that can be used to solve such equations. You will be delighted to know there is a group of applied problems that require rational equations to be solved, and we will also study these in this chapter.

"There is one more piece of information that I feel should be mentioned before beginning our study. As I said before, the form of a rational expression is similar to that of numbers in arithmetic called common fractions. We will use this similarity to develop the methods for operating on rational expressions. That is, rational expressions are reduced, added, multiplied, and so on, in much the same way that common fractions are. Now, if there are no questions, let's begin our study of rational expression."

SECTION 4-1. Negative Integer Exponents

1. A definition of b^{-n}, where n is a positive integer
2. Three useful theorems related to negative exponents
3. Applying the properties of exponents to negative exponents
4. Scientific notation as an application of exponents

To write a polynomial expression, only positive integers are needed for exponents. A rational expression, as we will see in Section 4-2, is a ratio of polynomials. Based on the definition of negative integer exponents we will examine in this section, we can write certain rational expressions using negative number exponents. Furthermore, we will use the five properties of exponents to simplify expressions that include positive, negative, and zero-number exponents.

A Definition of b^{-n}, Where n Is a Positive Integer

Many streets have lights that monitor traffic flow. Drivers agree to stop at a red light, and to continue when the light is green. We might say this agreement is "by definition", because no explanation is needed as to *why* these symbols are given their interpretations. The symbols have a purpose (control traffic flow), and everyone agrees to the interpretations (stop on red, go on green). As a consequence, the symbols are effective.

Through the years, many symbols used in mathematics have also acquired commonly accepted meanings. In the development of some mathematical concept, a symbol is needed to represent some aspect of the development. A particular symbol will then be selected, and the precise meaning of the symbol will be given "by definition". Anyone who uses the symbol must agree to the definition, and may not alter or abridge the meaning in any way. As an illustration, Definition 4.1 contains the meaning given to negative integer exponents.

Definition 4.1. Negative integer exponents
If b is a real number ($b \neq 0$) and n is a positive integer:

$$b^{-n} = \frac{1}{b^n}$$

Example 1. Write each expression with only positive exponents. Assume all variables are not 0.

a. 5^{-2} **b.** $2x^{-1}$ **c.** $-t^{-3}$ **d.** $3x^{-5}y^4$

Solution.

a. $5^{-2} = \frac{1}{5^2}$ or $\frac{1}{25}$ — $b^{-n} = \frac{1}{b^n}$

b. $2x^{-1} = 2 \cdot x^{-1} = 2 \cdot \frac{1}{x} = \frac{2}{x}$ — The negative exponent is not on the 2.

c. $-t^{-3} = \frac{-1}{t^3}$ — $-t^{-3}$ can be written as $-1 \cdot t^{-3}$

d. $3x^{-5}y^4 = \frac{3y^4}{x^5}$ — Only the x has a negative exponent.

Three Useful Theorems Related to Negative Exponents

Theorems 1–3 are formulas that can be shown to be true by using the equation in Definition 4.1. These theorems, along with the definition, are useful for writing expressions with negative exponents in terms of equivalent ones using only positive exponents.

> If a and b are nonzero real numbers and m and n are positive integers:
>
> $$b^{-n} = \frac{1}{b^n}$$
>
> **Theorem 1.** $\frac{1}{b^{-n}} = b^n$
>
> **Theorem 2.** $\left(\frac{a}{b}\right)^{-n} = \left(\frac{b}{a}\right)^n$
>
> **Theorem 3.** $a^m b^{-n} = \frac{a^m}{b^n}$

Example 2. Write each expression with only positive exponents. Assume all variables are not 0.

a. $\frac{1}{t^{-2}}$ **b.** $\left(\frac{u}{2}\right)^{-5}$ **c.** $2y^3z^{-2}$

Solution.

a. $\frac{1}{t^{-2}} = t^2$ — $\frac{1}{b^{-n}} = b^n$

b. $\left(\frac{u}{2}\right)^{-5} = \left(\frac{2}{u}\right)^5$ or $\frac{32}{u^5}$ — $\left(\frac{a}{b}\right)^{-n} = \left(\frac{b}{a}\right)^n$

c. $2y^3z^{-2} = \frac{2y^3}{z^2}$ — Write the z^2 in the denominator.

Applying the Properties of Exponents to Negative Exponents

It can be shown that the five properties of exponents given in Section 3-1 are valid for the above definition of b^{-n}. A summary of the properties and definitions of exponents is given below.

Five properties and two definitions of exponents
Let a and b be nonzero real numbers, with m and n integers.

Property 1. $a^m \cdot a^n = a^{m+n}$ — When multiplying like bases, add exponents.

Property 2. $(a^m)^n = a^{mn}$ — When raising a power to a power, multiply exponents.

Property 3. $\dfrac{a^m}{a^n} = a^{m-n}$ — When dividing like bases, subtract exponents.

Property 4. $(ab)^m = a^m \cdot b^m$ — When raising a product to a power, raise each factor to the power.

Property 5. $\left(\dfrac{a}{b}\right)^m = \dfrac{a^m}{b^m}$ — When raising a quotient to a power, raise each factor to the power.

Definition 3.2. If $a \neq 0$, then $a^0 = 1$.

Definition 4.1. If $b \neq 0$, then $b^{-n} = \dfrac{1}{b^n}$.

Example 3. Simplify each expression. Write answers with only positive exponents. Assume all variables are not 0.

a. $(2u^{-2})^{-3}$ **b.** $\left(\dfrac{p^{-3}}{3}\right)^{-2}$

Solution. **a.**

$(2u^{-2})^{-3}$	The given expression
$= 2^{-3}(u^{-2})^{-3}$	Raise each factor to the (-3) power.
$= 2^{-3}u^6$	$(u^{-2})^{-3} = u^{(-2)(-3)} = u^6$
$= \dfrac{u^6}{8}$	$2^{-3} = \dfrac{1}{2^3} = \dfrac{1}{8}$

b.

$\left(\dfrac{p^{-3}}{3}\right)^{-2}$	The given expression
$= \dfrac{(p^{-3})^{-2}}{3^{-2}}$	Raise each factor to the (-2) power.
$= 9p^6$	$(p^{-3})^{-2} = p^6$ and $\dfrac{1}{3^{-2}} = 9$

Example 4. Simplify each expression. Write answers with only positive exponents. Assume all variables are not 0.

a. $(-2r^3)^{-1}(2r^{-2})^{-3}$ **b.** $\dfrac{(-3s^2)^{-2}}{(3s)^{-3}}$

Solution. **Discussion.** Raising to powers has priority over multiplication and division. Therefore, all powers are simplified first before any products or quotients.

a. $(-2r^3)^{-1}(2r^{-2})^{-3}$ — The given expression

$= (-2)^{-1}(r^3)^{-1} \cdot 2^{-3}(r^{-2})^{-3}$ — Remove the parentheses.

$= (-2)^{-1}r^{-3} \cdot 2^{-3}r^6$ — $(r^3)^{-1} = r^{-3}$ and $(r^{-2})^{-3} = r^6$

$= \frac{-1}{2} \cdot r^3 \cdot \frac{1}{8}$ — $(-2)^{-1} = \frac{-1}{2}$ and $r^{-3} \cdot r^6 = r^3$

$= \frac{-r^3}{16}$ — The simplified form

b. $\frac{(-3s^2)^{-2}}{(3s)^{-3}}$ — The given expression

$= \frac{3^{-2}(s^2)^{-2}}{3^{-3}s^{-3}}$ — Since $(-3)^{-2} = 3^{-2}$, the exponent is even.

$= \frac{3^{-2}s^{-4}}{3^{-3}s^{-3}}$ — $(s^2)^{-2} = s^{-4}$

$= \frac{3}{s}$ — $3^{-2-(-3)} = 3$ and $\frac{1}{s^{-3-(-4)}} = \frac{1}{s}$

Scientific Notation as an Application of Exponents

Example **a** is the speed of light in meters per second, and example **b** is the diameter of a smallpox virus.

a. 300,000,000 meters per second

b. 0.00000025 meters

Working with such large or small numbers written in **ordinary notation** is frequently difficult. Scientists frequently write such numbers in scientific notation.

> **A definition of scientific notation**
>
> A number is written in **scientific notation** if it is written as
>
> $$d \times 10^k, \text{ where}$$
>
> **1.** d is a number between 1 and 10; that is, $1 \leq d < 10$
>
> **2.** k is an integer

To change x from ordinary notation to scientific notation:

Step 1. Relocate the decimal point in x so that exactly one nonzero digit is to the left of the decimal point. Call this number d.

Step 2. Count the number of places the decimal point was moved.

Step 3. Multiply d by 10^k, where k is the number of places counted in Step 2.

a. Make k positive if the decimal point was moved to the left.
b. Make k negative if the decimal point was moved to the right.
c. Make k equal to 0 if the decimal point was not moved.

Example 5. Write each number in scientific notation.

a. 5,280 **b.** 0.000128

Solution. **a. Step 1.** $d = 5.28$

Step 2. The decimal point was moved three places to the left.

Step 3. $k = 3$ and $5{,}280 = 5.28 \times 10^3$

b. Step 1. $d = 1.28$

Step 2. The decimal point was moved four places to the right.

Step 3. $k = -4$ and $0.000128 = 1.28 \times 10^{-4}$

To change a number written in scientific notation to one written in ordinary notation, the decimal point is moved right or left, depending on whether the exponent on the 10 is positive or negative, respectively.

Example 6. Write each number in ordinary notation.

a. 2.47×10^7 **b.** 1.06×10^{-5}

Solution. **a.** The exponent is positive $(7 > 0)$; thus, the decimal point is moved seven places to the right.

$$2.47 \times 10^7 = 24{,}700{,}000$$

b. The exponent is negative $(-5 < 0)$; thus, the decimal point is moved five places to the left.

$$1.06 \times 10^{-5} = 0.0000106$$

SECTION 4-1. Practice Exercises

In exercises **1–70**, simplify each expression. Write answers with only positive exponents. Assume all variables are not 0.

[Example 1] **1.** 2^{-4} **2.** 3^{-2} **3.** x^{-8}

4. x^{-5} **5.** $-p^{-6}$ **6.** $-p^{-7}$

7. $(-k)^{-4}$ **8.** $(-k)^{-9}$ **9.** $2m^{-2}n^{3}$

10. $7m^{-3}n^{2}$ **11.** $(-2)^{-3}x^{5}y^{-4}$ **12.** $(-3)^{-2}x^{3}y^{-6}$

[Example 2] **13.** $\dfrac{1}{y^{-4}}$ **14.** $\dfrac{-1}{y^{-7}}$ **15.** $\dfrac{-p^{4}q^{-5}}{r^{-1}}$

16. $\dfrac{p^{-6}q}{r^{-3}}$ **17.** $\left(\dfrac{-5}{x}\right)^{-2}$ **18.** $\left(\dfrac{-3}{x}\right)^{-4}$

19. $\left(\dfrac{2m}{n}\right)^{-3}$ **20.** $\left(\dfrac{-m}{3n}\right)^{-4}$ **21.** $-6a^{-1}b^{2}c^{-3}$

22. $-9^{-1}ab^{-2}c$ **23.** $\dfrac{(-3)^{-3}a^{2}b^{-2}}{c^{-1}}$ **24.** $\dfrac{a^{-3}b}{(-5)^{-2}c^{2}}$

25. $\left(\dfrac{2w}{v^{2}}\right)^{-4}$ **26.** $\left(\dfrac{5w^{2}}{v}\right)^{-3}$ **27.** $\left(\dfrac{-p^{2}}{3q^{3}}\right)^{-1}$

28. $\left(\dfrac{-p^{2}q^{3}}{10}\right)^{-2}$ **29.** $\left(\dfrac{5x}{yz^{2}}\right)^{-3}$ **30.** $\left(\dfrac{7x^{2}}{yz}\right)^{-2}$

[Example 3] **31.** $x^{-1}\cdot x^{-3}$ **32.** $x^{-4}\cdot x^{2}$ **33.** $t^{-2}\cdot t^{4}\cdot t$

34. $t^{-3}\cdot t^{7}\cdot t^{-2}$ **35.** $(-u^{-2})^{-1}$ **36.** $(-u^{-3})^{-2}$

37. $(v^{3})^{-4}$ **38.** $(v^{2})^{-5}$ **39.** $(-w^{-4})^{2}$

40. $(-w^{-3})^{5}$ **41.** $\dfrac{a^{-3}}{a^{-7}}$ **42.** $\dfrac{a^{-9}}{a^{-5}}$

43. $\dfrac{-b^{-2}}{b}$ **44.** $\dfrac{-b^{3}}{b^{-1}}$ **45.** $(-2m^{-2})^{-3}$

46. $(-3m^{4})^{-2}$ **47.** $(10s^{2}t^{-3})^{-5}$ **48.** $(5s^{-3}t^{4})^{-2}$

49. $\left(\dfrac{-k^{-1}}{6}\right)^{-2}$ **50.** $\left(\dfrac{-k^{-3}}{2}\right)^{-4}$ **51.** $\left(\dfrac{2p^{-2}}{q}\right)^{3}$

52. $\left(\dfrac{3p^{2}}{q^{-1}}\right)^{5}$ **53.** $\left(\dfrac{-u^{-2}}{w}\right)^{-4}$ **54.** $\left(\dfrac{-7u^{3}}{v^{-1}w^{2}}\right)^{-2}$

[Example 4] **55.** $(-2t^{2})(5t^{-4})$ **56.** $(-10t^{-5})(-13t^{8})$

57. $(3u)(-6u^{-3})(10u^{5})$ **58.** $(-7u^{-4})(-u^{9})(-12u^{-7})$

59. $(x^{-2}y^{3})(xy^{-2})^{-2}$ **60.** $(x^{2}y^{-3})(x^{-1}y^{2})^{-3}$

61. $(5z^{-2})^{-1}(-2z^{-3})^{2}$ **62.** $(-3z^{2})^{3}(-6z^{5})^{-1}$

63. $\dfrac{(p^{2}q^{-1})^{-3}}{(p^{2}q^{3})^{-1}}$ **64.** $\dfrac{(p^{-4}q^{2})^{-2}}{(p^{-2}q^{3})^{-3}}$

65. $(2ab^{2})^{-3}(3ab^{-3})^{2}$ **66.** $(2ab^{-5})^{3}(3a^{-3}b^{6})^{-1}$

67. $\dfrac{(-10x^{-2})^{2}}{(2^{-2}x^{3})^{-2}}$ **68.** $\dfrac{(-3^{-2}x^{3})^{-1}}{(6x^{-2})^{2}}$

69. $\dfrac{(-5m^{-1}n^{-1})^2(2m^4n)^{-3}}{(10m^3n)^{-3}}$

70. $\dfrac{(-3m^3)^{-2}(2mn^6)^{-1}}{(6mn^2)^{-3}}$

In exercises **71–80**, write each number in scientific notation.

[Example 5] **71.** 360

72. 1760

73. 0.0078

74. 0.091

75. 500,000

76. 70,000

77. 0.0000015

78. 0.000000063

79. 92,100,000

80. 452,000

In exercises **81–90**, write each number in ordinary notation.

[Example 6] **81.** 2.5×10^4

82. 3.9×10^6

83. 1.4×10^{-2}

84. 5.8×10^{-3}

85. 4.39×10^6

86. 8.07×10^8

87. 6.025×10^{11}

88. 4.107×10^9

89. 1.12×10^{-9}

90. 2.15×10^{-11}

SECTION 4-1. Ten Review Exercises

In exercises **1–3**, solve each equation.

1. $5(6u + 1) - 4(12u - 5) = 10$

2. $|2v - 9| = 13$

3. $12 - 5w = 2w^2$

In exercises **4–6**, solve and graph each inequality.

4. $3(4x + 3) - 4 < -9x - 9$

5. $4 \leq \dfrac{3y - 4}{2} \leq 7$

6. $|3b + 2| < 7$

In exercises **7–10**, simplify each expression.

7. $\dfrac{3^4a^5b^2}{3^3a^2b^2}$

8. $(2k)^3(-3k^2)^2$

9. $\left(\dfrac{-m^3}{n}\right)^5\left(\dfrac{-2n^2}{y^4}\right)^4$

10. $(-6u^3v)^2\left(\dfrac{-1}{2uv}\right)^3$

SECTION 4-1. Supplementary Exercises

In exercises **1–20**, simplify each expression. Assume a and b are positive integers.

1. $\dfrac{x^a}{x^{a-1}}$

2. $\dfrac{x^{a+b}}{x^{a-b}}$

3. $(x^a)^{-2}$

4. $(x^{-b})^{-3}$

5. $x^{1-a} \cdot x^{a+2}$

6. $x^{a-b} \cdot x^{a+b}$

7. $\dfrac{2^a \cdot 2^b}{2^{a+b}}$

8. $\dfrac{5^{a-b}}{5^a \cdot 5^{-b}}$

9. $\dfrac{(10^a)^2}{10^{2a-1}}$

10. $\dfrac{10^{b-1}}{10^{1+b}}$

11. $2^a \cdot 2^b \cdot 2^{a-b}$

12. $3^{-a} \cdot 3^{-b} \cdot 3^{a-b}$

13. $(x^{-a} \cdot y^b)^{-2}$

14. $(x^a \cdot y^{-b})^{-3}$

15. $(x^3y^{-2})^a$

16. $(x^{-1}y^3)^{-b}$

17. $\left(\dfrac{x^a}{y^b}\right)^{-3}$

18. $\left(\dfrac{x^{-2a}}{y^{-3b}}\right)^{-2}$

19. $\left(\dfrac{x^2}{y}\right)^{-a}$

20. $\left(\dfrac{x^{-3}}{y^2}\right)^{-2b}$

Example. Evaluate $\dfrac{2^{-3} + 2^{-1}}{2^{-2}}$.

Solution. **Discussion.** The numerator is a binomial. Therefore, the quotient property of exponents and the definition of a negative exponent cannot be applied directly.

$\dfrac{2^{-3} + 2^{-1}}{2^{-2}}$ — The given expression

$= 2^2(2^{-3} + 2^{-1})$ — Write $\dfrac{1}{2^{-2}}$ as 2^2.

$= 2^{-1} + 2$ — $2^2 \cdot 2^{-3} = 2^{-1}$ and $2^2 \cdot 2^{-1} = 2$

$= \dfrac{1}{2} + 2$ or $\dfrac{5}{2}$ — The value of the expression is $\dfrac{5}{2}$.

In exercises **21–28**, evaluate each expression.

21. $\dfrac{5^{-4} + 5^{-2}}{5^{-3}}$

22. $\dfrac{3^{-8} + 3^{-6}}{3^{-7}}$

23. $\dfrac{2^{-5}}{2^{-4} + 2^{-6}}$

24. $\dfrac{10^{-2}}{10^{-1} + 10^{-3}}$

25. $\dfrac{3^{-1} - 3^{-5}}{3^{-2}}$

26. $\dfrac{7^{-4} - 7^{-6}}{7^{-5}}$

27. $\dfrac{2^{-3} + 2^{-5}}{2^{-4} - 2^{-6}}$

28. $\dfrac{3^{-4} + 3^{-6}}{3^{-5} - 3^{-7}}$

In exercises **29–32**, simplify each expression and write answers with positive number exponents.

29. a. $3^{-4} \cdot 3^{-2}$
b. $3^{-4} + 3^{-2}$
c. $3^{-4} \div 3^{-2}$

30. a. $2^{-5} \cdot 2^{2}$
b. $2^{-5} + 2^{2}$
c. $2^{-5} \div 2^{2}$

31. a. $(2^{-2} \cdot 2^{-3}) \cdot 2^{-1}$
b. $(2^{-2} + 2^{-3}) \cdot 2^{-1}$
c. $(2^{-2} \cdot 2^{-3}) \div 2^{-1}$
d. $(2^{-2} + 2^{-3}) \div 2^{-1}$

32. a. $3^{-2} \cdot (3^{-1} \cdot 3^{-3})$
b. $3^{-2} \cdot (3^{-1} + 3^{-3})$
c. $3^{-2} \div (3^{-1} \cdot 3^{-3})$
d. $3^{-2} \div (3^{-1} + 3^{-3})$

Example. Simplify each expression. Leave answers in scientific notation.

a. $(3.1 \times 10^{7})(4.0 \times 10^{-11})$ **b.** $\dfrac{2.0 \times 10^{-7}}{5.0 \times 10^{-9}}$

Solution. **a.** $(3.1 \times 10^{7})(4.0 \times 10^{-11})$

$= 12.4 \times 10^{-4}$

$= (1.24 \times 10^{1}) \times 10^{-4}$

$= 1.24 \times 10^{-3}$

b. $\dfrac{2.0 \times 10^{-7}}{5.0 \times 10^{-9}}$

$= 0.4 \times 10^{2}$

$= (4.0 \times 10^{-1}) \times 10^{2}$

$= 4.0 \times 10^{1}$

In exercises **33–46**, simplify each expression. Leave answers in scientific notation.

33. $(3 \times 10^{4})(1.5 \times 10^{3})$

34. $(5 \times 10^{8})(1.2 \times 10^{-4})$

35. $(2.1 \times 10^{-7})(4 \times 10^{-1})$

36. $(7.5 \times 10^{-2})(2 \times 10^{-5})$

37. $\dfrac{7.2 \times 10^{-4}}{1.2 \times 10^{-7}}$

38. $\dfrac{8 \times 10^{5}}{2 \times 10^{9}}$

39. $\dfrac{2.5 \times 10^{-5}}{5 \times 10^{-3}}$

40. $\dfrac{1.6 \times 10^{-10}}{4 \times 10^{-6}}$

41. $(2 \times 10^{7})^{3}$

42. $(1.5 \times 10^{4})^{2}$

43. $(3 \times 10^{-4})^{4}$

44. $(2 \times 10^{-3})^{5}$

45. $\dfrac{(3.5 \times 10^{6})(1.6 \times 10^{-3})}{(7 \times 10^{-1})(4 \times 10^{5})}$

46. $\dfrac{(8.4 \times 10^{-5})(1.2 \times 10^{3})}{(2 \times 10^{-4})^{2}}$

In scientific work, the following powers of 10 are given names:

10^{-3}	is called *milli-*	10^{-3} meters = 1 millimeter (mm)
10^{-6}	is called *micro-*	10^{-6} meters = 1 micrometer (μm)
10^{-9}	is called *nano-*	10^{-9} meters = 1 nanometer (nm)
10^{-12}	is called *pico-*	10^{-12} meters = 1 picometer (pm)

In exercises **47–50**, change each measurement to the indicated unit.

47. 0.000065 m
a. mm
b. μm

48. 0.00000000051 m
a. nm
b. μm

49. 6.45×10^{-10} m
a. pm
b. nm

50. 4.081×10^{-8} m
a. μm
b. nm

SECTION 4-2. Rational Expressions, Defined and Reduced

1. A definition of a rational expression
2. Restricted values for rational expressions
3. The reduced form of a rational expression
4. Writing rational expressions with different denominators

In this section, we will define and reduce rational expressions. Also, since division by 0 is undefined, we will examine the concept of restricted values for some rational expressions. Furthermore, as we will see, a common denominator is needed to add and subtract rational expressions. Therefore, we will study the process of changing rational expressions to equivalent ones with different denominators.

A Definition of a Rational Expression

The expressions in examples **a** and **b** are polynomials in t.

a. $t^4 - 3t^3 + 6t - 2$ **b.** $t^4 - 16$

When the polynomials in **a** and **b** are written as a ratio, then a rational expression in t is formed, as shown in examples **c** and **d**.

c. $\dfrac{t^4 - 3t^3 + 6t - 2}{t^4 - 16}$

d. $\dfrac{t^4 - 16}{t^4 - 3t^3 + 6t - 2}$

(c and d) Rational expressions in t

Definition 4.2. Rational expressions

If P and Q are polynomials and $Q \neq 0$, then $\dfrac{P}{Q}$ is a **rational expression.**

Examples **e–h** illustrate some rational expressions.

e. $\dfrac{x + 3}{3x - 4}$ A rational expression in x

f. $\dfrac{x^2 - 9xy + 20y^2}{x^2 + 9y^2}$ A rational expression in x and y

g. $\dfrac{k^3 - 27}{10}$ A rational expression in k

h. $\dfrac{-5}{p^2 - pq - 2q^2}$ A rational expression in p and q

Restricted Values for Rational Expressions

In Definition 4.2, $\frac{P}{Q}$ is a rational expression provided $Q \neq 0$. Since Q is a polynomial, the variable or variables in Q must be restricted from taking on any value, or values, that will result in Q being 0. If any values do exist, then they are called **restricted values.**

Example 1. Find any restricted values for each expression.

a. $\dfrac{5t + 8}{3t - 2}$ b. $\dfrac{33}{u^2 - u - 12}$

Solution. a. **Discussion.** A restricted value is any value of t that makes the denominator equal to 0. To find any restricted values, we solve an equation formed by setting the denominator equal to 0.

$3t - 2 = 0$ Set the denominator equal to 0.

$t = \dfrac{2}{3}$ Solve for t.

For the expression $\dfrac{5t + 8}{3t - 2}$, $\dfrac{2}{3}$ is a restricted value. This can be written $\dfrac{5t + 8}{3t - 2}$ and $t \neq \dfrac{2}{3}$.

b. $u^2 - u - 12 = 0$ Set the denominator equal to 0.

$(u + 3)(u - 4) = 0$ Factor.

$u + 3 = 0$ or $u - 4 = 0$ Use the zero-product property.

$u = -3$ $u = 4$ Solve for u.

Thus, $\dfrac{33}{u^2 - u - 12}$ and $u \neq -3$ or 4.

Example 2. Find any restricted values for each expression.

a. $\dfrac{2a + b}{a - 2b}$ b. $\dfrac{x^2 - 4}{x^2 + 1}$

Solution. a. **Discussion.** The denominator is a binomial in a and b. Thus, a and b must be restricted so that $a - 2b$ cannot be 0. The number of combinations of values for a and b that must be restricted is infinite. Therefore, the restrictions are stated as a formula using the not-equal-to ($\neq$) relation.

$a - 2b = 0$ Set the denominator equal to 0.

$a = 2b$ Solve for a in terms of b.

Thus, $\dfrac{2a + b}{a - 2b}$, and $a \neq 2b$.

b. Discussion. For all values of x, x^2 is greater than or equal to 0. Therefore, there are no restrictions, because $x^2 + 1 \neq 0$ for all x.

The Reduced Form of a Rational Expression

Example **i** is a rational number and example **j** is a rational expression. If t in example **j** is replaced by 3 and the indicated operations carried out, then the rational number in example **i** is obtained.

i. $\dfrac{27}{99}$ **j.** $\dfrac{4t^2 - 9}{6t^2 + 13t + 6}$

Fractions should always be reduced. That is, since $\frac{a}{a} = 1$ (whenever $a \neq 0$), and $a \cdot 1 = a$ for any number a, a fraction such as $\frac{27}{99}$ can be reduced. The complete process is shown below:

$$\frac{27}{99} = \frac{9 \cdot 3}{9 \cdot 11} = \frac{9}{9} \cdot \frac{3}{11} = 1 \cdot \frac{3}{11} = \frac{3}{11}$$

Similarly, the rational expression in example **j** can be reduced.

j. $\dfrac{4t^2 - 9}{6t^2 + 13t + 6}$ The given expression

$= \dfrac{(2t + 3)(2t - 3)}{(2t + 3)(3t + 2)}$ Factor numerator and denominator.

$= \dfrac{2t - 3}{3t + 2}$ The common factor $(2t + 3)$ is changed to 1 and is removed from the expression.

It is important to recognize that $\dfrac{4t^2 - 9}{6t^2 + 13t + 6}$ and $\dfrac{2t - 3}{3t + 2}$ have the *same value* for all replacements of t, other than the restricted value on t. The reduced fraction has fewer symbols and operations; therefore, it would be easier to evaluate than the given fraction. As a consequence, the reduced fraction is the **preferred form.**

Example 3. Reduce each expression.

a. $\dfrac{3y^2 - 9y}{y^2 - 4y + 3}$ **b.** $\dfrac{4u^2 - 9v^2}{8u^3 - 27v^3}$

Solution. **Discussion.** In any problem involving the reduction of rational expressions, assume the variable, or variables, have been restricted so that the denominators are never 0.

a. $\dfrac{3y^2 - 9y}{y^2 - 4y + 3}$ The given expression

$= \dfrac{3y(y - 3)}{(y - 1)(y - 3)}$ Factor numerator and denominator.

$= \dfrac{3y}{y - 1}$ Remove the common factor $y - 3$.

b. $\dfrac{4u^2 - 9v^2}{8u^3 - 27v^3}$ The given expression

$= \dfrac{(2u + 3v)(2u - 3v)}{(2u - 3v)(4u^2 + 6uv + 9v^2)}$ Factor numerator and denominator.

$= \dfrac{2u + 3v}{4u^2 + 6uv + 9v^2}$ Remove the common factor $2u - 3v$.

Writing Rational Expressions with Different Denominators

Consider the indicated sum in example **k**.

k. $\dfrac{1}{t - 1} + \dfrac{t}{t + 2}$

The sum in example **k** cannot be simplified until both expressions are written with the same denominator. In Section 4-4, we will simplify sums such as the one in example **k**. To prepare for these problems, we will change some rational expressions to equivalent ones with different denominators. The procedure is the opposite to the one in which fractions are reduced. That is, instead of **removing common factors** in the numerator and denominator, we use the multiplication property of 1 to **multiply-in common factors.** The procedure is sometimes referred to as "building-up" a fraction.

Example 4. Write equivalent expressions with the indicated denominators.

a. $\dfrac{3}{2k^2}$ with denominator $12k^3$

b. $\dfrac{2u - v}{u + 2v}$ with denominator $2u^2 + uv - 6v^2$

Solution. **a. Discussion.** First divide $12k^3$ by $2k^2$ to determine what factors are missing. Then multiply numerator and denominator by the missing factors.

$\dfrac{12k^3}{2k^2} = 6k$ The missing factors are 6 and k.

$\dfrac{3}{2k^2} \cdot \dfrac{6k}{6k} = \dfrac{18k}{12k^3}$ Multiply by $6k$ over $6k$.

Thus, $\dfrac{3}{2k^2}$ and $\dfrac{18k}{12k^3}$ are equivalent expressions for all values of k, and $k \neq 0$.

b. $\dfrac{2u^2 + uv - 6v^2}{u + 2v} = \dfrac{(u + 2v)(2u - 3v)}{(u + 2v)}$ Factor the numerator.

$= (2u - 3v)$ The missing factor

$\dfrac{2u - v}{u + 2v} \cdot \dfrac{2u - 3v}{2u - 3v}$ Multiply numerator and denominator by $2u - 3v$.

$= \dfrac{4u^2 - 8uv + 3v^2}{2u^2 + uv - 6v^2}$ An equivalent expression with $(2u^2 + uv - 6v^2)$ as denominator.

SECTION 4-2. Practice Exercises

In exercises 1–20, find any restricted values for each expression. For any expression that has no restricted value, write "none".

[Examples 1 and 2]

1. $\dfrac{x + 5}{x - 3}$

2. $\dfrac{x - 7}{x + 8}$

3. $\dfrac{4y^2}{2y + 10}$

4. $\dfrac{5y^2}{3y - 9}$

5. $\dfrac{a^2 + 2a + 1}{7a}$

6. $\dfrac{a^2 - 4a + 4}{3a}$

7. $\dfrac{2z + 3}{8}$

8. $\dfrac{9z - 5}{6}$

9. $\dfrac{p - 7}{p^2 - 4p - 5}$

10. $\dfrac{p + 4}{p^2 - 7p + 6}$

11. $\dfrac{x + y}{x - y}$

12. $\dfrac{x - y}{x + y}$

13. $\dfrac{10a}{2a + b}$

14. $\dfrac{a - b}{a - 2b}$

15. $\dfrac{m + n}{m(5m - 3n)}$

16. $\dfrac{m^2 - n}{m(6m + 7n)}$

17. $\dfrac{x^2}{x^2 - xy - 2y^2}$

18. $\dfrac{16}{x^2 + 3xy - 4y^2}$

19. $\dfrac{k + 3}{k^2 + 25}$

20. $\dfrac{k^2 + 4k + 1}{49 + k^2}$

In exercises 21–44, reduce each expression. Assume the restricted values have been taken into consideration and have been eliminated as possible replacements for the variables.

[Example 3]

21. $\dfrac{24x^2y}{40xy^3}$

22. $\dfrac{60x^2y^3}{42x^3y^2}$

23. $\dfrac{4m^2 - 2m}{4m^2 - 1}$

24. $\dfrac{m^2 - 4}{2m - 4}$

25. $\dfrac{9p^2 - 4}{9p^2 - 12p + 4}$

26. $\dfrac{p^2 + 3p - 10}{5p^2 - 9p - 2}$

27. $\dfrac{x^2 - 6x + 9}{x^3 - 27}$

28. $\dfrac{8x^3 + 1}{4x^2 + 4x + 1}$

29. $\dfrac{5s - 2t}{2t - 5s}$

30. $\dfrac{7s - p}{p - 7s}$

31. $\dfrac{25 - b^2}{b - 5}$

32. $\dfrac{6 - 7b}{49b^2 - 36}$

33. $\dfrac{2m^2 + 9m - 35}{2m^2 + 19m + 35}$

34. $\dfrac{9m - 6m^2}{6m^2 - 5m - 6}$

35. $\dfrac{p^2 - 2p^3}{2p^3 + 2p - 1 - p^2}$

36. $\dfrac{pq - p + 3q - 3}{pq - 3q - p + 3}$

37. $\dfrac{a^3 - 5a^2 + a - 5}{a^2 - 4a - 5}$

38. $\dfrac{2a^3 - 3a^2}{2a^3 - 3a^2 + 2a - 3}$

39. $\dfrac{2ab^2 - 3a^2b}{9a^3b - 4ab^3}$

40. $\dfrac{75a^2b^2 - 12a^4}{6a^3 - 15a^2b}$

41. $\dfrac{27x^3 - 125}{9x^2 + 15x + 25}$

42. $\dfrac{50x^2 - 20x + 8}{125x^3 + 8}$

43. $\dfrac{p^2 - 4p + 4 - q^2}{p - q - 2}$

44. $\dfrac{3 - 2p + q}{9 - 4p^2 + 4pq - q^2}$

In exercises **45–60**, write equivalent expressions with the indicated denominators. Assume that no denominator is equal to 0.

[Example 4] **45.** $\dfrac{5y}{2x}$ with denominator $6xy$

46. $\dfrac{x}{3y}$ with denominator $12xy$

47. $\dfrac{3}{a - 1}$ with denominator $4a - 4$

48. $\dfrac{a}{a + 1}$ with denominator $10a + 10$

49. $\dfrac{2b}{3b + 1}$ with denominator $3b^2 + b$

50. $\dfrac{1}{2b - 5}$ with denominator $4b^2 - 10b$

51. $\dfrac{2}{2m + n}$ with denominator $10m^2n + 5mn^2$

52. $\dfrac{m}{m - 3n}$ with denominator $5m^2n - 15mn^2$

53. $\dfrac{1}{x - 2}$ with denominator $x^2 - 4$

54. $\dfrac{x}{3x + 1}$ with denominator $9x^2 - 1$

55. $\dfrac{5}{y + 3}$ with denominator $y^2 + 5y + 6$

56. $\dfrac{4}{2y - 5}$ with denominator $2y^2 - y - 10$

57. $\dfrac{2}{a + 2b}$ with denominator $a^2 - 3ab - 10b^2$

58. $\dfrac{a + b}{2a - b}$ with denominator $4a^2 - b^2$

59. $\dfrac{3}{x^2 + 2x + 4}$ with denominator $x^3 - 8$

60. $\dfrac{5}{3x + 1}$ with denominator $27x^3 + 1$

SECTION 4-2. Ten Review Exercises

1. Multiply $(3t^2)(7t)$.

2. Evaluate the product obtained in exercise **1** for $t = 2$.

3. Multiply $(3 + t^2)(7 + t)$.

4. Evaluate the product obtained in exercise **3** for $t = 2$ and compare with the answer in exercise 2.

5. Divide $\dfrac{36k^2}{6k}$.

6. Evaluate the quotient obtained in exercise **5** for $k = 5$.

7. Divide $\dfrac{36 - k^2}{6 - k}$.

8. Evaluate the quotient obtained in exercise **7** for $k = 5$ and compare with the answer in exercise **6**.

In exercises **9** and **10**, simplify each expression.

9. $(5uv)^2$

10. $(5 + uv)^2$

SECTION 4-2. Supplementary Exercises

Common *factors* can be removed to reduce a rational expression. Common *terms* cannot.

In exercises **1–8**:

(i) Evaluate expressions **a–d** for $x = 2$.

(ii) Compare the evaluations. Identify the expressions with the same value.

(iii) Reduce the expression in part **a**. State whether **b**, **c**, or **d** is the reduced form of **a**.

1. a. $\frac{x^2 + 7x}{x}$
   b. $x^2 + 7$
   c. $x + 7x$
   d. $x + 7$

2. a. $\frac{x^2 + 5x + 6}{x + 2}$
   b. $x + 5x + 3$
   c. $x^2 + 5 + 3$
   d. $x + 3$

3. a. $\frac{x^2 - 25}{x - 5}$
   b. $x - 5$
   c. $x + 5$
   d. $-5x$

4. a. $\frac{3x^2 - 27}{3}$
   b. $x^2 - 27$
   c. $3x^2 - 9$
   d. $x^2 - 9$

5. a. $\frac{x^3 + 1}{x + 1}$
   b. $x^2 - x + 1$
   c. x^2
   d. $x^2 + 1$

6. a. $\frac{x^2 + 6x + 8}{x^2 - 4x - 12}$
   b. $\frac{6x + 8}{-4x - 12}$
   c. $\frac{x + 4}{x - 6}$
   d. $2x - 4$

7. a. $\frac{2(x - 5)}{5 - x}$
   b. $-10x$
   c. -2
   d. x

8. a. $\frac{4x^2}{8x^3 - 72}$
   b. $\frac{1}{2x - 72}$
   c. $\frac{1}{8x - 18}$
   d. $\frac{x^2}{2(x^3 - 9)}$

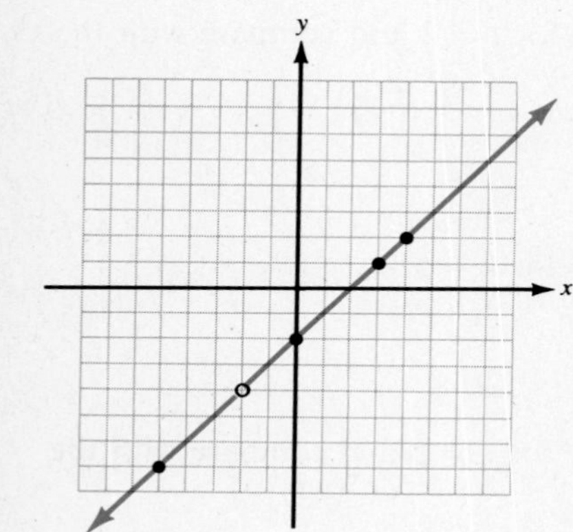

Figure 4-1. A graph of $y = \frac{x^2 - 4}{x + 2}$.

9. Figure 4-1 is a graph of $y = \frac{x^2 - 4}{x + 2}$.
   a. Use the graph to find y for $x = 3$ and $x = -5$.
   b. Use the rational expression in x to find y for $x = 3$ and $x = -5$.
   c. Use the graph to find x for $y = 2$ and $y = -2$.
   d. Reduce the rational expression in x. Use the reduced form to find y for $x = 3$ and $x = -5$.
   e. Is there a value for y if $x = -2$?
   f. Is there a value for x if $y = -4$?
   g. Are there any restricted values for the rational expression in x?
   h. Are there any restricted values for the reduced form found in part **d**?
   i. Use the results of parts **g** and **h** to explain the need for the "hole" in the graph at $(-2, -4)$.

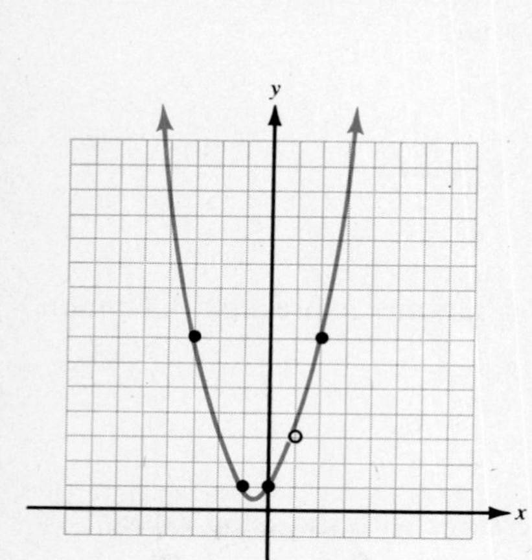

Figure 4-2. A graph of $y = \frac{x^3 - 1}{x - 1}$.

10. Figure 4-2 is a graph of $y = \frac{x^3 - 1}{x - 1}$.
   a. Use the graph to find y for $x = 2$ and $x = -1$.
   b. Use the rational expression in x to find y for $x = 2$ and $x = -1$.
   c. Use the graph to find x for $y = 7$ and $y = 1$. How many values of x are there for each value of y?
   d. Reduce the rational expression in x. Use the reduced form to find y for $x = 2$ and $x = -1$.
   e. Are there any values of y for $x = 1$?
   f. Are there any values of x for $y = 3$?
   g. Are there any restricted values for the rational expression in x?
   h. Are there any restricted values for the reduced form found in part **d**?

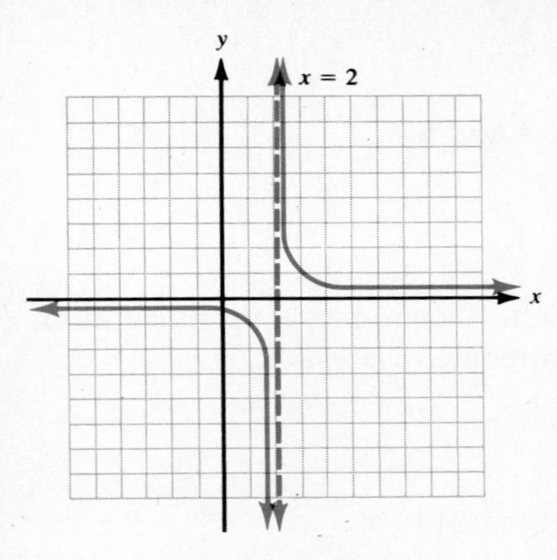

Figure 4-3. A graph of $y = \frac{1}{x-2}$.

i. Use the results of parts **g** and **h** to explain the need for the "hole" in the graph at (1, 3).

11. Figure 4-3 is a graph of $y = \dfrac{1}{x-2}$.

a. Find the restricted value for $\dfrac{1}{x-2}$.

b. What is drawn on the graph for the restricted value of part **a**?

c. Based on the graph, what happens to values of y as x takes on values "close" to 2 but always less than 2?

d. Based on the graph, what happens to values of y as x takes on values "close" to 2 but always greater than 2?

e. Based on the graph, can y ever take on the value of 0? Is this observation consistent with the expression $\dfrac{1}{x-2}$?

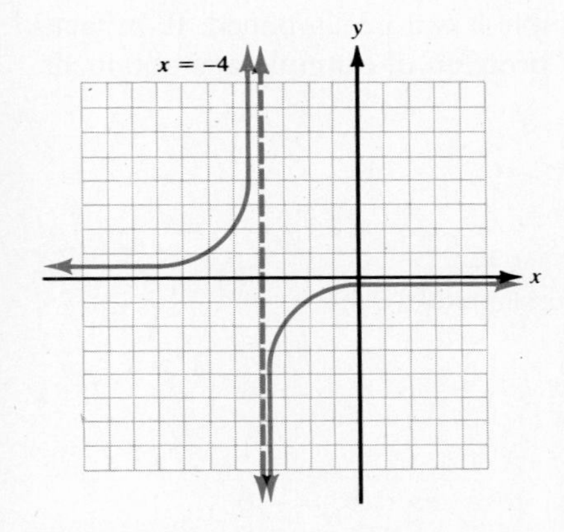

Figure 4-4. A graph of $y = \frac{-2}{x+4}$.

12. Figure 4-4 is a graph of $y = \dfrac{-2}{x+4}$.

a. Find the restricted value for $\dfrac{-2}{x+4}$.

b. What is drawn on the graph for the restricted value of part **a**?

c. Based on the graph, what happens to the values of y as x takes on values "close" to -4 but always greater than -4?

d. Based on the graph, what happens to the values of y as x takes on values "close" to -4 but always greater than -4?

e. Based on the graph, can y ever take on the value of 0? Is this observation consistent with the expression $\dfrac{-2}{x+4}$?

In exercises **13–18**, reduce each expression. Assume that a and b are integers and no denominator is 0.

13. $\dfrac{x^{2a} + 2x^a + 1}{x^a + 1}$

14. $\dfrac{x^{2a} - y^{2b}}{x^a - y^b}$

15. $\dfrac{x^{3a} - 1}{x^a - 1}$

16. $\dfrac{x^{a+b} + x^b y^b - x^a - y^b}{x^a + y^b}$

17. $\dfrac{y^{2a} - y^a - 6}{y^{2a} - 6y^a + 9}$

18. $\dfrac{x^{2a} - 5x^a}{x^{2a} - 25}$

SECTION 4-3. Multiplication and Division of Rational Expressions

1. Multiplying rational expressions
2. Dividing rational expressions
3. Combined operations

In this section, we will multiply and divide rational expressions. As we will see, the procedures for multiplying and dividing rational expressions are similar to the ones used to multiply and divide rational numbers.

Multiplying Rational Expressions

Consider the indicated products in examples **a** and **b**.

a. $\frac{10}{21} \cdot \frac{28}{25}$ **b.** $\frac{2t}{t^2 - 4} \cdot \frac{t^2 + t - 2}{t^2}$

To simplify the product in example **a**, we can write the product of the numerators over the product of the denominators and then reduce, if possible.

Divide 10 and 25 by 5.

a. $\frac{10}{21} \cdot \frac{28}{25} = \frac{10 \cdot 28}{21 \cdot 25} = \frac{2 \cdot 4}{3 \cdot 5} = \frac{8}{15}$ The simplified product

Divide 21 and 28 by 7.

In much the same way, the product in example **b** can be simplified. If, in fact, we replace t by 5 in this example, the number problem of example **a** is obtained.

b. $\frac{2t}{t^2 - 4} \cdot \frac{t^2 + t - 2}{t^2}$ The indicated product.

$= \frac{2t(t^2 + t - 2)}{(t^2 - 4)t^2}$ $\frac{\text{Multiply the numerators}}{\text{Multiply the denominators}}$

$= \frac{2t(t + 2)(t - 1)}{t^2(t + 2)(t - 2)}$ $t^2 + t - 2 = (t + 2)(t - 1)$, and $t^2 - 4 = (t + 2)(t - 2)$

$= \frac{2(t - 1)}{t(t - 2)}$ Remove the common factors t and $(t + 2)$.

The following is a general summary of the steps taken to simplify the product in example **b**:

> If $\frac{P}{Q}$ and $\frac{R}{S}$ are rational expressions, where $Q \neq 0$ and $S \neq 0$, then to simplify $\frac{P}{Q} \cdot \frac{R}{S}$:
>
> **Step 1.** Write $\frac{P \cdot R}{Q \cdot S}$. $\frac{\text{Multiply the numerators}}{\text{Multiply the denominators}}$
>
> **Step 2.** If possible, factor P, R, Q, and S.
>
> **Step 3.** Remove any factors common to the numerator and denominator. Unless otherwise instructed, leave the simplified expression in factored form.

To simplify the study in this section, the variables in all rational expressions have been restricted so that no denominator can be 0.

Example 1. Multiply $\frac{24a^3}{35b^2} \cdot \frac{49b^3}{60a^5}$.

Solution. **Discussion.** The expressions are all monomials and do not require factoring.

$$\frac{24a^3}{35b^2} \cdot \frac{49b^3}{60a^5} \qquad \frac{P}{Q} \cdot \frac{R}{S}$$

$$= \frac{24a^3 \cdot 49b^3}{35b^2 \cdot 60a^5} \qquad = \frac{P \cdot R}{Q \cdot S}$$

$$= \frac{2 \cdot 7b}{5 \cdot 5a^2}$$ Remove the common factors of 12, 7, b^2, and a^3.

$$= \frac{14b}{25a^2}$$ Simplify the indicated products of $2 \cdot 7$ and $5 \cdot 5$.

Example 2. Multiply $\dfrac{y^2 + 2y - 15}{4y^2 - 2y} \cdot \dfrac{1 - 2y}{5 + y}$.

Solution.

$$\frac{y^2 + 2y - 15}{4y^2 - 2y} \cdot \frac{1 - 2y}{5 + y}$$

$$= \frac{(y^2 + 2y - 15)(1 - 2y)}{(4y^2 - 2y)(5 + y)}$$

$$= \frac{(y + 5)(y - 3)(1 - 2y)}{2y(2y - 1)(y + 5)}$$

$$= \frac{(y - 3)(-1)(2y - 1)}{2y(2y - 1)}$$ Remove the $(y + 5)$ factor. Write $(1 - 2y)$ as $-1 \cdot (2y - 1)$.

$$= \frac{-1(y - 3)}{2y} \text{ or } \frac{3 - y}{2y}$$ Now remove the $(2y - 1)$ common factor.

Dividing Rational Expressions

Consider the indicated quotients in example **c** and **d**.

c. $\dfrac{999}{16} \div \dfrac{111}{2}$ **d.** $\dfrac{8u^3 - 1}{u^2 - 9} \div \dfrac{4u^2 + 2u + 1}{u - 3}$

The quotient in **c** can be simplified by first writing the division as a multiplication. The products of the numerators and denominators are then formed and the fraction is reduced.

Divide 999 and 111 by 111.

c. $$\frac{999}{16} \div \frac{111}{2} = \frac{999}{16} \cdot \frac{2}{111} = \frac{999 \cdot 2}{16 \cdot 111} = \frac{9 \cdot 1}{8 \cdot 1} = \frac{9}{8}$$

Divide 16 and 2 by 2.

Similarly, the quotient in example **d** can be simplified. Notice that the rational expressions in **d** become the fractions in **c** when u is replaced by 5.

d. $\dfrac{8u^3 - 1}{u^2 - 9} \div \dfrac{4u^2 + 2u + 1}{u - 3}$ The indicated quotient

$= \dfrac{8u^3 - 1}{u^2 - 9} \cdot \dfrac{u - 3}{4u^2 + 2u + 1}$ Multiply by the reciprocal of the divisor.

$= \dfrac{(8u^3 - 1)(u - 3)}{(u^2 - 9)(4u^2 + 2u + 1)}$ $\dfrac{\text{Multiply the numerators}}{\text{Multiply the denominators}}$

$= \dfrac{(2u - 1)(4u^2 + 2u + 1)(u - 3)}{(u + 3)(u - 3)(4u^2 + 2u + 1)}$ $8u^3 - 1 = (2u - 1)(4u^2 + 2u + 1)$; $u^2 - 9 = (u + 3)(u - 3)$

$= \dfrac{2u - 1}{u + 3}$ Remove the common factors of $(4u^2 + 2u + 1)$ and $(u - 3)$.

The following is a general summary of the steps taken to simplify the quotient in example **d**:

> If $\dfrac{P}{Q}$ and $\dfrac{R}{S}$ are rational expressions and $Q \neq 0$, $R \neq 0$, and $S \neq 0$, then to simplify $\dfrac{P}{Q} \div \dfrac{R}{S}$:
>
> **Step 1.** Write $\dfrac{P}{Q} \cdot \dfrac{S}{R}$. $\dfrac{S}{R}$ is the reciprocal of $\dfrac{R}{S}$
>
> **Step 2.** Now write $\dfrac{P \cdot S}{Q \cdot R}$. $\dfrac{\text{Multiply the numerators}}{\text{Multiply the denominators}}$
>
> **Step 3.** If possible, factor P, R, Q, and S.
>
> **Step 4.** Remove any factors common to the numerator and denominator.

Example 3. Find the indicated quotients.

a. $\dfrac{-40p^3}{57q^2} \div \dfrac{10p^3}{19q}$

b. $\dfrac{x^3 - 8}{6x^2 + 15x} \div \dfrac{x^2 - x - 2}{2x + 5}$

Solution. **a. Discussion.** The expressions are all monomials and do not require factoring.

$\dfrac{-40p^3}{57q^2} \div \dfrac{10p^3}{19q}$ $\dfrac{P}{Q} \div \dfrac{R}{S}$

$= \dfrac{-40p^3}{57q^2} \cdot \dfrac{19q}{10p^3}$ $= \dfrac{P}{Q} \cdot \dfrac{S}{R}$

$= \dfrac{-40p^3 \cdot 19q}{57q^2 \cdot 10p^3}$ $= \dfrac{P \cdot S}{Q \cdot R}$

$= \dfrac{-4}{3q}$ Remove the common factors 10, 19, p^3, and q.

b. $\dfrac{x^3 - 8}{6x^2 + 15x} \div \dfrac{x^2 - x - 2}{2x + 5}$

$$= \frac{x^3 - 8}{6x^2 + 15x} \cdot \frac{2x + 5}{x^2 - x - 2}$$

$$= \frac{(x^3 - 8)(2x + 5)}{(6x^2 + 15x)(x^2 - x - 2)}$$

$$= \frac{(x - 2)(x^2 + 2x + 4)(2x + 5)}{3x(2x + 5)(x + 1)(x - 2)}$$

$$= \frac{x^2 + 2x + 4}{3x(x + 1)}$$

Combined Operations

If two or more operations are indicated in the same expression, then the rule for order of operations must be observed when simplifying the expression.

Example 4. Simplify $\left(\dfrac{2k^2 - 3k + 1}{3k^2 + 2k - 5}\right)\left(\dfrac{2k^3 - k^2 + 2k - 1}{4k^2 - 1} \div \dfrac{k^2 + 1}{3k + 5}\right)$.

Solution. **Discussion.** The division is grouped by parentheses. Therefore, the quotient is computed first, and then multiplied by the first rational expression.

$$\left(\frac{2k^2 - 3k + 1}{3k^2 + 2k - 5}\right)\left(\frac{2k^3 - k^2 + 2k - 1}{4k^2 - 1} \cdot \frac{3k + 5}{k^2 + 1}\right)$$

$$= \left(\frac{2k^2 - 3k + 1}{3k^2 + 2k - 5}\right)\left(\frac{(k^2 + 1)(2k - 1)(3k + 5)}{(2k + 1)(2k - 1)(k^2 + 1)}\right)$$

$$= \left(\frac{2k^2 - 3k + 1}{3k^2 + 2k - 5}\right)\left(\frac{3k + 5}{2k + 1}\right)$$ Remove the common factors $(k^2 + 1)$ and $(2k - 1)$.

$$= \frac{(2k - 1)(k - 1)(3k + 5)}{(3k + 5)(k - 1)(2k + 1)}$$ $2k^2 - 3k + 1 = (2k - 1)(k - 1)$; $3k^2 + 2k - 5 = (3k + 5)(k - 1)$

$$= \frac{2k - 1}{2k + 1}$$ Remove the common factors $(k - 1)$ and $(3k + 5)$.

Consider again the three rational expressions in Example 4:

$$\left(\frac{2k^2 - 3k + 1}{3k^2 + 2k - 5}\right)\left(\frac{2k^3 - k^2 + 2k - 1}{4k^2 - 1} \div \frac{k^2 + 1}{3k + 5}\right)$$

Suppose we needed to evaluate this expression for $k = 5$. Instead of replacing k by 5 in this expression, we would simply evaluate the answer to the problem, namely $\dfrac{2k - 1}{2k + 1}$. Replacing k by 5:

$$\frac{2k - 1}{2k + 1} \text{ becomes } \frac{2(5) - 1}{2(5) + 1} = \frac{10 - 1}{10 + 1} = \frac{9}{11}$$

The value of the original expression would also simplify to $\frac{9}{11}$ when k is replaced by 5. That is, *in carrying out the indicated operations and reducing the fraction, we change the form of the expression, but not the values it represents for specified values of* k.

SECTION 4-3. Practice Exercises

In exercises **1–52**, do the indicated operations.

[Example 1]

1. $\frac{6a}{b^3} \cdot \frac{5b^2}{12a}$

2. $\frac{10a}{3b} \cdot \frac{9b}{5a^2}$

3. $\frac{uv^2}{3w^3} \cdot \frac{6w^4}{u^2v^3}$

4. $\frac{7u}{vw^2} \cdot \frac{v^2w^5}{14u}$

[Example 2]

5. $\frac{6}{k^2 - k} \cdot \frac{k^2 - 1}{30}$

6. $\frac{2k + 3}{5k} \cdot \frac{35k^3}{2k^2 + k - 3}$

7. $\frac{2r - 3s}{5r} \cdot \frac{10r^2}{4r^2 - 12rs + 9s^2}$

8. $\frac{9r^2 + 6rs + s^2}{6r^2s^2} \cdot \frac{3rs}{3r + s}$

9. $\frac{1}{x^2 - xy - 6y^2} \cdot (x^2 - 6xy + 9y^2)$

10. $(x^3 - 125y^3) \cdot \frac{1}{x^2 - 10xy + 25y^2}$

11. $\frac{39}{27a^3 - b^3} \cdot \frac{9a^2 + 3ab + b^2}{13}$

12. $\frac{15ab}{25a^2 - 16b^2} \cdot \frac{125a^3 - 64b^3}{60ab}$

13. $\frac{n - m}{7mn} \cdot \frac{35m^2n^2}{m^2 - 2mn + n^2}$

14. $\frac{65m^3n^2}{2m^2 + 7mn - 15n^2} \cdot \frac{3n - 2m}{13mn}$

15. $\frac{u^2 + uv - 2v^2}{3u^3v^2 - u^2v^3} \cdot \frac{u^2v^2}{u - v}$

16. $\frac{3u - v}{5u^2v^2} \cdot \frac{5u^3v^2 - 15u^2v^3}{6u^2 + uv - v^2}$

17. $\frac{4y^2 - x^2}{3x^2 + 6xy} \cdot \frac{6xy}{2xy - 4y^2}$

18. $\frac{-25y^2 + 20xy - 4x^2}{4x^2y - 10xy^2} \cdot \frac{20x^3y^2}{4x^3y - 10x^2y^2}$

19. $\frac{3r^2 + rs}{3r^2 - 5rs - 2s^2} \cdot \frac{r^2 - 4s^2}{rs + 2s^2}$

20. $\frac{rs + 3s^2}{2r^2 + 7rs + 3s^2} \cdot \frac{4r^2 - s^2}{4r^2 - 2rs}$

21. $\frac{9x^2 + 6x + 1 - y^2}{3x^3 - y - x^2y + 3x} \cdot \frac{x^2 + 1}{3x + y + 1}$

22. $\frac{x^3 + y^3 + x^2y + xy^2}{9 - x^2 - 2xy - y^2} \cdot \frac{x + y + 3}{x^2 + y^2}$

[Example 3]

23. $\frac{42r}{s^2t^3} \div \frac{6r^2}{s^3t^2}$

24. $\frac{st}{150r} \div \frac{s^3t^3}{90r^3}$

25. $\frac{u^3v^3}{12w^2} \div \frac{u^3v}{48w}$

26. $\frac{u^3v^3}{10w} \div \frac{u^4v^2}{60w^2}$

27. $\frac{6x^2y^2}{x^2 - y^2} \div \frac{3xy}{x + y}$

28. $\frac{x^3 - 8y^3}{8x^3y^3} \div \frac{x - 2y}{4xy}$

29. $\dfrac{a^3 + 27b^3}{3ab} \div \dfrac{a^2 - 3ab + 9b^2}{9a^2b^2}$

30. $\dfrac{3a^2 + 2ab - 9a - 6b}{5a^2b^2} \div \dfrac{3a + 2b}{15a^3b^3}$

31. $\dfrac{2k^2 + k - 15}{5k} \div (5 - 2k)$

32. $\dfrac{2k^4 + 7k^2 - 4}{2k} \div (k^2 + 4)$

33. $\dfrac{r^2 - 6rs + 9s^2}{49r^2s^2} \div \dfrac{r - 3s}{7rs}$

34. $\dfrac{100r^2 + 20rs + s^2}{15r^5s^4} \div \dfrac{10r + s}{3r^3s^3}$

35. $\dfrac{x^2 - 25}{10 + 13x - 3x^2} \div \dfrac{2x^2 + 7x - 15}{6x^2 - 5x - 6}$

36. $\dfrac{x^3 - 8}{4 - x^2} \div \dfrac{x^2 + 2x + 4}{x + 2}$

37. $\dfrac{y^3 + 3y^2 + y + 3}{2y^2} \div \dfrac{3y^2 + 3}{2y}$

38. $\dfrac{y + 3}{y^2 + 5} \div \dfrac{2y^2 + 6y}{3y^3 + 15y - y^2 - 5}$

39. $\dfrac{2z^3 + 12z - z^2 - 6}{6z^2 - 3z} \div \dfrac{z^4 - 36}{6z^2}$

40. $\dfrac{9z^4 - 1}{10z} \div \dfrac{3z^3 + z - 21z^2 - 7}{5z^2 - 35z}$

41. $\dfrac{m^2 - 4mn + 4n^2}{6m^2n - 12mn^2} \div \dfrac{m^2 - 4n^2}{3m^2n + 6mn^2}$

42. $\dfrac{2m^3n^2 + 3m^2n^3}{4m^2 + 12mn + 9n^2} \div \dfrac{6m^2n - 9mn^2}{4m^2 - 9n^2}$

43. $\dfrac{a^2 - b^2 + 2b - 1}{3a^2 + b + 3ab + a} \div \dfrac{a - b + 1}{3a + 1}$

44. $\dfrac{3a^3 + 8b + 12a + 2a^2b}{100 - a^2 + 2ab - b^2} \div \dfrac{a^2 + 4}{10 - a + b}$

[Example 4] **45.** $\left(\dfrac{u^2v - uv^2}{u^2 + 2uv + v^2} \div \dfrac{2u + v}{3uv}\right)\left(\dfrac{2u^2 + 3uv + v^2}{6u^2v - 6uv^2}\right)$

46. $\left(\dfrac{u^2 - 2uv + v^2}{3u - v} \div \dfrac{2u^3v^3}{5u + 25v}\right)\left(\dfrac{6u^3v^2 - 2u^2v^3}{10u^2 + 40uv - 50v^2}\right)$

47. $\left(\dfrac{k^2 - k - 2}{k^2 + 3k + 2} \cdot \dfrac{k^2 + 4k + 4}{k^2 + k - 2}\right) \div \left(\dfrac{k^2 + k - 6}{k^2 - k^3}\right)$

48. $\left(\dfrac{k^2 + 2k + 1}{k^2 + 4k} \cdot \dfrac{k^2 - k^3}{k^2 + 2k - 3}\right) \div \left(\dfrac{k^2 - 3k - 4}{k^2 - 16}\right)$

49. $\left[\left(\dfrac{x^2}{x^2 - 25} \cdot \dfrac{x^3 + 125}{3x^2}\right) \div \dfrac{x^2 - 5x + 25}{6x + 3}\right]\left[\dfrac{2x^2 - 10x}{4x^2 + 4x + 1}\right]$

50. $\left[\left(\dfrac{8x^3 - 1}{4x^3 + 2x^2} \cdot \dfrac{5x - 15}{6x^2 - 13x + 5}\right) \div \dfrac{15x^2}{2x^4 + x^3}\right]\left[\dfrac{6x - 10}{4x^2 + 2x + 1}\right]$

51. $\left(\dfrac{t^2 - 9}{5} \div \dfrac{3t^3 + 2t - 9t^2 - 6}{t}\right) \div \left(\dfrac{t^3 + 3t^2}{15t^2 + 10t}\right)$

52. $\left(\dfrac{25t^2 - 5t + 1}{10t} \div \dfrac{t^2 - 1}{10t^2 - 10t}\right) \div \left(\dfrac{125t^3 + 1}{t^3 - 5t + t^2 - 5}\right)$

SECTION 4-3. Ten Review Exercises

1. Simplify $(9 - 2x) - (3 - 5x)$.

2. Multiply $(9 - 2x)(3 - 5x)$.

3. Factor $10x^2 - 51x + 27$.

4. Divide $(10x^2 - 51x + 27) \div (3 - 5x)$.

5. Reduce $\dfrac{27 - 51x + 10x^2}{9 - 2x}$

In exercises **6–10**, simplify each expression. Assume all variables are not 0.

6. $\dfrac{(-ab^2)^3}{(2a^2b^3)^2}$

7. $\dfrac{(-a^{-1}b^2)^{-3}}{(2a^2b^{-3})^2}$

8. $(-3xy^2)^3(x^2y^2)$

9. $(-3x^{-1}y^2)^{-3}(x^2y^2)$

10. $\left(\dfrac{-3k^{-1}}{5}\right)^2\left(\dfrac{-10k}{9}\right)^2$

SECTION 4-3. Supplementary Exercises

In exercises **1–8**:

a. Evaluate the given expressions for the indicated value of x and find the indicated product or quotient.

b. Simplify the indicated product or quotient.

c. Evaluate the simplified product or quotient of part **b** and compare with the answer to part **a**.

1. $\dfrac{2x^2}{3} \cdot \dfrac{9}{4x}$; $x = 7$

2. $\dfrac{15}{2x^2} \cdot \dfrac{x^3}{5}$; $x = -3$

3. $\dfrac{x^2 + x - 2}{3x^3 - x^2} \cdot \dfrac{x^2}{x - 1}$; $x = 10$

4. $\dfrac{x^2}{x - 3} \cdot \dfrac{x^2 - 7x + 12}{x^2 - 4x}$; $x = -2$

5. $\dfrac{6x^2}{x^2 - 1} \div \dfrac{3x}{x + 1}$; $x = 6$

6. $\dfrac{5x + 10}{10} \div \dfrac{x^2 + 3x + 2}{2x^2 - 6x - 8}$; $x = 0$

7. $\dfrac{x^2 - 4x + 4}{6x^2 - 12x} \div \dfrac{x^2 - 4}{3x^2 + 6x}$; $x = -5$

8. $\dfrac{x + 3}{x - 2} \div \dfrac{9 - x^2}{2 - x}$; $x = 4$

In exercises **9–14**, find the value of k in each expression on the left that reduces it to the form of the expression on the right.

9. $\dfrac{21x^2}{20y^k} \cdot \dfrac{10y^3}{42x^3}$ becomes $\dfrac{1}{4x}$

10. $\dfrac{60x^2}{99y^2z^k} \div \dfrac{80yz}{55x}$ becomes $\dfrac{5x^3}{12y^3z^2}$

11. $\dfrac{x - 3}{x - 4} \cdot \dfrac{x^2 - k}{x^2 - 9}$ becomes $\dfrac{x + 4}{x + 3}$

12. $\dfrac{4x + 20}{x^2 + kx + 16} \cdot \dfrac{x^2 + 4x}{x^2 + 7x + 10}$ becomes $\dfrac{4x}{(x + 4)(x + 2)}$

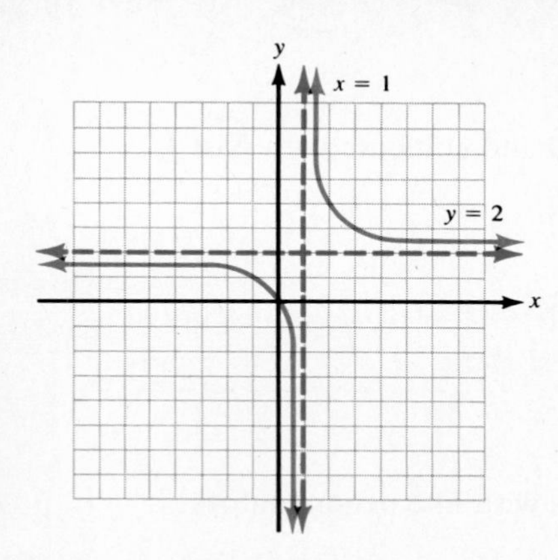

Figure 4-5. A graph of $y_1 = \frac{2x}{x-1}$.

13. $\dfrac{y^2 - 3y + 2}{5y^2 - 26y + 5} \div \dfrac{y^2 - y - 2}{5y^2 + ky - 5}$ becomes $\dfrac{(y-1)(y+5)}{(y-5)(y+1)}$

14. $\dfrac{x^3 - 8}{x + 5} \div \dfrac{x^2 + kx + 4}{x^2 - 25}$ becomes $(x-2)(x-5)$

In exercises **15–20**, do the indicated operations. Write answers in reduced form.

15. $\dfrac{a-b}{c-d} \cdot \dfrac{d-c}{a-b}$

16. $\dfrac{a-b}{c-d} \cdot \dfrac{d-c}{b-a}$

17. $\dfrac{x^2 - 49}{x^2 - 8x - 20} \cdot \dfrac{x+2}{7-x}$

18. $\dfrac{5x}{60 - 5x - 5x^2} \div \dfrac{2x^2}{2x^3 - 14x^2 + 24x}$

19. $\dfrac{x-1}{x-2} \cdot \dfrac{2-x}{x-3} \cdot \dfrac{3-x}{1-x}$

20. $\dfrac{ax - bx}{ax - ay} \cdot \dfrac{ax + ay}{ab - b^2} \div \dfrac{xy + y^2}{by - bx}$

21. Figure 4-5 is a graph of $y_1 = \dfrac{2x}{x-1}$ and Figure 4-6 is a graph of $y_2 = \dfrac{x+2}{x}$.

a. Use the graphs to find values for y_1 and y_2 if $x = 2$.
b. Use the values from part **a** to compute $y_1 \cdot y_2$.
c. Use the graphs to find values for y_1 and y_2 if $x = -1$.
d. Use the values from part **c** to compute $y_1 \cdot y_2$.
e. Multiply $\dfrac{2x}{x-1} \cdot \dfrac{x+2}{x}$.
f. Evaluate the product of part **e** for $x = 2$ and compare with part **b**.
g. Evaluate the product of part **e** for $x = -1$ and compare with part **d**.

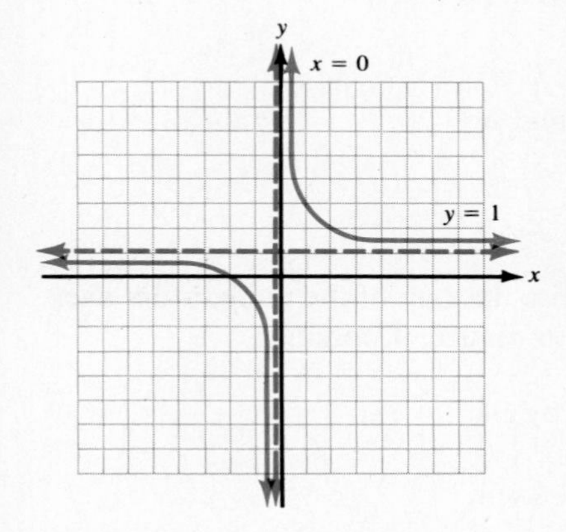

Figure 4-6. A graph of $y_2 = \frac{x+2}{x}$.

22. Figure 4-7 is a graph of $y_1 = \dfrac{x-1}{x+1}$ and Figure 4-8 is a graph of $y_2 = \dfrac{3x}{x^2 - 1}$.

a. Use the graphs to find values for y_1 and y_2 if $x = -2$.
b. Use the values from part **a** to compute $y_1 \div y_2$.
c. Divide $\dfrac{x-1}{x+1} \div \dfrac{3x}{x^2 - 1}$.
d. Evaluate the quotient of part **c** for $x = -2$ and compare with part **b**.
e. Use the graphs to find values for y_1 and y_2 if $x = 0$.
f. Use the values from **e** to compute $y_1 \cdot y_2$.

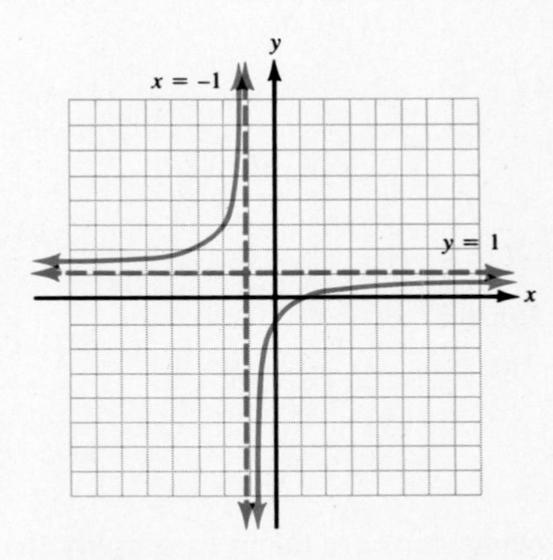

Figure 4-7. A graph of $y_1 = \frac{x-1}{x+1}$.

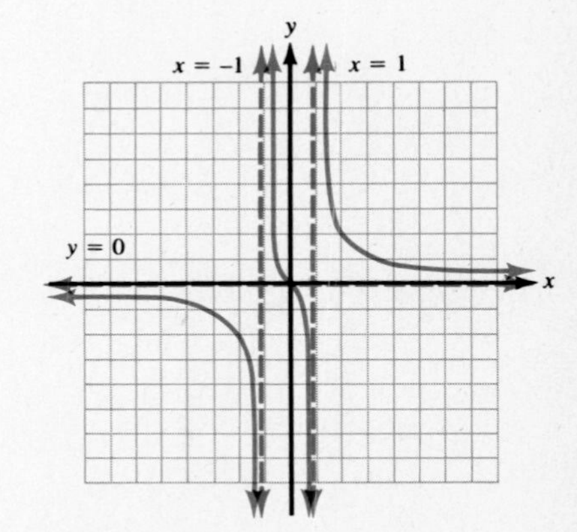

Figure 4-8. A graph of $y_2 = \frac{3x}{x^2 - 1}$.

g. Multiply $\dfrac{x-1}{x+1} \cdot \dfrac{3x}{x^2-1}$.

h. Evaluate the product of part **g** for $x = 0$ and compare with part **f**.

SECTION 4-4. Addition and Subtraction of Rational Expressions

1. Adding and subtracting rational expressions with like denominators
2. Finding a least common denominator (LCD)
3. Adding and subtracting rational expressions with unlike denominators

In this section, we will add and subtract rational expressions. As in Section 4-3, we will add and subtract rational numbers to develop a procedure for adding and subtracting rational expressions.

Adding and Subtracting Rational Expressions with Like Denominators

Consider the indicated sums in examples **a** and **b**.

a. $\dfrac{5}{98} + \dfrac{9}{98}$ **b.** $\dfrac{t^2-4t}{3t^2+5t-2} + \dfrac{3(t-2)}{3t^2+5t-2}$

To simplify the sum in example **a**, we can write the sum of the numerators over the common denominator (namely 98) and then reduce, if possible.

Divide 14 and 98 by 14.

a. $\dfrac{5}{98} + \dfrac{9}{98} = \dfrac{5+9}{98} = \dfrac{14}{98} = \dfrac{1}{7}$ The simplified sum

The sum in example **b** can be simplified in a similar way. If, in fact, we replace t by 5, the number problem of example **a** is obtained.

b. $\dfrac{t^2-4t}{3t^2+5t-2} + \dfrac{3(t-2)}{3t^2+5t-2}$ The given expression

$= \dfrac{t^2-4t+3(t-2)}{3t^2+5t-2}$ Combine the numerators over the common denominator.

$= \dfrac{t^2-t-6}{3t^2+5t-2}$ Simplify the numerator.

$= \dfrac{(t-3)(t+2)}{(3t-1)(t+2)}$ Factor numerator and denominator.

$= \dfrac{t-3}{3t-1}$ Remove the common factor $t+2$.

The following steps are taken to simplify the sum in example **b**.

If $\frac{P}{Q}$ and $\frac{R}{Q}$ are rational expressions and $Q \neq 0$, then to simplify $\frac{P}{Q} + \frac{R}{Q}$ or $\frac{P}{Q} - \frac{R}{Q}$:

Step 1. Write the indicated sum or difference of the numerators over the common denominator. $\frac{\text{Add (or subtract) the numerators}}{\text{The common denominator}}$

Step 2. Simplify the numerator.

Step 3. If possible, factor and reduce the fraction.

As in Section 4-3, the variables in all rational expressions have been restricted so that no denominator can be 0.

Example 1. Simplify each expression.

a. $\frac{31}{10ab} + \frac{7}{10ab}$ **b.** $\frac{5x^2 + 8x}{x^2 + 10x + 25} - \frac{2x^2 - 5x + 10}{x^2 + 10x + 25}$

Solution. **a.** $\frac{31}{10ab} + \frac{7}{10ab}$ Both expressions have the denominator $10ab$.

$= \frac{31 + 7}{10ab}$ $\frac{\text{Add the numerators}}{\text{The common denominator}}$

$= \frac{38}{10ab}$ Simplify the numerator.

$= \frac{19}{5ab}$ Reduce the fraction.

b. $\frac{5x^2 + 8x}{x^2 + 10x + 25} - \frac{2x^2 - 5x + 10}{x^2 + 10x + 25}$

$$= \frac{5x^2 + 8x - (2x^2 - 5x + 10)}{x^2 + 10x + 25}$$

$$= \frac{5x^2 + 8x - 2x^2 + 5x - 10}{x^2 + 10x + 25}$$

$$= \frac{3x^2 + 13x - 10}{x^2 + 10x + 25}$$

$$= \frac{(3x - 2)(x + 5)}{(x + 5)(x + 5)}$$

$$= \frac{3x - 2}{x + 5}$$

Finding a Least Common Denominator (LCD)

The expressions in examples **c** and **d** cannot be simplified as written.

c. $\frac{11}{35} - \frac{11}{84} + \frac{2}{60}$ **d.** $\frac{t-1}{5t-25} - \frac{t-1}{t^2-5t} + \frac{2}{5t}$

To combine rational expressions like these, it is necessary to first change them to equivalent expressions with a common denominator. The work is usually made easier if the least common multiple (LCM) of the denominators is used. In such a case, the LCM is called the **least common denominator (LCD).** The similarity between examples **c** and **d** can be seen by evaluating the rational expressions in **d** for $t = 12$.

To find the LCD of two or more rational expressions:

Step 1. Write each denominator as a product of primes.

Step 2. List each different factor that occurs in any denominator.

Step 3. Give each factor the largest exponent it has in any of the denominators.

Step 4. The LCD is the indicated product of the factors from Step 2 raised to the powers determined in Step 3.

Example 2. Find the LCD of each set of expressions.

a. 35, 84, and 60 **b.** $5t - 25$, $t^2 - 5t$, and $5t$

Solution. **a. Step 1.** $35 = 5 \cdot 7$

$84 = 2 \cdot 2 \cdot 3 \cdot 7 = 2^2 \cdot 3 \cdot 7$

$60 = 2 \cdot 2 \cdot 3 \cdot 5 = 2^2 \cdot 3 \cdot 5$

Step 2. The factors of the LCD are 2, 3, 5, and 7.

Step 3. The highest power of 2 is 2.
The highest power of 3, 5, and 7 is 1.

Step 4. The LCD is $2^2 \cdot 3 \cdot 5 \cdot 7$, or 420.

Notice that 420 is the smallest number that 35, 84, and 60 will divide with a remainder 0.

b. Step 1. $5t - 25 = 5(t - 5)$

$t^2 - 5t = t(t - 5)$

$5t$

Step 2. The factors of the LCD are 5, t, and $(t - 5)$.

Step 3. The highest power of 5, t, and $(t - 5)$ is 1.

Step 4. The LCD is $5t(t - 5)$.

Adding and Subtracting Rational Expressions with Unlike Denominators

Once the LCD of rational expressions with different denominators has been determined, then each expression is written as an equivalent fraction with the LCD as denominator before they are added or subtracted.

Example 3. Simplify $\frac{5}{4u} - \frac{1}{3t} + \frac{3t - 2u}{12tu}$.

Solution. **Discussion.** First determine the LCD of the three fractions. Then build up each fraction so that they are equivalent fractions with the LCD as denominator. Finally, combine and simplify the numerators over the LCD.

The LCD of $4u$, $3t$, and $12tu$ is $12tu$.

$$\frac{5}{4u} \cdot \frac{3t}{3t} - \frac{1}{3t} \cdot \frac{4u}{4u} + \frac{3t - 2u}{12tu}$$

$$= \frac{15t - 4u + 3t - 2u}{12tu}$$

$$= \frac{18t - 6u}{12tu}$$

$$= \frac{6(3t - u)}{12tu} = \frac{3t - u}{2tu}$$

Example 4. Simplify $\frac{t - 1}{5(t - 5)} - \frac{t - 1}{t(t - 5)} + \frac{2}{5t}$.

Solution. **Discussion.** From Example 2, the LCD is $5t(t - 5)$.

$$\frac{t - 1}{5(t - 5)} \cdot \frac{t}{t} - \frac{t - 1}{t(t - 5)} \cdot \frac{5}{5} + \frac{2}{5t} \cdot \frac{t - 5}{t - 5}$$

$$= \frac{t(t - 1) - 5(t - 1) + 2(t - 5)}{5t(t - 5)}$$ Combine the numerators over the LCD.

$$= \frac{t^2 - t - 5t + 5 + 2t - 10}{5t(t - 5)}$$ Remove parentheses in the numerator.

$$= \frac{t^2 - 4t - 5}{5t(t - 5)}$$ Simplify the numerator.

$$= \frac{(t + 1)(t - 5)}{5t(t - 5)} = \frac{t + 1}{5t}$$ Factor and reduce.

Example 5. Simplify $\frac{w}{w - 2} - \frac{w^2}{w^2 - 4} + \frac{w + 1}{w + 2}$.

Solution. Since $w^2 - 4 = (w - 2)(w + 2)$, the LCD is $(w - 2)(w + 2)$.

$$\frac{w}{w-2} - \frac{w^2}{(w-2)(w+2)} + \frac{w+1}{w+2}$$

$$= \frac{w}{w-2} \cdot \frac{w+2}{w+2} - \frac{w^2}{(w-2)(w+2)} + \frac{w+1}{w+2} \cdot \frac{w-2}{w-2}$$

$$= \frac{w(w+2) - w^2 + (w+1)(w-2)}{(w-2)(w+2)}$$

$$= \frac{w^2 + 2w - w^2 + w^2 - w - 2}{(w-2)(w+2)}$$

$$= \frac{w^2 + w - 2}{(w-2)(w+2)}$$

$$= \frac{(w+2)(w-1)}{(w-2)(w+2)} = \frac{w-1}{w-2}$$

SECTION 4-4. Practice Exercises

In exercises **1–20**, simplify each expression.

[Example 1]

1. $\frac{12}{5u} + \frac{3}{5u}$

2. $\frac{8}{3u^2} - \frac{2}{3u^2}$

3. $\frac{16}{9mn} - \frac{22}{9mn}$

4. $\frac{7}{12mn} + \frac{11}{12mn}$

5. $\frac{5x}{x+3} + \frac{15}{x+3}$

6. $\frac{3x}{x-3} - \frac{9}{x-3}$

7. $\frac{3a}{9a^2 - 4b^2} - \frac{2b}{9a^2 - 4b^2}$

8. $\frac{a}{a^2 - 25b^2} + \frac{5b}{a^2 - 25b^2}$

9. $\frac{y+2}{4y^2 - 1} + \frac{3y}{4y^2 - 1}$

10. $\frac{9y}{y^2 - 36} - \frac{5y + 24}{y^2 - 36}$

11. $\frac{p^2 - pq}{p^2 - q^2} + \frac{q^2 - pq}{p^2 - q^2}$

12. $\frac{p^2 + 5pq}{p^3 - q^3} - \frac{4pq + q^2}{p^3 - q^3}$

13. $\frac{2(k^2 + 1)}{k^2 - 7k + 12} - \frac{5(k + 1)}{k^2 - 7k + 12}$

14. $\frac{k^2 - 4k}{3k^2 + 5k - 2} + \frac{3(k - 2)}{3k^2 + 5k - 2}$

15. $\frac{a(2a - 1)}{ax + ay - x - y} + \frac{2a - 3}{ax + ay - x - y}$

16. $\frac{x^2 + 4xy}{ax + ay - x - y} - \frac{xy - 2y^2}{ax + ay - x - y}$

17. $\frac{t^2 + 5t}{t^3 - 1} - \frac{4t - 1}{t^3 - 1}$

18. $\frac{5t^2 + 13}{8t^3 + 27} - \frac{12t - 3t^2 - 5}{8t^3 + 27}$

19. $\frac{3(u - 2)}{3u^2 + 5u - 2} - \frac{3u}{3u^2 + 5u - 2} + \frac{u(u - 1)}{3u^2 + 5u - 2}$

20. $\frac{-2(2u - 1)}{u^2 - 7u + 12} + \frac{u(2u - 1)}{u^2 - 7u + 12} - \frac{5}{u^2 - 7u + 12}$

In exercises **21–34**, find the LCD of each set of expressions.

[Example 2]

21. $\frac{1}{6k^2}, \frac{1}{12k^4}$, and $\frac{1}{18}$

22. $\frac{1}{21}, \frac{1}{k^5}$, and $\frac{1}{6k^3}$

23. $\frac{1}{4m^2 + 6m}$ and $\frac{1}{6m + 9}$

24. $\frac{1}{18m + 6}$ and $\frac{1}{12m^2 + 4m}$

25. $\frac{1}{p - 5q}$ and $\frac{1}{5q - p}$

26. $\frac{1}{p^2 - pq - 2q^2}$ and $\frac{1}{2q^2 + pq - p^2}$

27. $\frac{1}{m^3 + 27n^3}$ and $\frac{1}{m^2 - 3mn + 9n^2}$

28. $\frac{1}{m^2 - 25n^2}$ and $\frac{1}{m + 5n}$

29. $\frac{1}{12x^2 - 6xy}, \frac{1}{6xy + 3y^2}$, and $\frac{1}{8x^2y - 4xy^2}$

30. $\frac{1}{3x^2 + 15xy}, \frac{1}{9xy + 45y^2}$, and $\frac{1}{x^2y + 5xy^2}$

31. $\frac{1}{u^2 + 2uv + v^2}$ and $\frac{1}{u^2 + 3uv + 2v^2}$

32. $\frac{1}{2u^2 + uv - v^2}$ and $\frac{1}{4u^2 - 4uv + v^2}$

33. $\frac{1}{3k^2 - 14k - 5}, \frac{1}{k^2 - 25}$, and $\frac{1}{3k^2 + 16k + 5}$

34. $\frac{1}{4k^2 - 9}, \frac{1}{2k^2 - 5k - 12}$, and $\frac{1}{2k^2 - 11k + 12}$

In exercises **35–72**, simplify each expression.

[Examples 3–5]

35. $\frac{3t}{8} + \frac{t}{12}$

36. $\frac{7t^2}{10} + \frac{2t^2}{15}$

37. $\frac{x + 4}{4y} + \frac{3x - 5}{5y}$

38. $\frac{3x + 1}{9y} - \frac{5 - 2x}{6y}$

39. $\frac{9r^2}{10st} - \frac{r^2}{2st} + \frac{3r^2}{5st}$

40. $\frac{r^2}{12st} + \frac{7r^2}{8st} - \frac{5r^2}{6st}$

41. $\frac{2}{3x^2} - \frac{1}{6x}$

42. $\frac{5}{4x} - \frac{3}{8x^2}$

43. $\frac{1}{10x} + \frac{3}{5y}$

44. $\frac{7}{12x} + \frac{5}{6y}$

45. $\frac{3}{10a} + \frac{a + 2}{2a} - \frac{2a + 4}{5a}$

46. $\frac{a - 1}{3a^2} + \frac{a - 1}{4a} - \frac{1}{6}$

47. $\frac{k - 3}{18k} - \frac{4k - 1}{24k} + \frac{1}{9}$

48. $\frac{2k + 1}{12k} + \frac{3k - 2}{18k} - \frac{1}{3}$

49. $\frac{m - 1}{m + 1} + \frac{1}{m}$

50. $\frac{1}{m} + \frac{1}{m - 1}$

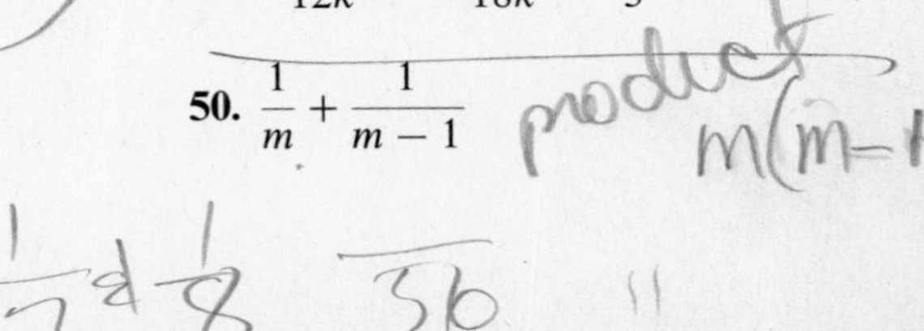

51. $\dfrac{2a}{3a-6} - \dfrac{a}{2a-4}$

52. $\dfrac{4}{5a+15} - \dfrac{1}{3a+9}$

53. $\dfrac{c+1}{c^2-cd} + \dfrac{d+1}{d^2-cd}$

54. $\dfrac{c}{3c-12} + \dfrac{d}{12-3d}$

55. $\dfrac{3}{p^2-4p} + \dfrac{p-3}{p^2-16}$

56. $\dfrac{3}{p^2-9} + \dfrac{3}{p^2-6p+9}$

57. $\dfrac{2}{u^2-5u+6} - \dfrac{2}{u^2-3u+2}$

58. $\dfrac{u}{u^2-4u+4} - \dfrac{1}{u^2-u-2}$

59. $\dfrac{5}{v^2-v-20} + \dfrac{v}{v^2+8v+16}$

60. $\dfrac{1}{6v^2-v-1} + \dfrac{2}{3v^2+7v+2}$

61. $\dfrac{-11}{5x-25} + \dfrac{4}{3x-15} - 2$

62. $\dfrac{7}{3x+9} - \dfrac{5}{4x+12} + 3$

63. $\dfrac{3}{z^2-4z} + 1 - \dfrac{3}{z^2+4z}$

64. $\dfrac{2}{z^2+z} - 3 + \dfrac{2}{z^2-z}$

65. $\dfrac{y-2}{y^2-36} + \dfrac{1}{2y+12} - \dfrac{1}{3y-18}$

66. $\dfrac{y}{4y^2-9} - \dfrac{1}{6y+9} + \dfrac{1}{4y-6}$

67. $\dfrac{y-2}{y^2+y} - \dfrac{4}{y+1} - \dfrac{3y}{1-y^2}$

68. $\dfrac{-y}{12-6y} - \dfrac{2}{3y+6} - \dfrac{y^2+12}{12y^2-48}$

69. $\dfrac{13x-12}{2x^3-54} + \dfrac{x}{x^2+3x+9} - \dfrac{1}{2x-6}$

70. $\dfrac{2}{6x+3} - \dfrac{x}{4x^2-2x+1} + \dfrac{4(x-1)}{24x^3+3}$

71. $\dfrac{2a}{a^2-2a+4} + \dfrac{40(1-a)}{5a^3+40} - \dfrac{3}{5a+10}$

72. $\dfrac{2(6a+1)}{81a^3-3} - \dfrac{4}{9a-3} + \dfrac{2a}{9a^2+3a+1}$

SECTION 4-4. Ten Review Exercises

1. State the opposite of $\dfrac{-3k}{5}$.

2. State the reciprocal of $\dfrac{-3k}{5}$, where $k \neq 0$.

In exercises 3–8, do the indicated operations. Write answers in reduced form.

3. $\dfrac{2^4-3}{3^2+1} + \dfrac{-3+4\cdot 5}{-6+4^2}$

4. $5(-17+5\cdot 2^2-3^2)+2\cdot 3^2$

5. $\dfrac{-1}{2}(4k^3-8k^2) - \dfrac{2}{3}(12k^2-6k^3) + \dfrac{3}{5}(20k-5)$

6. $(2a-3)^2-(a+5)(3a-7)-9(4-3a)$

7. $(4b^4 + 2b^3 - 16b^2 + 13b - 3) \div (2b^2 - 3b + 1)$

8. $(2z^2 - z + 1)(2z^2 + z - 3)$

In exercises **9** and **10** factor each expression.

9. $16x^2 + 9y^2 - 24xy$

10. $13m - 10 + 3m^2$

SECTION 4-4. Supplementary Exercises

In exercises **1–8**, show that expressions A and B are not equivalent by evaluating both expressions for the indicated value(s) of the variable(s) and comparing the results.

1. A: $\frac{1}{x} + \frac{1}{x+1}$ B: $\frac{2}{x(x+1)}$ $x = 5$

2. A: $\frac{1}{x+1} + \frac{1}{x-1}$ B: $\frac{2}{x}$ $x = -3$

3. A: $\frac{y}{x} - \frac{x}{y}$ B: $\frac{y-x}{xy}$ $x = 2,\ y = 4$

4. A: $\frac{x}{x+3} - \frac{5}{x+3}$ B: $x - 5$ $x = 7$

5. A: $\frac{2x}{x-7} + \frac{14}{7-x}$ B: $\frac{2x+14}{x-7}$ $x = -3$

6. A: $\frac{1}{x+8} - \frac{1}{x-2}$ B: $\frac{-10}{(x+8)(x-2)}$ $x = 0$

7. A: $\frac{x}{x^3+1} + \frac{1}{x^2+x}$ B: $\frac{x+1}{x^3+x^2+x+1}$ $x = 1$

8. A: $\frac{x}{x-4} - \frac{1}{2x-4}$ B: $\frac{2x-1}{2x-4}$ $x = -8$

In exercises **9–16**, find the value of k so that the LCD of both expressions is as indicated.

9. $\frac{1}{4x^2y}$ and $\frac{1}{6xy^k}$ The LCD is $12x^2y^3$.

10. $\frac{1}{10pq^3}$ and $\frac{1}{2p^kq}$ The LCD is $20pq^3$.

11. $\frac{1}{z^2-k}$ and $\frac{1}{z+1}$ The LCD is $z^2 - 1$.

12. $\frac{1}{b-2}$ and $\frac{1}{k}$ The LCD is $b^2 - 2b$.

13. $\frac{1}{x^2+kx+6}$ and $\frac{1}{x^2+2x-3}$ The LCD is $(x+2)(x+3)(x-1)$.

14. $\dfrac{1}{m^2 - 2m + k}$ and $\dfrac{1}{m^2 - 1}$ The LCD is $(m - 1)(m - 1)(m + 1)$.

15. $\dfrac{1}{4p - 5}$ and $\dfrac{1}{5 - kp}$ The LCD is $4p - 5$.

16. $\dfrac{1}{r^2 - r}$ and $\dfrac{1}{r^2 + kr}$ and $\dfrac{1}{r^2 + r - 2}$ The LCD is $r(r + 2)(r + 1)$.

In exercises **17–20**, do the indicated operations. Write answers in reduced form.

17. $\left(\dfrac{9}{s - 2} - \dfrac{3}{s}\right)\left(\dfrac{5s - 10}{6}\right)$

18. $\left(\dfrac{2p}{p^2 - 1} + \dfrac{-1}{p + 1}\right) \div \left(\dfrac{4}{p - 1}\right)$

19. $\left(\dfrac{5}{x^2 - 6x + 9} - \dfrac{4}{x^2 - 9}\right) \div \left(\dfrac{x + 27}{x^2 - 9}\right)$

20. $\left(\dfrac{3}{a - 2} - \dfrac{2}{3a} - \dfrac{18}{3a^2 - 6a}\right)\left(\dfrac{a^2 + a}{7}\right)$

SECTION 4-5. Combined Operations and Complex Fractions

1. Simplifying rational expressions with combined operations
2. A definition of a complex fraction
3. Two techniques for simplifying a complex fraction
4. Complex fractions written with negative exponents

Frequently in mathematics we see expressions that indicate two or more operations on rational expressions. Furthermore, many expressions used in higher mathematics and applied sciences are so-called complex fractions. We will study the techniques used to simplify these types of expressions in this section.

Simplifying Rational Expressions with Combined Operations

Example **a** is an expression that indicates three operations on rational expressions.

a. $\dfrac{1}{4(x + 2)} + \dfrac{x - 5}{x^2 - 5x + 6} \cdot \dfrac{x - 3}{x + 2} + \dfrac{3}{4(x - 2)}$

To simplify this expression, the rule for order of operations must be observed. That is, do the multiplication first, then do the additions in order from left to right.

Example 1. Simplify $\dfrac{1}{4(x + 2)} + \dfrac{x - 5}{x^2 - 5x + 6} \cdot \dfrac{x - 3}{x + 2} + \dfrac{3}{4(x - 2)}$.

Solution.

$$\frac{1}{4(x+2)} + \frac{(x-5)(x-3)}{(x-3)(x-2)(x+2)} + \frac{3}{4(x-2)}$$ Do the multiplication.

$$= \frac{1}{4(x+2)} + \frac{x-5}{(x-2)(x+2)} + \frac{3}{4(x-2)}$$ Change the common factor $(x-3)$ to 1.

$$= \frac{(x-2) + 4(x-5) + 3(x+2)}{4(x+2)(x-2)}$$ The LCD is $4(x+2)(x-2)$.

$$= \frac{x - 2 + 4x - 20 + 3x + 6}{4(x+2)(x-2)}$$ Remove parentheses in the numerator

$$= \frac{8x - 16}{4(x+2)(x-2)}$$ Simplify the numerator.

$$= \frac{8(x-2)}{4(x+2)(x-2)}$$ Factor the numerator.

$$= \frac{2}{x+2}$$ Reduce the fraction.

Example 2. Simplify $(u-3)(u+3)(2u-3)\left(\frac{u-1}{2u^2 - 9u + 9} + \frac{u+1}{2u^2 + 3u - 9}\right)$.

Solution. **Discussion.** The distributive property can be used to distribute the three binomials outside the parentheses to each of the rational expressions within the parentheses.

$$(u-3)(u+3)(2u-3)\left(\frac{u-1}{2u^2 - 9u + 9} + \frac{u+1}{2u^2 + 3u - 9}\right)$$

$$= \frac{(u-3)(u+3)(2u-3)(u-1)}{(2u-3)(u-3)} + \frac{(u-3)(u+3)(2u-3)(u+1)}{(2u-3)(u+3)}$$

$$= (u+3)(u-1) + (u-3)(u+1)$$

$$= u^2 + 2u - 3 + u^2 - 2u - 3$$

$$= 2u^2 - 6$$

To emphasize the relationship between the given expression and $2u^2 - 6$, suppose we needed to evaluate the given expression for any suitable replacement for u, say $u = 5$. The value of the expression would be the same as the value obtained by replacing u by 5 in $2u^2 - 6$; that is:

$$2(5)^2 - 6 = 50 - 6 = 44$$

A Definition of a Complex Fraction

Consider the expressions in examples **b** and **c**.

b. $\dfrac{2\frac{1}{4}}{3\frac{3}{8}}$ **c.** $\dfrac{t + \frac{t-1}{2t}}{(t+1) + \frac{t+1}{4t}}$

In example **b**, the ratio of mixed numbers can be rewritten:

$$\frac{2\frac{1}{4}}{3\frac{3}{8}} \text{ can be written as } \frac{2 + \frac{1}{4}}{3 + \frac{3}{8}}$$

In example **c**, the numerator and denominator are mixed expressions in t. If t is replaced by 2, the expression becomes the number expression in example **b**. These examples are illustrations of expressions that can be called complex fractions.

Definition 4.3. Complex fraction
A **complex fraction** is an expression in which either the numerator, or denominator, or both, contains a fraction.

Two Techniques for Simplifying a Complex Fraction

The division bar that separates the numerator and denominator of a complex fraction can be called the **principal,** or **main division.** Plan A is a procedure that can be used to simplify many complex fractions.

Plan A for simplifying a complex fraction

Step 1. Carry out the indicated operations above and below the principal division bar.

Step 2. Multiply the numerator by the reciprocal of the denominator.

Step 3. If possible, simplify the resulting product.

Plan A is used to simplify the expressions in examples **b** and **c**.

Example 3. Simplify $\dfrac{2 + \frac{1}{4}}{3 + \frac{3}{8}}$.

Solution.

$\dfrac{2 + \frac{1}{4}}{3 + \frac{3}{8}}$ The given expression

$= \dfrac{\frac{9}{4}}{\frac{27}{8}}$ $2 + \frac{1}{4} = \frac{2 \cdot 4 + 1}{4} = \frac{9}{4}$; $3 + \frac{3}{8} = \frac{3 \cdot 8 + 3}{8} = \frac{27}{8}$

$= \dfrac{9}{4} \cdot \dfrac{8}{27}$ Multiply by the reciprocal of $\frac{27}{8}$.

$= \dfrac{2}{3}$ Multiply and reduce.

The solution to Example 3 suggests techniques that can be used to simplify some complex fractions.

In the following equations, assume a, b, c, and d are nonzero expressions.

Technique A. $a + \frac{b}{c} = \frac{ac + b}{c}$

Technique B. $\frac{a}{b} + \frac{c}{d} = \frac{ad + bc}{bd}$, where b and d have no common factors

Technique C. $\frac{\frac{a}{b}}{\frac{c}{d}} = \frac{ad}{bc}$

Example 4. Simplify $\dfrac{t + \frac{t-1}{2t}}{(t+1) + \frac{t+1}{4t}}$.

Solution.

$$\frac{t + \frac{t-1}{2t}}{(t+1) + \frac{t+1}{4t}}$$ The given expression

$$= \frac{\frac{t(2t) + (t-1)}{2t}}{\frac{(t+1)4t + (t+1)}{4t}}$$ $a + \frac{b}{c} = \frac{ac+b}{c}$

$$= \frac{\frac{2t^2 + t - 1}{2t}}{\frac{4t^2 + 5t + 1}{4t}}$$ Simplify both numerators.

$$= \frac{(2t^2 + t - 1)4t}{2t(4t^2 + 5t + 1)}$$ Technique C

$$= \frac{2(2t - 1)}{4t + 1}$$ Factor and reduce.

Example **d** is also a complex fraction.

d. $\dfrac{\frac{u}{3v} + \frac{5}{6} - \frac{v}{2u}}{\frac{u}{v} + \frac{v}{6u} - \frac{5}{6}}$

The numerator and denominator each have three terms. The denominators of these six terms are monomials. The LCD of these six monomials is $6uv$.

Plan B is an easier procedure to use to simplify expressions such as the one in example **d**. The procedure uses the multiplication property of 1 (that is, $a \cdot 1 = a$), where 1 has the form of the LCD of the monomial denominators over itself.

Plan B for simplifying a complex fraction

Step 1. Determine the LCD of the rational expressions in the numerator and denominator of the complex fraction.

Step 2. Multiply the complex fraction by the LCD over itself and distribute to each term of the complex fraction. If possible, factor and reduce the resulting rational expression.

Example 5. Simplify $\dfrac{\frac{u}{3v}+\frac{5}{6}-\frac{v}{2u}}{\frac{u}{v}+\frac{v}{6u}-\frac{5}{6}}$.

Solution. **Step 1.** The LCD of $3v$, 6, $2u$, v, $6u$, and 6 is $6uv$.

Step 2.

$$\frac{\frac{u}{3v}+\frac{5}{6}-\frac{v}{2u}}{\frac{u}{v}+\frac{v}{6u}-\frac{5}{6}}\cdot\frac{6uv}{6uv}$$ Multiply by LCD over LCD.

$$=\frac{\frac{u}{3v}\cdot 6uv+\frac{5}{6}\cdot 6uv-\frac{v}{2u}\cdot 6uv}{\frac{u}{v}\cdot 6uv+\frac{v}{6u}\cdot 6uv-\frac{5}{6}\cdot 6uv}$$ Distributive property

$$=\frac{2u^2+5uv-3v^2}{6u^2+v^2-5uv}$$ Simplify all six products.

$$=\frac{(2u-v)(u+3v)}{(2u-v)(3u-v)}$$ Factor numerator and denominator.

$$=\frac{u+3v}{3u-v}$$ Reduce.

Complex Fractions Written with Negative Exponents

Examples **e** and **f** are not rational expressions because the numerators and denominators are not polynomials. However, when the definition of a negative integer exponent is used, then the expressions take on the form of complex fractions.

e. $\dfrac{x^{-1}+y^{-1}}{x^{-2}-y^{-2}}$ **f.** $\dfrac{8n^{-3}+27m^{-3}}{2n^{-1}+3m^{-1}}$

If the definition of a negative exponent is used to write these expressions as complex fractions, then Plan A or Plan B can be used to simplify them.

Example 6. Simplify each expression.

a. $\dfrac{x^{-1}+y^{-1}}{x^{-2}-y^{-2}}$ **b.** $\dfrac{8n^{-3}+27m^{-3}}{2n^{-1}+3m^{-1}}$

Solution. **a. Using Plan A**

$$\frac{x^{-1}+y^{-1}}{x^{-2}-y^{-2}} = \frac{\dfrac{1}{x}+\dfrac{1}{y}}{\dfrac{1}{x^2}-\dfrac{1}{y^2}} = \frac{\dfrac{y+x}{xy}}{\dfrac{y^2-x^2}{x^2y^2}} = \frac{y+x}{xy}\cdot\frac{x^2y^2}{y^2-x^2} = \frac{x^2y^2(y+x)}{xy(y+x)(y-x)} = \frac{xy}{y-x}$$

b. Using Plan B

$$\frac{8n^{-3}+27m^{-3}}{2n^{-1}+3m^{-1}} = \frac{\dfrac{8}{n^3}+\dfrac{27}{m^3}}{\dfrac{2}{n}+\dfrac{3}{m}} = \frac{\dfrac{8}{n^3}+\dfrac{27}{m^3}}{\dfrac{2}{n}+\dfrac{3}{m}}\cdot\frac{m^3n^3}{m^3n^3} = \frac{8m^3+27n^3}{2m^3n^2+3m^2n^3} = \frac{(2m+3n)(4m^2-6mn+9n^2)}{m^2n^2(2m+3n)} = \frac{4m^2-6mn+9n^2}{m^2n^2}$$

SECTION 4-5. Practice Exercises

In exercises **1–16**, simplify each expression.

[Example 1]

1. $\dfrac{a^2-2ab-2b^2}{a^2-ab-20b^2}-\dfrac{2ab+3b^2}{a^2-16b^2}\div\dfrac{a-5b}{a-4b}$

2. $\dfrac{3a^2b+ab^2}{2a^2-3ab+b^2}-\dfrac{3a^2b-ab^2}{a-b}\div\dfrac{6a^2-5ab+b^2}{a+3b}$

3. $\dfrac{1}{k^2+k}+\dfrac{5k}{3k+3}\cdot\dfrac{6}{5k-5}+\dfrac{1}{k^2-k}$

4. $\dfrac{1}{2k+4}-\dfrac{4}{5k+10}\cdot\dfrac{k+6}{8k-16}+\dfrac{1}{5k-10}$

5. $\dfrac{8}{15m}+\dfrac{2m-2}{3m^2+3m}\cdot\dfrac{m+1}{5m-5}-\dfrac{4}{15m}$

6. $\dfrac{m+3}{24m}-\dfrac{m+1}{24m}+\dfrac{3m-1}{6m^2+18m}\div\dfrac{12m-4}{m+3}$

7. $\dfrac{3u^2+5uv+2v^2}{u^2+5uv+6v^2}+\dfrac{2u+6v}{u^2-4v^2}\div\dfrac{u^2+6uv+9v^2}{u^2v-2uv^2}$

8. $\dfrac{u^2+2v^2}{u^2-uv-2v^2}-\dfrac{u^2-2uv}{u^2+2uv+v^2}\div\dfrac{u^2-4uv+4v^2}{3uv+3v^2}$

[Example 2]

9. $6ab\left(\dfrac{1}{2a}+\dfrac{1}{3b}\right)$

10. $10a^2b^2\left(\frac{1}{5a^2} - \frac{1}{2b^2}\right)$

11. $(12x^2 - 12y^2)\left(\frac{3}{4(x + y)} + \frac{2}{3(x - y)}\right)$

12. $(3x^3 + 24y^3)\left(\frac{2}{3x + 6y} - \frac{2(x + 2y)}{3x^2 - 6xy + 12y^2}\right)$

13. $(u + 2)(u - 2)^2\left(\frac{u + 1}{u^2 - 4} + \frac{u - 1}{u^2 - 4u + 4}\right)$

14. $(u^2 + 2)(u - 1)(u - 2)\left(\frac{1}{u^2 - 3u + 2} - \frac{u - 2}{u^3 + 2u - u^2 - 2}\right)$

15. $(r^2 + 1)(2r - 3s)(r + s)\left(\frac{r + 3s}{r^3 + r^2s + r + s} - \frac{2r - 9s}{2r^3 - 3r^2s + 2r - 3s}\right)$

16. $(r^2 + 4)(s^2 + 1)(r + 2s)\left(\frac{r^2}{r^3 + 2r^2s + 4r + 8s} - \frac{s^2}{2s^3 + rs^2 + r + 2s}\right)$

In exercises **17–26**, simplify each expression using Plan A.

[Examples 3 and 4] 17. $\dfrac{\frac{1}{3} + \frac{1}{4}}{1 - \frac{1}{12}}$

18. $\dfrac{\frac{1}{5} + \frac{1}{3}}{\frac{1}{15}}$

19. $\dfrac{\frac{4}{a} - a}{1 + \frac{a}{2}}$

20. $\dfrac{a - \frac{9}{a}}{\frac{a}{3} - 1}$

21. $\dfrac{\frac{u}{v} - \frac{v}{u}}{\frac{u}{v} - 1}$

22. $\dfrac{1 - \frac{2v}{u + v}}{1 - \frac{3v}{u + 2v}}$

23. $\dfrac{\frac{x}{3} - \frac{3}{x}}{3 - x}$

24. $\dfrac{x - 5}{\frac{5}{x} - \frac{x}{5}}$

25. $\dfrac{m + 1 - \frac{2}{m}}{m + 4 + \frac{4}{m}}$

26. $\dfrac{m - \frac{9}{2} - \frac{9}{m}}{m + \frac{13}{2} + \frac{15}{2m}}$

In exercises **27–34**, simplify each expression using Plan B.

[Example 5] 27. $\dfrac{\frac{1}{b - 3} + \frac{1}{b + 3}}{\frac{1}{b - 3} - \frac{1}{b + 3}}$

28. $\dfrac{\frac{4}{b^2 - 4} + 1}{\frac{2}{b - 2} + 1}$

29. $\dfrac{2 - \dfrac{5}{k} - \dfrac{12}{k^2}}{3 - \dfrac{10}{k} - \dfrac{8}{k^2}}$

30. $\dfrac{\dfrac{5}{k^2} - \dfrac{11}{k} + 2}{10 - \dfrac{3}{k} - \dfrac{1}{k^2}}$

31. $\dfrac{\dfrac{3}{9z^2 - 1}}{\dfrac{2}{3z + 1} + \dfrac{2}{3z - 1}}$

32. $\dfrac{\dfrac{4}{4z^2 - 9}}{\dfrac{3}{2z + 3} + \dfrac{3}{2z - 3}}$

33. $\dfrac{\dfrac{2}{2u + v} - \dfrac{3}{3u - v}}{\dfrac{6}{3u - v} - \dfrac{4}{2u + v}}$

34. $\dfrac{\dfrac{3}{3u + 2v} - \dfrac{4}{4u - v}}{\dfrac{1}{4u - v} + \dfrac{2}{3u + 2v}}$

In exercises **35–42**, simplify each expression using either Plan A or Plan B.

[Examples 3–5] 35. $\dfrac{\dfrac{1}{6} - \dfrac{1}{p}}{\dfrac{1}{36} - \dfrac{1}{p^2}}$

36. $\dfrac{\dfrac{1}{3} + \dfrac{1}{p}}{1 - \dfrac{1}{3p}}$

37. $\dfrac{x - \dfrac{1}{x}}{1 - x}$

38. $\dfrac{2x}{\dfrac{x}{30} - \dfrac{x}{40}}$

39. $\dfrac{\dfrac{n + 2}{n - 1} - \dfrac{n - 1}{n + 2}}{\dfrac{1}{n + 2} + \dfrac{1}{n - 1}}$

40. $\dfrac{\dfrac{n}{n - 1} - \dfrac{1}{n + 1}}{\dfrac{n}{n - 1} + \dfrac{1}{n + 1}}$

41. $\dfrac{s - 4 - \dfrac{9}{s - 4}}{s + 2 + \dfrac{9}{s - 4}}$

42. $\dfrac{s - 5 - \dfrac{8}{s + 2}}{s - 2 - \dfrac{5}{s + 2}}$

In exercises **43–56**, simplify each expression.

[Example 6] 43. $\dfrac{2^{-1} + 3^{-1}}{6^{-1}}$

44. $\dfrac{5^{-1} + 2^{-1}}{10^{-1}}$

45. $\dfrac{a^{-1} + 2}{2a + 1}$

46. $\dfrac{5a + 1}{5 + a^{-1}}$

47. $\dfrac{(mn)^{-1}}{m^{-1} + n^{-1}}$

48. $\dfrac{m^{-1} - n^{-1}}{(mn)^{-1}}$

49. $\dfrac{9^{-1}k - k^{-1}}{k + 3}$

50. $\dfrac{8^{-1}k - k^{-2}}{k^2 + 2k + 4}$

51. $\dfrac{u^{-1} + 4v^{-1}}{v^2 - 16u^2}$

52. $\dfrac{2v^{-1} - 3u^{-1}}{4u^2 - 12uv + 9v^2}$

53. $\dfrac{s^{-3} + r^{-3}}{r^{-1} + s^{-1}}$

54. $\dfrac{s^{-3} - r^{-3}}{s^{-1} - r^{-1}}$

55. $\dfrac{(x - 1)^{-1} - x^{-1}}{(x + 1)^{-1}}$

56. $\dfrac{(x - 1)^{-1}}{x^{-1} - (x + 1)^{-1}}$

SECTION 4-5. Ten Review Exercises

1. Use the commutative property of addition to rewrite $3t + (7 + t)$.
2. Use the associative property of addition to regroup the answer to exercise **1**.
3. Use the distributive property to change the answer to exercise **2** to two terms.

In exercises **4–7**, do the indicated operations. Write answers in reduced form.

4. $\dfrac{-1}{2}(4t^2 - 8t + 10) + \dfrac{3}{5}(20t^2 - 5t + 10)$
5. $(2x - 3)^2 + (x + 5)(3x + 7) - 9(4 - 3x)$
6. $\dfrac{2y}{y + 3} - \dfrac{1}{y - 3} + \dfrac{2y}{y^2 - 9}$
7. $(4z^4 + 2z^3 - 16z^2 - 3z + 15) \div (2z^2 - 3)$

In exercises **8–10**, factor each expression.

8. $18k^3 - 2k$
9. $81u^4 - 1$
10. $a^2 - 10ab + 25b^2$

SECTION 4-5. Supplementary Exercises

In exercises **1–18**, simplify each expression.

1. $1 - \dfrac{1}{1 + \dfrac{1}{a - 1}}$

2. $a + \dfrac{a + 1}{a - \dfrac{1}{a}}$

3. $\dfrac{1 - \dfrac{1}{k}}{1 + \dfrac{1}{1 - \dfrac{2}{k}}}$

4. $\dfrac{1 + \dfrac{1}{1 - \dfrac{2}{k}}}{1 - \dfrac{3}{1 - \dfrac{2}{k}}}$

5. $\dfrac{m + \dfrac{9}{m - 6}}{1 + \dfrac{3}{m - 6}}$

6. $\dfrac{m + \dfrac{25}{m + 10}}{1 - \dfrac{5}{m + 10}}$

7. $\dfrac{1 - \dfrac{a}{b}\left(1 - \dfrac{b}{a}\right)}{b\left(1 + \dfrac{a}{b}\right) - a\left(2 - \dfrac{b}{a}\right)}$

8. $\dfrac{\dfrac{a^2}{b^2} - \dfrac{1 - a^2}{1 - b^2}}{\dfrac{a - 1}{b + b^2} + \dfrac{a + 1}{b - b^2}}$

9. $\dfrac{9x - x^3}{\left(9 - \dfrac{6x}{x + 1}\right)\left(x + \dfrac{3 - x^2}{2}\right)}$

10. $\dfrac{\dfrac{-x}{4}}{\dfrac{1}{2} - \dfrac{1}{1 - \dfrac{x - 2}{x + 2}}}$

11. $\dfrac{1 + r^3}{1 - \dfrac{r}{1 + \dfrac{r}{1 - r}}}$

12. $\dfrac{r - 2}{3(1 - r) - \dfrac{1 - r^3}{r + \dfrac{r}{r + 1}}}$

13. $(a^{-1} - 2^{-1})^{-1}$

14. $(a^{-1} + b^{-1})^{-1}$

15. $\left(\dfrac{k^{-1} + 2^{-1}}{4^{-2} - k^{-2}}\right)^{-1}$

16. $\left(\dfrac{25^{-2} - k^{-2}}{5^{-1} - k^{-1}}\right)^{-1}$

17. $(m^{-1} + 2)^{-2}$

18. $(m^{-1} + 2^{-1})^{-2}$

The formula below relates the amount of pull (P) needed by a differential pulley to lift a weight (W):

$$P = \frac{WR - Wr}{2R}$$

In exercises **19–22**, find a value or an expression for P given the indicated values of W, R, and r.

19. $W = 100$, $R = \dfrac{3}{4}$, and $r = \dfrac{1}{3}$

20. $W = 1$, $R = \dfrac{1}{y}$, and $r = \dfrac{1}{y + 1}$

21. $W = \dfrac{1}{t}$, $R = 4$, and $r = 2$

22. $W = x$, $R = \dfrac{1}{y}$, and $r = \dfrac{1}{2y}$

One formula used in trigonometry is:

$$\text{Tan}(A - B) = \frac{\text{Tan } A - \text{Tan } B}{1 + (\text{Tan } A)(\text{Tan } B)}$$

In exercises **23–26**, use the formula to find an expression for $\text{Tan}(A - B)$ given the indicated values of Tan A and Tan B.

23. $\text{Tan } A = \dfrac{1}{2}$, $\text{Tan } B = \dfrac{-1}{3}$

24. $\text{Tan } A = \dfrac{1}{x}$, $\text{Tan } B = \dfrac{1}{y}$

25. $\text{Tan } A = 1$, $\text{Tan } B = \dfrac{1}{p^2}$

26. $\text{Tan } A = \dfrac{x}{x + 1}$, $\text{Tan } B = \dfrac{1}{x - 1}$

The following example illustrates a procedure called **decomposition of rational expressions.** The procedure writes a rational expression with a quadratic denominator as an indicated sum or difference of rational expressions with denominators of degree 1.

Example. Write $\dfrac{5x + 12}{x^2 + 5x + 6}$ as a sum of two rational expressions.

Solution. **Discussion.** Since $x^2 + 5x + 6$ factors as $(x + 2)(x + 3)$, $\dfrac{5x + 12}{x^2 + 5x + 6}$ can be written as $\dfrac{5x + 12}{(x + 2)(x + 3)}$. To decompose the fraction, we consider it to be the sum of two fractions with unknown numerators (say, A and B), and denominators $x + 2$ and $x + 3$, respectively. The task is to find A and B.

$$\frac{5x + 12}{(x + 2)(x + 3)} = \frac{A}{x + 2} + \frac{B}{x + 3}$$ Write as a sum of two rational expressions.

$$= \frac{A(x + 3) + B(x + 2)}{(x + 2)(x + 3)}$$ Combine the two fractions.

Since the denominators on the left side and right side are the same, we may conclude the numerators are equal, and write:

$$5x + 12 = A(x + 3) + B(x + 2)$$

To make $B(x + 2)$ equal to 0, replace x by -2 in the equation.

$$5(-2) + 12 = A(-2 + 3) + B(-2 + 2)$$

$$2 = A$$ Thus, $A = 2$.

To make $A(x + 3)$ equal to 0, replace x by -3 in the equation.

$$5(-3) + 12 = A(-3 + 3) + B(-3 + 2)$$

$$-3 = -B$$

$$B = 3$$ Thus, $B = 3$.

$$\frac{5x + 12}{x^2 + 5x + 6} = \frac{2}{x + 2} + \frac{3}{x + 3}$$

In exercises **27–30**, write each expression as a sum of two rational expressions.

27. $\dfrac{3x + 4}{(x + 1)(x + 2)}$

28. $\dfrac{4x + 7}{(x + 1)(x + 2)}$

29. $\dfrac{7x - 4}{x^2 + x - 6}$

30. $\dfrac{-3x - 14}{x^2 + x - 6}$

SECTION 4-6. Solving Rational Equations and Inequalities

1. The definition of a rational equation
2. A procedure for solving a rational equation
3. Restricted values and apparent solutions
4. A procedure for solving a rational inequality

Examples **a–c** are three equations. Examples **a** and **b** are equations that can be solved by techniques learned in earlier sections. Example **c** is a rational equation that can be solved by a technique learned in this section.

a. $5(3 - t) - 13 = 8(t + 4) - 7t$ A linear equation in t

b. $5k + 4 = 6k^2$ A quadratic equation in k

c. $\frac{x-5}{x+3} + \frac{x-2}{x-3} = \frac{x+2}{x+3} + \frac{x^2+x}{x^2-9}$ A rational equation in x

As we will see, some rational equations can be changed to equations that are either linear or quadratic in one variable.

The Definition of a Rational Equation

The following statement can be used as a definition of a rational equation:

> **Definition 4.4. Rational equation**
> A **rational equation** is one in which at least one term of the equation is a rational expression.

A Procedure for Solving a Rational Equation

The multiplication property of equality has previously been used to solve equations:

$$\text{If (1) } A = B \text{ and } C \neq 0, \text{ then (2) } AC = BC.$$

This property asserts that any solution of (1) will also be a solution of (2), provided $C \neq 0$. In other words, (1) and (2) are **equivalent equations.** For a given rational equation, we can determine the LCD of all the terms in the equation. If we restrict the variable in the equation so that the LCD may not be 0, then we can use the multiplication property of equality to multiply the given equation by the LCD.

> **To solve a rational equation:**
>
> **Step 1.** If necessary, factor each denominator in the equation.
>
> **Step 2.** Determine the LCD of the denominators.
>
> **Step 3.** Determine all values of the variable that would make the LCD zero.
>
> **Step 4.** Multiply both sides of the equation by the LCD.
>
> **Step 5.** Solve the equation obtained in Step 4.
>
> **Step 6.** Reject any solutions obtained in Step 5 that are restricted values and state the solution set.

Example 1. Solve $\frac{1}{2t} - \frac{1}{6} = \frac{1}{4t} - 1 - \frac{1}{t}$.

Solution.

Step 1. The denominators are $2t$, 6, $4t$, and t.

Step 2. The LCD is $12t$.

Step 3. If $12t = 0$, then $t = 0$. Therefore, for the given equation, $t \neq 0$.

Step 4. Assuming $12t \neq 0$:

$$12t\left(\frac{1}{2t} - \frac{1}{6}\right) = 12t\left(\frac{1}{4t} - 1 - \frac{1}{t}\right) \qquad \text{Multiply both sides by LCD.}$$

$$6 - 2t = 3 - 12t - 12 \qquad \text{A linear equation in } t$$

Step 5. $6 - 2t = -12t - 9$ — Combine like terms.

$6 + 10t = -9$ — Add $12t$ to both sides.

$10t = -15$ — Subtract 6 from both sides.

$t = \frac{-15}{10}$ — Divide both sides by 10.

$t = \frac{-3}{2}$ — Reduce the fraction.

Step 6. Since $-\frac{3}{2}$ is not a restricted value, it is accepted as a solution of the given equation. Thus, the solution set is $\{-\frac{3}{2}\}$.

Example 2. Solve $\frac{1}{x + 2} = \frac{3}{x^2 + 2x} - \frac{1}{x}$.

Solution. **Step 1.** The denominators are $(x + 2)$, $x(x + 2)$, and x.

Step 2. The LCD is $x(x + 2)$.

Step 3. If $x(x + 2) = 0$, then $x = 0$ or $x + 2 = 0$. Therefore, for the given equation, $x \neq 0$ and $x \neq -2$.

Step 4. Assuming $x(x + 2) \neq 0$:

$$x(x + 2)\left(\frac{1}{x + 2}\right) = x(x + 2)\left(\frac{3}{x(x + 2)} - \frac{1}{x}\right)$$

$$x = 3 - (x + 2) \qquad \text{A linear equation in } x$$

Step 5. $x = 3 - x - 2$ — Remove parentheses.

$2x = 1$ — Add x to both sides.

$x = \frac{1}{2}$ — Divide both sides by 2.

Step 6. Since $\frac{1}{2}$ is not a restricted value, it is accepted as a solution of the given equation. Thus, the solution set is $\{\frac{1}{2}\}$.

Restricted Values and Apparent Solutions

The multiplication property of equality yields an equivalent equation provided $C \neq 0$. In Examples 1 and 2 we "assumed" the LCD was not 0 when we multiplied both sides of the equation by it. The solutions we got in Step 5 were actually **apparent solutions** of the given equation. They were not accepted as solutions of the **given equation** until it was verified that they were not values that made the LCD expression 0. If an apparent solution does make the LCD zero, then it is a restricted value for at least one of the terms of the rational equation, and is therefore rejected.

> A solution of a rational equation cannot be a restricted value for any of the terms of the equation.

Example 3. Solve $\dfrac{u+1}{u-1} - \dfrac{2}{u^2-u} = \dfrac{4}{u}$.

Solution.

Step 1. The denominators are $(u-1)$, $u(u-1)$, and u.

Step 2. The LCD is $u(u-1)$.

Step 3. If $u(u-1)=0$, then $u=0$ or $u-1=0$. Therefore, for the given equation, $u \neq 0$ and $u \neq 1$.

Step 4. Assuming $u(u-1) \neq 0$:

$$u(u-1)\left(\frac{u+1}{u-1} - \frac{2}{u(u-1)}\right) = u(u-1)\left(\frac{4}{u}\right)$$

$$u(u+1) - 2 = 4(u-1)$$

Step 5.

$u^2 + u - 2 = 4u - 4$	A quadratic equation in u
$u^2 - 3u + 2 = 0$	Set equal to 0.
$(u-2)(u-1) = 0$	Factor the trinomial.
$u - 2 = 0$ or $u - 1 = 0$	The zero-product property
$u = 2$ $u = 1$	Solve both equations.

Step 6. The apparent solution 1 is a restricted value and is therefore rejected. The apparent solution 2 is not a restricted value and is accepted. Thus, the solution set is $\{2\}$.

Example 4. Solve $\dfrac{y+7}{2y-6} = 1 + \dfrac{5}{y-3}$.

Solution.

Step 1. The denominators are $2(y-3)$ and $(y-3)$.

Step 2. The LCD is $2(y-3)$.

Step 3. If $2(y-3)=0$, then $y-3=0$. Therefore, for the given equation, $y \neq 3$.

Step 4. Assuming $2(y-3) \neq 0$:

$$2(y-3)\left(\frac{y+7}{2(y-3)}\right) = 2(y-3)\left(1 + \frac{5}{y-3}\right)$$

$y + 7 = 2(y-3) + 2(5)$ A linear equation in y

Step 5. $y + 7 = 2y - 6 + 10$ Remove parentheses.

$y + 7 = 2y + 4$ Combine like terms.

$3 = y$ Add $-y$ and -4 to both sides.

Step 6. The apparent solution 3 is a restricted value and is therefore rejected. Since there are no other solutions, the solution set of the given equation is $\varnothing$.

A Procedure for Solving a Rational Inequality

If the equals sign in a rational equation is replaced by one of the forms of the inequality relation (that is, $<$, $\leq$, $>$, or $\geq$), then a rational inequality is formed.

Definition 4.5. Rational inequality
A **rational inequality** is one in which at least one term of the inequality is a rational expression.

Examples **e** and **f** illustrate rational inequalities.

e. $\dfrac{t+3}{t-5} > 0$ **f.** $\dfrac{4}{u+1} \leq 1$

Recall the multiplication property of inequality from Chapter 2:

If $a < b$ and **Case 1:** $c > 0$, then $ac < bc$. Order is unchanged.
Case 2: $c < 0$, then $ac > bc$. Order is reversed.

A given order relation is unchanged or reversed, depending on whether the multiplication involves a *positive* or *negative number, respectively.* As a consequence, if an inequality is multiplied by a **variable expression,** then some statement must be made about whether the expression represents a positive number or a negative number.

To illustrate, consider the inequality in example **f.** Suppose we want to multiply both sides of the inequality by $(u + 1)$. Such a multiplication can only be done by first stating whether $(u + 1)$ is positive or negative. The procedure could be shown as follows:

Case 1: $(u + 1) > 0$ and $(u+1)\left(\dfrac{4}{u+1}\right) \leq (u+1)(1)$

Order is unchanged.

Case 2: $(u + 1) < 0$ and $(u+1)\left(\dfrac{4}{u+1}\right) \geq (u+1)(1)$

Order is reversed.

The union of the solution sets of Cases 1 and 2 is then the solution set of the given inequality. Rather than using such an algebraic approach to solving the inequality, a geometric procedure using a number line is frequently easier. The details of the procedure are identified in the following six steps:

To solve a rational inequality with a number line:

Step 1. If necessary, make one side of the inequality 0.

Step 2. Write the nonzero side of the inequality as a single rational expression.

Step 3. Write the numerator and denominator in factored form.

Step 4. On a number line, plot the numbers that make any factor, on top or bottom, equal to 0. Use these numbers to divide the number line into separated regions.

Step 5. Pick a test point in each region and evaluate the given inequality with the coordinate of that point.

Step 6. Select as solutions the numbers on any intervals that make the inequality true.

Example 5. Solve and graph $\dfrac{t+3}{t-5} > 0$.

Solution. **Step 1.** If necessary, make one side of the inequality 0.

$$\frac{t+3}{t-5} > 0 \qquad \text{The right side is 0.}$$

Step 2. Write the nonzero side as a single expression.

$$\frac{t+3}{t-5} > 0 \qquad \text{The given ratio is a single expression.}$$

Step 3. Write top and bottom in factored form.

$$\frac{t+3}{t-5} > 0 \qquad \text{Both expressions are prime polynomials.}$$

Step 4. Plot any numbers that make each factor 0.
If $t + 3 = 0$, then $t = -3$.
If $t - 5 = 0$, then $t = 5$.
In Figure 4-9, -3 and 5 divide the number line into three separated regions.

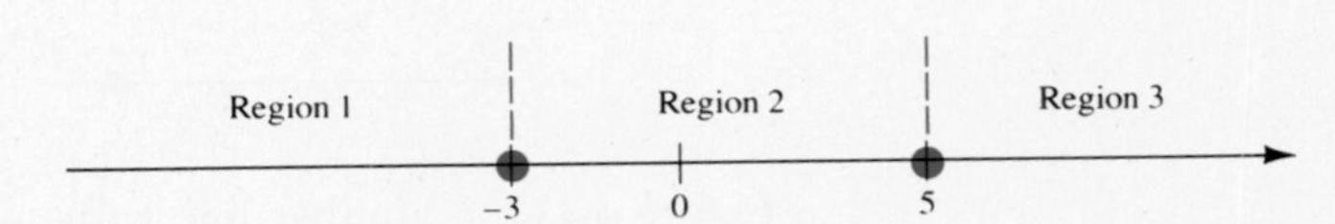

Figure 4-9. A number line with -3 and 5 plotted.

Step 5. Evaluate the inequality with the coordinate of a test point in each region.

In Region 1, select -4:

$\frac{-4+3}{-4-5} > 0$ and $\frac{-1}{-9} > 0$ is *true* — Region 1 contains solutions.

In Region 2, select 0:

$\frac{0+3}{0-5} > 0$ and $\frac{3}{-5} > 0$ is *false* — Region 2 does not contain solutions.

In Region 3, select 6:

$\frac{6+3}{6-5} > 0$ and $\frac{9}{1} > 0$ is *true* — Region 3 contains solutions.

Step 6. Region 1 contains solutions; thus, $t < -3$. Region 3 contains solutions; thus, $t > 5$. The solution set is $\{t \mid t < -3 \text{ or } t > 5\}$. A graph of the solution set is shown in Figure 4-10.

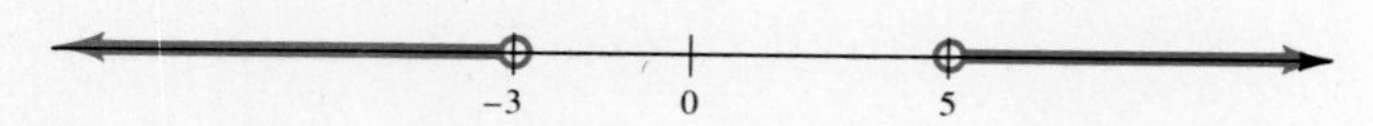

Figure 4-10. A graph of $\{t \mid t < -3 \text{ or } t > 5\}$.

Example 6. Solve and graph $\frac{4}{u+1} \leq 1$.

Solution.

Step 1. $\frac{4}{u+1} - 1 \leq 0$ — Subtract 1 from both sides.

Step 2. $\frac{4-(u+1)}{u+1} \leq 0$ — The LCD is $(u+1)$.

Step 3. $\frac{3-u}{u+1} \leq 0$ — Simplify the numerator.

Step 4. In Figure 4-11, the numbers 3 and -1 are plotted. — If $u = 3$, then $3 - u = 0$. If $u = -1$, then $u + 1 = 0$.

Step 5.

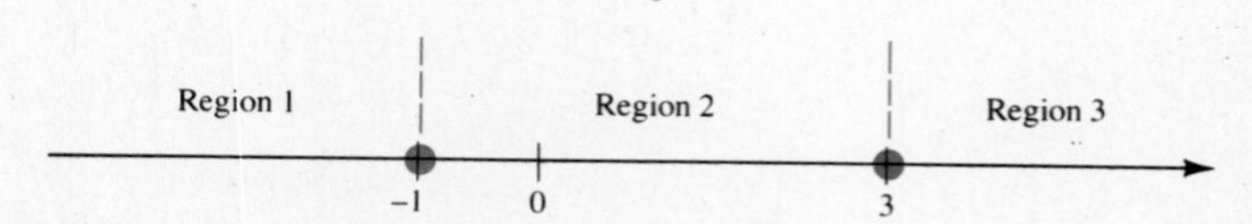

Figure 4-11. A number line with -1 and 3 plotted.

In Region 1, select -2:

$\frac{4}{-2+1} \leq 1$ and $-4 \leq 1$ is *true* — Region 1 contains solutions.

In Region 2, select 0:

$\frac{4}{0+1} \le 1$ and $4 \le 1$ is *false* — Region 2 does not contain solutions.

In Region 3, select 4:

$\frac{4}{4+1} \le 1$ and $\frac{4}{5} \le 1$ is *true* — Region 3 contains solutions.

Step 6. Regions 1 and 3 contain solutions. The inequality is $\le$; therefore, 3 is included in the set. However, -1 makes the denominator 0 and is not a solution. The solution set is $\{u \mid u < -1 \text{ or } u \ge 3\}$. A graph of the solution set is shown in Figure 4-12.

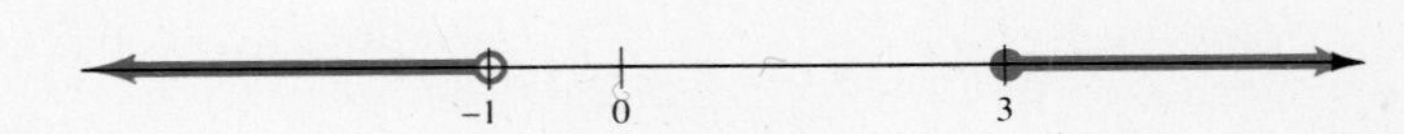

Figure 4-12. A graph of $\{u \mid u < -1 \text{ or } u \ge 3\}$.

Example 7. Solve and graph $\frac{3}{2x-5} < \frac{1}{x+4}$.

Solution.

Step 1. $\frac{3}{2x-5} - \frac{1}{x+4} < 0$ — Subtract $\frac{1}{x+4}$ from both sides.

Step 2. $\frac{3(x+4)-(2x-5)}{(2x-5)(x+4)} < 0$ — The LCD is $(2x-5)(x+4)$.

Step 3. $\frac{3x+12-2x+5}{(2x-5)(x+4)} < 0$ — Remove parentheses in numerator.

$\frac{x+17}{(2x-5)(x+4)} < 0$ — Simplify numerator.

Step 4. In Figure 4-13, the numbers -17, -4, and $\frac{5}{2}$ are plotted. — If $x = -17$, then $x + 17 = 0$. If $x = \frac{5}{2}$, then $2x - 5 = 0$. If $x = -4$, then $x + 4 = 0$.

Step 5.

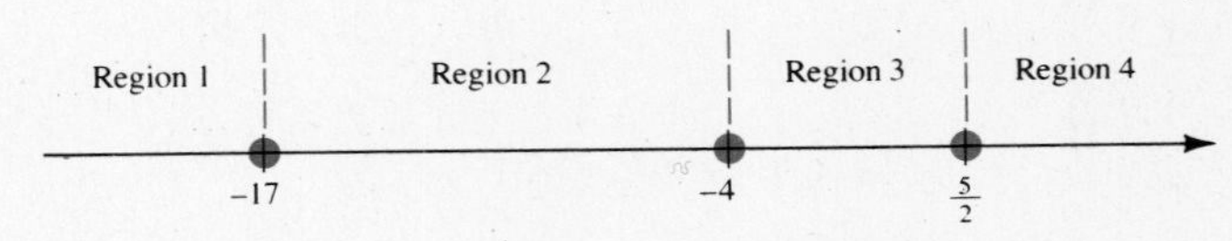

Figure 4-13. A number line with -17, -14, and $\frac{5}{2}$ plotted.

In Region 1, select -18:

$\frac{3}{2(-18)-5} < \frac{1}{-18+4}$ and $\frac{-3}{41} < \frac{-1}{14}$ is *true*

In Region 2, select -5:

$$\frac{3}{2(-5)-5} < \frac{1}{-5+4} \quad \text{and} \quad \frac{-1}{5} < -1 \text{ is } \textit{false}$$

In Region 3, select 0:

$$\frac{3}{2(0)-5} < \frac{1}{0+4} \quad \text{and} \quad \frac{-3}{5} < \frac{1}{4} \text{ is } \textit{true}$$

In Region 4, select 3:

$$\frac{3}{2(3)-5} < \frac{1}{3+4} \quad \text{and} \quad 3 < \frac{1}{7} \text{ is } \textit{false}$$

Step 6. Regions 1 and 3 contain solutions. The solution set is $\{x \mid x < -17 \text{ or } -4 < x < \frac{5}{2}\}$. A graph of the solution set is shown in Figure 4-14.

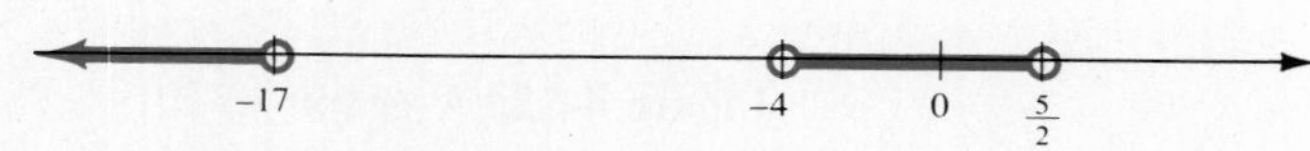

Figure 4-14. A graph of $\{x \mid x < -17 \text{ or } -4 < x < \frac{5}{2}\}$.

SECTION 4-6. Practice Exercises

In exercises **1–30**, solve each equation.

[Examples 1 and 2]

1. $\frac{2x}{9} - \frac{1}{2} = \frac{1}{6} - \frac{x}{9}$

2. $\frac{x}{10} + \frac{1}{5} = 1 - \frac{3x}{50}$

3. $\frac{5b}{4} + \frac{5}{3} = 1 - \frac{3b}{4} - \frac{13}{3}$

4. $\frac{25b}{9} + 1 + \frac{7b}{3} = \frac{2b}{3} - \frac{15}{9}$

5. $\frac{2}{5} - \frac{13}{15r} = \frac{1}{10} + \frac{14}{15r}$

6. $\frac{3}{10} - \frac{9}{20r} = \frac{9}{10r} - \frac{3}{20}$

7. $\frac{1}{3} + \frac{t+1}{4t} = \frac{1}{3t} - \frac{t+5}{6t}$

8. $\frac{2t+15}{3t} = \frac{1}{6} + \frac{5-t}{3t}$

9. $\frac{a+6}{a+2} = \frac{a+2}{a-1}$

10. $\frac{a-4}{a-2} = \frac{a}{a+1}$

11. $\frac{5t-4}{t+3} = 0$

12. $\frac{t-5}{t+6} = 0$

13. $\frac{2}{k} + \frac{3}{k^2-1} = \frac{2}{k+1}$

14. $\frac{1}{k-4} + \frac{4}{k^3-16k} = \frac{1}{k+4}$

15. $\frac{b}{2b+1} - \frac{1}{2b-1} = \frac{b+1}{2b+1} - \frac{8}{4b^2-1}$

16. $\dfrac{b}{b+4} + \dfrac{1}{b+1} = \dfrac{b+3}{b+4} - \dfrac{5}{b^2+5b+4}$

17. $\dfrac{1-4t}{t^2-1} - \dfrac{t}{1-t} = 1 + \dfrac{3}{t^2-1}$

18. $\dfrac{-t}{2-2t} - \dfrac{1}{2} - \dfrac{t-3}{t^2+t-2} = \dfrac{6}{2t+4}$

19. $\dfrac{2}{y^2-3y+2} = \dfrac{2}{y^2+y-6} - \dfrac{1}{y^2+2y-3}$

20. $\dfrac{1}{y^2+y-2} - \dfrac{3}{y^2-2y-8} = \dfrac{1}{y^2-5y+4}$

[Examples 3 and 4]

21. $\dfrac{2u+5}{4u} = \dfrac{5}{4u} - \dfrac{1}{3}$

22. $\dfrac{3u+2}{3u} = \dfrac{1}{3} + \dfrac{2}{3u}$

23. $\dfrac{b-4}{b-1} = \dfrac{b+3}{b-1}$

24. $\dfrac{b+4}{b-3} = \dfrac{b-5}{b-3}$

25. $\dfrac{p}{2p+4} = \dfrac{4}{p^2-4} - \dfrac{p}{2p-4}$

26. $\dfrac{p}{p+2} = \dfrac{6}{p^2+p-2} + \dfrac{1}{p-1}$

27. $\dfrac{z-1}{z-3} = \dfrac{4}{z^2-5z+6} + \dfrac{2z}{z-2}$

28. $1 + \dfrac{2}{z-1} = \dfrac{2}{z^2-z}$

29. $\dfrac{x^2-1}{x^2+10x+21} = \dfrac{2}{x+3}$

30. $\dfrac{x^2-14}{x^2+9x+20} = \dfrac{2}{x+4}$

In exercises **31–50**, solve and graph each inequality.

[Examples 5-7]

31. $\dfrac{p-3}{p+3} < 0$

32. $\dfrac{p+5}{p-5} > 0$

33. $\dfrac{3x+1}{x-3} \geq 0$

34. $\dfrac{2x-9}{x+2} \leq 0$

35. $\dfrac{4}{w} > 1$

36. $\dfrac{7}{w} < 2$

37. $\dfrac{10}{t+1} < 3$

38. $\dfrac{5}{2t-1} > 2$

39. $\dfrac{a}{a-3} \leq \dfrac{a}{a+3}$

40. $\dfrac{a}{a-5} \geq \dfrac{a}{a+2}$

41. $\dfrac{y-1}{y-6} \geq \dfrac{y+1}{y+2}$

42. $\dfrac{y-4}{y-10} \leq \dfrac{y+2}{y+6}$

43. $\dfrac{x+1}{x-5} < \dfrac{x}{x-3}$

44. $\dfrac{x}{x+6} > \dfrac{x+2}{x-2}$

45. $2 \leq \dfrac{t^2-2}{t-1}$

46. $1 > \dfrac{t^2-1}{4t-1}$

47. $\dfrac{2}{w+1} < 1 - \dfrac{w}{w+2}$

48. $\dfrac{-6}{w+3} > 2 - \dfrac{2w}{w-3}$

49. $\frac{1}{b} + \frac{1}{b+2} > \frac{6}{b^2 + 2b}$

50. $\frac{3}{b+4} < \frac{23}{b^2 + 4b} - \frac{2}{b}$

SECTION 4-6. Ten Review Exercises

In exercises **1–4**, solve each equation.

1. $5(3 - 4k) = 9 + 2(3 - k)$

2. $0 = |3x - 1| - 8$

3. $\frac{6m + 1}{8m} = \frac{13}{4} + \frac{5}{8m}$

4. $6t^2 - 2 = -11t$

In exercises **5–8**, solve and graph each inequality.

5. $3(u - 4) - 12 < 9u - 6(1 - 3u)$

6. $|3 - v| < 9$

7. $|2w - 5| \geq 13$

8. $\frac{x}{x-5} < \frac{x}{x+2}$

In exercises **9** and **10**, simplify each expression. Assume all variables represent positive numbers.

9. $(2^{-3}y^2)^{-4}(2^5y^{-3})^{-2}$

10. $\frac{(5^{-3}z^4)^{-2}}{(5z^{-1})^4}$

SECTION 4-6. Supplementary Exercises

In exercises **1–14**, solve each equation.

1. $\frac{x + 7}{x} = 0$

2. $\frac{2z + 3}{z} + \frac{z}{3} = 0$

3. $2 + \frac{3}{y - 2} - \frac{3}{y^2 - 2y} = 0$

4. $\frac{x}{x + 3} - \frac{2}{x + 1} = 0$

5. $\frac{1}{z + 1} - \frac{3}{z + 4} + \frac{5}{z^2 + 5z + 4} = 0$

6. $\frac{8}{4z^2 - 1} - \frac{1}{2z - 1} - \frac{1}{2z + 1} = 0$

7. $\frac{2}{a + 1} - \frac{2}{a} - \frac{3}{a^2 - 1} = 0$

8. $\frac{1}{a + 4} - \frac{1}{a - 4} - \frac{4}{a^3 - 16a} = 0$

9. $\frac{44}{x^2 - 9} - \frac{x}{x + 3} + 5 = 0$

10. $\frac{x - 10}{x^2 - 16} - \frac{x}{x - 4} - \frac{2}{3} = 0$

11. $\frac{y}{y - 2} - \frac{y}{2y + 4} - \frac{4}{4 - y^2} = 0$

12. $\frac{4}{y + 5} - \frac{y}{y - 5} - \frac{50}{25 - y^2} = 0$

13. $\frac{z}{2z + 4} + \frac{z}{2z - 4} - \frac{4}{z^2 - 4} = 0$

14. $\frac{z}{z + 2} - \frac{1}{z - 1} - \frac{6}{z^2 + z - 2} = 0$

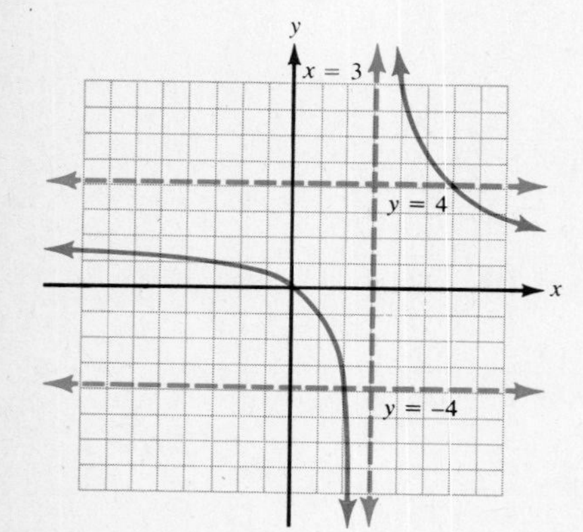

Figure 4-15. Graphs of $y = \frac{2x}{x-3}$, $y = 4$, and $y = -4$.

Example. Figure 4-15 contains graphs of $y = \dfrac{2x}{x-3}$, $y = 4$, and $y = -4$. Use these graphs to find the following solution sets of each inequality.

a. $\dfrac{2x}{x-3} > 4$ **b.** $\dfrac{2x}{x-3} < -4$

Solution.

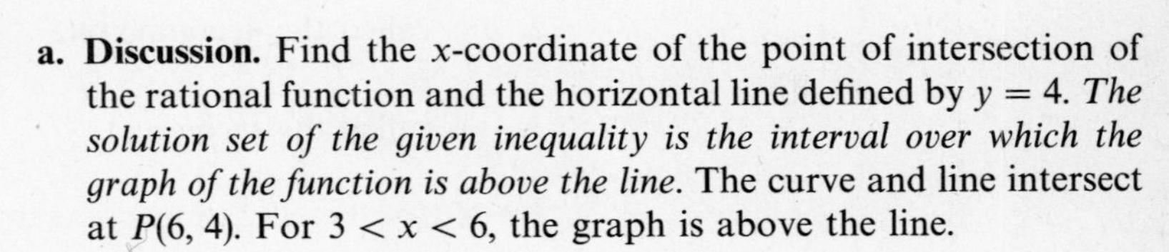

a. Discussion. Find the x-coordinate of the point of intersection of the rational function and the horizontal line defined by $y = 4$. *The solution set of the given inequality is the interval over which the graph of the function is above the line.* The curve and line intersect at $P(6, 4)$. For $3 < x < 6$, the graph is above the line.

The solution set is $\{x \mid 3 < x < 6\}$.

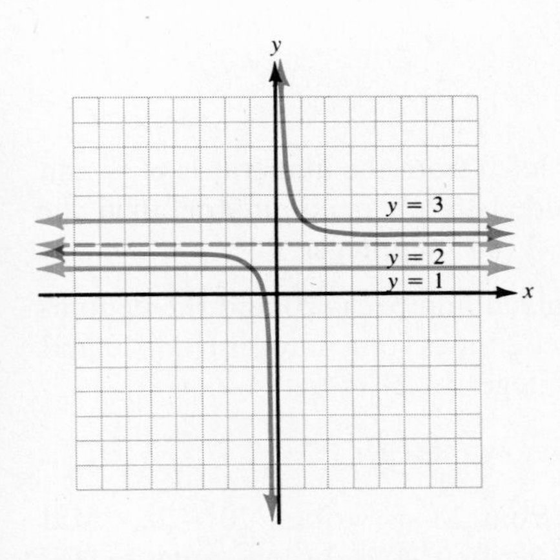

Figure 4-16. A graph of $y = \frac{2x+1}{x}$.

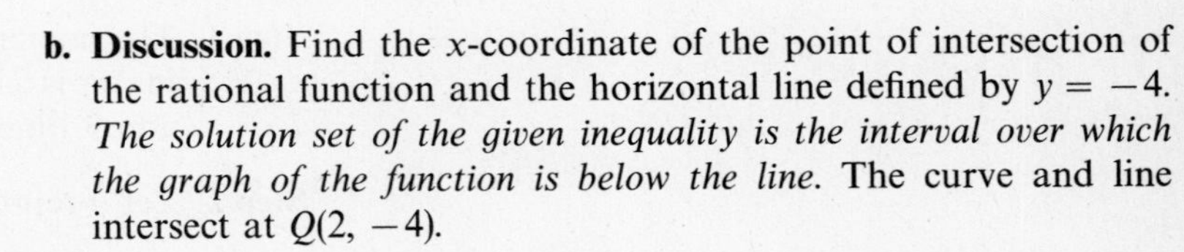

b. Discussion. Find the x-coordinate of the point of intersection of the rational function and the horizontal line defined by $y = -4$. *The solution set of the given inequality is the interval over which the graph of the function is below the line.* The curve and line intersect at $Q(2, -4)$.

For $2 < x < 3$, the graph is below the line.

The solution set is $\{x \mid 2 < x < 3\}$.

In exercises **15** and **16**, use the graph in Figure 4-16 to find the solution of each inequality. Verify your answers by solving the inequalities.

15. $\dfrac{2x+1}{x} > 3$

16. $\dfrac{2x+1}{x} < 1$

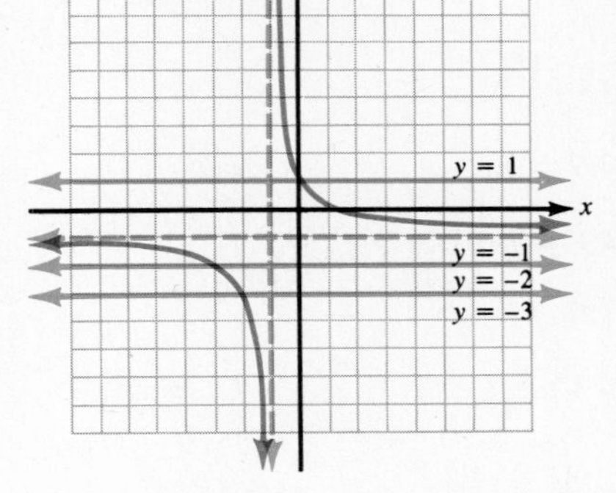

Figure 4-17. A graph of $y = \frac{1-x}{1+x}$.

In exercises **17–20**, use the graph in Figure 4-17 to find the solution of each inequality. Verify your answers by solving the inequalities.

17. $\dfrac{1-x}{1+x} > 1$

18. $\dfrac{1-x}{1+x} < -3$

19. $\dfrac{1-x}{1+x} \geq 0$

20. $\dfrac{1-x}{1+x} \leq -2$

SECTION 4-7. Applied Problems

1. Number problems
2. Work problems
3. Uniform motion problems
4. Literal equations with fractions

In this section, we will solve some applied problems using equations that have at least one rational expression. We can use the same procedure we used to solve rational equations in Section 4-6.

Number Problems

If $\frac{a}{b}$ is a rational number not equal to 0, then:

1. a is called the **numerator.**

2. b is called the **denominator.**

3. $\frac{b}{a}$ is called the **reciprocal.**

Example 1. If 2 times some integer is subtracted from the numerator of $\frac{20}{21}$ and 3 times the same integer is added to the denominator, then the rational number reduces to $\frac{7}{15}$. Find the integer.

Solution. **Step 1.** The numerator of the given number is 20 and the denominator is 21. We subtract 2 times some integer from 20 and add 3 times the same integer to 21 to get $\frac{7}{15}$.

Step 2. Let n represent the unknown integer.

Step 3. "Subtract two times n from 20" is written $20 - 2n$. "Add three times n to 21" is written $20 + 3n$. An equation for the problem is $\frac{20 - 2n}{21 + 3n} = \frac{7}{15}$.

Step 4. The LCD of $3(7 + n)$ and 15 is $15(7 + n)$.

$$15(7 + n)\left(\frac{20 - 2n}{3(7 + n)}\right) = \frac{7}{15} \cdot 15(7 + n)$$

$$5(20 - 2n) = 7(7 + n)$$

$$100 - 10n = 49 + 7n$$

$$51 = 17n$$

$$3 = n$$

Step 5. $\frac{20 - 2(3)}{21 + 3(3)} = \frac{14}{30} = \frac{7}{15}$, *check*

Step 6. The integer is 3.

Example 2. The sum of a rational number and 2 times its reciprocal is $\frac{17}{6}$. Find the number.

Solution. **Step 1.** The reciprocal of a number is 1 over the number.

Step 2. Let r represent the rational number.

Step 3. "Two times the reciprocal of r" is written $2 \cdot \frac{1}{r}$, or $\frac{2}{r}$.

$$r + \frac{2}{r} = \frac{17}{6}$$

Step 4.

$$6r\left(r + \frac{2}{r}\right) = 6r \cdot \frac{17}{6} \qquad \text{The LCD is } 6r.$$

$$6r^2 + 12 = 17r \qquad \text{A quadratic equation in } r$$

$$6r^2 - 17r + 12 = 0 \qquad \text{Write in standard form.}$$

$$(3r - 4)(2r - 3) = 0 \qquad \text{Factor the left side.}$$

$$3r - 4 = 0 \quad \text{or} \quad 2r - 3 = 0 \qquad \text{The zero-product property}$$

$$r = \frac{4}{3} \qquad r = \frac{3}{2} \qquad \text{Solve both equations.}$$

Step 5. If $r = \frac{4}{3}$, then $\frac{1}{r} = \frac{3}{4}$.

$$\frac{4}{3} + 2\left(\frac{3}{4}\right) = \frac{4}{3} + \frac{3}{2} = \frac{8}{6} + \frac{9}{6} = \frac{17}{6}, \ \textit{check}$$

If $r = \frac{3}{2}$, then $\frac{1}{r} = \frac{2}{3}$.

$$\frac{3}{2} + 2\left(\frac{2}{3}\right) = \frac{3}{2} + \frac{4}{3} = \frac{9}{6} + \frac{8}{6} = \frac{17}{6}, \ \textit{check}$$

Step 6. The rational number is $\frac{4}{3}$ or $\frac{3}{2}$.

Work Problems

The following principles are used to solve work problems.

Principle 1. If the amount of time needed to complete some piece of work is t, then the rate at which work is done is r, and:

$$r = \frac{1}{t}$$

Examples **a–c** illustrate the equation.

Work to be completed	Time needed to complete the work	Rate at which work is done
a. Build a wall	8 days	$\frac{1}{8}$ of wall per day
b. Fill a pool	25 hours	$\frac{1}{25}$ of pool filled per hour
c. Process some data	45 minutes	$\frac{1}{45}$ of data processed per minute

Principle 2. The product of the amount of time worked and the rate at which work is done must equal 1 to show that the job is completed. That is, $r \cdot t = 1$. The 1 represents "one job completed".

Principle 3. When two or more "machines" are used on a job, then the sum of the products of times worked and rates of work must equal 1. Thus, if t_A and r_A are the time and rate for machine A and t_B and r_B are the time and rate for machine B, then:

$$r_A \cdot t_A + r_B \cdot t_B = 1$$

Example 3. Charlotte Strauser is the accounting manager at a branch of a large banking institution. One of her responsibilities is doing a daily computer run of the checking accounts at this branch office.

Computer AK3027 can process the data in 30 minutes. Computer 2T588 can process the same data in 40 minutes, and both computers can work on the data simultaneously. Last Friday, Charlotte was using AK3027 for 10 minutes when 2T588 became available, and she used it to help finish the job. On this day, how long did it take to process all the data?

Solution. Let m represent the number of minutes the AK3027 worked on the data on this day. Then $(m - 10)$ represents the number of minutes the 2T588 computer worked.

Machines	Time to process data alone	Rate at which data is processed	Amount of work done on this day
AK3027	30 minutes	$\frac{1}{30}$ per minute	$\frac{1}{30} \cdot m$ or $\frac{m}{30}$
2T588	40 minutes	$\frac{1}{40}$ per minute	$\frac{1}{40}(m - 10)$ or $\frac{m - 10}{40}$

$$\begin{pmatrix}\text{Work done}\\ \text{by AK3027}\end{pmatrix} + \begin{pmatrix}\text{Work done}\\ \text{by 2T588}\end{pmatrix} = \begin{pmatrix}\text{One data}\\ \text{set processed}\end{pmatrix}$$

$$\frac{m}{30} + \frac{m - 10}{40} = 1$$

$$120\left(\frac{m}{30} + \frac{m - 10}{40}\right) = 120 \cdot 1$$

$$4m + 3(m - 10) = 120$$

$$7m = 150$$

$$m = 21\frac{3}{7} \text{ minutes}$$

Under the stated conditions, it took about $21\frac{3}{7}$ minutes to process the data on this day.

Example 4. Joe Bennett works at Della Bueno Winery. One of the large holding tanks at the winery is filled by inlet valve A in 8 hours. Outlet valve B can empty the tank in 20 hours and is connected to smaller holding tanks. Yesterday, Joe wanted to fill the large tank, but in the process he let valve B remain open during the first 2 hours of filling. At the end of the 2 hours, he closed valve B and filled the tank using only valve A. Find the total time valve A was open during the filling process.

Solution. **Discussion.** Valve B operates in opposition to valve A; thus, we assign a negative rate to the work it does. Let t represent the time valve A is open.

Machines	Time open	Rate	Amount of work
Valve A	t hours	$\frac{1}{8}$ per hour	$t \cdot \frac{1}{8}$ or $\frac{t}{8}$
Valve B	2 hours	$\frac{-1}{20}$ per hour	$2 \cdot \frac{-1}{20}$ or $\frac{-1}{10}$

$$\begin{pmatrix}\text{Work done}\\ \text{by A}\end{pmatrix} + \begin{pmatrix}\text{Work done}\\ \text{by B}\end{pmatrix} = \begin{pmatrix}\text{One holding}\\ \text{tank filled}\end{pmatrix}$$

$$\frac{t}{8} + \frac{-1}{10} = 1$$

$$40\left(\frac{t}{8} - \frac{1}{10}\right) = 40 \cdot 1$$

$$5t - 4 = 40$$

$$5t = 44$$

$$t = 8.8 \text{ hours}$$

Under the stated conditions, it took 8.8 hours to fill the tank.

Uniform Motion Problems

The equation $d = rt$ can be used to find the distance d that an object moves in a given time t when moving at a uniform rate r. When the distance and time are known values, then the rate can be found using Equation (1):

$$(1) \quad r = \frac{d}{t}$$

When the distance and rate are known values, then the time can be found using Equation (2):

$$(2) \quad t = \frac{d}{r}$$

Example 5. A model BL5T jet made a test run of 825 miles in 1.5 hours aided by a 50 mph tailwind. The same plane made a run of 675 miles in the same time against the 50 mph wind. Based on these data, what was the speed of the plane without the effect of the wind?

Solution. Let r represent the speed of the plane without the wind.

Model BL5T	Rate	Distance	Time
With wind	$(r + 50)$ mph	825 miles	$\frac{825}{r + 50}$ hours
Against wind	$(r - 50)$ mph	675 miles	$\frac{675}{r - 50}$ hours

$$\begin{pmatrix}\text{Time traveled}\\ \text{with wind}\end{pmatrix} = \begin{pmatrix}\text{Time traveled}\\ \text{against wind}\end{pmatrix}$$

$$\frac{825}{r + 50} = \frac{675}{r - 50}$$

$$825(r - 50) = 675(r + 50)$$

$$825r - 41{,}250 = 675r + 33{,}750$$

$$150r = 75{,}000$$

$$r = 500 \text{ mph}$$

Based on these data, the model BL5T averaged 500 mph without the wind.

Literal Equations with Fractions

Many literal equations in mathematics contain one or more terms that are rational expressions. To solve for one of the variables in the equation may require the technique for solving fractional equations. In the following examples, the variables are suitably restricted so that no denominator is 0.

Example 6. Solve $\frac{1}{R} = \frac{1}{r_1} + \frac{1}{r_2}$ for R.

A formula for calculating the total resistance (R) of two electrical resistors (r_1 and r_2) when connected in parallel

Solution.

$$Rr_1r_2\left(\frac{1}{R}\right) = Rr_1r_2\left(\frac{1}{r_1} + \frac{1}{r_2}\right)$$ The LCD is Rr_1r_2.

$$r_1r_2 = Rr_2 + Rr_1$$ Simplify both sides.

$$r_1r_2 = R(r_1 + r_2)$$ Factor the right side.

$$\frac{r_1r_2}{r_1 + r_2} = R$$ Divide by $(r_1 + r_2)$.

$$R = \frac{r_1r_2}{r_1 + r_2}$$ If $x = y$, then $y = x$.

Example 7. Solve $y = \frac{2x + 5}{x - 4}$ for x.

Solution. **Discussion.** Notice that there are two x-terms in the equation. To solve for x, we need to combine these terms on the same side of the equation. Then use the distributive property to change the two terms to one term.

$$(x - 4)y = (x - 4)\left(\frac{2x + 5}{x - 4}\right)$$ The LCD is $(x - 4)$.

$$xy - 4y = 2x + 5$$ Distribute y on the left side.

$$xy - 2x = 4y + 5$$ Combine x-terms on left side.

$$x(y - 2) = 4y + 5$$ Factor an x.

$$x = \frac{4y + 5}{y - 2}$$ Divide by $(y - 2)$.

SECTION 4-7. Practice Exercises

In exercises **1–36**, solve each word problem.

[Example 1] **1.** Two times a number is added to the numerator of $\frac{2}{9}$, and the same number is subtracted from the denominator. The resulting number is $\frac{4}{3}$. Find the number.

2. Three times a number is subtracted from the numerator of $\frac{24}{35}$. Two times the same number is added to the denominator. The resulting number is $\frac{1}{5}$. Find the number.

3. What number must be added to the numerator and denominator of $\frac{18}{23}$ to yield a fraction whose value is $\frac{5}{6}$?

4. What number must be substracted from the numerator and denominator of $\frac{26}{31}$ to yield a fraction whose value is $\frac{3}{4}$?

5. A number is added to the numerator of $\frac{1}{8}$, and 2 times the number is added to the denominator. The value of the new fraction is $\frac{3}{4}$. Find the number.

6. A number is added to the numerator of $\frac{9}{7}$, and 3 times the number is added to the denominator. The value of the new fraction is $\frac{3}{4}$. Find the number.

7. A number is added to the numerator and one-half of the number is added to the denominator of $\frac{15}{16}$. Find the number if the result is 1.

8. A number is added to the numerator and one-fourth of the number is added to denominator of $\frac{16}{13}$. The result is $\frac{8}{5}$. Find the number.

[Example 2]

9. The sum of the reciprocals of two consecutive integers is equal to 17 times the reciprocal of the product of the two numbers. Find both numbers.

10. Three times the reciprocal of one integer plus 2 times the reciprocal of the next consecutive integer is equal to 28 times the reciprocal of the product of the two numbers. Find both numbers.

11. The sum of the reciprocal of 3 times a number and $\frac{1}{30}$ is equal to the reciprocal of 5 times the number. Find the number.

12. The difference between $\frac{5}{84}$ and 4 times the reciprocal of a number is equal to the reciprocal of 6 times the number. Find the number.

13. One number is 4 times a second number. The difference in their reciprocals is $\frac{3}{4}$. Find the numbers.

14. The larger of two numbers is 3 times the smaller. Two times the reciprocal of the smaller number plus the reciprocal of the larger is $\frac{1}{3}$. Find the numbers.

15. The sum of a number and 5 times its reciprocal is 6. Find the number.

16. The difference between a number and 4 times its reciprocal is 3. Find the number.

[Examples 3 and 4]

17. Computer A can process a quantity of data in 6 hours, while computer B can process the same amount of data in 10 hours. How long will it take both machines, working together, to process the data?

18. One well can irrigate a field of alfalfa in 15 hours, and a second well can irrigate the same field in 20 hours. How long does it take to irrigate the field using both wells?

19. Fred can paint a fence in 12 hours. Fred and Al together can paint the fence in 8 hours. How long would it take Al to paint the fence alone?

20. A sheet metal worker can complete a furnace job in 10 days. Working with a helper, the same job takes 6 days. How long would it take the helper to do the job alone?

21. An inlet valve can fill a pool in 10 hours, and a hose can fill the same pool in 24 hours. After the inlet valve has been operating 2 hours, the hose is also turned on. How long must the inlet valve remain open to fill the pool?

22. A tub for watering horses can be drained in 20 minutes. The inlet valve can fill the empty tub in 12 minutes. If the drain is open for the first 5 minutes the tank is being filled, how long will it take the inlet valve to fill the tub?

23. Sleepy Hollow Golf Course has three machines to cut the grass. Mower A can do the job in 15 hours, mower B can do the job in 20 hours, and mower C can do the job in 30 hours. How long does it take the three mowers, working together, to do the job?

24. One typist can type a quantity of material in 100 minutes, and a second typist can type the same amount of material in 2 hours. On a similar quantity of material, the two have been typing for 30 minutes when a third typist helps to finish the job in 20 minutes. How long would it take the third typist to do the typing alone?

[Example 5] 25. The current in a river is 4 mph. A ski boat can go 48 miles down the river with the current in the same time that it can go 36 miles up the river against the current. Find the speed of the boat in still water.

26. The current in a large river is 2 mph. Julie can paddle her canoe for $4\frac{3}{4}$ miles with the current in the same time she can paddle for $2\frac{3}{4}$ miles against the current. How fast can Julie paddle her canoe in still water?

27. Moving at 55 mph, a car takes 2 hours longer to travel a certain distance than a train that averages 75 mph. Find the distance.

28. Edna and Frank travel south from Cleveland to a small town. On the way down, they average 45 mph. On the way back, they average 55 mph. If the total travel time is 2 hours, how far south of Cleveland is the town?

29. An airliner can travel 1,860 miles in a given time with a 40 mph tail wind, but it can travel only 1,620 miles in the same time when the wind is a head wind. Find the speed of the airliner when no wind is present.

30. A small private plane can fly at a speed of 280 mph with no wind. When traveling with a tail wind, the plane can travel 930 miles in a set time, but it can travel only 750 miles in the same time when the wind is a head wind. Find the speed of the wind.

31. Rick and Nancy leave the same house for a common destination. Rick averages 10 mph on his bicyle, and Nancy averages 40 mph in her car. If Nancy can leave the house 30 minutes after Rick and arrive at the same time as Rick, how far from home is the destination?

32. A police helicopter spotted a truck 5 miles north of an intersection on an interstate highway. Ten minutes later a highway patrol car, traveling at $1\frac{1}{2}$ times the speed of the truck, left the intersection to overtake the truck. If the patrol car overtook the truck 45 miles north of the intersection, how long did it take the car to catch the truck?

33. A boat travels 30 miles upstream in the same time that it travels 45 miles downstream. If the speed of the boat in still water is 10 mph, find the rate of the current.

34. A plane with a tail wind can travel 400 miles in the time it takes to travel 300 miles against the same wind. Find the speed of the wind if the speed of the plane in still air is 175 mph.

35. It takes one train 1 hour longer to travel 300 miles than it takes another train traveling at the same rate to go 250 miles. Find the time of each train.

36. Sally and Sam both jog at the same rate. Sally can jog 20 miles in $\frac{1}{2}$ hour less time than Sam can jog 24 miles. Find Sally's and Sam's rates.

In exercises **37–48**, solve each equation for the indicated variable.

[Examples 6 and 7]

37. $\frac{1}{F} = \frac{1}{a} + \frac{1}{b}$, for F

38. $\frac{1}{t} = \frac{2}{a} - \frac{2}{b}$, for a

39. $\frac{y + a}{y - b} = \frac{b}{a}$, for y

40. $\frac{x - a}{x - b} = \frac{b}{a}$, for x

41. $\frac{1}{x} - \frac{3}{y} = \frac{1}{2}$, for x

42. $\frac{3}{x} + \frac{2}{y} = \frac{1}{5}$, for y

43. $\frac{t - 2}{t + 2} = \frac{a + b}{a - b}$, for t

44. $\frac{1}{3t} + \frac{1}{2t} = \frac{a}{b}$, for t

45. $y = \frac{3x}{x + 3}$, for x

46. $y = \frac{-2x}{3x - 4}$, for x

47. $y = k\left(\frac{x + 1}{x - 5}\right)$, for x

48. $y = t\left(\frac{2x - 1}{2x + 1}\right)$, for x

SECTION 4-7. Ten Review Exercises

In exercises **1** and **2**, evaluate each expression for the given value.

1. $5(4k - 1)$, for $k = \frac{-5}{4}$

2. $2(4k - 3) - 14$, for $k = \frac{-5}{4}$

3. Solve $5(4k - 1) = 2(4k - 3) - 14$.

In exercises **4** and **5**, evaluate each expression for the given value.

4. $\frac{1}{x - 4} - \frac{1}{x + 4}$, for $x = \frac{1}{2}$

5. $\frac{4}{x^3 - 16x}$, for $x = \frac{1}{2}$

6. Solve $\frac{1}{x - 4} = \frac{1}{x + 4} + \frac{4}{x^3 - 16x}$.

In exercises **7–10**, factor each expression.

7. $27t^3 - u^3$

8. $27t^2 - 3u^2$

9. $27t^2 + 18tu + 3u^2$

10. $27t^2 + 3u^2$

SECTION 4-7. Supplementary Exercises

In exercises **1–12**, answer parts **a** and **b**.

1. a. Add 3 to the numerator of $\frac{5}{6}$.
b. Is the number of part **a** greater than, equal to, or less than $\frac{5}{6}$?

2. a. Add a number $p (p > 0)$ to the numerator of $\frac{x}{y}$, where $y \neq 0$.
b. Is the number of part **a** greater than, equal to, or less than $\frac{x}{y}$?

3. a. Subtract 5 from the numerator of $\frac{9}{10}$.
b. Is the number of part **a** greater than, equal to, or less than $\frac{9}{10}$?

4. a. Subtract a number $p (p > 0)$ from the numerator of $\frac{x}{y}$, where $y \neq 0$.
b. Is the number of part **a** greater than, equal to, or less than $\frac{x}{y}$?

5. a. Add 4 to the denominator of $\frac{3}{5}$.
b. Is the number of part **a** greater than, equal to, or less than $\frac{3}{5}$?

6. a. Add a number $p (p > 0)$ to the denominator of $\frac{x}{y}$, where $y \neq 0$.
b. Is the number of part **a** greater than, equal to, or less than $\frac{x}{y}$?

7. a. Subtract 2 from the denominator of $\frac{5}{8}$.
b. Is the number of part **a** greater than, equal to, or less than $\frac{5}{8}$?

8. a. Subtract a number $p (p > 0)$ from the denominator of $\frac{x}{y}$, where $y \neq 0$.
b. Is the number of part **a** greater than, equal to, or less than $\frac{x}{y}$?

9. a. Add 6 to the numerator and denominator of $\frac{1}{3}$.
b. Is the number of part **a** greater than, equal to, or less than $\frac{1}{3}$?

10. a. Add a number $p (p > 0)$ to the numerator and denominator of $\frac{x}{y}$, where $y \neq 0$.
b. Is the number of part **a** greater than, equal to, or less than $\frac{x}{y}$?

11. a. Add 1,000 to the numerator and denominator of $\frac{7}{8}$.
b. Is there any number p that can be added to the numerator and denominator of $\frac{7}{8}$ that will yield a number greater than 1?

12. a. Subtract 4 from the numerator and denominator of $\frac{11}{9}$.
b. Is there any number p that can be subtracted from the numerator and denominator of $\frac{11}{9}$ that will yield a number less than 1?

In exercises **13–16**, use the formula $R = \dfrac{r_1 r_2}{r_1 + r_2}$ to find R for the indicated values of r_1 and r_2.

13. $r_1 = 30, r_2 = 50$

14. $r_1 = 50, r_2 = 100$

15. $r_1 = 100, r_2 = 1{,}000$

16. $r_1 = 10, r_2 = 1{,}000$

In exercises **17–20**:

a. State the restrictions on x.

b. Solve the equation for x in terms of y.

c. State the restrictions on y based on the equation obtained in part **b**.

17. $y = \dfrac{5}{x - 1}$

18. $y = \dfrac{9}{x + 3}$

19. $y = \dfrac{2x + 5}{x}$

20. $y = \dfrac{4x - 1}{x}$

In exercises **21** and **22**, answer parts **a–c**.

21. a. Solve $\dfrac{1}{R} = \dfrac{1}{r_1} + \dfrac{1}{r_2} + \dfrac{1}{r_3}$ for R.
b. Use the result of part **a** to find R for $r_1 = 10$, $r_2 = 15$ and $r_3 = 20$.
c. Compare R to r_1, r_2, and r_3.

22. a. Solve $\dfrac{1}{R} = \dfrac{1}{r_1} + \dfrac{1}{r_2} + \dfrac{1}{r_3} + \dfrac{1}{r_4}$ for R.
b. Use the results of part **a** to find R for $r_1 = 10$, $r_2 = 20$, $r_3 = 5$ and $r_4 = 30$.
c. Compare R to r_1, r_2, r_3, and r_4.

23. On a trip from city A to city B, Brad Hexom averaged 60 mph. On the return trip, he averaged 40 mph.
a. Make a guess as to the average rate for the complete round trip.
b. If x represents the distance between the two cities, what is the distance for the round trip?
c. Write an algebraic expression for the time of travel from A to B $\left(t = \dfrac{d}{r}\right)$.
d. Write an algebraic expression for the time of travel from B to A.
e. Use the expressions in parts **c** and **d** to write an algebraic expression for the total time of travel.
f. Write an algebraic expression for the round-trip rate using total distance and total time.
g. Simplify the expression in part ***f***.
h. Compare the result found in part **g** with your guess in part **a**.
i. Would the average rate have been different if the cities were a different distance apart?

24. On a trip from city A to city B, a car averaged r_1 mph. On the return trip, the car averaged r_2 mph.
a. If x represents the distance between the two cities, what is the distance of the round trip?
b. Write an algebraic expression for the time of travel from A to B $\left(t = \dfrac{d}{r}\right)$.
c. Write an algebraic expression for the time of travel from B to A.
d. Use the expressions from parts **b** and **c** to write an algebraic expression for the total time of travel.
e. Write an algebraic expression for the round-trip rate using total distance and total time.
f. Simplify the expression obtained in part **e**.
g. Would the distance x between the two cities affect the rate expression obtained in part **f**?

Name

Date

Score

SECTION 4-5. Summary Exercises

Answer

In exercises **1** and **2**, simplify each expression.

1. $\dfrac{x^2 - 3x - 7}{x^2 - 4x + 4} + \dfrac{x + 1}{x^2 - 7x + 10} \cdot \dfrac{3x - 15}{x - 2}$

1. ______

2. $(m - 3)(m + 3)^2\left(\dfrac{m - 1}{m^2 + 6m + 9} + \dfrac{m + 1}{m^2 - 9}\right)$

2. ______

3. Simplify $\dfrac{\dfrac{1}{x} + \dfrac{1}{y}}{\dfrac{1}{xy}}$ using Plan A.

3. ______

4. Simplify $\dfrac{\dfrac{x}{y} - 2 - \dfrac{24y}{x}}{\dfrac{1}{y^2} - \dfrac{10}{xy} + \dfrac{24}{x^2}}$ using Plan B.

4. ______

In exercises **5–8**, simplify each expression.

5. $\dfrac{y - \dfrac{16}{y}}{4 - \dfrac{16}{y}}$

5. ____________________

6. $\dfrac{x + 5 - \dfrac{10}{x + 2}}{x + 2 - \dfrac{25}{x + 2}}$

6. ____________________

7. $\dfrac{1 - 2^{-1}n}{2^{-1}n - 1}$

7. ____________________

8. $\dfrac{6 + 7x^{-1} - 3x^{-2}}{9 + 3x^{-1} - 2x^{-2}}$

8. ____________________

Name

Date

Score

SECTION 4-7. Summary Exercises

Answer

In exercises **1–4**, solve each word problem.

1. An integer was added to the denominator of $\frac{-5}{2}$ and $\frac{1}{2}$ the integer was added to the numerator. The result was $\frac{1}{8}$. Find the integer.

1. ______

2. Norm's Rapid Print Shoppe had a large order of promotional material to print. His best machine could print the material in 40 minutes. His old machine could do the work in 60 minutes. Norm started the job using only the old machine. However, 5 minutes later he also started using the best machine. Under this plan of machine usage, how long did it take to complete the job?

2. ______

3. A canoeist can paddle at a rate of 8 mph in still water. She can travel 18 miles against a current in the same time that she can travel 30 miles with the current. Find the rate of the current.

3. ________________

4. Given: $y = \dfrac{5 - x}{2x + 1}$

a. State any restricted values for x.

4. a. ________________

b. Solve the equation for x in terms of y.

b. ________________

c. State any restricted values for y based on part **b**.

c. ________________

CHAPTER 4. Review Exercises

In exercises **1–8**, simplify each expression. Write answers with only positive exponents. Assume all variables are not 0.

1. $(-4)^{-3}$

2. $\left(\frac{3}{5}\right)^{-2}$

3. $-2^{-1}x^2y^{-3}$

4. $\frac{-5^{-2}}{a^{-1}b^3}$

5. $\frac{5^3 \cdot 5^{-1}}{5^{-4} \cdot 5^6}$

6. $\left(\frac{2a^{-2}b^3}{5^{-1}a^4b^{-2}}\right)^{-3}$

7. $\frac{(3^{-1}p^{-2}q)^{-3}}{(6^{-1}pq^{-3})^{-2}}$

8. $\left(\frac{6x^{-3}y}{-7x^2y^{-3}}\right)^0$

In exercises **9** and **10**, write each number in scientific notation.

9. 3,970,000

10. 0.0000196

In exercises **11** and **12**, write each number in ordinary notation.

11. 9.35×10^{-3}

12. 3.05×10^5

In exercises **13–16**, find any restricted values for each expression. For any expression that has no restricted values, write "none".

13. $\frac{5x - 1}{x - 5}$

14. $\frac{y^2 + 36}{y^2 - 36}$

15. $\frac{a^2 - ab + b^2}{a^2 + 100b^2}$

16. $\frac{5}{m^2 - 9mn + 20n^2}$

In exercises **17–20**, reduce each expression. Assume that no denominator is 0.

17. $\frac{18m^3n^2}{42mn^3}$

18. $\frac{x^2 + 10x + 25}{5x^2 + 26x + 5}$

19. $\frac{y^3 + 1}{y^2 - 1}$

20. $\frac{a^3 + a^2b - 3a - 3b}{a^3 - 15 + 5a^2 - 3a}$

In exercises **21–24**, write equivalent expressions with the indicated denominators.

21. $\frac{4}{3ab}$ with denominator $12a^2b^2$

22. $\frac{1}{x - 2y}$ with denominator $5x^2y - 10xy^2$

23. $\frac{m}{2m + 3}$ with denominator $4m^2 - 9$

24. $\frac{2p - 3}{p + 5}$ with denominator $p^2 - 2p - 35$

In exercises **25–34**, do the indicated operations. Write answers in reduced form.

25. $\dfrac{8x^3}{15y^5} \cdot \dfrac{5y^4}{4x}$

26. $\dfrac{7x^3}{16y} \div \dfrac{49x^5}{64y^2}$

27. $\dfrac{a^2 + 3a}{a^2 - 16} \div \dfrac{a^2 - 9}{a^2 - 7a + 12}$

28. $\dfrac{x^3 - 125}{x^3 - 25x} \cdot \dfrac{2x^3 + 16x^2}{x}$

29. $\left[\dfrac{u^4 - u^2 + 6u - 9}{9u^3 - 18u^2} \cdot \dfrac{9u^2}{9u^2 - 1}\right] \div \dfrac{u^2 - u + 3}{3u^2 - 5u - 2}$

30. $\left[\dfrac{4u^3 + 18u^2 - 10u}{u^2 - 81} \div \dfrac{15 - 7u - 2u^2}{2u^3 + 18u - 3u^2 - 27}\right] \cdot \dfrac{3u^2 + 27u}{12u^3 - 6u^2}$

31. $\dfrac{2}{15r} + \dfrac{8}{15r}$

32. $\dfrac{29}{24rs} - \dfrac{11}{24rs}$

33. $\dfrac{3k(k + 3)}{2k^2 - 7k - 15} - \dfrac{2(11k + 5)}{2k^2 - 7k - 15}$

34. $\dfrac{k(k - 4)}{k^3 + 8} + \dfrac{4(k - 1)}{k^3 + 8}$

In exercises **35–38**, find the LCD of each set of expressions.

35. $\dfrac{1}{3a^2b}$ and $\dfrac{1}{5ab^2}$

36. $\dfrac{1}{12x^2}$ and $\dfrac{1}{6x^2 + 6}$

37. $\dfrac{1}{y^2 + 7y + 10}$ and $\dfrac{2}{y^2 + 2y - 15}$

38. $\dfrac{1}{u^2 + 6uv + 9v^2}, \dfrac{1}{2u^2 + 5uv - 3v^2}$, and $\dfrac{1}{2u^2v + 6uv^2}$

In exercises **39–50**, do the indicated operations. Write answers in reduced form.

39. $\dfrac{a - 3}{3} + \dfrac{a + 2}{2} - \dfrac{a}{4}$

40. $1 + \dfrac{1}{b + 2} - \dfrac{1}{b - 2}$

41. $\dfrac{x + 1}{x^2 - x - 6} + \dfrac{x + 4}{x^2 + x - 2} - \dfrac{x - 4}{x^2 - 4x + 3}$

42. $\dfrac{15(5 - 2y)}{2y^3 + 250} - \dfrac{3}{2y + 10} + \dfrac{2y}{y^2 - 5y + 25}$

43. $\dfrac{3}{2x^2y^2} + \dfrac{2x - y}{6x^2 + 12xy} \cdot \dfrac{5x + 10y}{2x^2y^2 - xy^3} - \dfrac{1}{3x^2y^2}$

44. $\dfrac{a(a - 6)}{a^2 - 7a + 10} + \dfrac{4a - 4}{a^2 - 4a - 5} \div \dfrac{a^2 - 3a + 2}{2a + 2}$

45. $(a^3 - 8b^3)\left(\dfrac{1}{a - 2b} - \dfrac{a - 2b}{a^2 + 2ab + 4b^2}\right)$

46. $(r^3 + 2)(r - 1)^2\left(\dfrac{2(r + 1)}{r^4 - r^3 + 2r - 2} + \dfrac{1}{r^2 - 2r + 1}\right)$

47. $\dfrac{1 - \dfrac{2}{k} + \dfrac{1}{k^2}}{\dfrac{2}{k^2} - \dfrac{1}{k} - 1}$

48. $\dfrac{\dfrac{10}{y^2} - \dfrac{3}{xy} - \dfrac{1}{x^2}}{\dfrac{2x}{y} + 9 - \dfrac{5y}{x}}$

49. $\dfrac{2^{-1} + 7^{-1}}{14^{-1}}$

50. $\dfrac{u^2 - 9uv - 10v^2}{(10v)^{-1} - u^{-1}}$

In exercises **51–54**, solve each equation.

51. $\dfrac{3x}{5} + \dfrac{1}{10} = \dfrac{5}{2} + \dfrac{x}{5}$

52. $\dfrac{2}{5y} + \dfrac{3}{10} = \dfrac{5}{2y} + \dfrac{1}{5}$

53. $\dfrac{1}{a + 4} = \dfrac{3}{2a + 8} - \dfrac{1}{2}$

54. $\dfrac{y}{y - 1} = \dfrac{2y + 1}{y^2 + y - 2} - \dfrac{3}{y + 2}$

In exercises **55–58**, solve and graph each inequality.

55. $\dfrac{x}{x - 6} < 0$

56. $\dfrac{y - 3}{y + 4} \geq 0$

57. $\dfrac{p}{p + 1} > \dfrac{p}{p - 1}$

58. $\dfrac{q + 2}{q + 6} \leq \dfrac{q - 4}{q - 10}$

In exercises **59–62**, solve each word problem.

59. Three times a number is added to the numerator of $\frac{13}{17}$. Two times the same number is subtracted from the denominator. The resulting fraction has the value 2. Find the number.

60. If 2 times a number is subtracted from its reciprocal, the result is $\frac{41}{15}$. Find the number.

61. Georgia and Samantha need to braid and groom three horses for a horse show. Georgia can do the job in 4 hours and Samantha can do the job in 5 hours. One hour after Georgia starts working, Samantha arrives to help finish the job. How long does Georgia work on the project?

62. A pleasure boat can travel at 12 mph in still water. A trip of 15 miles against a river's current takes the same time as a trip of 21 miles with the current. Find the rate of the current.

In exercises **63** and **64**, solve each equation for the indicated variable.

63. $\dfrac{t + 3}{t - 1} = \dfrac{x + 2}{x - 2}$, for t

64. $y = \dfrac{5x - 3}{2x + 5}$, for x

CHAPTER

5

Roots, Radicals, and Complex Numbers

"Many things we do every day," Ms. Glaston said, "can be looked at in terms of an **operation.** To illustrate, consider the following actions as operations in which we do something." She then wrote on the board:

Operation

a. Put on a pair of shoes.

b. Start the engine of a car.

c. Climb up a ladder.

"Now," she continued, "I would like to hear some suggestions on what might be considered **inverse operations** of the ones I have listed. She then added to the examples on the board the following heading:

Operation	**Inverse operation**
a. Put on a pair of shoes.	
b. Start the engine of a car.	
c. Climb up a ladder.	

"I've got the first one," Bill Grant said. "I'd say the inverse operation would be taking off a pair of shoes. Is that right?"

"I agree with Bill," said Cindy Gries, "and I think turning off a car's engine and climbing down a ladder are inverse operations of examples **b** and **c**, respectively."

"Okay," Ms. Glaston said, "I think you have the idea. The concept that an inverse operation **reverses** the result of the first operation is shown in Bill's and Cindy's answers. Let's turn now to operations in mathematics." She then wrote on the board:

Mathematical operations	**Corresponding inverse operations**
a. Addition	
b. Multiplication (except by 0)	
c. Raising to powers	

Ester Washington raised her hand and said, "Subtraction and division are the obvious answers to the addition and multiplication operations. I would guess that finding the square root would be the inverse operation to squaring a number. But

I'm not sure what would be the inverse operation of cubing a number, or some other power."

"Good answers, Ester," Ms. Glaston replied. "In Beginning Algebra we studied the square roots of real numbers, and learned that the square root of a number was related to a squared number. In this chapter, we will find the inverse operations of raising numbers to powers 3, 4, and so on. The operation is called "extracting roots of numbers." As we will see, there are cube roots, fourth roots, and, in general, n-th roots of numbers. Also, in order to find certain roots of negative numbers, we will need to define a new set of numbers."

SECTION 5-1. Roots of Numbers

1. A review of the square root of numbers
2. Cube roots of numbers
3. Higher-order roots of numbers
4. Two properties of roots of numbers

Square roots of numbers are studied in most Beginning Algebra courses. In this section, we will first review the definition and some properties of square roots. We will then give a definition of higher-order roots, such as cube roots and fourth roots. Finally, we will study the properties of n-th roots in general.

A Review of the Square Root of a Number

The exponent 2 on a number is read **squared.** For example, t^2 is read "t-squared". To "unsquare" a number, we look for the number that was originally squared. This operation is read "square-root".

Definition 5.1. The square root of x
If x and y are numbers such that

$$y^2 = x$$

then y is a **square root of x.**

Examples **a–d** illustrate Definition 5.1.

	y	x	**y is a square root of x because $y^2 = x$**
a.	5	25	5 is a square root of 25 because $5^2 = 25$
b.	-7	49	-7 is a square root of 49 because $(-7)^2 = 49$
c.	$\frac{1}{10}$	$\frac{1}{100}$	$\frac{1}{10}$ is a square root of $\frac{1}{100}$ because $\left(\frac{1}{10}\right)^2 = \frac{1}{100}$
d.	$\frac{-2}{3}$	$\frac{4}{9}$	$\frac{-2}{3}$ is a square root of $\frac{4}{9}$ because $\left(\frac{-2}{3}\right)^2 = \frac{4}{9}$

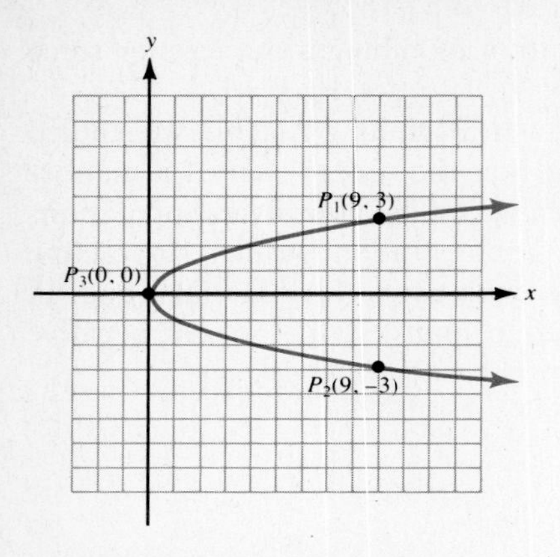

Figure 5-1. A graph of $y^2 = x$.

If x and y are both restricted to real numbers, then a graph of $y^2 = x$ will reveal three important facts related to the square-root operation. Figure 5-1 is a graph of $y^2 = x$.

Fact 1. *If* $x > 0$*, then there are two square roots, one positive and one negative.* In Figure 5-1, both $P_1(9, 3)$ and $P_2(9, -3)$ are on the graph. Thus, 3 and -3 are square roots of 9.

Fact 2. *If* $x = 0$*, then there is one square root, namely 0.* The only square root of 0 is 0.

Fact 3. *If* $x < 0$*, then there are no real-number square roots of* x. The graph in Figure 5-1 does not extend to the left of the y-axis. The square root of a negative number is not a real number. (We will study these square roots in Section 5-6.)

A symbol called a **radical sign** can be used to indicate the square root of a number.

If x is a positive real number, then:

1. $\sqrt{x}$ is the **positive number** whose square is x. $(\sqrt{x})^2 = x$
   Read $\sqrt{x}$ as "the square root of x".

2. $-\sqrt{x}$ is the **negative number** whose square is x. $(-\sqrt{x})^2 = x$
   Read $-\sqrt{x}$ as "the opposite (or negative) of the square root of x".

3. $\sqrt{-x}$ is not a real number.
   For example, $\sqrt{-2}$ and $\sqrt{-9}$ are not real numbers.
   The x in the symbol $\sqrt{x}$ is called the **radicand.**

If x is a positive real number, then $\sqrt{x}$ will name a positive real number that is either **rational** or **irrational.** If x can be written as a ratio of squares of positive integers (called **perfect squares**) that have no common factor other than 1, then $\sqrt{x}$ will be a **rational number.** The squares of the integers from 1–30 are listed below.

Squares of integers 1–30				
$1^2 = 1$	$7^2 = 49$	$13^2 = 169$	$19^2 = 361$	$25^2 = 625$
$2^2 = 4$	$8^2 = 64$	$14^2 = 196$	$20^2 = 400$	$26^2 = 676$
$3^2 = 9$	$9^2 = 81$	$15^2 = 225$	$21^2 = 441$	$27^2 = 729$
$4^2 = 16$	$10^2 = 100$	$16^2 = 256$	$22^2 = 484$	$28^2 = 784$
$5^2 = 25$	$11^2 = 121$	$17^2 = 289$	$23^2 = 529$	$29^2 = 841$
$6^2 = 36$	$12^2 = 144$	$18^2 = 324$	$24^2 = 576$	$30^2 = 900$

Example 1. Simplify each expression.

a. $\sqrt{256}$ **b.** $\sqrt{\frac{49}{16}}$ **c.** $\sqrt{11\frac{1}{9}}$

Solution.

a. $\sqrt{256} = 16$ 256 is a perfect square

b. $\sqrt{\frac{49}{16}} = \frac{7}{4}$ 49 and 16 are perfect squares

c. $\sqrt{11\frac{1}{9}} = \sqrt{\frac{100}{9}}$ Change to an improper fraction.

$= \frac{10}{3}$ 100 and 9 are perfect squares

If x cannot be written as a ratio of perfect squares, then $\sqrt{x}$ will be an **irrational number.** A decimal approximation of $\sqrt{x}$ can be obtained with a table or calculator. Appendix Table 1 can be used to get three decimal-place approximations. An electronic calculator can be used to get more than three decimal-place approximations.

Example 2. Use Appendix Table 1 or a calculator to find a decimal approximation of each expression.

a. $\sqrt{12}$ **b.** $\sqrt{57}$

Solution. **a.** $\sqrt{12} \approx$ 3.464, using Appendix Table 1; 3.464101615, using a calculator

b. $\sqrt{57} \approx$ 7.550, using Appendix Table 1; 7.549834435, using a calculator

Cube Roots of Numbers

The exponent 3 on a number is read **cubed.** For example, t^3 is read "t-cubed". To "uncube" a number, we look for the number that was originally cubed. This operation is read "cube-root".

> **Definition 5.2. The cube root of x**
> If x and y are numbers such that
> $$y^3 = x$$
> then y is the **cube root of x.**

Examples **e–h** illustrate Definition 5.2.

	y	x	y is a cube root of x because $y^3 = x$
e.	2	8	2 is a cube root of 8 because $2^3 = 8$
f.	-3	-27	-3 is a cube root of -27 because $(-3)^3 = -27$
g.	$\frac{1}{10}$	$\frac{1}{1,000}$	$\frac{1}{10}$ is a cube root of $\frac{1}{1,000}$ because $\left(\frac{1}{10}\right)^3 = \frac{1}{1,000}$
h.	$\frac{-5}{6}$	$\frac{-125}{216}$	$\frac{-5}{6}$ is a cube root of $\frac{-125}{216}$ because $\left(\frac{-5}{6}\right)^3 = \frac{-125}{216}$

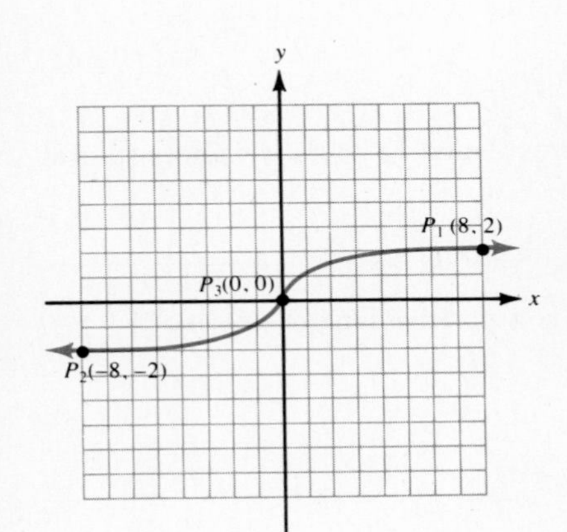

Figure 5-2. A graph of $y^3 = x$.

If x and y are real numbers, then a graph of $y^3 = x$ will reveal three important facts related to the cube-root operator. Figure 5.2 is a graph of $y^3 = x$.

Fact 1. *If* $x > 0$, *then there is one positive-number cube root of* x.
In Figure 5-2, $P_1(8, 2)$ is on the graph. Thus, 2 is the cube root of 8.

Fact 2. *If* $x < 0$, *then there is one negative-number cube root of* x.
In Figure 5-2, $P_2(-8, -2)$ is on the graph. Thus, -2 is the cube root of -8.

Fact 3. *If* $x = 0$, *then the cube root is 0.*
In Figure 5-2, $P_3(0, 0)$ is on the graph.

The radical sign can also be used to indicate the cube root of a number.

If x and y are real numbers such that $y^3 = x$, then $y = \sqrt[3]{x}$, which is read "the cube root of x". The 3 in the radical sign is called a **root index.** The x is called the **radicand.**

If x is any real number, then $\sqrt[3]{x}$ will be a real number that is either rational or irrational. If x can be written as a ratio of cubes of integers (called **perfect cubes**) that have no common factor other than 1, then $\sqrt[3]{x}$ will be a **rational number.** The cubes of the integers from 1–10 are listed below.

Cubes of integers 1–10				
$1^3 = 1$ $2^3 = 8$	$3^3 = 27$ $4^3 = 64$	$5^3 = 125$ $6^3 = 216$	$7^3 = 343$ $8^3 = 512$	$9^3 = 729$ $10^3 = 1{,}000$

Example 3. Simplify each expression.

a. $\sqrt[3]{512}$ **b.** $\sqrt[3]{-1{,}000}$ **c.** $\sqrt[3]{\frac{27}{8}}$

Solution. **a.** $\sqrt[3]{512} = 8$ 512 is a perfect cube

b. $\sqrt[3]{-1{,}000} = -10$ If $x < 0$, then $\sqrt[3]{x} < 0$.

c. $\sqrt[3]{\frac{27}{8}} = \frac{3}{2}$ 27 and 8 are perfect cubes

Because of the various ways in which calculators obtain decimal approximations of irrational number cube roots, we will wait to discuss these problems in Section 5-2.

Higher-Order Roots of Numbers

The definition of square and cube roots is extended to include **indices** (plural for index) greater than 3.

Definition 5.3. The n-th root of x, where n is a positive integer and $n > 1$
If x and y are numbers such that

$$y^n = x$$

then y is the **n-th root of x.**

The radical sign with a root index of n can be used to indicate the n-th root:

$\sqrt[n]{x}$ indicates "the n-th root of x"

If $n = 2$, then $\sqrt[2]{x}$ is written $\sqrt{x}$. The index 2 is not written.

Examples i–l illustrate some n-th roots that are rational numbers.

i. $\sqrt[4]{81} = 3$, because $3^4 = 81$

j. $\sqrt[4]{\dfrac{1}{16}} = \dfrac{1}{2}$, because $\left(\dfrac{1}{2}\right)^4 = \dfrac{1}{16}$

k. $\sqrt[5]{100{,}000} = 10$ because $10^5 = 100{,}000$

l. $\sqrt[5]{-32} = -2$, because $(-2)^5 = -32$

The following table contains a summary of square and cube roots of a real number x:

Case 1: $x > 0$	**Case 2: $x = 0$**	**Case 3: $x < 0$**
x has two square roots: $\sqrt{x} > 0$ and $-\sqrt{x} < 0$	x has one square root: $\sqrt{0} = 0$	x has no real-number square root
x has one cube root: $\sqrt[3]{x} > 0$	x has one cube root: $\sqrt[3]{0} = 0$	x has one cube root: $\sqrt[3]{x} < 0$

The summary in the table can now be extended to n-th roots of x. Now, however, we make the distinction based on the root index n. If n is even $(2, 4, 6, \ldots)$, we get the entries in the first row of the table. If n is odd $(3, 5, 7, \ldots)$, we get the entries in the second row.

If	**Case 1: $x > 0$**	**Case 2: $x = 0$**	**Case 3: $x < 0$**
n is even: $(2, 4, 6, \ldots)$	x has two n-th roots: $\sqrt[n]{x} > 0$ and $-\sqrt[n]{x} < 0$	x has one n-th root: $\sqrt[n]{0} = 0$	x has no real-number n-th root
n is odd: $(3, 5, 7, \ldots)$	x has one n-th root: $\sqrt[n]{x} > 0$	x has one n-th root: $\sqrt[n]{0} = 0$	x has one n-th root: $\sqrt[n]{x} < 0$

Example 4. Simplify each expression. If a root is not a real number, then write "not a real number". For these examples, any root that is a real number is rational.

a. $\sqrt[4]{625}$ **b.** $\sqrt[4]{-16}$ **c.** $-\sqrt[4]{16}$

d. $\sqrt[5]{243}$ **e.** $\sqrt[5]{-243}$ **f.** $-\sqrt[5]{-243}$

Solution.

a. $\sqrt[4]{625} = 5$ — $5^4 = 625$

b. $\sqrt[4]{-16}$ is not a real number — An even root of a negative number is not real.

c. $-\sqrt[4]{16} = -1 \cdot \sqrt[4]{16}$ — $-a$ can be written as $-1 \cdot a$

$= -1 \cdot 2$ — $2^4 = 16$

$= -2$ — $(-2)^4$ is also equal to 16

d. $\sqrt[5]{243} = 3$ — $3^5 = 243$

e. $\sqrt[5]{-243} = -3$ — $(-3)^5 = -243$

f. $-\sqrt[5]{-243} = -1 \cdot \sqrt[5]{-243}$ — $-a$ can be written as $-1 \cdot a$

$= -1 \cdot (-3)$ — $\sqrt[5]{-243} = -3$

$= 3$ — Simplify the product.

Two Properties of Roots of Numbers

Addition and subtraction are **inverse operations.** That is, if y is added to x, and then y is subtracted from the sum, the result is always x. Or, if y is subtracted from x, and then y is added to the difference, the result is always x.

To illustrate, let y be 8 and x be 13.

$$13 + 8 = 21 \quad \text{and} \quad 21 - 8 = 13$$

$$13 - 8 = 5 \quad \text{and} \quad 5 + 8 = 13$$

Multiplication and division are also inverse operations, except y cannot be 0. It may appear that raising to powers and extracting roots are inverse operations as well; it turns out that they are if the extraction of roots is done first, and the raising to powers is done second. Examples **m–o** illustrate.

m. $(\sqrt{36})^2 = 6^2 = 36$ Start with 36, end with 36.

n. $(\sqrt[3]{-64})^3 = (-4)^3 = -64$ Start with -64, end with -64.

o. $(\sqrt[4]{10{,}000})^4 = 10^4 = 10{,}000$ Start with 10,000, end with 10,000.

However, if raising to powers is done first, as shown by putting the exponent on the radicand, then the final number may, or may not, match the beginning radicand. Examples **p** and **q** illustrate.

p. $\sqrt{7^2} = \sqrt{49} = 7$ Start with 7, end with 7.

q. $\sqrt{(-7)^2} = \sqrt{49} = 7$ Start with -7, end with 7.

In example **p**, the final number matched the given radicand. In example **q**, the final number did not match, but was equal to the absolute value of the given radicand (that is, $7 = |-7|$).

Examples **r** and **s** illustrate that an odd-root index always yields the given radicand, whether the radicand is positive or negative.

r. $\sqrt[3]{(-5)^3} = \sqrt[3]{-125} = -5$ Start with -5, end with -5.

s. $\sqrt[3]{6^3} = \sqrt[3]{216} = 6$ Start with 6, end with 6.

Properties 1 and 2 are general statements of the previous development.

Two properties of *n*-th roots

Let x be a real number and n be an integer greater than 1.

Property 1. $(\sqrt[n]{x})^n = x$, for any x and n

Property 2. **Case 1:** $\sqrt[n]{x^n} = x$, for any x and n odd

Case 2: $\sqrt[n]{x^n} = x$, for $x \geq 0$ and n even

Case 3: $\sqrt[n]{x^n} = |x|$, for $x < 0$ and n even

Example **q** above illustrates Case 3 of Property 2.

Example 5. Simplify each expression.

a. $(\sqrt[3]{5})^3$ **b.** $(\sqrt[4]{20})^4$ **c.** $\sqrt[3]{(-7)^3}$

d. $\sqrt{1.3^2}$ **e.** $(\sqrt[5]{-0.2})^5$ **f.** $\sqrt[4]{(-3)^4}$

Solution.

a. $(\sqrt[3]{5})^3 = 5$ — Based on Property 1

b. $(\sqrt[4]{20})^4 = 20$ — Based on Property 1

c. $\sqrt[3]{(-7)^3} = -7$ — Based on Case 1 of Property 2

d. $\sqrt{1.3^2} = 1.3$ — Based on Case 2 of Property 2

e. $(\sqrt[5]{-0.2})^5 = -0.2$ — Based on Property 1

f. $\sqrt[4]{(-3)^4} = |-3|$ — Based on Case 3 of Property 2

$= 3$

SECTION 5-1. Practice Exercises

In exercises **1–12**, simplify each expression.

[Example 1]

1. $\sqrt{25}$ **2.** $\sqrt{81}$ **3.** $\sqrt{\frac{100}{121}}$

4. $\sqrt{\frac{361}{144}}$ **5.** $-\sqrt{16}$ **6.** $-\sqrt{49}$

7. $\sqrt{-64}$ **8.** $\sqrt{-9}$ **9.** $\sqrt{2\frac{14}{25}}$

10. $\sqrt{2\frac{7}{9}}$ **11.** $\sqrt{0}$ **12.** $\sqrt{-1}$

In exercises **13–20**, use Appendix Table 1 or a calculator to find a decimal approximation to three places for each number.

[Example 2]

13. $\sqrt{8}$ **14.** $\sqrt{10}$ **15.** $-\sqrt{24}$

16. $-\sqrt{22}$ **17.** $\sqrt{77}$ **18.** $\sqrt{65}$

19. $-\sqrt{99}$ **20.** $-\sqrt{80}$

In exercises **21–64**, simplify each expression.

[Example 3]

21. $\sqrt[3]{-64}$ **22.** $\sqrt[3]{-125}$ **23.** $\sqrt[3]{216}$

24. $\sqrt[3]{343}$ **25.** $-\sqrt[3]{-1}$ **26.** $-\sqrt[3]{-8}$

27. $\sqrt[3]{\frac{729}{125}}$ **28.** $\sqrt[3]{\frac{27}{512}}$ **29.** $\sqrt[3]{\frac{-343}{1{,}000}}$

30. $\sqrt[3]{\frac{-64}{1{,}331}}$ **31.** $\sqrt[3]{0.008}$ **32.** $\sqrt[3]{0.027}$

[Example 4]

33. $\sqrt[4]{81}$ **34.** $\sqrt[4]{625}$ **35.** $\sqrt[6]{64}$

36. $\sqrt[6]{729}$ **37.** $\sqrt[5]{-1{,}024}$ **38.** $\sqrt[5]{-32}$

39. $\sqrt[4]{-625}$ **40.** $\sqrt[4]{-81}$ **41.** $\sqrt[5]{-1}$

42. $\sqrt[5]{0}$ **43.** $-\sqrt[4]{1{,}296}$ **44.** $-\sqrt[4]{2{,}401}$

[Example 5]

45. $(\sqrt[5]{11})^5$ **46.** $(\sqrt[4]{3})^4$ **47.** $\sqrt[4]{(-12)^4}$

48. $\sqrt[6]{(-20)^6}$
49. $\sqrt[3]{(-8)^3}$
50. $\sqrt[5]{(-5)^5}$
51. $\sqrt[8]{(6)^8}$
52. $\sqrt[8]{(7)^8}$
53. $(\sqrt[7]{19})^7$
54. $(\sqrt[7]{11})^7$
55. $\sqrt[6]{(0.5)^6}$
56. $\sqrt[4]{(0.7)^4}$
57. $\sqrt{(-4.1)^2}$
58. $\sqrt{(-3.7)^2}$
59. $\sqrt[8]{\left(\frac{-3}{4}\right)^8}$
60. $\sqrt[10]{\left(\frac{-2}{5}\right)^{10}}$
61. $\sqrt[5]{(-13)^5}$
62. $\sqrt[3]{(-19)^3}$
63. $(\sqrt[4]{31})^4$
64. $(\sqrt[4]{29})^4$

SECTION 5-1. Ten Review Exercises

In exercises **1–4**, simplify each expression.

1. $(8t + 3) + (5t - 4)$
2. $(8t + 3)(5t - 4)$
3. $\left(\frac{5m}{2}\right)^2$
4. $\left(\frac{5 - m}{2}\right)^2$

In exercises **5–8**, solve each equation.

5. $10k + 7(5k - 2) = 2(5k + 2) + 3(15k - 4)$
6. $|3u + 2| = 10$
7. $t^2 = 4t + 60$
8. $\frac{2 + 3m}{3m} = \frac{1}{3} + \frac{2}{3m}$

In exercises **9** and **10**, solve and graph each inequality.

9. $|5x - 4| < 16$
10. $\frac{y + 2}{2} > \frac{y - 3}{4} - \frac{y - 1}{4}$

SECTION 5-1. Supplementary Exercises

In exercises **1–24**, simplify each expression. If the radical does not always name a real number, then write "not always a real number". Assume any variable represents a real number that is not necessarily positive.

1. $\sqrt[3]{x^3}$
2. $\sqrt[5]{x^5}$
3. $\sqrt[4]{x^4}$
4. $\sqrt{x^2}$
5. $-\sqrt{x^2}$
6. $-\sqrt[4]{x^4}$
7. $(\sqrt[5]{x})^5$
8. $(\sqrt[7]{x})^7$
9. $\sqrt{-9x^2}$
10. $\sqrt{-25x^4}$
11. $\sqrt[3]{x^6y^3}$
12. $\sqrt[3]{x^3y^9}$
13. $\sqrt[4]{x^{12}y^4}$
14. $\sqrt[4]{x^4y^8}$
15. $\sqrt{-27x^3}$
16. $\sqrt{-8x^9}$
17. $\sqrt[4]{(-7x)^4}$
18. $\sqrt[6]{(-3x)^6}$

19. $(\sqrt[5]{8x^2})^5$ **20.** $(\sqrt[3]{25x})^3$ **21.** $\sqrt[6]{x^6}$

22. $\sqrt[4]{x^4}$ **23.** $\sqrt{16x^2}$ **24.** $\sqrt{9x^2}$

In exercises **25** and **26**, answer parts **a–e**.

25. a. Complete the following table for $y^4 = x$:

x					
y	-2	-1	0	1	2

b. Plot the points on the axes below and draw a smooth curve through the points.

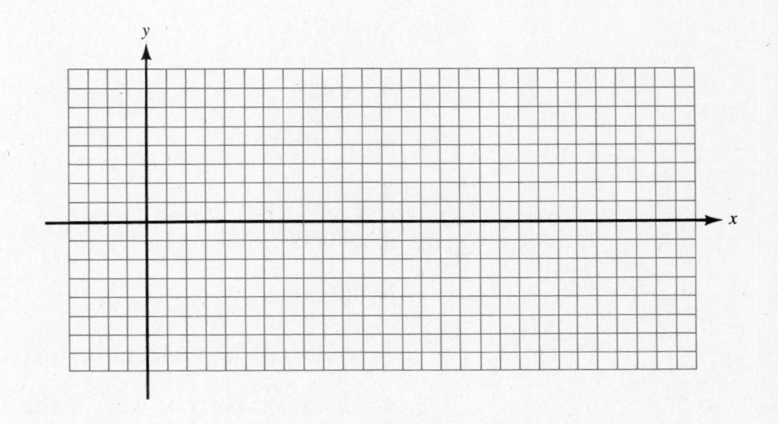

c. If $x > 0$, then how many values of y are paired with each value of x?
d. If $x < 0$, then how many values of y are paired with each value of x?
e. For $x = 0$, how many values of y are there on the graph?

26. a. Complete the following table for $y^5 = x$:

x					
y	-2	-1	0	1	2

b. Plot the points on the axes below and draw a smooth curve through the points.

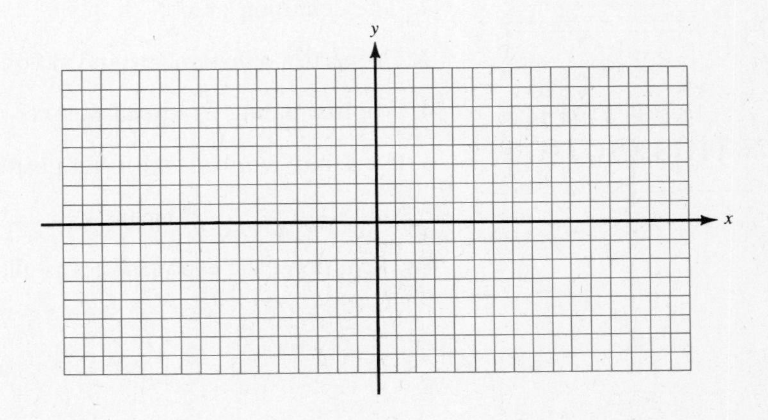

c. If $x > 0$, then how many values of y are paired with each value of x?
d. If $x < 0$, then how many values of y are paired with each value of x?
e. For $x = 0$, how many values of y are there on the graph?

The scientific calculator has the ability to find the square of a number or the square root of a number. Some calculators have a key for each operation; that is, the square of number on the display can be found by pressing $\boxed{x^2}$. The square root of the number on the display can be found by pressing $\boxed{\sqrt{\ }}$. Most calculators have these two operations "share" a key. Usually, the square root is found by pressing the $\boxed{\sqrt{\ }}$ key. The square is found by pressing an "INV" or "second function" key. Typically, $\boxed{\text{INV}}\boxed{\sqrt{\ }}$ will find the square ($\boxed{x^2}$) of a number.

In exercises **27–32**, evaluate parts **a** and **b** with a calculator.

27. a. $\sqrt{32^2}$
b. $(\sqrt{32})^2$

28. a. $\sqrt{14^2}$
b. $(\sqrt{14})^2$

29. a. $\sqrt{(-43)^2}$
b. $\sqrt{(-21)^2}$

30. a. $\sqrt{(-123)^2}$
b. $\sqrt{(-105)^2}$

31. a. $\sqrt{(0.3)^2}$
b. $(\sqrt{0.3})^2$

32. a. $\sqrt{(-0.5)^2}$
b. $\sqrt{(-0.7)^2}$

In exercises **33** and **34**, answer parts **a–c.**

33. a. Using Property 1, what is the value of $(\sqrt{-5})^2$?
b. Use a calculator to evaluate $(\sqrt{-5})^2$.
c. Did the calculator in part **b** give the same answer obtained in part **a**?

34. a. Using Property 1, what is the value of $(\sqrt{-0.3})^2$?
b. Use a calculator to evaluate $(\sqrt{-0.3})^2$.
c. Did the calculator in part **b** give the same answer obtained in part **a**?

SECTION 5-2. Rational-Number Exponents

1. The definition of $x^{1/n}$
2. The definition of $x^{m/n}$
3. Properties of exponents and rational-number exponents
4. Approximating irrational-number roots with a calculator (Optional)

If x is any number and n is a positive integer, then:

$$x^n \text{ means } x \cdot x \cdot x \ldots x \qquad n\text{-factors of } x$$

By definitions, the exponent n was allowed to be 0 or a negative integer. If $x \neq 0$, then:

$$x^0 = 1 \qquad \text{and} \qquad x^{-n} = \frac{1}{x^n}$$

In this section, we will give further definitions for expressions

$$x^{1/n} \quad \text{and} \quad x^{m/n}$$

where m is any integer, n is a positive integer, and x is a real number that must be restricted to positive numbers when n is even. This section lays much of the groundwork for material to be covered in Chapter 9.

The Definition of $x^{1/n}$

The expressions in examples **a** and **b** have **rational-number exponents.**

a. $25^{1/2}$ **b.** $(-64)^{1/3}$

The equation in Definition 5.4 identifies the way these exponential expressions can be written as equivalent radical expressions.

Definition 5.4. Equivalent exponential and radical expressions
If x is a real number and n is an integer greater than 1, then

Exponential form $\quad x^{1/n} = \sqrt[n]{x} \quad$ Radical form

wherever $\sqrt[n]{x}$ names a real number.

Using the equation in Definition 5.4 on the expressions in examples **a** and **b**:

a. $25^{1/2} = \sqrt{25}$ **b.** $(-64)^{1/3} = \sqrt[3]{-64}$

(i) The base of the exponent becomes the radicand.

(ii) The denominator of the fraction exponent becomes the root index.

Since $\sqrt{25} = 5$, we can assert that $25^{1/2}$ is another way to write 5. Similarly, $\sqrt[3]{-64} = -4$; therefore, $(-64)^{1/3}$ is another form of -4.

The equation of Definition 5.4 cannot be used whenever $\sqrt[n]{x}$ is not a real number. Recall that $\sqrt[n]{x}$ does not name a real number whenever n is even and $x < 0$. For example, $\sqrt{-9}$ and $\sqrt[4]{-16}$ are not real numbers. As a consequence of the restriction stated in Definition 5.4, we will not use the exponential form on radicals that do not name real numbers. Specifically, if the root index is even when the radicand is negative, do not write $\sqrt[n]{x}$ as $x^{1/n}$.

Example 1. Write each expression in radical form and simplify, if possible.

a. $10^{1/2}$ **b.** $(-15)^{1/5}$ **c.** $125^{1/3}$ **d.** $-16^{1/4}$

Solution.

a. $10^{1/2} = \sqrt{10}$ — Cannot be simplified

b. $(-15)^{1/5} = \sqrt[5]{-15}$ — An odd root of a negative number is real.

c. $125^{1/3} = \sqrt[3]{125}$ — The root index is 3.

$= 5$ — 125 is a perfect cube

d. $-16^{1/4} = -\sqrt[4]{16}$ — The base is 16, not (-16).

$= -2$ — $2^4 = 16$

Example 2. Write each expression in exponential form.

a. $\sqrt{39}$ **b.** $\sqrt[3]{101}$ **c.** $-\sqrt[4]{2}$ **d.** $\sqrt[5]{-21}$

Solution.

a. $\sqrt{39} = 39^{1/2}$ — The root index is 2.

b. $\sqrt[3]{101} = 101^{1/3}$ — The root index is 3.

c. $-\sqrt[4]{2} = -2^{1/4}$ — The base is 2, not (-2).

d. $\sqrt[5]{-21} = (-21)^{1/5}$ — Put -21 in parentheses.

The Definition of $x^{m/n}$

The expressions in examples **c** and **d** have rational-number exponents in which the numerators are not 1.

c. $36^{3/2}$ **d.** $(-27)^{2/3}$

The equation in Definition 5.5 identifies the way these exponential expressions can be written as equivalent radical expressions.

> **Definition 5.5. Equivalent exponential and radical expressions**
>
> **Case 1.** If x is any real number, m is any integer, and n is a positive integer, then whenever $\sqrt[n]{x}$ names a real number
>
> Exponential form $\quad x^{m/n} = (\sqrt[n]{x})^m \quad$ Radical form
>
> *Note:* Cannot be used if $x < 0$ and n is even.
>
> **Case 2.** If $x > 0$, m is any integer, and n is a positive integer, then:
>
> $x^{m/n}$ can be written as $(\sqrt[n]{x})^m$ or $\sqrt[n]{x^m}$

The base in example **c** is positive $(36 > 0)$, so we can use Case 2 of Definition 5.5:

$36^{3/2}$ can be written as $(\sqrt{36})^3 = 6^3 = 216$ or $\sqrt{36^3} = \sqrt{46{,}656} = 216$

The base in example **d** is negative $(-27 < 0)$, but n is 3, an odd number. Therefore, $(-27)^{2/3}$ is a real number.

$(-27)^{2/3} = (\sqrt[3]{-27})^2$ — Extract the root first.

$= (-3)^2$ — 27 is a perfect cube

$= 9$ — Now square.

Example 3. Write each expression in radical form and simplify, if possible.

a. $6^{3/4}$ **b.** $(-8)^{5/3}$ **c.** $-49^{3/2}$

Solution.

a. $6^{3/4} = (\sqrt[4]{6})^3$ — $x^{m/n} = (\sqrt[n]{x})^m$

or $\sqrt[4]{6^3}$ — Case 2 applies because $6 > 0$.

$= \sqrt[4]{216}$ — Simplify the radicand.

b. $(-8)^{5/3} = (\sqrt[3]{-8})^5$ — The base is negative, but n is odd.

$= (-2)^5$ — 8 is a perfect cube

$= -32$ — Simplify the power.

c. $-49^{3/2} = -(\sqrt{49})^3$ — The base is 49, not -49.

$= -(7)^3$ — Simplify the square root.

$= -343$ — Simplify the power.

In Definition 5.6, m can be any integer while n is restricted to only positive integers. If $m < 0$, then $\frac{m}{n} < 0$.

Definition 5.6. Negative rational-number exponent

If $x \neq 0$ and $\frac{m}{n} > 0$, then:

$$x^{-m/n} = \frac{1}{x^{m/n}}$$

Example 4. Simplify each expression.

a. $81^{-1/4}$ **b.** $(-32)^{-3/5}$

Solution. **Discussion.** Use Definition 5.6 to write the numbers with positive exponents. Then simplify the resulting expressions, if possible.

a. $81^{-1/4} = \frac{1}{81^{1/4}}$ — $x^{-m/n} = \frac{1}{x^{m/n}}$

$= \frac{1}{\sqrt[4]{81}}$ — Write as a radical.

$= \frac{1}{3}$ — $\sqrt[4]{81} = \sqrt[4]{3^4} = 3$

b. $(-32)^{-3/5} = \frac{1}{(-32)^{3/5}}$

$= \frac{1}{(\sqrt[5]{-32})^3}$

$= \frac{1}{(-2)^3}$

$= \frac{1}{-8}$ or $\frac{-1}{8}$

Properties of Exponents and Rational-Number Exponents

With the definition given to $x^{m/n}$, it can be shown that the five properties of exponents are valid for rational-number exponents. In the following equations, assume r and s are rational numbers and x and y are restricted so that x^r, x^s, and y^r are always real numbers.

Properties of exponents	Examples
1. $x^r \cdot x^s = x^{r+s}$	$t^{1/2} \cdot t^{1/3} = t^{1/2+1/3} = t^{5/6}$
2. $(x^r)^s = x^{rs}$	$(t^{3/2})^{1/3} = t^{3/2 \cdot 1/3} = t^{1/2}$
3. $\dfrac{x^r}{x^s} = x^{r-s}$	$\dfrac{t^{1/2}}{t^{1/3}} = t^{1/2-1/3} = t^{1/6}$
4. $(xy)^r = x^r \cdot y^r$	$(49t)^{1/2} = 49^{1/2} \cdot t^{1/2} = 7t^{1/2}$
5. $\left(\dfrac{x}{y}\right)^r = \dfrac{x^r}{y^r}, y \neq 0$	$\left(\dfrac{t^2}{125}\right)^{1/3} = \dfrac{(t^2)^{1/3}}{125^{1/3}} = \dfrac{t^{2/3}}{5}$

Example 5. Simplify each expression. Assume all variables are positive numbers.

a. $(1{,}000u^6)^{4/3}$ **b.** $\left(\dfrac{36}{k^8}\right)^{-1/2}$

Solution. **a.**

$(1{,}000u^6)^{4/3}$ The given expression

$= 1{,}000^{4/3} \cdot (u^6)^{4/3}$ Raise each factor to $\frac{4}{3}$ power.

$= (\sqrt[3]{1{,}000})^4 \cdot u^8$ Multiply exponents.

$= 10{,}000u^8$ $(\sqrt[3]{1{,}000})^4 = 10^4 = 10{,}000$

b.

$\left(\dfrac{36}{k^8}\right)^{-1/2}$ The given expression

$= \left(\dfrac{k^8}{36}\right)^{1/2}$ $\left(\dfrac{x}{y}\right)^{-n} = \left(\dfrac{y}{x}\right)^n$

$= \dfrac{(k^8)^{1/2}}{36^{1/2}}$ Raise each factor to $\frac{1}{2}$ power.

$= \dfrac{k^4}{6}$ $36^{1/2} = \sqrt{36} = 6$

Example 6. Simplify each expression. Assume all variables are positive numbers.

a. $(x^{1/2} + 5)(2x^{1/2} - 3)$ **b.** $\dfrac{10t^{7/3} + 2t^{4/3}}{2t^{1/3}}$

Solution. **a.**

$(x^{1/2} + 5)(2x^{1/2} - 3)$ The given expression

$= 2x^{1/2}x^{1/2} - 3x^{1/2} + 10x^{1/2} - 15$ Use the FOIL method.

$= 2x + 7x^{1/2} - 15$ $x^{1/2} \cdot x^{1/2} = x^1 = x$

b.

$\dfrac{10t^{7/3} + 2t^{4/3}}{2t^{1/3}}$ The given expression

$= \dfrac{10t^{7/3}}{2t^{1/3}} + \dfrac{2t^{4/3}}{2t^{1/3}}$ Write each term of the numerator over the denominator.

$= 5t^2 + t$ Subtract exponents on t.

Approximating Irrational-Number Roots with a Calculator (Optional)

Expressions such as $\sqrt[3]{5}$ and $\sqrt[4]{2}$ name irrational numbers. The **exact form** of such numbers uses the radical sign. However, it is frequently necessary to get a decimal approximation for such numbers. A scientific calculator with a $\boxed{y^x}$ key can be used to obtain such approximations.

Example 7. Use a calculator to approximate each root to three decimal places.

a. $\sqrt[3]{12}$ **b.** $(\sqrt[4]{8})^3$

Solution. **Discussion.** A scientific calculator has two functions that can be used to approximate a number with a rational number exponent. The functions are x^y (or y^x) and $x^{1/y}$. Some calculators have individual keys for these functions, but others require using an "inverse", or "2nd" function key in combination with the $\boxed{x^y}$ and $\boxed{x^{1/y}}$ keys. The specific key-stroke sequences may differ from the ones in this example, depending on the brand of calculator.

a. $\sqrt[3]{12} = 12^{1/3}$ Change to exponential form.

Operation	Display
Enter $\boxed{12}$	12.
Press $\boxed{x^{1/y}}$	12.
Enter $\boxed{3}$	3.
Press $\boxed{=}$	2.289428485

$\sqrt[3]{12} \approx 2.289$

b. $(\sqrt[4]{8})^3 = 8^{3/4}$ Change to exponential form.

Operation	Display
Enter $\boxed{8}$	8.
Press $\boxed{x^y}$	8.
Press $\boxed{(}$	[01 0.
Enter $\boxed{3}$	3.
Press $\boxed{\div}$	3.
Enter $\boxed{4}$	4.
Press $\boxed{)}$	0.75
Press $\boxed{=}$	4.756828460

$(\sqrt[4]{8})^3 \approx 4.757$

SECTION 5-2. Practice Exercises

In exercises **1–14**, write each expression in radical form and simplify, if possible.

[Example 1] **1.** $4^{1/2}$ **2.** $100^{1/2}$

3. $21^{1/2}$ 4. $29^{1/2}$

5. $1{,}000^{1/3}$ 6. $343^{1/3}$

7. $(-10)^{1/3}$ 8. $(-17)^{1/3}$

9. $-5^{1/6}$ 10. $-12^{1/6}$

11. $256^{1/4}$ 12. $81^{1/4}$

13. $0.000001^{1/6}$ 14. $0.000064^{1/6}$

In exercises **15–26**, write each expression in exponential form.

[Example 2]

15. $\sqrt{26}$ 16. $\sqrt{29}$ 17. $\sqrt[3]{71}$

18. $\sqrt[3]{65}$ 19. $\sqrt[3]{-21}$ 20. $\sqrt[3]{-34}$

21. $\sqrt[4]{14}$ 22. $\sqrt[4]{18}$ 23. $-\sqrt[4]{7}$

24. $-\sqrt[4]{6}$ 25. $-\sqrt[7]{-44}$ 26. $-\sqrt[7]{-51}$

In exercises **27–42**, write each expression in radical form and simplify, if possible.

[Example 3]

27. $16^{3/4}$ 28. $16^{5/4}$

29. $(-27)^{4/3}$ 30. $(-27)^{2/3}$

31. $-25^{3/2}$ 32. $-16^{3/2}$

33. $4^{7/2}$ 34. $9^{5/2}$

35. $(-216)^{4/3}$ 36. $(-1{,}000)^{4/3}$

37. $15^{0/4}$ 38. $22^{0/3}$

39. $12^{2/5}$ 40. $14^{2/7}$

41. $\left(\frac{27}{8}\right)^{2/3}$ 42. $\left(\frac{1{,}000}{343}\right)^{2/3}$

In exercises **43–96**, simplify each expression. Assume all variables are positive numbers.

[Example 4]

43. $25^{-1/2}$ 44. $16^{-1/2}$

45. $1{,}000^{-1/3}$ 46. $64^{-1/3}$

47. $49^{-3/2}$ 48. $25^{-3/2}$

49. $(-216)^{-2/3}$ 50. $(-27)^{-2/3}$

51. $9^{-3/2}$ 52. $100^{-3/2}$

53. $(-8)^{-4/3}$ 54. $(-125)^{-4/3}$

55. $(-1)^{-5/7}$ 56. $(-1)^{-4/7}$

57. $16^{-0/4}$ 58. $8^{-0/3}$

59. $-100^{-5/2}$ 60. $-9^{-5/2}$

61. $\left(\frac{-27}{8}\right)^{-2/3}$ 62. $\left(\frac{-1{,}000}{343}\right)^{-2/3}$

[Example 5]

63. $2^{1/2} \cdot 2^{1/3}$ 64. $10^{3/5} \cdot 10^{2/3}$

65. $\frac{3^{2/3}}{3^{1/2}}$

66. $\frac{12^{1/2}}{12^{1/3}}$

67. $(6^{1/2})^3$

68. $(6^4)^{1/3}$

69. $\left(\frac{5}{9}\right)^{1/2}$

70. $\left(\frac{3}{16}\right)^{1/4}$

71. $(8x)^{1/3}$

72. $(625x)^{1/4}$

73. $(81x^2y^4)^{1/4}$

74. $(16x^8y^4)^{3/2}$

75. $\left(\frac{25}{x^6}\right)^{-1/2}$

76. $\left(\frac{-343}{x^6y^3}\right)^{-1/3}$

77. $\frac{10^{1/3} \cdot 10^{1/4}}{10^{-1/4}}$

78. $\frac{3^{3/4} \cdot 3^{-2/3}}{3^{-1/4}}$

79. $\frac{6^{1/2} \cdot 5^{2/3}}{6^{1/3} \cdot 5^{1/3}}$

80. $\frac{7^{3/4} \cdot 3^{3/5}}{7^{1/3} \cdot 3^{2/5}}$

81. $(8^{1/2})^{2/3} \cdot (8^{-1/2})^3$

82. $(5^{1/3})^{1/2} \cdot (5^{-1/4})^3$

[Example 6] 83. $(x^{1/2} + 5)(x^{1/2} - 5)$

84. $(x^{1/2} - 3)(x^{1/2} + 3)$

85. $(y^{1/2} + 6)(y^{1/2} + 3)$

86. $(y^{1/2} - 2)(y^{1/2} + 7)$

87. $(3t^{1/4} + 10)^2$

88. $(5t^{1/3} - 6)^2$

89. $(x^{1/3} + 2^{1/3})(x^{2/3} - 2^{1/3}x^{1/3} + 2^{2/3})$

90. $(x^{1/3} - 5^{1/3})(x^{2/3} + 5^{1/3}x^{1/3} + 5^{2/3})$

91. $\frac{27s^{9/4} - 30s^{5/4} + 9s^{1/4}}{3s^{1/4}}$

92. $\frac{25s^{12/5} - 35s^{7/5} - 10s^{2/5}}{5s^{2/5}}$

93. $\frac{16u^{16/5} - 2u^{1/5}}{2u^{1/5}}$

94. $\frac{81u^{10/3} + 3u^{1/3}}{3u^{1/3}}$

95. $\frac{2^{9/4}x^{7/3} - 2^{5/4}x^{5/3}}{2^{1/4}x^{2/3}}$

96. $\frac{7^{11/5}x^{15/4} + 7^{6/5}x^{11/4}}{7^{1/5}x^{3/4}}$

In exercises **97–106**, use a calculator to approximate each root to three decimal places. (Optional)

[Example 7] 97. $\sqrt[3]{10}$

98. $\sqrt[3]{15}$

99. $\sqrt[4]{20}$

100. $\sqrt[4]{33}$

101. $(\sqrt[4]{35})^3$

102. $(\sqrt[5]{72})^3$

103. $(\sqrt[3]{6})^5$

104. $(\sqrt[3]{21})^4$

105. $\sqrt{8^5}$

106. $\sqrt{12^3}$

SECTION 5-2. Ten Review Exercises

In exercises **1–8**, do the indicated operations and simplify.

1. $(-3t^2u)^2(-5u^2)^3$

2. $\dfrac{(-5xy^2)^2}{(2x^3y)^3}$

3. $(3a + 2)(3a - 2)(2a + 5)$

4. $(4b^4 - 21b^2 + 9) \div (2b^2 + 3b - 3)$

5. $3(m^2 - 2mn + 5n^2) - (2m^2 + 5mn - n^2) + 4(3mn - 4n^2 + m^2)$

6. $\dfrac{x^3 - 8y^3}{x^3 + 8y^3} \cdot \dfrac{x + 2y}{x^2 + 2xy + 4y^2}$

7. $\dfrac{3p^2 + 2pq - 9p - 6q}{6p^2q} \div \dfrac{3p + 2q}{24p^2q^2}$

8. $\dfrac{1}{9} + \dfrac{k - 3}{18k} - \dfrac{4k - 1}{24k}$

9. Evaluate $3(5z - 8) - 7z - 8 + 7(4z + 8)$ for $z = \frac{-2}{3}$.

10. Solve $3(5z - 8) - 7z = 8 - 7(4z + 8)$.

SECTION 5-2. Supplementary Exercises

In exercises **1–30**, simplify each expression, if possible. If the expression does not represent a real number, then write "not a real number".

1. $(4^{-2/3})^{-1/2}$

2. $(25^{-3/4})^{-1/3}$

3. $(-16)^{1/2}$

4. $(-49)^{1/2}$

5. $(-125)^{-1/3}$

6. $(-8)^{-1/3}$

7. $\dfrac{\sqrt{2} \cdot \sqrt[3]{2}}{\sqrt[4]{2}}$

8. $\dfrac{\sqrt[3]{3} \cdot \sqrt[4]{3}}{\sqrt{3}}$

9. $\dfrac{\sqrt[5]{-7}}{\sqrt[3]{-7}}$

10. $\dfrac{\sqrt[3]{-13}}{\sqrt[7]{-13}}$

11. $(49^{3/2})^{-1/2}$

12. $(100^{5/2})^{-1/2}$

13. $\sqrt{(-5)^2}$

14. $\sqrt{(-8)^2}$

15. $(\sqrt{-5})^2$

16. $(\sqrt{-8})^2$

17. $\sqrt{\sqrt{x}}$

18. $\sqrt[3]{\sqrt[3]{x}}$

19. $\sqrt[3]{\sqrt{y}}$

20. $\sqrt{\sqrt[3]{y}}$

21. $\sqrt[5]{\sqrt[4]{z}}$

22. $\sqrt[4]{\sqrt[6]{z}}$

23. $\sqrt[6]{\sqrt[3]{t^2}}$

24. $\sqrt[4]{\sqrt[3]{t^2}}$

25. $\dfrac{1}{36^{-1/2}}$

26. $\dfrac{1}{81^{-1/2}}$

27. $\dfrac{1}{(-27)^{-2/3}}$ **28.** $\dfrac{1}{(-125)^{-2/3}}$

29. $\dfrac{1}{9^{-5/2}}$ **30.** $\dfrac{1}{16^{-3/4}}$

In exercises **31–36**:

a. Simplify by multiplying.

b. Remove as a common factor the term with the rational-number exponent.

31. a. $p^{1/4}(p^{3/4} + 1)$
b. $p^1 + p^{3/4}$

32. a. $p^{2/5}(p^{3/5} - 1)$
b. $p^1 - p^{3/5}$

33. a. $x^{3/2}(x^{1/2} - 1)$
b. $x^3 - x^{5/2}$

34. a. $x^{5/4}(x^{3/4} + 1)$
b. $x^2 + x^{7/4}$

35. a. $m^{4/3}(m^{5/3} + m^{2/3})$
b. $m^2 + m^{3/4}$

36. a. $m^{4/5}(m^{6/5} - m^{2/5})$
b. $m^2 - m^{3/5}$

SECTION 5-3. The Product and Quotient Properties of Radicals

1. The product property of radicals
2. The quotient property of radicals
3. Multiplying and dividing radicals with different indices

Based on Definition 5.4, stated in Section 5-2, we can write the radicals in examples **a** and **b** with rational-number exponents.

a. $\sqrt{10}$ can be written as $10^{1/2}$

b. $\sqrt[3]{-7}$ can be written as $(-7)^{1/3}$

One of the benefits of this relationship is that we can use some of the properties of exponents to develop corresponding properties of radicals. We will study two properties of radicals in this section.

The Product Property of Radicals

In examples **c** and **d**, numbers written as radicals are multiplied. As shown, it is possible to write each of the multiplications with one radical symbol. The procedure is to first write these numbers in exponential form, then use a property of exponents to write the product with one exponent.

c.	**d.**	
$\sqrt{5} \cdot \sqrt{7}$	$\sqrt[3]{3} \cdot \sqrt[3]{4}$	Two indicated products
$= 5^{1/2} \cdot 7^{1/2}$	$= 3^{1/3} \cdot 4^{1/3}$	Write in exponential form.
$= (5 \cdot 7)^{1/2}$	$= (3 \cdot 4)^{1/3}$	$x^n \cdot y^n = (x \cdot y)^n$
$= 35^{1/2}$	$= 12^{1/3}$	Simplify the products.
$= \sqrt{35}$	$= \sqrt[3]{12}$	Write in radical form.

The following equations are general statements of examples **c** and **d**:

The product property of radicals

If x and y are real numbers and n is an integer greater than 1, then

$$\sqrt[n]{x} \cdot \sqrt[n]{y} = \sqrt[n]{xy} \quad \text{and} \quad \sqrt[n]{xy} = \sqrt[n]{x} \cdot \sqrt[n]{y}$$

provided $x \geq 0$ and $y \geq 0$ whenever n is even.

Example 1. Write $\sqrt[4]{2} \cdot \sqrt[4]{3} \cdot \sqrt[4]{5}$ as a single radical.

Solution. **Discussion.** The root index on all three radicals is 4. Thus, the radicands can all be multiplied.

$= \sqrt[4]{2} \cdot \sqrt[4]{3} \cdot \sqrt[4]{5}$ The given expression

$= \sqrt[4]{2 \cdot 3 \cdot 5}$ Multiply the radicands.

$= \sqrt[4]{30}$ Simplify the product.

Example 2. Write each expression as a single radical and simplify, if possible.

a. $\sqrt[3]{6t^2} \cdot \sqrt[3]{36t}$

b. $\sqrt[4]{2x^3} \cdot \sqrt[4]{4x^3} \cdot \sqrt[4]{2x^2}$

Solution. **a.** $\sqrt[3]{6t^2} \cdot \sqrt[3]{36t} = \sqrt[3]{216t^3}$ Multiply under one radical.

$= 6t$ $\sqrt[3]{216} = 6$ and $\sqrt[3]{t^3} = t$

b. $\sqrt[4]{2x^3} \cdot \sqrt[4]{4x^3} \cdot \sqrt[4]{2x^2}$ The given expression

$= \sqrt[4]{16x^8}$ Multiply under one radical.

$= 2x^2$ $\sqrt[4]{16} = 2$ and $\sqrt[4]{x^8} = x^2$

Notice that in part **a** of Example 2 it was not necessary to put t in absolute value bars because the root index (namely, 3) is odd. In part **b** of Example 2, it was not necessary to put x^2 in absolute value bars because x^2 is nonnegative (that is, $x^2 \geq 0$) for all x.

The Quotient Property of Radicals

In examples **e** and **f**, numbers written as radicals are divided. As shown, it is possible to write each of the divisions with one radical symbol. The procedure is similar to the one used to write an indicated product of two or more radicals with the same root index under one radical.

e. $\dfrac{\sqrt[3]{-30}}{\sqrt[3]{3}}$ **f.** $\dfrac{\sqrt[4]{72}}{\sqrt[4]{6}}$ Two indicated quotients

$= \dfrac{(-30)^{1/3}}{3^{1/3}}$ $= \dfrac{72^{1/4}}{6^{1/4}}$ Write in exponential form.

$= \left(\dfrac{-30}{3}\right)^{1/3}$ $= \left(\dfrac{72}{6}\right)^{1/4}$ $\dfrac{x^n}{y^n} = \left(\dfrac{x}{y}\right)^n$

$= (-10)^{1/3}$ $= 12^{1/4}$ Simplify the quotients.

$= \sqrt[3]{-10}$ $= \sqrt[4]{12}$ Write in radical form.

The following equations are general statements of examples **e** and **f**:

> **The quotient property of radicals**
> If x and y are real numbers ($y \neq 0$) and n is an integer greater than 1, then
>
> $$\frac{\sqrt[n]{x}}{\sqrt[n]{y}} = \sqrt[n]{\frac{x}{y}} \quad \text{and} \quad \sqrt[n]{\frac{x}{y}} = \frac{\sqrt[n]{x}}{\sqrt[n]{y}}$$
>
> provided $x \geq 0$ and $y > 0$ whenever n is even.

Example 3. Write $\dfrac{\sqrt[4]{7} \cdot \sqrt[4]{15}}{\sqrt[4]{21}}$ as a single radical.

Solution. **Discussion.** The root index on all three radicals is 4. Thus, the radicands can be written with one radical.

$\dfrac{\sqrt[4]{7} \cdot \sqrt[4]{15}}{\sqrt[4]{21}}$ The given expression

$= \sqrt[4]{\dfrac{7 \cdot 15}{21}}$ Use one radical.

$= \sqrt[4]{5}$ Reduce the fraction.

Example 4. Write each expression as a single radical and simplify, if possible.

a. $\dfrac{\sqrt[3]{-54u}}{\sqrt[3]{2u^4}}$; $u \neq 0$ **b.** $\dfrac{\sqrt[4]{7v^7}}{\sqrt[4]{112v^3}}$; $v \neq 0$

Solution. **a.** $\dfrac{\sqrt[3]{-54u}}{\sqrt[3]{2u^4}} = \sqrt[3]{\dfrac{-54u}{2u^4}}$ Divide under one radical.

$= \sqrt[3]{\dfrac{-27}{u^3}}$ Simplify the quotient.

$= \dfrac{-3}{u}$ The index is odd.

b. $\dfrac{\sqrt[4]{7v^7}}{\sqrt[4]{112v^3}} = \sqrt[4]{\dfrac{7v^7}{112v^3}}$ Divide under one radical.

$= \sqrt[4]{\dfrac{v^4}{16}}$ Simplify the fraction.

$= \dfrac{|v|}{2}$ $\sqrt[4]{v^4} = |v|$. The index is even.

Multiplying and Dividing Radicals with Different Indices

The product and quotient properties can be used on radicals with the same indices. They cannot be used directly on radical expressions such as those in examples **g** and **h**.

g. $\sqrt{3} \cdot \sqrt[3]{2}$ The indices are 2 and 3.

h. $\dfrac{\sqrt[4]{2}}{\sqrt{2}}$ The indices are 4 and 2.

The expressions in these examples can be written as one radical provided the given radicals are written as equivalent ones with the same root index. The power of a power property of exponents can be used to make these changes.

g. $\sqrt{3} \cdot \sqrt[3]{2} = 3^{1/2} \cdot 2^{1/3}$ Write in exponential form.

$= (3^{1/2})^{3/3} \cdot (2^{1/3})^{2/2}$ The LCD of $\frac{1}{2}$ and $\frac{1}{3}$ is 6.

$= 3^{3/6} \cdot 2^{2/6}$ Multiply exponents.

$= (3^3)^{1/6} \cdot (2^2)^{1/6}$ Write both to the $\frac{1}{6}$ power.

$= \sqrt[6]{3^3 \cdot 2^2}$ Write in radical form.

$= \sqrt[6]{108}$ Simplify the product.

The numbers in example **h** can similarly be written under one radical.

h. $\dfrac{\sqrt[4]{12}}{\sqrt{2}} = \dfrac{12^{1/4}}{2^{1/2}} = \dfrac{12^{1/4}}{2^{2/4}} = \dfrac{12^{1/4}}{4^{1/4}} = \left(\dfrac{12}{4}\right)^{1/4} = 3^{1/4} = \sqrt[4]{3}$

In general, supposing $x > 0$ and n, m, and k are integers greater than 1:

$\sqrt[n]{x^m}$ can be written $x^{m/n}$ Change to exponential form.

$= x^{mk/nk}$ Multiply m and n by k.

$= \sqrt[nk]{x^{mk}}$ Change back to radical form.

These steps can also be reversed to **reduce a root index** when a root index and radicand have a common factor.

Changing the index on a radical

If x is a real number, with m any integer, and n and k integers greater than 1, then:

(1) $\sqrt[n]{x^m} = \sqrt[nk]{x^{mk}}$ — Multiply the index by k. Multiply any exponent on x by k.

(2) $\sqrt[nk]{x^{mk}} = \sqrt[n]{x^m}$ — Divide the index by k. Divide any exponent on x by k.

provided $x \geq 0$ whenever n and nk are even.

Example 5. Use equation (1) or (2) to change each radical to the indicated index.

a. $\sqrt[4]{100}$, to index 2

b. $\sqrt[9]{-216}$, to index 3

c. $\sqrt[3]{7}$, to index 6

d. $\sqrt{5t}$, $t > 0$, to index 8

Solution. a. **Discussion.** To reduce the index 4 to 2, we must divide by 2. Thus, we need to write the radicand as a number to the power 2.

$\sqrt[4]{100} = \sqrt[4]{10^2}$ $\quad 100 = 10^2$

$= \sqrt{10}$ $\quad$ Divide index and exponent by 2.

b. $\sqrt[9]{-216} = \sqrt[9]{(-6)^3}$ $\quad -216 = (-6)^3$

$= \sqrt[3]{-6}$ $\quad$ Divide index and exponent by 3.

c. $\sqrt[3]{7} = \sqrt[6]{7^2}$ $\quad$ Multiply index and exponent by 2.

$= \sqrt[6]{49}$ $\quad$ Simplify the power.

d. $\sqrt{5t} = \sqrt[8]{(5t)^4}$ $\quad$ Multiply index and exponent by 4.

$= \sqrt[8]{625t^4}$ $\quad$ Simplify the power.

SECTION 5-3. Practice Exercises

In exercises **1–44**, write each expression as a single radical and simplify if possible. Assume all variables represent positive real numbers.

[Example 1]

1. $\sqrt[4]{5}\sqrt[4]{3}$
2. $\sqrt[4]{7}\sqrt[4]{10}$
3. $\sqrt[6]{2y^2}\sqrt[6]{3y^3}$
4. $\sqrt[6]{13y}\sqrt[6]{3y^4}$
5. $\sqrt[9]{8}\sqrt[9]{-5}\sqrt[9]{2}$
6. $\sqrt[7]{10}\sqrt[7]{12}\sqrt[7]{-3}$
7. $\sqrt[5]{5x}\sqrt[5]{-6x^3}$
8. $\sqrt[3]{-7x}\sqrt[3]{2x}$

[Example 2]

9. $\sqrt[3]{25}\sqrt[3]{5}$
10. $\sqrt[3]{10}\sqrt[3]{100}$
11. $\sqrt{12}\sqrt{3}$
12. $\sqrt{2}\sqrt{8}$
13. $(-\sqrt{5})(-\sqrt{20})$
14. $(-\sqrt{72})(-\sqrt{2})$
15. $\sqrt[3]{40}\sqrt[3]{25}$
16. $\sqrt[3]{108}\sqrt[3]{2}$
17. $\sqrt[4]{4x}\sqrt[4]{4x^3}$
18. $\sqrt[4]{27y^2}\sqrt[4]{3y^2}$
19. $\sqrt[5]{\dfrac{t^2}{27}}\sqrt[5]{\dfrac{t^3}{9}}$
20. $\sqrt[5]{\dfrac{4}{t^3}}\sqrt[5]{\dfrac{8}{t^2}}$
21. $\sqrt[3]{2t^2}\sqrt[3]{25t^2}\sqrt[3]{20t^2}$
22. $\sqrt[3]{5t}\sqrt[3]{t^4}\sqrt[3]{25t}$
23. $\sqrt[4]{\dfrac{10}{y^2}}\sqrt[4]{\dfrac{3}{y^3}}\sqrt[4]{\dfrac{2}{y^3}}$
24. $\sqrt[4]{\dfrac{5}{y}}\sqrt[4]{\dfrac{3}{y^2}}\sqrt[4]{\dfrac{2}{y}}$

[Example 3]

25. $\dfrac{\sqrt{24}}{\sqrt{8}}$
26. $\dfrac{\sqrt{10}}{\sqrt{2}}$
27. $\dfrac{\sqrt[3]{72}}{\sqrt[3]{12}}$
28. $\dfrac{\sqrt[3]{120}}{\sqrt[3]{10}}$
29. $\dfrac{\sqrt[4]{10x^3}}{\sqrt[4]{10x^2}}$
30. $\dfrac{\sqrt[4]{3x^2}}{\sqrt[4]{3x}}$

31. $\dfrac{\sqrt[6]{15}\sqrt[6]{14}}{\sqrt[6]{6}}$

32. $\dfrac{\sqrt[6]{20}\sqrt[6]{9}}{\sqrt[6]{15}}$

33. $\dfrac{\sqrt[5]{-16p^2}\sqrt[5]{3p}}{\sqrt[5]{8p^4}}$

34. $\dfrac{\sqrt[5]{22p}\sqrt[5]{5p}}{\sqrt[5]{11p^4}}$

[Example 4] 35. $\dfrac{\sqrt[3]{10}}{\sqrt[3]{80}}$

36. $\dfrac{\sqrt[3]{3}}{\sqrt[3]{375}}$

37. $\dfrac{-\sqrt[5]{96}}{\sqrt[5]{3}}$

38. $\dfrac{-\sqrt[5]{200{,}000}}{\sqrt[5]{2}}$

39. $\dfrac{\sqrt[4]{80x^2}}{\sqrt[4]{5x^6}}$

40. $\dfrac{\sqrt[4]{7x^5}}{\sqrt[4]{567x}}$

41. $\dfrac{\sqrt[3]{3y^5}\sqrt[3]{2y^3}}{\sqrt[3]{384y^2}}$

42. $\dfrac{\sqrt[3]{5y^4}\sqrt[3]{3y^3}}{\sqrt[3]{3{,}240y}}$

43. $\sqrt[3]{-128m^{11}} \div \sqrt[3]{-16m^2}$

44. $\sqrt[3]{54m^5} \div \sqrt[3]{2m^2}$

In exercises **45–60**, change each radical to the indicated index.

[Example 5] 45. $\sqrt[4]{9}$, to index 2

46. $\sqrt[4]{25}$, to index 2

47. $\sqrt[9]{-343}$, to index 3

48. $\sqrt[9]{-216}$, to index 3

49. $\sqrt{5}$, to index 6

50. $\sqrt{3}$, to index 6

51. $\sqrt[3]{2w^2}$, to index 12

52. $\sqrt[3]{7w}$, to index 9

53. $\sqrt[4]{121v^2}$, to index 2

54. $\sqrt[4]{196v^2w^2}$, to index 2

55. $\sqrt[8]{625}$, to index 2

56. $\sqrt[8]{81}$, to index 2

57. $-\sqrt[9]{27k^3}$, to index 3

58. $-\sqrt[9]{125k^6}$, to index 3

59. $\sqrt[3]{x^2y}$, to index 15

60. $\sqrt[3]{xyz^2}$, to index 12

SECTION 5-3. Ten Review Exercises

1. Factor $4x^2 - 4x - 15$.

2. Evaluate $4x^2 - 4x - 15$ for $x = \dfrac{5}{2}$.

3. Solve $4x^2 = 4x + 15$.

4. Reduce $\dfrac{4x^2 - 4x - 15}{2x + 3}$.

5. Evaluate $\dfrac{4x^2 - 4x - 15}{2x - 5}$ for $x = \dfrac{-3}{2}$.

6. State the square roots of 144.

7. Simplify $\sqrt{144}$.

8. Solve $t^2 = 144$.

9. Simplify $\sqrt[3]{27}$.

10. State the cube roots of 27.

SECTION 5-3. Supplementary Exercises

In exercises **1–12**, write each expression as a single radical and simplify, if possible. Assume all radicands represent positive real numbers.

1. $\dfrac{\sqrt{3x-6}}{\sqrt{108}}$

2. $\dfrac{\sqrt{5x+15}}{\sqrt{20}}$

3. $\dfrac{\sqrt[3]{2x^2-2y^2}}{\sqrt[3]{250}}$

4. $\dfrac{\sqrt[3]{10x-40y}}{\sqrt[3]{270}}$

5. $\dfrac{\sqrt[4]{x^2-5x+6}}{\sqrt[4]{x-3}}$

6. $\dfrac{\sqrt[4]{x^2-x-2}}{\sqrt[4]{x-2}}$

7. $\sqrt[3]{(x+5)^2}\sqrt[3]{x+5}$

8. $\sqrt[3]{(x-1)}\sqrt[3]{(x-1)^2}$

9. $\dfrac{\sqrt[3]{8y^3-1}}{\sqrt[3]{2y-1}}$

10. $\dfrac{\sqrt[4]{x^3-y^3}}{\sqrt[4]{x-y}}$

11. $\sqrt[3]{t^2-4t+4}\sqrt[3]{t-2}$

12. $\sqrt[3]{t+3}\sqrt[3]{t^2+6t+9}$

In exercises **13–22**, change each radical to a common root index and write the indicated product as a single radical. Assume all variables represent positive real numbers.

13. $\sqrt{3}\cdot\sqrt[3]{5}$

14. $\sqrt[3]{4}\cdot\sqrt{5}$

15. $\sqrt[5]{7x}\cdot\sqrt{2y}$

16. $\sqrt[4]{3x}\cdot\sqrt[3]{2y^2}$

17. $\sqrt{u}\cdot\sqrt[4]{2u}$

18. $\sqrt[4]{3u}\cdot\sqrt{5u}$

19. $\sqrt[3]{m^2n}\cdot\sqrt[6]{7mn^2}$

20. $\sqrt[6]{5m}\cdot\sqrt[3]{2n^2}$

21. $\sqrt{2t}\cdot\sqrt[3]{2t}\cdot\sqrt[6]{2t}$

22. $\sqrt[3]{3t}\cdot\sqrt[6]{3t}\cdot\sqrt{3t}$

Example. Use a calculator to approximate the left and right sides of $\sqrt[3]{10}\cdot\sqrt[3]{7}=\sqrt[3]{70}$, and compare the approximations.

Solution.

Left side	**Right side**
$\sqrt[3]{10}\cdot\sqrt[3]{7}$	$\sqrt[3]{70}$
$=10^{1/3}\cdot 7^{1/3}$	$=70^{1/3}$
$=(2.154434\ldots)(1.912931\ldots)$	$=4.121285\ldots$
$=4.121285\ldots$	

The approximations are the same.

In exercises **23–26**, use a calculator to verify each equation.

23. $\sqrt[3]{9}\cdot\sqrt[3]{5}=\sqrt[3]{45}$

24. $\sqrt[4]{15}\cdot\sqrt[4]{8}=\sqrt[4]{120}$

25. $\sqrt{3}\cdot\sqrt[3]{5}=\sqrt[6]{3^3\cdot 5^2}$

26. $\sqrt[3]{2}\cdot\sqrt{5}=\sqrt[6]{2^2\cdot 5^3}$

Joule's Law
The heat generated (H) in a conductor by an electric current is proportional to the resistance (R) of the conductor, the time (T) during which the current flows, and the square of the strength of the current (I^2). (H is in calories per second, R is in ohms, I is in amperes, and T is in seconds.)

Use Joule's Law to approximate the current, given the formula:

$$I = 5 \cdot \sqrt{\frac{H}{6RT}}$$

In exercises **27–30**, find the current I.

27. $H = 36{,}000$, $R = 6$, and $T = 10$

28. $H = 43{,}560$, $R = 12$ and $T = 5$

29. $H = 1{,}800$, $R = \sqrt{6}$, and $T = \sqrt{24}$

30. $H = 5{,}292$, $R = \sqrt{54}$, and $T = \sqrt{6}$

Charles' Law
At a constant pressure, the volume of a gas is proportional to its absolute temperature.

Based on this law, one of the adiabolic compression formulae for gasses is

$$V_2 = V_1\left(\frac{P_1}{P_2}\right)^{0.71}$$

where V_1 and P_1 are the first volume and pressure and P_2 is a new pressure that causes a new volume V_2. (V is in cubic feet and P in pounds per square inch).

In exercises **31–34**, approximate V_2 to two decimal places.

31. $V_1 = 165$, $P_1 = 15$, and $P_2 = 80$

32. $V_1 = 120$, $P_1 = 20$, and $P_2 = 125$

33. $V_1 = 100$, $P_1 = x$, and $P_2 = 3x$

34. $V_1 = 100$, $P_1 = 2x$, and $P_2 = 5x$

SECTION 5-4. Simplified Forms of Radicals

1. A description of a simplified radical
2. Simplifying radicals based on Criterion 1
3. Simplifying radicals based on Criterion 2
4. Simplifying radicals based on Criterion 3
5. Simplifying radicals based on Criterion 4
6. Simplifying radicals based on Criterion 5

A number can always be written in more than one way. For example:

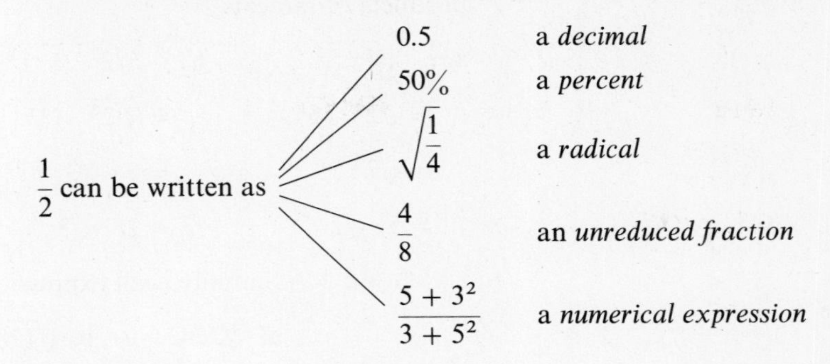

Depending on what the number is being used for, there may be a **preferred form** for the way in which $\frac{1}{2}$ should be written. In general, radical expressions also have preferred forms for the way in which they should be written. In this section, we will examine, and then apply, the five criteria that are the generally accepted preferred forms for radicals.

A Description of a Simplified Radical

The instruction to simplify a given radical usually means to change it to a preferred, but equivalent, form. Criteria 1–5 identify the form that is generally accepted as the simplified (or preferred) form of a radical expression.

A simplified radical expression must satisfy the following criteria:

Criterion 1. When written as a product of prime factors, the radicand has no factor with an exponent greater than or equal to the root index.

Criterion 2. If possible, the radicand is written as a positive number.

Criterion 3. The radicand is not a fraction.

Criterion 4. The denominator of a fraction does not contain any radicals.

Criterion 5. The root index is reduced to the smallest possible value.

We will first examine radicals that are not simplified because of one or more of the criteria given above. We will then determine a course of action to change these radicals to the simplified form without changing the value of the expression.

Simplifying Radicals Based on Criterion 1

The radicals in examples **a** and **b** are not simplified based on Criterion 1.

a. $\sqrt{50}$
- $50 = 2 \cdot 5 \cdot 5 = 2 \cdot 5^2$
- The exponent on 5 matches the root index.

b. $\sqrt[3]{24t^3}$
- $24t^3 = 2 \cdot 2 \cdot 2 \cdot 3 \cdot t^3 = 2^3 \cdot 3 \cdot t^3$
- The exponents on 2 and t match the root index.

The product property of radicals can be used to rewrite the given radicals as products of radicals.

a.	**b.**	
$\sqrt{2 \cdot 5^2}$	$\sqrt[3]{2^3 \cdot 3 \cdot t^3}$	The given radicals
$= \sqrt{2} \cdot \sqrt{5^2}$	$= \sqrt[3]{2^3} \cdot \sqrt[3]{3} \cdot \sqrt[3]{t^3}$	Product property of radicals
$= \sqrt{2} \cdot 5$	$= 2 \cdot \sqrt[3]{3} \cdot t$	
$= 5 \cdot \sqrt{2}$	$= 2t \cdot \sqrt[3]{3}$	The preferred form

Example 1. Simplify each expression.

a. $\sqrt[5]{224}$ **b.** $\sqrt{128a^2}$

Solution. **a.** $\sqrt[5]{224} = \sqrt[5]{2^5 \cdot 7}$ Prime factor 224

$= \sqrt[5]{2^5} \cdot \sqrt[5]{7}$ Product property of radicals

$= 2 \cdot \sqrt[5]{7}$ The simplified form

b. Discussion. Look for factors of $128a^2$ to powers greater than or equal to 2, the root index.

$\sqrt{128a^2} = \sqrt{64a^2 \cdot 2}$ $64a^2 = (8a)^2$

$= \sqrt{(8a)^2} \cdot \sqrt{2}$

$= 8|a| \cdot \sqrt{2}$

Simplifying Radicals Based on Criterion 2

The radicals in examples **c** and **d** are not simplified based on Criterion 2.

c. $\sqrt[3]{-5}$
- $-5 < 0$. Thus, the radicand is not positive.
- The root index is 3. Thus, the radical is a real number.

d. $\sqrt[5]{-12t^2}$
- For any $t \neq 0$, $-12t^2 < 0$.
- The root index is 5. Thus, the radical is a real number.

If n is an *odd integer* and $x > 0$, then:

$\sqrt[n]{-x} = \sqrt[n]{-1 \cdot x}$ Write $-x$ as $-1 \cdot x$.

$= \sqrt[n]{-1} \cdot \sqrt[n]{x}$ The product property of radicals

$= -1 \cdot \sqrt[n]{x}$ Any odd root of -1 is -1.

$= -\sqrt[n]{x}$ "The opposite of the n-th root of x"

Using this property on examples **c** and **d**:

c. $\sqrt[3]{-5} = -\sqrt[3]{5}$ **d.** $\sqrt[5]{-12t^2} = -\sqrt[5]{12t^2}$

Example 2. Simplify $\sqrt[3]{-135}$.

Solution. **Discussion.** The radicand is not simplified based on Criteria 1 and 2:

$-135 = -1 \cdot 27 \cdot 5$ and $27 = 3^3$

$-135 < 0$, a negative radicand

$\sqrt[3]{-135} = -\sqrt[3]{135}$ Make the radicand positive.

$= -3\sqrt[3]{5}$ $\sqrt[3]{135} = \sqrt[3]{27 \cdot 5} = 3\sqrt[3]{5}$

Simplifying Radicals Based on Criterion 3

The radicals in examples **e** and **f** are not simplified based on Criterion 3. In both examples, the radicand is a fraction.

e. $\sqrt[3]{\dfrac{5t}{64}}$ **f.** $\sqrt[4]{\dfrac{10}{k^8}}$; $k \neq 0$

The quotient property can be used to rewrite the given radicals as quotients of radicals. Then, if possible, the radicals in the numerator and denominator should be simplified.

e. $\sqrt[3]{\dfrac{5t}{64}}$ **f.** $\sqrt[4]{\dfrac{10}{k^8}}$ The given radicals

$= \dfrac{\sqrt[3]{5t}}{\sqrt[3]{64}}$ $= \dfrac{\sqrt[4]{10}}{\sqrt[4]{k^8}}$ The quotient property of radicals

$= \dfrac{\sqrt[3]{5t}}{4}$ $= \dfrac{\sqrt[4]{10}}{k^2}$ Simplify the denominator radicals.

Example 3. Simplify $\sqrt[3]{\dfrac{-2}{27x^3}}$; $x \neq 0$

Solution. $\sqrt[3]{\dfrac{-2}{27x^3}}$ The given radical

$= \dfrac{\sqrt[3]{-2}}{\sqrt[3]{27x^3}}$ The quotient property of radicals

$= \dfrac{\sqrt[3]{-2}}{3x}$ $\sqrt[3]{27x^3} = \sqrt[3]{(3x)^3} = 3x$

$= \dfrac{-\sqrt[3]{2}}{3x}$ Write the numerator with a positive radicand.

Simplifying Radicals Based on Criterion 4

The radicals in examples **g** and **h** are not simplified based on Criterion 4. In both examples, the denominator is a radical expression.

g. $\dfrac{10}{\sqrt[3]{25}}$ **h.** $\dfrac{6}{\sqrt[4]{2}}$

The multiplication property of 1 can be used to rewrite each fraction as an equivalent one without a denominator that has a radical.

g. $\dfrac{10}{\sqrt[3]{25}}$ The given radical expression

$= \dfrac{10}{\sqrt[3]{25}} \cdot \dfrac{\sqrt[3]{5}}{\sqrt[3]{5}}$ 125 is the smallest perfect cube of which 25 is a factor

$= \dfrac{10\sqrt[3]{5}}{5}$ $\sqrt[3]{25} \cdot \sqrt[3]{5} = \sqrt[3]{125} = 5$

$= 2\sqrt[3]{5}$ Reduce the fraction.

h. $\dfrac{6}{\sqrt[4]{2}}$ The given radical expression

$= \dfrac{6}{\sqrt[4]{2}} \cdot \dfrac{\sqrt[4]{8}}{\sqrt[4]{8}}$ $16 = 2^4$ is the smallest fourth power of which 2 is a factor

$= \dfrac{6\sqrt[4]{8}}{2}$ $\sqrt[4]{2} \cdot \sqrt[4]{8} = \sqrt[4]{16} = 2$

$= 3\sqrt[4]{8}$ Reduce the fraction.

Example 4. Simplify each expression.

a. $\dfrac{1}{\sqrt[3]{7}}$ **b.** $\dfrac{6}{\sqrt[5]{16x^3}}$; $x \neq 0$

Solution. **a. Discussion.** The root index is 3. The given radicand has one factor of 7. Therefore, we need two more.

$\dfrac{1}{\sqrt[3]{7}} \cdot \dfrac{\sqrt[3]{49}}{\sqrt[3]{49}}$ Multiply numerator and denominator by $\sqrt[3]{7^2} = \sqrt[3]{49}$.

$= \dfrac{\sqrt[3]{49}}{7}$ $\sqrt[3]{343} = 7$

b. Discussion. The root index is 5. Therefore, each factor of the given radicand must be multiplied by additional ones so that each has an exponent of 5.

Factors present in radicand	Number of factors needed	Product of given factors and factors needed
$16 = 2^4$	2^1	$2^4 \cdot 2^1 = 2^5$
x^3	x^2	$x^3 \cdot x^2 = x^5$

$\dfrac{6}{\sqrt[5]{16x^3}} \cdot \dfrac{\sqrt[5]{2x^2}}{\sqrt[5]{2x^2}}$ Multiply numerator and denominator by $\sqrt[5]{2x^2}$.

$= \dfrac{6\sqrt[5]{2x^2}}{2x}$ $\sqrt[5]{32x^5} = 2x$

$= \dfrac{3\sqrt[5]{2x^2}}{x}$ Reduce the fraction.

Simplifying Radicals Based on Criterion 5

The radicals in examples **i** and **j** are not simplified based on Criterion 5. In both examples, the root index and the exponent on the radicand have a common factor.

i. $\sqrt[4]{9} = \sqrt[4]{3^2}$ $\qquad 9 = 3^2$

j. $\sqrt[9]{125t^6} = \sqrt[9]{(5t^2)^3}$ $\qquad 125 = 5^3$ and $t^6 = (t^2)^3$

The equations for changing the index on a radical given in Section 5-3 can be used to remove the common factors. Specifically:

$\sqrt[nk]{x^{mk}} = \sqrt[n]{x^m}$ $\qquad$ Divide index and exponent by k.

i. $\sqrt[4]{3^2} = \sqrt{3}$ $\qquad$ Divide index and exponent by 2.

j. $\sqrt[9]{(5t^2)^3} = \sqrt[3]{5t^2}$ $\qquad$ Divide index and exponent by 3.

Example 5. Simplify $\sqrt[8]{81u^4}$; $u > 0$

Solution. Since $81 = 3 \cdot 3 \cdot 3 \cdot 3 = 3^4$:

$\sqrt[8]{81u^4} = \sqrt[8]{3^4u^4} = \sqrt{3u}$ $\qquad$ Divide index and exponents by 4.

SECTION 5-4. Practice Exercises

In exercises **1–60**, simplify each expression. Assume all variables are positive real numbers.

[Example 1]

1. $\sqrt{45}$ **2.** $\sqrt{75}$

3. $-\sqrt{98}$ **4.** $-\sqrt{28}$

5. $\sqrt[3]{250}$ **6.** $\sqrt[3]{56}$

7. $-\sqrt[5]{128a^6b^8}$ **8.** $-\sqrt[5]{64a^{10}b^6}$

9. $\sqrt[4]{20x^8y^5}$ **10.** $\sqrt[4]{5x^{16}y^7}$

11. $-\sqrt[3]{270m^3n^7}$ **12.** $\sqrt[3]{432m^2n^8}$

[Example 2]

13. $\sqrt[3]{-10}$ **14.** $\sqrt[3]{-7}$

15. $\sqrt[5]{-3t^2}$ **16.** $\sqrt[7]{-8t^5}$

17. $\sqrt[5]{-96}$ **18.** $\sqrt[3]{-486}$

19. $\sqrt[3]{-500p^7q^3}$ **20.** $\sqrt[3]{-320p^6q^5}$

21. $\sqrt[3]{\dfrac{-240}{x^3}}$ **22.** $\sqrt[3]{\dfrac{-160}{x^9}}$

[Example 3]

23. $\sqrt{\dfrac{7}{25}}$ **24.** $\sqrt{\dfrac{13}{100}}$

25. $\sqrt[3]{\dfrac{21}{1,000}}$

26. $\sqrt[3]{\dfrac{9}{64}}$

27. $\sqrt[3]{\dfrac{9}{8x^3}}$

28. $\sqrt[3]{\dfrac{4}{343x^6}}$

29. $\sqrt{\dfrac{5t^6}{324}}$

30. $\sqrt{\dfrac{10t^{10}}{121}}$

31. $\sqrt{\dfrac{47}{81y^{12}}}$

32. $\sqrt{\dfrac{17}{10,000y^4}}$

[Example 4] 33. $\dfrac{3}{\sqrt{6}}$

34. $\dfrac{2}{\sqrt{14}}$

35. $\dfrac{-3}{\sqrt{15}}$

36. $\dfrac{-5}{\sqrt{30}}$

37. $\dfrac{10}{\sqrt[3]{2}}$

38. $\dfrac{6}{\sqrt[3]{3}}$

39. $\dfrac{-21}{\sqrt[3]{9}}$

40. $\dfrac{-20}{\sqrt[3]{100}}$

41. $\dfrac{10}{\sqrt[4]{125}}$

42. $\dfrac{35}{\sqrt[4]{7}}$

43. $\dfrac{20}{\sqrt[3]{2x^2}}$

44. $\dfrac{10}{\sqrt[3]{25x}}$

45. $\dfrac{-1}{\sqrt[4]{xy^3}}$

46. $\dfrac{-1}{\sqrt[4]{x^2y^3}}$

47. $\dfrac{4a}{\sqrt[5]{8ab^3}}$

48. $\dfrac{6b}{\sqrt[5]{4a^2b^4}}$

[Example 5] 49. $\sqrt[4]{49}$

50. $\sqrt[4]{121}$

51. $\sqrt[6]{27}$

52. $\sqrt[6]{1,000}$

53. $-\sqrt[4]{x^2y^2}$

54. $-\sqrt[4]{4x^2}$

55. $-\sqrt[6]{4x^2}$

56. $-\sqrt[6]{x^2y^4}$

57. $\sqrt[8]{100x^2}$

58. $\sqrt[8]{49x^2y^4}$

59. $\sqrt[8]{16}$

60. $\sqrt[8]{64}$

SECTION 5-4. Ten Review Exercises

In exercises **1** and **2**, evaluate each expression.

1. $2t^2 + 15t + 7$, for $t = -3$

2. $5t + 8$, for $t = -3$

3. Simplify $(2t^2 + 15t + 7)(5t + 8)$

4. Evaluate the answer obtained in exercise **3** for $t = -3$.

5. Simplify $(10t^3 + 91t^2 + 155t + 56) \div (5t + 8)$.

6. Evaluate the answer obtained in exercise **5** for $t = -7$.

In exercises **7–10**, simplify each expression.

7. $(3k^2)^2$

8. $(3 - k^2)^2$

9. $\dfrac{9p^2q^2}{3pq}$

10. $\dfrac{9 - p^2q^2}{3 - pq}$

SECTION 5-4. Supplementary Exercises

In exercises **1–30**:

a. State which criterion is (are) not satisfied.

b. Simplify the given radical. Assume all variables represent positive real numbers.

1. $\sqrt[3]{200}$
2. $\sqrt[3]{800}$
3. $\dfrac{1}{\sqrt[3]{-64}}$
4. $\dfrac{1}{\sqrt[3]{-27}}$
5. $\sqrt[3]{-44}$
6. $\sqrt[5]{-121}$
7. $\sqrt[3]{\dfrac{28}{27}}$
8. $\sqrt[3]{\dfrac{36}{125}}$
9. $\sqrt{0.09}$
10. $\sqrt{0.04}$
11. $-\sqrt[6]{8}$
12. $-\sqrt[4]{49}$
13. $\dfrac{6}{\sqrt[3]{5}}$
14. $\dfrac{3}{\sqrt[3]{11}}$
15. $\sqrt[3]{\dfrac{x^5}{y^6}}$
16. $\sqrt[4]{\dfrac{x^3}{y^4}}$
17. $\dfrac{-10}{\sqrt[4]{x^3y^3}}$
18. $\dfrac{-18}{\sqrt[5]{x^3y}}$
19. $\sqrt[5]{-21p^6q^{10}}$
20. $\sqrt[5]{-12p^{15}q^9}$
21. $\sqrt[3]{\dfrac{1}{343}}$
22. $\sqrt[4]{\dfrac{1}{256}}$
23. $\sqrt{24} \div \sqrt{2}$
24. $\sqrt{150} \div \sqrt{3}$
25. $\sqrt{\dfrac{243x^3}{y^2}}$
26. $\sqrt{\dfrac{432x^2}{y^4}}$
27. $\dfrac{\sqrt[3]{-5}}{\sqrt[3]{4}}$
28. $\dfrac{\sqrt[3]{6}}{\sqrt[3]{-49}}$
29. $\sqrt[6]{\dfrac{a^3}{b^6}}$
30. $\sqrt[8]{\dfrac{a^{16}}{b^8}}$

In exercises **31–40**, simplify each expression. Assume all radicands represent positive real numbers.

31. $\sqrt{50(x^2 + 1)^3}$
32. $\sqrt{12(x^4 + 12x^2 + 36)}$
33. $\sqrt[3]{(y - 1)^4}$
34. $\sqrt[3]{(y + 2)^9}$
35. $\sqrt[3]{\dfrac{x - 1}{(x + 1)^2}}$
36. $\sqrt[3]{\dfrac{x + 3}{(x - 3)^2}}$
37. $\sqrt[4]{\dfrac{1}{(x^2 + 4x + 4)^2}}$
38. $\sqrt[4]{\dfrac{1}{(x^2 + 6x + 9)^2}}$
39. $\sqrt[6]{8(x + 5)^3}$
40. $\sqrt[8]{16(x - 7)^2}$

SECTION 5-5. Operations with Radicals

1. Adding and subtracting radicals
2. Multiplying expressions with radicals
3. The definition of irrational conjugates
4. Dividing expressions with radicals
5. Simplifying an indicated quotient involving a radical

In Section 5-4, we examined the simplified (or preferred) forms for radicals. In this section, we will perform some operations on numbers and expressions that include one or more radicals. In any example or exercise, the answer must be written in terms of only simplified radicals.

Adding and Subtracting Radicals

The numerical expressions in examples **a** and **b** are a binomial and a trinomial, respectively.

a. $3\sqrt{6} + 5\sqrt{6}$ **b.** $3\sqrt[3]{7} - 9\sqrt[3]{7} + \sqrt[3]{7}$

Each term of these expressions contains a radical as a factor, and the radical is an irrational number. The order of operations requires that the root and multiplication operations be performed before any indicated addition or subtraction. However, since the same radical appears in each term, the distributive property can be used to "factor out the radicals", and the additions and subtractions can then be performed.

a. $3\sqrt{6} + 5\sqrt{6}$ **b.** $4\sqrt[3]{7} - 9\sqrt[3]{7} + \sqrt[3]{7}$

$= (3 + 5)\sqrt{6}$ $= (4 - 9 + 1)\sqrt[3]{7}$ Factor radicals.

$= 8\sqrt{6}$ $= -4\sqrt[3]{7}$ Simplify within parentheses.

The given binomial and trinomial have both been changed to a monomial. Furthermore, the numbers represented by the monomials have the same values as the given expressions. Any approximation of either expression is more simply obtained by using the monomials. The instructions for exercises illustrated by examples **a** and **b** and will therefore be to simplify.

Example 1. Simplify each expression.

a. $2\sqrt{7} + 3\sqrt{10} + 3\sqrt{7} - \sqrt{10}$

b. $4\sqrt[3]{5} + \sqrt{5} - 9\sqrt{5} - 3\sqrt[3]{5} + 6\sqrt{5}$

Solution. **a. Discussion.** This expression contains two $\sqrt{7}$ terms and two $\sqrt{10}$ terms. These can be combined to one term each.

$2\sqrt{7} + 3\sqrt{10} + 3\sqrt{7} - \sqrt{10}$

$= (2 + 3)\sqrt{7} + (3 - 1)\sqrt{10}$ Factor the radicals.

$= 5\sqrt{7} + 2\sqrt{10}$ The simplified answer

b. Discussion. This expression contains two $\sqrt[3]{5}$ terms and three $\sqrt{5}$ terms. These can be combined to one term each.

$$4\sqrt[3]{5} + \sqrt{5} - 9\sqrt{5} - 3\sqrt[3]{5} + 6\sqrt{5}$$

$$= (4 - 3)\sqrt[3]{5} + (1 - 9 + 6)\sqrt{5}$$

$$= \sqrt[3]{5} + (-2)\sqrt{5}$$

$$= \sqrt[3]{5} - 2\sqrt{5} \qquad \text{The simplified answer}$$

Notice that the simplified answers cannot be changed to monomials. *The order of operations prevents us from combining these terms, because the roots and multiplications must be done before the indicated addition or subtraction.*

Example 2. Simplify $3\sqrt{28} - \sqrt{343} - 2\sqrt{7} + 2\sqrt{252}$.

Solution. **Discussion.** As written, none of the radicands are alike and the terms cannot be combined. However, the radicals can—with the exception of $\sqrt{7}$—be simplified. If any radicals simplify to the same radicand, then they can be combined.

$$3\sqrt{28} - \sqrt{343} - 2\sqrt{7} + 2\sqrt{252}$$

$$= 3(2\sqrt{7}) - 7\sqrt{7} - 2\sqrt{7} + 2(6\sqrt{7})$$

$$= (6 - 7 - 2 + 12)\sqrt{7}$$

$$= 9\sqrt{7}$$

Multiplying Expressions with Radicals

The numerical expressions in examples **c** and **d** indicate a product of a monomial with a binomial and a trinomial, respectively.

Monomial

c. $\sqrt{6}(3\sqrt{2} + 5\sqrt{3})$ — the part in parentheses is labelled **Binomial**

d. $\sqrt[3]{10}(\sqrt[3]{25} - 3\sqrt[3]{4} + 2\sqrt[3]{100})$ — the part in parentheses is labelled **Trinomial**

The distributive property can be used to distribute the monomials to each term of the grouped expressions. The multiplication property of radicals can then be used to multiply the radicands. If possible, the radicands must then be simplified.

c.

$$\sqrt{6}(3\sqrt{2} + 5\sqrt{3})$$

$$= 3\sqrt{12} + 5\sqrt{18}$$

$$= 3(2\sqrt{3}) + 5(3\sqrt{2})$$

$$= 6\sqrt{3} + 15\sqrt{2}$$

d.

$$\sqrt[3]{10}(\sqrt[3]{25} - 3\sqrt[3]{4} + 2\sqrt[3]{100})$$

$$= \sqrt[3]{250} - 3\sqrt[3]{40} + 2\sqrt[3]{1000}$$

$$= 5\sqrt[3]{2} - 3(2\sqrt[3]{5}) + 2(10)$$

$$= 5\sqrt[3]{2} - 6\sqrt[3]{5} + 20$$

There are no terms in either expression that can be combined.

Example 3. Simplify each expression.

a. $(2 + 3\sqrt{5})(4 - \sqrt{5})$ **b.** $(3 - 2\sqrt{10})^2$

Solution. **a. Discussion.** To multiply the two binomials, we can follow the pattern used in the FOIL method.

$$(2 + 3\sqrt{5})(4 - \sqrt{5})$$

$$= 2 \cdot 4 - 2 \cdot \sqrt{5} + 3\sqrt{5} \cdot 4 - 3\sqrt{5} \cdot \sqrt{5}$$

$$= 8 - 2\sqrt{5} + 12\sqrt{5} - 15$$

$$= -7 + 10\sqrt{5}$$

Put the radical term on the right.

b. Discussion. To square the binomial we can use:

$$(x - y)^2 = x^2 - 2xy + y^2$$

$$(3 - 2\sqrt{10})^2 = 3^2 - 2(3)(2\sqrt{10}) + (2\sqrt{10})^2$$

$$= 9 - 12\sqrt{10} + 40$$

$$= 49 - 12\sqrt{10}$$

The Definition of Irrational Conjugates

The expressions in examples **e** and **f** are not simplified because both have a radical term in the denominator.

e. $\dfrac{14}{3 + \sqrt{2}}$ **f.** $\dfrac{5}{7 - 2\sqrt{6}}$

Notice that the denominators are binomials. Therefore, to eliminate the irrational number in the denominator (called **rationalizing the denominator**), we need to multiply by the **conjugate of the denominator.**

The conjugate of a binomial involving square roots
If x and y are positive numbers, and either $\sqrt{x}$, or $\sqrt{y}$, or both are irrational numbers, then the conjugate of $\sqrt{x} + \sqrt{y}$ is $\sqrt{x} - \sqrt{y}$.

Example 4. For each binomial:

a. $3 + \sqrt{2}$ **b.** $7 - 2\sqrt{6}$

(i) State the conjugate.

(ii) Compute the product of each binomial and its conjugate to verify that the product is a rational number.

Solution. **a. (i)** The conjugate of $3 + \sqrt{2}$ is $3 - \sqrt{2}$.

(ii) To compute the product we can use:

$$(x + y)(x - y) = x^2 - y^2$$

$$(3 + \sqrt{2})(3 - \sqrt{2}) = (3)^2 - (\sqrt{2})^2$$

$$= 9 - 2$$

$$= 7\text{, a rational number}$$

b. (i) The conjugate of $7 - 2\sqrt{6}$ is $7 + 2\sqrt{6}$.

$$\text{(ii)}\ (7 + 2\sqrt{6})(7 - 2\sqrt{6}) = (7)^2 - (2\sqrt{6})^2$$
$$= 49 - 24$$
$$= 25, \text{ a rational number}$$

Dividing Expressions with Radicals

Examples **e** and **f** are repeated below.

e. $\dfrac{14}{3 + \sqrt{2}}$ **f.** $\dfrac{5}{7 - 2\sqrt{6}}$

These expressions can be considered division problems, in which the binomials are the divisors. In this case, we are looking for numbers u and v, such that the following equalities will be true:

e. $14 \div (3 + \sqrt{2}) = u$, such that $14 = u(3 + \sqrt{2})$

f. $5 \div (7 - 2\sqrt{6}) = v$, such that $5 = v(7 - 2\sqrt{6})$

The values for u *and* v *can be found by multiplying the given expressions by the conjugates of the denominators.*

e. $\dfrac{14}{3 + \sqrt{2}} \cdot \dfrac{3 - \sqrt{2}}{3 - \sqrt{2}}$

$= \dfrac{14(3 - \sqrt{2})}{9 - 2}$

$= \dfrac{14(3 - \sqrt{2})}{7}$

$= 2(3 - \sqrt{2})$

Thus, $u = 6 - 2\sqrt{2}$.

f. $\dfrac{5}{7 - 2\sqrt{6}} \cdot \dfrac{7 + 2\sqrt{6}}{7 + 2\sqrt{6}}$

$= \dfrac{5(7 + 2\sqrt{6})}{49 - 24}$

$= \dfrac{5(7 + 2\sqrt{6})}{25}$

$= \dfrac{7 + 2\sqrt{6}}{5}$

Thus, $v = \dfrac{7 + 2\sqrt{6}}{5}$.

The results of multiplying the expressions in examples **e** and **f** by the conjugates are numbers with **rational denominators** (specifically, 1 for **e** and 5 for **f**). As a consequence, when a division problem involves an irrational number divisor, the instructions given are usually "rationalize the denominator (divisor)".

Example 5. Rationalize the denominator of $\dfrac{\sqrt{10}}{2\sqrt{5} - \sqrt{2}}$.

Solution. **Discussion.** The conjugate of $2\sqrt{5} - \sqrt{2}$ is $2\sqrt{5} + \sqrt{2}$.

$\dfrac{\sqrt{10}}{2\sqrt{5} - \sqrt{2}} \cdot \dfrac{2\sqrt{5} + \sqrt{2}}{2\sqrt{5} + \sqrt{2}}$ Multiply by $\dfrac{\text{conjugate}}{\text{conjugate}}$.

$= \dfrac{2(5\sqrt{2}) + 2\sqrt{5}}{(2\sqrt{5})^2 - (\sqrt{2})^2}$ $(x - y)(x + y) = x^2 - y^2$

$= \dfrac{2(5\sqrt{2} + \sqrt{5})}{18}$ Factor a 2 in the numerator.

$= \dfrac{5\sqrt{2} + \sqrt{5}}{9}$ Reduce the fraction.

Simplifying an Indicated Quotient Involving a Radical

In Chapter 6, we will solve quadratic equations with a formula. Expressions like these in examples **g** and **h** are frequently obtained as solutions when the formula is used.

g. $\dfrac{-8 + \sqrt{48}}{20}$ **h.** $\dfrac{6 - \sqrt{45}}{12}$

To change the numbers in examples **g** and **h** to **preferred form,** there are frequently two tasks to perform:

Task 1. Simplify the radical.

Task 2. Reduce the fraction.

Example 6. Simplify $\dfrac{-8 + \sqrt{48}}{20}$.

Solution.

Task 1. $\dfrac{-8 + \sqrt{48}}{20} = \dfrac{-8 + 4\sqrt{3}}{20}$ Simplify the radical.

Task 2. $= \dfrac{4(-2 + \sqrt{3})}{20}$ Factor the numerator.

$= \dfrac{-2 + \sqrt{3}}{5}$ Reduce the fraction.

SECTION 5-5. Practice Exercises

In exercises **1–62**, simplify each expression. Assume all variables are positive real numbers.

[Example 1]

1. $7\sqrt{10} + 8\sqrt{10}$

2. $3\sqrt{7} + 9\sqrt{7}$

3. $6\sqrt[4]{3} - 4\sqrt[4]{3}$

4. $9\sqrt[5]{2} - 3\sqrt[5]{2}$

5. $7\sqrt{5} - 3\sqrt{5} + \sqrt{5}$

6. $-2\sqrt{11} + 10\sqrt{11} - \sqrt{11}$

7. $\sqrt[3]{9} - 5\sqrt[3]{9} + 3\sqrt[3]{9} - 4\sqrt[3]{9}$

8. $2\sqrt[3]{20} + 3\sqrt[3]{20} - \sqrt[3]{20} - 8\sqrt[3]{20}$

9. $7\sqrt[4]{5} + 3\sqrt{6} - 3\sqrt[4]{5} - 7\sqrt{6}$

10. $3\sqrt{5} - 2\sqrt[4]{3} + 15\sqrt{5} - 17\sqrt[4]{3}$

11. $6\sqrt[5]{3} - 2\sqrt{3} + 15\sqrt[5]{3} - 17\sqrt{3}$

12. $9\sqrt[7]{5} + 3\sqrt{5} - 2\sqrt[7]{5} + 11\sqrt{5}$

13. $-5\sqrt[3]{2} - \sqrt{2} + 3\sqrt{2} + 7\sqrt[3]{2}$

14. $-10\sqrt{2} + 6\sqrt[3]{2} - 2\sqrt{2} + 8\sqrt[3]{2}$

[Example 2]

15. $3\sqrt{50} + \sqrt{72}$

16. $2\sqrt{27} + 3\sqrt{75}$

17. $\sqrt{125} - 2\sqrt{45}$

18. $3\sqrt{150} - \sqrt{216}$

19. $\sqrt[3]{81} - \sqrt[3]{3}$

20. $\sqrt[3]{16} - \sqrt[3]{2}$

21. $\sqrt[3]{80} + \sqrt[3]{270}$

22. $\sqrt[3]{640} - \sqrt[3]{2{,}160}$

23. $2\sqrt{200} - 4\sqrt{98} + 3\sqrt{8}$

24. $4\sqrt{108} + 3\sqrt{3} - 7\sqrt{48}$

25. $\sqrt[3]{343} + \sqrt[3]{135} - \sqrt[3]{5}$

26. $\sqrt[3]{54} + \sqrt[3]{16} - \sqrt[3]{27}$

27. $-\sqrt{5} - \sqrt{100} + 2\sqrt{20}$

28. $4\sqrt{12} + \sqrt{81} - \sqrt{27}$

29. $\sqrt[3]{375} - \sqrt[3]{72} - \sqrt[3]{192}$

30. $\sqrt[3]{108} - \sqrt[3]{54} + \sqrt[3]{250}$

31. $\sqrt{250} + 2\sqrt{80} - \sqrt{245}$

32. $3\sqrt{90} + \sqrt{500} - 2\sqrt{250}$

33. $\sqrt{49x} - 6\sqrt{18y} - \sqrt{4x} + 3\sqrt{50y}$

34. $\sqrt{108x} + \sqrt{200y} - 4\sqrt{12x} - \sqrt{50y}$

35. $3\sqrt{\frac{3}{2}} - \sqrt{\frac{2}{3}}$

36. $2\sqrt{\frac{3}{5}} - \sqrt{\frac{5}{3}}$

37. $\sqrt{\frac{5}{6}} - 2\sqrt{\frac{6}{5}}$

38. $4\sqrt{\frac{1}{6}} + \sqrt{\frac{2}{3}}$

[Example 3] 39. $\sqrt{2}(3\sqrt{2} + \sqrt{5})$

40. $\sqrt{3}(3\sqrt{2} - \sqrt{3})$

41. $\sqrt[3]{3}(2\sqrt[3]{2} - \sqrt[3]{9})$

42. $\sqrt[3]{2}(5\sqrt[3]{4} + 2\sqrt[3]{5})$

43. $\sqrt[4]{8}(\sqrt[4]{10} + 3\sqrt[4]{2})$

44. $\sqrt[4]{27}(\sqrt[4]{30} - 2\sqrt[4]{3})$

45. $\sqrt{14}(5\sqrt{2} - 2\sqrt{7} + \sqrt{14})$

46. $\sqrt{15}(2\sqrt{3} + \sqrt{5} - 2\sqrt{15})$

47. $(3 - 2\sqrt{2})(1 + \sqrt{2})$

48. $(1 + 4\sqrt{3})(3 - 2\sqrt{3})$

49. $(\sqrt{2} + 3)(\sqrt{3} - 5)$

50. $(\sqrt{5} + 2)(3\sqrt{7} - 1)$

51. $(2\sqrt{2} - \sqrt{3})(\sqrt{2} - 5\sqrt{3})$

52. $(3\sqrt{3} + 2\sqrt{5})(\sqrt{3} - 3\sqrt{5})$

53. $(-\sqrt{10} - 2\sqrt{6})(2\sqrt{10} + \sqrt{6})$

54. $(-\sqrt{15} + 3\sqrt{5})(\sqrt{15} + 2\sqrt{5})$

55. $(3 + \sqrt{5})^2$

56. $(6 + \sqrt{7})^2$

57. $(5 - 2\sqrt{7})^2$

58. $(2 - 5\sqrt{2})^2$

59. $(\sqrt{3} + \sqrt{2})^2$

60. $(\sqrt{11} + \sqrt{3})^2$

61. $(2\sqrt{10} - 4\sqrt{3})^2$

62. $(3\sqrt{5} - 2\sqrt{3})^2$

In exercises **63–70**, for each binomial:

a. State the conjugate.

b. Compute the product of the binomial and its conjugate.

[Example 4] 63. $3 + \sqrt{10}$

64. $2 + \sqrt{5}$

65. $8 - \sqrt{2}$

66. $4 - \sqrt{3}$

67. $-1 + 2\sqrt{5}$

68. $-9 + 4\sqrt{2}$

69. $-3\sqrt{7} - 4\sqrt{2}$

70. $-5\sqrt{3} - 7\sqrt{2}$

In exercises **71–86**, rationalize the denominator of each expression.

[Example 5] 71. $\dfrac{3 - \sqrt{6}}{\sqrt{3}}$

72. $\dfrac{4 + \sqrt{10}}{\sqrt{2}}$

73. $\dfrac{-\sqrt{10} + \sqrt{15}}{\sqrt{5}}$

74. $\dfrac{-\sqrt{20} - \sqrt{15}}{\sqrt{10}}$

75. $\dfrac{\sqrt{35} - 7}{2\sqrt{14}}$

76. $\dfrac{\sqrt{6} + 9}{5\sqrt{3}}$

77. $\dfrac{1}{2 + \sqrt{3}}$

78. $\dfrac{1}{3 + \sqrt{10}}$

79. $\dfrac{1}{-5 + 2\sqrt{6}}$

80. $\dfrac{1}{-4 + 2\sqrt{15}}$

81. $\dfrac{3}{\sqrt{2} - 2\sqrt{5}}$

82. $\dfrac{-3}{\sqrt{6} + 3\sqrt{7}}$

83. $\dfrac{\sqrt{5} + \sqrt{2}}{\sqrt{5} - \sqrt{2}}$

84. $\dfrac{\sqrt{7} - \sqrt{3}}{\sqrt{7} + \sqrt{3}}$

85. $\dfrac{15}{-\sqrt{7} + \sqrt{2}}$

86. $\dfrac{-12}{-\sqrt{10} - \sqrt{6}}$

In exercises **87–96**, simplify each expression.

[Example 6] **87.** $\dfrac{2 + \sqrt{24}}{2}$

88. $\dfrac{4 - \sqrt{8}}{2}$

89. $\dfrac{6 - \sqrt{27}}{3}$

90. $\dfrac{-8 + \sqrt{48}}{4}$

91. $\dfrac{-15 + \sqrt{125}}{-10}$

92. $\dfrac{2 - \sqrt{52}}{-10}$

93. $\dfrac{4 - \sqrt{200}}{14}$

94. $\dfrac{-21 + \sqrt{98}}{14}$

95. $\dfrac{-35 - \sqrt{147}}{-21}$

96. $\dfrac{12 - \sqrt{72}}{-21}$

SECTION 5-5. Ten Review Exercises

In exercises **1–8**, simplify each expression.

1. $-5(3^2) - 6^2 \div (2 \cdot 3) + 7^2$

2. $(-6)^2 - 2^6 + \sqrt{9} \cdot \sqrt{16} - 5(-3)$

3. $5(-17 + 5 \cdot 4 - 3^2) + 6^2 \div 2$

4. $\dfrac{\sqrt{9} \cdot \sqrt{25}}{1 - 2^2} + \dfrac{1 + \sqrt{16}}{2^4 - 11}$

5. $(3xy^2)(-7x^3)$

6. $(ab^2)(2ab)^2$

7. $(-2t)^3(3t)^2$

8. $\left(\dfrac{-3u}{2}\right)^2\left(\dfrac{-2u^2}{5}\right)^3$

In exercises **9** and **10**, solve and graph each inequality.

9. $3x - 1 < 8$

10. $|3x - 1| < 8$

SECTION 5-5. Supplementary Exercises

In exercises **1–12**, simplify each expression. Assume all variables are positive real numbers.

1. $7\sqrt[4]{5} + \sqrt{5} - 3\sqrt[4]{5} - 10\sqrt[3]{5} + 2\sqrt{5} - \sqrt[3]{5}$

2. $3\sqrt{5} - 2\sqrt[4]{5} - 9\sqrt{5} + 8\sqrt[3]{5} + 3\sqrt[4]{5} - 5\sqrt[3]{5}$

3. $\sqrt[3]{75}(7\sqrt[3]{5} - 8\sqrt[3]{9})$

4. $\sqrt[3]{98}(14\sqrt[3]{4} + 10\sqrt[3]{7})$

5. $(\sqrt{7 + \sqrt{13}})(\sqrt{7 - \sqrt{13}})$

6. $(\sqrt{8 - \sqrt{15}})(\sqrt{8 + \sqrt{15}})$

7. $(\sqrt{11 - \sqrt{10}})(\sqrt{11 + \sqrt{10}})$

8. $(\sqrt{15 + \sqrt{14}})(\sqrt{15 - \sqrt{14}})$

9. $\dfrac{2\sqrt{2}}{\sqrt{5} + \sqrt{3}}$

10. $\dfrac{8}{9 - 3\sqrt{7}}$

11. $\dfrac{\sqrt[3]{2x} + \sqrt[3]{4x^2}}{\sqrt[3]{4x}}$

12. $\dfrac{\sqrt[3]{9x^2} - \sqrt[3]{3x}}{\sqrt[3]{3x}}$

In exercises **13–20**, simplify each expression. Assume all radicands are positive real numbers.

13. $\sqrt{x}(\sqrt{x} + \sqrt{xy})$

14. $(\sqrt{a} - \sqrt{b})^2$

15. $\dfrac{2y}{\sqrt{y}}$

16. $\dfrac{\sqrt{m} - \sqrt{n}}{\sqrt{m}}$

17. $\dfrac{\sqrt{p} + \sqrt{q}}{\sqrt{p} - \sqrt{q}}$

18. $\dfrac{1}{\sqrt{s} + \sqrt{t}}$

19. $\dfrac{\sqrt{x + 1}}{\sqrt{x + 1} + 1}$

20. $\dfrac{\sqrt{x + 2}}{\sqrt{x + 2} + 2}$

Example. Rationalize the denominator of $\dfrac{1}{\sqrt[3]{2} + 1}$.

Solution. The denominator contains two terms, one of which is a cube root. Using $(x + y)(x^2 - xy + y^2) = x^3 + y^3$, with $x = \sqrt[3]{2}$ and $y = 1$, the **rationalizing factor** is the trinomial $\sqrt[3]{4} - \sqrt[3]{2} + 1$. Thus,

$$\frac{1}{\sqrt[3]{2} + 1} \cdot \frac{\sqrt[3]{4} - \sqrt[3]{2} + 1}{\sqrt[3]{4} - \sqrt[3]{2} + 1} = \frac{\sqrt[3]{4} - \sqrt[3]{2} + 1}{(\sqrt[3]{2})^3 + 1^3}$$

$$= \frac{\sqrt[3]{4} - \sqrt[3]{2} + 1}{3}$$

In exercises **21–24**, rationalize the denominator of each expression.

21. $\dfrac{1}{\sqrt[3]{3} + 1}$

22. $\dfrac{1}{\sqrt[3]{5} - 1}$

23. $\dfrac{12}{\sqrt[3]{25} - \sqrt[3]{5} + 1}$

24. $\dfrac{-15}{\sqrt[3]{36} + \sqrt[3]{6} + 1}$

In exercises **25–28**, rationalize the numerator of each expression.

25. $\dfrac{4-\sqrt{15}}{7}$ **26.** $\dfrac{7-\sqrt{13}}{2}$

27. $\dfrac{\sqrt{11}-\sqrt{5}}{\sqrt{5}}$ **28.** $\dfrac{\sqrt{7}-\sqrt{2}}{\sqrt{2}}$

In exercises **29** and **30**, simplify each expression.

29. $(\sqrt{x+3}+\sqrt{x-3})^2$ **30.** $(\sqrt{x-2}-\sqrt{x+2})^2$

SECTION 5-6. Complex Numbers

1. The definition of the imaginary unit

2. The definition of complex numbers

3. The definition of equality of complex numbers

4. Definitions for adding and subtracting complex numbers

There are times when we need to solve equations such as $t^2 = -9$. From the form of this equation we know that t cannot be replaced by any real number to make the equation true. This follows from the fact that if t is any real number, then $t^2 \geq 0$. Thus, there is no real number solution to the equation.

In this section, we will define numbers that are solutions to equations like the one above. We will also define equality for these numbers, and give definitions for adding and subtracting them.

The Definition of the Imaginary Unit

One of the first persons to work on the problem of solving equations such as $t^2 = -9$ was the German mathematician Karl Friedrich Gauss (1777–1855). Gauss decided that a set of numbers that are not real numbers was needed for solutions to such equations. He therefore made the following definition for a number whose square is negative.

Definition 5.7. The imaginary unit *i*
The imaginary unit is i, and i is the number such that:

$$i^2 = -1 \quad \text{and} \quad i = \sqrt{-1}$$

The imaginary unit i can be used to write the square root of a negative radicand in terms of a number with a positive radicand.

Definition 5.8. The *i*-form of a number
If x is a positive number, then $\sqrt{-x} = i\sqrt{x}$.

Example 1. Write each expression in the i-form and simplify.

a. $\sqrt{-9}$ **b.** $\sqrt{-40}$ **c.** $-\sqrt{\frac{-5}{64}}$

Solution. **Discussion.** First, use i to write each number with a positive radicand. Then, if possible, simplify the radical.

a. $\sqrt{-9} = i\sqrt{9}$ If $x > 0$, then $\sqrt{-x} = i\sqrt{x}$.

$= i3$ or $3i$ The preferred form is $3i$.

b. $\sqrt{-40} = i\sqrt{40}$

$= i2\sqrt{10}$ or $2i\sqrt{10}$ The preferred form is $2i\sqrt{10}$.

c. $-\sqrt{\frac{-5}{64}} = -i\sqrt{\frac{5}{64}}$

$= \frac{-i\sqrt{5}}{8}$ $\sqrt{\frac{5}{64}} = \frac{\sqrt{5}}{\sqrt{64}} = \frac{\sqrt{5}}{8}$

The product and quotient properties of n-th roots applies only to square roots of nonnegative radicands. If $x < 0$, or $y < 0$, or both are negative, then the square roots of x and y can be correspondingly written in the i-form and the properties can then be applied.

Example 2. Simplify each expression.

a. $\sqrt{-6} \cdot \sqrt{-15}$ **b.** $\frac{\sqrt{-56}}{\sqrt{2}}$

Solution. **a.** $\sqrt{-6} \cdot \sqrt{-15}$ The given expression

$= i\sqrt{6} \cdot i\sqrt{15}$ Written in the i-forms

$= i^2\sqrt{90}$ Multiply the radicands.

$= -3\sqrt{10}$ $i^2 = -1$ and $\sqrt{90} = 3\sqrt{10}$

b. $\frac{\sqrt{-56}}{\sqrt{2}}$ The given expression

$= \frac{i\sqrt{56}}{\sqrt{2}}$ Write $\sqrt{-56}$ in the i-form.

$= i\sqrt{28}$ $\frac{\sqrt{56}}{\sqrt{2}} = \sqrt{\frac{56}{2}} = \sqrt{28}$

$= 2i\sqrt{7}$ $\sqrt{28} = \sqrt{4 \cdot 7} = 2\sqrt{7}$

The Definition of Complex Numbers

The imaginary unit i is used to define a set of numbers for which the set of real numbers is a proper subset.

Definition 5.9. The set of complex numbers
If x and y are real numbers and $i = \sqrt{-1}$, then a number that can be written in the form $x + yi$ is called a **complex number.** The $x + yi$ form is called the **standard form:**

$$x + yi$$

If $y = 0$, then $x + yi = x$, a *real number*.

If $x = 0$, then $x + yi = yi$, an *imaginary number*.

Example 3. Write each expression in standard form and simplify.

a. $-3 + \sqrt{-100}$ **b.** $\sqrt{18} - \sqrt{-75}$

Solution. **Discussion.** In standard form, the real-number part is written first and the imaginary (or i-part) is written second.

a. $-3 + \sqrt{-100} = -3 + 10i$ The $x + yi$ form

b. $\sqrt{18} - \sqrt{-75} = \sqrt{18} - i\sqrt{75}$ Written in standard form

$= 3\sqrt{2} - 5i\sqrt{3}$ Simplify the radicals.

The i-part of the complex number in part **b** of Example 3 can be written in the following ways:

$i\sqrt{75}$ can be written as $i5\sqrt{3}$, $5i\sqrt{3}$, $5\sqrt{3}i$

In this text the $5i\sqrt{3}$ form will be used as the *preferred form*

The Definition of Equality of Complex Numbers

Complex numbers consist of two parts. The complex numbers in examples **a** and **b** are used to illustrate.

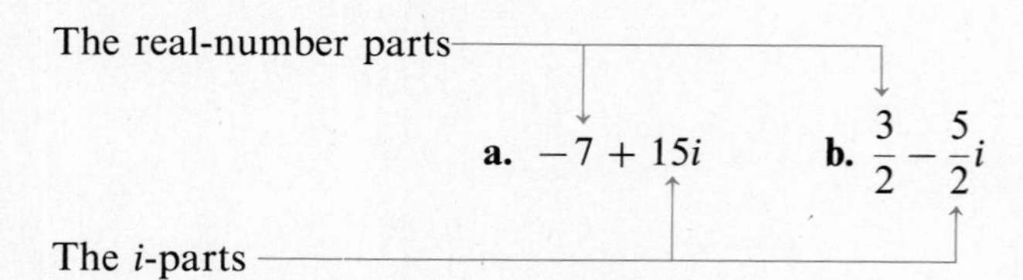

Based on Definition 5.10, two complex numbers are equal if, and only if, the real parts and the i-parts of the two numbers are equal.

Definition 5.10. Equality of complex numbers
If $x + yi$ and $a + bi$ are complex numbers, then

$$x + yi = a + bi$$

if, and only if:

(1) $x = a$ and (2) $y = b$

Example 4. Solve $-3u - 6i = 15 - 2vi$ for u and v.

Solution. **Discussion.** The equation states that two complex numbers are equal. By Definition 5.10, the numbers will be equal provided the real parts and the i-parts are equal.

Equating the real parts	**Equating the i-parts**
$-3u = 15$	$-6 = -2v$
$u = -5$	$3 = v$

Thus, if u is -5 and v is 3, the two complex numbers will be equal.

Definitions for Adding and Subtracting Complex Numbers

Examples **c** and **d** are an indicated sum and difference, respectively, of complex numbers.

c. $(7 - 2i) + (-4 - 9i)$ **d.** $(-3 + 5i) - (1 + 8i)$

Definition 5.11 contains equations that change both the indicated sum and difference to one complex number.

Definition 5.11. Adding and subtracting complex numbers
If $a + bi$ and $c + di$ are complex numbers, then:

$$(1) \quad (a + bi) + (c + di) = (a + c) + (b + d)i$$

$$(2) \quad (a + bi) - (c + di) = (a - c) + (b - d)i$$

Based on equations (1) and (2), two complex numbers are added or subtracted by combining the real parts and the i-parts. Using examples **c** and **d** to illustrate:

c. $(7 - 2i) + (-4 - 9i)$: $7 + (-4) = 3$; $-2 + (-9) = -11$; $\Rightarrow 3 - 11i$

d. $(-3 + 5i) - (1 + 8i)$: $-3 - 1 = -4$; $5 - 8 = -3$; $\Rightarrow -4 - 3i$

Example 5. Simplify $(\sqrt{8} + \sqrt{-50}) - (\sqrt{32} - \sqrt{-162})$.

Solution. **Discussion.** The numbers are first changed to standard form and the radicals simplified.

$$(\sqrt{8} + i\sqrt{50}) - (\sqrt{32} - i\sqrt{162})$$
$$= (2\sqrt{2} + 5i\sqrt{2}) - (4\sqrt{2} - 9i\sqrt{2})$$
$$= -2\sqrt{2} + 14i\sqrt{2}$$

SECTION 5-6. Practice Exercises

In exercises **1–40**, write each expression in the *i*-form and simplify.

[Example 1]

1. $\sqrt{-25}$ **2.** $\sqrt{-49}$ **3.** $\sqrt{-144}$

4. $\sqrt{-100}$ **5.** $-\sqrt{-196}$ **6.** $-\sqrt{-324}$

7. $\sqrt{-12}$ **8.** $\sqrt{-50}$ **9.** $-\sqrt{-98}$

10. $-\sqrt{-128}$ **11.** $\sqrt{\frac{-4}{49}}$ **12.** $\sqrt{\frac{-9}{64}}$

13. $\sqrt{\frac{-5}{36}}$ **14.** $\sqrt{\frac{-2}{9}}$ **15.** $\sqrt{\frac{-3}{7}}$

16. $\sqrt{\frac{-7}{10}}$ **17.** $-\sqrt{\frac{-4}{5}}$ **18.** $-\sqrt{\frac{-25}{3}}$

19. $\sqrt{\frac{-2}{45}}$ **20.** $\sqrt{\frac{-8}{75}}$

[Example 2]

21. $\sqrt{-2}\sqrt{-3}$ **22.** $\sqrt{-7}\sqrt{-2}$ **23.** $\sqrt{-6}\sqrt{-10}$

24. $\sqrt{-15}\sqrt{-10}$ **25.** $3\sqrt{-11}\sqrt{-7}$ **26.** $8\sqrt{-5}\sqrt{-13}$

27. $(4\sqrt{-10})(3\sqrt{-15})$ **28.** $(6\sqrt{-35})(9\sqrt{-14})$ **29.** $\frac{\sqrt{-10}}{\sqrt{5}}$

30. $\frac{\sqrt{-6}}{\sqrt{2}}$ **31.** $\frac{\sqrt{-45}}{\sqrt{5}}$ **32.** $\frac{\sqrt{-20}}{\sqrt{5}}$

33. $\frac{\sqrt{-490}}{\sqrt{-2}}$ **34.** $\frac{\sqrt{-540}}{\sqrt{-3}}$ **35.** $\frac{\sqrt{-336}}{\sqrt{-7}}$

36. $\frac{\sqrt{-693}}{\sqrt{-11}}$ **37.** $\frac{\sqrt{-8}\cdot\sqrt{-5}}{\sqrt{-64}}$ **38.** $\frac{\sqrt{-63}\cdot\sqrt{-2}}{\sqrt{-144}}$

39. $\frac{-\sqrt{-500}}{\sqrt{-50}\cdot\sqrt{3}}$ **40.** $\frac{-\sqrt{-450}}{\sqrt{-90}\cdot\sqrt{2}}$

In exercises **41–56**, write each expression in standard form and simplify.

[Example 3]

41. $10+\sqrt{-81}$ **42.** $5+\sqrt{-16}$ **43.** $-3-\sqrt{-121}$

44. $-8-\sqrt{-225}$ **45.** $12-\sqrt{-75}$ **46.** $1-\sqrt{-12}$

47. $\sqrt{64}+\sqrt{-121}$ **48.** $\sqrt{25}+\sqrt{-169}$ **49.** $\sqrt{8}+\sqrt{-8}$

50. $\sqrt{27}-\sqrt{-125}$ **51.** $\frac{7-\sqrt{-100}}{3}$ **52.** $\frac{4+\sqrt{-64}}{5}$

53. $\frac{-4-\sqrt{-68}}{3}$ **54.** $\frac{-12-\sqrt{-150}}{13}$ **55.** $\frac{-\sqrt{20}+\sqrt{-45}}{5}$

56. $\frac{\sqrt{300}-\sqrt{-90}}{7}$

In exercises **57–70**, solve each equation for x and y.

[Example 4] **57.** $x + yi = 7 + 2i$

58. $x + yi = -3 - 10i$

59. $x + yi = 6i - 10$

60. $x + yi = 9i + 1$

61. $x + yi = 7i$

62. $x + yi = 15$

63. $x + yi = 3\sqrt{2} + 7i\sqrt{2}$

64. $x + yi = \sqrt{5} - 3i\sqrt{5}$

65. $4x - 5yi = 6 + 6i$

66. $5x + 8yi = 8 - 12i$

67. $3x - 8i = -15 + 2yi$

68. $2x + 14i = 12 - 7yi$

69. $x\sqrt{2} + yi\sqrt{3} = -4 + 12i$

70. $x\sqrt{3} + yi\sqrt{5} = 9 - 20i$

In exercises **71–90**, simplify each expression.

[Example 5] **71.** $(3 + 2i) + (4 + i)$

72. $(7 + 3i) + (2 + 5i)$

73. $(1 - 3i) + (2 - 4i)$

74. $(8 - 6i) + (9 - i)$

75. $(7 - 7i) - (-2 + 3i)$

76. $(-2 - 2i) - (3 - 8i)$

77. $(-13 - 9i) - (6 - 10i)$

78. $(8 + 2i) - (14 - 7i)$

79. $(7 + 4i) + (3 - 4i)$

80. $(-11 - i) + (11 - 5i)$

81. $(17 + \sqrt{-64}) + (-11 - \sqrt{-16})$

82. $(-16 + \sqrt{-100}) + (20 - \sqrt{-121})$

83. $(-2 - \sqrt{-4}) - (5 - \sqrt{-25})$

84. $(21 - \sqrt{-169}) - (18 - \sqrt{-225})$

85. $(-\sqrt{20} - \sqrt{-125}) + (\sqrt{45} - \sqrt{-20})$

86. $(\sqrt{108} - \sqrt{-48}) + (\sqrt{300} - \sqrt{-147})$

87. $(\sqrt{24} + \sqrt{-54}) - (2\sqrt{6} - \sqrt{-150})$

88. $(2\sqrt{20} - \sqrt{-50}) - (\sqrt{80} + \sqrt{-32})$

89. $(4\sqrt{8} - 5\sqrt{-6}) - (-\sqrt{32} - \sqrt{-150})$

90. $(3\sqrt{6} + \sqrt{-18}) - (\sqrt{54} - 3\sqrt{-2})$

SECTION 5-6. Ten Review Exercises

In exercises **1–4**, solve each equation.

1. $\frac{2}{5}(5k + 3) = \frac{29}{45} - \frac{1}{9}(4 - 27k)$

2. $|2x + 3| = 3$

3. $2z^2 + 9 = 9z$

4. $\frac{2m + 4}{m + 3} = \frac{2m + 7}{m + 4}$

5. Solve $A = \frac{1}{2}h(a + b)$ for b.

In exercises **6–10**, solve and graph each inequality.

6. $3(4t + 3) < 4 - 9(t + 1)$

7. $3 \le 4b + 5 \le 11$

8. $5x - 1 < 3x - 21$ or $7x - 2 > 4x - 17$

9. $|2y - 5| < 9$

10. $\dfrac{2z - 9}{2z + 3} \leq 0$

SECTION 5-6. Supplementary Exercises

In exercises **1–12**, simplify each expression.

1. $(1 - 3i) + (4 + 5i) - (2 - i)$

2. $(-3 + 5i) - (7 - i) + (1 - 2i)$

3. $(10 - 7i) - (-2 - 6i) + (5) - (4 + 3i) + (4i)$

4. $(5 - 5i) - (3 + 3i) - (8i) + (2 + 2i) + (-4)$

5. $(1 - \sqrt{-1}) - (3 + \sqrt{-4}) + (-1 + \sqrt{-4}) - (5 - \sqrt{-16})$

6. $(4 + \sqrt{-9}) + (-2 - \sqrt{-4}) - (5 + \sqrt{-36}) - (-1 - \sqrt{-49})$

7. $\sqrt{-21 - 4} + \sqrt{40 - 4} - \sqrt{-45 - 4}$

8. $\sqrt{20 - 4} + \sqrt{-77 - 4} - \sqrt{5 - 4}$

9. $\sqrt{125 - 4} + \sqrt{-140 - 4} + \sqrt{-165 - 4}$

10. $\sqrt{-5 - 4} - \sqrt{-60 - 4} + \sqrt{104 - 4}$

11. $(6\sqrt{72} + 5\sqrt{-5}) - (-\sqrt{32} - 2\sqrt{-125})$

12. $(3\sqrt{3} + \sqrt{-18}) + (\sqrt{108} - 3\sqrt{-18})$

In exercises **13–20**, solve each equation for x and y.

13. $(8 + yi) + (x - i) = 11 - 3i$

14. $(2 - yi) + (x + 6i) = 6 + 5i$

15. $(-6 - yi) + (x - 4i) = -6 - 5i$

16. $(-4 + yi) + (x - i) = -3 - i$

17. $(3x + yi) - (x - (y - 2)i) = 4 + 4i$

18. $(x - 2yi) + ((x + 2) + yi) = 8 - 2i$

19. $(x + (y - 3)i) - ((x - 2) - yi) = (x - 1) + (y + 5)i$

20. $(x + (y + 1)i) - ((x - 1) - yi) = (x - 4) + (3y + 4)i$

In exercises **21–24**, z is the complex number $a + bi$. Find the expressions for a and b so that each equation is true.

21. $z + (x + yi) = 0$

22. $z + (x + yi) = 2x + 2yi$

23. $z + (x + yi) = -2x$

24. $z + (x + yi) = -3yi$

Definition 5.12. The absolute value (or modulus) of a complex number
If a and b are real numbers and i is the imaginary unit, then:

$$|a + bi| = \sqrt{a^2 + b^2}$$

In exercises **25–30**, find the modulus of each complex number.

25. $|6 + 8i|$ **26.** $|4 - 3i|$ **27.** $|7i|$

28. $|10i|$ **29.** $|5 + 5i|$ **30.** $|3 - 3i|$

SECTION 5-7. Multiplication and Division of Complex Numbers

1. Multiplying complex numbers
2. The definition of a complex conjugate
3. Dividing complex numbers
4. Combined operations

In Section 5-6, we studied definitions for equality of complex numbers, adding complex numbers, and subtracting complex numbers. In this section, we will study definitions for multiplying and dividing complex numbers.

Multiplying Complex Numbers

Example **a** is the indicated product of two complex numbers. The product of these numbers is also a complex number.

a. $(5 + 7i)(2 - 3i)$

The standard form of a complex number is a binomial. Therefore, to simplify this product four products must be found. The FOIL method for multiplying binomials can be used:

F L

a. $(5 + 7i)(2 - 3i)$

I

O

F: $5(2) = 10$, a *real number*

O: $5(-3i) = -15i$, an *imaginary number*

I: $7i(2) = 14i$, an *imaginary number*

L: $7i(-3i) = -21i^2$, and $i^2 = -1$

$= 21$, a *real number*

The two real and imaginary numbers can now be combined to yield a complex number in standard form:

$$10 - 15i + 14i + 21$$

$$= 31 - i$$ The product of $(5 + 7i)(2 - 3i)$

Definition 5.13. Multiplying complex numbers
If $a + bi$ and $c + di$ are complex numbers, then:

$$(3) \quad (a + bi)(c + di) = \underbrace{(ac - bd)}_{\text{Real part}} + \underbrace{(ad + bc)}_{\text{i-part}}i$$

Example 1. Find the indicated products and simplify.

a. $-4i(7 - 13i)$ **b.** $(9 - \sqrt{-20})(-1 + \sqrt{-245})$

Solution. **a. Discussion.** The real part of the left number is 0. To simplify this product, we distribute the $-4i$ to both parts of the other complex number.

$$\begin{aligned} -4i(7 - 13i) &= -28i + 52i^2 && \text{The distributive property} \\ &= -28i + 52(-1) && \text{Replace } i^2 \text{ by } -1. \\ &= -52 - 28i && \text{Written in standard form} \end{aligned}$$

b. Discussion. First write both complex numbers in standard form with simplified radicals.

$$\begin{aligned} &(9 - \sqrt{-20})(-1 + \sqrt{-245}) \\ &= (9 - 2i\sqrt{5})(-1 + 7i\sqrt{5}) && \text{In standard form} \\ &= -9 + 63i\sqrt{5} + 2i\sqrt{5} - 14i^2(5) && \sqrt{5} \cdot \sqrt{5} = 5 \\ &= -9 + 65i\sqrt{5} + 70 && -14i^2(5) = 70 \\ &= 61 + 65i\sqrt{5} && \text{The simplified product} \end{aligned}$$

The Definition of a Complex Conjugate

Example **b** is an indicated product of an irrational number and its **conjugate.** The product of these numbers is a **rational number.**

b. $(-3 + \sqrt{5})(-3 - \sqrt{5}) = (-3)^2 - (\sqrt{5})^2 = 9 - 5 = 4$

Example **c** is an indicated product of a complex number and its **conjugate.** The product of these numbers is a **real number.**

c. $(-3 + 5i)(-3 - 5i) = (-3)^2 - (5i)^2 = 9 - (-25) = 34$

The conjugate of an irrational number and conjugate of a complex number have similar forms. The product of an irrational number and its conjugate is a rational number. The product of a complex number and its conjugate is a real number.

Definition 5.14. Conjugate of a complex number
The conjugate of $x + yi$ is $x - yi$.

Example 2. Given: $\frac{-3}{2} - \frac{5}{2}i$

a. State the conjugate.

b. Compute the product of the number and its conjugate.

Solution. **a.** The conjugate of $\frac{-3}{2} - \frac{5}{2}i$ is $\frac{-3}{2} + \frac{5}{2}i$.

Same real parts

Same i-parts

Different signs

b.

$$\left(\frac{-3}{2} - \frac{5}{2}i\right)\left(\frac{-3}{2} + \frac{5}{2}i\right) \qquad (x - yi)(x + yi)$$

$$= \left(\frac{-3}{2}\right)^2 - \left(\frac{5}{2}i\right)^2 \qquad = x^2 - (yi)^2$$

$$= \frac{9}{4} - \frac{-25}{4} \qquad \frac{25}{4}i^2 = \frac{-25}{4}$$

$$= \frac{34}{4} = \frac{17}{2} \qquad \text{Combine and reduce.}$$

Dividing Complex Numbers

Examples **d** and **e** show indicated quotients of complex numbers.

d. $\frac{1 - 3i}{2i}$ **e.** $\frac{17}{-3 + 5i}$

In these examples, we are looking for numbers u and v, such that the following equalities will be true:

d. $(1 - 3i) \div 2i = u$, such that $1 - 3i = u(2i)$

e. $17 \div (-3 + 5i) = v$, such that $17 = v(-3 + 5i)$

The values for u and v can be found by multiplying the given expressions by the conjugates of the denominators.

d.

$$\frac{1 - 3i}{2i} \cdot \frac{-2i}{-2i}$$

$$= \frac{-2i + 6i^2}{-4i^2}$$

$$= \frac{-6 - 2i}{4}$$

$$= \frac{-3 - i}{2}$$

e.

$$\frac{17}{-3 + 5i} \cdot \frac{-3 - 5i}{-3 - 5i}$$

$$= \frac{17(-3 - 5i)}{9 - 25i^2}$$

$$= \frac{17(-3 - 5i)}{34}$$

$$= \frac{-3 - 5i}{2} \quad \text{or} \quad \frac{-3}{2} - \frac{5}{2}i$$

Definition 5.15. Dividing complex numbers
If $c + di \neq 0 + 0i$, then:

$$\frac{a + bi}{c + di} = \frac{a + bi}{c + di} \cdot \frac{c - di}{c - di} = \frac{ac + bd}{c^2 + d^2} + \frac{bc - ad}{c^2 + d^2}i$$

Example 3. Find the quotient of $\dfrac{-6 + 10i}{1 - i}$.

Solution. **Discussion.** Multiply by $\dfrac{\text{conjugate}}{\text{conjugate}}$, where "conjugate" is the conjugate of the divisor (or denominator) of the given expression.

$$\frac{-6 + 10i}{1 - i} \cdot \frac{1 + i}{1 + i}$$ The conjugate of $1 - i$ is $1 + i$.

$$= \frac{-6 - 6i + 10i + 10i^2}{1 - i^2}$$ $(1 - i)(1 + i) = 1^2 - i^2$

$$= \frac{-16 + 4i}{2}$$ $-6 + 10i^2 = -6 - 10 = -16$

$$= -8 + 2i$$ Simplify the fraction.

Combined Operations

When simplifying expressions with more than one operation indicated, the order of operations must be followed. This rule also applies to expressions with complex numbers.

Example 4. Simplify $(10 - 3i)^2 - (-5 + 2i)(3 + 7i)$.

Solution. **Discussion.** Do the power first. Then simplify the indicated product. Finally, combine any real parts and i-parts, and write the answer in the standard form of a complex number.

$$(10 - 3i)^2 - (-5 + 2i)(3 + 7i)$$
$$= 100 - 60i + 9i^2 - (-5 + 2i)(3 + 7i)$$
$$= 100 - 60i - 9 - (-15 - 35i + 6i + 14i^2)$$
$$= 100 - 60i - 9 + 15 + 35i - 6i + 14$$
$$= 120 - 31i$$

SECTION 5-7. Practice Exercises

In exercises **1–36**, find the indicated products and simplify.

[Example 1]

1. $-6(9i)$

2. $-7(5i)$

3. $8(2 - 5i)$

4. $9(-4 + 7i)$

5. $-11(-7 + 13i)$

6. $-6(13 - 12i)$

7. $5i(-4 + 3i)$

8. $3i(8 + 2i)$

9. $-i(15 + 9i)$

10. $-i(17 - 6i)$

11. $(3 + 2i)(2 + 3i)$

12. $(7 + i)(1 + 7i)$

13. $(1 - 5i)(4 - 7i)$

14. $(4 - 9i)(6 - i)$

15. $(-3 + 10i)(-6 - 2i)$

16. $(-5 - 4i)(-1 + 3i)$

17. $(2 - \sqrt{-72})(1 + \sqrt{-8})$

18. $(-3 + \sqrt{-200})(-1 - \sqrt{-18})$

19. $(-1 + \sqrt{-27})(-3 - \sqrt{-48})$

20. $(-2 - \sqrt{-75})(2 + \sqrt{-12})$

21. $(5 + \sqrt{-2})(5 - \sqrt{-2})$

22. $(7 + \sqrt{-3})(7 - \sqrt{-3})$

23. $(-\sqrt{24} - \sqrt{-100})(-\sqrt{24} + \sqrt{-100})$

24. $(-\sqrt{18} + \sqrt{-121})(-\sqrt{18} - \sqrt{-121})$

25. $(-\sqrt{2} + \sqrt{-18})(3\sqrt{2} - 2\sqrt{-18})$

26. $(2\sqrt{5} - \sqrt{-50})(-\sqrt{5} - 3\sqrt{-50})$

27. $(4 + 3i)^2$

28. $(7 + i)^2$

29. $(2 - 5i)^2$

30. $(10 - 3i)^2$

31. $(\sqrt{3} + \sqrt{-27})^2$

32. $(-\sqrt{2} - \sqrt{-20})^2$

33. $(2\sqrt{5} - \sqrt{-75})^2$

34. $(-3\sqrt{7} + \sqrt{-98})^2$

35. $\left(\frac{-1}{2} + \frac{\sqrt{3}}{2}i\right)^2$

36. $\left(\frac{-5}{2} - \frac{\sqrt{3}}{2}i\right)^2$

In exercises **37–46**, for each complex number:

a. State the conjugate.

b. Compute the product of the number and its conjugate.

[Example 2]

37. $10 - 3i$

38. $8 - 5i$

39. $-7 + 4i$

40. $-3 + 9i$

41. $6 - 6i$

42. $3 - 3i$

43. $\sqrt{10} + \sqrt{-81}$

44. $\sqrt{6} + \sqrt{-49}$

45. $-\sqrt{-24} + 3\sqrt{5}$

46. $\sqrt{-72} - 4\sqrt{3}$

In exercises **47–84**, find the indicated quotients and simplify.

[Example 3]

47. $\frac{2 + 10i}{-2i}$

48. $\frac{6 - 9i}{3i}$

49. $\frac{-1 - 3i}{i\sqrt{5}}$

50. $\frac{4 + i}{-i\sqrt{5}}$

51. $\frac{12 - 20i}{-2i\sqrt{3}}$

52. $\frac{-15 + 24i}{3i\sqrt{2}}$

53. $\dfrac{1}{1+i}$

54. $\dfrac{1}{-1-i}$

55. $\dfrac{3}{-2+2i}$

56. $\dfrac{5}{2-2i}$

57. $\dfrac{-8i}{3-3i}$

58. $\dfrac{-2i}{-3+3i}$

59. $\dfrac{1-i}{1+i}$

60. $\dfrac{1+i}{1-i}$

61. $\dfrac{1+2i}{2+3i}$

62. $\dfrac{2+3i}{3+2i}$

63. $\dfrac{4-8i}{-1-i}$

64. $\dfrac{-6+10i}{-1+i}$

65. $\dfrac{9-15i}{-1-3i}$

66. $\dfrac{-21+30i}{3+i}$

67. $\dfrac{-\sqrt{3}-i\sqrt{2}}{\sqrt{3}-i\sqrt{2}}$

68. $\dfrac{-\sqrt{2}-i\sqrt{3}}{\sqrt{2}-i\sqrt{3}}$

69. $\dfrac{1+2i\sqrt{3}}{3-i\sqrt{3}}$

70. $\dfrac{4-2i\sqrt{2}}{3+i\sqrt{2}}$

71. $\dfrac{10+\sqrt{-24}}{-1-\sqrt{-54}}$

72. $\dfrac{-3-\sqrt{-216}}{2+\sqrt{-150}}$

[Example 4] **73.** $(2+i)^2 - 4(2+i)$

74. $(3-i)^2 - 5(1+2i)$

75. $(5+3i)^2 + 4i(7-i)$

76. $(8+2i)^2 + 6i(5-i)$

77. $\dfrac{(5+10i)+(-3-7i)}{-3+2i}$

78. $\dfrac{(-12-7i)+(16+5i)}{1-3i}$

79. $\dfrac{(-15-25i)-(5+20i)}{4+2i}$

80. $\dfrac{(27+16i)-(30-2i)}{1+i}$

81. $\dfrac{(-2+i)^2}{-2-i} + \dfrac{4i}{-2-i}$

82. $\dfrac{(1-3i)^2}{1+3i} + \dfrac{6i}{1+3i}$

83. $\dfrac{1+2i}{1-2i} - \dfrac{1-2i}{1+2i}$

84. $\dfrac{3-2i}{3+2i} - \dfrac{3+2i}{3-2i}$

SECTION 5-7. Ten Review Exercises

1. Simplify $2(k+6) - 14 + 6(k+3)$.

2. Solve $2(k+6) = 14 - 6(k+3)$.

3. Factor $4t^4 + 7t^2 - 36$.

4. Solve $4t^4 = 36 - 7t^2$ over the complex numbers.

5. Factor $10u - 8 + 3u^2$.

6. Solve and graph $3u^2 + 10u - 8 < 0$.

7. Simplify $\dfrac{x+2}{x+4} + \dfrac{x-4}{x-5} - 2$.

8. Solve $\dfrac{x+2}{x+4} + \dfrac{x-4}{x-5} = 2$.

In exercises **9** and **10**, solve each equation for y.

9. $3y - x = by + 5$

10. $x = \dfrac{3y}{2y+3}$

SECTION 5-7. Supplementary Exercises

In exercises **1–20**, use the definition $i^2 = -1$ to simplify each power.

1. i^3 **2.** i^4 **3.** i^5

4. i^6 **5.** i^7 **6.** i^8

7. i^9 **8.** i^{10} **9.** i^{11}

10. i^{12} **11.** i^{20} **12.** i^{28}

13. i^{30} **14.** i^{31} **15.** i^{35}

16. i^{39} **17.** i^{48} **18.** i^{50}

19. i^{101} **20.** i^{110}

In exercises **21–28**, simplify each expression.

21. $5i(-4i)(1 - 5i)$ **22.** $2i(3 - 4i)(5 + 6i)$

23. $(1 + i)(2 - i)(3 - i)$ **24.** $(1 - 2i)(3 + i)(2 + i)$

25. $(1 + i)^3$ **26.** $(2 + i)^3$

27. $(\sqrt{3} - i)^3$ **28.** $(\sqrt{2} + i\sqrt{3})^3$

In exercises **29–32**, evaluate each expression for the indicated values.

29. $y^2 - 6y + 10$
- **a.** $y = 5$
- **b.** $y = 2 + \sqrt{3}$
- **c.** $y = 3 - i$

30. $y^2 - 4y + 9$
- **a.** $y = 2$
- **b.** $y = 2 + \sqrt{5}$
- **c.** $y = 2 - i\sqrt{5}$

31. $z^2 + iz + 6$
- **a.** $z = -2$
- **b.** $z = \sqrt{2}$
- **c.** $z = 2i$

32. $z^2 - 2iz - 2$
- **a.** $z = 2$
- **b.** $z = -\sqrt{2}$
- **c.** $z = 1 + i$

In exercises **33–38**, solve each equation for z where z is a complex number of the form $a + bi$.

33. $6z = 12 - 6i$

34. $5z = 15 - 20i$

35. $2iz = 8 + 10i$

36. $3iz = 12 + 18i$

37. $3z - iz = 20$

38. $z + iz = 5$

Name

Date

Score

SECTION 5-1. Summary Exercises

Answer

In exercises **1–17**, simplify each expression.

1. $-\sqrt{169}$

1. ______________________

2. $\sqrt[3]{\dfrac{-125}{216}}$

2. ______________________

3. $\sqrt[4]{256}$

3. ______________________

4. $-\sqrt[5]{\dfrac{-1}{243}}$

4. ______________________

5. $\sqrt[4]{(-10)^4}$

5. ______________________

6. $\sqrt[5]{(-9)^5}$

6. ______________________

7. $(\sqrt[3]{-24})^3$

7. ______________________

8. Use Appendix Table 1 or a calculator to find to 3 places a decimal approximation of $\sqrt{50}$.

8. ______________________

Name

Date

Score

SECTION 5-2. Summary Exercises

Answer

In exercises **1** and **2**, write each expression in radical form and simplify, if possible.

1. a. $(-26)^{1/3}$ **1. a.** ______________

b. $625^{1/4}$ **b.** ______________

2. a. $31^{4/3}$ **2. a.** ______________

b. $64^{2/3}$ **b.** ______________

3. Write each expression in exponential form.

a. $\sqrt[3]{-5}$ **3. a.** ______________

b. $\sqrt[8]{3}$ **b.** ______________

In exercises **4–7**, simplify each expression.

4. a. $9^{-1/2}$ **4. a.** ______________

b. $16^{-3/4}$ **b.** ______________

5. $\dfrac{5^{2/3} \cdot 5^{3/4}}{5^{1/12}}$

5. ______________________

6. $(-125x^3y^6)^{1/3}$

6. ______________________

7. $(4z^{1/2} + 1)(3z^{1/2} - 2)$

7. ______________________

8. Use a calculator to approximate $(\sqrt[5]{83})^3$ to three decimal places.

8. ______________________

Name

Date

Score

SECTION 5-3. Summary Exercises

Answer

In exercises **1–7**, write each expression as a single radical and simplify, if possible. Assume all variables represent positive real numbers.

1. $\sqrt[5]{3p} \cdot \sqrt[5]{-12p^3}$

1. ____________________

2. $(-\sqrt[4]{162})(\sqrt[4]{8})$

2. ____________________

3. $\sqrt[3]{12x^2} \cdot \sqrt[3]{36x^2} \cdot \sqrt[3]{4x^2}$

3. ____________________

4. $\dfrac{\sqrt[6]{264}}{\sqrt[6]{8}}$

4. ____________________

5. 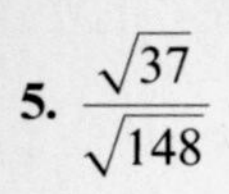
$\dfrac{\sqrt{37}}{\sqrt{148}}$

5. ______________________

6. $\dfrac{\sqrt[5]{224x^7}}{\sqrt[5]{7x^2}}$

6. ______________________

7. $\dfrac{\sqrt[3]{6x^2}\,\sqrt[3]{-2x^2}}{\sqrt[3]{-324x}}$

7. ______________________

8. Change $\sqrt[8]{64x^2}$ to an index of 4.

8. ______________________

Name

Date

Score

SECTION 5-4. Summary Exercises

Answer

In exercises **1–8**, simplify each expression. Assume all variables are positive real numbers.

1. $-\sqrt[3]{750}$

1. ______

2. $\sqrt[3]{-13}$

2. ______

3. $\sqrt[5]{-64}$

3. ______

4. $\sqrt[4]{\dfrac{5}{81}}$

4. ______

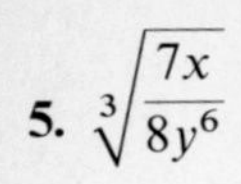

5. $\sqrt[3]{\dfrac{7x}{8y^6}}$ 5. ____________

6. $\dfrac{9}{\sqrt[4]{27}}$ 6. ____________

7. $\dfrac{10}{\sqrt[5]{4a^3b^4}}$ 7. ____________

8. $\sqrt[8]{25x^4}$ 8. ____________

Name

Date

Score

SECTION 5-5. Summary Exercises

Answer

In exercises **1–4**, simplify each expression.

1. $5\sqrt[4]{8} + 2\sqrt[3]{6} - \sqrt[4]{8} + 7\sqrt[3]{6}$

1. ____________

2. $\sqrt{175} - \sqrt{112} - 2\sqrt{28}$

2. ____________

3. $(9 - 2\sqrt{5})(4 + 3\sqrt{5})$

3. ____________

4. $(2\sqrt{3} - \sqrt{5})^2$

4. ____________

5. Given: $-\sqrt{7} + 2\sqrt{5}$

a. State the conjugate.

5. a. ____________________

b. Compute the product of the binomial and its conjugate.

b. ____________________

In exercises **6** and **7**, rationalize each denominator.

6. $\dfrac{13 - 4\sqrt{5}}{\sqrt{10}}$

6. ____________________

7. $\dfrac{-15}{\sqrt{5} - 5\sqrt{2}}$

7. ____________________

8. Simplify $\dfrac{8 - \sqrt{80}}{4}$.

8. ____________________

Name

Date

Score

SECTION 5-6. Summary Exercises

Answer

In exercises **1** and **2**, write each expression in the *i*-form and simplify.

1. $\sqrt{-405}$ **1.** ____________

2. $\sqrt{\dfrac{-15}{16}}$ **2.** ____________

In exercises **3** and **4**, simplify each expression.

3. $4\sqrt{-6}\sqrt{-30}$ **3.** ____________

4. $\dfrac{\sqrt{-275}}{\sqrt{-11}}$ **4.** ____________

5. Write $\sqrt{4} - \sqrt{-25}$ in standard form.

5. ______________________

6. Solve $4x + yi = 3i\sqrt{2} + 10$ for x and y.

6. ______________________

In exercises **7** and **8**, simplify each expression.

7. $(17 - 12i) - (32 - 19i)$

7. ______________________

8. $(8 - \sqrt{-81}) + (12 - \sqrt{-144})$

8. ______________________

Name

Date

Score

SECTION 5-7. Summary Exercises

Answer

In exercises **1–3**, find the indicated products and simplify.

1. $-7i(3-6i)$

1. ______

2. $(-2-\sqrt{-25})(8+\sqrt{-4})$

2. ______

3. $(\sqrt{5}+\sqrt{-18})^2$

3. ______

4. Given: $-4-9i$

a. State the conjugate.

4. a. ______

b. Compute the product of the number and its conjugate.

b. ______

In exercises **5–8**, do the indicated operations and simplify.

5. $\dfrac{10 - 7i}{-3i}$

5. ______________________

6. $\dfrac{5 - i}{3 + 2i}$

6. ______________________

7. $(1 - 3i)^2 - 5(1 - 3i) + 6$

7. ______________________

8. $\dfrac{(3 - 2i)^2}{4 + 3i}$

8. ______________________

CHAPTER 5. Review Exercises

In exercises **1–4**, simplify each expression.

1. $\sqrt{196}$

2. $-\sqrt{\frac{121}{64}}$

3. $\sqrt{1\frac{9}{16}}$

4. $-\sqrt{12^2}$

In exercises **5** and **6**, use Appendix Table 1 or a calculator to find a decimal approximation of each expression.

5. $\sqrt{23}$

6. $\sqrt{71}$

In exercises **7–18**, simplify each expression. If a root is not a real number, then write "not a real number".

7. $\sqrt[3]{\frac{-125}{8}}$

8. $-\sqrt[4]{256}$

9. $\sqrt[4]{\frac{81}{16}}$

10. $\sqrt[5]{-243}$

11. $\sqrt{-3^2}$

12. $\sqrt[4]{-81}$

13. $(\sqrt[3]{41})^3$

14. $\sqrt[5]{(-4)^5}$

15. $\sqrt[4]{(-7)^4}$

16. $(\sqrt[6]{0.8})^6$

17. $(\sqrt{15})^2$

18. $\sqrt{13^2}$

In exercises **19–22**, write each expression in radical form and simplify, if possible.

19. $121^{1/2}$

20. $1{,}296^{1/4}$

21. $12^{1/3}$

22. $0.06^{1/5}$

In exercises **23–26**, write each expression in exponential form.

23. $\sqrt[3]{-18}$

24. $-\sqrt[5]{2}$

25. $\sqrt{17}$

26. $\sqrt[4]{5x^2}$

In exercises **27–30**, write each expression in radical form and simplify, if possible.

27. $(-27)^{2/3}$

28. $\left(\frac{-125}{8}\right)^{4/3}$

29. $x^{3/4}$; $x > 0$

30. $(2a)^{3/5}$

In exercises **31** and **32**, simplify each expression.

31. $196^{-1/2}$

32. $2{,}401^{-3/4}$

In exercises **33–38**, simplify each expression. Assume all variables represent positive real numbers.

33. $(343x^6)^{1/3}$

34. $\dfrac{x^{2/3}y^{-3/4}}{x^{-1/3}y^{1/4}}$

35. $\left(\dfrac{a^2b^4}{100}\right)^{-1/2}$

36. $(x^{1/2} + 2y^{1/2})^2$

37. $\dfrac{t^{4/3} - t^{1/3}}{t^{1/3}}$

38. $\dfrac{54x^{11/4} + 42x^{7/4}}{6x^{3/4}}$

In exercises **39** and **40**, use a calculator to approximate each root to three decimal places.

39. $\sqrt[5]{41}$

40. $\sqrt[4]{153}$

In exercises **41–50**, write each expression as a single radical and simplify, if possible. Assume all variables represent positive real numbers.

41. $\sqrt{7}\sqrt{22}$

42. $\sqrt[3]{5}\sqrt[3]{2}\sqrt[3]{11}$

43. $\sqrt[3]{4}\sqrt[3]{16}$

44. $\sqrt[4]{3}\sqrt[4]{9}\sqrt[4]{3}$

45. $\sqrt[5]{2x^2}\sqrt[5]{4x}\sqrt[5]{8x^2}$

46. $\sqrt[6]{9y}\sqrt[6]{3y^3}\sqrt[6]{27y^2}$

47. $\dfrac{\sqrt{128x}}{\sqrt{2}}$

48. $\dfrac{\sqrt[3]{6}}{\sqrt[3]{750}}$

49. $\dfrac{\sqrt[4]{7}\sqrt[4]{27}}{\sqrt[4]{3{,}024}}$

50. $\dfrac{\sqrt[5]{729}}{\sqrt[5]{-1}\sqrt[5]{3}}$

In exercises **51** and **52**, change each radical to the indicated index.

51. $\sqrt[6]{144}$, to index 3

52. $\sqrt[4]{9w}$, to index 8

In exercises **53–62**, simplify each expression. Assume all variables are positive real numbers.

53. $\sqrt{252}$

54. $\sqrt[3]{80}$

55. $\sqrt[3]{-375x^4}$

56. $\sqrt[4]{48a^4b^9}$

57. $\dfrac{5}{\sqrt{10}}$

58. $\dfrac{3}{\sqrt[3]{9x}}$

59. $\dfrac{14a}{\sqrt[4]{49a^3}}$

60. $\sqrt[3]{\dfrac{17}{1{,}000x^6}}$

61. $\sqrt[4]{36}$

62. $\sqrt[6]{216x^3}$

In exercises **63–72**, simplify each expression. Assume all variables are positive real numbers.

63. $4\sqrt{2} + 3\sqrt{3} - \sqrt{2} + 2\sqrt{3}$

64. $2\sqrt{20} - \sqrt{54} + \sqrt{600} - 3\sqrt{5}$

65. $8\sqrt[3]{2x} + 2\sqrt{2x} - \sqrt[3]{250x} - 5\sqrt{8x}$

66. $4\sqrt[3]{5y^2} - 3\sqrt[3]{5y} + \sqrt[3]{40y} - \sqrt[3]{135y^2}$

67. $\sqrt{7}(3\sqrt{7} - \sqrt{2})$

68. $\sqrt[3]{6}(2\sqrt[3]{18} + \sqrt[3]{12})$

69. $(10 + 3\sqrt{5})(3 - 2\sqrt{5})$

70. $(2 - \sqrt[3]{4})(5 + \sqrt[3]{2})$

71. $(3 - \sqrt{7})^2$

72. $(2\sqrt{6} + 5\sqrt{2})^2$

In exercises **73** and **74**, for each binomial:

a. State the conjugate.

b. Compute the product of the number and its conjugate.

73. $9 - 2\sqrt{3}$

74. $-2\sqrt{5} + 5\sqrt{2}$

In exercises **75–78**, rationalize the denominator of each expression.

75. $\dfrac{7 - \sqrt{2}}{\sqrt{3}}$

76. $\dfrac{10}{4 - \sqrt{11}}$

77. $\dfrac{-6 + \sqrt{2}}{\sqrt{7} + \sqrt{5}}$

78. $\dfrac{2\sqrt{6} + \sqrt{3}}{2\sqrt{6} - \sqrt{3}}$

In exercises **79** and **80**, simplify each expression.

79. $\dfrac{12 + \sqrt{32}}{8}$

80. $\dfrac{-10 - \sqrt{75}}{15}$

In exercises **81–84**, write each expression in the *i*-form and simplify.

81. $\sqrt{-36}$

82. $-\sqrt{-120}$

83. $-\sqrt{\dfrac{-7}{25}}$

84. $\sqrt{\dfrac{-4x}{75}}$; $x > 0$

In exercises **85–88**, simplify each expression.

85. $\sqrt{-10} \cdot \sqrt{-3}$

86. $\sqrt{-35} \cdot \sqrt{-14}$

87. $\dfrac{\sqrt{-252}}{\sqrt{7}}$

88. $\dfrac{\sqrt{336}}{\sqrt{-3}}$

In exercises **89–92**, write each expression in standard $a + bi$ form.

89. $-\sqrt{27} + \sqrt{-27}$

90. $-5 - \sqrt{-180}$

91. $\dfrac{-4 + \sqrt{-28}}{10}$

92. $\dfrac{\sqrt{18} - \sqrt{-72}}{3}$

In exercises **93–96**, solve each equation for x and y.

93. $3x + 5yi = -9 - 25i$

94. $2x - 3yi = -3 + 5yi$

95. $x\sqrt{6} - 3yi\sqrt{2} = 4 + 6i$

96. $-2x + 5yi = 4\sqrt{3}$

In exercises **97–102**, simplify each expression.

97. $(-6 + 9i) + (8 - 11i)$

98. $(-7 - 2i) - (-13 + i)$

99. $(21 - \sqrt{-121}) - (17 + \sqrt{-4})$

100. $(-11 + \sqrt{-36}) + (7 + \sqrt{-81})$

101. $(\sqrt{20} - \sqrt{-18}) - (\sqrt{125} + \sqrt{-98}) + (-\sqrt{500} + \sqrt{-200})$

102. $(-\sqrt{24} + 2\sqrt{-3}) + (\sqrt{216} - \sqrt{-108}) - (4\sqrt{6} - \sqrt{-75})$

In exercises **103–108**, find the indicated products and simplify.

103. $6i(-7 + 9i)$

104. $(-3 + i)(8 - 5i)$

105. $\left(\frac{8}{17} + \frac{15}{17}i\right)\left(\frac{8}{17} - \frac{15}{17}i\right)$

106. $(-4 + \sqrt{-24})(-3 - \sqrt{-6})$

107. $(\sqrt{3} - 2i\sqrt{7})(2\sqrt{3} + 5i\sqrt{7})$

108. $(9 + 5i)^2$

In exercises **109** and **110**, for each complex number:

a. State the conjugate.

b. Compute the product of the number and its conjugate.

109. $-9 - 13i$

110. $3\sqrt{5} - \sqrt{-50}$

In exercises **111–114**, find the indicated quotients and simplify.

111. $\frac{14 - 35i}{-7i}$

112. $\frac{-3 + i}{-2 - 2i}$

113. $\frac{\sqrt{6} + i\sqrt{3}}{2\sqrt{6} - 3i\sqrt{3}}$

114. $\frac{-\sqrt{12} - \sqrt{-27}}{\sqrt{75} + \sqrt{-3}}$

CHAPTER

6

Quadratic Equations and Inequalities

Before the day's lesson had begun, Kim Kimiecik raised her hand and asked, "Ms. Glaston, is there a formula to figure out how long it takes something to fall a certain distance when dropped from a very high point?"

"Yes, Kim, there is," Ms. Glaston replied. "If air resistance is neglected, then the following equation can be used." She wrote on the board:

$d = 16t^2$ — t is the *time* an object falls in seconds; d is the corresponding *distance* it falls in feet

"By the way, Kim," Ms. Glaston continued, "what makes you interested in such an equation?"

"Well," Kim replied, "recently I started lessons in skydiving. Last week I had my first jump from a plane. It was really a fantastic experience, and more than a little frightening. And, even though these first jumps are made with the chute being opened for me, eventually I'll be eligible to make jumps in which I will decide when to open the chute. That's when I got to thinking about the time and distance relationship. Could you show me how to use the formula for a few values of d?"

"Certainly, Kim," Ms. Glaston replied, "it will give us all some practice solving quadratic equations." She then wrote on the board:

(1) If $d = 3{,}600$ feet, then $3{,}600 = 16t^2$ and $t =$ ______ or ______.

(2) If $d = 6{,}400$ feet, then $6{,}400 = 16t^2$ and $t =$ ______ or ______.

(3) If $d = 10{,}000$ feet, then $10{,}000 = 16t^2$ and $t =$ ______ or ______.

"Now," Ms. Glaston asked, "how do we solve these equations?"

"By factoring," Shelby White answered quickly, "and then by using the zero-product property. But first you need to make one side of the equation equal to 0."

"Very good," Ms. Glaston nodded approvingly. "I'll push the chalk and you all tell me what to write." After about 5 minutes, she had written the following display on the board:

Equation (1)	**Equation (2)**	**Equation (3)**
$16t^2 - 3{,}600 = 0$	$16t^2 - 6{,}400 = 0$	$16t^2 - 10{,}000 = 0$
$16(t^2 - 225) = 0$	$16(t^2 - 400) = 0$	$16(t^2 - 625) = 0$
$16(t + 15)(t - 15) = 0$	$16(t + 20)(t - 20) = 0$	$16(t + 25)(t - 25) = 0$
$t + 15 = 0$ or $t - 15 = 0$	$t + 20 = 0$ or $t - 20 = 0$	$t + 25 = 0$ or $t - 25 = 0$
$t = -15$ or $t = 15$	$t = -20$ or $t = 20$	$t = -25$ or $t = 25$

"Since t represents the time an object falls," Ms. Glaston continued, "we would reject the negative-number solutions. Therefore, we can see that it takes about 15 seconds to fall 3,600 feet, 20 seconds to fall 6,400 feet, and 25 seconds to fall 10,000 feet."

"Ms. Glaston," Marty Rosen said, "you were either real lucky, or you cheated somehow. The distances you selected in all three equations gave us expressions that we could factor. What if we wanted to find t for 2,137 feet? I'll bet that equation won't factor so easily."

"You're right, Marty," Ms. Glaston replied, "I did pick numbers that would give us equations that would easily factor. However, in the chapter we are starting today, we will learn other techniques for solving quadratic equations. These techniques do not require factoring the expression in the equation. Therefore, we will be able to find the corresponding times for distances such as 2,137 feet. But first, let's review factoring and the zero-product property method for solving quadratic equations."

SECTION 6-1. Solving Quadratic Equations by Factoring and the Square-Root Theorem

1. The definition of a quadratic equation in x
2. Solving quadratic equations by factoring
3. Solving equations of the form $X^2 = k$
4. Solving literal equations that are quadratic
5. Quadratic equations with solutions that are not real numbers

In Section 3-7, we solved quadratic equations in one variable as an application of factoring. The procedure used the zero-product property after the polynomial in the equation was written in factored form. In this section, we will review the material in Section 3-7, and then solve certain quadratic equations with the square-root theorem.

The Definition of a Quadratic Equation in x

The following definition previously appeared as Definition 3.5:

Definition 6.1. The standard form of a quadratic equation in x
A quadratic equation in x is one that can be written in the form

$$ax^2 + bx + c = 0 \quad \text{or} \quad 0 = ax^2 + bx + c$$

where a, b, and c are real numbers and $a > 0$.

A solution of a quadratic equation is any complex number that makes the equation true.

Example 1. Verify that $\frac{3}{5}$ and -4 are solutions of $12 - 17t = 5t^2$.

Solution.

$12 - 17\left(\frac{3}{5}\right) = 5\left(\frac{3}{5}\right)^2$	First replace t by $\frac{3}{5}$.
$\frac{60}{5} - \frac{51}{5} = 5 \cdot \frac{9}{25}$	Write 12 as $\frac{60}{5}$.
$\frac{9}{5} = \frac{9}{5}$, true	Thus, $\frac{3}{5}$ is a solution.
$12 - 17(-4) = 5(-4)^2$	Now replace t by (-4).
$12 - (-68) = 5(16)$	$(-4)^2 = (-4)(-4) = 16$
$80 = 80$, *true*	Thus, -4 is a solution.

Example 2. Verify that $2 + \sqrt{6}$ is a solution of $u^2 - 4u = 2$.

Solution.

$(2 + \sqrt{6})^2 - 4(2 + \sqrt{6}) = 2$	Replace u by $2 + \sqrt{6}$.
$4 + 4\sqrt{6} + 6 - 8 - 4\sqrt{6} = 2$	$(2 + \sqrt{6})^2 = 2^2 + 2(2)\sqrt{6} + (\sqrt{6})^2$
$2 = 2$, *true*	Thus, $2 + \sqrt{6}$ is a solution.

Solving Quadratic Equations by Factoring

Examples **a** and **b** are quadratic equations in k and w, respectively. By manipulating the expressions in the equations, they can be changed to the standard form given in Definition 6.1.

a. $(k + 2)^2 = 3(k + 8)$ can be written as $k^2 + k - 20 = 0$

b. $\frac{7 + 2w^2}{2} = \frac{7w^2 - 11w}{3}$ can be written as $0 = 8w^2 - 22w - 21$

The polynomials in these equations can be factored using only integers. As a consequence, the zero-product property can be used to solve these equations.

> **Zero-product property**
> If a and b are numbers and $a \cdot b = 0$, then $a = 0$ or $b = 0$ or a and b are both 0.

Example 3. Solve $(k + 2)^2 = 3(k + 8)$.

Solution. **Discussion.** First write the equation in standard form, then factor and apply the zero-product property.

$(k + 2)^2 = 3(k + 8)$	The given equation
$k^2 + 4k + 4 = 3k + 24$	Remove parentheses.
$k^2 + k - 20 = 0$	Subtract $3k$ and 24.
$(k + 5)(k - 4) = 0$	Factor the trinomial.
$k + 5 = 0$ or $k - 4 = 0$	Use the zero-product property.
$k = -5$ or $k = 4$	Solve for k.

The check is omitted. The solution set is $\{-5, 4\}$.

Example 4. Solve $\dfrac{7 + 2w^2}{2} = \dfrac{7w^2 - 11w}{3}$.

Solution.

$\dfrac{7 + 2w^2}{2} = \dfrac{7w^2 - 11w}{3}$	The given equation
$6\left(\dfrac{7 + 2w^2}{2}\right) = 6\left(\dfrac{7w^2 - 11w}{3}\right)$	The LCM of 2 and 3 is 6.
$3(7 + 2w^2) = 2(7w^2 - 11w)$	$6 \div 2 = 3$ and $6 \div 3 = 2$
$21 + 6w^2 = 14w^2 - 22w$	Remove parentheses.
$0 = 8w^2 - 22w - 21$	Write in standard form.
$0 = (4w + 3)(2w - 7)$	Factor the trinomial.
$4w + 3 = 0$ or $2w - 7 = 0$	The zero-product property
$w = \dfrac{-3}{4}$ or $w = \dfrac{7}{2}$	Solve for w.

The check is omitted. The solution set is $\left\{\dfrac{-3}{4}, \dfrac{7}{2}\right\}$.

Solving Equations of the Form $X^2 = k$

Examples **c** and **d** are quadratic equations in k and z, respectively.

c. $k^2 - 10 = 0$ These equations cannot be factored using only integers.

d. $9z^2 - 20 = 0$

Equations such as these can be readily solved by using the square-root theorem.

> **The square-root theorem**
> If X is an algebraic expression, k is a real number, and $X^2 = k$, then:
>
> (1) $X = \sqrt{k}$ or (2) $X = -\sqrt{k}$

Equations (1) and (2) are simply statements of the fact that any nonzero real number has two square roots:

Since $6^2 = (-6)^2 = 36$, 6 and -6 are square roots of 36.
Since $(3i)^2 = (-3i)^2 = -9$, $3i$ and $-3i$ are square roots of -9.

Example 5. Solve $k^2 - 10 = 0$.

Solution. **Discussion.** First write the equation in the form $X^2 = k$. Then apply the square-root theorem.

$k^2 = 10$	Add 10 to both sides.
$k = \sqrt{10}$ or $k = -\sqrt{10}$	Apply the square-root theorem.

The check is omitted. The solution set is $\{\sqrt{10}, -\sqrt{10}\}$.

Example 6. Solve $(2u - 1)^2 = 13$.

Solution. $(2u - 1)^2 = 13$ In the square-root theorem, X is $(2u - 1)$.

$$2u - 1 = \sqrt{13} \quad \text{or} \quad 2u - 1 = -\sqrt{13}$$

$$2u = 1 + \sqrt{13} \quad \text{or} \quad 2u = 1 - \sqrt{13}$$

$$u = \frac{1 + \sqrt{13}}{2} \quad \text{or} \quad u = \frac{1 - \sqrt{13}}{2}$$

The check is omitted. The solution set is $\left\{\frac{1 + \sqrt{13}}{2}, \frac{1 - \sqrt{13}}{2}\right\}$.

A $\pm$ symbol, which is read "plus-minus", can be used to write the two solutions of Example 6 as a single ratio:

$$\frac{1 \pm \sqrt{13}}{2} \quad \text{means} \quad \frac{1 + \sqrt{13}}{2} \quad \text{or} \quad \frac{1 - \sqrt{13}}{2}$$

Solving Literal Equations That Are Quadratic

Examples **e** and **f** are literal equations in that each one has more than one variable.

e. $2x^2 - 7xy - 15y^2 = 0$ An equation in x and y

f. $6at + 2t^2 = bt + 3ab$ An equation in a, b, and t

Equations such as these can frequently be solved for one of the variables by factoring and using the zero-product property.

Example 7. Solve $2x^2 - 7xy - 15y^2 = 0$ for y in terms of x.

Solution.

$2x^2 - 7xy - 15y^2 = 0$ The given equation

$(2x + 3y)(x - 5y) = 0$ Factor.

$2x + 3y = 0$ or $x - 5y = 0$ The zero-product property

$y = \frac{-2x}{3}$ or $y = \frac{x}{5}$ Solve for y.

Thus, $y = \frac{-2x}{3}$ or $y = \frac{x}{5}$.

Example 8. Solve $6at + 2t^2 = bt + 3ab$ for t in terms of a and b.

Solution.

$$2t^2 + 6at - bt - 3ab = 0$$

$$2t(t + 3a) - b(t + 3a) = 0$$

$$(t + 3a)(2t - b) = 0$$

$$t + 3a = 0 \quad \text{or} \quad 2t - b = 0$$

$$t = -3a \quad \text{or} \quad t = \frac{b}{2}$$

Thus, $t = -3a$ or $t = \frac{b}{2}$.

Quadratic Equations with Solutions That Are Not Real Numbers

All quadratic equations do not have real-number solutions. For example, consider the equation in example **g.**

g. $u^2 + 25 = 0$ The sum of u^2 and 25 is 0.

The square of any nonzero real number is positive. Therefore, for example, we cannot find a real number whose square when added to 25 will equal 0, as in example **g.** The solutions of this equation are imaginary numbers.

Example 9. Solve $u^2 + 25 = 0$.

Solution.

$u^2 + 25 = 0$ The given equation

$u^2 = -25$ Subtract 25 from both sides.

$u = \sqrt{-25}$ or $u = -\sqrt{-25}$ The square-root theorem

$u = 5i$ or $u = -5i$ $\sqrt{-25} = i\sqrt{25} = 5i$

Check: If $u = 5i$:

$(5i)^2 + 25 = 0$

$25i^2 + 25 = 0$

$-25 + 25 = 0$, *true*

If $u = -5i$:

$(-5i)^2 + 25 = 0$

$25i^2 + 25 = 0$

$-25 + 25 = 0$, *true*

Thus, the solution set is $\{5i, -5i\}$.

SECTION 6-1. Practice Exercises

In exercises **1–12**, determine whether each number is a solution of the given equation.

[Examples 1 and 2]

1. $x^2 + x = 30$
a. 6
b. 5

2. $x^2 = 3x + 28$
a. -4
b. -3

3. $3y^2 = 2 - y$
a. $\frac{2}{3}$
b. -1

4. $3y^2 = 6 + 17y$
a. 6
b. $\frac{-1}{3}$

5. $x^2 + 4x + 2 = 0$
a. $-2 - \sqrt{2}$
b. $2 - \sqrt{2}$

6. $x^2 - 6x - 3 = 0$
a. $3 + 3\sqrt{2}$
b. $3 + 2\sqrt{3}$

7. $z^2 + 30 = -12z$
a. $-6 - \sqrt{6}$
b. $-6 + \sqrt{6}$

8. $z^2 + 2 = -4z$
a. $-2 - \sqrt{2}$
b. $-2 + \sqrt{2}$

9. $p^2 = -2p - 2$
a. $1 - i$
b. $-1 + i$

10. $p^2 = 2p - 3$
a. $1 + i\sqrt{2}$
b. $-1 - i\sqrt{2}$

11. $(x + 2)^2 = 11$
a. $-2 - \sqrt{11}$
b. $-2 + \sqrt{11}$

12. $(x - 4)^2 = 7$
a. $4 + \sqrt{7}$
b. $4 - \sqrt{7}$

In exercises **13–58**, solve each equation.

[Examples 3 and 4]

13. $(4t + 1)(t - 6) = 0$

14. $(9t - 2)(t + 3) = 0$

15. $y(3y - 8) = 0$

16. $(6y + 11)y = 0$

17. $23(k + 1)(k - 6) = 0$

18. $14(k - 10)(k - 3) = 0$

19. $0 = m(4 - m)(5m + 1)$

20. $0 = m(2m + 1)(6 - 5m)$

21. $2x^2 = 5x$

22. $10x = 3x^2$

23. $x^3 - 5x^2 + 4x = 0$

24. $x^3 - 11x^2 + 10x = 0$

25. $2 = (q - 3)(q - 4)$

26. $5 = (q + 3)(q - 1)$

27. $\dfrac{x^2 - 1}{x} = \dfrac{x + 1}{2}$

28. $\dfrac{x - 1}{2} = \dfrac{x^2 - 1}{x}$

29. $(8x^2 - 14x - 4)(x - 9) = 0$

30. $(3x^2 + 45x - 300)(x + 3) = 0$

[Example 5]

31. $y^2 = 121$

32. $y^2 = 144$

33. $s^2 - 21 = 0$

34. $s^2 - 15 = 0$

35. $m^2 - 8 = 0$

36. $m^2 - 12 = 0$

37. $u^2 - 6 = 75$

38. $u^2 + 2 = 38$

39. $w^2 = -25$

40. $w^2 = -81$

41. $25x^2 - 1 = 0$

42. $49x^2 - 1 = 0$

43. $2y^2 - 9 = 0$

44. $3y^2 - 64 = 0$

45. $3v^2 = 5$

46. $7v^2 = 2$

47. $\dfrac{2}{5} = 4z^2$

48. $\dfrac{3}{7} = 6z^2$

[Example 6]

49. $(x + 1)^2 = 49$

50. $(x - 2)^2 = 16$

51. $(2p - 3)^2 = 25$

52. $(2p + 1)^2 = 81$

53. $(t - 8)^2 = 6$

54. $(t + 12)^2 = 3$

55. $(3a - 1)^2 = 18$

56. $(5a - 2)^2 = 12$

57. $\left(\dfrac{z + 3}{2}\right)^2 = 7$

58. $\left(\dfrac{z - 1}{5}\right)^2 = 3$

In exercises **59–74**, solve each equation for x in terms of the other variables.

[Example 7]

59. $(x + 2y)(x - 5y) = 0$

60. $(x - 7y)(x + y) = 0$

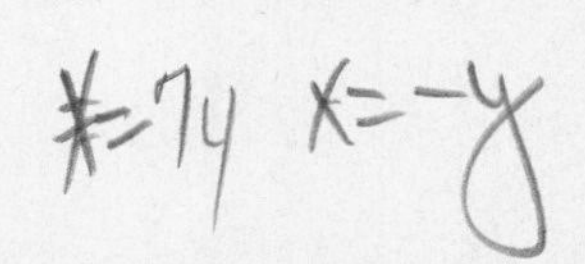

61. $(4x + 3a)(2x - a) = 0$

62. $(3x - a)(5x + 6a) = 0$

63. $x^2 - 4xq - 12q^2 = 0$

64. $x^2 + 2xq - 15q^2 = 0$

65. $14x^2 + 17xm = 6m^2$

66. $20x^2 + 11xm = 3m^2$

[Example 8] **67.** $x^2 - 2xb + ax - 2ab = 0$

68. $x^2 + bx - 3ax - 3ab = 0$

69. $x^2 - 3ax + 4bx = 12ab$

70. $x^2 + 3ax - 5bx = 15ab$

71. $8x^2 + 3ab = 2bx + 12ax$

72. $15x^2 + 2ab = 10bx + 3ax$

73. $35ax + 10ab = -42x^2 - 12bx$

74. $18bx + 3ab = -24x^2 - 4ax$

In exercises **75–90**, solve each equation.

[Example 9] **75.** $p^2 = -4$

76. $p^2 = -25$

77. $w^2 = -10$

78. $w^2 = -19$

79. $v^2 + 50 = 0$

80. $v^2 + 72 = 0$

81. $(m - 6)^2 = -100$

82. $(m + 2)^2 = -225$

83. $4y^2 + 1 = 0$

84. $16y^2 + 1 = 0$

85. $3w^2 = -1$

86. $5w^2 = -1$

87. $7t^2 = -3$

88. $3t^2 = -10$

89. $(3t - 2)^2 = -12$

90. $(4t + 1)^2 = -20$

SECTION 6-1. Ten Review Exercises

In exercises **1** and **2**, evaluate each expression for $t = -4$.

1. $20t - 3(t + 4)$

2. $4(t - 9) + (t - 24)$

3. Solve $20t - 3(t + 4) = 4(t - 9) + (t - 24)$.

4. Solve and graph $20t - 4(t - 9) \geq t + 3(t + 4) - 24$.

5. Evaluate $\left|\frac{2u - 3}{5}\right|$ for $u = 19$ and $u = -16$.

6. Solve $\left|\frac{2u - 3}{5}\right| = 7$.

7. Solve and graph $\left|\frac{2u - 3}{5}\right| < 7$.

In exercises **8–10**, simplify each expression.

8. $\sqrt[3]{-108t^3}$

9. $\sqrt[3]{\frac{1}{25k}}$; $k \neq 0$

10. $\sqrt[6]{4}$

SECTION 6-1. Supplementary Exercises

In exercises **1–10**, solve each equation for x.

1. $(ax + b)(cx + d) = 0;\ a \neq 0,\ c \neq 0$

2. $ax(bx + c) = 0;\ a \neq 0,\ b \neq 0$

3. $(ax + b)^2 = c^2;\ a \neq 0,\ c > 0$

4. $(ax)^2 = b^2;\ a \neq 0,\ b > 0$

5. $(ax + b)^2 = 0;\ a \neq 0$

6. $(ax)^2 = 0;\ a \neq 0$

7. $x^2 + ax - bx - ab = 0$

8. $x^2 - ax - bx + ab = 0$

9. $x^2 + 2xy + y^2 = k$

10. $x^2 - 4xy + 4y^2 = k$

In exercises **11–14**, use the graphs in each figure to answer parts **a–d.**

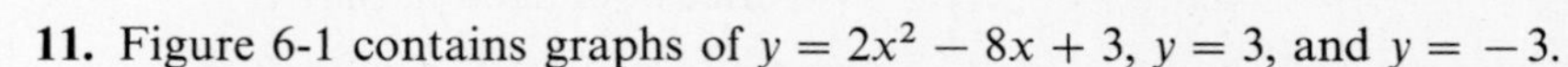

11. Figure 6-1 contains graphs of $y = 2x^2 - 8x + 3$, $y = 3$, and $y = -3$.

a. From the graph, read the values of x where the line $y = 3$ crosses the parabola.

b. Solve $3 = 2x^2 - 8x + 3$ by factoring. Compare the solutions with the x-values obtained in part **a**.

c. From the graph, read the values of x where the line $y = -3$ crosses the parabola.

d. Solve $-3 = 2x^2 - 8x + 3$ by factoring. Compare the solutions with the x-values obtained in part **c**.

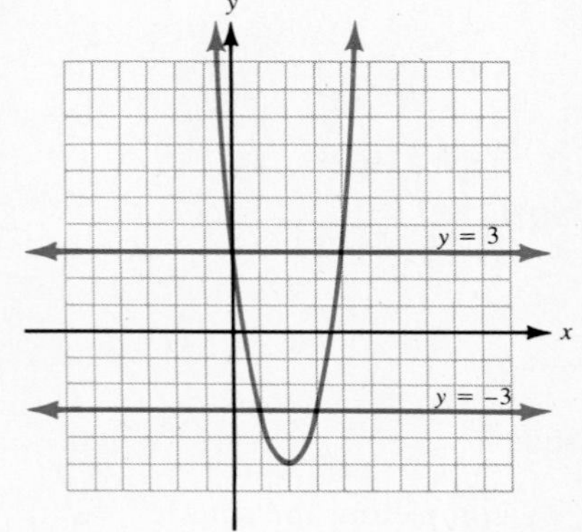

Figure 6-1. A graph of $y = 2x^2 - 8x + 3$.

12. Figure 6-2 contains graphs of $y = 2x^2 - 4x - 2$, $y = -2$, and $y = 4$.

a. From the graph, read the values of x where the line $y = -2$ crosses the parabola.

b. Solve $-2 = 2x^2 - 4x - 2$ by factoring. Compare the solutions with the x-values obtained in part **a**.

c. From the graph, read the values of x where the line $y = 4$ crosses the parabola.

d. Solve $4 = 2x^2 - 4x - 2$ by factoring. Compare the solutions with the x-values obtained in part **c**.

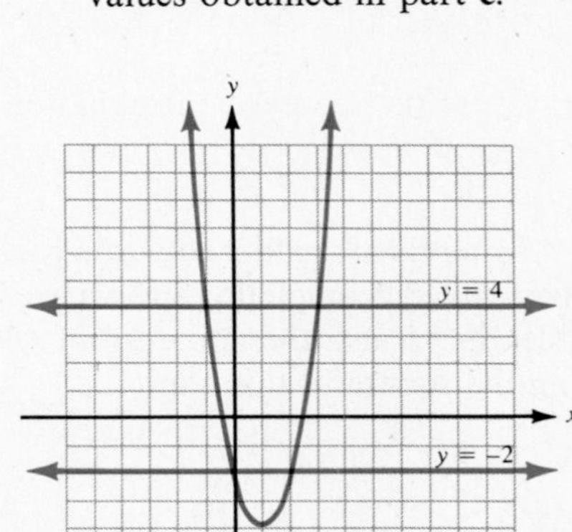

Figure 6-2. A graph of $y = 2x^2 - 4x - 2$.

Figure 6-3. A graph of $y = -2x^2 - 12x - 10$.

13. Figure 6-3 contains graphs of $y = -2x^2 - 12x - 10$, $y = 0$, and $y = 6$.

a. From the graph, read the values of x where the line $y = 0$ crosses the parabola.

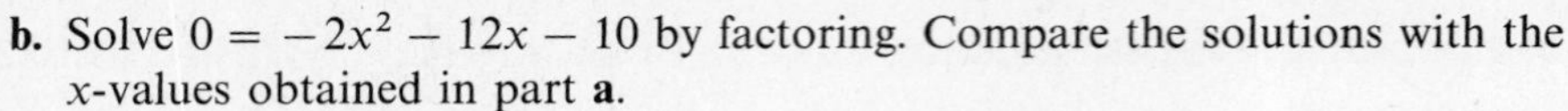

b. Solve $0 = -2x^2 - 12x - 10$ by factoring. Compare the solutions with the x-values obtained in part **a**.

c. From the graph, read the values of x where the line $y = 6$ crosses the parabola.

d. Solve $6 = -2x^2 - 12x - 10$ by factoring. Compare the solutions with the x-values obtained in part **c**.

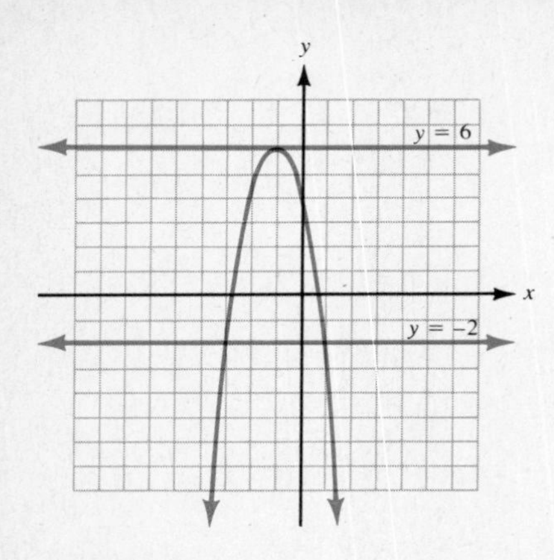

Figure 6-4. A graph of $y = -2x^2 - 4x + 4$.

14. Figure 6-4 contains graphs of $y = -2x^2 - 4x + 4$, $y = 6$, and $y = -2$.

a. From the graph, read the values of x where the line $y = 6$ crosses the parabola.

b. Solve $6 = -2x^2 - 4x + 4$ by factoring. Compare the solutions with the x-values obtained in part **a**.

c. From the graph, read the values of x where the line $y = -2$ crosses the parabola.

d. Solve $-2 = -2x^2 - 4x + 4$ by factoring. Compare the solutions with the x values obtained in part **c**.

SECTION 6-2. Solving Quadratic Equations by Completing the Square

1. Changing $x^2 + 2xy$ to a perfect square trinomial
2. Solving $x^2 + bx + c = 0$ by completing the square
3. Solving $ax^2 + bx + c = 0$, where $a \neq 0$ or 1, by completing the square
4. Quadratic equations with solutions that are not real numbers

As we have seen in Section 6-1, not all quadratic equations can be solved by factoring and the zero-product property if we are restricted only to integers. However, the square-root theorem can solve any equation of the form:

$$ax^2 + bx + c = 0$$

However, to use the theorem, the equation must be written in the form:

$$X^2 = k$$

The $ax^2 + bx$ terms of the first equation can be changed to the square of a binomial by a process called "completing the square". The equation then takes on the form that enables us to use the square-root theorem. We will study the process of completing the square in this section.

Changing $x^2 + 2xy$ to a Perfect-Square Trinomial

Consider the binomials in examples **a** and **b**.

a. $p^2 + 6p$ A binomial in p

b. $q^2 - 20q$ A binomial in q

Suppose we want to change these binomials to **perfect-square trinomials,** so that the factored forms would be the **squares of binomials.** To accomplish this task, we need to add terms to both binomials. The problem can be illustrated with a diagram by letting a □ represent the needed third terms and a △ the missing terms in the squared binomials.

$$\left.\begin{array}{l}\textbf{a. } p^2 + 6p + \square = (p + \triangle)^2 \\ \textbf{b. } \underbrace{q^2 - 20q + \square}_{\textbf{Perfect-square trinomials}} = (q - \triangle)^2\end{array}\right\} \textbf{ Squares of binomials}$$

Recall the general formulas for the multiplied and factored forms of such expressions:

Multiplied forms		Factored forms
$x^2 + 2xy + y^2$	$=$	$(x + y)^2$
$x^2 - 2xy + y^2$	$=$	$(x - y)^2$

Comparing the general forms with examples **a** and **b**, we need to find y's to replace the △'s, and y^2-values to replace the □'s. *In both examples, the y's are one-half the coefficients of the linear terms.*

In example **a**, $y = \frac{1}{2} \cdot 6 = 3$, and $y^2 = 3^2 = 9$.

In example **b**, $y = \frac{1}{2} \cdot 20 = 10$, and $y^2 = 10^2 = 100$.

Therefore, completing the squares on p and q:

a. $p^2 + 6p + 9$ can be written $(p + 3)^2$

b. $q^2 - 20q + 100$ can be written $(q - 10)^2$

> To complete the squares of expresssions with the form
>
> $$x^2 + 2yx + \square \quad \text{or} \quad x^2 - 2yx + \square$$
>
> add $\left(\frac{1}{2} \cdot 2y\right)^2 = y^2$ to the expressions.

Example 1. Given: $k^2 + 18k$

a. Find the number to add to form a perfect-square trinomial.

b. Write the resulting trinomial as the square of a binomial.

Solution. **a.** For the given binomial, $2y = 18$ and $y = 9$. Thus, $y^2 = 9^2 = 81$. $k^2 + 18k + 81$ is a perfect-square trinomial

b. $k^2 + 18k + 81 = (k + 9)^2$

Example 2. Given: $n^2 - 7n$

a. Find the number to add to form a perfect-square trinomial.

b. Write the resulting trinomial as the square of a binomial.

Solution. **a.** For the given binomial, $2y = 7$ and $y = \frac{7}{2}$.

$$\text{Thus, } y^2 = \left(\frac{7}{2}\right)^2 = \frac{49}{4}.$$

$$n^2 - 7n + \frac{49}{4} \text{ is a perfect-square trinomial}$$

b. $n^2 - 7n + \frac{49}{4} = \left(n - \frac{7}{2}\right)^2$

Solving $x^2 + bx + c = 0$ by Completing the Square

The binomials in examples **a** and **b** are used to write the equations in examples **c** and **d**.

c. $p^2 + 6p = 3$ A quadratic equation in p

d. $q^2 - 20q = -82$ A quadratic equation in q

As we have seen, when 9 is added to $p^2 + 6p$ and 100 is added to $q^2 - 20q$, the resulting expressions are perfect-square trinomials. The addition property of equality can be used to add these numbers to both sides of the appropriate given equations. Then, by factoring the trinomials, the square-root theorem can be used to solve the equations.

c.
$$\begin{aligned} p^2 + 6p \quad &= 3 \\ +9 \quad &\quad +9 \\ \hline p^2 + 6p + 9 &= 12 \\ (p+3)^2 &= 12 \\ p + 3 &= \pm\sqrt{12} \\ p &= -3 \pm \sqrt{12} \\ p &= -3 \pm 2\sqrt{3} \end{aligned}$$

The solution set is $\{-3 \pm 2\sqrt{3}\}$.

d.
$$\begin{aligned} q^2 - 20q \quad &= -82 \\ +100 \quad &\quad +100 \\ \hline q^2 - 20q + 100 &= 18 \\ (q-10)^2 &= 18 \\ q - 10 &= \pm\sqrt{18} \\ q &= 10 \pm \sqrt{18} \\ q &= 10 \pm 3\sqrt{2} \end{aligned}$$

The solution set is $\{10 \pm 3\sqrt{2}\}$.

Example 3. Solve $z^2 - 3z = 1$ by completing the square.

Solution.

$z^2 - 3z = 1$ The given equation

$z^2 - 3z + \frac{9}{4} = 1 + \frac{9}{4}$ Add $\left(\frac{1}{2} \cdot 3\right)^2 = \frac{9}{4}$ to both sides.

$\left(z - \frac{3}{2}\right)^2 = \frac{13}{4}$ Simplify the right side.

$z - \frac{3}{2} = \pm\sqrt{\frac{13}{4}}$ Use the square-root theorem.

$z = \frac{3}{2} \pm \frac{\sqrt{13}}{2}$ Simplify the radical.

$z = \frac{3 \pm \sqrt{13}}{2}$ Simplify the expression.

The solution set is $\left\{\frac{3 \pm \sqrt{13}}{2}\right\}$.

Solving $ax^2 + bx + c = 0$, Where $a \neq 0$ or 1, by Completing the Square

The coefficient of the squared term in example **e** is not 1, as were the coefficients of the squared terms in examples **c** and **d**.

e. $4t^2 + 16t = 3$ A quadratic equation in t

We can use the multiplication property of equality to make the coefficient of the t^2-term 1, and then proceed as we did in Example 3.

e. $\frac{1}{4}(4t^2 + 16t) = \frac{1}{4} \cdot 3$ Multiply both sides by $\frac{1}{4}$.

$t^2 + 4t = \frac{3}{4}$ Simplify both sides.

$t^2 + 4t + 4 = \frac{3}{4} + 4$ Add $\left(\frac{1}{2} \cdot 4\right)^2 = 4$ to both sides.

$(t + 2)^2 = \frac{19}{4}$ Factor left side and simplify right.

$t + 2 = \pm\sqrt{\frac{19}{4}}$ Use the square-root theorem.

$t = -2 \pm \frac{\sqrt{19}}{2}$ Solve for t.

or $t = \frac{-4 \pm \sqrt{19}}{2}$ The preferred form for solutions

The solution set is $\left\{\frac{-4 \pm \sqrt{19}}{2}\right\}$.

The above example suggests the following five-step procedure:

To solve $ax^2 + bx + c = 0$, where $a > 0$ by the method of completing the square:

Step 1. If $a \neq 1$, divide each term by a.

Step 2. Move the constant term to the opposite side of the equation.

Step 3. Add the square of one-half the coefficient of the x-term to both sides of the equation.

Step 4. Write the left side as the square of a binomial and simplify the right side.

Step 5. Use the square-root theorem to solve the equation of Step 4.

Example 4. Solve $2x^2 - 3x - 1 = 0$ by completing the square.

Solution. **Step 1.** $x^2 - \frac{3}{2}x - \frac{1}{2} = 0$ Divide each term by 2.

Step 2. $x^2 - \frac{3}{2}x = \frac{1}{2}$ Add $\frac{1}{2}$ to both sides.

Step 3. $x^2 - \frac{3}{2}x + \frac{9}{16} = \frac{1}{2} + \frac{9}{16}$ Add $\left(\frac{1}{2} \cdot \frac{3}{2}\right)^2 = \frac{9}{16}$ to both sides.

Step 4. $\left(x - \frac{3}{4}\right)^2 = \frac{17}{16}$ Factor left side, simplify right.

Step 5. $x - \frac{3}{4} = \pm\sqrt{\frac{17}{16}}$ Use the square-root theorem.

$x = \frac{3}{4} \pm \frac{\sqrt{17}}{4}$ Solve for x.

or, $x = \frac{3 \pm \sqrt{17}}{4}$ The preferred form

The solution set is $\left\{\frac{3 \pm \sqrt{17}}{4}\right\}$.

Quadratic Equations with Solutions That Are Not Real Numbers

The method of solving by completing the square can also be used on quadratic equations whose solutions are not real numbers.

Example 5. Solve $2m^2 - 2m + 1 = 0$ by completing the square.

Solution. **Step 1.** $m^2 - m + \frac{1}{2} = 0$

Step 2. $m^2 - m = \frac{-1}{2}$

Step 3. $m^2 - m + \frac{1}{4} = \frac{-1}{2} + \frac{1}{4}$

Step 4. $\left(m - \frac{1}{2}\right)^2 = \frac{-1}{4}$

Step 5. $m - \frac{1}{2} = \pm\sqrt{\frac{-1}{4}}$

$m - \frac{1}{2} = \pm\frac{i}{2}$

$m = \frac{1 \pm i}{2}$

The solution set is $\left\{\frac{1 \pm i}{2}\right\}$.

SECTION 6-2. Practice Exercises

In exercises **1–18**, for each expression:

a. Find the number to add to form a perfect-square trinomial.

b. Write the resulting trinomial as the square of a binomial.

[Example 1]

1. $x^2 + 16x$ **2.** $x^2 - 4x$

3. $y^2 - 10y$ **4.** $y^2 + 20y$

5. $z^2 + 24z$ **6.** $z^2 - 14z$

7. $w^2 - 2w$ **8.** $w^2 + 26w$

[Example 2]

9. $p^2 - 3p$ **10.** $p^2 - 5p$

11. $t^2 + t$ **12.** $t^2 - t$

13. $y^2 + 9y$ **14.** $y^2 + 11y$

15. $x^2 - \frac{4}{3}x$ **16.** $x^2 - \frac{2}{5}x$

17. $z^2 - \frac{5}{6}z$ **18.** $z^2 - \frac{3}{7}z$

In exercises **19–60**, solve each equation by completing the square.

[Example 3]

19. $t^2 + 6t + 7 = 0$ **20.** $t^2 + 10t + 20 = 0$

21. $y^2 - 8y = 8$ **22.** $y^2 - 12y = 4$

23. $z^2 - 2z - 10 = 0$ **24.** $z^2 + 4z - 8 = 0$

25. $m^2 = -14m - 17$ **26.** $m^2 = -16m - 16$

27. $t^2 + t - 3 = 0$ **28.** $t^2 - t - 5 = 0$

29. $v^2 - 14 = 4v$ **30.** $v^2 + 6v = 11$

31. $w^2 - 5w = -2$ **32.** $w^2 - 3w + 1 = 0$

33. $x^2 - 9 = 3x$ **34.** $x^2 - 23 = 5x$

35. $a^2 = 6a + 16$ **36.** $a^2 = 5 - 4a$

37. $y^2 + 3y + \frac{3}{2} = 0$ **38.** $y^2 + y - \frac{1}{2} = 0$

[Example 4]

39. $4u^2 = 12u - 1$ **40.** $10u + 1 = 2u^2$

41. $5z^2 - 1 = 2z$ **42.** $3z^2 + 3 = -8z$

43. $7p + 6 = 3p^2$ **44.** $4p - 15 = -4p^2$

45. $2(6 - x^2) = 5x$ **46.** $x(2x - 1) = 1$

47. $2y^2 = 8y + 5$ **48.** $5 = 6y + 3y^2$

49. $6z = 4 - 9z^2$ **50.** $12z = 9z^2 - 1$

[Example 5]

51. $m^2 + 10 = 2m$

52. $m^2 + 4m = -8$

53. $u^2 + 16 = 4u$

54. $u^2 + 6u = -27$

55. $x^2 + 7 = 3x$

56. $x^2 + 7 = 5x$

57. $w^2 = 6w - 13$

58. $w^2 = -4w - 5$

59. $t^2 + 2t + 3 = 0$

60. $t^2 + 12 = 6t$

SECTION 6-2. Ten Review Exercises

1. Evaluate $3(t^2 - 2t + 7) - 2(t^2 - 3t + 8)$ for $t = -3$.
2. Simplify $3(t^2 - 2t + 7) - 2(t^2 - 3t + 8)$.
3. Evaluate the expression of exercise **2** for $t = -3$ and compare with that in exercise **1**.
4. Evaluate $(6u)^3(-3u)^{-2}$ for $u = 2$.
5. Simplify $(6u)^3(-3u)^{-2}$.
6. Evaluate the expression of exercise **2** for $u = 2$ and compare with that in exercise **4**.
7. For what values of x will $\sqrt{x - 3}$ be real numbers?
8. For what values of x will $\sqrt[3]{x + 1}$ be real numbers?

In exercises **9** and **10**, simplify each expression. Do *not* assume all variables are positive real numbers.

9. $\sqrt{75x^4y^2}$

10. $\sqrt[3]{-250x^6y^3}$

SECTION 6-2. Supplementary Exercises

In exercises **1–10**, solve each equation for x by completing the square.

1. $x^2 + 2bx = 0$

2. $x^2 - 4bx = 0$

3. $x^2 - bx = 0$

4. $x^2 + bx = 0$

5. $x^2 + 2bx + c = 0$

6. $x^2 - 4bx + c = 0$

7. $ax^2 + abx + ac = 0$; $a \neq 0$.

8. $ax^2 + abx + a = 0$; $a \neq 0$.

9. $x^2 - xy + y^2 = 0$; $y > 0$.

10. $x^2 + 3xy + y^2 = 0$; $y > 0$.

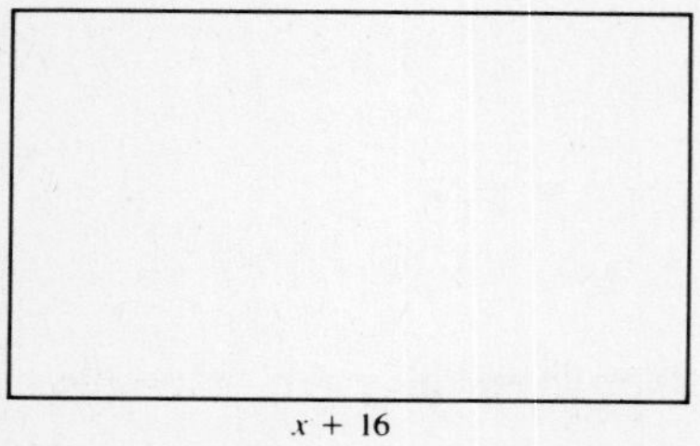

$x + 16$

Figure 6-5. A rectangle x by $x + 16$.

Figure 6-5 is a rectangle with width x and length $x + 16$. The area of the rectangle is A, and:

$$A = x(x + 16) = x^2 + 16x$$

How much area would have to be added to yield a square with side $x + k$, where k is a rational number?

To determine k, consider the same rectangle in Figure 6-6, in which the line divides the area into two parts. One part is the square x by x. The other part is the rectangle x by 16.

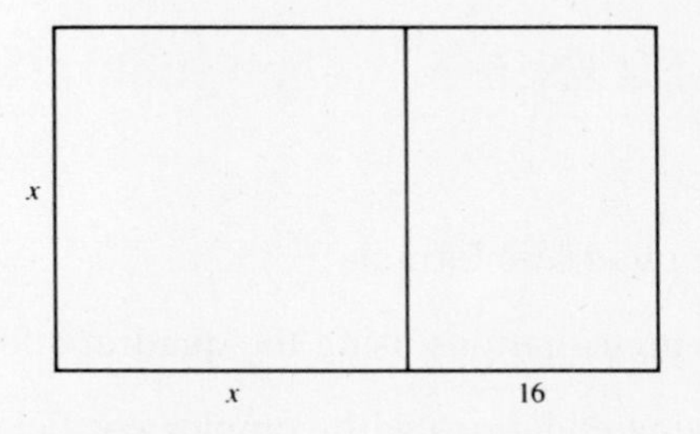

Figure 6-6. The rectangle divided into two parts.

Figure 6-7. The rectangle divided into three parts.

In Figure 6-7, the rectangle is further divided into the two rectangles x by 8 and x by 8. The two pieces x by 8 units are labeled B and C.

In Figure 6-8, the area labeled C is moved as shown. To **complete the square,** the area labeled D must be added. The dimensions of D are 8×8. Thus, the area is 64.

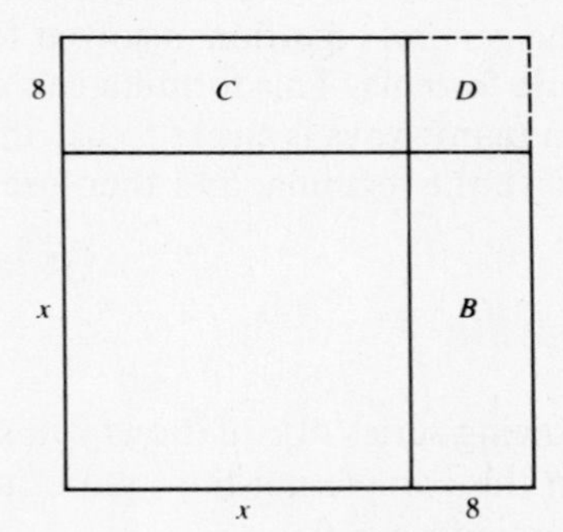

Figure 6-8. If D is added to the area, the result is a square.

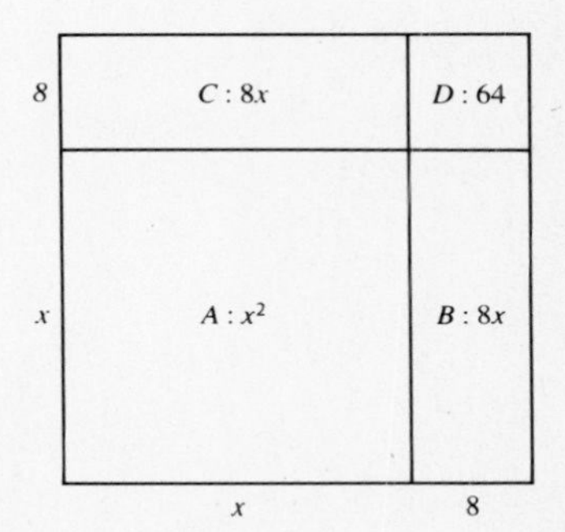

Figure 6-9. The area of the square is $(x + 8)^2 = x^2 + 16x + 64$.

In Figure 6-9, the four regions are labeled A, B, C, and D. The areas of each region are shown. *Notice:* $(x + 8)^2 = x^2 + 16x + 64$. Thus, $k = 8$.

In exercises **11–18**, the given expressions represent the areas of rectangles. Refer to Figure 6-9 to identify the regions A, B, and C.

a. Find the area of region A.

b. Find the area of region B.

c. Find the area of region C.

d. Determine the value of k to form a square region from the given rectangles.

11. $x(x + 4)$

12. $x(x + 8)$

13. $x^2 + 6x$

14. $x^2 + 10x$

15. $x^2 + 3x$

16. $x^2 + 5x$

17. $x^2 + \frac{7}{2}x$

18. $x^2 + \frac{9}{2}x$

SECTION 6-3. The Quadratic Formula

1. The quadratic formula
2. Solving equations using the quadratic formula
3. Solving equations with complex coefficients
4. The discriminant

According to Definition 6-1, a quadratic equation in x is one that can be written in the form

$$ax^2 + bx + c = 0$$

where a, b, and c are real numbers and $a > 0$. In Section 6-2, we studied a method that can be used to solve any such equation. The process requires completing the square in x, and then using the square-root theorem.

When the general equation is solved for x, then the resulting equation is called the **quadratic formula.** This formula can also be used to solve any quadratic equation, and in many ways is easier to use than completing the square. In this section, we will derive the formula, and then use it to solve several equations.

The Quadratic Formula

The following series of equations will solve $ax^2 + bx + c = 0$ for x by using the six steps in the completing-the-square method. In each equation, a, b, and c are real numbers and $a > 0$.

Step 1. $x^2 + \frac{b}{a}x + \frac{c}{a} = 0$ — Divide each term by a.

Step 2. $x^2 + \frac{b}{a}x = \frac{-c}{a}$ — Subtract $\frac{c}{a}$ from both sides.

Step 3. $x^2 + \frac{b}{a}x + \frac{b^2}{4a^2} = \frac{-c}{a} + \frac{b^2}{4a^2}$ — Add $\left(\frac{1}{2}\cdot\frac{b}{a}\right)^2 = \frac{b^2}{4a^2}$ to both sides.

Step 4. $\left(x + \frac{b}{2a}\right)^2 = \frac{b^2 - 4ac}{4a^2}$ — $\frac{-c}{a}\cdot\frac{4a}{4a} = \frac{-4ac}{4a^2}$

Step 5. $x + \frac{b}{2a} = \pm\sqrt{\frac{b^2 - 4ac}{4a^2}}$ — Use the square-root theorem.

$x = \frac{-b}{2a} \pm \frac{\sqrt{b^2 - 4ac}}{2a}$ — $\sqrt{4a^2} = 2a$, since $a > 0$

$x = \frac{-b \pm \sqrt{b^2 - 4ac}}{2a}$ — The preferred form

The quadratic formula

If $ax^2 + bx + c = 0$, where a, b, and c are real numbers and $a > 0$, then:

$$x = \frac{-b + \sqrt{b^2 - 4ac}}{2a} \quad \text{or} \quad x = \frac{-b - \sqrt{b^2 - 4ac}}{2a}$$

The solution set is $\left\{\frac{-b \pm \sqrt{b^2 - 4ac}}{2a}\right\}$.

Solving Equations Using the Quadratic Formula

The following six steps provide a procedure for solving quadratic equations using the formula:

To solve $ax^2 + bx + c = 0$ with the quadratic formula:

Step 1. If necessary, write the equation in standard form, with $a > 0$.

Step 2. Identify the values of a, b, and c.

Step 3. Replace a, b, and c in $\frac{-b \pm \sqrt{b^2 - 4ac}}{2a}$ with the values listed in Step 2.

Step 4. If possible, simplify the radical.

Step 5. If possible, factor the numerator and reduce the fraction.

Step 6. State the solution set.

Example 1. Solve $t + 6 = 12t^2$.

Solution.

Step 1. $0 = 12t^2 - t - 6$ In standard form with $a > 0$

Step 2. $0 = 12t^2 + (-1)\ t + (-6)$

$a = 12$, $b = -1$, and $c = -6$

Step 3. $t = \frac{-(-1) \pm \sqrt{(-1)^2 - 4(12)(-6)}}{2(12)}$ Note that $-b$ is $-(-1)$.

Step 4. $t = \frac{1 \pm \sqrt{289}}{24}$ $(-1)^2 - 4(12)(-6) = 289$

$= \frac{1 \pm 17}{24}$ $\sqrt{289} = 17$

Step 5. The radical yielded an integer. Therefore, we write two equations and simplify both.

$$t = \frac{1 + 17}{24} \quad \text{or} \quad t = \frac{1 - 17}{24}$$

$$= \frac{18}{24} \text{ or } \frac{3}{4} \qquad = \frac{-16}{24} \text{ or } \frac{-2}{3}$$

The solution set is $\left\{\frac{3}{4}, \frac{-2}{3}\right\}$.

Example 2. **a.** Solve $9u^2 - 5 = 6u$.

b. Approximate the solutions to one decimal place.

Solution. **a. Step 1.** $9u^2 - 6u - 5 = 0$

Step 2. $a = 9$, $b = -6$ and $c = -5$

Step 3. $u = \dfrac{-(-6) \pm \sqrt{(-6)^2 - 4(9)(-5)}}{2(9)}$

Step 4. $u = \dfrac{6 \pm \sqrt{216}}{18}$ $\qquad (-6)^2 - 4(9)(-5) = 216$

$= \dfrac{6 \pm 6\sqrt{6}}{18}$ $\qquad \sqrt{216} = \sqrt{36 \cdot 6} = 6\sqrt{6}$

Step 5. $u = \dfrac{6(1 \pm \sqrt{6})}{18}$ $\qquad$ Factor a 6 in the numerator.

$u = \dfrac{1 \pm \sqrt{6}}{3}$ $\qquad$ Reduce the fraction.

Step 6. The solution set is $\left\{\dfrac{1 \pm \sqrt{6}}{3}\right\}$.

b. Using a calculator:

$$\frac{1+\sqrt{6}}{3} = \frac{1 + 2.449\ldots}{3} \qquad \frac{1-\sqrt{6}}{3} = \frac{1 - 2.449\ldots}{3}$$

$$= 1.149\ldots \qquad = -0.483\ldots$$

$$\approx 1.1 \qquad \approx -0.5$$

In Example 2, the solutions are irrational numbers, because $\sqrt{6}$ is irrational. Notice also that the solutions are **irrational conjugates**—that is, $\dfrac{1+\sqrt{6}}{3}$ and $\dfrac{1-\sqrt{6}}{3}$. *If* a, b *and* c *are rational numbers, then any irrational number solutions of* $ax^2 + bx + c = 0$ *will occur as conjugate pairs.*

Example 3. Solve $4v = 7 + v^2$.

Solution. **Step 1.** $0 = v^2 - 4v + 7$

Step 2. $a = 1$, $b = -4$, and $c = 7$

Step 3. $v = \dfrac{-(-4) \pm \sqrt{(-4)^2 - 4(1)(7)}}{2(1)}$

Step 4. $v = \dfrac{4 \pm \sqrt{-12}}{2}$ $\qquad (-4)^2 - 4(1)(7) = -12$

$= \dfrac{4 \pm 2i\sqrt{3}}{2}$ $\qquad \sqrt{-12} = \sqrt{-1 \cdot 4 \cdot 3} = 2i\sqrt{3}$

Step 5. $v = \dfrac{2(2 \pm i\sqrt{3})}{2}$ Factor a 2 in the numerator.

$= 2 \pm i\sqrt{3}$ Reduce the fraction.

Step 6. The solution set is $\{2 \pm i\sqrt{3}\}$.

In Example 3, the solutions are imaginary numbers. Notice also that the solutions are **complex conjugates**—that is, $2 + i\sqrt{3}$ and $2 - i\sqrt{3}$. If a, b, and c are rational numbers, then any imaginary-number solutions of $ax^2 + bx + c = 0$ will occur as conjugate pairs.

Solving Equations with Complex Coefficients

In Definition 6-1, a, b, and c in $ax^2 + bx + c = 0$ are identified as real numbers. Also, in the formula for solving such equations they are identified as being real numbers. However, as shown in Examples 4 and 5, the quadratic formula can be used to solve a quadratic equation in which these numbers are imaginary.

Example 4. Solve $-6iz = 11 - 2z^2$.

Solution. **Step 1.** $2z^2 - 6iz - 11 = 0$ $-6i$ is an imaginary number

Step 2. $a = 2$, $b = -6i$, and $c = -11$

Step 3. $z = \dfrac{-(-6i) \pm \sqrt{(-6i)^2 - 4(2)(-11)}}{2(2)}$

Step 4. $z = \dfrac{6i \pm \sqrt{52}}{4}$ $(-6i)^2 - 4(2)(-11) = 52$

$= \dfrac{6i \pm 2\sqrt{13}}{4}$

Step 5. $z = \dfrac{2(3i \pm \sqrt{13})}{4}$

$= \dfrac{3i \pm \sqrt{13}}{2}$

Step 6. The solution set is $\left\{\dfrac{3i \pm \sqrt{13}}{2}\right\}$.

Example 5. Solve $-2ik^2 = 3i + k$.

Solution. **Step 1.** $0 = 2ik^2 + k + 3i$ $2i$ and $3i$ are imaginary numbers

Step 2. $a = 2i$, $b = 1$, and $c = 3i$

Step 3. $k = \dfrac{-1 \pm \sqrt{1^2 - 4(2i)(3i)}}{2(2i)}$

Step 4. $k = \dfrac{-1 \pm \sqrt{25}}{4i}$ $1^2 - 4(2i)(3i) = 1 - (-24) = 25$

$= \dfrac{-1 \pm 5}{4i}$

Step 5. Write two equations and simplify both.

$$k = \frac{-1 + 5}{4i} \quad \text{or} \quad k = \frac{-1 - 5}{4i}$$

$$k = \frac{4}{4i} \cdot \frac{i}{i} \quad \text{or} \quad k = \frac{-6}{4i} \cdot \frac{i}{i}$$

$$k = -i \qquad k = \frac{3i}{2}$$

Step 6. The solution set is $\left\{-i, \frac{3i}{2}\right\}$.

The Discriminant

In Examples 1–3, a, b, and c in the equations are real numbers. As seen in the solution sets of these equations, the solutions of $ax^2 + bx + c = 0$ can be real or imaginary numbers. When a computer solves a quadratic equation, it first determines whether the solutions will be real or imaginary numbers. If the determination shows that the solutions will be real numbers, one series of steps in the program is followed; otherwise, another series of steps is used. The radicand in the quadratic formula can be used to make that determination.

Definition 6.2. The discriminant
If a, b, and c are real numbers and $a \neq 0$, then for the equation $ax^2 + bx + c = 0$, the expression $b^2 - 4ac$ is called the **discriminant.**

Example 6. Find the value of the discriminant for $m = 35 - 6m^2$.

Solution. **Discussion.** Before trying to identify a, b, and c, first write the equation in standard form.

$6m^2 + m - 35 = 0$ Add $6m^2 - 35$ to both sides.

Now, $a = 6$, $b = 1$, and $c = -35$.
$b^2 - 4ac$ becomes $1^2 - 4(6)(-35) = 841$
For the given equation, the value of the discriminant is 841.

Most of the quadratic equations we will solve in this book have rational numbers for a, b, and c. As a consequence, the use of the discriminant to determine the kinds of solutions that a given quadratic equation will have is based on these numbers being rational.

To use the discriminant to determine the nature of the solutions:
If a, b, and c are rational numbers and $a \neq 0$, then the solutions of $ax^2 + bx + c = 0$ will be:

1. *Rational and equal,* if $b^2 - 4ac = 0$ One real-number solution
2. *Real and unequal,* if $b^2 - 4ac > 0$ Two real-number solutions
   a. If $b^2 - 4ac$ is a *perfect square,* then the solutions will be *rational numbers.*
   b. If $b^2 - 4ac$ is not a *perfect square,* then the solutions will be *irrational numbers* (that is, **irrational conjugates**).
3. *Imaginary and unequal,* if $b^2 - 4ac < 0$ Two imaginary-number solutions

   The solutions will be *complex conjugates.*

In Example 6, the value of the discriminant was 841. Since $841 > 0$, from 2 above we know the solutions of $6m^2 + m - 35 = 0$ are *real and unequal.* Furthermore, $841 = 29^2$, a perfect square. Thus, from 2a, the solutions are *rational numbers.*

Example 7. Determine the nature of the solutions of $\frac{q^2 + 1}{3} = \frac{4q + 1}{8}$.

Solution.

$24\left(\frac{q^2 + 1}{3}\right) = 24\left(\frac{4q + 1}{8}\right)$ — Multiply both sides by 24.

$8(q^2 + 1) = 3(4q + 1)$ — Reduce the fractions.

$8q^2 + 8 = 12q + 3$ — Use the distributive property.

$8q^2 - 12q + 5 = 0$ — Write in standard form.

$a = 8$, $b = -12$, and $c = 5$

$(-12)^2 - 4(8)(5)$ — Replace in $b^2 - 4ac$.

$= 144 - 160$ — Simplify.

$= -16$ — The value of the discriminant is -16.

Since $-16 < 0$, we know from 3 above that the solutions are *imaginary and unequal.*

SECTION 6-3. Practice Exercises

In exercises **1–12**, solve each equation using the quadratic formula.

[Example 1]

1. $2x^2 + 11x + 15 = 0$ **2.** $3x^2 + 8x + 4 = 0$

3. $4z^2 + 17z = 15$ **4.** $2z^2 - 5 = 3z$

5. $3p^2 - 5p = 0$ **6.** $7p^2 + 2p = 0$

7. $16 = 9t^2$ **8.** $25 - 4t^2 = 0$

9. $30 = 3m(2m + 1)$ **10.** $3 = 2m(4m + 5)$

11. $27y = 18 - 5y^2$ **12.** $4y^2 = 17y + 15$

In exercises **13–26**, for each equation:

a. Solve using the quadratic formula.

b. Approximate the solutions to one decimal place.

[Example 2]

13. $2t + 2 = t^2$ **14.** $t^2 + 1 = 4t$

15. $0 = u^2 + 4 + 6u$ **16.** $0 = 10u + u^2 + 23$

17. $v^2 - 4v = 14$ **18.** $v^2 = 11 - 6v$

19. $8t = 4t^2 + 1$ **20.** $7 + 4t^2 = 12t$

21. $9x^2 - 44 + 6x = 0$ **22.** $12x - 16 + 9x^2 = 0$

23. $4 + 20v - 25v^2 = 0$

24. $-25v^2 - 6 + 30v = 0$

25. $z^2 - 2\sqrt{2}z = 1$

26. $z^2 = 4\sqrt{3}z - 7$

In exercises **27–50**, solve each equation.

[Example 3] 27. $t^2 + 13 = 6t$

28. $4t - 13 = t^2$

29. $0 = 8y + 17 + y^2$

30. $40 - 12y = -y^2$

31. $-z^2 - 6z - 14 = 0$

32. $2z - z^2 - 7 = 0$

33. $5x(6 - 5x) = 11$

34. $10x = 3(3 + x^2)$

35. $7 = y(2 - 3y)$

36. $2y(y + 1) = -11$

37. $4 + t^2 = 2\sqrt{3}t$

38. $6 + t^2 = 2\sqrt{2}t$

[Examples 4 and 5] 39. $u^2 + 2iu + 5 = 0$

40. $u^2 + 2iu + 7 = 0$

41. $v^2 - 3iv + 4 = 0$

42. $v^2 - 4iv + 5 = 0$

43. $2w^2 - iw + 1 = 0$

44. $3w^2 + 2iw + 1 = 0$

45. $3p^2 - ip = -4$

46. $7 - 5ip = -2p^2$

47. $2iy^2 = -6y - 3i$

48. $2i + 3iy^2 = 5y$

49. $10t^2 = 16it$

50. $12it = 9t^2$

In exercises **51–60**, find the value of the discriminant for each equation.

[Example 6] 51. $2x^2 - x - 5 = 0$

52. $6x^2 + 2x - 7 = 0$

53. $7y = 3y^2 + 9$

54. $y = -3 - 5y^2$

55. $4z^2 = 12z - 9$

56. $0 = 60z + 36z^2 + 25$

57. $\dfrac{t+2}{t} = \dfrac{2t-1}{5}$

58. $\dfrac{-3}{3t+1} = \dfrac{t}{t-1}$

59. $1 - \dfrac{1}{u} = \dfrac{u+3}{2}$

60. $3 - \dfrac{2}{u} = \dfrac{u}{u-2}$

In exercises **61–80**, determine the nature of the solutions of each equation. For solutions that are real and unequal, state whether the solutions are rational or irrational numbers.

[Example 7] 61. $6x^2 = 7 - 19x$

62. $13x + 45 = 2x^2$

63. $y = y^2 + 5$

64. $2y + 4 + 3y^2 = 0$

65. $144z^2 + 25 = 120z$

66. $-9z^2 = 42z + 49$

67. $2t + 2 = t^2$

68. $5t^2 + t = 3$

69. $3u - 1 = 4u^2$

70. $7 + 5u + 2u^2 = 0$

71. $20w(w + 1) = 4w(w - 1) - 9$

72. $2w(w - 10) = 2(2 - w^2) - 29$

73. $6v^2 = 4 + 23v$

74. $17v + 28 = 3v^2$

75. $3x(x - 3) = 2(1 - 8x)$

76. $2x(x + 1) = 5(1 - x)$

77. $\dfrac{2y}{y+1} = 1 - \dfrac{3}{y}$

78. $2 - \dfrac{1}{y} = \dfrac{-2y}{y-2}$

79. $\dfrac{6z+1}{z} = \dfrac{29-2z}{z+3}$

80. $2z(z+8) = 3(z-6) - 2$

SECTION 6-3. Ten Review Exercises

In exercises **1** and **2**, evaluate each expression for $t = -4$.

1. $2t^2 + 10t + 3$

2. $23 + 3t - t^2$

3. Based on the answers to exercises **1** and **2**, is -4 a solution of $2t^2 + 10t + 3 = 23 + 3t - t^2$?

4. Find both solutions of $2t^2 + 10t + 3 = 23 + 3t - t^2$.

5. Based on the solutions of the equation in exercise 4, what value would you expect for the discriminant of the equation?

In exercises **6–10**, do the indicated operations.

6. $(k^3 - k^2 + 2k + 3)(k^2 + k - 1)$

7. $(3p^2 - 1)^2 + (2p + 3)(3p - 1) - (3p^2 - 1)(3p^2 + 1)$

8. $(8x^3 - 6x^2y - 5xy^2 + 3y^3) \div (4x + 3y)$

9. $\dfrac{20t^3}{4t^3 - 10t^2} \cdot \dfrac{-25 + 20t - 4t^2}{4t^2 - 10t}$

10. $\dfrac{m^2 - n^2 + 2n - 1}{3m^2 + n + 3mn + m} \div \dfrac{m - n + 1}{3m + 1}$

SECTION 6-3. Supplementary Exercises

The **sum-and-product-of-solutions theorem** identifies a relationship between the sum and the product of the solutions of $ax^2 + bx + c = 0$ and the real numbers a, b, and c.

The sum-and-product-of-solutions theorem

If r and s are solutions of $ax^2 + bx + c = 0$ and $a > 0$, then:

Condition 1. $r + s = \dfrac{-b}{a}$ — The sum of the solutions equals the opposite of b over a.

Condition 2. $r \cdot s = \dfrac{c}{a}$ — The product of the solutions equals c over a.

Example. Determine whether $5 \pm 2\sqrt{3}$ are solutions of $10t = 13 + t^2$.

Solution. **Step 1.** If necessary, write the equation in standard form with $a > 0$, and identify a, b, and c.

$0 = t^2 - 10t + 13$ Write in standard form.

$a = 1$, $b = -10$, and $c = 13$

Step 2. Form the ratios $\frac{-b}{a}$ and $\frac{c}{a}$.

$\frac{-(-10)}{1} = 10$ The sum of the solutions is 10.

$\frac{13}{1} = 13$ The product of the solutions is 13.

Step 3. Compute the sum and product of the apparent solutions. Let $r = 5 + 2\sqrt{3}$ and $s = 5 - 2\sqrt{3}$.

$r + s$ becomes:

$(5 + 2\sqrt{3}) + (5 - 2\sqrt{3})$

$= 10$

$r \cdot s$ becomes:

$(5 + 2\sqrt{3})(5 - 2\sqrt{3})$

$= 5^2 - (2\sqrt{3})^2$

$= 25 - 12 = 13$

Step 4. Compare $r + s$ with $\frac{-b}{a}$ and $r \cdot s$ with $\frac{c}{a}$.

$r + s$ and $\frac{-b}{a}$ both equal 10, *check*

$r \cdot s$ and $\frac{c}{a}$ both equal 13, *check*

Thus, $5 \pm 2\sqrt{3}$ are the solutions of $10t = 13 + t^2$.

In exercises **1–12**, determine whether each pair of numbers are solutions of the given equation by using the sum-and-product-of-solutions theorem.

1. $4x^2 + 4x - 15 = 0$; $\frac{3}{2}$ and $\frac{-5}{2}$

2. $5x^2 + 14x = 3$; -3 and $\frac{1}{5}$

3. $y^2 - 10y + 22 = 0$; $5 \pm \sqrt{3}$

4. $y^2 + 4y + 2 = 0$; $-2 \pm \sqrt{2}$

5. $z^2 = 6z - 10$; $3 \pm i$

6. $4z + z^2 + 13 = 0$; $-2 \pm 3i$

7. $4w^2 = 20w - 25$; $\frac{5}{2}$ and $\frac{5}{2}$

8. $24w + 9 = -16w^2$; $\frac{-3}{4}$ and $\frac{-3}{4}$

9. $6x + 19 = 9x^2$; $\frac{1 \pm 2\sqrt{5}}{3}$

10. $2x^2 = 10x - 11$; $\frac{5 \pm \sqrt{3}}{2}$

11. $25y^2 + 30y + 11 = 0$; $\frac{-3 \pm i\sqrt{2}}{5}$

12. $3 = -3y^2 - 2y$; $\frac{-1 \pm 2i\sqrt{2}}{3}$

A graph of $y = ax^2 + bx + c$ will show the kinds of solutions for the corresponding equation $ax^2 + bx + c = 0$.

Case 1. If the graph of $y = ax^2 + bx + c$ intersects the x-axis at two different points, then $ax^2 + bx + c = 0$ will have two real roots. (See Figure 6-10.)

Case 2. If the graph of $y = ax^2 + bx + c$ intersects the x-axis at exactly one point, then $ax^2 + bx + c = 0$ will have one real root. (See Figure 6-11.)

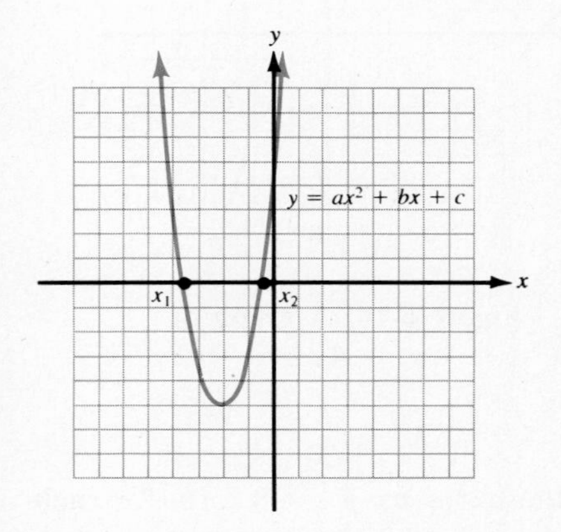

Figure 6-10. x_1 and x_2 are roots of $ax^2 + bx + c = 0$.

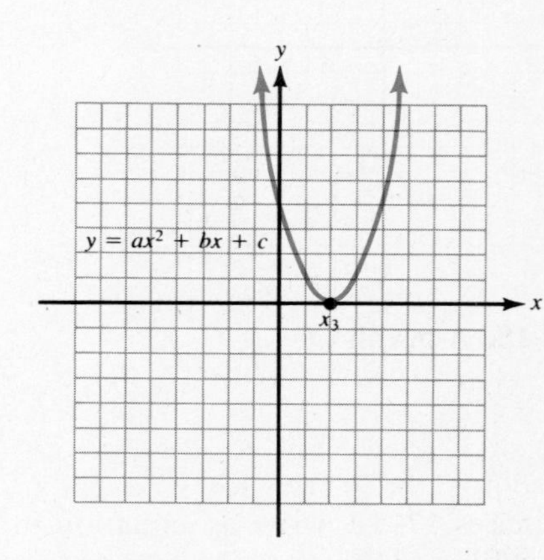

Figure 6-11. x_3 is the only root of $ax^2 + bx + c = 0$.

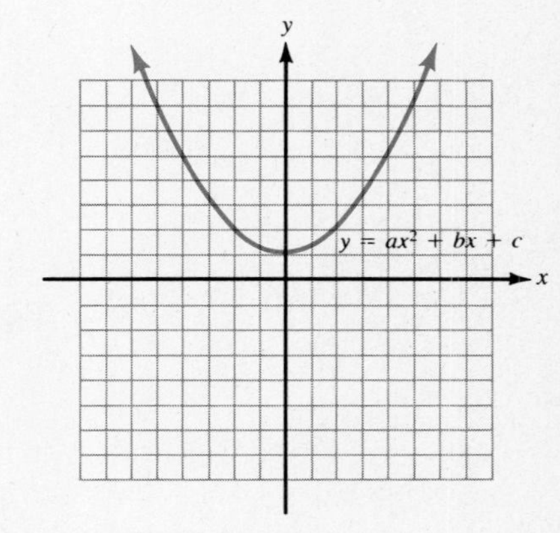

Figure 6-12. No real roots of $ax^2 + bx + c = 0$.

Case 3. If the graph of $y = ax^2 + bx + c$ does not intersect the x-axis at any points, then $ax^2 + bx + c = 0$ will have two imaginary roots. (See Figure 6-12.)

In exercises **13–16**, use the graphs of the quadratic functions in each figure.

a. Evaluate $b^2 - 4ac$ to verify the kinds of solutions for the corresponding quadratic equation.

b. Solve the equation and state the solution set.

13. $x^2 - 4x - 8 = 0$

14. $-x^2 - 2x + 2 = 0$

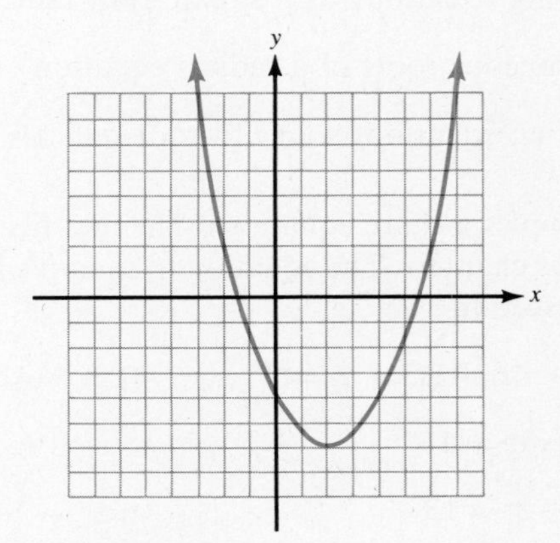

Figure 6-13. A graph of $2y = x^2 - 4x - 8$.

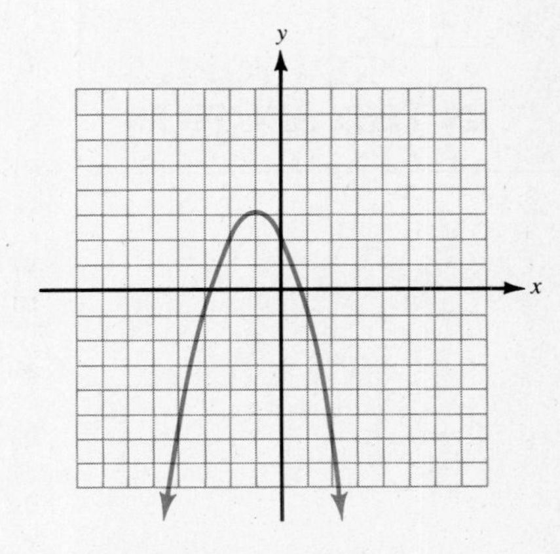

Figure 6-14. A graph of $y = -x^2 - 2x + 2$.

15. $-x^2 + 6x - 9 = 0$

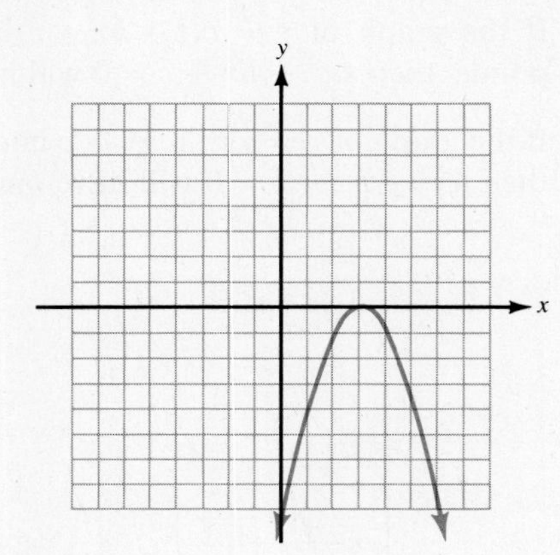

Figure 6-15. A graph of $y = -x^2 + 6x - 9$.

16. $-x^2 + 4x - 7 = 0$

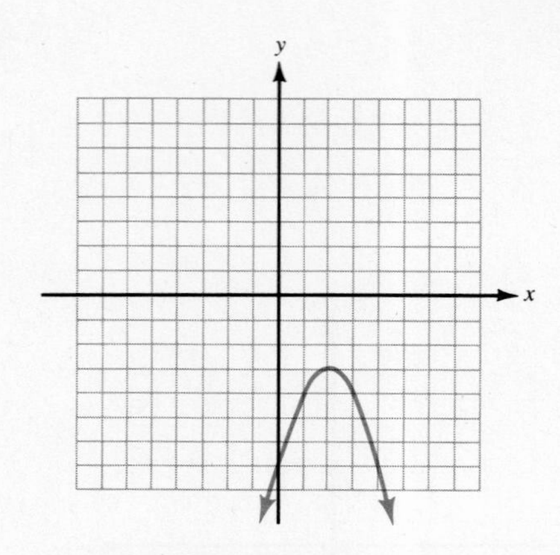

Figure 6-16. A graph of $y = -x^2 + 4x - 7$.

In exercises **17–20**, write an equation in the $ax^2 + bx + c = 0$ form for each pair of solutions. (*Hint:* use the sum-and-product-of-solutions theorem.)

17. $-6, \frac{3}{2}$

18. $5\sqrt{10}, -5\sqrt{10}$

19. $3 + 2\sqrt{2}, 3 - 2\sqrt{2}$

20. $-4 + 3i, -4 - 3i$

SECTION 6-4. Equations with Radicals

KEY TOPICS IN THIS SECTION

1. A definition of a radical equation
2. A procedure for solving radical equations
3. Solving equations with square-root radicals
4. Extraneous roots of a radical equation
5. Solving equations with *n*th-root radicals

Examples **a–f** are equations. The first five are equations that we have solved in previous chapters. The equation in example **f** is one that we will learn how to solve in this section.

a. $5(3 - 4t) - 2(3 - t) = 9$ — A linear equation in t

b. $3x + 5y = 15$ — A literal equation in x and y

c. $|2u + 3| = 13$ — An absolute value equation in u

d. $w^2 + 8w = 33$ — A quadratic equation in w

e. $\frac{5}{4y} - \frac{1}{3} = \frac{5 + 2y}{4y}$ A rational equation in y

f. $z - 4 = \sqrt{5z - 20}$ A radical equation in z

A Definition of a Radical Equation

Frequently we need to solve equations that contain one or more terms in which a variable is part of a radicand.

> **Definition 6.3. Radical equations**
> A **radical equation** is one that has at least one term with a variable under a radical.

The $\sqrt{5z - 20}$ term in example **f** makes the equation radical.

A Procedure for Solving Radical Equations

The procedure for solving a radical equation requires eliminating all radical terms. The following theorem can be used for this purpose:

> **Raising both sides of an equation to a power**
> If n is a positive integer greater than 1, then any solutions of $X = Y$ are also solutions of $X^n = Y^n$.

The following equation is based on one of the properties of nth-roots:

$$(\sqrt[n]{x})^n = x$$ The nth-root of x to the power n is x.

When this property is used together with the raising-to-powers theorem, then we can systematically remove radical terms from equations. For the simplest of equations, consider examples **a** and **b**.

a. $\sqrt{t} = 5$ The square root of t is 5.

b. $\sqrt[3]{u} = -6$ The cube root of u is -6.

Since $(\sqrt{t})^2 = t$, we can **square both sides** of example **a** to remove the radical.

$(\sqrt{t})^2 = 5^2$ Square both sides of the equation.

$t = 25$ The radical is eliminated.

Check: $\sqrt{25} = 5$ 5 is the **positive** square root of 25

Since $(\sqrt[3]{u})^3 = u$, we can **cube both sides** of example **b** to remove the radical.

$(\sqrt[3]{u})^3 = (-6)^3$ Cube both sides of the equation.

$u = -216$ The radical is eliminated.

Check: $\sqrt[3]{-216} = -6$ The cube root of a negative number is negative.

Examples **a** and **b** suggest the following five-step procedure for solving radical equations:

To solve a radical equation:

Step 1. If necessary, isolate a radical term on one side of the equation.

Step 2. Raise both sides of the equation to the nth power, where n is the root index.

Step 3. Simplify both sides of the equation.

Step 4. If necessary, repeat Steps 1–3, until all radical terms have been eliminated. Solve the final equation that has no radical terms.

Step 5. Check any solutions of Step 4 in the original equation. Reject any solution that does not check.

Solving Equations with Square-Root Radicals

The five-step procedure is used to solve the equations in Examples 1 and 2.

Example 1. Solve $5 + \sqrt{2t - 1} = 12$.

Solution. **Step 1.** Isolate the radical term on the left side.

$$\sqrt{2t - 1} = 7 \qquad \text{Subtract 5 from both sides.}$$

Step 2. Square both sides of the equation.

$$(\sqrt{2t - 1})^2 = 7^2 \qquad \text{The root index is 2.}$$

Step 3. Simplify both sides of the equation.

$$2t - 1 = 49 \qquad (\sqrt{x})^2 = x$$

Step 4. Solve this equation for t.

$$2t = 50 \qquad \text{Add 1 to both sides.}$$

$$t = 25 \qquad \text{Divide both sides by 2.}$$

Step 5. Check 25 in the given equation.

$$5 + \sqrt{2(25) - 1} = 12 \qquad \text{Replace } t \text{ by 25.}$$

$$5 + \sqrt{49} = 12 \qquad \text{Simplify the radicand.}$$

$$5 + 7 = 12, \textit{ true}$$

The solution set is $\{25\}$.

Example 2. Solve $k + 3 = 7 + \sqrt{5k - 20}$.

Solution. **Step 1.** $k - 4 = \sqrt{5k - 20}$ Subtract 7 from both sides.

Step 2. $(k - 4)^2 = (\sqrt{5k - 20})^2$ The root index is 2.

Step 3. $k^2 - 8k + 16 = 5k - 20$ $(x - y)^2 = x^2 - 2xy + y^2$

Step 4. $k^2 - 13k + 36 = 0$ — A quadratic equation in k

$(k - 9)(k - 4) = 0$ — Factor the trinomial.

$k - 9 = 0$ or $k - 4 = 0$ — Zero-product property.

$k = 9$ $\quad$ $k = 4$ — Solve both equations.

Step 5. If $k = 9$:

$$9 + 3 = 7 + \sqrt{(5(9) - 20}$$

$$12 = 7 + \sqrt{25}, \textit{ true}$$

If $k = 4$:

$$4 + 3 = 7 + \sqrt{5(4) - 20}$$

$$7 = 7 + \sqrt{0}, \textit{ true}$$

The solution set is $\{9, 4\}$.

Extraneous Roots of a Radical Equation

The raising-to-powers theorem contains a statement that is not present in the addition and multiplication properties of equality:

If $X = Y$, then $X + Z = Y + Z$.
If $X = Y$ and $Z \neq 0$, then $X \cdot Z = Y \cdot Z$.
} The solution sets of the pairs of equations are the same.

We say the addition and multiplication properties yield **equivalent equations,** because the solution sets are the same. The raising-to-powers theorem guarantees that no solutions are lost when an equation is raised to some power n. The problem is that the new equation may have some solutions that are not solutions of the given equation. These "extra" solutions are called **extraneous.**

Definition 6.4. Extraneous solutions
If c is a real number and c is a solution of $X^n = Y^n$ but not a solution of $X = Y$, then c is an **extraneous solution** of $X = Y$.

Example 3. Solve $16 + \sqrt{2z - 1} = 11$.

Solution. **Step 1.** $\sqrt{2z - 1} = -5$

Step 2. $(\sqrt{2z - 1})^2 = (-5)^2$

Step 3. $2z - 1 = 25$

Step 4. $2z = 26$

$z = 13$

Step 5. If $z = 13$:

$$16 + \sqrt{2(13) - 1} = 11$$

$$16 + \sqrt{25} = 11$$

$$16 + 5 = 11, \textit{ false}$$

Since 13 does not check in the original equation, it is rejected as an extraneous solution. There are no other solutions to check from the squared equation. Therefore, the solution set is $\varnothing$.

Example 4. Solve $u + 5 = 2\sqrt{u+4} + 4$.

Solution. **Step 1.** $u + 1 = 2\sqrt{u+4}$

Step 2. $(u+1)^2 = (2\sqrt{u+4})^2$

Step 3. $u^2 + 2u + 1 = 4(u+4)$

Step 4. $u^2 - 2u - 15 = 0$

$$(u-5)(u+3) = 0$$

$$u - 5 = 0 \quad \text{or} \quad u + 3 = 0$$

$$u = 5 \quad \text{or} \quad u = -3$$

Step 5. If $u = 5$:

$$5 + 5 = 2\sqrt{5+4} + 4$$
$$10 = 2\sqrt{9} + 4$$
$$10 = 6 + 4, \textit{ true}$$

If $u = -3$:

$$-3 + 5 = 2\sqrt{-3+4} + 4$$
$$2 = 2\sqrt{1} + 4$$
$$2 = 2 + 4, \textit{ false}$$

The apparent solution -3 is extraneous. Thus, the solution set is $\{5\}$.

Solving Equations with *n*th-root Radicals

In Example 5, the root index is 3. Therefore, both sides of the equation are cubed to remove the radical. In Example 6, the root index is 4. Therefore, both sides of the equation are raised to the fourth power to remove the radical.

Example 5. Solve $p = 2 + \sqrt[3]{p^3 - 7p^2 + 14p}$.

Solution. **Step 1.** $p - 2 = \sqrt[3]{p^3 - 7p^2 + 14p}$

Step 2. $(p-2)^3 = (\sqrt[3]{p^3 - 7p^2 + 14p})^3$

Step 3. $(p-2)^2(p-2) = p^3 - 7p^2 + 14p$

Step 4. $p^3 - 6p^2 + 12p - 8 = p^3 - 7p^2 + 14p$

$$p^2 - 2p - 8 = 0$$

$$(p-4)(p+2) = 0$$

$$p - 4 = 0 \quad \text{or} \quad p + 2 = 0$$

$$p = 4 \qquad p = -2$$

Step 5. If $p = 4$:

$$4 = 2 + \sqrt[3]{8}$$
$$4 = 2 + 2, \textit{ true}$$

If $p = -2$:

$$-2 = 2 + \sqrt[3]{-64}$$
$$-2 = 2 + (-4), \textit{ true}$$

Both solutions check. Therefore, the solution set is $\{4, -2\}$.

Example 6. Solve $\sqrt[4]{6 + 10x} - 4 = 0$.

Solution. **Step 1.** $\sqrt[4]{6 + 10x} = 4$

Step 2. $(\sqrt[4]{6 + 10x})^4 = 4^4$

Step 3. $6 + 10x = 256$

Step 4. $10x = 250$

$x = 25$

Step 5. If $x = 25$:

$\sqrt[4]{6 + 10(25)} - 4 = 0$

$\sqrt[4]{256} - 4 = 0$

$4 - 4 = 0$ True

The solution set is $\{25\}$.

SECTION 6-4. Practice Exercises

In exercises **1–68**, solve each equation. If no solutions check, write $\varnothing$.

[Example 1]

1. $\sqrt{x} = 9$

2. $\sqrt{x} = 4$

3. $\sqrt{5w} = 15$

4. $\sqrt{3w} = 6$

5. $\sqrt{2y - 1} = 3$

6. $\sqrt{5y - 9} = 4$

7. $\sqrt{9 - 3z} = 3$

8. $\sqrt{16 + 7z} = 4$

9. $0 = \sqrt{1 - 7x} - 6$

10. $0 = \sqrt{1 - 4x} - 3$

11. $\sqrt{2t + 1} = \sqrt{t + 4}$

12. $\sqrt{5t - 2} = \sqrt{3t + 4}$

13. $\sqrt{3v - 1} = 2\sqrt{v - 7}$

14. $\sqrt{12 + 3v} = 3\sqrt{4 - v}$

[Example 2]

15. $w = \sqrt{3w + 13} - 5$

16. $w + 4 = \sqrt{9w + 28}$

17. $2 = x + \sqrt{6 - 3x}$

18. $x = \sqrt{7x + 21} - 3$

19. $y - 3 = \sqrt{2y - 6}$

20. $2 + \sqrt{4y - 8} = y$

21. $\sqrt{3z - 2} = 3z - 2$

22. $\sqrt{2z - 1} = 2z - 1$

23. $1 - 6m = -\sqrt{18m - 5}$

24. $-\sqrt{6 - 5m} = 2m - 3$

[Example 3]

25. $\sqrt{k + 1} + 2 = 0$

26. $5 + \sqrt{2k - 1} = 0$

27. $5 + \sqrt{9m - 2} = 0$

28. $\sqrt{6m - 2} + 2 = 0$

29. $10 + \sqrt{x + 1} = 2$

30. $15 = \sqrt{x - 1} + 18$

31. $1 - \sqrt{3t + 7} = 6$

32. $15 = 2 - \sqrt{5t - 1}$

[Example 4]

33. $\sqrt{2y - 1} + 2 = y$

34. $\sqrt{6 + 2y} + 1 = y$

35. $\sqrt{p - 1} + p = 3$

36. $p = \sqrt{p + 8} - 2$

37. $\sqrt{z + 8} + 2 = -z$

38. $\sqrt{2z - 3} = 3 - z$

39. $0 = 1 + \sqrt{2u + 1} - u$

40. $5 + \sqrt{4u + 5} - 2u = 0$

41. $w = \sqrt{w + 2}$

42. $w - \sqrt{w + 6} = 0$

43. $\sqrt{2x^2 + 3x - 5} - x = 1$

44. $2x = \sqrt{5x^2 + 2x - 2} - 1$

45. $2 + \sqrt{2x^2 - 4x} = x$

46. $x = \sqrt{2x^2 - 6x + 3}$

[Example 5 and 6] 47. $\sqrt[3]{3x - 1} = 5$

48. $\sqrt[3]{5 - 2x} = 3$

49. $4 + \sqrt[3]{3 + y} = 1$

50. $10 + \sqrt[3]{4 + 3y} = 8$

51. $\sqrt[3]{12z + 4} = \sqrt[3]{15z - 11}$

52. $\sqrt[3]{x^2 - 19} - \sqrt[3]{x^2 - x - 7} = 0$

53. $\sqrt[4]{1 - 3t} = 2$

54. $\sqrt[4]{6 - 25t} = 3$

55. $\sqrt[4]{2u^2 + 5u + 6} - \sqrt[4]{2u^2 + 6u - 4} = 0$

56. $\sqrt[4]{u^2 + 17} = \sqrt[4]{u^2 + 2u + 1}$

57. $\sqrt[4]{v + 1} + 3 = 0$

58. $5 + \sqrt[4]{3 - v} = 0$

59. $\sqrt[5]{k + 1} = -2$

60. $5 = 7 - \sqrt[5]{2k}$

61. $m + 1 = \sqrt[3]{m^3 + 3m^2}$

62. $m - 1 = \sqrt[3]{m^3 - 3m^2}$

63. $w = 2 + \sqrt[3]{w^3 - 6w^2}$

64. $w = \sqrt[3]{w^3 + 6w^2} - 2$

65. $2x = 1 + \sqrt[3]{8x^3 - 12x^2}$

66. $\sqrt[3]{8x^3 + 12x^2} = 2x + 1$

67. $\sqrt[3]{x^3 + 16x - 3} - x - 1 = 0$

68. $\sqrt[3]{x^3 + 4x^2 - 2} - x - 2 = 0$

SECTION 6-4. Ten Review Exercises

In exercises **1–4**, simplify each expression.

1. $\sqrt{8} + \sqrt{75}$

2. $\sqrt{8} \cdot \sqrt{75}$

3. $\dfrac{1}{2\sqrt{3}}$

4. $\dfrac{1}{2 + \sqrt{3}}$

In exercises **5** and **6**, write each expression using only positive exponents.

5. $2t^{-1}$

6. $\dfrac{2^{-1} + t^{-1}}{(2t)^{-1}}$

In exercises **7** and **8**, solve each equation.

7. $9k - 5 = 11$

8. $9k^2 - 5 = 11$

In exercises **9** and **10**, factor each expression.

9. $6t^2 - 2at - ab + 3bt$

10. $a^4b^4 - 16$

SECTION 6-4. Supplementary Exercises

In exercises **1–20**, solve each equation.

1. $\sqrt{n + 3} + \sqrt{n - 3} = 3$

2. $\sqrt{n + 3} - \sqrt{n - 3} = 3$

3. $\sqrt{t + 5} - \sqrt{t - 5} = 5$

4. $\sqrt{t + 5} + \sqrt{t - 5} = 5$

5. $\sqrt{u+5} = 5 - \sqrt{u}$

6. $\sqrt{u} = \sqrt{u+12} - 2$

7. $\sqrt{3v-2} - 2 = \sqrt{v}$

8. $\sqrt{v+6} = \sqrt{5v-1} - 3$

9. $3 = \sqrt{w+13} - \sqrt{w-2}$

10. $\sqrt{3w+6} - \sqrt{3w-26} = 2$

11. $\sqrt{3x+1} - \sqrt{x-7} = \sqrt{x}$

12. $\sqrt{5x+4} - \sqrt{2x+1} = \sqrt{x-3}$

13. $\sqrt{3y+18} = \sqrt{y} + \sqrt{2y+6}$

14. $\sqrt{16y-5} = \sqrt{y+1} + \sqrt{y}$

15. $\sqrt{x^2 - \sqrt{3x^2+6}} = x - 1$

16. $\sqrt{x^2 + \sqrt{16x+1}} = x + 1$

17. $\sqrt{m^2 - \sqrt{2m-2}} = 2 - m$

18. $\sqrt{2m - 5 + \sqrt{3m+1}} - 3 = 0$

19. $\sqrt{t+6} = \sqrt{t+2} + \sqrt{8t-1}$

20. $\sqrt{8t+25} = \sqrt{2t+8} + \sqrt{2t+5}$

In exercises **21–24**, answer parts **a–d**. Assume a and b are real numbers.

21. $\sqrt{x} = a$
- **a.** Solve for x.
- **b.** Can a be a positive number?
- **c.** Can a be a negative number?
- **d.** Can a be 0?

22. $\sqrt[3]{x} = a$
- **a.** Solve for x.
- **b.** Can a be a positive number?
- **c.** Can a be a negative number?
- **d.** Can a be 0?

23. $\sqrt{x} + b = a$
- **a.** Solve for x.
- **b.** Can $a > b$?
- **c.** Can $b > a$?
- **d.** Can $a = b$?

24. $\sqrt[3]{x} + b = a$
- **a.** Solve for x.
- **b.** Can $a > b$?
- **c.** Can $b > a$?
- **d.** Can $a = b$?

In exercises **25** and **26**, answer parts **a–e**. Assume a and b are real numbers.

25. $\sqrt{x+a} = b$
- **a.** Solve for x.
- **b.** Can $x + a > 0$?
- **c.** Can $x + a < 0$?
- **d.** Can $b > a$?
- **e.** Can $a > b$?

26. $\sqrt{ax} = b$
- **a.** Solve for x.
- **b.** Can $a > 0$ and $b > 0$?
- **c.** Can $a > 0$ and $b < 0$?
- **d.** Can $a < 0$ and $b < 0$?
- **e.** Can $a < 0$ and $b > 0$?

SECTION 6-5. Quadratic Inequalities

KEY TOPICS IN THIS SECTION

1. A definition of a quadratic inequality in x
2. Classifications of quadratic inequalities based on solution sets
3. A graphical technique for solving a quadratic inequality
4. Approximating irrational numbers k_1 and k_2

The equations in examples **a** and **b** are quadratic.

a. $t^2 - 8t - 33 = 0$ **b.** $3u^2 + 7u - 6 = 0$

If the equal signs are replaced by $<$, $>$, $\leq$, or $\geq$, then a quadratic inequality is formed, as in examples **a*** and **b***.

a*. $t^2 - 8t - 33 < 0$ A "less than" quadratic inequality in t

b*. $3u^2 + 7u - 6 > 0$ A "greater than" quadratic inequality in u

In this section, we will study a procedure for finding the solution sets of quadratic inequalities.

A Definition of a Quadratic Inequality in x

The following identifies the standard forms of quadratic inequalities:

Definition 6.5. Quadratic inequalities in x
If a, b, and c are real numbers and $a \neq 0$, then a quadratic inequality in x can be written in one of the following forms:

1. $ax^2 + bx + c < 0$ **2.** $ax^2 + bx + c > 0$

3. $ax^2 + bx + c \leq 0$ **4.** $ax^2 + bx + c \geq 0$

1 and 2 are called **simple inequalities,** and 3 and 4 are called **compound inequalities.**

A *solution* of a quadratic inequality is any real-number replacement for the variable that makes a true inequality. The *solution set* is the set of all solutions.

Example 1. Determine whether 7 is a solution of $t^2 - 8t - 33 < 0$.

Solution. **Discussion.** Replace t by 7 in the given inequality and simplify. If the resulting inequality is *true,* then 7 is a *solution.* If the inequality is *false,* then 7 is *not a solution.*

$7^2 - 8(7) - 33 < 0$ Replace t by 7.

$49 - 56 - 33 < 0$ Simplify.

$-40 < 0$, *true* A true inequality

Therefore, 7 is a solution of $t^2 - 8t - 33 < 0$.

Example 2. Determine whether -2 is a solution of $3u^2 + 7u - 6 > 0$.

Solution. $3(-2)^2 + 7(-2) - 6 > 0$ Replace u by -2.

$12 + (-14) - 6 > 0$ Simplify.

$-8 > 0$, *false* A false inequality

Therefore, -2 is not a solution of $3u^2 + 7u - 6 > 0$.

Classifications of Quadratic Inequalities Based on Solution Sets

In Section 6-3, we used $b^2 - 4ac$ to determine the nature of the solutions of $ax^2 + bx + c = 0$. Recall that there were three possibilities for solution sets.

Possibility 1. One real-number solution if $b^2 - 4ac = 0$

Possibility 2. Two real-number solutions if $b^2 - 4ac > 0$

Possibility 3. Two imaginary-number solutions if $b^2 - 4ac < 0$

We are now using $ax^2 + bx + c$ to form quadratic inequalities. The solution sets of the inequalities depend in part on whether $b^2 - 4ac$ is 0, positive, or negative. If $b^2 - 4ac > 0$ (Possibility 2), then the solution sets can be found using a number line and the two real-number solutions of the corresponding quadratic equations. (We will study these inequalities in the following discussion. The inequalities formed by equations with discriminants from Possibilities 1 and 3 are discussed in the Supplementary Exercises of this section, p. 430.)

A Graphical Technique for Solving a Quadratic Inequality

Consider again the equations in examples **a** and **b**.

a. $t^2 - 8t - 33 = 0$ **b.** $3u^2 + 7u - 6 = 0$

Both equations have two real-number solutions.

a. $(t + 3)(t - 11) = 0$ **b.** $(u + 3)(3u - 2) = 0$

The solution set is $\{-3, 11\}$. The solution set is $\left\{-3, \frac{2}{3}\right\}$.

The solutions to these equations are plotted in Figure 6-17.

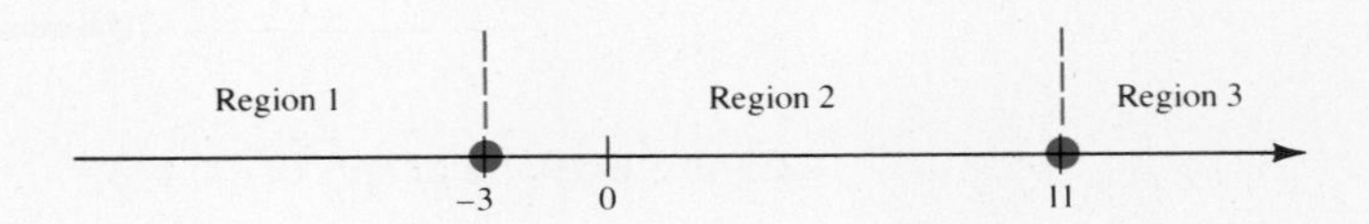

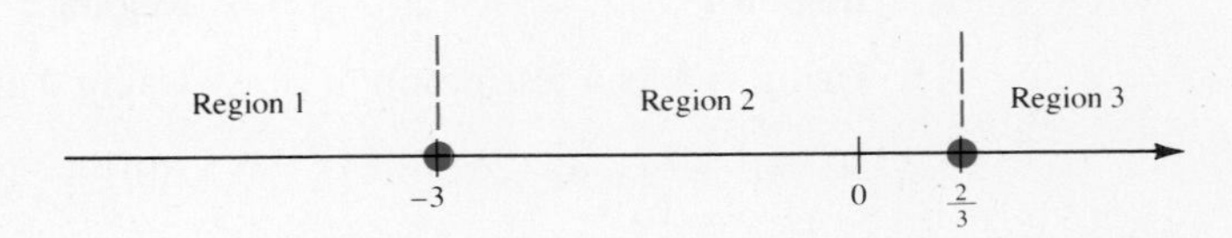

Figure 6-17. Graphs of $\{-3, 11\}$ and $\left\{-3, \frac{2}{3}\right\}$.

As shown, the solutions divide the number lines on which they are plotted into three regions.

Region 1. The points to the left of the graph of the smaller solution

Region 2. The points between the graphs of the solutions

Region 3. The points to the right of the graph of the larger solution

It can be shown that only two possibilities exist for the solution sets of $<$ or $>$ inequalities formed using the expressions $t^2 - 8t - 33$ and $3u^2 + 7u - 6$.

Possibility 1. The coordinates of points in Regions 1 and 3.

Possibility 2. The coordinates of points in Region 2.

To determine which possibility exists for a given inequality, we simply evaluate the inequality for any number in each of the three regions. To illustrate, consider the inequalities in examples **a*** and **b***.

a*. $t^2 - 8t - 33 < 0$

Region 1	**Region 2**	**Region 3**
Using -4 as a test point:	Using 0 as a test point:	Using 12 as a test point:
$(-4)^2 - 8(-4) - 33 < 0$	$0^2 - 8(0) - 33 < 0$	$12^2 - 8(12) - 33 < 0$
$15 < 0$, *false*	$-33 < 0$, *true*	$15 < 0$, *false*

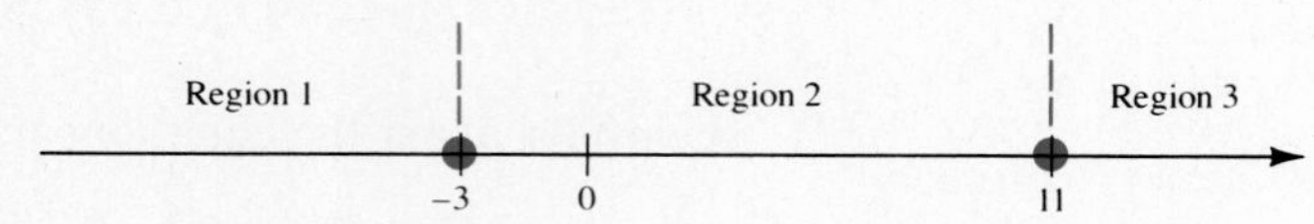

Figure 6-18. A graph of $\{-3, 11\}$.

Thus, the points between -3 and 11 have coordinates that are solutions of the given inequality. The solution set is $\{t \mid -3 < t < 11\}$, and a graph is shown in Figure 6-19.

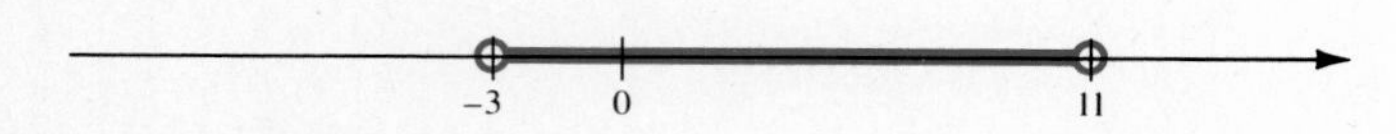

Figure 6-19. A graph of $\{t \mid -3 < t < 11\}$.

b*. $3u^2 + 7u - 6 > 0$

Region 1	**Region 2**	**Region 3**
Using -4 as a test point:	Using 0 as a test point:	Using 1 as a test point:
$3(-4)^2 + 7(-4) - 6 > 0$	$3(0)^2 + 7(0) - 6 > 0$	$3(1)^2 + 7(1) - 6 > 0$
$14 > 0$, *true*	$-6 > 0$, *false*	$4 > 0$, *true*

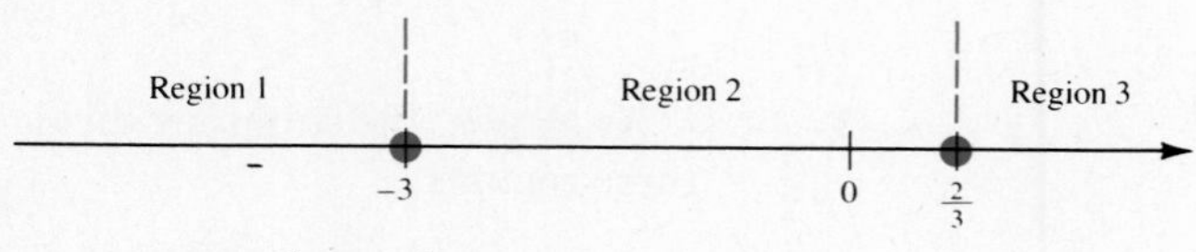

Figure 6-20. A graph of $\left\{-3, \frac{2}{3}\right\}$.

Thus, the points to the left of -3 and to the right of $\frac{2}{3}$ have coordinates that are solutions of the given inequality. The solution set is $\{u \mid u < -3 \text{ or } u > \frac{2}{3}\}$, and a graph is shown in Figure 6-21.

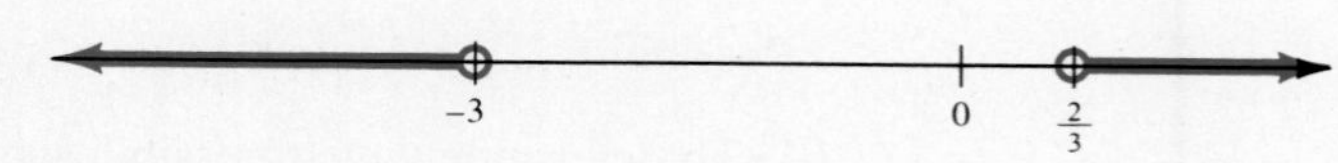

Figure 6-21. A graph of $\left\{u \mid u < -3 \text{ or } u > \frac{2}{3}\right\}$.

Examples **a*** and **b*** suggest the following five-step procedure for solving quadratic inequalities whose corresponding quadratic equations have two real-number solutions:

> If $ax^2 + bx + c = 0$ has two real number solutions, then to find the solution set of an inequality formed using $ax^2 + bx + c$:
>
> **Step 1.** If necessary, write the inequality in standard form based on Definition 6-5.
>
> **Step 2.** Find k_1 and k_2, the real-number solutions of $ax^2 + bx + c = 0$.
>
> **Step 3.** Plot k_1 and k_2 on a number line.
>
> **Step 4.** Evaluate the given inequality with the coordinate of a test point in each of the three regions determined by k_1 and k_2.
>
> **Step 5.** State the solution set of the given inequality based on the results obtained in Step 4.

Example 3. Solve and graph $2x^2 + 9x > 5$.

Solution.

Step 1. $2x^2 + 9x - 5 > 0$ — In standard form

Step 2. $2x^2 + 9x - 5 = 0$ — A corresponding equation

$(x + 5)(2x - 1) = 0$ — Factor the trinomial.

$x + 5 = 0$ or $2x - 1 = 0$ — The zero-product property

$x = -5$ or $x = \frac{1}{2}$ — Solve for x.

Thus, $k_1 = -5$ and $k_2 = \frac{1}{2}$. — Let k_1 be the smaller number.

Step 3. — Plot -5 and $\frac{1}{2}$.

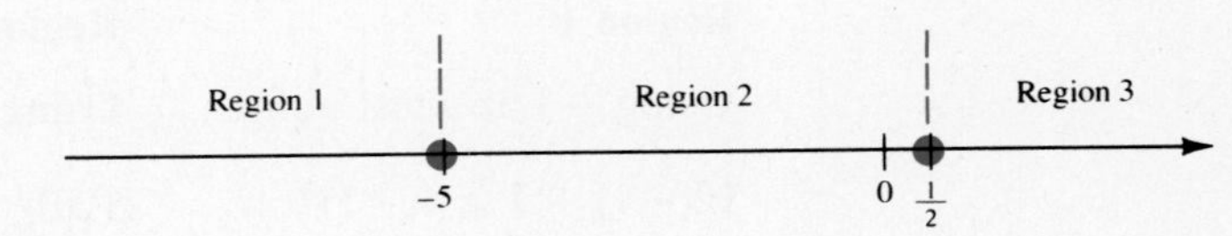

Figure 6-22. A number line with k_1 and k_2 plotted.

Step 4.

Region 1	**Region 2**	**Region 3**
Using -6 as a test point:	Using 0 as a test point:	Using 1 as a test point:
$2(-6)^2 + 9(-6) > 5$	$2(0)^2 + 9(0) > 5$	$2(1)^2 + 9(1) > 5$
$72 + (-54) > 5$, *true*	$0 > 5$, *false*	$2 + 9 > 5$, *true*

Step 5. The solution set is $\{x|x < -5 \text{ or } x > \frac{1}{2}\}$. A graph of the solution set is shown in Figure 6-23.

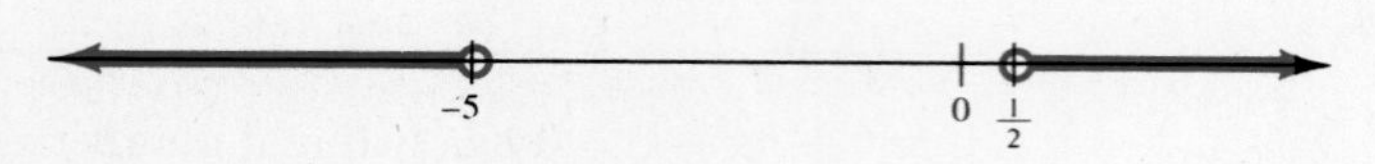

Since the inequality is only less than, hollow dots are used.

Figure 6-23. A graph of $\left\{x|x < -5 \text{ or } x > \frac{1}{2}\right\}$.

Example 4. Solve and graph $12y + 7 \geq 4y^2$.

Solution.

Step 1. $0 \geq 4y^2 - 12y - 7$ — If $a \geq b$,

$4y^2 - 12y - 7 \leq 0$ — then $b \leq a$.

Step 2. $4y^2 - 12y - 7 = 0$

$(2y + 1)(2y - 7) = 0$ — Factor the trinomial.

$2y + 1 = 0 \quad \text{or} \quad 2y - 7 = 0$ — The zero-product property.

$y = \frac{-1}{2} \quad \text{or} \quad y = \frac{7}{2}$ — Solve for y.

Thus, $k_1 = \frac{-1}{2}$ and $k_2 = \frac{7}{2}$. — Let k_1 be the smaller number.

Step 3.

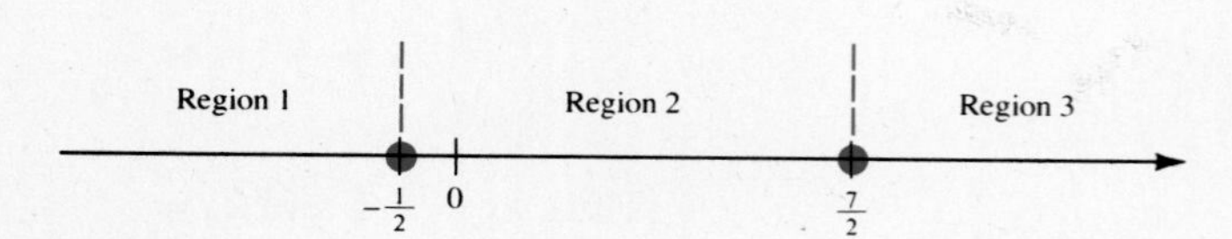

Plot $\frac{-1}{2}$ and $\frac{7}{2}$.

Figure 6-24. A number line with k_1 and k_2 plotted.

Step 4.

Region 1	**Region 2**	**Region 3**
Using -1 as a test point:	Using 0 as a test point:	Using 4 as a test point:
$12(-1) + 7 \geq 4(-1)^2$	$12(0) + 7 \geq 4(0)^2$	$12(4) + 7 \geq 4(4)^2$
$-5 \geq 4$, *false*	$7 \geq 0$, *true*	$55 \geq 64$, *false*

Step 5. The solution set is $\{y|\frac{-1}{2} \leq y \leq \frac{7}{2}\}$. A graph of the solution set in shown in Figure 6-25.

Since the inequality is compound, the dots are solid.

Figure 6-25. A graph of $\left\{y\middle|-\frac{1}{2} \leq y \leq \frac{7}{2}\right\}$.

Approximating Irrational Numbers k_1 and k_2

If the solutions of $ax^2 + bx + c = 0$ are irrational numbers, then an approximation is used to help pick the test points in Regions 1, 2, and 3.

Example 5. Solve and graph $t^2 + 4 \geq 6t$.

Solution. **Step 1.** $t^2 - 6t + 4 \geq 0$ — Write in standard form.

Step 2. $t^2 - 6t + 4 = 0$ — A corresponding equation.

$$t = \frac{6 \pm \sqrt{36 - 4(1)(4)}}{2}$$ The quadratic formula

$$= \frac{6 \pm 2\sqrt{5}}{2}$$ $\sqrt{36 - 16} = \sqrt{20} = 2\sqrt{5}$

$$= 3 \pm \sqrt{5}$$ Factor a 2 and reduce.

Step 3. $k_1 = 3 - \sqrt{5} \approx 0.8$

$k_2 = 3 + \sqrt{5} \approx 5.2$

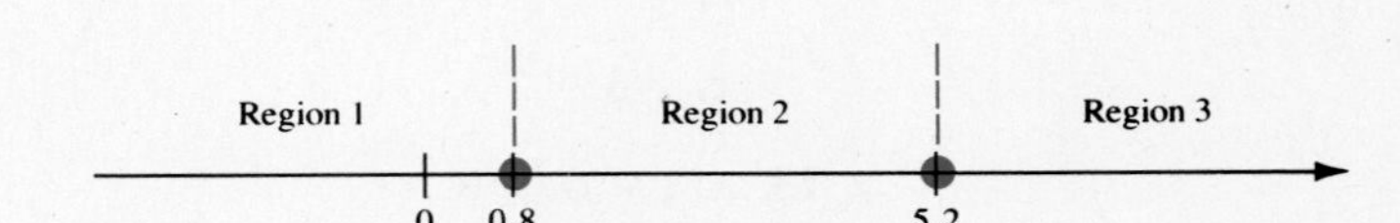

Figure 6-26. Approximate locations of k_1 and k_2.

Step 4.

Region 1	**Region 2**	**Region 3**
Using 0 as a test point:	Using 2 as a test point:	Using 6 as a test point:
$0^2 + 4 \geq 6(0)$	$2^2 + 4 \geq 6(2)$	$6^2 + 4 \geq 6(6)$
$4 \geq 0$, *true*	$4 + 4 \geq 12$, *false*	$36 + 4 \geq 36$, *true*

Step 5. The solution set is $\{t \mid t \leq 3 - \sqrt{5} \text{ or } t \geq 3 + \sqrt{5}\}$. A graph of the solution set is shown in Figure 6-27.

Figure 6-27. A graph of $\{t \mid t \leq 3 - \sqrt{5} \text{ or } t \geq 3 + \sqrt{5}\}$.

The exact values of k_1 and k_2 are used to graph the solution set.

SECTION 6-5. Practice Exercises

In exercises **1–14**, determine whether the given values are solutions of each inequality.

[Examples 1 and 2]

1. $x^2 - 9x + 8 < 0$
a. 2
b. 7

2. $x^2 + 12x + 20 < 0$
a. -9
b. -3

3. $2x^2 + 13x > 45$
a. -10
b. 2

4. $2x^2 + x > 21$
a. -4
b. 2

5. $3x^2 \leq 24 - 14x$
a. 5
b. -5

6. $3x^2 \leq 14 - x$
a. 3
b. -2

7. $x^2 + 4x \geq -4$
a. 0
b. $\frac{1}{2}$

8. $9x^2 \geq 6x - 1$
a. 0
b. $\frac{1}{3}$

9. $x^2 < 2x - 5$
a. $\frac{-3}{5}$
b. 3

10. $x^2 < x - 3$
a. -10
b. $\frac{8}{3}$

11. $(x + 5)(x - 2) < 0$
a. 0
b. $\frac{7}{5}$
c. $\sqrt{3}$

12. $(x + 3)(x - 4) < 0$
a. $\frac{5}{2}$
b. 0
c. $\sqrt{11}$

13. $x(x - 6) \geq 0$
a. -3
b. $\frac{1}{3}$
c. $1 - \sqrt{2}$

14. $x(x + 2) \leq 0$
a. -1
b. $\frac{1}{9}$
c. $2 - \sqrt{2}$

In exercises **15–56**, solve and graph each inequality. State any irrational number solution in exact form.

[Examples 3 and 4]

15. $(x + 5)(x - 2) < 0$

16. $(x + 1)(x - 8) < 0$

17. $(x - 1)(x - 10) > 0$

18. $(x - 2)(x - 9) > 0$

19. $(2x - 5)(x + 6) \leq 0$

20. $(3x + 1)(x - 9) \leq 0$

21. $(5x + 9)(x - 3) \geq 0$

22. $(2x + 9)(x - 2) \geq 0$

23. $x^2 + 3x - 10 < 0$

24. $x^2 - x - 12 < 0$

25. $x^2 - 4x > 12$

26. $x^2 - 5x > 14$

27. $10 > 7x - x^2$

28. $8 > 9x - x^2$

29. $18 + 2x^2 \leq 13x$

30. $2x^2 + 7x \leq 30$

31. $28x \geq -3x^2 - 9$

32. $5 \geq -5x^2 - 26x$

33. $2x^2 + x < 28$

34. $2x^2 < -6 - 13x$

35. $3x^2 > 14x - 8$

36. $3x^2 + 5x > 8$

37. $4x^2 \leq 9$

38. $100x^2 \leq 81$

39. $25 \leq 16x^2$

40. $36 \leq x^2$

41. $x^2 < 12x$

42. $x^2 < -8x$

[Example 5]

43. $15x < x^2$

44. $x^2 + 7x > 0$

45. $x^2 - 2x \leq 1$

46. $x^2 + 6 \leq 6x$

47. $x^2 + 19 \geq 10x$

48. $x^2 - 4x \geq 6$

49. $x^2 + 6x + 4 < 0$

50. $x^2 < 12 - 2x$

51. $x^2 < 5$

52. $x^2 > 7$

53. $x^2 - 1 > 4x$

54. $x^2 \geq 10x - 23$

55. $x^2 + 29 > -12x$

56. $x^2 + 14 < -10x$

SECTION 6-5. Ten Review Exercises

In exercises **1–6**, solve each equation.

1. $(2t + 3)(2t - 1) = (4t + 3)(t - 2)$

2. $(2t + 3)(t - 2) = (4t + 3)(t - 2)$

3. $7k - 3 = 11$

4. $|7k - 3| = 11$

5. $\sqrt{3m - 1} + 4 = 0$

6. $\sqrt[3]{3m - 1} + 4 = 0$

In **7–10**, simplify each expression.

7. $(5uv)^2$

8. $(5u + v)^2$

9. $\dfrac{16u^4v^4}{2uv}$

10. $\dfrac{16 - u^4v^4}{2 - uv}$

SECTION 6-5. Supplementary Exercises

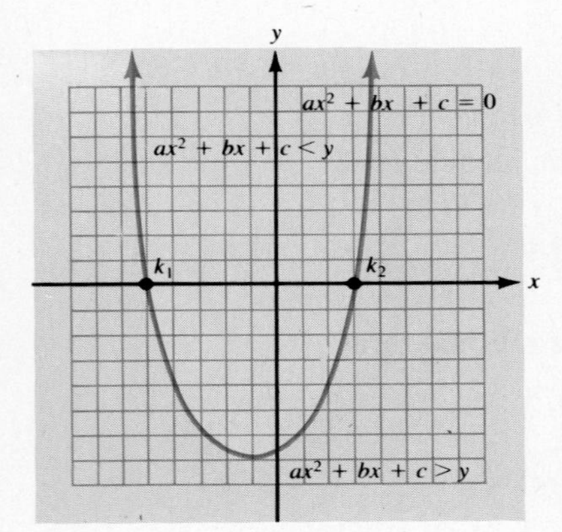

Figure 6-28. $b^2 - 4ac > 0$.

Figure 6-28 is a graph of $y = ax^2 + bx + c$. For this equation, $a > 0$ and $b^2 - 4ac > 0$. The curve intersects the x-axis at points with coordinates k_1 and k_2, the solutions of:

$$ax^2 + bx + c = 0$$

Note 1. For all points in the interior of the curve:

$$y > ax^2 + bx + c \quad \text{or} \quad ax^2 + bx + c < y$$

In particular, if $y = 0$, then all points on the x-axis between k_1 and k_2 have coordinates that satisfy $ax^2 + bx + c < 0$.

Note 2. For all points in the exterior of the curve:

$$y < ax^2 + bx + c, \quad \text{or} \quad ax^2 + bx + c > y$$

In particular, if $y = 0$, then all points on the x-axis to the left of k_1 or to the right of k_2 have coordinates that satisfy $ax^2 + bx + c > 0$.

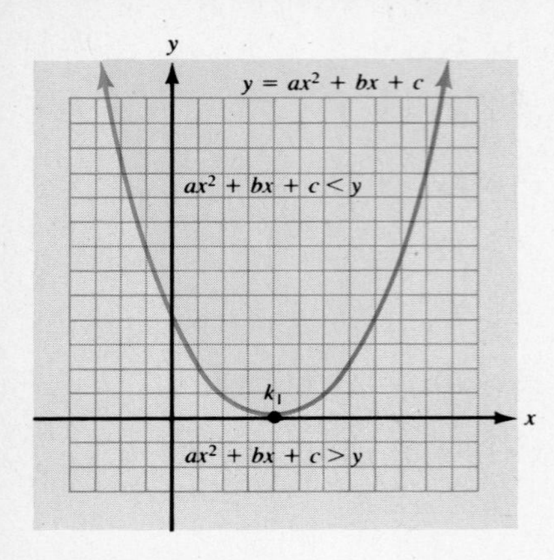

Figure 6-29. $b^2 - 4ac = 0$.

Figure 6-29 is a graph of $y = ax^2 + bx + c$. For this equation, $a > 0$ and $b^2 - 4ac = 0$. The curve intersects the x-axis at one point with coordinate k_1, the only solution of:

$$ax^2 + bx + c = 0$$

Note 3. No points of the x-axis are in the interior of the curve.

Thus, $ax^2 + bx + c < 0$ has no solutions.

Note 4. All points of the x-axis except k_1 are in the exterior of the curve.

Thus, $ax^2 + bx + c > 0$ has all real numbers as solutions except k_1.

Figure 6-30 is a graph of $y = ax^2 + bx + c$. For this equation, $a > 0$ and $b^2 - 4ac < 0$. The curve does not intersect the x-axis at any point. The solutions of $ax^2 + bx + c = 0$ are imaginary numbers.

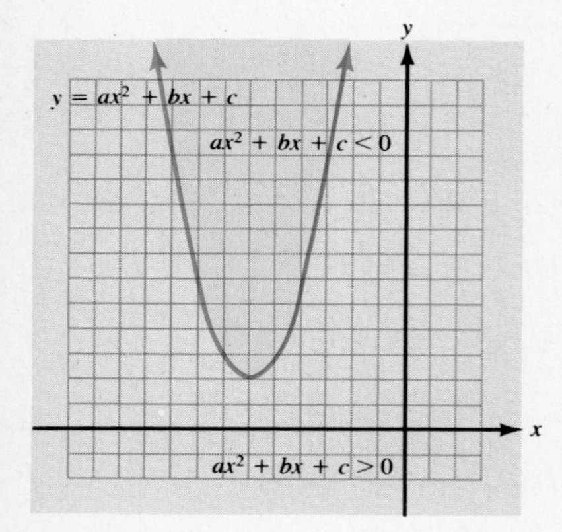

Figure 6-30. $b^2 - 4ac < 0$.

Note 5. No points of the x-axis are in the interior of the curve.

Thus, $ax^2 + bx + c < 0$ has no solutions.

Note 6. All points of the x-axis are in the exterior of the curve.

Thus, $ax^2 + bx + c > 0$ has all real numbers as solutions.

In exercises **1–10**, refer to the graphs in Figures 6-28, 6-29, and 6-30.

a. If necessary, write each inequality in one of the standard forms of Definition 6.5, with $a > 0$.

b. Evaluate $b^2 - 4ac$.

c. Determine the solution set of each inequality.

1. $x^2 - 4x + 5 > 0$

2. $x^2 - 6x + 12 > 0$

3. $x^2 + 2x + 4 < 0$

4. $x^2 + 10x + 26 < 0$

5. $x^2 > 6x - 9$

6. $0 < x^2 + 16x + 64$

7. $x^2 + 8x + 16 \le 0$

8. $x^2 \le 4x - 4$

9. $4x^2 - 4x + 3 \ge 0$

10. $9x^2 + 12x + 5 < 0$

Example. Solve $\dfrac{5}{x + 1} < 2$.

Solution. **Step 1.** Make the right side of the inequality 0.

$$\frac{5}{x + 1} - 2 < 0 \qquad \text{Add } -2 \text{ to both sides.}$$

Step 2. Write the left side as one rational term.

$$\frac{5}{x + 1} - \frac{2(x + 1)}{x + 1} < 0 \qquad \text{The LCD is } x + 1.$$

$$\frac{3 - 2x}{x + 1} < 0$$

Step 3. Find the values of x that make the numerator and denominator 0.

$$3 - 2x = 0 \qquad x + 1 = 0$$

$$x = \frac{3}{2} \qquad x = -1$$

Step 4. Divide a number line into separate regions, using the values of x from Step 3.

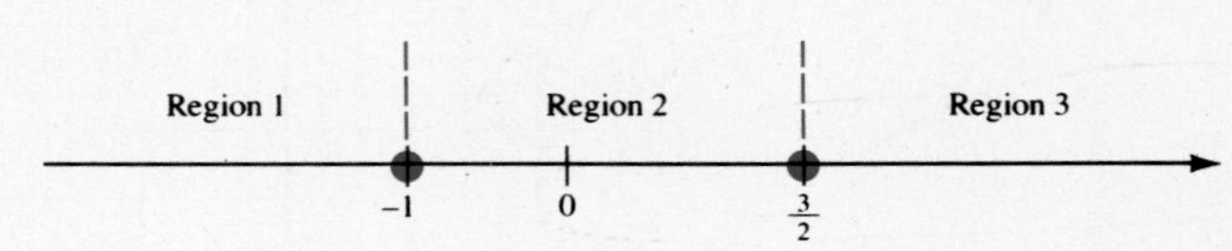

Figure 6-31. A graph of $\left\{-1, \frac{3}{2}\right\}$.

Step 5. Evaluate the given inequality with the coordinate of a test point in each of the regions found in Step 4.

Region 1

Using -2 as a test point:

$$\frac{5}{-2+1} < 2$$

$-5 < 2$, *true*

Region 2

Using 0 as a test point:

$\frac{5}{0+1} < 2$, *false*

Region 3

Using 2 as a test point:

$$\frac{5}{2+1} < 2$$

$\frac{5}{3} < 2$, *true*

Step 6. Select the region (or regions) that contain solutions.

$$\left\{x \middle| x < -1 \quad \text{or} \quad x > \frac{3}{2}\right\}$$

In exercises **11–20**, solve each inequality.

11. $\frac{x+1}{x-2} < 0$

12. $\frac{2x+6}{x-3} < 0$

13. $\frac{3-x}{x+4} > 0$

14. $\frac{x}{x+5} > 0$

15. $\frac{4x}{x-1} - 5 < 0$

16. $\frac{2x}{x-5} - 4 < 0$

17. $\frac{21}{2x-1} \geq -3$

18. $\frac{10}{2x-5} \geq -1$

19. $\frac{x^2}{x+4} \geq x - 1$

20. $\frac{x^2}{x-6} \leq x + 2$

In exercises **21–30**, use the graph of $y = f(x)$ to find the values of x such that:

a. $f(x) > 0$
b. $f(x) < 0$
c. $f(x) \geq 0$
d. $f(x) \leq 0$

21.

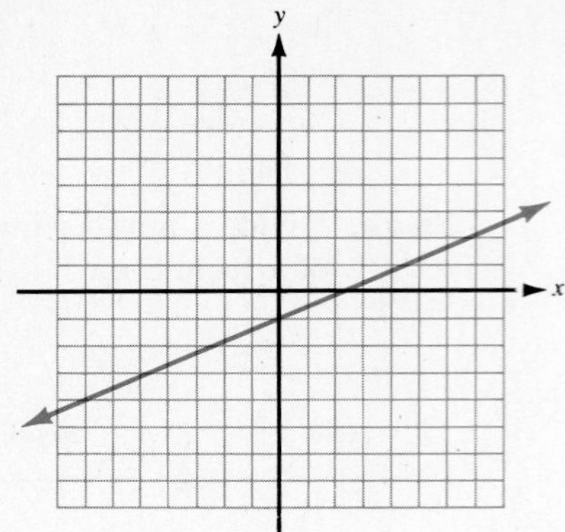

22.

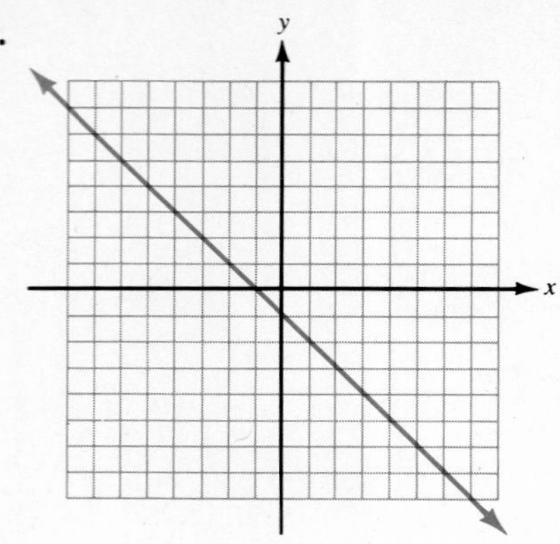

23.

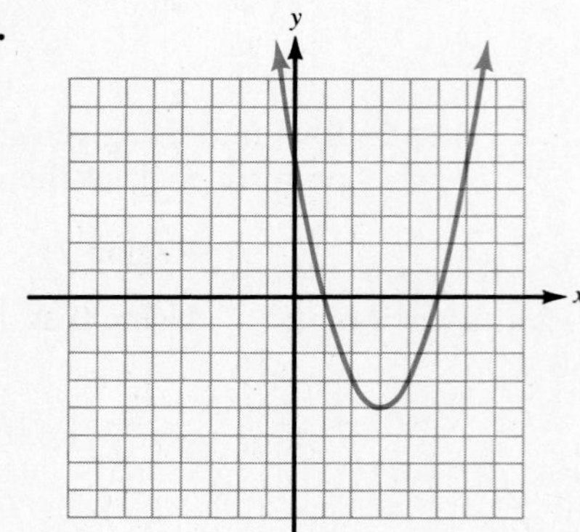

24.

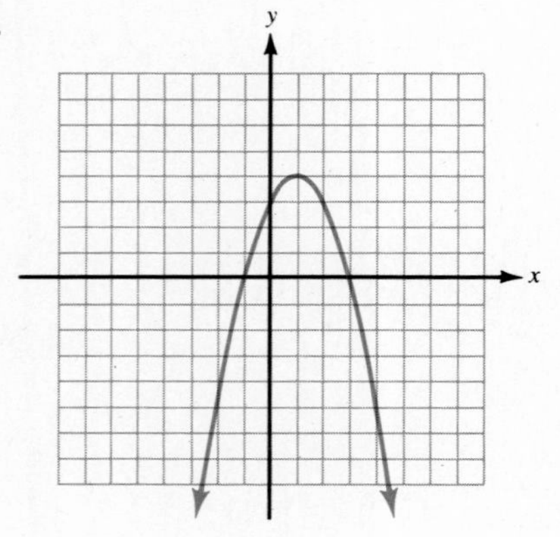

25.

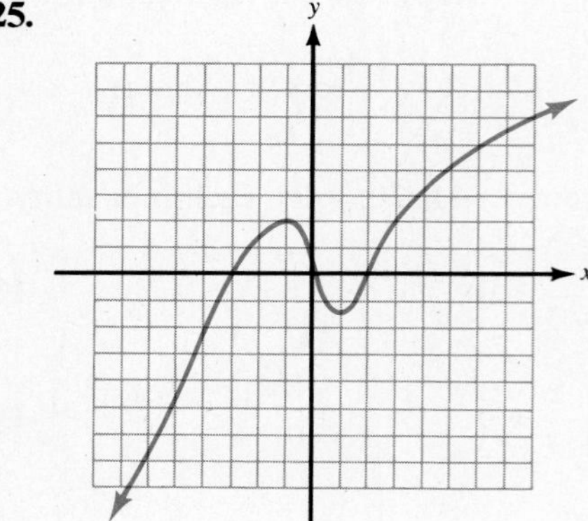

26.

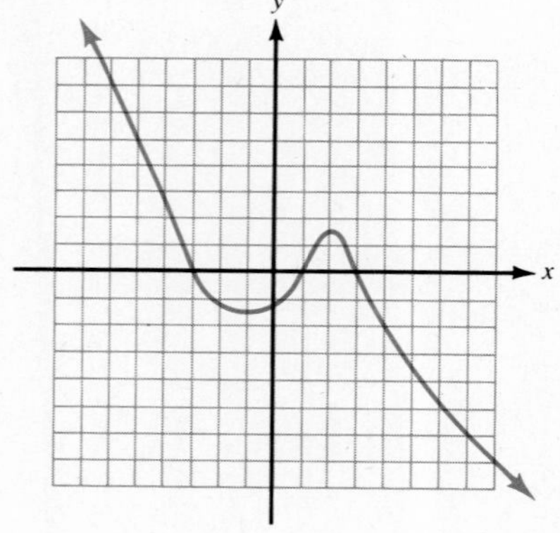

27.

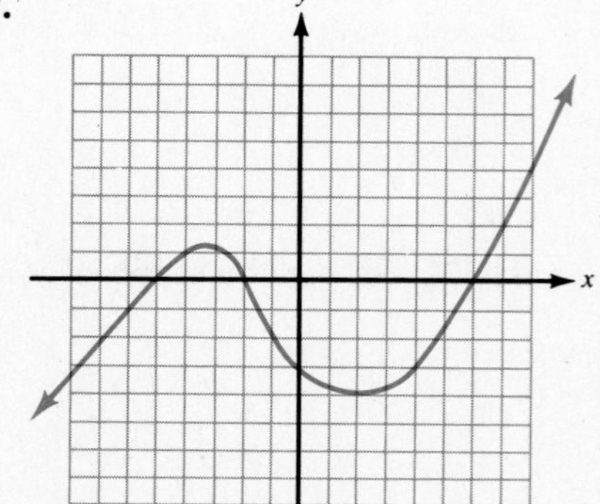

28.

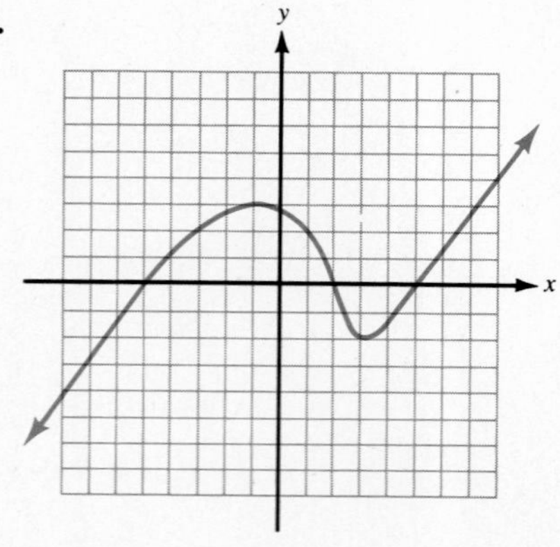

29.

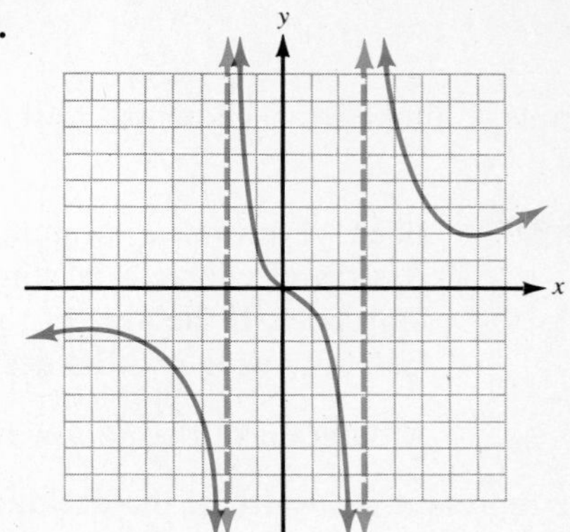

30.

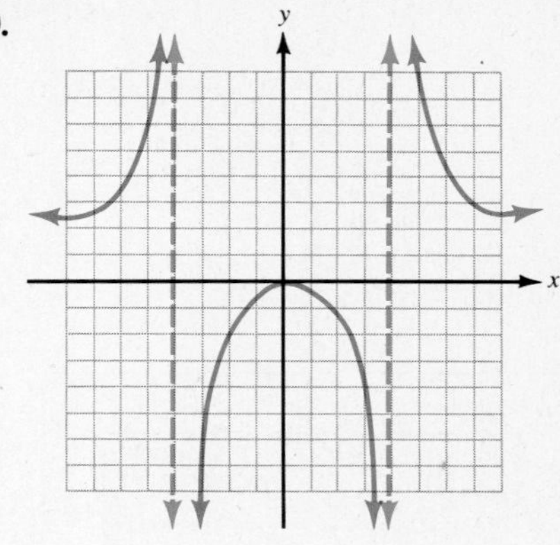

SECTION 6-6. Applications

1. Literal quadratic equations
2. Geometry problems
3. Work problems
4. Motion problems

In this section, we will study a few types of applied problems that require solving quadratic equations to answer.

Literal Quadratic Equations

Many equations contain two or more variables of degree 2, as in examples **a–c**. (We will study some of these equations in Chapter 11.)

a. $a^2 + b^2 = c^2$ — The Pythagorean Theorem

b. $(x - 3)^2 + (y - 2)^2 = 16$ — The center-radius equation of a circle

c. $4x^2 + 9y^2 = 36$ — An equation of an ellipse

We frequently need to solve such equations for one of the variables in terms of the other variables. The process is illustrated in Example 1.

Example 1. Solve $4x^2 + 9y^2 = 36$ for y in terms of x.

Solution.

$$4x^2 + 9y^2 = 36 \quad \text{The given equation}$$

$$9y^2 = 36 - 4x^2 \quad \text{Subtract } 4x^2 \text{ from both sides.}$$

$$y^2 = \frac{36 - 4x^2}{9} \quad \text{Divide both sides by 9.}$$

$$y = \pm\sqrt{\frac{36 - 4x^2}{9}} \quad \text{Use the square-root theorem.}$$

$$y = \pm\sqrt{\frac{4}{9}(9 - x^2)} \quad \text{Factor a 4 in the numerator.}$$

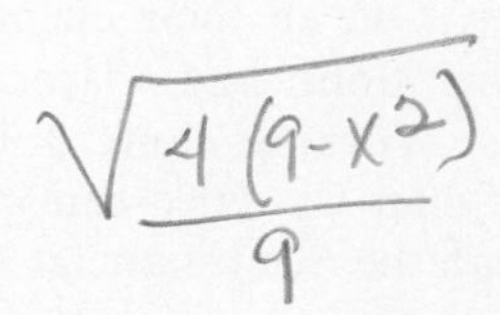

$$y = \frac{\pm 2\sqrt{9 - x^2}}{3} \quad \text{Simplify the radical.}$$

Geometry Problems

Example 2 illustrates a geometry problem that requires a quadratic equation to solve.

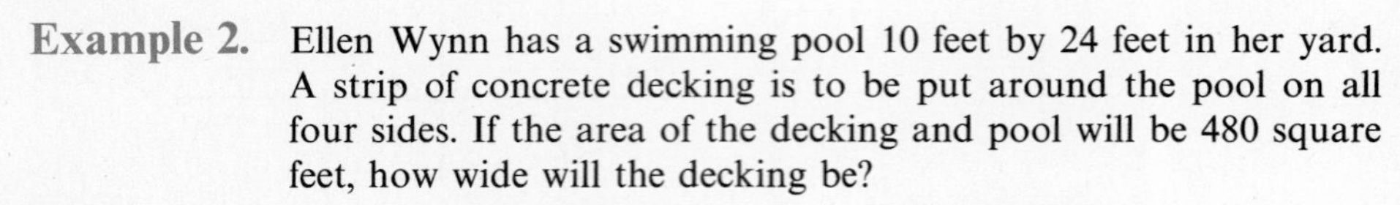

Example 2. Ellen Wynn has a swimming pool 10 feet by 24 feet in her yard. A strip of concrete decking is to be put around the pool on all four sides. If the area of the decking and pool will be 480 square feet, how wide will the decking be?

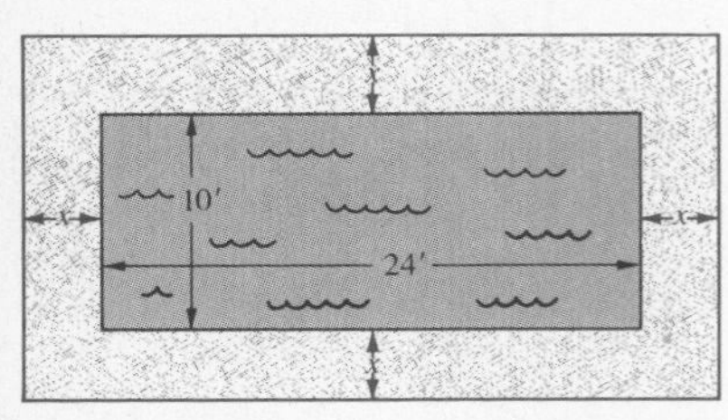

Figure 6-32. A diagram of the pool with the strip of decking.

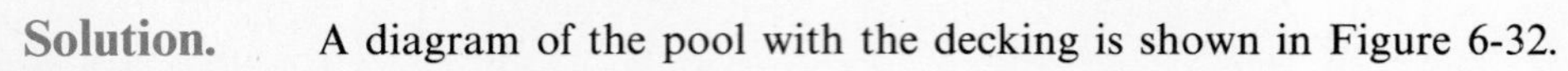

Solution. A diagram of the pool with the decking is shown in Figure 6-32.

Let x represent the width of the decking.

$10 + 2x$ represents the width of pool plus decking

$24 + 2x$ represents the length of pool plus decking

$$\text{Area} = (\text{width})(\text{length})$$

$$480 = (10 + 2x)(24 + 2x)$$

$$480 = 240 + 68x + 4x^2$$

$$4x^2 + 68x - 240 = 0 \qquad \text{Write in standard form.}$$

$$x^2 + 17x - 60 = 0 \qquad \text{Multiply both sides by } \frac{1}{4}.$$

$$(x + 20)(x - 3) = 0 \qquad \text{Factor the trinomial.}$$

$$x + 20 = 0 \quad \text{or} \quad x - 3 = 0 \qquad \text{The zero-product property}$$

$$x = -20, \textit{ reject} \quad x = 3 \qquad \text{Solve for } x.$$

Since x represents a distance (the width of the decking), the apparent solution -20 is rejected. Thus, the width of the decking must be 3 feet.

Check: $10 + 2(3) = 16$ feet wide
$24 + 2(3) = 30$ feet long
Area $= (16)(30) = 480 \text{ ft}^2$

Work Problems

The following symbols are used to solve a work problem:

1. t represents the *time* a "machine" works on a certain job
2. r represents the fractional part of the job completed by the "machine" for each unit of time the "machine" works (r stands for the *rate* at which the job is completed)
3. $r \cdot t$ represents the part of the *total job* completed by that "machine"
4. If more than one "machine" is used to complete a job, then the sum of the products of the r's and t's must equal 1.

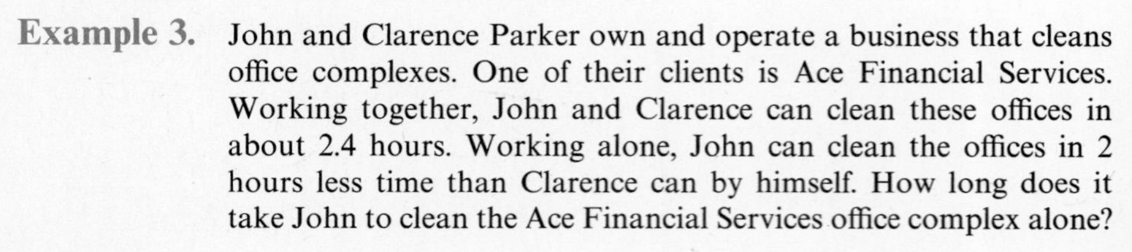

Example 3. John and Clarence Parker own and operate a business that cleans office complexes. One of their clients is Ace Financial Services. Working together, John and Clarence can clean these offices in about 2.4 hours. Working alone, John can clean the offices in 2 hours less time than Clarence can by himself. How long does it take John to clean the Ace Financial Services office complex alone?

Solution. The information for this problem is given in the following table:

	Time (in hours) to do the job alone	Fractional part of work done in 1 hour	Fractional part of total job done in 2.4 hours
John	t hours	$\frac{1}{t}$ of job	$2.4\left(\frac{1}{t}\right)$ or $\frac{2.4}{t}$
Clarence	$(t + 2)$ hours	$\frac{1}{t+2}$ of job	$2.4\left(\frac{1}{t+2}\right)$ or $\frac{2.4}{t+2}$

Working together, at the end of 2.4 hours the job is completed. Therefore, an equation for the problem is:

$$\frac{2.4}{t} + \frac{2.4}{t+2} = 1 \qquad \boxed{\text{Work done by John}} + \boxed{\text{Work done by Clarence}} = \boxed{\text{One completed job}}$$

$$t(t+2)\left(\frac{2.4}{t} + \frac{2.4}{t+2}\right) = t(t+2) \qquad \text{Multiply both sides by } t(t+2).$$

$$2.4(t+2) + 2.4t = t(t+2) \qquad \text{Distribute on the left side.}$$

$$2.4t + 4.8 + 2.4t = t^2 + 2t \qquad \text{Remove the parentheses.}$$

$$t^2 - 2.8t - 4.8 = 0 \qquad \text{Write in standard form.}$$

$$5t^2 - 14t - 24 = 0 \qquad \text{Multiply both sides by 5.}$$

$$(5t+6)(t-4) = 0 \qquad \text{Factor the trinomial.}$$

$$5t + 6 = 0 \quad \text{or} \quad t - 4 = 0 \qquad \text{The zero-product property}$$

$$t = \frac{-6}{5}, \textit{ reject} \qquad t = 4 \qquad \text{Solve the equations for } t.$$

Since t represents time, $\frac{-6}{5}$ is rejected, while 4 is accepted as a possible solution. If $t = 4$, then $t + 2 = 6$.

$$\frac{2.4}{4} + \frac{2.4}{6} = 1$$

$$0.6 + 0.4 = 1, \textit{ true}$$

Thus, it takes John 4 hours to clean the office complex alone, and Clarence 6 hours.

Motion Problems

The following symbols are used to solve a motion problem:

1. d represents the *distance* an object travels in some specified time
2. r represents the *rate* at which the object is traveling
3. t represents the *time* during which the motion is measured
4. $d = r \cdot t$ is the formula that relates the quantities

5. $r = \dfrac{d}{t}$ is the formula solved for r in terms of d and t

6. $t = \dfrac{d}{r}$ is the formula solved for t in terms of d and r

Example 4. Debbie Bacus flew her plane from her home in Virginia to her parents' home in Maryland, a distance of 480 miles. On the trip to Maryland, she flew against a steady headwind of 40 mph. On the return trip to Virginia, the same wind was now a tail wind. If the time for the roundtrip was 5 hours, how fast can Debbie fly her plane with no wind interference?

Solution. Let r represent the rate of the plane with no wind interference.

	Distance	Rate	Time
Against the wind	480 miles	$(r - 40)$ mph	$\dfrac{480}{r - 40}$ hours
With the wind	480 miles	$(r + 40)$ mph	$\dfrac{480}{r + 40}$ hours

The time for the roundtrip was 5 hours. Therefore, an equation for the problem is:

$$\frac{480}{r-40} + \frac{480}{r+40} = 5 \qquad \boxed{\text{Time out}} + \boxed{\text{Time back}} = \boxed{\text{Total time}}$$

$$(r-40)(r+40)\left(\frac{480}{r-40} + \frac{480}{r+40}\right) = 5(r-40)(r+40)$$

$$480(r+40) + 480(r-40) = 5(r-40)(r+40)$$

$$480r + 19{,}200 + 480r - 19{,}200 = 5r^2 - 8{,}000$$

$$5r^2 - 960r - 8{,}000 = 0$$

$$r^2 - 192r - 1{,}600 = 0$$

$$(r - 200)(r + 8) = 0$$

$$r - 200 = 0 \quad \text{or} \quad r + 8 = 0$$

$$r = 200 \quad \text{or} \quad r = -8, \textit{ reject}$$

The check is omitted. The speed of Debbie's plane with no wind interference is 200 mph.

SECTION 6-6. Practice Exercises

[Example 1] In exercises **1–10**, solve each equation for the indicated variable. Simplify the answers.

1. $x^2 + y^2 = 9$, for y

2. $x^2 + 4y^2 = 4$, for y

3. $9x^2 + y^2 = 9$, for x

4. $9x^2 + 4y^2 = 36$, for x

5. $x^2 - y^2 = 16$, for y

6. $4x^2 - 25y^2 = 100$, for y

7. $s = at^2 + vt$, for t

8. $s = 2t^2 - 5t$, for t

9. $A = 4\pi r^2$, for r

10. $A = 2\pi rh + 2\pi r^2$, for r

In exercises **11–34**, solve each word problem.

[Example 2] **11.** A flower garden is 4 feet wide and 30 feet long. A concrete mowing strip is placed around the edge of the flower bed. If the area of the flower bed and mowing strip is 192 square feet, how wide is the mowing strip?

12. Sheila has a bedspread that is 8 feet wide and 10 feet long. She sewed a fancy edge down the sides and across the bottom. The bedspread now covers 144 square feet. How wide is the edge that Sheila sewed to the bedspread?

13. The length of a rectangle is 1 foot less than 2 times the width. If the area of the rectangle is 120 square feet, find the length and width.

14. The length of a rectangle is 2 feet more than 3 times the width. If the area of the rectangle is 85 square feet, find the length and width.

15. A flower bed in Central Park in New York City has the shape of an isosceles triangle. The base of the triangle is 6 meters longer than the height. If the area of the triangle is 108 square meters, find the lengths of the base and height of the flower bed.

16. A sign used to give directions has the shape of a right triangle. The bottom edge of the sign is the longer leg of the triangle. This edge is 3 feet more than 2 times the vertical edge. If the area of the sign is 22 square feet, find the lengths of the two legs of the sign.

17. The legs of a right triangle are equal in length. If the length of the hypotenuse is 10 feet, approximate the lengths of the legs to one decimal place.

18. The length of the longer leg of a right triangle is 1 meter more than 2 times the length of the shorter leg. If the length of the hypotenuse is 12 meters, approximate the lengths of the legs to one decimal place.

[Example 3] **19.** Harold and Fred can type the company's quarterly report in 2 hours when they work together. Working alone, it takes Fred 3 hours longer than it takes Harold. How long would it take Harold to type the report? How long would it take Fred?

20. Working together, two trucks (large truck A and small truck B) can haul a quantity of gravel from the pit to a road job in 15 hours. It takes truck B 16 hours longer than it takes truck A to haul the gravel alone. How long does it take truck A to haul the gravel? How long does it take truck B?

21. When two valves are both used, a large holding tank can be filled with gasoline in 8 hours. The smaller valve (valve X) takes 12 hours longer than the larger valve (valve Y) to fill the tank when the valves are used alone. How long does it take valve X to fill the tank? How long does it take valve Y?

22. When two hoses of different sizes are used, an area of grass can be watered in 6 hours. Using the smaller hose alone, it takes 5 hours longer than when the larger hose is used alone. How long does it take to water the area using only the larger hose? How long does it take using only the smaller hose?

23. Machine A and machine B can process a quantity of data in 12 minutes when both are used. It takes machine B 7 minutes longer than machine A to process the data alone. How long does it take machine A to process the data? How long does it take machine B?

24. Bill and John can assemble an electronic component in 30 minutes when they work together. It takes John 32 more minutes than Bill to assemble the component alone. How long does it take Bill to assemble the component? How long does it take John?

25. Every year Barbara Bulas and Stacey Matthews get together to paint the summer cottage they own jointly in northern Tennessee. If they work together, they can paint the cabin in 8 hours. The women estimate that it would take Stacey about 3 hours longer than it would take Barbara to paint it alone. Based on the 3-hour estimate, calculate to one decimal place the approximate number of hours it would take Barbara and Stacey, each working alone, to paint the cottage.

26. The Rocking Horse T Ranch has several hundred feet of white board fence that needs repairs and touch-up painting every spring. It takes Pecos and Cheyenne 6 days to complete the task when they work together. The men estimate it would take Cheyenne about 2 days longer than it would take Pecos to do the job working alone. To one decimal place, calculate the approximate number of days it would take Pecos and Cheyenne, each working alone, to complete the repair job.

[Example 4] 27. A small plane makes a trip from city D to city C, a distance of 600 miles, against a wind of 25 mph. It makes the return trip from city C to city D with the same wind. If the round trip takes 7 hours, find the speed of the plane with no wind.

28. A large military cargo plane made a flight of 4,950 miles against a prevailing head wind. It made the return flight with the same wind as a tail wind. If the plane is capable of averaging 500 mph with no wind, find the speed of the wind on this trip if the total flying time for the trip was 20 hours.

29. A baseball is thrown vertically upward with an initial speed of 128 feet per second. Its distance d above the ground at time t is approximated with the equation:

$$d = 128t - 16t^2$$

Find the time it takes for the ball to reach a height of 256 feet.

30. A missile is fired vertically upward with an initial speed of 560 feet per second. Its distance d above the ground at time t is approximated with the equation:

$$d = 560t - 16t^2$$

Find the time it takes the missile to attain a height of 4,416 feet on its way up.

31. Ships A and B meet in the middle of the Atlantic ocean. Ship A is heading south and ship B is heading east. Ship A is traveling at a rate of speed that is 6 mph faster than ship B. Three hours after they meet, the distance between the ships is 90 miles. Find the rates of ship A and ship B in miles per hour.

32. A commercial airliner and a private plane cross paths, with the airliner heading north and the private plane heading west. The speed of the airliner is 280 mph faster than the private plane. One hour after they pass, the two planes are 680 miles apart. Find the speeds of the airliner and the private plane.

33. The owner of Atwater River Rentals is experimenting with a new paddle design. He canoes up a river against a current of 5 mph and returns with the current. The 12-mile round trip took a total of 1.6 hours. Find the rate of the canoe in still water.

34. Ken and Charlie have to hike a level trail 6 miles long and camp overnight at a lake. So that Charlie can hike in fast and fish the first day, Ken will carry both packs in and Charlie will carry both packs out. Ken's rate carrying both packs is $1\frac{1}{2}$ mph slower than normal. His rate without a pack is 1 mph faster than normal. Find his normal rate if his total travel time, in and out, is 14 hours.

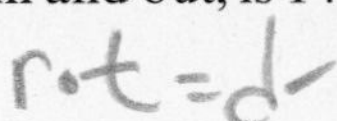

SECTION 6-6. Ten Review Exercises

1. Evaluate $8t + 1$ for $t = 6$.

2. Based on the answer to exercise **1**, is 6 a solution of $\sqrt{8t + 1} = t + 1$?

3. Solve $\sqrt{8t + 1} = t + 1$.

4. Did both apparent solutions of exercise **3** check, or was one extraneous?

5. Evaluate $u^3 - 7u^2 + 10u$ for $u = -4$.

6. Based on the answer to exercise **5**, is -4 a solution of $\sqrt[3]{u^3 - 7u^2 + 10u} = u - 2$?

7. Solve $\sqrt[3]{u^3 - 7u^2 + 10u} = u - 2$.

8. Did both apparent solutions of exercise **7** check, or was one extraneous?

In exercises **9** and **10**, factor each expression.

9. $9k^2 + 49 - 42k$

10. $81u^4 - 625$

SECTION 6-6. Supplementary Exercises

If a moving object is **uniformly accelerated,** then the distance d that the object moves during the time it is accelerated can be approximated with the formula:

$$d = at^2 + vt$$

a is the *acceleration* in units per time^2

v is *velocity* prior to acceleration in units per time

Example. Sid Goldstein is traveling down I-550 at 44 feet per second (about 30 mph). He accelerates the car 3 feet per second^2. If during acceleration the car travels 600 feet, then approximately how many seconds (to one decimal place) did Sid accelerate?

Solution. **Discussion.** Use $s = at^2 + vt$ with $a = 3$ feet per second^2, $d = 600$ feet, and $v = 44$ feet per second.

$$600 = 3t^2 + 44t \qquad \text{Replacing } d, a, \text{ and } v$$

$$0 = 3t^2 + 44t - 600 \qquad \text{Write in standard form.}$$

$$t = \frac{-44 \pm \sqrt{1936 - 4(3)(-600)}}{6} \qquad \text{The quadratic formula}$$

$$t = \frac{-44 \pm 95.582\ldots}{6} \qquad \sqrt{9136} \approx 95.582\ldots$$

$$t \approx 8.6 \quad \text{or} \quad t \approx -23.3, \textit{ reject}$$

Thus, Sid accelerated his car approximately 8.6 seconds. At the end of acceleration, Sid was going about 48 mph.

In exercises **1–6**, find the value of t to one decimal place.

1. A car was traveling on a highway at a speed of 60 feet per second. The driver accelerated the car 2 feet per second^2. If the car traveled 500 feet during acceleration, approximately how long was the car accelerated?

2. A truck was traveling down the interstate at a speed of 75 feet per second. The driver accelerated the truck 3 feet per second^2. If the truck traveled 1,000 feet during acceleration, approximately how long was the truck accelerated?

3. A bicyclist was riding on a level road at a speed of 15 feet per second. The road reached a grade that uniformly açcelerated the bike 1.5 feet per second^2. If the length of the grade was 2,500 feet, approximately how long was the bike accelerated?

4. A bicyclist was riding on a level road at a speed of 20 feet per second. The road reached a grade that uniformly accelerated the bike 2.5 feet per second^2. If the length of the grade was 1,200 feet, approximately how long was the bike accelerated?

5. A car was traveling on a highway at a speed of 100 feet per second. The driver applied the brakes, uniformly slowing the car 5 feet per second. If the car traveled 480 feet during deceleration, approximately how long were the brakes applied? (*Note:* $a = -5$.)

6. A truck was traveling on an interstate at a speed of 88 feet per second. The driver slowed the truck uniformly at a rate of 2 feet per second^2. If the truck traveled 800 feet during deceleration, approximately how long was the truck slowed? (*Note:* $a = -2$.)

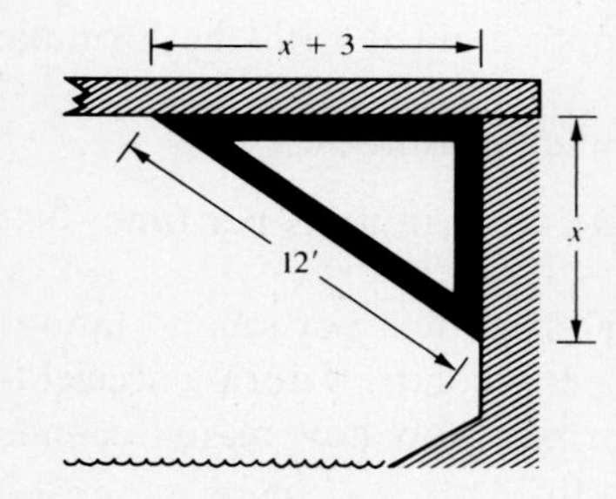

Figure 6-33. A support for a bridge.

Example. A support for a bridge is in the shape of a right triangle, as shown in Figure 6-33. The longer leg is 3 feet more than the shorter leg, and the length of the hypotenuse is 12 feet. To one decimal place, approximate the lengths of the legs.

Solution. Let x represent the length of the shorter leg and $x + 3$ the length of the longer leg.

$$x^2 + (x + 3)^2 = 12^2 \qquad (\text{leg})^2 + (\text{leg})^2 = (\text{hypotenuse})^2$$

$$2x^2 + 6x - 135 = 0 \qquad \text{Write in standard form.}$$

$$x = \frac{-6 \pm \sqrt{36 - 4(2)(-135)}}{4} \qquad \text{The quadratic formula}$$

$$x \approx \frac{-6 \pm 33.406\ldots}{4} \qquad \sqrt{1116} \approx 33.406\ldots$$

$$x \approx 6.851\ldots \quad \text{or} \quad x \approx -9.851\ldots, \textit{ reject}$$

$$x \approx 6.9, \text{ to one decimal place}$$

$$x + 3 \approx 9.9, \text{ to one decimal place}$$

The legs of the support are approximately 6.9 feet and 9.9 feet.

In exercises **7–12**, approximate the lengths of the unknown sides to one decimal place.

7. The longer leg is 2 feet more than the shorter leg, and the hypotenuse is 7 feet. Find the lengths of both legs.

8. The shorter leg is 6 feet less than the longer leg, and the hypotenuse is 20 feet. Find the lengths of both legs.

9. The hypotenuse is twice as long as the shorter leg, and the longer leg is 6 feet. Find the lengths of the shorter leg and the hypotenuse.

10. The hypotenuse is 3 times the length of the shorter leg, and the longer leg is 10 feet. Find the lengths of the shorter leg and the hypotenuse.

11. The longer leg is 3 inches more than the shorter leg, and the hypotenuse is 1 inch more than 2 times the length of the shorter leg. Find the lengths of all three sides.

12. The shorter leg is 5 inches less than the longer leg, and the hypotenuse is 3 inches more than the longer leg. Find the lengths of all three sides.

In exercises **13–18**, solve each equation for x. First write each equation in the form $ax^2 + bx + c = 0$ and use the quadratic formula.

13. $y = x^2 + 2x$

14. $y = 2x^2 - x$

15. $y = x^2 - 2x - 2$

16. $y = x^2 - 10x + 24$

17. $y = x^2 + 4x + 4$

18. $y = x^2 - 20x + 100$

SECTION 6-7. Equations That Are Quadratic in Form (Optional)

1. Solving equations of degree 4
2. Solving equations of degree 6
3. Solving equations with binomial factors
4. Solving equations with rational-number exponents
5. Solving rational equations that are quadratic in form (Supplementary Exercises)

Many equations that are not quadratic can be solved using one of the techniques used to solve quadratic equations. We will solve some of these equations in this section.

Solving Equations of Degree 4

An equation of degree 4 may be changed to one of degree 2 in a different variable by making a suitable substitution, as illustrated in Example 1.

Example 1. Solve $4t^4 + 31t^2 - 8 = 0$.

Solution. **Discussion.** By letting another variable, say x, stand for t^2, x^2 then can stand for t^4. The result is an equation that is quadratic in x.

$4x^2 + 31x - 8 = 0$ — Replace t^2 by x.

$(4x - 1)(x + 8) = 0$ — Factor.

$4x - 1 = 0$ or $x + 8 = 0$ — The zero-product property

$x = \frac{1}{4}$ or $x = -8$ — Solve for x.

$t^2 = \frac{1}{4}$ or $t^2 = -8$ — Replace x by t^2.

$t = \pm\sqrt{\frac{1}{4}}$ or $t = \pm\sqrt{-8}$ — The square-root theorem

$t = \pm\frac{1}{2}$ $\quad t = \pm 2i\sqrt{2}$ — Simplify the radicals.

The solution set is $\left\{\frac{-1}{2}, \frac{1}{2}, -2i\sqrt{2}, 2i\sqrt{2}\right\}$.

Solving Equations of Degree 6

An equation of degree 6 may be changed to one of degree 2 in a different variable by making a suitable substitution, as illustrated in Example 2.

Example 2. Solve $z^6 + 117z^3 - 1{,}000 = 0$.

Solution. Replacing z^3 by x and $z^6 = (z^3)^2$ by x^2:

$x^2 + 117x - 1{,}000 = 0$ — A quadratic equation in x

$(x + 125)(x - 8) = 0$ — Factor.

$x + 125 = 0$ or $x - 8 = 0$ — The zero-product property

$z^3 + 125 = 0$ $\quad z^3 - 8 = 0$ — Replace x by z^3

$(z + 5)(z^2 - 5z + 25) = 0$ or $(z - 2)(z^2 + 2z + 4) = 0$

$z + 5 = 0$ or $z^2 - 5z + 25 = 0$ or $z - 2 = 0$ or $z^2 + 2z + 4 = 0$

$z = -5$ or $z = \frac{5 \pm \sqrt{25 - 100}}{2}$ or $z = 2$ or $z = \frac{-2 \pm \sqrt{4 - 16}}{2}$

or $z = \frac{5 \pm 5i\sqrt{3}}{2}$ or $z = -1 \pm i\sqrt{3}$

The solution set is $\left\{-5, \frac{5 \pm 5i\sqrt{3}}{2}, 2, -1 \pm i\sqrt{3}\right\}$.

Solving Equations with Binomial Factors

An equation may contain a grouped expression, such as a binomial, to a power 2. To facilitate solving the equation, the grouped expression can temporarily be replaced by a single variable, as illustrated in Example 3.

Example 3. Solve $2(m^2 - 3m)^2 = 11(m^2 - 3m) - 12$.

Solution. **Discussion.** Replace $(m^2 - 3m)$ by x, then $(m^2 - 3m)^2$ can be replaced by x^2.

$2x^2 = 11x - 12$ A quadratic equation in x

$2x^2 - 11x + 12 = 0$ Set equal to 0.

$(2x - 3)(x - 4) = 0$ Factor.

$2x - 3 = 0$ or $x - 4 = 0$

$x = \frac{3}{2}$ or $x = 4$ Solve for x.

$m^2 - 3m = \frac{3}{2}$ or $m^2 - 3m = 4$ Replace x by $m^2 - 3m$.

$2m^2 - 6m - 3 = 0$ or $m^2 - 3m - 4 = 0$ Solve for m.

$m = \frac{6 \pm \sqrt{36 + 24}}{4}$ or $(m - 4)(m + 1) = 0$

$m = \frac{6 \pm 2\sqrt{15}}{4}$ or $m - 4 = 0$ or $m + 1 = 0$

$m = \frac{3 \pm \sqrt{15}}{2}$ or $m = 4$ or $m = -1$

The solution set is $\left\{\frac{3 \pm \sqrt{15}}{2}, 4, -1\right\}$.

Solving Equations with Rational-Number Exponents

In advanced mathematics courses, we sometimes work with equations with variables raised to a rational-number power such as $\frac{2}{3}$ or $\frac{2}{5}$. If such an equation must be solved, the techniques learned in this chapter may be appropriate, provided a temporary replacement is made for the variable by another variable of degree 1. The details are illustrated in Example 4.

Example 4. Solve $2p^{2/3} = 15 - 7p^{1/3}$.

Solution. **Discussion.** By letting $x = p^{1/3}$, and $x^2 = (p^{1/3})^2 = p^{2/3}$, the given equation will then be quadratic in x.

$$2x^2 = 15 - 7x \quad \text{Replace } p^{1/3} \text{ by } x.$$

$$2x^2 + 7x - 15 = 0 \quad \text{Set equal to 0.}$$

$$(2x - 3)(x + 5) = 0 \quad \text{Factor.}$$

$$2x - 3 = 0 \quad \text{or} \quad x + 5 = 0 \quad \text{The zero-product property}$$

$$x = \frac{3}{2} \quad \text{or} \quad x = -5 \quad \text{Solve for } x.$$

$$p^{1/3} = \frac{3}{2} \quad \text{or} \quad p^{1/3} = -5 \quad \text{Replace } x \text{ by } p^{1/3}.$$

$$(p^{1/3})^3 = \left(\frac{3}{2}\right)^3 \quad \text{or} \quad (p^{1/3})^3 = (-5)^3$$

$$p = \frac{27}{8} \quad \text{or} \quad p = -125$$

The solution set is $\left\{\frac{27}{8}, -125\right\}$.

SECTION 6-7. Practice Exercises

In exercises **1–30**, solve each equation.

[Example 1]

1. $y^4 - 13y^2 + 36 = 0$

2. $y^4 - 25y^2 + 144 = 0$

3. $p^4 - 27p^2 + 50 = 0$

4. $p^4 - 19p^2 + 48 = 0$

5. $z^4 - 13z^2 + 40 = 0$

6. $z^4 - 15z^2 + 36 = 0$

7. $m^4 + 21m^2 - 100 = 0$

8. $m^4 + 15m^2 - 16 = 0$

9. $4t^4 + 23t^2 - 6 = 0$

10. $9t^4 + 98t^2 - 11 = 0$

[Example 2]

11. $y^6 - 26y^3 - 27 = 0$

12. $y^6 + 19y^3 - 216 = 0$

13. $x^6 + 56x^3 - 512 = 0$

14. $x^6 - 124x^3 - 125 = 0$

15. $216n^6 + 19n^3 - 1 = 0$

16. $512n^6 + 72n^3 + 1 = 0$

[Example 3]

17. $2(y^2 - y)^2 = 6 + (y^2 - y)$

18. $7(y^2 - 3y) = 2(y^2 - 3y)^2 - 4$

19. $(u^2 + 5u)^2 = 4(u^2 + 5u) + 12$

20. $(u^2 - 2u)^2 = 32 + 4(u^2 - 2u)$

21. $2(2v^2 + v)^2 = 15 + (2v^2 + v)$

22. $2 + 3(2v^2 - 3v)^2 = 7(2v^2 - 3v)$

[Example 4]

23. $2m^{2/3} - 9m^{1/3} = 5$

24. $5m^{1/3} + 3 + 2m^{2/3} = 0$

25. $3z^{2/3} + 7z^{1/3} - 6 = 0$

26. $10z^{2/3} = z^{1/3} + 3$

27. $3y^{1/5} = 2y^{2/5} - 2$

28. $3y^{2/5} + 2 = 5y^{1/5}$

29. $p^{2/3} = 4p^{1/3}$

30. $p^{1/2} = 2p^{1/4}$

SECTION 6-7. Ten Review Exercises

In exercises **1–8**, simplify each expression. Assume all variables represent positive real numbers.

1. $(9y^6t^2)^{1/2}$

2. $\left(\frac{u^6}{8}\right)^{2/3}$

3. $v^{1/3}(v^{5/3} - 2v^{2/3} + 5v^{-1/3})$

4. $\frac{72x^{-3/2}}{12x^{1/2}}$

5. $(3y + 2)^2$

6. $(3y^{1/2} + 2)^2$

7. $\frac{4z^2 - 25}{2z + 5}$

8. $\frac{4z - 25}{2\sqrt{z} + 5}$

In exercises **9** and **10**, solve each word problem.

9. The difference between 5 times an odd integer and 3 times the next consecutive odd integer is 16. Find both integers.

10. The base of a triangle is 1 meter less than 2 times the altitude to that base. If the area of the triangle is 60 square meters, find the lengths of the base and altitude.

SECTION 6-7. Supplementary Exercises

Solving Rational Equations That Are Quadratic in Form

Some of the expressions in the equations in examples **a** and **b** are rational.

a. $\frac{4}{t^4} = 225 - \frac{11}{t^2}$, where $t \neq 0$ — A rational equation in t

b. $\frac{5}{(n-1)^2} = \frac{13}{n-1} - 6$, where $n \neq 1$ — A rational equation in n

By making suitable substitutions for $\frac{1}{t^2}$ and $\frac{1}{n-1}$, a quadratic equation in another variable can be formed. To illustrate, in example **a**, replace $\frac{1}{t^2}$ by x and $\frac{1}{t^4}$ by x^2. In example **b**, replace $\frac{1}{n-1}$ by x and $\frac{1}{(n-1)^2}$ by x^2.

In exercises **1–12**, solve each equation.

1. $\frac{3}{u^2} - 4 + \frac{4}{u} = 0$

2. $\frac{2}{u^2} + \frac{5}{u} = 3$

3. $\frac{3}{v^2} + \frac{1}{v} = 4$

4. $\frac{6}{v^2} - \frac{1}{v} = 2$

5. $\frac{4}{p^4} + 36 = \frac{25}{p^2}$

6. $\frac{36}{p^4} - \frac{85}{p^2} = -9$

7. $\frac{2}{(w-1)^2} = 5 - \frac{3}{(w-1)}$

8. $\frac{7}{(w+2)^2} + \frac{4}{(w+2)} = 3$

9. $\frac{6}{(z+1)^2} = \frac{1}{(z+1)} + 1$

10. $1 = \frac{8}{(z-2)^2} + \frac{2}{(z-2)}$

11. $\frac{3}{(y-3)^2} - \frac{1}{(y-3)} = 2$

12. $\frac{42}{(y+4)^2} + 4 = \frac{31}{(y+4)}$

In exercises **13–20**, use an appropriate substitution to write a quadratic equation in x. Use the quadratic formula to solve for x. Replace x by the given variable and solve.

13. $r^4 - 4r^2 + 1 = 0$

14. $r^4 - 6r^2 + 7 = 0$

15. $4u^4 = 8u^2 + 1$

16. $6u^2 - 1 = 2u^4$

17. $v^4 + 7v^2 + 1 = 0$

18. $4v^2 - 11 = -4v^4$

19. $18 - 10z^2 + z^4 = 0$

20. $-14z^2 + 39 + z^4 = 0$

Name

Date

Score

SECTION 6-2. Summary Exercises

Answer

In exercises **1** and **2**, for each expression:

a. Find the numbers to add to form perfect square trinomials.

b. Write the resulting trinomials as squares of binomials.

1. $w^2 + 20w$

1. a. ______

b. ______

2. $w^2 + \frac{5}{3}w$

2. a. ______

b. ______

In exercises **3–8**, solve each equation by completing the square.

3. $m^2 + 2m - 4 = 0$

3. ______

4. $x^2 - 8x + 2 = 0$

4. ______

5. $y^2 = 3y + 5$

5. ____________________

6. $z^2 + 7z + 2 = 0$

6. ____________________

7. $2w^2 - 5w = 10$

7. ____________________

8. $x^2 + 22 = 4x$

8. ____________________

Name

Date

Score

SECTION 6-4. Summary Exercises

Answer

In exercises **1–8**, solve each equation.

1. $\sqrt{2y} = 36$

1. ______

2. $\sqrt{3x + 1} = 8$

2. ______

3. $\sqrt{p + 10} = \sqrt{3p}$

3. ______

4. $w - \sqrt{w - 1} = 3$

4. ______

5. $\sqrt{4t + 1} = -5$

5. ______________________

6. $3 + \sqrt{x^2 - 2x - 23} = x$

6. ______________________

7. $\sqrt[3]{z} = -8$

7. ______________________

8. $\sqrt[3]{q^3 - 4q^2 + 3q} = q - 1$

8. ______________________

Name

Date

Score

SECTION 6-7 Summary Exercises

Answer

In exercises **1–4**, solve each equation.

1. $s^4 - 85s^2 + 324 = 0$

1. ____________________

2. $p^6 + 7p^3 - 8 = 0$

2. ____________________

3. $(2x^2 - 7x)^2 = 12 + (2x^2 - 7x)$

3. ____________________

4. $x^{1/3} + 6 = x^{2/3}$

4. ____________________

CHAPTER 6. Review Exercises

In exercises **1** and **2**, determine whether each number is a solution of the given equation.

1. $2x^2 = x + 6$
  a. $\frac{2}{3}$
  b. 2

2. $x^2 + 8x + 11 = 0$
  a. $-4 + \sqrt{5}$
  b. $4\sqrt{5}$

In exercises **3–14**, solve each equation.

3. $(3u + 2)(2u - 3) = 0$

4. $7v + 3v^2 = 0$

5. $x^2 = 40 + 3x$

6. $13y + 6 + 2y^2 = 0$

7. $x^3 - 9x = 0$

8. $(x + 4)(x^2 - 5x + 6) = 0$

9. $p^2 = 225$

10. $w^2 = 18$

11. $2z^2 - 25 = 0$

12. $3t^2 = \frac{1}{4}$

13. $(2y - 1)^2 = 81$

14. $(5m + 2)^2 = 11$

In exercises **15** and **16**, solve each equation for x in terms of the other variables.

15. $2x^2 + xy - y^2 = 0$

16. $3x^2 + 3px = xm + pm$

In exercises **17** and **18**, solve each equation.

17. $v^2 + 24 = 0$

18. $(2p + 5)^2 = -4$

In exercises **19** and **20**, for each expression:

a. Find the number to add to form a perfect square trinomial.

b. Write the resulting trinomial as the square of a binomial.

19. $u^2 + 10u$

20. $w^2 + 7w$

In exercises **21** and **22**, solve each equation by completing the square.

21. $s^2 - 12s = 4$

22. $m^2 = m + 4$

In exercises **23–28**, solve each equation using the quadratic formula.

23. $7z + 3 = 6z^2$

24. $30 + 19w = 4w^2$

25. $m^2 + 10m + 1 = 0$

26. $2p(p - 1) = -5$

27. $2ix^2 + 3x + 5i = 0$

28. $5p^2 - 3ip = -1$

In exercises **29** and **30**, solve each equation using the quadratic formula and approximate the solutions to one decimal place.

29. $x^2 + 4x + 1 = 0$

30. $m^2 - 6m = 3$

In exercises **31** and **32**, for each equation:

a. Find the value of the discriminant.

b. Determine the nature of the roots.

31. $v^2 = \frac{3}{2} - \frac{5}{2}v$

32. $x^2 = 4(1 - x)$

In exercises **33–38**, solve each equation. If there are no solutions, write $\varnothing$.

33. $\sqrt{2z - 1} = 5$

34. $\sqrt{3y} - \sqrt{y + 10} = 0$

35. $w + 2 = \sqrt{7w + 14}$

36. $\sqrt{2x + 3} + 5 = 0$

37. $\sqrt[3]{k^3 - 8} = k - 2$

38. $\sqrt[4]{t + 2} - 3 = 0$

In exercises **39–46**, solve and graph each inequality.

39. $(x + 4)(x - 5) < 0$

40. $(2y - 3)(6y + 1) \geq 0$

41. $p^2 \geq 4$

42. $w^2 < 4w$

43. $x^2 + 4x + 2 \leq 0$

44. $q^2 > 6q + 11$

In exercises **45** and **46**, solve each equation for the indicated variable.

45. $V = \frac{1}{3}\pi r^2 h$, for r

46. $9x^2 - 16y^2 = 144$, for y

In exercises **47–49**, solve each word problem.

47. The length of a rectangle is 3 feet more than 2 times the width. If the width is increased by 1 foot and the length by 2 feet, the area of the rectangle is 90 square feet. Find the original length and width.

48. A tank has two outlet valves, labeled F and S. When both are opened, a full tank can be emptied in 12 hours. When S is used alone, it takes 10 hours longer than when F is used alone. How long does it take valve F to empty a full tank? How long does it take valve S?

49. A small motor boat travels 75 miles up a river against a current of 5 mph. It then makes the return trip with the current. If the trip up and back takes 8 hours, find the rate the boat can travel in still water.

CHAPTER

7 First-Degree Equations in Two Variables

The following equations were on the board when the students arrived for class:

1. $5F° - 9C° = 160°$
2. $C° = \frac{5}{9}(F° - 32°)$
3. $F° = \frac{5}{9}C° + 32°$

$F°$ stands for a temperature measured in Fahrenheit degrees.

$C°$ stands for the corresponding temperature measured in Celsius degrees.

"A bank down the street," Ms. Glaston began, "has an electronic sign that gives the current outside temperature. The sign shows the temperature in Fahrenheit degrees for about 30 seconds, and then shows the same temperature in Celsius degrees for about 30 seconds. The recorded temperatures are the same; therefore, the different numbers are the result of differences in the scales used by the two methods. Does anyone know what the standards are for these two temperature scales?"

"I believe the freezing and boiling points of water are used for both scales," Tyson Lubin said. "In the Fahrenheit scale, water freezes at 32° and boils at 212°. In the Celsius scale, water freezes at 0° and boils at 100°."

"Isn't that confusing," Sheila Kearney asked, "to have two different scales for measuring temperature? Furthermore, the two scales don't even have the same spacing between degrees. There are 180 divisions between the freezing and boiling points of water in the Fahrenheit scale, but only 100 divisions in the Celsius scale."

"Yes, it could cause some problems Sheila," Ms. Glaston replied, "but we need to keep in mind the relationship between the two scales as indicated by the formula. With these equations, we can convert a temperature in one scale to the corresponding temperature in the other." She then turned to the board and wrote:

To find $F°$, given $C°$, use $F° = \frac{9}{5}C° + 32°$.

If $C° = 0°$, then $F° = \frac{9}{5} \cdot 0° + 32° = 32°$.

If $C° = 100°$, then $F° = \frac{9}{5} \cdot 100° + 32° = 180° + 32° = 212°$.

To find C°, given F°, use $C^\circ = \frac{5}{9}(F^\circ - 32^\circ)$.

$$\text{If } F^\circ = 32^\circ, \text{ then } C^\circ = \frac{5}{9}(32^\circ - 32^\circ) = 0^\circ.$$

$$\text{If } F^\circ = 212^\circ, \text{ then } C^\circ = \frac{5}{9}(212^\circ - 32^\circ) = \frac{5}{9}(180^\circ) = 100^\circ.$$

"As you can see," Ms. Glaston continued, "I have used two forms of the formula to verify the comparable temperatures that Tyson gave us earlier. These are but two pairs of values from a virtually infinite number of possible pairs that could be listed. To make a record of these pairings in a compact way, in mathematics we use **ordered-pair notation.** For example, if we agree to list a Fahrenheit temperature first and the corresponding Celsius temperature second, then the ordered pairs for this temperature relationship would have the following form." She again turned to the board and wrote:

(F°, C°) As examples: $(32^\circ, 0^\circ)$ and $(212^\circ, 100^\circ)$.

"Ms. Glaston," Eric Hauben interrupted, "since there are so many ordered pairs possible, couldn't we use a graph of some kind to show them all?"

"I'm so glad you brought that up, Eric," Ms. Glaston replied. "I just happen to have such a graph." She then put an acetate on the projector to show the graph in Figure 7-1.

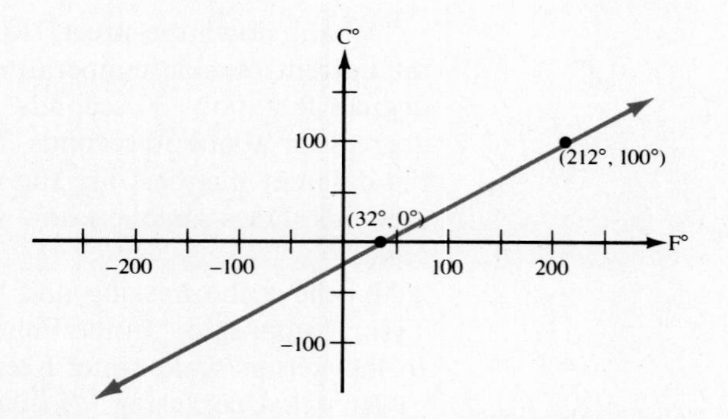

Figure 7-1. A graph of the Fahrenheit and Celsius relationship.

"In this brief discussion," Ms. Glaston continued, "we have looked at several concepts that we will be studying in detail in this chapter. First, we have used equations in two variables, not one variable. Second, I have introduced ordered-pair notation as a way of writing solutions to equations in two variables. Third, the graph on the screen is a geometric representation of all the solutions in the relationship stated algebraically by the equations. In other words, we have almost casually moved from the one-dimensional orientation of equations in one variable and one-number-line graphs, to a two-dimensional orientation of equations in two variables and a two-number-line scheme for graphing. At this time, we will back up and go through this transition at a much slower pace. However, this discussion should provide some indication of the direction of study we will be following in this chapter."

SECTION 7-1. The Rectangular Coordinate System, Distance Formula, and Midpoint Formula

1. The rectangular coordinate system
2. Plotting points in a rectangular coordinate system
3. The distance formula
4. The midpoint formula

In this chapter, we will begin a study of equations in two variables that continues for the next four chapters. As we will see, a solution of an equation in two variables is an ordered pair of numbers. To illustrate, (3, 7) is a solution of the equations in examples **a** and **b.**

a. $8x - 3y = 3$ **b.** $2x^2 + y^2 = 67$

$8(3) - 3(7) = 3$ ⟵ Replace x by 3. Replace y by 7. ⟶ $2(3)^2 + 7^2 = 67$

$24 - 21 = 3$, *true* $18 + 49 = 67$, *true*

A geometric model for graphing ordered pairs of numbers uses two number lines. One number line is used to graph the first number in an ordered pair, and the second number line is used to graph the second number. We will study the details of the two-axis system in this section.

The Rectangular Coordinate System

Many two-variable equations in mathematics are written in terms of x and y. A number replacement for x is typically listed first in an **ordered pair,** and the corresponding number replacement for y is listed second. To plot (or graph) such an ordered pair of numbers, two number lines (called axes) are drawn perpendicular to each other, as in Figure 7-2.

As shown, the **horizontal number line** is labeled the ***x*-axis.** The **vertical number line** is then labeled the ***y*-axis.** Notice the following features of this coordinate system:

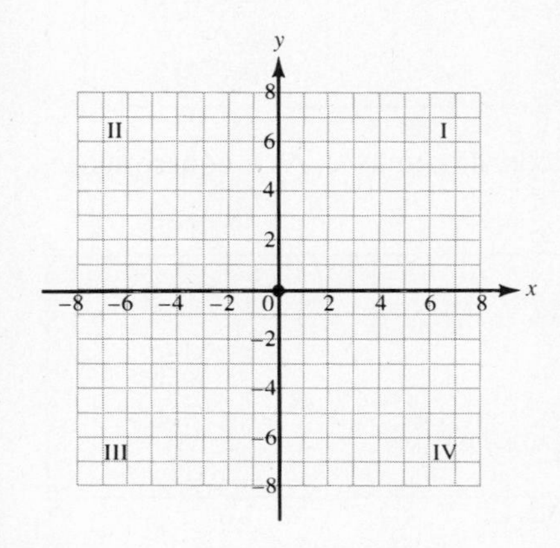

Figure 7-2. An xy-coordinate system.

Feature 1. The point labeled O is called the **origin** of the system. The ordered pair (0, 0) are the coordinates of O, written $O(0, 0)$.

Feature 2. The positive numbers on the x-axis are graphed to the right of O. The positive numbers on the y-axis are graphed above O.

Feature 3. The axes divide the plane on which they are drawn into four quarters, called **quadrants.**

Feature 4. The quadrants are labeled I, II, III, and IV beginning in the upper right quadrant and moving counterclockwise.

Plotting Points in a Rectangular Coordinate System

For every point in a plane on which xy-axes are drawn, such as in Figure 7-2, one and only one ordered pair of numbers can be assigned as **coordinates.** Furthermore, for every ordered pair of real numbers, there is exactly one point in the coordinate system that can be assigned as the **graph.** In mathematical terms, we say there is a **one-to-one correspondence** between the set of ordered pairs of real numbers and the set of points in the plane of an xy-coordinate system.

> If a and b are real numbers, then to plot (a, b) in an xy-coordinate system:
>
> **Step 1.** Locate the point with coordinate a on the x-axis. Imagine a vertical line is drawn through this point.
>
> **Step 2.** Locate the point with coordinate b on the y-axis. Imagine a horizontal line is drawn through this point.
>
> **Step 3.** Locate the point of intersection of the vertical line from Step 1 and the horizontal line from Step 2. The point of intersection is the graph of (a, b).

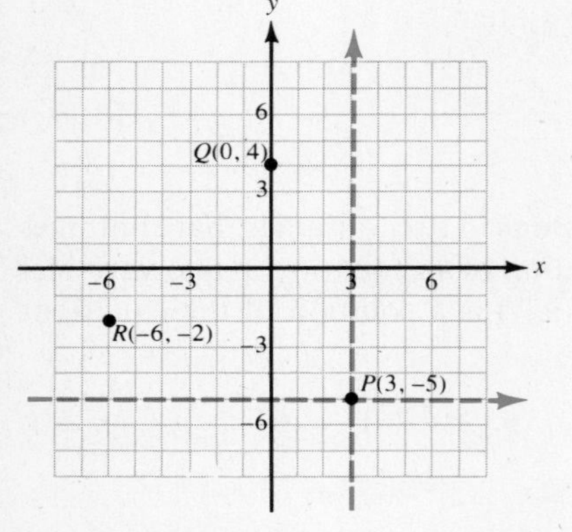

Figure 7-3. Graphs of P, Q, and R.

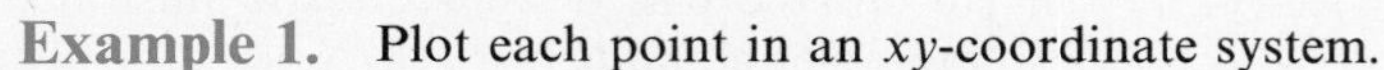

Example 1. Plot each point in an xy-coordinate system.

a. $P(3, -5)$ **b.** $Q(0, 4)$ **c.** $R(-6, -2)$

Solution. **a.** In Figure 7-3, a vertical line is drawn through 3 on the x-axis and a horizontal line is drawn through -5 on the y-axis. As shown, the lines intersect at $P(3, -5)$.

b. With $x = 0$ and $y = 4$, the point $Q(0, 4)$ is graphed in Figure 7-3.

c. With $x = -6$ and $y = -2$, the point $R(-6, -2)$ is graphed in Figure 7-3.

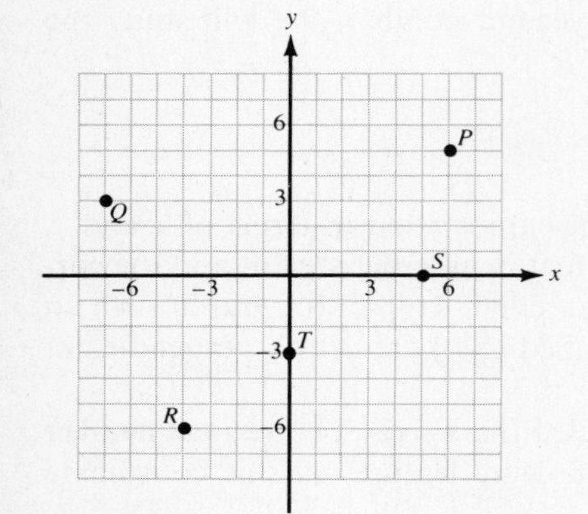

Figure 7-4. Graphs of points P through T.

Example 2. State the coordinates of P, Q, R, S, and T in Figure 7-4.

Solution.

Point	x-Coordinate	y-Coordinate	Ordered pair
P	6	5	(6, 5)
Q	-7	3	$(-7, 3)$
R	-4	-6	$(-4, -6)$
S	5	0	(5, 0)
T	0	-3	$(0, -3)$

In Example 2, points S and T lie one of the coordinate axes. As a consequence, one of the coordinates is 0.

> If a and b are real numbers, then:
>
> **1.** $P(a, 0)$ is a point on the x-axis.
>
> **2.** $Q(0, b)$ is a point on the y-axis.

The Distance Formula

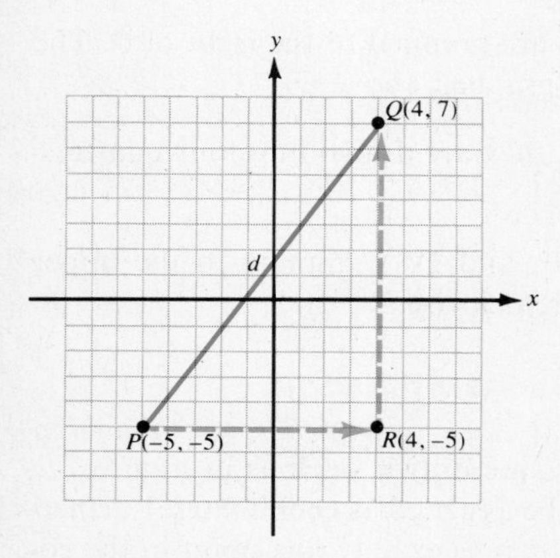

Figure 7-5. The distance between P and Q is d.

In Figure 7-5, points $P(-5, -5)$ and $Q(4, 7)$ are plotted. Suppose we want to calculate the distance between these points. The distance, labeled d, is the length of the line segment with endpoints P and Q. To compute d, we use the following procedure:

Step 1. Along a horizontal line through P, count the number of units to R, where R is on a vertical line through Q. As shown, the coordinates of R are

(4, −5). The distance is the difference in the x-coordinates of P and R:

$4-(-5)=9$ Subtract the x-coordinates.

Step 2. Along the vertical line through Q and R, count the number of units from R to Q. The distance is the difference in the y-coordinates of Q and R:

$7-(-5)=12$ Subtract the y-coordinates.

Step 3. Triangle PQR is a right triangle with the right angle at R. By the Pythagorean Theorem, the sum of the squares of the lengths of segments PR and QR is equal to the square of d.

$d^2 = 9^2 + 12^2$ $(\text{hypotenuse})^2 = (\text{leg})^2 + (\text{leg})^2$

$d^2 = 81 + 144$

$d^2 = 225$

$d = \sqrt{225} = 15$

The distance between P and Q is 15. The general formula based on the example above is known as the **distance formula.**

The distance formula

If $P(x_1, y_1)$ and $Q(x_2, y_2)$ are two points in an xy-coordinate system, then the distance d between P and Q is:

$$d = \sqrt{(x_2 - x_1)^2 + (y_2 - y_1)^2}$$

Example 3. Find the distance between $P(-4, 2)$ and $Q(2, 8)$.

Solution. With $(x_1, y_1) = (-4, 2)$ and $(x_2, y_2) = (2, 8)$:

$d = \sqrt{(2-(-4))^2 + (8-2)^2}$ $d = \sqrt{(x_2 - x_1)^2 + (y_2 - y_1)^2}$

$= \sqrt{36 + 36}$ Square the differences.

$= \sqrt{72}$ Sum the squares.

$= 6\sqrt{2}$ $\sqrt{72} = \sqrt{36 \cdot 2} = 6\sqrt{2}$

Thus, the distance between P and Q is $6\sqrt{2}$ units.

Example 4. If the distance between $P(2, -1)$ and $Q(x, 3)$ is 5 units, find the possible values of x.

Solution. **Discussion.** Since d is given, the distance formula can be used to find x. There are two possible locations of Q that are 5 units from P (see Figure 7-6).

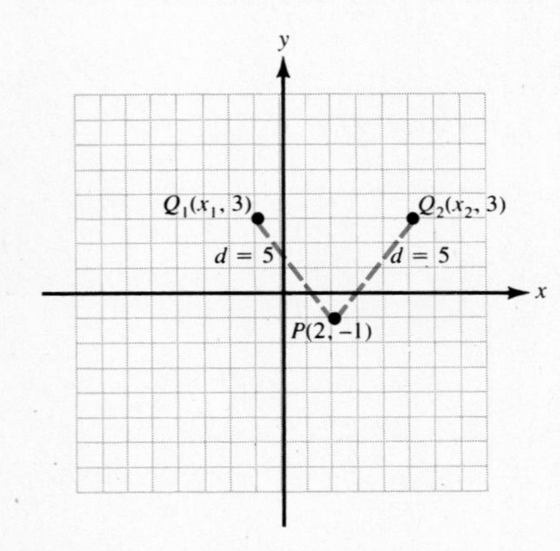

Figure 7-6. Q_1 and Q_2 are both 5 units from P.

$5 = \sqrt{(x-2)^2 + (3-(-1))^2}$

$25 = x^2 - 4x + 4 + 16$

$0 = x^2 - 4x - 5$

$0 = (x-5)(x+1)$

$x - 5 = 0$ or $x + 1 = 0$

$x = 5$ or $x = -1$

Thus, $Q_1(-1, 3)$ and $Q_2(5, 3)$ are two points 5 units from $P(2, -1)$.

The Midpoint Formula

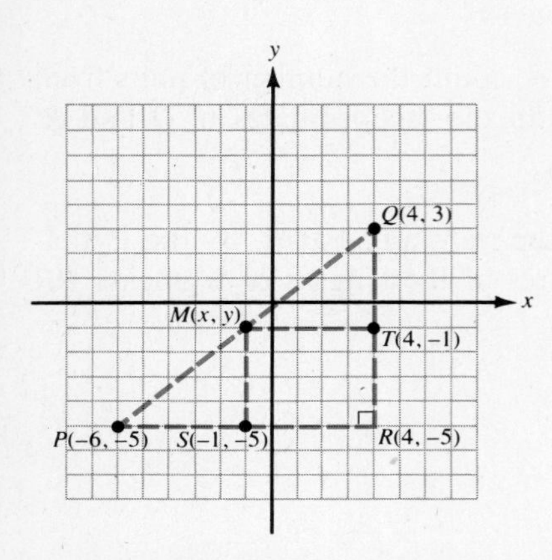

Figure 7-7. M is the midpoint of line segment PQ.

In geometry, we learn that the **midpoint** of a line segment is a point of the segment that is the same distance from the ends of the segment. In Figure 7-7, points $P(-6, -5)$ and $Q(4, 3)$ are plotted. Suppose we want the coordinates of M, the midpoint of line segment PQ. To find (x, y), the coordinates of M, we use the following procedure:

Step 1. Draw a horizontal line through P and a vertical line through Q. These lines intersect at $R(4, -5)$ as shown.

Step 2. The midpoint S of the segment with endpoints P and R has coordinates $(-1, -5)$:

$$x\text{-coordinate of } S = \frac{4 + (-6)}{2} = \frac{-2}{2} = -1$$

Step 3. The midpoint T of the segment with endpoints Q and R has coordinates $(4, -1)$:

$$y\text{-coordinate of } T = \frac{3 + (-5)}{2} = \frac{-2}{2} = -1$$

Step 4. A vertical line through S and a horizontal line through T intersect the line segment with endpoints P and Q at $M(x, y)$. The x-coordinate of M is the x-coordinate of S, and the y-coordinate of M is the y-coordinate of T. Thus, for point M:

$$(x, y) = (-1, -1)$$

The general formula based on the above example is known as the **midpoint formula.**

> **The midpoint formula**
>
> If $P(x_1, y_1)$ and $Q(x_2, y_2)$ are points in an xy-coordinate system, then the midpoint of the line segment with endpoints P and Q is $M(x, y)$, where:
>
> $$x = \frac{x_1 + x_2}{2} \quad \text{and} \quad y = \frac{y_1 + y_2}{2}$$

Example 5. Find the midpoint of the line segment with endpoints $P(-3, 8)$ and $Q(6, -1)$.

Solution. With $(x_1, y_1) = (-3, 8)$ and $(x_2, y_2) = (6, -1)$:

$$x\text{-coordinate of } M = \frac{-3 + 6}{2} \qquad x = \frac{x_1 + x_2}{2}$$

$$= \frac{3}{2}$$

$$y\text{-coordinate of } M = \frac{8 + (-1)}{2} \qquad y = \frac{y_1 + y_2}{2}$$

$$= \frac{7}{2}$$

Thus, $M\left(\frac{3}{2}, \frac{7}{2}\right)$ is the midpoint of the line segment with endpoints $P(-3, 8)$ and $Q(6, -1)$.

Example 6. If $M(7, 3)$ is the midpoint of the line segment with endpoints $P(x_1, 9)$ and $Q(13, y_2)$, find x_1 and y_2.

Solution. With $(x_1, y_1) = (x_1, 9)$, $(x_2, y_2) = (13, y_2)$, and $(7, 3)$ the midpoint:

$$7 = \frac{x_1 + 13}{2} \qquad x = \frac{x_1 + x_2}{2}$$

$$14 = x_1 + 13 \qquad \text{Multiply both sides by 2.}$$

$$1 = x_1 \qquad P(1, 9) \text{ is one endpoint.}$$

$$3 = \frac{9 + y_2}{2} \qquad y = \frac{y_1 + y_2}{2}$$

$$6 = 9 + y_2 \qquad \text{Multiply both sides by 2.}$$

$$-3 = y_2 \qquad Q(13, -3) \text{ is the other endpoint.}$$

SECTION 7-1. Practice Exercises

In exercises **1–8**, plot each set of points in an *xy*-coordinate system.

[Example 1]

1. $A(-4, 4)$, $B(0, 8)$, $C(-2), 6)$ **2.** $D(1, -1)$, $E(-6, -2)$, $F(8, 0)$

3. $G(-4, -3)$, $H(-1, 0)$, $I(2, 3)$ **4.** $J(5, 3)$, $K(0, -2)$, $L(-1, -3)$

5. $M(5, 2)$, $N(-1, 2)$, $O(-3, 2)$ **6.** $P(4, -3)$, $Q(2, -3)$, $R(-1, -3)$

7. $S(-4, -2)$, $T(-4, 0)$, $U(-4, 5)$ **8.** $V(1, -7)$, $W(1, -2)$, $X(1, 0)$

In exercises **9–18**, state the coordinates of each point plotted in Figure 7-8.

[Example 2]

9. *A* **10.** *B* **11.** *C* **12.** *D* **13.** *E*

14. *F* **15.** *G* **16.** *H* **17.** *I* **18.** *J*

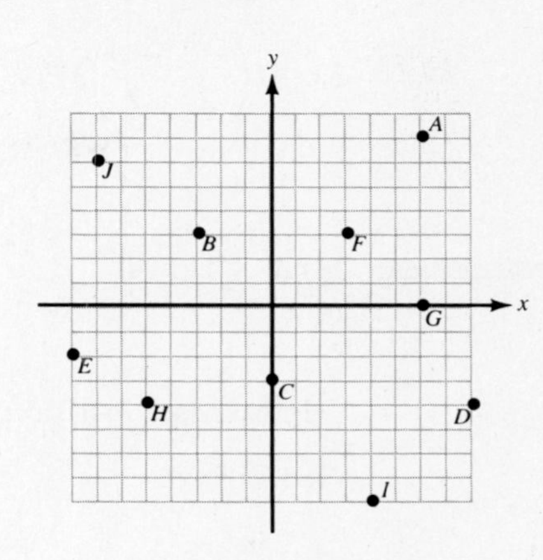

Figure 7-8.

In exercises **19–28**, find the distance between P and Q.

[Example 3]

19. $P(2, 1)$, $Q(5, 5)$

20. $P(2, 3)$, $Q(6, 6)$

21. $P(-2, 6)$, $Q(4, -2)$

22. $P(-6, -2)$, $Q(2, 4)$

23. $P(8, 3)$, $Q(-1, 3)$

24. $P(-5, 6)$, $Q(-5, -1)$

25. $P(6, 4)$, $Q(-1, 5)$

26. $P(5, 1)$, $Q(1, -1)$

27. $P(1, -7)$, $Q(-5, -4)$

28. $P(-9, 10)$, $Q(-7, 4)$

In exercises **29–36**, find the possible value(s) of each unknown coordinate for the given distance d between the points.

[Example 4]

29. $P(-1, 1)$, $Q(x, -3)$; $d = 5$

30. $P(1, -2)$, $Q(x, 2)$; $d = 5$

31. $P(-3, 2)$, $Q(5, y)$; $d = 10$

32. $P(-6, -3)$, $Q(2, y)$; $d = 10$

33. $P(6, y)$, $Q(6, -1)$; $d = 4$

34. $P(x, -3)$, $Q(-2, -3)$; $d = 2$

35. $P(6, 3)$, $Q(-1, y)$; $d = 5\sqrt{2}$

36. $P(-5, -2)$, $Q(1, y)$; $d = 3\sqrt{5}$

In exercises **37–46**, find the midpoints of each line segment with the given endpoints.

[Example 5]

37. $P(-8, 1)$, $Q(-2, -3)$

38. $P(-2, -9)$, $Q(-6, -1)$

39. $A(5, -8)$, $B(-9, 0)$

40. $A(-6, -4)$, $B(0, 6)$

41. $M(4, -2)$, $N(8, -2)$

42. $M(10, 2)$, $N(10, -4)$

43. $S(14, -17)$, $T(-17, 15)$

44. $S(18, 11)$, $T(-2, -10)$

45. $U(-12, 5)$, $V(15, -3)$

46. $U(9, -12)$, $V(-12, -10)$

In exercises **47–54**, find the values of x and y given the midpoint M and endpoints P and Q.

[Example 6]

47. $M(3, -1)$, $P(9, y)$, $Q(x, -8)$

48. $M(-5, -1)$, $P(-3, y)$, $Q(x, -7)$

49. $M(-2, 3)$, $P(x, 4)$, $Q(0, y)$

50. $M(5, 1)$, $P(x, 0)$, $Q(2, y)$

51. $M(-11, -1)$, $P(-3, y)$, $Q(x, -12)$

52. $M(2, -18)$, $P(10, y)$, $Q(x, -19)$

53. $M\left(\frac{-3}{2}, \frac{-9}{2}\right)$, $P(x, -5)$, $Q(-6, y)$

54. $M\left(\frac{7}{2}, \frac{-9}{2}\right)$, $P(x, -9)$, $Q(1, y)$

SECTION 7-1. Ten Review Exercises

In exercises **1–8**, do the indicated operations.

1. $(7t^2)^2$

2. $(7 + t^2)^2$

3. $\dfrac{25u^4}{5u^2}$

4. $\dfrac{25 - u^4}{5 - u^2}$

5. $(2pq)(10pq)$

6. $(2p + q)(10p + q)$

7. $\dfrac{(2m^2n)(3mn^3)}{6mn}$

8. $\dfrac{2m^2n + 3mn^3}{6mn}$

In exercises **9** and **10**, solve each equation.

9. $17 - 3(2z - 5) = 2$

10. $17 - 3\,|2z - 5| = 2$

SECTION 7-1. Supplementary Exercises

In exercises **1–4**, P, Q, R, and S are vertices of parallelograms.

a. Plot the points in an xy-coordinate system.

b. Find the perimeter of the figure.

c. Find the length of each diagonal.

d. Find the midpoint of each diagonal.

1. $P(-4, -2)$, $Q(-1, 2)$, $R(5, 2)$, $S(2, -2)$

2. $P(1, 1)$, $Q(-2, 5)$, $R(4, 5)$, $S(7, 1)$

3. $P(-5, -2)$, $Q(-1, 3)$, $R(7, 5)$, $S(3, 0)$

4. $P(-8, 1)$, $Q(-4, 6)$, $R(4, 2)$, $S(0, -3)$

In exercises **5–10**, determine whether P, Q, and R are **colinear** (lie on the same line). If they are colinear, then:

$$d(P, Q) \quad + \quad d(Q, R) \quad = d(P, R)$$

(distance from P to Q) + (distance from Q to R) = (distance from P to R).

5. $P(-6, 4)$, $Q(-3, 0)$, $R(3, -8)$

6. $P(-2, -4)$, $Q(2, -1)$, $R(10, 5)$

7. $P(-5, -2)$, $Q(-1, 2)$, $R(3, 4)$

8. $P(-6, 1)$, $Q(-1, 2)$, $R(6, -8)$

9. $P(4, 3)$, $Q(2, 3)$, $R(-6, 3)$

10. $P(-2, -5)$, $Q(-2, 1)$, $R(-2, 3)$

In exercises **11–14**, points P, Q, and R are vertices of a triangle.

a. Plot the points in an xy-system.

b. Find the lengths of the three sides.

c. Use the Pythagorean Theorem ($a^2 + b^2 = c^2$) and the lengths obtained in part **b** to determine whether the triangle is a right triangle.

d. Determine the midpoints of the three sides.

e. Determine whether the triangle whose vertices are the midpoints of part **d** is a right triangle.

11. $P(-3, 2)$, $Q(3, 2)$, $R(-3, 6)$

12. $P(0, 6)$, $Q(6, 6)$, $R(6, -2)$

13. $P(4, 0)$, $Q(-2, -6)$, $R(0, 4)$

14. $P(-3, 2)$, $Q(1, 6)$, $R(3, -4)$

SECTION 7-2. Linear Equations in Two Variables

1. A definition of linear equations in x and y
2. Solutions of linear equations in x and y
3. Graphs of linear equations in x and y
4. Graphs of horizontal and vertical lines
5. Slopes of lines
6. Graphing lines using a point and the slope

In everyday life, we frequently encounter pairs of variable quantities that are in some way *related to each other*. By related, we mean that when a specific value of one of the quantities is known, we can then determine the corresponding value of the second quantity. The details of a relationship can frequently be given by a **formula,** and it is often possible to write the formula as a mathematical equation. The equations used by Ms. Glaston to introduce this chapter are examples of such formulas. In this section, we will begin our study of pairs of quantities that have a so-called "linear relationship".

A Definition of Linear Equations in x and y

The simplest relationship that can exist between two quantities x and y is one that can be described by an equation in which both variables have degree 1. Such an equation is called **linear in x and y.**

Definition 7.1. A linear equation in x and y
If a, b and c are real numbers and a and b are not both 0, then a linear equation in x and y is one that can be written in the form:

$$(1) \quad ax + by = c$$

Equation (1) is called the **standard form** of a linear equation.

In the introduction to this chapter, Ms. Glaston used the equation that shows the relationship between temperatures measured in Fahrenheit and Celsius scales. Notice that the equation qualifies as linear in the variables $F°$ and $C°$:

$$5F° - 9C° = 160$$

a has the value 5
b has the value -9
c has the value 160

Examples **a–d** are equations that are *not* linear in x and y.

a. $x^2 + 2y = 5$ — The exponent on x is 2, not 1.

b. $xy = 3$ — x and y are both factors of the same term

c. $\sqrt{x} + 3\sqrt{y} = 10$ — x and y are under radical signs

d. $\frac{4}{x} + \frac{9}{y} = 1$ — x and y are denominators of terms

Solutions of Linear Equations in x and y

A linear equation in x and y becomes either true or false when the variables are replaced by numbers. Since the equation has two variables, two numbers are needed. To indicate which number is to be used to replace which variable, we use **ordered-pair notation.**

> **Definition 7.2. A solution of a linear equation in x and y**
> If d and e are real numbers, then the ordered pair (d, e) is a solution of a linear equation in x and y provided the equation is true when x is replaced by d and y is replaced by e.

Example 1. Determine whether the given ordered pairs are solutions of $2x + 3y = -4$.

a. $(-5, 2)$ **b.** $(8, -7)$

Solution.

a. $2(-5) + 3(2) = -4$ Replace x by -5 and y by 2.

$-10 + 6 = -4$ Multiply first.

$-4 = -4$, *true* Thus, $(-5, 2)$ is a solution.

b. $2(8) + 3(-7) = -4$ Replace x by 8 and y by -7.

$16 + (-21) = -4$ Multiply first.

$-5 = -4$, *false* Thus, $(8, -7)$ is not a solution.

If a value for x or y is given, then the corresponding value of the ordered-pair solution can be found.

Example 2. Find the solutions of $2x + 3y = -4$ for:

a. $x = 7$ **b.** $y = -3$

Solution.

a. $2(7) + 3y = -4$ Replace x by 7.

$14 + 3y = -4$ Solve for y.

$y = -6$ Thus, $(7, -6)$ is a solution.

b. $2x + 3(-3) = -4$ Replace y by -3.

$2x + (-9) = -4$ Solve for x.

$x = \frac{5}{2}$ Thus, $\left(\frac{5}{2}, -3\right)$ is a solution.

Graphs of Linear Equations in x and y

Examples **e** and **f** are linear equations in x and y.

e. $x + 2y = 10$ **f.** $3x - y = 6$

The number of ordered pairs that are solutions of these equations is infinite. A geometric representation of all the solutions can be obtained using the rectangular coordinate system studied in Section 7-1.

For any linear equation in x and y, a graph of the solutions lies on a straight line. For a given equation, a line that has points whose coordinates are solutions of the equation is called a **graph of the equation.** The instruction "graph the equation $ax + by = c$" means to locate such a line.

Example 3. Graph $x + 2y = 10$.

Solution. **Discussion.** Through two different points, one and only one line can be drawn. However, we will find three solutions and use the third point to check our work.

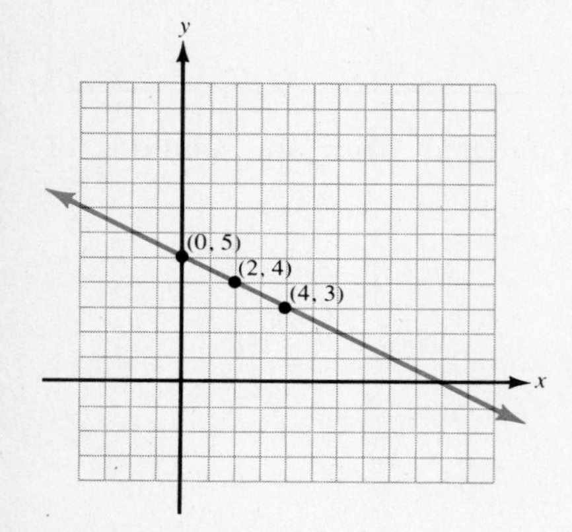

Figure 7-9. A graph of $x + 2y = 10$.

Step 1. Pick any three numbers for x (or y), then find the corresponding values for the other variable.

Pick $x = 0$: $0 + 2y = 10$

$y = 5$ (0, 5) is a solution

Pick $x = 2$: $2 + 2y = 10$

$y = 4$ (2, 4) is a solution

Pick $x = 4$: $4 + 2y = 10$

$y = 3$ (4, 3) is a solution

Step 2. Plot the points with coordinates that are the solutions obtained in Step 1.
In Figure 7-9, the three ordered pairs are graphed.

Step 3. Draw a straight line through the three points.
If the points do not lie on a line, then a mistake has been made and the work of Steps 1 and 2 must be checked.

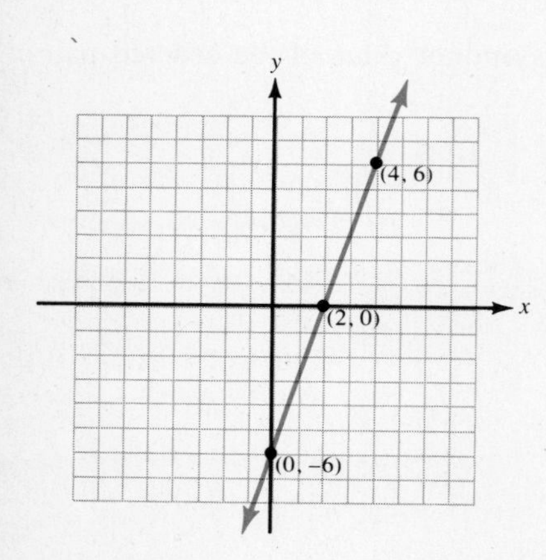

Figure 7-10. A graph of $3x - y = 6$.

Example 4. Graph $3x - y = 6$.

Solution. **Step 1.** Pick $x = 0$: $3(0) - y = 6$

$y = -6$ (0, −6) is a solution

Pick $x = 2$: $3(2) - y = 6$

$y = 0$ (2, 0) is a solution

Pick $x = 4$: $3(4) - y = 6$

$y = 6$ (4, 6) is a solution

Step 2. In Figure 7-10, the points with coordinates $(0, -6)$, $(2, 0)$, and $(4, 6)$ are plotted.

Step 3. The line through these points is a graph of the given equation.

Graphs of Horizontal and Vertical Lines

In Definition 7.1, the statement is made that in the equation

$$ax + by = c \qquad \text{A linear equation in } x \text{ and } y$$

a and b are not both 0. If a is 0, then the equation has a **horizontal line** for a graph. (A horizontal line is parallel to the x-axis.) If b is 0, then the equation has a **vertical line** for a graph. (A vertical line is parallel to the y-axis.)

Figure 7-11. A graph of $y = 3$.

Example 5. Graph $y = 3$.

Solution. **Discussion.** The given equation can be written as:

$0 \cdot x + y = 3$ The form $ax + by = c$, with $a = 0$

If $x = -4$, then $0(-4) + y = 3$. $(-4, 3)$ is a solution

If $x = 0$, then $0(0) + y = 3$. $(0, 3)$ is a solution

If $x = 5$, then $0(5) + y = 3$. $(5, 3)$ is a solution

The solutions are graphed in Figure 7-11. The horizontal line through the points is a graph of $y = 3$.

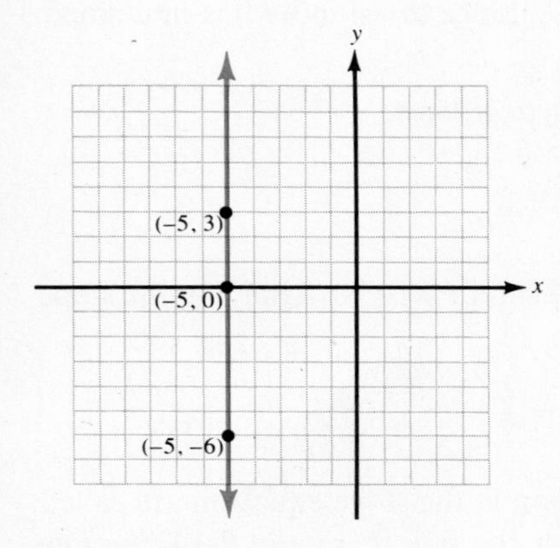

Figure 7-12. A graph of $x = -5$.

Example 6. Graph $x = -5$.

Solution. **Discussion.** The given equation can be written as:

$x + 0 \cdot y = -5$ The $ax + by = c$ form, with $b = 0$

If $y = -6$, then $x + 0(-6) = -5$. $(-5, -6)$ is a solution

If $y = 0$, then $x + 0(0) = -5$. $(-5, 0)$ is a solution

If $y = 3$, then $x + 0(3) = -5$. $(-5, 3)$ is a solution

The solutions are graphed in Figure 7-12. The vertical line through the points is a graph of $x = -5$.

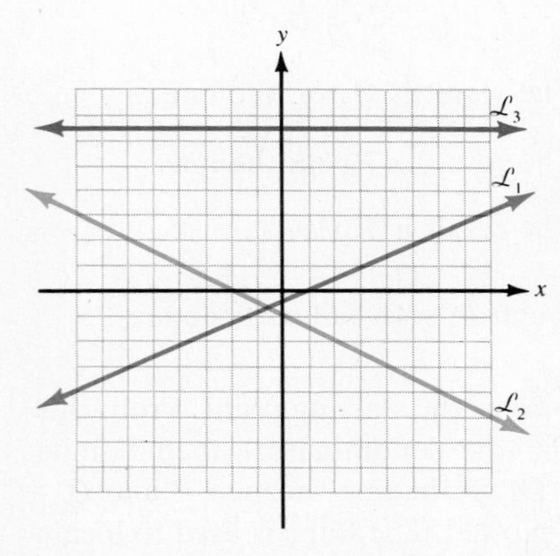

Figure 7-13. Three lines $\mathscr{L}_1$, $\mathscr{L}_2$ and $\mathscr{L}_3$.

Slopes of Lines

In Figure 7-13, three lines are drawn. To each nonvertical line, a real number can be given that describes the **slope** of the line. Intuitively, the slope of a line is the amount of inclination of the line relative to the positively directed x-axis. For the three lines in Figure 7-13:

$\mathscr{L}_1$ is inclined *upward* to the right.

$\mathscr{L}_2$ is inclined *downward* to the right.

$\mathscr{L}_3$ is neither inclined upward nor downward, but is *horizontal.*

The slope of a line can be determined when the coordinates of two points on the line are known.

Definition 7.3. The slope of a line
If a line $\mathscr{L}$ passes through $P_1(x_1, y_1)$ and $P_2(x_2, y_2)$, and $x_1 \neq x_2$, then the slope of $\mathscr{L}$ is m, and:

$$(2) \quad m = \frac{y_2 - y_1}{x_2 - x_1} \quad \text{or} \quad m = \frac{y_1 - y_2}{x_1 - x_2}$$

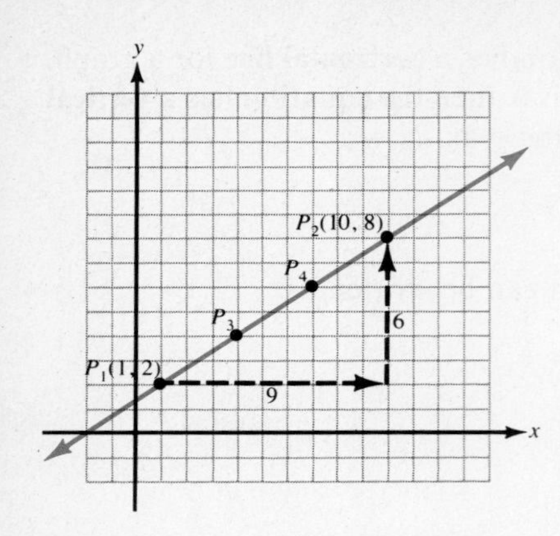

Figure 7-14. A line through $P_1(1, 2)$ and $P_2(10, 8)$ with slope $\frac{2}{3}$.

Example 7. Find the slope of the line that passes through $P_1(-5, 7)$ and $P_2(9, -1)$.

Solution. With $(x_1, y_1) = (-5, 7)$ and $(x_2, y_2) = (9, -1)$:

$$m = \frac{-1-7}{9-(-5)} \qquad \text{Using equation (2), } m = \frac{y_2 - y_1}{x_2 - x_1}.$$

$$= \frac{-8}{14} \qquad \text{Simplify top and bottom.}$$

$$= \frac{-4}{7} \qquad \text{Reduce the fraction.}$$

In the definition of slope, the restriction $x_1 \neq x_2$ is given. If $x_1 = x_2$, then P_1 and P_2 are on a vertical line. If equation (2) is used for points on a vertical line, then the denominator of the expression is 0. Since division by 0 is undefined, we say:

"Slope is undefined for vertical lines."

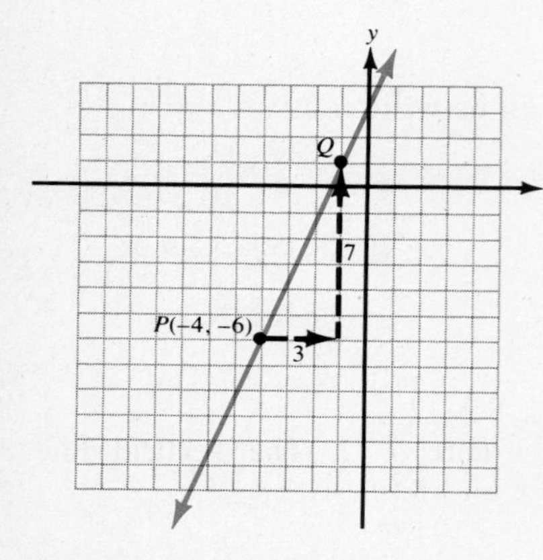

Figure 7-15. A line through $P(-4, -6)$ with slope $\frac{7}{3}$.

Graphing Lines Using a Point and the Slope

In Figure 7-14, a line is drawn through $P_1(1, 2)$ and $P_2(10, 8)$. Applying the slope equation:

$$m = \frac{8-2}{10-1} = \frac{6}{9} = \frac{2}{3}$$

The numerator and denominator of the expression in the slope equation are called the **rise** and **run,** respectively, of the line. Using the line in Figure 7-14, *the line rises 2 units for every 3-unit run.* To illustrate:

Starting at P_1:

1. Move 3 units to the right and 2 units up. The result is P_3 on the line.

2. Move 3 units to the right and 2 units up. The result is P_4 on the line.

3. Move 3 units to the right and 2 units up. The result is $P_2(10, 8)$.

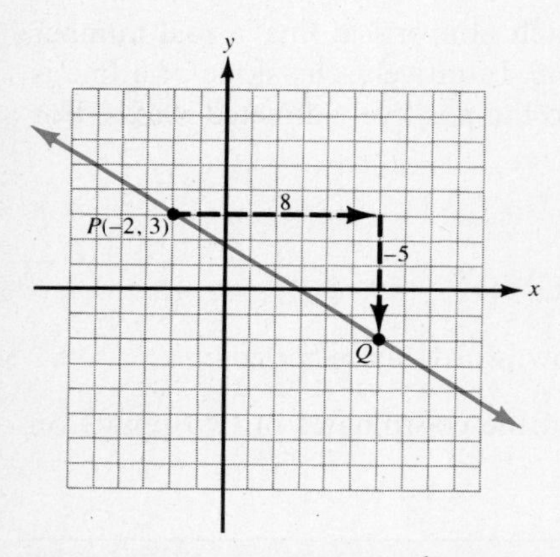

Figure 7-16. A line through $P(-2, 3)$ with slope $\frac{-5}{8}$.

Example 8. Graph the line that passes through $P(-4, -6)$ with slope $\frac{7}{3}$.

Solution. **Discussion.** Locate the given point as the starting point. From this point, move 3 units to the right (the denominator), then 7 units up (the numerator) to point Q. Draw the line through P and Q.

In Figure 7-15, $P(-4, -6)$ is plotted. The slope is used to locate Q. The line through these points is a graph of the line.

If the slope is negative, then the "rise" is down. However, the run is still to the right.

Example 9. Graph the line that passes through $P(-2, 3)$ with slope $\frac{-5}{8}$.

Solution. In Figure 7-16, locate $P(-2, 3)$. From this point, move 8 units to the right (the run), then 5 units down (the rise) to locate Q. The line through P and Q is a graph of the line.

SECTION 7-2. Practice Exercises

In exercises **1–10**, determine whether the given ordered pairs are solutions of each equation.

[Example 1]

1. $5x + y = 7$
a. $(1, 2)$
b. $(2, -2)$
c. $(-3, 22)$

2. $2x + 7y = 10$
a. $(5, 0)$
b. $(1, 1)$
c. $(-2, 2)$

3. $x - 4y = 9$
a. $(-3, -3)$
b. $(9, 0)$
c. $(1, 2)$

4. $4x - y = 12$
a. $(2, -4)$
b. $(0, -12)$
c. $(-3, 0)$

5. $3u + 4v = 20$
a. $(5, 1)$
b. $(4, 2)$
c. $(-4, 8)$

6. $4u + 7v = 15$
a. $(5, -1)$
b. $(-5, 5)$
c. $(2, 1)$

7. $\frac{1}{2}u - \frac{1}{3}v = 4$
a. $(8, 0)$
b. $(0, -12)$
c. $(2, -9)$

8. $\frac{3}{4}u - \frac{1}{5}v = 2$
a. $(0, -10)$
b. $(-4, -25)$
c. $(8, 20)$

9. $\frac{2}{3}m + \frac{3}{2}n = 6$
a. $(12, 0)$
b. $(0, 4)$
c. $(27, -8)$

10. $\frac{4}{5}m - \frac{1}{2}n = 10$
a. $(15, 0)$
b. $(0, -20)$
c. $(15, 4)$

In exercises **11–20**, find the value of the missing component so that each ordered pair is a solution of the given equation.

[Example 2]

11. $x - 4y = 10$
a. $(2, \square)$
b. $(\square, -1)$

12. $3x + 5y = 9$
a. $(-2, \square)$
b. $(\square, 0)$

13. $4x + 7y = -8$
a. $(\square, -4)$
b. $(-2, \square)$

14. $10x - 3y = -5$
a. $(\square, 5)$
b. $\left(\frac{-1}{2}, \square\right)$

15. $2m - 3n = 1$
a. $(-4, \square)$
b. $(\square, 3)$

16. $3m + 10n = 7$
a. $(-1, \square)$
b. $(\square, -2)$

17. $\frac{-1}{2}u + \frac{1}{3}v = 12$
a. $(\square, 18)$
b. $(-30, \square)$

18. $\frac{-4}{3}u + \frac{3}{4}v = 9$
a. $(\square, 28)$
b. $(-9, \square)$

19. $\frac{3}{10}r - \frac{2}{5}s = -15$
a. (30, □)
b. (□, −15)

20. $\frac{-7}{10}r + \frac{3}{5}s = -10$
a. (10, □)
b. (□, 30)

In exercises **21–50**, graph each equation.

[Examples 3 and 4]

21. $x + 2y = 8$

22. $3x + y = 6$

23. $2x - y = 6$

24. $x - 4y = 8$

25. $2x - 3y = 3$

26. $5x - 2y = 12$

27. $3x + 2y = 8$

28. $2x + 5y = 25$

29. $x + y = 5$

30. $x + y = -2$

31. $x - y = -4$

32. $x - y = 6$

33. $5x - 2y = -5$

34. $7x - 2y = 3$

35. $3x + 4y = -10$

36. $3x + 4y = 14$

37. $x - y = 0$

38. $x + y = 0$

[Examples 5 and 6]

39. $y = 6$

40. $y = 4$

41. $x = 5$

42. $x = 7$

43. $y = -2$

44. $y = -6$

45. $x = \frac{-3}{2}$

46. $x = \frac{-5}{2}$

47. $x + 1 = 0$

48. $y - 3 = 0$

49. $3y - 12 = 0$

50. $2x + 10 = 0$

In exercises **51–66**, find the slope of the line that passes through each pair of points. For any vertical line, write "undefined".

[Example 7]

51. $P_1(4, 5)$, $P_2(8, 12)$

52. $P_1(2, 3)$, $P_2(8, 5)$

53. $P_1(-2, 7)$, $P_2(3, 4)$

54. $P_1(-10, 5)$, $P_2(-3, 3)$

55. $P_1(6, -1)$, $P_2(7, 4)$

56. $P_1(0, -4)$, $P_2(1, 5)$

57. $P_1(-4, -3)$, $P_2(-3, -5)$

58. $P_1(-8, -3)$, $P_2(-7, -13)$

59. $P_1(9, -6)$, $P_2(1, -9)$

60. $P_1(-6, 9)$, $P_2(4, 16)$

61. $P_1(-7, 5)$, $P_2(7, 5)$

62. $P_1(-2, 2)$, $P_2(8, 2)$

63. $P_1(-7, 5)$, $P_2(-7, -5)$

64. $P_1(-2, 2)$, $P_2(-2, -2)$

65. $P_1\left(8, \frac{4}{3}\right)$, $P_2\left(9, \frac{2}{3}\right)$

66. $P_1\left(-3, \frac{9}{2}\right)$, $P_2(-2, 2)$

In exercises **67–80**, graph the line that passes through each point with the given slope.

[Examples 8 and 9]

67. $P(3, 1)$; slope $= \frac{1}{2}$

68. $P(1, 4)$; slope $= \frac{1}{3}$

69. $P(-4, 0)$; slope $= \frac{-2}{3}$

70. $P(2, 0)$; slope $= \frac{-3}{2}$

71. $P(5, -2)$; slope $= 1$

72. $P(-4, 3)$; slope $= 1$

73. $P(-2, -3)$; slope $= -2$

74. $P(-1, -4)$; slope $= -3$

75. $P(3, -5)$; slope $= 0$

76. $P(6, -2)$; slope $= 0$

77. $P(1, -4)$; slope $= \frac{5}{3}$

78. $P(-2, 1)$; slope $= \frac{-5}{2}$

79. $P(0, 5)$; slope $= \frac{-7}{2}$

80. $P(0, -6)$; slope $= \frac{9}{5}$

SECTION 7-2. Ten Review Exercises

1. Evaluate $2(3t - 4) - 3(t - 1) - 6 + 5(2t - 3)$ for $t = 2$.

2. Simplify $2(3t - 4) - 3(t - 1) - 6 + 5(2t - 3)$.

3. Solve $2(3t - 4) - 3(t - 1) = 6 - 5(2t - 3)$.

4. Solve and graph $6 - 5(2t - 3) > 2(3t - 4) - 3(t - 1)$.

In exercises **5** and **6**, simplify each expression.

5. $\sqrt{128t^6}$; $t > 0$

6. $\sqrt[3]{128t^6}$

In exercises **7** and **8**, rationalize the denominator of each expression.

7. $\frac{1}{\sqrt{12}}$

8. $\frac{1}{\sqrt[3]{24}}$

In exercises **9** and **10**, simplify each expression.

9. $2\sqrt{45} - \sqrt{125}$

10. $3\sqrt[3]{16t} + 2\sqrt[3]{54t}$

SECTION 7-2. Supplementary Exercises

In exercises **1–6**, an equation and ordered pairs A, B, and C are given. Two of the ordered pairs are solutions and one is not.

a. Determine which ordered pair is *not* a solution.

b. Change the y-value of the ordered pair of part **a** to make it a solution.

1. $x + y = 1$; $A(-7, 8)$, $B(-2, 3)$, $C(3, 3)$

2. $x - y = 4$; $A(-2, -6)$, $B(1, 3)$, $C(5, 1)$

3. $2x - y = 5$; $A(-1, -17)$, $B(0, -5)$, $C(4, 3)$

4. $3x + y = 6$; $A(3, -3)$, $B(2, 2)$, $C(4, -6)$

5. $4x - 3y = 8$; $A(2, -2)$, $B(-4, -8)$, $C(5, 4)$

6. $2x - 3y = 8$; $A(-8, -8)$, $B(4, 0)$, $C(7, -2)$

In exercises **7–12**, a line $\mathscr{L}$ has the given slope and passes through points P, Q, and R.

a. Find the x-coordinate of Q.

b. Find the y-coordinate of R.

7. $P(2, -2)$, $Q(x, -6)$, $R(5, y)$; slope $= \dfrac{2}{3}$

8. $P(10, 2)$, $Q(x, -4)$, $R(0, y)$; slope $= \dfrac{2}{5}$

9. $P(-3, 6)$, $Q(x, 12)$, $R(7, y)$; slope $= \dfrac{-6}{5}$

10. $P(6, -3)$, $Q(x, 4)$, $R(-14, y)$; slope $= \dfrac{-7}{10}$

11. $P\left(4, \dfrac{-14}{5}\right)$, $Q\left(x, \dfrac{-42}{5}\right)$, $R(-8, y)$; slope $= \dfrac{-7}{10}$

12. $P\left(-9, \dfrac{-7}{5}\right)$, $Q\left(x, \dfrac{11}{5}\right)$, $R(-4, y)$; slope $= \dfrac{8}{5}$

Example. Find the value of a such that the ordered pair $(3a, -2a)$ is a solution of $2x + 7y = 16$.

Solution. Replacing x by $3a$ and y by $-2a$, and solving for a:

$$2(3a) + 7(-2a) = 16$$
$$6a - 14a = 16$$
$$-8a = 16$$
$$a = -2$$

Thus, when $a = -2$ the ordered pair $(3a, -2a)$ is a solution of $2x + 7y = 16$.

In exercises **13–20**, find the value of a such that the given ordered pair is a solution of each equation.

13. $x - 3y = -15$; $(a, 2a)$

14. $2x + y = 35$; $(3a, a)$

15. $4x + 5y = -68$; $(-2a, 5a)$

16. $7x - 2y = 25$; $(-3a, 2a)$

17. $\dfrac{1}{2}x - \dfrac{3}{2}y = 22$; $(10a, -4a)$

18. $\dfrac{2}{3}x + \dfrac{5}{3}y = 6$; $(6a, -3a)$

19. $10x + 9y = 4$; $\left(\dfrac{1}{2}a, \dfrac{1}{3}a\right)$

20. $8x - 3y = 6$; $\left(\dfrac{3}{2}a, \dfrac{4}{3}a\right)$

SECTION 7-3. The Slope-Intercept and Point-Slope Equations of a Line

1. The definition of the intercepts of a line
2. The slope-intercept equation of a line
3. Parallel and perpendicular lines
4. The point-slope equation of a line

The standard form of a linear equation in x and y is $ax + by = c$. There are two different forms of such an equation. We will study these equations in this section, together with the primary uses for them.

The Definition of the Intercepts of a Line

In many of the remaining sections of this text, we will study other kinds of equations in x and y besides linear ones. We will also learn techniques for graphing the solution sets of these equations. The techniques generally require finding any points where the graphs intersect the x- and y-axes. To locate such points we use the x- and y-intercepts of the equations.

> **Definition 7.4.** ***x*-intercept and *y*-intercept**
> If $P_1(a, 0)$ and $P_2(0, b)$ are points of a nonvertical and nonhorizontal line $\mathscr{L}$, then a is the ***x*-intercept** and b is the ***y*-intercept.**

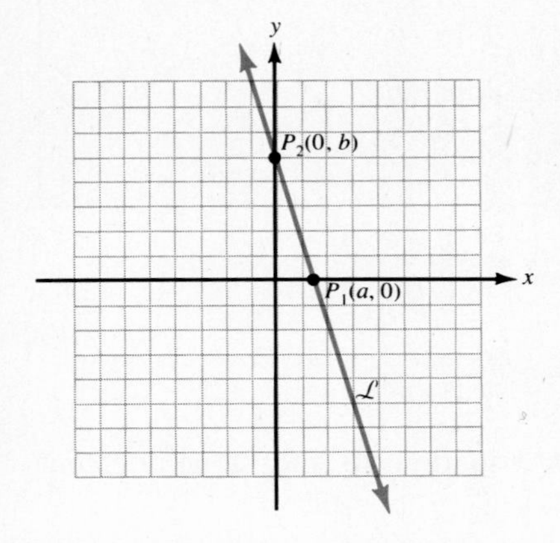

Figure 7-17. A line $\mathscr{L}$ with intercepts a and b.

In Figure 7-17, a nonhorizontal and nonvertical line $\mathscr{L}$ is drawn. The line intersects the x-axis at $P_1(a, 0)$; therefore, a is the x-intercept. The line also intersects the y-axis at $P_2(0, b)$; therefore, b is the y-intercept.

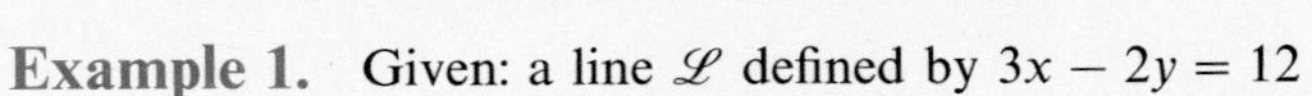

Example 1. Given: a line $\mathscr{L}$ defined by $3x - 2y = 12$

a. Find the x-intercept.

b. Find the y-intercept.

Solution. **a. Discussion.** To find the x-intercept, replace y by 0 and solve for x.

$3x - 2(0) = 12$ Replace y by 0.

$3x = 12$ Solve for x.

$x = 4$ The x-intercept is 4.

b. Discussion. To find the y-intercept, replace x by 0 and solve for y.

$3(0) - 2y = 12$ Replace x by 0.

$-2y = 12$ Solve for y.

$y = -6$ The y-intercept is -6.

The Slope-Intercept Equation of a Line

The equations in examples **a** and **b** are linear in x and y.

a. $5x + 4y = 12$ **b.** $2x - 3y = 21$

The following steps solve these equations for y in terms of x:

a. $4y = -5x + 12$ **b.** $-3y = -2x + 21$

$y = \frac{-5}{4}x + 3$ $y = \frac{2}{3}x - 7$

When written in this form, the following facts can be shown:

Fact 1. The coefficients of x are the slopes of the graphs of the equations.

For example **a**, the slope of the line is $\frac{-5}{4}$.

For example **b**, the slope of the line is $\frac{2}{3}$.

Fact 2. The constant terms are the y-intercepts.

For example **a**, the y-intercept is 3.

For example **b**, the y-intercept is -7.

The general-form equation illustrated by these examples is called the slope-intercept equation of a line.

The slope-intercept equation of a line
If $\mathscr{L}$ is a line defined by $ax + by = c$ and $b \neq 0$, then the **slope-intercept equation** of $\mathscr{L}$ is:

$$y = mx + b$$

m is the slope of $\mathscr{L}$

b is the y-intercept

Example 2. Given: a line defined by $4x + 7y = 35$

a. Find the slope.

b. Find the y-intercept.

Solution. **Discussion.** Change the given equation to the slope-intercept form.

$4x + 7y = 35$ The given equation

$7y = -4x + 35$ Subtract $4x$ from both sides.

$y = \frac{-4}{7}x + 5$ Divide each term by 7.

a. The slope is $\frac{-4}{7}$. The slope is the x-coefficient.

b. The y-intercept is 5. The y-intercept is the constant term.

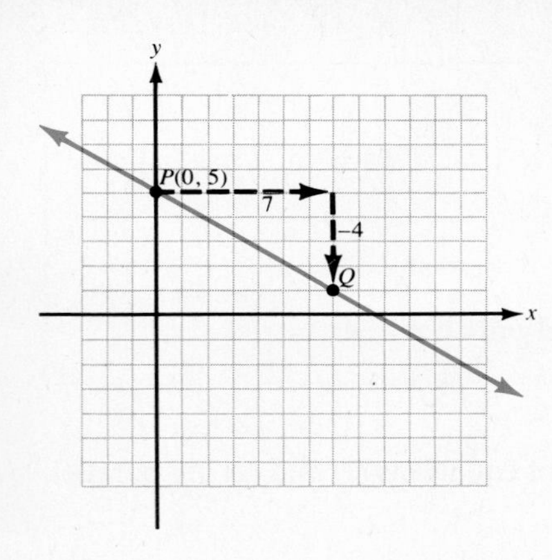

Figure 7-18. A graph of $4x + 7y = 35$.

The slope-intercept form of an equation is useful for graphing lines, as illustrated in Examples 3 and 4.

Example 3. Graph $4x + 7y = 35$.

Solution. From Example 2, the slope-intercept form is $y = \frac{-4}{7}x + 5$.

Step 1. Locate $P(0, b)$.
The y-intercept is 5. Thus, (0, 5) is the point where the line intersects the y-axis.

Step 2. Use m to locate Q.
The slope is $\frac{-4}{7}$. From P, count 7 units to the right and 4 units down to point Q in Figure 7-18. The line through these points is a graph of the equation.

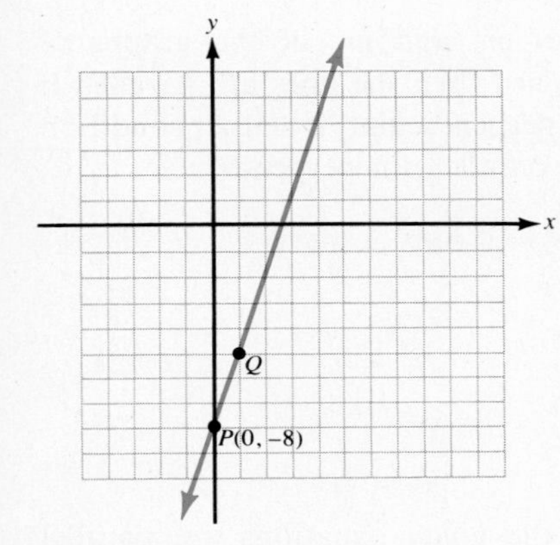

Figure 7-19. A graph of $3x - y = 8$.

Example 4. Graph $3x - y = 8$.

Solution.

$3x - y = 8$ The given equation

$y = 3x - 8$ The $y = mx + b$ form

The slope is 3, and the y-intercept is -8.

Step 1. In Figure 7-19, $P(0, -8)$ is plotted.

Step 2. From P count 1 unit to the right and 3 units up to Q. The line through these points is a graph of the equation.

Parallel and Perpendicular Lines

In Figure 7-20a, $\mathscr{L}_1$ and $\mathscr{L}_2$ are **parallel lines,** and in 7-20b, $\mathscr{L}_3$ and $\mathscr{L}_4$ are **perpendicular lines.** For lines that are nonhorizontal and nonvertical, a special relationship exists between the slopes of lines that are parallel, and those that are perpendicular.

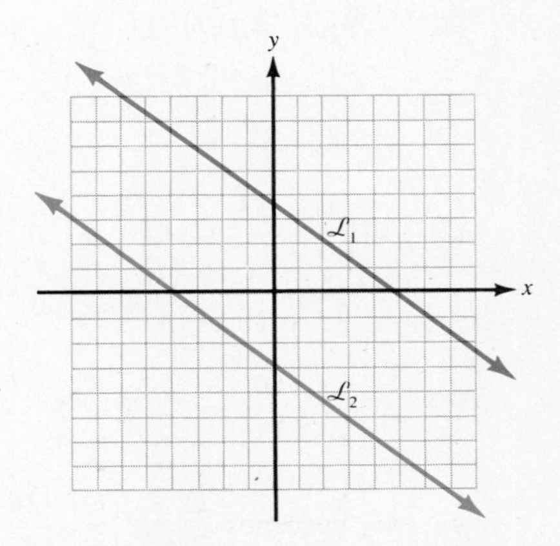

Figure 7-20a. $\mathscr{L}_1$ and $\mathscr{L}_2$ are parallel.

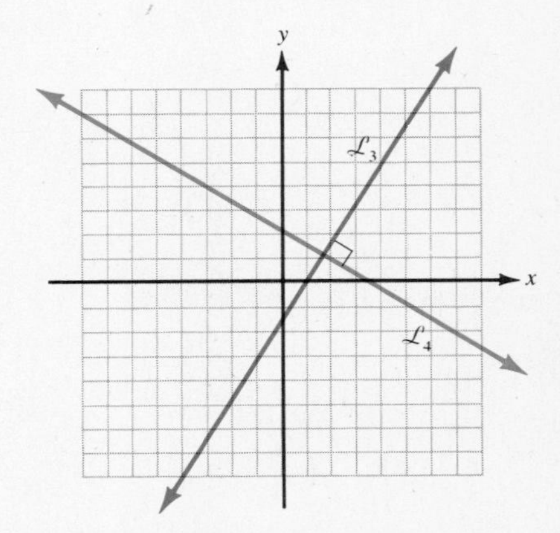

Figure 7-20b. $\mathscr{L}_3$ and $\mathscr{L}_4$ are perpendicular.

Parallel and perpendicular lines

If $\mathscr{L}_1$ and $\mathscr{L}_2$ are nonhorizontal and nonvertical lines with slopes m_1 and m_2, respectively, then:

$\mathscr{L}_1$ and $\mathscr{L}_2$ are **parallel** if and only if $m_1 = m_2$.

$\mathscr{L}_1$ and $\mathscr{L}_2$ are **perpendicular** if and only if $m_1 \cdot m_2 = -1$.

Same slope

Example 5. Determine whether the graphs of equations (1) and (2) are parallel, perpendicular, or neither:

(1) $7x + 5y = 10$

(2) $y = \frac{-7}{5}x + 9$

Solution. **Discussion.** Compare the slopes m_1 and m_2 of the graphs of (1) and (2), respectively. If $m_1 = m_2$, then the lines are parallel. If $m_1 \cdot m_2 = -1$, then the lines are perpendicular. If neither condition holds, then the lines are neither parallel nor perpendicular.

(1) $5y = -7x + 10$ Write in $y = mx + b$ form.

$y = \frac{-7}{5}x + 2$ $\quad m_1 = \frac{-7}{5}$

(2) $y = \frac{-7}{5}x + 9$ $\quad m_2 = \frac{-7}{5}$

Since $m_1 = m_2$, the graphs of the given equations are parallel lines.

Example 6. Determine whether the graphs of equations (1) and (2) are parallel, perpendicular, or neither:

(1) $2x - 7y = 35$

(2) $\frac{y + 5}{7} = \frac{2 - x}{2}$

Solution. (1) $-7y = -2x + 35$ Write in $y = mx + b$ form.

$y = \frac{2}{7}x - 5$ $\quad m_1 = \frac{2}{7}$

(2) $14\left(\frac{y + 5}{7}\right) = 14\left(\frac{2 - x}{2}\right)$ Multiply both sides by 14.

$2y + 10 = 14 - 7x$ Remove the parentheses.

$y = \frac{-7}{2}x + 2$ $\quad m_2 = \frac{-7}{2}$

$m_1 \cdot m_2$ becomes $\frac{2}{7} \cdot \frac{-7}{2} = \frac{-14}{14} = -1$

Thus, the graphs of the given equations are perpendicular lines.

The statement regarding parallel and perpendicular lines exclude $\mathscr{L}_1$ and $\mathscr{L}_2$ from being horizontal or vertical lines. However, the following summary statements can be made regarding any horizontal and vertical lines:

> **Statement 1.** If $\mathscr{L}_1$ and $\mathscr{L}_2$ are horizontal, then they are parallel to each other.
>
> **Statement 2.** If $\mathscr{L}_1$ and $\mathscr{L}_2$ are vertical, then they are parallel to each other.
>
> **Statement 3.** If $\mathscr{L}_1$ is horizontal and $\mathscr{L}_2$ is vertical, then they are perpendicular to each other.

The Point-Slope Equation of a Line

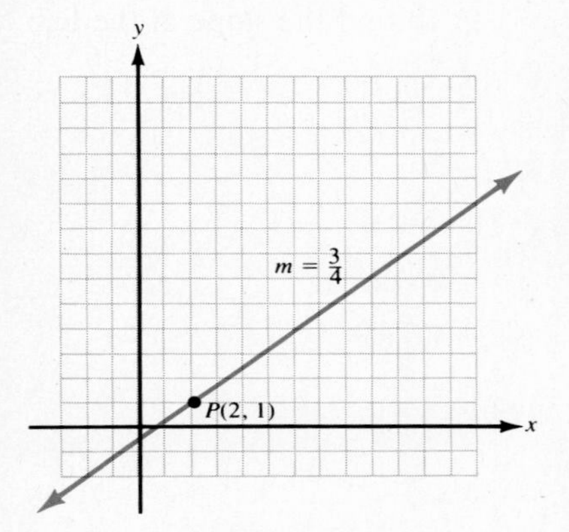

Figure 7-21. A line through $P(2, 1)$ with slope $\frac{3}{4}$.

In Figure 7-21, the line with slope $\frac{3}{4}$ and passing through $P(2, 1)$ is shown. Suppose that we want an equation that defines this line. That is, *we want an equation* $ax + by = c$ *for which the coordinates of every point on the line are solutions.*

To write such an equation, we use the fact that the slope of a line is the same, regardless of what two points are used to compute it. Thus, we let $Q(x, y)$ be a **variable point** on the line, in that Q represents every point on the line except P. Now, using the slope formula:

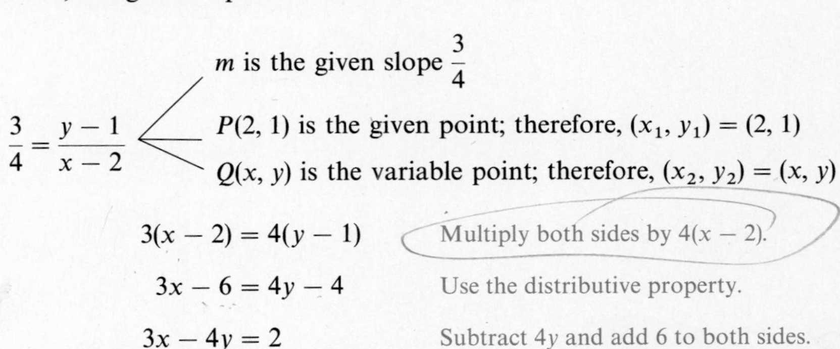

$$\frac{3}{4} = \frac{y-1}{x-2}$$

m is the given slope $\frac{3}{4}$

$P(2, 1)$ is the given point; therefore, $(x_1, y_1) = (2, 1)$

$Q(x, y)$ is the variable point; therefore, $(x_2, y_2) = (x, y)$

$3(x-2) = 4(y-1)$ Multiply both sides by $4(x-2)$.

$3x - 6 = 4y - 4$ Use the distributive property.

$3x - 4y = 2$ Subtract $4y$ and add 6 to both sides.

An equation of the line is therefore $3x - 4y = 2$. Notice that we were able to determine this equation by knowing a *point on the line* and the *slope of the line.* The general formula is known as the point-slope equation of a line.

> **The point-slope equation of a line**
>
> If $\mathscr{L}$ is a nonvertical line with slope m and passing through $P(x_1, y_1)$, then the **point-slope equation of** $\mathscr{L}$ is:
>
> $$y - y_1 = m(x - x_1)$$
>
> The given point (points to x_1, y_1) — The given slope (points to m)

Example 7. Write in the $ax + by = c$ form an equation for the line that passes through $P_1(-4, -7)$ and $P_2(3, -2)$

Solution. **Discussion.** Use the coordinates of P_1 and P_2 to compute m.

$$m = \frac{-7-(-2)}{-4-3} = \frac{-5}{-7} = \frac{5}{7} \qquad m = \frac{y_1 - y_2}{x_1 - x_2}$$

Arbitrarily using $(3, -2)$ for (x_1, y_1):

$y - (-2) = \frac{5}{7}(x - 3)$	The point-slope equation
$7(y + 2) = 5(x - 3)$	Multiply both sides by 7.
$7y + 14 = 5x - 15$	Use the distributive property.
$29 = 5x - 7y$	Subtract $7y$ and add 15 to both sides.
$5x - 7y = 29$	The $ax + by = c$ form, with $a > 0$

Example 8. Write an equation for the line that passes through $P(4, -5)$ and is perpendicular to the graph of $x - 3y = 9$.

Solution. **Discussion.** Write $x - 3y = 9$ in the $y = mx + b$ form to find the slope of the graph. Use $m_1 \cdot m_2 = -1$ to find the slope of the line whose equation we want.

$x - 3y = 9$	The given equation
$y = \frac{1}{3}x - 3$	The slope of the line is $\frac{1}{3}$.
$m_1 \cdot \frac{1}{3} = -1$	Slopes of perpendicular lines
$m_1 = -3$	The slope of unknown line
$y - (-5) = -3(x - 4)$	$m = -3$ and $(x_1, y_1) = (4, -5)$
$y + 5 = -3x + 12$	Use the distributive property.
$3x + y = 7$	The $ax + by = c$ form, with $a > 0$

SECTION 7-3. Practice Exercises

In exercises **1–14**, for each equation:

a. Find the x-intercept.

b. Find the y-intercept.

[Example 1]

1. $x + 4y = 4$

2. $3x + y = 6$

3. $5x - y = 15$

4. $x - 2y = 2$

5. $2x + 5y = 10$

6. $5x + 4y = 20$

7. $9x - 8y = -72$

8. $10x - 3y = -30$

9. $2y + 8 = 0$

10. $-5y + 20 = 0$

11. $-3x + 12 = 0$

12. $4x + 32 = 0$

13. $7x - 15y = 0$

14. $-5x + 8y = 0$

In exercises **15–36**, for each equation:

a. Find the slope, if it is defined.

b. Find the y-intercept, if one exists.

c. Graph the equation.

[Examples 2–4]

15. $5x + y = 3$

16. $-2x + y = 4$

17. $3x - 4y = 12$

18. $4x - 5y = 10$

19. $10x + 9y = -9$

20. $8x - 7y = -14$

21. $x + y = 3$

22. $x - y = 4$

23. $3x - y = 0$

24. $x + 3y = 0$

25. $y - 4 = 0$

26. $3y + 9 = 0$

27. $2(y + 2) = 8 - 3x$

28. $3y + 1 = 5(3x + 2)$

29. $x = \frac{3}{2}y - 6$

30. $x = 8 - \frac{4}{3}y$

31. $2x + 5 = 0$

32. $0 = x - 8$

33. $\frac{2y + x}{2} = \frac{x - 18}{3}$

34. $\frac{y - 1}{5} = \frac{x + 2}{2}$

35. $\frac{x + y + 1}{4} = \frac{2x + 5}{5}$

36. $\frac{y - x + 1}{4} = \frac{7 - x}{7}$

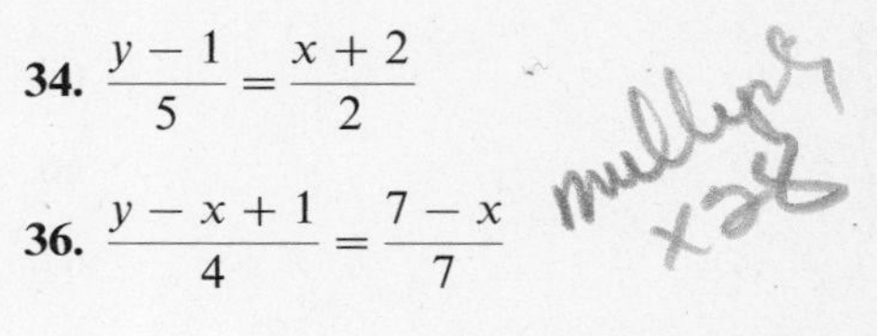

In exercises **37–52**, determine whether the graphs of each pair of equations are parallel, perpendicular, or neither.

[Examples 5 and 6]

37. (A) $2x + y = 5$
(B) $6x + 3y = 10$

38. (A) $x - 6y = 3$
(B) $-3x + 18y = 10$

39. (A) $x - 2y = 10$
(B) $2x + y = 7$

40. (A) $3x + 5y = 10$
(B) $10x - 6y = -9$

41. (A) $3x + 7y = 14$
(B) $3x + 8y = 16$

42. (A) $7x - 2y = 6$
(B) $5x - 2y = 10$

43. (A) $2y + 5 = 0$
(B) $y - 7 = 0$

44. (A) $3y - 2 = 0$
(B) $y = 14$

45. (A) $y = 3$
(B) $x = 5$

46. (A) $x = -4$
(B) $y = -2$

47. (A) $x + \frac{1}{3}y = 4$
(B) $3x + y = -2$

48. (A) $\frac{-2}{3}x + y = 10$
(B) $x - \frac{2}{3}y = 4$

49. (A) $\frac{3}{5}x - y = 6$
(B) $x + \frac{3}{5}y = 1$

50. (A) $-x + \frac{2}{7}y = 12$
(B) $\frac{2}{7}x + y = 3$

51. (A) $x + 3y = 0$
(B) $3x + y = 0$

52. (A) $2x - 5y = 0$
(B) $5x - 2y = 0$

In exercises **53–72**, write an equation for each line in the:

a. Point-slope form

b. $ax + by = c$ form

[Example 7] **53.** Slope 3 and containing (2, 4)

54. Slope -5 and containing $(7, -2)$

55. Slope $\frac{2}{3}$ and containing $(0, -4)$

56. Slope $\frac{7}{4}$ and containing $(-3, 0)$

57. Slope $\frac{-5}{2}$ and containing $(-10, 6)$

58. Slope $\frac{-1}{6}$ and containing $(-1, 5)$

59. Slope 0 and containing (3, 8)

60. Slope 0 and containing $(85, -2)$

61. Slope $\frac{1}{2}$ and containing $(-4, 2)$

62. Slope $\frac{3}{4}$ and containing $(6, -9)$

63. Passing through (0, 5) and $(-4, 0)$

64. Passing through $(-1, 4)$ and (5, 0)

65. Passing through $(2, -1)$ and $(-4, 3)$

66. Passing through $(-8, 10)$ and $(-4, 6)$

67. Passing through $(-3, -2)$ and (5, 5)

68. Passing through $(-4, -9)$ and $(-2, -1)$

69. Passing through $(12, -6)$ and $(-12, -6)$

70. Passing through $(-9, 5)$ and (9, 5)

71. Passing through $\left(\frac{1}{2}, \frac{2}{3}\right)$ and $\left(\frac{-1}{3}, \frac{-1}{2}\right)$

72. Passing through $\left(\frac{3}{5}, \frac{-1}{4}\right)$ and $\left(\frac{-3}{4}, \frac{2}{5}\right)$

In exercises **73–80**, write an equation in the $ax + by = c$ form for the line that contains the given point and is:

a. Parallel to the given line

b. Perpendicular to the given line

[Example 8] **73.** Containing $P(1, 4)$ and the line $2x - 3y = 5$

74. Containing $P(4, -2)$ and the line $5x - y = 9$

75. Containing $P(-6, 3)$ and the line $x + 2y = 10$

76. Containing $P(-2, -5)$ and the line $4x + 3y = 20$

77. Containing $P(0, 0)$ and the line $-2x + 6y = 5$

78. Containing $P(0, 0)$ and the line $-5x + 10y = 7$

79. Containing $P(3, -6)$ and the line $y = 8$

80. Containing $P(-5, 9)$ and the line $y = -10$

SECTION 7-3. Ten Review Exercises

In exercises **1–4**, simplify each expression.

1. $(3x + 2y) - (2x - 3y)$

2. $(3\sqrt{2} + 2\sqrt{3}) - (2\sqrt{2} - 3\sqrt{3})$

3. $2(4x^2 - 2x + 3) - 5(x^2 - x + 1)$

4. $2(4\sqrt[3]{5} - 2\sqrt{5} + 3) - 5(\sqrt[3]{5} - \sqrt{5} + 1)$

In exercises **5–10**, use the points $P(5, 7)$ and $Q(-10, -1)$.

5. Find the distance between P and Q.

6. Find the midpoint of the segment with P and Q as endpoints.

7. In which quadrant is the plot of Q?

8. Find the slope of the line that passes through P and Q.

9. Is the line that passes through P and Q parallel to the line defined by $8x - 15y = 16$?

10. Is the line that passes through P and Q perpendicular to the line defined by $15x + 8y = 16$?

SECTION 7-3. Supplementary Exercises

Jay Marlow has been contracted to haul boxes stored in a warehouse to a new location. There are two types of boxes in the warehouse based on size. The large boxes are called Type A and the small boxes are called Type B. Based on the shape of his truck, Jay can haul a certain number of these boxes on each trip.

In exercises **1–4**, the beginning inventory of Type A and Type B boxes in the warehouse is given. Furthermore, the number of Type A and Type B boxes that Jay hauls in each trip is given. In these exercises:

x represents the number of Type A boxes left in the warehouse after each trip

y represents the corresponding number of Type B boxes left

1. The beginning inventory consists of 24 Type A and 20 Type B boxes, and on each trip Jay transports 4 Type A and 3 Type B boxes.

a. Complete the following table:

	Trip 0	**Trip 1**	**Trip 2**	**Trip 3**	**Trip 4**
x	24		16		
y	20				8

b. Plot and draw a line through the five data points in part **a**.
c. Write in $y = mx + b$ form an equation for this relation.
d. After the last Type A box has been moved, how many Type B boxes are left in the warehouse?

2. The beginning inventory consists of 27 Type A and 20 Type B boxes, and on each trip Jay transports 3 Type A and 2 Type B boxes.

a. Complete the following table:

	Trip 0	Trip 1	Trip 2	Trip 3	Trip 4
x	27		21		
y	20				12

b. Plot and draw a line through the data points in part **a**.
c. Write in $y = mx + b$ form an equation for this relation.
d. After the last Type B box has been moved, how many Type A boxes are left in the warehouse?

3. The beginning inventory consists of 29 Type A and 29 Type B boxes, and on each trip Jay transports 4 Type A and 4 Type B boxes.

a. Complete the following table:

	Trip 0	Trip 1	Trip 2	Trip 3	Trip 4
x	29	25			
y	29			17	

b. Plot and draw a line through the data points in part **a**.
c. Write in $y = mx + b$ form an equation for this relation.
d. How many Type A and Type B boxes are left in the warehouse after the last full load has been moved?

4. The beginning inventory consists of 30 Type A and 17 Type B boxes, and on each trip Jay transports 2 Type A and 1 Type B boxes.

a. Complete the following table:

	Trip 0	Trip 1	Trip 2	Trip 3	Trip 4
x	30				22
y	17		15		

b. Plot and draw a line through the five data points in part **a**.
c. Write in $y = mx + b$ form an equation for this relation.
d. After the last Type A box has been moved, how many Type B boxes are left in the warehouse?

5. The beginning inventory consisted of 25 Type A and 37 Type B boxes. Let x and y represent the numbers of Type A and Type B boxes, respectively, that Jay transported on each load. Find the smallest possible values for x and y if after several trips there were 20 Type A and 22 Type B boxes left in the warehouse.

6. The beginning inventory consisted of 30 Type A and 20 Type B boxes. Let x and y represent the numbers of Type A and Type B boxes, respectively, that Jay transported on each load. Find the smallest possible values for x and y if after several trips there were 15 Type A and 2 Type B boxes left in the warehouse.

Lucy McCray has a small greenhouse in which she grows flowers as a hobby. Two flowers that she grows are daisies and pansies. Lucy has decided that the number of daisies (x) and the number of pansies (y) that she will grow each year will follow a formula based on the numbers grown in two consecutive years.

7. a. Complete the following table:

	Year 1	**Year 2**	**Year 3**	**Year 4**	**Year 5**
x	2	6		14	
y	2	5			14

b. Write in $y = mx + b$ form an equation for this formula.
c. How many pansies will she have the year she grows 30 daisies?

8. a. Complete the following table:

	Year 1	**Year 2**	**Year 3**	**Year 4**	**Year 5**
x	2	7		17	
y	2	5			14

b. Write in $y = mx + b$ form an equation for this formula.
c. How many pansies will she have the year she grows 42 daisies?

9. a. Complete the following table:

	Year 1	**Year 2**	**Year 3**	**Year 4**	**Year 5**
x	2	5			
y	1	3			

b. Write in $y = mx + b$ form an equation for this formula.
c. How many pansies will she have the year she grows 32 daisies?

10. a. Complete the following table:

	Year 1	**Year 2**	**Year 3**	**Year 4**	**Year 5**
x	1	4			
y	3	5			

b. Write in $y = mx + b$ form an equation for this formula.
c. How many pansies will she have the year she grows 28 daisies?

Two Polynesian cultures developed simultaneously on opposite ends of a large South Pacific island. The northern culture uses the x-*alop* as the standard unit of length. The southern culture uses the y-*aplot* as the standard unit of length. To compare these standards, the leaders of the two cultures measure the lengths of a fallen tree and a large rock.

In exercises **11–14**, find a formula to convert x-alops to y-aplots, based on the results of the given measurements.

11. Tree: 10 x-alops and 12 y-aplots
Rock: 5 x-alops and 6 y-aplots

12. Tree: 14 x-alops and 10 y-aplots
Rock: 7 x-alops and 5 y-aplots

13. Tree: 21 x-alops and 15 y-aplots
Rock: 7 x-alops and 5 y-aplots

14. Tree: 18 x-alops and 24 y-aplots
Rock: 3 x-alops and 4 y-aplots

SECTION 7-4. Graphs of Linear Inequalities in Two Variables

1. A definition of a linear inequality in x and y
2. A solution of a linear inequality in x and y
3. A procedure for graphing a linear inequality in x and y
4. Graphing linear inequalities with horizontal or vertical boundary lines.

In Figure 7-22a, point P is a graph of $x = 3$. The half-line to the left of P contains points whose coordinates are solutions of $x < 3$, as shown in Figure 7-22b. The half-line to the right of P contains points whose coordinates are solutions of $x > 3$, as shown in Figure 7-22c.

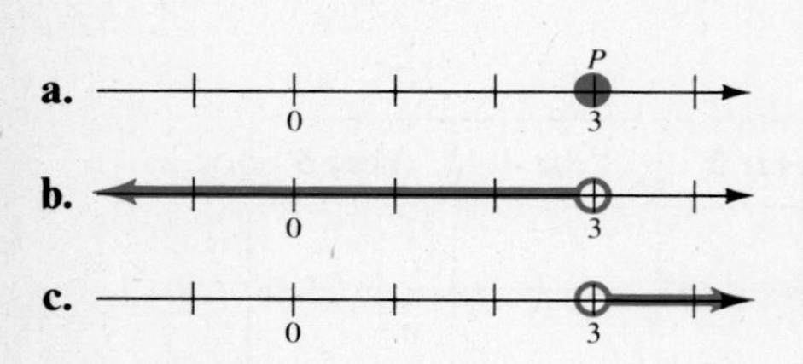

Figure 7-22. **a.** A graph of $x = 3$.
b. A graph of $x < 3$.
c. A graph of $x > 3$.

The equation and inequalities graphed in Figure 7-22 are linear in one variable. Notice that P, the graph of $x = 3$, divides the line into **two half-lines.** Now consider line $\mathscr{L}$ in Figure 7-23, which is a graph of $x + y = 3$. Notice that $\mathscr{L}$ divides the plane on which it is drawn into **two half-planes.** In this section, we will learn how these half-planes can be used to graph the solutions of linear inequalities in x and y.

A Definition of a Linear Inequality in x and y

The inequalities in Definition 7.5 identify the standard forms of linear inequalities in x and y.

Definition 7.5. Linear inequalities in x and y
If a, b, and c are real numbers and a and b are not both 0, then the standard form of a linear inequality in x and y is:

$$ax + by < c \quad \text{or} \quad ax + by > c \quad \text{or} \quad ax + by \leq c \quad \text{or} \quad ax + by \geq c$$

Examples **a** and **b** illustrate two linear inequalities in x and y.

a. $3x - 5y > 10$ A "greater than" linear inequality

b. $2x + y \leq 4$ A "less than or equal to" linear inequality

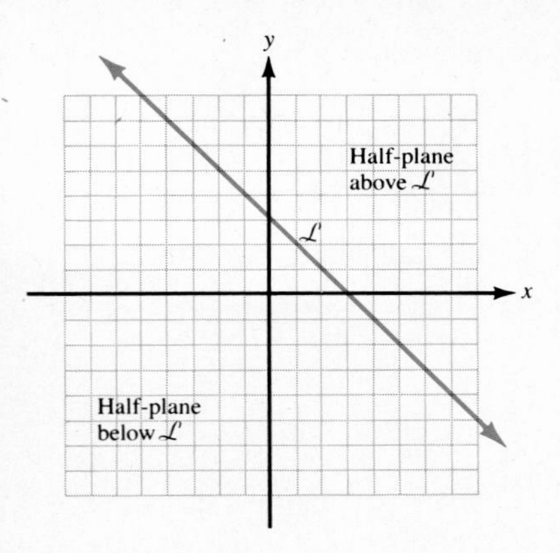

Figure 7-23. A graph of $x + y = 3$.

A Solution of a Linear Inequality in x and y

An ordered pair (d, e) is a solution of a linear inequality in x and y if the inequality is true when x is replaced by d and y is replaced by e.

Example 1. Determine whether $(-2, -5)$ is a solution of $7x - 4y < 12$.

Solution.

$7(-2) - 4(-5) < 12$	Replace x by -2 and y by -5.
$-14 - (-20) < 12$	Do the indicated multiplications.
$6 < 12$, *true*	Thus, $(-2, -5)$ is a solution.

Example 2. Determine whether $(13, 8)$ is a solution of $-3x + 5y \geq 4$.

Solution.

$-3(13) + 5(8) \geq 4$	Replace x by 13 and y by 8.
$-39 + 40 \geq 4$	Do the indicated multiplications.
$1 \geq 4$, *false*	Thus, $(13, 8)$ is not a solution.

A Procedure for Graphing a Linear Inequality in x and y

The number of solutions of a linear inequality in x and y is infinite. Therefore, a graph is one way to display the solution set of such an inequality. In an xy-coordinate system, a graph of a linear inequality is a **half-plane of points** whose coordinates are solutions. To indicate the solution set, the half-plane is *shaded*.

To locate the graph of a linear inequality in x and y:

Step 1. Graph the line $ax + by = c$.
- **a.** Use a *dotted line* if the inequality is $<$ or $>$.
- **b.** Use a *solid line* if the inequality is $\leq$ or $\geq$.

Step 2. Select a **test point** $P(d, e)$ anywhere in the plane, except on the line obtained in Step 1.
- **a.** If the line does not pass through the origin, pick $P(0, 0)$.
- **b.** If the line passes through the origin, pick $P(d, 0)$ or $P(0, e)$.

Step 3. Replace x by d and y by e in the given inequality and simplify.

Step 4. If (d, e) is a solution, then shade the half-plane containing P. If (d, e) is not a solution, then shade the half-plane that does not contain P.

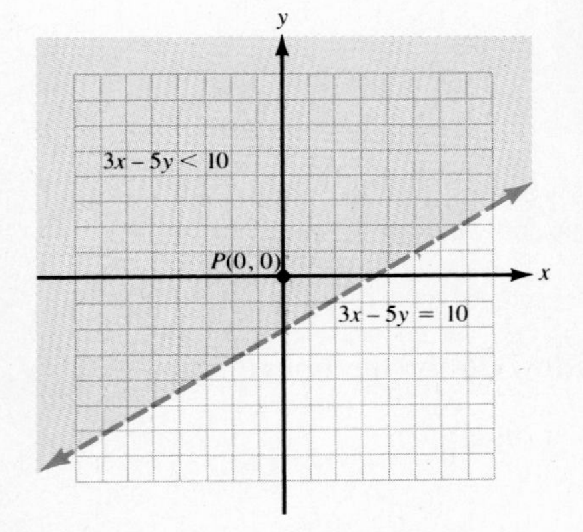

Figure 7-24. A graph of $3x - 5y < 10$.

The line (dotted or solid) in Step 1 is the **boundary,** or **edge,** to the half-plane that is the graph of the solution set of the given inequality.

Example 3. Graph $3x - 5y < 10$.

Solution.

Step 1. $3x - 5y = 10$	Write the corresponding equation.
$-5y = -3x + 10$	Write in $y = mx + b$ form.
$y = \frac{3}{5}x - 2$	$m = \frac{3}{5}$ and y-intercept is -2

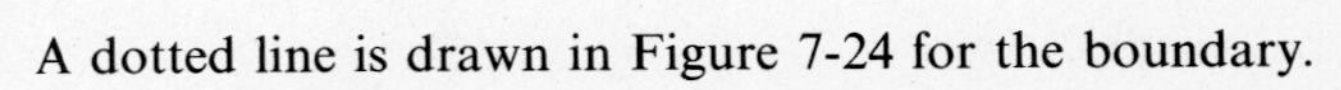

A dotted line is drawn in Figure 7-24 for the boundary.

Step 2. $P(0, 0)$ is used as a test point.

Step 3. $3(0) - 5(0) < 10$ — Replace x by 0 and y by 0.

$0 - 0 < 10$, *true* — (0, 0) is a solution

Step 4. Since (0, 0) is a solution, the half-plane containing P is shaded as shown in Figure 7-24.

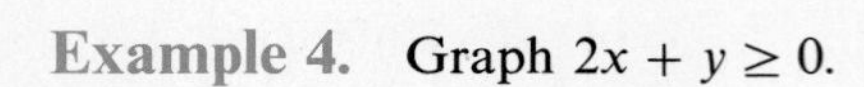

Example 4. Graph $2x + y \geq 0$.

Solution. **Step 1.** $2x + y = 0$ — Write the corresponding equation.

$y = -2x$ — Write in $y = mx + b$ form.

$y = -2x + 0$ — $m = -2$ and y-intercept is 0

A solid line is drawn in Figure 7-25 for the boundary.

Step 2. $P(0, 6)$ is used as a test point.

Step 3. $2(0) + 6 \geq 0$ — Replace x by 0 and y by 6.

$0 + 6 \geq 0$, *true* — (0, 6) is a solution

Step 4. Since (0, 6) is a solution, the half-plane containing P is shaded as shown in Figure 7-25.

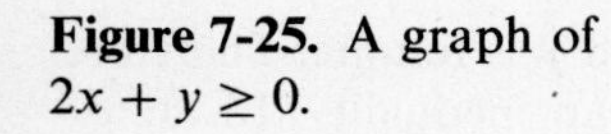

Figure 7-25. A graph of $2x + y \geq 0$.

Graphing Linear Inequalities with Horizontal or Vertical Boundary Lines

The inequalities in examples **c** and **d** can be written in one of the standard forms of a linear inequality in x and y.

c. $y + 2 < 0$ can be written as $0 \cdot x + y < -2$

d. $x - 3 \geq 0$ can be written as $x + 0 \cdot y \geq 3$

As linear inequalities in x and y, the graphs of the solution sets of these inequalities are shaded half-planes. The boundary lines are a horizontal line (example **c**) and a vertical line (example **d**).

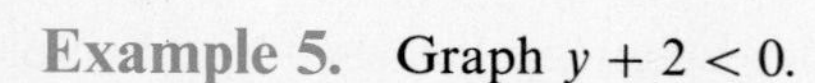

Example 5. Graph $y + 2 < 0$.

Solution. **Discussion.**

Step 1. $y + 2 = 0$ — Write the corresponding equation.

$y = -2$ — A horizontal line with y-intercept -2

A dotted line is drawn in Figure 7-26 for the boundary.

Step 2. $P(0, 0)$ is used as a test point.

Step 3. $0 + 2 < 0$ — Replace y by 0.

$2 < 0$, *false* — (0, 0) is not a solution

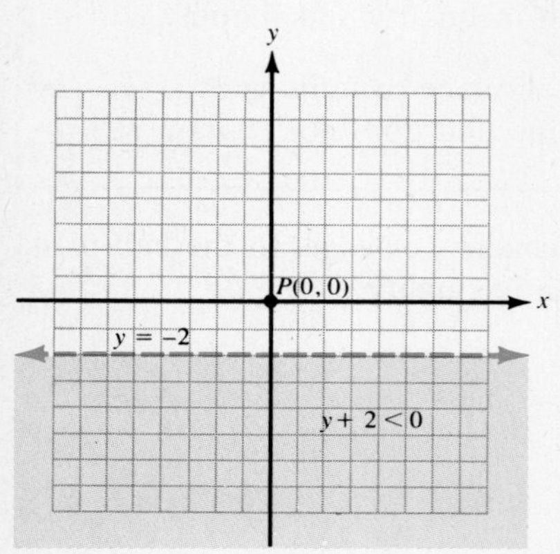

Figure 7-26. A graph of $y + 2 < 0$.

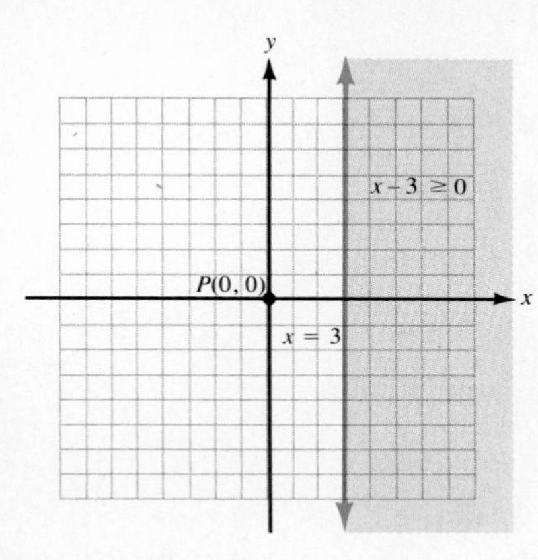

Figure 7-27. A graph of $x - 3 \geq 0$.

Step 4. Since (0, 0) is not a solution, the half-plane not containing P is shaded as shown in Figure 7-26.

Example 6. Graph $x - 3 \geq 0$.

Solution.

Step 1. $x - 3 = 0$ — Write the corresponding equation.

$x = 3$ — A vertical line with x-intercept 3

A solid line is drawn in Figure 7-27 for the boundary.

Step 2. $P(0, 0)$ is used as a test point.

Step 3. $0 - 3 \geq 0$ — Replace x by 0.

$-3 \geq 0$, *false* — (0, 0) is not a solution

Step 4. Since (0, 0) is not a solution, the half-plane not containing P is shaded as shown in Figure 7-27.

SECTION 7-4. Practice Exercises

In exercises **1–10**, determine whether the given ordered pairs are solutions of each inequality.

[Examples 1 and 2]

1. $3x + 5y > 10$
a. (1, 2)
b. (3, 1)

2. $2x - 3y > 5$
a. (5, −1)
b. (1, −2)

3. $x - 8y < 4$
a. (10, 2)
b. (4, 0)

4. $5x + 2y < 10$
a. (−3, 10)
b. (2, 0)

5. $-10x + 7y \geq 0$
a. (3, 6)
b. (0, −2)

6. $-6x - 5y \geq 0$
a. (−4, 3)
b. (−1, −2)

7. $-2x - 9y \leq 8$
a. $\left(\frac{1}{2}, -2\right)$
b. $\left(-6, \frac{1}{3}\right)$

8. $-x + 6y \leq 13$
a. $\left(-10, \frac{3}{2}\right)$
b. (−5, 2)

9. $y > 11$
a. (−20, 12)
b. (50, 13)

10. $x \leq 3$
a. (0, 100)
b. (3, −90)

In exercises **11–50**, graph each inequality.

[Examples 3–6]

11. $y > \frac{3}{4}x - 6$

12. $y > \frac{5}{2}x - 7$

13. $y < \frac{-1}{3}x + 5$

14. $y < \frac{-3}{2}x + 3$

15. $y \geq 5x + 1$

16. $y \leq -3x - 2$

17. $y + 3 < 0$

18. $5y - 10 > 0$

19. $2x - 6 \geq 0$

20. $5x + 15 < 0$

21. $x + 3y < 12$

22. $2x + y < 6$

23. $3x - y > 6$

24. $x - 3y > 2$

25. $x + 2y \geq -4$

26. $5x + 4y \geq -16$

27. $3x - 5y \geq 15$

28. $4x - y \geq 8$

29. $x + y > 2$

30. $x + 3y > 3$

31. $3x - 7y < -21$

32. $5x - 6y < -24$

33. $5x + y \leq -5$

34. $3x + 2y < -14$

35. $3x - 2y \geq 12$

36. $6x - y \geq 8$

37. $x + 2y > 0$

38. $3x + 5y > 0$

39. $-x + 6y < 0$

40. $-4x + 3y < 0$

41. $4x - y \leq 0$

42. $2x - 5y \leq 0$

43. $y - 2 \leq 0$

44. $3x + 12 \leq 0$

45. $4 - x \geq 0$

46. $x + 3 < 0$

47. $y < 0$

48. $x > 0$

49. $x \leq 0$

50. $y \geq 0$

SECTION 7-4. Ten Review Exercises

1. Factor $3x^2 - 8 - 10x$.

2. Solve $10x + 8 = 3x^2$.

3. a. Determine the constant to add to $t^2 - 10t$ to form a perfect-square trinomial.

b. Write the trinomial of part **a** as the square of a binomial.

4. Solve $t^2 - 10t + 13 = 0$ by completing the square.

In exercises **5–10**, do the indicated operations.

5. $(2u - 3)(u + 5)$

6. $(2k^{1/2} - 3)(k^{1/2} + 5)$

7. $(2\sqrt{6} - 3)(\sqrt{6} + 5)$

8. $(2 - 3i)(1 + 5i)$

9. $\left(\frac{2}{z} - 3\right)\left(\frac{1}{z} + 5\right)$

10. $(2\sqrt[3]{36} - 3)(\sqrt[3]{6} + 5)$

SECTION 7-4. Supplementary Exercises

In exercises **1–14**, write an inequality in one of the following forms:

(1) $ax + by < c$ (2) $ax + by > c$

(3) $ax + by \leq c$ (4) $ax + by \geq c$

Example. The graph of a solution set is a half-plane containing $O(0, 0)$. The edge of the half-plane contains $P_1(-6, -3)$ and $P_2(8, 2)$, and the edge is not included in the solution set.

Solution. **Step 1.** Write in $ax + by = c$ form an equation for the edge of the half-plane.

$$m = \frac{2-(-3)}{8-(-6)}$$ $(x_1, y_1) = (-6, -3)$ and $(x_2, y_2) = (8, 2)$.

$$\frac{5}{14}$$ Simplify.

$$y - 2 = \frac{5}{14}(x - 8)$$ $m = \frac{5}{14}$ and $(x_2, y_2) = (8, 2)$

$$5x - 14y = 12$$ Write in the $ax + by = c$ form.

Step 2. Determine whether the inequality is $<$ or $>$.
Replace x by 0 and y by 0 in the equation of Step 1, since $O(0, 0)$ is given as a solution, and determine which order relation will give a true statement.

$$5(0) - 14(0) \quad 12$$

$$0 < 12, \textit{ true}$$ Insert the $<$ relation.

Step 3. Write an appropriate inequality using the results of Steps 1 and 2.
The edge is not included in the solution set. Thus, an inequality for the given conditions is: $5x - 14y < 12$.

1. The edge of the half-plane that is the graph of the solution set passes through $P_1(0, -5)$ and $P_2(4, 0)$. The coordinates of the origin are a solution of the inequality, and the boundary is not included in the solution set.

2. The edge of the half-plane that is the graph of the solution set passes through $P_1(-2, 0)$ and $P_2(0, 5)$. The coordinates of the origin are a solution of the inequality, and the boundary is not included in the solution set.

3. The edge of the half-plane that is the graph of the solution set passes through $P_1(1, 1)$ and $P_2(3, -1)$. The coordinates of the origin are not a solution of the inequality, and the boundary is included in the solution set.

4. The edge of the half-plane that is the graph of the solution set passes through $P_1(1, -1)$ and $P_2(5, 2)$. The coordinates of the origin are not a solution of the inequality, and the boundary is included in the solution set.

5. The edge of the half-plane that is the graph of the solution set passes through $P_1(-10, 4)$ and $P_2(9, 4)$. The coordinates of the origin are a solution of the inequality, and the boundary is not included in the solution set.

6. The edge of the half-plane that is the graph of the solution set passes through $P_1(-8, 10)$ and $P_2(-8, -6)$. The coordinates of the origin are not a solution of the inequality, and the boundary is included in the solution set.

7. The edge of the half-plane that is the graph of the solution set passes through $P_1(0, 0)$ and $P_2(6, 8)$. The point $P_3(0, 5)$ is in the half-plane. The boundary is included in the solution set.

8. The edge of the half-plane that is the graph of the solution set passes through $P_1(0, 0)$ and $P_2(5, -4)$. The point $P_3(10, 0)$ is in the half-plane. The boundary is not included in the solution set.

9.

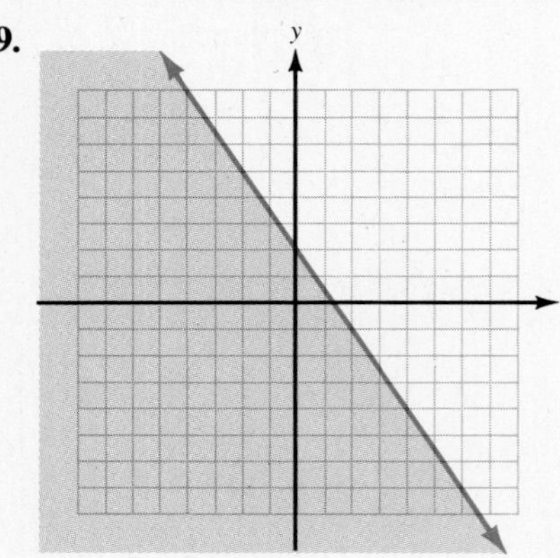

10.

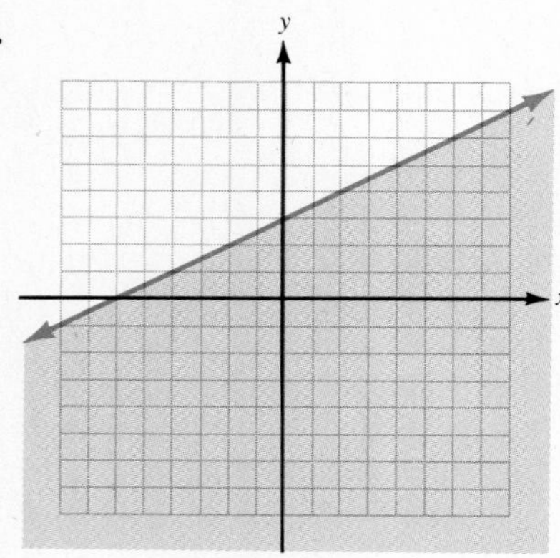

11.

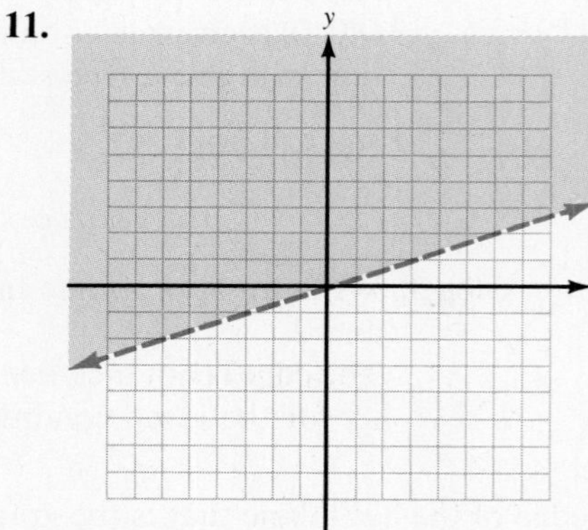

12.

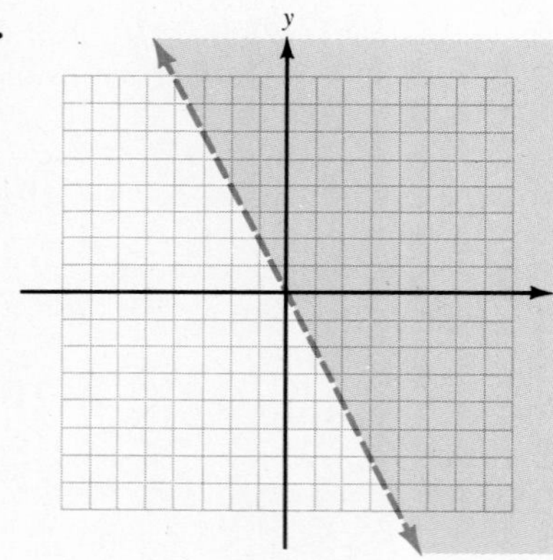

13.

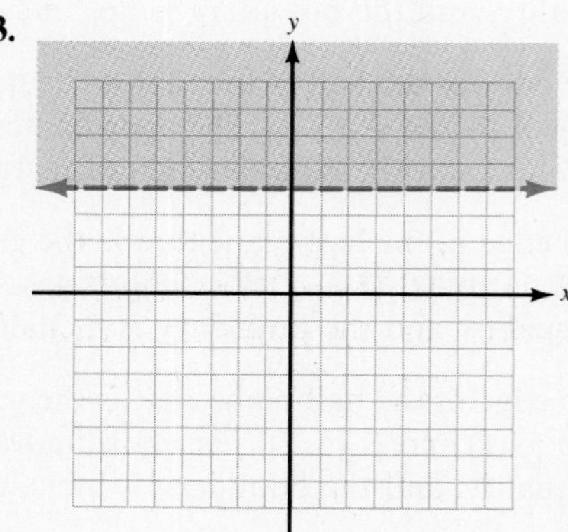

14.

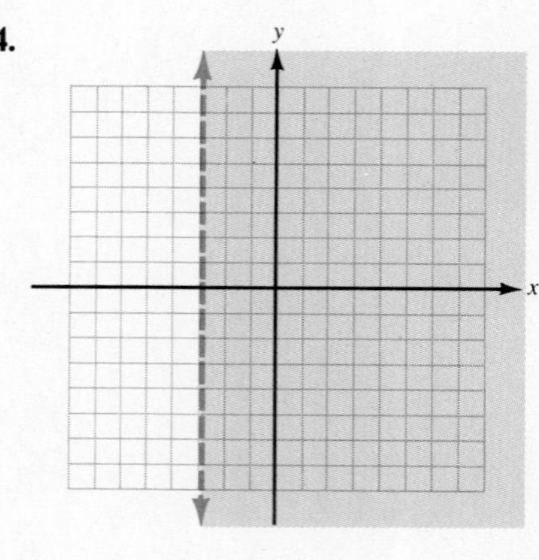

SECTION 7-5. Linear Functions and Their Inverses

1. A definition of a relation
2. A definition of a function
3. A definition of a linear function in x and y
4. The inverse of a linear function in x and y
5. Graphing a linear function and its inverse on the same set of axes.

The concept of relations and functions is very important in mathematics. In this section, we will use the linear equation in x and y to begin the study of relations and functions. We will study other types of relations and functions throughout the remainder of the text.

A Definition of a Relation

Many quantities can be **paired** in a meaningful way. Examples **a–e** identify five such quantities.

Quantity A	Quantity B
a. The *time* during which a vehicle moves at 50 mph	**a*.** The corresponding *distance* the vehicle moves
b. The length of an object in *inches*	**b*.** The corresponding length in *centimeters*
c. The temperature in *Fahrenheit*°	**c*.** The corresponding temperature in *Celsius*°
d. The *radius* of a circle	**d*.** The corresponding *area* of the circle
e. The *amount* of a purchase	**e*.** The corresponding *sales tax* on the purchase

For each example, a given value for a Quantity A item is paired with a corresponding value for a Quantity B item. In mathematics, such a pairing of numbers is called a relation.

> **Definition 7.6. Relation**
> A **relation** is a set of ordered pairs.

Many relations are defined by equations or inequalities that can be thought of as **rules of correspondence.** That is, a given rule specifies how to determine the value of a Quantity B item to pair with a given value of a Quantity A item. To illustrate, consider examples **a** and **a***.

The following formula can be used to describe the relationship between time and distance:

$$d = r \cdot t \qquad \text{distance traveled} = \text{rate} \cdot (\text{time of travel})$$

If the vehicle is moving at 50 mph, then $r = 50$. The rule of correspondence for the relation becomes:

$$d = 50t \begin{cases} \text{for } t = 1, & d = 50 \cdot 1 = 50 \\ \text{for } t = 3, & d = 50 \cdot 3 = 150 \\ \text{for } t = 7.5, & d = 50 \cdot 7.5 = 375 \end{cases}$$

These three paired values can be written as (1, 50), (3, 150), and (7.5, 375).

Example 1. The equation that relates centimeters (c) to inches (i) is:

$c = 2.54i$

Find the ordered pairs of the form (i, c) for each value of i.

a. 5 **b.** 12 **c.** 30.5

Solution. **Discussion.** Replace i in the equation $c = 2.54i$ for each of the given values, then find the corresponding values of c.

a. $c = 2.54(5)$ Replace i by 5.

$= 12.7$ (5, 12.7) is in the relation

b. $c = 2.54(12)$ Replace i by 12.

$= 30.48$ (12, 30.48) is in the relation

c. $c = 2.54(30.5)$ Replace i by 30.5.

$= 77.47$ (30.5, 77.47) is in the relation

The terms domain and range are important in the study of relations.

Definition 7.7. The domain and range of a relation
The set of all *first elements* in the ordered pairs of a relation is the **domain.** The set of all *second elements* in the ordered pairs of a relation is the **range.**

Example 2. A relation consists of the following set of ordered pairs: $\{(-2, -8), (-1, -5), (0, -2), (1, 1), (2, 4), (3, 7), (4, 10)\}$

a. Identify the domain.

b. Identify the range.

Solution. **a. Discussion.** The domain is the set of first elements: $\{-2, -1, 0, 1, 2, 3, 4\}$.

b. Discussion. The range is the set of second elements: $\{-8, -5, -2, 1, 4, 7, 10\}$.

A Definition of a Function

Many relations are defined in such a way that each number in the domain is paired with exactly one number in the range. These relations are called functions.

> **Definition 7.8. Function**
> A **function** is a relation in which each domain element is paired with exactly one range element. Lower-case letters such as f, g, and h are used to represent functions.

Examples **a–e** at the beginning of this section are all functions.

For a given quantity A		There is exactly one quantity B
a. time t in hours	$\longrightarrow$	**a*.** one distance d in miles
b. measurement in inches i	$\longrightarrow$	**b*.** one measurement in centimeters c
c. temperature in Fahrenheit°	$\longrightarrow$	**c*.** one temperature in Celsius°
d. radius r of circle	$\longrightarrow$	**d*.** one area A of circle
e. amount a of purchase	$\longrightarrow$	**e*.** one sales tax t on purchase

As examples of relations that are not functions, consider the relations defined in examples **f** and **g**.

f. $x = |y|$ *For a given positive number* x, *there are two numbers* y *whose absolute values are x.*

If $x = 3$, then $y = 3$ or $y = -3$.
If $x = 10$, then $y = 10$ or $y = -10$. } *Each* $x > 0$ *is paired with two* y*'s.*

g. $x = y^2$ *For a given positive number* x, *there are two numbers* y *whose squares are* x.

If $x = 25$, then $y = 5$ or $y = -5$.
If $x = 169$, then $y = 13$ or $y = -13$ } *Each* $x > 0$ *is paired with two* y*'s.*

Example 3. Determine whether each set of ordered pairs defines a function.

a. $\{(-3, 6), (-2, -1), (-1, -2), (0, -3), (1, -2), (2, 1), (3, 6)\}$

b. $\{(2, -3), (1, -2), (1, 0), (2, 1), (3, 2), (4, 3)\}$

Solution. **a.** Each domain element $\{-3, -2, -1, 0, 1, 2, 3\}$ is paired with exactly one range element $\{6, -1, -2, -3, 1\}$. Therefore, the given set of ordered pairs defines a function.

b. The following pairings of domain elements 2 and 1 occur in the set of ordered pairs:

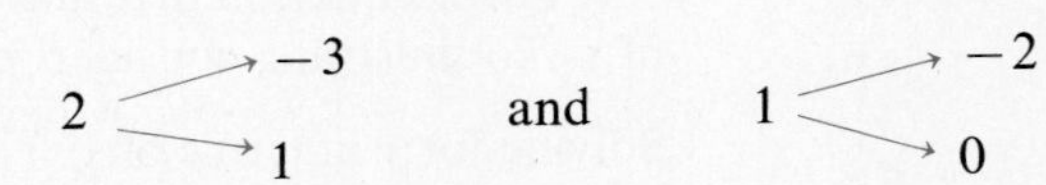

Since 2 and 1 are both paired with two different range elements, the set of ordered pairs does not define a function.

A Definition of a Linear Function in x and y

A linear equation in x and y is one that can be written in the form:

$$ax + by = c \qquad a \text{ and } b \text{ are not both } 0$$

The solution set of such an equation is a set of ordered pairs of real numbers. As a consequence, every linear equation in x and y defines a relation. If $ax + by = c$ can be written in the slope-intercept form, then the relation is a function.

Definition 7.9. A linear function in x and y
If m and b are constants, then a linear function in x and y can be defined by an equation of the form:

$$y = mx + b$$

The domain of any linear function is the set of real numbers, written "all x".

The only linear equations in x and y that are not functions are those that have vertical lines for graphs. Example **h** is such an equation. For this equation, two solutions are given that show that the same x is paired with two different y's. Therefore, the equation does not define a function.

h. $x + 0 \cdot y = 5$ $\quad$ $(5, -3)$ and $(5, 8)$ are solutions, and 5 is paired with -3 and 8

Example 4. Find each ordered pair in the function $y = \frac{2}{3}x - 5$.

a. $x = 6$ $\quad$ **b.** $x = -9$

Solution.

a. $y = \frac{2}{3}(6) - 5$ $\quad$ Replace x by 6.

$= 4 - 5$ $\quad$ Multiply first.

$= -1$ $\quad$ Thus, $(6, -1)$ is in the function.

b. $y = \frac{2}{3}(-9) - 5$ $\quad$ Replace x by -9.

$= -6 - 5$ $\quad$ Multiply first.

$= -11$ $\quad$ Thus, $(-9, -11)$ is in the function.

The Inverse of a Linear Function in x and y

A linear equation in x and y can be solved for y in terms of x, or for x in terms of y. To illustrate, consider $3x + 2y = 6$.

Solving for y in terms of x:

$3x + 2y = 6$

$2y = -3x + 6$

$y = \frac{-3}{2}x + 3 \quad (1)$

Equation (1) defines a function in x.

Solving for x in terms of y:

$3x + 2y = 6$

$3x = -2y + 6$

$x = \frac{-2}{3}y + 2 \quad (2)$

Equation (2) defines a function in y

In (1), replace x by -2, 0, and 4:

$$y = \frac{-3}{2}(-2) + 3 = 6; (-2, 6)$$

$$y = \frac{-3}{2}(0) + 3 = 3; (0, 3)$$

$$y = \frac{-3}{2}(4) + 3 = -3; (4, -3)$$

In (2), replace y by 6, 3, and -3:

$$x = \frac{-2}{3}(6) + 2 = -2; (6, -2)$$

$$x = \frac{-2}{3}(3) + 2 = 0; (3, 0)$$

$$x = \frac{-2}{3}(-3) + 2 = 4; (-3, 4)$$

Notice that the ordered pairs in the left column are *reversed* in the ordered pairs in the right column.

$(-2, 6)$	$(6, -2)$
$(0, 3)$	$(3, 0)$
$(4, -3)$	$(-3, 4)$

In many cases, the interchanging of the roles of the variables in a function has some advantages. As a consequence, mathematicians have added this dimension to the study of functions. However, to keep the role of the variables in the functions the same (that is, x as independent and y as dependent), the variables are interchanged in the equation that defines the function of x. The new equation in which the variables have been interchanged is called the **inverse function.** If f is used for the given linear function, then f^{-1}, which is read "inverse function", is the inverse of f. (*Note:* f^{-1} does not represent $\frac{1}{f}$ where (-1) is interpreted as an exponent.)

Definition 7.10. The inverse of a linear function in x and y
If f is defined by $y = mx + b$, then f^{-1} is defined by $x = my + b$.
For example, $y = 5x + 9$. Therefore, $x = 5y + 9$.
If $m \neq 0$, then the domain of f^{-1} is the set of real numbers.

Example 5. Find f^{-1}, for f defined by $y = 3x - 15$.

Solution. **Discussion.** To find an equation that defines f^{-1}, replace x by y and y by x in the equation that defines f.

f: $y = 3x - 15$	The given equation
f^{-1}: $x = 3y - 15$	An equation that defines f^{-1}
$x + 15 = 3y$	Add 15 to both sides.
$\frac{1}{3}x + 5 = y$	Divide each term by 3.
$y = \frac{1}{3}x + 5$	Written in the $y = mx + b$ form

Example 6. Find f^{-1}, for f defined by $y = \frac{-3}{5}x + 4$.

Solution.

f: $y = \frac{-3}{5}x + 4$ — The given equation

f^{-1}: $x = \frac{-3}{5}y + 4$ — An equation that defines f^{-1}

$5x = -3y + 20$ — Multiply both sides by 5.

$3y = -5x + 20$ — Subtract $5x$ and add $3y$ to both sides.

$y = \frac{-5}{3}x + \frac{20}{3}$ — Written in the $y = mx + b$ form

Graphing a Linear Function and Its Inverse on the Same Set of Axes

In Examples 7 and 8, the equations that define f^{-1} are determined, and then f and f^{-1} are both graphed on the same pair of axes.

Example 7. For f defined by $y = \frac{2}{3}x - 2$:

a. Find an equation that defines f^{-1}.

b. Graph f and f^{-1} on the same set of axes.

Solution.

a. f: $y = \frac{2}{3}x - 2$ — The given equation

f^{-1}: $x = \frac{2}{3}y - 2$ — An equation that defines f^{-1}

$x + 2 = \frac{2}{3}y$ — Add 2 to both sides.

$\frac{3}{2}x + 3 = y$ — Multiply each term by $\frac{3}{2}$.

$y = \frac{3}{2}x + 3$ — Written in the $y = mx + b$ form

b. $y = \frac{2}{3}x - 2$ — y-intercept is -2; slope is $\frac{2}{3}$

$y = \frac{3}{2}x + 3$ — y-intercept is 3; slope is $\frac{3}{2}$

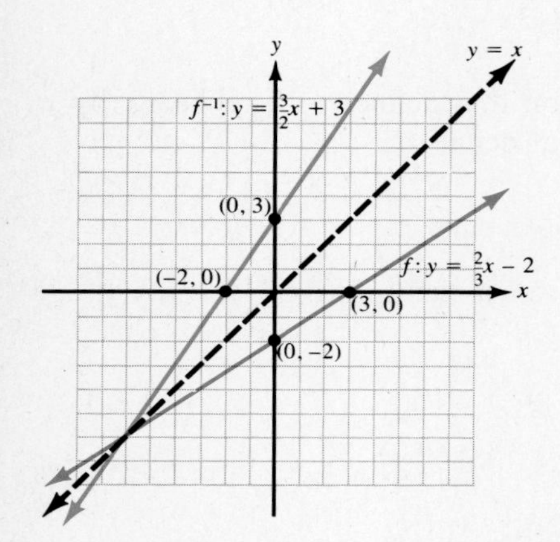

Figure 7-28. Graphs of f and f^{-1}.

The graphs of f and f^{-1} are shown in Figure 7-28. The line defined by $y = x$ is also shown. Notice that the graphs of f and f^{-1} would be the same line if the grid on which the graphs are drawn is folded along this line. Thus, for each (x, y) plotted on

f, the ordered pair (y, x) is located on f^{-1}. For the four points shown on the graphs:

$(0, -2)$ is on f and $(-2, 0)$ is on f^{-1}

$(3, 0)$ is on f and $(0, 3)$ is on f^{-1}

Example 8. For f defined by $y = \frac{-5}{3}x + 5$:

a. Find an equation that defines f^{-1}.

b. Graph f and f^{-1} on the same set of axes.

Solution. **a.** f: $y = \frac{-5}{3}x + 5$ The given equation

f^{-1}: $x = \frac{-5}{3}y + 5$ An equation that defines f^{-1}

$3x = -5y + 15$ Multiply both sides by 3.

$5y = -3x + 15$ Subtract $3x$ and add $5y$ to both sides.

$y = \frac{-3}{5}x + 3$ Written in the $y = mx + b$ form

b. $y = \frac{-5}{3}x + 5$ — y-intercept is 5; slope is $\frac{-5}{3}$

$y = \frac{-3}{5}x + 3$ — y-intercept is 3; slope is $\frac{-3}{5}$

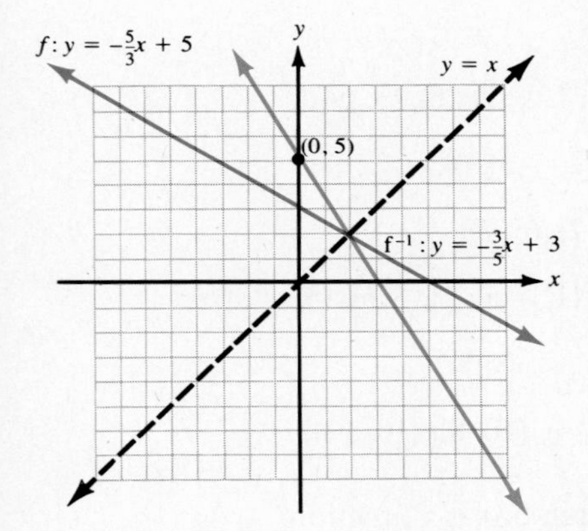

Figure 7-29. Graphs of f and f^{-1}.

Graphs of f and f^{-1} are shown in Figure 7-29. The line $y = x$ is also shown as a dotted line.

SECTION 7-5. Practice Exercises

In exercises **1–10**, find the ordered pairs of the given equation for each value.

[Example 1] The *circumference* C of a circle is related to the *diameter* d by $C = \pi d$, where $\pi \approx 3.14$. Write the ordered pairs in the form (d, C).

1. a. $d = 4$ **b.** $d = 2.5$ **2. a.** $d = 7$ **b.** $d = 4.8$

The total *price* p of an item, including sales tax, is related to the *cost* c by $p = 1.06c$. Write the ordered pairs in the form (c, p). If necessary, round p to the nearest cent.

3. a. $c = \$10$ **b.** $c = \$5.95$ **4. a.** $c = \$30$ **b.** $c = \$14.49$

The *sales price* S of an item, after a 20% discount, is related to the *original price* P by $S = 0.8P$. Write ordered pairs in the form (P, S).

5. a. $P = \$12$ **b.** $P = \$72.50$ **6. a.** $P = \$149$ **b.** $P = \$18.75$

The value of the *last term* L of a particular arithmetic progression is related to the *number* N of the term by $L = 12 + (N - 1)4$. Write ordered pairs in the form (N, L).

7. a. $N = 6$ **b.** $N = 102$ **8. a.** $N = 9$ **b.** $N = 91$

The *discriminant* D associated with a particular quadratic is related to the *constant* C by $D = 36 - 4C$. Write ordered pairs in the form (C, D).

9. a. $C = 4$ **b.** $C = 0$ **10. a.** $C = 9$ **b.** $C = 10$

In exercises **11–20**, identify the domain and range of each relation.

[Example 2] **11.** $\{(-2, 3), (-1, 4), (0, 5), (1, 6), (2, 7)\}$

12. $\{(-2, -7), (-1, -5), (0, -3), (1, -1), (2, 1)\}$

13. $\{(-4, 18), (-2, 6), (0, 2), (2, 6), (4, 18)\}$

14. $\{(-2, 0), (-1, 6), (0, 8), (1, 6), (2, 0)\}$

15. $\{(-3, -44), (-2, -6), (-1, 8), (0, 10), (1, 12), (2, 26), (3, 64)\}$

16. $\{(-3, 32), (-2, 13), (-1, 6), (0, 5), (1, 4), (2, -3), (3, -22)\}$

17. $\{(-6, 2), (-4, 2), (-2, 2), (0, 2), (2, 2), (4, 2), (6, 2)\}$

18. $\{(-5, -5), (-3, -5), (-1, -5), (0, -5), (1, -5), (3, -5), (5, -5)\}$

19. $\{(-8, -10), (-8, -5), (-8, 0), (-8, 5), (-8, 10)\}$

20. $\{(10, -12), (10, -8), (10, -5), (10, 0), (10, 4), (10, 8), (10, 12)\}$

In exercises **21–28**, determine whether each relation is a function.

[Example 3] **21.** $\{(-3, 80), (-2, 35), (-1, 8), (0, -1), (1, 8), (2, 35), (3, 80)\}$

22. $\{(-5, -14), (-3, -10), (-1, -6), (0, -4), (1, -2), (3, 2)\}$

23. $\{(-3, 1), (-3, -1), (-1, 2), (-1, -2), (0, 3), (0, -3)\}$

24. $\{(0, 4), (0, -4), (1, \sqrt{10}), (1, -\sqrt{2}), (2, \sqrt{3}), (2, -\sqrt{3})\}$

25. $\{(-3, 6), (-2, 6), (-1, 6), (0, 6), (1, 6), (2, 6), (3, 6)\}$

26. $\{(-1, 5), (-1, 3), (-1, 1), (-1, -1), (-1, -3), (-1, -5)\}$

27. $\{(9, -9), (-9, -9), (10, -9), (-10, -9)\}$

28. $\{(2, 0), (-2, 0), (4, 0), (-4, 0)\}$

In exercises **29–36**, find the ordered pairs of the given function for each value in the domain.

[Example 4] **29.** $y = 3x - 5$
a. $x = -4$
b. $x = 0$

30. $y = 7x - 2$
a. $x = -6$
b. $x = 0$

31. $y = 10 - 3x$
a. $x = -6$
b. $x = -11$

32. $y = 5 - 4x$
a. $x = -13$
b. $x = 15$

33. $y = 8$
a. $x = 4$
b. $x = \frac{-1}{2}$

34. $y = -16$
a. $x = -3$
b. $x = \frac{2}{3}$

35. $y = \frac{2}{3}x + \frac{1}{4}$
a. $x = -6$
b. $x = \frac{9}{4}$

36. $y = \frac{4}{5}x - \frac{1}{3}$
a. $x = 10$
b. $x = \frac{-5}{8}$

In exercises **37–48**, write in $y = mx + b$ form an equation for f^{-1} for each function f.

[Examples 5 and 6]

37. $f\colon y = 8x$

38. $f\colon y = \frac{x}{2}$

39. $f\colon y = x - 3$

40. $f\colon y = x + 10$

41. $f\colon y = 8x - 4$

42. $f\colon y = 3x + 6$

43. $f\colon y = \frac{-2}{3}x + 12$

44. $f\colon y = \frac{-5}{4}x + 20$

45. $f\colon y = \frac{x}{2} - 9$

46. $f\colon y = \frac{x}{6} + 2$

47. $f\colon y = \frac{2x + 5}{5}$

48. $f\colon y = \frac{4x - 1}{3}$

In exercises **49–60**, for each function f:

a. Graph f.

b. Write in $y = mx + b$ form an equation that defines f^{-1}.

c. Graph f^{-1} on the same axes as f.

[Examples 7 and 8]

49. $f\colon y = 2x + 6$

50. $f\colon y = 4x - 8$

51. $f\colon y = \frac{-1}{3}x - 2$

52. $f\colon y = \frac{-1}{5}x + 1$

53. $f\colon y = \frac{6}{5}x$

54. $f\colon y = \frac{4}{3}x$

55. $f\colon y = \frac{3}{4}x - 6$

56. $f\colon y = \frac{3}{5}x + 6$

57. $f\colon y = \frac{-7}{2}x + 7$

58. $f\colon y = \frac{-5}{3}x - 5$

59. $f\colon y = x - \dfrac{1}{2}$

60. $f\colon y = -x + \dfrac{3}{2}$

SECTION 7-5. Ten Review Exercises

In exercises **1–10**, do the indicated operations.

1. $(5xy)(2xy)$

2. $(5x + y)(2x + y)$

3. $(5xy)^2$

4. $(5x - y)^2$

5. $(5xy)(2xy)^{-1}$

6. $5xy(2x - y)(2x + y)$

7. $\dfrac{(-6t^2u^{-3})^{-2}}{-3t^{-5}u^4}$

8. $\dfrac{15t^4u - 35t^3u^2 - 40t^2u^3}{5t^2u}$

9. $\dfrac{t^3 - 125u^3}{t^2 - 10tu + 25u^2}$

10. $\dfrac{10t^3 + 33t^2u - 22tu^2 + 3u^3}{5t - u}$

SECTION 7-5. Supplementary Exercises

The graph of the inverse is a reflection of the function across the line $y = x$ (see Fig. 7-28). Each point on the inverse is the same distance away from $y = x$ as its corresponding point on the function, but the two points are on opposite sides of the line $y = x$ unless both are on the line.

Example. Graph f^{-1} by using the graph of f.

Solution. Graph $y = x$. Draw two lines perpendicular to $y = x$. Mark the distance from $y = x$ to the points on the opposite side of $y = x$. Draw a straight line through these points.

In exercises **1–8**, graph f^{-1} by using the graphs of f.

1.

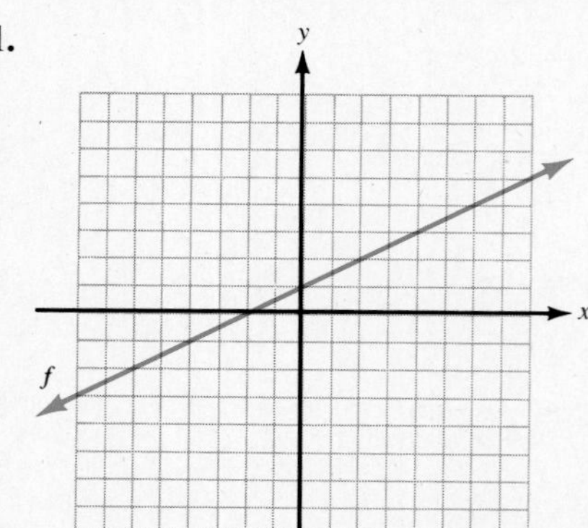

2.

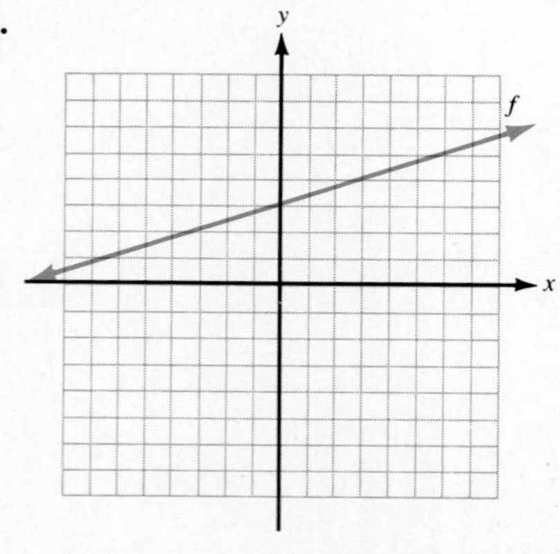

3.

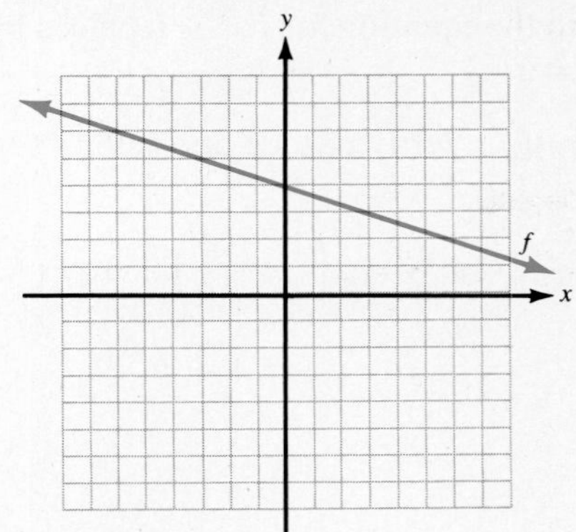

4.

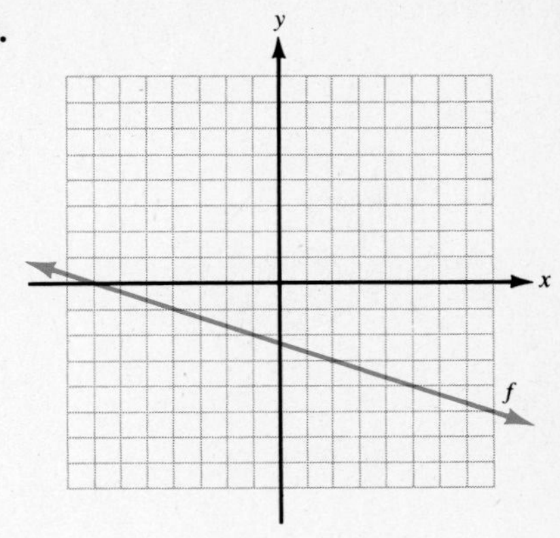

5.

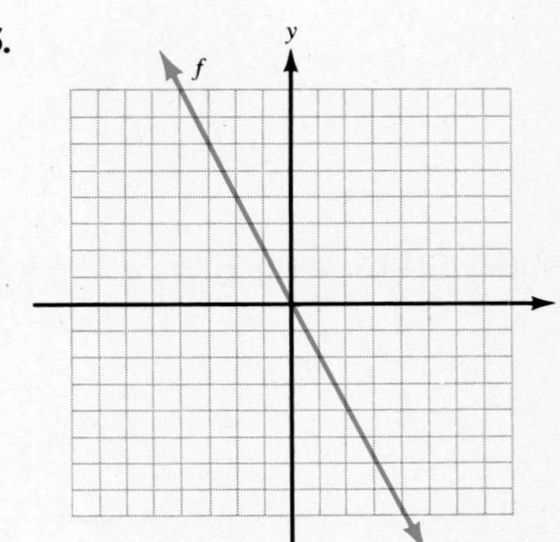

6.

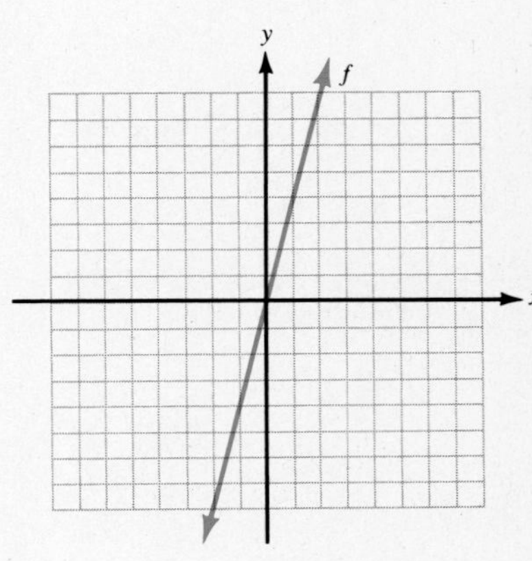

7.

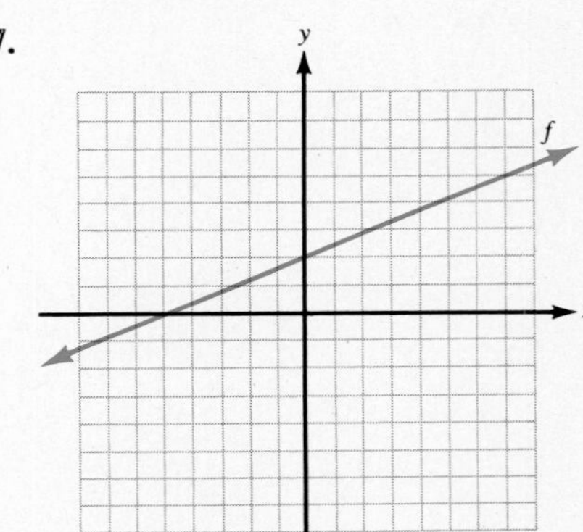

8.

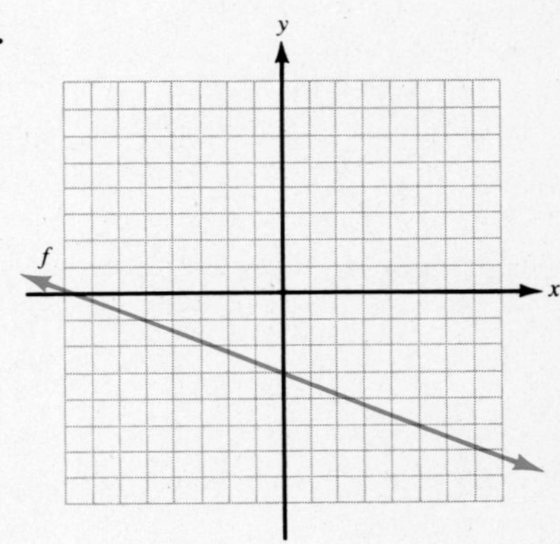

In exercises **9–12**, for each function f:

a. Write an equation for f^{-1} in the $y = mx + b$ form.

b. Write an equation for the inverse obtained in part **a** and compare with the given equation.

9. $f: y = 3x + 1$

10. $f: y = 4x + 2$

11. $f: y = \frac{4}{3}x - 5$

12. $f: y = \frac{5}{2}x - 3$

Suppose f is defined by $y = m_1x + b_1$ and f^{-1} is defined by $y = m_2x + b_2$

a. If x in the equation for f is replaced by $m_2x + b_2$, then the simplified equation is $y = x$.

b. If x in the equation for f^{-1} is replaced by $m_1x + b_1$, then the simplified equation is $y = x$.

Example. Verify parts **a** and **b** for f: $y = 2x + 3$ and f^{-1}: $y = \frac{1}{2}x - \frac{3}{2}$.

Solution. **a.** $y = 2\left(\frac{1}{2}x - \frac{3}{2}\right) + 3$ — Replace x in f by $\frac{1}{2}x - \frac{3}{2}$.

$y = x - 3 + 3$ — Distribute the 2.

$y = x$ — The simplified equation

b. $y = \frac{1}{2}(2x + 3) - \frac{3}{2}$ — Replace x in f^{-1} by $2x + 3$.

$y = x + \frac{3}{2} - \frac{3}{2}$ — Distribute the $\frac{1}{2}$.

$y = x$ — The simplified equation

In exercises **13–16**, verify parts **a** and **b** for each pair of f and f^{-1}.

13. f: $y = 5x - 2$
f^{-1}: $y = \frac{1}{5}x + \frac{2}{5}$

14. f: $y = 6x + 4$
f^{-1} $y = \frac{1}{6}x - \frac{2}{3}$

15. f: $y = \frac{1}{2}x + 3$
f^{-1}: $y = 2x - 6$

16. f: $y = \frac{1}{4}x - 5$
f^{-1}: $y = 4x + 20$

Name

Date

Score

SECTION 7-5. Summary Exercises

Answer

1. The *number of feet d* that a car travels over a *time period t* while moving at 60 mph is found using $d = 88t$. Find the number of feet traveled if the time:

a. $t = 7$ seconds

1. a. __________

b. $t = 4.5$ seconds

b. __________

In exercises **2** and **3**, identify the domain and range of each relation.

2. $\{(-2, 4), (6, 3), (10, 0), (11, -3)\}$

2. Domain: __________

Range: __________

3. $\{(5, 4), (3, 2), (1, 0), (-1, -2)\}$

3. Domain: __________

Range: __________

In exercises **4** and **5**, determine whether each relation is a function.

4. $\{(0, 0), (3, 2), (-6, 7), (4, -1)\}$

4. __________

5. $\{(11, 23), (-10, 8), (9, 4), (11, 32)\}$

5. ____________________

In exercises **6** and **7**, for each function f find f^{-1}, and write equations in the $y = mx + b$ form.

6. $f\colon y = \frac{1}{4}x - 1$

6. ____________________

7. $f\colon y = -2x + 18$

7. ____________________

8. a. Graph $f\colon y = \frac{-3}{2}x + 8.$

b. Write in $y = mx + b$ form an equation that defines f^{-1}.

c. Graph f^{-1} on the same axes as f.

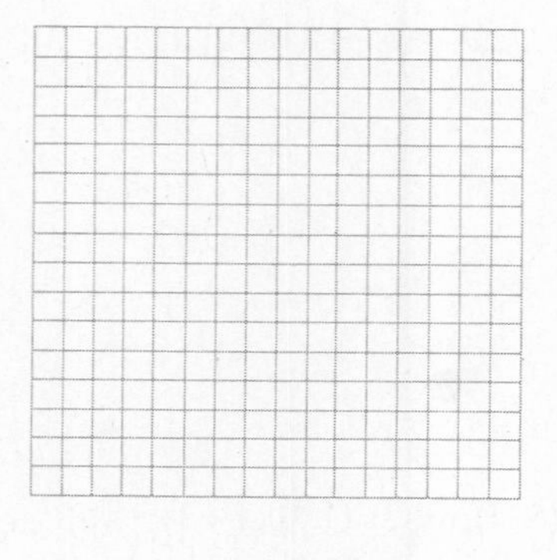

CHAPTER 7 Review Exercises

In exercises **1** and **2**, find the distance between P and Q.

1. $P(-3, -2)$, $Q(5, 4)$

2. $P(6, -2)$, $Q(-3, -5)$

In exercises **3** and **4**, find the possible values of x and y for the given distance d between the points.

3. $P(x, 2)$, $Q(-4, 4)$: $d = 2\sqrt{5}$

4. $P(-6, 8)$, $Q(0, y)$: $d = 10$

In exercises **5** and **6**, find the midpoints of each line segment with endpoints P and Q.

5. $P(-10, 12)$, $Q(-2, -4)$

6. $P(7, -3)$, $Q(6, -6)$

In exercises **7** and **8**, find the values of x and y given the midpoint M and endpoints P and Q.

7. $M(-2, 0)$, $P(x, 1)$, $Q(-6, y)$

8. $M(2, 1)$, $P(5, y)$, $Q(x, -5)$

In exercises **9** and **10**, determine whether the given ordered pairs are solutions of each equation.

9. $6x - 5y = 17$
- **a.** $(-3, -7)$
- **b.** $(-2, -6)$
- **c.** $(2, -1)$

10. $3u + 7v = -13$
- **a.** $(5, -4)$
- **b.** $(-9, 2)$
- **c.** $(3, -3)$

In exercises **11** and **12**, find the value of the missing point so that each ordered pair is a solution of the given equation.

11. $2x + 9y = 13$
- **a.** $(2, \square)$
- **b.** $(\square, -1)$

12. $\frac{1}{4}u - \frac{1}{3}v = 7$
- **a.** $(\square, -15)$
- **b.** $(32, \square)$

In exercises **13–16**, graph each equation.

13. $5x + 3y = 12$

14. $4x - 7y = 35$

15. $y - 6 = 0$

16. $x + 4 = 0$

In exercises **17–20**, find the slope of the line that passes through each pair of points. For any vertical line, write "undefined".

17. $P_1(-7, 4)$, $P_2(3, 10)$

18. $P_1(5, 8)$, $P_2(-4, 8)$

19. $P_1(13, 10)$, $P_2(13, -5)$

20. $P_1(3, -15)$, $P_2(-3, 9)$

In exercises **21** and **22**, graph the line that passes through P with slope m.

21. $P(1, 6)$, $m = \frac{-3}{4}$

22. $P(-5, -3)$, $m = 3$

In exercises **23** and **24**, for each equation:

a. Find the x-intercept.

b. Find the y-intercept.

23. $6x - 9y = 18$

24. $x - 1 = \frac{1}{3}(3 - 2y)$

In exercises **25** and **26**, for each equation:

a. Find the slope.

b. Find the y-intercept.

25. $8x + 3y = 6$

26. $\frac{y - x + 2}{2} = \frac{x + 1}{3}$

In exercises **27** and **28**, graph each equation.

27. $y = \frac{1}{2}x - 4$

28. $y = \frac{-7}{4}x + 5$

In exercises **29** and **30**, determine whether the graphs of each pair of equations are parallel, perpendicular, or neither.

29. (A) $10x - 9y = 18$

(B) $\frac{x + 3}{6} = \frac{3y - x - 1}{4}$

30. (A) $5x - 6y = -6$

(B) $\frac{4 + 2x - y}{2} = \frac{8x + 5}{5}$

In exercises **31–38**, write an equation for each line in the $ax + by = c$ form.

31. Containing $(8, -8)$ and with slope -2

32. Containing $(-4, 6)$ and with slope $\frac{3}{4}$

33. Containing $(3, -5)$ and $(-7, 1)$

34. Containing $(4, -9)$ and $(0, 0)$

35. Containing $(0, -10)$ and parallel to $4x + 10y = 7$

36. Containing $(-9, 6)$ and perpendicular to $15x - 10y = -12$

37. With slope $\frac{10}{7}$ and y-intercept -5

38. With slope $\frac{-3}{5}$ and y-intercept 8

In exercises **39** and **40**, determine whether the given ordered pairs are solutions of each inequality.

39. $2x + 5y > -12$
a. $(3, -3)$
b. $(-6, 0)$

40. $4x - 7y \leq 16$
a. $(-2, -4)$
b. $(4, 0)$

In exercises **41–44**, graph each inequality.

41. $x - 2y < -2$

42. $5x + y \geq -6$

43. $2y - 6 > 0$

44. $4x - 20 \leq 0$

In exercises **45–48**, for each set of ordered pairs:

a. Identify the domain.

b. Identify the range.

c. Determine whether it is a function.

45. $\{(-3, 0), (7, 2), (4, -5), (1, 0), (-10, -9)\}$

46. $\{(8, 2), (-6, 3), (5, 1), (-8, 2)\}$

47. $\{(4, 2), (-6, 2), (0, 2), (8, 2)\}$

48. $\{(-2, -1), (0, -1), (2, 3), (4, 3), (6, 3)\}$

In exercises **49** and **50**, find the value in the range of the given function for each value in the domain.

49. $y = \frac{3}{2}x - 12$

a. $x = 6$
b. $x = -12$

50. $y = \frac{-1}{4}x + 5$

a. $x = 16$
b. $x = -4$

In exercises **51** and **52**, write in $y = mx + b$ form an equation for f^{-1} for each function f.

51. $f\colon y = 6x + 18$

52. $f\colon y = \frac{-3}{5}x + 6$

In exercises **53** and **54**, for each function f:

a. Graph f.

b. Write in $y = mx + b$ form an equation that defines f^{-1}.

c. Graph f^{-1} on the same axes as f.

53. $f\colon y = \frac{-1}{2}x + 2$

54. $f\colon y = 4x - 3$

In exercises **55–60**, for each pair of points:

a. Plot the points on an x-y axis.

b. Find the distance between the points.

c. Find the midpoint of the line segment joining the points.

d. Find the slope of the line passing through these points.

e. Write an equation in $ax + by = c$ form for the line passing through the points.

f. Write an inequality for the half-plane that has the line as boundary and the point $O(0, 0)$ in the half-plane.

55. $P(-5, 3)$, $Q(3, -7)$

56. $P(-3, 6)$, $Q(5, -4)$

57. $P(-5, -1)$, $Q(2, 1)$

58. $P(-2, -3)$, $Q(7, 0)$

59. $P(4, -8)$, $Q(4, 2)$

60. $P(-3, -2)$, $Q(7, -2)$

CHAPTER

8 Graphs of Other Functions

Ms. Glaston was carrying a box when she entered the classroom. "During the last few sessions," she began, "we have been studying relationships that can be described by linear equations in two variables." She then wrote on the board:

(1) $ax + by = c$ The standard-form equation

(2) $y = mx + b$ The slope-intercept equation

(3) $y - y_1 = m(x - x_1)$ The point-slope equation

"Not all relationships between variable quantities," she continued, "can be described by linear equations in two variables. For example, consider this box on the desk. At this time, I will tell you that the lengths of all the edges of this box are the same. There are quantities related to this box that have certain relationships to each other. Furthermore, these relationships cannot be described by linear equations. Can anyone suggest one of these relationships?"

After a lengthy silence, Sam Gore said, "I'm not sure this is what you're looking for, but I think the volume of a box has something to do with the dimensions of the box. That is, how long it is, how wide it is, and how high it is."

"Okay," Ms. Glaston replied, "Sam has suggested that once the dimensions of a box are known, one can use an equation, or formula, to calculate the volume. For this box, all the edges are the same length, so the length, width, and height are the same value. Suppose we let s represent the length of an edge, and V the volume of the box. It can be shown that V and s are related in a way that is described by the following equation." She then wrote on the board:

$V = s^3$ The volume is the cube of the length of a side.

"Since s is raised to the third power," Ms. Glaston continued, "this equation is not linear."

"If the volume is related to the length of a side," Annie Todd said, "then the outside area of the box must also be related to a side."

"It would seem so, Annie," Ms. Glaston replied. Then, pointing to the box, she added, "Each face of this box is a square. Therefore, the area of each face is s^2. The box has six of these squares, so the total area of the box is six times the area of each face. Let's use A for the total area." She then wrote on the board:

$A = 6s^2$ The area of the box is six times the square of the length of a side.

"Since s is raised to the second power," Ms. Glaston continued, "this equation is not linear."

"How about the weight of the box?" Bob Smith asked. "Is there a formula for computing that?"

"Well," Ms. Glaston replied, "the weight depends not only on the dimensions of the box, but also on what is in the box, and what the box is made of. Physicists use the concept of density to describe the weight, or mass, of an object in terms of a unit volume. For example, if we knew that the density of this box is 5 pounds per cubic foot, then the following formula could be written."

$W = 5s^3$ Weight equals 5 pounds per ft^3 times the volume.

"There are other quantities whose relationships we could examine," Ms. Glaston continued, "but I think we have some idea of the fact that quantities can be related by equations that are not linear in two variables. In the chapter we begin studying today, we will look at a few of the types that commonly occur. As we will also see, many of these relations are described by equations that are functions."

SECTION 8-1. The Absolute Value Function

1. The definition of the absolute value function
2. Graphing $y = a|x|$
3. Graphing $y = |x - b|$
4. Graphing $y = |x| + c$
5. Graphing $y = a|x - b| + c$

In previous sections, the absolute value operator has been used for several purposes:

a. To take the absolute value of a number $|8| = 8$ and $|-2| = 2$

b. To write an absolute value equation $|2t - 3| = 7$

c. To write an absolute value inequality $|x - 1| < 5$ and $|x + 5| > 9$

In this section, we will define the absolute value function and study a procedure for graphing such functions.

The Definition of the Absolute Value Function

Figure 8-1 is a graph of $y = x$. The points with coordinates $(-6, -6)$, $(-3, -3)$, $(0, 0)$, $(3, 3)$, and $(6, 6)$ are plotted on the graph. Suppose now that x in the equation is placed in absolute value bars to define the relation:

$$y = |x|$$

What effect would this have on the graph shown in Figure 8-1? Consider the effect of the absolute value operator on the five points shown on $y = x$:

x	$y = \|x\|$	Point on graph
-6	$y = \|-6\| = 6$	$(-6, 6)$
-3	$y = \|-3\| = 3$	$(-3, 3)$
0	$y = \|0\| = 0$	$V(0, 0)$
3	$y = \|3\| = 3$	$(3, 3)$
6	$y = \|6\| = 6$	$(6, 6)$

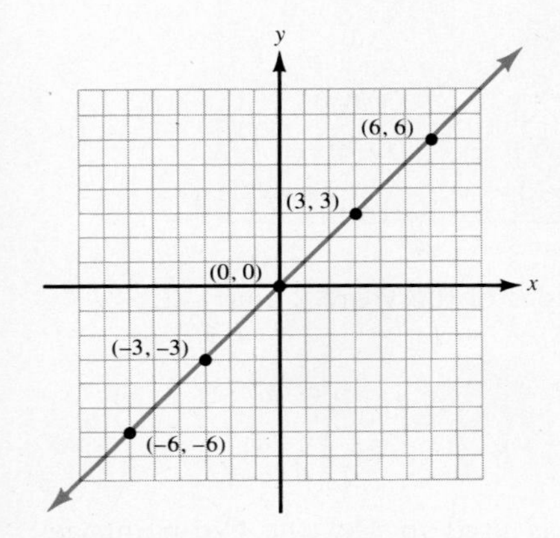

Figure 8-1. A graph of $y = x$.

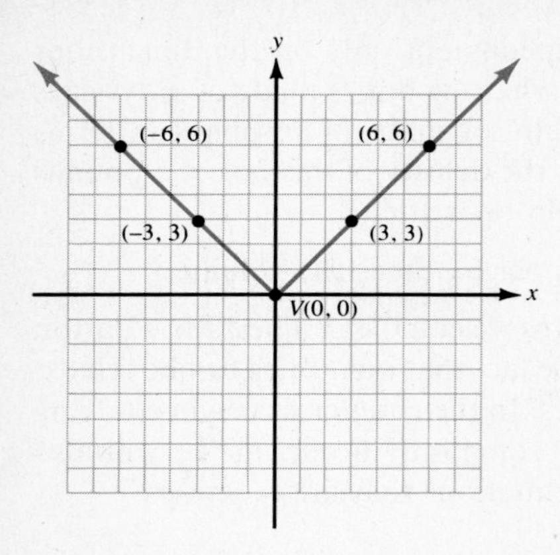

Figure 8-2. A graph of $y = |x|$.

In Figure 8-2, these points are plotted to give shape to a graph of $y = |x|$. Notice that the line in Figure 8-1 has been "broken" and part of it has been directed upward to form a V. This V-shape is typical of a graph of a function in which x is put inside absolute value bars. The point of the V is called the **vertex.** By using constants a, b, and c, the slopes of the sides of the V can be changed and the location of the V can be shifted to the left or to the right, and upward or downward.

Definition 8.1. The absolute value function
If a, b, and c are real numbers and $a \neq 0$, then an absolute value function can be defined by an equation of the form:

$$y = a|x - b| + c$$

The domain is the set of real numbers, written "all x".

Graphing $y = a|x|$

In Figure 8-2, the slopes of the sides of the V are 1 and -1. The positive 1 is the slope of the side to the right of the vertex, and the negative 1 the slope of the side to the left. In examples **a** and **b**, the absolute value of x is multiplied by nonzero constants.

a. $y = 2|x|$ **b.** $y = \frac{-2}{3}|x|$

The effect of these constants on the graph in Figure 8-2 is to change the slope of the sides. Furthermore, in example **b** the negative constant changes the direction of the V from upward to downward.

Example 1. Graph $y = 2|x|$.

Solution. The following table of values is used to plot the five points in Figure 8-3:

x	$y = 2\|x\|$	Ordered pair
-6	$y = 2\|-6\| = 12$	$(-6, 12)$
-3	$y = 2\|-3\| = 6$	$(-3, 6)$
0	$y = 2\|0\| = 0$	$V(0, 0)$
3	$y = 2\|3\| = 6$	$(3, 6)$
6	$y = 2\|6\| = 12$	$(6, 12)$

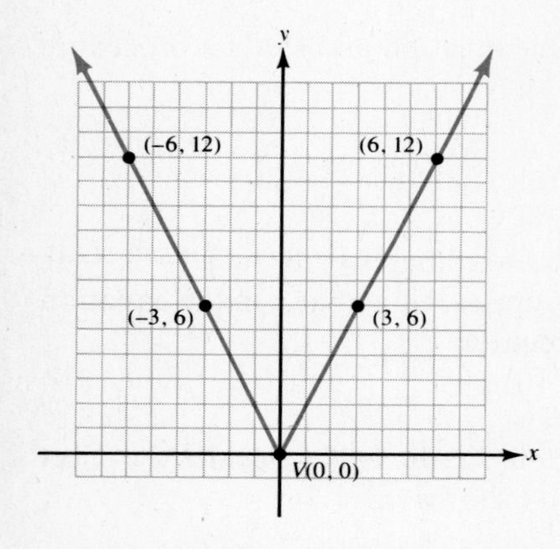

Figure 8-3. A graph of $y = 2|x|$.

Notice that the slopes of the sides of the V are 2 and -2.

Example 2. Graph $y = \frac{-2}{3}|x|$.

Solution. The following table of values is used to plot the five points in Figure 8-4:

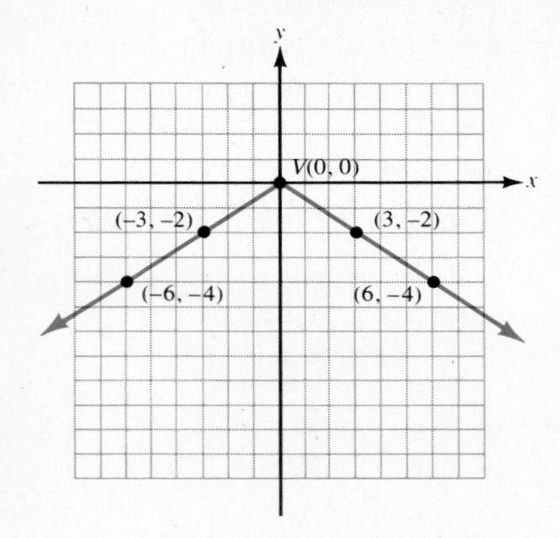

Figure 8-4. A graph of $y = \frac{-2}{3}|x|$.

x	$y = \frac{-2}{3}\lvert x\rvert$	Ordered pair
-6	$y = \frac{-2}{3}\lvert -6\rvert = -4$	$(-6, -4)$
-3	$y = \frac{-2}{3}\lvert -3\rvert = -2$	$(-3, -2)$
0	$y = \frac{-2}{3}\lvert 0\rvert = 0$	$V(0, 0)$
3	$y = \frac{-2}{3}\lvert 3\rvert = -2$	$(3, -2)$
6	$y = \frac{-2}{3}\lvert 6\rvert = -4$	$(6, -4)$

Notice that the slopes of the sides of the V are $\frac{-2}{3}$ and $\frac{2}{3}$. Furthermore, because $\frac{-2}{3} < 0$, the V is directed downward.

If $y = a|x|$, where a is a nonzero constant, then a graph of the function is a V.

1. The slopes of the sides of the V are a and $-a$.
2. The V is directed upward if $a > 0$.
3. The V is directed downward if $a < 0$.

Graphing $y = |x - b|$

In Figure 8-2, the vertex is $V(0, 0)$. The equation for the function can be written $y = |x - 0|$, which indicates a 0 is subtracted from x within the absolute value bars. In examples **c** and **d**, nonzero constants are subtracted from x within the absolute value bars.

c. $y = |x - 2|$ **d.** $y = |x - (-3)|$, or $y = |x + 3|$

The effect of these constants on the graph in Figure 8-2 is to shift the vertex to the right, or to the left, depending on the sign of the constant.

Example 3. Graph $y = |x - 2|$.

Solution. The following table of values is used to plot the five points in Figure 8-5:

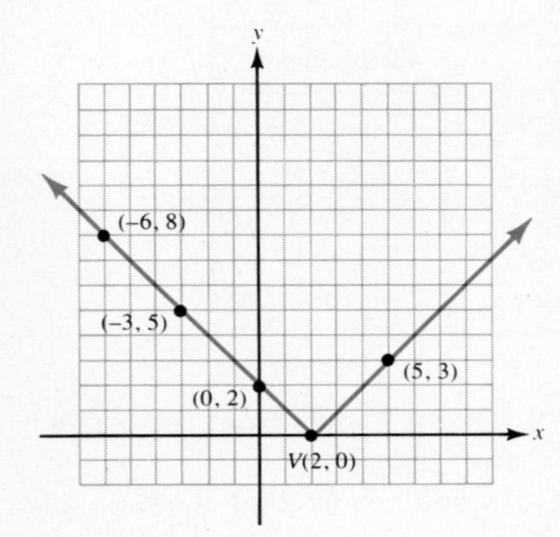

Figure 8-5. A graph of $y = |x - 2|$.

x	$y = \lvert x - 2\rvert$	Ordered pair
-6	$y = \lvert -6 - 2\rvert = 8$	$(-6, 8)$
-3	$y = \lvert -3 - 2\rvert = 5$	$(-3, 5)$
0	$y = \lvert 0 - 2\rvert = 2$	$(0, 2)$
2	$y = \lvert 2 - 2\rvert = 0$	$V(2, 0)$
5	$y = \lvert 5 - 2\rvert = 3$	$(5, 3)$

Notice that the 2 in $x - 2$ shifted the vertex *to the right* of the origin 2 units.

Example 4. Graph $y = |x + 3|$.

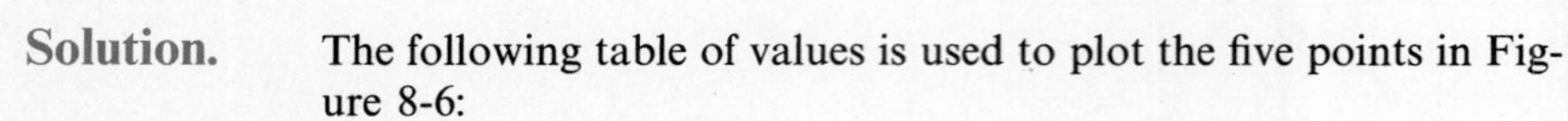

Solution. The following table of values is used to plot the five points in Figure 8-6:

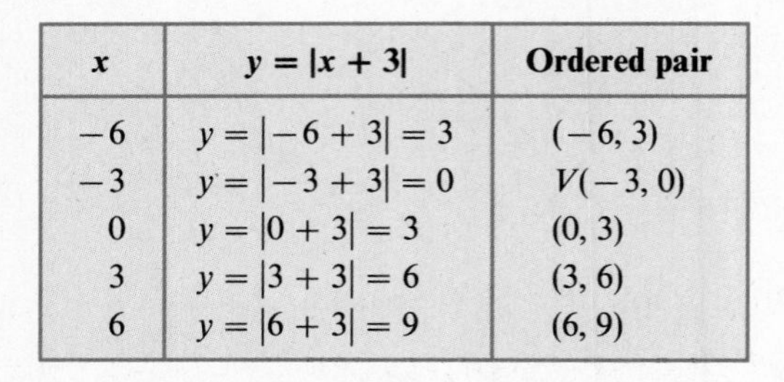

x	$y = \lvert x + 3 \rvert$	Ordered pair
-6	$y = \lvert -6 + 3 \rvert = 3$	$(-6, 3)$
-3	$y = \lvert -3 + 3 \rvert = 0$	$V(-3, 0)$
0	$y = \lvert 0 + 3 \rvert = 3$	$(0, 3)$
3	$y = \lvert 3 + 3 \rvert = 6$	$(3, 6)$
6	$y = \lvert 6 + 3 \rvert = 9$	$(6, 9)$

Notice that the -3 in $x - (-3)$ shifted the vertex *to the left* of the origin 3 units.

> If $y = |x - b|$, where b is a nonzero constant, then a graph of the function is a V whose vertex is $V(b, 0)$.

Graphing $y = |x| + c$

In Figure 8-2, the vertex is $V(0, 0)$. The equation for the function can be written $y = |x| + 0$, which indicates a 0 added to the absolute value of x. In examples **e** and **f**, nonzero constants are added to the absolute values of x.

e. $y = |x| + 3$ **f.** $y = |x| + (-2)$, or $y = |x| - 2$

The effect of these constants on the graph in Figure 8-2 is to shift the vertex up or down, depending on the sign of the constant.

Example 5. Graph $y = |x| + 3$.

Solution. The following table of values is used to plot the five points in Figure 8-7:

x	$y = \lvert x \rvert + 3$	Ordered pair
-6	$y = \lvert -6 \rvert + 3 = 9$	$(-6, 9)$
-3	$y = \lvert -3 \rvert + 3 = 6$	$(-3, 6)$
0	$y = \lvert 0 \rvert + 3 = 3$	$V(0, 3)$
3	$y = \lvert 3 \rvert + 3 = 6$	$(3, 6)$
6	$y = \lvert 6 \rvert + 3 = 9$	$(6, 9)$

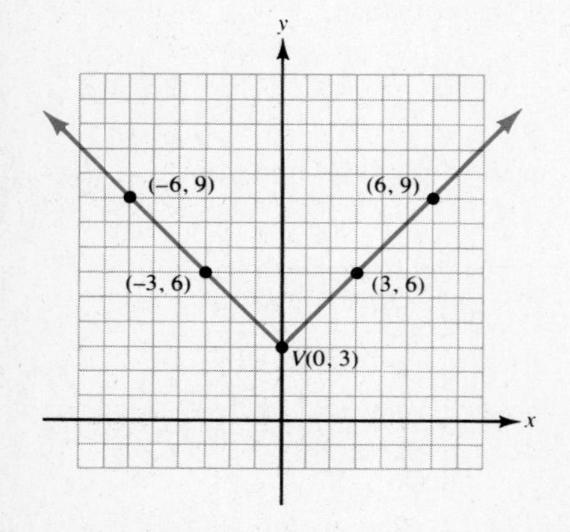

Figure 8-7. A graph of $y = |x| + 3$.

Notice that the 3 in the equation shifted the vertex up 3 units from the origin. *The shift is up because a positive 3 is added to* $|x|$.

An analysis similar to the one in Example 5 would show that a graph of $y = |x| - 2$ would be the graph in Figure 8-2 shifted down 2 units. Thus, the vertex would be $V(0, -2)$. *The shift would be down because* (-2) *is added to* $|x|$.

> If $y = |x| + c$, where c is a nonzero constant, then a graph of the function is a V whose vertex is $V(0, c)$.

Graphing $y = a|x - b| + c$

Examples **g** and **h** are equations that define absolute value functions. In these equations a, b, and c are all nonzero constants.

g. $y = \frac{3}{2}|x - 1| + 2$ **h.** $y = -2|x + 2| - 1$

The following steps can be used to graph these functions:

> **To graph $y = a|x - b| + c$:**
>
> **Step 1.** List the values of a, b, and c.
>
> **Step 2.** Locate the vertex $V(b, c)$.
>
> **Step 3.** Use a and $-a$ as the slopes of the sides to locate one or two points on each side of the V. Remember, if $a > 0$, the V is directed upward. If $a < 0$, the V is directed downward.
>
> **Step 4.** Draw the V using the points plotted in Steps 2 and 3.

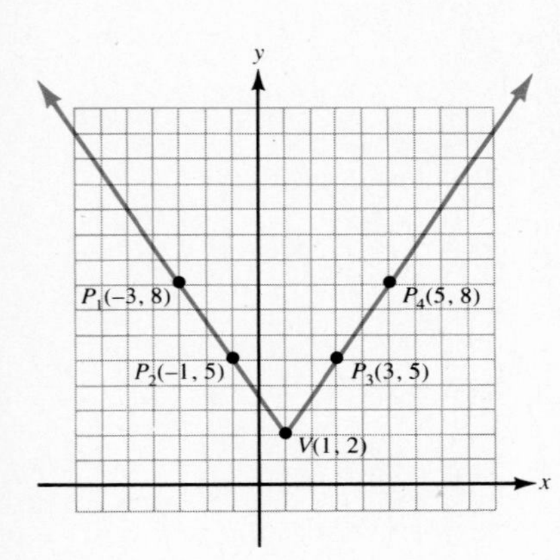

Figure 8-8. A graph of $y = \frac{3}{2}|x - 1| + 2$.

Example 6. Graph $y = \frac{3}{2}|x - 1| + 2$.

Solution.

Step 1. $a = \frac{3}{2}$, $b = 1$, and $c = 2$

Step 2. With $b = 1$ and $c = 2$, $V(1, 2)$ is the vertex and is plotted in Figure 8-8.

Step 3. Since $\frac{3}{2} > 0$, the V is directed upward. Using $-a = \frac{-3}{2}$, the points $P_1(-3, 8)$ and $P_2(-1, 5)$ are plotted in Figure 8-8. Using $a = \frac{3}{2}$, the points $P_3(3, 5)$ and $P_4(5, 8)$ are plotted in the figure.

Step 4. Using P_1, P_2, V, P_3 and P_4, the graph in Figure 8-8 is drawn.

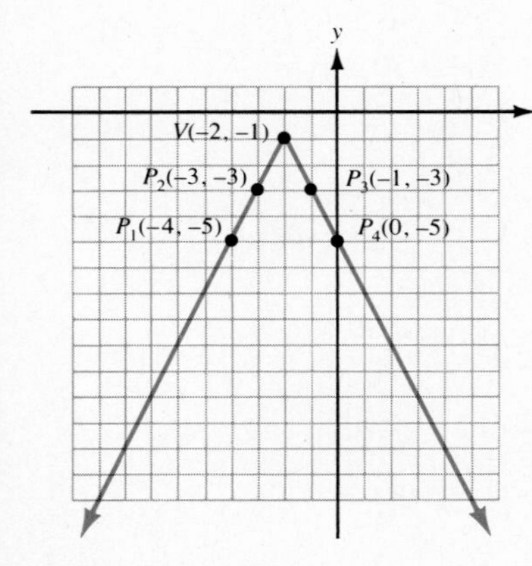

Figure 8-9. A graph of $y = -2|x + 2| - 1$.

Example 7. Graph $y = -2|x + 2| - 1$.

Solution.

Step 1. $a = -2$, $b = -2$, and $c = -1$

Step 2. With $b = -2$ and $c = -1$, the point $V(-2, -1)$ is plotted in Figure 8-9.

Step 3. Since $-2 < 0$, the V is directed downward. Using $-a = 2$, the points $P_1(-4, -5)$ and $P_2(-3, -3)$ are plotted. Using $a = -2$, the points $P_3(-1, -3)$ and $P_4(0, -5)$ are plotted.

Step 4. Using P_1, P_2, V, P_3, and P_4, the V in Figure 8-9 is drawn.

SECTION 8-1. Practice Exercises

In exercises **1–50**, graph each equation.

[Examples 1 and 2]

1. $y = 4|x|$

2. $y = 5|x|$

3. $y = \frac{1}{3}|x|$

4. $y = \frac{1}{2}|x|$

5. $y = -2|x|$

6. $y = -|x|$

7. $y = \frac{-3}{5}|x|$

8. $y = \frac{-4}{7}|x|$

[Examples 3 and 4]

9. $y = |x + 1|$

10. $y = |x + 3|$

11. $y = |x - 3|$

12. $y = |x - 5|$

13. $y = 2|x + 8|$

14. $y = \frac{2}{3}|x + 6|$

15. $y = -|x - 7|$

16. $y = -3|x - 2|$

17. $y = \frac{5}{3}|x - 1|$

18. $y = \frac{3}{5}|x - 4|$

19. $y = \frac{-1}{4}|x + 2|$

20. $y = \frac{-3}{2}|x + 3|$

[Example 5]

21. $y = |x| + 1$

22. $y = |x| + 5$

23. $y = 4 - |x|$

24. $y = 6 - 3|x|$

25. $y = 2|x| - 4$

26. $y = \frac{3}{4}|x| - 6$

27. $y = \frac{-1}{2}|x| + 8$

28. $y = \frac{-5}{3}|x| + 9$

[Examples 6 and 7]

29. $y = |x + 2| - 1$

30. $y = |x - 3| - 2$

31. $y = 8 - |x + 4|$

32. $y = 6 - |x - 1|$

33. $y = \frac{3}{2}|x - 2| + 1$

34. $y = \frac{2}{5}|x + 3| - 1$

35. $y = -2|x| + 5$

36. $y = -3|x| - 2$

37. $y = -4|x + 1| + 3$

38. $y = -3|x - 2| + 4$

39. $y = \frac{4}{3}|x - 3| + 1$

40. $y = \frac{5}{2}|x + 4| - 3$

41. $5y - 5 = 2|x|$

42. $4y - 12 = -3|x|$

43. $y = |2x - 8|$

44. $y = |3x + 9|$

45. $2y + 3|x| = 6$

46. $3y + 5|x| = 3$

47. $y = |3 - x| - 2$

48. $y = |5 + x| - 3$

49. $30 + 5y = 4|1 - x|$

50. $2y - 10 = -7|4 - x|$

SECTION 8-1. Ten Review Exercises

In exercises **1–10**, simplify each expression. Write exponential expressions with only positive exponents.

1. $\sqrt{\dfrac{3}{2}}$

2. $3\sqrt[3]{5} - \sqrt[3]{40}$

3. $\sqrt[4]{\dfrac{7}{27}}$

4. $\dfrac{18}{4 - \sqrt{7}}$

5. $\sqrt[6]{8t^3}$; $t > 0$

6. $(-u^2v^{-1})^3(2u^4v^{-2})^{-2}$

7. $\dfrac{(6a^{-3})^2}{(-3a)^{-3}}$

8. $(2k^{1/2} + 3)^2$

9. $(5m^{1/2} + 3)(5m^{1/2} - 3)$

10. $(-2x^{-2}y)^{-5}$

SECTION 8-1. Supplementary Exercises

In exercises **1–8**, write in the $y = a|x - b| + c$ form an equation for each graph.

1.

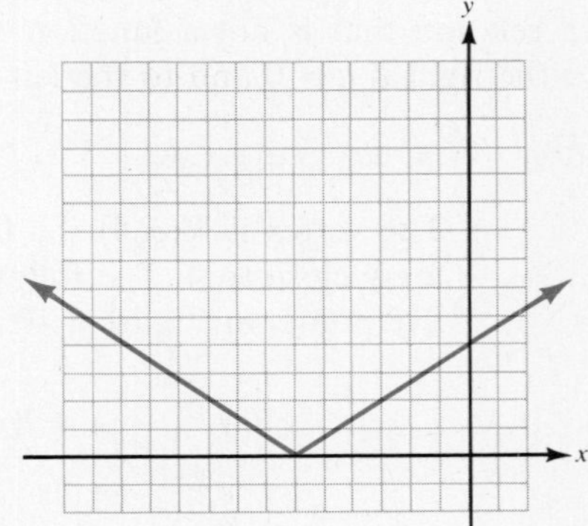

2.

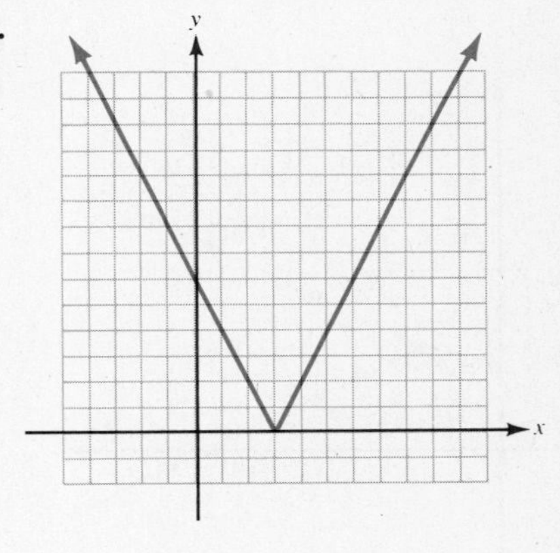

3.

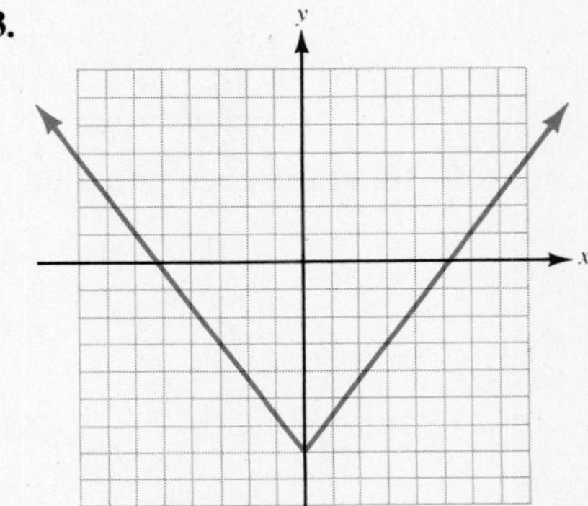

4.

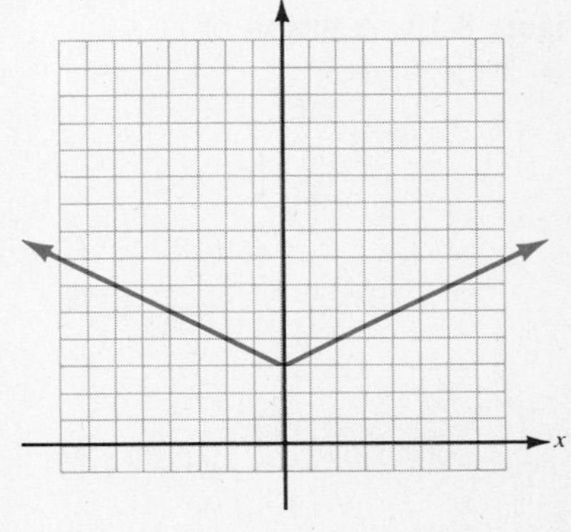

5.

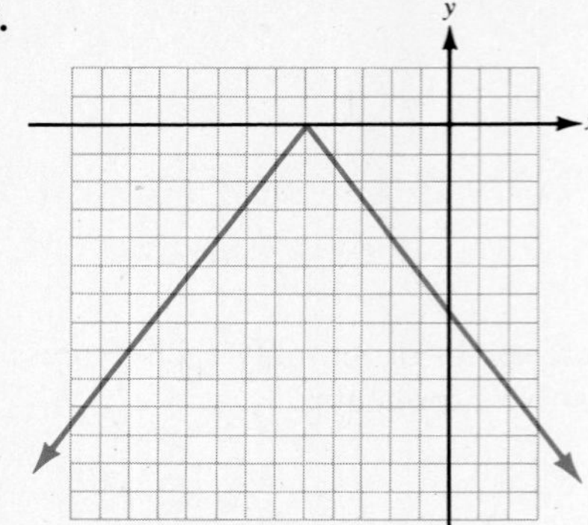

6.

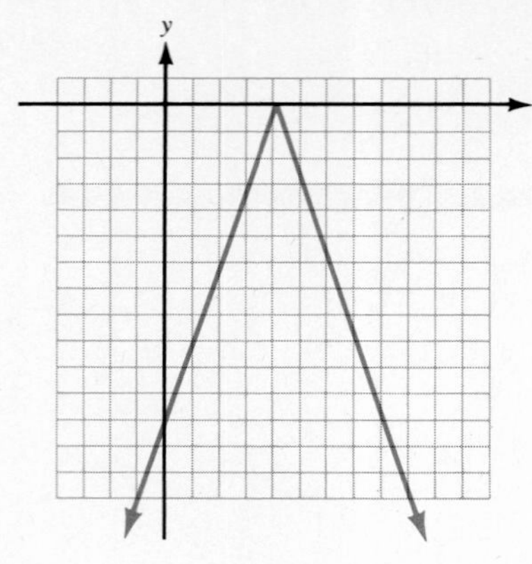

7.

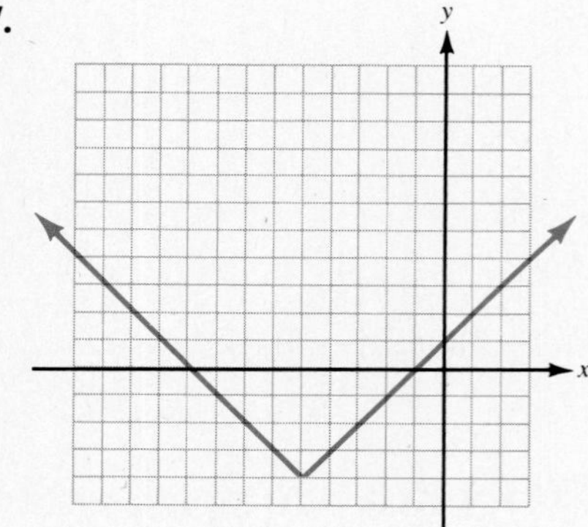

8.

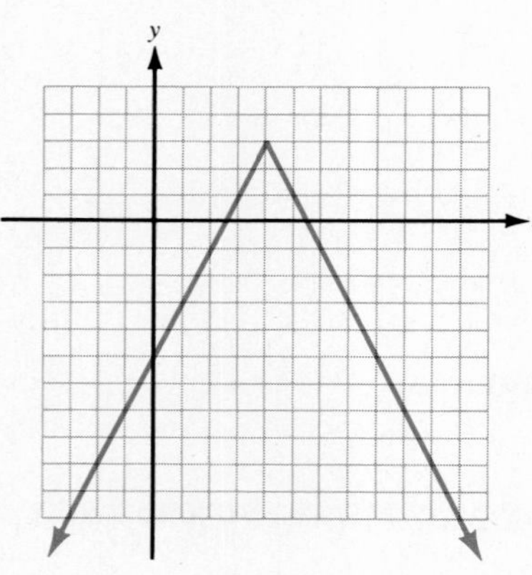

An equation written in the form

$$x = a|y - b| + c$$

defines a relation that is not a function. A graph of such a relation is a V that opens to the right if $a > 0$ and to the left if $a < 0$.

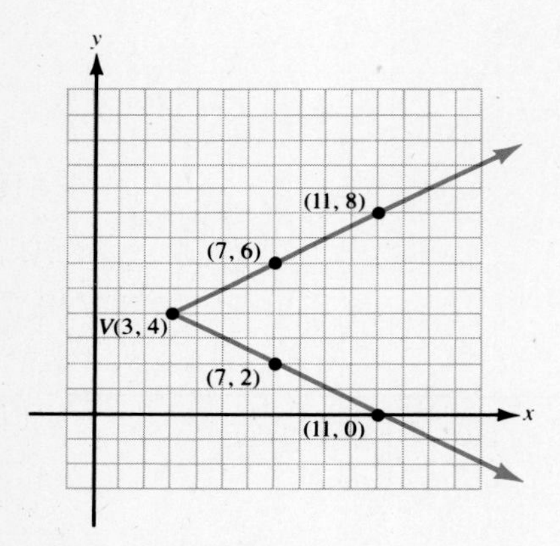

Figure 8-10. A graph of $x = 2|y - 4| + 3$.

Example. Graph $x = 2|y - 4| + 3$.

Solution. The vertex is $V(3, 4)$. To find more points on the V, select values for y close to 4. The following table of values is used to plot the five points in Figure 8-10:

y	$x = 2\|y - 4\| + 3$	Ordered pair
8	$x = 2\|8 - 4\| + 3 = 11$	(11, 8)
6	$x = 2\|6 - 4\| + 3 = 7$	(7, 6)
4	$x = 2\|4 - 4\| + 3 = 3$	$V(3, 4)$
2	$x = 2\|2 - 4\| + 3 = 7$	(7, 2)
0	$x = 2\|0 - 4\| + 3 = 11$	(11, 0)

In exercises **9–18**, grapn each equation.

9. $x = |y|$

10. $x = -|y|$

11. $x = |y - 2|$

12. $x = |y + 1|$

13. $x = -2|y + 3|$

14. $x = \frac{-2}{3}|y - 1|$

15. $x = \frac{5}{4}|y| - 2$

16. $x = \frac{-3}{4}|y| + 1$

17. $x = 3|y + 2| - 5$

18. $x = -|y - 4| + 3$

An inequality written in the form

$$y < a|x - b| + c \qquad \text{or} \qquad y > a|x - b| + c$$

defines an **absolute value inequality.** A graph of such a relation is either the *interior* or *exterior* of a V defined by a corresponding absolute value equation. Use a **test point** to determine whether the graph of a given inequality is the interior or exterior.

In exercises **19–24**, graph each inequality.

19. $y > 2|x| - 4$

20. $y > \frac{3}{2}|x| + 1$

21. $y < 8 - 3|x|$

22. $y < 6 - \frac{4}{5}|x|$

23. $y > \frac{1}{3}|x - 4| - 2$

24. $y < \frac{-3}{4}|x + 2| + 5$

SECTION 8-2. The Quadratic Function

1. The definition of the quadratic function
2. Graphing $y = ax^2$
3. Graphing $y = (x - h)^2$
4. Graphing $y = x^2 + k$
5. Graphing $y = a(x - h)^2 + k$
6. Graphing $y = ax^2 + bx + c$

In previous sections, we have solved quadratic equations by factoring, completing the square, and the quadratic formula. In this section, we will use the quadratic expression $ax^2 + bx + c$ to define the quadratic function. We will then study the graphs of such functions.

The Definition of the Quadratic Function

The square of any real number is greater than or equal to 0. As a consequence, when the equation $y = x^2$ is used to define a function, the smallest value that y can have is 0. This minimum value of y occurs at the point on the graph called the **vertex.** The following table of values is used to plot points in Figure 8-11 to give shape to a graph of $y = x^2$:

x	$y = x^2$	Ordered pair
-4	$y = (-4)^2 = 16$	$(-4, 16)$
-2	$y = (-2)^2 = 4$	$(-2, 4)$
0	$y = 0^2 = 0$	$V(0, 0)$
2	$y = 2^2 = 4$	$(2, 4)$
4	$y = 4^2 = 16$	$(4, 16)$

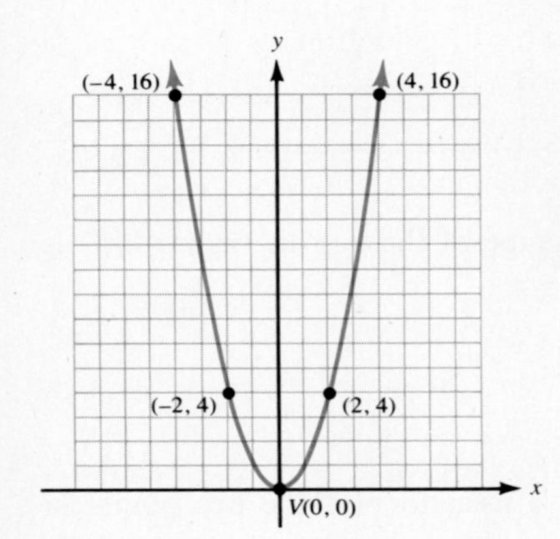

Figure 8-11. A graph of $y = x^2$.

The cup-shaped curve in Figure 8-11 is typical of graphs of quadratic functions. For this curve, the y-axis is the **axis of symmetry.** Therefore, if the grid is folded along this line, the two branches of the curve will coincide.

Definition 8.2. The quadratic function
If a, b, and c are real numbers and $a \neq 0$, then a quadratic function can be defined by an equation of the form:

$$y = ax^2 + bx + c$$

The domain is the set of real numbers, written "all x".

As in the case of the absolute value function, the constants a, b, and c in the equation of Definition 8.2 alter the shape and location of the cup in Figure 8-11. The details of how each constant affects the graph are discussed below.

Graphing $y = ax^2$

In the functions defined by the equations in examples **a** and **b**, the x^2 is multiplied by nonzero constants:

a. $y = 2x^2$ **b.** $y = \frac{-1}{3}x^2$

The effect of these constants on the graph in Figure 8-11 is to change the size of the cup. Furthermore, in example **b** the negative constant changes the direction of the cup from upward to downward.

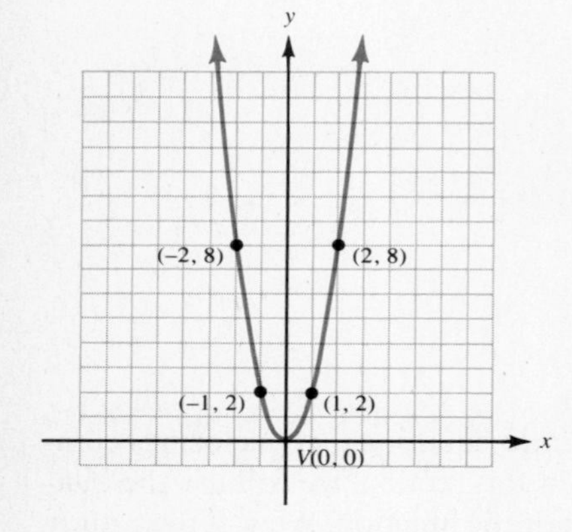

Figure 8-12. A graph of $y = 2x^2$.

Example 1. Graph $y = 2x^2$.

Solution. The following table of values is used to plot the five points in Figure 8-12:

x	$y = 2x^2$	Ordered pair
-2	$y = 2(-2)^2 = 8$	$(-2, 8)$
-1	$y = 2(-1)^2 = 2$	$(-1, 2)$
0	$y = 2(0)^2 = 0$	$V(0, 0)$
1	$y = 2(1)^2 = 2$	$(1, 2)$
2	$y = 2(2)^2 = 8$	$(2, 8)$

Compared to the cup in Figure 8-11, the shape of the cup in Figure 8-12 is narrower.

Example 2. Graph $y = \frac{-1}{3}x^2$.

Solution. The following table of values is used to plot the five points in Figure 8-13:

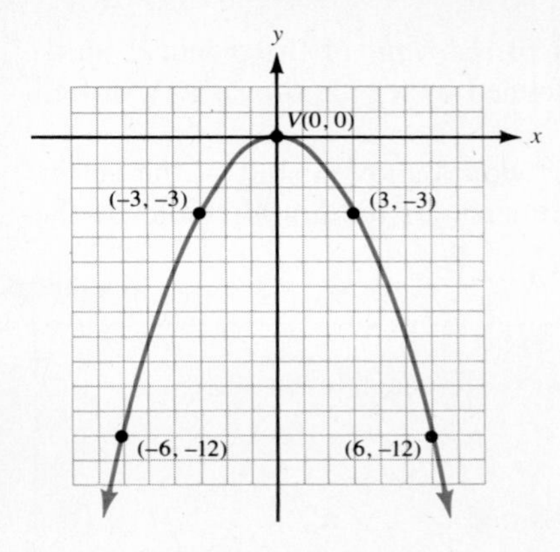

Figure 8-13. A graph of $y = \frac{-1}{3}x^2$.

x	$y = \frac{-1}{3}x^2$	Ordered pair
-6	$y = \frac{-1}{3}(-6)^2 = -12$	$(-6, -12)$
-3	$y = \frac{-1}{3}(-3)^2 = -3$	$(-3, -3)$
0	$y = \frac{-1}{3}(0)^2 = 0$	$V(0, 0)$
3	$y = \frac{-1}{3}(3^2) = -3$	$(3, -3)$
6	$y = \frac{-1}{3}(6)^2 = -12$	$(6, -12)$

Compared to the cup in Figure 8-11, the shape of the cup in Figure 8-13 is broader and directed downward.

> If $y = ax^2$, where a is a nonzero constant, then a graph of the function is shaped like a cup, and when compared to a graph of $y = x^2$ (where $a = 1$):
>
> **1.** The cup is "narrower" when $|a| > 1$.
>
> **2.** The cup is "broader" when $0 < |a| < 1$.
>
> **3.** The cup is directed upward if $a > 0$.
>
> **4.** The cup is directed downward if $a < 0$.

Graphing $y = (x - h)^2$

In Figure 8-11, the vertex is $V(0, 0)$. The equation for the function can be written $y = (x - 0)^2$, which indicates a 0 is subtracted from x before the value is squared. In examples **c** and **d**, nonzero constants are subtracted from x before the values are squared.

c. $y = (x - 2)^2$ **d.** $y = (x - (-3))^2$, or $y = (x + 3)^2$

The effect of these constants on the graph in Figure 8-11 is to shift the vertex to the right, or to the left of the origin.

Example 3. Graph $y = (x - 2)^2$.

Solution. The following table of values is used to plot the five points in Figure 8-14:

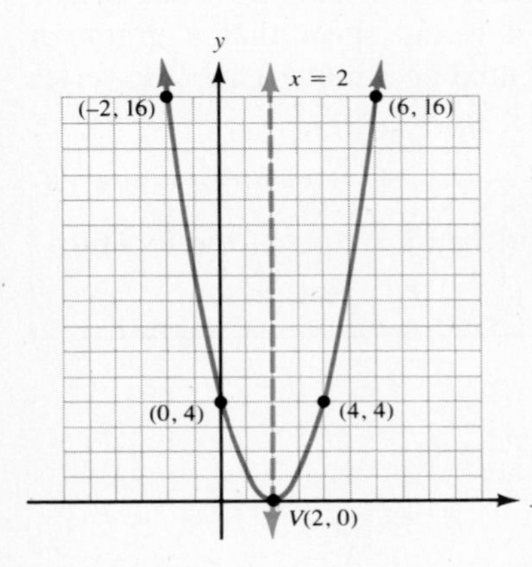

Figure 8-14. A graph of $y = (x - 2)^2$.

x	$y = (x - 2)^2$	Ordered pair
-2	$y = (-2 - 2)^2 = 16$	$(-2, 16)$
0	$y = (0 - 2)^2 = 4$	$(0, 4)$
2	$y = (2 - 2)^2 = 0$	$V(2, 0)$
4	$y = (4 - 2)^2 = 4$	$(4, 4)$
6	$y = (6 - 2)^2 = 16$	$(6, 16)$

Notice that the 2 in $(x - 2)$ shifted the vertex *to the right* of the origin 2 units. The axis of symmetry is now the vertical line defined by $x = 2$, shown as a dotted line in the figure.

In a similar way, a graph of $y = (x - (-3))^2$ would show a shift of the vertex *to the left* of the origin 3 units. The axis of symmetry for the graph would be the vertical line defined by $x = -3$.

> If $y = (x - h)^2$, then a graph of the function is shaped like a cup.
>
> **1.** The vertex is $V(h, 0)$.
>
> **2.** The axis of symmetry is the vertical line defined by $x = h$.

Graphing $y = x^2 + k$

In Figure 8-11 the vertex is $V(0, 0)$. The equation for the function can be written $y = x^2 + 0$, which indicates a 0 added to the square of x. In examples **e** and **f**, nonzero constants are added to the square of x.

e. $y = x^2 + (-4)$, or $y = x^2 - 4$ **f.** $y = x^2 + 3$

The effect of these constants on the graph in Figure 8-11 is to shift the vertex up, or down from the origin.

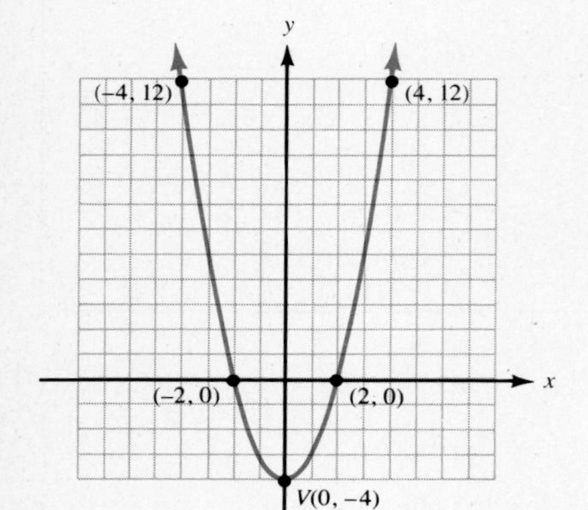

Figure 8-15. A graph of $y = x^2 - 4$.

Example 4. Graph $y = x^2 - 4$.

Solution. The following table of values is used to plot the five points in Figure 8-15:

x	$y = x^2 - 4$	Ordered pair
-4	$y = (-4)^2 - 4 = 12$	$(-4, 12)$
-2	$y = (-2)^2 - 4 = 0$	$(-2, 0)$
0	$y = (0)^2 - 4 = -4$	$V(0, -4)$
2	$y = 2^2 - 4 = 0$	$(2, 0)$
4	$y = 4^2 - 4 = 12$	$(4, 12)$

Notice that the -4 added to x^2 shifted the vertex *down* 4 units from the origin.

An analysis similar to the one in Example 4 would show that a graph of $y = x^2 + 3$ would be the graph in Figure 8-11 shifted *up* 3 units. Thus, the vertex would be $V(0, 3)$.

> If $y = x^2 + k$, where k is a nonzero constant, then a graph of the function has as vertex $V(0, k)$.

Graphing $y = a(x - h)^2 + k$

If the quadratic function defined by:

$$y = ax^2 + bx + c \quad \text{is written as} \quad y = a(x - h)^2 + k$$

then the graph of the function is in the shape of a cup, and:

1. Has vertex $V(h, k)$
2. The axis of symmetry is the line defined by $x = h$
3. Is directed upward if $a > 0$, and directed downward if $a < 0$

Example 5. Graph $y = \frac{1}{4}(x + 2)^2 - 3$.

Solution. With $a = \frac{1}{4}$, $h = -2$, and $k = -3$:

1. $V(-2, -3)$ is the vertex.
2. The line $x = -2$ is the axis of symmetry.
3. With $a > 0$, the curve opens upward.

Discussion. Since $a = \frac{1}{4}$, x will be replaced with values that differ from h by multiples of 4.

The following five points are plotted to give shape to the graph given in Figure 8-16:

x	$y = \frac{1}{4}(x + 2)^2 - 3$	Ordered pair
-10	$y = \frac{1}{4}(-10 + 2)^2 - 3 = 13$	$(-10, 13)$
-6	$y = \frac{1}{4}(-6 + 2)^2 - 3 = 1$	$(-6, 1)$
-2	$y = \frac{1}{4}(-2 + 2)^2 - 3 = -3$	$V(-2, -3)$
2	$y = \frac{1}{4}(2 + 2)^2 - 3 = 1$	$(2, 1)$
6	$y = \frac{1}{4}(6 + 2)^2 - 3 = 13$	$(6, 13)$

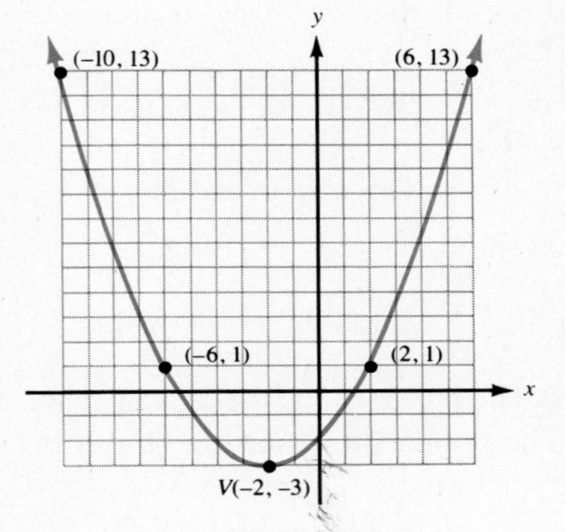

Figure 8-16. A graph of $y = \frac{1}{4}(x + 2)^2 - 3$.

Graphing $y = ax^2 + bx + c$

If a quadratic function is defined by an equation

$$y = ax^2 + bx + c$$

then to graph the function it is easier to first write the equation as:

$$y = a(x - h)^2 + k$$

The change may require completing the square in x. Use the following six-step procedure:

> **To change $y = ax^2 + bx + c$ to $y = a(x - h)^2 + k$:**
>
> **Step 1.** If $a \neq 1$, then multiply both sides of the equation by $\frac{1}{a}$.
>
> **Step 2.** If $\frac{c}{a} \neq 0$, then add $\frac{-c}{a}$ to both sides of the equation.
>
> **Step 3.** Add $\left(\frac{1}{2} \cdot \frac{b}{a}\right)^2$ to both sides of the equation.
>
> **Step 4.** Simplify the left side and write the right side as the square of a binomial.
>
> **Step 5.** Move the constant on the left side to the right side.
>
> **Step 6.** If $a \neq 1$, then multiply both sides by a.

Example 6. Graph $y = x^2 - 6x + 4$.

Solution.

Step 1.	$y = x^2 - 6x + 4$	In this equation, $a = 1$.
Step 2.	$y - 4 = x^2 - 6x$	Subtract 4 from both sides.
Step 3.	$y - 4 + 9 = x^2 - 6x + 9$	Add $\left(\frac{1}{2} \cdot 6\right)^2$ to both sides.
Step 4.	$y + 5 = (x - 3)^2$	Simplify the left side.
Step 5.	$y = (x - 3)^2 - 5$	Subtract 5 from both sides.
Step 6.	$y = (x - 3)^2 - 5$	The coefficient of y is 1.

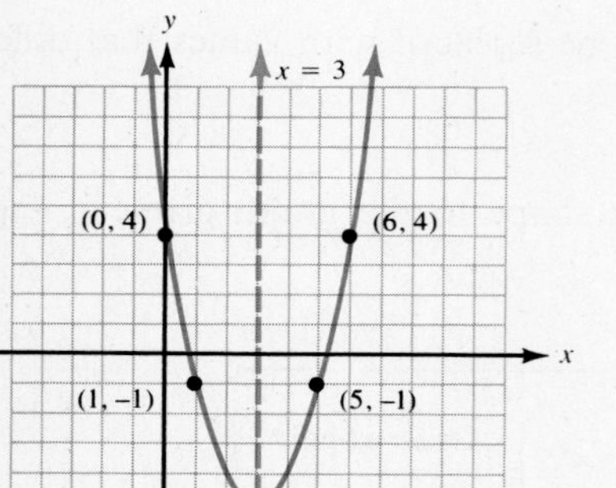

Figure 8-17. A graph of $y = x^2 - 6x + 4$.

With $h = 3$ and $k = -5$, $V(3, -5)$ is the vertex. The line $x = 3$ is the axis of symmetry. Since $1 > 0$, the curve opens upward. The points shown on the graph in Figure 8-17 give shape to the curve.

Example 7. Graph $y = 4x - \frac{1}{2}x^2$.

Solution.

Step 1.	$-2y = x^2 - 8x$	Multiply both sides by -2.
Step 2.	$-2y = x^2 - 8x$	The constant term is 0.
Step 3.	$-2y + 16 = x^2 - 8x + 16$	Add $\left(\frac{1}{2} \cdot 8\right)^2$ to both sides.
Step 4.	$-2y + 16 = (x - 4)^2$	Write $x^2 - 8x + 16$ as $(x - 4)^2$.
Step 5.	$-2y = (x - 4)^2 - 16$	Add -16 to both sides.
Step 6.	$y = \frac{-1}{2}(x - 4)^2 + 8$	Multiply both sides by $\frac{-1}{2}$.

Figure 8-18. A graph of $y = 4x - \frac{1}{2}x^2$.

With $h = 4$ and $k = 8$, $V(4, 8)$ is the vertex and $x = 4$ is the axis of symmetry. Since $\frac{-1}{2} < 0$, the curve opens downward. The indicated points are used to graph Figure 8-18.

SECTION 8-2. Practice Exercises

In exercises **1–50**, graph each function.

[Examples 1 and 2]

1. $y = 3x^2$

2. $y = 4x^2$

3. $y = \frac{1}{2}x^2$

4. $y = \frac{1}{3}x^2$

5. $y = -2x^2$

6. $y = -3x^2$

7. $y = \frac{-5}{4}x^2$

8. $y = \frac{-3}{4}x^2$

[Example 3]

9. $y = (x - 1)^2$

10. $y = (x - 3)^2$

11. $y = (x + 4)^2$

12. $y = (x + 2)^2$

13. $y = (x - 7)^2$

14. $y = (x - 5)^2$

15. $y = (7 - x)^2$

16. $y = (5 - x)^2$

[Example 4]

17. $y = x^2 + 2$

18. $y = x^2 + 1$

19. $y = x^2 - 1$

20. $y = x^2 - 3$

21. $y = -x^2 + 2$

22. $y = -x^2 - 3$

23. $y = -x^2 - 5$

24. $y = -x^2 + 6$

[Example 5]

25. $y = 2(x + 1)^2 - 3$

26. $y = 3(x + 2)^2 - 1$

27. $y = \frac{-1}{4}(x + 3)^2 + 4$

28. $y = \frac{-1}{2}(x + 4)^2 + 2$

29. $y = -(x - 4)^2 + 1$

30. $y = -(x - 1)^2 + 3$

31. $y = \frac{1}{2}(x - 2)^2 - 6$

32. $y = \frac{3}{4}(x - 4)^2 - 5$

33. $y = \frac{3}{2}(x - 1)^2 + 2$

34. $y = \frac{5}{3}(x + 3)^2 + 1$

35. $y = -2(x + 3)^2 - 1$

36. $y = \frac{-4}{3}(x - 2)^2 - 4$

[Examples 6 and 7]

37. $y = x^2 + 6x + 9$

38. $y = x^2 + 10x + 25$

39. $y = x^2 - 12x + 36$

40. $y = x^2 - 16x + 64$

41. $y = x^2 - 2x + 5$

42. $y = x^2 - 4x + 7$

43. $y = x^2 + 14x + 50$

44. $y = x^2 + 16x + 68$

45. $y = x^2 - 6x$

46. $y = x^2 + 4x$

47. $y = 4x^2 + 32x + 61$

48. $y = 2x^2 + 4x + 5$

49. $y = -5x^2 - 10x + 1$

50. $y = -3x^2 + 12x - 8$

SECTION 8-2. Ten Review Exercises

In exercises **1–3**, evaluate each expression for the indicated value.

1. $2(x - 1)$, for $x = 5$

2. $2|x - 1|$, for $x = -3$

3. $2(x - 1)^2$, for $x = -2$

In exercises **4–6**, solve each equation.

4. $2(x - 1) = 8$

5. $2|x - 1| = 8$

6. $2(x - 1)^2 = 8$

In exercises **7–10**, graph each function.

7. $y = 2(x - 1)$

8. $y = 2|x - 1|$

9. $y = 2(x - 1)^2$

10. $y = 2x^2 - 4x + 2$

SECTION 8-2. Supplementary Exercises

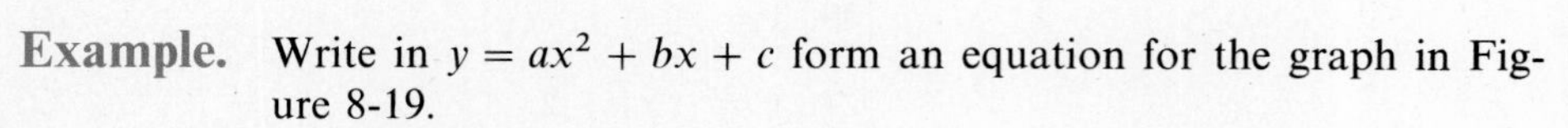

Example. Write in $y = ax^2 + bx + c$ form an equation for the graph in Figure 8-19.

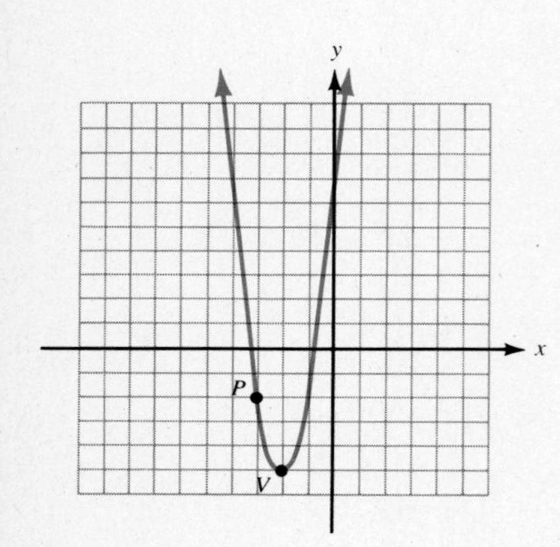

Figure 8-19.

Solution. From the graph, the following points are identified:

$V(-2, -5)$ and $P(-3, -2)$

In the $y = a(x - h)^2 + k$ form:

$h = -2$ and $k = -5$

Thus, to find a:

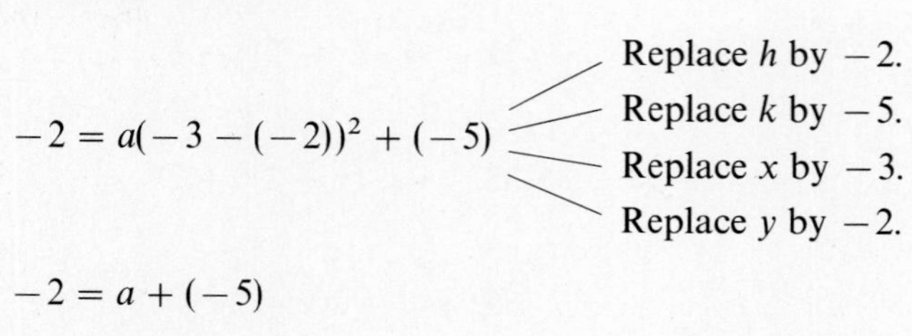

$$-2 = a(-3 - (-2))^2 + (-5)$$

Replace h by -2.
Replace k by -5.
Replace x by -3.
Replace y by -2.

$$-2 = a + (-5)$$

$$a = 3$$

Therefore: $y = 3(x + 2)^2 - 5$ Use the $y = a(x - h)^2 + k$ form.
$y = 3x^2 + 12x + 7$ An equation for the function

In exercises **1–8**, write an equation of the form $y = ax^2 + bx + c$ for the graph with the given vertex and point on the graph.

1. $V(2, 4)$, $P(3, 5)$

2. $V(1, 2)$, $P(0, 3)$

3. $V(0, 3)$, $P(2, -5)$

4. $V(0, -1)$, $P(-2, -3)$

5. $V(-2, 0)$, $P(0, -2)$

6. $V(5, 0)$, $P(7, 2)$

7. $V(-2, -3)$, $P(-1, 0)$

8. $V(4, -2)$, $P(3, 6)$

In geometry, a **tangent** to a curve is a line that intersects the curve at exactly one point. In exercises **9–12**, tangents are drawn to the graphs of four quadratic functions. Use the coordinates of the points given on the tangents to write equations of the lines in each figure in the $y = mx + b$ form.

9.

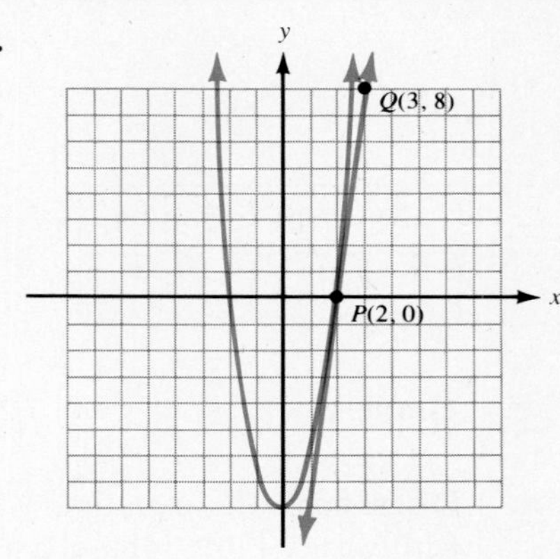

10.

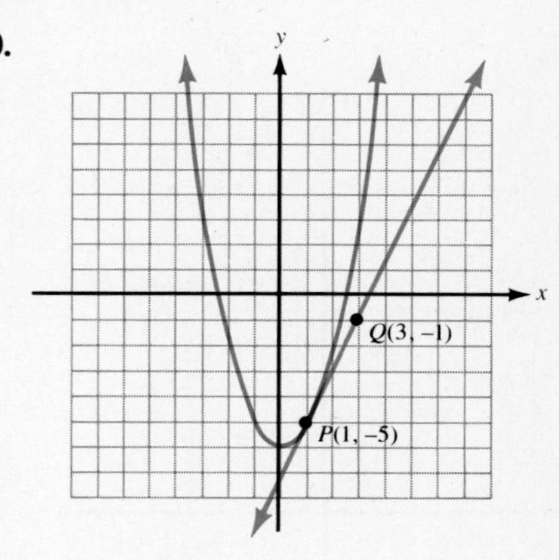

11.

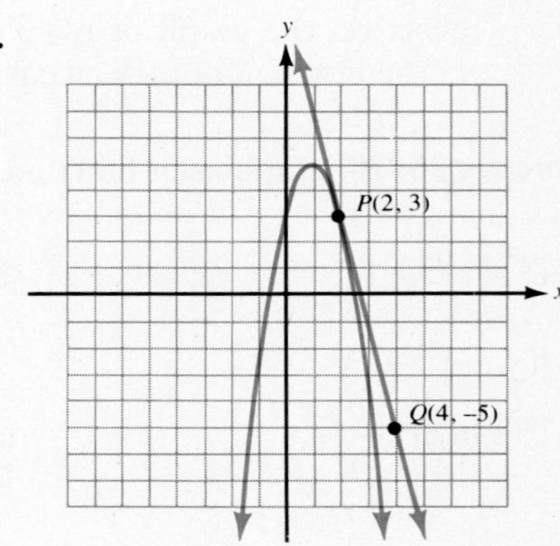

12.

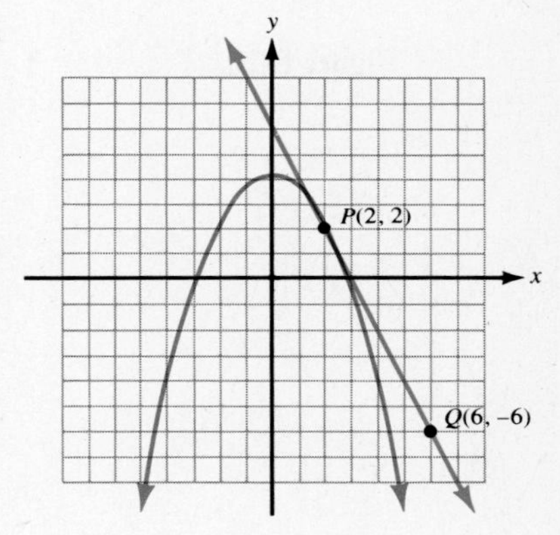

In exercises **13–20**, graph the solution set of each quadratic inequality. The solution set is either the region inside the curve, or outside the curve. Use a test point to determine the solution set of each inequality.

13. $y > x^2$

14. $y > \frac{1}{2}x^2 - 4$

15. $y < 2x^2 - 6$

16. $y < \frac{2}{3}x^2 - 3$

17. $y > 2 - x^2$

18. $y > 4 - \frac{3}{2}x^2$

19. $y < 4x - x^2$

20. $y < 6x + x^2$

In exercises **21–24**:

a. Complete the table using the given equation.

b. Plot the ordered pairs in an xy-coordinate system.

c. Draw a smooth curve through the points.

21. $y = \frac{1}{2}x^3$

x	-2	-1	0	1	2
y					

22. $y = \frac{1}{4}x^3$

x	-3	-2	0	2	3
y					

23. $y = \frac{-1}{4}x^3$

x	-3	-2	0	2	3
y					

24. $y = \frac{-1}{8}x^3$

x	-4	-2	0	2	4
y					

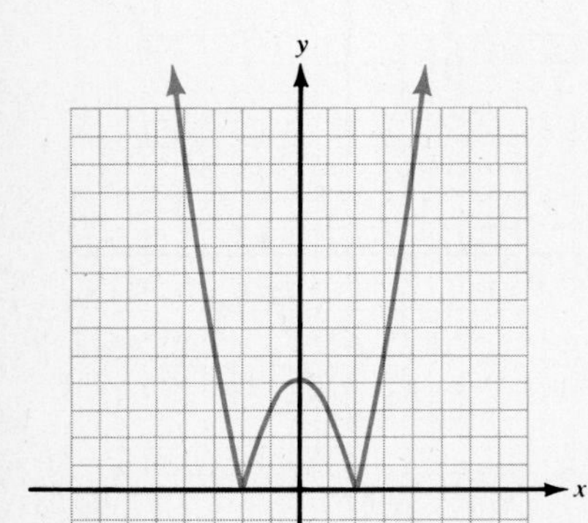

Figure 8-20.

Example. Graph $y = |x^2 - 4|$.

Solution. **Discussion.** The equation $y = x^2 - 4$ defines a quadratic function with vertex $V(0, -4)$ and axis of symmetry $x = 0$. However, the absolute value bars make any negative values of y nonnegative. The effect on the graph of $y = x^2 - 4$ is to reflect the negative portion of the graph about the x-axis. The result is the graph in Figure 8-20.

In exercises **25–30**, graph each function.

25. $y = |x^2 - 9|$

26. $y = \frac{1}{2}|x^2 - 16|$

27. $y = |(x - 1)^2 - 3|$

28. $y = |(x + 2)^2 - 4|$

29. $y = |2(x + 3)^2 - 1|$

30. $y = |3(x - 2)^2 - 5|$

SECTION 8-3. The Rational Function

1. The definition of the rational function
2. Determining the domain of a rational function
3. Some useful notation
4. Vertical asymptotes for graphs of rational functions
5. Horizontal asymptotes for graphs of rational functions
6. A procedure for sketching graphs of rational functions

In Chapter 4, we studied **rational expressions,** such as those in examples **a** and **b**.

a. $\frac{x + 1}{x - 1}$ **b.** $\frac{x^2 + 2x - 8}{x^2 - 9}$

When expressions such as these are set equal to 0, then **rational equations** are formed, such as those in examples **c** and **d**.

c. $\dfrac{x+1}{x-1} = 0$ **d.** $\dfrac{x^2+2x-8}{x^2-9} = 0$

If equations are formed in which the zeros in examples **c** and **d** are replaced by y, then **rational functions** are defined.

e. $y = \dfrac{x+1}{x-1}$ **f.** $y = \dfrac{x^2+2x-8}{x^2-9}$

In this section we will learn how to graph functions such as those in examples **e** and **f**.

The Definition of the Rational Function

In examples **e** and **f**, the numerators and denominators of the expressions in x are polynomials. The ratio of these polynomials is a rational expression in x, provided the bottom expressions are not 0.

Definition 8.3. The rational function in x
If P and Q are polynomials in x and $Q \neq 0$, then a rational function in x is one defined by:

$$y = \frac{P}{Q}$$

Determining the Domain of a Rational Function

The domain of a rational function is the set of real numbers, except any real numbers that make the denominator polynomial 0.

Example 1. Find the domain of $y = \dfrac{x+1}{x-1}$.

Solution. **Discussion.** The domain of this rational function is the set of real numbers, except any values of x for which $x - 1 = 0$.

$x - 1 = 0$ Set the denominator equal to 0.

$x = 1$ Solve for x.

Since 1 makes the denominator 0, the domain contains all real numbers except 1. This can be written as "all x, $x \neq 1$".

Example 2. Find the domain of $y = \dfrac{x^2+2x-8}{x^2-9}$.

Solution.

$x^2 - 9 = 0$ Set the denominator equal to 0.

$(x+3)(x-3) = 0$ Factor the difference of squares.

$x + 3 = 0$ or $x - 3 = 0$ Use the zero-product property.

$x = -3$ $x = 3$ Solve for x.

The domain is: all x, $x \neq -3, 3$.

Some Useful Notation

The domain of the function defined by $y = \dfrac{x+1}{x-1}$ contains all real numbers except 1. To graph this function, we would need to examine values of x "close to 1", but not equal to 1. The following notation can be used to describe this examination:

$x \longrightarrow 1^-$ "x approaches 1 with values less than 1"

$x \longrightarrow 1^+$ "x approaches 1 with values greater than 1"

Example 3. Complete each table for y, where $y = \dfrac{x+1}{x-1}$.

a. $x \longrightarrow 1^-$

x	0	0.5	0.9	0.99
y				

Table A

b. $x \longrightarrow 1^+$

x	2	1.5	1.1	1.01
y				

Table B

Solution. **Discussion.** In Table A, the values of x are progressively "closer to 1", but always less than 1. In Table B, the values of x are progressively "closer to 1", but always greater than 1.

a. $x = 0$ $\qquad$ $x = 0.5$

$$\frac{0+1}{0-1} = -1 \qquad \frac{0.5+1}{0.5-1} = -3$$

$x = 0.9$ $\qquad$ $x = 0.99$

$$\frac{0.9+1}{0.9-1} = -19 \qquad \frac{0.99+1}{0.99-1} = -199$$

Table A

$x \longrightarrow 1^-$

x	0	0.5	0.9	0.99
y	-1	-3	-19	-199

Notice that the y-values are *decreasing* for values of x closer to 1, but less than 1. This fact can be written as follows:

$y \to -\infty$ "the y-values become negatively infinite"

b. $x = 2$ $\qquad$ $x = 1.5$ $\qquad$ $x = 1.1$ $\qquad$ $x = 1.01$

$$\frac{2+1}{2-1} = 3 \qquad \frac{1.5+1}{1.5-1} = 5 \qquad \frac{1.1+1}{1.1-1} = 21 \qquad \frac{1.01+1}{1.01-1} = 201$$

Table B

$x \longrightarrow 1^+$

x	2	1.5	1.1	1.01
y	3	5	21	201

Notice that the y-values are *increasing* for values of x closer to 1, but greater than 1. This fact can be written as follows:

$y \longrightarrow +\infty$ "the y-values become positively infinite"

Vertical Asymptotes for Graphs of Rational Functions

From Example 3, the following analysis can be made for the function defined by $y = \dfrac{x+1}{x-1}$:

As $x \longrightarrow 1^-, y \longrightarrow -\infty$ "y decreases as x approaches 1 from below 1"

As $x \longrightarrow 1^+, y \longrightarrow +\infty$ "y increases as x approaches 1 from above 1"

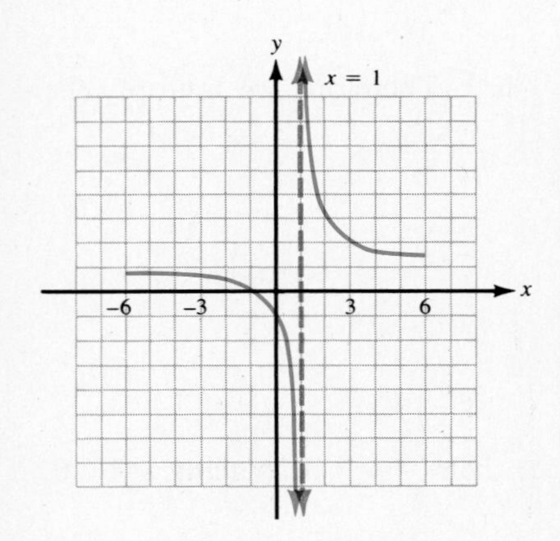

Figure 8-21. A graph of $y = \frac{x+1}{x-1}$, where $-6 \leq x \leq 6$.

In Figure 8-21, a partial graph of the function is shown for $-6 \leq x \leq 6$. The dotted vertical line $x = 1$ is a line that the graph *approaches, but never crosses,* as x takes on values closer and closer to 1, but always different from 1. The graphs of many rational functions have one or more vertical lines such as these, called asymptotes.

Definition 8.4. A vertical asymptote

If f is a rational function defined by $y = \dfrac{P}{Q}$, and a is a real number that makes Q equal to 0, but not P, then

$$x = a$$

is the equation of a **vertical asymptote** for a graph of f.

Example 4. Determine the equation of any vertical asymptotes for the graph of $y = \dfrac{2x-3}{x^2-x-42}$.

Solution.

$x^2 - x - 42 = 0$ Set the bottom polynomial equal to 0.

$(x-7)(x+6) = 0$ Factor.

$x - 7 = 0$ or $x + 6 = 0$ The zero-product property

$x = 7$ $x = -6$ Solve for x.

Since both 7 and -6 make the denominator 0, but not the numerator $[2(7) - 3 \neq 0$ and $2(-6) - 3 \neq 0]$, the vertical lines defined by $x = 7$ and $x = -6$ are vertical asymptotes for the graph of this function.

Horizontal Asymptotes for Graphs of Rational Functions

To complete a graph of $y = \dfrac{x+1}{x-1}$, we need to examine the effect on values of y as $|x|$ takes on larger and larger values. That is:

$x \longrightarrow +\infty$ "The values of x are becoming positively infinite."

$x \longrightarrow -\infty$ "The values of x are becoming negatively infinite."

To make such an analysis, we multiply the top and bottom polynomial by $\frac{1}{x^n}$, where n is the degree of the denominator polynomial. It can be shown that for any positive integer n:

As $x \longrightarrow +\infty$ or As $x \longrightarrow -\infty$: $\frac{1}{x^n} \to 0$

That is, as $|x|$ becomes larger and larger, the ratio $\frac{1}{x^n}$ becomes smaller and smaller, and approaches 0.

Example 5. Examine the values of y as $x \to +\infty$ or $x \to -\infty$, and $y = \frac{x+1}{x-1}$.

Solution. **Discussion.** The degree of $x - 1$ is 1. Therefore, we multiply the top and bottom by $\frac{1}{x}$.

$$\frac{x+1}{x-1} \cdot \frac{\frac{1}{x}}{\frac{1}{x}} = \frac{1 + \frac{1}{x}}{1 - \frac{1}{x}}$$

Now, as $x \to +\infty$ or $x \to -\infty$, then $\frac{1}{x} \to 0$. Replacing $\frac{1}{x}$ by 0, we get:

$$= \frac{1+0}{1-0}$$

$$= 1$$

Thus, as $x \to +\infty$ or $x \to -\infty$, $y \to 1$. That is, *the values of* y *get closer and closer to 1 as values of* x *become positively and negatively infinite.*

Example 6. Examine the values of y as $x \to +\infty$ or $x \to -\infty$ for $y = \frac{2x^2 + 1}{x^2 - 9}$.

Solution.

$$\frac{2x^2 + 1}{x^2 - 9} \cdot \frac{\frac{1}{x^2}}{\frac{1}{x^2}}$$

The degree of $x^2 - 9$ is 2.

$$= \frac{2 + \frac{1}{x^2}}{1 - \frac{9}{x^2}}$$

Write each term over x^2 and simplify.

$$= \frac{2+0}{1-0}$$

As $x \longrightarrow +\infty$ or $x \longrightarrow -\infty$, $\frac{1}{x^2} \longrightarrow 0$ and $\frac{9}{x^2} \longrightarrow 0$.

$$= 2$$

Thus, $y \longrightarrow 2$ as $|x| \longrightarrow \infty$.

That is, values of y get closer and closer to 2 as values of x become infinitely large.

Definition 8.5. A horizontal asymptote

If f is a rational function defined by $y = \frac{P}{Q}$, and b is a real number such that

$$y \longrightarrow b \quad \text{as} \quad |x| \longrightarrow \infty$$

then the line defined by $y = b$ is a **horizontal asymptote** for the graph of f.

From Example 5, $y \to 1$ as $x \to +\infty$ or $x \to -\infty$. Thus, the line defined by $y = 1$ is a horizontal asymptote for the graph of $y = \frac{x+1}{x-1}$. In Figure 8-22, a completed graph of the function is shown. The dotted line $y = 1$ is a horizontal asymptote for the graph. Thus, the curve gets closer and closer to this line as $x \to +\infty$ and $x \to -\infty$. As seen in the figure, as $x \to +\infty$, the curve approaches the line from above. As $x \to -\infty$, the curve approaches the line from below.

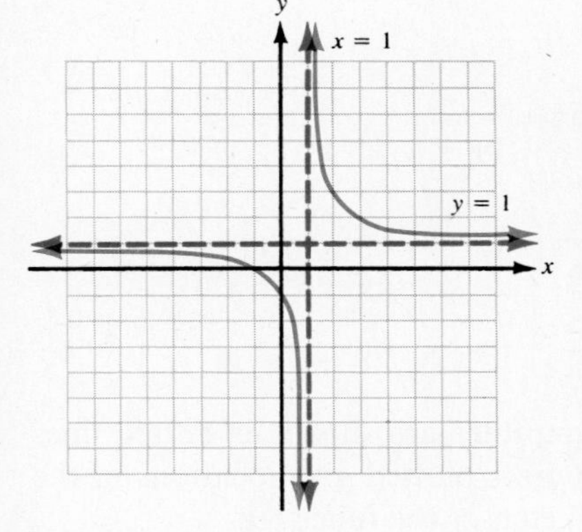

Figure 8-22. A graph of $y = \frac{x+1}{x-1}$.

A Procedure for Sketching Graphs of Rational Functions

The following six steps provide a procedure for graphing rational functions:

To sketch a graph of a rational function defined by $y = \frac{P}{Q}$:

Step 1. Locate any vertical asymptotes.

Step 2. Locate any horizontal asymptotes.

Step 3. If one exists, determine any y-intercept by evaluating $\frac{P}{Q}$ for $x = 0$.

Step 4. If any exist, determine the x-intercepts by setting $P = 0$ and solving.

Step 5. If necessary, plot a few more points on the curve.

Step 6. Connect any points found in Steps 3–5, but do not cross a vertical asymptote.

Example 7. Graph $y = \frac{3x}{x+3}$.

Solution. **Step 1.** $x + 3 = 0$ — Set the bottom polynomial equal to 0.

$x = -3$ — A vertical asymptote at $x = -3$.

Step 2. $\dfrac{3x}{x+3}\cdot\dfrac{\frac{1}{x}}{\frac{1}{x}}$ Multiply the top and bottom by $\frac{1}{x}$.

$= \dfrac{3}{1+\frac{3}{x}}$ Write each term over x and simplify.

$= \dfrac{3}{1+0}$ $\frac{3}{x} \longrightarrow 0$ as $|x| \longrightarrow \infty$

$= 3$ The line $y = 3$ is a horizontal asymptote.

Step 3. $\dfrac{3(0)}{0+3}$ Replace x by 0.

$= 0$ The y-intercept is 0.

Step 4. $3x = 0$ Set the top polynomial equal to 0.

$x = 0$ The x-intercept is 0.

Step 5. Determine a few more solutions for the function.

x	-6	3	6
y	6	1.5	2

Step 6. In Figure 8-23, the asymptotes are shown as dotted lines and labeled. The four points plotted give locations of the two curves that are a sketch of the function.

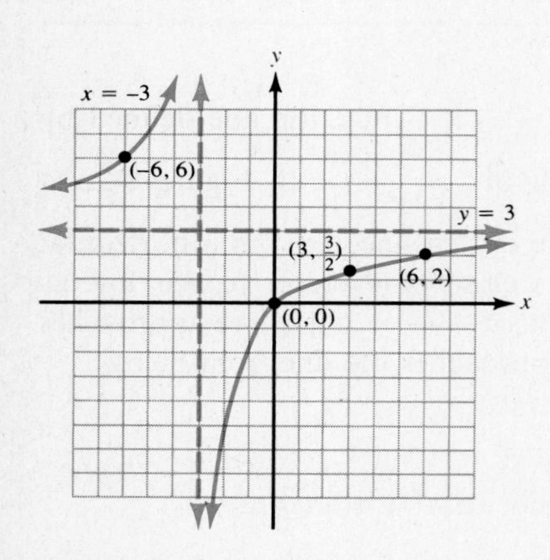

Figure 8-23. A graph of $y = \frac{3x}{x+3}$.

Example 8. Graph $y = \dfrac{x}{x^2 - 9}$.

Solution. **Step 1.** $x^2 - 9 = 0$

$(x+3)(x-3) = 0$

$x = -3$ or $x = 3$ Vertical asymptotes at $x = \pm 3$

Step 2. $\dfrac{x}{x^2-9}\cdot\dfrac{\frac{1}{x^2}}{\frac{1}{x^2}}$

$= \dfrac{\frac{1}{x}}{1-\frac{9}{x^2}}$

$= \dfrac{0}{1-0}$ $\frac{1}{x} \longrightarrow 0$ and $\frac{9}{x^2} \longrightarrow 0.$

$= 0$ Horizontal asymptote at $y = 0$

Step 3. $\dfrac{0}{0-9} = 0$ The y-intercept is 0.

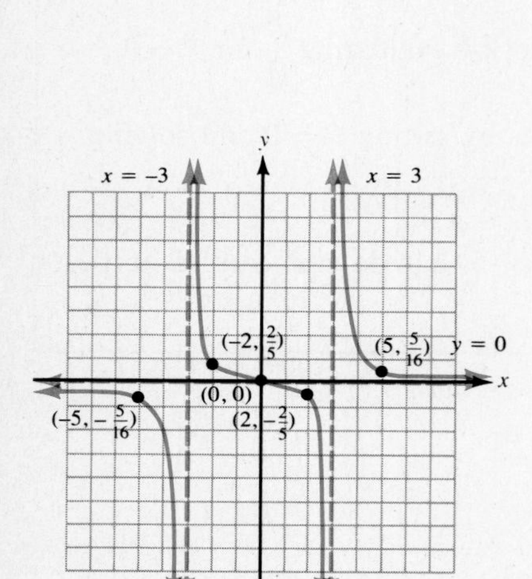

Figure 8-24. A graph of $y = \frac{x}{x^2-9}$.

Step 4. $x = 0$ The x-intercept is 0.

Step 5.

x	-5	-2	2	5
y	$\frac{-5}{16}$	$\frac{2}{5}$	$\frac{-2}{5}$	$\frac{5}{16}$

Determine a few more solutions for the function.

Step 6. In Figure 8-24, the asymptotes are shown as dotted lines and labeled. The five points plotted give locations of the curves that are a sketch of the function. Notice that the graph crosses the horizontal asymptote at (0, 0). However, it can never cross a vertical asymptote.

SECTION 8-3. Practice Exercises

In exercises **1–20**, find the domain of each rational function.

[Examples 1 and 2]

1. $y = \dfrac{x}{x + 7}$

2. $y = \dfrac{-x}{x + 12}$

3. $y = \dfrac{x^2 + 9x + 1}{15 - x}$

4. $y = \dfrac{x^2 + 10}{11 - x}$

5. $y = \dfrac{2x + 7}{x(x - 3)}$

6. $y = \dfrac{5x - 1}{x(x + 9)}$

7. $y = \dfrac{(x + 1)(x - 4)}{x^2 - 25}$

8. $y = \dfrac{(x + 7)^3}{x^2 - 64}$

9. $y = \dfrac{x - 5}{x^2 + 25}$

10. $y = \dfrac{x^2}{4x^2 + 1}$

11. $y = \dfrac{4}{x^2 - 7x + 10}$

12. $y = \dfrac{3 + x}{x^2 - 8x + 7}$

13. $y = \dfrac{x^2 + 49}{x^3 - 6x^2 - 16x}$

14. $y = \dfrac{x^2}{x^3 + x^2 - 20x}$

15. $y = \dfrac{x^2 + 11x + 10}{3x}$

16. $y = \dfrac{x^2 + 11x + 18}{x^2}$

17. $y = \dfrac{x^3 - 125}{12}$

18. $y = \dfrac{x^4 - 81}{3}$

19. $y = \dfrac{12}{x^3 - 8}$

20. $y = \dfrac{3}{x^4 - 16}$

In exercises **21–28**, complete each table.

[Example 3]

21. $y = \dfrac{x}{x - 3}$

a. $x \longrightarrow 3^-$

x	2	2.5	2.9	2.99
y				

b. $x \longrightarrow 3^+$

x	4	3.5	3.1	3.01
y				

22. $y = \dfrac{2x}{x-6}$

a. $x \longrightarrow 6^-$

x	5	5.5	5.9	5.99
y				

b. $x \longrightarrow 6^+$

x	7	6.5	6.1	6.01
y				

23. $y = \dfrac{x+4}{x+2}$

a. $x \longrightarrow -2^-$

x	-3	-2.5	-2.1	-2.01
y				

b. $x \longrightarrow -2^+$

x	-1	-1.5	-1.9	-1.99
y				

24. $y = \dfrac{x+10}{x+5}$

a. $x \longrightarrow -5^-$

x	-6	-5.5	-5.1	-5.01
y				

b. $x \longrightarrow -5^+$

x	-4	-4.5	-4.9	-4.99
y				

25. $y = \dfrac{x+3}{x}$

a. $x \longrightarrow 0^-$

x	-1	-0.5	-0.1	-0.01
y				

b. $x \longrightarrow 0^+$

x	1	0.5	0.1	0.01
y				

26. $y = \dfrac{x+6}{x}$

a. $x \longrightarrow 0^-$

x	-1	-0.5	-0.1	-0.01
y				

b. $x \longrightarrow 0^+$

x	1	0.5	0.1	0.01
y				

27. $y = \dfrac{1}{x - 5}$

a. $x \longrightarrow 5^-$

x	4	4.5	4.9	4.99
y				

b. $x \longrightarrow 5^+$

x	6	5.5	5.1	5.01
y				

28. $y = \dfrac{1}{x + 7}$

a. $x \longrightarrow -7^-$

x	-8	-7.5	-7.1	-7.01
y				

b. $x \longrightarrow -7^+$

x	-6	-6.5	-6.9	-6.99
y				

In exercises **29–40**, determine the equation of any vertical asymptotes for the graphs of each rational function.

[Example 4] **29.** $y = \dfrac{x + 9}{x - 8}$

30. $y = \dfrac{x - 3}{x + 6}$

31. $y = \dfrac{x^2 + 4x + 4}{x + 6}$

32. $y = \dfrac{x^2 - 7x + 10}{x - 4}$

33. $y = \dfrac{x}{x^2 - 7x - 18}$

34. $y = \dfrac{x^2}{x^2 + 3x - 28}$

35. $y = \dfrac{6 - x}{2x^2 - 7x - 15}$

36. $y = \dfrac{10 - x}{2x^2 - 3x - 14}$

37. $y = \dfrac{x - 5}{x^3 + 8}$

38. $y = \dfrac{2x - 1}{x^3 - 64}$

39. $y = \dfrac{x^2 + 1}{x^3 - 9x}$

40. $y = \dfrac{x^2 + x - 1}{4x^3 - x}$

In exercises **41–54**, for each function, examine the values of y as:

a. $x \longrightarrow +\infty$

b. $x \longrightarrow -\infty$

[Examples 5 and 6] **41.** $y = \dfrac{x + 3}{x - 10}$

42. $y = \dfrac{x - 6}{x + 8}$

43. $y = \dfrac{2x + 5}{x + 1}$

44. $y = \dfrac{5x - 3}{x - 2}$

45. $y = \dfrac{12}{7x - 3}$

46. $y = \dfrac{8}{6x + 5}$

47. $y = \dfrac{x + 11}{x^2 + 16}$

48. $y = \dfrac{x - 9}{x^2 + 49}$

49. $y = \dfrac{3x^2 + 4x}{2x^2 + 7}$

50. $y = \dfrac{6x^2 + 13}{5x^2 - 2x}$

51. $y = \dfrac{x^3 + 6x^2 + 15}{4x^3 + 1}$

52. $y = \dfrac{2x^3 + x - 7}{2x^3 + 3}$

53. $y = \dfrac{6x^2 + 4x - 1}{3x^2 - x + 2}$

54. $y = \dfrac{8x^2 - 5x + 10}{2x^2 + 3x - 7}$

In exercises **55–70**, graph each rational function.

[Examples 7 and 8] **55.** $y = \dfrac{x}{x - 5}$

56. $y = \dfrac{x}{x + 8}$

57. $y = \dfrac{x - 3}{x + 2}$

58. $y = \dfrac{x + 6}{x - 2}$

59. $y = \dfrac{x + 4}{x}$

60. $y = \dfrac{x - 7}{x}$

61. $y = \dfrac{x - 5}{x - 1}$

62. $y = \dfrac{x + 2}{x + 8}$

63. $y = \dfrac{x}{x^2 - 4}$

64. $y = \dfrac{x}{x^2 - 25}$

65. $y = \dfrac{x - 5}{x^2 - 1}$

66. $y = \dfrac{x + 2}{x^2 - 16}$

67. $y = \dfrac{x^2 + 1}{x^2 - 1}$

68. $y = \dfrac{x^2 + 4}{x^2 - 9}$

69. $y = \dfrac{3x^2}{x^2 + x - 2}$

70. $y = \dfrac{2x^2}{x^2 - x - 6}$

SECTION 8-3. Ten Review Exercises

In exercises **1–10**, do the indicated operations.

1. $\dfrac{3 + \sqrt{3}}{2 - 2\sqrt{3}}$

2. $\dfrac{3 + i\sqrt{3}}{2 - 2i\sqrt{3}}$

3. $(12 - \sqrt{8}) - (17 - \sqrt{50})$

4. $(12 - \sqrt{-8}) - (17 - \sqrt{-50})$

5. $(-2 + \sqrt{2})(5 - \sqrt{2})$

6. $(-2 + 2i)(5 - 3i)$

7. $\dfrac{1}{t-4} - \dfrac{1}{t+4} + \dfrac{4}{t^3 - 16t}$

8. $\dfrac{\dfrac{u}{v} - \dfrac{v}{u}}{\dfrac{u}{v} - 1}$

9. $\dfrac{x^2 - 3x - 7}{x^2 - 4x + 4} + \dfrac{x + 1}{x^2 - 7x + 10} \cdot \dfrac{3x - 15}{x - 2}$

10. $(8b^4 - 14b^3 + 2b^2 + 8b - 1) \div (2b^2 - 1)$

SECTION 8-3. Supplementary Exercises

In exercises **1–4**, the graph of the rational functions defined by the given equations do not have any vertical asymptotes. For each function, verify that there is no value of x that makes the denominator 0.

1. $y = \dfrac{x - 3}{10}$

2. $y = \dfrac{9 - x^2}{7}$

3. $y = \dfrac{4x^2 - 1}{x^2 + 4}$

4. $y = \dfrac{2 - x}{9 + x^2}$

In exercises **5–12**, the graphs of the rational functions defined by the given equations do not have vertical asymptotes. However, there is a "hole" in the curve for values of x that make the denominator 0. Write each function with a reduced rational expression and the corresponding restrictions on x.

5. $y = \dfrac{x^2 - 3x}{x}$

6. $y = \dfrac{x^3 - x^2 - 2x}{5x}$

7. $y = \dfrac{x^2 - 5x + 6}{x - 3}$

8. $y = \dfrac{x^3 - 8}{x - 2}$

9. $y = \dfrac{4x^2 - 9}{2x + 3}$

10. $y = \dfrac{25x^2 - 1}{5x - 1}$

11. $y = \dfrac{2x^3 + 3x^2 - 4x - 6}{x^2 - 2}$

12. $y = \dfrac{3x^3 - 6x^2 - x + 2}{3x^2 - 1}$

In exercises **13–16**, show that graphs of each rational function do not have horizontal asymptotes. (*Hint:* divide numerator by denominator and get a linear function of x plus a remainder. As $|x| \to \infty$, the remainder approaches 0 and y approaches the line defined by the linear function.)

13. $y = \dfrac{x^2 - 2x}{x + 1}$

14. $y = \dfrac{4x^2 + x}{2x - 1}$

15. $y = \dfrac{9x^2}{3x - 2}$

16. $y = \dfrac{10x^2}{5x + 3}$

SECTION 8-4. Functions and the Inverses of Functions

1. The $f(x)$ notation
2. The definition of the inverse of a function
3. Forming the inverse of a function
4. The definition of a one-to-one function
5. The vertical and horizontal line tests

In Section 7-5, we studied linear functions and the inverses of linear functions. In this section, we will extend the concepts studied in that section to other functions. As we will see, the inverses of some functions do not qualify as functions. However, there is a category of functions, called one-to-one, whose inverses are always functions.

The $f(x)$ Notation

Before beginning our study of inverse functions, we will first introduce some notation that is frequently used to represent y in an equation that defines a function.

> **The $f(x)$ notation**
> If f is a function and x is an element of the domain of f, then $f(x)$, which is read "f of x", is the corresponding element of the range of f.

Examples **a–c** are three equations that define functions using x and y. Examples **a*–c*** are equations that define the same functions using x and $f(x)$.

Functions defined using x and y	Same functions defined using x and $f(x)$
a. $y = 5x - 3$	**a*.** $f(x) = 5x - 3$
b. $y = x^2 - 4x - 5$	**b*.** $f(x) = x^2 - 4x - 5$
c. $y = \dfrac{3x}{x-3}$	**c*.** $f(x) = \dfrac{3x}{x-3}$

Example 1. If a function f is defined by $f(x) = 3x^2 + 5x - 2$, find:

a. $f(2)$ **b.** $f(-3)$ **c.** $f(x + a)$

Solution. **a. Discussion.** To find $f(2)$, replace x by 2 in the expression that defines $f(x)$.

$f(2) = 3(2)^2 + 5(2) - 2$ — $f(x) = 3x^2 + 5x - 2$

$= 12 + 10 - 2$ — Do the indicated operations.

$= 20$ — Thus, (2, 20) is a solution.

b. $f(-3) = 3(-3)^2 + 5(-3) - 2$ — Replace x by (-3).

$= 27 + (-15) - 2$ — Do the indicated operations.

$= 10$ — Thus, $(-3, 10)$ is a solution.

c. $f(x + a) = 3(x + a)^2 + 5(x + a) - 2$

$$= 3(x^2 + 2ax + a^2) + 5x + 5a - 2$$

$$= 3x^2 + 6ax + 3a^2 + 5x + 5a - 2$$

The Definition of the Inverse of a Function

The following statement can be used to identify the form of the ordered pairs of the inverse of a function:

Definition 8.6. If a and b are real numbers and (a, b) is an ordered pair in a function f, then (b, a) is an ordered pair in the inverse of f. The symbol f^{-1}, read "f inverse", can be used for the inverse of f.

Example 2. List the ordered pairs in f^{-1}, where

$$f = \{(-2, -12), (-1, -4), (0, -2), (1, 0), (2, 8)\}$$

Solution. **Discussion.** The inverse f^{-1} is the set of ordered pairs of f in which the components are reversed.

$$f^{-1} = \{(-12, -2), (-4, -1), (-2, 0), (0, 1), (8, 2)\}$$

Consider the domains and ranges of f and f^{-1} in Example 2:

Domain of f $= \{-2, -1, 0, 1, 2\} =$ **Range of f^{-1}**

Range of f $= \{-12, -4, -2, 0, 8\} =$ **Domain of f^{-1}**

By reversing the ordered pairs in f to form f^{-1}, the domain of f became the range of f^{-1}, and the range of f became the domain of f^{-1}.

In general, if f^{-1} is the inverse of f, then:

1. The domain of f is the range of f^{-1}.
2. The range of f is the domain of f^{-1}.

Example 3. A function f is defined by the equation $y = \sqrt{x - 3}$. The domain of f is all x, where $x \geq 3$. The range of f is all y, where $y \geq 0$.

a. State the domain of f^{-1}.

b. State the range of f^{-1}.

Solution. a. The domain of f^{-1} is all x, where $x \geq 0$.

b. The range of f^{-1} is all y, where $y \geq 3$.

Forming the Inverse of a Function

When a function is defined by an equation, then the inverse can be formed by interchanging the variables in the equation. *Such an interchanging of variables in the defining equation effectively reverses the ordered pairs in the function.* If the equation for f is written using the $f(x)$ notation, then it is usually easier to form an equation that defines f^{-1} by temporarily replacing $f(x)$ by y.

To write an equation that defines f^{-1} for a function f:

Step 1. Replace $f(x)$ by y in the equation that defines f.

Step 2. Replace x by y and y by x in the equation of Step 1.

Step 3. If possible, solve the equation in Step 2 for y.

Step 4. If the equation of Step 3 defines a function, replace y by $f^{-1}(x)$, read "f inverse of x".

Notice that Step 4 states that the $f^{-1}(x)$ notation can be used if f^{-1} is a function. *In the examples and Practice Exercises of this section, the inverses of the functions used will all be functions.* However, in the Supplementary Exercises a few functions are given whose inverses are relations that are not functions.

Example 4. If f is defined by $f(x) = 2x - 5$, then write an equation that defines f^{-1}, using $f^{-1}(x)$ notation.

Solution.

Step 1. $y = 2x - 5$ Replace $f(x)$ by y.

Step 2. $x = 2y - 5$ Replace x by y and y by x.

Step 3. $x + 5 = 2y$ Add 5 to both sides.

$\frac{x+5}{2} = y$ Divide both sides by 2.

$y = \frac{x+5}{2}$ Write y on the left side.

Step 4. $f^{-1}(x) = \frac{x+5}{2}$ The equation defines a function.

In Step 3 of Example 4, we got the equation:

$y = \frac{x+5}{2}$ An equation that defines f^{-1}

If x is replaced by any real number, then exactly one real-number value will be obtained for y. Therefore, the equation defines a function.

Example 5. If f is defined by $f(x) = \frac{x}{x+3}$:

a. Write an equation that defines f^{-1}, using $f^{-1}(x)$ notation.

b. Compute $f^{-1}(2)$.

Solution.

a. Step 1. $y = \frac{x}{x+3}$ Replace $f(x)$ by y.

Step 2. $x = \frac{y}{y+3}$ Replace x by y and y by x.

Step 3. $x(y + 3) = y$ — Multiply both sides by $(y + 3)$.

$xy + 3x = y$ — Use the distributive property.

$3x = y - xy$ — Subtract xy from both sides.

$3x = (1 - x)y$ — Factor the y.

$\frac{3x}{1 - x} = y$ — Divide both sides by $(1 - x)$.

$y = \frac{3x}{1 - x}$ — Write the y on the left side.

Step 4. $f^{-1}(x) = \frac{3x}{1 - x}$ — The equation defines a function.

b. $f^{-1}(2) = \frac{3(2)}{1 - 2}$ — Replace x by 2.

$= -6$ — Simplify.

Thus, $(2, -6)$ is a solution of f^{-1}.

Example 6. If f is defined by $f(x) = x^3 - 7$:

a. Compute $f(2)$.

b. Write an equation that defines f^{-1}, using $f^{-1}(x)$ notation.

c. Compute $f^{-1}(1)$.

Solution. **a.** $f(2) = 2^3 - 7$ — Replace x by 2 in $x^3 - 7$.

$= 1$ — Thus, $(2, 1)$ is a solution of f.

b. $y = x^3 - 7$ — Replace $f(x)$ by y.

$x = y^3 - 7$ — Replace x by y and y by x.

$x + 7 = y^3$ — Add 7 to both sides.

$\sqrt[3]{x + 7} = y$ — Take the cube root of both sides.

$y = \sqrt[3]{x + 7}$ — Write the y on the left side.

$f^{-1}(x) = \sqrt[3]{x + 7}$ — Replace y by $f^{-1}(x)$.

c. $f^{-1}(1) = \sqrt[3]{1 + 7}$ — Replace x by 1.

$= 2$ — Thus, $(1, 2)$ is a solution of f^{-1}.

The Definition of a One-to-One Function

The inverse of any function can be formed by interchanging the variables in an equation that defines the function. It should be pointed out, however, that f^{-1} may not be a function. Definition 8.7 identifies a type of function whose inverse is also a function.

Definition 8.7. One-to-one functions
A function f is one-to-one if and only if:

1. For each x in the domain, there is exactly one y in the range. (This statement guarantees that f is a function.)
2. For each y in the range, there is exactly one x in the domain. (This statement guarantees that f is one-to-one.)

Example 7. Determine whether each function is one-to-one.

a. $f = \{(-3, -26), (-2, -7), (-1, 0), (0, 1), (1, 2), (2, 9), (3, 28)\}$

b. $g = \{(-3, 10), (-2, 5), (-1, 2), (0, 1), (1, 2), (2, 5), (3, 10)\}$

Solution. **Discussion.** Check the ordered pairs in f and g to see whether any second element is paired with two different first elements. If the answer to the search is *no,* then the function is one-to-one. If the answer is *yes,* then the function is not one-to-one.

a. In f, each range element, $\{-26, -7, 0, 1, 2, 9, 28\}$, is paired with exactly one domain element. Thus, f is one-to-one.

b. In g, the range elements are $\{10, 5, 2, 1\}$. As seen:

$$10 \begin{cases} \nearrow -3 \\ \searrow 3 \end{cases} \quad \text{and} \quad 5 \begin{cases} \nearrow -2 \\ \searrow 2 \end{cases} \quad \text{and} \quad 2 \begin{cases} \nearrow -1 \\ \searrow 1 \end{cases}$$

Since these range elements are each paired with two different domain elements, g is not one-to-one.

In Example 7, g is not one-to-one. As a consequence, g^{-1} is not a function. Notice that $g^{-1} = \{(10, -3), (5, -2), (2, -1), (1, 0), (2, 1), (5, 2), (10, 3)\}$ is not a function because 10, 5, and 2 in the domain are each paired with two numbers in the range.

The Vertical and Horizontal Line Tests

For many relations defined by equations, it is difficult to determine whether or not the relation is a function. Similarly, for many functions defined by equations, it is difficult to determine whether or not the function is one-to-one. If a graph of a relation or function can be studied, then the vertical and horizontal line tests can be used to make these decisions.

The vertical and horizontal line tests

Part A. If r is a relation and any *vertical line* drawn in the plane of a graph of r can intersect the graph in at most *one point,* then r is a function.

Part B. If f is a function and any *horizontal line* drawn in the plane of a graph of f can intersect the graph in at most *one point,* then f is one-to-one.

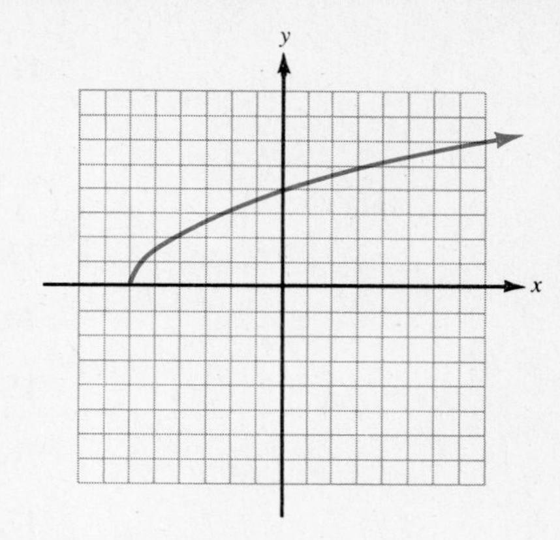

Figure 8-25. Graphs of two relations.

Example 8. Consider the relations graphed in Figure 8-25.

(i) Determine whether the relation is a function.

(ii) If the relation is a function, determine whether it is one-to-one.

Solution. **a.** Apply Part A of the line tests.

(i) Move from left to right over the graph and mentally draw *vertical lines.* Since no vertical line can intersect the graph in more than one point, the relation in this figure is a function.

(ii) Apply Part B of the line tests. Move from bottom to top over the graph and mentally draw *horizontal lines.* Since any horizontal line drawn between $y = -2$ and $y = 2$ will intersect the graph in three points, the function is not one-to-one.

b. Apply Part A of the line tests.

(i) The graph passes the vertical line test. Therefore, the relation is a function.

(ii) Apply Part B of the line tests. The graph also passes the horizontal line test. Therefore, the function is one-to-one.

SECTION 8-4. Practice Exercises

In exercises **1–12**, for each function f, find:

a. $f(-3)$ **b.** $f(0)$ **c.** $f(5)$ **d.** $f(x + a)$

[Example 1]

1. $f(x) = 4x - 5$

2. $f(x) = x^2 + 4$

3. $f(x) = x^3 - x + 2$

4. $f(x) = \dfrac{x + 3}{x - 3}$

5. $f(x) = 9 - x^2$

6. $f(x) = \sqrt{x + 4}$

7. $f(x) = \sqrt{36 - x^2}$

8. $f(x) = \sqrt{x^2 + 16}$

9. $f(x) = \dfrac{x}{x + 4}$

10. $f(x) = \sqrt[3]{9x}$

11. $f(x) = \frac{-1}{3}x + 10$

12. $f(x) = 2$

In exercises **13–20**, for each function f, list the ordered pairs in f^{-1}.

[Example 2] **13.** $f = \{(1, 2), (2, -1), (3, -4), (4, -7), (5, -10)\}$

14. $f = \{(-6, -2), (-3, 0), (0, 2), (3, 4), (6, 6)\}$

15. $f = \{(-2, 8), (-1, 5), (0, 4), (1, -5), (2, -8)\}$

16. $f = \{(-2, 0), (-1, 3), (0, 4), (1, 5), (2, 8)\}$

17. $f = \{(-3, 0), (-2, 1), (1, 2), (6, 3), (13, 4)\}$

18. $f = \{(4, 0), (3, 1), (0, 2), (-5, 3), (-12, 4)\}$

19. $f = \left\{\left(-3, \frac{1}{9}\right), \left(-2, \frac{1}{4}\right), (-1, 1), \left(\frac{-1}{2}, 4\right), \left(\frac{-1}{3}, 9\right)\right\}$

20. $f = \left\{\left(\frac{1}{3}, 81\right), \left(\frac{1}{2}, 16\right), (1, 1), \left(2, \frac{1}{16}\right), \left(3, \frac{1}{81}\right)\right\}$

In exercises **21–26**, the domain and range of a function f are given. For each function:

a. State the domain of f^{-1}.

b. State the range of f^{-1}.

[Example 3] **21.** $y = \sqrt{x + 2}$ Domain: all x, where $x \geq -2$
Range: all y, where $y \geq 0$

22. $y = \sqrt{x + 6}$ Domain: all x, where $x \geq -6$
Range: all y, where $y \geq 0$

23. $y = |x + 2|$ Domain: all x
Range: all y, where $y \geq 0$

24. $y = |6 - x|$ Domain: all x
Range: all y, where $y \geq 0$

25. $y = \frac{1}{x + 1}$ Domain: all x, where $x \neq -1$
Range: all y, where $y \neq 0$

26. $y = \frac{3}{x - 4}$ Domain: all x, where $x \neq 4$
Range: all y, where $y \neq 0$

In exercises **27–32**, write an equation that defines f^{-1}, using $f^{-1}(x)$ notation.

[Example 4] **27.** $f(x) = 3x - 10$

28. $f(x) = 5x + 1$

29. $f(x) = \frac{-2}{5}x + 4$

30. $f(x) = \frac{4}{3}x - 7$

31. $f(x) = \frac{2}{3}x - \frac{8}{3}$

32. $f(x) = \frac{-5}{7}x + \frac{10}{7}$

In exercises **33–40**, for each function:

a. Write an equation that defines f^{-1}, using $f^{-1}(x)$ notation.

b. Compute $f^{-1}(x)$ for the given value of x.

[Example 5] **33.** $f(x) = \dfrac{x}{x-8}$; $f^{-1}(-1)$

34. $f(x) = \dfrac{x}{2x-3}$; $f^{-1}(2)$

35. $f(x) = \dfrac{x+3}{x-1}$; $f^{-1}(2)$

36. $f(x) = \dfrac{x+5}{x-4}$; $f^{-1}(10)$

37. $f(x) = \dfrac{x+7}{x}$; $f^{-1}(2)$

38. $f(x) = \dfrac{x-4}{x}$; $f^{-1}\left(\dfrac{1}{2}\right)$

39. $f(x) = \dfrac{x+6}{2-x}$; $f^{-1}(-9)$

40. $f(x) = \dfrac{x-3}{5-x}$; $f^{-1}(1)$

In exercises **41–48**, for each function:

a. Compute $f(k)$ for the value of k.

b. Write an equation that defines f^{-1}, using $f^{-1}(x)$ notation.

c. Compute $f^{-1}(\ell)$ for the value of ℓ.

[Example 6] **41.** $f(x) = x^3 + 2$; $k = -1$ and $\ell = 1$

42. $f(x) = x^3 + 7$; $k = -2$ and $\ell = -1$

43. $f(x) = (x+1)^3$; $k = 2$ and $\ell = 27$

44. $f(x) = (x-2)^3$; $k = 0$ and $\ell = -8$

45. $f(x) = \sqrt[3]{x+1}$; $k = 7$ and $\ell = -2$

46. $f(x) = \sqrt[3]{x-2}$; $k = 29$ and $\ell = -3$

47. $f(x) = 1 - x^5$; $k = -2$ and $\ell = -31$

48. $f(x) = x^5 + 4$; $k = 1$ and $\ell = 3$

In exercises **49–56**, determine whether each function is one-to-one.

[Example 7] **49.** $f = \{(6, 2), (4, 3), (-2, -1), (-5, 0)\}$

50. $f = \{(8, -2), (5, -3), (-4, 1), (-6, 8)\}$

51. $f = \{(-4, 4), (-3, 3), (-1, 1), (0, 0), (1, 1)\}$

52. $f = \{(-2, 4), (-1, 1), (0, 0), (1, 1), (2, 4)\}$

53. $f = \{(6, 2), (7, 2), (10, 2)\}$

54. $f = \{(-3, 5), (0, 5), (2, 5)\}$

55. $f = \{(-9, 6), (-7, 4), (-5, 2), (-2, -2), (0, -4), (2, -6)\}$

56. $f = \{(8, -10), (4, -5), (2, -1), (1, 1), (-2, 3), (-5, 6)\}$

In exercises **57–64**, for the relation graphed in each figure:

a. Determine whether the relation is a function.

b. If the relation is a function, determine whether it is one-to-one.

57.

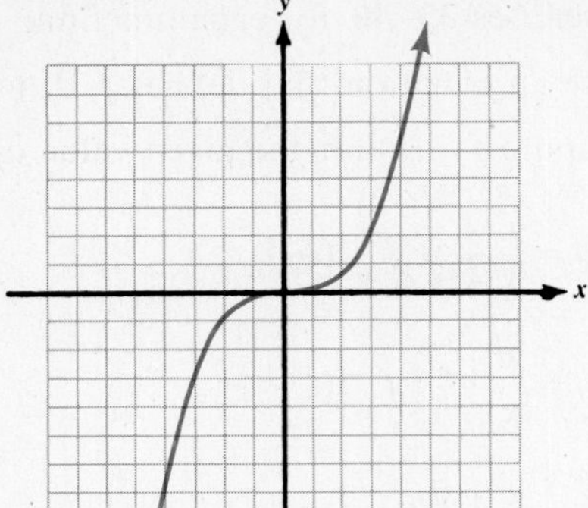

58.

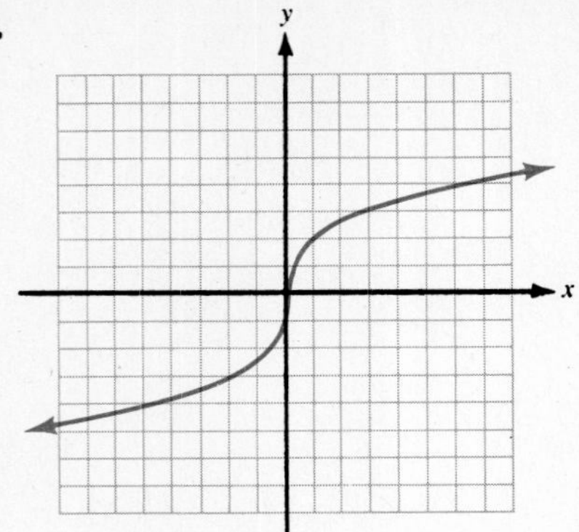

59.

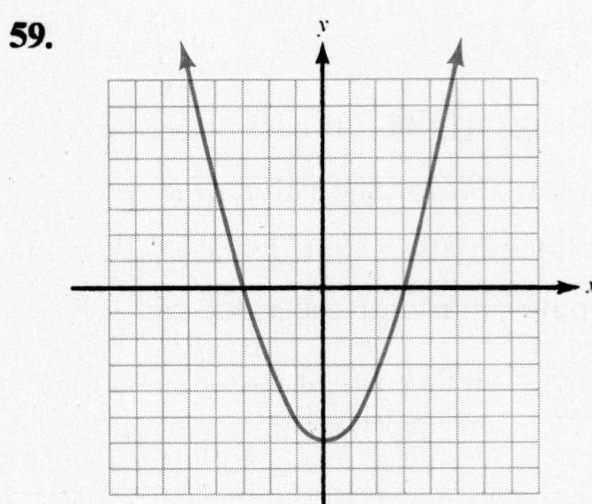

60.

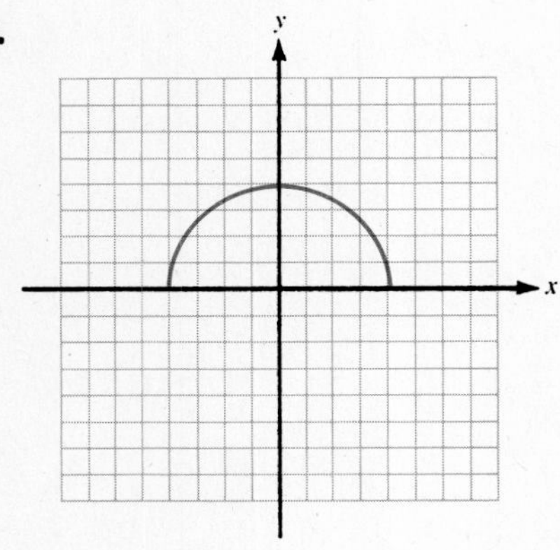

61.

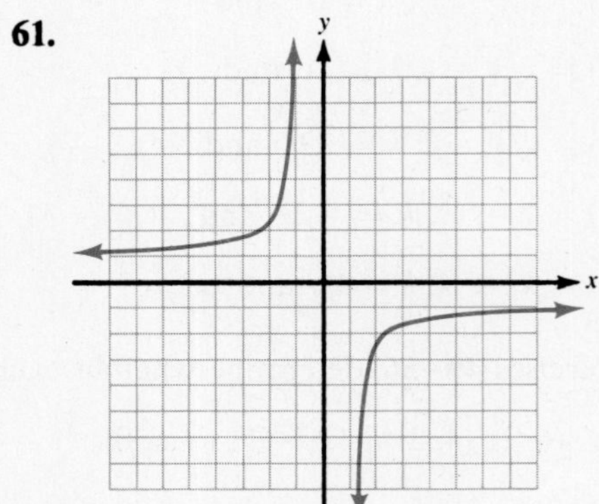

62.

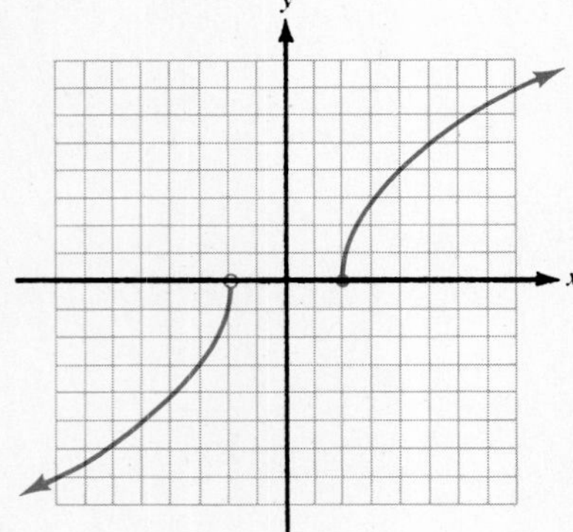

63.

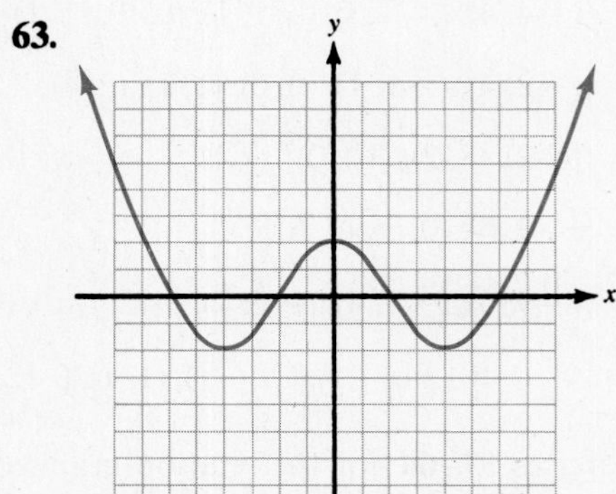

64.

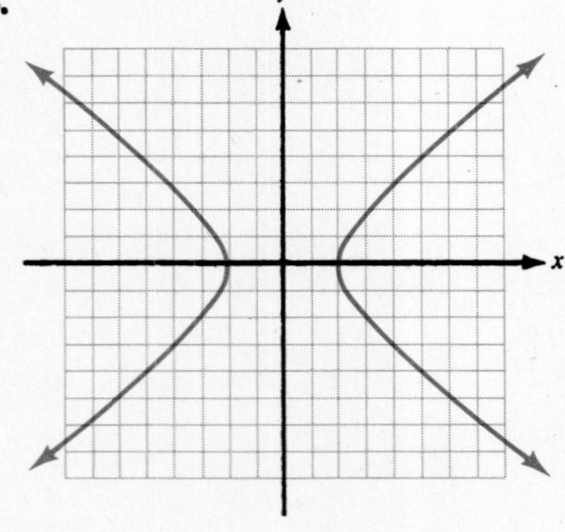

SECTION 8-4. Ten Review Exercises

1. Factor $x^2 - 4x - 5$.

2. Solve $x^2 - 4x - 5 = 0$.

3. For the function defined by $y = x^2 - 4x - 5$, find the values of x where the curve intersects the x-axis.

4. For the function defined by $y = \dfrac{3}{x^2 - 4x - 5}$, find equations of the vertical asymptotes for a graph of the function.

In exercises **5–10**, use $P(-7, 4)$ and $Q(8, -4)$.

5. Find the distance between P and Q.

6. Find the coordinates of M, the midpoint of the segment with P and Q as endpoints.

7. Find the slope of the line that passes through P and Q.

8. Find the slope of the line perpendicular to the line that passes through P and Q.

9. Write in $ax + by = c$ form an equation of the line that passes through P and Q.

10. Is the line defined by $15x + 8y = 12$ parallel to the line that passes through P and Q?

SECTION 8-4. Supplementary Exercises

In exercises **1–16**, in **a** find $f^{-1}(x)$ and then in **b** and **c**, find the indicated values.

1. $f(x) = 4x - 3$
   a. $f^{-1}(x)$
   b. $f^{-1}(1)$
   c. $f^{-1}(-3)$

2. $f(x) = \dfrac{x + 4}{2}$
   a. $f^{-1}(x)$
   b. $f^{-1}(0)$
   c. $f^{-1}(2)$

3. $f(x) = x^3 - 2$
   a. $f^{-1}(x)$
   b. $f^{-1}(-1)$
   c. $f^{-1}(6)$

4. $f(x) = 1 - x^3$
   a. $f^{-1}(x)$
   b. $f^{-1}(2)$
   c. $f^{-1}(28)$

5. $f(x) = \dfrac{2x}{x + 5}$; $x \neq -5$
   a. $f^{-1}(x)$
   b. $f^{-1}(3)$
   c. $f^{-1}(7)$

6. $f(x) = \dfrac{x - 2}{2x + 3}$; $x \neq \dfrac{-3}{2}$
   a. $f^{-1}(x)$
   b. $f^{-1}(1)$
   c. $f^{-1}(0)$

7. $f(x) = \sqrt{x};\ x \geq 0$
a. $f^{-1}(x)$
b. $f^{-1}(5)$
c. $f^{-1}(4)$

8. $f(x) = -\sqrt{x};\ x \geq 0$
a. $f^{-1}(x)$
b. $f^{-1}(-6)$
c. $f^{-1}(-4)$

9. $f(x) = \sqrt{2x - 1};\ x \geq \frac{1}{2}$
a. $f^{-1}(x)$
b. $f^{-1}(25)$
c. $f^{-1}(1)$

10. $f(x) = \sqrt{3x + 12};\ x \geq -4$
a. $f^{-1}(x)$
b. $f^{-1}(6)$
c. $f^{-1}(3)$

11. $f(x) = \sqrt{6 - 3x};\ x \leq 2$
a. $f^{-1}(x)$
b. $f^{-1}(3)$
c. $f^{-1}(0)$

12. $f(x) = \sqrt{10 - 2x};\ x \leq 5$
a. $f^{-1}(x)$
b. $f^{-1}(0)$
c. $f^{-1}(6)$

13. $f(x) = x^2 + 1;\ x \geq 0$
a. $f^{-1}(x)$
b. $f^{-1}(101)$
c. $f^{-1}(50)$

14. $f(x) = x^2 - 5;\ x \leq 0$
a. $f^{-1}(x)$
b. $f^{-1}(4)$
c. $f^{-1}(20)$

15. $f(x) = 6 - x^4;\ x \leq 0$
a. $f^{-1}(x)$
b. $f^{-1}(5)$
c. $f^{-1}(-75)$

16. $f(x) = -2 - x^4;\ x \geq 0$
a. $f^{-1}(x)$
b. $f^{-1}(-18)$
c. $f^{-1}(-2)$

In exercises **17–26**:

a. Determine whether the relation defined by the given equation is a function.

b. If it is a function, determine whether it is one-to-one.

17. $y = 4x - 8$

18. $y = -2x + 5$

19. $y = 4x^2$

20. $y = -9x^2$

21. $y^2 = x^2 + 1$

22. $y^2 = 9 - x^2$

23. $y = x^4$

24. $y = x^6$

25. $y = x^5$

26. $y = x^7$

In exercises **27–30**, the functions defined by the given equations are not one-to-one. For the given domains, state a subset of the domain over which the function is one-to-one. The answers are not unique.

27. $y = x^2$; all x

28. $y = 2x^2 - 1$; all x

29. $y = \sqrt{9 - x^2};\ -3 \leq x \leq 3$

30. $y = \sqrt[4]{36 - x^2};\ -6 \leq x \leq 6$

Name

Date

Score

SECTION 8-1. Summary Exercises

In exercises **1–6**, graph each function.

1. $y = \frac{3}{2}|x|$

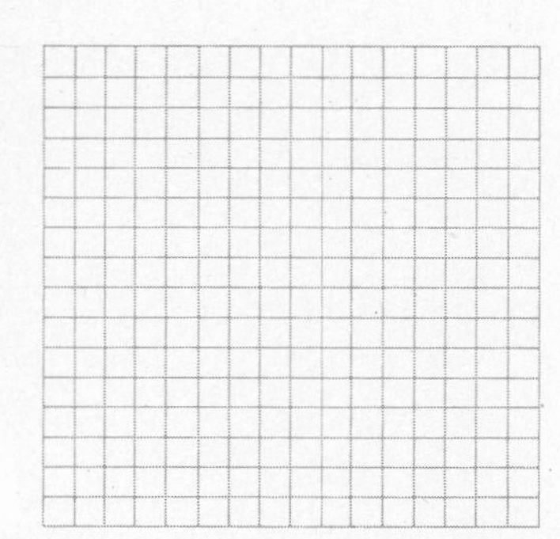

2. $y = \frac{-1}{4}|x|$

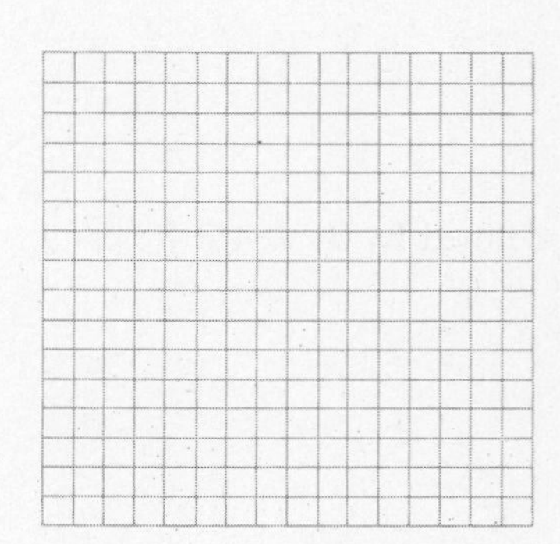

3. $y = |x + 4| - 3$

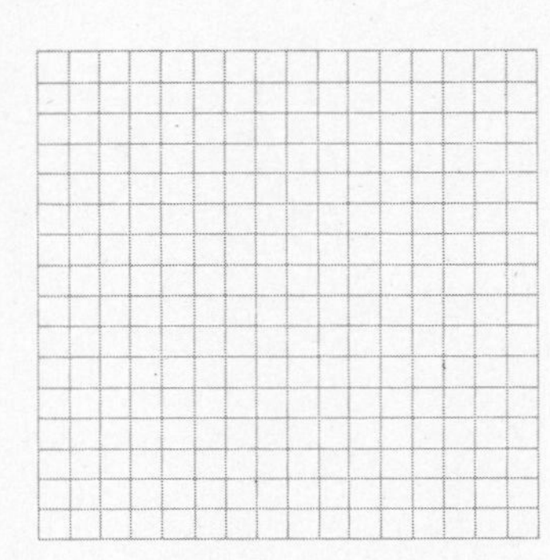

4. $y = -|x - 2| + 4$

5. $y = 2|x - 5|$

6. $y = -3|x + 1| + 8$

In exercises **7** and **8**, refer to the function defined by $3y = 4|x + 5| - 9$.

7. Identify the coordinates of the vertex.

7. $V(\quad,\quad)$

8. Identify the slopes of the side.

8. ____________________

CHAPTER 8. Review Exercises

In exercises **1–12**, graph each equation.

1. $y = \frac{-4}{5}|x|$

2. $y = |x - 7|$

3. $y = |x| - 6$

4. $y = 3|x + 1|$

5. $y = |x - 2| - 3$

6. $y = -|x| + 4$

7. $y = \frac{-3}{2}x^2$

8. $y = (x - 3)^2$

9. $y = x^2 + 2$

10. $y = 3(x - 1)^2$

11. $y = x^2 + 6x$

12. $y = 2x^2 - 8x + 7$

In exercises **13–16**, for each function:

a. Determine the domain of the function.

b. Find the equations of the vertical asymptotes.

c. Determine the nature of y as $x \to +\infty$ and $x \to -\infty$.

13. $y = \frac{2x - 5}{x + 10}$

14. $y = \frac{2}{x - 4}$

15. $y = \frac{x^2 + 1}{4x^2 - 25}$

16. $y = \frac{x}{x^2 + 1}$

In exercises **17** and **18**, graph each function.

17. $y = \frac{x + 8}{x + 1}$

18. $y = \frac{4x^2}{x^2 + 2x - 8}$

In exercises **19** and **20**, for each function f, find:

a. $f(-2)$ **b.** $f(3)$ **c.** $f(x + a)$

19. $f(x) = \frac{x + 2}{x - 4}$

20. $f(x) = x^2 + 3x$

In exercises **21** and **22**, for each function f, list the ordered pairs in f^{-1}.

21. $f = \{(-5, -6), (0, -4), (5, -2), (10, 0)\}$

22. $f = \left\{(-4, -16), (-2, 4), (0, 0), \left(1, \frac{1}{4}\right), \left(3, \frac{3}{2}\right)\right\}$

In exercises **23** and **24**, for each function f, state the domain and range of f^{-1}.

23. $y = x^2 - 5$ Domain: all x
Range: all y, where $y \geq -5$

24. $y = \frac{3x}{x - 3}$ Domain: all x, where $x \neq 3$
Range: all y, where $y \neq 3$.

In exercises **25–30**, write an equation that defines f^{-1}, using $f^{-1}(x)$ notation.

25. $f(x) = \dfrac{2x}{x - 1}$

26. $f(x) = -2x^3$

27. $f(x) = \sqrt{1 - x}$

28. $f(x) = \sqrt[3]{2x + 5}$

29. $f(x) = \dfrac{1}{x^3 + 1}$

30. $f(x) = \dfrac{-3}{2}x + 4$

CHAPTER

9

Systems of Linear Equations and Inequalities

Ms. Glaston was carrying a small box when she entered the classroom. "Good morning, everyone," she said. "In this box, I have some blue and white marbles." She then shook the box and the sound of colliding objects could be heard throughout the room.

"The total number of marbles in the box," she continued, "is 100. Now, who will tell me how many blue and how many white marbles there are in this box." She held the box up over her head, as though this action would somehow help the students determine the answer.

Nobody said anything for almost a minute. Finally, Monica Stutzman said, "I don't think I can answer your question, Ms. Glaston. There are many ways in which the numbers of blue and white marbles could add up to 100."

"You're right, Monica," Ms. Glaston replied. "In fact, we can use two variables and write an equation that expresses this relationship." She then turned on the projector to show the following display on the screen:

Let x represent the number of blue marbles in the box.

Let y represent the number of white marbles in the box.

(1) $x + y = 100$ The total number of marbles is 100.

"That's a linear equation in two variables," Bob Smith observed. "Would it help to graph it?"

Ms. Glaston uncovered the rest of the acetate to show the graph in Figure 9-1 (see p. 576). "Any point with whole-number coordinates on this line," Ms. Glaston said, "would be a solution of equation (1). For example, (25, 75) and (65, 35) represent the possibilities of 25 blue and 75 white marbles, and 65 blue and 35 white marbles, respectively."

"Suppose," she continued, "that I told you that the number of white marbles is 4 more than 3 times the number of blue marbles. Would this additional information enable you to calculate the numbers of blue and white marbles in the box?"

"We could use the information to write another equation in x and y," Carrie Mattaine said. "Then we could also graph this equation, and see if it crosses the line in the figure."

"I am so happy you suggested that, Carrie," Ms. Glaston replied. "In anticipation that someone would say what Carrie just suggested, I prepared this dis-

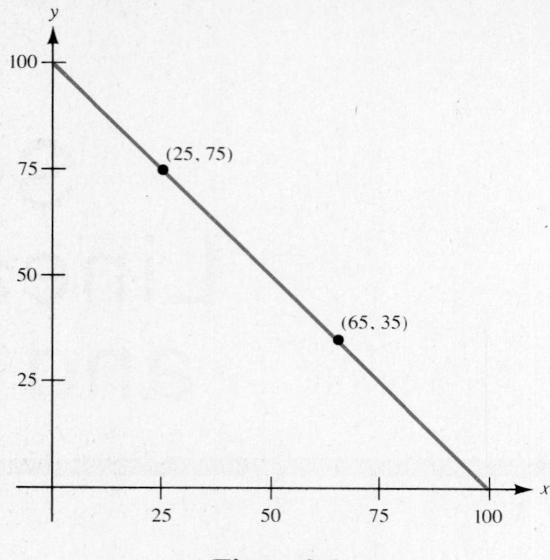

Figure 9-1.

play." She then put an acetate on the projector that showed the following display on the screen (Figure 9-2):

x represents the number of blue marbles

y represents the number of white marbles

(1) $x + y = 100$	The total number of marbles is 100.
(2) $y = 3x + 4$	The number of white marbles is 4 more than 3 times the number of blue.

"As you can see from the graphs," she continued, "the lines do cross at a point I labeled (x_1, y_1). Since this ordered pair is a solution of the equations I have labeled (1) and (2), it satisfies the two statements I made about the numbers of blue and white marbles in this box. Notice that two equations were needed to restrict x and y to just one pair of values. The two equations form a system of equations, and (x_1, y_1) is the solution of the system."

"It looks like (x_1, y_1) is about (25, 75)," Dan Coleman said. "It's hard to tell from the graph."

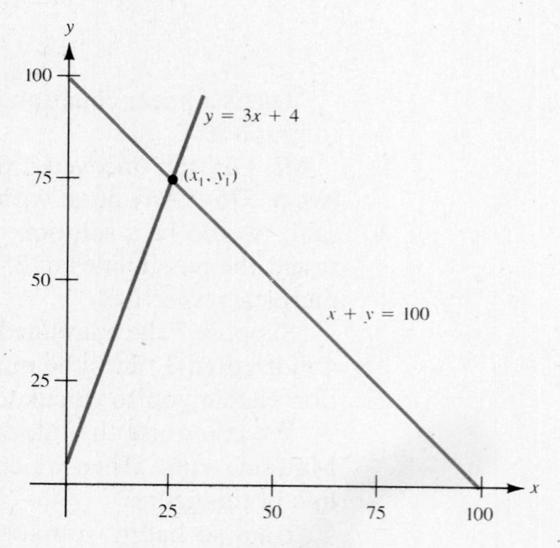

Figure 9-2.

"That's close, Dan," Ms. Glaston replied. "Actually, the ordered pair is (24, 76)." She then turned to the board and wrote:

(1) $24 + 76 = 100$ (2) $76 = 3(24) + 4$

$100 = 100$, *true* $76 = 72 + 4$, *true*

"As Dan has pointed out," she continued, "it is difficult to accurately read the solution from the graph. As a consequence, in the chapter we are beginning today we will learn how to find solutions of systems of equations by algebraic methods. Furthermore, we will apply such algebraic methods to solve systems of equations that have more than two equations and more than two variables."

SECTION 9-1. Systems of Linear Equations in Two Variables

1. A description of a system of linear equations in x and y
2. A solution of a system of linear equations in x and y
3. Solving a system by the graphical method
4. Classifications of systems of equations
5. Solving a system by the substitution method
6. Solving a system by the addition method

In this section, we will study systems of linear equations in x and y, including the graphical, substitution, and addition methods for solving such systems.

A Description of a System of Linear Equations in x and y

If a, b, and c are real numbers, and a and b are not both 0, then $ax + by = c$ is a linear equation in x and y. If two such equations are paired, then the pair of equations is called a **system of equations.** Since the system contains linear equations, we would call it a **system of linear equations in x and y.**

A system of linear equations in x and y can be written as:
(1) $a_1x + b_1y = c_1$, where a_1 and b_1 are not both 0.
(2) $a_2x + b_2y = c_2$, where a_2 and b_2 are not both 0.

Examples **a** and **b** illustrate systems of linear equations in x and y.

a. (1) $2x + y = 11$
(2) $x - 2y = -7$

b. (1) $5x - 4y = -28$
(2) $3x + 7y = 2$

Variables other than x and y can be used for the equations of a system, but the variables must be the same for both equations.

A Solution of a System of Linear Equations in x and y

A solution of a system of linear equations in x and y is identified in Definition 9.1.

Definition 9.1.
If (x_1, y_1) is a solution of equations (1) and (2), then it is a solution of the system:

$$(1)\quad a_1x + b_1y = c_1$$
$$(2)\quad a_2x + b_2y = c_2$$

Example 1. Determine whether $(-5, 8)$ is a solution of the system:

$$(1)\quad 4x + 3y = 4$$
$$(2)\quad 3x - 8y = -79$$

Solution. (1) $4(-5) + 3(8) = 4$ (2) $3(-5) - 8(8) = -79$

$-20 + 24 = 4$, *true* $-15 - 64 = -79$, *true*

Since $(-5, 8)$ is a solution of both equations, it is a solution of the system.

Solving a System by the Graphical Method

It may be possible to determine the solution of a system of linear equations in x and y by graphing the equations. If accurately drawn graphs of the lines intersect at a corner of one of the grids, then the solution can be "read" from the graph. Specifically, the point of intersection has coordinates that are the solution of the system. In this section, the systems solved by this graphical procedure contain equations whose graphs intersect at points with integer coordinates.

Example 2. Given: (1) $x - 2y = 8$
(2) $x + y = 2$

Solve by the graphical method.

Solution. (1) $x - 2y = 8$

$$-2y = -x + 8$$

$$y = \frac{1}{2}x - 4$$

slope is $\frac{1}{2}$; y-intercept is -4

(2) $x + y = 2$

$$y = -x + 2$$

slope is -1; y-intercept is 2

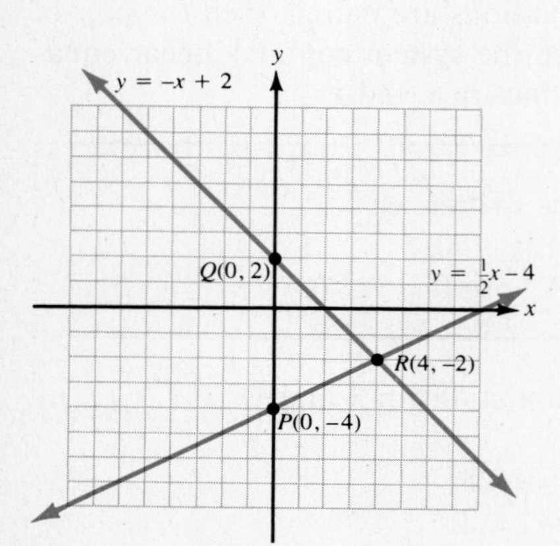

Figure 9-3. Graph of the system (1) $x - 2y = 8$ and (2) $x + y = 2$.

In Figure 9-3, the lines are graphed. $R(4, -2)$ is the point of intersection. The solution set is $\{(4, -2)\}$.

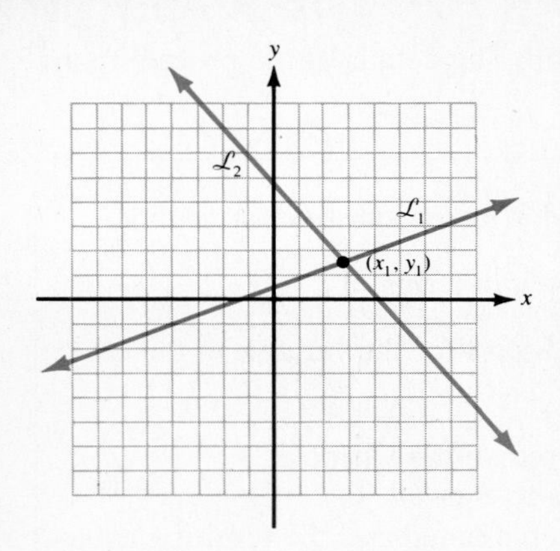

a. $\mathscr{L}_1$ and $\mathscr{L}_2$ intersect.

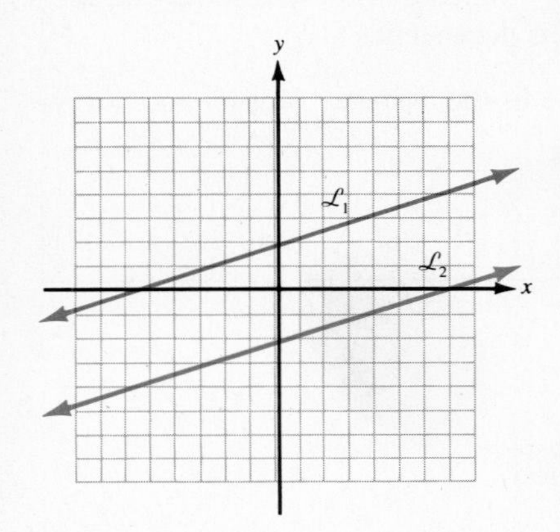

b. $\mathscr{L}_1$ and $\mathscr{L}_2$ are parallel.

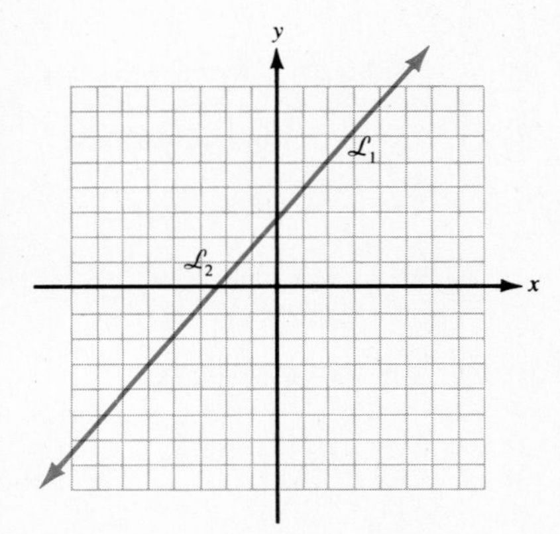

c. $\mathscr{L}_1$ and $\mathscr{L}_2$ are same line.

Figure 9-4. Three possible relationships for $\mathscr{L}_1$ and $\mathscr{L}_2$.

Classifications of Systems of Equations

In Example 2, the graphs of the equations of the system intersected in one point. Therefore, the coordinates of that point were the solution of the system. All systems of linear equations in x and y do not have one ordered pair as a solution. A system may have no solution, or it may have an infinite number of ordered pairs that are solutions. Thus, *there are three possibilities for solution sets of systems of linear equations*. These possibilities are illustrated graphically by the lines defined by the equations of the system.

Let $\mathscr{L}_1$ and $\mathscr{L}_2$ be the graphs of (1) $a_1x + b_1y = c_1$ and (2) $a_2x + b_2y = c_2$. If $\mathscr{L}_1$ and $\mathscr{L}_2$ are drawn on the same coordinate system, then $\mathscr{L}_1$ and $\mathscr{L}_2$ will have one of three possible relationships. These possibilities are shown in Figure 9-4.

For each of the three possibilities, a classification is given for the system composed of the corresponding equations.

Classification 1. The system is *independent* and *consistent*. The graphs of (1) and (2) are two different lines that intersect in exactly one point (Figure 9-4a.) The solution set is $\{(x_1, y_1)\}$.

Classification 2. The system is *inconsistent*. The graphs of (1) and (2) are two different lines that are parallel (Figure 9-4b). The solution set is $\varnothing$.

Classification 3. The system is *dependent*. The graphs of (1) and (2) are the same line (Figure 9-4c.). The solution set contains an infinite number of ordered pairs that lie on a line defined by either equation of the system.

Example 3. Given: (1) $3x - 4y = 20$
(2) $-9x + 12y = 48$

Classify the system as independent and consistent, inconsistent, or dependent.

Solution. **Discussion.** Write (1) and (2) in the $y = mx + b$ form.

1. If the slopes are different, then the system is independent and consistent.
2. If the slopes are the same and the y-intercepts are different, then the system is inconsistent.
3. If the slopes and y-intercepts are the same, then the system is dependent.

(1) $-4y = -3x + 20$ (2) $12y = 9x + 48$

$$y = \frac{3}{4}x - 5 \qquad y = \frac{3}{4}x + 4$$

$$m_1 = \frac{3}{4} \text{ and } b_1 = -5 \qquad m_2 = \frac{3}{4} \text{ and } b_2 = 4$$

Since $m_1 = m_2$ and $b_1 \neq b_2$, the graphs of these equations are parallel lines. Thus, the solution set is $\varnothing$ and the system is inconsistent.

Solving a System by the Substitution Method

One of the basic algebraic techniques for solving a system of linear equations in x and y is the **substitution method.** This technique eliminates one of the variables in equation (1) or (2) by means of a substitution. The details of the procedure are given in the following six steps:

To solve a system of linear equations in x and y by the substitution method:

Step 1. Solve (1) or (2) for x in terms of y, or y in terms of x.

Step 2. Substitute the expression obtained in Step 1 for the variable in the other equation of the system.

Step 3. Solve the equation of Step 2 for the remaining variable.
Case 1. If a root is obtained, then go to Steps 4–6.
Case 2. If the second variable is also eliminated and the resulting equation is *false*, then the system is **inconsistent.**
Case 3. If the second variable is also eliminated and the resulting equation is *true*, then the system is **dependent.**

Step 4. Find the value of the second variable in the system.

Step 5. Check the solution in (1) and (2).

Step 6. State the solution set.

Example 4. Given: (1) $3x + y = 6$
(2) $9x - 2y = -7$

Solve by the substitution method.

Solution. **Discussion.** The y-term in (1) has a coefficient of 1. We therefore solve for y in this equation.

Step 1. (1) $y = 6 - 3x$ — Subtract $3x$ from both sides.

Step 2. (2) $9x - 2(6 - 3x) = -7$ — Replace y in (2) by $6 - 3x$.

Step 3. (2) $9x - 12 + 6x = -7$ — Remove parentheses.

$15x = 5$ — Add 12 to both sides.

$x = \frac{1}{3}$ — The x-component is $\frac{1}{3}$.

Step 4. (1) $3\left(\frac{1}{3}\right) + y = 6$ — Replace x by $\frac{1}{3}$ in (1).

$y = 5$ — The y-component is 5.

Step 5. Checking $\left(\frac{1}{3}, 5\right)$ in (1) and (2):

(1) $3\left(\frac{1}{3}\right) + 5 = 6$ (2) $9\left(\frac{1}{3}\right) - 2(5) = -7$

$1 + 5 = 6$, *true* $3 - 10 = -7$, *true*

Step 6. The solution set is $\left\{\left(\frac{1}{3}, 5\right)\right\}$.

Example 5. Given: (1) $4x - 32y = 65$
(2) $x - 8y = -10$

Solve by the substitution method.

Solution. **Discussion.** The x-term in (2) has a coefficient of 1. We therefore solve for x in this equation.

Step 1. (2) $x = 8y - 10$ Add $8y$ to both sides.

Step 2. (1) $4(8y - 10) - 32y = 65$ Replace x in (1).

Step 3. (1) $32y - 40 - 32y = 65$ Remove parentheses.

$-40 = 65$, *false* The y-terms combine to 0.

Since the resulting number statement is false, the solution set of the system is $\varnothing$.

Solving a System by the Addition Method

Another basic algebraic technique for solving a system of linear equations is the **addition method.** This technique eliminates one of the variables by adding the equations. The details of the procedure are given in the following six steps:

To solve a system of linear equations in x and y by the addition method:

Step 1. If necessary, multiply (1) and (2) by nonzero constants so that the coefficients of one of the variables are equal, but opposite in sign.

Step 2. Add (1) and (2).

Step 3. *Case 1.* If a variable remains, then solve the equation and proceed to Steps 4–6.
Case 2. If both variables are eliminated and the equation is *false,* then the system is **inconsistent.**
Case 3. If both variables are eliminated and the equation is *true,* then the system is **dependent.**

Step 4. Use (1) or (2) to find the value of the other variable.

Step 5. Check the solution in (1) and (2).

Step 6. State the solution set.

Example 6. Given: (1) $3x + 5y = 24$
(2) $4x - 7y = -50$

Solve by the addition method.

Solution. **Discussion.** To eliminate the y-terms when (1) and (2) are added, first change the coefficients to 35 and -35, respectively. Notice that 35 is the least common multiple of 5 and 7.

Step 1. (1) $7(3x + 5y = 24) \longrightarrow$ (1) $21x + 35y = 168$
(2) $5(4x - 7y = -50) \longrightarrow$ (2) $20x - 35y = -250$

Step 2. (1) $21x + 35y = 168$
(2) $20x - 35y = -250$ (+)

$41x = -82$

Step 3. $x = -2$ — The system is independent and consistent.

Step 4. (1) $3(-2) + 5y = 24$ — Replace x by -2 in (1).

$-6 + 5y = 24$

$5y = 30$

$y = 6$ — $(-2, 6)$ is the apparent solution

Step 5. (1) $3(-2)+5(6)=24$ (2) $4(-2)-7(6)=-50$

$-6+30=24$, *true* $-8-42=-50$, *true*

Step 6. The solution set is $\{(-2, 6)\}$.

Example 7. Given: (1) $-4x + 3y = 9$
(2) $12x - 9y = -27$

Solve by the addition method.

Solution. **Step 1.** Multiply (1) by 3.

(1) $3(-4x + 3y = 9) \longrightarrow$ (1) $-12x + 9y = 27$

Step 2. (1) $-12x + 9y = 27$
(2) $12x - 9y = -27$ (+)

$0 = 0$, *true*

Step 3. Both x and y are eliminated and the resulting equation is true. Thus, the system is dependent. The solution set can be written as $\{(x, y) | -4x + 3y = 9\}$. Any solution of the equation is a solution of the system. A particular solution *depends* on which value of x (or y) is arbitrarily selected.

SECTION 9-1. Practice Exercises

In exercises **1–10** determine whether the given ordered pair is a solution of each system.

[Example 1]

1. (1) $2x - 3y = 8$
(2) $7x + y = -41$
$(-5, -6)$

2. (1) $4x + y = 27$
(2) $x - 5y = -9$
$(6, 3)$

3. (1) $2b = 19 - 3a$
(2) $3b = 6 - 2a$
$(9, -4)$

4. (1) $5b = 4a + 56$
(2) $6b = 5a + 69$
$(-9, 4)$

5. (1) $10m + 9n = 8$
(2) $12m - 15n = 0$
$\left(\frac{1}{2}, \frac{1}{3}\right)$

6. (1) $10m + 12n = -23$
(2) $20m - 24n = -16$
$\left(\frac{-3}{2}, \frac{-2}{3}\right)$

7. (1) $5u - 3v = 7$
(2) $10u + 9v = -6$

$\left(\frac{3}{5}, \frac{-4}{3}\right)$

8. (1) $6u + 6v = -14$
(2) $10u - 12v = -27$

$\left(\frac{-5}{2}, \frac{1}{6}\right)$

9. (1) $3x - 2y = 8\sqrt{3}$
(2) $8x + 5y = 11\sqrt{3}$
$(2\sqrt{3}, -\sqrt{3})$

10. (1) $\sqrt{5}x - 2\sqrt{3}y = 3$
(2) $2\sqrt{5}x + \sqrt{3}y = 36$
$(3\sqrt{5}, 2\sqrt{3})$

In exercises **11–20**, solve each system by the graphical method.

[Example 2]

11. (1) $x + 2y = 4$
(2) $3x - 2y = 4$

12. (1) $-2x + y = 2$
(2) $x + y = -4$
(−2, −2)

13. (1) $x + 2y = 2$
(2) $x - y = 5$

14. (1) $x + 2y = 10$
(2) $3x - 2y = -2$

15. (1) $x + 4y = 20$
(2) $3x - 4y = -4$

16. (1) $2x - 3y = 9$
(2) $x + 3y = -18$

17. (1) $2x - 5y = 25$
(2) $x = 5$

18. (1) $x + 3y = 12$
(2) $x + 6 = 0$

19. (1) $5x + 4y = 12$
(2) $y = -7$

20. (1) $6x - 7y = 28$
(2) $y - 2 = 0$

In Exercises **21–30**, classify each system as independent and consistent, inconsistent, or dependent.

[Example 3]

21. (1) $x - 3y = 9$
(2) $2x - 6y = 0$

22. (1) $2x + y = -3$
(2) $6x + 3y = 12$

23. (1) $4x + 3y = 18$
(2) $12x + 9y = 54$

24. (1) $3x - 2y = 8$
(2) $6x - 4y = 16$

25. (1) $2x - 6y = 12$
(2) $4x - y = 2$

26. (1) $7x - 3y = 27$
(2) $14x + 9y = 45$

27. (1) $2(2x - 1) = 10y$
(2) $2(x - y) = 3y + 1$

28. (1) $x = 2(3 - 2y)$
(2) $3(x + 2y) = 6(3 - y)$

29. (1) $3x - 5 = -3(x + y)$
(2) $3(x + 3y) = 18 - 15x$

30. (1) $x - 8 = 4(y - x)$
(2) $2(x - 4) = 8(y - x)$

In exercises **31–60**, solve each system by the substitution or addition method. Write "inconsistent" for a system with no solution and "dependent" for a system with more than one solution.

[Examples 4–7]

31. (1) $5x + y = 14$
(2) $y = -1$

32. (1) $x - 3y = 14$
(2) $y = -4$

33. (1) $3a - 7b = -26$
(2) $a + 6b = 8$

34. (1) $4a + 9b = 42$
(2) $-2a + b = 12$

35. (1) $7m - 3n = 34$
(2) $5m + 9n = 28$

36. (1) $10m + 2n = -19$
(2) $9m - 8n = -22$

37. (1) $4u + 5v = -12$
(2) $2u = -1$

38. (1) $3u - 10v = -41$
(2) $3u = -1$

39. (1) $2r + 3s = 30$
(2) $3r - 5s = -50$

40. (1) $-4r + 5s = 28$
(2) $6r - 6s = -42$

41. (1) $-5x + 10y = -56$
(2) $x - 2y = 0$

42. (1) $9x + y = 36$
(2) $y = -9x$

43. (1) $8a - b + 58 = 0$
(2) $-7a + 4b - 82 = 0$

44. (1) $10a + b - 42 = 0$
(2) $12a - 6b - 108 = 0$

45. (1) $2m + 3n - 15 = 0$
(2) $15n + 10m - 75 = 0$

46. (1) $4m - 7n + 32 = 0$
(2) $28n - 16m - 128 = 0$

47. (1) $5u - 6v = -59$
(2) $3u + 5v = 42$

48. (1) $4u + 9v = 7$
(2) $7u - 4v = 32$

49. (1) $9x + 6y = 2$
(2) $6x - 9y = 10$

50. (1) $15x - 5y = 16$
(2) $5x - 15y = 32$

51. (1) $r + \frac{14}{3}s = -3$
(2) $9r + 20s = -16$

52. (1) $r + \frac{9}{2}s = \frac{-9}{2}$
(2) $14r + 30s = -19$

53. (1) $4a + 6b = 28$
(2) $a + \frac{3}{2}b = 6$

54. (1) $6a - 9b = -36$
(2) $\frac{2}{3}a - b = 4$

55. (1) $21m - 8n = -57$
(2) $-35m + 12n = 96$

56. (1) $3m - 20n = 18$
(2) $-8m + 15n = -25$

57. (1) $15x - 2y + 17 = 0$
(2) $6x + 7y - 40 = 0$

58. (1) $4x - 9y - 38 = 0$
(2) $6x - 5y - 23 = 0$

59. (1) $21u + 20v + 13 = 0$
(2) $15u - 8v + 26 = 0$

60. (1) $8u - 9v - 18 = 0$
(2) $10u + 12v - 7 = 0$

SECTION 9-1. Ten Review Exercises

1. Factor $x^2 - 2x - 8$.

2. Solve $x^2 = 8 + 2x$.

In exercises **3–6**, refer to $y = x^2 - 2x - 8$.

3. Write the equation in the form $y = a(x - h)^2 + k$.

4. Identify the coordinates of the vertex of the graph.

5. Write an equation for the axis of symmetry.

6. Graph the function.

In exercises **7–10**, simplify each expression.

7. $(2t - 3)^2 - (2t + 1)(2t + 9)$

8. $\frac{2}{3}(6u - 3) + \frac{4}{5}(15u - 5) - \frac{2}{7}(21u + 14) + 15$

9. $\frac{1}{w - 4} - \frac{1}{w + 4} + \frac{32}{w^3 - 16w}$

10. $\frac{y^2 - 3y - 7}{y^2 - 4y + 4} + \frac{y + 1}{y^2 - 7y + 10} \cdot \frac{3y - 15}{y - 2}$

SECTION 9-1. Supplementary Exercises

In exercises **1–6**, solve each system.

1. (1) $3x + 5y = 9$
(2) $x - 3y = -11$
(3) $2x + y = -1$

2. (1) $x + 3y = 9$
(2) $x - 2y = 4$
(3) $2x - 3y = 9$

3. (1) $5x - 2y = 2$
(2) $x - 2y = -6$
(3) $2x = y$

4. (1) $3x + y = -1$
(2) $3x + 2y = 4$
(3) $5x = -2y$

5. (1) $3x + y = -2$
(2) $15x + y = 4$
(3) $7x = -y$

6. (1) $9x - 5y = -20$
(2) $9x + 5y = -25$
(3) $x = 5y$

In exercises **7–12**, solve. (*Hint:* Replace $\frac{1}{x}$ by u and $\frac{1}{y}$ by v. Solve equations (1) and (2) for u and v. Then use the values for u and v to solve for x and y.)

7. (1) $\frac{4}{x} - \frac{3}{y} = -1$
(2) $\frac{2}{x} + \frac{1}{y} = 7$

8. (1) $\frac{1}{x} + \frac{2}{y} = 16$
(2) $\frac{5}{x} - \frac{3}{y} = 2$

9. (1) $\frac{2}{x} - \frac{3}{y} = -6$
(2) $\frac{-5}{x} + \frac{2}{y} = 37$

10. (1) $\frac{-3}{x} + \frac{4}{y} = -37$
(2) $\frac{7}{x} - \frac{3}{y} = 23$

11. (1) $\frac{9}{x} + \frac{4}{y} = 2$
(2) $\frac{12}{x} - \frac{8}{y} = 6$

12. (1) $\frac{25}{x} - \frac{2}{y} = -6$
(2) $\frac{-35}{x} + \frac{8}{y} = 11$

In exercises **13–18**, write a system of equations for each graph in the form:

(1) $a_1x + b_1y = c_1$

(2) $a_2x + b_2y = c_2$

13.

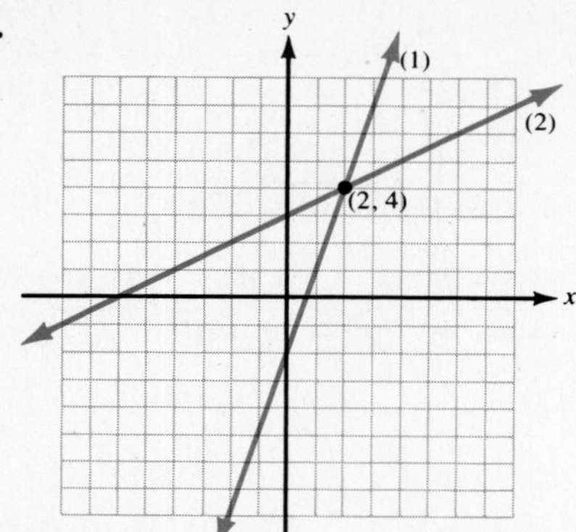

14.

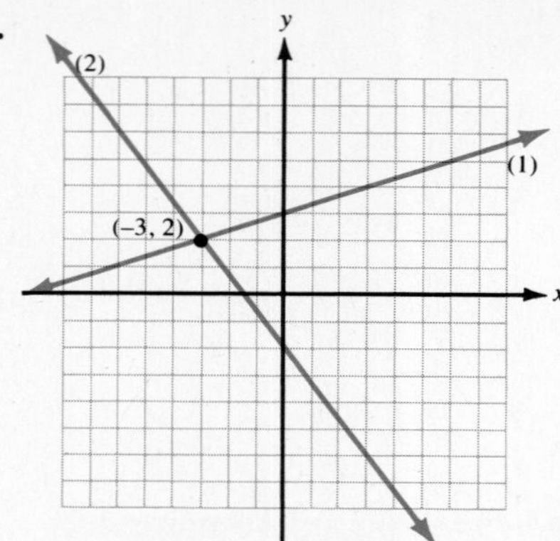

15.

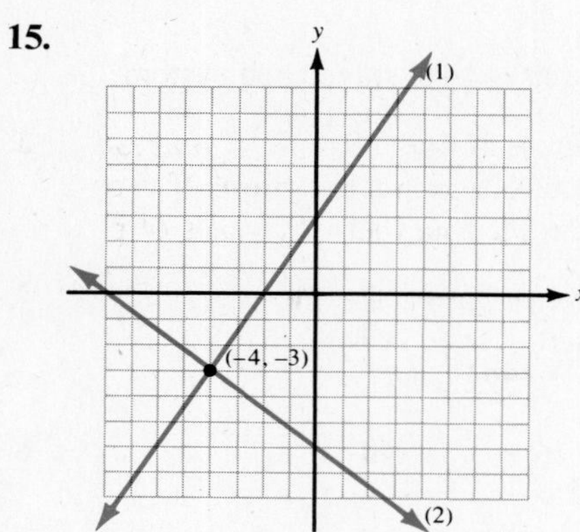

16.

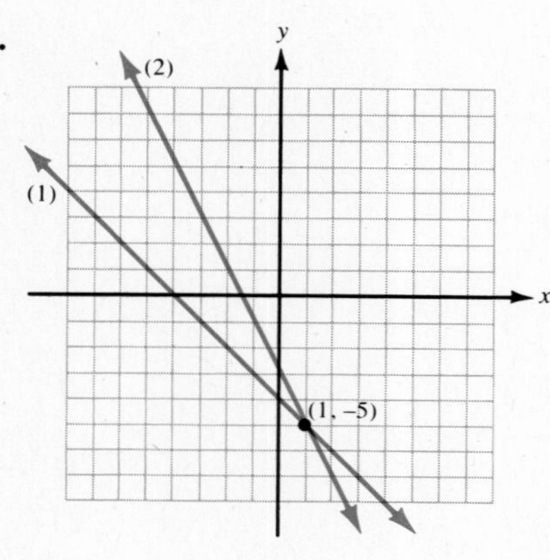

17.

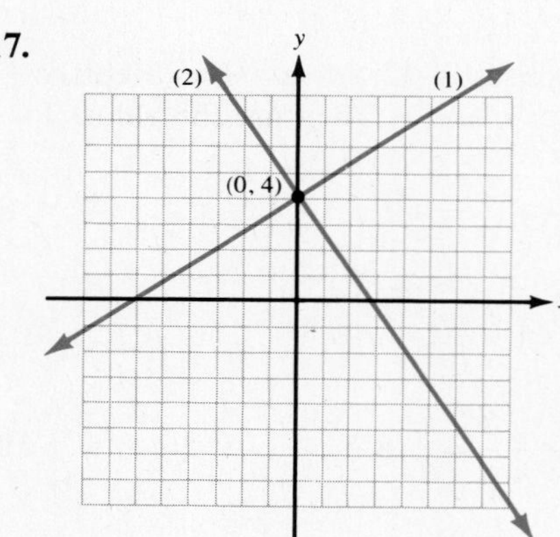

18.

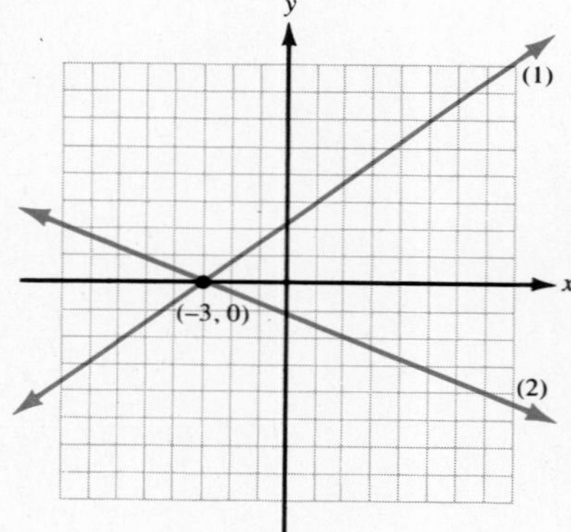

In exercises **19–22**, the systems of equations are nonlinear. Check each ordered pair to determine whether it is a solution.

19. (1) $3x^2 + y^2 = 28$ **a.** $(3, -1)$ **b.** $(1, -5)$
(2) $4x - 2y = 14$

20. (1) $x^2 + 3y^2 = 7$ **a.** $(-2, 1)$ **b.** $(2, -1)$
(2) $x^2 - 2y^2 = 2$

21. (1) $2x^2 - 4y^2 = -64$ **a.** $(-2, 2)$ **b.** $(0, -4)$
(2) $3x^2 - y = 4$

22. (1) $2x^2 + 4y^2 = 12$ **a.** $(2, 2)$ **b.** $(2, -1)$
(2) $3x - y = 7$

SECTION 9-2. Systems of Linear Equations in Three Variables

KEY TOPICS IN THIS SECTION

1. A system of linear equations in x, y, and z
2. A solution of a system in x, y, and z
3. Classifying systems of linear equations in x, y, and z
4. Solving a system by the substitution or addition method

The substitution and addition methods we studied in Section 9-1 can be used to solve systems consisting of three linear equations in three variables.

A System of Linear Equations in x, y, and z

If a cz-term is added to the general form equation of a line, then a linear equation in x, y, and z is formed. A graph of such an equation is a **plane.** The floor, ceiling, and walls of a room are examples of planes.

Linear equation in x and y:	$ax + by = c$	A graph is a line.
Linear equation in x, y, and z:	$ax + by + cz = d$	A graph is a plane.

When three linear equations in x, y, and z are grouped, the result is a system of equations.

A system of linear equations in x, y and z can be written as:

(1) $a_1x + b_1y + c_1z = d_1$, where a_1, b_1, and c_1 are not all 0

(2) $a_2x + b_2y + c_2z = d_2$, where a_2, b_2, and c_2 are not all 0

(3) $a_3x + b_3y + c_3z = d_3$, where a_3, b_3, and c_3 are not all 0

Examples **a** and **b** illustrate systems of linear equations in x, y, and z.

a. (1) $x - 2y + 3z = 14$
(2) $2x + y - 4z = -12$
(3) $3x + 3y - 5z = -18$

b. (1) $2x + 3y = -3$
(2) $5y - 3z = -10$
(3) $x - y + 4z = 16$

A Solution of a System in x, y, and z

A solution of a system of linear equations in x, y, and z is identified in Definition 9.2.

Definition 9.2.
If (x_1, y_1, z_1) is a solution of equations (1), (2), and (3), then it is a solution of the system:

(1) $a_1x + b_1y + c_1z = d_1$
(2) $a_2x + b_2y + c_2z = d_2$
(3) $a_3x + b_3y + c_3z = d_3$

(x_1, y_1, z_1) is called an **ordered triple.**

Example 1. Determine whether $(1, -2, 3)$ is a solution of the system in example **a**.

Solution. **Discussion.** By Definition 9.2, the ordered triple is a solution of the system if it is a solution of (1), (2), and (3).

(1) $1 - 2(-2) + 3(3) = 14$ Check $(1, -2, 3)$ in (1).
$1 - (-4) + 9 = 14$, *true* Yes, a solution of (1)

(2) $2(1) + (-2) - 4(3) = -12$ Check $(1, -2, 3)$ in (2).
$2 + (-2) - 12 = -12$, *true* Yes, a solution of (2)

(3) $3(1) + 3(-2) - 5(3) = -18$ Check $(1, -2, 3)$ in (3).
$3 + (-6) - 15 = -18$, *true* Yes, a solution of (3)

Since $(1, -2, 3)$ is a solution of equations (1), (2), and (3), it is a solution of the system.

Classifying Systems of Linear Equations in x, y, and z

A graph of equations (1), (2), or (3) is a plane in an xyz-coordinate system. If all three planes are drawn in the same coordinate system, then the planes can be related as follows:

Case 1. The planes can intersect in exactly one point, as in Figure 9-5. A system composed of such equations is classified *independent* and *consistent*, and has one ordered triple as a solution.

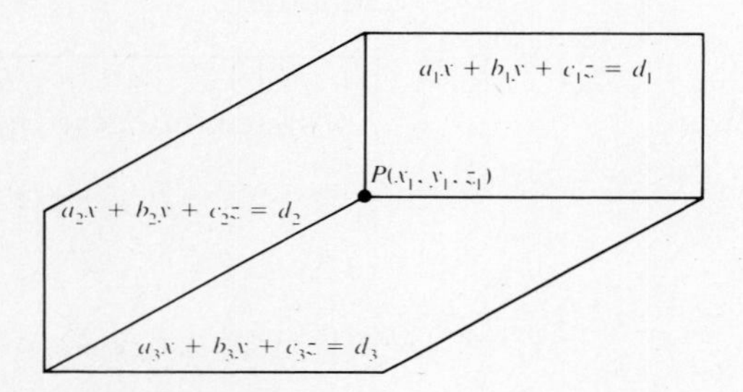

Figure 9-5. The planes intersect at $P(x_1, y_1, z_1)$.

Case 2. The planes do not have any one point that is common to all three planes. Figure 9-6 illustrates one such possibility, in that the intersections of two different pairs of planes are parallel lines. A system composed of such equations is *inconsistent*.

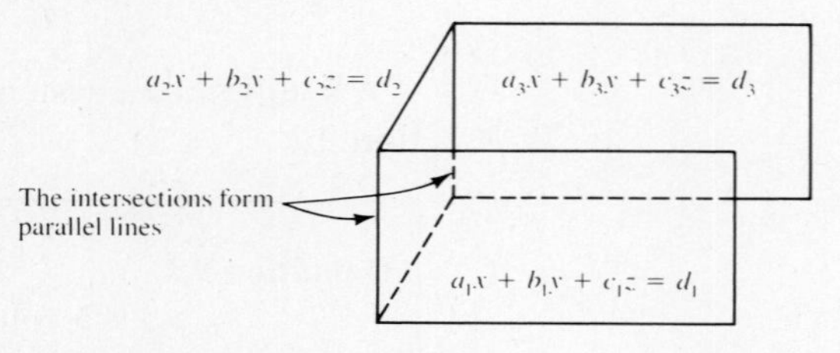

Figure 9-6. The planes intersect in two parallel lines.

Case 3. The planes can intersect in an infinite number of points. Figure 9-7 illustrates one such possibility, in that the intersection of the three planes is a line. Thus, the coordinate of any point on this line is a solution of the system. A system composed of such equations is *dependent*.

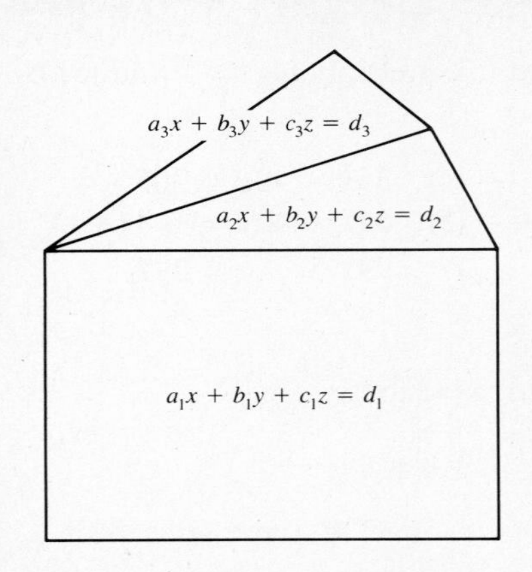

Figure 9-7. The planes intersect in a line of points.

The substitution and addition methods we used to solve systems in Section 9-1 can also be used to solve the systems in this section. If the process eliminates one variable at a time yielding one ordered triple as a solution, then the system is independent and consistent. However, if all the variables are eliminated and the result is a number statement, then the system is either inconsistent or dependent.

> If in the process of solving a system of linear equations in x, y, and z the three variables are eliminated and a number statement is obtained, then:
> 1. The system is **inconsistent** and the solution set is $\emptyset$ if the number statement is *false*.
> 2. The system is **dependent** and the solution set contains an infinite number of solutions if the number statement is *true*.

Solving a System by the Substitution or Addition Method

One technique for solving a system composed of (1), (2), and (3) is to reduce the system to two equations in the same variables. Specifically, form equations (4) and (5) by eliminating the same variable from any of the following combinations:

Combination 1
- Use equations (1) and (2) to form (4).
- Use equations (1) and (3) to form (5).

Combination 2
- Use equations (1) and (2) to form (4).
- Use equations (2) and (3) to form (5).

Combination 3
- Use equations (1) and (3) to form (4).
- Use equations (2) and (3) to form (5).

Example 2. Given:

$$\begin{aligned} (1)\quad 3x + 2y - 4z &= -29 \\ (2)\quad 2x - 5y + 2z &= 45 \\ (3)\quad 5x + 3y + 4z &= 28 \end{aligned}$$

Solve by the substitution or addition method.

Solution. **Discussion.** By inspecting the coefficients of x, y, and z, it appears that the coefficients of z make this variable the easiest one to eliminate to form equations (4) and (5).

Step 1. Multiply (2) by 2 and add to (1) to form (4).

$$(2)\quad 2(2x - 5y + 2z = 45) \longrightarrow \begin{aligned} (2)\quad 4x - 10y + 4z &= 90 \\ (1)\quad 3x + 2y - 4z &= -29\ (+) \\ \hline (4)\quad 7x - 8y \qquad &= 61 \end{aligned}$$

Step 2. Add (1) and (3) to form (5).

$$\begin{aligned} (1)\quad 3x + 2y - 4z &= -29 \\ (3)\quad 5x + 3y + 4z &= 28\ (+) \\ \hline (5)\quad 8x + 5y \qquad &= -1 \end{aligned}$$

We now have the following system in x and y:

$$\begin{aligned} (4)\quad 7x - 8y &= 61 \\ (5)\quad 8x + 5y &= -1 \end{aligned}$$

Step 3. To eliminate y in (4) and (5), multiply (4) by 5 and (5) by 8, and add.

$$\begin{array}{llcll} (4) & 5(7x - 8y = 61) & \longrightarrow & (4) & 35x - 40y = 305 \\ (5) & 8(8x + 5y = -1) & \longrightarrow & (5) & 64x + 40y = -8\ (+) \\ \hline & & & & 99x \qquad = 297 \\ & & & & x = 3 \end{array}$$

Step 4. Replace x by 3 in (4) and solve for y.

(4) $7(3) - 8y = 61$ — Replace x by 3 in (4)

$-8y = 40$ — Subtract 21 from both sides.

$y = -5$ — The value of y_1 is -5.

Step 5. Replace x by 3 and y by -5 in (1), (2), or (3).

(1) $3(3) + 2(-5) - 4z = -29$ — Using equation (1)

$-4z = -28$ — Add 1 to both sides.

$z = 7$ — The value of z_1 is 7.

The check is omitted. The solution set is $\{(3, -5, 7)\}$.

Example 3. Given:

$$\begin{array}{ll} (1) & 2x + 3y \qquad = -3 \\ (2) & \qquad 5y - 3z = -10 \\ (3) & x - y + 4z = 16 \end{array}$$

Solve by the substitution or addition method.

Solution. **Discussion.** Equation (1) has no z-term and equation (2) has no x-term. Either of these equations can be used for (4). If (1) is used, then eliminate the z-term using (2) and (3). If (2) is used, then eliminate the x-term using (1) and (3).

Step 1. Multiply (3) by (-2) and add to (1) to eliminate x.

$$\begin{array}{llcll} (3) & -2(x - y + 4z = 16) & \longrightarrow & (3) & -2x + 2y - 8z = -32 \\ & & & (1) & 2x + 3y \qquad = -3\ (+) \\ \hline & & & (5) & 5y - 8z = -35 \end{array}$$

We now have the following system in y and z:

$$\left.\begin{array}{ll} (4) & 5y - 3z = -10 \\ (5) & 5y - 8z = -35 \end{array}\right.$$

Step 2. Multiply (4) by (-1) and add to (5).

$$\begin{array}{llcll} (4) & -1(5y - 3z = -10) & \longrightarrow & (4) & -5y + 3z = 10 \\ & & & (5) & 5y - 8z = -35\ (+) \\ \hline & & & & -5z = -25 \\ & & & & z = 5 \end{array}$$

Step 3. Replace z by 5 in (2).

(2) $5y - 3(5) = -10$ — Replace z by 5.

$y = 1$ — The value of y_1 is 1.

Step 4. Replace y by 1 in (1).

$$(1)\quad 2x + 3(1) = -3 \qquad \text{Replace } y \text{ by } 1.$$
$$x = -3 \qquad \text{The value of } x_1 \text{ is } -3.$$

The check is omitted. The solution set is $\{(-3, 1, 5)\}$.

Example 4. Given:

$$\begin{aligned} (1)\quad -2x - y + 4z &= 15 \\ (2)\quad 5x + y - 3z &= 12 \\ (3)\quad 4x + 2y - 8z &= -32 \end{aligned}$$

Solve by the substitution or addition method.

Solution. **Discussion.** The easiest variable to eliminate first is y, using (1) and (2), and then (1) and (3).

Step 1. Add (1) and (2).

$$\begin{aligned} (1)\quad -2x - y + 4z &= 15 \\ (2)\quad 5x + y - 3z &= 12\ (+) \\ \hline (4)\quad 3x + z &= 27 \end{aligned}$$

Step 2. Multiply (1) by 2 and add to (3).

$$(1)\quad 2(-2x - y + 4z = 15) \longrightarrow \begin{aligned} (1)\quad -4x - 2y + 8z &= 30 \\ (3)\quad 4x + 2y - 8z &= -32\ (+) \\ \hline 0 &= -2, \textit{false} \end{aligned}$$

All three variables were simultaneously eliminated and the resulting number statement is false. Therefore, the given system is inconsistent and the solution set is $\varnothing$.

SECTION 9-2. Practice Exercises

In exercises **1–6**, determine whether the given ordered triple is a solution of each system.

[Example 1]

1. (1) $x + 2y + 3z = 9$
(2) $2x - y - z = 2$
(3) $3x - 2y + z = 11$

$(2, -1, 3)$

2. (1) $x - y + z = -3$
(2) $2x + y - z = 12$
(3) $3x - 2y - z = 9$

$(3, 2, -4)$

3. (1) $x + y - z = 0$
(2) $x - 2y + 3z = -18$
(3) $3x - y + 2z = -25$

$(-5, 2, -3)$

4. (1) $2x - y + 3z = 16$
(2) $3x - y + 2z = 18$
(3) $6x + 8y - 3z = 0$

$(4, -2, 2)$

5. (1) $4x - 9y = 8$
(2) $12x + 8z = -6$
(3) $8x + 3y - 4z = 8$

$\left(\frac{1}{2}, \frac{-2}{3}, \frac{-3}{2}\right)$

6. (1) $6x + 5y - 8z = -5$
(2) $15y + 2z = 10$
(3) $-9x + 16z = 14$

$\left(\frac{-2}{3}, \frac{3}{5}, \frac{1}{2}\right)$

In exercises **7–26**, solve each system by the substitution or addition method. Write "inconsistent" for a system with no solution and "dependent" for a system with more than one solution.

[Examples 2–4]

7. (1) $x + y + z = 4$
(2) $2x + 3y - z = 14$
(3) $3x - y - z = 4$

8. (1) $2x - 2y + z = -3$
(2) $5x - 3y - 2z = -17$
(3) $3x - 5y + 5z = 2$

9. (1) $2x + 3y - 2z = 3$
(2) $-3x + y - z = 7$
(3) $-4x + 2y + 3z = 30$

10. (1) $3x + 2y + z = 0$
(2) $x + 4y - z = 0$
(3) $2x - 3y + 2z = 2$

11. (1) $2x - y - 3z = 19$
(2) $5x - 2y + 2z = 19$
(3) $3x + y + 4z = 3$

12. (1) $2x + 5y + 5z = -22$
(2) $x - 2y - 3z = 16$
(3) $x + 3y + 2z = -14$

13. (1) $3x - 2y + 4z = 7$
(2) $5x + y - z = -10$
(3) $-9x + 6y - 12z = -20$

14. (1) $x + 3y - 2z = 5$
(2) $-2x - y + 7z = 4$
(3) $-5x - 15y + 10z = -20$

15. (1) $x - 2z = 15$
(2) $x - 2y - 3z = 1$
(3) $2x + 5y = 56$

16. (1) $2x + 3y = 9$
(2) $-4y + 5z = 44$
(3) $x - 6y - 2z = -4$

17. (1) $2x + z = -6$
(2) $x + 2y = 4$
(3) $y + 2z = 26$

18. (1) $x + y + z = 0$
(2) $2x - z = -17$
(3) $-x + 2y = -6$

19. (1) $2x - 6y + 10z = 24$
(2) $-x + 3y - 5z = -12$
(3) $10x - 30y + 50z = 120$

20. (1) $6x + 3y - 15z = 27$
(2) $-2x - y + 5z = -9$
(3) $8x + 4y - 20z = 36$

21. (1) $4x - 15y + 3z = 0$
(2) $-2x + 5y - 9z = -3$
(3) $10x - 25y - 6z = -2$

22. (1) $-12x + 8y - 5z = 8$
(2) $3x - 20y - 5z = -18$
(3) $-9x + 4y + 5z = 8$

23. (1) $-6x + 2y - 5z = 2$
(2) $3x - 12y + 20z = 4$
(3) $15x + 20y - 30z = -18$

24. (1) $12x + 14y - 2z = 0$
(2) $-6x + 20y + 36z = 39$
(3) $18x + 8y - 4z = 5$

25. (1) $2x + y - z = \frac{-7}{6}$
(2) $x - y - z = \frac{-10}{6}$
(3) $3x + y - z = \frac{-10}{6}$

26. (1) $x - 2y + z = \frac{-7}{6}$
(2) $x + y - 2z = \frac{-4}{6}$
(3) $2x - y + 3z = \frac{1}{6}$

SECTION 9-2. Ten Review Exercises

In exercises **1–10**, let f be the linear function containing $P(-4, 2)$ and $Q(4, -8)$.

1. State the domain of f.

2. Find the slope of the line that passes through P and Q.

3. Write in $f(x) = mx + b$ form an equation that defines f.

4. Find the x-intercept of a graph of f.

5. Write in $g(x) = mx + b$ form an equation of a line that contains $R(1, 3)$ and is perpendicular to a graph of f.

6. Write an equation that defines f^{-1} using $f^{-1}(x)$ notation.

7. Write an equation that defines g^{-1} using $g^{-1}(x)$ notation.

8. Find the point of intersection of the graphs of f and f^{-1}.

9. Find the distance between P and Q.

10. Find the midpoint of the segment with endpoints P and Q.

SECTION 9-2. Supplementary Exercises

If a system of equations has fewer equations than variables (such as two equations in three variables), then it is **underdetermined.** Such a system is dependent. It may, however, be possible to write a general form for the solutions using one variable in the system.

Example. Given: (1) $x + 2y - z = 4$
(2) $2x - y + z = 7$

Write the solutions in terms of x.

Solution. **Step 1.** Eliminate the z-terms in (1) and (2).

$$\begin{array}{lrl} (1) & x + 2y - z = & 4 \\ (2) & 2x - y + z = & 7\,(+) \\ \hline (3) & 3x + y \quad\;\; = & 11 \end{array}$$

Step 2. Solve (3) for y in terms of x:

(3) $y = 11 - 3x$

Step 3. Eliminate the y-terms in (1) and (2).

$$\begin{array}{lrl} (1) & x + 2y - z = & 4 \\ (2) & 4x - 2y + 2z = & 14\,(+) \\ \hline (4) & 5x \quad\;\; + z = & 18 \end{array}$$

Step 4. Solve (4) for z in terms of x.

(4) $z = 18 - 5x$

Step 5. Write the solution set in terms of x.

$$\{(x, 11 - 3x, 18 - 5x)\}$$

In exercises **1–4**, write the solutions of each system in terms of x.

1. (1) $3x + y + z = 6$
(2) $x - y + z = 2$

2. (1) $5x + 2y - z = -4$
(2) $2x - 2y - z = 2$

3. (1) $x + 2y - 4z = 7$
(2) $3x - 2y + 2z = 0$

4. (1) $6x - 3y + 8z = 10$
(2) $x - y + 2z = -5$

In exercises **5–8**, write the solutions of each system in terms of z.

5. (1) $4x + 3y - z = 8$
(2) $x - y + 2z = -4$

6. (1) $3x - 2y + z = 0$
(2) $2x + 4y - 3z = 1$

7. (1) $x - y + 4z = 10$
(2) $3x + 2y + z = -4$

8. (1) $x + y - z = 7$
(2) $2x + 3y + z = 2$

Tina Smith works in a fast food outlet called The Burger Shack. She noticed one day that the cash register was printing the number of items and the total cost of the sale, but was not printing the cost of each item.

In exercises **9** and **10**, find the cost of one of each item.

9. (1) 3 burgers, 1 order of fries, and 2 drinks cost \$10.00
(2) 2 burgers, 1 order of fries, and 3 drinks cost \$8.50
(3) 4 burgers, 2 orders of fries, and 1 drink cost \$12.00

10. (1) 3 burgers, 2 orders of fries, and 3 drinks cost \$9.75
(2) 2 burgers, 1 order of fries, and 3 drinks cost \$7.50
(3) 4 burgers, 3 orders of fries, and 1 drink cost \$9.50

SECTION 9-3. Applications

1. Number problems
2. Mixture problems
3. Geometry problems
4. Motion problems
5. Age problems

In this section, we will solve some applied problems using systems of linear equations in x and y, or systems of linear equations in x, y, and z.

Number Problems

A number problem includes two statements about two numbers, or three statements about three numbers. Each statement yields a linear equation in two, or three variables. These equations form a system of linear equations, and the solution of the system will be the unknown numbers.

Example 1. The sum of 3 times one integer and 2 times another is -7. When the second integer is subtracted from 4 times the first, the difference is 20. Find both integers.

Solution. Let x represent the first integer and y the second integer. "The sum of 3 times one integer and 2 times another is -7."

(1) $3x + 2y = -7$

"When the second integer is subtracted from 4 times the first, the difference is 20."

(2) $4x - y = 20$

Writing (1) and (2) as a system:

(1)	$3x + 2y = -7$	Multiply (2) by 2 and add to (1) to eliminate the y-terms.
(2)	$4x - y = 20$	

(1)	$3x + 2y = -7$	
(2)	$8x - 2y = 40\ (+)$	$2(4x - y = 20) \longrightarrow 8x - 2y = 40$
	$11x = 33$	
	$x = 3$	The first integer is 3.
(1)	$3(3) + 2y = -7$	Replace x by 3 in (1).
	$y = -8$	The second integer is -8.

Check: (1) $3(3) + 2(-8) = -7$

$9 + (-16) = -7$, *true*

(2) $4(3) - (-8) = 20$

$12 - (-8) = 20$, *true*

Thus, the integers are 3 and -8.

Mixture Problems

A mixture problem includes two or three statements about quantities that are combined. Frequently, the information indicates *how much* of the various quantities are being combined, and *some value* associated with the mixture.

Example 2. Jain Simmons, Joe Bennett, and Shannon Schanze are neighbors that live in the Horizon's West Townhouse complex. Recently, they shopped together at the local supermarket. All three bought the same brands of frozen carrots, broccoli, and corn. Jain paid \$6.22 for one package of carrots, three broccoli, and five corn. Joe paid \$7.11 for six packages of carrots, one broccoli, and four corn. Shannon paid \$9.67 for five packages of carrots, two broccoli, and eight corn. Find the costs of one package of carrots, one package of broccoli, and one package of corn.

Solution. Let x be the cost of one package of carrots, y be the cost of one package of broccoli, and z be the cost of one package of corn.

(1)	$x + 3y + 5z = 622$	The combination Jain bought
(2)	$6x + y + 4z = 711$	The combination Joe bought
(3)	$5x + 2y + 8z = 967$	The combination Shannon bought

Use (1) and (2), and also (2) and (3) to eliminate the y-terms:

(1)	$x + 3y + 5z = 622$	Multiply (2) by -3 and add to (1).
(2)	$-18x - 3y - 12z = -2{,}133\ (+)$	
(4)	$-17x - 7z = -1{,}511$	

(2)	$-12x - 2y - 8z = -1{,}422$	Multiply (2) by -2 and add to (3).
(3)	$5x + 2y + 8z = 967\ (+)$	
(5)	$-7x = -455$	The y and z-terms are eliminated.
	$x = 65$	The carrots cost \$0.65.

(4) $-17(65) - 7z = -1{,}511$ Replace x by 65 in (4).

$z = 58$ The corn cost \$0.58.

(1) $65 + 3y + 5(58) = 622$ Replace x and z in (1).

$y = 89$ The broccoli cost \$0.89.

The check is omitted. The following values are asserted:

Carrots cost \$0.65 per package.
Broccoli cost \$0.89 per package.
Corn cost \$0.58 per package.

Geometry Problems

Frequently, some information is given regarding a graph of some equation, or graphs of two or more equations. A system of equations may be formed from the given information, and the solution of the system can be used to write an equation for the graph.

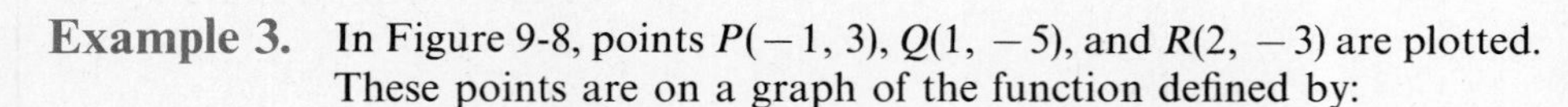

Example 3. In Figure 9-8, points $P(-1, 3)$, $Q(1, -5)$, and $R(2, -3)$ are plotted. These points are on a graph of the function defined by:

$$y = ax^2 + bx + c$$

Find the values for a, b, and c.

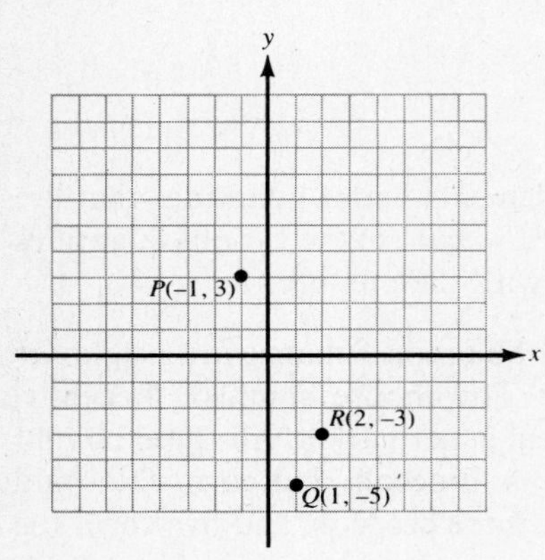

Figure 9-8. Points P, Q, and R.

Solution. **Discussion.** Replace x and y by the coordinates of P, Q, and R to form a system of three equations in a, b, and c.

Using $P(-1, 3)$: (1) $3 = a - b + c$ or $a - b + c = 3$

Using $Q(1, -5)$: (2) $-5 = a + b + c$ or $a + b + c = -5$

Using $R(2, -3)$: (3) $-3 = 4a + 2b + c$ or $4a + 2b + c = -3$

Multiply (1) by -1 and add the equation to (2), and then to (3), to eliminate the c-terms.

(1) $(-1)(a - b + c = 3) \longrightarrow$ (1) $-a + b - c = -3$

(1) $-a + b - c = -3$
(2) $a + b + c = -5\ (+)$
(4) $2b = -8$

(1) $-a + b - c = -3$
(3) $4a + 2b + c = -3\ (+)$
(5) $3a + 3b = -6$

$b = -4$

Replace b by -4 in (5) and solve for a.

(5) $3a + 3(-4) = -6$

$a = 2$

Replace a by 2 and b by -4 in (1).

(1) $2 - (-4) + c = 3$

$c = -3$

Thus, $a = 2$, $b = -4$, and $c = -3$, and an equation for the function is $y = 2x^2 - 4x - 3$. To check, replace x and y by the coordinates of P, Q, and R to verify they are solutions. The check is omitted.

Motion Problems

A motion problem usually includes a statement about two moving objects, or about an object that moves at two different rates, or about an object that moves two different distances in some specified time. The formula used in any of these problems is:

$$d = r \cdot t \qquad \text{distance} = (\text{rate})(\text{time})$$

Example 4. Netta Gilboa owns a cosmetic firm that has offices in Highland Park and Sandusky. On a recent round trip between the two cities, she flew the company plane from Sandusky to Highland Park in 4 hours against a wind that slowed the ground speed of the plane. The return trip to Sandusky was made in 3 hours with the same wind increasing the ground speed of the plane. If the distance between the two cities is 960 miles, find the ground speed of the plane with no wind, and the speed of the wind on this trip.

Solution. Let x represent the speed of the plane with no wind. Let y represent the speed of the wind on this trip.

Trip	Rate	Time	Distance
From Sandusky to Highland Park	$(x - y)$ mph	4 hours	$4(x - y)$ miles
From Highland Park to Sandusky	$(x + y)$ mph	3 hours	$3(x + y)$ miles

(1) $960 = 4(x - y)$ From Sandusky to Highland Park
(2) $960 = 3(x + y)$ From Highland Park to Sandusky

(1) $240 = x - y$ Divide both sides by 4.
(2) $320 = x + y\ (+)$ Divide both sides by 3.
$560 = 2x$ Add (1) and (2).

$280 = x$ The speed of the plane with no wind

(1) $240 = 280 - y$ Replace x by 280 in (1).

$y = 40$ The speed of the wind

On this trip, the speed of the plane without wind was 280 mph and the speed of the wind was 40 mph.

Age Problems

In an age problem, two or more statements are given about two or more people or objects. The statements can be used to write linear equations about the ages of the items in the problem. These equations can be used to form a system of equations, and the solution of the system can be used to determine the unknown ages.

Example 5. Sharon and Bob Gazdacko have a son Craig. The sum of their current ages is 92 years. The difference between Bob's and Sharon's ages is 7 years. If Bob's age 7 years ago is multiplied by 2, the product is the same as the sum of the ages of Sharon and Craig in 10 years. Find the ages of Sharon, Bob, and Craig.

Solution. The information is shown in the following table:

Person	Current ages	Ages at some other time
Sharon	x years	10 years from now: $x + 10$
Bob	y years	7 years ago: $y - 7$
Craig	z years	10 years from now: $z + 10$

(1) $x + y + z = 92$ — The sum of current ages is 92.

(2) $y - x = 7$ — Bob's age minus Sharon's is 7.

(3) $2(y - 7) = (x + 10) + (z + 10)$ — Twice Bob's age 7 years ago equals the sum of Sharon's and Craig's ages 10 years from now.

Writing (1), (2), and (3) as a system:

(1) $x + y + z = 92$
(2) $-x + y = 7$
(3) $-x + 2y - z = 34$

Let (2) be equation (4). Then add (1) and (3) to eliminate the z-terms.

(4) $-x + y = 7$
(5) $3y = 126$ — Both x and z-terms are eliminated.

$y = 42$ — The current age of Bob

(2) $-x + 42 = 7$ — Replace y by 42 in (2).

$x = 35$ — The current age of Sharon

(1) $35 + 42 + z = 92$ — Replace x and y in (1) by 35 and 42.

$z = 15$ — The current age of Craig

The check is omitted. The current ages of Sharon, Bob, and Craig are, respectively, 35 years, 42 years, and 15 years.

SECTION 9-3. Practice Exercises

In exercises **1–16**, solve each word problem.

[Example 1]

1. The sum of 4 times the smaller of two numbers and 3 times the larger is 68. The difference between 3 times the smaller number and 2 times the larger is 0. Find both numbers.

2. The larger of two numbers is 1 more than 2 times the smaller. The difference between 5 times the smaller number and 3 times the larger is -10. Find both numbers.

3. The smaller of two numbers is 7 less than $\frac{1}{3}$ the larger. The sum of 3 times the smaller and 4 times the larger is 9. Find both numbers.

4. Two times the smaller of two numbers is 4 less than 7 times the larger. The difference between 3 times the smaller number and 10 times the larger is -7. Find both numbers.

5. The sum of 8 times the smaller of two numbers and 6 times the larger is 14. When 3 times the larger number is subtracted from 10 times the smaller, the difference is 0. Find both numbers.

6. The smaller of two numbers is 1 less than 2 times the larger. The sum of 6 times the smaller and 4 times the larger is -18. Find both numbers.

7. The sum of three numbers is 3. The sum of the smallest and 2 times the largest is 8. The difference between 2 times the smallest and 6 times the middle-sized number is -14. Find all three numbers.

8. The sum of 3 times the smallest, 2 times the middle-sized, and the largest number is 5. The difference between 2 times the smallest number and 3 times the largest is -19. The sum of the middle-sized number and 2 times the largest is 13. Find all three numbers.

[Example 2] **9.** Charlie and Rick stopped at a roadside fruit stand on the way home from work. Charlie bought 5 pounds of apples and 8 pounds of oranges for a total of $4.09. Rick bought 3 pounds of the same apples and 5 pounds of the same oranges for $2.50. What was the price per pound for the apples and oranges?

10. A local nursery had two kinds of lawn seed for sale. Annette bought a mix consisting of 20 pounds of the more expensive brand and 10 pounds of the cheaper brand for $56.00. Pete bought 8 pounds of the more expensive brand and 12 pounds of the cheaper brand for $33.60. What was the price per pound for each brand of lawn seed?

11. Edna has $2,500 invested in a mutual fund and $6,000 in a time certificate. On these two investments, she earns $770 a year in interest. Lisa has $4,200 in the same mutual fund and $4,800 in a similar time certificate. Lisa earns $792 a year in interest. Find the annual interest rate on each type of investment.

12. Richard has a $6,000 loan from Apex Finance Company and a $2,500 loan from the Civil Employees Credit Union. He pays $1,405 annual interest on these loans. Scott has an $8,500 loan from Apex Finance Company at the same interest rate as Richard's loan, and a $1,200 loan from the credit union with the same interest rate as Richard's. Scott pays $1,686 annual interest on these loans. What are the annual interest rates on the loans at Apex Finance Company and Civil Employees Credit Union, respectively?

13. Storage tanks A and B in a chemical company's warehouse hold a cleaning solvent with different percents of a caustic chemical. When 10 gallons from tank A are mixed with 20 gallons from tank B, the mixture contains 30% solvent. When 15 gallons from tank A are mixed with 10 gallons from tank B, the mixture contains 26% solvent. Find the percent concentration of solvent in tanks A and B, respectively.

14. A hospital storeroom has two flasks with different percent concentrations of alcohol. Nurse Ellie mixed 5 liters from flask A and 10 liters from flask B and got a mixture that was 28% alcohol. Nurse Sheila mixed 15 liters from flask A with 5 liters from flask B and got a mixture that was 35.5% alcohol. Find the percent concentration of alcohol in flasks A and B, respectively.

15. A market has three kinds of nuts for sale, labeled X, Y, and Z. Nancy paid a total of $7.95 for 2 pounds of X, 1 pound of Y, and 3 pounds of Z. Mitzi paid a total of $4.90 for 1 pound of X, 2 pounds of Y, and 1 pound of Z. Zelda paid a total of $6.85 for 2 pounds of X, 3 pounds of Y, and 1 pound of Z. Find the costs of 1 pound each of X, Y, and Z.

16. A feed-grain dealer wants to mix oats, barley, and corn to make a grain mix for horses and cattle. Based on the current retail prices of these grains, 1 pound of oats, 1 pound of barley, and 1 pound of corn together sell for a total of 31 cents. A mix of 3 pounds of oats, 1 pound of barley, and 2 pounds of corn sells for a total of 61 cents. A mix of 2 pounds of oats, 2 pounds of barley, and 3 pounds of corn sells for a total of 74 cents. Find the current retail prices for 1 pound of oats, 1 pound of barley, and 1 pound of corn, respectively.

[Example 3] In exercises **17–20**, the given points are on the graph of $y = ax^2 + bx + c$. Find the values of a, b, and c.

17. $P_1(1, 1), P_2(2, -4), P_3(-2, 4)$

18. $P_1(-3, 9), P_2(-1, 5), P_3(1, -7)$

19. $P_1(4, 3), P_2(2, -1), P_3(3, 0)$

20. $P_1(3, 4), P_2(1, 6), P_3(5, -2)$

In exercises **21–24**, the given points are on the graph of a circle defined by $x^2 + y^2 + ax + by = c$. Find the values of a, b, and c.

21. $P_1(3, -1), P_2(1, 1), P_3(-1, -1)$

22. $P_1(6, 1), P_2(3, 4), P_3(0, 1)$

23. $P_1(4, 1), P_2(-5, 4), P_3(2, -3)$

24. $P_1(-6, -6), P_2(3, -3), P_3(1, 1)$

In exercises **25–38**, solve each word problem.

[Example 4] **25.** Amelia Hart is flying her plane in a wind. If Amelia flies in the direction the wind is blowing, she can travel 675 miles in 3 hours. If she flies in the opposite direction of the wind, she can travel only 620 miles in 4 hours. Find the speed of the plane with no wind, and find the speed of the wind.

26. Fritz Chrysler is cruising his jet boat on a large river. In $3\frac{1}{2}$ hours, he travels 126 miles upriver against the current. On the return trip, in 2 hours he is 14 miles from the starting point. Find the cruising speed of Fritz's boat and the speed of the current in the river.

27. To travel to Denver, Kate has two options available.

Option 1: Travel 2 hours by bus to city A, then fly to Denver in 3 hours.

Option 2: Fly to city B in 4 hours, then travel to Denver by bus in 1 more hour.

Option 1 covers 1,350 miles and option 2 covers 1,725 miles. Assuming the rates of the buses are the same and the rates of the plane are the same in options 1 and 2, find the rates of the bus and the plane, respectively.

28. Fred Quillen wants to visit his son in New Orleans. He has two options available for making the trip.

Option 1: Travel by train for $2\frac{1}{2}$ hours to city C and then 5 hours by bus to New Orleans, for a total distance of 375 miles.

Option 2: Travel by bus for 3 hours to city D and then $4\frac{1}{2}$ hours by train to New Orleans, for a total distance of 405 miles.

Assuming the rates of the buses are the same and the rates of the trains are the same in options 1 and 2, find the rates of the bus and the plane, respectively.

29. Sue needs to make a trip to New York City. She has two options for making the trip.

Option 1: Travel by car at an average speed of 45 mph to Metropolitan Airport, then travel by plane at an average speed of 350 mph to New York City, a total of 1,490 miles.

Option 2: Travel by car at an average speed of 60 mph to a small landing field, then fly by commuter plane at an average speed of 250 mph to New York City, a total of 1,120 miles.

How many hours must Sue travel by car and plane, respectively, assuming the trip takes the same amount of time with either option?

30. The president of a major corporation must make a trip to corporate headquarters in Los Angeles. She has two travel plans available.

Option 1: Travel by car at an average speed of 55 mph for a time t_1, then by a corporation jet at an average speed of 580 mph for a time t_2 to Los Angeles, a total of 3,590 miles.

Option 2: Travel by helicopter at an average speed of 120 mph for a time t_1, then by commercial airliner at an average speed of 450 mph for a time t_2 to Los Angeles, a total of 2,940 miles.

Find t_1 and t_2.

[Example 5] **31.** Desiree has a brother David who is 7 years older than she is. The sum of their ages now is 9 years less than twice the sum of their ages 8 years ago. Find the ages of Desiree and David.

32. Beth is 25 years younger than her mother Clarise. The difference between 6 times Beth's age now and 2 times Clarise's age now is 6 years. Find the ages of Beth and Clarise.

33. Henry has a favorite Uncle George who is 12 years older than he is. Two times Henry's age 2 years ago is the same as George's age in 4 years. How old are Henry and George?

34. Sandi has a friend named Elke who lives in Sweden. Elke is 8 years older than Sandi. Five times Sandi's age 4 years ago is the same as 2 times Elke's age 6 years from now. Find the ages of Sandi and Elke.

35. Linda has two horses, Par Bleu and Zanetto. Two times Par Bleu's age now is 1 year less than Zanetto's age now. Nine times Par Bleu's age 2 years ago is the same as 3 times Zanetto's age 2 years ago. How old is each horse?

36. Pam and Dan have a large maple tree and a hickory tree in their yard. Three times the current age of the hickory tree is 7 years less than the current age of the maple tree. Five times the age of the hickory tree in 7 years is the same as 2 times the age of the maple tree 2 years ago. How old is each tree?

37. Gladys has three pieces of antique jewelry: a gold locket, a ruby pin, and a diamond ring. The sum of the current ages of the three pieces is 310 years. The sum of 2 times the age of the locket 5 years ago and the age of the pin 10 years ago is 240 years. The difference between 3 times the locket's age in 3 years and 2 times the ring's age 8 years ago is 0. Find the ages of the locket, the pin, and the ring.

38. Jean and Bill Coffee have three children. Kerry and Jerry are Jean's from a previous marriage, and Glenn is their mutual son. The sum of the current ages of the children is 41 years. Kerry's age 5 years ago is the same as 3 times Glenn's age in 2 years. Two times Kerry's age in 1 year is the same as 3 times Jerry's age 4 years ago. Find the ages of the three children.

SECTION 9-3. Ten Review Exercises

1. Simplify $\frac{x^2 - x - 6}{x^2 - 4}$.

2. State the restricted values for $\frac{x^2 + 4}{x^2 - 4}$.

3. Simplify $\frac{x}{x + 2} + \frac{4}{x - 2} - \frac{3x + 14}{x^2 - 4}$.

4. Solve $\frac{x}{x + 2} + \frac{4}{x - 2} = \frac{3x + 14}{x^2 - 4}$.

In exercises **5–10**, refer to the function defined by $y = \frac{x - 3}{x - 2}$.

5. Write the equation of any vertical asymptotes for the graph.

6. State the domain of the function.

7. Write the equation of any horizontal asymptotes for the graph.

8. State the range of the function.

9. Graph the function.

10. Write an equation that defines the inverse of the function.

SECTION 9-3. Supplementary Exercises

In exercises **1–4**, for each system let x represent the first of two integers and y the second.

a. Write in words the interpretation of equation (1).

b. Write in words the interpretation of equation (2).

c. Solve the system.

1. (1) $x + y = 42$
(2) $2x - 3y = 9$

2. (1) $2x + y = 50$
(2) $x + 3y = 55$

3. (1) $x - 2y = 9$
(2) $3x + 4y = 7$

4. (1) $5x - 3y = 7$
(2) $4x - y = -7$

In Exercises **5–8**, the ordered pairs are solutions to:

$$Ax^2 + By^2 = C$$

Find A and B.

5. $\left(-1, \frac{\sqrt{3}}{2}\right), \left(\sqrt{3}, \frac{1}{2}\right); C = 4$

6. $(1, -2\sqrt{3}), (-\sqrt{3}, 2); C = 16$

7. $\left(1, \frac{3\sqrt{3}}{2}\right), \left(\frac{4\sqrt{2}}{3}, -1\right); C = 36$

8. $\left(-3, \frac{8}{5}\right), \left(\frac{5\sqrt{3}}{2}, -1\right); C = 100$

SECTION 9-4. Solving Systems of Linear Inequalities

KEY TOPICS IN THIS SECTION

1. A definition of a system of linear inequalities in x and y
2. A solution of a system of linear inequalities in x and y
3. Solving a system of linear inequalities in x and y
4. Solving systems of linear inequalities in x and y with no solutions

A linear inequality in x and y can be written in one of the following forms:

(1) $ax + by < c$ or (2) $ax + by > c$ or

(3) $ax + by \leq c$ or (4) $ax + by \geq c$

In this section, we will study **systems of linear inequalities in x and y** formed by taking two or more of the inequalities defined by (1), (2), (3), or (4).

A Definition of a System of Linear Inequalities in x and y

If two or more linear inequalities in x and y are grouped, then a system of inequalities is formed.

A system of two linear inequalities in x and y can be written as:

(1) $a_1x + b_1y < c_1$, where a_1 and b_1 are not both 0

(2) $a_2x + b_2y < c_2$, where a_2 and b_2 are not both 0

A system can also be formed by replacing $<$ in (1), or (2), or both (1) and (2) by $>$, $\leq$, or $\geq$. Examples **a** and **b** illustrate two such systems.

a. (1) $3x + 5y < 30$
(2) $4x - 3y \geq 6$

b. (1) $x - 2y > -4$
(2) $2x + y < 4$

A Solution of a System of Linear Inequalities in x and y

A solution of a system of linear inequalities in x and y is identified in Definition 9.3.

> **Definition 9.3.**
> If (x_1, y_1) is a solution of inequalities (1) and (2), then it is a solution of the system:
>
> (1) $a_1x + b_1y < c_1$
> (2) $a_2x + b_2y < c_2$
>
> The definition applies if $<$ is replaced by $>$, $\leq$, or $\geq$.

Example 1. Determine whether (2, 3) is a solution of the system:

(1) $4x - y \leq 5$
(2) $2x + 7y > 20$

Solution.

(1) $4(2) - 3 \leq 5$ — Replace x by 2 and y by 3 in (1).

$8 - 3 \leq 5$, *true* — Thus, (2, 3) is a solution of (1).

(2) $2(2) + 7(3) > 20$ — Replace x by 2 and y by 3 in (2).

$4 + 21 > 20$, *true* — Thus, (2, 3) is a solution of (2).

Since (2, 3) is a solution of both (1) and (2), it is a solution of the system.

Solving a System of Linear Inequalities in x and y

In Figure 9-9, the graphs of (1) $x + y = 4$ and (2) $5x - 2y = 6$ are drawn. The two lines intersect at $P(2, 2)$. Notice that the plane on which the lines are drawn is divided into four regions, labeled I, II, III, and IV. When the $=$ signs in (1) and (2) are changed to $<$ or $>$, then the solution sets of the resulting systems of inequalities will be the coordinates of the points in one of the four regions. To determine which region contains the solution set, the coordinates of a test point can be checked in the inequalities of the system. *The region whose coordinates make both inequalities true contains the solution set.*

Figure 9-9. Graph of the system (1) $x + y = 4$ and (2) $5x - 2y = 6$.

To illustrate, suppose the following system is formed using the graph in Figure 9-9:

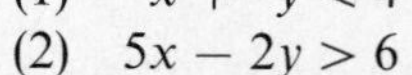

(1) $x + y < 4$
(2) $5x - 2y > 6$

In general, try to pick a test point on an axis in each of the four regions.

In I, pick (6, 0).

(1) $6 + 0 < 4$, *false*

(2) $30 - 0 > 6$, *true*

Since (1) is false, Region I does not contain the solution set.

In II, pick (0, 6).

(1) $0 + 6 < 4$, *false*

(2) $0 - 12 > 6$, *false*

Since (1) and (2) are false, Region II does not contain the solution set.

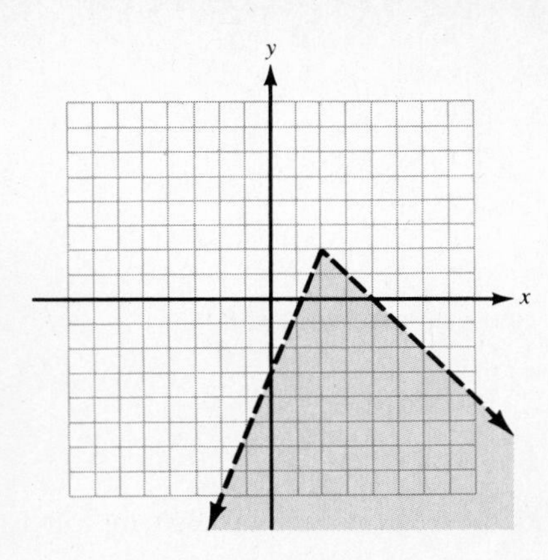

Figure 9-10. The solution set is shaded.

In III, pick (0, 0).

(1) $0 + 0 < 4$, *true*

(2) $0 - 0 > 6$, *false*

Since (2) is false, Region III does not contain the solution set.

In IV, pick (0, −6).

(1) $0 - \ 6 < 4$, *true*

(2) $0 + 12 > 6$, *true*

Since (1) and (2) are both true, Region IV contains the solution set.

Once the solution set of a system of inequalities is identified, we usually shade that region. In Figure 9-10, the solution set for the above system is shaded. Since the inequalities in the system do not include the equal sign, the boundaries are dotted lines.

Example 2. Given: (1) $2x + \ y < 4$
(2) $x - 2y \leq -4$

Graph the solution set.

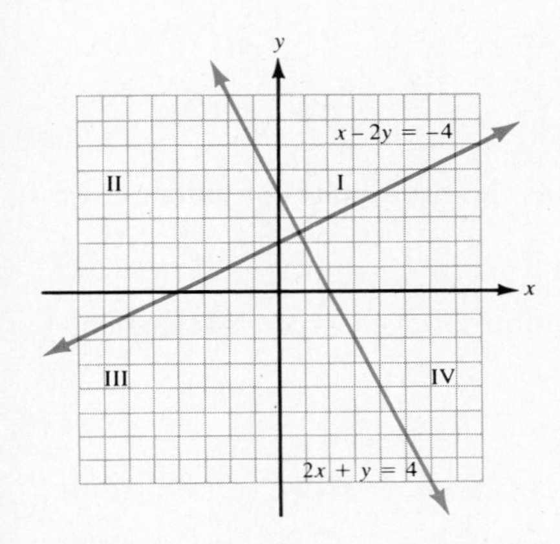

Figure 9-11. Regions I, II, III, and IV.

Solution. **Step 1.** Graph the lines defined by (1) and (2) in which the inequalities are replaced with equal signs.

(1) $2x + y = 4$

$y = -2x + 4$ — y-intercept is 4; slope is -2

(2) $x - 2y = -4$

$y = \frac{1}{2}x + 2$ — y-intercept is 2; slope is $\frac{1}{2}$

The lines are drawn and the regions are labeled in Figure 9-11.

Step 2. Check the coordinates of a test point in each region until one is found that makes both inequalities true. In Region I, pick (0, 8).

(1) $0 + 8 < 4$, *false*

Since (1) is false, do not bother to check (2). In Region II, pick (−8, 0).

(1) $-16 + 0 < 4$, *true*

(2) $-8 - 0 \leq -4$, *true*

Since (1) and (2) are true, Region II contains the solution set.

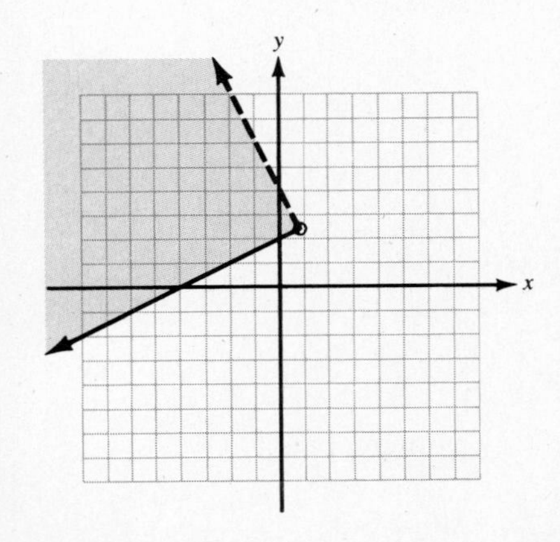

Figure 9-12. The solution set is shaded.

Step 3. Shade the region that contains the solution set. In Figure 9-12, the solution set of the system is shaded. Since (1) is <, the line for this inequality is dotted. Since (2) includes the equals relation, the line for this inequality is solid.

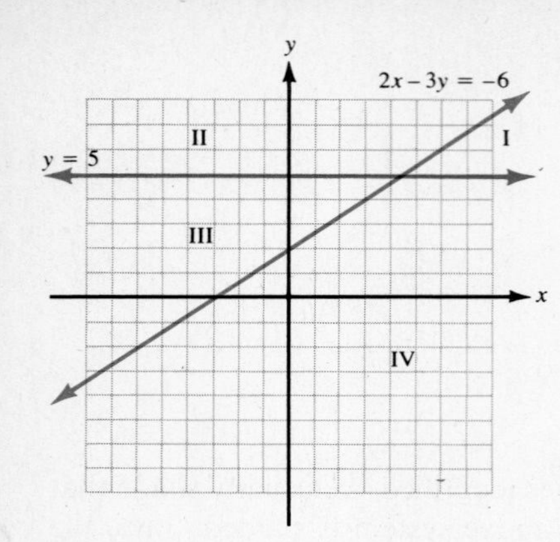

Figure 9-13. Regions I, II, III, and IV.

Example 3. Given: (1) $2x - 3y \leq -6$
(2) $y \leq 5$

Graph the solution set.

Solution. **Step 1.** (1) $2x - 3y = -6$

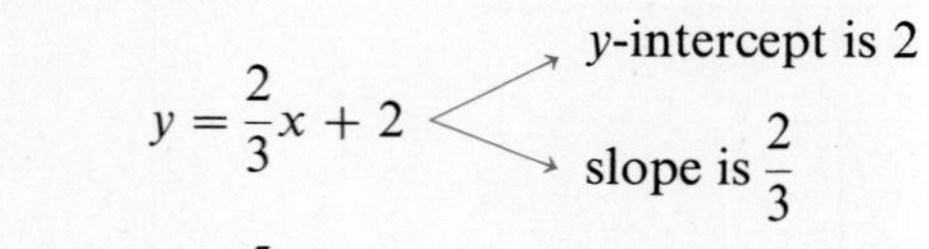

(2) $y = 5$

The lines are drawn and the regions are labeled in Figure 9-13.

Step 2. **Discussion.** Since (2) asserts $y \leq 5$, we may conclude that the solution set must be below the horizontal line. In Region III, pick $(-8, 0)$.

(1) $-16 - 0 \leq -6$, *true*

(2) $0 \leq 5$, *true*

The solution set contains the coordinates of points in Region III.

Step 3. In Figure 9-14, the solution set of the system is shaded.

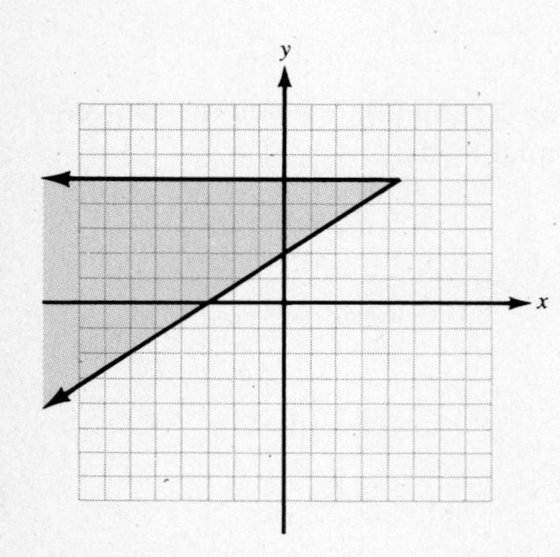

Figure 9-14. The solution set is shaded.

Solving Systems of Linear Inequalities in x and y with No Solutions

Not all systems of linear inequalities have solutions. The inequalities may be defined in such a way that no ordered pair of real numbers satisfies the inequalities at the same time.

Example 4. Given: (1) $3x - y > 5$
(2) $3x - y < -4$

Graph the solution set.

Solution. **Step 1.** (1) $3x - y = 5$

$y = 3x - 5$ — y-intercept is -5; slope is 3

(2) $3x - y = -4$

$y = 3x + 4$ — y-intercept is 4; slope is 3

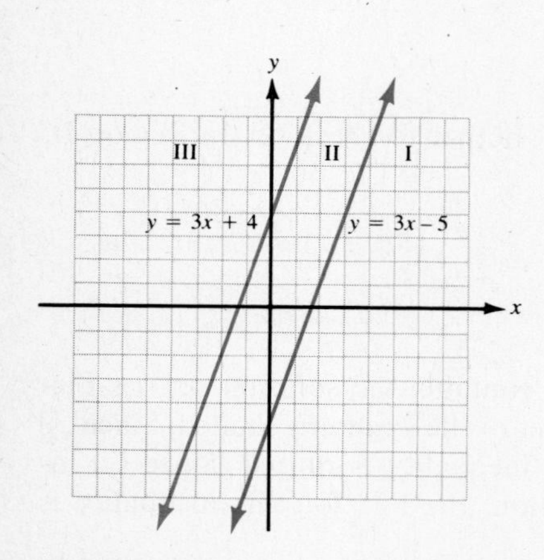

Figure 9-15. Regions I, II, and III.

The lines are graphed in Figure 9-15 as parallel lines that divide the plane into three regions.

Step 2. In Region I, pick (6, 0).

(1) $18 - 0 > 5$, *true*

(2) $18 - 0 < -4$, *false*

Region I is not the solution set.

In Region II, pick (0, 0).

(1) $0 - 0 > 5$, *false*

Region II is not the solution set.

In Region III, pick $(-6, 0)$.

(1) $0 - 18 - 0 > 5$, *false*

Region III is not the solution set.

Step 3. No region contains points with coordinates that simultaneously satisfy both inequalities. Thus, the solution set is $\varnothing$.

SECTION 9-4. Practice Exercises

In exercises **1–10**, determine whether each ordered pair is a solution of the given system.

[Example 1]

1. (1) $x + 2y \leq 5$
(2) $3x - 4y > 8$
a. $(0, -3)$ **b.** $(2, -2)$

2. (1) $5x - 4y < 10$
(2) $2x + 7y > 20$
a. $(2, 3)$ **b.** $(5, 4)$

3. (1) $10x - 3y \geq 2$
(2) $9x - 3y \leq 2$
a. $(-1, -4)$ **b.** $(2, 6)$

4. (1) $x - 6y < 20$
(2) $2x - 7y \geq 20$
a. $(10, -2)$ **b.** $(9, -1)$

5. (1) $3x + 4y < 0$
(2) $-x + 5y < 0$
a. $(-3, 1)$ **b.** $(5, -2)$

6. (1) $2x - 7y < -15$
(2) $x - 9y > -30$
a. $(6, 3)$ **b.** $(4, 5)$

7. (1) $2x + 9y \leq 4$
(2) $6x + 15y \geq -8$
a. $\left(\frac{1}{2}, \frac{1}{3}\right)$ **b.** $\left(\frac{-1}{2}, \frac{-1}{3}\right)$

8. (1) $4x - 3y \geq -4$
(2) $10x - 9y \leq 9$
a. $\left(\frac{3}{2}, \frac{2}{3}\right)$ **b.** $\left(\frac{-3}{2}, \frac{-2}{3}\right)$

9. (1) $5x + 12y < 6$
(2) $10x + 6y > -10$
a. $\left(\frac{3}{5}, \frac{1}{6}\right)$ **b.** $\left(\frac{-3}{5}, \frac{-1}{6}\right)$

10. (1) $35x + 18y > -2$
(2) $25x + 24y \leq 10$
a. $\left(\frac{-2}{5}, \frac{5}{6}\right)$ **b.** $\left(\frac{2}{5}, \frac{-5}{6}\right)$

In exercises **11–40**, graph the solution set of each system. If a given solution set is empty, write $\varnothing$.

[Examples 2–4]

11. (1) $x - y > -1$
(2) $x + y < 1$

12. (1) $x - y > -2$
(2) $x + 2y > -2$

13. (1) $2x - y < -2$
(2) $2x + 5y > 10$

14. (1) $x + 2y < 4$
(2) $x - y < 0$

15. (1) $2x - y < 6$
(2) $x + y \geq 0$

16. (1) $x + y < 2$
(2) $3x - 2y \geq -9$

17. (1) $4x - y \leq -4$
(2) $3x + y < -3$

18. (1) $x - y > 2$
(2) $x + y \geq 0$

19. (1) $4x - y > -4$
(2) $4x - y < 4$

20. (1) $x + 2y < 8$
(2) $x + 2y > 0$

21. (1) $4x + y \geq -4$
(2) $4x + y \leq 16$

22. (1) $3x - y \geq -3$
(2) $3x - y \leq 6$

23. (1) $2x + 3y \leq 6$
(2) $2x + 3y \geq 12$

24. (1) $5x - 2y < 10$
(2) $5x - 2y > 15$

25. (1) $2x - y > 0$
(2) $4x + 3y \leq 20$

26. (1) $3x + 2y < 16$
(2) $x - y \leq 2$

27. (1) $3x - y \geq 2$
(2) $x - 3y \leq -2$

28. (1) $x - y > -4$
(2) $x - 2y > -6$

29. (1) $3x + 2y > 8$
(2) $x + 2y < 4$

30. (1) $x + y > -1$
(2) $2x - y < 4$

31. (1) $2x - 3y \leq 2$
(2) $y < 2$

32. (1) $x + 2y < -3$
(2) $y \geq -1$

33. (1) $5x - 3y \leq -3$
(2) $x > -3$

34. (1) $x \geq 1$
(2) $2x + y < 4$

35. (1) $x \geq 0$
(2) $y < 3$

36. (1) $y \leq 0$
(2) $x \geq -1$

37. (1) $x \geq 3$
(2) $y > 2$

38. (1) $x < -2$
(2) $y \leq -1$

39. (1) $x - y > 0$
(2) $x + y > 0$

40. (1) $3x + y > 0$
(2) $3x - y < 0$

SECTION 9-4. Ten Review Exercises

In exercises **1–5**, solve each equation.

1. $3(1 - 4k) + 7 = 8k + 12 - 4(k + 2)$

2. $|3t - 2| = 14$

3. $20y^2 = 23y - 6$

4. $\dfrac{z-5}{z+3} + \dfrac{z-2}{z-3} = \dfrac{z+2}{z+3} + \dfrac{z^2+3}{z^2-9}$

5. $\sqrt{u+2} - \sqrt{u-3} = 1$

In exercises **6–10**, solve and graph each inequality.

6. $7t + 5 \leq 40$ and $2t \geq t + 1$

7. $0 < 6 - 9v < 24$

8. $|x + 5| \geq 8$

9. $|8 - 5y| < 13$

10. $z^2 \leq 25$

SECTION 9-4. Supplementary Exercises

In exercises **1–6**, graph the solution set of each system.

1. (1) $x - y < 4$
(2) $x + y < 4$
(3) $x > 0$

2. (1) $x - 2y > 2$
(2) $3x + 4y < 24$
(3) $y > 0$

3. (1) $x + y < 8$
(2) $x - y > 0$
(3) $y > 0$

4. (1) $3x - 2y > 0$
(2) $x + y < 7$
(3) $y > 0$

5. (1) $x + y < 0$
(2) $x + 8 > 0$
(3) $y + 1 > 0$

6. (1) $x - y > 0$
(2) $y + 3 > 0$
(3) $x - 7 < 0$

In exercises **7–12**, write a system of inequalities that defines the shaded region in each figure.

7.

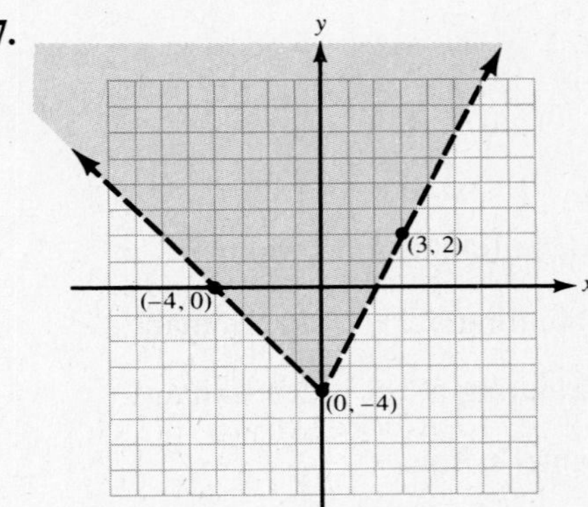

8.

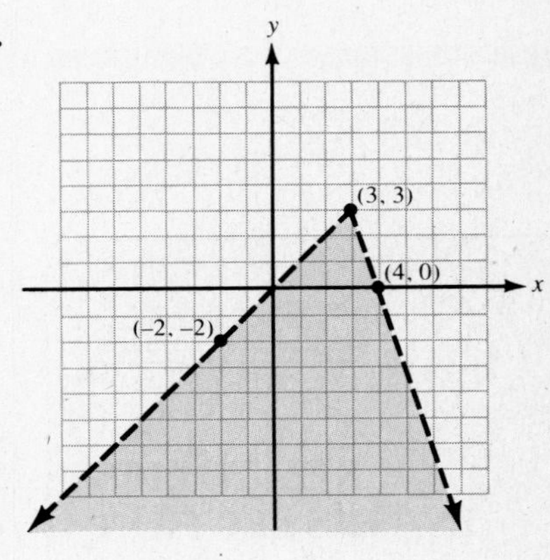

9.

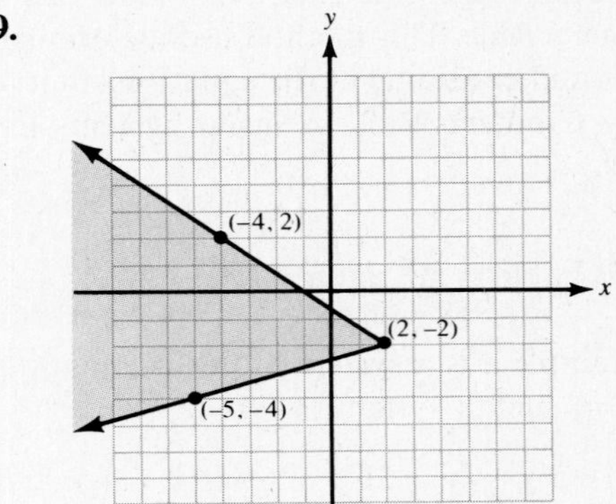

10.

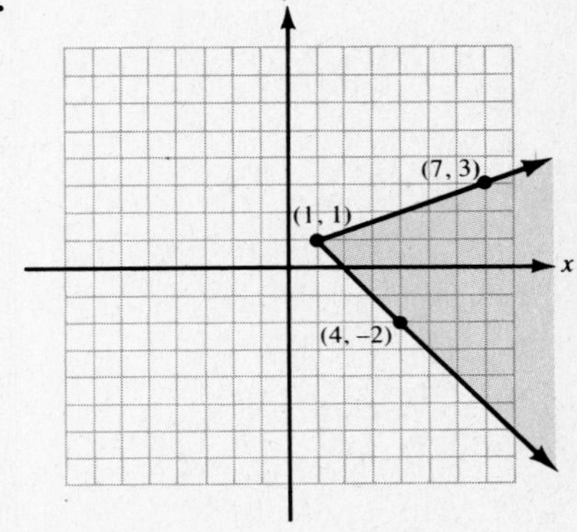

11.

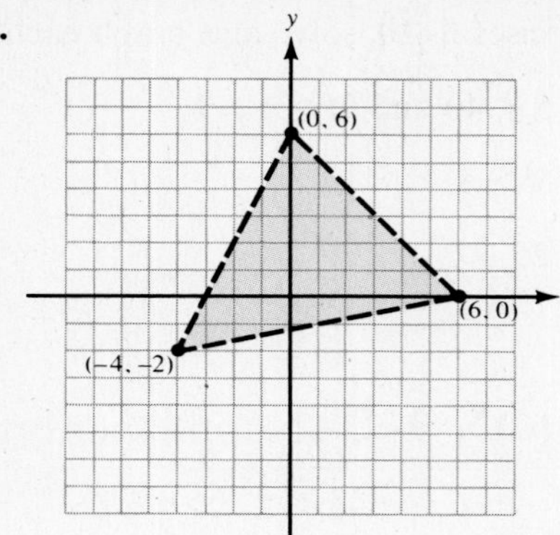

12.

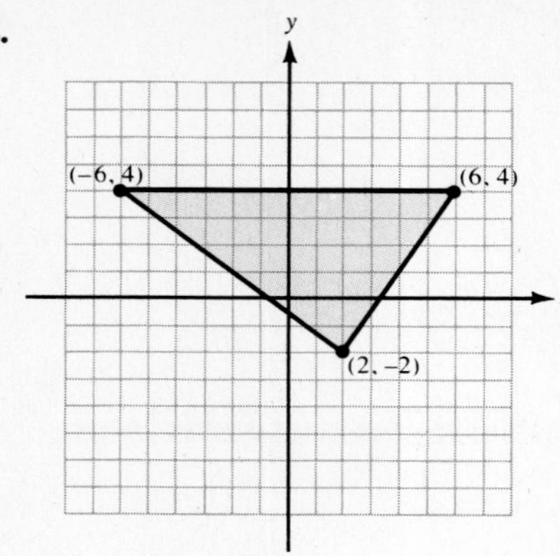

In exercises **13–20**, identify the quadrant or quadrants that contain the solutions for each system.

13. (1) $x > 0$
(2) $y < 0$

14. (1) $x > 0$
(2) $y > 0$

15. (1) $x < 0$
(2) $y < 0$

16. (1) $x < 0$
(2) $y > 0$

17. (1) $x < 0$
(2) $y \neq 0$

18. (1) $x > 0$
(2) $y \neq 0$

19. (1) $x \neq 0$
(2) $y > 0$

20. (1) $x \neq 0$
(2) $y < 0$

SECTION 9-5. Determinants and Cramer's Rule (Optional)

1. A definition of a determinant
2. Evaluating a 2 × 2 determinant
3. Evaluating a 3 × 3 determinant
4. Cramer's Rule
5. Inconsistent and dependent systems

Another algebraic method for solving systems of linear equations is called Cramer's Rule. This method uses determinants that are computed using the coefficients and constants in the equations that form the system. In this section, we will apply Cramer's Rule to linear systems in x and y, and linear systems in x, y, and z.

A Definition of a Determinant

Example **a** is a system of linear equations in x and y. Example **b** is the general form of such a system.

a. (1) $5x + 2y = 29$
(2) $2x + 3y = 16$

b. (1) $a_1x + b_1y = c_1$
(2) $a_2x + b_2y = c_2$

Both systems will now be solved for x using the addition method. To solve for x, the y-terms will be eliminated. Because the coefficients and constants in example **b** are written in terms of letters, the products cannot be simplified as they are in example **a**. However, the procedures are the same in other respects.

In **a**: multiply (1) by 3 and (2) by (-2) and add.

In **b**: multiply (1) by b_2 and (2) by $(-b_1)$ and add.

a. (1) $15x + 6y = 87$
(2) $-4x - 6y = -32\ (+)$
$11x = 55$

b. (1) $a_1b_2x + b_1b_2y = b_2c_1$
(2) $-a_2b_1x - b_1b_2y = -b_1c_2\ (+)$
$(a_1b_2 - a_2b_1)x = (b_2c_1 - b_1c_2)$

Now divide both sides of the resulting equations by the coefficients of x. In example **b**, we must assume that $a_1b_2 - a_2b_1 \neq 0$.

a. $x = \dfrac{55}{11} = 5$ **b.** $x = \dfrac{b_2c_1 - b_1c_2}{a_1b_2 - a_2b_1}$

The result of example **b** can be viewed as a formula for finding the solution (if one exists) of x in any system of linear equations in x and y. The differences in the products can more readily be applied to the a's, b's, and c's in the system by writing the numbers in rows and columns.

b. $x = \dfrac{b_2c_1 - b_1c_2}{a_1b_2 - a_2b_1}$ can be written as $x = \dfrac{\begin{vmatrix} c_1 & b_1 \\ c_2 & b_2 \end{vmatrix}}{\begin{vmatrix} a_1 & b_1 \\ a_2 & b_2 \end{vmatrix}}$

In the left equation, we see a ratio of differences in two pairs of products. In the right equation, we see a ratio of **arrays** of numbers in two rows and two columns each. The vertical bars indicate that the arrays can be evaluated by simplifying the products and differences indicated in the equation on the left. The numbers obtained by doing these operations are called the determinants of the arrays. The following definition asserts that a determinant can be calculated for arrays of real numbers in n-rows and n-columns, where n is some positive integer.

Definition 9.4. Determinant
A **determinant** is a real number assigned to an array of real numbers in n-rows and n-columns. The **order of a determinant** is $n \times n$, which is read "n by n".

Evaluating a 2 × 2 Determinant

Equation (1) can be used to evaluate a 2×2 determinant.

Definition 9.5.
If a_1, a_2, b_1, and b_2 are real numbers, then:

$$(1)\quad \begin{vmatrix} a_1 & b_1 \\ a_2 & b_2 \end{vmatrix} = a_1b_2 - a_2b_1$$

Example 1. Given: $\begin{vmatrix} 5 & 2 \\ 2 & 3 \end{vmatrix}$

Evaluate.

Solution. **Discussion.** To evaluate the given expression means to find the real number (or determinant) that the array represents.

$\begin{vmatrix} 5 & 2 \\ 2 & 3 \end{vmatrix} = 5 \cdot 3 - 2 \cdot 2$ $\qquad \begin{vmatrix} a_1 & b_1 \\ a_2 & b_2 \end{vmatrix} = a_1b_2 - a_2b_1$

$= 15 - 4$ $\qquad$ Multiply first.

$= 11$ $\qquad$ The determinant is 11.

Evaluating a 3 × 3 Determinant

Definition 9.6.
If a_1, a_2, a_3, b_1, b_2, b_3, c_1, c_2, and c_3 are real numbers, then:

$$(2) \quad \begin{vmatrix} a_1 & b_1 & c_1 \\ a_2 & b_2 & c_2 \\ a_3 & b_3 & c_3 \end{vmatrix} = a_1b_2c_3 + a_2b_3c_1 + a_3b_1c_2 - a_3b_2c_1 - a_2b_1c_3 - a_1b_3c_2$$

To find the determinant of an array of three rows and three columns, we must compute six products of three numbers each; that is, $a_1b_2c_3, a_2b_3c_1, \ldots, a_1b_3c_2$. Three of the products are added, and the other three are then subtracted. The following arrangement can be used to do the operations in the proper order:

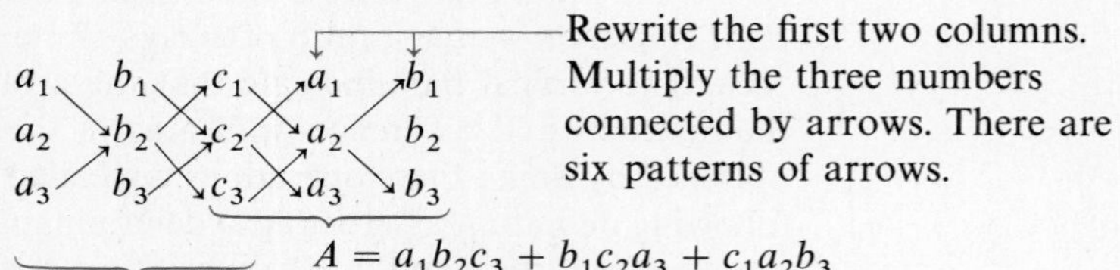

$A = a_1b_2c_3 + b_1c_2a_3 + c_1a_2b_3$

$B = a_3b_2c_1 + b_3c_2a_1 + c_3a_2b_1$

The determinant is $A - B$.

Example 2. Given: $\begin{vmatrix} 2 & -4 & 6 \\ 3 & 9 & 1 \\ -5 & 0 & -2 \end{vmatrix}$

Evaluate.

Solution. Rewrite the first two columns.

$$\begin{matrix} 2 & -4 & 6 & 2 & -4 \\ 3 & 9 & 1 & 3 & 9 \\ -5 & 0 & -2 & -5 & 0 \end{matrix}$$

$A = (2)(9)(-2) + (-4)(1)(-5) + (6)(3)(0)$

$= -36 + 20 + 0 = -16$

$B = (-5)(9)(6) + (0)(1)(2) + (-2)(3)(-4)$

$= -270 + 0 + 24 = -246$

With $A = -16$ and $B = -246$:

$A - B = -16 - (-246) = 230$

The given determinant is 230.

Cramer's Rule

Consider the system stated earlier as example **b**.

b. (1) $a_1x + b_1y = c_1$
(2) $a_2x + b_2y = c_2$

The general form of a system of linear equations in x and y

It was shown that for this system:

$$x = \frac{b_2c_1 - b_1c_2}{a_1b_2 - a_2b_1}, \text{ which can be written as } x = \frac{\begin{vmatrix} c_1 & b_1 \\ c_2 & b_2 \end{vmatrix}}{\begin{vmatrix} a_1 & b_1 \\ a_2 & b_2 \end{vmatrix}}$$

It can similarly be shown that for this system:

$$y = \frac{a_1c_2 - a_2c_1}{a_1b_2 - a_2b_1}, \text{ which can be written as } y = \frac{\begin{vmatrix} a_1 & c_1 \\ a_2 & c_2 \end{vmatrix}}{\begin{vmatrix} a_1 & b_1 \\ a_2 & b_2 \end{vmatrix}}$$

To write these ratios of determinants in compact form, the following symbols are used:

$$D = \begin{vmatrix} a_1 & b_1 \\ a_2 & b_2 \end{vmatrix} \qquad N_x = \begin{vmatrix} c_1 & b_1 \\ c_2 & b_2 \end{vmatrix} \qquad N_y = \begin{vmatrix} a_1 & c_1 \\ a_2 & c_2 \end{vmatrix}$$

Using these symbols:

$$x = \frac{N_x}{D} \quad \text{and} \quad y = \frac{N_y}{D}$$

These equations for the solutions of a system of linear equations in x and y are called **Cramer's Rule** for such a two-variable system.

Cramer's Rule
For the system of linear equations:

(1) $a_1x + b_1y = c_1$
(2) $a_2x + b_2y = c_2$

$x = \dfrac{N_x}{D}$ and $y = \dfrac{N_y}{D}$, provided $D \neq 0$

Example 3. Given: (1) $3x - 4y = 25$
(2) $2x + 7y = -22$

Solve using Cramer's rule.

Solution. **Discussion.** Use the coefficients and constants to compute D, N_x, and N_y.

$$D = \begin{vmatrix} 3 & -4 \\ 2 & 7 \end{vmatrix} = 21 - (-8) \qquad D = \begin{vmatrix} a_1 & b_1 \\ a_2 & b_2 \end{vmatrix}$$

$$= 29$$

$$N_x = \begin{vmatrix} 25 & -4 \\ -22 & 7 \end{vmatrix} = 175 - 88 \qquad N_x = \begin{vmatrix} c_1 & b_1 \\ c_2 & b_2 \end{vmatrix}$$

$$= 87$$

$$N_y = \begin{vmatrix} 3 & 25 \\ 2 & -22 \end{vmatrix} = -66 - 50 \qquad N_y = \begin{vmatrix} a_1 & c_1 \\ a_2 & c_2 \end{vmatrix}$$

$$= -116$$

$$x = \frac{87}{29} = 3 \qquad x = \frac{N_x}{D}$$

$$y = \frac{-116}{29} = -4 \qquad y = \frac{N_y}{D}$$

The check is omitted. The solution set is $\{(3, -4)\}$.

Consider now the system in example **c.**

c. (1) $a_1x + b_1y + c_1z = d_1$
(2) $a_2x + b_2y + c_2z = d_2$
(3) $a_3x + b_3y + c_3z = d_3$

The general form of a system of linear equations in x, y, and z.

It can be shown that the solution for this system can be found using 3×3 determinants. The elements of the determinants are the coefficients and constants in the equations of the system. The following symbols are used to represent the determinants:

$$D = \begin{vmatrix} a_1 & b_1 & c_1 \\ a_2 & b_2 & c_2 \\ a_3 & b_3 & c_3 \end{vmatrix} \qquad N_x = \begin{vmatrix} d_1 & b_1 & c_1 \\ d_2 & b_2 & c_2 \\ d_3 & b_3 & c_3 \end{vmatrix}$$

$$N_y = \begin{vmatrix} a_1 & d_1 & c_1 \\ a_2 & d_2 & c_2 \\ a_3 & d_3 & c_3 \end{vmatrix} \qquad N_z = \begin{vmatrix} a_1 & b_1 & d_1 \\ a_2 & b_2 & d_2 \\ a_3 & b_3 & d_3 \end{vmatrix}$$

Cramer's Rule

For the system of linear equations:

(1) $a_1x + b_1y + c_1z = d_1$
(2) $a_2x + b_2y + c_2z = d_2$
(3) $a_3x + b_3y + c_3z = d_3$

$x = \dfrac{N_x}{D}$; $y = \dfrac{N_y}{D}$; $z = \dfrac{N_z}{D}$, provided $D \neq 0$

Example 4. Given: (1) $x - 2y + 3z = 5$
(2) $2x + 3y + z = 6$
(3) $-4x - y - 2z = 3$

Solve using Cramer's Rule.

Solution. **Discussion.** First find the values of D, N_x, N_y, and N_z.

$$D = \begin{vmatrix} 1 & -2 & 3 \\ 2 & 3 & 1 \\ -4 & -1 & -2 \end{vmatrix} = [(1)(3)(-2) + (-2)(1)(-4) + (3)(2)(-1)] - [(-4)(3)(3) + (-1)(1)(1) + (-2)(2)(-2)]$$
$$= (-4) - (-29) = 25$$

$$N_x = \begin{vmatrix} 5 & -2 & 3 \\ 6 & 3 & 1 \\ 3 & -1 & -2 \end{vmatrix} = [(5)(3)(-2) + (-2)(1)(3) + (3)(6)(-1)] - [(3)(3)(3) + (-1)(1)(5) + (-2)(6)(-2)]$$
$$= (-54) - 46 = -100$$

$$N_y = \begin{vmatrix} 1 & 5 & 3 \\ 2 & 6 & 1 \\ -4 & 3 & -2 \end{vmatrix} = [(1)(6)(-2) + (5)(1)(-4) + (3)(2)(3)] - [(-4)(6)(3) + (3)(1)(1) + (-2)(2)(5)]$$
$$= (-14) - (-89) = 75$$

$$N_z = \begin{vmatrix} 1 & -2 & 5 \\ 2 & 3 & 6 \\ -4 & -1 & 3 \end{vmatrix} = [(1)(3)(3) + (-2)(6)(-4) + (5)(2)(-1)] - [(-4)(3)(5) + (-1)(6)(1) + (3)(2)(-2)]$$
$$= 47 - (-78) = 125$$

Solving for x, y, and z:

$$x = \frac{-100}{25} = -4 \qquad x = \frac{N_x}{D}$$

$$y = \frac{75}{25} = 3 \qquad y = \frac{N_y}{D}$$

$$z = \frac{125}{25} = 5 \qquad z = \frac{N_z}{D}$$

The check is omitted. The solution set is $\{(-4, 3, 5)\}$.

Inconsistent and Dependent Systems

In the statement of Cramer's Rule, the determinant D is restricted from being 0. If $D = 0$, then the associated system of equations is either inconsistent or dependent.

If in applying Cramer's Rule, $D = 0$, then:

Conclusion 1. The system is **inconsistent**

a. if $N_x \neq 0$, or $N_y \neq 0$ for a system in x and y.

b. if $N_x \neq 0$, or $N_y \neq 0$, or $N_z \neq 0$ for a system in x, y, and z.

Conclusion 2. The system is **dependent**

a. if $N_x = N_y = 0$ for a system in x and y.

b. if $N_x = N_y = N_z = 0$ for a system in x, y, and z.

Example 5. Given: (1) $7x - 5y = 10$
(2) $-14x + 10y = 17$

Solve using Cramer's Rule.

Solution. $D = \begin{vmatrix} 7 & -5 \\ -14 & 10 \end{vmatrix} = 70 - 70 = 0$ → The system is inconsistent. or The system is dependent.

$$N_x = \begin{vmatrix} 10 & -5 \\ 17 & 10 \end{vmatrix} = 100 - (-85) = 185$$

Since $N_x \neq 0$, the system is inconsistent and the solution set is $\varnothing$.

SECTION 9-5. Practice Exercises

In exercises **1–20**, evaluate each determinant.

[Example 1]

1. $\begin{vmatrix} 5 & 2 \\ 3 & 4 \end{vmatrix}$ **2.** $\begin{vmatrix} 6 & -1 \\ 5 & 2 \end{vmatrix}$ **3.** $\begin{vmatrix} -4 & 3 \\ 8 & -6 \end{vmatrix}$

4. $\begin{vmatrix} -6 & 3 \\ 4 & -2 \end{vmatrix}$ **5.** $\begin{vmatrix} -24 & 36 \\ -\frac{1}{6} & \frac{3}{8} \end{vmatrix}$ **6.** $\begin{vmatrix} 54 & -63 \\ \frac{5}{9} & -\frac{2}{9} \end{vmatrix}$

7. $\begin{vmatrix} 0.3 & 1.5 \\ -0.2 & 2.4 \end{vmatrix}$ **8.** $\begin{vmatrix} -0.5 & 0.8 \\ 0.6 & -0.4 \end{vmatrix}$ **9.** $\begin{vmatrix} -\frac{1}{2} & \frac{3}{2} \\ -\frac{1}{3} & \frac{4}{3} \end{vmatrix}$

10. $\begin{vmatrix} \frac{3}{4} & -\frac{2}{5} \\ \frac{1}{4} & -\frac{4}{5} \end{vmatrix}$ **11.** $\begin{vmatrix} \sqrt{3} & -\sqrt{2} \\ 2\sqrt{3} & -\sqrt{2} \end{vmatrix}$ **12.** $\begin{vmatrix} -3\sqrt{6} & 2\sqrt{2} \\ \sqrt{6} & -\sqrt{2} \end{vmatrix}$

[Example 2]

13. $\begin{vmatrix} 1 & -3 & 2 \\ 2 & -2 & -1 \\ 3 & -1 & 1 \end{vmatrix}$ **14.** $\begin{vmatrix} 2 & 1 & 5 \\ -3 & 2 & 3 \\ 5 & -1 & 1 \end{vmatrix}$ **15.** $\begin{vmatrix} 4 & 1 & 3 \\ 2 & -6 & -5 \\ -3 & 4 & 2 \end{vmatrix}$

16. $\begin{vmatrix} 3 & 2 & 2 \\ -1 & 0 & -2 \\ 9 & 2 & 8 \end{vmatrix}$ **17.** $\begin{vmatrix} 9 & 6 & 3 \\ -6 & 5 & 0 \\ 7 & -6 & 0 \end{vmatrix}$ **18.** $\begin{vmatrix} -12 & 15 & -4 \\ 9 & -8 & 0 \\ -8 & 7 & 0 \end{vmatrix}$

19. $\begin{vmatrix} 10 & 0 & -9 \\ 13 & -5 & 25 \\ -9 & 0 & 8 \end{vmatrix}$

20. $\begin{vmatrix} -5 & 0 & 3 \\ 39 & 6 & -42 \\ 8 & 0 & -5 \end{vmatrix}$

In exercises **21–40**, solve each system using Cramer's Rule.

[Example 3]

21. (1) $2x - 5y = 31$
(2) $5x + 4y = -5$

22. (1) $7x + 10y = -8$
(2) $9x - 4y = -44$

23. (1) $4x + 11y = 53$
(2) $-12x + 3y = 93$

24. (1) $-3x + 7y = 8$
(2) $-7x + 9y = -18$

25. (1) $6x - 13y = 60$
(2) $-2x + 11y = -20$

26. (1) $12x + 5y = -40$
(2) $9x - 4y = 32$

27. (1) $6x - 4y = 3$
(2) $9x + 8y = 8$

28. (1) $8x + 5y = 6$
(2) $2x - 10y = -3$

29. (1) $10x + 9y = 12$
(2) $-4x - 15y = -1$

30. (1) $6x + 14y = 3$
(2) $21x - 10y = -19$

[Example 4]

31. (1) $4x - 2y - z = 7$
(2) $4x + 2y - z = 3$
(3) $-x + 2y + z = -1$

32. (1) $x - 3y - 6z = 13$
(2) $2x + 5y - z = 4$
(3) $3x + 2y - 7z = 17$

33. (1) $x + 3y - 3z = 12$
(2) $5x - 8y - z = -55$
(3) $2x + 5y + 4z = 19$

34. (1) $x + y - z = 9$
(2) $3x - 2y + 3z = -3$
(3) $2x - 3y + z = 3$

35. (1) $x + y - 3z = 2$
(2) $x - y + 2z = 0$
(3) $2x + 2y + 3z = -14$

36. (1) $2x + 3y - 2z = -16$
(2) $x - 4y + z = 20$
(3) $3x - 2y - z = 8$

37. (1) $2x + y - 2z = 10$
(2) $4x - 5z = 17$
(3) $6x + 2y = 22$

38. (1) $x - 2y + z = 11$
(2) $4x - 5y = 30$
(3) $3y + 4z = 8$

39. (1) $x - 2y + z = 5$
(2) $3x + y - 2z = -9$
(3) $2x - y + 2z = 4$

40. (1) $x - 2y + 3z = -9$
(2) $x - y - 2z = 7$
(3) $3x + y - z = 10$

In exercises **41–50**, solve each system using Cramer's Rule. Write "inconsistent" for any system with no solutions and "dependent" for any system with more than one solution.

[Example 5]

41. (1) $9x + 4y = 27$
(2) $27x + 12y = 62$

42. (1) $-11x + 8y = 28$
(2) $22x - 16y = -42$

43. (1) $5x - 2y = 12$
(2) $-10x + 4y = -24$

44. (1) $6x - 4y = 20$
(2) $24x - 16y = 5$

45. (1) $5x - 4y = -15$
(2) $10x - 2y = 0$

46. (1) $2x + y = -2$
(2) $-10x + 3y = -14$

47. (1) $x + y - z = 10$
(2) $-2x + 5z = 7$
(3) $-3x - 3y + 3z = 15$

48. (1) $4x - 2y + 6z = 7$
(2) $-2x + y - 3z = -8$
(3) $-12x + 6y - 18z = -24$

49. (1) $4x + 9y - 5z = 48$
(2) $x - 4y + 3z = -28$
(3) $8x + 18y - 10z = 96$

50. (1) $5x - y + z = -13$
(2) $-15x + 3y - 3z = 39$
(3) $4x + 2y - z = 7$

SECTION 9-5. Ten Review Exercises

In exercises **1–10**, do the indicated operations and simplify.

1. $-4(2 - 7) + 10(2^2 - 3^2) - 6(-5)$

2. $\dfrac{(3a^{-2}b^{-1})^{-2}}{(6ab^{-2})^{-1}}$

3. $\left(\dfrac{-3u^2v}{5}\right)^2\left(\dfrac{-10uv^2}{3}\right)\left(\dfrac{u^3v}{6}\right)$

4. $4(3x^2 - 2x - 1) - (4x^2 - 10 - 3x) - 2(3 - x + 2x^2)$

5. $3(7a^2 - 6ab + 10b^2) + 2(9ab - 4a^2 - 5b^2) - (10a^2 - 20b^2 - ab)$

6. $\dfrac{10m^3n + 5mn^2}{10m^3 + 2m^2n + 5mn + n^2}$

7. $\dfrac{2}{2t + 3} - \dfrac{18}{4t^2 - 9} + \dfrac{3}{2t - 3}$

8. $\left(\dfrac{4u^2 - 9}{2u^2 + 2u} \div \dfrac{6u^2 - 9u}{u + 1}\right)\left(\dfrac{18u^3 + 12u^2}{9u^2 + 12u + 4}\right)$

9. $\left(\dfrac{8}{27}\right)^{2/3}$

10. $4^{-3/2}$

SECTION 9-5. Supplementary Exercises

A procedure for evaluating determinants of order 3×3 or larger involves the minor and cofactor of an element in a determinant.

Definition 9.7.
If a is an element in an $n \times n$ determinant, then the **minor** of a is M, and M is the determinant obtained by deleting the row and column containing a.

Definition 9.8.
If a is an element in an $n \times n$ determinant, then the **cofactor** of a is the product of M and $(+1)$ or (-1). A $(+1)$ is used if the sum of the row number and column number containing a is even, and a (-1) if the sum is odd.

Example. Given: $\begin{vmatrix} 3 & 1 & 2 \\ -1 & 4 & 8 \\ 7 & -2 & -3 \end{vmatrix}$

a. State the minor of 8 in the determinant.

b. State the cofactor of 8 in the determinant.

Solution. **a.** The minor of 8 is:

$$\begin{vmatrix} 3 & 1 \\ 7 & -2 \end{vmatrix}$$

Mentally cross out the row and column containing 8:

$$\begin{vmatrix} 3 & 1 & 2 \\ -1 & 4 & 8 \\ 7 & -2 & -3 \end{vmatrix}$$

b. The 8 is in row two and column three. Thus, 8 is in a (-1) position, because $2 + 3 = 5$ and 5 is an odd number. The cofactor of 8 is:

$$(-1)\begin{vmatrix} 3 & 1 \\ 7 & -2 \end{vmatrix} = (-1)(-6 - 7) = 13$$

Equation (3) can be used to evaluate the 3×3 determinant by the **expansion method.** This equation shows the expansion about the elements in the first column. Similar equations can be written for expansions of the determinant about the elements in any row or any column:

$$\begin{vmatrix} a_1 & b_1 & c_1 \\ a_2 & b_2 & c_2 \\ a_3 & b_3 & c_3 \end{vmatrix} = \underbrace{a_1(+1)\begin{vmatrix} b_2 & c_2 \\ b_3 & c_3 \end{vmatrix}}_{\textbf{Cofactor of } a_1} + \underbrace{a_2(-1)\begin{vmatrix} b_1 & c_1 \\ b_3 & c_3 \end{vmatrix}}_{\textbf{Cofactor of } a_2} + \underbrace{a_3(+1)\begin{vmatrix} b_1 & c_1 \\ b_2 & c_2 \end{vmatrix}}_{\textbf{Cofactor of } a_3}$$

Example. Given: $\begin{vmatrix} 1 & 2 & -3 \\ 4 & 5 & 6 \\ 7 & -8 & 9 \end{vmatrix}$

Evaluate.

Solution. Expanding about the elements in column one:

$$\begin{vmatrix} 1 & 2 & -3 \\ 4 & 5 & 6 \\ 7 & -8 & 9 \end{vmatrix} = 1(+1)\begin{vmatrix} 5 & 6 \\ -8 & 9 \end{vmatrix} + 4(-1)\begin{vmatrix} 2 & -3 \\ -8 & 9 \end{vmatrix} + 7(+1)\begin{vmatrix} 2 & -3 \\ 5 & 6 \end{vmatrix}$$

$$= 1(45 + 48) - 4(18 - 24) + 7(12 + 15)$$

$$= 93 - (-24) + 189$$

$$= 306$$

In exercises **1–10**, evaluate each determinant using the expansion method.

1. $\begin{vmatrix} 4 & -3 & 2 \\ 2 & 1 & -1 \\ 3 & 4 & 5 \end{vmatrix}$

2. $\begin{vmatrix} -2 & 5 & 2 \\ 4 & -2 & 3 \\ -3 & 6 & 9 \end{vmatrix}$

3. $\begin{vmatrix} -1 & 4 & -5 \\ -2 & -3 & 2 \\ 3 & -2 & -4 \end{vmatrix}$

4. $\begin{vmatrix} 1 & 1 & 2 \\ -5 & 5 & -7 \\ 3 & 3 & 1 \end{vmatrix}$

5. $\begin{vmatrix} 2 & 4 & -1 \\ 1 & -6 & 1 \\ 3 & -2 & 5 \end{vmatrix}$

6. $\begin{vmatrix} -4 & 7 & 2 \\ 3 & -1 & 1 \\ -4 & 7 & 2 \end{vmatrix}$

7. $\begin{vmatrix} 9 & -1 & 9 \\ -2 & 4 & -2 \\ 5 & 6 & 5 \end{vmatrix}$

8. $\begin{vmatrix} 5 & -9 & 2 \\ 0 & 8 & 11 \\ 3 & 4 & -7 \end{vmatrix}$

9. $\begin{vmatrix} -2 & 2 & -5 \\ 0 & 7 & 4 \\ 3 & -6 & -12 \end{vmatrix}$

10. $\begin{vmatrix} -2 & 0 & 3 \\ 1 & -4 & 2 \\ -2 & 1 & 3 \end{vmatrix}$

The following property of determinants can be used to change the elements in a given row (or column) to exactly one nonzero element before expanding the determinant:

> If a constant multiple of the elements in any row (or column) of a determinant is added to the corresponding elements in any other row (or column), then the determinant is unchanged.

Example. Given: $\begin{vmatrix} 1 & 2 & -2 \\ 3 & 10 & -9 \\ -4 & -6 & 13 \end{vmatrix}$

Evaluate.

Solution. **Discussion.** Before using the expansion equation, the property of determinants stated above will be used to change the 3 and -4 in column one to 0's.

Step 1. Multiply the elements in row one by (-3) and add the products to the corresponding elements in row two.

$$\begin{vmatrix} 1 & 2 & -2 \\ 0 & 4 & -3 \\ -4 & -6 & 13 \end{vmatrix} \quad \begin{aligned} 1(-3) + 3 &= -3 + 3 = 0 \\ 2(-3) + 10 &= -6 + 10 = 4 \\ (-2)(-3) + (-9) &= 6 + (-9) = -3 \end{aligned}$$

Step 2. Multiply the elements in row one by 4 and add the products to the corresponding elements in row three.

$$\begin{vmatrix} 1 & 2 & -2 \\ 0 & 4 & -3 \\ 0 & 2 & 5 \end{vmatrix} \quad \begin{aligned} 1(4) + (-4) &= 4 + (-4) = 0 \\ 2(4) + (-6) &= 8 + (-6) = 2 \\ -2(4) + 13 &= -8 + 13 = 5 \end{aligned}$$

Step 3. Expand the determinant about the elements in column one.

$$\begin{vmatrix} 1 & 2 & -2 \\ 0 & 4 & -3 \\ 0 & 2 & 5 \end{vmatrix} = 1(+1)\begin{vmatrix} 4 & -3 \\ 2 & 5 \end{vmatrix} + 0 + 0$$

$$= 1(20 - (-6)) = 26$$

In exercises **11–20**, first change a row (or column) so that it contains exactly one nonzero value, and then evaluate the determinant using the expansion method.

11. $\begin{vmatrix} 1 & 3 & -5 \\ 2 & 8 & -7 \\ -3 & -5 & 12 \end{vmatrix}$

12. $\begin{vmatrix} 1 & -2 & 3 \\ -4 & 10 & -15 \\ 5 & -6 & 12 \end{vmatrix}$

13. $\begin{vmatrix} -2 & 1 & 4 \\ 5 & -1 & -6 \\ 4 & -2 & -5 \end{vmatrix}$

14. $\begin{vmatrix} 6 & 1 & 4 \\ -9 & -2 & 10 \\ 20 & 3 & -13 \end{vmatrix}$

15. $\begin{vmatrix} 3 & -3 & 1 \\ -13 & 13 & -5 \\ 15 & -22 & 6 \end{vmatrix}$

16. $\begin{vmatrix} 2 & -6 & 1 \\ 10 & -39 & 7 \\ 3 & 12 & -2 \end{vmatrix}$

17. $\begin{vmatrix} -4 & -9 & -15 \\ 1 & 2 & 3 \\ -3 & -1 & -5 \end{vmatrix}$

18. $\begin{vmatrix} 9 & -36 & 40 \\ 1 & -4 & 5 \\ 6 & -25 & 25 \end{vmatrix}$

19. $\begin{vmatrix} -15 & -2 & 10 \\ 6 & 1 & -8 \\ -2 & -1 & 10 \end{vmatrix}$

20. $\begin{vmatrix} -40 & 4 & 21 \\ -9 & 1 & 5 \\ 60 & -6 & -28 \end{vmatrix}$

SECTION 9-6. Using Matrices to Solve a System of Equations (Optional)

KEY TOPICS IN THIS SECTION

1. A definition of a matrix
2. An augmented matrix
3. Equivalent augmented matrices
4. Elementary row operations
5. Using matrices to solve a system of linear equations
6. Inconsistent and dependent systems

In this section, we will study another algebraic method for solving systems of linear equations. This method uses matrices (the plural of matrix).

A Definition of a Matrix

Definition 9.9. Matrix
A **matrix** is a rectangular array of numbers (or other entries) in rows and columns. The numbers in the array are called the **elements** of the matrix.

To show that a given array of numbers is a matrix, a pair of brackets encloses the array. Examples **a** and **b** are two matrices.

a. $\begin{bmatrix} 2 & 5 \\ 3 & -7 \end{bmatrix}$ A matrix with two rows and two columns, written 2×2, which is read "2 by 2".

b. $\begin{bmatrix} 1 & 2 & 4 & 24 \\ 2 & -1 & -3 & -22 \\ 3 & 0 & 5 & 19 \end{bmatrix}$ A matrix with three rows and four columns, written 3×4, which is read "3 by 4".

An Augmented Matrix

The coefficients and constants of a system of linear equations can be used to write two matrices. The **coefficient matrix** is formed using the coefficients in the system. The **augmented matrix** is formed using the coefficients and constants.

Example 1. Given: (1) $2x - y = -6$
(2) $3x - 5y = -16$

Write the augmented matrix.

Solution.

$$\left[\begin{array}{cc|c} 2 & -1 & -6 \\ 3 & -5 & -16 \end{array}\right]$$

Write the coefficients to the left of the bar and the constants to the right.

Example 2. Given:

$$\left[\begin{array}{ccc|c} 1 & 3 & 0 & 7 \\ 0 & 1 & 5 & -3 \\ 2 & -1 & 3 & 8 \end{array}\right]$$

Write the associated system.

Solution. **Discussion.** The elements in column one are the coefficients of x, the elements in column two are the coefficients of y, and the elements in column three are the coefficients of z.

$$\begin{aligned} &(1) \quad x + 3y = 7 \\ &(2) \quad y + 5z = -3 \\ &(3) \quad 2x - y + 3z = 8 \end{aligned}$$

Don't write the z-term in (1).
Don't write the x-term in (2).

Equivalent Augmented Matrices

If two systems of linear equations have the same solution sets, then the systems are equivalent. Similarly, the augmented matrices of equivalent systems are also equivalent.

Definition 9.10. Equivalent augmented matrices
Two augmented matrices A and B are **equivalent,** written $A \equiv B$, if and only if the associated systems of linear equations have the same solution sets.

Example 3. Given that the associated systems of linear equations have exactly one solution, determine whether or not $A \equiv B$.

Matrix A

$$\left[\begin{array}{ccc|c} 1 & -1 & -1 & -10 \\ 2 & -1 & 1 & -6 \\ 3 & -5 & -8 & -55 \end{array}\right]$$

Matrix B

$$\left[\begin{array}{ccc|c} 1 & 0 & 0 & -2 \\ 0 & 1 & 0 & 5 \\ 0 & 0 & 1 & 3 \end{array}\right]$$

Solution. **Discussion.** To show that $A \equiv B$, verify that the solution of B is also the solution of the system associated with A.

System associated with A

$$\begin{aligned} &(1) \quad x - y - z = -10 \\ &(2) \quad 2x - y + z = -6 \\ &(3) \quad 3x - 5y - 8z = -55 \end{aligned}$$

System associated with B

$$\begin{aligned} &(1) \quad x = -2 \\ &(2) \quad y = 5 \\ &(3) \quad z = 3 \end{aligned}$$

The solution set is $\{(-2, 5, 3)\}$.

Checking $(-2, 5, 3)$ in (1), (2), and (3) of A:

(1) $-2 - 5 - 3 = -10$, *true*

(2) $2(-2) - 5 + 3 = -6$

$-4 - 5 + 3 = -6$, *true*

(3) $3(-2) - 5(5) - 8(3) = -55$

$-6 - 25 - 24 = -55$, *true*

Thus $A \equiv B$.

Elementary Row Operations

In Example 3, the augmented matrices A and B were shown to be equivalent. The solution of the system of equations associated with B was easy to see, because the coefficient matrix had ones along the diagonal from upper left to lower right, and 0's for every other element. There are three **elementary row operations** that we can use to change a given augmented matrix to an equivalent one in the form of B. Such a matrix is said to be **reduced.** Examples **e** and **f** are, respectively, 2×3 and 3×4 reduced augmented matrices.

e. $\left[\begin{array}{cc|c} 1 & 0 & k_1 \\ 0 & 1 & k_2 \end{array}\right]$ k_1 and k_2 are real numbers

f. $\left[\begin{array}{ccc|c} 1 & 0 & 0 & k_1 \\ 0 & 1 & 0 & k_2 \\ 0 & 0 & 1 & k_3 \end{array}\right]$ k_1, k_2, and k_3 are real numbers

Elementary row operations for an augmented matrix	Corresponding operations to the associated system of equations
1. Interchange any two rows.	**1*.** Interchange any two equations.
2. Multiply the elements in any row by a nonzero constant.	**2*.** Multiply both sides of an equation by a nonzero constant.
3. A nonzero multiple of the elements of any row can be added to the corresponding elements in any other row.	**3*.** The terms of any equation can be multiplied by a nonzero constant and the products added to the corresponding terms of another equation.

These elementary row operations can be used to change one augmented matrix to an equivalent one that is reduced.

Using Matrices to Solve a System of Linear Equations

In Examples 4 and 5, the augmented matrices associated with the given systems of equations are reduced by using the elementary row operations.

Example 4. Given: (1) $5x + 15y = 115$
(2) $8x + 22y = 170$

Solve using matrices.

Solution. $\left[\begin{array}{cc|c} 5 & 15 & 115 \\ 8 & 22 & 170 \end{array}\right]$ The given augmented matrix

Step 1. *If necessary, change the element in row one and column one to 1.*

$\equiv \left[\begin{array}{cc|c} 1 & 3 & 23 \\ 8 & 22 & 170 \end{array}\right]$ Multiply row one by $\frac{1}{5}$.

Step 2. *Change the remaining elements in column one to 0's.* Multiply the elements in row one by -8 and add the products to the corresponding elements in row two.

$\equiv \left[\begin{array}{cc|c} 1 & 3 & 23 \\ 0 & -2 & -14 \end{array}\right]$ $\begin{array}{l} 1(-8) + 8 = 0 \\ 3(-8) + 22 = -2 \\ 23(-8) + 170 = -14 \end{array}$

Step 3. *Change the element in row two and column two to 1.*

$\equiv \left[\begin{array}{cc|c} 1 & 3 & 23 \\ 0 & 1 & 7 \end{array}\right]$ Multiply row two by $\frac{-1}{2}$.

Step 4. *Change the remaining elements in column two to 0's.* Multiply row two by -3 and add to row one.

$\equiv \left[\begin{array}{cc|c} 1 & 0 & 2 \\ 0 & 1 & 7 \end{array}\right]$ $\begin{array}{l} 1(-3) + 3 = 0 \\ 7(-3) + 23 = 2 \end{array}$

Step 5. *State the solution set of the system associated with the matrix of Step 4.*

(1) $x \quad = 2$
(2) $\quad y = 7$ The solution set is $\{(2, 7)\}$.

Example 5. Given: (1) $4x + 5y + 13z = 58$
(2) $x + 2y + 6z = 26$
(3) $2x + 5y + 16z = 69$

Solve using matrices.

Solution. $\left[\begin{array}{ccc|c} 4 & 5 & 13 & 58 \\ 1 & 2 & 6 & 26 \\ 2 & 5 & 16 & 69 \end{array}\right]$ The given augmented matrix

$\equiv \left[\begin{array}{ccc|c} 1 & 2 & 6 & 26 \\ 4 & 5 & 13 & 58 \\ 2 & 5 & 16 & 69 \end{array}\right]$ Interchange rows one and two.
Multiply row one by -4 and add to row two.
Multiply row one by -2 and add to row three.

Second row **Third row**

$\left.\begin{array}{r} -4(1) + 4 = 0 \\ -4(2) + 5 = -3 \\ -4(6) + 13 = -11 \\ -4(26) + 58 = -46 \end{array}\right\} \equiv \left[\begin{array}{ccc|c} 1 & 2 & 6 & 26 \\ 0 & -3 & -11 & -46 \\ 0 & 1 & 4 & 17 \end{array}\right] \left\{\begin{array}{l} -2(1) + 2 = 0 \\ -2(2) + 5 = 1 \\ -2(6) + 16 = 4 \\ -2(26) + 69 = 17 \end{array}\right.$

$\equiv \left[\begin{array}{ccc|c} 1 & 2 & 6 & 26 \\ 0 & 1 & 4 & 17 \\ 0 & -3 & -11 & -46 \end{array}\right]$ Interchange rows two and three.
Multiply row two by -2 and add to row one.
Multiply row two by 3 and add to row three.

First row **Third row**

$$\left.\begin{array}{r} -2(0)+1=1 \\ -2(1)+2=0 \\ -2(4)+6=-2 \\ -2(17)+26=-8 \end{array}\right\} \equiv \left[\begin{array}{rrr|r} 1 & 0 & -2 & -8 \\ 0 & 1 & 4 & 17 \\ 0 & 0 & 1 & 5 \end{array}\right] \left\{\begin{array}{r} 3(0)+0=0 \\ 3(1)+(-3)=0 \\ 3(4)+(-11)=1 \\ 3(17)+(-46)=5 \end{array}\right.$$

Notice that there is a 1 in row three and column three. Multiply row three by 2 and add to row one. Multiply row three by -4 and add to row two.

First row **Second row**

$$\left.\begin{array}{r} 2(0)+1=1 \\ 2(0)+0=0 \\ 2(1)+(-2)=0 \\ 2(5)+(-8)=2 \end{array}\right\} \equiv \left[\begin{array}{rrr|r} 1 & 0 & 0 & 2 \\ 0 & 1 & 0 & -3 \\ 0 & 0 & 1 & 5 \end{array}\right] \left\{\begin{array}{r} -4(0)+0=0 \\ -4(0)+1=1 \\ -4(1)+4=0 \\ -4(5)+17=-3 \end{array}\right.$$

The system associated with this last matrix is:

$$\begin{array}{llcr} (1) & x & = & 2 \\ (2) & y & = & -3 \\ (3) & z & = & 5 \end{array}$$

The solution set is $\{(2, -3, 5)\}$.

Inconsistent and Dependent Systems

In the process of solving a system of linear equations using matrices, there are two results that indicate the system is inconsistent or dependent.

When solving a system of linear equations using augmented matrices:

1. If any row has only 0's to the left of the vertical bar and a nonzero constant to the right side, then the associated system is **inconsistent.**
2. If any row has only 0's to the left and right sides of the vertical bar, then the associated system is **dependent.**

SECTION 9-6. Practice Exercises

In exercises **1–4**, write the augmented matrix of each linear system.

[Example 1]

1. (1) $-4x + 5y = 30$
(2) $7x - 9y = -54$

2. (1) $2x - 7y = -29$
(2) $6x + y = -21$

3. (1) $x + 3y + 2z = 2$
(2) $2y - 4z = -12$
(3) $2x + 3z = 1$

4. (1) $2x + z = 10$
(2) $3y - 7z = 6$
(3) $-4x + 2y = -16$

In exercises **5–8**, write the associated linear system for each augmented matrix.

[Example 2]

5. $\left[\begin{array}{rr|r} 1 & 3 & 5 \\ 2 & -3 & 7 \end{array}\right]$

x+3y=5
2x-3y=7

6. $\left[\begin{array}{rr|r} -3 & 5 & -6 \\ 1 & -4 & 4 \end{array}\right]$

-3x+5y=-6
x-4y=4

7. $\left[\begin{array}{ccc|c} 3 & 2 & 0 & -5 \\ 0 & -3 & 4 & 6 \\ 1 & 0 & -8 & 10 \end{array}\right]$

8. $\left[\begin{array}{ccc|c} -3 & 0 & 5 & -6 \\ 4 & -3 & 0 & 7 \\ -2 & 4 & 9 & -8 \end{array}\right]$

In exercises **9** and **10**, given that the associated systems of linear equations have exactly one solution, determine whether or not $A \equiv B$.

[Example 3]

9. $A\left[\begin{array}{cc|c} -2 & 3 & 23 \\ 3 & 5 & 13 \end{array}\right]$ $\quad B\left[\begin{array}{cc|c} 1 & 0 & -4 \\ 0 & 1 & 5 \end{array}\right]$

10. $A\left[\begin{array}{ccc|c} 1 & -1 & 3 & 23 \\ 2 & 5 & -1 & -10 \\ 3 & 2 & 0 & 5 \end{array}\right]$ $\quad B\left[\begin{array}{ccc|c} 1 & 0 & 0 & 3 \\ 0 & 1 & 0 & -2 \\ 0 & 0 & 1 & 6 \end{array}\right]$

In exercises **11–24**, solve each system of linear equations using matrices.

[Examples 4 and 5]

11. (1) $x - 4y = 11$
(2) $3x + 2y = 5$

12. (1) $2x + 7y = -1$
(2) $x - 5y = -9$

13. (1) $2x - 4y = 22$
(2) $5x - 3y = -1$

14. (1) $3x + 6y = -15$
(2) $-4x - 3y = -10$

15. (1) $2x + 6y = 9$
(2) $6x - 12y = -13$

16. (1) $3x - 6y = -7$
(2) $-9x + 4y = 0$

17. (1) $4x - y - z = 5$
(2) $x - y + 2z = 2$
(3) $2x + y - z = 9$

18. (1) $x + 2y + z = 4$
(2) $3x + 4y + 6z = 18$
(3) $2x + 3y + 3z = 9$

19. (1) $x - 3y + 2z = -6$
(2) $5x - y + 8z = 2$
(3) $3x - 2y - 3z = 14$

20. (1) $x + y + z = 1$
(2) $2x - 3y + 4z = 6$
(3) $2x + 5y - 2z = 8$

21. (1) $x - 2y + 3z = 1$
(2) $x + y - z = 6$
(3) $2x - y + z = -9$

22. (1) $x + y + 2z = -1$
(2) $2x - y + 2z = 0$
(3) $x + 3y - 6z = 7$

23. (1) $x + y - 2z = 1$
(2) $2x + y - z = 8$
(3) $3x - 2y + z = 1$

24. (1) $2x - 3y + z = -2$
(2) $x + y - 3z = -15$
(3) $3x + y - 2z = 4$

In exercises **25–30**, solve each system of linear equations using matrices. Write "inconsistent" for any system with no solution and "dependent" for any system with more than one solution.

[Example 6]

25. (1) $4x - 12y = -5$
(2) $x - 3y = 5$

26. (1) $3x + 2y = -1$
(2) $15x + 10y = -5$

27. (1) $2x - 2y + 2z = 10$
(2) $3x + y + z = -5$
(3) $x - y + z = 5$

28. (1) $3x + 2y - 4z = 4$
(2) $-x + 2y - 3z = -7$
(3) $-6x - 4y + 8z = -8$

29. (1) $9x + 3y - 6z = 15$
(2) $3x + y - 2z = 8$
(3) $x - 4y + z = -20$

30. (1) $8x - 2y + 10z = 0$
(2) $x - 8y + 25z = 12$
(3) $4x - y + 5z = 2$

SECTION 9-6. Ten Review Exercises

In exercises **1–6**, simplify each expression. Assume variables represent positive real numbers.

1. $\dfrac{1}{\sqrt{2x}}$ **2.** $\dfrac{1}{\sqrt{x+2}}$ **3.** $\dfrac{1}{2\sqrt{x}}$

4. $\dfrac{1}{x\sqrt{2}}$ **5.** $\dfrac{1}{x-\sqrt{2}}$ **6.** $\dfrac{1}{\sqrt{x}+\sqrt{2}}$

In exercises **7–10**, use the points $P_1(-5, 4)$ and $P_2(7, -1)$.

7. Compute the distance between P_1 and P_2.

8. Compute the slope of the line that passes through P_1 and P_2.

9. Write an equation of the line that passes through P_1 and P_2 in $y = mx + b$ form.

10. Find the x-coordinate of the line that passes through P_1 and P_2 where it crosses the x-axis.

SECTION 9-6. Supplementary Exercises

In exercises **1–12**, state in words what operation was performed on A to transform it to the equivalent matrix B.

1. $A\left[\begin{array}{cc|c} 3 & -12 & 27 \\ -2 & 5 & 3 \end{array}\right]$ $B\left[\begin{array}{cc|c} 1 & -4 & 9 \\ -2 & 5 & 3 \end{array}\right]$

2. $A\left[\begin{array}{cc|c} -2 & 8 & 10 \\ 4 & 1 & 3 \end{array}\right]$ $B\left[\begin{array}{cc|c} 1 & -4 & -5 \\ 4 & 1 & 3 \end{array}\right]$

3. $A\left[\begin{array}{cc|c} 5 & 3 & 7 \\ 4 & 6 & 1 \end{array}\right]$ $B\left[\begin{array}{cc|c} 1 & -3 & 6 \\ 4 & 6 & 1 \end{array}\right]$

4. $A\left[\begin{array}{cc|c} -3 & 2 & 5 \\ 4 & -1 & 7 \end{array}\right]$ $B\left[\begin{array}{cc|c} 1 & 1 & 12 \\ 4 & -1 & 7 \end{array}\right]$

5. $A\left[\begin{array}{cc|c} 1 & -1 & 3 \\ 2 & 5 & -2 \end{array}\right]$ $B\left[\begin{array}{cc|c} 1 & -1 & 3 \\ 0 & 7 & -8 \end{array}\right]$

6. $A\left[\begin{array}{cc|c} 1 & 0 & 5 \\ -3 & 2 & -2 \end{array}\right]$ $B\left[\begin{array}{cc|c} 1 & 0 & 5 \\ 0 & 2 & 13 \end{array}\right]$

7. $A\left[\begin{array}{ccc|c} 4 & 2 & 1 & 5 \\ 3 & -6 & 7 & -3 \\ 7 & 4 & 8 & 1 \end{array}\right]$ $B\left[\begin{array}{ccc|c} 4 & 2 & 1 & 5 \\ 6 & -12 & 14 & -6 \\ 7 & 4 & 8 & 1 \end{array}\right]$

8. $A\left[\begin{array}{ccc|c} 2 & 10 & 6 & 4 \\ 1 & -6 & 3 & 5 \\ 3 & 5 & -2 & -8 \end{array}\right]$ $B\left[\begin{array}{ccc|c} 1 & 5 & 3 & 2 \\ 1 & -6 & 3 & 5 \\ 3 & 5 & -2 & -8 \end{array}\right]$

9. $A\left[\begin{array}{rrr|r} -2 & 5 & 1 & 9 \\ 4 & 1 & -3 & 2 \\ 1 & 6 & 5 & -1 \end{array}\right]$ $B\left[\begin{array}{rrr|r} 1 & 6 & 5 & -1 \\ 4 & 1 & -3 & 2 \\ -2 & 5 & 1 & 9 \end{array}\right]$

10. $A\left[\begin{array}{rrr|r} 1 & -4 & -9 & 11 \\ 2 & 6 & 3 & -4 \\ 1 & -2 & 6 & 1 \end{array}\right]$ $B\left[\begin{array}{rrr|r} 1 & -4 & -9 & 11 \\ 1 & -2 & 6 & 1 \\ 2 & 6 & 3 & -4 \end{array}\right]$

11. $A\left[\begin{array}{rrr|r} 7 & 4 & 1 & 0 \\ 5 & 3 & 2 & 6 \\ 2 & -5 & 3 & -2 \end{array}\right]$ $B\left[\begin{array}{rrr|r} 1 & 19 & -8 & 6 \\ 5 & 3 & 2 & 6 \\ 2 & -5 & 3 & -2 \end{array}\right]$

12. $A\left[\begin{array}{rrr|r} -5 & 8 & 1 & 4 \\ 3 & -2 & 3 & 0 \\ 2 & 5 & -1 & -2 \end{array}\right]$ $B\left[\begin{array}{rrr|r} 1 & 4 & 7 & 4 \\ 3 & -2 & 3 & 0 \\ 2 & 5 & -1 & -2 \end{array}\right]$

In exercises **13–20**, solve for x, y, and z. In each system, a and b are real numbers.

13. (1) $5x - y + 2z = 13a$
(2) $x + 2y - z = -a$
(3) $x - y + z = 4a$

14. (1) $3x + 2y - z = -9a$
(2) $x - 3y + 2z = -2a$
(3) $2x - y + z = -5a$

15. (1) $x + 2y - z = 2b$
(2) $2x - 3y + z = -b$
(3) $-x + y - 3z = -8b$

16. (1) $3x - y + 2z = 4b$
(2) $-x + 2y - 3z = -9b$
(3) $3x - 2y - z = -9b$

17. (1) $3x + 5y = 6a + 15b$
(2) $4x - 7y = 8a - 21b$

18. (1) $2x - 5y = 2a + 15b$
(2) $5x + 4y = 5a - 12b$

19. (1) $x + 2y = 3a - b$
(2) $2x - y = a + 3b$

20. (1) $3x - 2y = 4b - a$
(2) $2x + 5y = 12a + 9b$

SECTION 9-7. Linear Programming (Optional)

1. A definition of a function in two variables
2. The domain of a function in two variables
3. Restricting the domain with constraints
4. Maximum and minimum values of a function in two variables

In Chapter 7, we studied functions of one variable. These functions define a dependent variable y whose values are determined by an independent variable x. We frequently identified the relationship with a *rule of correspondence*, and wrote:

$$y = f(x) \qquad \text{y is a function of x}$$

In this section, we will study a special example of a dependent variable z whose values are determined by two independent variables, x and y. We will identify the relationships with rules of correspondence, and write:

$$z = f(x, y) \qquad \text{z is a function of x and y}$$

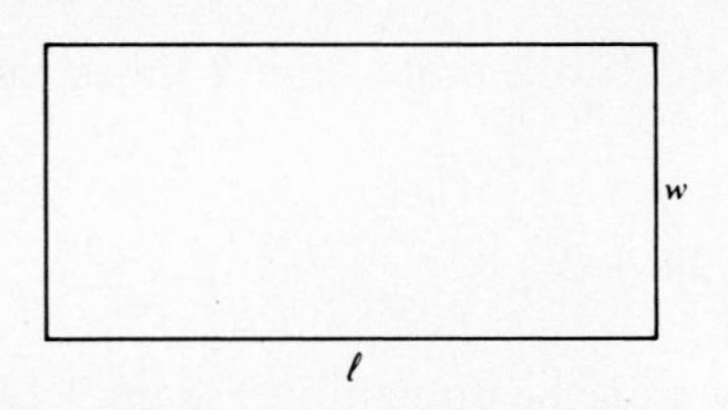

Figure 9-16. A rectangle with length ℓ and width w.

A Definition of a Function in Two Variables

In Figure 9-16, a rectangle is drawn with length ℓ and width w. The perimeter P and area A both *depend* on ℓ and w. Their relationships are shown in equations (1) and (2).

(1) $P = 2\ell + 2w$ The perimeter is the sum of 2 times the length and 2 times the width.

(2) $A = \ell \cdot w$ The area is the product of the length and width.

In mathematical terms, we say that P and A are functions of ℓ and w. This means that each value of P or A depends on arbitrarily selected values of ℓ and w. For any given ordered pair of values of ℓ and w, there is exactly one value for P and one value for A. Using function notation, the equations could be written as follows:

(1) $f(\ell, w) = 2\ell + 2w$ The perimeter is a function of ℓ and w.

(2) $g(\ell, w) = \ell \cdot w$ The area is a function of ℓ and w.

Definition 9.11. Functions of two variables
If z is a variable related to x and y, and for each ordered pair of values for x and y there is one and only one value for z, then z is a function of x and y. The relationship can be stated by the equation:

$z = f(x, y)$ z is a function of x and y

The Domain of a Function in Two Variables

Consider again the rectangle in Figure 9-16. Suppose no **constraints,** or restrictions, are placed on ℓ and w other than that they must be nonnegative numbers. Any ordered pair of nonnegative numbers (ℓ, w) will yield corresponding values for P and A, as shown in examples **a** and **b**.

a. If $(\ell, w) = (10, 7)$; then $P = 2(10) + 2(7) = 34$

$A = (10)(7) = 70$

b. If $(\ell, w) = (15, 8)$; then $P = 2(15) + 2(8) = 46$

$A = (15)(8) = 120$

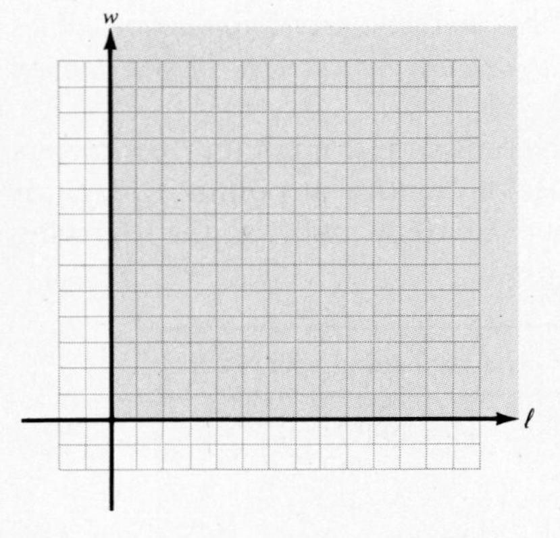

Figure 9-17. The domain of the perimeter and area functions.

The domain can be represented geometrically as the area in Quadrant I of an ℓw-coordinate system, as shown in Figure 9-17.

Restricting the Domain with Constraints

Now suppose that two constraints are placed on the possible values of length and width:

Constraint 1. The sum of the length and width must be less than or equal to 20 units. Written as an inequality:

(1) $\ell + w \leq 20$

Constraint 2. The length must be less than or equal to 2 more than 2 times the width. Written as an inequality:

$$(2) \quad \ell \leq 2w + 2$$

When the graphs of these inequalities are drawn on the domain in Figure 9-17, we get the region shown in Figure 9-18. The constraints put on ℓ and w yield a **restricted domain** bounded by the axes and the two inequalities.

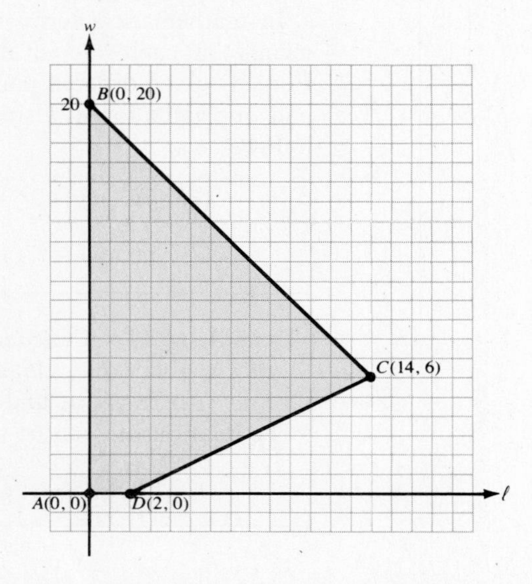

Figure 9-18. The restricted domain of the perimeter and area functions.

The figure is called a **convex polygon,** because it has no *indentations.* The points $A(0, 0)$, $B(0, 20)$, $C(14, 6)$, and $D(2, 0)$ are vertices of the polygon. With this restricted domain, we can now replace ℓ and w by any pair of values within the shaded region or on the edges of the polygon.

Maximum and Minimum Values of a Function in Two Variables

A function of two variables with a domain such as the one in Figure 9-18 has a maximum value and a minimum value. *These maximum and minimum values occur for values of the independent variables at one of the vertices of the convex polygon.*

As a consequence, we can assert that the maximum and minimum perimeters and areas for the rectangle will occur for values of ℓ and w at points A, B, C, or D. The coordinates of these points are used to evaluate P and A in the following table:

Point	Corresponding value of P	Corresponding value of A
$A(0, 0)$	$P = 2(0) + 2(0) = 0$	$A = (0)(0) = 0$
$B(0, 20)$	$P = 2(0) + 2(20) = 40$	$B = (0)(20) = 0$
$C(14, 6)$	$P = 2(14) + 2(6) = 40$	$C = (14)(6) = 84$
$D(2, 0)$	$P = 2(2) + 2(0) = 4$	$D = (2)(0) = 0$

From the results in the table, we can make the following conclusions:

1. The maximum perimeter occurs when $\ell = 0$ and $w = 20$, or $\ell = 14$ and $w = 6$.
2. The minimum perimeter occurs when $\ell = 0$ and $w = 0$, or $\ell = 2$ and $w = 0$.
3. The maximum area occurs when $\ell = 14$ and $w = 6$.
4. The minimum area occurs when either $\ell = 0$, $w = 0$, or both are 0.

The branch of mathematics that solves maximum and minimum values of functions of two variables by the technique shown above is called **linear programming.** The equation that defines the dependent variable in terms of the two independent variables is called the **objective function.** The region that represents the domain of the function is called the **feasible set.** The procedure used to solve such problems is outlined in Example 1.

Example 1. Given: $z = 18x + 12y$

Find the maximum value of z subject to the following constraints:

(1) $x + 3y \leq 36$
(2) $10x + 7y \leq 130$
(3) $x \geq 0$
(4) $y \geq 0$

Solution.

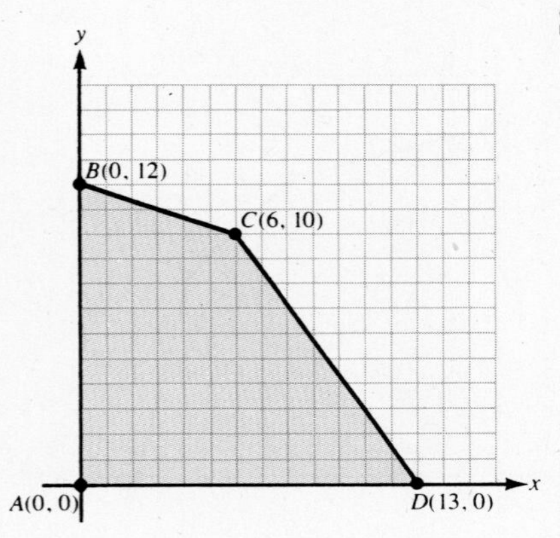

Figure 9-19. A graph of the feasible set.

Step 1. *Use the inequalities to graph the domain of the function.* In Figure 9-19, the line through points B and C is the boundary to (1). The line through points C and D is the boundary to (2). Inequalities (3) and (4) restrict the domain to Quadrant I.

Step 2. *Determine the coordinates of the vertices of the region from Step* 1.
$A(0, 0)$ is the point of intersection of the boundaries of (3) and (4).
$B(0, 12)$ is the point of intersection of the boundaries of (1) and (3).
$C(6, 10)$ is the point of intersection of the boundaries (1) and (2).
$D(13, 0)$ is the point of intersection of the boundaries of (2) and (4).

Step 3. *Evaluate the objective function for the values of the independent variables at each of the vertices.*

$A(0, 0)$: $z = 18(0) + 12(0) = 0$

$B(0, 12)$: $z = 18(0) + 12(12) = 144$

$C(6, 10)$: $z = 18(6) + 12(10) = 228$

$D(13, 0)$: $z = 18(13) + 12(0) = 234$

Step 4. *Use the results of Step* 3 *to answer the problem.*
A maximum value for z is needed. Of the four values computed in Step 3, the maximum is 234, which occurred for $x = 13$ and $y = 0$.

SECTION 9-7. Practice Exercises

In exercises **1–4**, for each feasible set and objective function z:

a. State the ordered pair (x, y) that yields the maximum value of z.

b. State the maximum value of z.

1. $z = 2x - y$

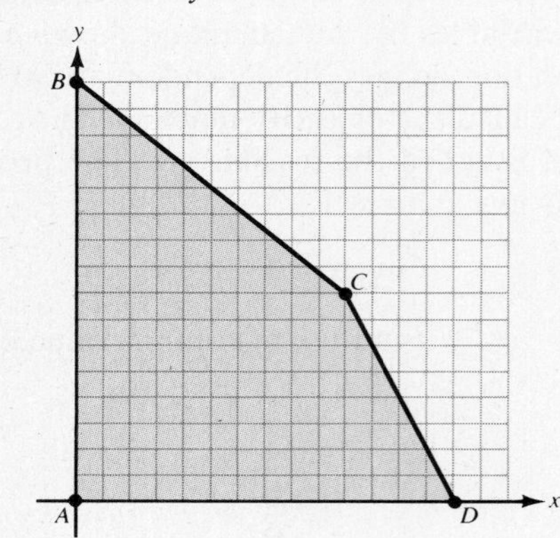

2. $z = y - x$

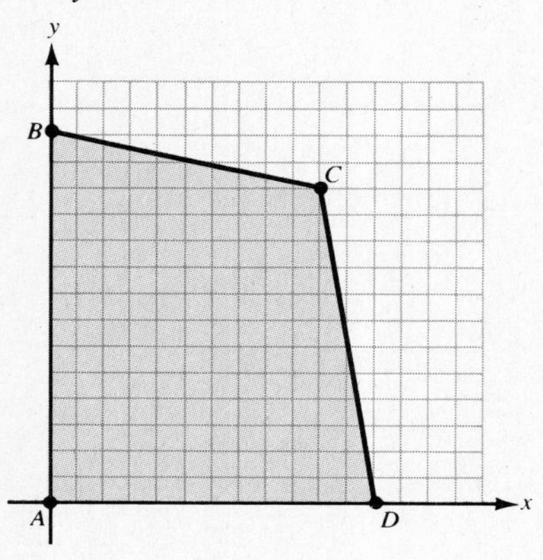

3. $z = 4x + 2y$

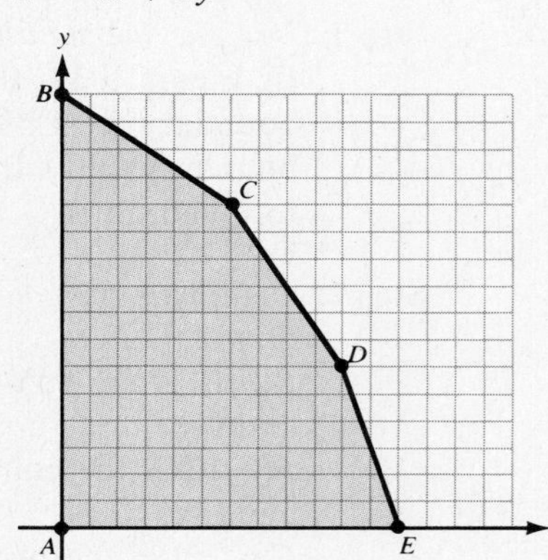

4. $z = 2x + 3y$

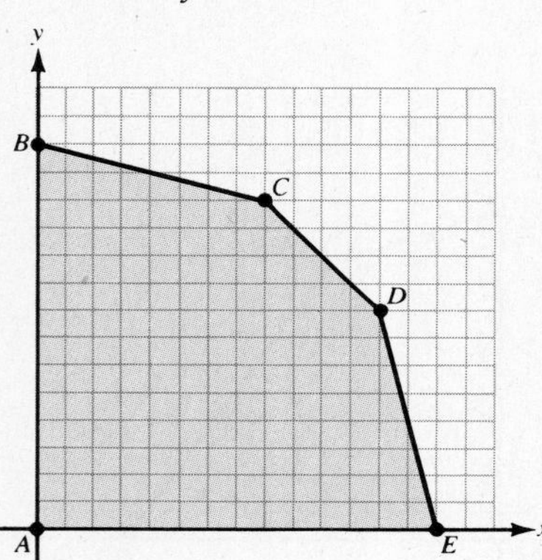

In exercises **5–12**, find the maximum value of z subject to the given constraints.

5. $z = 5x + 2y$
(1) $4x + 5y \leq 20$
(2) $x \geq 0$
(3) $y \geq 0$

6. $z = 4x + 3y$
(1) $3x + 2y \leq 6$
(2) $x \geq 0$
(3) $y \geq 0$

7. $z = \frac{1}{2}x + 2y$
(1) $x + 2y \leq 16$
(2) $x \leq 6$
(3) $x \geq 0$
(4) $y \geq 0$

8. $z = \frac{1}{5}x - 3y$
(1) $x + y \leq 8$
(2) $x \leq 5$
(3) $x \geq 0$
(4) $y \geq 0$

9. $z = 6x + 3y$
(1) $x + 2y \leq 6$
(2) $2x - 3y \geq -2$
(3) $x \geq 0$
(4) $y \geq 0$

10. $z = 3x + 4y$
(1) $x - y \geq -1$
(2) $x + 2y \leq 14$
(3) $x \geq 0$
(4) $y \geq 0$

11. $z = 4x - \frac{1}{5}y$

(1) $5x - y \leq 10$

(2) $2y - 5x \leq 10$

(3) $5x + y \leq 20$

(4) $x \geq 0$

(5) $y \geq 0$

12. $z = 5x + 2y$

(1) $x - y \leq 1$

(2) $x - 3y \geq -8$

(3) $2y + 3x \leq 23$

(4) $x \geq 0$

(5) $y \geq 0$

SECTION 9-7. Ten Review Exercises

In exercises **1–7**, given: $3(3t + 1) - 2t - 5(t - 2) - 19$

1. Simplify the expression.

2. Set the expression of exercise **1** equal to 3 and solve for t.

3. Set the expression of exercise **1** less than 4 and state the solutions of the inequality.

4. Set the absolute value of the expression of exercise **1** equal to 5 and solve for t.

5. Set the absolute value of the expression of exercise **1** greater than 6 and state the solutions of the inequality.

6. Set the expression of exercise **1** equal to u. Identify by name the function defined by the equation.

7. For a graph of the function of exercise **6**:
a. Identify the slope.
b. Identify the u-intercept.

In exercises **8–10**, given: $3x^2 + 2x - 8$

8. Write the expression in factored form.

9. Find the solution set when the expression is set equal to 0.

10. Identify by name the function formed when the expression is set equal to y, and name the graph of the function.

SECTION 9-7. Supplementary Exercises

In exercises **1–4**, solve each word problem.

1. Donna Gabrielli has \$5,000 to invest in savings accounts at Bank A and Bank B. Bank A offers 5% interest on money invested in savings accounts, and Bank B offers 6% interest. Donna must invest at least \$2,500 in Bank A and \$1,000 in Bank B. She wants to invest no more than $\frac{2}{5}$ of the total amount in Bank B. To get maximum interest, how much should Donna invest in Bank A and Bank B?

2. Tamara Davis has $10,000 to invest in two money market accounts, called Fund A and Fund B. Fund A yields a 7.5% annual return, and Fund B yields an annual return of 8.0%. Tamara must invest at least $2,000 in Fund A and $3,000 in Fund B. She wants to invest no more than 60% of the total in Fund B. To get maximum return on her investment, how much should Tamara invest in Fund A and Fund B?

3. A store called the Candy Corner wants to mix two kinds of candy for sampler packages. The proprietors have 150 pounds of red candy and 240 pounds of green candy to mix. A mixture of $\frac{1}{4}$ of a pound of red candy and $\frac{3}{4}$ of a pound of green candy sells for $0.50. A mixture of $\frac{1}{2}$ of a pound of each kind sells for $0.90. How many packages should be mixed to get maximum income?

4. The Stamp House wants to sell two kinds of cancelled stamps, some from North America and some from South America. The proprietors have 72 North American stamps and 120 South American stamps. A package of three North American stamps and nine South American stamps sells for $1.00. A package of six of each kind of stamp sells for $0.80. How many packages of each type should be made to get maximum income?

In exercises **5–8**, find the *minimum* value of z subject to the given constraints.

5. $z = 3x - 2y$
   (1) $x + y \geq 5$
   (2) $x \geq 2$
   (3) $y \geq 1$

6. $z = 10x - y$
   (1) $2x + y \geq 10$
   (2) $x \geq 1$
   (3) $y \geq 4$

7. $z = x + 2y$
   (1) $x + y \geq 8$
   (2) $3x + y \geq 0$
   (3) $5y - x \geq 10$

8. $z = 2x + 7y$
   (1) $x + 2y \geq 8$
   (2) $3x - y \geq 3$
   (3) $2x - 3y \leq 9$

Name

Date

Score

SECTION 9-5. Summary Exercises

Answer

In exercises **1–4**, evaluate each determinant.

1. $\begin{vmatrix} 5 & 7 \\ 2 & 9 \end{vmatrix}$

1. ______________

2. $\begin{vmatrix} -3 & -11 \\ 18 & 12 \end{vmatrix}$

2. ______________

3. $\begin{vmatrix} -15 & 9 \\ -5 & 3 \end{vmatrix}$

3. ______________

4. $\begin{vmatrix} 2 & 1 & -2 \\ 10 & 3 & -9 \\ -6 & -4 & 13 \end{vmatrix}$

4. ______________

In exercises **5–8**, solve each system using Cramer's Rule. Write "inconsistent" for any system with no solution and "dependent" for any system with more than one solution.

5. (1) $2x + 3y = 5$
(2) $x - 2y = -8$

5. ______________

6. (1) $2x - 2y = 7$
(2) $-3x + 3y = 1$

6. ______________

7. (1) $2x + y = 8$
(2) $8x - 5y = 5$

7. ______________

8. (1) $4x - y + 3z = 3$
(2) $3x + 2y - z = 4$
(3) $x + y - 2z = -1$

8. ______________

CHAPTER 9. Review Exercises

In exercises **1** and **2**, determine whether the given ordered pair is a solution of each system.

1. (1) $5x + 2y = -10$
(2) $9x - 8y = -75$

$(-4, 5)$

2. (1) $2a - 3b = -5$
(2) $4a + 15b = 4$

$\left(\frac{-3}{2}, \frac{2}{3}\right)$

In exercises **3–8**, solve each system. Write "inconsistent" for a system with no solution and "dependent" for a system with more than one solution.

3. (1) $m - 7n = -45$
(2) $3m + 10n = 51$

4. (1) $3u + 7v = 1$
(2) $-2u + 9v = -28$

5. (1) $14s + 11t = 24$
(2) $7s + \frac{11}{2}t = 10$

6. (1) $4a + 9b = 6$
(2) $-6a - 15b = -11$

7. (1) $3x + 2y = -10$
(2) $y = 6x$

8. (1) $2x - 9y = -4$
(2) $x = 3y$

In exercises **9** and **10**, determine whether the given ordered triple is a solution of each system.

9. $9x - 10y + 13z = 62$; $(-2, -3, 4)$

10. $12x + 15y - 8z = 15$; $\left(\frac{3}{4}, \frac{-2}{5}, \frac{-3}{2}\right)$

In exercises **11** and **12**, determine whether the given ordered triple is a solution of each system.

11. (1) $6x - 9y + 5z = -6$
(2) $30x + 27y - 15z = -6$
(3) $-16x + 33y + 40z = 54$

$\left(\frac{-1}{2}, \frac{2}{3}, \frac{3}{5}\right)$

12. (1) $x + 2y + 7z = -13$
(2) $-5x + 10z = -20$
(3) $9y - 8z = 70$

$(-2, 5, -3)$

In exercises **13** and **14**, solve each system by the substitution or addition method. Write "inconsistent" for a system with no solution and "dependent" for a system with more than one solution.

13. (1) $4x + 2y - 3z = -4$
(2) $14x + 5y - 6z = -11$
(3) $8x - 3y + 9z = 2$

14. (1) $9x - 2y + 20z = -3$
(2) $15x - 4y + 32z = -5$
(3) $-21x + 6y - 48z = 8$

In exercises **15–20**, solve each word problem.

15. The sum of 8 times the larger of two numbers and 6 times the smaller is 18. The difference between 4 times the larger and 9 times the smaller is 1. Find both numbers.

16. Shirley and Helen are sisters. The sum of 2 times Shirley's age and 3 times Helen's age is 147 years. The difference between 3 times Shirley's age 4 years ago and Helen's age 7 years from now is 20 years. How old is each person?

17. Vic bowls in a Friday night league at Sunrise Bowling Lanes. Last week the sum of the scores for the three games Vic bowled was 575. The sum of the scores of the first two games was 131 more than the score of the third game. The difference between 3 times the score of the second game and 2 times the score of the third game was 60. Find the scores of each of the three games.

18. Jill and Sid Newman bought a small plane to travel to and from a summer home in the mountains. Last week they made the 900-mile trip against a headwind in 5 hours. They made the return trip with the same wind as a tailwind in 3 hours. Find the speed of the plane in still air, and the speed of the wind.

19. Find an equation in the $y = ax^2 + bx + c$ form for a function if the points $P(-1, -3)$, $Q(-2, 3)$ and $R(-3, 5)$ are on a graph of that function.

20. Container A has a 10% solution of antiseptic. Container B has a 30% solution of antiseptic. When the two containers are mixed, 50 liters of a 28% solution are produced. Find the amount of solution in container A and in container B, respectively.

In exercises **21** and **22**, determine whether each ordered pair is a solution of the given system.

21. (1) $2x - 5y \leq 4$
(2) $x + 6y < 0$
a. $(-3, -2)$ **b.** $(7, 3)$

22. (1) $14x + 15y \geq -1$
(2) $20x - 3y < 10$
a. $\left(\frac{1}{2}, \frac{-1}{3}\right)$ **b.** $\left(\frac{-3}{2}, \frac{4}{3}\right)$

In exercises **23–26**, graph the solution set of each system.

23. (1) $5x - 3y + 15 > 0$
(2) $x + y + 2 > 0$

24. (1) $5x + y + 5 \leq 0$
(2) $x \leq -3$

25. (1) $3x - 5y > -20$
(2) $3x - 5y < 0$

26. (1) $x - 4 \leq 0$
(2) $y + 1 > 0$

In exercises **27–30**, evaluate each determinant.

27. $\begin{vmatrix} 32 & -24 \\ \frac{7}{6} & \frac{5}{8} \end{vmatrix}$

28. $\begin{vmatrix} -\frac{3}{8} & \frac{9}{8} \\ \frac{3}{2} & -\frac{5}{2} \end{vmatrix}$

29. $\begin{vmatrix} -2 & 4 & 3 \\ 0 & -5 & 2 \\ 3 & 6 & -4 \end{vmatrix}$

30. $\begin{vmatrix} 4 & -5 & 6 \\ -3 & 2 & 7 \\ -1 & 3 & 9 \end{vmatrix}$

In exercises **31–34**, solve each system using Cramer's Rule. Write "inconsistent" for any system with no solutions and "dependent" for any system with more than one solution.

31. (1) $2x - 3y = 10$
(2) $7x + 6y = 24$

32. (1) $-8x + 5y = -7$
(2) $12x - 15y = 12$

33. (1) $x - 3y + 2z = 3$
(2) $2x + 4y - z = -4$
(3) $3x - 5y + z = -7$

34. (1) $2x - y - 3z = 5$
(2) $x + 4y + 5z = 0$
(3) $3x - 6y + 2z = 23$

In exercises **35** and **36**, write the augmented matrix of each linear system.

35. (1) $-10x + 3y = 5$
(2) $7x - 6y = 16$

36. (1) $x - 3y + 5z = 10$
(2) $3x \quad + 4z = 15$
(3) $6y - z = 8$

In exercises **37** and **38**, write the associated linear system for each augmented matrix.

37. $\left[\begin{array}{ccc|c} 1 & -6 & 4 & -8 \\ 2 & 0 & -5 & 3 \\ 3 & 4 & -1 & 7 \end{array}\right]$

38. $\left[\begin{array}{ccc|c} 1 & \frac{3}{4} & -\frac{2}{3} & 8 \\ 0 & 1 & \frac{5}{3} & 2 \\ 0 & 0 & 1 & -6 \end{array}\right]$

In exercises **39** and **40**, given that the associated systems of linear equations have exactly one solution, determine whether $A \equiv B$.

39. $A\left[\begin{array}{cc|c} 10 & -16 & 10 \\ -15 & 8 & -11 \end{array}\right]$ $\quad B\left[\begin{array}{cc|c} 1 & 0 & \frac{3}{5} \\ 0 & 1 & -\frac{1}{4} \end{array}\right]$

40. $A\left[\begin{array}{ccc|c} 3 & 8 & 9 & -5 \\ -2 & 4 & 3 & 5 \\ 5 & -10 & -12 & -11 \end{array}\right]$ $\quad B\left[\begin{array}{ccc|c} 1 & 0 & 0 & -2 \\ 0 & 1 & 0 & \frac{1}{2} \\ 0 & 0 & 1 & -\frac{1}{3} \end{array}\right]$

In exercises **41–44**, solve each system of linear equations using matrices. Write "inconsistent" for any system with no solution and "dependent" for any system with more than one solution.

41. (1) $x - 2y = -18$
(2) $5x + 7y = 29$

42. (1) $3x - 12y = -14$
(2) $-9x + 8y = 21$

43. (1) $-x - 4y - 5z = -18$
(2) $3x + 13y + 17z = 61$
(3) $2x + 6y + 7z = 26$

44. (1) $x + 2y + 7z = -19$
(2) $-3x - 3y - 7z = 7$
(3) $2x + 3y + 9z = -20$

CHAPTER

10

Exponential and Logarithmic Functions

Ms. Glaston held a small cellophane packet above her head. In the packet was a decal displaying a panther, the school's mascot. "Today," Ms. Glaston began, "someone in this class may win this beautiful school decal to put on his or her car. To win this valuable prize (for which I paid 75 cents at the bookstore), someone must make a correct estimation for the following activity."

She then placed on the desk a stack of cardboard cut in 8-inch squares. "Each piece of cardboard in this stack," Ms. Glaston continued, "is approximately 0.1 inches thick." She selected one piece of the cardboard and placed it directly in front of her on the desk.

"Now," she said, "how high is the stack of cardboard that I just placed in front of me?"

"Are you trying to trick us, Ms. Glaston?" Carrie Maittain asked. "There's only one piece of cardboard in the stack, so the height must be 0.1 inches."

"Good, Carrie," Ms. Glaston replied with a smile, "maybe you'll be the one to win the decal." She now placed a second piece of cardboard on top of the first one. "How high is the stack now?" she asked.

"You doubled the pieces of cardboard," Bob Smith replied, "so you must have doubled the height. The height must therefore be 0.2 inches. But there must be a trick somewhere, because your problems are never this easy."

"Don't be so suspicious, Bob," Ms. Glaston replied, and placed two more pieces on the stack. "Now how high is the stack?" she asked.

"Well, 0.2 inches doubled must be 0.4 inches," Kim Kimiecik replied. "Does this mean you're going to add four pieces to the stack next time?"

"Exactly, Kim," Ms. Glaston said, and added four more pieces to the pile. "As you all know," she continued, "the stack is now 0.8 inches high. Suppose that I had enough pieces of cardboard to continue this doubling process just 30 times. The height of the cardboard would have the following value." She then wrote on the board:

$$(0.1 \text{ inches})(2^{30})$$

"The winner of this decal," she said, "will be the one that can guess to the nearest foot the height of the stack of cardboard."

"It won't be a foot high," Steve Drummond said, "so I'll guess a foot."

"No, it'll be more than a foot, Steve," Terry Retchless said, "but it'll be less than 10 feet. So, I'll guess 10 feet."

"Ten feet is too much, Terry," Kate Derrick added. "I'll guess half that much, say 5 feet."

A considerable amount of discussion took place for the next several minutes, until someone suggested that a calculator could settle the arguments and give an accurate value for the expression.

Ms. Glaston then interrupted and closed the time for any further guesses.

"Well," she said, "I guess I can keep this decal as a prize for the next class, because nobody came close to the answer. With a calculator, we can evaluate the expression on the board, and then change the inches to feet, and then feet to miles. Such an exercise would show the height of cardboard to be approximately 1,695 miles. The factor 2^{30} in the expression is the reason for such a large number. It appears to be a small and insignificant number. As a consequence, it is commonly underestimated. The expression 2^{30} is one of the values of the exponential function defined by $y = 2^x$. Today we begin a study of these functions, and their inverse functions, which are called logarithms. As you will soon see, these functions have characteristics that make them quite different from other functions we have previously studied."

SECTION 10-1. The Exponential Function

1. A definition of an exponential function
2. Graphing an exponential function
3. Some properties of the exponential function
4. Solving some exponential equations

In this section, we will study the exponential function. As we will see, a graph of an exponential function has a shape that is quite different from the functions studied in Chapters 7 and 8.

A Definition of an Exponential Function

In previous chapters, we have used definitions to assign values to expressions such as 3^{-1}, 5^0, $16^{3/4}$, and $49^{-1/2}$. The exponents have always been rational numbers. Furthermore, certain restrictions have been enforced when a base is negative. For example, $(-8)^{2/3}$ is a real number because 3 is odd, while $(-4)^{1/2}$ is not because 2 is even.

To define an exponential function, we want to also give values to bases with irrational-number exponents, such as $2^{\sqrt{5}}$. It can be shown that $2^{\sqrt{5}}$ can be approximated by rational numbers a and b, such that $a < \sqrt{5} < b$. The following inequalities illustrate how $2^{\sqrt{5}}$ can be "squeezed" between 2^a and 2^b, as the values of a and b more closely approximate $\sqrt{5}$:

$a < \sqrt{5} < b$	$2^a < 2^{\sqrt{5}} < 2^b$
$2.2 < \sqrt{5} < 2.3$	$2^{2.2} < 2^{\sqrt{5}} < 2^{2.3}$
$2.23 < \sqrt{5} < 2.24$	$2^{2.23} < 2^{\sqrt{5}} < 2^{2.24}$
$2.236 < \sqrt{5} < 2.237$	$2^{2.236} < 2^{\sqrt{5}} < 2^{2.237}$

The approximation could be continued by increasing the number of digits used to approximate $\sqrt{5}$. Using a calculator on the last inequality, to eight decimal places:

$$4.71089116 < 2^{\sqrt{5}} < 4.71415763$$

Definition 10.1. The exponential function

If f is an **exponential function,** then f can be defined by an equation written in the form:

$$f(x) = b^x \qquad \text{Where } b > 0 \text{ and } b \neq 1$$

The domain is the set of real numbers, written "all x".

Example 1. If $f(x) = 8^x$, find each function value.

a. $f(2)$ **b.** $f(-1)$ **c.** $f\left(\frac{1}{3}\right)$ **d.** $f\left(\frac{4}{3}\right)$

Solution.

a. $f(2) = 8^2 = 64$

b. $f(-1) = 8^{-1} = \frac{1}{8}$ $\qquad b^{-n} = \frac{1}{b^n}$

c. $f\left(\frac{1}{3}\right) = 8^{1/3} = \sqrt[3]{8} = 2$ $\qquad b^{1/n} = \sqrt[n]{b}$

d. $f\left(\frac{4}{3}\right) = 8^{4/3} = (\sqrt[3]{8})^4 = 16$ $\qquad b^{m/n} = (\sqrt[n]{b})^m$

Graphing an Exponential Function

In Definition 10.1, the base b is positive ($b > 0$) and not equal to 1 ($b \neq 1$). There are two possible sets of values for b.

Possibility 1. $0 < b < 1$ $\qquad$ b is between 0 and 1

Possibility 2. $b > 1$ $\qquad$ b is greater than 1

For possibility 1, a graph of $f(x) = b^x$ has several general characteristics that are illustrated by the function in Example 2.

Example 2. Graph $f(x) = \left(\frac{1}{2}\right)^x$.

Solution. **Discussion.** The following table contains seven solutions for the function. A smooth curve through the plots of these ordered pairs yields a graph of the function shown in Figure 10-1.

x	-3	-2	-1	0	1	2	3
$f(x) = \left(\frac{1}{2}\right)^x$	8	4	2	1	$\frac{1}{2}$	$\frac{1}{4}$	$\frac{1}{8}$

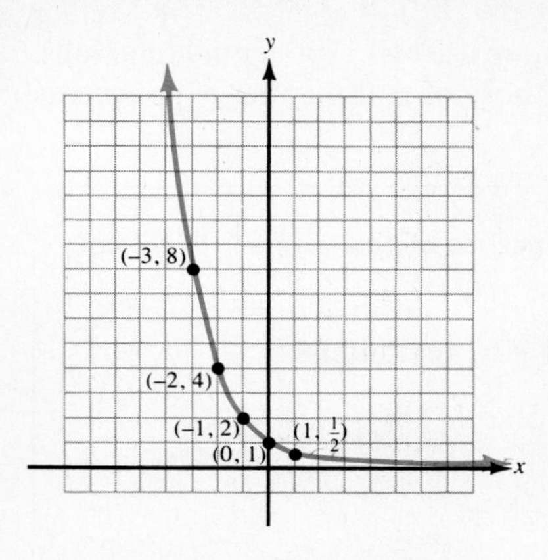

Figure 10-1. A graph of $y = (\frac{1}{2})^x$.

The graph in Figure 10-1 reveals the following characteristics of exponential functions with bases between 0 and 1:

Characteristic 1. For all x, $f(x) > 0$. Thus, the range of f is $f(x) > 0$.

Characteristic 2. As $x \to +\infty$, $f(x) \to 0^+$. Therefore, the x-axis is a horizontal asymptote for the graph.

Characteristic 3. As $x \to -\infty$, $f(x) \to +\infty$.

Characteristic 4. $P(0, 1)$ is a point on the graph, and the y-intercept is 1.

Characteristic 5. The graph passes the vertical and horizontal line tests of a one-to-one function.

For possibility 2, a graph of $f(x) = b^x$ has several general characteristics that are illustrated by the function in Example 3.

Example 3. Graph $f(x) = 2^x$.

Solution. **Discussion.** The following table contains seven solutions for the function. A smooth curve through the plots of these ordered pairs yields a graph of the function shown in Figure 10-2.

x	-3	-2	-1	0	1	2	3
$f(x) = 2^x$	$\frac{1}{8}$	$\frac{1}{4}$	$\frac{1}{2}$	1	2	4	8

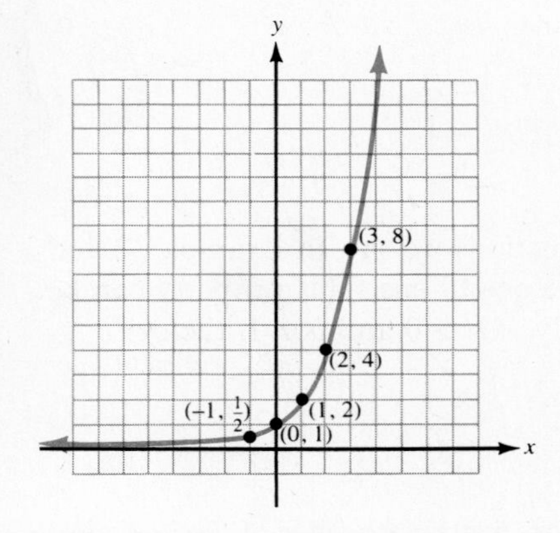

Figure 10-2. A graph of $y = 2^x$.

The graph in Figure 10-2 reveals the following characteristics of exponential functions with bases greater than 1:

Characteristic 1. For all x, $f(x) > 0$. Thus, the range of f is $f(x) > 0$.

Characteristic 2. As $x \to +\infty$, $f(x) \to +\infty$.

Characteristic 3. As $x \to -\infty$, $f(x) \to 0^+$. Therefore, the x-axis is a horizontal asymptote for the graph.

Characteristic 4. $P(0, 1)$ is a point on the graph, and the y-intercept is 1.

Characteristic 5. The graph passes the vertical and horizontal line tests of a one-to-one function.

Some Properties of the Exponential Function

Properties of real-number exponents are frequently used to simplify expressions such as $x^3 \cdot x^2$ and $(x^5)^4$. As examples **a–c** illustrate, these properties of exponents are also properties of exponential functions.

a. $5^{2x} \cdot 5^x = 5^{2x+x} = 5^{3x}$ Add the exponents.

b. $(10^3)^{2x} = 10^{3(2x)} = 10^{6x}$ Multiply the exponents.

c. $\frac{3^{4x}}{3^x} = 3^{4x-x} = 3^{3x}$ Subtract the exponents.

To write general statements of these examples, we use two exponential functions. The parallelism between these properties with those of real-number exponents is also shown.

Properties of exponential functions	**Properties of real-number exponents**
Let $f(x) = b^x$ and $f(y) = b^y$ be exponential functions.	Let b be a positive number and x and y be real numbers.
1. $f(x) \cdot f(y) = f(x + y)$	$b^x \cdot b^y = b^{x+y}$
2. $\dfrac{f(x)}{f(y)} = f(x - y)$	$\dfrac{b^x}{b^y} = b^{x-y}$
3. $[f(x)]^y = f(x \cdot y)$	$(b^x)^y = b^{xy}$

Example 4. Simplify $f(x) = 3^{2x} \cdot 27^{-x}$.

Solution. **Discussion.** For the properties of exponential functions to be used, the bases must be the same. We must therefore write 27 as a power of 3.

$f(x) = 3^{2x} \cdot (3^3)^{-x}$ — Write 27 as 3^3.

$= 3^{2x} \cdot 3^{-3x}$ — Multiply the exponents 3 and $-x$.

$= 3^{2x+(-3x)}$ — Add the exponents $2x$ and $-3x$.

$= 3^{-x}$ or $\left(\frac{1}{3}\right)^x$ — Simplify.

Solving Some Exponential Equations

The characteristics of graphs of exponential functions we saw in Examples 2 and 3 include the statement that these functions are **one-to-one.** This property can be used to solve exponential equations. Specifically, if $b > 0$ and $b \neq 1$, then:

$$b^x = b^y \qquad \text{if and only if} \qquad x = y$$

Example 5. Solve $5^{x+1} = 125^x$.

Solution. **Discussion.** To solve this equation, write both sides of the equation as power terms with the same base. Then set the exponents equal and solve for x.

$5^{x+1} = (5^3)^x$ — Write 125 as 5^3.

$5^{x+1} = 5^{3x}$ — Multiply the exponents.

$x + 1 = 3x$ — If $b^x = b^y$, then $x = y$.

$x = \frac{1}{2}$ — The solution is $\frac{1}{2}$.

Thus, $5^{3/2} = 125^{1/2}$. — Replace x by $\frac{1}{2}$ and simplify.

The solution set is $\left\{\frac{1}{2}\right\}$.

Example 6. Solve $10^{x^2-4} = 1$.

Solution. **Discussion.** To write both sides of the equation with the same base, we write 1 as 10^0.

$10^{x^2-4} = 10^0$	Write 1 as 10^0.
$x^2 - 4 = 0$	If $b^x = b^y$, then $x = y$.
$(x+2)(x-2) = 0$	Factor as the difference of squares.
$x = -2$ or $x = 2$	Solve for x.

The solution set is $\{-2, 2\}$.

SECTION 10-1. Practice Exercises

In exercises **1–8**, find each value for the indicated function.

[Example 1]

1. $f(x) = 5^x$
- **a.** $f(3)$
- **b.** $f(-1)$
- **c.** $f(-2)$
- **d.** $f(0)$

2. $f(x) = 7^x$
- **a.** $f(2)$
- **b.** $f(-2)$
- **c.** $f(0)$
- **d.** $f(1)$

3. $g(x) = 4^x$
- **a.** $g(-1)$
- **b.** $g(3)$
- **c.** $g\left(\frac{1}{2}\right)$
- **d.** $g\left(\frac{-1}{2}\right)$

4. $g(x) = 9^x$
- **a.** $g(-2)$
- **b.** $g\left(\frac{1}{2}\right)$
- **c.** $g(3)$
- **d.** $g\left(\frac{-1}{2}\right)$

5. $h(x) = \left(\frac{1}{3}\right)^x$
- **a.** $h(2)$
- **b.** $h(-1)$
- **c.** $h(0)$
- **d.** $h(-2)$

6. $h(x) = \left(\frac{1}{4}\right)^x$
- **a.** $h(0)$
- **b.** $h(-2)$
- **c.** $h\left(\frac{1}{2}\right)$
- **d.** $h\left(\frac{-3}{2}\right)$

7. $f(x) = 16^x$
- **a.** $f(0)$
- **b.** $f(2)$
- **c.** $f\left(\frac{-1}{4}\right)$
- **d.** $f\left(\frac{3}{4}\right)$

8. $f(x) = \left(\frac{1}{16}\right)^x$
- **a.** $f(-1)$
- **b.** $f\left(\frac{1}{2}\right)$
- **c.** $f\left(\frac{-3}{4}\right)$
- **d.** $f\left(\frac{3}{2}\right)$

In exercises **9–16**, graph each function.

[Examples 2 and 3]

9. $y = 3^x$

10. $y = \left(\frac{1}{3}\right)^x$

11. $y = 4^x$

12. $y = \left(\frac{1}{4}\right)^x$

13. $y = \left(\frac{2}{3}\right)^x$

14. $y = \left(\frac{3}{2}\right)^x$

15. $y = \left(\frac{3}{4}\right)^x$

16. $y = \left(\frac{4}{3}\right)^x$

In exercises **17–24**, simplify each expression.

[Example 4]

17. $10^{3x} \cdot 100^x$

18. $3^{-2x} \cdot 81^{2x}$

19. $25^{-x} \cdot 125^{2x}$

20. $64^{2x} \cdot 32^{-3x}$

21. $343^{2-x} \cdot 49^{x+1}$

22. $216^{3-x} \cdot 36^{2x-1}$

23. $\dfrac{121^{3x+5}}{11^{6x+10}}$

24. $\dfrac{625^{x-3}}{125^{2x-4}}$

In exercises **25–54**, solve each equation.

[Examples 5 and 6]

25. $5^x = 5^{2x-3}$

26. $10^{3x-1} = 10^{x-3}$

27. $3^x = 243$

28. $2^x = 64$

29. $7^x = \dfrac{1}{49}$

30. $10^x = \dfrac{1}{10{,}000}$

31. $5^{2x} = 125$

32. $4^x = 8$

33. $9^x = \dfrac{1}{81}$

34. $49^x = \dfrac{1}{343}$

35. $125^{1-x} = \left(\dfrac{1}{25}\right)^x$

36. $216^{2-x} = \left(\dfrac{1}{36}\right)^x$

37. $2^{x^2} = 8^3$

38. $3^{x^2} = 9^2$

39. $27^{2x} = 81$

40. $125^{2x} = 625$

41. $8^x = \left(\dfrac{1}{2}\right)^{x-2}$

42. $343^x = \left(\dfrac{1}{49}\right)^{2-x}$

43. $13^{x^2-4} = 1$

44. $15^{9x^2-1} = 1$

45. $(\sqrt{2})^{x+2} = 8$

46. $(\sqrt{3})^{x-6} = \dfrac{1}{9}$

47. $25 \cdot 5^{2x} = 625$

48. $27 \cdot 3^{x+1} = 243$

49. $2^x \cdot 4^x = 64$

50. $2^x \cdot 4^{x-1} = 128$

51. $3^{-x} \cdot 9^x \cdot 3 = 81$

52. $4 \cdot 16^{x+1} \cdot 4^{-x} = 256$

53. $\dfrac{1}{49} \cdot 7^{2x+1} = 7 \cdot 7^x$

54. $\dfrac{1}{10^{x-2}} \cdot 100^x = 10^{x-1} \cdot 10^{2x-1}$

SECTION 10-1. Ten Review Exercises

In exercises **1–10**, simplify each expression. Assume all variables represent positive real numbers.

1. $t^3 \cdot t^2$

2. $5^{3t} \cdot 5^{2t}$

3. $\dfrac{u^5}{u^2}$; $u \neq 0$

4. $\dfrac{10^{5u}}{10^{2u}}$

5. $(v^3)^2$

6. $(7^{3v})^2$

7. $4\sqrt{5} - 2\sqrt{5} + \sqrt{5}$

8. $4k^{1/2} - 2k^{1/2} + k^{1/2}$

9. $(3\sqrt{2} + 5)^2$

10. $(3z^{1/2} + 5)^2$

SECTION 10-1. Supplementary Exercises

For $b > 0$ and $b \neq 1$, a graph of

$$y = b^{x-h}$$

is a graph of $y = b^x$ shifted horizontally $|h|$ units. The shift is to the right if $h > 0$ and to the left if $h < 0$.

In exercises **1–4**, graph each function.

1. $y = 2^{x-1}$

2. $y = 3^{x-2}$

3. $y = \left(\frac{1}{3}\right)^{x+2}$

4. $y = \left(\frac{1}{2}\right)^{x+1}$

For $b > 0$ and $b \neq 1$, a graph of

$$y = b^x + k$$

is a graph of $y = b^x$ shifted vertically $|k|$ units. The shift is upward if $k > 0$ and downward if $k < 0$.

In exercises **5–8**, graph each function.

5. $y = 2^x - 1$

6. $y = 3^x - 2$

7. $y = \left(\frac{1}{3}\right)^x + 2$

8. $y = \left(\frac{1}{2}\right)^x + 1$

In exercises **9–12**, graph each function.

9. $y = 2^{x-1} - 1$

10. $y = 3^{x-2} - 2$

11. $y = \left(\frac{1}{3}\right)^{x+2} + 2$

12. $y = \left(\frac{1}{2}\right)^{x+1} + 1$

In exercises **13–16**, use the following functions:

$$f \text{ is defined by } f(x) = 2^x$$

$$g \text{ is defined by } g(x) = x^2$$

13. a. Compute $f(4) - f(3)$.
b. Compute $g(4) - g(3)$.
c. Are the answers to parts **a** and **b** the same?

14. a. Compute $f(6) - f(4)$.
b. Compute $g(6) - g(4)$.
c. Are the answers to parts **a** and **b** the same?

15. a. Compute $f(x + 1) - f(x)$. (*Note:* $2^{x+1} = 2 \cdot 2^x$)
b. Compute $g(x + 1) - g(x)$.

16. a. Compute $f(x - 2) - f(x)$. $\left(\textit{Note: } 2^{x-2} = \frac{2^x}{2^2}\right)$
b. Compute $g(x - 2) - g(x)$.

SECTION 10-2. The Logarithmic Function

1. A definition of logarithmic functions
2. Finding y, when $y = \log_b x$
3. Finding x, when $y = \log_b x$
4. Finding b, when $y = \log_b x$
5. Graphs of logarithmic functions

As we saw in Section 10-1, exponential functions defined by $f(x) = b^x$, where $b > 0$ and $b \neq 1$, are one-to-one. As a consequence, the inverse of any exponential function is also a function. In this section, we will study the inverses of exponential functions.

A Definition of Logarithmic Functions

To write an equation that defines the inverse of an exponential function, we will first replace $f(x)$ by y. The equation of the inverse function can then be written by replacing x by y and y by x in the equation that defines the function.

Definition 10.2. The inverse of an exponential function
If f is a function defined by $y = b^x$, where $b > 0$ and $b \neq 1$, then f^{-1} is the function defined by $x = b^y$. The domain of f^{-1} is all x, where $x > 0$. The range of f^{-1} is all y.

Examples **a** and **b** illustrate two exponential functions and their inverses.

Exponential function	**Inverse of the exponential function**
a. $y = \left(\frac{1}{2}\right)^x$	**a*.** $x = \left(\frac{1}{2}\right)^y$
b. $y = 2^x$	**b*.** $x = 2^y$

To solve equations **a*** and **b*** for y in terms of x, we use the term **logarithm,** which is abbreviated "log".

For $b > 0$ and $b \neq 0$, y any real number, and $x > 0$:

(1) $x = b^y$ if and only if (2) $y = \log_b x$

Read $\log_b x$ as "log x base b".

Using this notation on equations **a*** and **b***:

Exponential form	**Logarithmic form**
a*. $x = \left(\frac{1}{2}\right)^y$	**a**.** $y = \log_{1/2} x$
b*. $x = 2^y$	**b**.** $y = \log_2 x$

Keep in mind that b, y, and x in equations (1) and (2) represent the same quantities:

b is a **base**

y is an **exponent**

x is equal to b^y

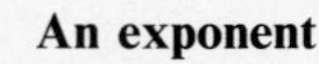

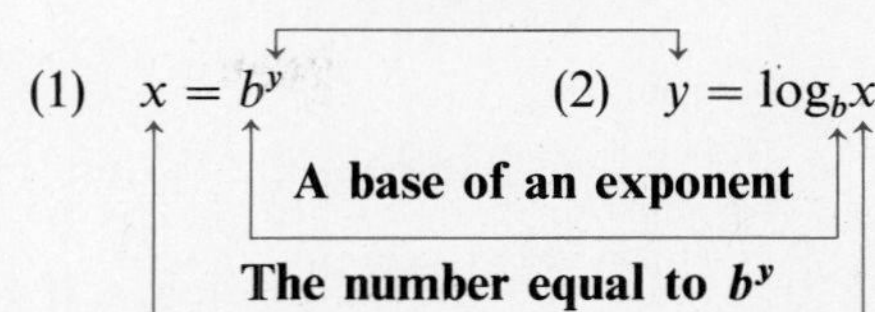

Example 1. Write $8 = 2^3$ in logarithmic form.

Solution. **Discussion.** The given equation is written in the form (1) $x = b^y$. We want to write it in the form (2) $y = \log_b x$.

$$8 = 2^3 \qquad \text{can be written} \qquad 3 = \log_2 8$$

Example 2. Write $-3 = \log_4 \frac{1}{64}$ in exponential form.

Solution. **Discussion.** The given equation is written in the form (2) $y = \log_b x$. We want to write it in the form (1) $x = b^y$.

$$-3 = \log_4 \frac{1}{64} \qquad \text{can be written} \qquad \frac{1}{64} = 4^{-3}$$

Finding y, When $y = \text{Log}_b x$

If $b > 0$ and $b \neq 1$, then $b^x = b^y$ if and only if $x = y$. This fact can be used to find y and $y = \log_b x$ for some values of b and x.

Example 3. Solve $y = \log_5 125$ for y.

Solution. **Step 1.** Write the equation in exponential form.

$125 = 5^y$ $y = \log_b x$ if and only if $x = b^y$

Step 2. Write both sides as power terms with the same base.

$5^3 = 5^y$ Write 125 as 5^3.

Step 3. Set the exponents equal and solve for y.

$3 = y$ Thus, $3 = \log_5 125$.

Finding x, When $y = \text{Log}_b x$

If b and y are known, then the exponential form of (2) $y = \log_b x$ can sometimes be used to find the value of x.

Example 4. Solve $\frac{-2}{3} = \log_8 x$ for x.

Solution. **Step 1.** Write the equation in exponential form.

$x = 8^{-2/3}$ $y = \log_b x$ if and only if $x = b^y$

Step 2. Simplify the right side.

$x = \frac{1}{4}$ Thus, $\frac{-2}{3} = \log_8 \frac{1}{4}$.

Finding b, When $y = \text{Log}_b x$

In (2), $y = \log_b x$. If x and y are known, then it may be possible to find the value of b.

Example 5. Solve $5 = \log_b 243$ for b.

Solution. **Step 1.** Write the equation in exponential form.

$243 = b^5$ $y = \log_b x$ if and only if $x = b^y$

Step 2. Write the left side as a power term with the same exponent as the power term on the right side.

$3^5 = b^5$ Write 243 as 3^5.

Step 3. Set the bases of the power terms equal.

$3 = b$ Thus, $5 = \log_3 243$.

Graphs of Logarithmic Functions

A graph of $y = \log_b x$ can be made using the exponential form of the equation. By selecting values for y and finding the corresponding values for x, a graph of the function can be sketched.

Example 6. Graph $y = \log_2 x$.

Solution. **Discussion.** Write the equation as $x = 2^y$. The following table contains seven solutions for the function. A smooth curve through the plots of these ordered pairs yields a graph of the function shown in Figure 10-3.

y	-3	-2	-1	0	1	2	3
$x = 2^y$	$\frac{1}{8}$	$\frac{1}{4}$	$\frac{1}{2}$	1	2	4	8

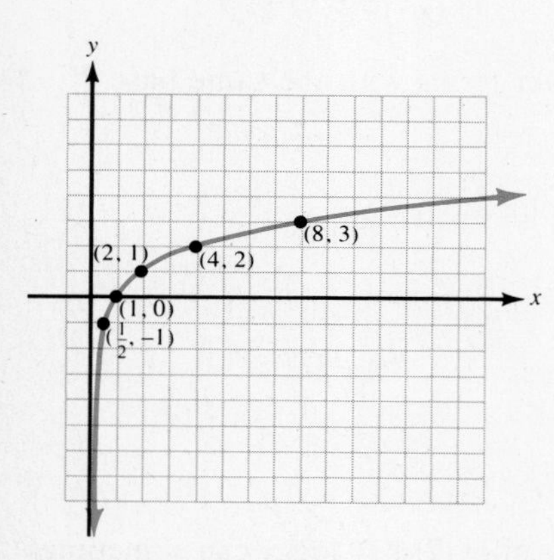

Figure 10-3. A graph of $y = \log_2 x$.

The graph in Figure 10-3 reveals the following characteristics of logarithmic functions base b, where $b > 1$.

Characteristic 1. For all values of y, $x > 0$. Thus, the domain is all x, where $x > 0$.

Characteristic 2. Since y can be positive, negative, or 0, the range is all y.

Characteristic 3. $P(1, 0)$ is a point on the graph. Thus the x-intercept is 1.

Characteristic 4. As $x \to 0^+$, $y \to -\infty$. Therefore, the y-axis is an asymptote for the graph.

Characteristic 5. The graph passes the vertical and horizontal line tests of a one-to-one function.

SECTION 10-2. Practice Exercises

In exercises **1–22**, write each equation in logarithmic form.

[Example 1]

1. $10^2 = 100$ **2.** $15^2 = 225$ **3.** $35^{1/5} = 2$

4. $27^{1/3} = 3$ **5.** $7^{-1} = \frac{1}{7}$ **6.** $5^{-2} = \frac{1}{25}$

7. $10^1 = 10$ **8.** $8^1 = 8$ **9.** $16^{3/4} = 8$

10. $216^{2/3} = 36$ **11.** $20^0 = 1$ **12.** $2^0 = 1$

13. $8^{-2/3} = \frac{1}{4}$ **14.** $9^{-3/2} = \frac{1}{27}$ **15.** $4^{2.5} = 32$

16. $25^{1.5} = 125$ **17.** $10^{0.4771} = 3$ **18.** $10^{0.6990} = 5$

19. $10^{3.6990} = 5{,}000$ **20.** $10^{2.4771} = 300$ **21.** $10^{-2.5229} = 0.003$

22. $10^{-0.3010} = 0.5$

In exercises **23–42**, write each equation in exponential form.

[Example 2]

23. $\log_{10} 1{,}000 = 3$ **24.** $\log_2 16 = 4$ **25.** $\log_5 \frac{1}{25} = -2$

26. $\log_3 \frac{1}{81} = -4$ **27.** $\log_8 8 = 1$ **28.** $\log_{20} 20 = 1$

29. $\log_{12} 1 = 0$ **30.** $\log_{10} 1 = 0$ **31.** $\log_{1/2} 64 = -6$

32. $\log_{1/5} 125 = -3$ **33.** $\log_{216} 1{,}296 = \frac{4}{3}$ **34.** $\log_4 8 = \frac{3}{2}$

35. $\log_8 4 = \frac{2}{3}$ **36.** $\log_{25} 5 = \frac{1}{2}$ **37.** $\log_9 \frac{1}{27} = \frac{-3}{2}$

38. $\log_4 \frac{1}{32} = \frac{-5}{2}$ **39.** $\log_{10} 6 = 0.7782$ **40.** $\log_{10} 8 = 0.9031$

41. $\log_{10} 400 = 2.6021$ **42.** $\log_{10} 7{,}000 = 3.8451$

In exercises **43–82**, find the value of each unknown.

[Examples 3–5]

43. $y = \log_2 64$ **44.** $y = \log_7 343$ **45.** $\log_5 x = 4$

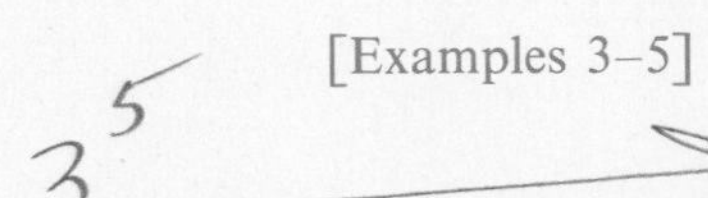

46. $\log_3 x = 5$ **47.** $\log_b 10{,}000 = 4$ **48.** $\log_b 216 = 3$

49. $y = \log_3 \frac{1}{27}$ **50.** $y = \log_2 \frac{1}{32}$ **51.** $\log_8 x = \frac{4}{3}$

52. $\log_9 x = \frac{3}{2}$ **53.** $\log_b 125 = \frac{3}{2}$ **54.** $\log_b 7 = \frac{1}{2}$

55. $y = \log_{12} 12$ **56.** $y = \log_{23} 23$ **57.** $\log_{20} x = -1$

58. $\log_4 x = \frac{-1}{2}$ **59.** $\log_b 0.01 = -2$ **60.** $\log_b 0.008 = -3$

61. $y = \log_{1/2}64$ **62.** $y = \log_{0.1}100$ **63.** $\log_2 x = -3$

64. $\log_6 x = -2$ **65.** $\log_b 9 = -2$ **66.** $\log_b 625 = -4$

67. $y = \log_7 1$ **68.** $y = \log_{13} 1$ **69.** $y = \log_2 4\sqrt{2}$

70. $y = \log_3 27\sqrt{3}$ **71.** $\log_{25} x = \frac{-3}{2}$ **72.** $\log_{100} x = \frac{-5}{2}$

73. $\log_b \frac{1}{128} = -7$ **74.** $\log_b \frac{8}{27} = -3$ **75.** $y = (\log_5 25)(\log_2 16)$

76. $y = \left(\log_2 \frac{1}{4}\right)\left(\log_5 \frac{1}{5}\right)$ **77.** $\log_{10} 10^{2x} = 4.3010$ **78.** $\log_{10} 10^{3x} = 1.4781$

79. $\log_b 10^{1.35} = 1.35$ **80.** $\log_b 5^{-0.63} = -0.63$ **81.** $y = 10^{\log_{10} 7.25}$

82. $y = 5^{\log_5 1.75}$

In exercises **83–90**, graph each equation.

[Example 6] **83.** $y = \log_3 x$ **84.** $y = \log_{1/3} x$ **85.** $y = \log_{1/4} x$

86. $y = \log_4 x$ **87.** $y = \log_{3/2} x$ **88.** $y = \log_{2/3} x$

89. $y = \log_{10} x$ **90.** $y = \log_{1/10} x$

SECTION 10-2. Ten Review Exercises

In exercises **1–6**, solve each equation.

1. $x^2 = 64$ **2.** $2^x = 64$ **3.** $x^3 = 27$

4. $3^x = 27$ **5.** $x^4 = 625$ **6.** $5^x = 625$

In exercises **7–10**, factor each expression.

7. $t^3 - 27$ **8.** $2t^3 - t^2 + 6t - 3$

9. $u^4 - 16$ **10.** $100 - x^2 + 2xy - y^2$

SECTION 10-2. Supplementary Exercises

For $b > 0$ and $b \neq 1$, a graph of

$$y = \log_b(x - h)$$

is a graph of $y = \log_b x$ shifted horizontally $|h|$ units. The shift is to the right if $h > 0$ and to the left if $h < 0$.

In exercises **1–4**, graph each function.

1. $y = \log_2(x - 1)$ **2.** $y = \log_3(x - 2)$

3. $y = \log_{1/3}(x + 2)$ **4.** $y = \log_{1/2}(x + 1)$

For $b > 0$ and $b \neq 1$, a graph of

$$y = \log_b x + k$$

is a graph of $y = \log_b x$ shifted vertically $|k|$ units. The shift is upward if $k > 0$ and downward if $k < 0$.

In exercises **5–8**, graph each function.

5. $y = \log_2 x - 1$

6. $y = \log_3 x - 2$

7. $y = \log_{1/3} x + 2$

8. $y = \log_{1/2} x + 1$

In exercises **9–12**, graph each function.

9. $y = \log_2(x - 1) - 1$

10. $y = \log_3(x - 2) - 2$

11. $y = \log_{1/3}(x + 2) + 2$

12. $y = \log_{1/2}(x + 1) + 1$

In exercises **13** and **14**, use the following functions:

f is defined by $f(x) = \sqrt{x}$

g is defined by $g(x) = \log_2 x$

13. **a.** Compute $f(64) - f(16)$
b. Compute $g(64) - g(16)$
c. Are the answers to parts **a** and **b** the same?

14. **a.** Compute $f(256) - f(4)$
b. Compute $g(256) - g(4)$
c. Are the answers to parts **a** and **b** the same?

SECTION 10-3. Properties of Logarithms

1. Properties of logarithms
2. Using the properties to change the form of $\log_b N$
3. Using the properties to change to the form $\log_b N$
4. Change-of-base theorem

There are several properties of exponents that can change the form of an exponential expression without changing its value. Since many properties of logarithms parallel the properties of exponents, the properties of logarithms can similarly change the form of a logarithmic expression without changing its value. We will study these properties in this section.

Properties of Logarithms

Example **a** illustrates the product-of-powers property of exponents.

a. $10^5 \cdot 10^3 = 10^{5+3} = 10^8$ When multiplying like bases, add the exponents.

Example **b** illustrates a property regarding the logarithm of a product.

b. $\log_{10}(5 \cdot 3) = \log_{10}5 + \log_{10}3$ — The logarithm of a product equals the sum of the logarithms.

In this section's Supplementary Exercises, the product property of exponents is used to verify the logarithm of a product property illustrated in example **b** (see p. 671). Similarly, the other properties of logarithms listed below can be verified using properties of exponents. The verifications are left as exercises.

In the following properties x, y, b, and r are real numbers, where $x > 0$, $y > 0$, and $b > 0$ and $b \neq 1$:

Property of logarithms	Example of property
1. $\log_b xy = \log_b x + \log_b y$	**1.** $\log_3(2 \cdot 5) = \log_3 2 + \log_3 5$
2. $\log_b \frac{x}{y} = \log_b x - \log_b y$	**2.** $\log_3 \frac{5}{2} = \log_3 5 - \log_3 2$
3. $\log_b x^r = r \log_b x$	**3.** $\log_3 2^5 = 5 \log_3 2$
4. $\log_b \frac{1}{x} = -\log_b x$	**4.** $\log_3 \frac{1}{5} = -\log_3 5$
5. $\log_b b = 1$	**5.** $\log_3 3 = 1$
6. $\log_b 1 = 0$	**6.** $\log_3 1 = 0$

Using the Properties to Change the Form of $\text{Log}_b N$

In Examples 1–5, the properties of logarithms are used to write expressions of the form $\log_b N$ to another equivalent form. Assume that variables are suitably restricted so that logarithms of only positive numbers are indicated.

Example 1. Write each expression as a sum of logarithms.

a. $\log_2 5t$ **b.** $\log_2(x^2 - 25)$

Solution. **Discussion.** In **a,** we assume $t > 0$. In **b,** we assume $x^2 - 25 > 0$.

a. $\log_2 5t = \log_2 5 + \log_2 t$ — $\log_b xy = \log_b x + \log_b y$

b. $\log_2(x^2 - 25) = \log_2(x + 5)(x - 5)$

$= \log_2(x + 5) + \log_2(x - 5)$

Example 2. Write each expression as a difference of logarithms.

a. $\log_2 \frac{t}{10}$ **b.** $\log_2 \frac{y}{y^2 + 1}$

Solution. **a.** $\log_2 \frac{t}{10} = \log_2 t - \log_2 10$ — $\log_b \frac{x}{y} = \log_b x - \log_b y$

b. $\log_2 \frac{y}{y^2 + 1} = \log_2 y - \log_2(y^2 + 1)$

Example 3. Write each expression as a product of a number and a logarithm.

a. $\log_5 3^4$ **b.** $\log_5 \sqrt{t^2 + 1}$

Solution. **a.** $\log_5 3^4 = 4 \log_5 3$ $\quad \log_b x^r = r \log_b x$

b. $\log_5 \sqrt{t^2 + 1} = \log_5 (t^2 + 1)^{1/2}$

$$= \frac{1}{2} \log_5 (t^2 + 1)$$

Example 4. Write $\log_3 \frac{1}{5}$ as the opposite of a logarithm.

Solution. $\log_3 \frac{1}{5} = -\log_3 5$ $\quad \log_b \frac{1}{x} = -\log_b x$

Example 5. Simplify each expression.

a. $\log_7 7$ **b.** $\log_7 1$

Solution. **a. Discussion.** Since $b^1 = b$ for any real number, $\log_7 7 = 1$.

b. Discussion. Since $b^0 = 1$ for $b \neq 0$, $\log_7 1 = 0$

Using the Properties to Change to the Form $\text{Log}_b N$

In Examples 6 and 7, the properties of logarithms are used to change some expressions to the form $\log_b N$. As before, when necessary assume that variables are suitably restricted so that logarithms of only positive numbers are indicated.

Example 6. Write each expression in the form $\log_b N$.

a. $\log_2 10 + \log_2 t$ **b.** $\log_2 x - \log_2 (x^2 + 5)$

Solution. **a.** $\log_2 10 + \log_2 t = \log_2 10t$ $\quad \log_b x + \log_b y = \log_b xy$

b. $\log_2 x - \log_2 (x^2 + 5) = \log_2 \frac{x}{x^2 + 5}$ $\quad \log_b x - \log_b y = \log_b \frac{x}{y}$

Example 7. Write $2 \log_3 5 + 4 \log_3 k$ in the form $\log_b N$.

Solution. $2 \log_3 5 + 4 \log_3 k$ $\quad$ The given expression

$= \log_3 5^2 + \log_3 k^4$ $\quad r \log_b x = \log_b x^r$

$= \log_3 25k^4$ $\quad$ In the form $\log_b N$

Change-of-Base Theorem

Any positive number different from 1 can be used as the base of a logarithm. In practice, as we will see in Section 10-4, only 10 and an irrational number e are commonly used as bases. With a table or a calculator, the logarithm of a positive number for base 10 and base e can be approximated. The **change-of-base theorem** can be used to write an indicated logarithm of a number to any suitable base as a ratio of logarithms with base 10 or base e.

Change-of-base theorem
If $x > 0$, $b > 0$ and $b \neq 1$, and $k > 0$ and $k \neq 1$, then:

$$\log_b x = \frac{\log_k x}{\log_k b}$$

Example 8. Write $\log_2 70$ in terms of base 10 logarithms.

Solution. $\log_2 70 = \dfrac{\log_{10} 70}{\log_{10} 2}$ $\quad \log_b x = \dfrac{\log_{10} x}{\log_{10} b}$

Example 9. Write $\log_5 340$ in terms of base e logarithms.

Solution. $\log_5 340 = \dfrac{\log_e 340}{\log_e 5}$ $\quad \log_b x = \dfrac{\log_e x}{\log_e b}$

SECTION 10-3. Practice Exercises

In exercises **1–30**, rewrite each expression in an equivalent or simplified form by using one or more of the properties of logarithms. Assume all variables are positive real numbers.

[Examples 1–5]

1. $\log_{10} 2(3)$ **2.** $\log_{10} 5(6)$ **3.** $\log_2 \dfrac{5}{8}$ **4.** $\log_2 \dfrac{3}{10}$

5. $\log_3 7^2$ **6.** $\log_3 5^4$ **7.** $\log_{10} \dfrac{1}{9}$ **8.** $\log_{10} \dfrac{1}{20}$

9. $\log_2 2$ **10.** $\log_8 8$ **11.** $\log_\pi 1$ **12.** $\log_{1/3} 1$

13. $\log_3 7(10)$ **14.** $\log_3 \dfrac{7}{10}$ **15.** $\log_3 10^7$ **16.** $\log_3 \sqrt[7]{10}$

17. $\log_{10} \dfrac{3}{7}$ **18.** $\log_{10} 3(7)$ **19.** $\log_{10} \sqrt[3]{7}$ **20.** $\log_{10} 3^7$

21. $\log_2 \dfrac{6(5)}{11}$ **22.** $\log_2 \dfrac{6}{5(11)}$ **23.** $\log_3 \sqrt{5x}$ **24.** $\log_3 \dfrac{\sqrt{5}}{x}$

25. $\log_5 10x^2$ **26.** $\log_5 x^2 y^3$ **27.** $\log_{10} \dfrac{\sqrt[3]{2}}{x^2}$ **28.** $\log_{10} \dfrac{xy}{\sqrt{5}}$

29. $\log_2 \dfrac{1}{7x}$ **30.** $\log_2 \dfrac{1}{3x^3}$

In exercises **31–60**, write each expression in the form $\log_b N$, if possible. Assume all variables are positive real numbers.

[Examples 6 and 7]

31. $\log_{10} 2 + \log_{10} 3$ **32.** $\log_{10} 7 + \log_{10} 5 + \log_{10} 3$

33. $\log_3 5 + \log_3 t$ **34.** $\log_3 7 + \log_3 x + \log_3 y$

35. $\log_2 7 - \log_2 10$ **36.** $\log_2 5 - \log_2 6$

37. $\log_5 6 + \log_5 t - \log_5 7$

38. $\log_5 3 - (\log_5 4 + \log_5 u)$

39. $2 \log_3 5$

40. $5 \log_3 2$

41. $\frac{1}{2} \log_{10} 6$

42. $\frac{1}{2} \log_{10} 8$

43. $2(\log_5 3 + \log_5 x)$

44. $3(\log_5 2 + \log_5 x)$

45. $3(\log_2 y - \log_2 5)$

46. $2(\log_2 x - \log_2 y)$

47. $\frac{1}{2} \log_3 10 + \log_3 x$

48. $2 \log_3 5 + 3 \log_3 y$

49. $\frac{1}{3} \log_{10} 5 - 2 \log_{10} 3$

50. $\frac{1}{2} \log_{10} 3 - 5 \log_{10} 2$

51. $\frac{1}{2}(\log_2 3 + \log_2 x) - \log_2 10$

52. $\frac{1}{2} \log_2 6 - 2(\log_2 10 + \log y)$

53. $(\log_3 2 + \log_3 x) - (\log_3 11 + \log_3 y)$

54. $(\log_3 10 + \log_3 y) - (\log_3 7 + \log_3 x)$

55. $\frac{1}{2}(\log_5 10 + \log_5 y) - (\log_5 3 + \log_5 x)$

56. $3 \log_5 2 - (\log_5 3 + \log_5 x + \log_5 y)$

57. $(3 \log_2 t + \log_2 10) - (2 \log_2 u + \log_2 7)$

58. $\left(\frac{1}{2} \log_2 t + \log_2 5\right) - (3 \log_2 u + \log_2 9)$

59. $(\log_3 4 + \log_3 k)(\log_3 10 + \log_3 k)$

60. $(\log_3 10 + 2 \log_3 k)(\log_3 20 + 3 \log_3 k)$

In exercises **61–68**, write each logarithm in terms of:

a. Base 10 logarithms

b. Base e logarithms

[Examples 8 and 9]

61. $\log_9 5$

62. $\log_7 4$

63. $\log_2 23$

64. $\log_3 29$

65. $\log_5 \frac{1}{2}$

66. $\log_{11} \frac{1}{5}$

67. $\log_{2/3} 17$

68. $\log_{4/3} 31$

SECTION 10-3. Ten Review Exercises

In exercises **1** and **2**, evaluate each expression for the indicated variable.

1. $t^2 - 2t - 8$, for $t = 4$
2. $t^2 - 8 - 2t$, for $t = -2$
3. Solve $t^2 = 2t + 8$
4. Write $t^2 - 2t - 8$ in the form $(t - h)^2 + k$.
5. Graph $y = x^2 - 2x - 8$.

In exercises **6–10**, do the indicated operations.

6. $(3 + 2y)(2 - y)$

7. $(3 + 2\sqrt{5})(2 - \sqrt{5})$

8. $(3 + 2i)(2 - i)$

9. $\dfrac{3 + 2\sqrt{5}}{2 - \sqrt{5}}$

10. $\dfrac{3 + 2i}{2 - i}$

SECTION 10-3. Supplementary Exercises

In exercises **1–20**, assume that for some base b, the following logarithms are correct to two decimal places:

$$\log_b 2 \approx 0.30 \qquad \log_b 3 \approx 0.48 \qquad \log_b 5 \approx 0.70$$

Approximate each logarithm to two decimal places.

Example. **a.** $\log_b \frac{5}{2}$ **b.** $\log_b 81$

Solution. **a.** $\log_b \frac{5}{2} = \log_b 5 - \log_b 2$

$= 0.70 - 0.30$

$= 0.40$

b. $\log_b 81 = \log_b 3^4$

$= 4 \log_b 3$

$= 4(0.48)$

$= 1.92$

1. $\log_b 6$

2. $\log_b 15$

3. $\log_b 0.4$

4. $\log_b 0.6$

5. $\log_b 4$

6. $\log_b 9$

7. $\log_b 125$

8. $\log_b 32$

9. $\log_b \sqrt{3}$

10. $\log_b \sqrt[3]{2}$

11. $\log_b 30$

12. $\log_b 1.2$

13. $\log_b \frac{1}{2}$

14. $\log_b \frac{1}{3}$

15. $\log_b 2(5)$

16. $\log_b \frac{5}{2}$

17. $\log_b \sqrt{8}$

18. $\log_b \sqrt{27}$

19. $\log_b \sqrt[3]{225}$

20. $\log_b \sqrt[3]{128}$

In exercises **21–28**, use the definition of a logarithm to show that the left and right expressions are not the same number.

21. $(\log_2 4)(\log_2 8) \neq \log_2 32$

22. $(\log_5 125)\left(\log_5 \frac{1}{5}\right) \neq \log_5(25)$

23. $\dfrac{\log_3 27}{\log_3 3} \neq \log_3 9$

24. $\dfrac{\log_{10} 100}{\log_{10} 100} \neq \log_{10} 1$

25. $\log_2(2 \cdot 4^2) \neq 2 \log_2(2 \cdot 4)$

26. $\log_3(3^4 \cdot 3) \neq 4 \log_3(3 \cdot 3)$

27. $\log_7 7 \neq 7^1$

28. $\log_{1/3} \frac{1}{3} \neq \left(\frac{1}{3}\right)^{-1}$

In exercises **29–36**, simplify each expression. Assume all variables are positive real numbers.

29. $\dfrac{\log_2 5 + \log_2 3}{\log_2 6 + \log_2 10}$

30. $\dfrac{\log_2 7 + \log_2 x}{\log_2 10 + \log_2 y}$

31. $(\log_{10} 2 + \log_{10} x)(\log_{10} 3 + \log_{10} y)$

32. $\log_{10} 4 - \log_{10} x)(\log_{10} y - \log_{10} 5)$

33. $\dfrac{8 \log_2 x}{2 \log_2 10}$

34. $\dfrac{3 \log_2 y}{15 \log_2 3}$

35. $(2 \log_3 6 + \log_3 x - \log_3 5)\left(\dfrac{1}{2} \log_3 10 + \log_3 y - \log_3 7\right)$

36. $(3 \log_3 2 + \log_3 x + \log_3 3)(2 \log_3 5 + \log_3 y + \log_3 3)$

Example. Verify that $\log_b(x \cdot y) = \log_b x + \log_b y$, where x and y are positive real numbers, and $b > 0$ and $b \neq 1$.

Solution. Let $M = \log_b x$ and $N = \log_b y$.

$b^M = x$ and $b^N = y$	Write in exponential form.
$b^M \cdot b^N = x \cdot y$	Multiply the equations.
$b^{M+N} = x \cdot y$	Add the exponents.
$M + N = \log_b(x \cdot y)$	Write in logarithmic form.
$\log_b x + \log_b y = \log_b(x \cdot y)$	Replace M by $\log_b x$ and N by $\log_b y$.

In exercises **37–41**, verify each property of logarithms.

37. $\log_b \dfrac{x}{y} = \log_b x - \log_b y$ (See Example above.)

38. $\log_b x^y = y \log_b x$ (Let $\log_b x = A$.)

39. $\log_b \dfrac{1}{x} = -\log_b x$ (Write $\dfrac{1}{x}$ as x^{-1}.)

40. $\log_b b = 1$ (Write in exponential form.)

41. $\log_b 1 = 0$ (Write in exponential form.)

SECTION 10-4. Common and Natural Logarithms

1. A definition of common logarithms
2. Finding $\log x$, for $1 \leq x \leq 10$
3. Finding $\log x$, for $0 < x < 1$ or $x > 10$
4. A definition of antilog x
5. A definition of natural logarithms
6. Finding $\ln x$ and antiln x

The positive numbers 10 and e (an irrational number) are frequently used as bases for logarithms. Tables and calculators can be used to approximate logarithms of positive numbers using these bases. We will study both bases in this section.

A Definition of Common Logarithms

Logarithms based on 10 are called **common logarithms.**

> **Definition 10.3. Common logarithms**
> If $b = 10$, then $\log_b x$ is written $\log x$. That is:
>
> $$y = \log x \quad \text{if and only if} \quad x = 10^y$$

Example 1. Write each equation in logarithmic form.

a. $10{,}000 = 10^4$ **b.** $0.01 = 10^{-2}$

Solution. **a.** $10{,}000 = 10^4$ can be written $4 = \log 10{,}000$

b. $0.01 = 10^{-2}$ can be written $-2 = \log 0.01$

Example 2. Write each equation in exponential form.

a. $\log 5.38 = 0.7308$ **b.** $\log 0.193 = -0.7144$

Solution. **a.** $\log 5.38 = 0.7308$ can be written $5.38 = 10^{0.7308}$

b. $\log 0.193 = -0.7144$ can be written $0.193 = 10^{-0.7144}$

As we will see, log 5.38 is equal to 0.7308, to four decimal places. Thus, it would be more accurate to write:

$$\log 5.38 \approx 0.7308$$

However, the popular convention is to acknowledge that the logarithms of most numbers using base 10 are **irrational numbers.** Thus, any stated value must be an approximation, and an equals sign is used even though the "approximately equal to" symbol would be more correct.

Finding log x, for $1 \leq x \leq 10$

In Figure 10-4, a graph of $y = \log x$ for $1 \leq x \leq 10$ is shown. Notice the following details in the graph:

Detail 1. If $x = 1$, then $y = 0$. That is, $\log 1 = 0$. $10^0 = 1$

Detail 2. If $x = 10$, then $y = 1$. That is, $\log 10 = 1$. $10^1 = 10$

Detail 3. If $1 < x < 10$, then $0 < y < 1$, or $0 < \log x < 1$. $0 < 10^y < 1$

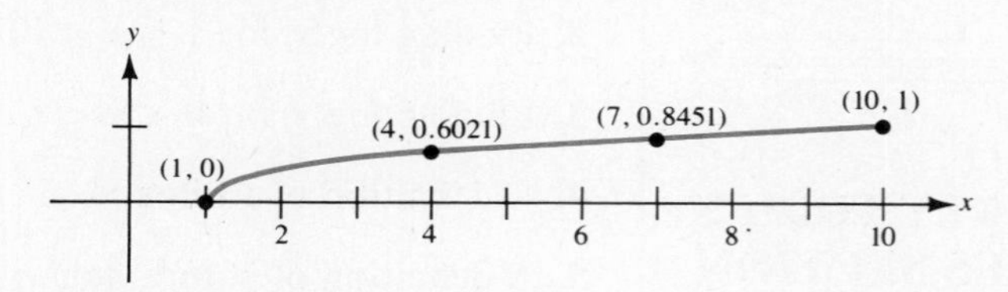

Figure 10-4. A graph of $y = \log x$ and $1 \leq x \leq 10$.

Appendix E contains a four-digit approximation of the base 10 logarithms of three-digit numbers from 1.00 through 9.99. A portion of this table is shown below.

Table 10-1. A Portion of Appendix E.

x	0	1	2	3	4	5	6	7	8	9
1.0	0.0000	0.0043	0.0086	0.0128	0.0170	0.0212	0.0253	0.0294	0.0334	0.0374
1.1	0.0414	0.0453	0.0492	0.0531	0.0569	0.0607	0.0645	0.0682	0.0719	0.0755
1.2	0.0792	0.0828	0.0864	0.0899	0.0934	0.0969	0.1004	0.1038	0.1072	0.1106
1.3	0.1139	0.1173	0.1206	0.1239	0.1271	0.1303	0.1335	0.1367	0.1399	0.1430
1.4	0.1461	0.1492	0.1523	0.1553	0.1584	0.1614	0.1644	0.1673	0.1703	0.1732
1.5	0.1761	0.1790	0.1818	0.1847	0.1875	0.1903	0.1931	0.1959	0.1987	0.2014
1.6	0.2041	0.2068	0.2095	0.2122	0.2148	0.2175	0.2201	0.2227	0.2253	0.2279
1.7	0.2304	0.2330	0.2355	0.2380	0.2405	0.2430	0.2455	0.2480	0.2504	0.2529
1.8	0.2553	0.2577	0.2601	0.2625	0.2648	0.2672	0.2695	0.2718	0.2742	0.2765
1.9	0.2788	0.2810	0.2833	0.2856	0.2878	0.2900	0.2923	0.2945	0.2967	0.2989
2.0	0.3010	0.3032	0.3054	0.3075	0.3096	0.3118	0.3139	0.3160	0.3181	0.3201
2.1	0.3222	0.3243	0.3263	0.3284	0.3304	0.3324	0.3345	0.3365	0.3385	0.3404
2.2	0.3424	0.3444	0.3464	0.3483	0.3502	0.3522	0.3541	0.3560	0.3579	0.3598
2.3	0.3617	0.3636	0.3655	0.3674	0.3692	0.3711	0.3729	0.3747	0.3766	0.3784
2.4	0.3802	0.3820	0.3838	0.3856	0.3874	0.3892	0.3909	0.3927	0.3945	0.3962
2.5	0.3979	0.3997	0.4014	0.4031	0.4048	0.4065	0.4082	0.4099	0.4116	0.4133
2.6	0.4150	0.4166	0.4183	0.4200	0.4216	0.4232	0.4249	0.4265	0.4281	0.4298
2.7	0.4314	0.4330	0.4346	0.4362	0.4378	0.4393	0.4409	0.4425	0.4440	0.4456
2.8	0.4472	0.4487	0.4502	0.4518	0.4533	0.4548	0.4564	0.4579	0.4594	0.4609
2.9	0.4624	0.4639	0.4654	0.4669	0.4683	0.4698	0.4713	0.4728	0.4742	0.4757

Example 3. Find log 2.48.

Solution.

Table method

Locate 2.4 under the column headed x on Table 10-1. Now move to the right along this row to the entry under the column headed 8. Read 0.3945.
log 2.48 = 0.3945

Calculator method

Enter 2.48 in the calculator. Press [log]. The display shows 0.394451681. To nine decimal places:
log 2.48 = 0.394451681

Keep in mind that log x is an exponent on 10 that yields x. Therefore, based on the results of Example 3, the following approximations can be written:

$$10^{0.3945} \approx 2.48 \qquad \text{and} \qquad 10^{0.394451681} \approx 2.48$$

Finding log x, for $0 < x < 1$ or $x > 10$

Appendix E can also be used to approximate logarithms of positive numbers not listed in the table by first writing them in scientific notation. By definition, x is written in scientific notation when it is written as:

$$x = N \times 10^k, \quad \text{where } 1 \leq N < 10 \text{ and } k \text{ is an integer.}$$

Example 4. Find log 24,800.

Solution.

Table method

$24{,}800 = 2.48 \times 10^4$	In scientific notation
$\log(2.48 \times 10^4) = \log 2.48 + \log 10^4$	$\log x \cdot y = \log x + \log y$
$= 0.3945 + 4$	$\log 10^4 = 4 \log 10 = 4$
$= 4.3945$	$\log 24{,}800 = 4.3945$

Calculator method

Enter 24,800 in the calculator and press [log]. The display shows: — Scientific notation is not needed.

4.394451681 — $\log 24{,}800 = 4.394451681$

Example 5. Find log 0.00248.

Solution.

Table method

$0.00248 = 2.48 \times 10^{-3}$	In scientific notation
$\log 0.00248 = \log 2.48 + \log 10^{-3}$	$\log x \cdot y = \log x + \log y$
$= 0.3945 + (-3)$	$\log 10^{-3} = -3 \log 10 = -3$
$= -2.6055$	$\log 0.00248 = -2.6055$

Calculator method

Enter 0.00248 in the calculator and press [log].

$\log 0.00248 = -2.605548319.$

When $\log x$ is written in the form $\log N + \log 10^k$, the term $\log N$ is called the **mantissa** and k is called the **characteristic.** In Examples 4 and 5, the mantissa is 0.3945. In Example 4, the characteristic is 4 and in Example 5 the characteristic is -3.

A Definition of Antilog x

If $y = \log x$, then by definition, $10^y = x$. If y is known, then x is 10 raised to the power y. For example, consider examples **a–d**.

	$y = \log x$	Given value of y	$x = 10^y$
a.	$5 = \log x$	5	$x = 10^5 = 100{,}000$
b.	$-2 = \log x$	-2	$x = 10^{-2} = 0.01$
c.	$0.4082 = \log x$	0.4082	$x = 10^{0.4082} = 2.56$
d.	$-1.0742 = \log x$	-1.0742	$x = 10^{-1.0742} = 0.0843$

In examples **a–d**, the values of x are called the **antilogarithms of y,** written "antilog y".

> **Definition 10.4.**
> If $y = \log x$, where x is a positive number, then:
>
> $$x = \text{antilog } y \qquad \text{and} \qquad x = 10^y$$

There are three commonly used ways to indicate an antilogarithm.

If $y = \log x$, then → $x = \text{antilog } y$.
→ $x = 10^y$.
→ $x = \text{Inv log } y$.

Read Inv log y as "inverse log y", which emphasizes that the logarithm function is the inverse of the exponential function.

Example 6. Find antilog 2.6493.

Solution.

Table method

$\text{antilog } 2.6493 = \text{antilog}(0.6493 + 2)$	Separate mantissa and characteristic.
$= \text{antilog } 0.6493 \times \text{antilog } 2$	Write as a product.
$= 4.46 \times 10^2$	$\log 4.46 = 0.6493$
$= 446$	$\log 446 = 2.6493$

Calculator method

Enter 2.6493 in the calculator and press $\boxed{10^x}$ key.

To three significant figures: antilog 2.6493 = 446

If $x = \text{antilog } y$ and $y < 0$, then a positive integer k (1, 2, 3, 4, . . .) must be added and subtracted to y to give a mantissa between 0 and 1. When y is changed to this form, Appendix E can be used to find $\log N$. The corresponding k is the exponent on 10. Examples **e–g** illustrate three choices of k and the corresponding changes on the form of y.

Antilog y	**k**	**Antilog $[(y + k) - k]$**
e. antilog(-0.6308)	1	$\text{antilog}[(-0.6308 + 1) - 1] = \text{antilog}(0.3692 - 1)$
f. antilog(-1.3726)	2	$\text{antilog}[(-1.3726 + 2) - 2] = \text{antilog}(0.6274 - 2)$
g. antilog(-2.1261)	3	$\text{antilog}[(-2.1261 + 3) - 3] = \text{antilog}(0.8739 - 3)$

Example 7. Find antilog(-2.1261).

Solution.

Table method

$\text{antilog}(-2.1261) = \text{antilog}(0.8739 - 3)$	Add and subtract 3.
$= \text{antilog } 0.8739 \times 10^{-3}$	The charactistic is -3.
$= 7.48 \times 10^{-3}$	$\log 7.48 = 0.8739$
$= 0.00748$	$\log 0.00748 = -2.1261$

Calculator method

Enter -2.1261 in the calculator and press $\boxed{10^x}$.

To three significant digits: $\text{antilog}(-2.1261) = 0.00748$

A Definition of Natural Logarithms

In many areas of applied mathematics, logarithms of numbers with base e are used. The number e is irrational. It is denoted e in honor of the Swiss mathematician Leonhard Euler (1707–1783). With high-speed computers, the approximation of e has been calculated to thousands of decimal places. For our purposes however:

$$e \approx 2.718$$

Definition 10.5.
If $b = e$, then $\log_e x$ is written "ln x". That is:

$$y = \ln x \qquad \text{if and only if} \qquad x = e^y$$

Values for $y = e^x$ and $y = e^{-x}$ for $0 \le x \le 10.0$ are given in Appendix F. Graphs of $y = e^x$ and $y = \ln x$ are shown in Figure 10.5.

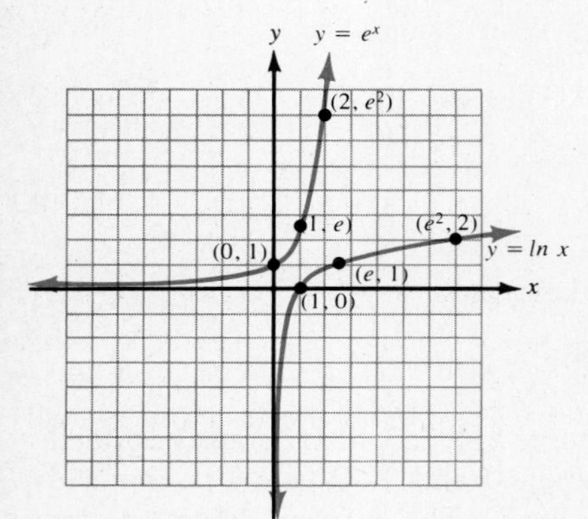

Figure 10-5. Graphs of $y = e^x$ and $y = \ln x$.

Finding Ln x and Antiln x

There are two commonly used methods for finding ln x and $x > 0$.

Method 1. Use a calculator with an $\boxed{\ln}$ key.

Method 2. Use a table of common logarithms and a **conversion factor 2.303.**

The conversion factor 2.303 is obtained by using the change-of-base theorem stated in Section 10-3 (see pp. 667–68).

$$\ln x = \frac{\log x}{\log e} \qquad \log_b x = \frac{\log x}{\log b}, \text{ and } b = e$$

$$= \frac{\log x}{0.43429} \qquad \text{To five places, } \log e = 0.43429.$$

$$\approx 2.303 \log x \qquad \text{To four significant digits}$$

Comparing methods 1 and 2, the calculator method is by far the easier one, and is demonstrated in Examples 8 and 9. The table method is illustrated in the Supplementary Exercises of this section (see p. 679).

Example 8. Find ln 420 to four decimal places.

Solution.

Calculator method

Enter 420 in the calculator and press $\boxed{\ln}$. The display shows:

$$6.040254711 = 6.0403, \text{ to four decimal places}$$

Thus, $e^{6.0403} \approx 420$.

Example 9. Find antiln 3.7416 to two significant digits.

Solution. **Calculator method**

Enter 3.7416 in the calculator and press $\boxed{e^x}$. The display shows:

$$42.16540086 = 42, \text{ to two significant digits}$$

Thus, $\ln 42 \approx 3.7416$.

SECTION 10-4. Practice Exercises

In exercises **1–10**, write each equation in logarithmic form.

[Example 1]

1. $10^2 = 100$ **2.** $10^3 = 1{,}000$

3. $10^{-5} = 0.00001$ **4.** $10^{-1} = 0.1$

5. $10^6 = 1{,}000{,}000$ **6.** $10^5 = 100{,}000$

7. $10^{1.5} = 31.6$ **8.** $10^{3.2} = 1{,}580$

9. $10^{-2.1} = 0.0079$ **10.** $10^{-0.8} = 0.158$

In exercises **11–20**, write each equation in exponential form.

[Example 2]

11. $\log 15 = 1.1761$ **12.** $\log 32 = 1.5051$

13. $\log 1.48 = 0.1703$ **14.** $\log 8.15 = 0.9112$

15. $\log 0.695 = -0.1580$ **16.** $\log 0.075 = -1.1249$

17. $\log 4500 = 3.6532$ **18.** $\log 125{,}000 = 5.0969$

19. $\log 0.0009 = -3.0458$ **20.** $\log 0.0056 = -2.2518$

In exercises **21–40**, find $\log x$. Write answers to four decimal places.

[Examples 3–5]

21. log 4.95 **22.** log 8.03 **23.** log 1.92

24. log 6.78 **25.** log 75 **26.** log 43

27. log 147 **28.** log 324 **29.** log 6,300

30. log 9,060 **31.** log 37,100 **32.** log 64,700

33. log 0.56 **34.** log 0.08 **35.** log 0.0234

36. log 0.0421 **37.** log 0.0083 **38.** log 0.00107

39. log 0.000092 **40.** log 0.00000539

In exercises **41–60**, find antilog y. Write answers to three significant digits.

[Examples 6 and 7]

41. antilog 0.5740 **42.** antilog 0.7945 **43.** antilog 0.9624

44. antilog 0.1703 **45.** antilog 1.8976 **46.** antilog 1.5563

47. antilog 2.7110 **48.** antilog 2.9143 **49.** antilog 4.2765

50. antilog 4.8041 **51.** antilog(−0.1180) **52.** antilog(−0.8416)

53. antilog(−2.3820) **54.** antilog(−2.1325) **55.** antilog(−4.0200)

56. antilog(−4.6216) **57.** antilog(−1.5143) **58.** antilog(−1.2197)

59. antilog(−3.1599) **60.** antilog(−3.5287)

In exercises **61–80**, approximate ln x. Write answers to four decimal places.

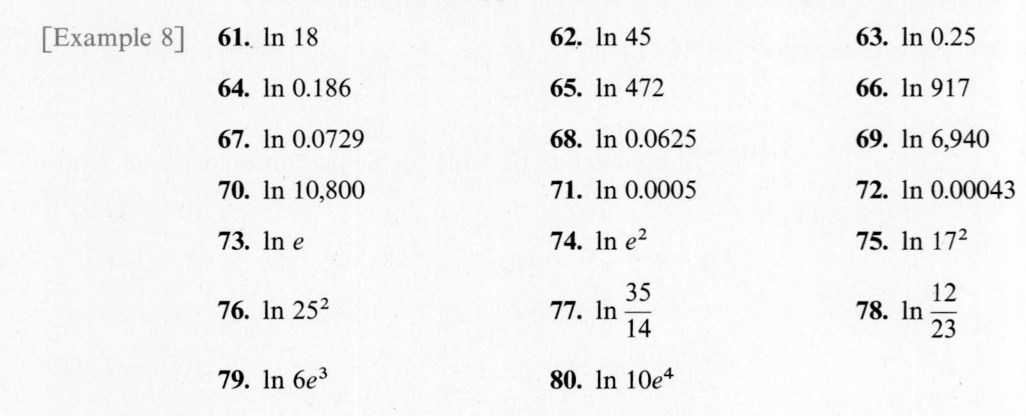

[Example 8] **61.** ln 18 **62.** ln 45 **63.** ln 0.25

64. ln 0.186 **65.** ln 472 **66.** ln 917

67. ln 0.0729 **68.** ln 0.0625 **69.** ln 6,940

70. ln 10,800 **71.** ln 0.0005 **72.** ln 0.00043

73. $\ln e$ **74.** $\ln e^2$ **75.** $\ln 17^2$

76. $\ln 25^2$ **77.** $\ln \frac{35}{14}$ **78.** $\ln \frac{12}{23}$

79. $\ln 6e^3$ **80.** $\ln 10e^4$

In exercises **81–100**, approximate antiln y. Write answers to three significant digits.

[Example 9] **81.** antiln 1.25 **82.** antiln 0.72 **83.** antiln 0.18

84. antiln 2.21 **85.** antiln 2.48 **86.** antiln 4.51

87. antiln 11.50 **88.** antiln 9.62 **89.** antiln(−0.35)

90. antiln(−0.90) **91.** antiln(−1.43) **92.** antiln(−1.63)

93. antiln(−2.04) **94.** antiln(−3.60) **95.** antiln(−3.75)

96. antiln(−5.89) **97.** antiln 0 **98.** antiln 1

99. antiln e **100.** antiln 10

SECTION 10-4. Ten Review Exercises

In exercises **1–10**, simplify each expression.

1. $(-x^2y)^{-2}(x^{-3}y^2)^{-1}$ **2.** $(6x^{-1}y)^2(3x^2y)^{-2}$

3. $2\sqrt{3} - \sqrt{75}$ **4.** $(2\sqrt{3})(\sqrt{75})$

5. $\frac{\sqrt{75}}{2\sqrt{3}}$ **6.** $(\sqrt{75})(2\sqrt{3})^2$

7. $(2u + 3)^2 - (u + 2)(4u - 1)$ **8.** $\frac{4u^2 + 12u + 9}{4u^2 - 9}$

9. $(2 + 3i)^2 - (1 + 2i)(4 - i)$

10. $\dfrac{2 + 3i}{1 + 2i}$

SECTION 10-4. Supplementary Exercises

Example. Find $\log_5 28$.

Solution.

$\log_5 28$ — The given problem

$= \dfrac{\log_{10} 28}{\log_{10} 5}$ — $\log_b x = \dfrac{\log_k x}{\log_k b}$

$= \dfrac{1.447158\ldots}{0.698970\ldots}$ — Do not round the logs.

$= 2.070415\ldots$ — Do the indicated division.

≈ 2.07 — Round to two places.

In exercises **1–20**, use the change-of-base theorem to approximate each logarithm to three significant digits.

1. $\log_2 43$ **2.** $\log_2 225$ **3.** $\log_3 0.7$

4. $\log_3 0.012$ **5.** $\log_5 810$ **6.** $\log_5 64$

7. $\log_6 0.09$ **8.** $\log_6 0.173$ **9.** $\log_8 5.3$

10. $\log_8 49.1$ **11.** $\log_9 8{,}400$ **12.** $\log_9 10{,}600$

13. $\log_{12} 2{,}500{,}000$ **14.** $\log_{12} 6{,}750{,}000$ **15.** $\log_{1/2} 13$

16. $\log_{1/2} 45$ **17.** $\log_{1/3} 0.2$ **18.** $\log_{1/3} 0.08$

19. $\log_{1/5} 63.5$ **20.** $\log_{1/5} 182$

In exercises **21–26**, use four digits and Appendix E to approximate each logarithm. Write $\ln x$ as $2.303 \log x$.

21. ln 109 **22.** ln 473 **23.** ln 8.83

24. ln 1.95 **25.** ln 0.108 **26.** ln 0.806

In exercises **27–32**, use antilogs to approximate each antiln to three significant digits. Write antiln y as antilog $0.4343\, y$.

27. antiln 3.19 **28.** antiln 7.52 **29.** antiln 0.23

30. antiln 0.784 **31.** antiln(-2.74) **32.** antiln(-1.59)

Common logarithms were frequently used to do computations before the hand-held calculator became so readily available.

Example. Compute $\sqrt[3]{5{,}830}$ to three significant digits.

Solution. Let $x = \sqrt[3]{5{,}830}$.

$\log x = \log \sqrt[3]{5{,}830}$	Take the log of both sides.
$= \frac{1}{3} \log 5{,}830$	$\log x^r = r \log x$
$= \frac{1}{3}(3.7657)$	$\log 5{,}830 = 3.7657$
$= 1.2552$	Do the indicated multiplication.
$x = \text{antilog } 1.2552$	$\log 1.80 = 0.2553$
$= 18.0$	

Thus, $\sqrt[3]{5{,}830} = 18.0$ to three significant digits.

In exercises **33–44**, use logarithms to do the indicated operations. Write answers correct to three significant digits.

33. $\sqrt{295}$ **34.** $\sqrt{488}$ **35.** $\sqrt[3]{5{,}480}$

36. $\sqrt[3]{7{,}320}$ **37.** $\sqrt[5]{28}$ **38.** $\sqrt[4]{42}$

39. (4,800)(0.0096) **40.** (0.0715)(1,650) **41.** (243)(0.054)

42. (908)(0.29) **43.** (0.869)(3,590) **44.** (74,900)(0.0415)

SECTION 10-5. Exponential and Logarithmic Equations

1. Some properties of exponential and logarithmic functions reviewed
2. A procedure to solve exponential equations
3. A procedure to solve logarithmic equations

Examples **a** and **b** are exponential equations that can be solved by the procedure studied in Section 10-1 (see pp. 656–57).

a.

$$8^{x+1} = 32^{x-1}$$
$$(2^3)^{x+1} = (2^5)^{x-1}$$
$$2^{3x+3} = 2^{5x-5}$$
$$3x + 3 = 5x - 5$$
$$8 = 2x$$
$$4 = x$$

The solution set is $\{4\}$.

b.

$$\left(\frac{1}{9}\right)^x = 27^{x-1}$$
$$(3^{-2})^x = (3^3)^{x-1}$$
$$3^{-2x} = 3^{3x-3}$$
$$-2x = 3x - 3$$
$$3 = 5x$$
$$\frac{3}{5} = x$$

The solution set is $\left\{\frac{3}{5}\right\}$.

The procedure demonstrated was possible because the expressions on both sides of the equation could be written in terms of a common base. In example **a**, the common base was 2, and in example **b**, the common base was 3.

Examples **c** and **d** are exponential equations in which the common base to use is not so easily identified. Consequently, we will study another procedure in this section for solving such equations.

c. $20^{t+1} = 5^{3t-1}$ **d.** $6^{u^2+2} = 1{,}512$

Some Properties of Exponential and Logarithmic Functions Reviewed

The following properties follow from the fact that the exponential and logarithmic functions are one-to-one. These properties will be used to solve exponential and logarithmic equations.

If $b > 0$ and $b \neq 1$, then for real numbers x and y:

Property 1. If $x = y$, then $b^x = b^y$.

Property 2. If $b^x = b^y$, then $x = y$.

Property 3. If $x > 0$, $y > 0$, and $x = y$, then $\log_b x = \log_b y$.

Property 4. If $x > 0$, $y > 0$, and $\log_b x = \log_b y$, then $x = y$.

Examples **e–h** illustrate these properties.

e. If $3t = t - 6$, then $10^{3t} = 10^{t-6}$. Property 1

f. If $e^{t^2} = e^{t+6}$, then $t^2 = t + 6$. Property 2

g. If $20^{t+1} = 5^{3t-1}$, then $\log 20^{t+1} = \log 5^{3t-1}$. Property 3

h. If $\ln t^2 = \ln(t - 2)$ and $t > 2$, then $t^2 = t - 2$. Property 4

A Procedure to Solve Exponential Equations

The procedure illustrated in Examples 1 and 2 can be used to solve many exponential equations. In each example, a solution is given in **exact form** using logarithms, and in **approximate form** correct to three significant digits.

Example 1. Solve $3^{2x-1} = 168$.

Solution. **Discussion.** Property 3 can be used to solve this equation. The phrase "take the log of both sides" expresses the application of the property. To obtain an approximation for the solution, we will use base 10.

$$\log 3^{2x-1} = \log 168$$ If $x = y$, then $\log x = \log y$.

$$(2x - 1)\log 3 = \log 168$$ $\log x^r = r \log x$

$$2x \log 3 - \log 3 = \log 168$$ The distributive property

$$x \log 9 = \log 168 + \log 3$$ $2x \log 3 = x \log 3^2$

$x \log 9 = \log 504$	$\log x + \log y = \log xy$
$x = \dfrac{\log 504}{\log 9}$	Exact form of solution
$x \approx \dfrac{2.702\ldots}{0.954\ldots}$	Do not round the logs.
$x \approx 2.832\ldots$	Approximate form
$x \approx 2.83$	To three significant digits

Example 2. Solve $40^{x-1} = 2^{3x+2}$.

Solution. **Discussion.** We will use base e to solve this equation.

$\ln 40^{x-1} = \ln 2^{3x+2}$	If $x = y$, then $\ln x = \ln y$.
$(x - 1) \ln 40 = (3x + 2) \ln 2$	$\ln x^r = r \ln x$
$x \ln 40 - \ln 40 = 3x \ln 2 + 2 \ln 2$	The distributive property
$x \ln 40 - \ln 40 = x \ln 8 + \ln 4$	$3x \ln 2 = x \ln 2^3 = x \ln 8$
$x \ln 40 - x \ln 8 = \ln 4 + \ln 40$	Combine x-terms on left.
$x(\ln 40 - \ln 8) = \ln 160$	$\ln 4 + \ln 40 = \ln 4(40)$
$x \ln 5 = \ln 160$	$\ln 40 - \ln 8 = \ln \dfrac{40}{8}$
$x = \dfrac{\ln 160}{\ln 5}$	Exact form of solution
$x \approx \dfrac{5.075\ldots}{1.609\ldots}$	Do not round the ln's.
$x \approx 3.15$	To three significant digits

A Procedure to Solve Logarithmic Equations

An equation in which the logarithm of an expression containing a variable occurs is called a **logarithmic equation.** The procedure illustrated in Examples 3 and 4 is used to solve many logarithmic equations. The goal is to change, if necessary, the equations to the form $\log x = \log y$ or $\ln x = \ln y$, and then use Property 4.

Example 3. Solve $\log(t - 2) + \log 6 = \log 2 + \log(14 - t)$.

Solution. **Discussion.** Use properties of logarithms to write both sides in the form $\log x = \log y$. Then apply Property 4 to write $x = y$, and solve.

$\log 6(t - 2) = \log 2(14 - t)$	$\log x + \log y = \log xy$
$6(t - 2) = 2(14 - t)$	If $\log x = \log y$, then $x = y$.
$6t - 12 = 28 - 2t$	The distributive property
$8t = 40$	Combine t-terms on left side.
$t = 5$	The apparent root is 5.

Check: $\log(5-2) + \log 6 = \log 2 + \log(14-5)$

$$\log 3 + \log 6 = \log 2 + \log 9$$

$$\log 18 = \log 18, \textit{ true}$$

The solution set is $\{5\}$.

Example 4. Solve $\ln 2 + 2\ln x = \ln(3-5x)$.

Solution.

$\ln 2 + \ln x^2 = \ln(3-5x)$	$r \ln x = \ln x^r$
$\ln 2x^2 = \ln(3-5x)$	In the form $\ln x = \ln y$
$2x^2 = 3 - 5x$	If $\ln x = \ln y$, then $x = y$.
$2x^2 + 5x - 3 = 0$	Solve the quadratic equation.
$(2x-1)(x+3) = 0$	Factor.
$2x - 1 = 0$ or $x + 3 = 0$	The zero-product property
$x = \frac{1}{2}$ or $x = -3$	The apparent solutions are $\frac{1}{2}$ and -3.

Check: $x = \frac{1}{2}$

$$\ln 2 + 2\ln\frac{1}{2} = \ln\left(3 - 5\left(\frac{1}{2}\right)\right)$$

$$\ln 2 + \ln\frac{1}{4} = \ln\left(\frac{6}{2} - \frac{5}{2}\right)$$

$$\ln\frac{1}{2} = \ln\frac{1}{2}, \textit{ true}$$

$x = -3$

$$\ln 2 + 2\ln(-3) = \ln(3 - 5(-3))$$

Reject -3, since $\ln(-3)$ is not defined.

The solution set is $\left\{\frac{1}{2}\right\}$.

SECTION 10-5. Practice Exercises

In exercises **1–30**, solve each equation. State answers in:

a. Exact form

b. Approximate form to three significant digits

[Examples 1 and 2]

1. $2^x = 12$

2. $5^x = 85$

3. $6^{2x} = 15$

4. $12^{2x} = 375$

5. $8^{x+1} = 32$

6. $9^{x+1} = 243$

7. $7^{x-1} = 19$

8. $5^{x-1} = 27$

9. $2^{2x+1} = 34$

10. $3^{2x-1} = 93$

11. $5^{3x-1} = 480$

12. $4^{3x-1} = 812$

13. $8^x = 24^{x-1}$

14. $9^x = 45^{x-1}$

15. $54^x = 3^{2x+1}$

16. $36^x = 2^{2x+1}$

17. $8^{x+1} = 12^{x-1}$

18. $4^{x+2} = 5^{x+1}$

19. $2^{2x+1} = 0.04$
20. $3^{2x-1} = 0.05$
21. $0.5^{x-2} = 0.2^{x+1}$
22. $0.3^{x+2} = 0.18^{1-x}$
23. $2^{x^2} = 48$
24. $3^{x^2} = 150$
25. $6^{x^2+1} = 90$
26. $12^{x^2+1} = 9{,}600$
27. $4^{\sqrt{x}} = 75$
28. $8^{\sqrt{x}} = 120$
29. $5^{\sqrt{x-1}} = 380$
30. $7^{\sqrt{x-2}} = 240$

In exercises **31–60**, solve each equation.

[Examples 3 and 4]

31. $2 \log x - \log 9 = \log 1$
32. $2 \log x - \log 64 = \log 1$
33. $3 \log x = 0$
34. $4 \log x = 0$
35. $\log x - \log(x - 1) = \log 2$
36. $\log x = \log(x - 5) + \log 6$
37. $\ln(x + 2) + \ln 3 = \ln 6 + \ln(x - 1)$
38. $\ln 11 + \ln(x - 3) = \ln(2x + 1) + \ln 2$
39. $\log(2x - 1) - \log(x + 3) = \log 3 - \log 2$
40. $\log(3x + 2) - \log(2x - 3) = \log 20 - \log 9$
41. $\ln 2 + \ln(x + 1) + \ln 3 = \ln 8 + \ln x$
42. $\ln 5 + \ln(x - 1) + \ln 2 = \ln 2x + \ln 4$
43. $2 \log(x - 3) = \log 25$
44. $2 \log(x - 1) = \log 9$
45. $\log_3(2x + 3) + \log_3 x = \log_3(3x - 1) + \log_3 4$
46. $\log_3 x + \log_3(3x + 1) = \log_3(2x - 1) + \log_3 6$
47. $\log_7 x - \log_7 6 = \log_7(3x - 2) - \log_7(6x + 1)$
48. $\log_7 2x - \log_7 3 = \log_7(5x - 1) - \log_7(2x + 1)$
49. $\log x + \log(x + 5) = \log 2 + \log(2x + 3)$
50. $\log 2x + \log(x + 4) = \log(x + 1) + \log 5 + \log 3$
51. $\ln x + \ln(x + 3) = \ln 4$
52. $\ln(x - 9) + \ln x = \ln 10$
53. $2 \log x - \log(x - 1.6) = 1$
54. $2 \log x - \log(x - 0.9) = 1$
55. $\frac{1}{2}[\log(x + 6) - \log(x - 2)] = \log 3$
56. $\frac{1}{2}[\log(x + 5) - \log(x - 1)] = \log 2$
57. $\log(x^2 + 21) - \log x = 1$
58. $\log(x^2 + 2{,}400) - \log x = 2$
59. $(2 \log x - 1)(2 \log x - 2) = 0$
60. $(\log x - 2)(\log x - 3) = 0$

SECTION 10-5. Ten Review Exercises

1. Simplify $m(m + 3) - m(1 - m) - 2(m^2 + 5)$.
2. Solve $m(m + 3) = 2(m^2 + 5) + m(1 - m)$.

3. Factor $x^2 + 5bx - 3ax - 15ab$.

4. Solve $3ax + 15ab = 5bx + x^2$ for x.

5. Simplify $\dfrac{2}{t+1} - \dfrac{3}{t^2-1} - \dfrac{2}{t}$.

6. Solve $\dfrac{3}{t^2-1} = \dfrac{2}{t+1} - \dfrac{2}{t}$.

In exercises **7–10**, solve and graph each inequality.

7. $|8 - x| < 3$

8. $|2x + 5| \geq 3$

9. $\dfrac{x+3}{x-10} > 0$

10. $2x^2 - x - 6 < 0$

SECTION 10-5. Supplementary Exercises

The *time* (in number of days) that it takes a carton of milk to become sour depends in part on the *temperature* (in degrees Fahrenheit) at which it is stored. Table 10-2 lists the approximate times for a few temperatures. If the paired data are plotted on an xy-coordinate system, then the points lie on a curve. However, when the following (x, y) points are plotted, the points lie on a line, where:

$$x = \log(\text{number of days})$$

$$y = \log(\text{temperature})$$

Table 10-2. Time versus temperature data on milk.

Number of days	Storage temperature
24.0	32.0 F°
11.0	40.0 F°
5.5	45.0 F°
2.0	50.0 F°
1.0	60.0 F°
0.5	70.0 F°

An equation relating the time versus temperature data in Table 10-2 can be written in the general form

$$y = bx^m$$

where b is the y-intercept and m is the slope of the line in the [log(number of days), log(temperature)] data. To find b and m, we use equations (1) and (2) on any two pair of data points in Table 10-2.

$$(1)\quad m = \frac{\log y_2 - \log y_1}{\log x_2 - \log x_1} \qquad \text{and} \qquad (2)\quad b = \frac{y_1}{{x_1}^m}$$

Example. Find the equation for the time versus temperature data in Table 10-2.

Solution. Arbitrarily picking the data points (1 day, 60 F°) and (0.5 day, 70 F°):

$$m = \frac{\log 70 - \log 60}{\log 0.5 - \log 1} \qquad m = \frac{\log y_2 - \log y_1}{\log x_2 - \log x_1}$$

$$= \frac{0.06694\ldots}{-0.30102\ldots} \qquad \text{Do not round.}$$

$$= -0.22239\ldots \qquad \text{Compute the quotient.}$$

$$\approx -0.2 \qquad \text{Round to one decimal place.}$$

$$b = \frac{60}{1^{-0.2}} \qquad b = \frac{y_1}{x_1{}^m}$$

$$= 60 \qquad \text{Simplify.}$$

Thus, an equation for the time versus temperature data is:

$$y = 60x^{-0.2}$$

x is the time in days

y is the temperature in F°

In exercises **1–8**, find the equation in the form $y = bx^m$ for each set of points. Write b and m to two significant digits.

1.

x	y
0.50	0.25
1.00	2.00
2.00	16.00
3.00	54.00
4.00	128.00
5.00	250.00

2.

x	y
0.50	0.14
1.00	0.20
2.00	0.28
3.00	0.35
4.00	0.40
5.00	0.45

3.

x	y
0.50	0.13
1.00	0.50
2.00	2.00
3.00	4.50
4.00	8.00
5.00	12.50

4.

x	y
0.5	0.05
1.0	0.4
2.0	3.2
3.0	10.8
4.0	25.6
5.0	50.0

5.

x	y
1.1	1.52
1.8	5.22
2.2	8.61
3.0	18.7

6.

x	y
1.5	3.65
1.9	4.01
2.5	4.47
3.4	5.06

7.

x	y
0.9	0.97
1.2	1.06
1.7	1.17
3.2	1.42

8.

x	y
1.4	6.44
2.5	12.91
3.2	17.36
4.6	26.84

9. (A)–(J) are ten properties of exponents and logarithms.

(A) $b^x \cdot b^y = b^{x+y}$

(B) $\log_b xy = \log_b x + \log_b y$

(C) $\dfrac{b^x}{b^y} = b^{x-y}$

(D) $\log_b \dfrac{x}{y} = \log_b x - \log_b y$

(E) $(b^x)^y = b^{xy}$ (F) $\log_b x^y = y \log_b x$

(G) $\left(\frac{a}{b}\right)^x = \frac{a^x}{b^x}$ (H) $(ab)^x = a^x \cdot b^x$

(I) $\log_b x = \log_b y$ if and only if $x = y$

(J) $x = b^y$ if and only if $\log_b x = y$

In the following set of equations, identify the property that justifies each step:

Step	Property
$\log y - \log y_1 = m(\log x - \log x_1)$	The point-slope equation
a. $\log \frac{y}{y_1} = m(\log x - \log x_1)$	**a.** ____________
b. $\log \frac{y}{y_1} = m \log \frac{x}{x_1}$	**b.** ____________
c. $\log \frac{y}{y_1} = \log\left(\frac{x}{x_1}\right)^m$	**c.** ____________
d. $\frac{y}{y_1} = \left(\frac{x}{x_1}\right)^m$	**d.** ____________
e. $\frac{y}{y_1} = \frac{x^m}{x_1{}^m}$	**e.** ____________
$y = \frac{y_1}{x_1{}^m} \cdot x^m$	Multiplication of equality

SECTION 10-6. Applications

1. Literal exponential equations
2. Hydrogen ion concentration and pH
3. Computing compound interest
4. Population growth and radioactive decay

The applied problems in this section can all be solved using logarithms. We will look at an example of each type to illustrate the procedures for solving such problems.

Literal Exponential Equations

Literal equations (or formulas) contain two or more variables. Such equations are frequently solved for one of the variables in terms of the other variable (or variables) in the equation.

Example 1. Solve $A = P(1 + r)^{2t}$ for t.

Solution. **Discussion.** In the given equation, t is a factor of the exponent on $(1 + r)$. To solve the equation for t, we need to take the log of both sides, then use the properties of logarithms to write t as a coefficient of $(1 + r)$.

$$\log A = \log P(1 + r)^{2t} \qquad \text{Take the log of both sides.}$$

$$\log A = \log P + \log(1 + r)^{2t} \qquad \log x \cdot y = \log x + \log y$$

$$\log A - \log P = 2t \log(1 + r) \qquad \log x^r = r \log x$$

$$\frac{\log A - \log P}{2 \log(1 + r)} = t \qquad \text{Solve for } t.$$

Hydrogen Ion Concentration and pH

A chemical solution is classified as acid, neutral, or alkaline depending on the concentration of hydrogen ions, written "$[H^+]$", in the solution.

Value of $[H^+]$	Classification of solution
About 10^{-7} moles per liter	Neutral
More than 10^{-7} moles per liter	Acid
Less than 10^{-7} moles per liter	Base (or alkaline)

An expression that measures $[H^+]$ is pH, which is defined by:

$$\text{pH} = -\log[H^+]$$

Example 2. If $[H^+] = 4.6 \times 10^{-4}$, find the pH.

Solution.

$$\text{pH} = -\log(4.6 \times 10^{-4}) \qquad \text{pH} = -\log [H^+]$$

$$= -[\log 4.6 + \log 10^{-4}] \qquad \log x \cdot y = \log x + \log y$$

$$= -[0.6628 - 4] \qquad \log 10^{-4} = -4 \log 10 = -4$$

$$= 3.3372$$

Usually, pH is rounded to one decimal place. Thus, the answer to Example 2 would be written:

$$\text{pH} = 3.3$$

Example 3. Find $[H^+]$ for a certain brand of tomatoes with pH = 4.2.

Solution.

$$\text{pH} = -\log[H^+] \qquad \text{By definition}$$

$$[H^+] = \text{antilog}(-\text{pH}) \qquad \text{Solve for } [H^+].$$

$$[H^+] = \text{antilog}(-4.2) \qquad \text{Replace pH by 4.2.}$$

$$= \text{antilog}(0.8 - 5) \qquad \text{Change to positive mantissa.}$$

$$= 6.31 \times 10^{-5} \qquad \log 6.31 = 0.8000$$

Thus, the hydrogen ion concentration in the juice of these tomatoes is about 6.31×10^{-5} moles per liter, a slightly acidic solution.

Computing Compound Interest

Individuals and institutions frequently invest money in accounts or funds that earn returns based on the amount of money invested. These investment accounts are usually designed to calculate the interest by one of two methods:

1. The simple interest method
2. The compound interest method

The compound interest method adds the interest to the amount invested before computing the interest for the subsequent time periods. By this method of computing interest, we can say that *interest earns interest*. The following formula is used when interest is compounded:

$$A = P\left(1 + \frac{r}{n}\right)^{nt}$$

P is the principal, that is, the amount of money originally invested; r is the annual rate at which interest is earned; and n is the number of times interest is compounded each year. For example:

If interest is computed semiannually, then $n = 2$.

If interest is computed quarterly, then $n = 4$.

If interest is computed monthly, then $n = 12$.

The variable t stands for the number of years for which the amount in the account is to be computed, and A is the amount of money in the account (principal plus interest) after t years.

Example 4. Ethel and Harley Keele invested \$10,000 in a savings plan that yields 8% annual interest and compounds interest quarterly. How much have they saved in 5 years?

Solution. **Discussion.** For the given plan:

$r = 0.08$ — Annual interest rate is 8%.

$n = 4$ — Interest is compounded quarterly.

$P = \$10{,}000$ — The Keeles invested \$10,000.

$t = 5$ — A is computed after 5 years.

Using logarithms

$$\log A = \log 10{,}000\left(1 + \frac{0.08}{4}\right)^{4(5)}$$

$$\log A = \log 10{,}000 + 20 \log 1.02$$

$$\log A = 4 + 0.1720\ldots$$

$$A = \text{antilog } 4.172\ldots = 10^{4.172\ldots}$$

$$A \approx 14{,}859.473\ldots$$

Using a calculator

$$A = 10{,}000\left(1 + \frac{0.08}{4}\right)^{4(5)}$$

$$A = 10{,}000(1.02)^{20}$$

$$A = 10{,}000(1.4859\ldots)$$

$$A = 14{,}859.473\ldots$$

To the nearest cent, $A = \$14{,}859.47$

Population Growth and Radioactive Decay

The expression "exponential growth" is frequently used to describe a population or a quantity that is exhibiting unusually large increases as time progresses. An equation that can frequently be used to define such relationships has the following general form:

$$P = P_0 e^{kt}$$

Here P is the size of the population at some time t, while P_0 is the size that the population had at some beginning time, usually written $t = 0$. As before, e is the irrational number base of the natural logarithm, k is a constant that describes the rate of increase (or rate of decrease) for the population, and t is the time interval over which P is computed. Depending on k, the time t can be measured in minutes, hours, days, or years.

Example 5. Debra Long is studying certain bacteria growing in a culture in her laboratory. At time $t = 0$, she observed 100 bacteria, and the rate of growth appeared to be $k = 0.2$ when t was measured in hours. Compute the number of bacteria there will be in 6 hours.

Solution. **Discussion.** For the given problem:

$P_0 = 100$ — The number of bacteria at $t = 0$

$k = 0.2$ — The rate at which the bacteria were increasing

$t = 6$ — The time at which the population will be counted

Using logarithms

$$P = 100e^{(0.2)(6)}$$
$$\ln P = \ln 100 + \ln e^{1.2}$$
$$\ln P = 4.605\ldots + 1.2$$
$$\ln P = 5.805\ldots$$
$$P = \text{antiln } 5.805\ldots = e^{5.805\ldots}$$
$$P \approx 332$$

Using a calculator

$$P = 100e^{(0.2)(6)}$$
$$P = 100e^{1.2}$$
$$P = 100(3.320\ldots)$$
$$P = 332.011\ldots$$
$$P \approx 332$$

Thus, there will be about 330 bacteria in the culture in 6 hours.

The amount by which a radioactive substance decays can be defined by an exponential equation with the following general form

$$y = y_0 e^{-kt}$$

where y is the amount of substance present at time t, and y_0 is the amount present at time $t = 0$. Since k is a positive rate of growth, $-k$ is a corresponding rate of decay, and t is the time at which y is computed. By definition, the half-life of a radioactive substance is the time it takes for y to equal $0.5y_0$.

Example 6. Find, to three significant digits, the half-life of a radioactive material that decays according to the equation $y = y_0 e^{-0.05t}$, where t is measured in years.

Solution. **Discussion.** For the given problem;

y_0 — The amount of material present at $t = 0$

$k = -0.05$ The constant rate of decay

t The time it takes for $y = \frac{1}{2}y_0$

Using logarithms	Using a calculator
$0.5y_0 = y_0e^{-0.05t}$	$0.5y_0 = y_0e^{-0.05t}$
$0.5 = e^{-0.05t}$	$0.5 = e^{-0.05t}$
$\ln 0.5 = \ln e^{-0.05t}$	$\ln 0.5 = -0.05t$
$\ln 0.5 = -0.05t$	$-0.693\ldots = -0.05t$
$-0.693\ldots = -0.05t$	$t = 13.86\ldots$
$t \approx 14$	$t \approx 14$

Thus, the half-life of the given material is about 14 years.

SECTION 10-6. Practice Exercises

In exercises **1–10**, solve each equation for the indicated variable.

[Example 1]

1. $S = Pa^{rt}$, for r

2. $S = Pb^{rt}$, for t

3. $P = Ae^{-rt}$, for t

4. $P = Ae^{-rt}$, for r

5. $A = B(1 + i)^{-kt}$, for k

6. $A = B(1 + i)^{kt}$, for t

7. $B = k2^{-t/h}$, for h

8. $B = k2^{-t/h}$, for t

9. $XY = ke^{t/r}$, for r

10. $XY = ke^{t/r}$, for t

In exercises **11–18**, find the pH for each value of $[H^+]$.

[Example 2]

11. $[H^+] = 2.7 \times 10^{-6}$

12. $[H^+] = 3.6 \times 10^{-4}$

13. $[H^+] = 1.4 \times 10^{-9}$

14. $[H^+] = 5.4 \times 10^{-8}$

15. $[H^+] = 7.5 \times 10^{-3}$

16. $[H^+] = 9.1 \times 10^{-2}$

17. $[H^+] = 4.7 \times 10^{-10}$

18. $[H^+] = 8.3 \times 10^{-12}$

In exercises **19–26**, find $[H^+]$ for each value of pH.

[Example 3]

19. pH = 2.8

20. pH = 3.6

21. pH = 4.5

22. pH = 5.8

23. pH = 7.2

24. pH = 6.4

25. pH = 9.1

26. pH = 8.7

In exercises **27–46**, solve each word problem.

[Example 4]

27. Betty Jones invested \$10,000 in a savings certificate that compounds interest semiannually. The annual interest rate is 10%, and the time the certificate must be held is 3 years. How much is the certificate worth at the end of the 3 years?

28. Tom Keen purchased a $2,500 savings certificate that compounds interest monthly. The annual interest rate is 12% and the certificate must be held 5 years. How much is the certificate worth at the end of the 5 years?

29. John Glascow invested $1,000 in a mutual fund that pays 14% annual interest, compounded semiannually. How much money is in the fund at the end of 10 years?

30. Anne Smythe invested $7,500 in a mutual fund that pays 12% interest, compounded monthly. How much money is in the fund at the end of 6 years?

31. Fred and Eileen Dinsmore want to have $10,000 in a fund to take a trip in 5 years. They have the opportunity to invest some money in a mutual fund that pays 8% annual interest compounded semiannually. To the nearest dollar, how much should they invest to have $10,000 at the end of 5 years?

32. Guy Norris needs $4,000 in 3 years. He can invest in a savings plan that pays 12% annual interest compounded quarterly. To the nearest dollar, how much does Guy need to invest to have $4,000 in the plan in 3 years?

33. Lucille Bowers has $6,000 to invest in a savings plan that pays 8% annual interest, compounded quarterly. How long, to the nearest year, must Lucille leave the money in the plan to have it build to $11,000?

34. George Hanover wants to double the $8,000 he has to invest. A time certificate pays 6% annual interest, compounded six times a year. To the nearest year, how long must George keep the certificate before it is worth $16,000?

[Example 5] 35. Desiree Johnson is studying bacteria growing in a laboratory culture. At time $t = 0$, she observes 50 bacteria and estimates the rate of growth at $k = 0.25$, measuring time in hours. How many bacteria will be present in 10 hours?

36. Dave Joblowski is studying the same bacteria as Desiree, but in a culture that appears to slow the rate of growth to $k = 0.10$. How many bacteria will Dave have in his culture in 10 hours if $P_0 = 50$?

37. A small community in Iowa had a population of 2,580 in 1980. The rate of growth is about $k = 0.04$ for t measured in years. Find the projected population of this city in the year 2,000.

38. A city in Illinois had a population of 36,475 in 1980. The rate of growth is about $k = 0.03$ for t measured in years. Find the predicted population of this city for 1990.

39. A large eastern city had a population of 692,490 in 1980. For this city, the rate of population change is $k = -0.005$ when t is measured in years. (The negative k indicates a *decline* in population.) Find the projected population for this city in the year 2000.

40. The population in a large urban area was 397,615 in 1980. For this area, the rate of population change is $k = -0.012$ when t is measured in years. Find the projected population of this area in the year 2,000.

[Example 6] 41. Find, to three significant digits, the half-life of a radioactive material that decays according to the equation $y = y_0e^{-0.08t}$.

42. Find, to three significant digits, the half-life of a radioactive material that decays according to the equation $y = y_0e^{-0.01}t$, if t is measured in years.

43. For a certain radioactive substance, $k = 0.025$. If $y_0 = 200$ milligrams, find y if $t = 20$ years.

44. For a certain radioactive substance, $k = 0.04$. If $y_0 = 100$ milligrams, find y if $t = 30$ years.

45. A certain radioactive substance decays from 200 milligrams to 100 milligrams in 120 years. Find the value of k to three decimal places for this substance.

46. A certain radioactive substance decays from 150 milligrams to 75 milligrams in 75 years. Find the value of k to three decimal places for this substance.

SECTION 10-6. Ten Review Exercises

In exercises **1–8**, refer to the function defined by $f(x) = x^2 - 6x + 5$.

1. Identify the name of a graph of the function.

2. Write the equation in the form $f(x) = a(x - h)^2 + k$.

3. Identify the coordinates of the vertex.

4. Write an equation of the axis of symmetry.

5. Identify the x-intercepts.

6. Identify the y-intercept.

7. State the domain of the function.

8. State the range of the function.

In exercises **9** and **10**, use $P(-5, 7)$ and $Q(1, -1)$.

9. Find the slope of the line that passes through P and Q.

10. Write in $f(x) = mx + b$ an equation of the line that passes through P and Q.

SECTION 10-6. Supplementary Exercises

Table 10-3 (p. 694) contains the past and projected future population figures for gray whales. If the following (x, y) ordered pairs are plotted, the points lie on a line where:

$$x = \text{year number}$$

$$y = \ln \text{ of the population}$$

An equation that relates the year number versus population data in Table 10-3 can be written in the general form

$$y = be^{mx}$$

where b is the y-intercept and m is the slope of the line in the [(year number), ln(population)] data. To find b and m, we use equations (3) and (4) on any two

Table 10-3. Year number versus population data on whales.

Year	Year number	Population
1988	0	20,000
1989	1	20,500
1990	2	21,010
1991	3	21,540
1992	4	22,080
1993	5	22,630
1994	6	23,190

pair of data points in Table 10-3:

$$(3)\quad m = \frac{\ln y_2 - \ln y_1}{x_2 - x_1} \qquad \text{and} \qquad (4)\ b = \frac{y_1}{e^{mx_1}}$$

Example. Find the equation for the year number versus population data in Table 10-3.

Solution. Arbitrarily picking the data points [(year 1, 20,500) and (year 2, 21,010)]:

$m = \dfrac{\ln 21{,}010 - \ln 20{,}500}{2 - 1}$ $\qquad m = \dfrac{\ln y_2 - \ln y_1}{x_2 - x_1}$

$= \dfrac{9.951\ldots - 9.928\ldots}{1}$ $\qquad$ Do not round.

$= 0.0245\ldots$ $\qquad$ Simplify.

≈ 0.025 $\qquad$ To two significant digits

(About a 2.5% growth factor)

$b = \dfrac{20{,}500}{e^{(0.025)(1)}}$ $\qquad b = \dfrac{y_1}{e^{mx_1}}$

$\approx 19{,}993.85\ldots$ $\qquad$ Do the division.

$\approx 20{,}000$ $\qquad$ Round to two significant digits.

Thus, $y = 20{,}000e^{0.025x}$.

In exercises **1–8**, find the equation in the form $y = be^{mx}$ for each set of points. Write b and m to two significant digits.

1.

x	y
0.50	2.47
1.00	4.08
2.00	11.08
3.00	30.13
4.00	81.90
5.00	222.62

2.

x	y
0.50	2.72
1.00	7.39
2.00	54.60
3.00	403.43
4.00	2980.96
5.00	22026.47

3.

x	y
0.50	2.57
1.00	3.30
2.00	5.44
3.00	8.96
4.00	14.78
5.00	24.36

4.

x	y
0.50	6.8
1.00	18.5
2.00	136.
3.00	1008.
4.00	7452.

5.

x	y
0.5	3.0
1.5	1.12
2.0	0.68
2.5	0.41

6.

x	y
0.5	1.10
1.5	0.15
2.0	0.055
2.5	0.02

7.

x	y
1	5.67
1.2	6.02
1.6	6.79
2.0	7.65
2.8	9.73

8.

x	y
2.1	6.35
2.3	7.15
2.8	9.66
3.3	13.04
3.7	16.57

Name

Date

Score

SECTION 10-3. Summary Exercises

Answer

In exercises **1–4**, rewrite each expression in an equivalent or simplified form by using one or more of the properties of logarithms. Assume all variables are real positive numbers.

1. $\log_5 3x$ **1.** ______________

2. $\log_4 \frac{6}{p}$ **2.** ______________

3. $\log_4 \sqrt[3]{xy}$ **3.** ______________

4. $\log_{10} \frac{2}{xy}$ **4.** ______________

In exercises **5** and **6**, write each expression in the form $\log_b N$. Assume all variables are positive real numbers.

5. $3 \log_3 x - \log_3 16$

5. ______________________

6. $\frac{1}{3}(\log_2 x + 2 \log_2 y)$

6. ______________________

7. Write $\log_8 5$ in terms of a base 10 logarithm.

7. ______________________

8. Write $\log_3 \left(\frac{1}{2}\right)$ in terms of a base e logarithm.

8. ______________________

Name

Date

Score

SECTION 10-4. Summary Exercises

Answer

1. Write $10^{-3} = 0.001$ in logarithmic form.

1. ______________

2. Write $\log_{10} 2.75 = 0.4393$ in exponential form.

2. ______________

In exercises **3** and **4**, find log x. Write answers to four decimal places.

3. log 53,900

3. ______________

4. log 0.0731

4. ______________

5. Find antilog 3.3139. Write answer to three significant digits. **5.** ____________

In exercises **6** and **7**, approximate ln x. Write answers to four decimal places.

6. ln 25.8 **6.** ____________

7. ln 0.00121 **7.** ____________

8. Find antiln(-3.52). Write answer to three significant digits. **8.** ____________

CHAPTER 10. Review Exercises

In exercises **1** and **2**, find each value for $f(x)$.

1. $f(x) = 10^x$

a. $f(-2)$

b. $f(0)$

2. $f(x) = \left(\frac{1}{4}\right)^x$

a. $f\left(\frac{1}{2}\right)$

b. $f(-2)$

In exercises **3** and **4**, graph each function.

3. $y = 5^x$

4. $y = \left(\frac{1}{5}\right)^x$

In exercises **5** and **6**, simplify each expression.

5. $(625)^x\left(\frac{1}{5}\right)^{-2x}$

6. $\left(\frac{1}{7}\right)^{x-1} 49^{3x}$

In exercises **7–12**, solve each equation.

7. $9^{2x} = 27$

8. $10^x = 1$

9. $36^{1-x} = \left(\frac{1}{6}\right)^{x+1}$

10. $15^{9x^2-16} = 1$

11. $125x^3 = 8$

12. $2\sqrt[4]{x} = 3$

In exercises **13–16**, write each equation in logarithmic form.

13. $5^4 = 625$

14. $\left(\frac{1}{36}\right)^{-3/2} = 216$

15. $10^0 = 1$

16. $2^1 = 2$

In exercises **17–20**, write each equation in exponential form.

17. $\log_3 243 = 5$

18. $\log_2 \frac{1}{16} = -4$

19. $\log_{10} 493 = 2.6928$

20. $\log_{10} 0.068 = -1.1675$

In exercises **21–26**, find the value of each unknown.

21. $y = \log_{3/2} \frac{81}{16}$

22. $y = \log_{1/5} 3{,}125$

23. $\log_8 x = \frac{-4}{3}$

24. $\log_{1/36} x = \frac{-1}{2}$

25. $\log_b 0.001 = -3$

26. $\log_b \frac{1}{243} = \frac{-5}{2}$

In exercises **27** and **28**, graph each equation.

27. $y = \log_6 x$

28. $y = \log_{1/6} x$

In exercises **29–32**, rewrite each expression in an equivalent or simplified form by using one or more of the properties of logarithms. Assume all variables are positive real numbers.

29. $\log_{10} \frac{3x}{5}$

30. $\log_{10} 7x^2$

31. $\log_3 \sqrt{2xy}$

32. $\log_3 \sqrt[3]{5y^2}$

In exercises **33–36**, write each expression in the form $\log_b N$. Assume all variables are positive real numbers.

33. $\frac{1}{2}(\log_2 5 + \log_2 3)$

34. $2(\log_2 x - \log_2 5)$

35. $3 \log_{10} 2 + 2 \log_{10} y$

36. $\frac{1}{2} \log_{10} 6 - 2 \log_{10} 10$

In exercises **37** and **38**, write each expression in terms of:

a. Base 10 logarithms

b. Base *e* logarithms

37. $\log_8 93$

38. $\log_2 3.9$

In exercises **39–42**, assume that for some base *b*, the following logarithms are correct to one decimal place:

$$\log_2 5 \approx 2.3 \qquad \log_2 10 \approx 3.3 \qquad \log_2 15 \approx 3.9$$

Approximate each logarithm to one decimal place.

39. $\log_2 50$

40. $\log_2 3$

41. $\log_2 \sqrt[3]{10}$

42. $\log_2 \sqrt{75}$

In exercises **43** and **44**, write each equation in logarithmic form.

43. $10^{2.8} = 631$

44. $10^{0.9036} = 8.01$

In exercises **45** and **46**, write each equation in exponential form.

45. $\log 7{,}840 = 3.8943$

46. $\log 0.0493 = -1.3072$

In exercises **47–50**, find $\log x$. Write answers to four decimal places.

47. $\log 28$

48. $\log 52{,}700$

49. $\log 0.0124$

50. $\log 0.000882$

In exercises **51–54**, find antilog *y*. Write answers to three significant digits.

51. antilog 2.8669

52. antilog 5.6702

53. antilog(−1.4828)

54. antilog(−3.0878)

In exercises **55–58**, approximate $\ln x$. Write answers to four decimal places.

55. $\ln 142$

56. $\ln 0.075$

57. $\ln 8e^2$

58. $\ln \frac{10}{e^2}$

In exercises **59–62**, approximate $\text{antiln } x$. Write answers to three significant digits.

59. antiln 0.72

60. antiln 3.39

61. antiln(-0.26)

62. antiln(-2.85)

In exercises **63–66**, solve each equation. State answers to three significant digits in:

a. Exact form

b. Approximate form

63. $12^x = 1{,}500$

64. $6^{2x+1} = 276$

65. $2^{x+1} = 12^{1-x}$

66. $15^{x^2} = 6{,}900$

In exercises **67–70**, solve each equation.

67. $2 \log x = 196$

68. $\log(4x - 2) - \log(x + 3) = \log 2$

69. $\log(x - 2) + \log 6 = \log 2 + \log(14 - x)$

70. $\log x + \log(x + 5) = \log 6 + \log(x + 1)$

In exercises **71** and **72**, solve each equation for the indicated variable.

71. $I = Ae^{kt}$, for t

72. $MN = ke^{-rt}$, for t

In exercises **73** and **74**, find the pH for each value of $[H^+]$.

73. $[H^+] = 3.7 \times 10^{-2}$

74. $[H^+] = 4.2 \times 10^{-11}$

In exercises **75** and **76**, find $[H^+]$ for each value of pH.

75. pH = 3.3

76. pH = 8.6

In exercises **77–79**, solve each word problem.

77. Susan Winters invested \$15,000 in a corporation bond that pays 8% annual interest, compounded quarterly. If the bond matures in 10 years, how much will Susan get for the bond?

78. A rodent population is increasing in a certain rural area according to the equation $P = P_0 e^{0.16t}$. If P_0 is estimated at 150,000 and t is measured in years, find P to three significant digits in 3 years.

79. Find to three significant digits the half-life of a radioactive material that decays according to the equation $y = y_0 e^{-0.004t}$, if t is measured in years.

CHAPTER

11

Quadratic Relations

"Good morning, everyone," Ms. Glaston said. "Today we're going to study a figure called a **right circular cone.**" She then turned on the projector, showing Figure 11-1.

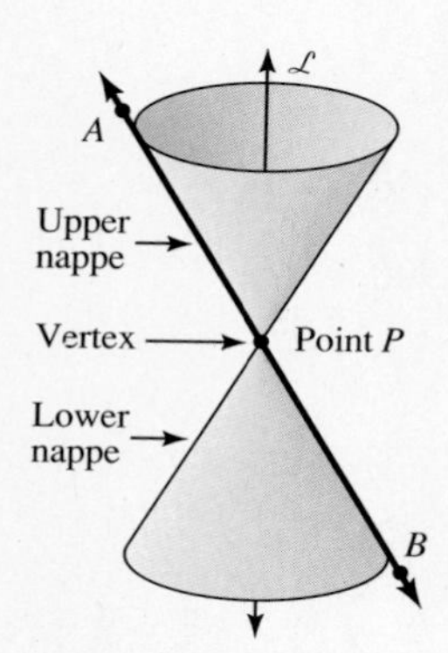

Figure 11-1. A cone.

"Note the parts of the cone identified in the figure." Ms. Glaston continued. Line $\mathscr{L}$ is the **axis** of the cone. The line that contains A and B is called a **generator** of the cone. Imagine this line rotated about point P in such a way that A and B follow a precise circular path. The result would be that the line would sweep out the cone in its entirety.

"Now suppose," she continued, "that a plane is allowed to intersect the cone. The figure formed by the points of the cone that intersect the plane belongs to a group of figures called **conics.** For example, suppose the plane is held in such a way that it is parallel to the line containing A and B. Such a position would cause the plane to intersect either the top or bottom **nappe** of the cone. The points of intersection would form a curve called a **parabola.**" She then placed an acetate on the projector showing Figure 11-2.

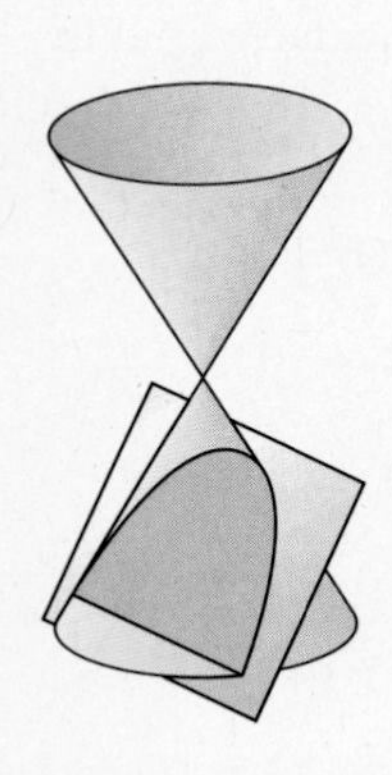

Figure 11-2. Intersecting a cone with a plane to get a parabola.

"That curve," Lynne Pickens said, "looks like the ones we got when we graphed quadratic functions."

"Exactly, Lynne," Ms. Glaston replied. "As we will see in this chapter, the parabola can in some cases be defined by an equation that has the form of the quadratic function."

"Now," she continued, "suppose the plane is held in such a way that it cuts through either the top or bottom nappe. The points of intersection would form a curve called an **ellipse.**" She then placed an acetate on the projector showing Figure 11-3.

Ms. Glaston now placed an acetate on the projector showing Figure 11-4. "To get the intersection shown in this figure," she continued, "the plane is held so that

Figure 11-3. The points of intersection form an ellipse.

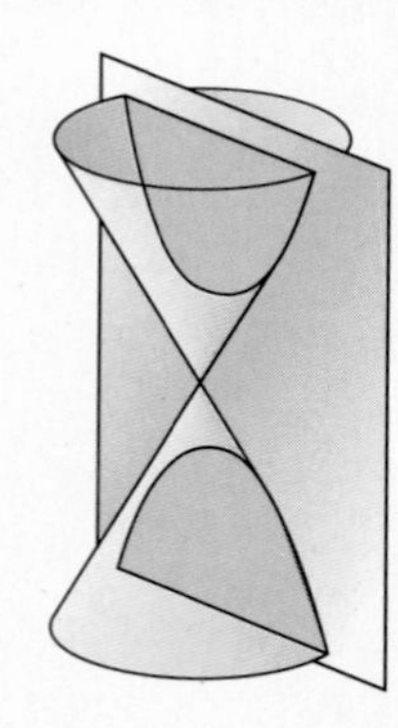

Figure 11-4. The points of intersection form a hyperbola.

it is parallel to the axis of the cone. Such a position would cause the plane to intersect both nappes of the cone. The figure formed is called a **hyperbola.** Each of these figures—parabola, ellipse, and hyperbola—can be defined by equations in x and y using some fixed points and the distance formula. We will examine the exact details of these relationships in the chapter we are beginning today.

SECTION 11-1. Parabolas

1. A definition of a parabola
2. Equations of parabolas with vertex $V(0, 0)$ and axis of symmetry the y-axis
3. Equations of parabolas with vertex $V(h, k)$ and axis of symmetry the line $x = h$

In this section, we will study conics called parabolas. As shown in Figure 11-2, a parabola is formed when a plane is held parallel to a generator of the right circular cone, and the plane intersects one of the nappes.

A Definition of a Parabola

The following statement is a definition of a parabola:

> **Definition 11.1. Parabola**
> A **parabola** is the set of points in a plane that are the same distance from a fixed point and a fixed line in the plane. The point F is called the **focus,** and the line $\mathscr{L}$ is called the **directrix.**

In Figure 11-5, a focus F and directrix $\mathscr{L}$ are drawn. When the points in the plane that are equally distant from F and $\mathscr{L}$ are determined, they all lie on the parabola shown in the figure. As indicated, P is one of the points on the parabola.

Figure 11-5 also shows V, the **vertex** of the parabola. The vertex is the "turning point" on the curve. The line through F and V and perpendicular to $\mathscr{L}$ is called the **axis of symmetry.** The left and right branches of the parabola will coincide if the graph is folded along the axis of symmetry.

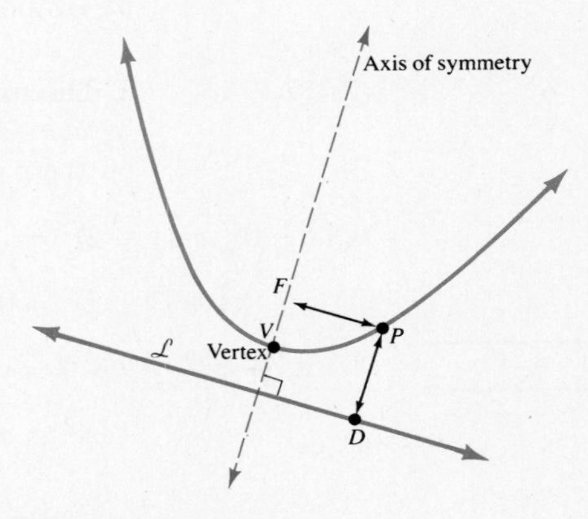

Figure 11-5. Point P is on the parabola.

The following symbols will be used:

1. $d(P, F)$ represents *the distance between* P *and* F
2. $d(P, D)$ represents *the distance between* P *and* D

If P is a point on the parabola, then by Definition 11.1:

$$d(P, F) = d(P, D)$$

Equations of Parabolas with Vertex $V(0, 0)$ and Axis of Symmetry the y-Axis

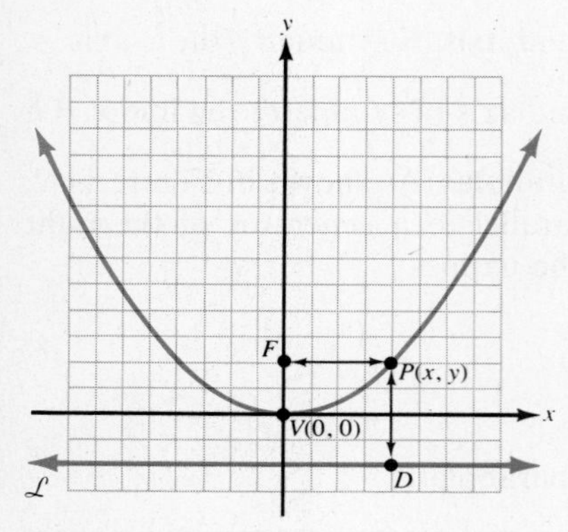

Figure 11-6. Parabola with $V(0, 0)$ and axis of symmetry $x = 0$.

To write an equation that defines a parabola, the focus and directrix are placed in an xy-coordinate system. The distance formula can then be used to write an equation that defines the parabola. The simplest location of F and $\mathscr{L}$ would be to position them so that the vertex of the parabola would be $V(0, 0)$, and the axis of symmetry would be an axis of the coordinate system. One such location of F and $\mathscr{L}$ is shown in Figure 11-6. There are four possible locations of $\mathscr{L}$ relative to F when the axis of symmetry is either a vertical or horizontal line.

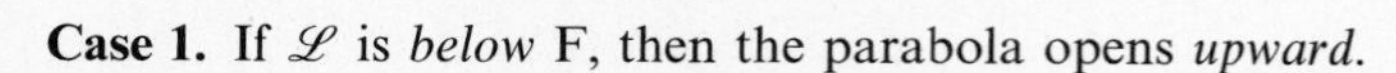

Case 1. If $\mathscr{L}$ is *below* F, then the parabola opens *upward.*

Case 2. If $\mathscr{L}$ is *above* F, then the parabola opens *downward.*

Case 3. If $\mathscr{L}$ is *to the left of* F, then the parabola opens *to the right.*

Case 4. If $\mathscr{L}$ is *to the right of* F, then the parabola opens *to the left.*

Cases 1 and 2 are discussed below. Cases 3 and 4 are studied in this section's Supplementary Exercises (see p. 718).

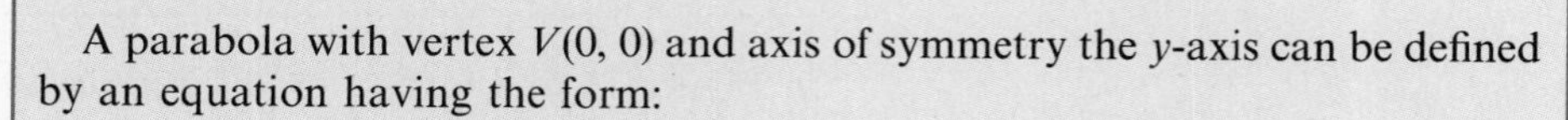

> A parabola with vertex $V(0, 0)$ and axis of symmetry the y-axis can be defined by an equation having the form:
>
> $y = ax^2$, where a is a nonzero real number

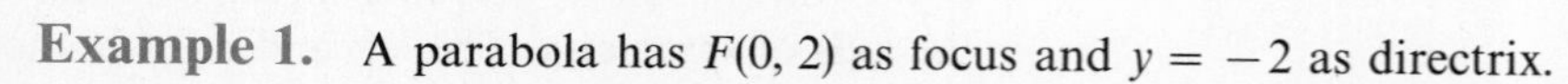

Example 1. A parabola has $F(0, 2)$ as focus and $y = -2$ as directrix.

a. Write an equation that defines the parabola in the form $y = ax^2$.

b. Graph the parabola.

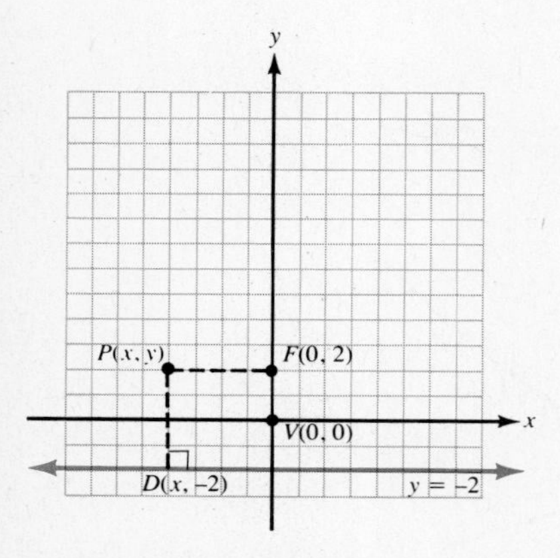

Figure 11-7. Focus at $F(0, 2)$ and directrix $y = -2$.

Solution. **a. Discussion.** In Figure 11-7, $F(0, 2)$ is plotted and the line defined by $y = -2$ is drawn. The vertex is shown at $V(0, 0)$. A point $P(x, y)$ on the parabola is also shown.

$$\sqrt{(x-0)^2 + (y-2)^2} = \sqrt{(x-x)^2 + (y-(-2))^2} \qquad d(P, F) = d(P, D)$$

$$x^2 + (y-2)^2 = 0^2 + (y+2)^2 \qquad \text{Square both sides.}$$

$$x^2 + y^2 - 4y + 4 = y^2 + 4y + 4 \qquad \text{Square the binomials.}$$

$$x^2 = 8y \qquad \text{Add } 4y \text{ to both sides.}$$

$$y = \frac{1}{8}x^2$$

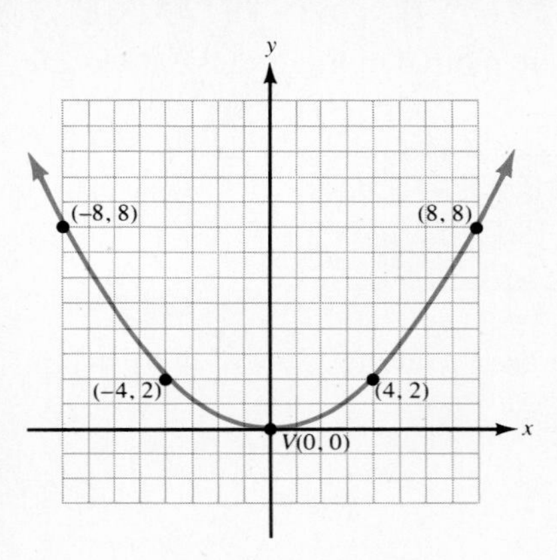

Figure 11-8. A graph of $y = \frac{1}{8}x^2$.

b. The following ordered pairs are plotted in Figure 11-8 to locate the parabola:

x	−8	−4	0	4	8
$y = \frac{1}{8}x^2$	8	2	0	2	8

Example 2. Graph $y = -2x^2$.

Solution. The following ordered pairs are plotted in Figure 11-9 to locate the parabola:

x	−2	−1	0	1	2
$y = -2x^2$	−8	−2	0	−2	−8

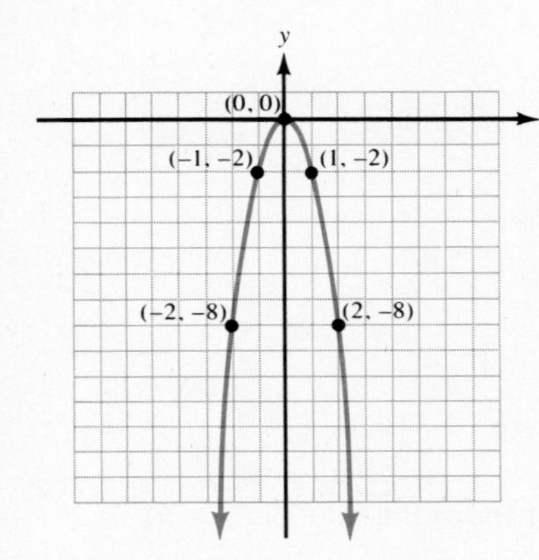

Figure 11-9. A graph of $y = -2x^2$.

Equations of Parabolas with Vertex $V(h, k)$ and Axis of Symmetry the Line $x = h$

The equation $y = ax^2$ for a parabola with vertex $V(0, 0)$ and focus on the y-axis is quadratic in x. We studied this equation in Chapter 8 as one of the forms of quadratic functions (see p. 534). At that time, we learned that a graph of the function defined by

$$y = ax^2 + bx + c, \quad \text{where } a, b, \text{ and } c \text{ are real numbers and } a \neq 0$$

has the cup-shaped appearance of a parabola. This relationship between quadratic functions and parabolas with foci on vertical lines and directrices (plural of directrix) that are horizontal lines is correct. We can thus use quadratic functions to define parabolas with the same cup-like graphs as those of quadratic functions.

If $V(h, k)$ is the vertex of a parabola with focus on the line $x = h$ and a horizontal directrix, then:

1. The general form of the equation of the parabola is:

$$y = ax^2 + bx + c, \quad \text{where } a, b, \text{ and } c \text{ are real numbers and } a \neq 0$$

2. For graphing purposes, the equation can be written in the form:

$$y = a(x - h)^2 + k$$

- The vertex is $V(h, k)$.
- The parabola opens *upward* if $a > 0$.
- The parabola opens *downward* if $a < 0$.
- The axis of symmetry is $x = h$.

Example 3. Graph $y = 2(x - 3)^2 - 4$.

Solution. **Discussion.** This equation defines a parabola with vertex $V(3, -4)$ and axis of symmetry $x = 3$. Since $a = 2$, the parabola opens upward.

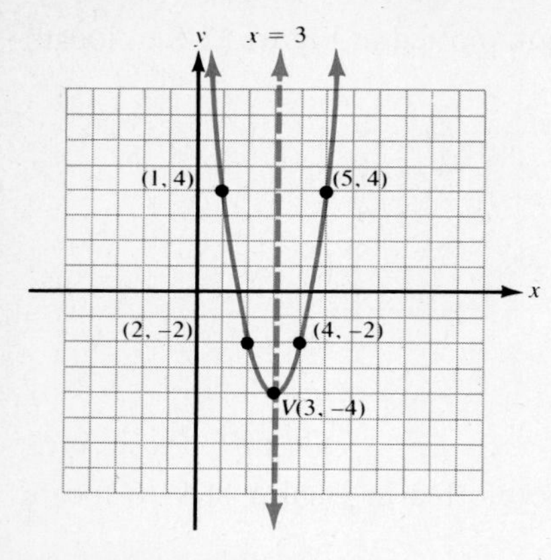

Figure 11-10. A graph of $y = 2(x - 3)^2 - 4$.

The following ordered pairs are plotted in Figure 11-10 to locate the parabola:

x	1	2	3	4	5
$y = 2(x - 3)^2 - 4$	4	−2	−4	−2	4

Example 4 reviews the six steps that can be used to write $y = ax^2 + bx + c$ in the form $y = a(x - h)^2 + k$.

Example 4. Graph $y = \frac{1}{2}x^2 - 4x + 3$.

Solution. **Step 1.** If $a \neq 1$, then multiply both sides by $\frac{1}{a}$.

$$2y = x^2 - 8x + 6 \qquad \text{Multiply both sides by 2.}$$

Step 2. If $\frac{c}{a} \neq 0$, then add $\frac{-c}{a}$ to both sides.

$$2y - 6 = x^2 - 8x \qquad \text{Add } -6 \text{ to both sides.}$$

Step 3. Add $\left(\frac{1}{2} \cdot \frac{b}{a}\right)^2$ to both sides.

$$2y - 6 + 16 = x^2 - 8x + 16 \qquad \text{Add } \left(\frac{1}{2}(-8)\right)^2 = 16.$$

Step 4. Simplify the left and write the right side as $(x - h)^2$.

$$2y + 10 = (x - 4)^2$$

Step 5. Move the constant on the left to the right side.

$$2y = (x - 4)^2 - 10 \qquad \text{Add } -10 \text{ to both sides.}$$

Step 6. If $a \neq 1$, then multiply both sides by a.

$$y = \frac{1}{2}(x - 4)^2 - 5 \qquad \text{Multiply both sides by } \frac{1}{2}.$$

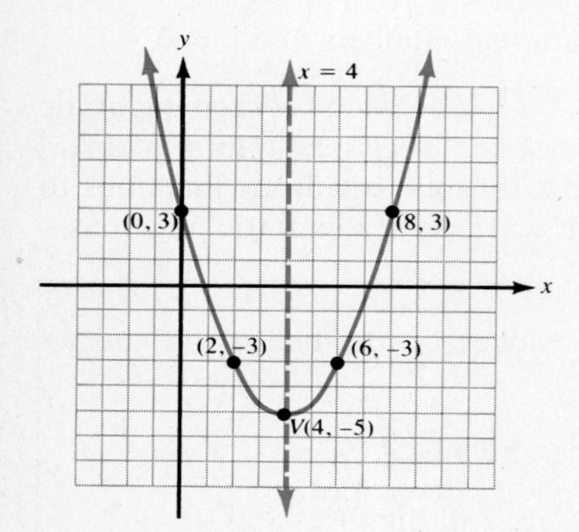

Figure 11-11. A graph of $y = \frac{1}{2}x^2 - 4x + 3$.

In this form, we can see that $h = 4$ and $k = -5$. Therefore, $V(4, -5)$ is the vertex and $x = 4$ is the axis of symmetry. Since $\frac{1}{2} > 0$, the parabola opens upward. The points plotted in Figure 11-11 are used to locate the parabola.

SECTION 11-1. Practice Exercises

In exercises **1–6**, for each combination of focus, directrix, or vertex:

a. Write an equation that defines the parabola in the form $y = ax^2$.

b. Graph the parabola.

[Example 1]

1. $F(0, 1)$ and $y = -1$
2. $F(0, 3)$ and $y = -3$
3. $F(0, -2)$ and $V(0, 0)$
4. $F(0, -4)$ and $V(0, 0)$
5. $V(0, 0)$ and $y = \frac{3}{4}$
6. $V(0, 0)$ and $y = \frac{-1}{2}$

In exercises **7–34**, graph each equation.

[Example 2]

7. $y = 2x^2$
8. $y = \frac{1}{2}x^2$
9. $y = \frac{-1}{3}x^2$
10. $y = -3x^2$
11. $y = \frac{2}{3}x^2$
12. $y = \frac{3}{2}x^2$

[Example 3]

13. $y = (x - 1)^2$
14. $y = (x - 3)^2$
15. $y = (x + 5)^2$
16. $y = (x + 2)^2$
17. $y = \frac{3}{2}(x + 1)^2$
18. $y = \frac{4}{3}(x + 3)^2$
19. $y = -x^2 + 1$
20. $y = -2x^2 + 6$
21. $y = 2(x + 4)^2 - 8$
22. $y = 3(x + 2)^2 - 9$

[Example 4]

23. $y = x^2 - 8x + 12$
24. $y = -x^2 - 2x + 5$
25. $y = \frac{1}{2}x^2 - 4x + 10$
26. $y = 3x^2 + 30x + 81$
27. $2y = 3x^2 - 6x - 1$
28. $3y = -x^2 - 4x + 5$
29. $y = 2x^2 - 12x + 9$
30. $y = -2x^2 - 10x - 18$
31. $y = -x^2 + 4x - 7$
32. $y = x^2 + 2x - 3$
33. $3y = x^2 - 2x + 7$
34. $4y = -x^2 + 6x - 5$

SECTION 11-1. Ten Review Exercises

In exercises **1–8**, solve each equation.

1. $u(u + 3) - 2(u^2 + 5) = u(1 - u)$
2. $5|v - 8| = 15$
3. $\frac{2}{t + 1} = \frac{2}{t} + \frac{3}{t^2 - 1}$
4. $6a(a + 3) = 5a - 6$
5. $\sqrt{b} = 1 + \sqrt{b - 7}$
6. $x - 2 = \sqrt[3]{x^3 - 6x^2}$
7. $5^{3x+1} = 480$
8. $\log x - \log 2 = \log(x - 1)$

In exercises **9** and **10**, solve and graph each inequality.

9. $2x^2 - x < 6$
10. $|x + 5| > 3$

SECTION 11-1. Supplementary Exercises

If a parabola is determined by a directrix that is a vertical line and a focus that lies on a horizontal line, then the parabola opens outward to the right or outward to the left. Such a parabola is defined by an equation with the form:

$$x = ay^2 + by + c$$

In the following example, a six-step procedure similar to the one in Example 4 is given for graphing such quadratic relations.

Example. Graph $x = 2y^2 - 8y + 4$.

Solution. **Step 1.** $\frac{1}{2}x = y^2 - 4y + 2$ — Multiply both sides by $\frac{1}{2}$.

Step 2. $\frac{1}{2}x - 2 = y^2 - 4y$ — Add -2 to both sides.

Step 3. $\frac{1}{2}x - 2 + 4 = y^2 - 4y + 4$ — Add 4 to both sides.

Step 4. $\frac{1}{2}x + 2 = (y - 2)^2$ — Simplify the left, factor the right.

Step 5. $\frac{1}{2}x = (y - 2)^2 - 2$ — Add -2 to both sides.

Step 6. $x = 2(y - 2)^2 - 4$ — Multiply both sides by 2.

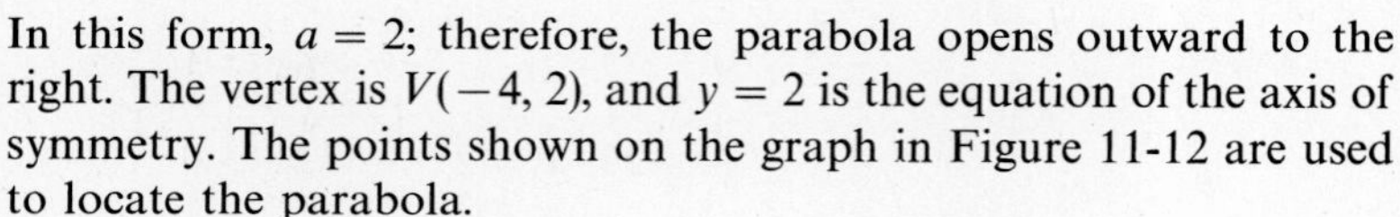

In this form, $a = 2$; therefore, the parabola opens outward to the right. The vertex is $V(-4, 2)$, and $y = 2$ is the equation of the axis of symmetry. The points shown on the graph in Figure 11-12 are used to locate the parabola.

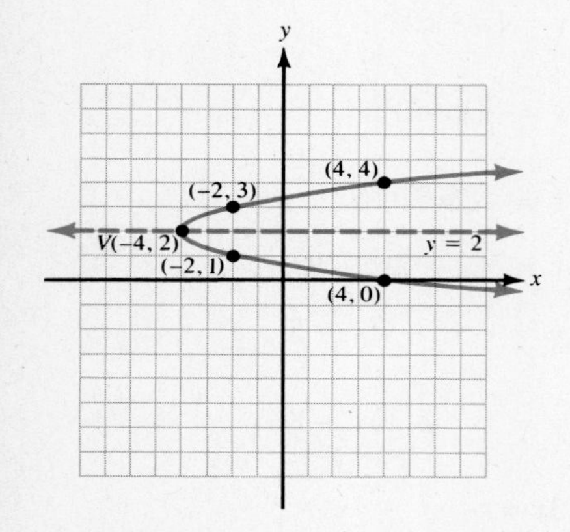

Figure 11-12. A graph of $x = 2y^2 - 8y + 4$.

In exercises **1–8**, graph each equation.

1. $x = y^2$

2. $x = 2y^2$

3. $x = -(y - 2)^2$

4. $x = -(y + 3)^2$

5. $x = 3y^2 - 12$

6. $x = 16 - 4y^2$

7. $x = -3y^2 + 18y - 23$

8. $2x = -y^2 - 4y - 2$

A graph of a function defined by

$$f(x) = a\sqrt{x - h}$$

is a semi-parabola. The vertex is $V(h, 0)$ and, depending on whether $a > 0$ or $a < 0$, a graph is the top half or bottom half, respectively, of a graph of a parabola that opens outward to the right.

In exercises **9–12**, graph each function.

9. $f(x) = 3\sqrt{x}$

10. $f(x) = 2\sqrt{x - 1}$

11. $f(x) = -2\sqrt{x + 2}$

12. $f(x) = \frac{-1}{2}\sqrt{x + 4}$

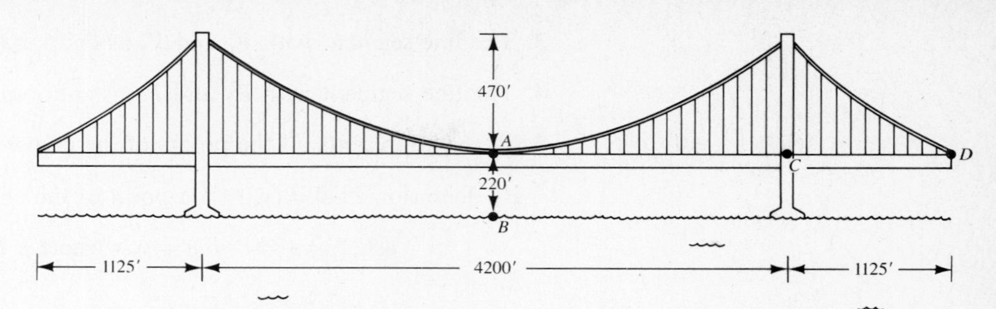

Figure 11-13. The Golden Gate Bridge.

The Golden Gate Bridge spanning the entrance to San Francisco Bay has two large cables that support the roadway. Because the stress on the cables is evenly distributed, their path is in the shape of a parabola.

In exercises **13–16**, find an equation in the form $y = a(x - h)^2 + k$ for the center cable if the origin of a coordinate system is placed at each point.

13. Point A

14. Point B

15. Point C

16. Point D

SECTION 11-2. Ellipses and Circles

1. A definition of an ellipse
2. The standard form equation of an ellipse with center (0, 0)
3. A procedure for graphing an ellipse
4. The center-radius equation of a circle
5. Finding the center and radius of a circle

In this section, we will study conics called ellipses and circles. As shown in Figure 11-3, an ellipse is formed when a plane intersects one, and only one of the nappes of the right circular cone.

A Definition of an Ellipse

Definition 11.2. Ellipse
An **ellipse** is the set of all points in a plane the sum of whose distances from two fixed points in the plane is a constant. The two fixed points, F_1 and F_2, are the **foci** (plural of focus) of the ellipse.

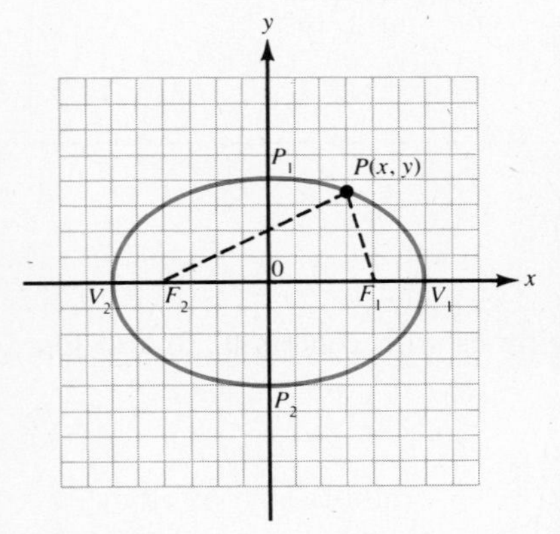

Figure 11-14. An ellipse with foci F_1 and F_2.

In Figure 11-14, an ellipse is drawn with foci F_1 and F_2. The following terminology is used for this figure:

1. Point O is the **center of the ellipse.**
2. V_1 and V_2 are the **vertices** (plural of vertex).

3. The line segment with V_1 and V_2 as endpoints is the **major axis.**

4. The line segment with P_1 and P_2 as endpoints is the **minor axis.**

5. $P(x, y)$ represents all the points of the ellipse.

By definition 11-2, $P(x, y)$ is a point of the ellipse, if and only if:

$$d(P, F_1) + d(P, F_2) = k, \quad \text{where } k \text{ is a positive number}$$

The Standard Form Equation of an Ellipse with Center (0, 0)

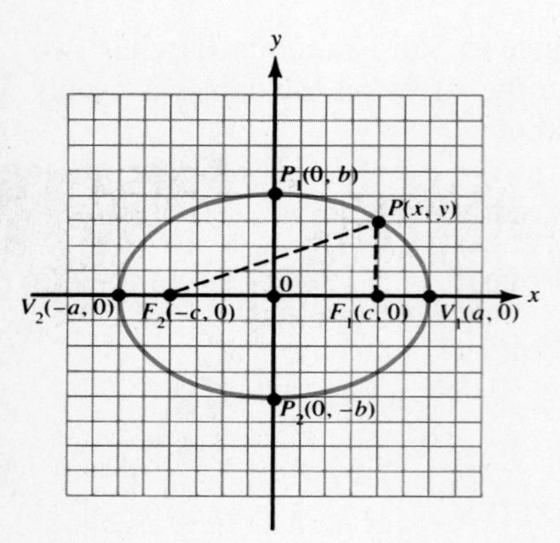

Figure 11-15a. Foci on x-axis.

To write an equation that defines an ellipse, F_1 and F_2 are placed in an xy-coordinate system. The distance formula can then be used to write an equation that defines the ellipse.

The simplest way to locate F_1 and F_2 is to position them so that 0, the center of the ellipse, coincides with the origin and F_1 and F_2 are on one of the axes. In Figure 11-15a, the foci are shown on the x-axis at $F_1(c, 0)$ and $F_2(-c, 0)$, where $c > 0$. The vertices are therefore at $V_1(a, 0)$ and $V_2(-a, 0)$, where $a > 0$. The endpoints of the minor axis are $P_1(0, b)$ and $P_2(0, -b)$, where $b > 0$. Notice that $a > b > 0$. Furthermore, both *the* x*-axis and* y*-axis are lines of symmetry for the graph.*

In Figure 11-15b, the foci are shown on the y-axis at $F_1(0, c)$ and $F_2(0, -c)$. The coordinates of V_1, V_2, P_1, and P_2 are correspondingly reversed from the ones in Figure 11-15a. When the distance formula is used to obtain equations for these ellipses, we get the so-called *standard form equation of an ellipse with center* (0, 0) *and foci on a coordinate axis.*

Figure 11-15b. Foci on y-axis.

The standard form equation of an ellipse with center (0, 0)

Foci on x-axis	**Foci on y-axis**
1. $F_1(c, 0)$ and $F_2(-c, 0)$ are foci.	**1.** $F_1(0, c)$ and $F_2(0, -c)$ are foci.
2. $V_1(a, 0)$ and $V_2(-a, 0)$ are vertices.	**2.** $V_1(0, a)$ and $V_2(0, -a)$ are vertices.
3. $P_1(0, b)$ and $P_2(0, -b)$ are endpoints of the minor axis.	**3.** $P_1(b, 0)$ and $P_2(-b, 0)$ are endpoints of the minor axis.
$\dfrac{x^2}{a^2} + \dfrac{y^2}{b^2} = 1$	$\dfrac{y^2}{a^2} + \dfrac{x^2}{b^2} = 1$

where $a > b > 0$, and $a^2 - b^2 = c^2$

A Procedure for Graphing an Ellipse

The equations in examples **a** and **b** define ellipses with centers (0, 0) and foci along the coordinate axes.

a. $4x^2 + 9y^2 = 144$ Foci are on the x-axis.

b. $4x^2 + y^2 = 64$ Foci are on the y-axis.

Example 1. Graph $4x^2 + 9y^2 = 144$.

Solution. **Step 1.** Write the equation in the standard form.

$$\frac{4x^2}{144} + \frac{9y^2}{144} = \frac{144}{144} \qquad \text{Divide each term by 144.}$$

$$\frac{x^2}{36} + \frac{y^2}{16} = 1 \qquad \text{Simplify.}$$

Step 2. Identify the values of a and b.

$a^2 = 36$ and $a = 6$ Remember that $a^2 > b^2$.

$b^2 = 16$ and $b = 4$

Step 3. Plot V_1, V_2, P_1, and P_2. In Figure 11-16, $V_1(6, 0)$, $V_2(-6, 0)$, $P_1(0, 4)$ and $P_2(0, -4)$ are plotted. These points determine the horizontal and vertical extents of the graph.

Step 4. Pick a value of x between 0 and a for ellipses with foci on the x-axis, and a value of x between 0 and b for ones with foci on the y-axis, and find the corresponding values for y

$4(3)^2 + 9y^2 = 144$ Replace x by 3.

$y \approx \pm 3.5$ Approximate to one decimal place.

Step 5. Use symmetry with respect to the x- and y-axes to plot additional points on the ellipse. With $x = \pm 3$ and $y \approx \pm 3.5$, the four points in Figure 11-16 are approximately located to give shape to the ellipse shown.

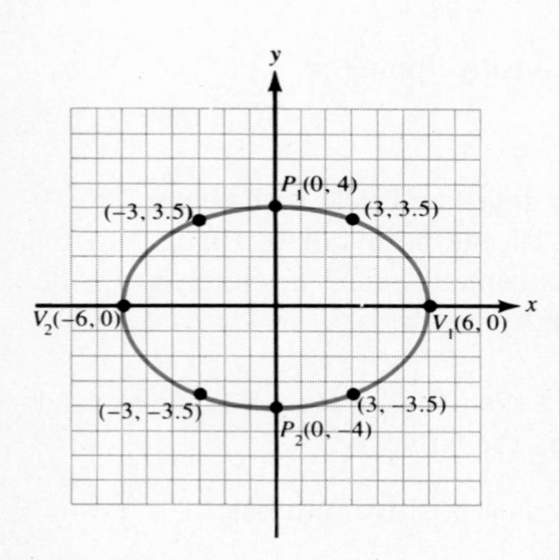

Figure 11-16. A graph of $4x^2 + 9y^2 = 144$.

Example 2. Graph $4x^2 + y^2 = 64$.

Solution. **Step 1.** $\frac{4x^2}{64} + \frac{y^2}{64} = \frac{64}{64}$ Divide each term by 64.

$$\frac{x^2}{16} + \frac{y^2}{64} = 1 \qquad \text{Simplify.}$$

Step 2. $a^2 = 64$ and $a = 8$ Remember that $a^2 > b^2$.

$b^2 = 16$ and $b = 4$

Step 3. In Figure 11-17, $V_1(0, 8)$, $V_2(0, -8)$, $P_1(4, 0)$, and $P_2(-4, 0)$ are plotted. These points determine the vertical and horizontal extents of the graph.

Step 4. $4(3)^2 + y^2 = 64$ Replace x by 3.

$y \approx \pm 5.3$ Approximate $\sqrt{28}$.

Step 5. With $x = \pm 3$ and $y \approx \pm 5.3$, the four points in Figure 11-17 are plotted to give shape to the ellipse shown.

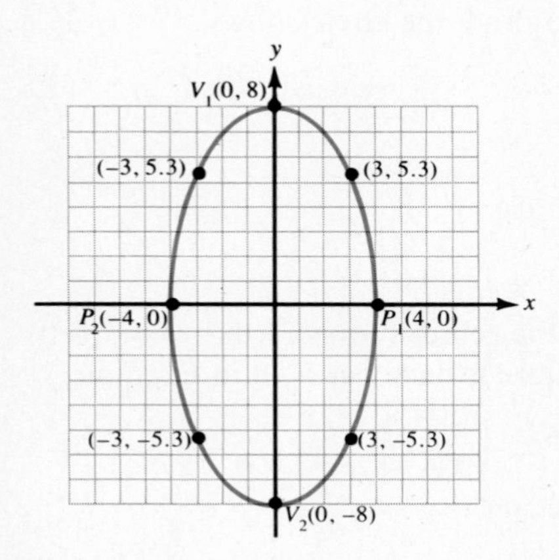

Figure 11-17. A graph of $4x^2 + y^2 = 64$.

The Center-Radius Equation of a Circle

The basic shape of an ellipse is oval. The amount of elongation to the oval is determined by the distance between the foci. Comparing the ellipses in Figures 11-16 and 11-17, the foci are farther apart in the latter figure because the oval is more elongated. The closer the foci are together, the more circular the shape. In fact, *if* F_1 *and* F_2 *are the same point, then the ellipse is a circle.* The following statement is consequently used to define a circle:

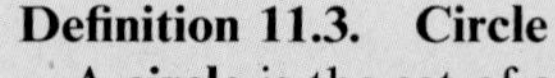

Definition 11.3. Circle
A **circle** is the set of all points in a plane that is a fixed distance from a given point in the plane. The fixed distance is called the **radius,** and the fixed point is called the **center.**

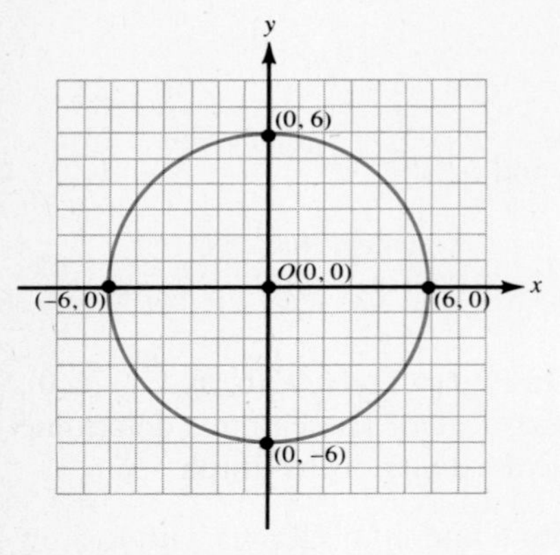

Figure 11-18. A graph of $x^2 + y^2 = 36$.

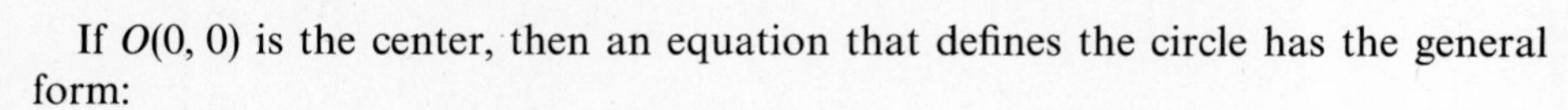

If $O(0, 0)$ is the center, then an equation that defines the circle has the general form:

$$x^2 + y^2 = r^2, \quad \text{where } r \text{ is a positive number}$$

Example 3. Graph $x^2 + y^2 = 36$.

Solution. **Discussion.** The given equation defines a circle with center $O(0, 0)$ and radius $r = 6$. In Figure 11-18, the points $(6, 0)$, $(0, 6)$, $(-6, 0)$, and $(0, -6)$ are plotted. An instrument called a compass can be used to accurately draw the circle shown.

If $O(h, k)$ is the center and r is the radius, then the distance formula can be used to obtain the **center-radius equation** that defines the circle:

$$(x - h)^2 + (y - k)^2 = r^2, \quad \text{where } r \text{ is a positive number}$$

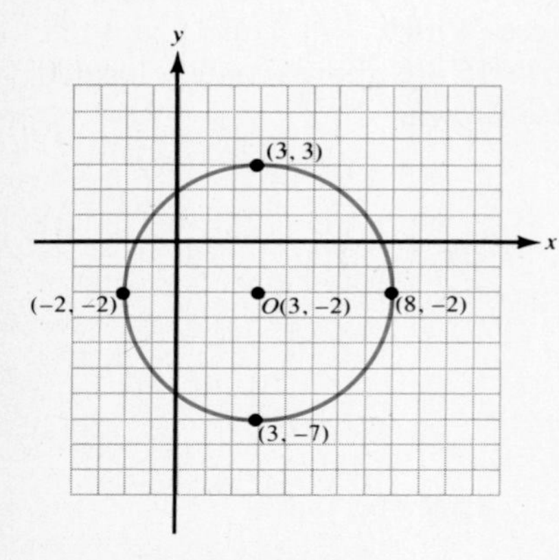

Figure 11-19. A graph of $(x - 3)^2 + (y + 2)^2 = 25$.

Example 4. Graph $(x - 3)^2 + (y + 2)^2 = 25$.

Solution. **Discussion.** Comparing the given equation with the center-radius form:

$$h = 3,\ k = -2, \text{ and } r = 5$$

In Figure 11-19, the point $O(3, -2)$ is plotted. A compass set at $r = 5$ units can then be used to draw the circle shown.

Finding the Center and Radius of a Circle

A graph of the equation in example **c** is a circle.

c. $x^2 + y^2 + 8x - 4y = 16$

To graph the circle, rewrite the equation in the center-radius form by completing the square in x and y. The four-step procedure is demonstrated in Example 5.

Example 5. Graph $x^2 + y^2 + 8x - 4y = 16$.

Solution. **Step 1.** Separate the x- and y-terms.

$$x^2 + 8x \quad + y^2 - 4y \quad = 16$$

Step 2. Complete the squares in x and y.

$$x^2 + 8x + 16 + y^2 - 4y + 4 = 16 + 16 + 4$$

Step 3. Write in the form $(x - h)^2 + (y - k)^2 = r^2$.

$$(x + 4)^2 + (y - 2)^2 = 36 \qquad h = -4, k = 2, \text{ and } r = 6$$

Step 4. Graph the circle from Step 3. In Figure 11-20, $O(-4, 2)$ is plotted. A compass set at $r = 6$ units can then be used to graph the circle shown.

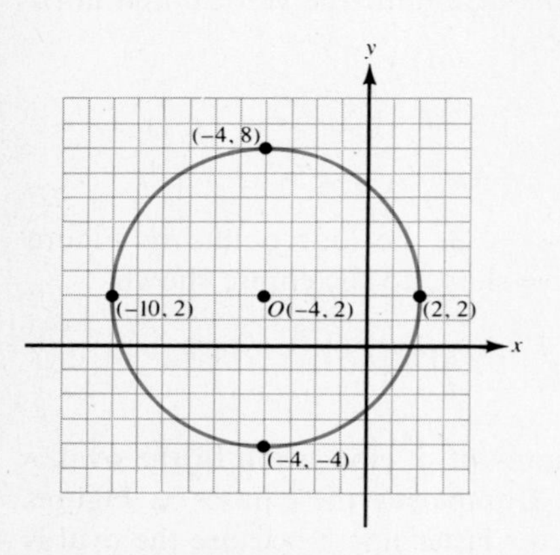

Figure 11-20. A graph of $x^2 + y^2 + 8x - 4y = 16$.

SECTION 11-2. Practice Exercises

In exercises **1–20**, graph each equation.

[Examples 1 and 2]

1. $x^2 + 4y^2 = 16$

2. $4x^2 + y^2 = 16$

3. $9x^2 + y^2 = 144$

4. $x^2 + 9y^2 = 144$

5. $4x^2 + 9y^2 = 36$

6. $9x^2 + 4y^2 = 36$

7. $x^2 + 9y^2 = 9$

8. $16x^2 + 9y^2 = 144$

9. $25x^2 + 16y^2 = 400$

10. $4x^2 + 25y^2 = 100$

11. $9x^2 + 25y^2 = 225$

12. $25x^2 + 4y^2 = 100$

13. $36x^2 + y^2 = 36$

14. $36x^2 + 25y^2 = 900$

15. $81x^2 + 4y^2 = 324$

16. $9x^2 + 100y^2 = 900$

17. $4x^2 + 9y^2 = 9$

18. $25x^2 + 16y^2 = 25$

19. $64x^2 + 4y^2 = 225$

20. $x^2 + 25y^2 = 9$

In exercises **21–28** graph each equation.

[Examples 3 and 4]

21. $x^2 + y^2 = 4$

22. $x^2 + y^2 = 64$

23. $x^2 + (y - 2)^2 = 9$

24. $x^2 + (y + 3)^2 = 16$

25. $(x + 1)^2 + y^2 = 25$

26. $(x - 1)^2 + y^2 = 36$

27. $(x - 2)^2 + (y - 5)^2 = 49$

28. $(x - 3)^2 + (y + 4)^2 = 9$

In exercises **29–40**, for each equation of a circle:

a. State the coordinates of the center and the radius.

b. Graph.

[Example 5]

29. $x^2 + y^2 - 10x + 9 = 0$

30. $x^2 + y^2 + 4x = 60$

31. $x^2 + y^2 + 6y = 16$

32. $x^2 + y^2 - 8y + 7 = 0$

33. $x^2 + y^2 - 2x + 4y = 31$

34. $x^2 + y^2 - 6x - 6y = 7$

35. $x^2 + y^2 + 4x - 2y = 95$

36. $x^2 + y^2 + 2x - 6y = 6$

37. $x^2 + y^2 + 8x + 10y + 37 = 0$

38. $x^2 + y^2 + 6x + 2y + 1 = 0$

39. $4x^2 + 4y^2 - 12x + 20y = 2$

40. $4x^2 + 4y^2 - 28x - 36y + 30 = 0$

SECTION 11-2. Ten Review Exercises

In exercises **1** and **2**, simplify each expression. Use only positive exponents.

1. $\left(\frac{3}{t}\right)^{-3}\left(\frac{6^{-1}}{t^{-4}}\right)^{-1}$

2. $(2u^{-2})^{-2}(8^{-1}u)^{-2}$

In exercises **3–6**, use f defined by $f(x) = \frac{x+3}{3x-2}$.

3. State the domain of f.

4. Write an equation that defines f^{-1} using $f^{-1}(x)$ notation.

5. State the domain of f^{-1}.

6. State the range of f^{-1}.

In exercises **7–10**, use the equation $4x - 3y = 3$.

7. Write in the form $f(x) = mx + b$.

8. Identify the slope of a graph of the equation.

9. Identify the y-intercept.

10. Identify the x-intercept.

SECTION 11-2. Supplementary Exercises

A graph of equations (1) and (2) are semicircles.

$$(1)\ y = \sqrt{r^2 - x^2} \quad \text{and} \quad (2)\ y = -\sqrt{r^2 - x^2}$$

The centers of the corresponding circles are $O(0, 0)$, and the radii are $r > 0$.

In exercises **1–4**, graph each equation.

1. $y = \sqrt{25 - x^2}$

2. $y = \sqrt{36 - x^2}$

3. $y = -\sqrt{49 - x^2}$

4. $y = -\sqrt{64 - x^2}$

If the $=$ in the standard-form equation of an ellipse is replaced by $<$ or $>$, then the graph is, respectively, the *interior* or *exterior* of an ellipse.

In exercises **5–10**, graph each inequality.

5. $9x^2 + 16y^2 > 144$

6. $25x^2 + 16y^2 < 400$

7. $x^2 + 16y^2 > 16$

8. $9x^2 + y^2 > 9$

9. $25x^2 + 9y^2 \leq 225$

10. $16x^2 + 81y^2 \leq 1{,}296$

If the $=$ in the center-radius equation of a circle is replaced by $<$ or $>$, then the graph is, respectively, the *interior* or *exterior* of a circle.

In exercises **11–14**, graph each inequality.

11. $(x + 2)^2 + (y - 1)^2 < 16$

12. $(x - 1)^2 + (y - 3)^2 \leq 49$

13. $x^2 + y^2 - 6x - 6y \leq 7$

14. $x^2 + y^2 - 2x + 4y > 31$

SECTION 11-3. Hyperbolas

1. A definition of a hyperbola
2. The standard form equation of a hyperbola with center (0, 0)
3. A procedure for graphing a hyperbola
4. Hyperbolas defined by $xy = k$

In this section, we will study conics called hyperbolas. As shown in Figure 11-4, a hyperbola is formed when a plane parallel to the axis of a right circular cone intersects both nappes.

A Definition of a Hyperbola

Definition 11.4. Hyperbola
A **hyperbola** is the set of all points in a plane the difference of whose distances from two fixed points in the plane is a constant. The two fixed points, F_1 and F_2, are the foci of the hyperbola.

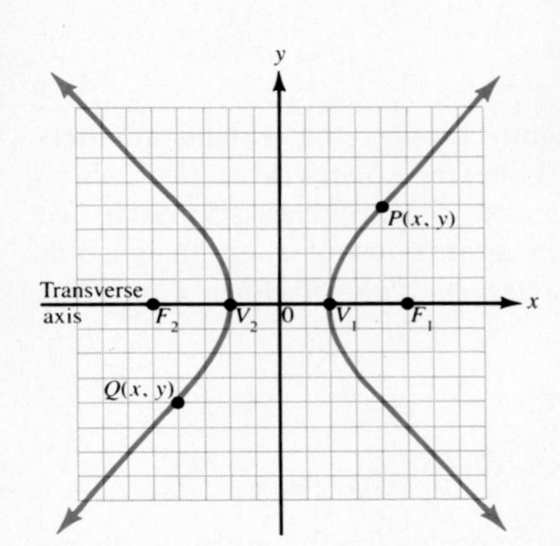

Figure 11-21. A hyperbola with foci F_1 and F_2.

In Figure 11-21, a hyperbola is drawn with foci F_1 and F_2. The following terminology is used for this figure:

1. Point O is the **center of the hyperbola.**
2. V_1 and V_2 are the **vertices.**
3. The line segment with V_1 and V_2 as endpoints is the **transverse axis.**
4. $P(x, y)$ represents all points on the right branch of the hyperbola, and $Q(x, y)$ represents all points on the left branch.

By Definition 11.4, $P(x, y)$ and $Q(x, y)$ are points on the hyperbola, if and only if:

$$\left.\begin{aligned} d(P, F_2) - d(P, F_1) = k \\ d(Q, F_1) - d(Q, F_2) = k \end{aligned}\right\} \text{ where } k \text{ is a positive number}$$

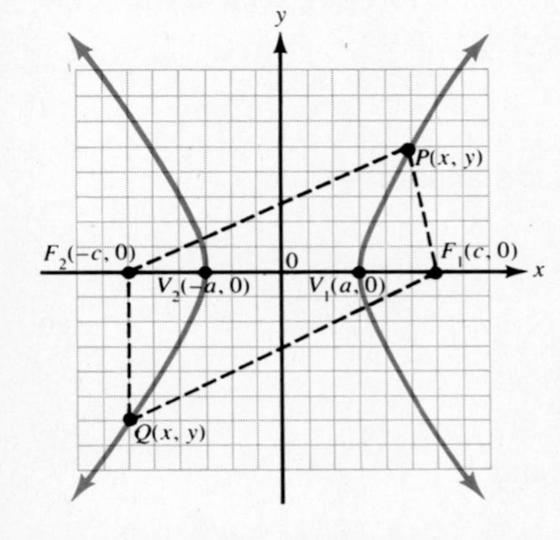

Figure 11-22a. A hyperbola with foci $F_1(c, 0)$ and $F_2(-c, 0)$.

The Standard Form Equation of a Hyperbola with Center (0, 0)

To write an equation that defines a hyperbola, F_1 and F_2 are placed in an xy-coordinate system. The distance formula can then be used to write an equation that defines the hyperbola.

The simplest way to locate F_1 and F_2 is to position them so that O, the center of the hyperbola, coincides with the origin and F_1 and F_2 are on one of the axes. In Figure 11-22a, the foci are shown on the x-axis at $F_1(c, 0)$ and $F_2(-c, 0)$, where $c > 0$. The vertices are at $V_1(a, 0)$ and $V_2(-a, 0)$, where $a > 0$. Notice that $c > a > 0$. Furthermore, *the* x- *and* y-*axes are lines of symmetry for the graph.*

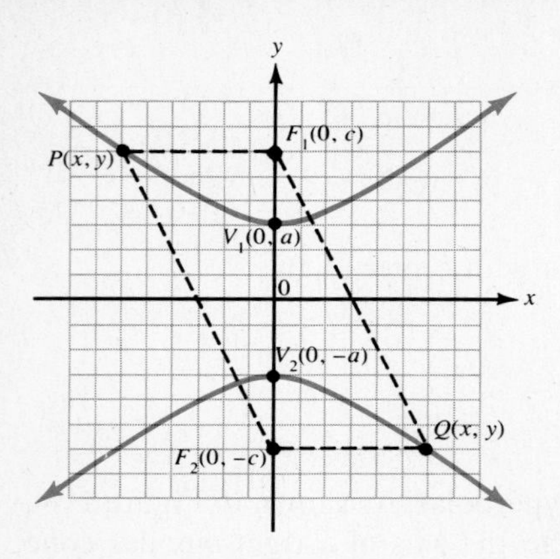

Figure 11-22b. A hyperbola with foci $F_1(0, c)$ and $F_2(O, -c)$.

In Figure 11-22b, the foci are shown on the y-axis at $F_1(0, c)$ and $F_2(0, -c)$. The vertices are now $V_1(0, a)$ and $V_2(0, -a)$, where again $c > a > 0$. Notice that the x- and y-axes are also lines of symmetry for this figure. When the distance formula is used to obtain the equations for the hyperbolae in these figures, we get the so-called *standard form equation of a hyperbola with center (0, 0) and foci on a coordinate axis.*

The standard form equation of a hyperbola with center (0, 0)

Foci on x-axis	**Foci on y-axis**
1. $F_1(c, 0)$ and $F_2(-c, 0)$ are foci.	**1.** $F_1(0, c)$ and $F_2(0, -c)$ are foci.
2. $V_1(a, 0)$ and $V_2(-a, 0)$ are vertices.	**2.** $V_1(0, a)$ and $V_2(0, -a)$ are vertices.
$\frac{x^2}{a^2} - \frac{y^2}{b^2} = 1$	$\frac{y^2}{a^2} - \frac{x^2}{b^2} = 1$

where $a > 0$, $b > 0$, and $a^2 + b^2 = c^2$

A Procedure for Graphing a Hyperbola

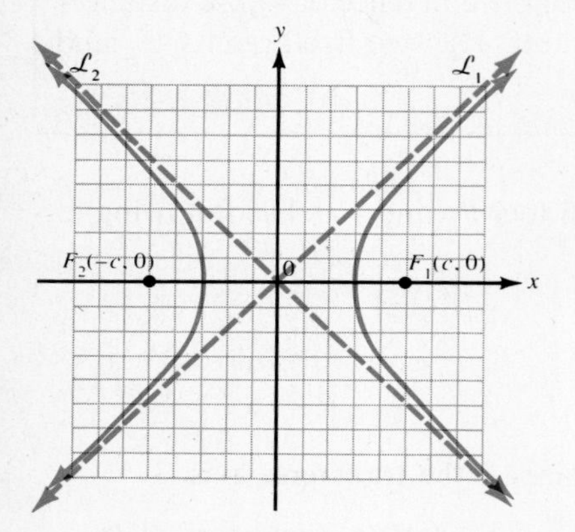

Figure 11-23. A hyperbola with foci $F_1(c, 0)$ and $F_2(-c, 0)$.

A graph of a hyperbola consists of two **disjoint** branches that extend infinitely far to the left and right (for foci on a horizontal line), or up and down (for foci on a vertical line). It can be shown that in either case, the branches approach two intersecting lines as **asymptotes.** Recall that an asymptote for a graph is a line that the graph gets closer and closer to for *extreme values of either* x *or* y. By extreme values we mean:

$$|x| \to \infty \qquad x \to +\infty \quad \text{or} \quad x \to -\infty$$

$$|y| \to \infty \qquad y \to +\infty \quad \text{or} \quad y \to -\infty$$

In Figure 11-23, the lines $\mathscr{L}_1$ and $\mathscr{L}_2$ are asymptotes for the given graph. As $x \to +\infty$ and $x \to -\infty$, the branches will get closer and closer to, but will never touch, these lines. Similarly, for a hyperbola with foci on the y-axis, lines can be drawn that are asymptotes for the graph.

In Examples 1 and 2, a six-step procedure is given for locating such asymptotes.

Example 1. Graph $4x^2 - y^2 = 16$.

Solution. **Step 1.** Write the equation in the standard form.

$$\frac{4x^2}{16} - \frac{y^2}{16} = \frac{16}{16} \qquad \text{Divide each term by 16.}$$

$$\frac{x^2}{4} - \frac{y^2}{16} = 1 \qquad \text{Simplify.}$$

Step 2. Identify the values of a and b.

$a^2 = 4$ and $a = 2$ — a^2 is always with the first term.

$b^2 = 16$ and $b = 4$

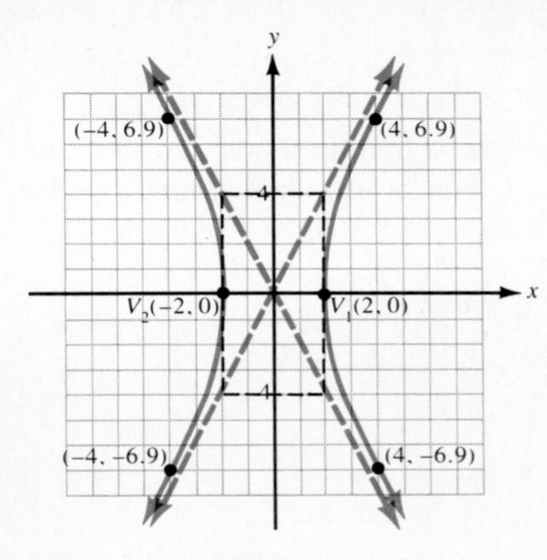

Figure 11-24. A graph of $4x^2 - y^2 = 16$.

Step 3. Plot $\pm a$ and $\pm b$ on the coordinate axes and construct a rectangle using dimension $2a$ by $2b$. In Figure 11-24, $V_1(2, 0)$ and $V_2(-2, 0)$ are shown as vertices, and 4 and -4 are plotted on the y-axis. The dotted lines through these points form a rectangle that is 4 units wide and 8 units high.

Step 4. Draw as asymptotes $\mathscr{L}_1$ and $\mathscr{L}_2$, the lines that contain the diagonals of the rectangle. In Figure 11-24, $\mathscr{L}_1$ and $\mathscr{L}_2$ are shown.

Step 5. Pick a suitable value for x or y, and find the corresponding values of the other variable. Let $x = 4$.

$4(4)^2 - y^2 = 16$ — Replace x by 4.

$y \approx \pm 6.9$ — Approximate to one decimal place.

Step 6. Use symmetry with respect to the x- and y-axes to plot additional points on the hyperbola. With $x = \pm 4$ and $y \approx \pm 6.9$, the four points in Figure 11-24 are approximately located to give shape to the hyperbola shown.

Example 2. Graph $25y^2 - 9x^2 = 225$.

Solution. **Step 1.** $\dfrac{y^2}{9} - \dfrac{x^2}{25} = 1$ — Divide each term by 225.

Step 2. $a^2 = 9$ and $a = 3$ — a^2 is always with the first term

$b^2 = 25$ and $b = 5$

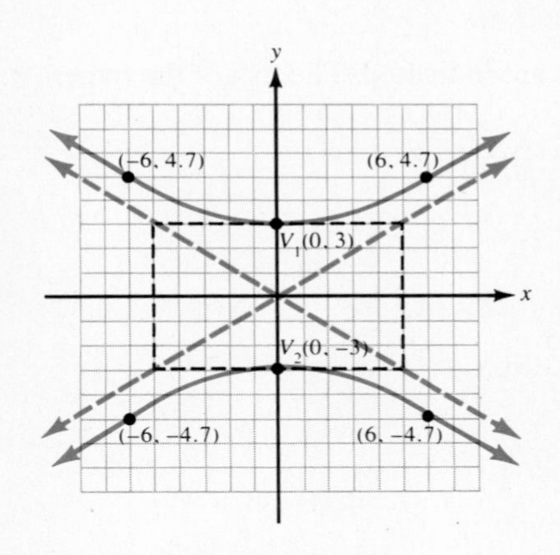

Figure 11-25. A graph of $25y^2 - 9x^2 = 225$.

Step 3. In Figure 11-25, $V_1(0, 3)$ and $V_2(0, -3)$ are shown as vertices, and 5 and -5 are plotted on the x-axis. The dotted lines through these points form a rectangle that is 10 units wide and 6 units high.

Step 4. $\mathscr{L}_1$ and $\mathscr{L}_2$ are drawn as asymptotes.

Step 5. Let $x = 6$.

$25y^2 - 9(6)^2 = 225$ — Replace x by 6.

$y \approx \pm 4.7$ — Approximate to one decimal place.

Step 6. With $x = \pm 6$ and $y \approx \pm 4.7$, the four points in Figure 11-25 are approximately located to give shape to the hyperbola shown.

Hyperbolas Defined by $xy = k$

The two branches of a hyperbola are also obtained as graphs of equations with the general form:

$$x \cdot y = k, \quad \text{where } k \text{ is a nonzero constant}$$

The foci and vertices of the hyperbola of such an equation lie on lines defined by either $y = x$ (provided $k > 0$), or $y = -x$ (provided $k < 0$). Examples 3 and 4 illustrate both cases.

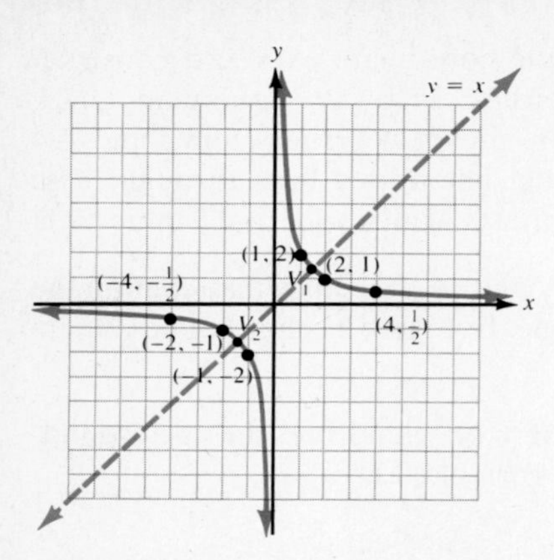

Figure 11-26. A graph of $xy = 2$.

Example 3. Graph $xy = 2$.

Solution. The product of x and y is positive. Therefore, $y > 0$ whenever $x > 0$, and $y < 0$ whenever $x < 0$. The points determined by the ordered pairs in the following table are plotted in Figure 11-26:

x	-4	-2	-1	1	2	4
$y = \frac{2}{x}$	$\frac{-1}{2}$	-1	-2	2	1	$\frac{1}{2}$

The line $y = -x$ and the vertices are indicated. The foci of the hyperbola are points on this line.

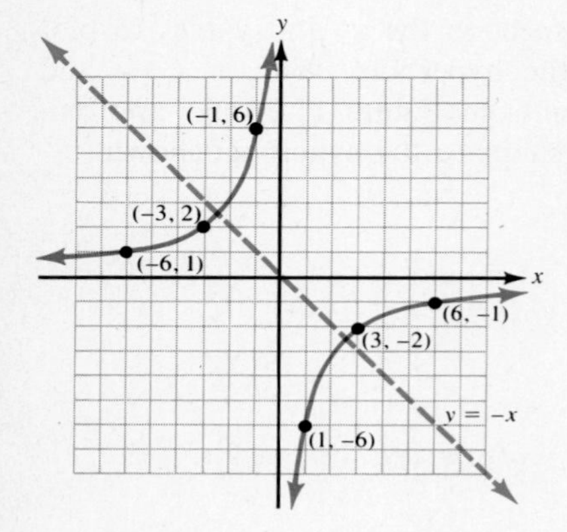

Figure 11-27. A graph of $xy = -6$.

Example 4. Graph $xy = -6$.

Solution. The product of x and y is negative. Therefore $y > 0$ whenever $x < 0$, and $y < 0$ whenever $x > 0$. The points determined by the ordered pairs in the following table are plotted in Figure 11-27:

x	6	3	1	-1	-3	-6
$y = \frac{-6}{x}$	-1	-2	-6	6	2	1

The line $y = x$ and the vertices are indicated. The foci of the hyperbola are points on this line.

SECTION 11-3. Practice Exercises

In exercises **1–24**, graph each equation.

[Examples 1 and 2]

1. $x^2 - y^2 = 4$

2. $y^2 - x^2 = 4$

3. $4y^2 - x^2 = 16$

4. $x^2 - 9y^2 = 9$

5. $x^2 - 16y^2 = 16$

6. $4x^2 - 9y^2 = 36$

7. $9y^2 - 4x^2 = 36$

8. $25y^2 - 4x^2 = 100$

9. $25x^2 - 16y^2 = 400$

10. $16y^2 - 25x^2 = 400$

11. $4y^2 - 49x^2 = 196$

12. $4x^2 - 49y^2 = 196$

13. $36x^2 - 25y^2 = 900$

14. $25x^2 - 36y^2 = 900$

15. $9y^2 - x^2 = 4$

16. $y^2 - 4x^2 = 9$

[Examples 3 and 4]

17. $xy = 4$

18. $xy = 8$

19. $xy = -5$

20. $xy = -6$

21. $xy = 10$

22. $xy = -10$

23. $2xy = 3$

24. $3xy = 2$

SECTION 11-3. Ten Review Exercises

In exercises **1–4**, do the indicated operations and simplify each expression.

1. $-5[3^2 - 2(18 - 5^2) - 17] + \dfrac{-3(-2 + 10)}{2^3 - 3^2}$

2. $\dfrac{8^{-1} + 12^{-1}}{96^{-1}}$

3. $(6t^4 + 13t^3 - 5t^2 - 9t + 10) \div (3t - 1)$

4. $\dfrac{30u}{u^2 - u - 30} + \dfrac{6u^2 - 3u}{u^2 - 5u - 6} \cdot \dfrac{2u^2 + 2u}{2u^2 + 9u - 5}$

In exercises **5–8**, solve each equation.

5. $|3 - x| - 6 = 0$

6. $-7t - 5(t + 3) = -15(t + 1)$

7. $1 + \dfrac{3}{2t + 2} = \dfrac{7}{t - 1} + \dfrac{2t^2 + 3}{2t^2 - 2}$

8. $x^2 = 4x - 53$

In exercises **9** and **10**, use f defined by $f(x) = \dfrac{x}{x + 2}$.

9. State the domain of f.

10. Write an equation for f^{-1} using $f^{-1}(x)$ notation.

SECTION 11-3. Supplementary Exercises

In exercises **1–8**, graph equations (1) and (2) on the same axes.

1. (1) $\dfrac{x^2}{9} + \dfrac{y^2}{4} = 1$
(2) $\dfrac{x^2}{9} - \dfrac{y^2}{4} = 1$

2. (1) $\dfrac{x^2}{25} - \dfrac{y^2}{9} = 1$
(2) $\dfrac{x^2}{25} + \dfrac{y^2}{9} = 1$

3. (1) $x^2 + y^2 = 9$
(2) $y^2 - x^2 = 9$

4. (1) $x^2 + y^2 = 16$
(2) $y^2 - x^2 = 16$

5. (1) $x^2 + y = 4$
(2) $x^2 + y^2 = 16$

6. (1) $y - x^2 = -6$
(2) $x^2 + y^2 = 36$

7. (1) $4x^2 - 9y^2 = 36$
(2) $4x^2 - 9y = 36$

8. (1) $25x^2 - 4y^2 = 100$
(2) $25x^2 - 4y = 20$

In exercises **9–16**, write an equation for the graph in each figure.

a. For an ellipse: $\frac{x^2}{a^2} + \frac{y^2}{b^2} = 1$ or $\frac{y^2}{a^2} + \frac{x^2}{b^2} = 1$

b. For a circle: $(x - h)^2 + (y - k)^2 = r^2$

c. For a hyperbola: $\frac{x^2}{a^2} - \frac{y^2}{b^2} = 1$ or $\frac{y^2}{a^2} - \frac{x^2}{b^2} = 1$

9.

10.

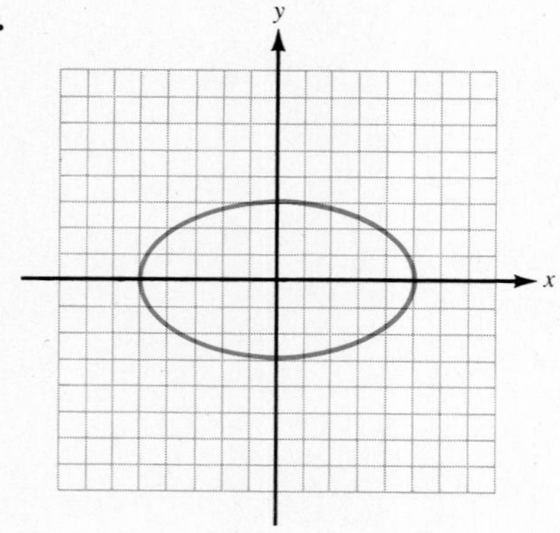

11.

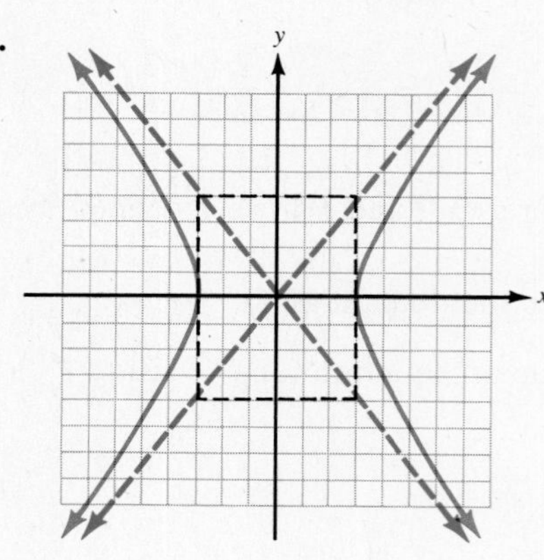

12.

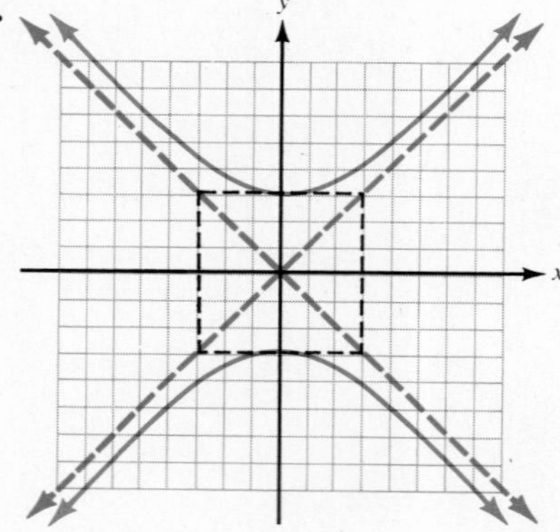

13.

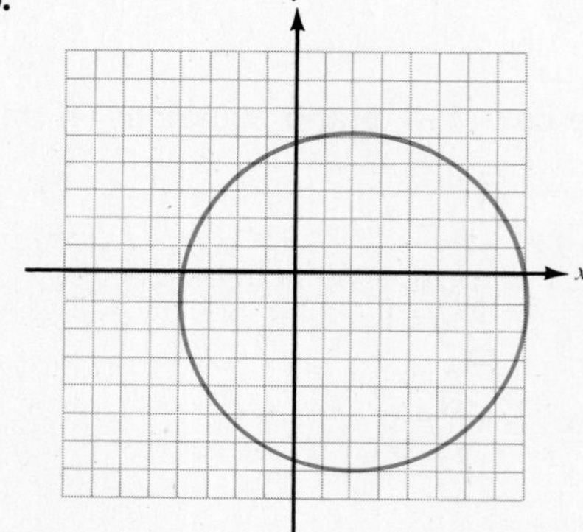

14.

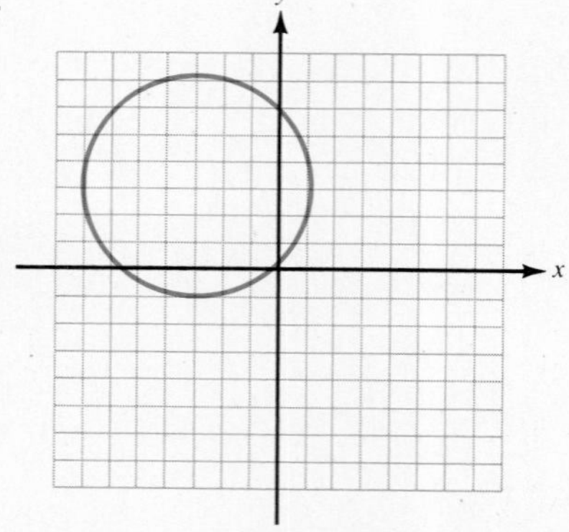

15.

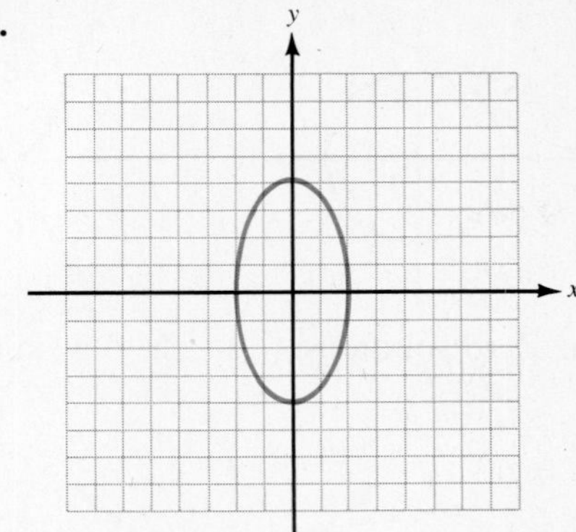

16.

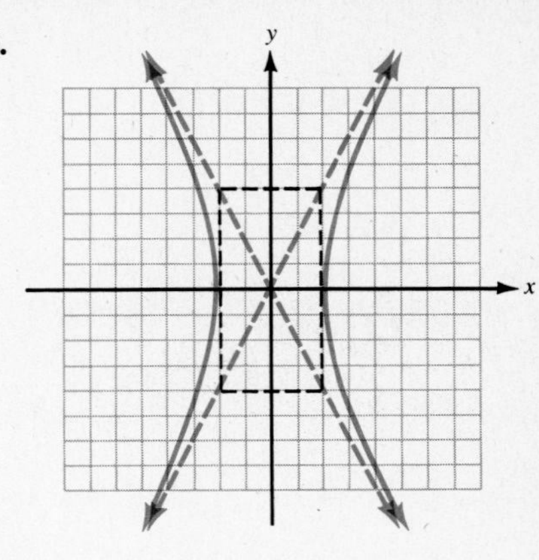

SECTION 11-4. Systems of Nonlinear Equations in x and y

1. A description of a system of nonlinear equations in x and y
2. Solving a nonlinear system by the substitution method
3. Solving a nonlinear system by the addition method
4. Some applications

In Chapter 9, we solved systems of linear equations in x and y. In this section, we will solve some systems of nonlinear equations in x and y.

A Description of a System of Nonlinear Equations in x and y

If two or more equations are grouped to form a system of equations, then the system is frequently called nonlinear when at least one of the equations is nonlinear. Examples **a–d** illustrate four such systems.

	Nonlinear system	Graph of the equation
a.	(1) $x + 2y = 10$	Line
	(2) $x^2 - y + 2 = 0$	Parabola
b.	(1) $x^2 + y^2 = 16$	Circle
	(2) $4x^2 - y^2 = 4$	Hyperbola
c.	(1) $x^2 + y^2 = 4$	Circle
	(2) $x^2 + 3y^2 = 3$	Ellipse
d.	(1) $x^2 - 4y = 0$	Parabola
	(2) $x^2 - y^2 = 3$	Hyperbola

The real solutions of a nonlinear system of two equations in two variables can be found (if only approximately) by graphing both equations on the same coordinate system. As shown in Figure 11-28 (p. 732), the coordinates of any point of intersection of the two graphs are a solution of the system.

A nonlinear system may have as solutions complex numbers that are not real. For example, $(3i, -4i)$ is a solution of the following system:

(1) $xy = 12$ $\qquad (3i)(-4i) = -12i^2 = -12(-1) = 12$

(2) $y^2 - 4x^2 = 20$ $\qquad (-4i)^2 - 4(3i)^2 = 16i^2 - 36i^2 = -20(-1) = 20$

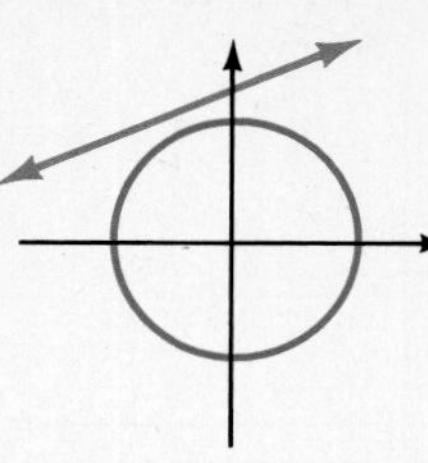

a. Zero solutions.

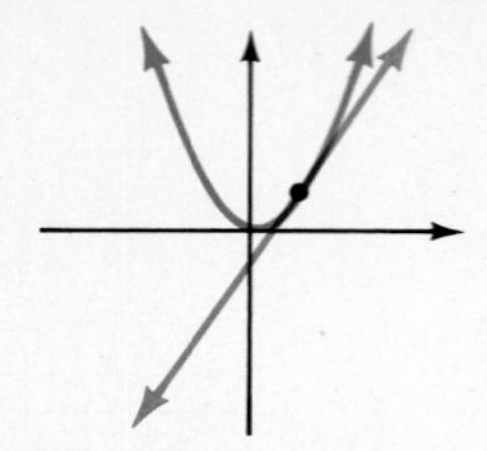

b. One solution.

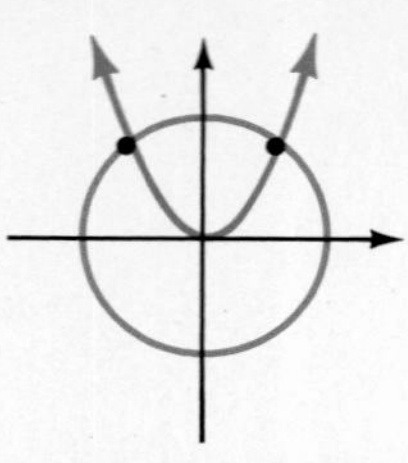

c. Two solutions.

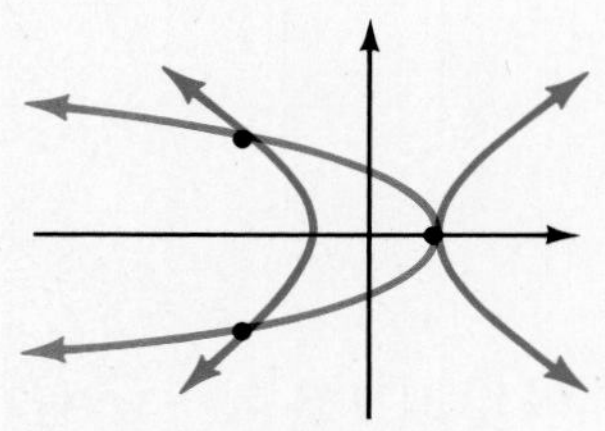

d. Three solutions.

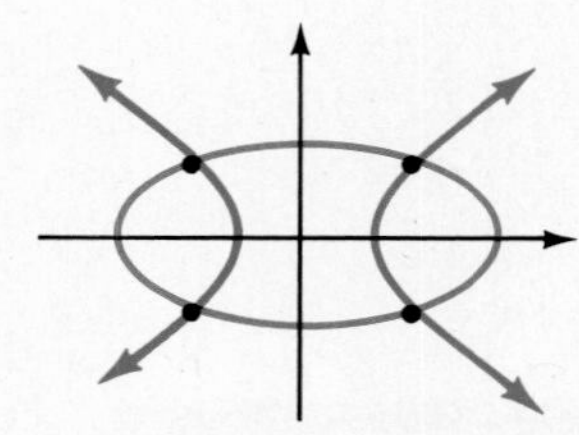

e. Four solutions.

Figure 11-28. The possible numbers of real solutions for nonlinear systems of two equations in two variables.

Such an ordered-pair solution would not appear as a point of intersection of the graphs of (1) and (2).

Solving a Nonlinear System by the Substitution Method

The substitution method can be used to solve some nonlinear systems of equations. This method is especially useful when one of the equations is linear. It can also be used when one of the equations can easily be solved for one of the variables.

Example 1. Given: (1) $x + 2y = 10$
(2) $x^2 - y = -2$

Solve by the substitution method.

Solution. **Discussion.** Equation (1) is linear and can be solved for either x or y. The expression for the solved variable can be substituted for it in (2).

(1) $x = 10 - 2y$ — Subtract $2y$.
(2) $(10 - 2y)^2 - y = -2$ — Replace x by $10 - 2y$.

$100 - 40y + 4y^2 - y = -2$ — $(x - y)^2 = x^2 - 2xy + y^2$

$4y^2 - 41y + 102 = 0$ — A quadratic equation

$(4y - 17)(y - 6) = 0$ — Factor.

$4y - 17 = 0$ or $y - 6 = 0$ — The zero-product property

$y = \dfrac{17}{4}$ $\quad$ $y = 6$ — Solve for y.

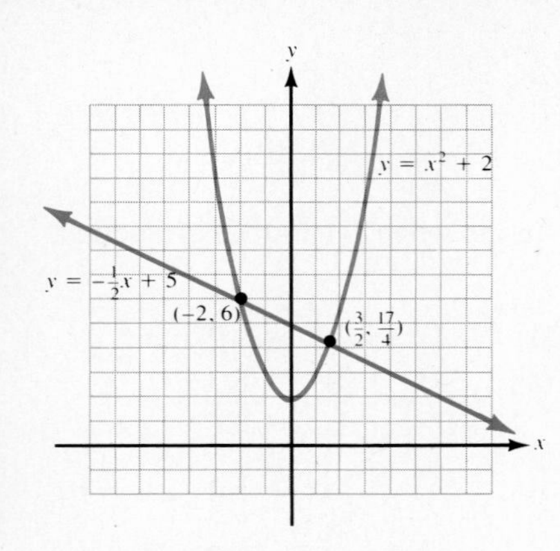

Figure 11-29. Graph of the system (1) $x + 2y = 10$ and (2) $x^2 - y = -2$.

$$x = 10 - 2\left(\frac{17}{4}\right) \quad \text{or} \quad x = 10 - 2(6) \qquad \text{Replace } y \text{ in (1).}$$

$$x = \frac{3}{2} \qquad\qquad x = -2$$

Thus, $(\frac{3}{2}, \frac{17}{4})$ and $(-2, 6)$ are solutions. The solutions are ordered pairs of real numbers. Therefore, the graphs of the equations of the system intersect in two points. The coordinates of these points are the solutions of the system, as shown in Figure 11-29.

Example 2. Given: (1) $x^2 + 9y^2 = 13$
(2) $xy = 2$

Solve by the substitution method.

Solution. **Discussion.** Equation (2) can easily be solved for x or y. The resulting expression can then be used in (1).

$$(2)\quad y = \frac{2}{x} \qquad \text{Solve (2) for } y.$$

$$(1)\quad x^2 + 9\left(\frac{2}{x}\right)^2 = 13 \qquad \text{Replace } y \text{ in (1).}$$

$$x^2 + \frac{36}{x^2} = 13 \qquad 9\left(\frac{2}{x}\right)^2 = 9\left(\frac{4}{x^2}\right) = \frac{36}{x^2}$$

$$x^4 + 36 = 13x^2 \qquad \text{Multiply by } x^2.$$

$$x^4 - 13x^2 + 36 = 0 \qquad \text{Set equal to 0.}$$

$$(x^2 - 9)(x^2 - 4) = 0 \qquad \text{Factor the trinomial.}$$

$$(x + 3)(x - 3)(x + 2)(x - 2) = 0 \qquad \text{Factor again.}$$

$$x + 3 = 0 \quad \text{or} \quad x - 3 = 0 \quad \text{or} \quad x + 2 = 0 \quad \text{or} \quad x - 2 = 0$$

$$x = -3 \qquad x = 3 \qquad x = -2 \qquad x = 2$$

Thus, $x = -3, 3, -2$, or 2. The corresponding values of y are obtained by using (2). The solution set is:

$$\left\{\left(-3, \frac{-2}{3}\right), \left(3, \frac{2}{3}\right), (-2, -1), (2, 1)\right\}$$

The four ordered pairs of real numbers are the intersections of the graphs of (1) and (2), as shown in Figure 11-30.

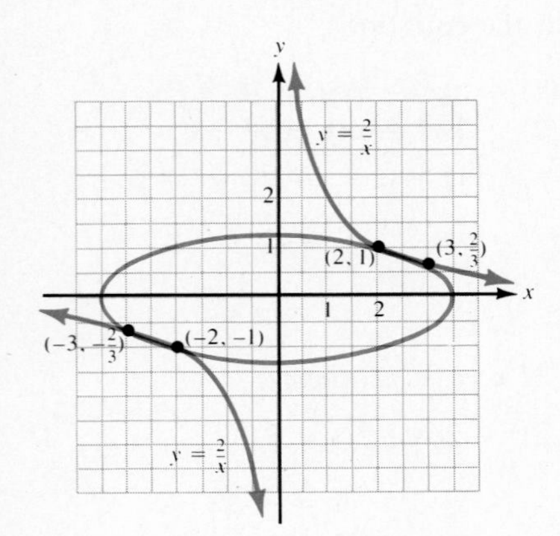

Figure 11-30. Graph of the system (1) $x^2 + 9y^2 = 13$ and (2) $xy = 2$.

Solving a Nonlinear System by the Addition Method

The addition method can be used to solve some nonlinear systems of equations. The technique is basically the same as the one used to solve some systems of linear equations. Use the multiplication property of equality to make the coefficients of one of the variables opposites. Then, when the equations are added, this variable will be eliminated.

Example 3. Given: (1) $2x^2 + y^2 = 1$
(2) $-3x^2 + 2y^2 = 30$

Solve by the addition method.

Solution. **Discussion.** To eliminate the y^2-terms when (1) and (2) are added, first multiply (1) by -2.

$$(1)\quad -2(2x^2 + y^2 = 1) \longrightarrow \begin{array}{ll} (1) & -4x^2 - 2y^2 = -2 \\ (2) & -3x^2 + 2y^2 = 30\ (+) \\ \hline & -7x^2 \qquad\quad = 28 \end{array}$$

$$x^2 = -4$$

$$x = \pm 2i$$

Replacing x^2 by -4 in (1):

(1) $2(-4) + y^2 = 1$ — Replace x^2 by -4.

$y^2 = 9$ — Add 8 to both sides.

$y = \pm 3$ — If $y^2 = k$, then $y = \pm\sqrt{k}$.

Thus, $(2i, 3)$, $(2i, -3)$, $(-2i, 3)$, and $(-2i, -3)$ are solutions.

Example 4. Given: (1) $4x^2 - 3xy = 2$
(2) $x^2 + 2xy = 6$

Solve by the addition method.

Solution. **Discussion.** To eliminate the xy-terms in (1) and (2), multiply (1) by 2 and (2) by 3, and then add the equations.

$$\begin{array}{ll} (1)\ 2(4x^2 - 3xy = 2) \longrightarrow & (1)\quad 8x^2 - 6xy = 4 \\ (2)\ \ 3(x^2 + 2xy = 6) \longrightarrow & (2)\quad 3x^2 + 6xy = 18 \\ \hline & \qquad 11x^2 \qquad\ = 22 \end{array}$$

$$x^2 = 2$$

$$x = \pm\sqrt{2}$$

If $x = \sqrt{2}$, then:

$(\sqrt{2})^2 + 2(\sqrt{2})y = 6$

$y = \sqrt{2}$

If $x = -\sqrt{2}$, then:

$(-\sqrt{2})^2 + 2(-\sqrt{2})y = 6$

$y = -\sqrt{2}$

The solution set is $\{(\sqrt{2}, \sqrt{2}), (-\sqrt{2}, -\sqrt{2})\}$.

Some Applications

Example 5 illustrates a problem that can be solved using systems of nonlinear equations.

Example 5. The sum of a number and 2 times another number is 14. The sum of the squares of the numbers is 73. Find both numbers.

Solution. Let x represent one number.
Let y represent the other number.

(1) $x + 2y = 14$	x plus two times y is 14
(2) $x^2 + y^2 = 73$	The sum of the squares is 73.

(1) $x = 14 - 2y$	Solve (1) for x.
(2) $(14 - 2y)^2 + y^2 = 73$	Replace x in (2).
$196 - 56y + 4y^2 + y^2 = 73$	Square the binomial.
$5y^2 - 56y + 123 = 0$	A quadratic equation
$(5y - 41)(y - 3) = 0$	Factor.
$y = \frac{41}{5}$ or $y = 3$	Solve for y.

If $y = \frac{41}{5}$, then: **If $y = 3$, then:**

$x = 14 - 2\left(\frac{41}{5}\right)$ or $x = 14 - 2(3)$

$x = \frac{-12}{5}$ $x = 8$

There are two pairs of numbers that satisfy (1) and (2): $\frac{-12}{5}$, $\frac{41}{5}$ and 8, 3.

SECTION 11-4. Practice Exercises

In exercises **1–30**, solve each system.

[Examples 1–4]

1. (1) $y = x^2 - 2x - 5$
(2) $x + y = 1$

2. (1) $x^2 = y + 5$
(2) $2x - y = 2$

3. (1) $x^2 + xy + 1 = 0$
(2) $y - x = 3$

4. (1) $x^2 + y^2 = 10$
(2) $2x + y = 5$

5. (1) $4x^2 + y^2 = 40$
(2) $2x + y = 8$

6. (1) $x^2 - y^2 = 12$
(2) $3y - x = 2$

7. (1) $7x - 3y = 10$
(2) $y^2 - x^2 = 20$

8. (1) $x = y^2 - 2y - 4$
(2) $x + y = 8$

9. (1) $x = 3y - 3$
(2) $x^2 + y^2 = 13$

10. (1) $x + 3y = 6$
(2) $x^2 + y^2 + 10x - 4y = -4$

11. (1) $3x + y = 5$
(2) $xy = -12$

12. (1) $2x - y = -7$
(2) $xy = 4$

13. (1) $2x + y = 4$
(2) $xy = -6$

14. (1) $xy = 5$
(2) $y - x = 4$

15. (1) $5x^2 - y^2 = 1$
(2) $xy = 2$

16. (1) $y^2 - x^2 = 5$
(2) $xy = -6$

17. (1) $3x + 2y = 5$
(2) $x^2 + 4xy = -35$

18. (1) $x^2 + xy = 60$
(2) $x - y = 2$

19. (1) $x^2 + 2y^2 = 1$
(2) $2x^2 - 3y^2 = 30$

20. (1) $4x^2 + 3y^2 = 11$
(2) $6x^2 + 5y^2 = 18$

21. (1) $5x^2 - 3y^2 = 5$
(2) $3x^2 - 2y^2 = -6$

22. (1) $2x^2 - 3y^2 = -6$
(2) $4x^2 - 5y^2 = 6$

23. (1) $4x^2 + 3y^2 = -106$
(2) $x^2 - y^2 = -23$

24. (1) $3x^2 - 5y^2 = 33$
(2) $2y^2 - x^2 = -15$

25. (1) $x^2 - 4y^2 = -16$
(2) $2x^2 + 2y^2 + 5y = 18$

26. (1) $4x^2 + y^2 = 25$
(2) $-2x + y^2 = 5$

27. (1) $x^2 + y^2 - 6x + 4y = 3$
(2) $x^2 + y^2 - 3x + 4y = 12$

28. (1) $x^2 - 2y^2 + 6x - 4y = -9$
(2) $x^2 - y^2 + 6x - 6y = -9$

(*Hint:* In exercises **29** and **30**, eliminate the constants on the right side.)

29. (1) $7x^2 - 9xy + 3y^2 = 1$
(2) $4x^2 - 3xy = -2$

30. (1) $3x^2 + 2xy = 24$
(2) $y^2 + xy = 7$

In exercises **31–40**, solve each word problem.

[Example 5] **31.** The sum of one number and 3 times another number is 26. The product of the numbers is 35. Find both numbers.

32. The difference between one number and 4 times another number is 4. The product of the numbers is 48. Find both numbers.

33. The distance around the floor of Linda's bedroom is 48 feet, and the area is 140 square feet. What are the dimensions of her bedroom?

34. The perimeter of the dining room in Cliff and Donna Simes' house is 54 feet. The sum of the squares of the length and width of the room is 369 square feet. What are the dimensions of the dining room?

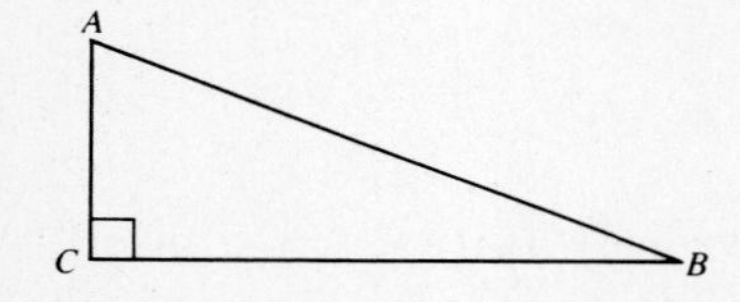

Figure 11-31. Triangle ABC.

35. In triangle ABC in Figure 11-31, angle C is a right angle. The length of side AB is $3\sqrt{13}$ meters. Find the lengths of sides AC and BC if side BC is 3 meters longer than side AC.

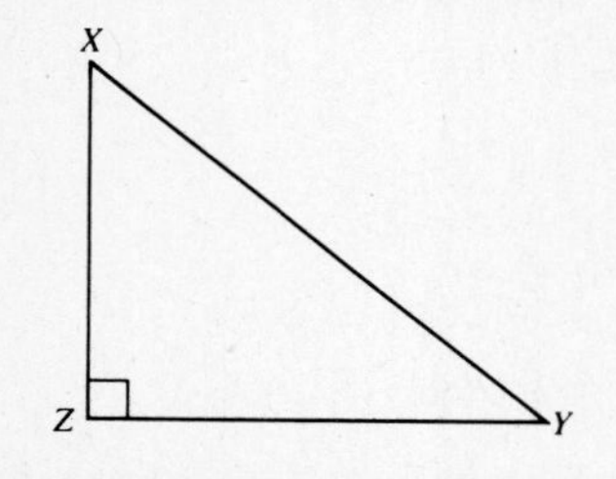

Figure 11-32. Triangle XYZ.

36. In triangle XYZ in Figure 11-32, angle Z is a right angle. The length of side XZ is $2\sqrt{51}$ yards. The length of side XY is 6 yards more than side YZ. Find the lengths of sides XY and YZ.

37. Maureen works as a waitress at the Stripped Tiger Restaurant. For a given number of workdays, she earned \$750 in wages from the restaurant (not counting tips). If the restaurant paid her \$5 a day less in wages, it would take her 5 more days to earn the same pay. How many days did Maureen work, and how much per day does she earn from the restaurant?

38. Clarence works part-time bagging groceries in a supermarket. For a given number of workdays, his gross earnings were \$660. If the market paid him \$3 less per day, it would take him 2 more days to earn the same amount. How many days did Clarence work during this pay period, and how much per day does he earn?

39. Two electrical resistors have resistances of x ohms and y ohms. (An *ohm* is a unit that measures electrical resistance.)

When hooked in series: $x + y = 10$ ohms

When hooked in parallel: $\frac{1}{x} + \frac{1}{y} = \frac{8}{5}$ ohms

Find the number of ohms in each resistor.

40. Two electrical resistors have resistances of x ohms and y ohms.

When hooked in series: $x + y = 24$ ohms

When hooked in parallel: $\frac{1}{x} + \frac{1}{y} = \frac{16}{3}$ ohms

Find the number of ohms in each resistor.

SECTION 11-4. Ten Review Exercises

In exercises **1–10**, simplify each expression.

1. $\sqrt[4]{25t^2}$; $t > 0$

2. $\dfrac{21}{\sqrt{10} + \sqrt{3}}$

3. $\sqrt[3]{56}$

4. $\dfrac{21}{\sqrt{3}}$

5. $3[(5y + 2)(5y - 2) - 1] - [(3y + 2)^2 + 2y(25y - 6)]$

6. $\dfrac{27u^3 - 1}{6u^2 + u - 1}$

7. $\dfrac{k + 3 - \dfrac{6}{k + 2}}{k + \dfrac{2k - 5}{k + 2}}$

8. $\dfrac{6 - \sqrt{45}}{9}$

9. $\dfrac{4 + \sqrt{-36}}{2}$

10. $\sqrt{\dfrac{1}{10}} + \sqrt{\dfrac{2}{5}} - \sqrt{\dfrac{5}{2}}$

SECTION 11-4. Supplementary Exercises

In exercises **1–6**, do the following:

a. Write equations (1) and (2) for the figures in the graphs.

b. Solve the system formed by (1) and (2). Approximate any real-number solutions.

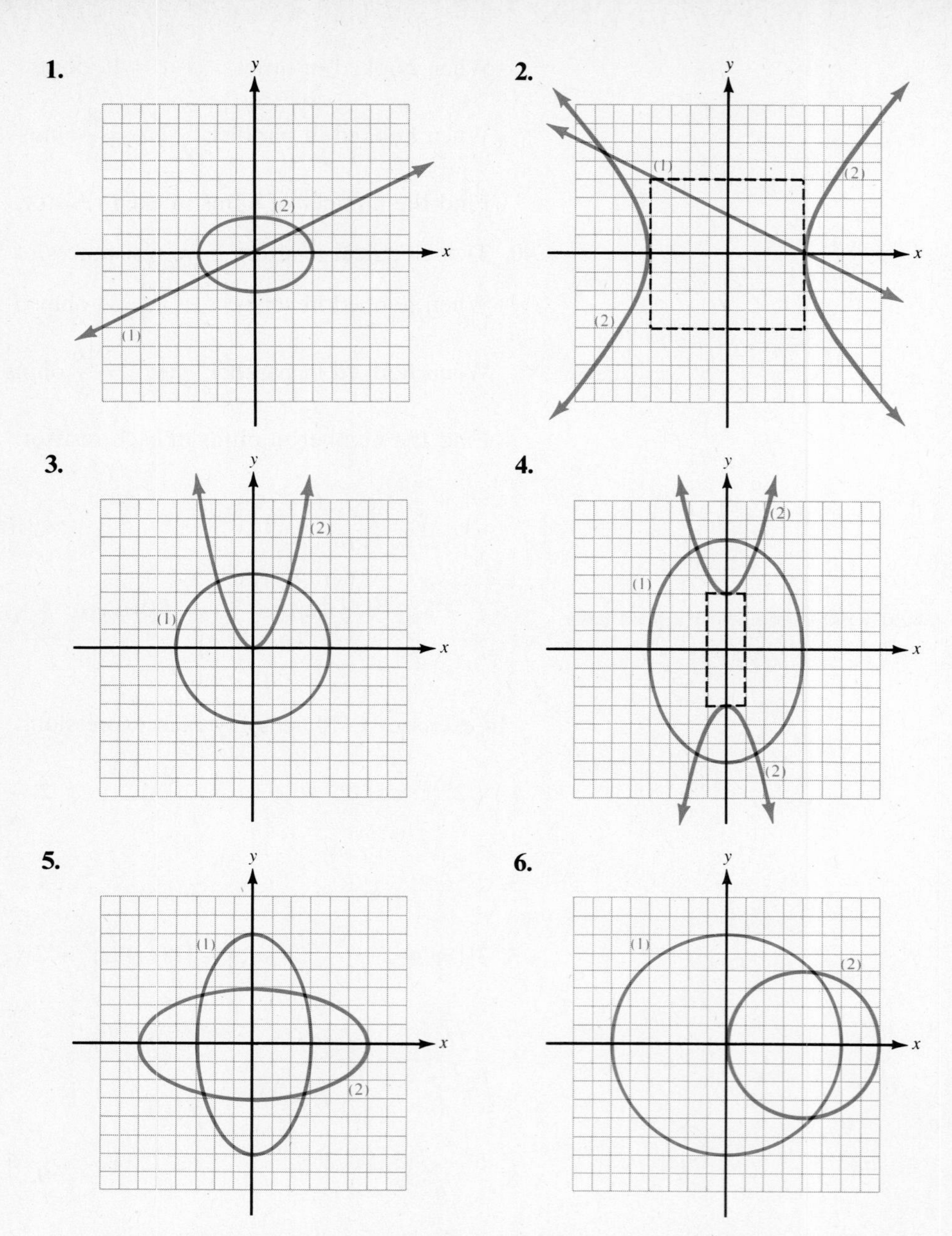

A bridge across a small river in Europe has an opening in the shape of a semi-ellipse. Large barges with rectangular tops need to pass under the bridge. As Figure 11-33 shows, the opening has a width w and a height h, and the barge has a width p and a height q.

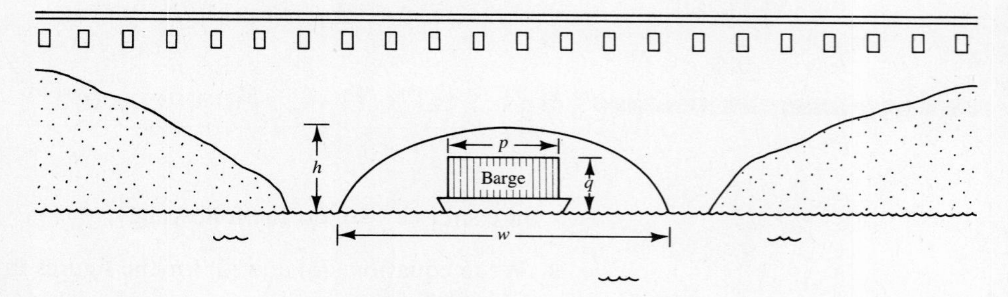

Figure 11-33. A bridge with a semi-elliptical opening.

In exercises **7–12**, determine whether the barge can pass under the bridge given each set of dimensions.

7. $w = 40$ ft, $h = 10$ ft
$p = 16$ ft, $q = 9.5$ ft

8. $w = 40$ ft, $h = 10$ ft
$p = 18$ ft, $q = 8.6$ ft

9. $w = 30$ ft, $h = 9$ ft
$p = 16$ ft, $q = 7$ ft

10. $w = 30$ ft, $h = 9$ ft
$p = 18$ ft, $q = 7.5$ ft

11. $w = 50$ ft, $h = 16$ ft
$p = 20$ ft, $q = 15$ ft

12. $w = 50$ ft, $h = 16$ ft
$p = 24$ ft, $q = 14.5$ ft

Name

Date

Score

SECTION 11-1. Summary Exercises

Answer

1. Write an equation that defines the parabola with focus $F(0, -6)$ and directrix $y = 6$ in the form $y = ax^2$.

1. ______________

In exercises **2–8**, graph each equation.

2. $y = \frac{-1}{9}x^2$

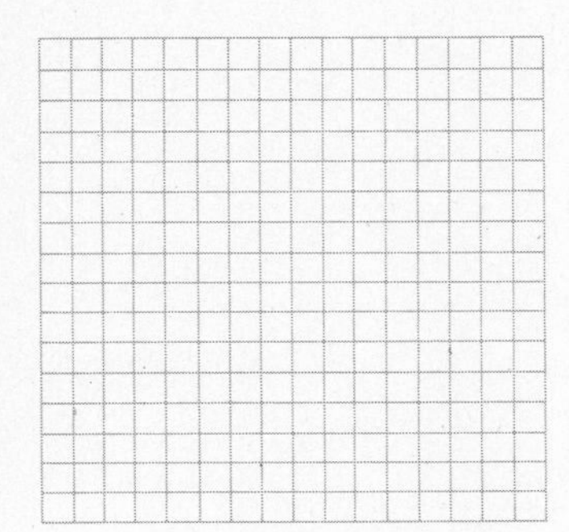

3. $y = (x + 3)^2$

4. $y = x^2 + 3$

5. $y = \frac{1}{2}(x - 2)^2$

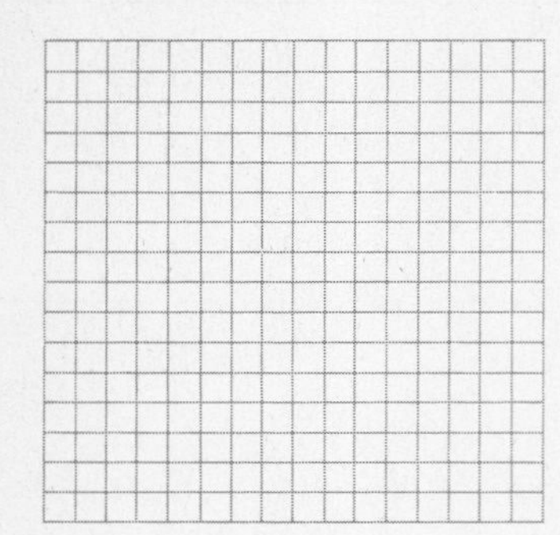

6. $y = \frac{-1}{2}x^2 + 3$

7. $y = -x^2 + 8x - 14$

8. $y = 2x^2 + 16x + 31$

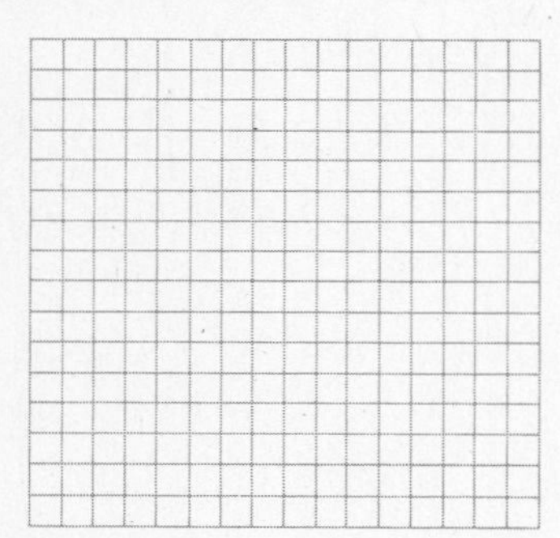

Name

Date

SECTION 11-2. Summary Exercises

Score

In exercises **1–6**, graph each equation.

1. $x^2 + 4y^2 = 64$

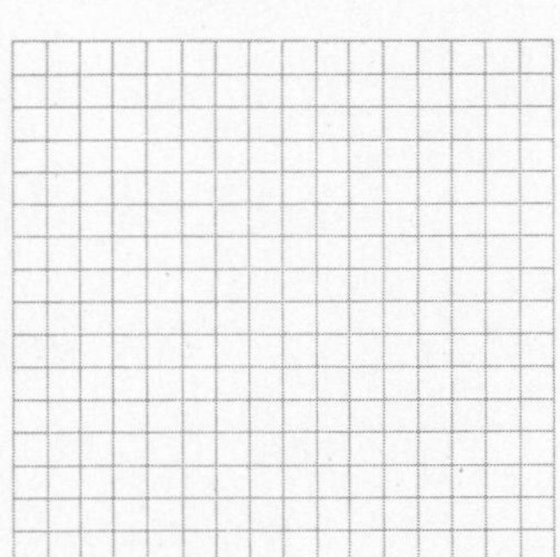

2. $x^2 + y^2 = 36$

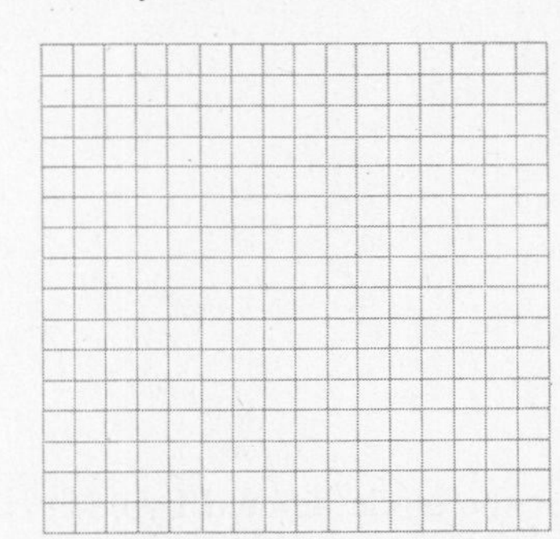

3. $(x - 2)^2 + (y + 1)^2 = 49$

4. $36x^2 + 4y^2 = 144$

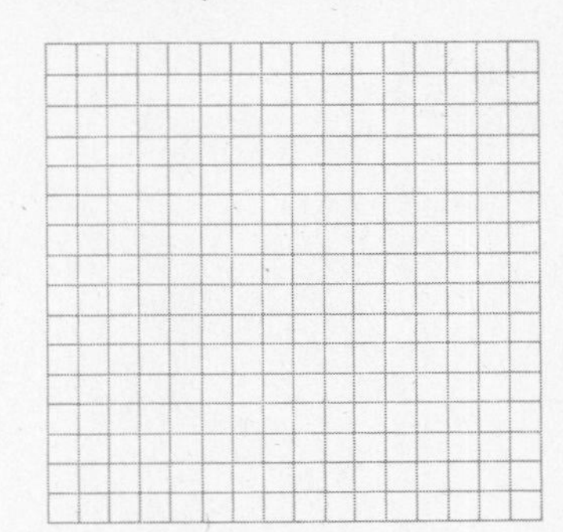

5. $9x^2 + 4y^2 = 144$

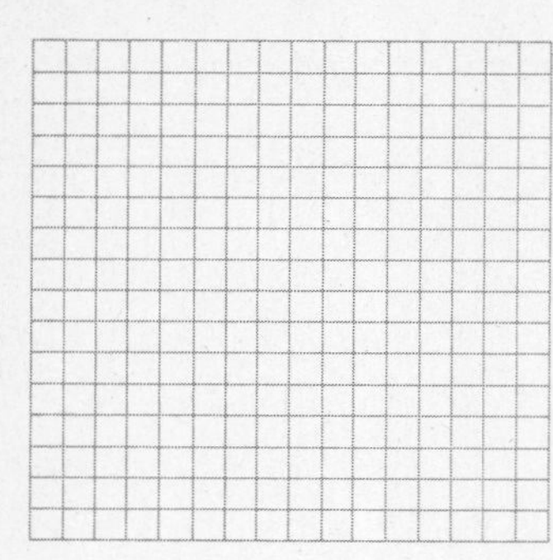

6. $x^2 + (y - 3)^2 = 16$

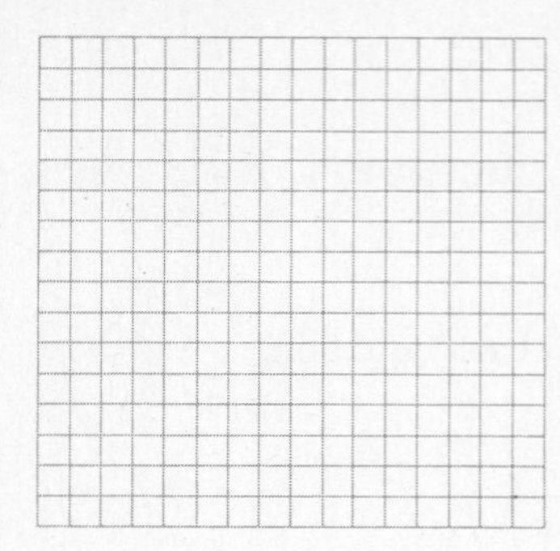

In exercises **7** and **8**, for the circle defined by each equation:

a. Find the center.

b. Find the radius.

7. $x^2 + y^2 - 6x + 2y = 6$

7. a. ______________________

b. ______________________

8. $x^2 + y^2 + 8x + 15 = 0$

8. a. ______________________

b. ______________________

Name

Date

Score

SECTION 11-3. Summary Exercises

In exercises **1–8**, graph each equation.

1. $\dfrac{x^2}{1} - \dfrac{y^2}{4} = 1$

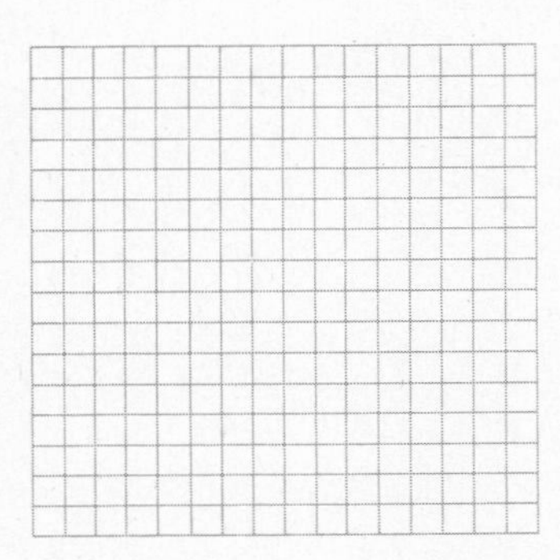

2. $\dfrac{y^2}{9} - \dfrac{x^2}{36} = 1$

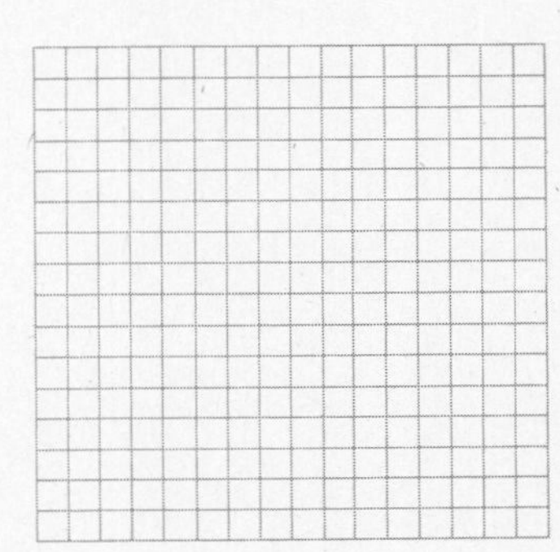

3. $x^2 - y^2 = 25$

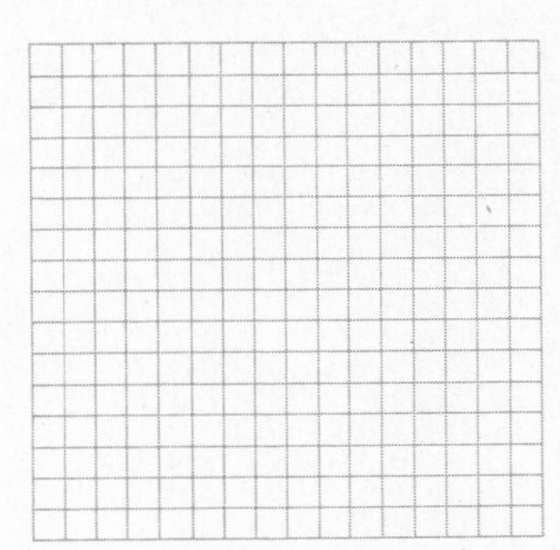

4. $9x^2 - 4y^2 = 36$

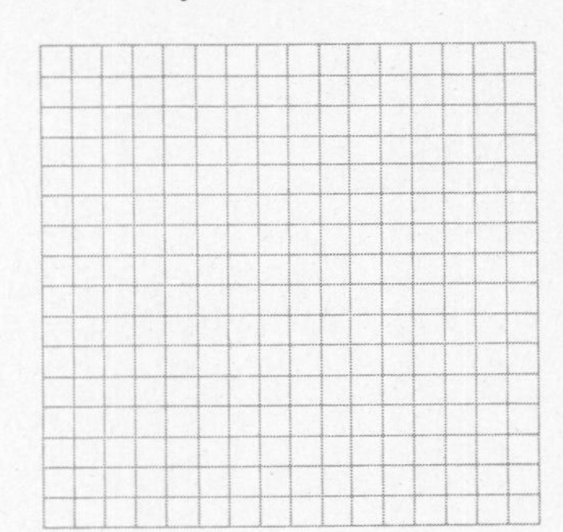

5. $36y^2 - 16x^2 = 576$

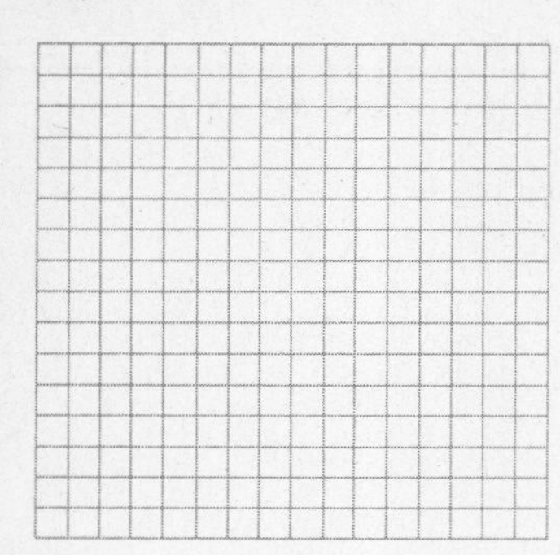

6. $y^2 - x^2 = 4$

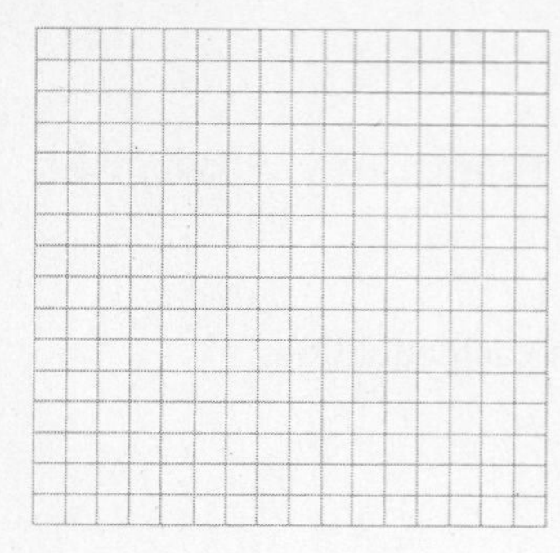

7. $xy = 12$

8. $x = \dfrac{-8}{y}$

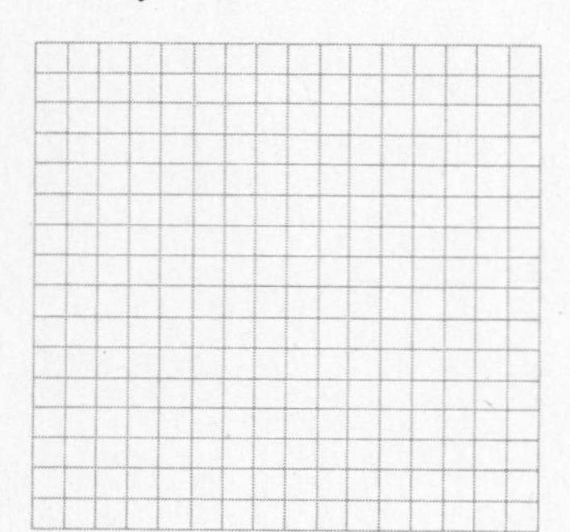

Name

Date

Score

SECTION 11-4. Summary Exercises

Answer

In exercises **1–4**, solve each system.

1. (1) $x^2 + y^2 = 10$
(2) $5x - y = 2$

1. ______

2. (1) $x^2 + 2y^2 = 18$
(2) $x^2 - y^2 = 9$

2. ______

3. (1) $x^2 + 4y^2 - 4 = 0$
(2) $4x^2 = 9y^2 + 36$

3. ____________________

4. (1) $x^2 + xy + y^2 = 1$
(2) $2x^2 - y^2 = 1$

4. ____________________

CHAPTER 11. Review Exercises

In exercises **1** and **2**, for each focus and directrix, write an equation that defines the parabola in the form $y = ax^2$.

1. $F(0, 9)$ and $y = -9$

2. $F\left(0, \frac{-2}{3}\right)$ and $y = \frac{2}{3}$

In exercises **3–8**, graph each equation.

3. $y = 2x^2 - 8$

4. $y = \frac{-3}{4}(x + 2)^2$

5. $2y = x^2 - 6x + 11$

6. $y = 4 - 8x - 4x^2$

7. $x^2 + (y - 3)^2 = 36$

8. $(x + 4)^2 + (y + 3)^2 = 81$

In exercises **9** and **10**, for each equation of a circle:

a. State the coordinates of the center and the radius.

b. Graph.

9. $x^2 + y^2 - 10x + 4y + 13 = 0$

10. $x^2 + y^2 + 8x - 10y = 8$

In exercises **11–20**, graph each equation.

11. $4x^2 + y^2 = 100$

12. $100x^2 + 9y^2 = 900$

13. $25x^2 + 16y^2 = 625$

14. $4x^2 + y^2 = 25$

15. $x^2 - 4y^2 = 16$

16. $y^2 - x^2 = 36$

17. $y^2 - 4x^2 = 9$

18. $25x^2 - 36y^2 = 36$

19. $xy = -16$

20. $y = \frac{10}{x}$

In exercises **21–24**, solve each system.

21. (1) $x - 2y = 3$
(2) $x^2 + y^2 = 5$

22. (1) $3x^2 + 4y^2 = 292$
(2) $x^2 + y^2 = 89$

23. (1) $4x^2 - y^2 = -7$
(2) $xy = 6$

24. (1) $x^2 - 2xy + y^2 = 16$
(2) $x^2 - 2xy - y^2 = 8$

In exercises **25** and **26**, solve each word problem.

25. The sum of 3 times one number and 2 times another number is 7. The product of the numbers is 2. Find both numbers.

26. The difference in the lengths of the legs of a right triangle is 4 inches. The length of the hypotenuse is $4\sqrt{5}$ inches. Find the lengths of the legs of the triangle.

APPENDIX

A

Sequences and Series

A Definition of a Sequence Function

The following statement is a definition of a sequence function:

> **Definition A.1. Sequence function**
> A **sequence** is a function whose domain is a set of consecutive positive integers, beginning with 1.

Sequence functions can be separated into two kinds based on the number of elements in the domain.

Finite sequence: one in which the domain is $\{1, 2, 3, \ldots, N\}$ for some positive N

Infinite sequence: one in which the domain is $\{1, 2, 3, 4, \ldots\}$

Examples **a** and **b** are finite and infinite sequences, respectively.

a. $\{(1, 1), (2, 4), (3, 9), (4, 16), (5, 25)\}$

b. $\{(1, 0), (2, 7), (3,26), (4, 63), \ldots\}$

Notation Used in Writing Sequences

The domain elements of any sequence function are positive integers. It is therefore common practice to list only the elements of the range. To form such a list, the general term of the range elements can be used. To emphasize that the domain contains only positive integers, an n is used instead of x.

> **Definition A.2.**
> The elements in the range of a sequence function are called **terms,** and they can be listed as follows:
>
> $$a_1, a_2, a_3, a_4, \ldots, a_n, \ldots$$
>
> a_1 is called the **first term,** a_2 is called the **second term,** and so on. The ***n*-th term** (also called the **general term**) is a_n. Symbols such as b_n, c_n, and so on can also be used to represent the general term.

Using the method described in Definition A.2, the functions in examples **a** and **b** can be written as follows:

a. $1, 4, 9, 16, 25$ $\qquad a_n = n^2$

b. $0, 7, 26, 63, \ldots, n^3 - 1, \ldots$ $\qquad a_n = n^3 - 1$

Example 1. List the first five terms of $a_n = 2n - 3$.

Solution. **Discussion.** Replace n in $2n - 3$ by 1, 2, 3, 4, and 5.

$$a_1 = 2(1) - 3 = -1 \qquad a_2 = 2(2) - 3 = 1$$

$$a_3 = 2(3) - 3 = 3 \qquad a_4 = 2(4) - 3 = 5$$

$$a_5 = 2(5) - 3 = 7$$

The first five terms in the sequence are -1, 1, 3, 5, and 7.

Example 2. List the first four terms and the general term of $c_n = \dfrac{n-1}{n+1}$.

Solution.

$$c_1 = \frac{1-1}{1+1} = 0 \qquad c_2 = \frac{2-1}{2+1} = \frac{1}{3}$$

$$c_3 = \frac{3-1}{3+1} = \frac{2}{4} = \frac{1}{2} \qquad c_4 = \frac{4-1}{4+1} = \frac{3}{5}$$

The terms of the sequence are $0, \frac{1}{3}, \frac{1}{2}, \frac{3}{5}, \ldots, \frac{n-1}{n+1}, \ldots$.

Arithmetic Sequences

Many sequences are such that after the first term, each term can be determined by adding the same nonzero constant to the previous term in the sequence. Such a sequence is called arithmetic.

> **Definition A.3. Arithmetic sequence**
>
> $$a_1, a_1 + d, a_1 + 2d, a_1 + 3d, \ldots, a_1 + (n-1)d, \ldots$$
>
> is called **arithmetic.** The d is a constant called the **common difference.**

Example 3. Find d for the sequence 5, 12, 19, 26, 33, ...

Solution. **Discussion.** To find d, subtract any two consecutive terms in the sequence, the previous term from the latter term.

$$\left.\begin{array}{lll} a_2 - a_1 & \text{becomes} & 12 - 5 = 7 \\ a_3 - a_2 & \text{becomes} & 19 - 12 = 7 \\ a_4 - a_3 & \text{becomes} & 26 - 19 = 7 \end{array}\right\} \text{Each difference is 7.}$$

Thus, d is 7.

Example 4. Find the number of terms in the sequence $10, \frac{28}{3}, \frac{26}{3}, \ldots, -20$.

Solution. **Discussion.** This sequence is arithmetic, with $a_1 = 10$, $a_n = -20$, and $d = \dfrac{-2}{3}\left(= \dfrac{28}{3} - 10\right)$.

$$-20 = 10 + (n-1)\left(\frac{-2}{3}\right) \qquad a_n = a_1 + (n-1)d$$

$$-30 = (n-1)\left(\frac{-2}{3}\right) \qquad \text{Subtract 10 from both sides.}$$

$$45 = n - 1 \qquad \text{Multiply both sides by } \frac{-3}{2}.$$

$$n = 46 \qquad \text{Add 1 to both sides.}$$

Thus, the given sequence has 46 terms.

Geometric Sequences

Many sequences are such that, after the first term, each term can be determined by multiplying the same nonzero constant to the previous term in the sequence. Such a sequence is called geometric.

Definition A.4. Geometric sequence

$$a_1, a_1r, a_1r^2, a_1r^3, \ldots, a_1r^{n-1}, \ldots$$

is called **geometric.** The r is a nonzero constant called the **common ratio.**

Example 5. Find r for the sequence 8, 12, 18, 27, ...

Solution. **Discussion.** To find r, divide any two consecutive terms in the sequence, the latter term by the previous term.

$$\left.\begin{aligned}\frac{12}{8} &= \frac{3}{2}\\ \frac{18}{12} &= \frac{3}{2}\\ \frac{27}{18} &= \frac{3}{2}\end{aligned}\right\} \text{Each ratio is } \frac{3}{2}.$$

Example 6. Find the tenth term of the sequence 96, 48, 24, 12, ...

Solution.

$$a_{10} = 96\left(\frac{1}{2}\right)^9$$

$$= \frac{3}{16} \qquad \text{The tenth term is } \frac{3}{16}.$$

A Definition of a Series

A listing of a set of numbers can be called a **sequence.** An indicated sum of these terms can be called a **series.**

> **Definition A.5.**
> A **series** is an indicated sum of the terms of a sequence.

The symbol s_n can be used to represent a series of n-terms. Examples **c** and **d** illustrate two sequences and the corresponding series.

	Sequences	Corresponding series
c.	1, 3, 5, 7, 9, 11, 13	$s_7 = 1 + 3 + 5 + 7 + 9 + 11 + 13$
d.	$\frac{-2}{3}, \frac{4}{9}, \frac{-8}{27}, \frac{16}{81}, \frac{-32}{243}$	$s_5 = \frac{-2}{3} + \frac{4}{9} - \frac{8}{27} + \frac{16}{81} - \frac{32}{243}$

A general term of a sequence can also be used to represent the terms of the corresponding series. In example **c,** $2n - 1$ is a general term for both the sequence and series. In example **d,** $(\frac{-2}{3})^n$ is a general term.

Notation Used in Writing a Series

When a general term of a series is known, S_n can be written using **sigma notation,** also called **summation notation.** The symbol Σ is used to mean **sum.**

> If a_n is a general term for a series, then:
>
> $$S_n = \sum_{i=1}^{n} a_i \qquad \textbf{Summation notation}$$
>
> $$= a_1 + a_2 + \cdots + a_n \qquad \textbf{Expanded form}$$
>
> The i is called the **index of summation.** Letters such as j and k can also be used as **indices** (plural of index). Thus:
>
> $$\sum_{i=1}^{n} a_i = \sum_{j=1}^{n} a_j = \sum_{k=1}^{n} a_k$$

Example 7. Find the indicated sum of the series $\sum_{i=1}^{4} (3i + 5)$.

Solution.

$$\sum_{i=1}^{4} (3i + 5) = (3 \cdot 1 + 5) + (3 \cdot 2 + 5) + (3 \cdot 3 + 5) + (3 \cdot 4 + 5)$$

$$= 8 + 11 + 14 + 17$$

$$= 50$$

The Sum of a Finite Arithmetic Sequence

The sum of an arithmetic series can be found with Equation (1).

The sum of a finite arithmetic series

If a_1 and a_n are the first and last terms, respectively, of n-terms of an arithmetic sequence, then:

$$(1) \quad S_n = \frac{n}{2}(a_1 + a_n)$$

$$S_n = \frac{n}{2}[2a_1 + (n-1)d] \quad \text{(Alternate formula)}$$

Example 8. Find S_n for the sequence 2, 6, 10, . . . , 158.

Solution. **Discussion.** The sequence is arithmetic, with $a_1 = 2$, $a_n = 158$, and $d = 4$. To use Equation (1), we must first find n.

$158 = 2 + (n-1)4$ — $S_n = a_1 + (n-1)d$

$n = 40$ — There are 40 terms in the sequence.

$S_{40} = \frac{40}{2}(2 + 158)$ — $S_n = \frac{n}{2}(a_1 + a_n)$

$= 3{,}200$ — The sum is 3200.

The Sum of a Finite Geometric Sequence

If a_n is the general term of a geometric sequence, then S_n is the indicated sum of the first n-terms of the sequence. Using summation notation, the sum can be written as follows:

$$S_n = \sum_{i=1}^{n} a_1 r^{i-1}$$

The value of S_n can be found with Equation (2).

The sum of a finite geometric series

If a_1 is the first term of a geometric sequence with common ratio r (where $r \neq 1$), then the sum of the first n-terms is S_n, and:

$$(2) \quad S_n = \frac{a_1(1 - r^n)}{1 - r}$$

Example 9. Find the sum of the first eight terms of the sequence 3, 6, 12, 24, ...

Solution. **Discussion.** This sequence is geometric, with $r = \frac{6}{3} = 2$ and $a_1 = 3$.

$S_8 = \frac{3(1 - 2^8)}{1 - 2}$ — $S_n = \frac{a_1(1 - r^n)}{1 - r}$

$= \frac{3(-255)}{-1}$ — $1 - 2^8 = 1 - 256 = -255$

$= 765$ — The sum of eight terms is 765.

A Definition of an Infinite Geometric Series

Definition A.6. Infinite geometric series
An **infinite geometric series** with first term a_1 and common ratio r can be written as:

$$a_1 + a_1r + a_1r^2 + a_1r^3 + \cdots + a_1r^{n-1} + \cdots$$

Using summation notation:

$$\sum_{i=1}^{\infty} a_1r^{i-1}$$

Examples **e** and **f** are infinite geometric series.

e. $1 + 2 + 4 + 8 + \cdots + 2^{n-1} + \ldots$ $\qquad a_1 = 1$ and $r = 2$

f. $1 + \frac{1}{2} + \frac{1}{4} + \frac{1}{8} + \cdots + \left(\frac{1}{2}\right)^{n-1} + \ldots$ $\qquad a_1 = 1$ and $r = \frac{1}{2}$

The Limit of a_n as n Becomes Infinitely Large

Notice the following trends in the terms of the series in examples **e** and **f**:

1. Each successive term in example **e** is larger than the preceeding terms in the series. That is, *as* n *increases,* 2^n *increases.*

n	5	8	12	20	26
2^n	32	256	4,096	1,048,576	67,108,864

As $n \to \infty$, 2^n becomes larger and larger, written "$2^n \to \infty$".

2. Each successive term in example **f** is smaller than the preceeding terms in the series. That is, *as* n *increases,* $(\frac{1}{2})^n$ *decreases.*

n	5	8	12	20	26
$\left(\frac{1}{2}\right)^n$	0.03125	0.00390625	0.000244141	0.000000954	0.000000015

As $n \to \infty$, $(\frac{1}{2})^n$ becomes smaller and smaller, written "$(\frac{1}{2})^n \to 0$". The term **limit** is used to express the fact that $(\frac{1}{2})^n$ gets closer and closer to 0, but never takes on the value 0, no matter how large n gets. In symbols:

$$\lim_{n\to\infty} \left(\frac{1}{2}\right)^n = 0$$

The limit, as n gets infinitely large, of $\left(\frac{1}{2}\right)^n$ is 0.

The following statement asserts that for any infinite geometric series in which $-1 < r < 1$, the limit of the n-th term is 0:

> If a_n is the n-th term of an infinite geometric series in which $-1 < r < 1$, then:
>
> $$\lim_{n \to \infty} a_n = 0$$

The Sum of an Infinite Geometric Series

The following formula was given for computing the sum of the first n-terms of a geometric series:

$$S_n = \frac{a_1(1 - r^n)}{1 - r}$$

The equation in the following statement can be used to find the sum of any infinite geometric series, provided $-1 < r < 1$:

> **The sum of an infinite geometric series, with $-1 < r < 1$**
> If a_1 is the first term of an infinite geometric series for which $-1 < r < 1$, then:
>
> $$S_\infty = a_1 + a_1 r + a_1 r^2 + \cdots + a_1 r^{n-1} + \ldots$$
> $$= \frac{a_1}{1 - r}$$

Example 10. Find the indicated sum of the series $5 + \frac{5}{3} + \frac{5}{9} + \cdots + 5\left(\frac{1}{3}\right)^{n-1} + \ldots$.

Solution. **Discussion.** The given series is geometric, with $a_1 = 5$ and $r = \frac{1}{3}$.

$$S_\infty = \frac{5}{1 - \frac{1}{3}} \qquad S_\infty = \frac{a_1}{1 - r}$$

$$= \frac{15}{2} \qquad \text{Multiply top and bottom by 3.}$$

Thus, the given series will have a "sum" as close to $\frac{15}{2}$ as required by simply adding a sufficient number of terms.

Example 11. Find $\sum_{i=1}^{\infty} 72\left(\frac{-3}{5}\right)^{i-1}$.

Solution. **Discussion.** The given series is geometric, with $a_1 = 72$ and $r = \frac{-3}{5}$.

$$S_\infty = \frac{72}{1 - \left(\frac{-3}{5}\right)} \qquad S_\infty = \frac{a_1}{1-r}$$

$$= \frac{72 \cdot 5}{8} \qquad \text{Multiply top and bottom by 5.}$$

$$= 45 \qquad \text{The sum of the series is 45.}$$

If r is the common ratio of an infinite geometric series and $r \leq -1$ or $r \geq 1$, then the series cannot have a finite number assigned as a sum. That is, there is no limit to the magnitude of the sum. For such a series we simply say "no sum".

Example 12. Find the indicated sum of the series

$$8 - 12 + 18 - 27 + \cdots + 8\left(\frac{-3}{2}\right)^{n-1} + \ldots.$$

Solution. **Discussion.** The given series is geometric, with $a_1 = 8$ and $r = \frac{-3}{2}$.

Since $\frac{-3}{2} \leq -1$, the given series has *no sum.*

Exercises

In exercises **1–10**, list the indicated number of terms of each sequence function.

[Example 1]

1. 5 terms; $a_n = 2n + 3$

2. 5 terms; $a_n = 10n - 4$

3. 5 terms; $c_n = n^3 + 8$

4. 5 terms; $c_n = 2n^3 - 9$

5. 5 terms; $a_n = 2^n + 1$

6. 5 terms; $a_n = 10 - 3^n$

7. 10 terms; $c_n = \frac{(-1)^n n}{n+1}$

8. 10 terms; $c_n = \frac{(-1)^n n^2}{2n+3}$

9. 10 terms; $a_n = \sqrt{n+1}$

10. 10 terms; $a_n = \sqrt{n^2+1}$

In exercises **11–22**, list the first four terms and the general term of each sequence function.

[Example 2]

11. $a_n = \frac{1-2n}{2+n}$

12. $a_n = \frac{n^2+1}{n}$

13. $b_n = n^2 + n$

14. $b_n = 2^n - 1$

15. $c_n = \frac{(-2)^n}{n^2}$

16. $c_n = \frac{(-3)^n}{n^3}$

17. $a_n = \sqrt{n^2 - 1}$

18. $a_n = \sqrt{n^3 + n}$

19. $b_n = \frac{\sqrt{n}}{n+1}$

20. $b_n = \frac{\sqrt[3]{n}}{n^2+1}$

21. $c_n = \log n$

22. $c_n = \log(2n + 1)$

In exercises **23–26**, find d for each sequence.

[Example 3]

23. 2, 5, 8, 11, . . .

24. −5, 8, 21, 34, . . .

25. $7, \frac{11}{2}, 4, \frac{5}{2}, \ldots$

26. $4, \frac{17}{5}, \frac{14}{5}, \frac{11}{5}, \ldots$

In exercises **27–32**, find the indicated term of each sequence.

27. The 16th term of $1, 6, 11, 16, \ldots$

28. The 21st term of $-10, -2, 6, 14, \ldots$

29. The 8th term of $1.5, 2.7, 3.9, 5.1, \ldots$

30. The 10th term of $0.6, 3.0, 5.4, 7.8, \ldots$

31. The 9th term of $10\sqrt{3}, 6\sqrt{3}, 2\sqrt{3}, \ldots$

32. The 14th term of $9\sqrt{2}, 7\sqrt{2}, 5\sqrt{2}, \ldots$

In exercises **33–38**, find the number of terms in each sequence.

[Example 4]

33. $1, 8, 15, \ldots, 141$

34. $-5, 4, 13, \ldots, 346$

35. $\frac{\pi}{5}, \frac{2\pi}{5}, \ldots, 6\pi$

36. $\frac{\pi}{6}, \frac{\pi}{3}, \frac{\pi}{2}, \ldots, 5\pi$

37. $\sqrt{3}, 5\sqrt{3}, \ldots, 137\sqrt{3}$

38. $30\sqrt{2}, 27\sqrt{2}, \ldots, -123\sqrt{2}$

In exercises **39–42**, find r for each sequence.

[Example 5]

39. $\frac{1}{54}, \frac{1}{27}, \frac{1}{9}, \frac{1}{3}, \ldots$

40. $\frac{1}{2}, 2, 8, 32, \ldots$

41. $\frac{1}{10}, \frac{-1}{2}, \frac{5}{2}, \frac{-25}{2}, \ldots$

42. $\frac{5}{36}, \frac{-5}{6}, 5, -30, \ldots$

In exercises **43–48**, find the indicated term of each sequence.

[Example 6]

43. The sixth term of $1, 3, 9, 27, \ldots$

44. The eighth term of $\frac{1}{8}, \frac{-1}{4}, \frac{1}{2}, -1, \ldots$

45. The sixth term of $100, 30, 9, 2.7, \ldots$

46. The ninth term of $64, 96, 144, \ldots$

47. The tenth term of $\frac{-1}{16}, \frac{1}{8}, \frac{-1}{4}, \ldots$

48. The sixth term of $-27x, 9x^2, -3x^3, \ldots$

In exercises **49–60**, find the indicated sums of each series.

[Example 7]

49. $\sum_{i=1}^{6} (2i + 3)$

50. $\sum_{i=1}^{6} (5i - 2)$

51. $\sum_{j=1}^{5} (j^2 + 1)$

52. $\sum_{j=1}^{5} (2j^2 - 3)$

53. $\sum_{k=1}^{8} (-1)^k(2^k - 1)$

54. $\sum_{k=1}^{8} (-1)^k(k^2 - 4)$

55. $\sum_{i=1}^{5} \frac{i(i+1)}{2}$ **56.** $\sum_{i=1}^{5} \frac{(2i-1)}{3}$ **57.** $\sum_{j=1}^{5} (j+1)(j+2)$

58. $\sum_{j=1}^{5} 2j(j-3)$ **59.** $\sum_{k=1}^{6} \log k$ **60.** $\sum_{k=1}^{6} \log(k+1)$

In exercises **61–64**, write each series using summation notation.

61. $2 + 4 + 8 + 16 + 32 + 64 + 128 + 256$; general term 2^n

62. $\frac{-1}{2} + \frac{1}{4} - \frac{1}{8} + \frac{1}{16} - \frac{1}{32} + \frac{1}{64} - \frac{1}{128} + \frac{1}{256}$; general term $\left(\frac{-1}{2}\right)^n$

63. $-1 + 4 - 7 + 10 - 13 + 16 - 19 + 22 - 25 + 28$; general term $(-1)^n(3n-2)$

64. $1 + 5 + 9 + 13 + 17 + 21 + 25 + 29 + 33 + 37$; general term $(4n-3)$

In exercises **65–70**, find S_n for each sequence.

[Example 8] **65.** $-8, -3, 2, \ldots, 87$ **66.** $3, 6, 9, \ldots, 75$

67. $0.3, 0.7, 1.1, \ldots, 7.9$ **68.** $0.1, 1.2, \ldots, 24.3$

69. $\frac{11}{2}, 5, \frac{9}{2}, \ldots, \frac{-43}{2}$ **70.** $1, \frac{4}{3}, \frac{5}{3}, \ldots, 8$

In exercises **71–76**, find S_n for each series.

71. $\sum_{i=1}^{15} (2i-1)$ **72.** $\sum_{i=1}^{20} (3i+2)$ **73.** $\sum_{j=1}^{12} (0.3j+1)$

74. $\sum_{j=1}^{16} (0.4j-5)$ **75.** $\sum_{i=1}^{21} (i\sqrt{5})$ **76.** $\sum_{i=1}^{15} (2i\sqrt{2})$

In exercises **77–80**, find S_n for each sequence.

[Example 9] **77.** S_6 for $1, 4, 16, 64, \ldots$ **78.** S_6 for $1, 5, 25, 125, \ldots$

79. S_7 for $1, -3, 9, -27, \ldots$ **80.** S_7 for $5, -10, 20, -40, \ldots$

In exercises **81–86**, find S_n for each series.

81. $\sum_{i=1}^{6} 2^i$ **82.** $\sum_{i=1}^{5} 5^i$ **83.** $\sum_{j=1}^{6} (-3)^j$

84. $\sum_{j=1}^{6} \left(\frac{-1}{5}\right)^j$ **85.** $\sum_{k=1}^{5} 10\left(\frac{3}{2}\right)^k$ **86.** $\sum_{k=1}^{5} 100\left(\frac{6}{5}\right)^k$

In exercises **87–111**, find the indicated sums of each series. If a given series has no sum, write "no sum".

[Examples 10–12] **87.** $2 + \frac{2}{3} + \frac{2}{9} + \cdots + 2\left(\frac{1}{3}\right)^{n-1} + \ldots$

88. $3 + \frac{9}{7} + \frac{27}{49} + \cdots + 3\left(\frac{3}{7}\right)^{n-1} + \ldots$

89. $-1 + \frac{1}{2} - \frac{1}{4} + \cdots + (-1)\left(\frac{-1}{2}\right)^{n-1} + \ldots$

90. $12 - 10 + \frac{25}{3} - \cdots + 12\left(\frac{-5}{6}\right)^{n-1} + \ldots$

91. $1 + \frac{4}{3} + \frac{16}{9} + \cdots + \left(\frac{4}{3}\right)^{n-1} + \ldots$

92. $3 - \frac{10}{3} + \frac{100}{27} - \cdots + \left(\frac{-10}{9}\right)^{n-1} + \ldots$

93. $-10 - 4 - \frac{8}{5} - \cdots - 10\left(\frac{2}{5}\right)^{n-1} - \ldots$

94. $-8 - 6 - \frac{9}{2} - \cdots - 8\left(\frac{3}{4}\right)^{n-1} - \ldots$

95. $\frac{32}{243} - \frac{16}{81} + \frac{8}{27} - \cdots - \frac{32}{243}\left(\frac{-3}{2}\right)^{n-1} + \ldots$

96. $\frac{81}{256} + \frac{9}{64} + \frac{3}{16} + \cdots + \frac{81}{256}\left(\frac{4}{3}\right)^{n-1} + \ldots$

97. $16 + 8\sqrt{2} + 8 + \cdots + 16\left(\frac{1}{\sqrt{2}}\right)^{n-1} + \ldots$

98. $9 + 3\sqrt{3} + 3 + \cdots + 9\left(\frac{1}{\sqrt{3}}\right)^{n-1} + \ldots$

99. $\sum_{i=1}^{\infty} 10\left(\frac{1}{6}\right)^{i-1}$

100. $\sum_{i=1}^{\infty} 4\left(\frac{5}{7}\right)^{i-1}$

101. $\sum_{j=1}^{\infty} 14\left(\frac{-1}{6}\right)^{j-1}$

102. $\sum_{j=1}^{\infty} 15\left(\frac{-2}{3}\right)^{j-1}$

103. $\sum_{k=1}^{\infty} 26(-0.3)^{k-1}$

104. $\sum_{k=1}^{\infty} 22(-0.1)^{k-1}$

105. $\sum_{i=1}^{\infty} 3(1.2)^{i-1}$

106. $\sum_{i=1}^{\infty} 4\left(\frac{9}{8}\right)^{i-1}$

107. $\sum_{i=1}^{\infty} 707(-0.01)^{i-1}$

108. $\sum_{i=1}^{\infty} 204(-0.02)^{i-1}$

109. $\sum_{j=1}^{\infty} -35\left(\frac{4}{9}\right)^{j-1}$

110. $\sum_{j=1}^{\infty} -18\left(\frac{5}{8}\right)^{j-1}$

APPENDIX

The Binomial Theorem

In mathematics, expressions such as

$$(x + 2)^2, (3t - 5)^3, \text{ and } (m + 4n)^4$$

are frequently encountered. Such binomials can be written in **expanded form** by raising them to the indicated powers. The Binomial Theorem is a formula that identifies the general form of each term of $(a + b)^n$, where n is an integer greater than or equal to 1.

The Expanded Form of Some Binomials

To obtain a possible pattern for the terms in the expanded form of $(a + b)^n$, the following expanded forms for $n = 1, 2, 3, 4$ and 5 are written. Let it be understood that $a \neq 0$ and $b \neq 0$.

$$(a + b)^1 = a^1 + b^1$$

$$(a + b)^2 = a^2 + 2ab + b^2$$

$$(a + b)^3 = a^3 + 3a^2b + 3ab^2 + b^3$$

$$(a + b)^4 = a^4 + 4a^3b + 6a^2b^2 + 4ab^3 + b^4$$

$$(a + b)^5 = a^5 + 5a^4b + 10a^3b^2 + 10a^2b^3 + 5ab^4 + b^5$$

The following observations can be made from these five expansions:

1. *The number of terms in the expanded form of* $(a + b)^n$ *is* $n + 1$. For example, for $(a + b)^5$, $n = 5$, and the number of terms is 6.
2. *The first and last terms are, respectively,* a^n *and* b^n. For example, for $(a + b)^5$, the first and last terms are, respectively, a^5 and b^5.
3. *The remaining terms have a factor of the form* a^ib^j, *where* $i + j = n$. For example, using the four interior terms of $(a + b)^5$:

Second term:	a^4b	$i = 4$ and $j = 1$, and $4 + 1 = 5$
Third term:	a^3b^2	$i = 3$ and $j = 2$, and $3 + 2 = 5$
Fourth term:	a^2b^3	$i = 2$ and $j = 3$, and $2 + 3 = 5$
Fifth term:	ab^4	$i = 1$ and $j = 4$, and $1 + 4 = 5$

4. *The coefficients follow an array called Pascal's Triangle.* To identify the pattern of the array, only the numerical coefficients of the expanded forms are listed. The coefficients for the a^n- and b^n-terms are 1; thus, each row begins and ends with a 1.

$$
\begin{array}{ccccccccccc}
 & & & & & 1 & & & & & \\
 & & & & 1 & & 1 & & & & \\
 & & & 1 & & 2 & & 1 & & & \\
 & & 1 & & 3 & & 3 & & 1 & & \\
 & 1 & & 4 & & 6 & & 4 & & 1 & \\
1 & & 5 & & 10 & & 10 & & 5 & & 1
\end{array}
$$

To write the next row in the array, notice that each entry in a given row (except for the first and last ones) is the sum of those entries immediately to the left and right in the row above it, so that the entries in the next row can be obtained as follows:

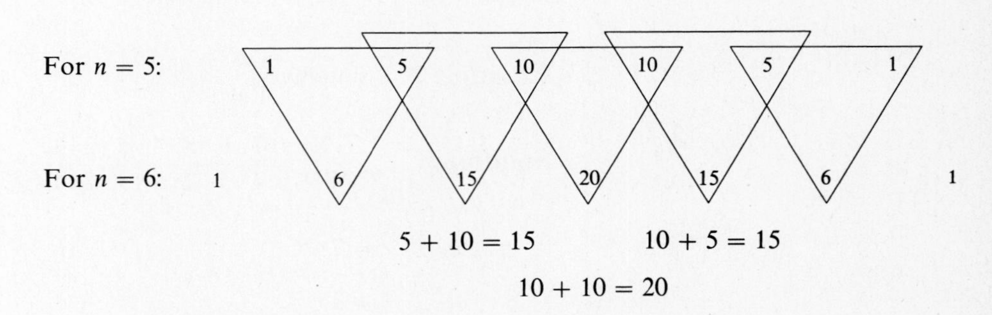

Example 1. Write $(t + 3)^4$ in expanded form.

Solution. **Factor 1.** The coefficient based on Pascal's Triangle

Factor 2. a^i, where $a = t$

Factor 3. b^j, where $b = 3$

From row five in Pascal's Triangle

$$
\begin{aligned}
(t + 3)^4 &= 1(t)^4 + 4(t)^3(3) + 6(t)^2(3)^2 + 4(t)(3)^3 + 1(3)^4 \\
&= t^4 + 12t^3 + 54t^2 + 108t + 81
\end{aligned}
$$

Example 2. Write $(u - 2)^5$ in expanded form.

Solution. **Discussion.** The given binomial is a subtraction of two terms. The expansion is easier if we first write it as $(u + (-2))^5$.

$$
\begin{aligned}
&(u + (-2))^5 \\
&= 1(u)^5 + 5(u)^4(-2) + 10(u)^3(-2)^2 + 10(u)^2(-2)^3 + 5(u)(-2)^4 + 1(-2)^5 \\
&= u^5 + (-10u^4) + 40u^3 + (-80u^2) + 80u + (-32) \\
&= u^5 - 10u^4 + 40u^3 - 80u^2 + 80u - 32
\end{aligned}
$$

Factorial Notation

To find the coefficients by writing out Pascal's Triangle can be very tedious for large values of n, and $(a + b)^n$. A more efficient way is to compute the coefficients by using a formula. One such formula uses the factorial of nonnegative integers.

Factorial notation
For any positive integer n:

$$n! = n(n-1)(n-2)\ldots(3)(2)(1)$$ Read $n!$ as "n-factorial"

By definition:

$$0! = 1$$

Examples **a** and **b** illustrate factorial notation.

a. $4! = 4 \cdot 3 \cdot 2 \cdot 1 = 24$ — 4 factorial is 24

b. $12! = 12 \cdot 11 \cdots 2 \cdot 1 = 479{,}001{,}600$ — 12 factorial is 479,001,600

Example 3. Simplify $\frac{7!}{4!3!}$.

Solution. $\frac{7!}{4!3!} = \frac{7 \cdot 6 \cdot 5 \cdot 4 \cdot 3 \cdot 2 \cdot 1}{4 \cdot 3 \cdot 2 \cdot 1 \cdot 3 \cdot 2 \cdot 1}$ — Write the factorial out.

$= 7 \cdot 5$ — Reduce out the $\frac{6 \cdot 4 \cdot 3 \cdot 2 \cdot 1}{4 \cdot 3 \cdot 2 \cdot 1 \cdot 3 \cdot 2 \cdot 1}$.

$= 35$

The Binomial Theorem

It can be shown that the numerical coefficients of $(a + b)^n$ follow a pattern that can be represented by a quotient of factorials. For example:

$$(a + b)^5 = a^5 + 5a^4b + 10a^3b^2 + 10a^2b^3 + 5ab^4 + b^5$$

$$= a^5 + \frac{5!}{4!1!}a^4b + \frac{5!}{3!2!}a^3b^2 + \frac{5!}{2!3!}a^2b^3 + \frac{5!}{1!4!}ab^4 + b^5$$

The coefficients of the a^5- and b^5-terms are not written using factorials because we know that they are equal to 1. To write a general expression for the other coefficients, we use the nonnegative integer r, where $0 \le r \le n$. In the expanded form of $(a + b)^n$, there are $n + 1$ terms. Using n, r, and factorial notation, a general expression can be written for the $(r + 1)$ term of $(a + b)^n$.

To find the $(r + 1)$ term in the expanded form of $(a + b)^n$:
If n and r are nonnegative integers and $r \le n$, then the form of the $(r + 1)$ term of $(a + b)^n$ is:

$$\frac{n!}{r!(n-r)!}a^{n-r}b^r$$

Example 4. Find the third term of $(3x + 1)^6$.

Solution. For the third term: $r + 1 = 3$, and $r = 2$
For the given binomial: $a = 3x$, $b = 1$, and $n = 6$

$$\frac{n!}{r!(n-r)!}a^{n-r}b^r \quad \text{becomes} \quad \frac{6!}{2!(6-2)!}(3x)^{6-2}(1)^2$$

$$= \frac{6 \cdot 5 \cdot 4!}{2 \cdot 1 \cdot 4!}(3x)^4(1)^2$$

$$= 15(81x^4)$$

$$= 1215x^4$$

An Alternate Form for the Binomial Theorem

The number $\frac{n!}{r!(n-r)!}$ is called a **binomial coefficient.** The following equation identifies a symbol that can be used to represent a binomial coefficient:

Binomial coefficient

If n and r are nonnegative integers and $r \leq n$, then:

$$\binom{n}{r} = \frac{n!}{r!(n-r)!}$$

$\binom{n}{r}$ can be read "the combination of n-things taken r at a time".

We can now write a compact formula for the $n + 1$ terms of $(a + b)^n$ by using the symbol $\binom{n}{r}$.

The Binomial Theorem

If n is a positive integer, then:

$$(a+b)^n = \binom{n}{0}a^n + \binom{n}{1}a^{n-1}b + \binom{n}{2}a^{n-2}b^2 + \cdots + \binom{n}{n-1}ab^{n-1} + \binom{n}{n}b^n$$

For any positive integer n:

$$\binom{n}{0} = 1 \qquad \binom{n}{1} = n \qquad \binom{n}{n-1} = n \qquad \binom{n}{n} = 1$$

Example 5. Write in simplified form the first three terms of the expanded form of $(3t^2 - 2)^6$.

Solution. **Discussion.** For the given binomial: $a = 3t^2$, $b = -2$, and $n = 6$

First term:	**Second term:**	**Third term:**
$\binom{6}{0}(3t^2)^6$	$\binom{6}{1}(3t^2)^5(-2)$	$\binom{6}{2}(3t^2)^4(-2)^2$
$= 1 \cdot 729t^{12}$	$= 6 \cdot 243t^{10}(-2)$	$= 15 \cdot 81t^8(4)$
$= 729t^{12}$	$= -2{,}916t^{10}$	$= 4{,}860t^8$

Thus, the first three terms are $729t^{12} - 2{,}916t^{10} + 4{,}860t^8$.

Exercises

In exercises **1–16**, write each expression in expanded form.

[Examples 1 and 2]

1. $(k + 1)^3$ **2.** $(k + 5)^3$ **3.** $(z - 2)^3$

4. $(2z - 1)^3$ **5.** $(y + 1)^4$ **6.** $(y + 7)^4$

7. $(x - 5)^4$ **8.** $(x - 1)^4$ **9.** $(2m + n)^4$

10. $(3m + 2n)^4$ **11.** $(3x - y)^5$ **12.** $(x - 4y)^5$

13. $(t^2 + 2)^6$ **14.** $(t^2 + 1)^6$ **15.** $(u^3 - 1)^6$

16. $(2u^3 - 3)^6$

In exercises **17–36**, simplify each expression.

[Example 3]

17. 4! **18.** 6! **19.** 7! **20.** 8!

21. $\frac{10!}{5!}$ **22.** $\frac{13!}{11!}$ **23.** $\frac{2!}{0!}$ **24.** $\frac{9!}{0!}$

25. $\frac{12!}{8!}$ **26.** $\frac{15!}{9!}$ **27.** $\frac{6!}{2!4!}$ **28.** $\frac{7!}{4!3!}$

29. $\frac{10!}{6!4!}$ **30.** $\frac{12!}{4!8!}$ **31.** $\frac{15!}{5!10!}$ **32.** $\frac{20!}{18!2!}$

33. $\frac{25!}{3!22!}$ **34.** $\frac{18!}{17!1!}$ **35.** $\frac{52!}{5!47!}$ **36.** $\frac{52!}{13!39!}$

In exercises **37–48**, find the indicated term of each expression.

[Example 4]

37. The second term of $(x + 2)^6$

38. The fifth term of $(x + 2)^6$

39. The third term of $(3t + 1)^7$

40. The fifth term of $(3t + 1)^7$

41. The fourth term of $(k^2 - 3)^8$

42. The sixth term of $(k^2 - 3)^8$

43. The seventh term of $(x - 2y)^9$

44. The third term of $(x - 2y)^9$

45. The second term of $(x^2 + 3y^2)^{10}$

46. The eighth term of $(x^2 + 3y^2)^{10}$

47. The third term of $(2x - y)^{10}$

48. The fourth term of $(a - 3b)^{10}$

In exercises **49–60**, write in simplified form the first three terms of the expanded form of each expression.

[Example 5] **49.** $(5k + 1)^6$ **50.** $(3k + 2)^6$ **51.** $(4z - 3)^8$

52. $(5z - 2)^8$ **53.** $(3u + 5)^7$ **54.** $(3u + 5)^9$

55. $(2v - 7)^9$ **56.** $(2v - 7)^7$ **57.** $(2m + n)^{10}$

58. $(m - 3n)^{10}$ **59.** $(2k^2 - 3)^6$ **60.** $(5k^2 + 4)^6$

APPENDIX

C

Linear Interpolation

Approximate values for logarithms of some numbers not listed in Appendix E can be found using **linear interpolation.** The procedure uses the relationship that exists between the lengths of corresponding sides of plane figures that are similar. That is, *the ratio of the lengths of corresponding sides of similar figures are equal.* To illustrate, in Figure C-1, triangle ABC is similar to triangle XYZ. The ratios of the lengths of the corresponding sides are equal.

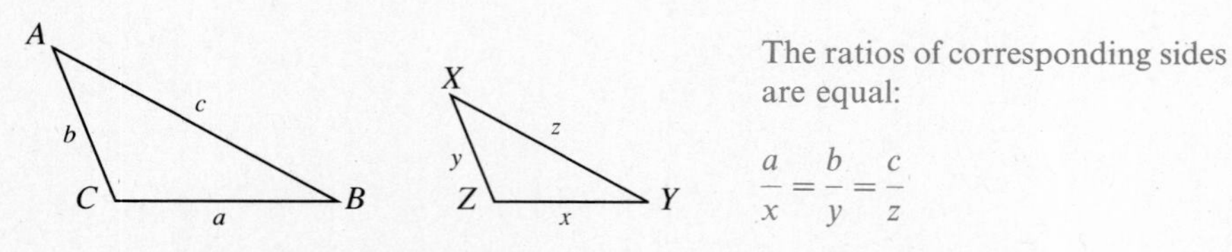

Figure C-1. Triangle ABC is similar to triangle XYZ.

To show how similar triangles are used to approximate the logarithms of numbers not given in the table, an enlargement of the graph of $y = \log x$ in the vicinity of $x = 5.864$ is shown in Figure C-2. A line connects the points on the graph of $x = 5.86$ and $x = 5.87$. The y-values for these values of x are obtained from Appendix E. These values are then used to find an approximate value for log 5.864.

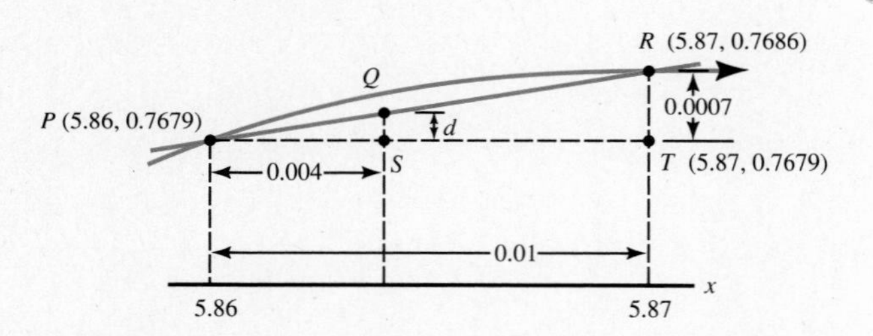

Figure C-2. A graph of $y = \log x$ around $x = 5.864$.

It can be shown that triangle PQS is similar to triangle PRT. As a consequence:

$$\frac{PS}{PT} = \frac{QS}{RT}$$

Substituting the values from Figure C-2:

$$\frac{0.004}{0.01} = \frac{d}{0.0007}$$

Solving for d:

$$\begin{aligned} d &= (0.4)(0.0007) \\ &= 0.00028 \\ &= 0.0003 \text{ to four decimal places} \end{aligned}$$

Adding d to 0.7679:

$$\log 5.864 \approx 0.7679 + 0.0003 = 0.7682$$

Thus, using the *line PR*, the approximate value for log 5.864 is 0.7682.

Example. Use Appendix E and linear interpolation to approximate log 4.356.

Solution. Since 4.356 is between $x_1 = 4.35$ and $x_2 = 4.36$, the logarithms for x_1 and x_2 are used to make the approximation. The work is organized by using the following array:

		x	$\log x$		
0.01		4.36	0.6395		0.0010
	0.006	4.356	?	d	
		4.35	0.6385		

Now writing the proportion:

$$\frac{0.006}{0.01} = \frac{d}{0.0010}$$

$$d = 0.0006$$

Adding d to 0.6385:

$$\log 4.356 \approx 0.6385 + 0.0006 = 0.6391$$

Practice Exercises

Use Appendix E and linear interpolation to approximate each logarithm.

1. log 2.145 **2.** log 3.625 **3.** log 3.912 **4.** log 4.058

5. log 4.103 **6.** log 5.833 **7.** log 6.747 **8.** log 7.386

9. log 8.056 **10.** log 9.555

APPENDIX

D

Significant Digits

The following statement identifies what is meant by a significant digit in a number:

> The **significant digits** of a number written in decimal form are the digits that indicate the precision of a measurement or a calculation.

The following cases stipulate what is generally agreed upon as identifying a significant digit in a number:

Case 1. All digits of an exact number are significant.

Case 2. For a number that is not exact, the significant digits are:

- **a.** all nonzero digits and
- **b.** the digit 0 if
  - **(i)** it is between nonzero digits, or
  - **(ii)** it is to the right of a nonzero digit and the number is written with a decimal point.

Examples **a–g** illustrate these rules:

Number	Number of significant digits	Reason
a. 1,250 pages	4	The units identify the number as exact.
b. 63,500	3	Only the nonzero digits are significant.
c. 63,500.	5	The decimal point makes all digits significant.
d. 60,305	5	The zeros are between nonzero digits.
e. 18.0	3	The zero follows a nonzero digit and the number has a decimal point.
f. 0.0025	2	Only the nonzero digits are significant.
g. 0.0500	3	The zeros follow a nonzero digit and the number has a decimal point.

APPENDIX

Table: Logarithms Base 10, Common Logarithms

x	0	1	2	3	4	5	6	7	8	9
1.0	0.0000	0.0043	0.0086	0.0128	0.0170	0.0212	0.0253	0.0294	0.0334	0.0374
1.1	0.0414	0.0453	0.0492	0.0531	0.0569	0.0607	0.0645	0.0682	0.0719	0.0755
1.2	0.0792	0.0828	0.0864	0.0899	0.0934	0.0969	0.1004	0.1038	0.1072	0.1106
1.3	0.1139	0.1173	0.1206	0.1239	0.1271	0.1303	0.1335	0.1367	0.1399	0.1430
1.4	0.1461	0.1492	0.1523	0.1553	0.1584	0.1614	0.1644	0.1673	0.1703	0.1732
1.5	0.1761	0.1790	0.1818	0.1847	0.1875	0.1903	0.1931	0.1959	0.1987	0.2014
1.6	0.2041	0.2068	0.2095	0.2122	0.2148	0.2175	0.2201	0.2227	0.2253	0.2279
1.7	0.2304	0.2330	0.2355	0.2380	0.2405	0.2430	0.2455	0.2480	0.2504	0.2529
1.8	0.2553	0.2577	0.2601	0.2625	0.2648	0.2672	0.2695	0.2718	0.2742	0.2765
1.9	0.2788	0.2810	0.2833	0.2856	0.2878	0.2900	0.2923	0.2945	0.2967	0.2989
2.0	0.3010	0.3032	0.3054	0.3075	0.3096	0.3118	0.3139	0.3160	0.3181	0.3201
2.1	0.3222	0.3243	0.3263	0.3284	0.3304	0.3324	0.3345	0.3365	0.3385	0.3404
2.2	0.3424	0.3444	0.3464	0.3483	0.3502	0.3522	0.3541	0.3560	0.3579	0.3598
2.3	0.3617	0.3636	0.3655	0.3674	0.3692	0.3711	0.3729	0.3747	0.3766	0.3784
2.4	0.3802	0.3820	0.3838	0.3856	0.3874	0.3892	0.3909	0.3927	0.3945	0.3962
2.5	0.3979	0.3997	0.4014	0.4031	0.4048	0.4065	0.4082	0.4099	0.4116	0.4133
2.6	0.4150	0.4166	0.4183	0.4200	0.4216	0.4232	0.4249	0.4265	0.4281	0.4298
2.7	0.4314	0.4330	0.4346	0.4362	0.4378	0.4393	0.4409	0.4425	0.4440	0.4456
2.8	0.4472	0.4487	0.4502	0.4518	0.4533	0.4548	0.4564	0.4579	0.4594	0.4609
2.9	0.4624	0.4639	0.4654	0.4669	0.4683	0.4698	0.4713	0.4728	0.4742	0.4757
3.0	0.4771	0.4786	0.4800	0.4814	0.4829	0.4843	0.4857	0.4871	0.4886	0.4900
3.1	0.4914	0.4928	0.4942	0.4955	0.4969	0.4983	0.4997	0.5011	0.5024	0.5038
3.2	0.5051	0.5065	0.5079	0.5092	0.5105	0.5119	0.5132	0.5145	0.5159	0.5172
3.3	0.5185	0.5198	0.5211	0.5224	0.5237	0.5250	0.5263	0.5276	0.5289	0.5302
3.4	0.5315	0.5328	0.5340	0.5353	0.5366	0.5378	0.5391	0.5403	0.5416	0.5428
3.5	0.5441	0.5453	0.5465	0.5478	0.5490	0.5502	0.5514	0.5527	0.5539	0.5551
3.6	0.5563	0.5575	0.5587	0.5599	0.5611	0.5623	0.5635	0.5647	0.5658	0.5670
3.7	0.5682	0.5694	0.5705	0.5717	0.5729	0.5740	0.5752	0.5763	0.5775	0.5786
3.8	0.5798	0.5809	0.5821	0.5832	0.5843	0.5855	0.5866	0.5877	0.5888	0.5899
3.9	0.5911	0.5922	0.5933	0.5944	0.5955	0.5966	0.5977	0.5988	0.5999	0.6010
4.0	0.6021	0.6031	0.6042	0.6053	0.6064	0.6075	0.6085	0.6096	0.6107	0.6117
4.1	0.6128	0.6138	0.6149	0.6160	0.6170	0.6180	0.6191	0.6201	0.6212	0.6222
4.2	0.6232	0.6243	0.6253	0.6263	0.6274	0.6284	0.6294	0.6304	0.6314	0.6325
4.3	0.6335	0.6345	0.6355	0.6365	0.6375	0.6385	0.6395	0.6405	0.6415	0.6425
4.4	0.6435	0.6444	0.6454	0.6464	0.6474	0.6484	0.6493	0.6503	0.6513	0.6522
x	0	1	2	3	4	5	6	7	8	9

TABLE: LOGARITHM BASE 10, COMMON LOGARITHMS (CONTINUED)

x	0	1	2	3	4	5	6	7	8	9
4.5	0.6532	0.6542	0.6551	0.6561	0.6571	0.6580	0.6590	0.6599	0.6609	0.6618
4.6	0.6628	0.6637	0.6646	0.6656	0.6665	0.6675	0.6684	0.6693	0.6702	0.6712
4.7	0.6721	0.6730	0.6739	0.6749	0.6758	0.6767	0.6776	0.6785	0.6794	0.6803
4.8	0.6812	0.6821	0.6830	0.6839	0.6848	0.6857	0.6866	0.6875	0.6884	0.6893
4.9	0.6902	0.6911	0.6920	0.6928	0.6937	0.6946	0.6955	0.6964	0.6972	0.6981
5.0	0.6990	0.6998	0.7007	0.7016	0.7024	0.7033	0.7042	0.7050	0.7059	0.7067
5.1	0.7076	0.7084	0.7093	0.7101	0.7110	0.7118	0.7126	0.7135	0.7143	0.7152
5.2	0.7160	0.7168	0.7177	0.7185	0.7193	0.7202	0.7210	0.7218	0.7226	0.7235
5.3	0.7243	0.7251	0.7259	0.7267	0.7275	0.7284	0.7292	0.7300	0.7308	0.7316
5.4	0.7324	0.7332	0.7340	0.7348	0.7356	0.7364	0.7372	0.7380	0.7388	0.7396
5.5	0.7404	0.7412	0.7419	0.7427	0.7435	0.7443	0.7451	0.7459	0.7466	0.7474
5.6	0.7482	0.7490	0.7497	0.7505	0.7513	0.7520	0.7528	0.7536	0.7543	0.7551
5.7	0.7559	0.7566	0.7574	0.7582	0.7589	0.7597	0.7604	0.7612	0.7619	0.7627
5.8	0.7634	0.7642	0.7649	0.7657	0.7664	0.7672	0.7679	0.7686	0.7694	0.7701
5.9	0.7709	0.7716	0.7723	0.7731	0.7738	0.7745	0.7752	0.7760	0.7767	0.7774
6.0	0.7782	0.7789	0.7796	0.7803	0.7810	0.7818	0.7825	0.7832	0.7839	0.7846
6.1	0.7853	0.7860	0.7868	0.7875	0.7882	0.7889	0.7896	0.7903	0.7910	0.7917
6.2	0.7924	0.7931	0.7938	0.7945	0.7952	0.7959	0.7966	0.7973	0.7980	0.7987
6.3	0.7993	0.8000	0.8007	0.8014	0.8021	0.8028	0.8035	0.8041	0.8048	0.8055
6.4	0.8062	0.8069	0.8075	0.8082	0.8089	0.8096	0.8102	0.8109	0.8116	0.8122
6.5	0.8129	0.8136	0.8142	0.8149	0.8156	0.8162	0.8169	0.8176	0.8182	0.8189
6.6	0.8195	0.8202	0.8209	0.8215	0.8222	0.8228	0.8235	0.8241	0.8248	0.8254
6.7	0.8261	0.8267	0.8274	0.8280	0.8287	0.8293	0.8299	0.8306	0.8312	0.8319
6.8	0.8325	0.8331	0.8338	0.8344	0.8351	0.8357	0.8363	0.8370	0.8376	0.8382
6.9	0.8388	0.8395	0.8401	0.8407	0.8414	0.8420	0.8426	0.8432	0.8439	0.8445
7.0	0.8451	0.8457	0.8463	0.8470	0.8476	0.8482	0.8488	0.8494	0.8500	0.8506
7.1	0.8513	0.8519	0.8525	0.8531	0.8537	0.8543	0.8549	0.8555	0.8561	0.8567
7.2	0.8573	0.8579	0.8585	0.8591	0.8597	0.8603	0.8609	0.8615	0.8621	0.8627
7.3	0.8633	0.8639	0.8645	0.8651	0.8657	0.8663	0.8669	0.8675	0.8681	0.8686
7.4	0.8692	0.8698	0.8704	0.8710	0.8716	0.8722	0.8727	0.8733	0.8739	0.8745
7.5	0.8751	0.8756	0.8762	0.8768	0.8774	0.8779	0.8785	0.8791	0.8797	0.8802
7.6	0.8808	0.8814	0.8820	0.8825	0.8831	0.8837	0.8842	0.8848	0.8854	0.8859
7.7	0.8865	0.8871	0.8876	0.8882	0.8887	0.8893	0.8899	0.8904	0.8910	0.8915
7.8	0.8921	0.8927	0.8938	0.8938	0.8943	0.8949	0.8954	0.8960	0.8965	0.8971
7.9	0.8976	0.8982	0.8993	0.8993	0.8998	0.9009	0.9009	0.9015	0.9020	0.9025
8.0	0.9031	0.9036	0.9042	0.9047	0.9053	0.9058	0.9063	0.9069	0.9074	0.9079
8.1	0.9085	0.9090	0.9096	0.9101	0.9106	0.9112	0.9117	0.9122	0.9128	0.9133
8.2	0.9138	0.9143	0.9149	0.9154	0.9159	0.9165	0.9170	0.9175	0.9180	0.9186
8.3	0.9191	0.9196	0.9201	0.9206	0.9212	0.9217	0.9222	0.9227	0.9232	0.9238
8.4	0.9243	0.9248	0.9253	0.9258	0.9263	0.9269	0.9274	0.9279	0.9284	0.9289
8.5	0.9294	0.9299	0.9304	0.9309	0.9315	0.9320	0.9325	0.9330	0.9335	0.9340
8.6	0.9345	0.9350	0.9355	0.9360	0.9365	0.9370	0.9375	0.9380	0.9385	0.9390
8.7	0.9395	0.9400	0.9405	0.9410	0.9415	0.9420	0.9425	0.9430	0.9435	0.9440
8.8	0.9445	0.9450	0.9455	0.9460	0.9465	0.9469	0.9474	0.9479	0.9484	0.9489
8.9	0.9494	0.9499	0.9504	0.9509	0.9513	0.9518	0.9528	0.9528	0.9533	0.9538
9.0	0.9542	0.9547	0.9552	0.9557	0.9562	0.9566	0.9571	0.9576	0.9581	0.9586
9.1	0.9590	0.9595	0.9600	0.9605	0.9609	0.9614	0.9619	0.9624	0.9628	0.9633
9.2	0.9638	0.9643	0.9647	0.9652	0.9657	0.9661	0.9666	0.9671	0.9675	0.9680
9.3	0.9685	0.9689	0.9694	0.9699	0.9703	0.9708	0.9713	0.9717	0.9722	0.9727
9.4	0.9731	0.9736	0.9741	0.9745	0.9750	0.9754	0.9759	0.9763	0.9768	0.9773
9.5	0.9777	0.9782	0.9786	0.9791	0.9795	0.9800	0.9805	0.9809	0.9814	0.9818
9.6	0.9823	0.9827	0.9832	0.9836	0.9841	0.9845	0.9850	0.9854	0.9859	0.9863
9.7	0.9868	0.9872	0.9877	0.9881	0.9886	0.9890	0.9894	0.9899	0.9903	0.9908
9.8	0.9912	0.9917	0.9921	0.9926	0.9930	0.9934	0.9939	0.9943	0.9948	0.9952
9.9	0.9956	0.9961	0.9965	0.9969	0.9974	0.9978	0.9983	0.9987	0.9991	0.9996
x	0	1	2	3	4	5	6	7	8	9

APPENDIX

Table: Exponential Function

x	e^x	e^{-x}	x	e^x	e^{-x}
0.00	1.0000	1.0000	3.0	20.086	0.0498
0.05	1.0513	0.9512	3.1	22.198	0.0450
0.10	1.1052	0.9048	3.2	24.533	0.0408
0.15	1.1618	0.8607	3.3	27.113	0.0369
0.20	1.2214	0.8187	3.4	29.964	0.0334
0.25	1.2840	0.7788	3.5	33.115	0.0302
0.30	1.3499	0.7408	3.6	36.598	0.0273
0.35	1.4191	0.7047	3.7	40.447	0.0247
0.40	1.4918	0.6703	3.8	44.701	0.0224
0.45	1.5683	0.6376	3.9	49.402	0.0202
0.50	1.6487	0.6065	4.0	54.598	0.0183
0.55	1.7333	0.5769	4.1	60.340	0.0166
0.60	1.8221	0.5488	4.2	66.686	0.0150
0.65	1.9155	0.5200	4.3	73.700	0.0136
0.70	2.0138	0.4966	4.4	81.451	0.0123
0.75	2.1170	0.4724	4.5	90.017	0.0111
0.80	2.2255	0.4493	4.6	99.484	0.0101
0.85	2.3396	0.4274	4.7	109.95	0.0091
0.90	2.4596	0.4066	4.8	121.51	0.0082
0.95	2.5857	0.3867	4.9	134.29	0.0074
1.0	2.7183	0.3679	5.0	148.41	0.0067
1.1	3.0042	0.3329	5.1	164.02	0.0061
1.2	3.3201	0.3012	5.2	181.27	0.0055
1.3	3.6693	0.2725	5.3	200.34	0.0050
1.4	4.0552	0.2466	5.4	221.41	0.0045
1.5	4.4817	0.2231	5.5	244.69	0.0041
1.6	4.9530	0.2019	5.6	270.43	0.0037
1.7	5.4739	0.1827	5.7	298.87	0.0033
1.8	6.0496	0.1653	5.8	330.30	0.0030
1.9	6.6859	0.1496	5.9	365.04	0.0027
2.0	7.3891	0.1353	6.0	403.43	0.0025
2.1	8.1662	0.1225	6.5	665.14	0.0015
2.2	9.0250	0.1108	7.0	1096.6	0.0009
2.3	9.9742	0.1003	7.5	1808.0	0.0006
2.4	11.023	0.0907	8.0	2981.0	0.0003
2.5	12.182	0.0821	8.5	4914.8	0.0002
2.6	13.464	0.0743	9.0	8103.1	0.0001
2.7	14.880	0.0672	9.5	13,360	0.00007
2.8	16.445	0.0608	10.0	22,026	0.00004
2.9	18.174	0.0550			

APPENDIX

G
Blank Graphs

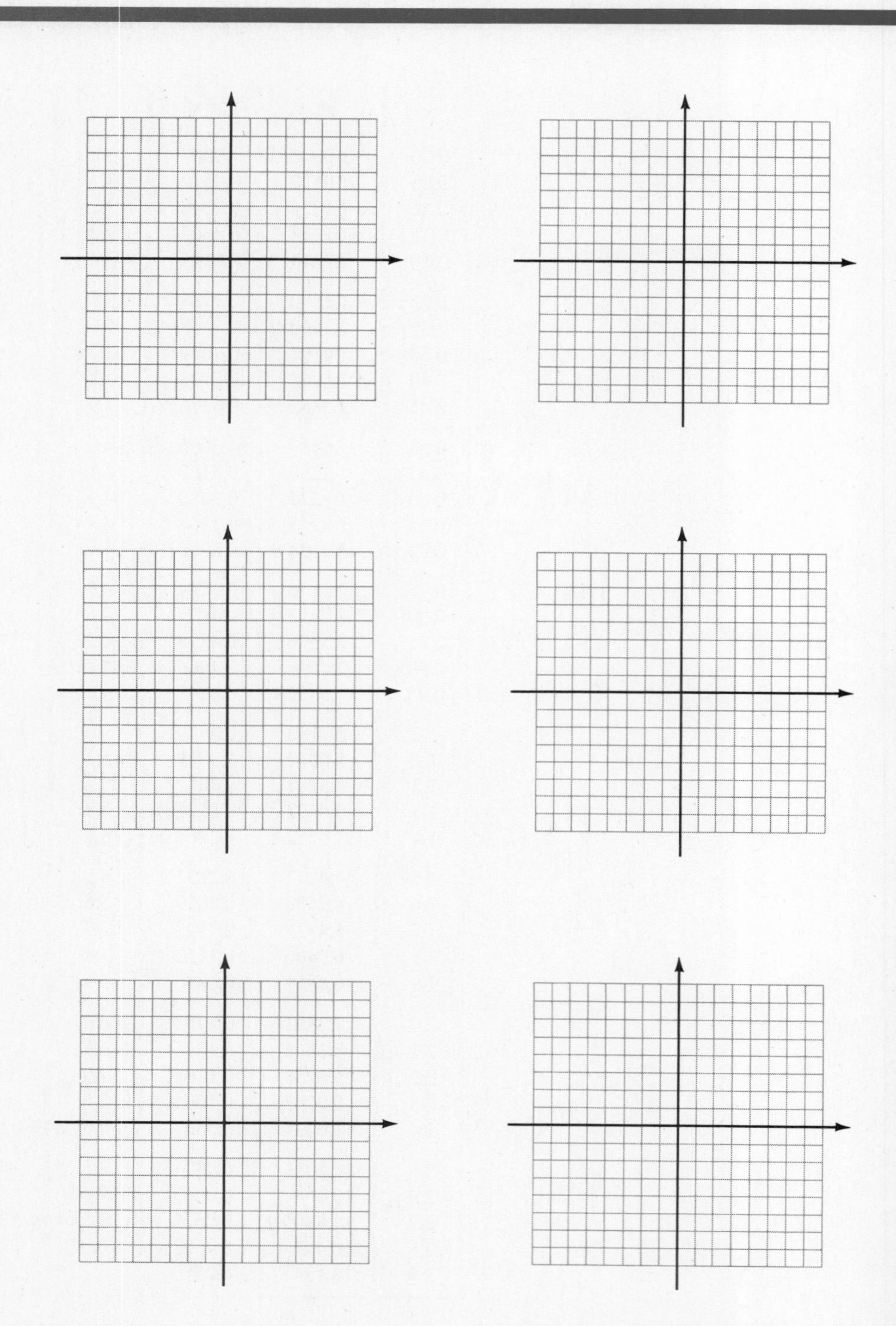

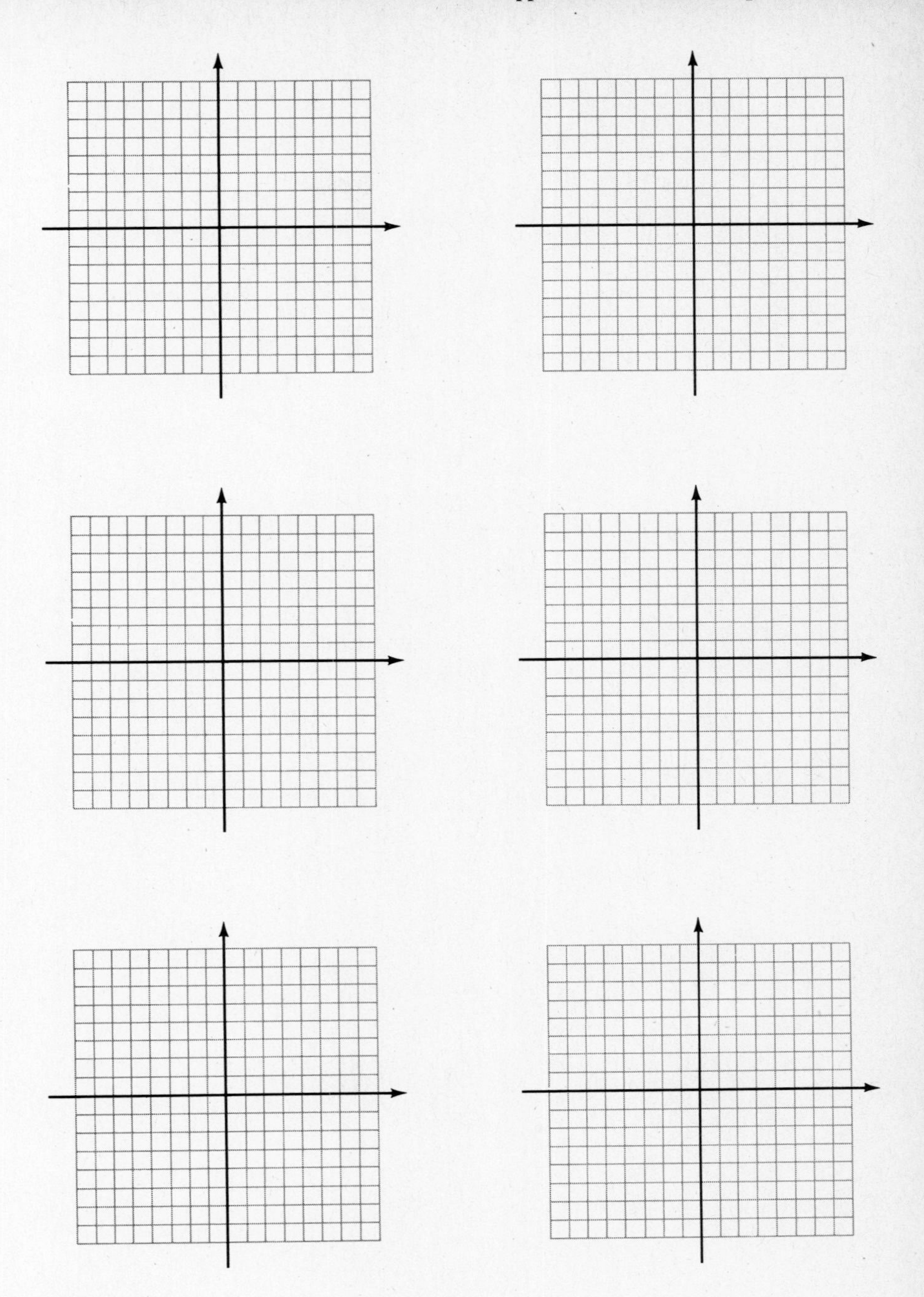

Answers For Odd-Numbered Exercises

Chapter 1 Essential Topics from Elementary Algebra

Section 1-1. Practice Exercises

1. $\frac{39}{1}$ or $\frac{78}{2}$, etc. **3.** $\frac{23}{5}$ **5.** $\frac{1}{4}$ **7.** $\frac{-8}{1}$ or $\frac{-16}{2}$, etc. **9.** $\frac{-13}{2}$ **11.** $\frac{-15}{4}$ **13.** $\frac{0}{1}$ or $\frac{0}{2}$, etc. **15.** $\frac{3}{16}$

	Whole numbers	Integers	Rational numbers	Irrational numbers	Real numbers
17.	X	X	X		X
19.			X		X
21.	X	X	X		X
23.				X	X
25.			X		X
27.			X		X
29.				X	X
31.			X		X
33.	X	X	X		X
35.	X	X	X		X

37. 5, 10, 15, 20, 25 **39.** 13, 26, 39, 52, 65 **41.** 19, 38, 57, 76, 95 **43.** 25, 50, 75, 100, 125 **45.** 39, 78, 117, 156, 195
47. 1, 2, 3, 4, 6, 12 **49.** 1, 3, 7, 21 **51.** 1, 5, 7, 35 **53.** 1, 43 **55.** 1, 2, 3, 6, 13, 26, 39, 78

57–69 Odd.
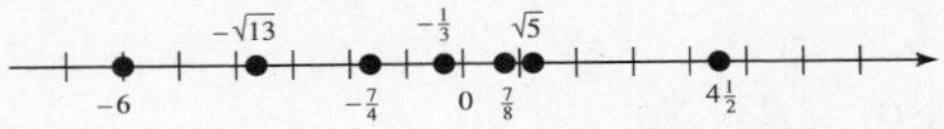

71 .> **73.** < **75.** < **77.** < **79.** >

81. Symmetric property of equality **83.** Transitive property of equality **85.** Antisymmetric property of inequality
87. Reflexive property of equality **89.** Transitive property of inequality

91.

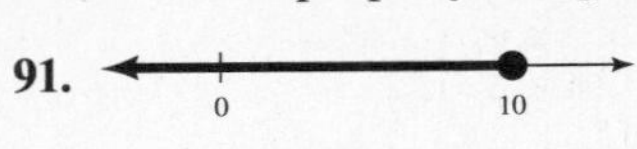

93.
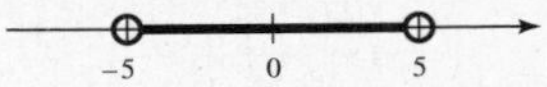

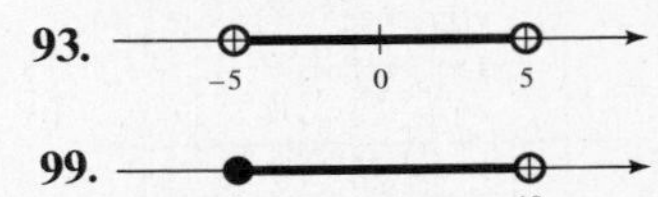

95.

97.
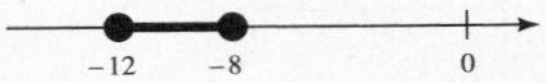

99.

101.
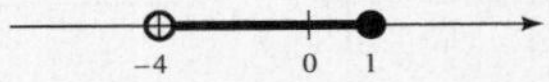

103. $\{x \mid x > -2\}$ **105.** $\{x \mid x \le 6\}$ **107.** $\{x \mid 2 \le x \le 7\}$ **109.** $\{x \mid -5 < x < 1\}$ **111.** $\{x \mid -20 < x \le -10\}$

Section 1-1. Supplementary Exercises

1. Never **3.** Always **5.** Never **7.** Sometimes **9.** Never **11.** Sometimes **13.** Never **15.** Sometimes
17. Sometimes **19.** Sometimes **21.** $2 \cdot 2 \cdot 3$ **23.** $2 \cdot 2 \cdot 7$ **25.** $3 \cdot 17$ **27.** $2 \cdot 2 \cdot 3 \cdot 5$ **29.** $2 \cdot 2 \cdot 3 \cdot 11$
31. $2 \cdot 2 \cdot 3 \cdot 29$ **33.** $\frac{7}{9}$ **35.** $\frac{4}{33}$ **37.** $\frac{3}{37}$ **39.** 30 **41.** 60 **43.** 150 **45.** 210 **47.** $x < -12$ **49.** $x \le 9$
51. $-10 < x < -3$ **53.** $-6 \le x < 5$ **55.** $x > 0$ **57.** $x \le 0$ **59.** $x \ge 0$

Section 1-2. Practice Exercises

1. 23 **3.** 39 **5.** 44 **7.** -12 **9.** $-x$ **11.** $a - b$ **13.** $-3a$ **15.** 36 **17.** -30 **19.** 13 **21.** -16
23. -45 **25.** 7 **27.** -30 **29.** 40 **31.** 58 **33.** -103 **35.** -6 **37.** 17 **39.** -16 **41.** -17 **43.** 10
45. 4 **47.** -4 **49.** 40 **51.** 30 **53.** -18 **55.** -21 **57.** 18 **59.** -60 **61.** 20 **63.** -7 **65.** -180

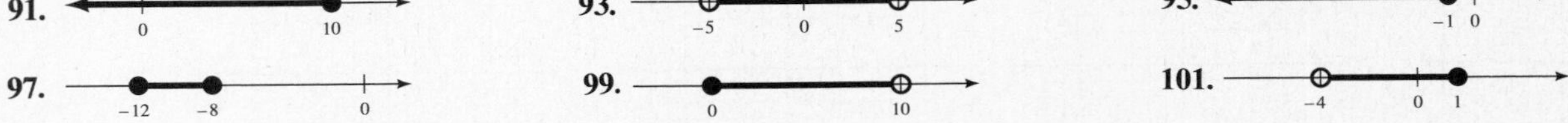

67. 0 **69.** $\frac{-2}{9}$ **71.** $\frac{3}{4}$ **73.** $\frac{1}{14}$ **75.** $\frac{25}{6}$ **77.** $\frac{1}{7}$ **79.** 6 **81.** -10 **83.** 4 **85.** -3 **87.** 4 **89.** $\frac{-1}{10}$

91. $\frac{2}{3}$ **93.** 20 **95.** -20 **97.** 125 **99.** $-1{,}000$ **101.** -16 **103.** 16 **105.** 1,728 **107.** 9 **109.** -4

111. 30 **113.** 193 **115.** 7 **117.** -44 **119.** 45 **121.** 27 **123.** 4 **125.** 2

Section 1-2. Supplementary Exercises

1. Closure under multiplication **3.** Commutative property of addition **5.** Associative property of multiplication
7. Identity element for addition **9.** Inverse elements under multiplication **11.** Distributive; multiplication over addition
13. Associative property of addition **15.** Commutative property of addition **17.** Closure under multiplication

19. Inverse elements under addition **21. a.** $\frac{-3}{8}$ **b.** $\frac{8}{3}$ **23. a.** $\frac{2}{9}$ **b.** $\frac{-9}{2}$ **25. a.** 7 **b.** $\frac{-1}{7}$

27. a. $\frac{-8}{3}$ **b.** $\frac{3}{8}$ **29.** 7 **31.** 8 **33.** 6 **35.** 21 **37.** 15 **39.** 6 **41.** -5 **43.** $15 - 7 \cdot 4 + 16 \div (4 \cdot 2)$

45. $(15 - 7) \cdot 4 + 16 \div (4 \cdot 2)$

Section 1-3. Practice Exercises

1. a. $5a^2bc^3$ **b.** 5 **c.** a^2bc^3 **3. a.** $\frac{-3}{4}xy^2$ **b.** $\frac{-3}{4}$ **c.** xy^2 **5. a.** $\frac{1}{3}l^3m^4n^7$ **b.** $\frac{1}{3}$ **c.** $l^3m^4n^7$

7. a. $-pq^2r^5$ **b.** -1 **c.** pq^2r^5 **9.** $8x$ **11.** yz **13.** $-6p^2 + 7p$ **15.** $5m^2 + 3m - 14$ **17.** $-12x^3y^5$
19. $16pq^2$ **21.** $24a^4b^2 - 6a^3b^4$ **23.** $-63s^8t^{13} + 14s^7t^{10} - 42s^6t^{13}$ **25.** $8x^2 - 2x$ **27.** $-19t^2 - 6t + 5$
29. $p^2 - 18p - 2$ **31.** $2u^3 - u^2 - u + 1$ **33.** $-x^3 + 2x^2 + 4x$ **35.** $-3y$ **37.** $-5t^3 - 13t^2 + 46t - 3$

39. $m^2 + 2mn + 2n^2$ **41. a.** -7 **b.** 35 **43. a.** 95 **b.** -10 **45.** 8 **47.** -144 **49.** $\frac{-1}{2}$

Section 1-3. Supplementary Exercises

1. 129 mi **3.** 150 mi **5.** 99 **7.** 50 **9.** $\frac{250}{3}$ o **11.** $\frac{9}{13}$ o **13.** 268 **15.** -520 **17.** $(-2t^2 + 18t - 50)$ t

19. $(-3t^2 + 20t - 46)$ t **21.** $4x^2 + 6$ **23.** $10x + 2$ **25.** $7x + 3y$

Section 1-4. Practice Exercises

1. a. (5, 28) **b.** $(-8, -50)$ **3. a.** (7, 14) **b.** $(-3, 24)$ **5. a.** (0, 0) **b.** $(-3, 1)$ **7. a.** (4, 5) **b.** $(2, -4)$
9. a. $\{-11, -3, 1, 7\}$ **b.** $\{-5, 2, 0, 3\}$ **11. a.** $\{-3, 5, 7, 21\}$ **b.** $\{-3, 0\}$ **13. a.** $\{0, 2, 8\}$ **b.** $\{3, 5, 7\}$
15. a. $\{-7\}$ **b.** $\{-8, -1, 0, 3, 12\}$ **17.** Yes **19.** Yes **21.** No **23.** No

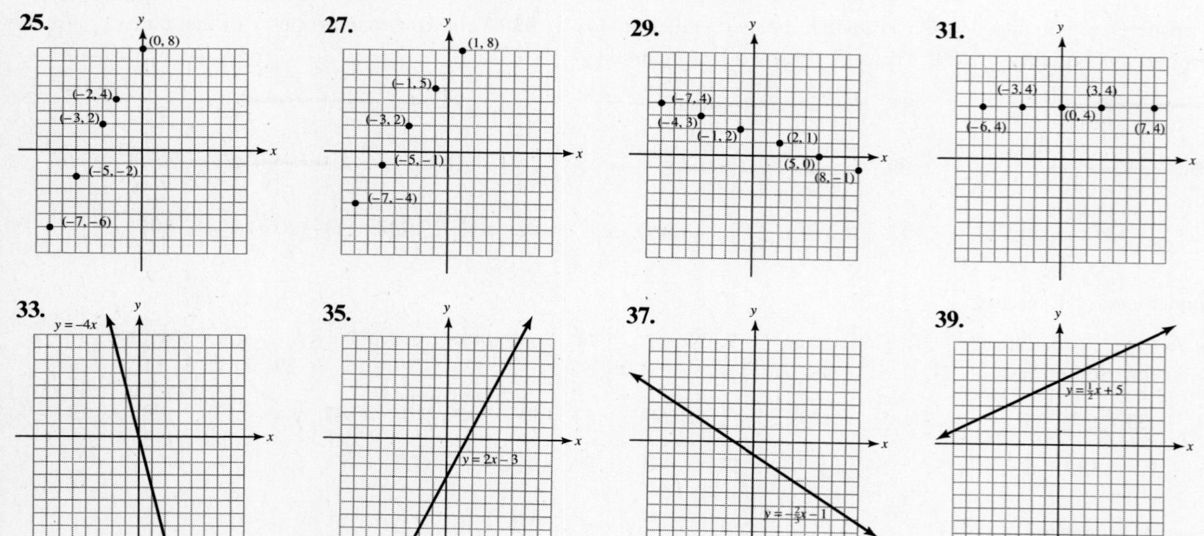

41. 5 **43.** $\frac{-3}{2}$ **45.** Undefined **47.** -1 **49.** 0

51.

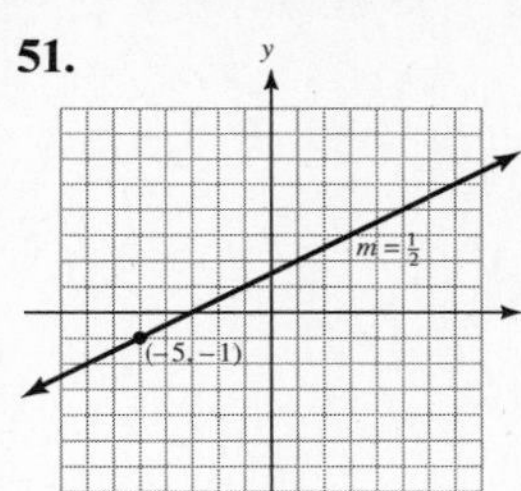

53.

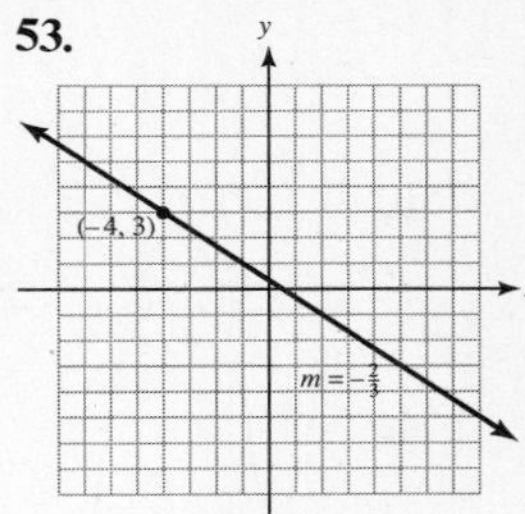

55.

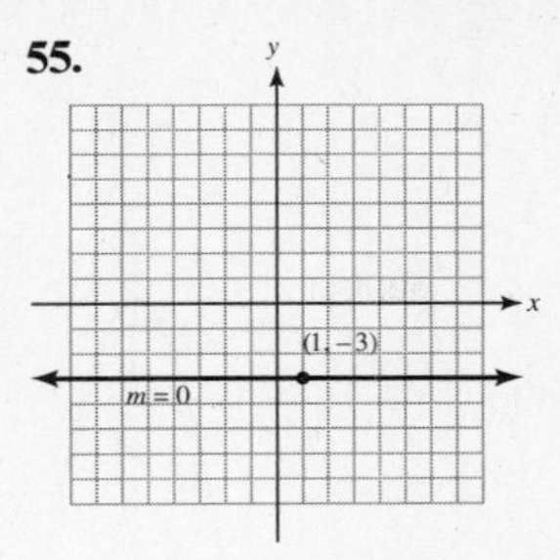

57.

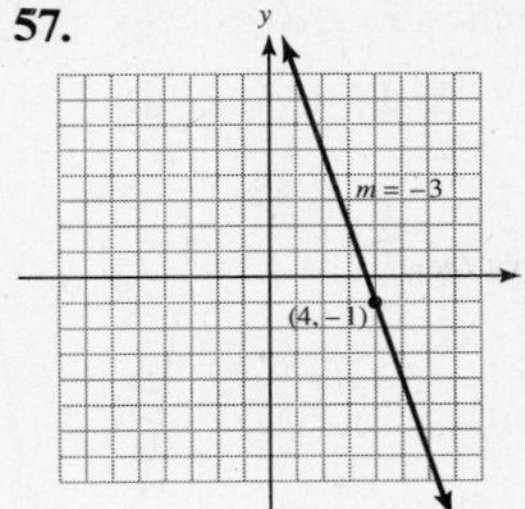

59. **a.** $y = -3x + 6$ **b.** -3 **c.** 6 **d.** 

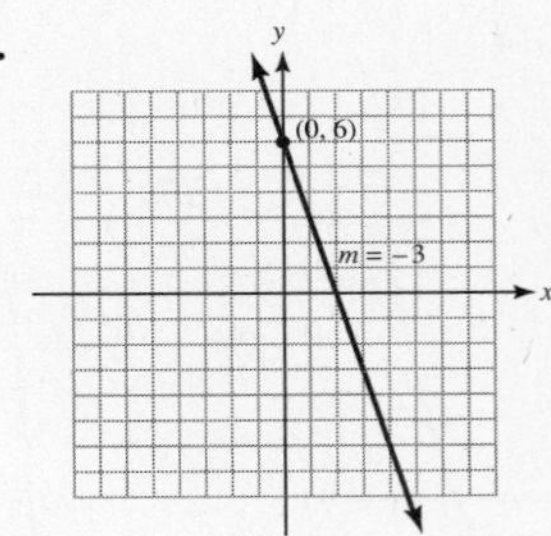

61. **a.** $y = \frac{1}{2}x - 4$ **b.** $\frac{1}{2}$ **c.** -4 **d.** 

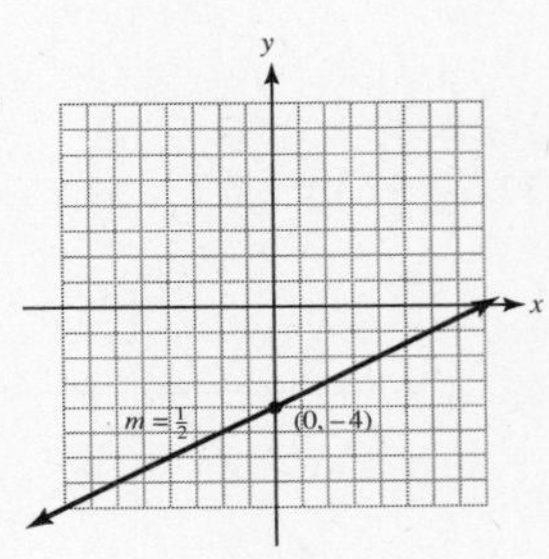

63. **a.** $y = \frac{-4}{3}x + 1$ **b.** $\frac{-4}{3}$ **c.** 1 **d.**

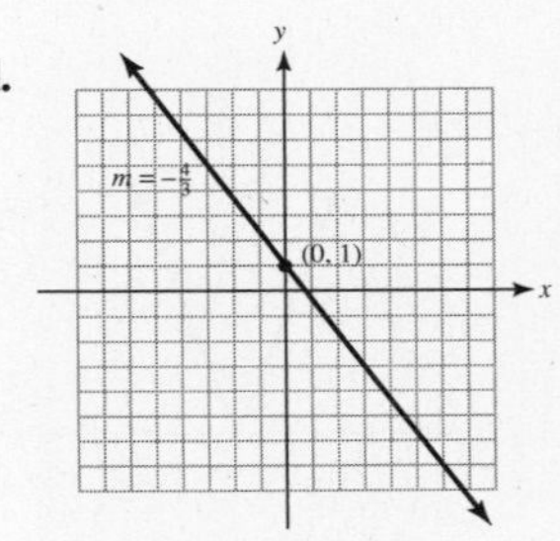

65. **a.** $y = \frac{7}{9}x - 3$ **b.** $\frac{7}{9}$ **c.** -3 **d.** 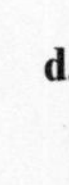

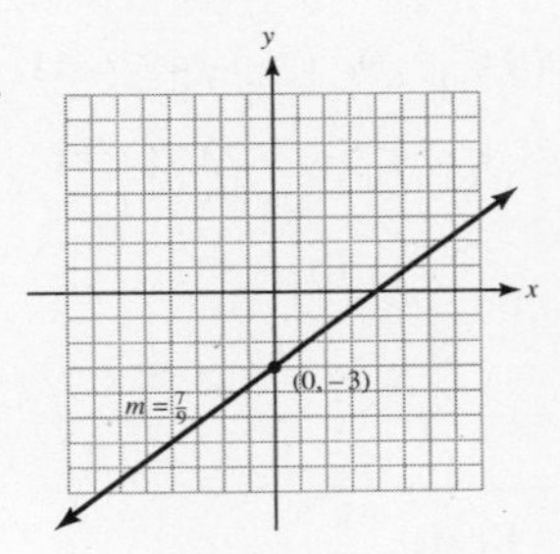

Section 1-4. Supplementary Exercises

1. $P(-6, 4)$; $Q(-2, 2)$; $R(4, -1)$ **3.** $\frac{-1}{2}$ **5.** 2 **7.** -2 **9.** $\frac{1}{2}$ **11.** -4 **13.** $x < 2$ **15.** $y < 1$ **17.** 2
19. $-6 < x < -2$ **21.** $x > -4$

Section 1-5. Practice Exercises

1. -4 or 4 **3.** 1 **5.** 0 or 2 **7.** 0 **9.** -4 or 0 **11.** 5

13.

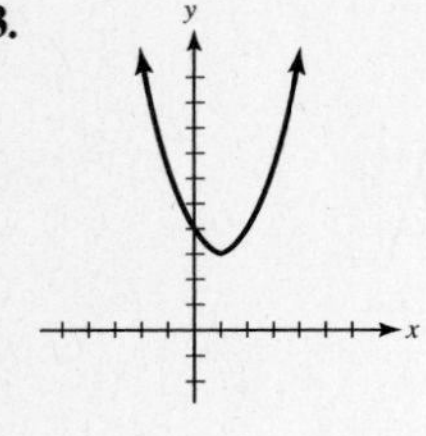

15.

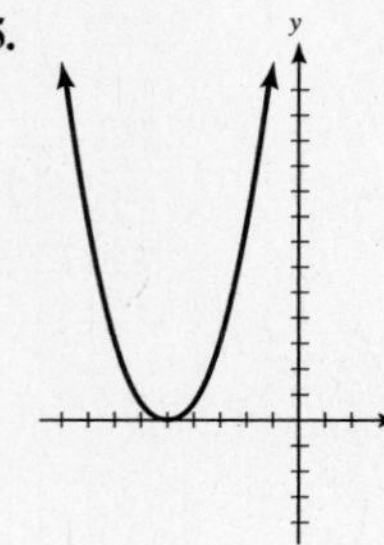

17.

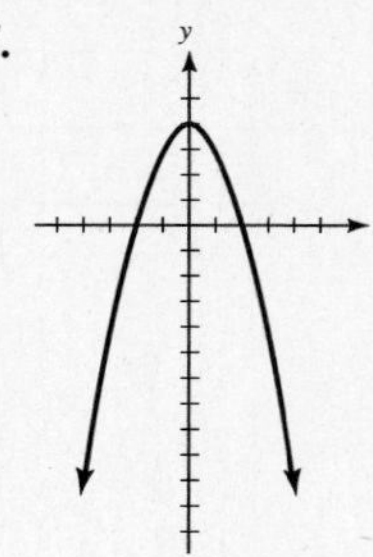

19.

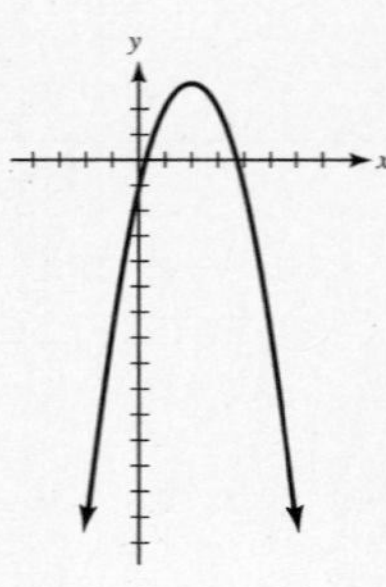

21.

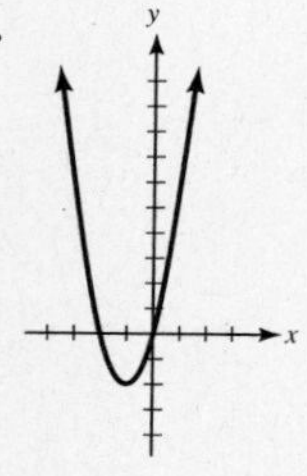

23.

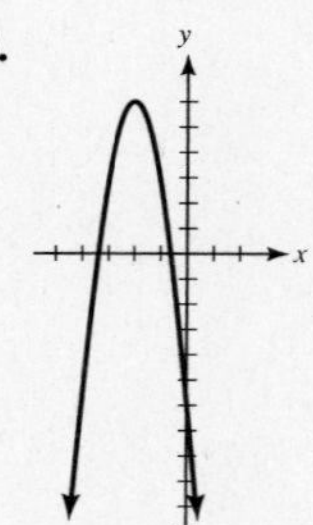

25.

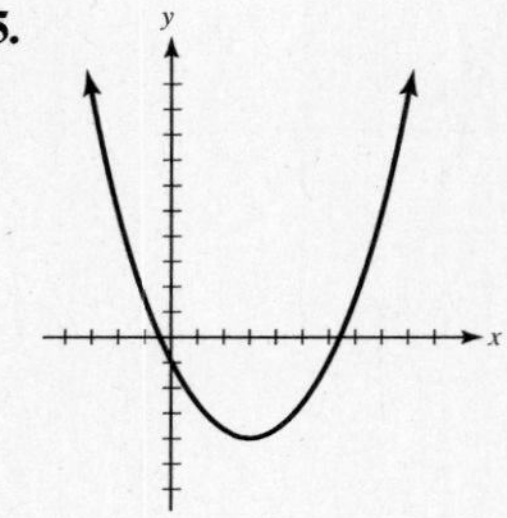

27.

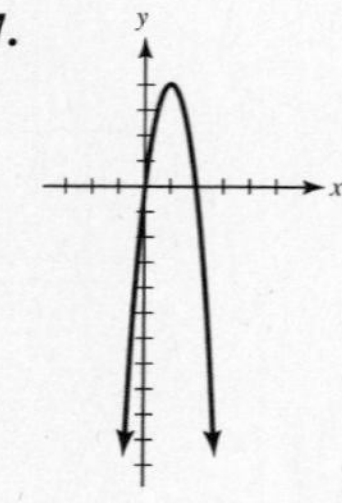

29. -2 or 8 **31.** 4 **33.** -4 or 0 **35.** 0 **37.** -5 or 7 **39.** -6

41.

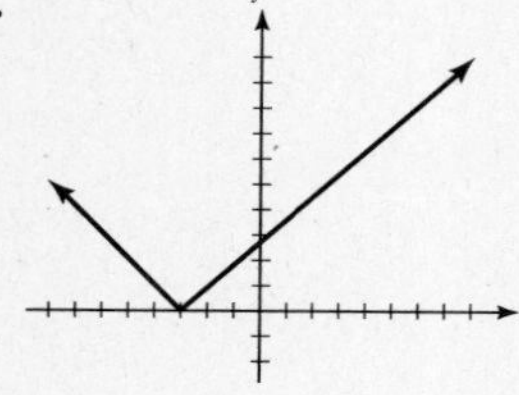

43.

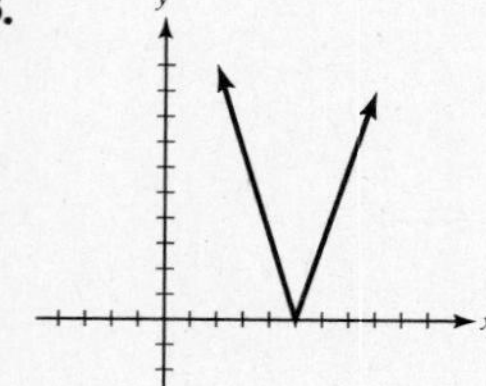

45.

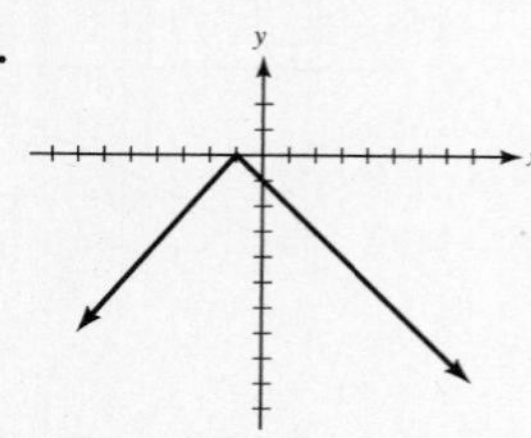

47.

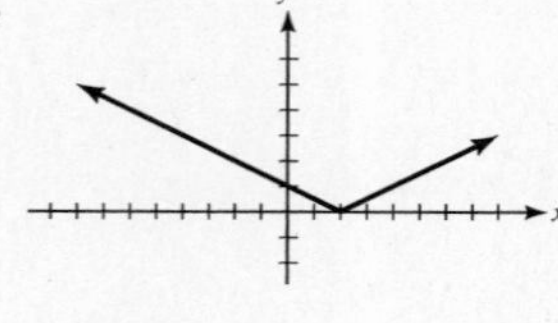

49.

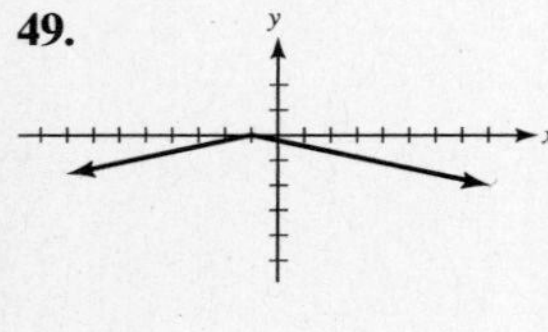

51.

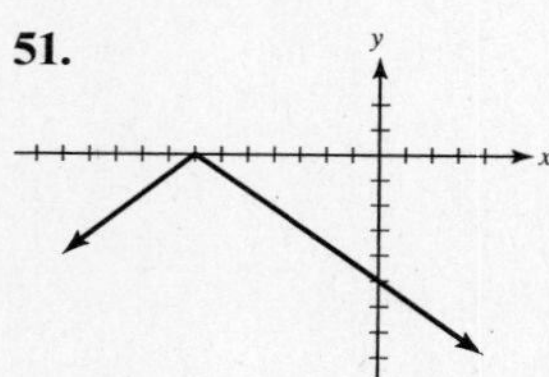

53.

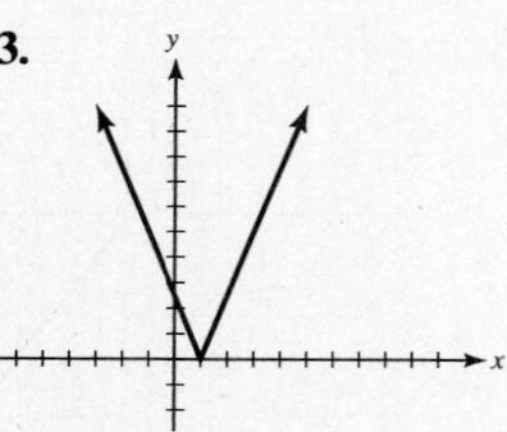

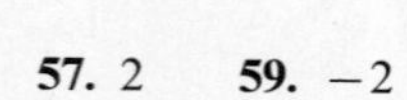

55. 2 **57.** 2 **59.** -2

Section 1-5. Supplementary Exercises

1. -1 **3.** -3 **5.** 1 or 5 **7.** $x < -5$ or $x > 11$ **9.** $y \geq -4$ **11.** -5 or 11 **13.** $-1 < x < 7$

15. $-3 < x < 9$ **17.** $\frac{-1}{2}$ **19.** -5 and 11 **21.** $1 < x < 5$ **23.** 1 or 5 **25.** $1 < x < 2$ or $4 < x < 5$

Chapter 1 Review Exercises

1. $\frac{22}{5}$

	Whole numbers	Integers	Rational numbers	Irrational numbers	Real numbers
3.			X		X
5.	X	X	X		X
7.			X		X

9. 90 **11.** 1, 2, 3, 5, 6, 10, 15, and 30 **13–17.** (number line: $-4\frac{1}{2}$, 0, $\frac{8}{5}$, 4) **19.** > **21.** <

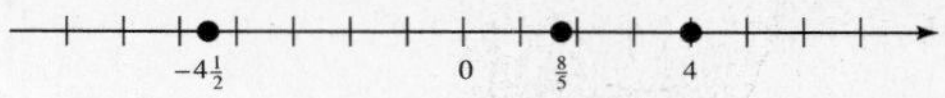

23. Symmetric property of equality **25.** Transitive property of inequality **27.** 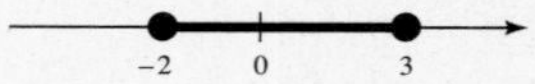 (number line: −2, 0, 3)

29. 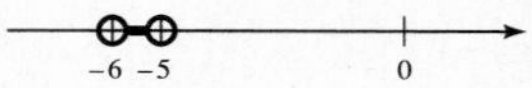 (number line: −6, −5, 0) **31.** $\{x \mid 0 \le x < 3\}$ **33.** 72 **35.** -12 **37.** $\frac{-7}{12}$ **39.** 80 **41.** -27 **43.** 343

45. -7 **47.** $\frac{3}{4}$ **49. a.** $-3xy^3z^5$ **b.** -3 **c.** xy^3z^5 **51.** $2x$ **53.** $21a^5b^2 + 3a^3b^3 - 27a^2b^4$

55. $4m^3 - 16m^2 + 37m - 10$ **57.** -90 **59. a.** $(-1, -10)$ **b.** (9,720) **61.** $\{3, 6, -2, -1\}$ **63.** No

65. **67.** -1 **69.** **71. a.** $\frac{1}{2}$ **b.** -3 **c.**

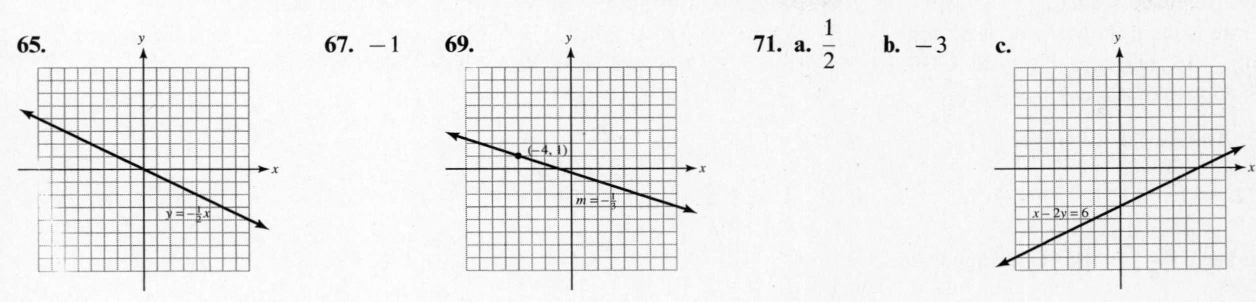

73. -3 or 1 **75.** **77.** -2 or 6 **79.** **81.** 3

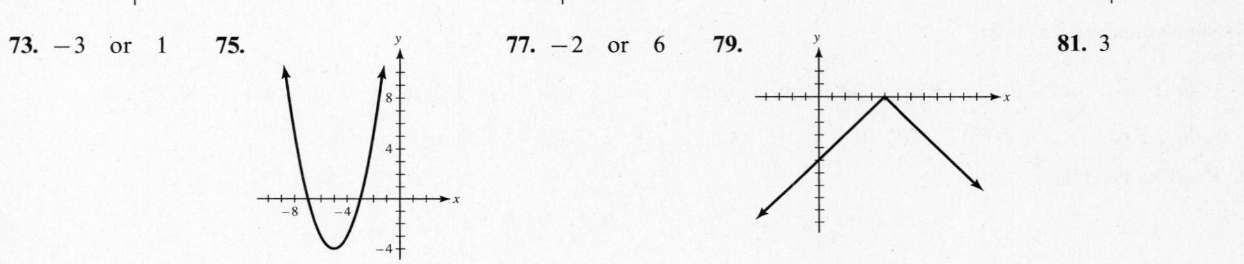

Chapter 2 Equations and Inequalities in One Variable

Section 2-1. Practice Exercises

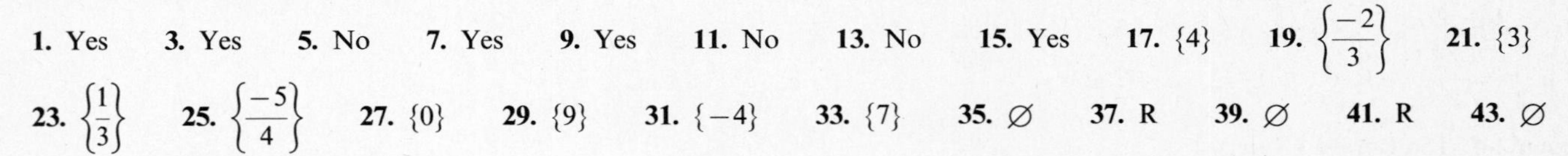

1. Yes **3.** Yes **5.** No **7.** Yes **9.** Yes **11.** No **13.** No **15.** Yes **17.** $\{4\}$ **19.** $\left\{\frac{-2}{3}\right\}$ **21.** $\{3\}$

23. $\left\{\frac{1}{3}\right\}$ **25.** $\left\{\frac{-5}{4}\right\}$ **27.** $\{0\}$ **29.** $\{9\}$ **31.** $\{-4\}$ **33.** $\{7\}$ **35.** $\varnothing$ **37.** R **39.** $\varnothing$ **41.** R **43.** $\varnothing$

Section 2-1. Ten Review Exercises

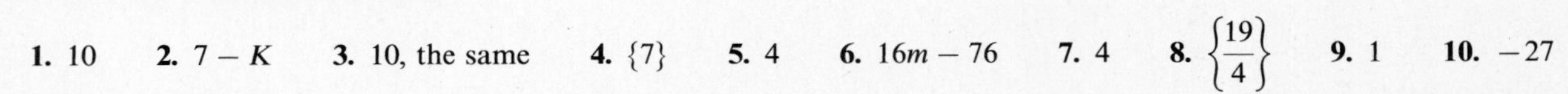

1. 10 **2.** $7 - K$ **3.** 10, the same **4.** $\{7\}$ **5.** 4 **6.** $16m - 76$ **7.** 4 **8.** $\left\{\frac{19}{4}\right\}$ **9.** 1 **10.** -27

Section 2-1. Supplementary Exercises

1. $\{2.0\}$ **3.** $\{3.125\}$ **5.** $\{0.0\}$ **7.** R **9.** 19 **11.** 5 **13.** 4 **15.** 3 **17.** $b = 7$ **19.** $b = 4$ **21.** $c = 9$

23. $\frac{a-6}{5}$ **25.** $\frac{16}{a}$ **27.** $\frac{7}{a-5}$ **29.** $\frac{3a+10}{2}$ **31. a.** 8 **b.** -6 **33. a.** -2 **b.** 6

Section 2-2. Practice Exercises

1. $xy - 3x$ **3.** $z^3 \div (x + y)$ or $\frac{z^3}{x+y}$ **5.** $(2x + 5)^2 - \frac{1}{4}x$ **7.** $(y + 12)z$ **9.** $11x + \frac{3}{x}$ **11.** The number is 26.

13. The number is -3. **15.** The number is 3. **17.** The number is 41. **19.** 20 ft by 35 ft **21.** 36 ft

23. 15 ft, 20 ft, 15 ft, 30 ft **25.** 72 yd by 90 yd

Section 2-2. Ten Review Exercises

1. Reflexive property of equality **2.** Antisymmetric property of inequality **3.** Symmetric property of equality

4. Transitive property of equality **5.** Transitive property of inequality **6.** $\frac{1}{3}$ **7.** $\frac{-1}{32}$ **8.** $\frac{-1}{6}$ **9.** 15 **10.** 1

Section 2-2. Supplementary Exercises

1. a.

8	5(8)	5(8) − 10	30	No
15	5(15)	5(15) − 10	65	No
12	5(12)	5(12) − 10	50	No
x	$5(x)$	$5(x) - 10$	40	

b. $5x - 10 = 40$

3. a.

7	7 + 9	2(7)	2(7 + 9)	2(7) + 2(7 + 9)	46	No
10	10 + 9	2(10)	2(10 + 9)	2(10) + 2(10 + 9)	58	No
15	15 + 9	2(15)	2(15 + 9)	2(15) + 2(15 + 9)	78	No
20	20 + 9	2(20)	2(20 + 9)	2(20) + 2(20 + 9)	98	No
x	$x + 9$	$2(x)$	$2(x + 9)$	$2(x) + 2(x + 9)$	110	

b. $2x + 2(x + 9) = 110$

5. 32 in by 38 in by 38 in **7.** Not enough information **9.** $\{-18\}$ **11.** $\{20\}$

Section 2-3. Practice Exercises

1. 1,300 general, 3,200 student **3.** \$8,800 in stocks, \$16,200 in bonds **5.** 8 ℓ from A, 12 ℓ from B
7. Train's rate is 60 mph; bus' rate is 52 mph. **9.** 52.5 mi **11. a.** 2 hr **b.** 112 mi **13. a.** 3 hr **b.** 1,820 mi
c. 1,380 mi **15.** Harry is 48 yr old; Gladys is 43 yr old. **17.** Carol is 24; Becky is 30; Darlene is 16.
19. Ring is 55 yr old; pin is 80 yr old.

Section 2-3. Ten Review Exercises

1. 210 **2.** 260 **3.** 1, 2, 3, 6, 7, 12, 14, 21, 42 **4.** 1, 2, 5, 10, 11, 22, 55, 110 **5.** 40 **6.** $\frac{-3}{2}$ **7.** $\{-9\}$ **8.** $\{0\}$

9. The number is $\frac{5}{14}$. **10.** Tim's age is 26.

Section 2-3. Supplementary Exercises

1. a. 1 gal **b.** 2 gal **c.** 4 gal **3. a.** $\frac{1}{2}$ gal **b.** $1\frac{1}{2}$ gal **5. a.** 3 gal **b.** 7.5 gal **7. a.** 6 gal **b.** 7.5 gal

9. a. 1 gal **b.** 2.5 gal **11.** 50% **13.** 30% **15.** 60% **17. a.** 20% **b.** Large amounts

Section 2-4. Practice Exercises

1. $a = \frac{7 - 3b}{2}$ **3.** $b = \frac{P}{ac}$ **5.** $x = \frac{4y - z + 13}{5}$ **7.** $m = \frac{y - b}{x}$ **9.** $a = \frac{D - b^2}{-4c}$ **11.** $b = a - 2m$ **13.** $x = \frac{c}{a - b}$

15. $d = \frac{s - 5}{n - 1}$ **17.** $m = \frac{b + 7}{a - 5}$ **19.** $y = \frac{2b - 5}{b^2 - 4}$ **21.** $h = \frac{3V}{\pi r^2}$ **23.** $w = \frac{p - 21}{2}$ **25.** $a = \frac{p - b - c}{2}$ **27.** $m = \frac{Fr}{v^2}$

29. $I = \frac{T}{1 + d}$ **31.** $y = \frac{x}{1 - x}$ **33.** $b^2 = c^2 - a^2$ **35.** $r^2 = \frac{GMN}{F}$ **37. a.** $y = \frac{8 - x}{2}$; **b.** 7; **c.** −1

39. a. $m = \frac{2E}{V^2}$ **b.** 6 **c.** 12 **41. a.** $r = \frac{s - a}{s}$ **b.** $\frac{1}{2}$ **c.** $\frac{2}{3}$ **43. a.** $a = \frac{y - k}{(x - h)^2}$ **b.** 3 **c.** −2

Section 2-4. Ten Review Exercises

1. 22 **2.** $\frac{1}{3}$ **3.** $\frac{-3}{4}$ **4.** 12 **5.** 0 **6.** $-16k + 32$ **7.** 0 **8.** $\left\{\frac{-1}{2}\right\}$ **9.** $x^2 = \frac{y - c}{a}$ **10.** $b = \frac{y - ax^2 - c}{x}$

Section 2-4. Supplementary Exercises

1. $n = \frac{x}{p}$ **3.** $P = \frac{x - m}{z}$ **5.** $r = \frac{mv^2}{F}$ **7.** $x = \frac{a - 2}{a^2 - 3}$ **9.** $a = \frac{128}{x - y + z}$ **11. a.** −1 **b.** 6 **c.** 7 **d.** −4

13. a. 2 **b.** −3 **c.** 3 **d.** −4 **15. a.** 9 **b.** 13.5 **c.** $\frac{1}{3}$ **d.** 4 **17.** $a + b + c + d$ **19.** $2a + 2b$

21. $2a + \pi b$

Section 2-5. Practice Exercises

1. $\{-6, 6\}$ **3.** $\{-\sqrt{3}, \sqrt{3}\}$ **5.** $\varnothing$ **7.** $\{0\}$ **9.** $\left\{\frac{-1}{5}, \frac{1}{5}\right\}$ **11.** $\{-72, 72\}$ **13.** $\{-14, 12\}$ **15.** $\{4, 8\}$

17. $\{-3, 4\}$ **19.** $\left\{0, \frac{-3}{2}\right\}$ **21.** $\{-7, 8\}$ **23.** $\{6\}$ **25.** $\left\{\frac{-19}{5}, 3\right\}$ **27.** $\{-14\}$ **29.** $\varnothing$ **31.** $\{-15, -3\}$

33. $\left\{\frac{-5}{9}, \frac{-1}{9}\right\}$ **35.** $\{-1, 9\}$ **37.** $\{2, 4\}$ **39.** $\left\{\frac{1}{3}, 5\right\}$ **41.** $\left\{\frac{-2}{3}, \frac{5}{2}\right\}$ **43.** $\{-4\}$ **45.** $\left\{\frac{-1}{7}, 5\right\}$

Section 2-5. Ten Review Exercises

1. 90 **2.** 1, 2, 3, 6, 11, 22, 33, 66 **3.** $\frac{-2}{3}$ **4.** -4 **5.** $\frac{-7}{8}$ **6.** $\frac{-4}{3}$ **7.** $\{-3\}$ **8.** $\left\{\frac{76}{57}\right\}$ **9.** $\left\{7, \frac{-11}{3}\right\}$
10. $\varnothing$

Section 2-5. Supplementary Exercises

1. $\{-9, 15\}$ **3.** $\{-11, 3\}$ **5.** $\left\{\frac{-11}{2}, \frac{13}{2}\right\}$ **7.** $\{-51, -9\}$ **9.** R **11.** $\left\{\frac{7}{12}\right\}$ **13.** $\left\{\frac{15}{14}\right\}$ **15.** $\{-6, 6\}$
17. $\left\{\frac{-1}{2}, \frac{7}{2}\right\}$ **19.** $\left\{\frac{-c}{a}, \frac{c}{a}\right\}$ **21.** $\left\{\frac{-b-c}{a}, \frac{-b+c}{a}\right\}$ **23.** $\{b - c, b + c\}$ **25. a.** $\{-11, 5\}$ **b.** $\{-7, 1\}$ **c.** $\varnothing$
27. a. $\{-10, 6\}$ **b.** $\{-6, 2\}$ **c.** $\{-2\}$ **29.** Always true **31.** Sometimes true **33.** Sometimes true
35. Sometimes true

Section 2-6. Practice Exercises

1. $y \geq 6$ **3.** $p > -12$ **5.** $x < 2$

7. $a < 0$ **9.** $b \leq \frac{1}{2}$ **11.** $y < 1$

13. $z > 3$ **15.** $t > \frac{1}{3}$ **17.** $u \geq 0$

19. $z < \frac{3}{2}$ **21.** $y > 1$ **23.** $x > 2$

25. $q \leq \frac{-4}{3}$ **27.** $p \leq 2$ **29.** $m \geq -2$

31. $\{\ldots, -1, 0, 1, 2\}$ **33.** $\{18, 19, 20, 21, \ldots\}$ **35.** 14 or more weeks **37.** $0 < w < 5$ in **39.** $0 < w \leq 10$ cm

41. $4 < s \leq 7$ cm **43.** 3, 4, 5, 6, 7, 8, or 9 units

Section 2-6. Ten Review Exercises

1. 3 **2.** 1 **3.** 2 **4.** -11 **5.** $-24t - 18$ **6.** 30 **7.** $\left\{\frac{-3}{4}\right\}$ **8.** $t < \frac{-3}{4}$

9. $\left\{\frac{1}{4}, \frac{3}{4}\right\}$ **10.** $\left\{\frac{-4}{5}, \frac{8}{3}\right\}$

Section 2-6. Supplementary Exercises

1. R **3.** $\varnothing$ **5.** R **7.** $\varnothing$ **9.** $x < c - b$ **11.** $x < \frac{c}{a}$ **13.** $x > \frac{c-b}{a}$ **15.** $x > \frac{cd-b}{a}$ **17.** $x < \frac{b-d}{a-c}$
19. a. $x < 6$ **b.** $x < -3$ **21. a.** $x \leq -3$ **b.** $x \leq 3$

Section 2-7. Practice Exercises

1. a. (i) T; (ii) T **b.** (i) T; (ii) F **3. a.** (i) T; (ii) F **b.** (i) F; (ii) F **5. a.** (i) T; (ii) F **b.** (i) T; (ii) T

7. a. (i) T; (ii) F **b.** (i) T; (ii) F **9. a.** (i) F; (ii) F **b.** (i) T; (ii) T **11.** $\{-5, 4\}$

13. -5 or $y > 1$ **15.** $z < 1$ or $z > 5$

17. $t \leq -5$ or $t \geq -2$ **19.** R **21.** $w \geq 5$

23. 1 or $p \leq -6$ **25.** $q \leq -2$ or $q > 6$

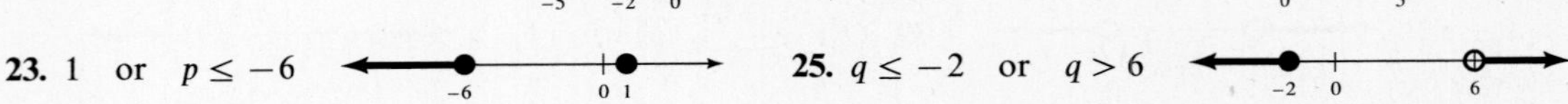

27. -1 or $x \geq 3$ (−1 0 3) **29.** $y < -5$ or $y > -1$ (−5 −1 0)

31. $1 < t < 4$ (0 1 4) **33.** $-10 \leq u \leq -2$ (−10 −2 0)

35. $-5 < v < 5$ (−5 0 5) **37.** 5 (0 5)

39. $-1 < q \leq 3$ (−1 0 3) **41.** $1 < x < 3$ (0 1 3)

43. $-3 < y < -1$ (−3 −1 0) **45.** $-5 \leq z < 2$ (−5 0 2)

47. $\frac{-3}{4} < t \leq \frac{1}{4}$ ($-\frac{3}{4}$ 0 $\frac{1}{4}$) **49.** 0 or $u < -3$ or $u > 3$ (−3 0 3)

51. R **53.** R **55.** $\varnothing$

Section 2-7. Ten Review Exercises

1. $3u - 2$ **2.** -17 **3.** $\left\{\frac{2}{3}\right\}$ **4.** $u > \frac{2}{3}$ (0 $\frac{2}{3}$)

5. $u < -1$ or $u > 2$ (−1 0 2) **6.** $\{-6, 7\}$ **7.** -30 **8.** -84 **9.** -75 **10.** $\frac{-1}{2}$

Section 2-7. Supplementary Exercises

1. $-4 \leq x \leq 5$ or $x > 12$ (−4 0 5 12) **3.** 6 or $y < 5$ or $y > 10$ (0 5 6 10)

5. $1 \leq z < 4$ (0 1 4) **7.** $\{x | x < -1$ or $x > 4\}$ **9.** $\{x | x < -4$ or $x > 1\}$ **11.** $\{x | 1 < x < 6\}$

13. $\{x | -6 < x < -1\}$ **15. a.** $8 > 4t$ **b.** $\{t | t < 2\}$ **c.** $8 < 4t$ **d.** $\{t | t > 2\}$ **e.** Not the same
17. $\{x | x < 0$ or $x > 1\}$ **19.** $\{x | x < 1$ or $x > 2\}$ **21.** $\{x | x < -2$ or $x > 0\}$

Section 2-8. Practice Exercises

1. $\{x | -3 < x < 3\}$ (−3 0 3) **3.** $\{y | -0.75 \leq y \leq 0.75\}$ (−0.75 0 0.75)

5. $\{z | -2 \leq z \leq 4\}$ (−2 0 4) **7.** $\{t | -6 < t < -2\}$ (−6 −2 0)

9. $\{u | -11 < u < 1\}$ (−11 0 1) **11.** $\{v | -1 \leq v \leq 4\}$ (−1 0 4)

13. $\left\{k \,\middle|\, -1 < k < \frac{9}{4}\right\}$ (−1 0 $\frac{9}{4}$) **15.** $\left\{p \,\middle|\, \frac{-4}{3} \leq p \leq 0\right\}$ ($-\frac{4}{3}$ 0)

17. $\left\{q \,\middle|\, \frac{-1}{2} \leq q \leq \frac{9}{2}\right\}$ ($-\frac{1}{2}$ 0 $\frac{9}{2}$) **19.** $\left\{x \,\middle|\, \frac{5 - \sqrt{2}}{2} < x < \frac{5 + \sqrt{2}}{2}\right\}$ (0 $\frac{5-\sqrt{2}}{2}$ $\frac{5+\sqrt{2}}{2}$)

21. $\{y | y \leq -6$ or $y \geq 6\}$ (−6 0 6) **23.** $\{z | z < -7$ or $z > 13\}$ (−7 0 13)

25. $\{t | t \leq -4$ or $t \geq 0\}$ (−4 0) **27.** $\{u | u < -17$ or $u > -3\}$ (−17 −3 0)

29. $\{v | v < -1$ or $v > 4\}$ (−1 0 4) **31.** $\{w | w < -1$ or $w > 9\}$ (−1 0 9)

33. $\left\{x \,\middle|\, x \leq \frac{1}{5} \text{ or } x \geq 1\right\}$ (0 $\frac{1}{5}$ 1) **35.** $\left\{y \,\middle|\, y < \frac{12}{5} \text{ or } y > \frac{13}{5}\right\}$ (0 $\frac{12}{5}$ $\frac{13}{5}$)

37. $\{z | z < 5 - \pi$ or $z > 5 + \pi\}$ (0 $5-\pi$ $5+\pi$) **39.** $\left\{t \,\middle|\, t < \frac{-14}{3} \text{ or } t > 4\right\}$ ($-\frac{14}{3}$ 0 4)

41. $|x| < 2$ **43.** $|x| \geq 1$ **45.** $|x + 1| < 2$ **47.** $|x - 2| \geq 2$ **49.** $|x + 7| < 3$

Section 2-8. Ten Review Exercises

1. and **2.**

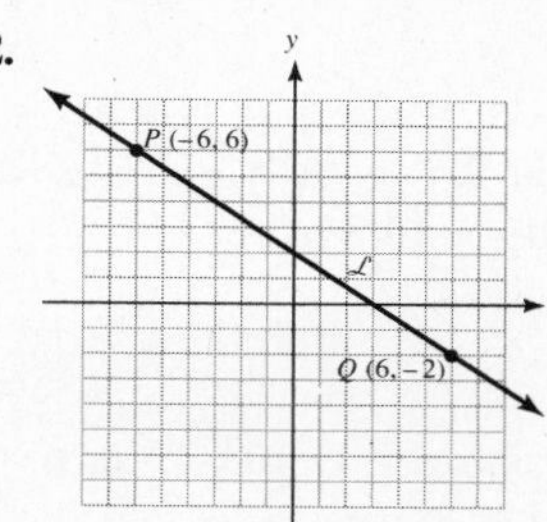

3. $\frac{-2}{3}$ **4.** $\frac{-2}{3}$ **5.** 2 **6.** 3 **7.** $y = \frac{-2}{3}x + 2$ **8.** $\{6\}$ **9.** $\{-3, 6\}$

10. $\{x \mid -3 < x < 6\}$

Section 2-8. Supplementary Exercises

1. $\left\{x \middle| -4 < x < \frac{8}{3}\right\}$ **3.** $\left\{y \middle| y < \frac{7}{6} \text{ or } y > \frac{4}{3}\right\}$

5. $\left\{z \middle| -\frac{13}{2} \le z \le \frac{11}{2}\right\}$ **7.** $\{t \mid t \le -7 \text{ or } t \ge 13\}$

9. $\left\{u \middle| \frac{-17}{2} < u < \frac{19}{2}\right\}$ **11.** $\{x \mid -6 < x < 10\}$ **13.** $\{2\}$ **15.** $\{x \mid x < -9 \text{ or } x > 3\}$

17. $\{x \mid x < -5 \text{ or } x > -1\}$ **19.** $\varnothing$ **21.** R **23.** All real numbers except 10 **25.** $\varnothing$ **27.** R **29.** $\{3\}$
31. $13 \le 5 + 8$ **33.** $8 \le 17 + 9$

Chapter 2 Review Exercises

1. $\{-3\}$ **3.** $\left\{\frac{-1}{2}\right\}$ **5.** R **7.** The number is -12. **9.** \$7,500 **11.** $\frac{4}{5}$ hr or 48 min **13.** 6 yr and 29 yr old

15. $a = \frac{s - vt}{t^2}$ **17.** $y = \frac{k - 4}{m - 2}$ **19.** $\{-20, 20\}$ **21.** $\{5, 11\}$ **23.** $\{-35, 0\}$

25. $\{y \mid y \ge 6\}$ **27.** $\{b \mid b \le 8\}$ **29.** Any integer less than 10

31. a. True **b.** False **33.** $\{x \mid x < 2 \text{ or } x > 4\}$

35. $\{x \mid x > 7\}$ **37.** $\{x \mid -16 < x < 10\}$

39. $\left\{x \middle| -1 < x < \frac{21}{5}\right\}$ **41.** $\varnothing$ **43.** $|x| < 4$

Chapter 3 Polynomial Expressions and Equations

Section 3-1. Practice Exercises

1. x^6 **3.** y^{10} **5.** 3^{12} **7.** a^8b^3 **9.** x^6 **11.** 5^{18} **13.** y^9z^8 **15.** a^{15} **17.** x^8 **19.** $\frac{1}{y^5}$ **21.** a^3 **23.** z^9

25. $\frac{10^6}{x^2}$ **27.** 2^5t^5 **29.** -3^7u^7 **31.** $4v^4w^6$ **33.** $256a^4b^8$ **35.** $\frac{x^6}{10^6}$ **37.** $\frac{y^6}{27}$ **39.** $\frac{a^4b^3}{16}$ **41.** $\frac{z^7}{32}$ **43.** 1

45. 1 **47.** -1 **49.** -1 **51.** $-27x^5y^5$ **53.** $80a^9$ **55.** $-108p^{11}q^7$ **57.** $\frac{a}{8b^2}$ **59.** $\frac{t^5}{12}$ **61.** $64u^{10}$

63. a^5b^3

Section 3-1. Ten Review Exercises

1. 0 **2.** $\{7\}$ **3.** $\{t \mid t > 7\}$ **4.** $\{t \mid t < 4 \text{ or } t > 7\}$

5. $\left\{5, \frac{-27}{5}\right\}$ **6.** $\left\{u \middle| \frac{-27}{5} < u < 5\right\}$ **7.** -6 **8.** $\{-11, -1\}$ **9.** $5x^2 - 3x + 1$ **10.** 12

Section 3-1. Supplementary Exercises

1. x^{2a} **3.** x^{8a} **5.** x^{2b} **7.** x^a **9.** x^{ab} **11. a.** 3125 **b.** 150 **c.** 4,000 **13. a.** $9z^2$ **b.** $41z^2$ **c.** $400z^4$ **15. a.** 96 **b.** 6 **c.** 16 **d.** $4x^2$ **e.** 16 **f.** They are the same. **21. b.** $3.377699719 \times 10^{12}$ in **c.** 52,309,654.65 mi

Section 3-2. Practice Exercises

1. a. $-x^2 + 5x + 10$ **b.** Two **c.** -1 **d.** 10 **e.** Trinomial **3. a.** $100x^8$ **b.** Eight **c.** 100 **d.** 0 **e.** Monomial **5. a.** $12x^3 - x^2 + 7x - 3$ **b.** Three **c.** 12 **d.** -3 **e.** Polynomial of four terms **7. a.** $-25x^8 + 36$ **b.** Eight **c.** -25 **d.** 36 **e.** Binomial **9. a.** $ax^3 + bx^2 + abx + 5$ **b.** Three **c.** a **d.** 5 **e.** Polynomial of four terms **11. a.** 3, 3, 3 **b.** $x^2y + 3xz^2 - 2y^2z$ **c.** $-2y^2z + x^2y + 3xz^2$ **d.** $3xz^2 - 2y^2z + x^2y$ **13. a.** 3, 5, 7, 5 **b.** $-x^4z^3 + 2x^3y^2 + xyz + 5y^3z^2$ **c.** $5y^3z^2 + 2x^3y^2 + xyz - x^4z^3$ **d.** $-x^4z^3 + 5y^3z^2 + xyz + 2x^3y^2$ **15. a.** 4, 2, 6, 3 **b.** $x^5y + 4x^3z - y^2 - z^3$ **c.** $-y^2 + x^5y + 4x^3z - z^3$ **d.** $-z^3 + 4x^3z + x^5y - y^2$ **17. a.** 3, 4, 4, 0, 4 **b.** $x^2y^2 + x^2z^2 + xyz - y^2z^2 + 2$ **c.** $x^2y^2 - y^2z^2 + xyz + x^2z^2 + 2$ **d.** $x^2z^2 - y^2z^2 + xyz + x^2y^2 + 2$ **19.** $3x^2 - 6x - 9$ **21.** $-2x^2 - 9x + 27$ **23.** $c^3 + 5c^2 - 4$ **25.** $x^3 + 17x^2 - 1$ **27.** $7x^2 - 16xy - 11y^2$ **29.** $11x^2 - y^2$ **31.** $-a^3 + 2a^2b + 2ab^2 + 3b^3$ **33.** $4x^3y^3 + x^2y^2 - 4xy$ **35.** $2x^2 + 2x - 2$ **37.** $-y^2 - 4y - 3$ **39.** $2x^3 - x^2$ **41.** $z^4 + z^2 - z - 4$ **43.** $5a^2 + 2ab$ **45.** $13x + 27$ **47.** $-6y^2 + y + 11$ **49.** $-5a^2 - 4a - 6$ **51.** $11c^3 - 6c^2 + 26c + 32$ **53.** $t^2 - tu - u^2$ **55.** $-7z^3 + 10z^2 - z + 7$ **57.** $2a^2 + 6b^2 - 15$ **59.** $0.04k^2 + 0.92k + 0.04$

Section 3-2. Ten Review Exercises

1. 14 **2.** $\frac{3}{2}$ **3.** -2 **4.** $\left\{\frac{-3}{5}\right\}$ **5.** $\left\{-3, \frac{5}{3}\right\}$ **6.** $y = \frac{5x + 8}{3 - k}$ **7.** $a > -5$

8. $x \leq -1$ or $x \geq 9$ **9.** $\frac{-6x^5y^4}{5}$ **10.** $\frac{2x}{3}$

Section 3-2. Supplementary Exercises

1. $(a + 9)x^2 + (b + 5)x + 10$ **3.** $7x^2 + (b + 4)x + (a + c)$ **5.** $(a + d + g)x^2 + (b - e + h)x + (c + f - i)$ **7. a.** $5x^2 + x + 1$ **b.** 131 **9. a.** $-x^2 + 9x - 3$ **b.** -39 **11.** $13x^2 - 2x + 4$ **13.** $23x - 7$ **15.** -11 **17.** 4 **19.** 2 **21.** -1

Section 3-3. Practice Exercises

1. $6x^3$ **3.** $-21m^5n^7$ **5.** $-10x^3y^4z^6$ **7.** $100p^9q^{10}$ **9.** $-72a^8b^6c^9$ **11.** $x^3 + 2x^2 + 3x + 6$ **13.** $9m^3 - m - 2$ **15.** $6r^3 + 5r^2 - 11r + 35$ **17.** $y^5 + 3y^4 + y^2 - y - 2$ **19.** $4z^4 - 5z^2 + 4z - 3$ **21.** $a^5 + 6a^2b^3 + ab^4 - 3b^5$ **23.** $t^9 - 343$ **25.** $x^5 - x^4 + 3x^3 - 9x^2 + x + 5$ **27.** $5y^5 + 15y^4 + 47y^3 - 54$ **29.** $u^3 + u^2 - u^2v + uv^2 - u^2 - v^3$ **31.** $x^3 - 5x^2 + 2x - 10$ **33.** $3x^2 + 4xy - 2y^2$ **35.** $2a^3 - 3a^2 - 10a + 15$ **37.** $6m^4 - 5m^2n^2 - 2n^4$ **39.** $t^5 + 10t^3 - 6t^2 - 60$ **41.** $2s^4 + 7s^2t + 3t^2$ **43.** $x^2 - 9$ **45.** $y^2 - 4y + 4$ **47.** $4r^2 - s^2$ **49.** $4a^4 + 12a^2 + 9$ **51.** $t^4 - 16$ **53.** $s^2t^2 - 18st + 81$ **55.** $u^4v^2 - 144$ **57.** $m^8n^4 + 6m^4n^2 + 9$ **59.** $16w^{10} - 81z^4$ **61.** $5a^3 + a^2 + 7a + 1$ **63.** $9x^4 - 5x^2$ **65.** $50m^2 + 2n^2$ **67.** $-22t$ **69.** $u^4 + 2u^2$

Section 3-3. Ten Review Exercises

1. 57 **2.** $11k + 2$ **3.** 57 **4.** $\left\{\frac{-2}{11}\right\}$ **5.** Rational, real **6.** $3x^6 + x^4 + 5x - 2$ **7.** Six **8.** 3 **9.** -2 **10.** 196

Section 3-3. Supplementary Exercises

1. $\frac{-m^9n^{11}}{2}$ **3.** $6m^8 - 10m^6 - 7m^4 + 5m^2 + 2$ **5.** $81m^4 - 18m^2n + n^2$ **7.** $27x^4 + 3x^2 - 10$ **9.** $9x^3 + 4x^2 + 3x^2y - 15xy + 8y^2$ **11.** 672 **13.** 2,783 **15.** 6,384 **17. a.** 9, 8 **b.** 72 **c.** $2x^3 + 3x^2 - 2x - 3$ **d.** 72 **e.** The values are the same. **19. a.** 7, 19 **b.** 133 **c.** $x^3 + 8$ **d.** 133 **e.** The values are the same. **21.** 196 **23.** 625 **25.** 396 **27.** 1,599 **29.** $2x^{2a} + 8x^a + 15$ **31.** $2x^{2a} - x^ay^b - 21y^{2b}$ **33.** $9x^{2a} - 12x^ay^b + 4y^{2b}$ **35.** $2x^{3a} + 5x^{2a} - 4x^a - 3$

Section 3-4. Practice Exercises

1. $8x^2$ **3.** $13t^7u^2$ **5.** $9a^3c^2$ **7.** $-2u^3v$ **9.** $4p^2q$ **11.** $3x^2 - 2x + 1$ **13.** $15m^3 - 3m^2n + 2mn^2 - 1$ **15.** $11s^2 - 14st + 6t^2$ **17.** $21a^3 - 5a^2b + 3ab^2 - 4b^3$ **19.** $10x^3 - xyz + 3y^2z^2$ **21.** $x^2 + 2x - 3$ **23.** $3y^2 + y - 2$

25. $a^2 - 2ab + 3b^2$ **27.** $3k^2 + 5k - 1 + \dfrac{-2}{2k^2 - k + 3}$ **29.** $8x^2 - 8x + 3$ **31.** $y^4 - 2y^3 + 4y^2 - 8y + 16$

33. $32a^5 + 48a^4 + 72a^3 + 108a^2 + 162a + 243$ **35.** $p^2 - 3$ **37.** $32z^4 - 16z^2 - 8z + 8 + \dfrac{8z}{2z^2 + 1}$

39. $b^4 - b^3 - b^2 - 13b + 15 + \dfrac{11b}{b^2 + b + 2}$

Section 3-4. Ten Review Exercises

1. a. 50 **b.** 56 **2.** $5t^2 - 3t - 4$ **3.** 106; Same values **4.** $-t^2 + 5t - 6$ **5.** -6; Same values

6. $6t^4 - 5t^3 - 17t^2 + 21t - 5$ **7.** 2,800; Same values **8.** $\left\{\dfrac{5}{6}\right\}$ **9.** $\{-7, 8\}$ **10.** $\{y \mid y > 2\}$ 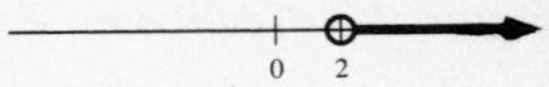

Section 3-4. Supplementary Exercises

1. a. 156, 13 **b.** $x + 2$ **c.** 12 **3. a.** 12,549, 123 **b.** $x^2 + 2 + \dfrac{3}{x^2 + 2x + 3}$ **c.** $102\dfrac{3}{123}$ or $102\dfrac{1}{41}$ **5.** $x^2 - 1$

7. $x^2 - 4$ **9.** $x^3 + 2x^2 + x + 2$ **11.** $3x^3 - x^2 + 5x - 7$ **13.** $6x^4 - x^2 + 3x - 2$ **15.** $4x^4 - 9x^2 + 7$

17. $2x^3 - 6x^2 - 50x - 42$ **19.** $4x^3 - 2x^2 - 36x + 18 + \dfrac{1}{x + \frac{1}{2}}$

Section 3-5. Practice Exercises

1. $x^3y^2(x - 2y)$ **3.** $2k^3(k^2 + 7k - 3)$ **5.** $3x^2y^2(2x^4 - 7x^2y + y^2)$ **7.** $5pq^2r^3(2p - q + 5r - 3)$ **9.** $(x + 18)(x - 2)$
11. $(y - 5)(y - 2)$ **13.** $(p - 7)(p - 6)$ **15.** $(w - 14)(w + 2)$ **17.** $(z + 6)(z + 15)$ **19.** $5pq(p^2 + 5q)(p + 4q)$
21. $2x^4(x + 2)(x + 1)$ **23.** $5t(t - 9)(t - 1)$ **25.** $12y^2(y - 3)(y + 1)$ **27.** $(x - 9)(x + 9)$ **29.** $(4 - 3y)(4 + 3y)$
31. $(a - 7b)(a + 7b)$ **33.** $7(pq - \pi)(pq + \pi)$ **35.** $3rs(r - 8s)(r + 8s)$ **37.** $(x + 10)^2$ **39.** $(9a - b)^2$ **41.** $(3k^2 - 10)^2$
43. $2(8r^2 + \pi)^2$ **45.** $3pq(p + 3q)^2$ **47.** $(y - 9)(y^2 + 9y + 81)$ **49.** $(7t + u)(49t^2 - 7tu + u^2)$
51. $(3x + 5)(9x^2 - 15x + 25)$ **53.** $(a + 10b)(a^2 - 10ab + 100b^2)$ **55.** $5xy(5x - y)(25x^2 + 5xy + y^2)$
57. $5k^2(2k + 1)(4k^2 - 2k + 1)$ **59.** $2(a^3b^2c - 2)(a^6b^4c^2 + 2a^3b^2c + 4)$

Section 3-5. Ten Review Exercises

1. $3x^6 + x^4 + 5x - 2$ **2.** Six **3.** 3 **4.** -2 **5.** 2 **6.** $4t - 14$ **7.** 2; Same values **8.** 0 **9.** $\{5\}$
10. $-3k + 15$

Section 3-5. Supplementary Exercises

1. $(k^2 - 14)(k^2 - 2)$ **3.** $2y^2(y^2 + 2)(y^2 + 1)$ **5.** $(2x + 1)(2x - 1)(4x^2 + 1)$ **7.** $(b^2 + 2)(b^4 - 2b^2 + 4)$
9. $(x - 5)^2(x + 5)^2$ **11.** $(1 - 4y^2)(1 + 4y^2 + 16y^4)$ **13.** $2(2x^2 - 9y)(2x^2 + 9y)$ **15.** $(7 - p - q)(7 + p + q)$
17. $(x + a + 1)^2$ **19.** $(m - n - 3)(m - n - 4)$ **21.** $4xy$ **23.** $(x^a - 5)(x^a - 3)$ **25.** $(x^a + 8y^b)(x^a - y^b)$
27. $(x^{a+b} - 7)(x^{a+b} - 3)$ **29.** $(4p^a - 1)(16p^{2a} + 4p^a + 1)$ **31.** $(5m^{2a} - 1)^2$ **33.** $(x^a - 2)(x^a + 1)(x^a - 6)(x^a - 3)$

Section 3-6. Practice Exercises

1. $(2t + 1)(t^2 + 3)$ **3.** $(3u + 2)(2u^2 + 9)$ **5.** $(3m^2 + n)(m + 4n)$ **7.** $(5x + 7y)(3xy + 4)$ **9.** $(2x + 3)(x^2 + 1)$
11. $(a + 3)(a^2 - 5)$ **13.** $(x - 5)(5y + 1)$ **15.** $(10a - 3)(5b - 1)$ **17.** $(y + 2)(2x^2 + x + 3)$ **19.** $(3z + 2)(z^4 - z^2 - 1)$
21. $(a - 3b)(3a^4 - a^2b^2 + b^4)$ **23.** $(m^2 - 2n)(m^2 + mn - n^2)$ **25.** $(5a + b - 1)(5a - b - 1)$ **27.** $(p - q + 3)(p - q - 3)$
29. $(10 + x - 9y)(10 - x + 9y)$ **31.** $[1 + 2(m^2 + n)][1 - 2(m^2 + n)]$ **33.** $(2p + 1)(4p + 5)$ **35.** $(2m + 3)(m - 4)$
37. $(4a - b)(a - 3b)$ **39.** $(5z - 1)(2z + 3)$ **41.** $(5k - 2)(k + 3)$ **43.** $(3t - 2u)(5t + u)$ **45.** $(6r + 5)(r - 3)$
47. $(2x - y)(2x + 9y)$ **49.** $(12a - b)(3a + b)$

Section 3-6. Ten Review Exercises

1. 0 **2.** Yes **3.** 0 **4.** Yes **5.** $(3t + 2)(t - 5)$ **6.** $2t^2(3t + 2)(t - 5)$ **7.** $-35a^4b^2$ **8.** $-2x^5y^5$ **9.** $\dfrac{-i^5j^8k^9}{81}$

10. $-7s^2$

Section 3-6. Supplementary Exercises

1. $(4n - 3)(3m + 1)$ **3.** $(6s + 1)(2s - 5)$ **5.** $(x + y)(2x^2 + 3xy + y^2)$ **7.** $(5q^2 - 1)(pq + 4)$
9. $(15p - q - r + s)(15p - q + r - s)$ **11. a.** $x^2 + 5x - 5x - 25$ **b.** $(x - 5)(x + 5)$ **13. a.** $100p^2 + 90pq - 90pq - 81q^2$
b. $(10p - 9q)(10p + 9q)$ **15. a.** $x^2 + 5x + 5x + 25$ **b.** $(x + 5)^2$ **17. a.** $49p^2 - 42pq - 42pq + 36q^2$ **b.** $(7p - 6q)^2$
19. a. $a^3 + a^2b - a^2b - ab^2 + ab^2 + b^3$ **b.** $(a + b)(a^2 - ab + b^2)$
21. a. $27m^3 + 9m - 9m - 3m^2 + 3m^2 + 1$ **b.** $(3m + 1)(9m^2 - 3m + 1)$

Section 3-7. Practice Exercises

1. $\{7\}$ **3.** $\{-2, 3\}$ **5.** $\left\{-5, \frac{1}{2}\right\}$ **7.** $\left\{0, \frac{7}{3}\right\}$ **9.** $\left\{\frac{-1}{3}, 0, \frac{1}{4}\right\}$ **11.** $\{-2, 12\}$ **13.** $\left\{0, \frac{3}{4}\right\}$ **15.** $\{-20, 5\}$

17. $\{1, 25\}$ **19.** $\left\{\frac{-2}{5}, \frac{2}{5}\right\}$ **21.** $\left\{\frac{-3}{4}, 2\right\}$ **23.** $\left\{-2, \frac{-3}{5}\right\}$ **25.** $\{0, 5\}$ **27.** $\{-11, 3\}$ **29.** $\left\{\frac{-4}{5}\right\}$

31. -3 and 4 **33.** -12 and -4, or 4 and 12 **35.** 6 and 9, or -12 and $\frac{-9}{2}$ **37.** 7 in, 10 in **39.** 9 cm

41. 20 ft, 22 ft

Section 3-7. Ten Review Exercises

1. $2t^3 - 5t^2 - 8t + 15$ **2.** 0 **3.** $2t^2 + t - 5$ **4.** 1 **5.** $3m(m^2 + 4m - 3)$ **6.** $3u(u + 2)^2$ **7.** $(u + 5)(u - 8)$
8. $(x + 5)(x^2 - 5x + 25)$ **9.** $(4y - 3)^2$ **10.** $(2z + 3)(z - 6)$

Section 3-7. Supplementary Exercises

1. $y = -x$ or $y = \frac{x}{3}$ **3.** $t = \frac{-s}{4}$ or $t = 2s$ **5.** $y = -6x$ or $y = 6x$ **7.** $a = -2c$ or $a = \frac{-3}{b}$

9. $\{-2, -1, 1, 2\}$ **11.** $\{-4, -3, 3, 4\}$ **13.** $\left\{-1, \frac{-2}{3}, \frac{2}{3}, 1\right\}$ **15. a.** $-2, 4$ **c.** $-1, 3$

17. a. $-1, 3$ **c.** 1 **e.** No

Chapter 3 Review Exercises

1. $-x^3y^5z^6$ **3.** $-9m^2n^2$ **5.** $\frac{9s^4}{t^8}$ **7.** $108t^{16}$ **9.** 1 **11. a.** $8x^5 + 5x^4 - x^3 + 3x^2 + 2x - 10$ **b.** Five **c.** 8
d. -10 **e.** Polynomial of six terms **13.** $x^3 + 2x^2 + 3x - 3$ **15.** $2m^2 + 2mn + 3n^2$ **17.** $2y^2 + 48y + 5$
19. $180x^{11}y^{10}z^3$ **21.** $x^3 + x^2 + xy + y^2 - y^3$ **23.** $100x^4 - 9$ **25.** $16k^6 - 8k^3 + 1$ **27.** $3x^2 - 13xy + 21y^2$
29. $-4x^2y$ **31.** $2m^2 - 5mn + n^2$ **33.** $2x^3 - x^2 - 6x + 3$ **35.** $2xy(x - y + 4)$ **37.** $(p - 4p)(p - 6q)$
39. $(4m - 3)(m + 2)$ **41.** $-3(k^2 + 5)(k^2 - 3)$ **43.** $(10x^2 - 3y^2)(10x^2 + 3y^2)$ **45.** $(1 - 10mn)^2$
47. $(5c - d)(25c^2 + 5cd + d^2)$ **49.** $(10k - 1)(100k^2 + 10k + 1)$ **51.** $(3a - 5b)(2c - 3)$ **53.** $(x + 3y + 13)(x + 3y - 13)$
55. $(r^2 + 3s - 8)(r^2 - 3s - 8)$ **57.** $(6a + 1)(3a - 2)$ **59.** $(4x - 3)(3x + 5)$ **61.** $\left\{0, \frac{-2}{3}\right\}$ **63.** $\{-5, 6\}$ **65.** $\left\{\frac{-1}{7}, \frac{1}{2}\right\}$

67. 4, 6, 8

Chapter 4 Rational Expressions and Equations

Section 4-1. Practice Exercises

1. $\frac{1}{16}$ **3.** $\frac{1}{x^8}$ **5.** $\frac{-1}{p^6}$ **7.** $\frac{1}{k^4}$ **9.** $\frac{2n^3}{m^2}$ **11.** $\frac{-x^5}{8y^4}$ **13.** y^4 **15.** $\frac{-p^4r}{q^5}$ **17.** $\frac{x^2}{25}$ **19.** $\frac{n^3}{8m^3}$ **21.** $\frac{-6b^2}{ac^3}$

23. $\frac{-a^2c}{27b^2}$ **25.** $\frac{v^8}{16w^4}$ **27.** $\frac{-3q^3}{p^2}$ **29.** $\frac{y^3z^6}{125x^3}$ **31.** $\frac{1}{x^4}$ **33.** t^3 **35.** $-u^2$ **37.** $\frac{1}{v^{12}}$ **39.** $\frac{1}{w^8}$ **41.** a^4

43. $\frac{-1}{b^3}$ **45.** $\frac{-m^6}{8}$ **47.** $\frac{t^{15}}{10^5s^{10}}$ **49.** $36k^2$ **51.** $\frac{8}{p^6q^3}$ **53.** u^8w^4 **55.** $\frac{-10}{t^2}$ **57.** $-180u^3$ **59.** $\frac{y^7}{x^4}$ **61.** $\frac{4}{5z^4}$

63. $\frac{q^6}{p^4}$ **65.** $\frac{9}{8ab^{12}}$ **67.** $\frac{25x^2}{4}$ **69.** $\frac{3125}{m^5n^2}$ **71.** 3.6×10^2 **73.** 7.8×10^{-3} **75.** 5.0×10^5 **77.** 1.5×10^{-6}

79. 9.21×10^7 **81.** 25,000 **83.** 0.014 **85.** 4,390,000 **87.** 602,500,000,000 **89.** 0.00000000112

Section 4-1. Ten Review Exercises

1. $\left\{\frac{5}{6}\right\}$ **2.** $\{-2, 11\}$ **3.** $\left\{-4, \frac{3}{2}\right\}$ **4.** $\left\{x \middle| x < \frac{-2}{3}\right\}$

5. $\{y | 4 \le y \le 6\}$ **6.** $\left\{b \middle| -3 < b < \frac{5}{3}\right\}$ **7.** $3a^3$ **8.** $72k^7$

9. $\frac{-16m^{15}n^3}{y^{16}}$ **10.** $\frac{-9u^3}{2v}$

Section 4-1. Supplementary Exercises

1. x **3.** x^{-2a} **5.** x^3 **7.** 1 **9.** 10 **11.** 2^{2a} **13.** $x^{2a}y^{-2b}$ **15.** $x^{3a}y^{-2a}$ **17.** $\frac{y^{3b}}{x^{3a}}$ **19.** $\frac{y^a}{x^{2a}}$ **21.** $\frac{26}{5}$

23. $\frac{2}{5}$ **25.** $\frac{80}{27}$ **27.** $\frac{10}{3}$ **29. a.** $\frac{1}{729}$ **b.** $\frac{10}{81}$ **c.** $\frac{1}{9}$ **31. a.** $\frac{1}{64}$ **b.** $\frac{3}{16}$ **c.** $\frac{1}{16}$ **d.** $\frac{3}{4}$ **33.** 4.5×10^7

35. 8.4×10^{-8} **37.** 6.0×10^3 **39.** 5.0×10^{-3} **41.** 8.0×10^{21} **43.** 8.1×10^{-15} **45.** 2.0×10^{-2}

47. a. 0.065 mm **b.** 65 μm **49. a.** 645 pm **b.** 0.645 nm

Section 4-2. Practice Exercises

1. 3 **3.** -5 **5.** 0 **7.** None **9.** $-1, 5$ **11.** $x \neq y$ **13.** $a \neq \frac{-b}{2}$ **15.** $0, 5m \neq 3n$ **17.** $x \neq -y, x \neq 2y$

19. None **21.** $\frac{3x}{5y^2}$ **23.** $\frac{2m}{2m+1}$ **25.** $\frac{3p+2}{3p-2}$ **27.** $\frac{x-3}{x^2+3x+9}$ **29.** -1 **31.** $-5-b$ **33.** $\frac{2m-5}{2m+5}$

35. $\frac{-p^2}{p^2+1}$ **37.** $\frac{a^2+1}{a+1}$ **39.** $\frac{-1}{3a+2b}$ **41.** $3x-5$ **43.** $p-2+q$ **45.** $\frac{15y^2}{6xy}$ **47.** $\frac{12}{4a-4}$ **49.** $\frac{2b^2}{3b^2+b}$

51. $\frac{10mn}{10m^2n+5mn^2}$ **53.** $\frac{x+2}{x^2-4}$ **55.** $\frac{5y+10}{y^2+5y+6}$ **57.** $\frac{2a-10b}{a^2-3ab-10b^2}$ **59.** $\frac{3x-6}{x^3-8}$

Section 4-2. Ten Review Exercises

1. $21t^3$ **2.** 168 **3.** $21+3t+7t^2+t^3$ **4.** 63, not the same **5.** $6k$ **6.** 30 **7.** $6+k$ **8.** 11, not the same
9. $25u^2v^2$ **10.** $25+10uv+u^2v^2$

Section 4-2. Supplementary Exercises

1. (i) a. 9 **b.** 11 **c.** 16 **d.** 9 **(ii)** a and d **(iii)** $x+7$; d **3. (i) a.** 7 **b.** -3 **c.** 7 **d.** -10
(ii) a and c **(iii)** $x+5$; c **5. (i) a.** 3 **b.** 3 **c.** 4 **d.** 5 **(ii)** a and b **(iii)** x^2-x+1; b **7. (i) a.** -2
b. -20 **c.** -2 **d.** 2 **(ii)** a and c **(iii)** -2; c **9. a.** $y=1, y=-7$ **b.** $y=1, y=-7$ **c.** $x=4$, no value
d. $x-2$, 1, 7 **e.** No **f.** No **g.** $x \neq -2$ **h.** $x \neq -2$ **i.** Restricted value at $x=-2$ **11. a.** $x \neq 2$
b. Vertical dotted line **c.** y becomes very negative **d.** y becomes very positive **e.** No, yes **13.** x^a+1

15. $x^{2a}+x^a+1$ **17.** $\frac{y^a+2}{y^a-3}$

Section 4-3. Practice Exercises

1. $\frac{5}{2b}$ **3.** $\frac{2w}{uv}$ **5.** $\frac{k+1}{5k}$ **7.** $\frac{2r}{2r-3s}$ **9.** $\frac{x-3y}{x+2y}$ **11.** $\frac{3}{3a-b}$ **13.** $\frac{-5mn}{m-n}$ **15.** $\frac{u+2v}{3u-v}$ **17.** -1 **19.** $\frac{r}{s}$

21. $\frac{3x-y+1}{3x-y}$ **23.** $\frac{7s}{rt}$ **25.** $\frac{4v^2}{w}$ **27.** $\frac{2xy}{x-y}$ **29.** $3ab(a+3b)$ **31.** $\frac{-(k+3)}{5k}$ **33.** $\frac{r-3s}{7rs}$ **35.** -1

37. $\frac{y+3}{3y}$ **39.** $\frac{2z}{z^2-6}$ **41.** $\frac{1}{2}$ **43.** $\frac{a+b-1}{a+b}$ **45.** $\frac{uv}{2(u+v)}$ **47.** $\frac{-k^2}{k+3}$ **49.** $\frac{2x}{2x+1}$ **51.** $\frac{3t+2}{3t^2+2}$

Section 4-3. Ten Review Exercises

1. $6+3x$ **2.** $27-51x+10x^2$ **3.** $(2x-9)(5x-3)$ **4.** $-(2x-9)$ **5.** $3-5x$ **6.** $\frac{-1}{4a}$ **7.** $\frac{-1}{4a}$ **8.** $-27x^5y^6$

9. $\frac{-x^5}{27y^4}$ **10.** $\frac{4}{9}$

Section 4-3. Supplementary Exercises

1. a. $\frac{21}{2}$ **b.** $\frac{3x}{2}$ **c.** $\frac{21}{2}$ **3. a.** $\frac{12}{29}$ **b.** $\frac{x+2}{3x-1}$ **c.** $\frac{12}{29}$ **5. a.** $\frac{12}{5}$ **b.** $\frac{2x}{x-1}$ **c.** $\frac{12}{5}$ **7. a.** $\frac{1}{2}$ **b.** $\frac{1}{2}$ **c.** $\frac{1}{2}$

9. $k=3$ **11.** $k=16$ **13.** $k=24$ **15.** -1 **17.** $\frac{-(x+7)}{x-10}$ **19.** -1 **21. a.** 4, 2 **b.** 8 **c.** 1, -1 **d.** 1

e. $\frac{2(x+2)}{x-1}$ **f.** 8, the same **g.** -1, the same

Section 4-4. Practice Exercises

1. $\frac{3}{u}$ **3.** $\frac{-2}{3mn}$ **5.** 5 **7.** $\frac{1}{3a+2b}$ **9.** $\frac{2}{2y-1}$ **11.** $\frac{p-q}{p+q}$ **13.** $\frac{2k+1}{k-4}$ **15.** $\frac{2a+3}{x+y}$ **17.** $\frac{1}{t-1}$ **19.** $\frac{u-3}{3u-1}$

21. $36k^4$ **23.** $6m(2m+3)$ **25.** $p-5q$ **27.** $(m+3n)(m^2-3mn+9n^2)$ **29.** $12xy(2x-y)(2x+y)$ **31.** $(u+v)^2(u+2v)$

33. $(k+5)(k-5)(3k+1)$ **35.** $\frac{11t}{24}$ **37.** $\frac{17x}{20y}$ **39.** $\frac{r^2}{st}$ **41.** $\frac{4-x}{6x^2}$ **43.** $\frac{y+6x}{10xy}$ **45.** $\frac{a+5}{10a}$ **47.** $\frac{-1}{8k}$

49. $\frac{m^2+1}{m(m+1)}$ **51.** $\frac{a}{6(a-2)}$ **53.** $\frac{-1}{cd}$ **55.** $\frac{p^2+12}{p(p-4)(p+4)}$ **57.** $\frac{4}{(u-1)(u-2)(u-3)}$ **59.** $\frac{v^2+20}{(v-5)(v+4)^2}$

61. $\frac{-30x+137}{15(x-5)}$ **63.** $\frac{z^3-16z+24}{z(z-4)(z+4)}$ **65.** $\frac{7}{6(y+6)}$ **67.** $\frac{y+2}{y(y-1)(y+1)}$ **69.** $\frac{x+7}{2(x^2+3x+9)}$ **71.** $\frac{7}{5(a+2)}$

Section 4-4. Ten Review Exercises

1. $\frac{3k}{5}$ **2.** $\frac{-5}{3k}$ **3.** 3 **4.** -12 **5.** $2k^3-4k^2+12k-3$ **6.** a^2+7a+8 **7.** $2b^2+4b-3$

8. $4z^4-5z^2+4z-3$ **9.** $(4x-3y)^2$ **10.** $(3m-2)(m+5)$

Section 4-4. Supplementary Exercises

1. a. $\frac{11}{30}$ **b.** $\frac{1}{15}$ **3. a.** $\frac{3}{2}$ **b.** $\frac{1}{4}$ **5. a.** 2 **b.** $\frac{-4}{5}$ **7. a.** 1 **b.** $\frac{1}{2}$ **9.** $k=3$ **11.** $k=1$ **13.** $k=5$

15. $k=4$ **17.** $\frac{5(s+1)}{s}$ **19.** $\frac{1}{x-3}$

Section 4-5. Practice Exercises

1. $\frac{a+b}{a+4b}$ **3.** $\frac{2}{k-1}$ **5.** $\frac{2}{5m}$ **7.** $\frac{3u+v}{u+3v}$ **9.** $2a+3b$ **11.** $17x-y$ **13.** $2u^2-4$ **15.** $10rs$ **17.** $\frac{7}{11}$

19. $\frac{2(2-a)}{a}$ **21.** $\frac{u+v}{u}$ **23.** $\frac{-x-3}{3x}$ **25.** $\frac{m-1}{m+2}$ **27.** $\frac{b}{3}$ **29.** $\frac{2k+3}{3k+2}$ **31.** $\frac{1}{4z}$ **33.** $\frac{-1}{2}$ **35.** $\frac{6p}{p+6}$

37. $\frac{-(x+1)}{x}$ **39.** 3 **41.** $\frac{s-7}{s-1}$ **43.** 5 **45.** $\frac{1}{a}$ **47.** $\frac{1}{m+n}$ **49.** $\frac{k-3}{9k}$ **51.** $\frac{1}{uv(v-4u)}$ **53.** $\frac{r^2-rs+s^2}{r^2s^2}$

55. $\frac{x+1}{x(x-1)}$

Section 4-5. Ten Review Exercises

1. $(7+t)+3t$ **2.** $7+(t+3t)$ **3.** $7+4t$ **4.** $10t^2+5t+1$ **5.** $7x^2+37x+8$ **6.** $\frac{2y+1}{y+3}$ **7.** $2z^2+z-5$

8. $2k(3k-1)(3k+1)$ **9.** $(3u-1)(3u+1)(9u^2+1)$ **10.** $(a-5b)^2$

Section 4-5. Supplementary Exercises

1. $\frac{1}{a}$ **3.** $\frac{k-2}{2k}$ **5.** $m-3$ **7.** $\frac{1}{b}$ **9.** $\frac{2x}{3}$ **11.** $1+r$ **13.** $\frac{2a}{2-a}$ **15.** $\frac{(k-4)(k+4)}{8k(k+2)}$ **17.** $\frac{m^2}{4m^2+4m+1}$

19. $27\frac{7}{9}$ **21.** $\frac{1}{4t}$ **23.** 1 **25.** $\frac{p^2-1}{p^2+1}$ **27.** $\frac{1}{x+1}+\frac{2}{x+2}$ **29.** $\frac{5}{x+3}+\frac{2}{x-2}$

Section 4-6. Practice Exercises

1. $\{3\}$ **3.** $\left\{\frac{-5}{2}\right\}$ **5.** $\{6\}$ **7.** $\{-1\}$ **9.** $\{10\}$ **11.** $\left\{\frac{4}{5}\right\}$ **13.** $\left\{\frac{2}{5}\right\}$ **15.** $\{2\}$ **17.** $\left\{\frac{-1}{3}\right\}$ **19.** $\{-6\}$

21. $\varnothing$ **23.** $\varnothing$ **25.** $\varnothing$ **27.** $\{1\}$ **29.** $\{5\}$ **31.** $-3<p<3$

33. $x\le\frac{-1}{3}$ or $x>3$

35. $0<w<4$

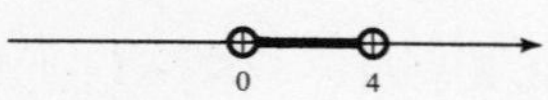

37. $t<-1$ or $t>\frac{7}{3}$

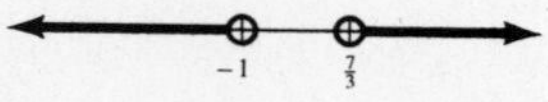

39. $a<-3$ or $0\le a<3$

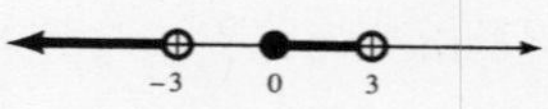

41. $-2 < y \leq \frac{-2}{3}$ or $y > 6$  **43.** $x < 1$ or $3 < x < 5$

45. $0 \leq t < 1$ or $t \geq 2$ **47.** $-2 < w < -1$

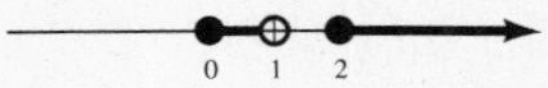

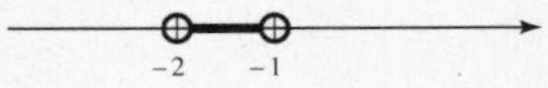

49. $-2 < b < 0$ or $b > 2$

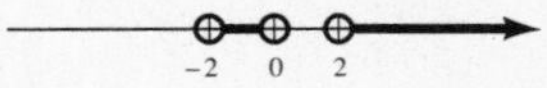

Section 4-6. Ten Review Exercises

1. $\{0\}$ **2.** $\left\{\frac{-7}{3}, 3\right\}$ **3.** $\left\{\frac{-1}{5}\right\}$ **4.** $\left\{-2, \frac{1}{6}\right\}$ **5.** $u > \frac{-3}{4}$

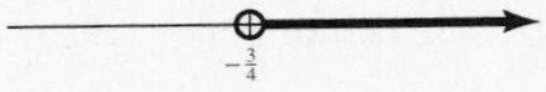

6. $-6 < v < 12$ **7.** $w \leq -4$ or $w \geq 9$

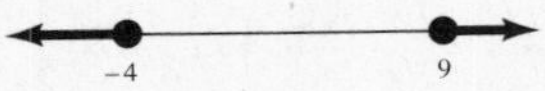

8. $x < -2$ or $0 < x < 5$ **9.** $\frac{4}{y^2}$ **10.** $\frac{25}{z^4}$

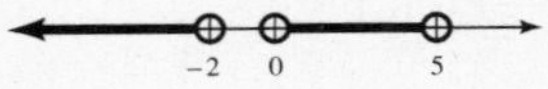

Section 4-6. Supplementary Exercises

1. $\{-7\}$ **3.** $\left\{\frac{3}{2}, -1\right\}$ **5.** $\{3\}$ **7.** $\left\{\frac{2}{5}\right\}$ **9.** $\left\{-1, \frac{1}{4}\right\}$ **11.** $\{-4\}$ **13.** $\varnothing$ **15.** $0 < x < 1$ **17.** $-1 < x < 0$
19. $-1 < x \leq 1$

Section 4-7. Practice Exercises

1. 3 **3.** 7 **5.** -10 **7.** 2 **9.** 8 and 9 **11.** -4 **13.** 1 and 4 **15.** 1 or 5 **17.** $3\frac{3}{4}$ hr **19.** 24 hr

21. $\frac{96}{17}$ hr **23.** $6\frac{2}{3}$ hr **25.** 28 mph **27.** 412.5 mi **29.** 580 mph **31.** $6\frac{2}{3}$ mi **33.** 2 mph **35.** 5 hr and 6 hr

37. $F = \frac{ab}{a + b}$ **39.** $y = \frac{a^2 + b^2}{b - a}$ **41.** $x = \frac{2y}{y + 6}$ **43.** $t = \frac{-2a}{b}$ **45.** $x = \frac{3y}{3 - y}$ **47.** $x = \frac{5y + k}{y - k}$

Section 4-7. Ten Review Exercises

1. -30 **2.** -30 **3.** $\left\{\frac{-5}{4}\right\}$ **4.** $\frac{-32}{63}$ **5.** $\frac{-32}{63}$ **6.** $\left\{\frac{1}{2}\right\}$ **7.** $(3t - u)(9t^2 + 3tu + u^2)$ **8.** $3(3t - u)(3t + u)$
9. $3(3t + u)^2$ **10.** $3(9t^2 + u^2)$

Section 4-7. Supplementary Exercises

1. a. $\frac{4}{3}$ **b.** Greater **3. a.** $\frac{2}{5}$ **b.** Less **5. a.** $\frac{1}{3}$ **b.** Less **7. a.** $\frac{5}{6}$ **b.** Greater **9. a.** $\frac{7}{9}$ **b.** Greater

11. a. $\frac{1{,}007}{1{,}008}$ **b.** No **13.** $18\frac{3}{4}$ **15.** $90\frac{10}{11}$ **17. a.** $x \neq 1$ **b.** $x = \frac{y + 5}{y}$ **c.** $y \neq 0$ **19. a.** $x \neq 0$

b. $x = \frac{5}{y - 2}$ **c.** $y \neq 2$ **21. a.** $R = \frac{r_1 r_2 r_3}{r_1 r_2 + r_1 r_3 + r_2 r_3}$ **b.** $\frac{60}{13}$ **c.** R is less than r_1 or r_2 or r_3 **23. b.** $2x$

c. $\frac{x}{60}$ **d.** $\frac{x}{40}$ **e.** $\frac{x}{60} + \frac{x}{40}$ **f.** $\frac{2x}{\frac{x}{60} + \frac{x}{40}}$ **g.** 48 mph **i.** No

Chapter 4 Review Exercises

1. $\frac{-1}{64}$ **3.** $\frac{-x^2}{2y^3}$ **5.** 1 **7.** $\frac{3p^8}{4q^9}$ **9.** 3.97×10^6 **11.** 0.00935 **13.** $x \neq 5$ **15.** None **17.** $\frac{3m^2}{7n}$

19. $\frac{y^2 - y + 1}{y - 1}$ **21.** $\frac{16ab}{12a^2b^2}$ **23.** $\frac{2m^2 - 3m}{4m^2 - 9}$ **25.** $\frac{2x^2}{3y}$ **27.** $\frac{a}{a + 4}$ **29.** $\frac{u^2 + u - 3}{3u - 1}$ **31.** $\frac{2}{3r}$ **33.** $\frac{3k + 2}{2k + 3}$

35. $15a^2b^2$ **37.** $(y - 3)(y + 2)(y + 5)$ **39.** $\frac{7a}{12}$ **41.** $\frac{x^2 + 3x - 5}{(x - 3)(x + 2)(x - 1)}$ **43.** $\frac{2}{x^2y^2}$ **45.** $6ab$ **47.** $\frac{1 - k}{2 + k}$ **49.** 9

51. $\{6\}$ **53.** $\{-3\}$ **55.** $0 < x < 6$ **57.** $p < -1$ or $0 < p < 1$

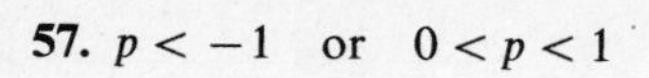

59. 3 **61.** $2\frac{2}{3}$ hr **63.** $t = x - 1$

Chapter 5 Roots, Radicals, and Complex Numbers

Section 5-1. Practice Exercises

1. 5 **3.** $\frac{10}{11}$ **5.** -4 **7.** Not a real number **9.** $\frac{8}{5}$ **11.** 0 **13.** 2.828 **15.** -4.899 **17.** 8.775

19. -9.950 **21.** -4 **23.** 6 **25.** 1 **27.** $\frac{9}{5}$ **29.** $\frac{-7}{10}$ **31.** 0.2 **33.** 3 **35.** 2 **37.** -4

39. Not a real number **41.** -1 **43.** -6 **45.** 11 **47.** 12 **49.** -8 **51.** 6 **53.** 19 **55.** 0.5 **57.** 4.1

59. $\frac{3}{4}$ **61.** -13 **63.** 31

Section 5-1. Ten Review Exercises

1. $13t - 1$ **2.** $40t^2 - 17t - 12$ **3.** $\frac{25m^2}{4}$ **4.** $\frac{25 - 10m + m^2}{4}$ **5.** $\left\{\frac{-3}{5}\right\}$ **6.** $\left\{-4, \frac{8}{3}\right\}$ **7.** $\{-6, 10\}$ **8.** $\varnothing$

9. $\frac{-12}{5} < x < 4$ **10.** $y > -3$

Section 5-1. Supplementary Exercises

1. x **3.** $|x|$ **5.** $-|x|$ **7.** x **9.** Not a real number **11.** x^2y **13.** $|x^3y|$ **15.** Not always a real number

17. $7|x|$ **19.** $8x^2$ **21.** $|x|$ **23.** $4|x|$ **25. a.** 16, 1, 0, 1, 16 **b.**

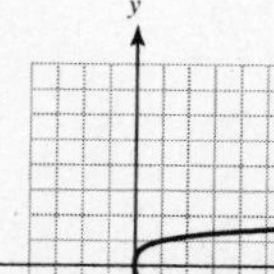

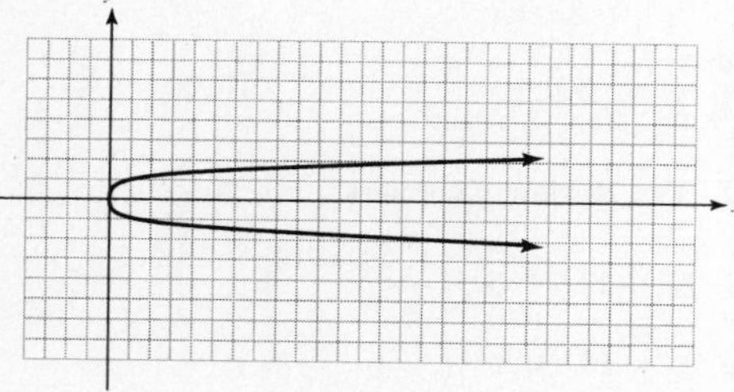

c. 2; **d.** None **e.** 1 **27. a.** 32 **b.** 32 **29. a.** 43 **b.** 21 **31. a.** 0.3 **b.** 0.3 **33. a.** -5
b. Calculator displays "ERROR" **c.** No

Section 5-2. Practice Exercises

1. $\sqrt{4} = 2$ **3.** $\sqrt{21}$ **5.** $\sqrt[3]{1{,}000} = 10$ **7.** $\sqrt[3]{-10}$ **9.** $-\sqrt[6]{5}$ **11.** $\sqrt[4]{256} = 4$ **13.** $\sqrt[6]{0.000001} = 0.1$ **15.** $26^{1/2}$
17. $71^{1/3}$ **19.** $(-21)^{1/3}$ **21.** $14^{1/4}$ **23.** $-7^{1/4}$ **25.** $-(-44)^{1/7}$ **27.** $(\sqrt[4]{16})^3 = 8$ **29.** $(\sqrt[3]{-27})^4 = 81$

31. $-(\sqrt{25})^3 = -125$ **33.** $(\sqrt{4})^7 = 128$ **35.** $(\sqrt[3]{-216})^4 = 1{,}296$ **37.** $(\sqrt[4]{15})^0 = 1$ **39.** $\sqrt[5]{144}$ **41.** $\left(\sqrt[3]{\frac{27}{8}}\right)^2 = \frac{9}{4}$

43. $\frac{1}{5}$ **45.** $\frac{1}{10}$ **47.** $\frac{1}{343}$ **49.** $\frac{1}{36}$ **51.** $\frac{1}{27}$ **53.** $\frac{1}{16}$ **55.** -1 **57.** 1 **59.** $\frac{-1}{100{,}000}$ **61.** $\frac{4}{9}$ **63.** $2^{5/6}$

65. $3^{1/6}$ **67.** $6^{3/2}$ **69.** $\frac{5^{1/2}}{3}$ **71.** $2x^{1/3}$ **73.** $3x^{1/2y}$ **75.** $\frac{x^3}{5}$ **77.** $10^{5/6}$ **79.** $6^{1/6} \cdot 5^{1/3}$ **81.** $\frac{1}{8^{7/6}}$

83. $x - 25$ **85.** $y + 9y^{1/2} + 18$ **87.** $9t^{1/2} + 60t^{1/4} + 100$ **89.** $x + 2$ **91.** $9s^2 - 10s + 3$ **93.** $8u^3 - 1$
95. $4x^{5/3} - 2x$ **97.** 2.154 **99.** 2.115 **101.** 14.390 **103.** 19.812 **105.** 181.019

Section 5-2. Ten Review Exercises

1. $1125t^4u^8$ **2.** $\frac{25y}{8x^7}$ **3.** $18a^3 + 45a^2 - 8a - 20$ **4.** $2b^2 - 3b - 3$ **5.** $5m^2 + mn$ **6.** $\frac{x - 2y}{x^2 - 2xy + 4y^2}$ **7.** $4q(p - 3)$

8. $\frac{1}{8k}$ **9.** 0 **10.** $\left\{\frac{-2}{3}\right\}$

Section 5-2. Supplementary Exercises

1. $4^{1/3}$ **3.** Not a real number **5.** $\frac{-1}{5}$ **7.** $2^{7/12}$ **9.** $\frac{1}{7^{2/15}}$ **11.** $\frac{1}{7^{3/2}}$ **13.** 5 **15.** -5 **17.** $x^{1/4}$
19. $y^{1/6}$ **21.** $z^{1/20}$ **23.** $t^{1/9}$ **25.** 6 **27.** 9 **29.** 243 **31. a.** $p + p^{1/4}$ **b.** $p^{3/4}(p^{1/4} + 1)$ **33. a.** $x^2 - x^{3/2}$
b. $x^{5/2}(x^{1/2} - 1)$ **35. a.** $m^3 + m^2$ **b.** $m^{3/4}(m^{5/4} + 1)$

Section 5-3. Practice Exercises

1. $\sqrt[4]{15}$ **3.** $\sqrt[6]{6y^5}$ **5.** $\sqrt[9]{-80}$ **7.** $\sqrt[5]{-30x^4}$ **9.** 5 **11.** 6 **13.** 10 **15.** 10 **17.** $2x$ **19.** $\frac{t}{3}$ **21.** $10t^2$
23. $\frac{\sqrt[4]{60}}{y^2}$ **25.** $\sqrt{3}$ **27.** $\sqrt[3]{6}$ **29.** $\sqrt[4]{x}$ **31.** $\sqrt[6]{35}$ **33.** $\sqrt[5]{\frac{-6}{p}}$ **35.** $\frac{1}{2}$ **37.** -2 **39.** $\frac{2}{x}$ **41.** $\frac{y^2}{4}$ **43.** $2m^3$
45. $\sqrt{3}$ **47.** $\sqrt[3]{-7}$ **49.** $\sqrt[6]{125}$ **51.** $\sqrt[12]{16w^8}$ **53.** $\sqrt{11v}$ **55.** $\sqrt{5}$ **57.** $-\sqrt[3]{3k}$ **59.** $\sqrt[15]{x^{10}y^5}$

Section 5-3. Ten Review Exercises

1. $(2x - 5)(2x + 3)$ **2.** 0 **3.** $\left\{\frac{-3}{2}, \frac{5}{2}\right\}$ **4.** $2x - 5$ **5.** 0 **6.** $-12, 12$ **7.** 12 **8.** $\{-12, 12\}$ **9.** 3 **10.** 3

Section 5-3. Supplementary Exercises

1. $\frac{\sqrt{x-2}}{6}$ **3.** $\frac{\sqrt[3]{x^2 - y^2}}{5}$ **5.** $\sqrt[4]{x-2}$ **7.** $x + 5$ **9.** $\sqrt[3]{4y^2 + 2y + 1}$ **11.** $t - 2$ **13.** $\sqrt[6]{675}$ **15.** $\sqrt[10]{1{,}568x^2y^5}$
17. $\sqrt[4]{2u^3}$ **19.** $\sqrt[6]{7m^5n^4}$ **21.** $2t$ **23.** 3.557 **25.** 2.962 **27.** 50 **29.** 25 **31.** 50.27 **33.** 45.84

Section 5-4. Practice Exercises

1. $3\sqrt{5}$ **3.** $-7\sqrt{2}$ **5.** $5\sqrt[3]{2}$ **7.** $-2ab\sqrt[5]{4ab^3}$ **9.** $x^2y\sqrt[4]{20y}$ **11.** $-3mn^2\sqrt[3]{10n}$ **13.** $-\sqrt[3]{10}$ **15.** $-\sqrt[3]{3t^2}$
17. $-2\sqrt[3]{3}$ **19.** $-5p^2q\sqrt[3]{4p}$ **21.** $\frac{-2\sqrt[3]{30}}{x}$ **23.** $\frac{\sqrt{7}}{5}$ **25.** $\frac{\sqrt[3]{21}}{10}$ **27.** $\frac{\sqrt[3]{9}}{2x}$ **29.** $\frac{t^3\sqrt{5}}{18}$ **31.** $\frac{\sqrt{47}}{9y^6}$ **33.** $\frac{\sqrt{6}}{2}$
35. $\frac{-\sqrt{15}}{5}$ **37.** $5\sqrt[3]{4}$ **39.** $-7\sqrt[3]{3}$ **41.** $2\sqrt[4]{5}$ **43.** $\frac{10\sqrt[3]{4x}}{x}$ **45.** $\frac{-\sqrt[4]{x^3y}}{xy}$ **47.** $\frac{2\sqrt[5]{4a^4b}}{b}$ **49.** $\sqrt{7}$ **51.** $\sqrt{3}$
53. $-\sqrt{xy}$ **55.** $-\sqrt[3]{2x}$ **57.** $\sqrt[4]{10x}$ **59.** $\sqrt{2}$

Section 5-4. Ten Review Exercises

1. -20 **2.** -7 **3.** $10t^3 + 91t^2 + 155t + 56$ **4.** 140 **5.** $2t^2 + 15t + 7$ **6.** 0 **7.** $9k^4$ **8.** $9 - 6k^2 + k^4$
9. $3pq$ **10.** $3 + pq$

Section 5-4. Supplementary Exercises

1. a. 1 **b.** $2\sqrt[3]{25}$ **3. a.** 1, 2, 4 **b.** $\frac{-1}{4}$ **5. a.** 2 **b.** $-\sqrt[3]{44}$ **7. a.** 3 **b.** $\frac{\sqrt[3]{28}}{3}$ **9. a.** 1 **b.** 0.3
11. a. 5 **b.** $-\sqrt{2}$ **13. a.** 4 **b.** $\frac{6\sqrt[3]{25}}{5}$ **15. a.** 1, 3 **b.** $\frac{x\sqrt[3]{x^2}}{y^2}$ **17. a.** 4 **b.** $\frac{-10\sqrt[4]{xy}}{xy}$ **19. a.** 1, 2
b. $-pq^2\sqrt[3]{21p}$ **21.** $\frac{1}{7}$ **23.** $2\sqrt{3}$ **25.** $\frac{9x\sqrt{3x}}{y}$ **27.** $\frac{-\sqrt[3]{10}}{2}$ **29.** $\frac{\sqrt{a}}{b}$ **31.** $5(x^2 + 1)\sqrt{2(x^2 + 1)}$
33. $(y - 1)\sqrt[3]{y - 1}$ **35.** $\frac{\sqrt[3]{x^2 - 1}}{x + 1}$ **37.** $\frac{\sqrt{x + 2}}{x + 2}$ **39.** $\sqrt{2(x + 5)}$

Section 5-5. Practice Exercises

1. $15\sqrt{10}$ **3.** $2\sqrt[4]{3}$ **5.** $5\sqrt{5}$ **7.** $-5\sqrt[3]{9}$ **9.** $4\sqrt[4]{5} - 4\sqrt{6}$ **11.** $21\sqrt[3]{3} - 19\sqrt{3}$ **13.** $2\sqrt[3]{2} + 2\sqrt{2}$ **15.** $21\sqrt{2}$
17. $-\sqrt{5}$ **19.** $2\sqrt[3]{3}$ **21.** $5\sqrt[3]{10}$ **23.** $-2\sqrt{2}$ **25.** $2\sqrt[3]{5} + 7$ **27.** $3\sqrt{5} - 10$ **29.** $\sqrt[3]{3} - 2\sqrt[3]{9}$ **31.** $5\sqrt{10} + \sqrt{5}$
33. $5\sqrt{x} - 3\sqrt{2y}$ **35.** $\frac{7\sqrt{6}}{6}$ **37.** $\frac{-7\sqrt{30}}{30}$ **39.** $6 + \sqrt{10}$ **41.** $2\sqrt[3]{6} - 3$ **43.** $2\sqrt[4]{5} + 6$ **45.** $10\sqrt{7} - 14\sqrt{2} + 14$
47. $-1 + \sqrt{2}$ **49.** $\sqrt{6} - 5\sqrt{2} + 3\sqrt{3} - 15$ **51.** $19 - 11\sqrt{6}$ **53.** $-32 - 10\sqrt{15}$ **55.** $14 + 6\sqrt{5}$ **57.** $53 - 20\sqrt{7}$
59. $5 + 2\sqrt{6}$ **61.** $88 - 16\sqrt{30}$ **63. a.** $3 - \sqrt{10}$ **b.** -1 **65. a.** $8 + \sqrt{2}$ **b.** 62 **67. a.** $-1 - 2\sqrt{5}$
b. -19 **69. a.** $-3\sqrt{7} + 4\sqrt{2}$ **b.** 31 **71.** $\sqrt{3} - \sqrt{2}$ **73.** $-\sqrt{2} + \sqrt{3}$ **75.** $\frac{\sqrt{10} - \sqrt{14}}{14}$ **77.** $2 - \sqrt{3}$

79. $-5-2\sqrt{6}$ **81.** $\dfrac{-\sqrt{2}-2\sqrt{5}}{6}$ **83.** $\dfrac{7+2\sqrt{10}}{3}$ **85.** $-3(\sqrt{7}+\sqrt{2})$ **87.** $1+\sqrt{6}$ **89.** $2-\sqrt{3}$ **91.** $\dfrac{3-\sqrt{5}}{2}$
93. $\dfrac{2-5\sqrt{2}}{7}$ **95.** $\dfrac{5+\sqrt{3}}{3}$

Section 5-5. Ten Review Exercises

1. -2 **2.** -1 **3.** -12 **4.** -4 **5.** $-21x^4y^2$ **6.** $4a^3b^4$ **7.** $-72t^5$ **8.** $\dfrac{-18u^8}{125}$

9. $x<3$ 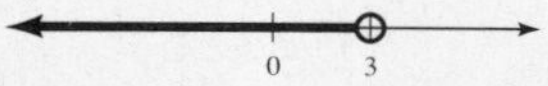

10. $\dfrac{-7}{3}<x<3$ 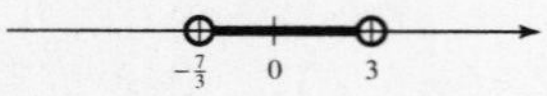

Section 5-5. Supplementary Exercises

1. $4\sqrt[4]{5}-11\sqrt[3]{5}+3\sqrt{5}$ **3.** $35\sqrt[3]{3}-24\sqrt[3]{25}$ **5.** 6 **7.** 1 **9.** $\sqrt{10}-\sqrt{6}$ **11.** $\dfrac{\sqrt[3]{4}+2\sqrt[3]{x}}{2}$ **13.** $x+x\sqrt{y}$
15. $2\sqrt{y}$ **17.** $\dfrac{p+q+2\sqrt{pq}}{p-q}$ **19.** $\dfrac{x+1-\sqrt{x+1}}{x}$ **21.** $\dfrac{\sqrt[3]{9}-\sqrt[3]{3}+1}{4}$ **23.** $2\sqrt[3]{5}+2$ **25.** $\dfrac{1}{28+7\sqrt{15}}$
27. $\dfrac{6}{5+\sqrt{55}}$ **29.** $2x+2\sqrt{x^2-9}$

Section 5-6. Practice Exercises

1. $5i$ **3.** $12i$ **5.** $-14i$ **7.** $2i\sqrt{3}$ **9.** $-7i\sqrt{2}$ **11.** $\dfrac{2}{7}i$ **13.** $\dfrac{i\sqrt{5}}{6}$ **15.** $\dfrac{i\sqrt{21}}{7}$ **17.** $\dfrac{-2i\sqrt{5}}{5}$ **19.** $\dfrac{i\sqrt{10}}{15}$
21. $-\sqrt{6}$ **23.** $-2\sqrt{15}$ **25.** $-3\sqrt{77}$ **27.** $-60\sqrt{6}$ **29.** $i\sqrt{2}$ **31.** $3i$ **33.** $7\sqrt{5}$ **35.** $4\sqrt{3}$ **37.** $\dfrac{i\sqrt{10}}{4}$
39. $\dfrac{-\sqrt{30}}{3}$ **41.** $10+9i$ **43.** $-3-11i$ **45.** $12-5i\sqrt{3}$ **47.** $8+11i$ **49.** $2\sqrt{2}+2i\sqrt{2}$ **51.** $\dfrac{7}{3}-\dfrac{10i}{3}$
53. $\dfrac{-4}{3}-\dfrac{2i\sqrt{17}}{3}$ **55.** $\dfrac{-2\sqrt{5}}{5}+\dfrac{3i\sqrt{5}}{5}$ **57.** $x=7;\ y=2$ **59.** $x=-10;\ y=6$ **61.** $x=0;\ y=7$
63. $x=3\sqrt{2};\ y=7\sqrt{2}$ **65.** $x=\dfrac{3}{2};\ y=\dfrac{-6}{5}$ **67.** $x=-5;\ y=-4$ **69.** $x=-2\sqrt{2};\ y=4\sqrt{3}$ **71.** $7+3i$
73. $3-7i$ **75.** $9-10i$ **77.** $-19+i$ **79.** 10 **81.** $6+4i$ **83.** $-7+3i$ **85.** $\sqrt{5}-7i\sqrt{5}$ **87.** $8i\sqrt{6}$
89. $12\sqrt{2}$

Section 5-6. Ten Review Exercises

1. $\{1\}$ **2.** $\{0, 3\}$ **3.** $\left\{\dfrac{3}{2}, 3\right\}$ **4.** $\{-5\}$ **5.** $b=\dfrac{2A}{h}-a$ **6.** $t<\dfrac{-2}{3}$ **7.** $\dfrac{-1}{2}\le b\le\dfrac{3}{2}$ 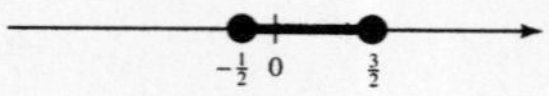

8. $x<-10$ or $x>-5$ 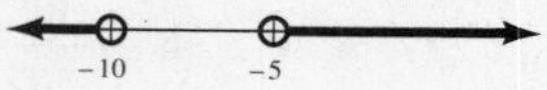

9. $-2<y<7$ 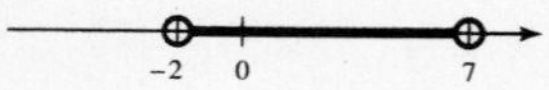

10. $\dfrac{-3}{2}<z\le\dfrac{9}{2}$ 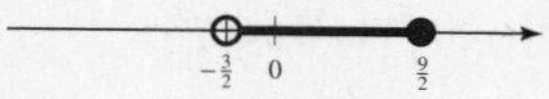

Section 5-6. Supplementary Exercises

1. $3+3i$ **3.** 13 **5.** $-8+7i$ **7.** $6-2i$ **9.** $11+25i$ **11.** $40\sqrt{2}+15i\sqrt{5}$ **13.** $x=3;\ y=-2$
15. $x=0;\ y=1$ **17.** $x=2;\ y=3$ **19.** $x=3;\ y=8$ **21.** $a=-x;\ b=-y$ **23.** $a=-3x;\ b=-y$ **25.** 10
27. 7 **29.** $5\sqrt{2}$

Section 5-7. Practice Exercises

1. $-54i$ **3.** $16-40i$ **5.** $77-143i$ **7.** $-15-20i$ **9.** $9-15i$ **11.** $13i$ **13.** $-31-27i$ **15.** $38-54i$
17. $26-2i\sqrt{2}$ **19.** $39-5i\sqrt{3}$ **21.** 27 **23.** 124 **25.** $30+30i$ **27.** $7+24i$ **29.** $-21-20i$ **31.** $-24+18i$
33. $-55-20i\sqrt{15}$ **35.** $\dfrac{-1}{2}-\dfrac{i\sqrt{3}}{2}$ **37. a.** $10+3i$ **b.** 109 **39. a.** $-7-4i$ **b.** 65 **41. a.** $6+6i$ **b.** 72

43. a. $\sqrt{10}-9i$ **b.** 91 **45. a.** $3\sqrt{5}+2i\sqrt{6}$ **b.** 69 **47.** $-5+i$ **49.** $\dfrac{3\sqrt{5}}{5}+\dfrac{i\sqrt{5}}{5}$ **51.** $\dfrac{10\sqrt{3}}{3}+\dfrac{6i\sqrt{3}}{3}$

53. $\frac{1}{2} - \frac{1}{2}i$ **55.** $\frac{-3}{4} - \frac{3}{4}i$ **57.** $\frac{4}{3} - \frac{4}{3}i$ **59.** $-i$ **61.** $\frac{8}{13} + \frac{1}{13}i$ **63.** $2 + 6i$ **65.** $\frac{18}{5} + \frac{21}{5}i$ **67.** -1

69. $\frac{-1}{4} + \frac{7i\sqrt{3}}{12}$ **71.** $\frac{-\sqrt{15}}{4} + \frac{1}{4}i$ **73.** -5 **75.** $20 - 2i$ **77.** $-i$ **79.** $\frac{-17}{2} - \frac{14}{2}i$ **81.** $\frac{-6}{5} + \frac{3}{5}i$ **83.** $\frac{8i}{5}$

Section 5-7. Ten Review Exercises

1. $8k + 16$ **2.** $\{-2\}$ **3.** $(t^2 + 4)(4t^2 - 9)$ **4.** $\left\{\frac{-3}{2}, \frac{3}{2}, -2i, 2i\right\}$ **5.** $(3u - 2)(u + 4)$

6. $-4 < u < \frac{2}{3}$ (number line: -4, 0, $\frac{2}{3}$) **7.** $\frac{14 - x}{(x + 4)(x - 5)}$ **8.** $\{14\}$ **9.** $y = \frac{x + 5}{3 - b}$

10. $y = \frac{-3x}{2x - 3}$ or $y = \frac{3x}{3 - 2x}$

Section 5-7. Supplementary Exercises

1. $-i$ **3.** i **5.** $-i$ **7.** i **9.** $-i$ **11.** 1 **13.** -1 **15.** $-i$ **17.** 1 **19.** i **21.** $20 - 100i$ **23.** 10 **25.** $-2 + 2i$ **27.** $-8i$ **29. a.** 5 **b.** $5 - 2\sqrt{3}$ **c.** 0 **31. a.** $10 - 2i$ **b.** $8 + i\sqrt{2}$ **c.** 0 **33.** $2 - i$ **35.** $5 - 4i$ **37.** $15 + 5i$

Chapter 5 Review Exercises

1. 14 **3.** $\frac{5}{4}$ **5.** 4.796 **7.** $\frac{-5}{2}$ **9.** $\frac{3}{2}$ **11.** Not a real number **13.** 41 **15.** 7 **17.** 15 **19.** 11

21. $\sqrt[3]{12}$ **23.** $(-18)^{1/3}$ **25.** $17^{1/2}$ **27.** 9 **29.** $\sqrt[4]{x^3}$ **31.** $\frac{1}{14}$ **33.** $7x^2$ **35.** $\frac{10}{ab^2}$ **37.** $t - 1$ **39.** 2.102

41. $\sqrt{154}$ **43.** 4 **45.** $2x\sqrt[5]{2}$ **47.** $8\sqrt{x}$ **49.** $\frac{1}{2}$ **51.** $\sqrt[3]{12}$ **53.** $6\sqrt{7}$ **55.** $-5x\sqrt[3]{3x}$ **57.** $\frac{\sqrt{10}}{2}$ **59.** $2\sqrt[4]{49a}$

61. $\sqrt{6}$ **63.** $3\sqrt{2} + 5\sqrt{3}$ **65.** $3\sqrt[3]{2x} - 8\sqrt{2x}$ **67.** $21 - \sqrt{14}$ **69.** $-11\sqrt{5}$ **71.** $16 - 6\sqrt{7}$ **73. a.** $a + 2\sqrt{3}$ **b.** 69 **75.** $\frac{7\sqrt{3} - \sqrt{6}}{3}$ **77.** $\frac{-6\sqrt{7} + \sqrt{14} + 6\sqrt{5} - \sqrt{10}}{2}$ **79.** $\frac{3 + \sqrt{2}}{2}$ **81.** $6i$ **83.** $\frac{-i\sqrt{7}}{5}$ **85.** $-\sqrt{30}$

87. $6i$ **89.** $-3\sqrt{3} + 3i\sqrt{3}$ **91.** $\frac{-2}{5} + \frac{i\sqrt{7}}{5}$ **93.** $x = -3$; $y = -5$ **95.** $x = \frac{2\sqrt{6}}{3}$; $y = -\sqrt{2}$ **97.** $2 - 2i$

99. $4 - 3i$ **101.** $-13\sqrt{5}$ **103.** $-54 - 42i$ **105.** 1 **107.** $76 + i\sqrt{21}$ **109. a.** $-9 + 13i$ **b.** 250 **111.** $5 + 2i$

113. $\frac{1}{17} + \frac{5i\sqrt{2}}{17}$

Chapter 6 Quadratic Equations and Inequalities

Section 6-1. Practice Exercises

1. a. No **b.** Yes **3. a.** Yes **b.** Yes **5. a.** Yes **b.** No **7. a.** Yes **b.** Yes **9. a.** No **b.** Yes

11. a. Yes **b.** Yes **13.** $\left\{\frac{-1}{4}, 6\right\}$ **15.** $\left\{0, \frac{8}{3}\right\}$ **17.** $\{-1, 6\}$ **19.** $\left\{\frac{-1}{5}, 0, 4\right\}$ **21.** $\left\{0, \frac{5}{2}\right\}$ **23.** $\{0, 1, 4\}$

25. $\{2, 5\}$ **27.** $\{-1, 2\}$ **29.** $\left\{\frac{-1}{4}, 2, 9\right\}$ **31.** $\{-11, 11\}$ **33.** $\{-\sqrt{21}, \sqrt{21}\}$ **35.** $\{-2\sqrt{2}, 2\sqrt{2}\}$ **37.** $\{-9, 9\}$

39. $\{-5i, 5i\}$ **41.** $\left\{\frac{-1}{5}, \frac{1}{5}\right\}$ **43.** $\left\{\frac{-3\sqrt{2}}{2}, \frac{3\sqrt{2}}{2}\right\}$ **45.** $\left\{\frac{-\sqrt{15}}{3}, \frac{\sqrt{15}}{3}\right\}$ **47.** $\left\{\frac{-\sqrt{10}}{10}, \frac{\sqrt{10}}{10}\right\}$ **49.** $\{-8, 6\}$

51. $\{-1, 4\}$ **53.** $\{8 - \sqrt{6}, 8 + \sqrt{6}\}$ **55.** $\left\{\frac{1 - 3\sqrt{2}}{3}, \frac{1 + 3\sqrt{2}}{3}\right\}$ **57.** $\{-3 - 2\sqrt{7}, -3 + 2\sqrt{7}\}$

59. $x = -2y$ or $x = 5y$ **61.** $x = \frac{-3a}{4}$ or $x = \frac{a}{2}$ **63.** $x = -2q$ or $x = 6q$ **65.** $x = \frac{-3m}{2}$ or $x = \frac{2m}{7}$

67. $x = -a$ or $x = 2b$ **69.** $x = -4b$ or $x = 3a$ **71.** $x = \frac{3a}{2}$ or $x = \frac{b}{4}$ **73.** $x = \frac{-5a}{6}$ or $x = \frac{-2b}{7}$

75. $\{-2i, 2i\}$ **77.** $\{-i\sqrt{10}, i\sqrt{10}\}$ **79.** $\{-5i\sqrt{2}, 5i\sqrt{2}\}$ **81.** $\{6 - 10i, 6 + 10i\}$ **83.** $\left\{\frac{-i}{2}, \frac{i}{2}\right\}$ **85.** $\left\{\frac{-i\sqrt{3}}{3}, \frac{i\sqrt{3}}{3}\right\}$

87. $\left\{\frac{-i\sqrt{21}}{7}, \frac{i\sqrt{21}}{7}\right\}$ **89.** $\left\{\frac{2 - 2i\sqrt{3}}{3}, \frac{2 + 2i\sqrt{3}}{3}\right\}$

Section 6-1. Ten Review Exercises

1. -80 **2.** 80 **3.** $\{-4\}$ **4.** $t \geq -4$ 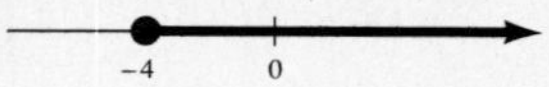

5. 7 and 7 **6.** $\{-16, 19\}$

7. $-16 < u < 19$ 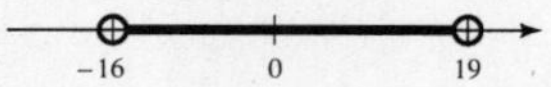

8. $-3t\sqrt[3]{4}$ **9.** $\dfrac{\sqrt[3]{5k^2}}{5k}$ **10.** $\sqrt[3]{2}$

Section 6-1. Supplementary Exercises

1. $\left\{\dfrac{-b}{a}, \dfrac{-d}{c}\right\}$ **3.** $\left\{\dfrac{-b \pm c}{a}\right\}$ **5.** $\left\{\dfrac{-b}{a}\right\}$ **7.** $\{-a, b\}$ **9.** $\{-y \pm \sqrt{k}\}$ **11. a.** 0 and 4 **b.** $\{0, 4\}$ **c.** 1 and 3 **d.** $\{1, 3\}$ **13. a.** -5 and -1 **b.** $\{-5, -1\}$ **c.** -4 and -2 **d.** $\{-4, -2\}$

Section 6-2. Practice Exercises

1. a. 64 **b.** $(x + 8)^2$ **3. a.** 25 **b.** $(y - 5)^2$ **5. a.** 144 **b.** $(z + 12)^2$ **7. a.** 1 **b.** $(w - 1)^2$ **9. a.** $\dfrac{9}{4}$ **b.** $\left(p - \dfrac{3}{2}\right)^2$ **11. a.** $\dfrac{1}{4}$ **b.** $\left(t + \dfrac{1}{2}\right)^2$ **13. a.** $\dfrac{81}{4}$ **b.** $\left(y + \dfrac{9}{2}\right)^2$ **15. a.** $\dfrac{4}{9}$ **b.** $\left(x - \dfrac{2}{3}\right)^2$ **17. a.** $\dfrac{25}{144}$ **b.** $\left(z - \dfrac{5}{12}\right)^2$ **19.** $\{-3 \pm \sqrt{2}\}$ **21.** $\{4 \pm 2\sqrt{6}\}$ **23.** $\{1 \pm \sqrt{11}\}$ **25.** $\{-7 \pm 4\sqrt{2}\}$ **27.** $\left\{\dfrac{-1 \pm \sqrt{13}}{2}\right\}$ **29.** $\{2 \pm 3\sqrt{2}\}$ **31.** $\left\{\dfrac{5 \pm \sqrt{17}}{2}\right\}$ **33.** $\left\{\dfrac{3 \pm 3\sqrt{5}}{2}\right\}$ **35.** $\{-2, 8\}$ **37.** $\left\{\dfrac{-3 \pm \sqrt{3}}{2}\right\}$ **39.** $\left\{\dfrac{3 \pm 2\sqrt{2}}{2}\right\}$ **41.** $\left\{\dfrac{1 \pm \sqrt{6}}{5}\right\}$ **43.** $\left\{\dfrac{-2}{3}, 3\right\}$ **45.** $\left\{-4, \dfrac{3}{2}\right\}$ **47.** $\left\{\dfrac{4 \pm \sqrt{26}}{2}\right\}$ **49.** $\left\{\dfrac{-1 \pm \sqrt{5}}{3}\right\}$ **51.** $\{1 \pm 3i\}$ **53.** $\{2 \pm 2i\sqrt{3}\}$ **55.** $\left\{\dfrac{3 \pm i\sqrt{19}}{2}\right\}$ **57.** $\{3 \pm 2i\}$ **59.** $\{-1 \pm i\sqrt{2}\}$

Section 6-2. Ten Review Exercises

1. 14 **2.** $t^2 + 5$ **3.** 14 **4.** 48 **5.** $24u$ **6.** 48 **7.** All x, where $x \geq 3$ **8.** All x **9.** $5x^2|y|\sqrt{3}$ **10.** $-5x^2y\sqrt[3]{2}$

Section 6-2. Supplementary Exercises

1. $\{0, -2b\}$ **3.** $\{0, b\}$ **5.** $\{-b \pm \sqrt{b^2 - c}\}$ **7.** $\left\{\dfrac{-b \pm \sqrt{b^2 - 4c}}{2}\right\}$ **9.** $\left\{\dfrac{y \pm yi\sqrt{3}}{2}\right\}$ **11. a.** x^2 **b.** $2x$ **c.** $2x$ **d.** 2 **13. a.** x^2 **b.** $3x$ **c.** $3x$ **d.** 3 **15. a.** x^2 **b.** $\dfrac{3x}{2}$ **c.** $\dfrac{3x}{2}$ **d.** $\dfrac{3}{2}$ **17. a.** x^2 **b.** $\dfrac{7x}{4}$ **c.** $\dfrac{7x}{4}$ **d.** $\dfrac{7}{4}$

Section 6-3. Practice Exercises

1. $\left\{-3, \dfrac{-5}{2}\right\}$ **3.** $\left\{-5, \dfrac{3}{4}\right\}$ **5.** $\left\{0, \dfrac{5}{3}\right\}$ **7.** $\left\{\dfrac{-4}{3}, \dfrac{4}{3}\right\}$ **9.** $\left\{\dfrac{-5}{2}, 2\right\}$ **11.** $\left\{-6, \dfrac{3}{5}\right\}$ **13. a.** $\{1 \pm \sqrt{3}\}$ **b.** 0.7; 2.7 **15. a.** $\{-3 \pm \sqrt{5}\}$ **b.** -0.8; -5.2 **17. a.** $\{2 \pm 3\sqrt{2}\}$ **b.** 6.2; -2.2 **19. a.** $\left\{\dfrac{2 \pm \sqrt{3}}{2}\right\}$ **b.** 1.9; 0.1 **21. a.** $\left\{\dfrac{-1 \pm 3\sqrt{5}}{3}\right\}$ **b.** 1.9; -2.6 **23. a.** $\left\{\dfrac{2 \pm 2\sqrt{2}}{5}\right\}$ **b.** 1.0; **c.** -0.2 **25. a.** $\{\sqrt{2} \pm \sqrt{3}\}$ **b.** 3.1; -0.3 **27.** $\{3 \pm 2i\}$ **29.** $\{-4 \pm i\}$ **31.** $\{-3 \pm i\sqrt{5}\}$ **33.** $\left\{\dfrac{3 \pm i\sqrt{2}}{5}\right\}$ **35.** $\left\{\dfrac{1 \pm 2i\sqrt{5}}{3}\right\}$ **37.** $\{\sqrt{3} \pm i\}$ **39.** $\{-i \pm i\sqrt{6}\}$ **41.** $\{-i, 4i\}$ **43.** $\left\{\dfrac{-i}{2}, i\right\}$ **45.** $\left\{-i, \dfrac{4i}{3}\right\}$ **47.** $\left\{\dfrac{-3 \pm \sqrt{15}}{2i}\right\}$ **49.** $\left\{0, \dfrac{8i}{5}\right\}$ **51.** 41 **53.** -59 **55.** 0 **57.** 116 **59.** -7 **61.** Rational and unequal **63.** Imaginary and unequal **65.** Rational and equal **67.** Irrational and unequal **69.** Imaginary and unequal **71.** Rational and equal **73.** Rational and unequal **75.** Irrational and unequal **77.** Imaginary and unequal **79.** Rational and unequal

Section 6-3. Ten Review Exercises

1. -5 **2.** -5 **3.** Yes **4.** $\left\{-4, \frac{5}{3}\right\}$ **5.** A perfect square **6.** $k^5 + 6k^2 + k - 3$ **7.** $7p - 1$ **8.** $2x^2 - 3xy + y^2$
9. -5 **10.** $\frac{m + n - 1}{m + n}$

Section 6-3. Supplementary Exercises

1. Yes **3.** Yes **5.** Yes **7.** Yes **9.** Yes **11.** Yes **13. a.** Two real roots **b.** $\{2 \pm 2\sqrt{3}\}$
15. a. One real root **b.** $\{3\}$ **17.** $2x^2 + 9x - 18 = 0$ **19.** $x^2 - 6x + 1 = 0$

Section 6-4. Practice Exercises

1. $\{81\}$ **3.** $\{45\}$ **5.** $\{5\}$ **7.** $\{0\}$ **9.** $\{-5\}$ **11.** $\{3\}$ **13.** $\{27\}$ **15.** $\{-4, -3\}$ **17.** $\{-1, 2\}$ **19.** $\{3, 5\}$
21. $\left\{\frac{2}{3}, 1\right\}$ **23.** $\left\{\frac{1}{3}, \frac{1}{2}\right\}$ **25.** $\varnothing$ **27.** $\varnothing$ **29.** $\varnothing$ **31.** $\varnothing$ **33.** $\{5\}$ **35.** $\{2\}$ **37.** $\{-4\}$ **39.** $\{4\}$
41. $\{2\}$ **43.** $\{2\}$ **45.** $\{2\}$ **47.** $\{42\}$ **49.** $\{-30\}$ **51.** $\{5\}$ **53.** $\{-5\}$ **55.** $\{10\}$ **57.** $\varnothing$ **59.** $\{-33\}$
61. $\left\{\frac{-1}{3}\right\}$ **63.** $\left\{\frac{2}{3}\right\}$ **65.** $\left\{\frac{1}{6}\right\}$ **67.** $\left\{\frac{1}{3}, 4\right\}$

Section 6-4. Ten Review Exercises

1. $2\sqrt{2} + 5\sqrt{3}$ **2.** $10\sqrt{6}$ **3.** $\frac{\sqrt{3}}{6}$ **4.** $2 - \sqrt{3}$ **5.** $\frac{2}{t}$ **6.** $t + 2$ **7.** $\left\{\frac{16}{9}\right\}$ **8.** $\left\{\frac{-4}{3}, \frac{4}{3}\right\}$ **9.** $(3t - a)(2t + b)$
10. $(a^2b^2 + 4)(ab + 2)(ab - 2)$

Section 6-4. Supplementary Exercises

1. $\left\{\frac{13}{4}\right\}$ **3.** $\varnothing$ **5.** $\{4\}$ **7.** $\{9\}$ **9.** $\{3\}$ **11.** $\{16\}$ **13.** $\{3\}$ **15.** $\{5\}$ **17.** $\left\{1, \frac{9}{8}\right\}$ **19.** $\left\{\frac{1}{4}\right\}$
21. a. $x = a^2$ **b.** Yes **c.** No **d.** Yes **23. a.** $x = (a - b)^2$ **b.** Yes **c.** No **d.** Yes **25. a.** $x = b^2 - a$
b. Yes **c.** No **d.** Yes **e.** Yes

Section 6-5. Practice Exercises

1. a. Yes **b.** Yes **3. a.** Yes **b.** No **5. a.** No **b.** Yes **7. a.** Yes **b.** Yes **9. a.** No **b.** No
11. a. Yes **b.** Yes **c.** Yes **13. a.** Yes **b.** No **c.** Yes **15.** $-5 < x < 2$

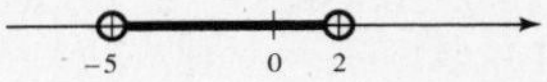

17. $x < 1$ or $x > 10$ 
19. $-6 \le x \le \frac{5}{2}$ 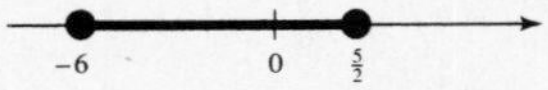
21. $x \le \frac{-9}{5}$ or $x \ge 3$ 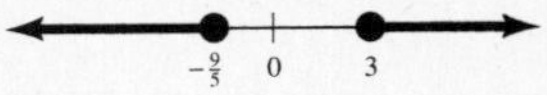
23. $-5 < x < 2$
25. $x < -2$ or $x > 6$ 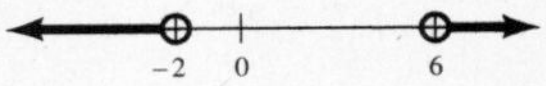
27. $x < 2$ or $x > 5$ 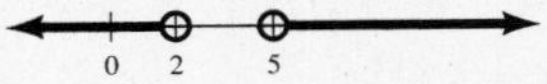
29. $2 \le x \le \frac{9}{2}$ 
31. $x \le -9$ or $x \ge \frac{-1}{3}$ 
33. $-4 < x < \frac{7}{2}$
35. $x < \frac{2}{3}$ or $x > 4$ 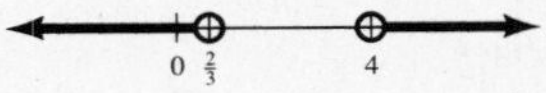
37. $\frac{-3}{2} \le x \le \frac{3}{2}$ 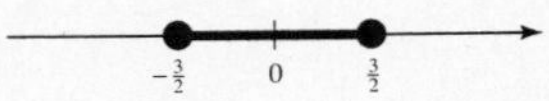
39. $x \le \frac{-5}{4}$ or $x \ge \frac{5}{4}$ 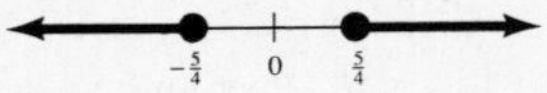
41. $0 < x < 12$
43. $x < 0$ or $x > 15$ 
45. $1 - \sqrt{2} \le x \le 1 + \sqrt{2}$ 
47. $x \le 5 - \sqrt{6}$ or $x \ge 5 + \sqrt{6}$ 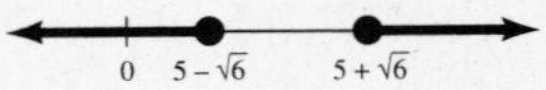
49. $-3 - \sqrt{5} < x < -3 + \sqrt{5}$
51. $-\sqrt{5} < x < \sqrt{5}$

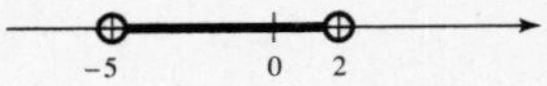

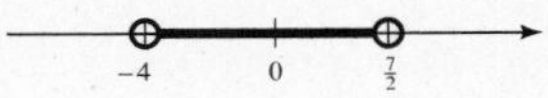

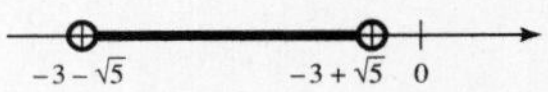

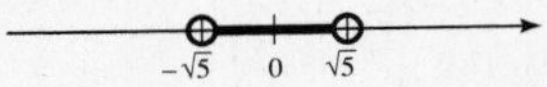

53. $x < 2 < \sqrt{5}$ or $x > 2 + \sqrt{5}$

$2 - \sqrt{5}$ 0 $2 + \sqrt{5}$

55. $x < -6 - \sqrt{7}$ or $x > -6 + \sqrt{7}$

$-6 - \sqrt{7}$ $-6 + \sqrt{7}$ 0

Section 6-5. Ten Review Exercises

1. $\left\{\frac{-1}{3}\right\}$ **2.** $\{0, 2\}$ **3.** $\{2\}$ **4.** $\left\{\frac{-8}{7}, 2\right\}$ **5.** $\varnothing$ **6.** $\{-21\}$ **7.** $25u^2v^2$ **8.** $25u^2 + 10uv + v^2$ **9.** $8u^3v^3$
10. $(4 + u^2v^2)(2 + uv)$

Section 6-5. Supplementary Exercises

1. R **3.** $\varnothing$ **5.** $R - \{3\}$ **7.** $\{-4\}$ **9.** R **11.** $-1 < x < 2$ **13.** $-4 < x < 3$ **15.** $x < 1$ or $x > 5$
17. $x \le -3$ or $x > \frac{1}{2}$ **19.** $-4 < x \le \frac{4}{3}$ **21. a.** $x > 2$ **b.** $x < 2$ **c.** $x \ge 2$ **d.** $x \le 2$
23. a. $x < 1$ or $x > 5$ **b.** $1 < x < 5$ **c.** $x \le 1$ or $x \ge 5$ **d.** $1 \le x \le 5$ **25. a.** $-3 < x < 0$ or $x > 2$
b. $x < -3$ or $0 < x < 2$ **c.** $-3 \le x \le 0$ or $x \ge 2$ **d.** $x \le -3$ or $0 \le x \le 2$ **27. a.** $-5 < x < -2$ or $x > 6$
b. $x < -5$ or $-2 < x < 6$ **c.** $-5 \le x \le -2$ or $x \ge 6$ **d.** $x \le -5$ or $-2 \le x \le 6$
29. a. $-2 < x < 0$ or $x > 3$ **b.** $x < -2$ or $0 < x < 3$ **c.** $-2 < x \le 0$ or $x > 3$ **d.** $x < -2$ or $0 \le x < 3$

Section 6-6. Practice Exercises

1. $y = \pm\sqrt{9 - x^2}$ **3.** $x = \frac{\pm\sqrt{9 - y^2}}{3}$ **5.** $y = \pm\sqrt{x^2 - 16}$ **7.** $t = \frac{-v \pm \sqrt{v^2 + 4as}}{2a}$ **9.** $r = \pm\frac{1}{2}\sqrt{\frac{A}{\pi}}$ **11.** 1 ft wide
13. 8 ft by 15 ft **15.** 12 m and 18 m **17.** 7.1 ft **19.** Harold takes 3 hr; Fred takes 6 hr
21. Valve A takes 12 hr; valve B takes 24 hr **23.** Machine A takes 21 min; machine B takes 28 min
25. Barbara takes about 14.6 hr; Stacy takes about 17.6 hr **27.** 175 mph **29.** 4 sec
31. 24 mph for ship A; 18 mph for ship B **33.** 10 mph

Section 6-6. Ten Review Exercises

1. 49 **2.** Yes **3.** $\{0, 6\}$ **4.** Both numbers are solutions **5.** -216 **6.** Yes **7.** $\{-4, 2\}$
8. Both numbers are solutions **9.** $(3k - 7)^2$ **10.** $(9u^2 + 25)(3u + 5)(3u - 5)$

Section 6-6. Supplementary Exercises

1. 6.8 sec **3.** 36.1 sec **5.** 8.0 sec **7.** 3.8 and 5.8 ft **9.** 3.5 and 6.9 ft **11.** 2.6, 5.6, and 6.1 in
13. $x = -1 \pm \sqrt{1 + y}$ **15.** $x = \pm\sqrt{3 + y}$ **17.** $x = -2 \pm \sqrt{y}$

Section 6-7. Practice Exercises

1. $\{\pm 3, \pm 2\}$ **3.** $\{\pm\sqrt{2}, \pm 5\}$ **5.** $\{\pm\sqrt{5}, \pm 2\sqrt{2}\}$ **7.** $\{\pm 5i, \pm 2\}$ **9.** $\left\{\pm\frac{1}{2}, \pm i\sqrt{6}\right\}$
11. $\left\{-1, 3, \frac{-1 \pm i\sqrt{3}}{2}, \frac{-3 \pm 3i\sqrt{3}}{2}\right\}$ **13.** $\{-4, 2, 2 \pm 2i\sqrt{3}, -1 \pm i\sqrt{3}\}$ **15.** $\left\{\frac{-1}{2}, \frac{1}{3}, \frac{-1 \pm i\sqrt{3}}{6}, \frac{1 \pm i\sqrt{3}}{4}\right\}$
17. $\left\{-1, 2, \frac{1 \pm i\sqrt{5}}{2}\right\}$ **19.** $\left\{-6, 1, \frac{-5 \pm \sqrt{17}}{2}\right\}$ **21.** $\left\{\frac{-3}{2}, 1, \frac{-1 \pm i\sqrt{19}}{4}\right\}$ **23.** $\left\{\frac{-1}{8}, 125\right\}$ **25.** $\left\{-27, \frac{8}{27}\right\}$
27. $\left\{\frac{-1}{32}, 32\right\}$ **29.** $\{0, 64\}$

Section 6-7. Ten Review Exercises

1. $3y^3t$ **2.** $\frac{u^4}{4}$ **3.** $v^2 - 2v + 5$ **4.** $\frac{6}{x^2}$ **5.** $9y^2 + 12y + 4$ **6.** $9y + 12y^{1/2} + 4$ **7.** $2z - 5$ **8.** $2\sqrt{z} - 5$
9. The integers are 11 and 13. **10.** 8 m and 15 m

Section 6-7. Supplementary Exercises

1. $\left\{\frac{-1}{2}, \frac{3}{2}\right\}$ **3.** $\left\{\frac{-3}{4}, 1\right\}$ **5.** $\left\{\pm\frac{1}{2}, \pm\frac{2}{3}\right\}$ **7.** $\left\{\frac{3}{5}, 2\right\}$ **9.** $\{-4, 1\}$ **11.** $\left\{\frac{3}{2}, 4\right\}$ **13.** $\{\pm\sqrt{2 \pm \sqrt{3}}\}$
15. $\left\{\pm\sqrt{\frac{2 \pm \sqrt{5}}{2}}\right\}$ **17.** $\left\{\pm\sqrt{\frac{-7 \pm 3\sqrt{5}}{2}}\right\}$ **19.** $\{\pm\sqrt{5 \pm \sqrt{7}}\}$

Chapter 6 Review Exercises

1. a. No **b.** Yes **3.** $\left\{\frac{-2}{3}, \frac{3}{2}\right\}$ **5.** $\{-5, 8\}$ **7.** $\{-3, 0, 3\}$ **9.** $\{-15, 15\}$ **11.** $\left\{\pm\frac{5\sqrt{2}}{2}\right\}$ **13.** $\{-4, 5\}$

15. $x = \frac{y}{2}$ or $x = -y$ **17.** $\{\pm 2i\sqrt{6}\}$ **19. a.** 25 **b.** $(u + 5)^2$ **21.** $\{6 \pm 2\sqrt{10}\}$ **23.** $\left\{\frac{-1}{3}, \frac{3}{2}\right\}$

25. $\{-5 \pm 2\sqrt{6}\}$ **27.** $\left\{-i, \frac{5i}{2}\right\}$ **29.** -0.3 and -3.7 **31. a.** 49 **b.** Rational and unequal **33.** $\{13\}$

35. $\{-2, 5\}$ **37.** $\{0, 2\}$ **39.** $-4 < x < 5$

41. $p \leq -2$ or $p \geq 2$ **43.** $-2 - \sqrt{2} \leq x \leq -2 + \sqrt{2}$

45. $r = \pm\sqrt{\frac{3V}{\pi h}}$ **47.** 5 ft by 13 ft **49.** 20 mph

Chapter 7 First-Degree Equations in Two Variables

Section 7-1. Practice Exercises

1.

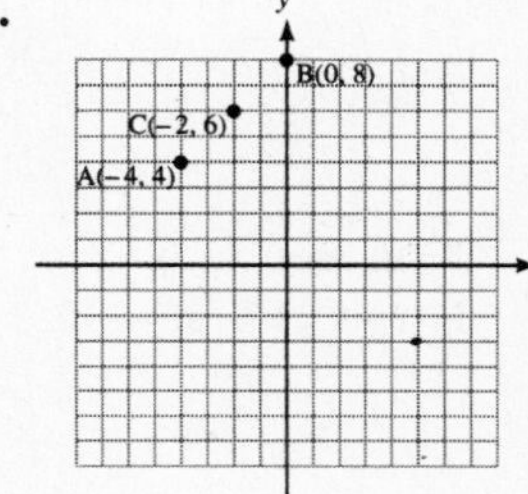

3.

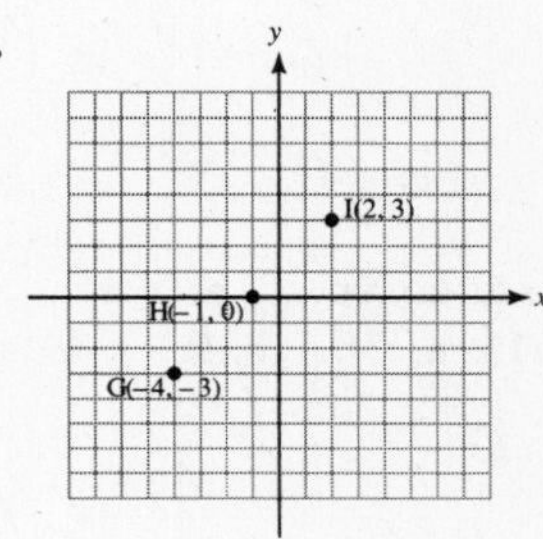

5.

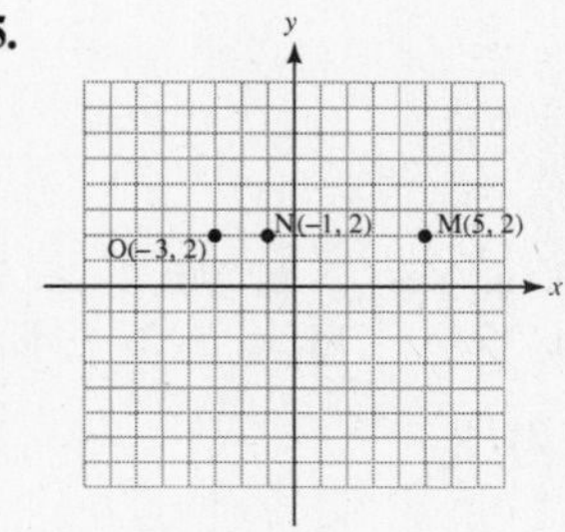

7.

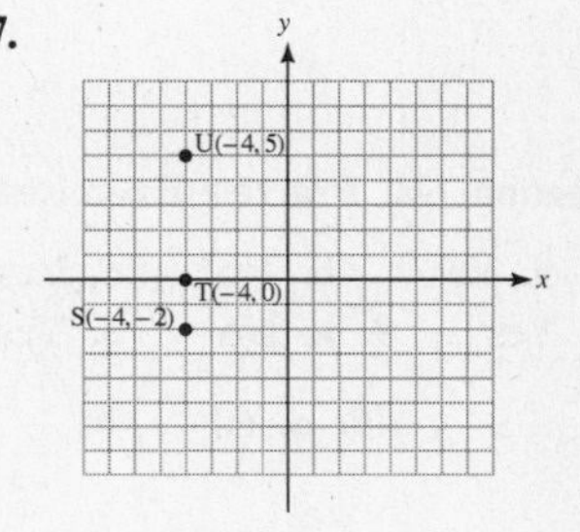

9. (6, 7) **11.** $(0, -3)$ **13.** $(-8, -2)$ **15.** (6, 0) **17.** $(4, -8)$ **19.** 5 **21.** 10 **23.** 9 **25.** $5\sqrt{2}$ **27.** $3\sqrt{5}$
29. 2 and -4 **31.** -4 and 8 **33.** -5 and 3 **35.** 2 and 4 **37.** $(-5, -1)$ **39.** $(-2, -4)$ **41.** $(6, -2)$
43. $\left(\frac{-3}{2}, -1\right)$ **45.** $\left(\frac{3}{2}, 1\right)$ **47.** $x = -3, y = 6$ **49.** $x = -4, y = 2$ **51.** $x = -19, y = 10$ **53.** $x = 3, y = -4$

Section 7-1. Ten Review Exercises

1. $49t^4$ **2.** $49 + 14t^2 + t^4$ **3.** $5u^2$ **4.** $5 + u^2$ **5.** $20p^2q^2$ **6.** $20p^2 + 12pq + q^2$ **7.** m^2n^3 **8.** $\frac{2m + 3n^2}{6}$

9. $\{5\}$ **10.** $\{0, 5\}$

Section 7-1. Supplementary Exercises

1. a.

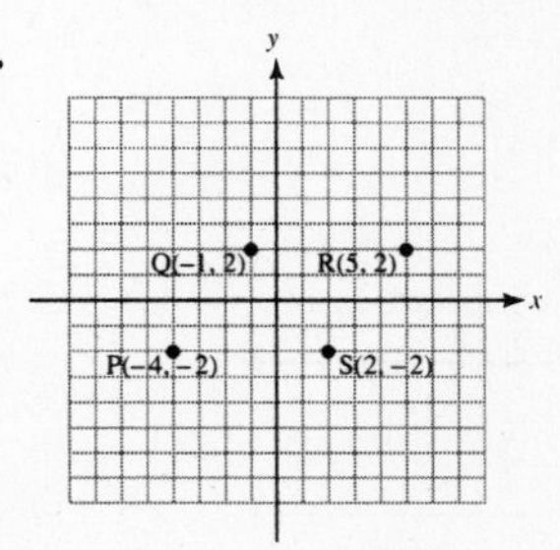

b. 22 **c.** $\sqrt{97}$ and 5 **d.** $\left(\frac{1}{2}, 0\right)$ **3. a.**

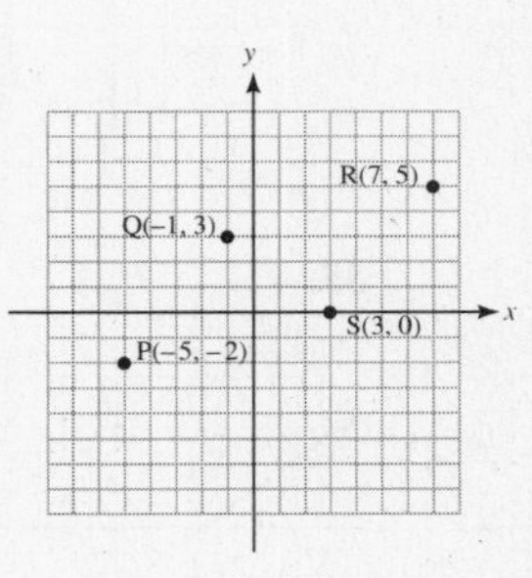

b. $4\sqrt{17} + 2\sqrt{41}$ **c.** $\sqrt{193}$ and 5 **d.** $\left(1, \frac{3}{2}\right)$ **5.** Yes **7.** No **9.** Yes

11. a.

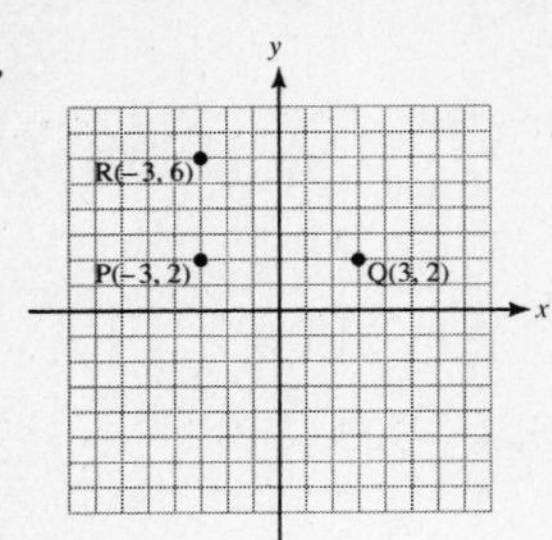

b. 4, 6 and $2\sqrt{13}$ **c.** Yes **d.** $(-3, 4)$, $(0, 2)$, and $(0, 4)$ **e.** Yes

13. a.

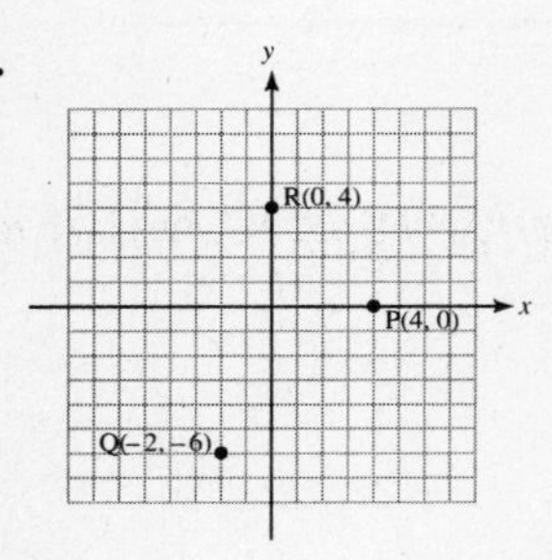

b. $4\sqrt{2}$, $6\sqrt{2}$ and $2\sqrt{26}$ **c.** Yes **d.** $(2, 2)$, $(1, -3)$ and $(-1, -1)$ **e.** Yes

Section 7-2. Practice Exercises

1. a. Yes **b.** No **c.** Yes **3. a.** Yes **b.** Yes **c.** No **5. a.** No **b.** Yes **c.** Yes **7. a.** Yes **b.** Yes **c.** Yes **9. a.** No **b.** Yes **c.** Yes **11. a.** -2 **b.** 6 **13. a.** 5 **b.** 0 **15. a.** -3 **b.** 5 **17. a.** -12 **b.** -9 **19. a.** 60 **b.** -70

21.

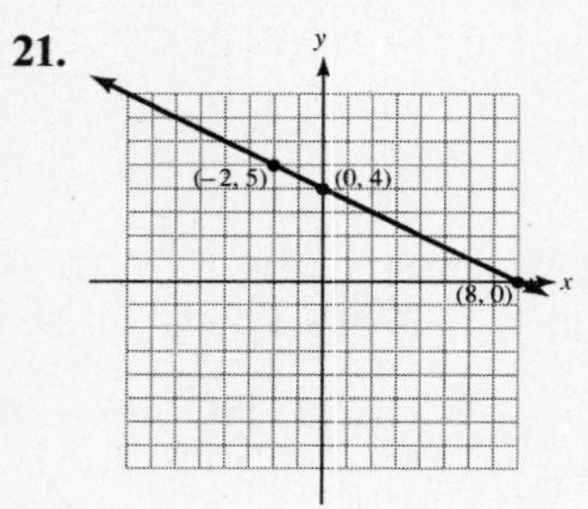

23.

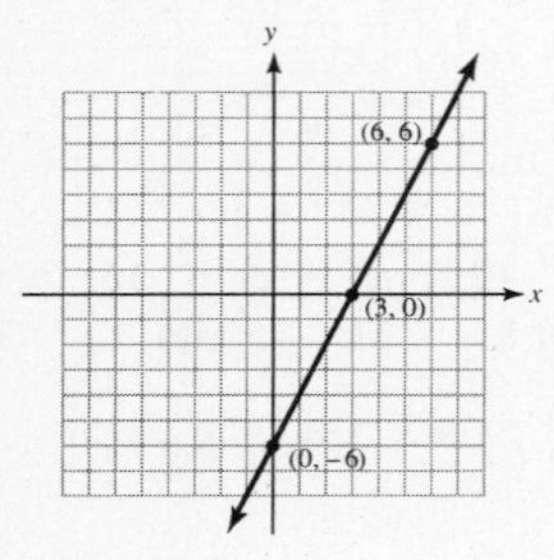

25.

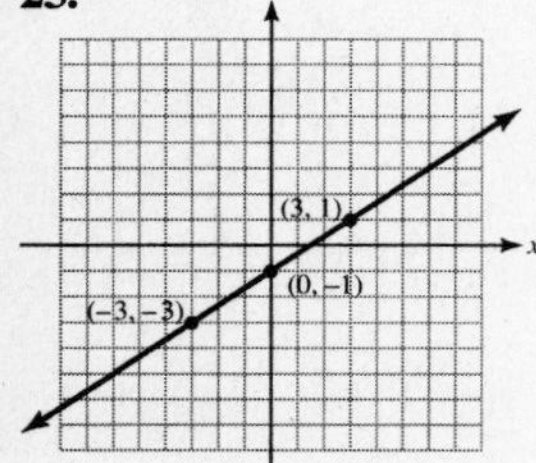

27.

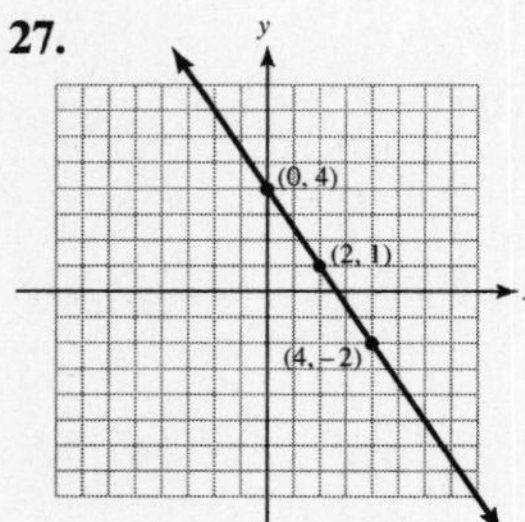

29.

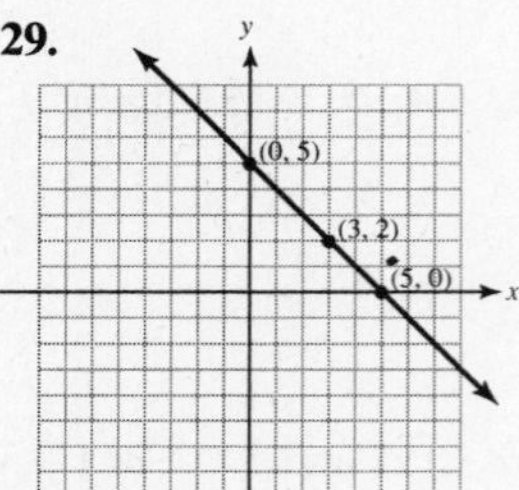

31.

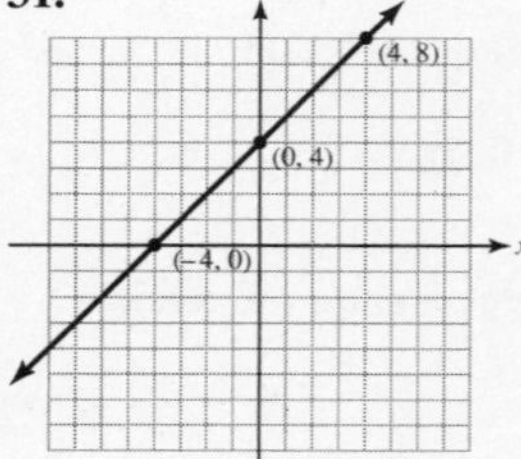

33.

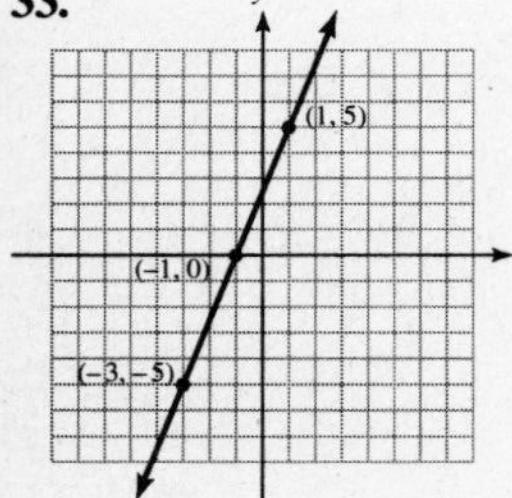

35.

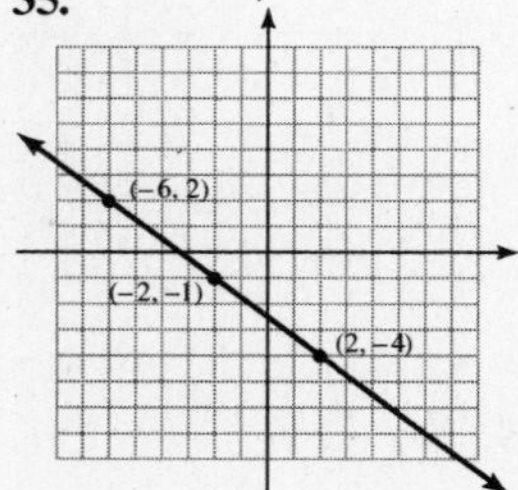

37.

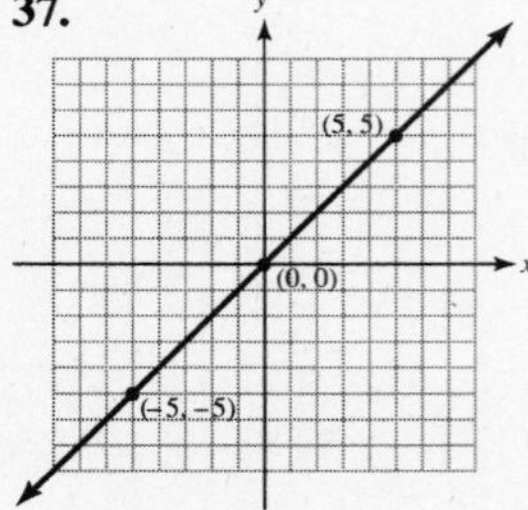

39.

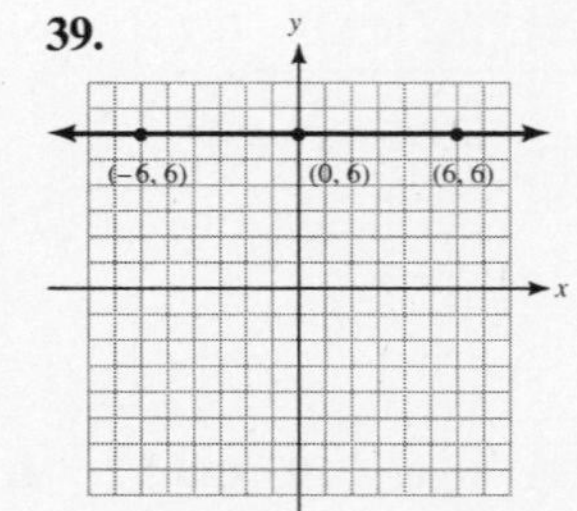

41.

43.

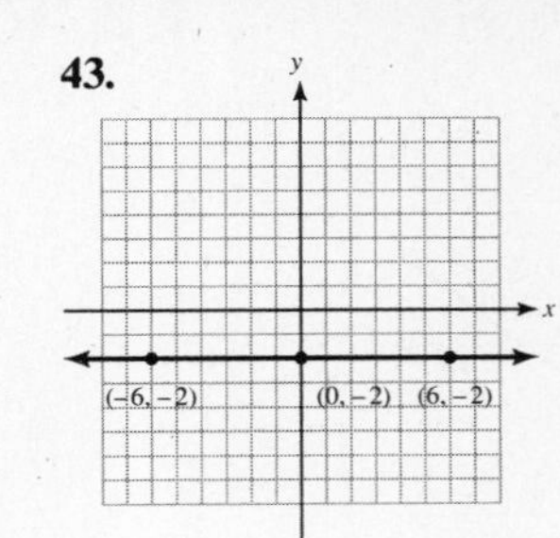

45.

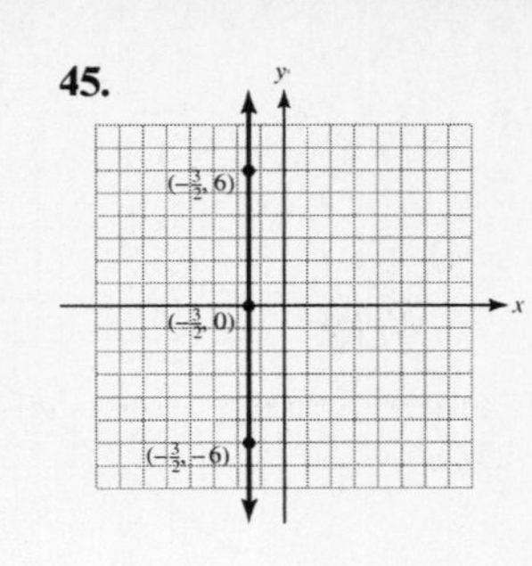

47.

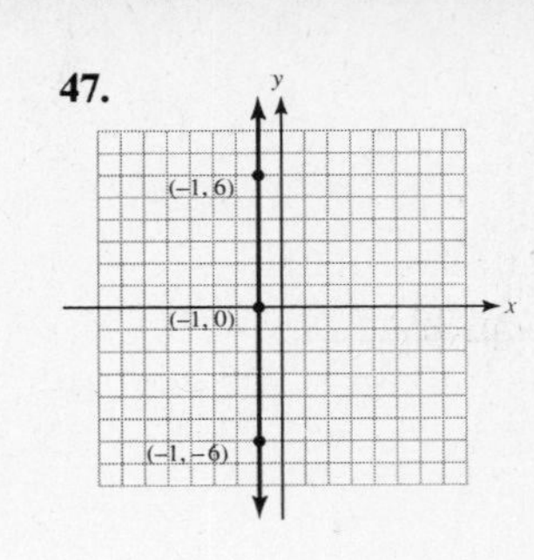

49.

51. $\frac{7}{4}$ **53.** $\frac{-3}{5}$ **55.** 5 **57.** -2 **59.** $\frac{3}{8}$ **61.** 0 **63.** Undefined **65.** $\frac{-2}{3}$

67.

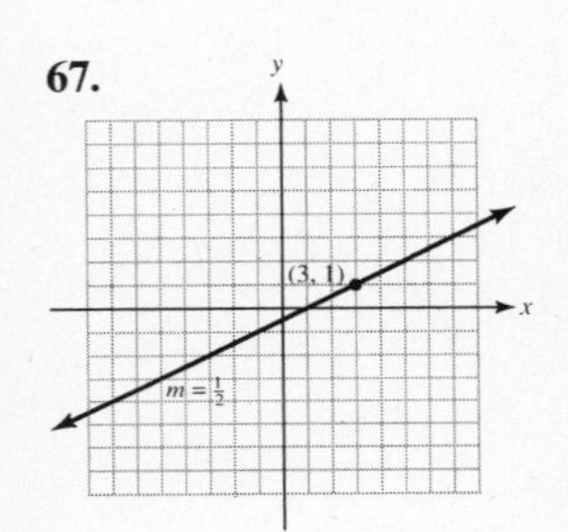

69.

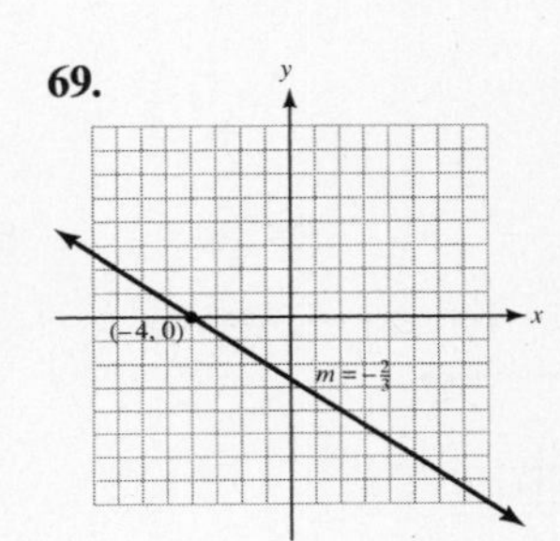

71.

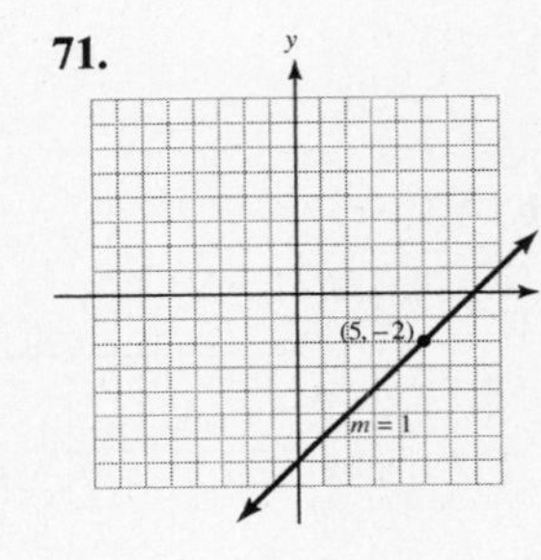

73.

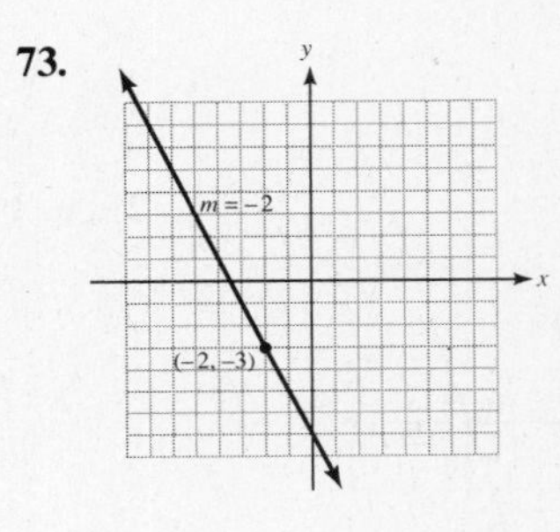

75.

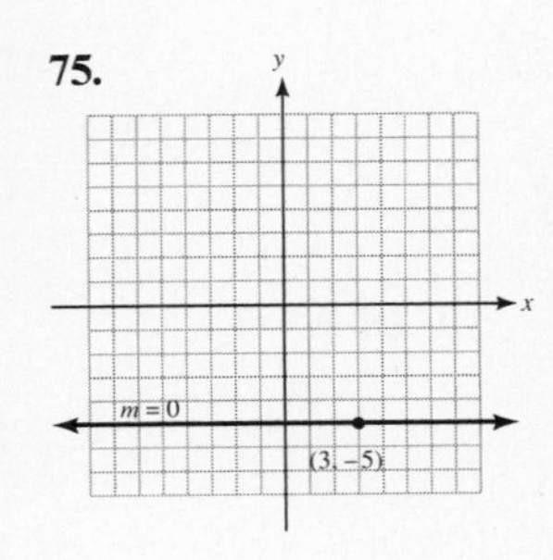

77.

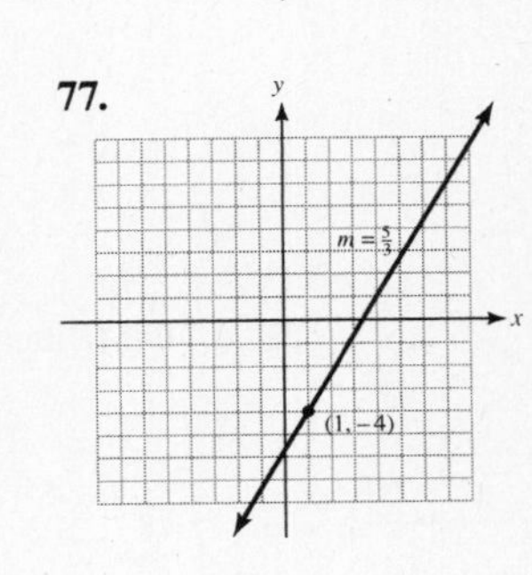

79.

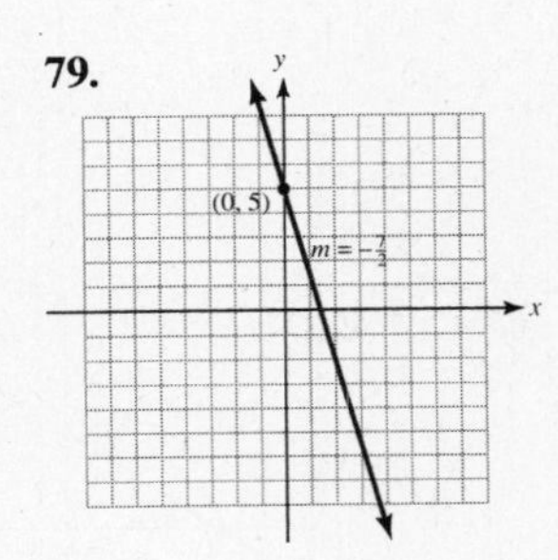

Section 7-2. Ten Review Exercises

1. 0 **2.** $13t - 26$ **3.** $\{2\}$ **4.** $t < 2$ **5.** $8t\sqrt[3]{2}$ **6.** $4t\sqrt[3]{2}$ **7.** $\frac{\sqrt{3}}{6}$ **8.** $\frac{\sqrt[3]{9}}{6}$ **9.** $\sqrt{5}$

10. $12\sqrt[3]{2t}$

Section 7-2. Supplementary Exercises

1. a. $C(3, 3)$ **b.** $C(3, -2)$ **3. a.** $A(-1, -17)$ **b.** $A(-1, -7)$ **5. a.** $A(2, -2)$ **b.** $A(2, 0)$ **7. a.** -4 **b.** 0

9. a. -8 **b.** -6 **11. a.** 12 **b.** $\frac{28}{5}$ **13.** 3 **15.** -4 **17.** 2 **19.** $\frac{1}{2}$

Section 7-3. Practice Exercises

1. a. 4 **b.** 1 **3. a.** 3 **b.** -15 **5. a.** 5 **b.** 2 **7. a.** -8 **b.** 9 **9. a.** No x-intercept **b.** -4

11. a. -4 **b.** No y-intercept **13. a.** 0 **b.** 0 **15. a.** -5 **b.** 3

c.

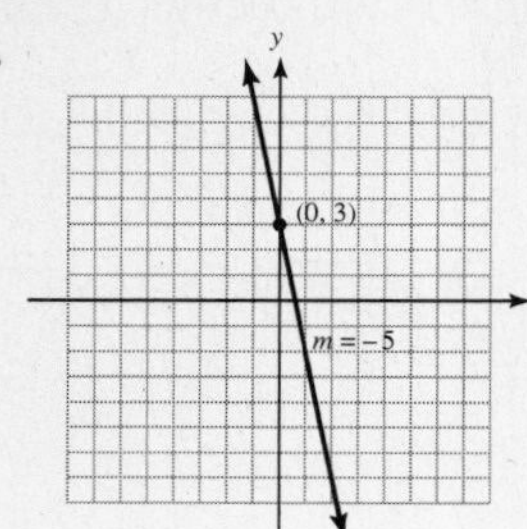

17. a. $\frac{3}{4}$ **b.** -3 **c.**

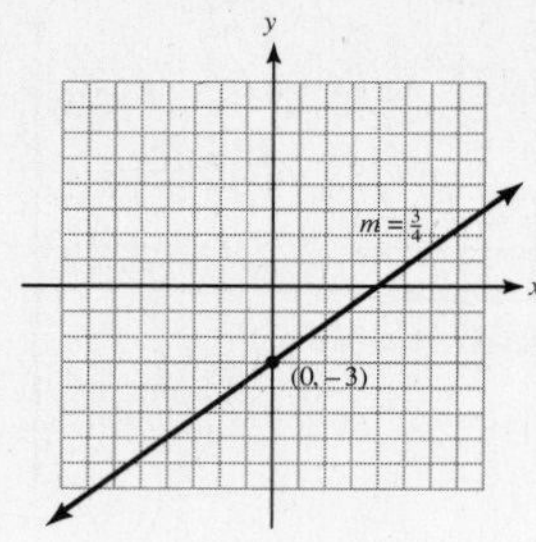

19. a. $\frac{-10}{9}$ **b.** -1 **c.**

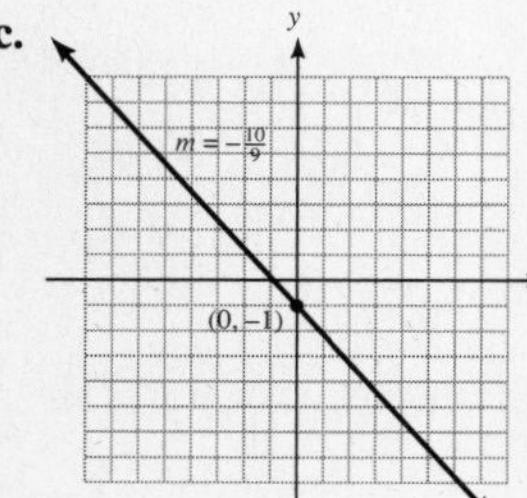

21. a. -1 **b.** 3 **c.**

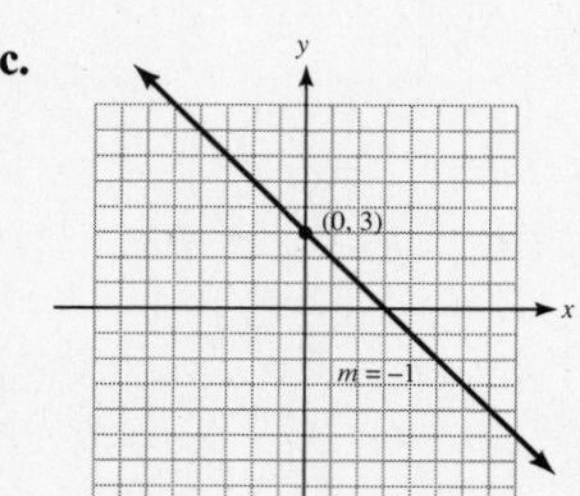

23. a. 3 **b.** 0 **c.**

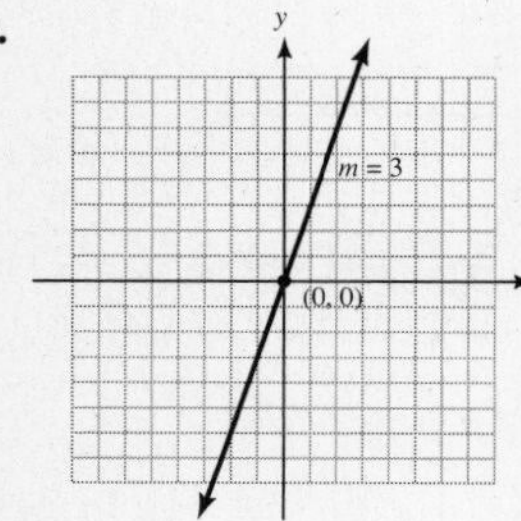

25. a. 0 **b.** 4 **c.**

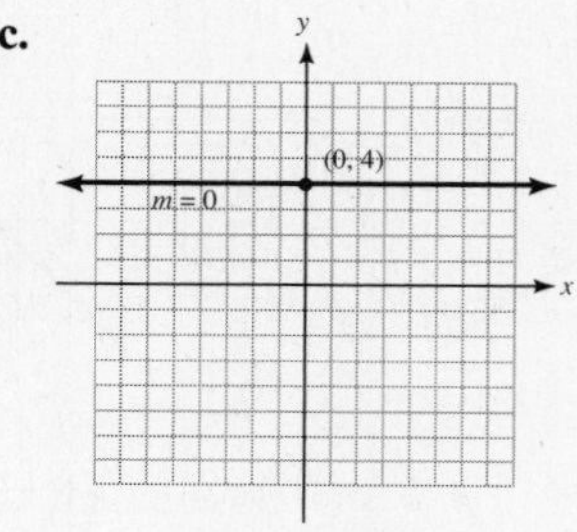

27. a. $\frac{-3}{2}$ **b.** 2 **c.**

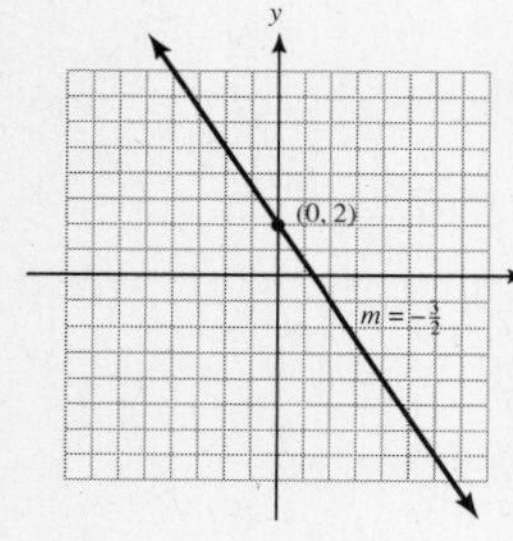

29. a. $\frac{2}{3}$ **b.** 4 **c.**

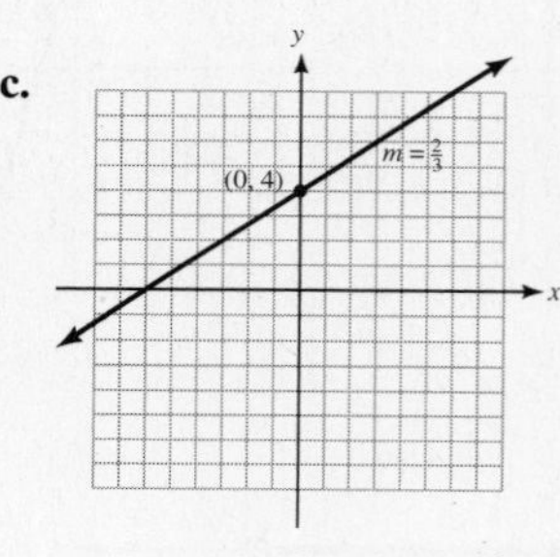

31. a. Undefined **b.** None **c.**

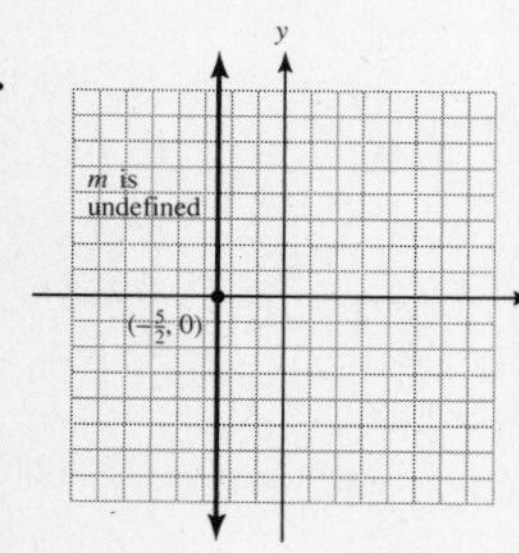

33. a. $\frac{-1}{6}$ **b.** -6 **c.**

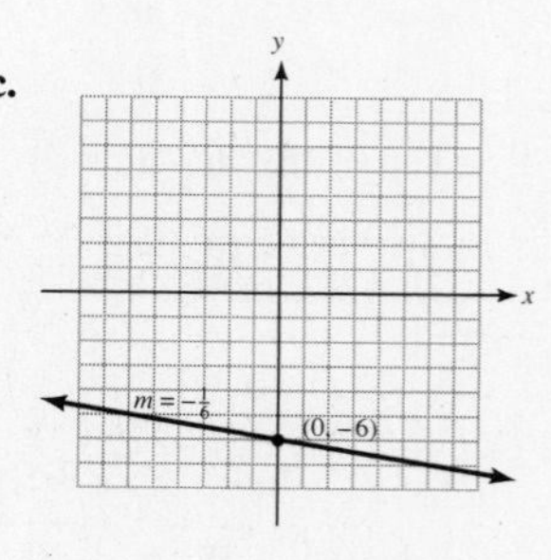

35. a. $\frac{3}{5}$ **b.** 3

c.

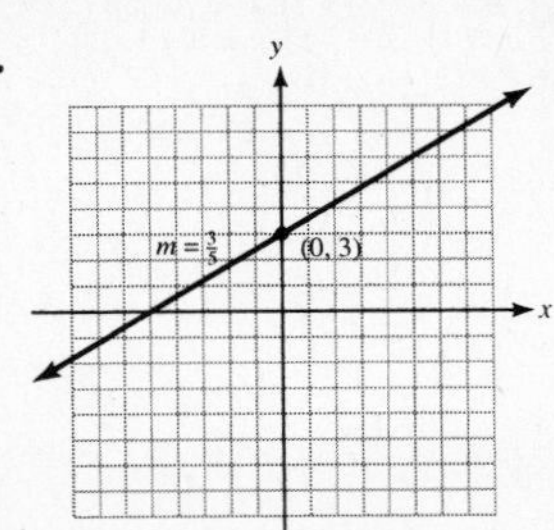

37. Parallel **39.** Perpendicular **41.** Neither **43.** Parallel **45.** Perpendicular

47. Parallel **49.** Perpendicular **51.** Neither **53. a.** $y - 4 = 3(x - 2)$ **b.** $3x - y = 2$ **55. a.** $y + 4 = \frac{2}{3}(x - 0)$ **b.** $2x - 3y = 12$ **57. a.** $y - 6 = \frac{-5}{2}(x + 10)$ **b.** $5x + 2y = -38$ **59. a.** $y - 8 = 0(x - 3)$ **b.** $y = 8$ **61. a.** $y - 2 = \frac{1}{2}(x + 4)$ **b.** $x - 2y = -8$ **63. a.** $y - 5 = \frac{5}{4}(x - 0)$ **b.** $5x - 4y = -20$ **65. a.** $y + 1 = \frac{-2}{3}(x - 2)$ **b.** $2x + 3y = 1$ **67. a.** $y + 2 = \frac{7}{8}(x + 3)$ **b.** $7x - 8y = -5$ **69. a.** $y + 6 = 0(x - 12)$ **b.** $y = -6$ **71. a.** $y - \frac{2}{3} = \frac{7}{5}\left(x - \frac{1}{2}\right)$ **b.** $42x - 30y = 1$ **73. a.** $2x - 3y = -10$ **b.** $3x + 2y = 11$ **75. a.** $x + 2y = 0$ **b.** $2x - y = -15$ **77. a.** $x - 3y = 0$ **b.** $3x + y = 0$ **79. a.** $y = -6$ **b.** $x = 3$

Section 7-3. Ten Review Exercises

1. $x + 5y$ **2.** $\sqrt{2} + 5\sqrt{3}$ **3.** $3x^2 + x + 1$ **4.** $3\sqrt[3]{5} + \sqrt{5} + 1$ **5.** 17 **6.** $\left(-\frac{5}{2}, 3\right)$ **7.** III **8.** $\frac{8}{15}$ **9.** Yes **10.** Yes

Section 7-3. Supplementary Exercises

1. a.

	Trip 0	Trip 1	Trip 2	Trip 3	Trip 4
x	24	20	16	12	8
y	20	17	14	11	8

b.

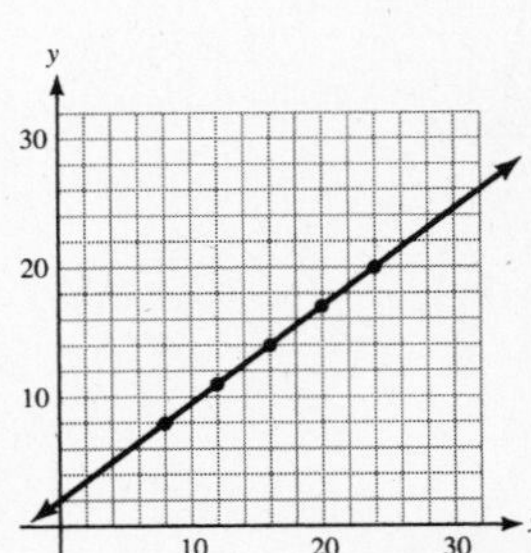

c. $y = \frac{3}{4}x + 2$ **d.** 2

3. a.

	Trip 0	Trip 1	Trip 2	Trip 3	Trip 4
x	29	25	21	17	13
y	29	25	21	17	13

b.

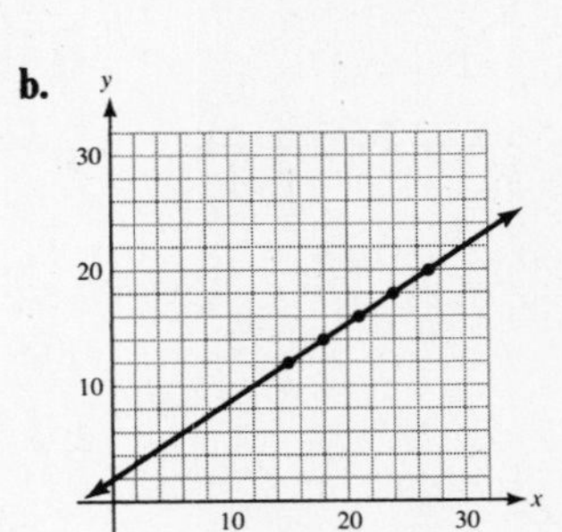

c. $y = x$ **d.** 0

5. $x = 13$ and $y = 1$ **7. a.**

	Year 1	Year 2	Year 3	Year 4	Year 5
x	2	6	10	14	18
y	2	5	8	11	14

b. $y = \frac{3}{4}x + \frac{1}{2}$ **c.** 23

9. a.

	Year 1	Year 2	Year 3	Year 4	Year 5
x	2	5	8	11	14
y	1	3	5	7	9

b. $y = \frac{2}{3}x - \frac{1}{3}$ **c.** 21

11. $y = \frac{6}{5}x$ **13.** $y = \frac{5}{7}x$

15. No, (0, 0) or 0 x-alops equals 0 y-aplots

Section 7-4. Practice Exercises

1. a. Yes **b.** Yes **3. a.** Yes **b.** No **5. a.** Yes **b.** No **7. a.** No **b.** No **9. a.** Yes **b.** Yes

11.

13.

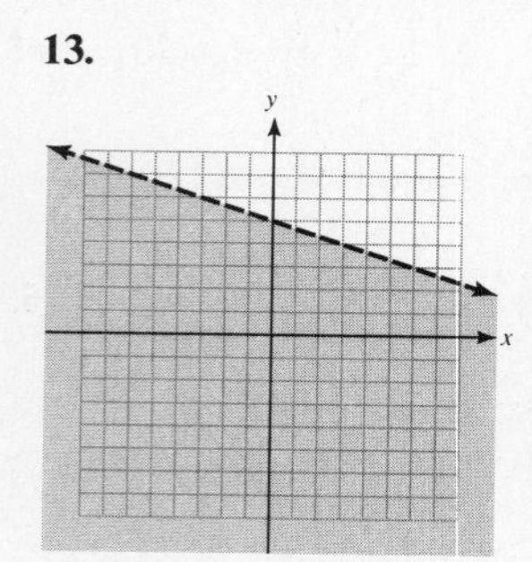

15.

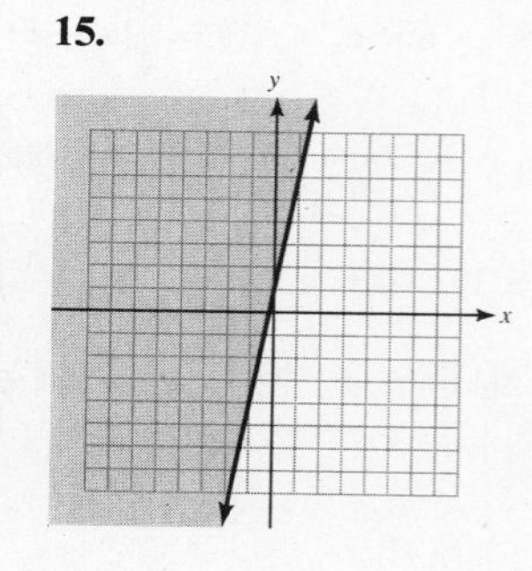

17.

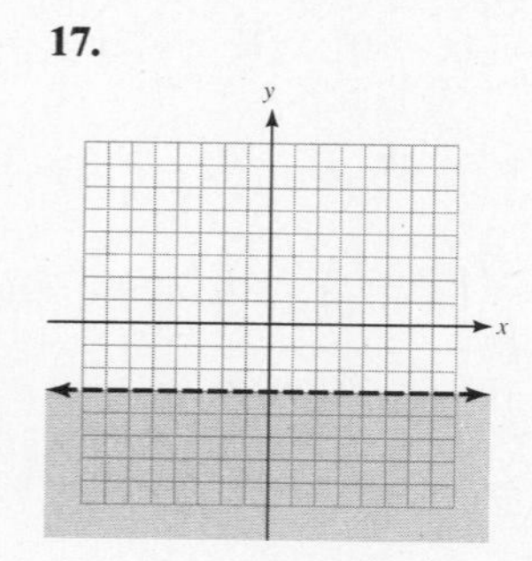

19.

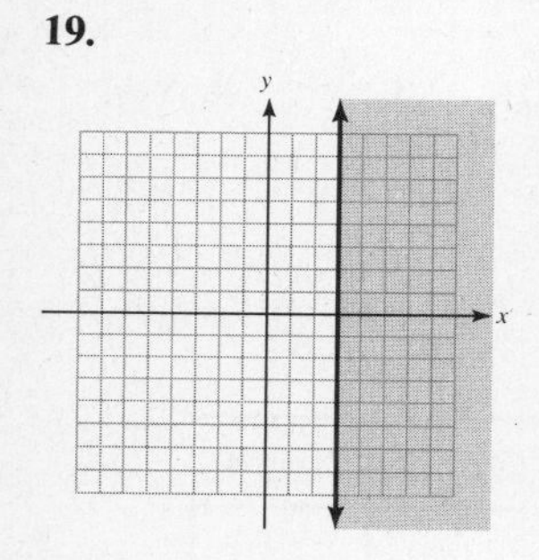

21.

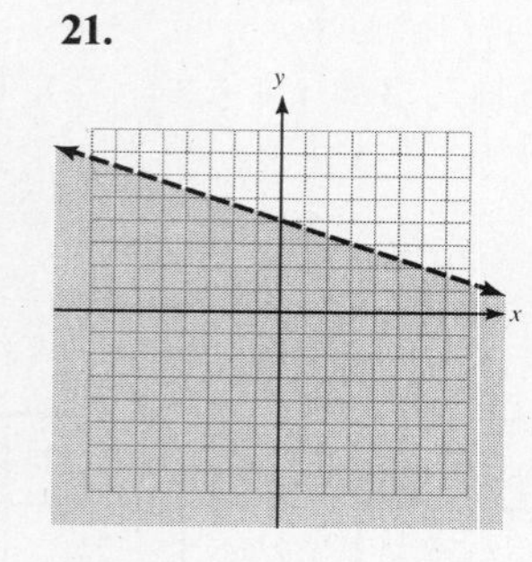

23.

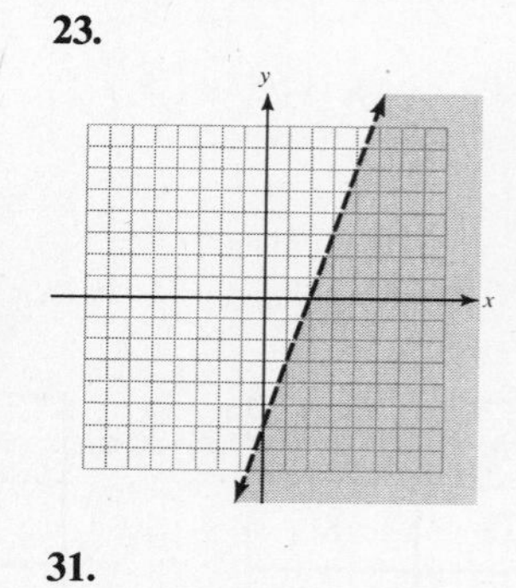

25.

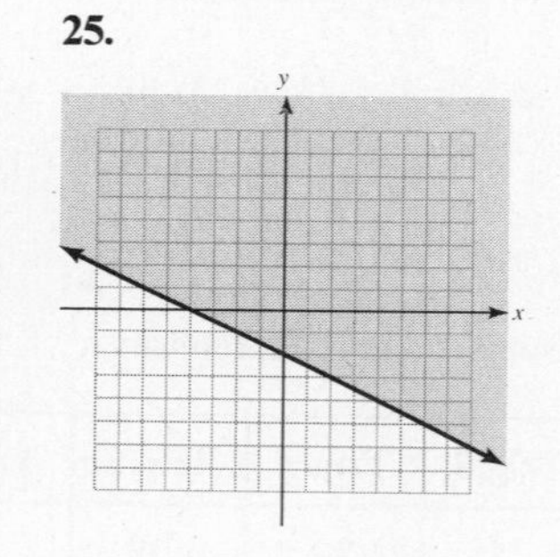

27.

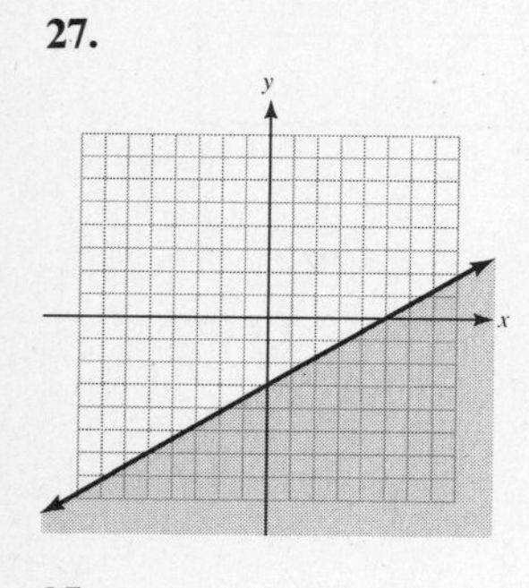

29.

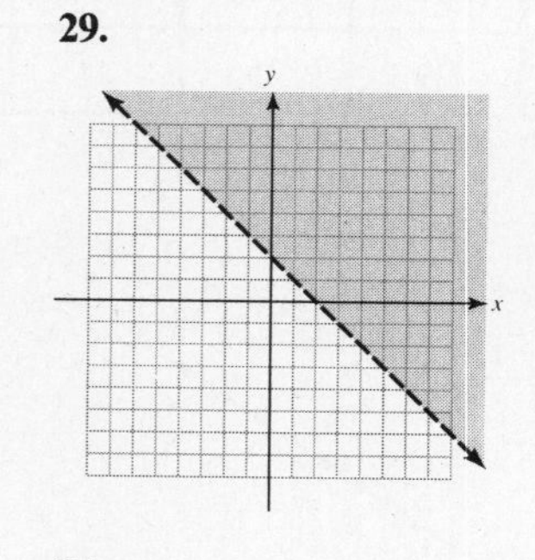

31.

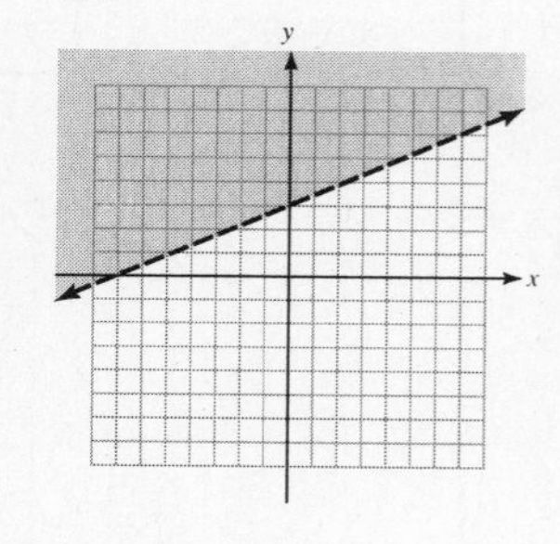

33.

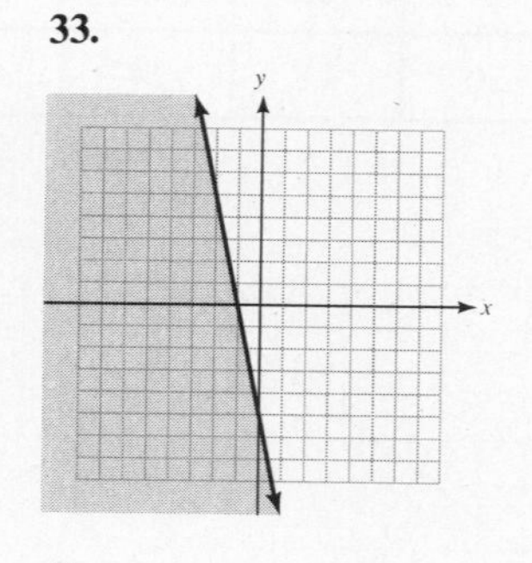

35.

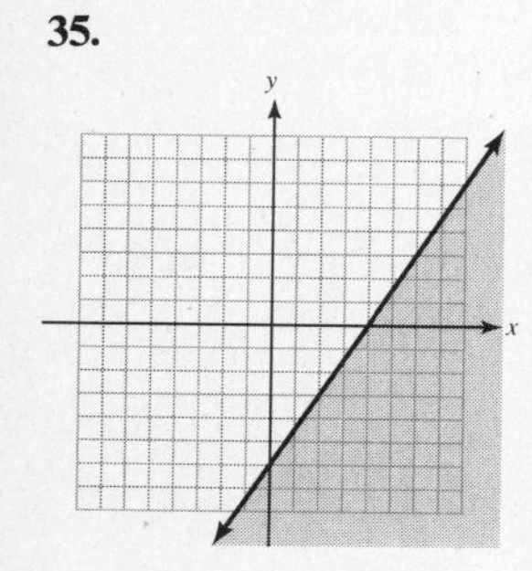

37.

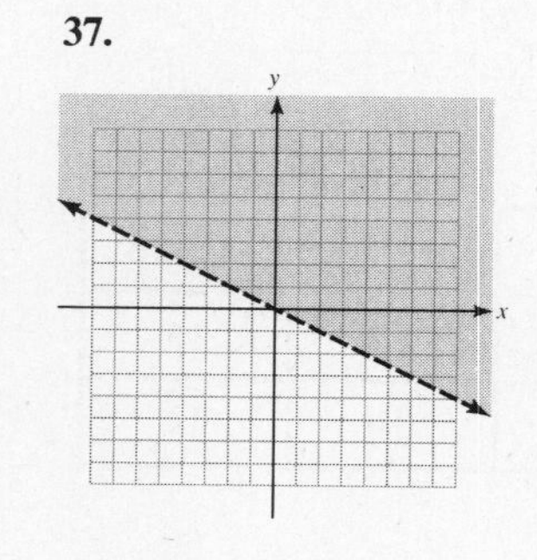

39.

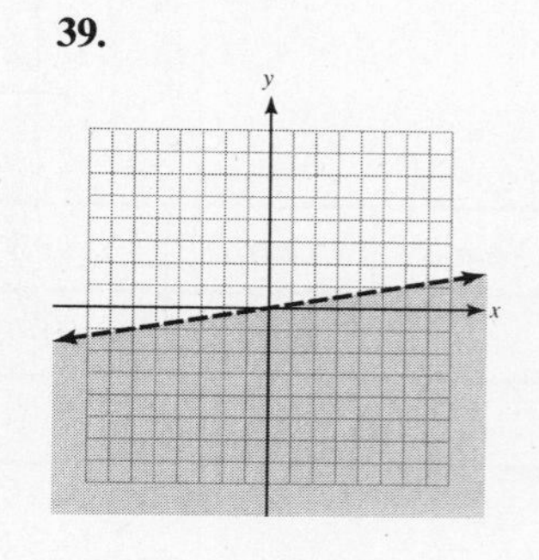

41.

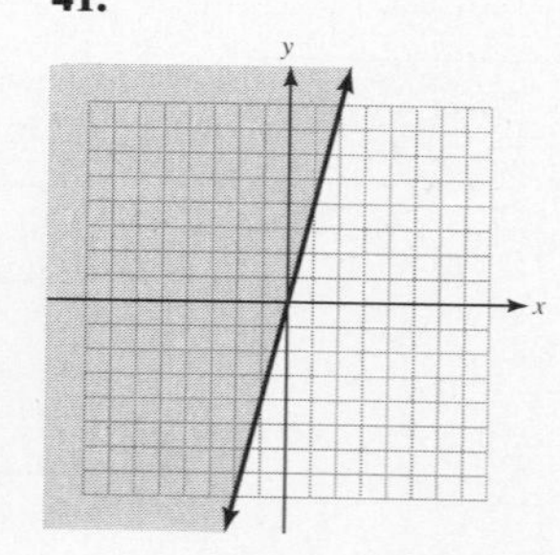

43.

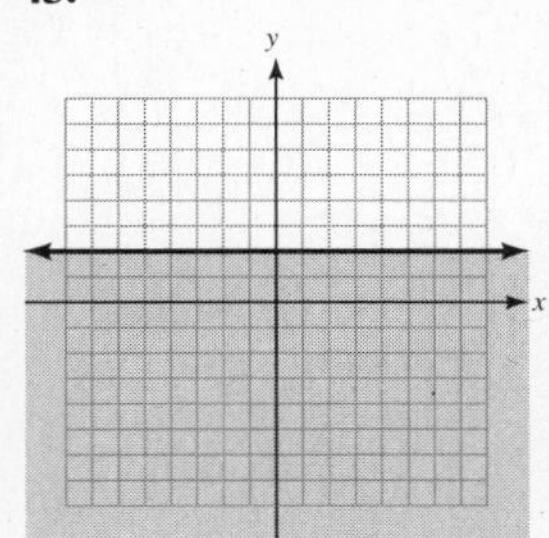

45.

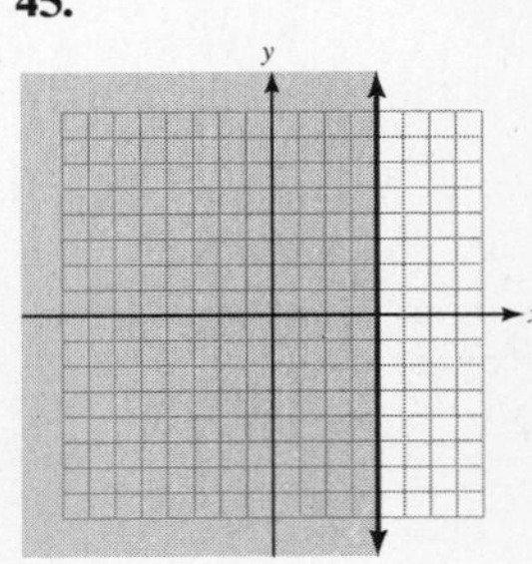

47.

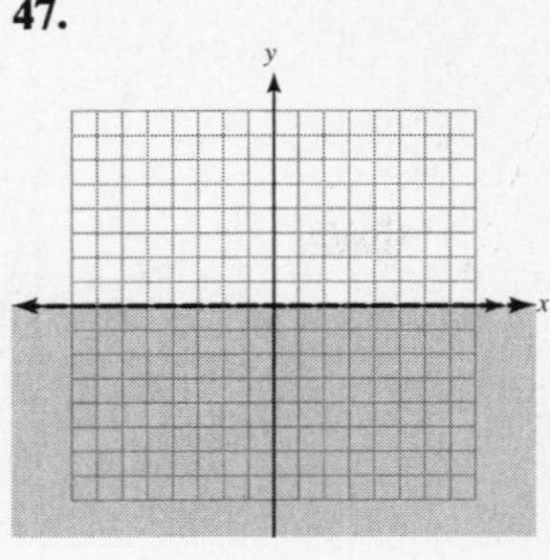

49.

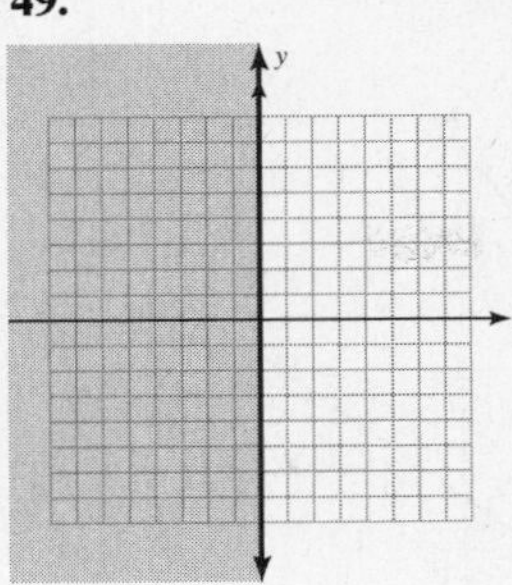

Section 7-4. Ten Review Exercises

1. $(3x + 2)(x - 4)$ **2.** $\left\{\frac{-2}{3}, 4\right\}$ **3. a.** 25 **b.** $(t - 5)^2$ **4.** $\{5 \pm 2\sqrt{3}\}$ **5.** $2u^2 + 7u - 15$ **6.** $2k + 7k^{1/2} - 15$

7. $-3 + 7\sqrt{6}$ **8.** $17 + 7i$ **9.** $\frac{2}{z^2} + \frac{7}{z} - 15$ **10.** $-3 - 3\sqrt[3]{6} + 10\sqrt[3]{36}$

Section 7-4. Supplementary Exercises

1. $5x - 4y < 20$ **3.** $x + y \geq 2$ **5.** $y < 4$ **7.** $4x - 3y \leq 0$ **9.** $3x + 2y \leq 2$ **11.** $x - 3y < 0$ **13.** $y > 4$

Section 7-5. Practice Exercises

1. a. (4, 12.56) **b.** (2.5, 7.85) **3. a.** ($10, $10.60) **b.** ($5.95, $6.31) **5. a.** ($12, $9.60) **b.** ($72.50, $58)
7. a. (6, 32) **b.** (102, 416) **9. a.** (4, 20) **b.** (0, 36) **11.** Domain = $\{-2, -1, 0, 1, 2\}$; Range = $\{3, 4, 5, 6, 7\}$
13. Domain = $\{-4, -2, 0, 2, 4\}$; Range = $\{18, 6, 2\}$
15. Domain = $\{-3, -2, -1, 0, 1, 2, 3\}$; Range = $\{-44, -6, 8, 10, 12, 26, 64\}$
17. Domain = $\{-6, -4, -2, 0, 2, 4, 6\}$; Range = $\{2\}$ **19.** Domain = $\{-8\}$; Range = $\{-10, -5, 0, 5, 10\}$ **21.** Function
23. Not a function **25.** Function **27.** Function **29. a.** $(-4, -17)$ **b.** $(0, -5)$ **31. a.** $(-6, 28)$ **b.** $(-11, 43)$
33. a. (4, 8) **b.** $\left(\frac{-1}{2}, 8\right)$ **35. a.** $\left(-6, \frac{-15}{4}\right)$ **b.** $\left(\frac{9}{4}, \frac{7}{4}\right)$ **37.** $y = \frac{1}{8}x$ **39.** $y = x + 3$ **41.** $y = \frac{1}{8}x + \frac{1}{2}$
43. $y = \frac{-3}{2}x + 18$ **45.** $y = 2x + 18$ **47.** $y = \frac{5}{2}x - \frac{5}{2}$ **49. a.** and **c.** **b.** $y = \frac{1}{2}x - 3$

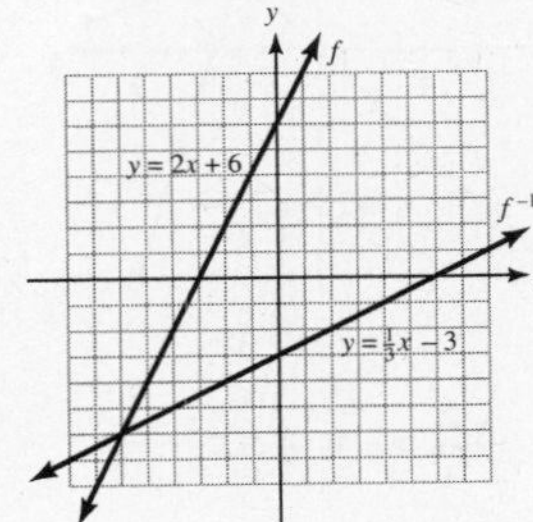

51. a. and **c.** **b.** $y = -3x - 6$

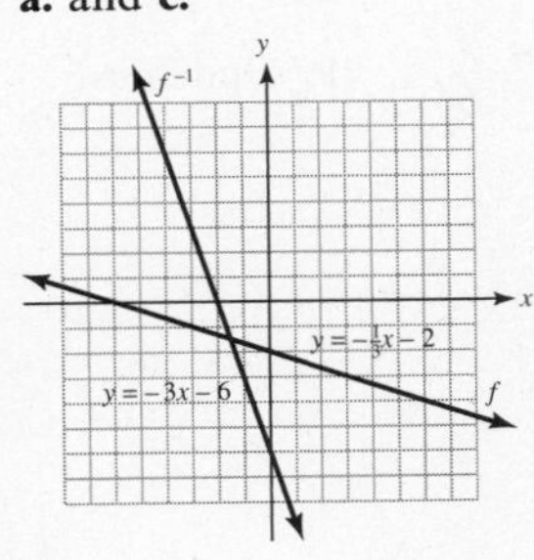

53. a. and **c.** **b.** $y = \frac{5}{6}x$

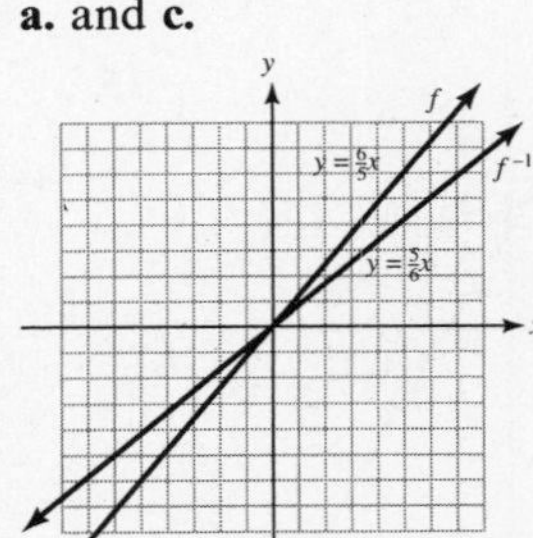

55. a. and **c.**

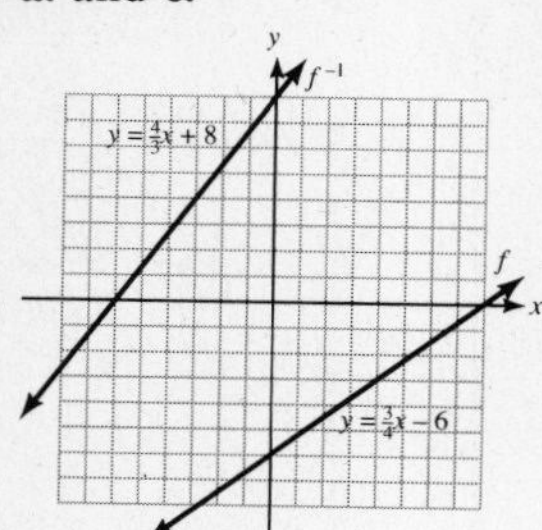

b. $y = \frac{4}{3}x + 8$

57. a. and **c.**

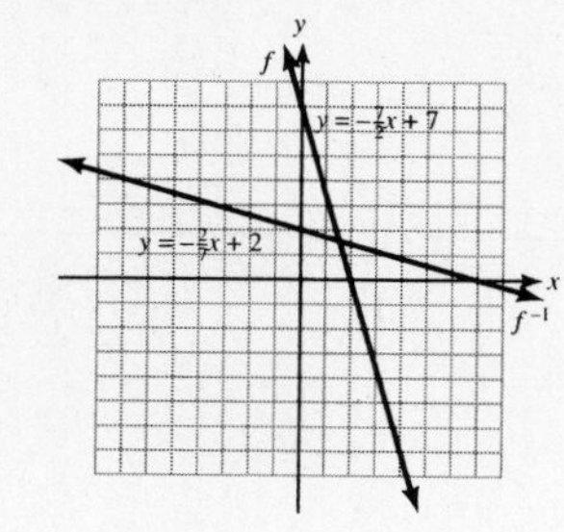

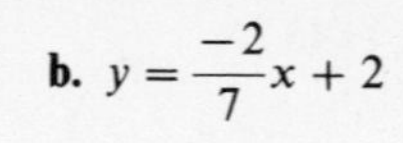

b. $y = \frac{-2}{7}x + 2$

59. a. and **c.**

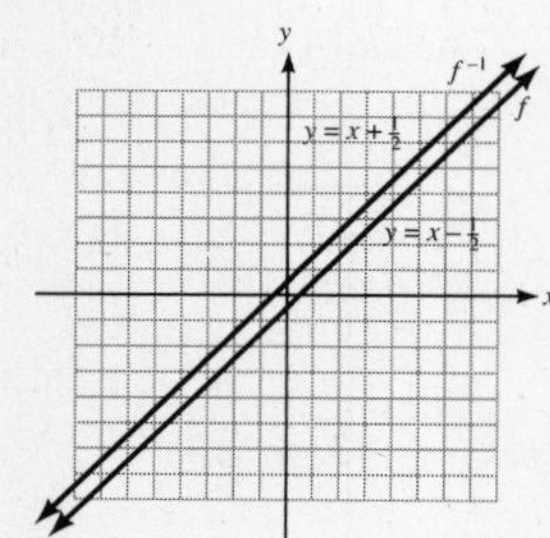

b. $y = x + \frac{1}{2}$

Section 7-5. Ten Review Exercises

1. $10x^2y^2$ **2.** $10x^2 + 7xy + y^2$ **3.** $25x^2y^2$ **4.** $25x^2 - 10xy + y^2$ **5.** $\frac{5}{2}$ **6.** $20x^3y - 5xy^3$ **7.** $\frac{-tu^2}{108}$
8. $3t^2 - 7tu - 8u^2$ **9.** $t - 5u$ **10.** $2t^2 + 7tu - 3u^2$

Section 7-5. Supplementary Exercises

1.

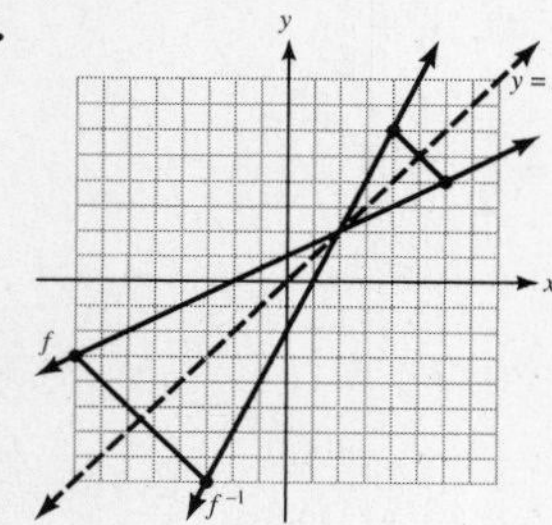

3.

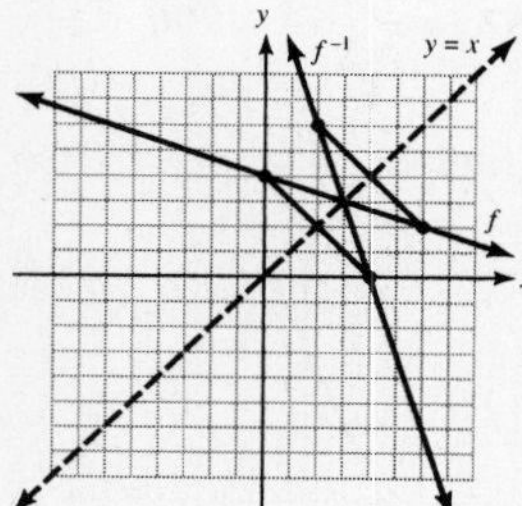

5.

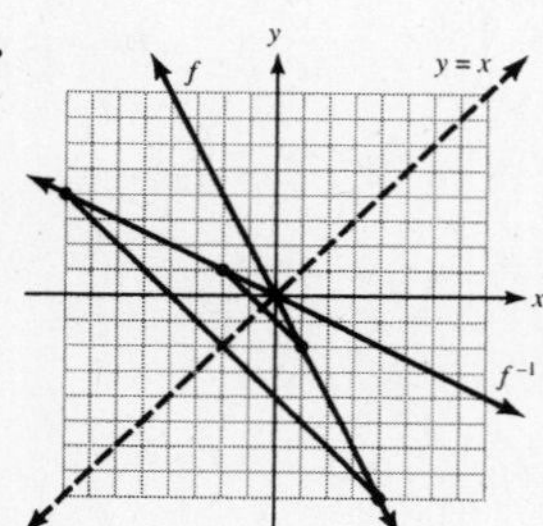

7.

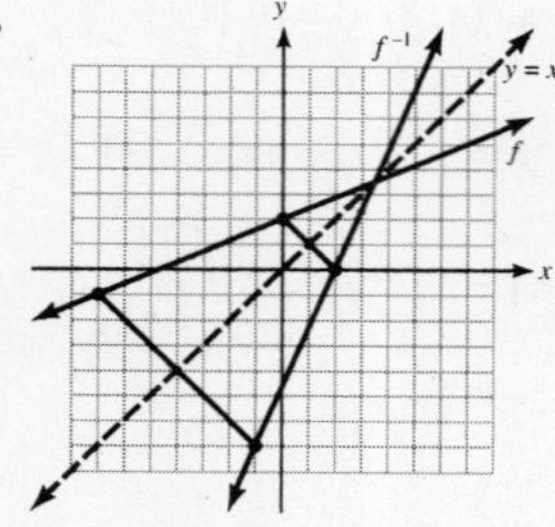

9. a. $y = \frac{1}{3}x - \frac{1}{3}$ **b.** $y = 3x + 1$ **11. a.** $y = \frac{3}{4}x + \frac{15}{4}$ **b.** $y = \frac{4}{3}x - 5$

Chapter 7 Review Exercises

1. 10 **3.** $x = 0$ or $x = -8$ **5.** $(-6, 4)$ **7.** $x = 2, y = -1$ **9. a.** Yes **b.** No **c.** Yes

11. a. (2, 1) **b.** (11, −1) **13.**

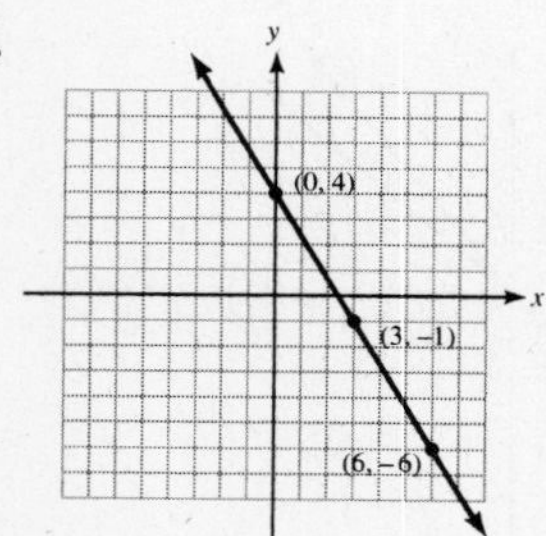

15.

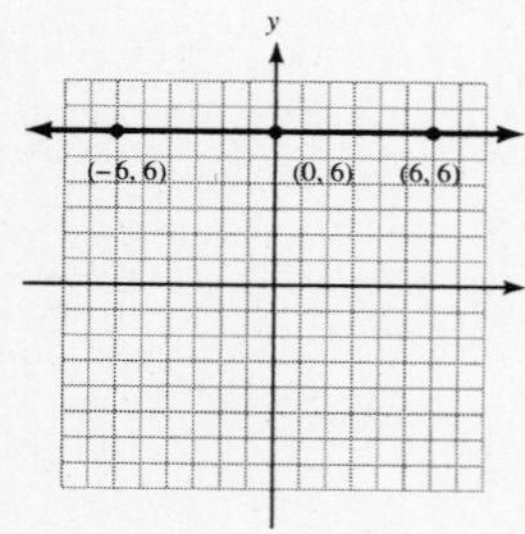

17. $\frac{3}{5}$ **19.** Undefined

21.

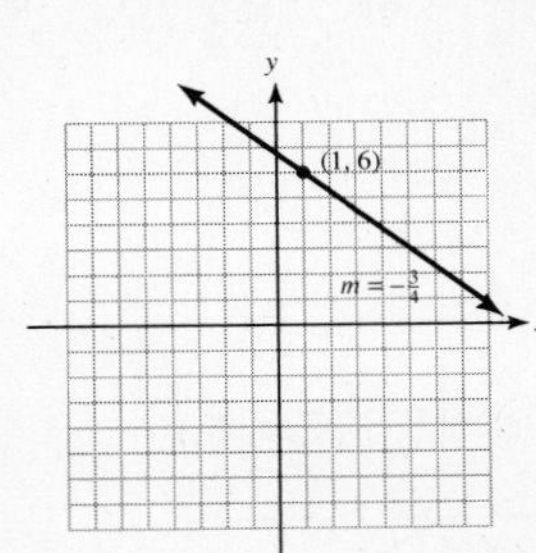

23. a. 3 **b.** -2 **25. a.** $\frac{-8}{3}$ **b.** 2 **27.**

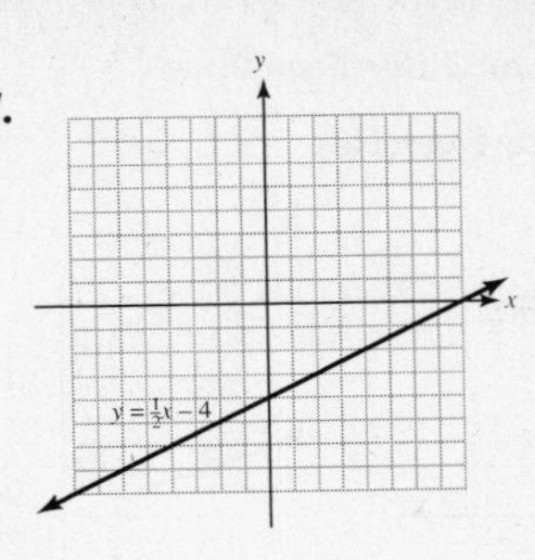

29. Parallel

31. $2x + y = 8$ **33.** $3x + 5y = -16$ **35.** $2x + 5y = -50$ **37.** $10x - 7y = 35$ **39. a.** Yes **b.** No

41.

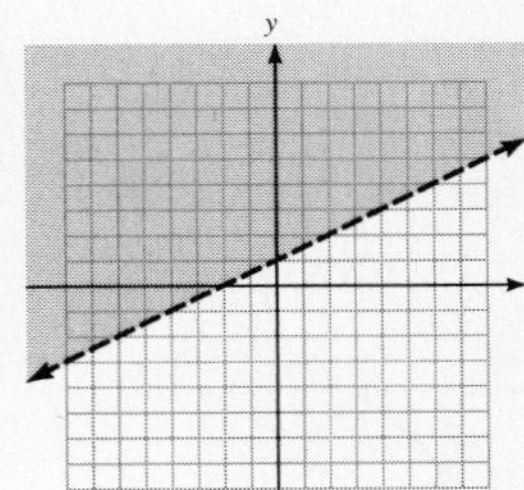

43.

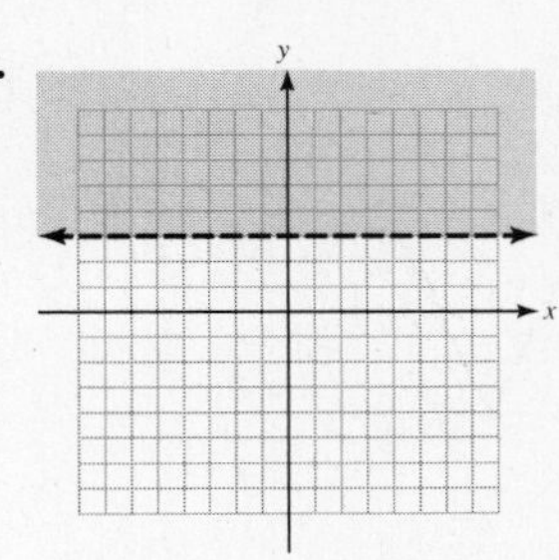

45. a. $\{-3, 7, 4, 1, -10\}$ **b.** $\{0, 2, -5, -9\}$ **c.** Function **47. a.** $\{4, -6, 0, 8\}$ **b.** $\{2\}$ **c.** Function **49. a.** -3 **b.** -30 **51.** $y = \frac{1}{6}x - 3$

53. **a** and **c.**

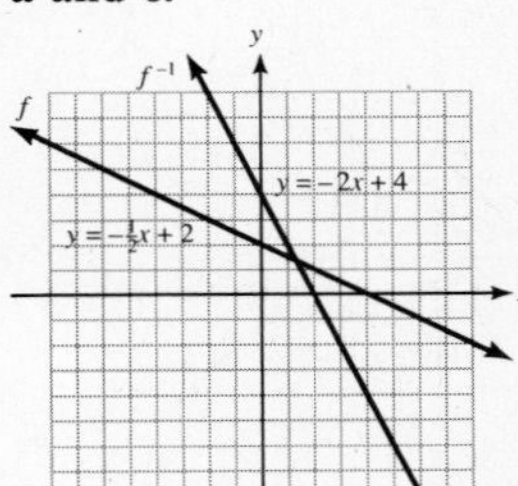

b. $y = -2x + 4$ **55. a.**

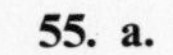

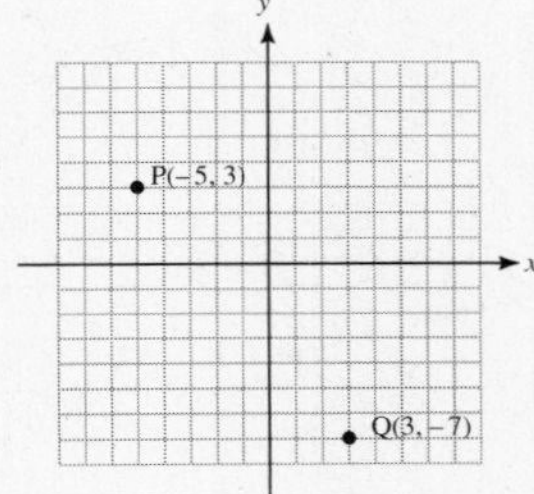

b. $2\sqrt{41}$ **c.** $(-1, -2)$ **d.** $\frac{-5}{4}$ **e.** $5x + 4y = -13$ **f.** $5x + 4y \geq -13$ **57. a.**

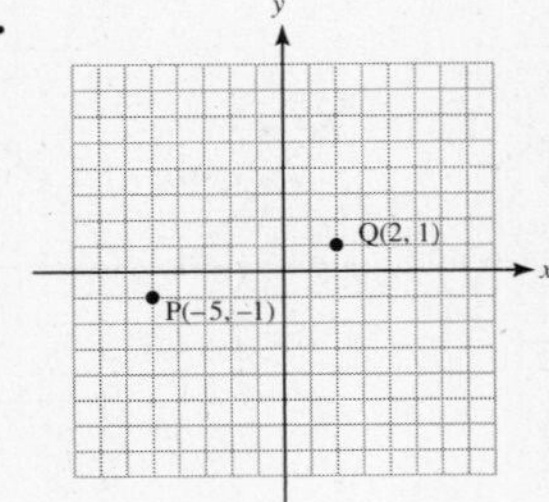

b. $\sqrt{53}$ **c.** $\left(\frac{-3}{2}, 0\right)$ **d.** $\frac{2}{7}$ **e.** $2x - 7y = -3$ **f.** $2x - 7y \geq -3$ **59. a.**

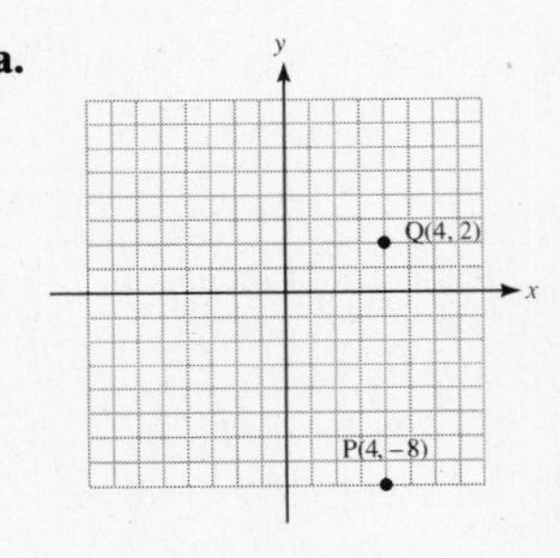

b. 10 **c.** $(4, -3)$ **d.** Undefined **e.** $x = 4$ **f.** $x \leq 4$

Chapter 8 Graphs of Other Functions

Section 8-1. Practice Exercises

1.

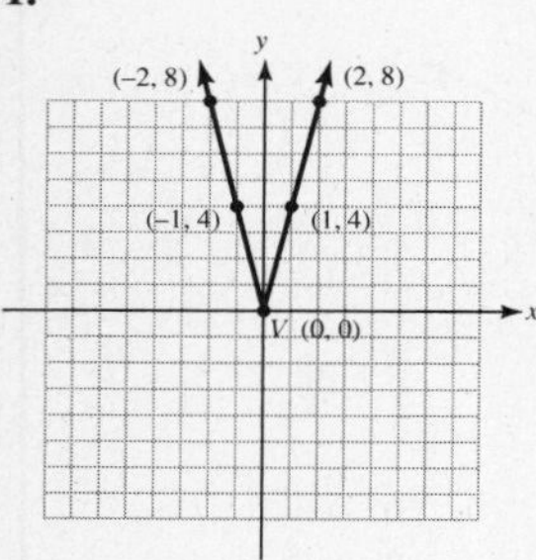

3.

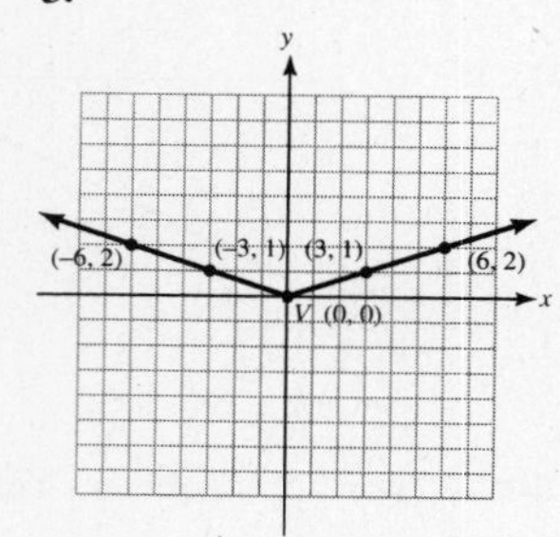

5.

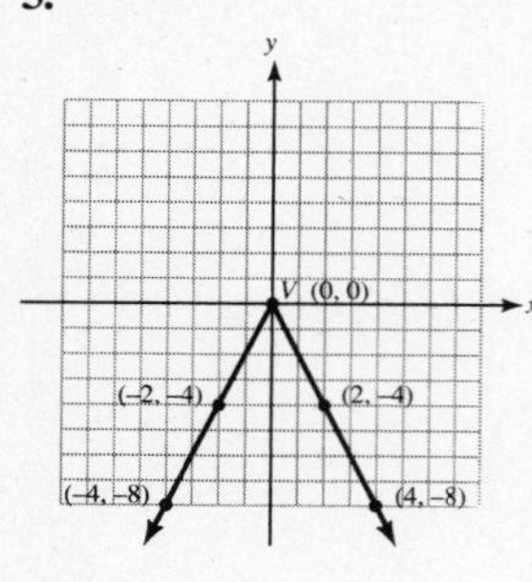

7.

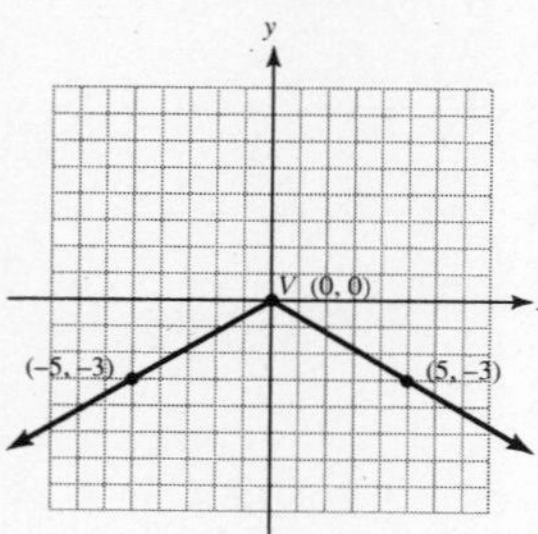

9.

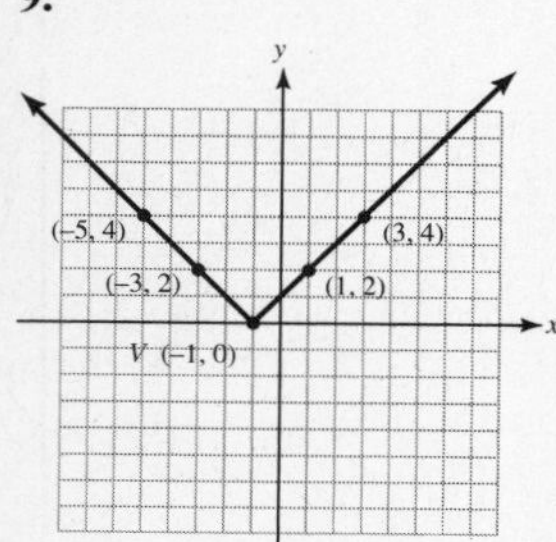

11.

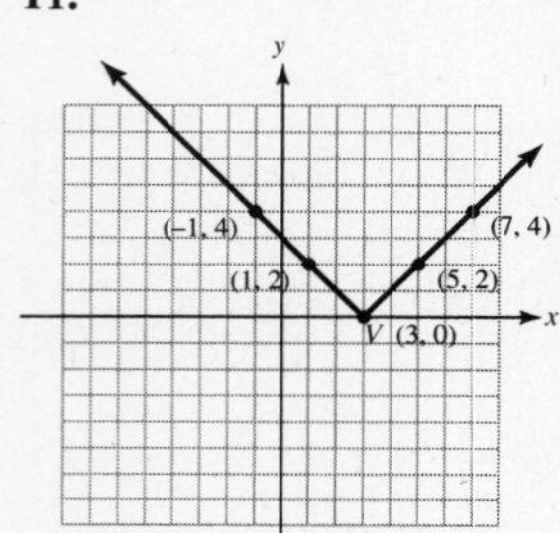

13.

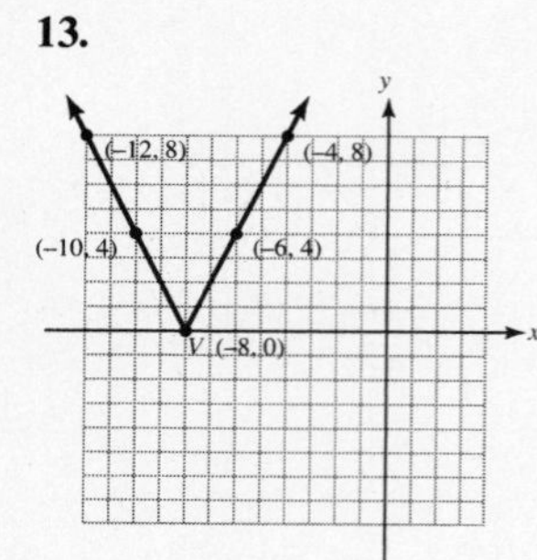

15.

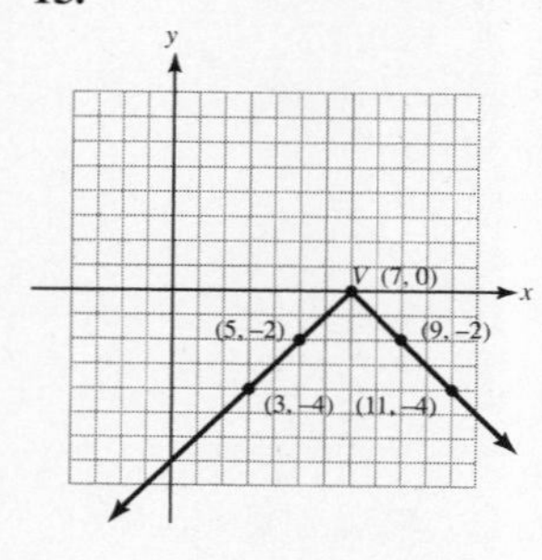

17.

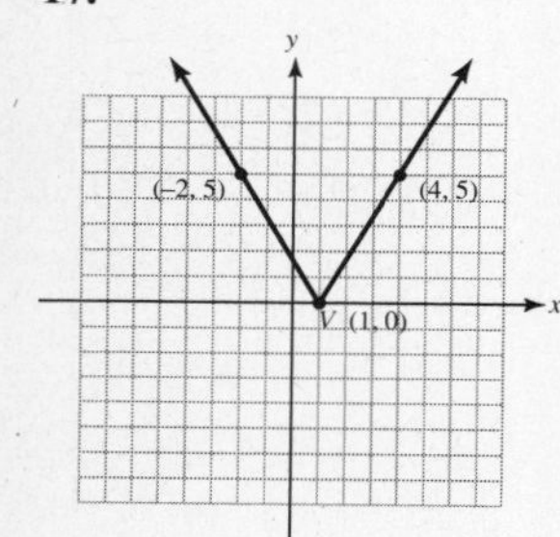

19.

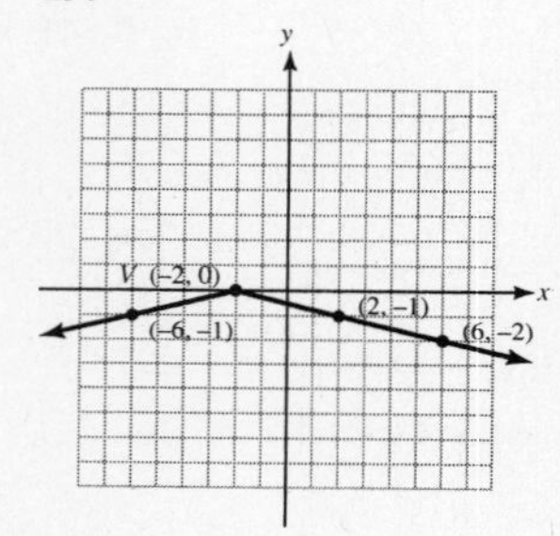

21.

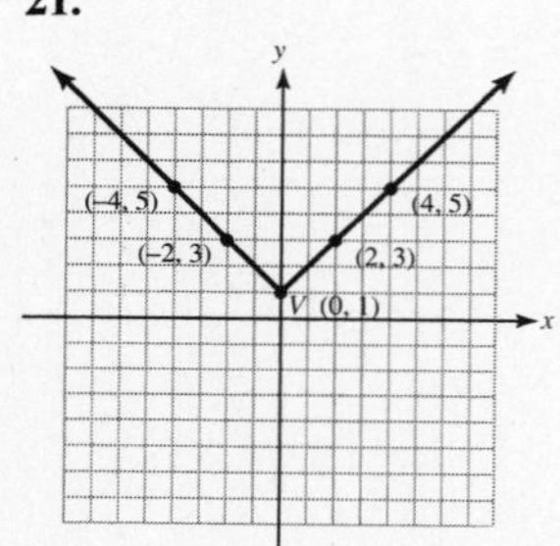

23.

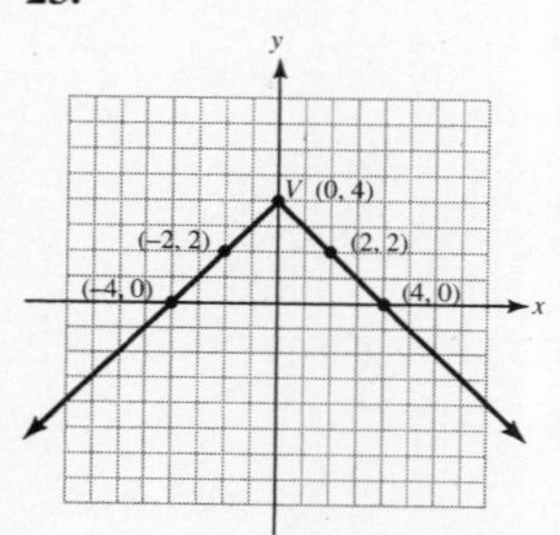

25.

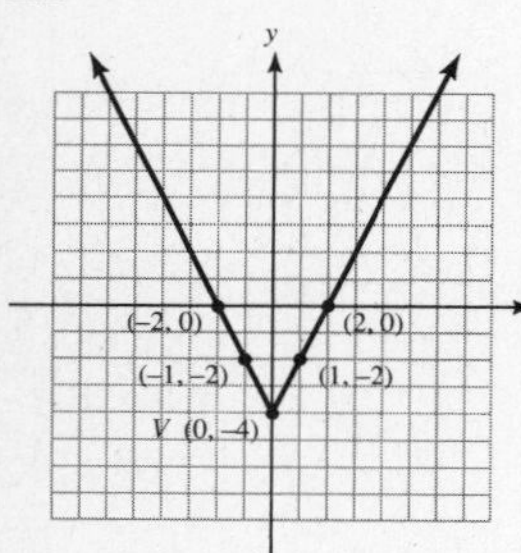

27.

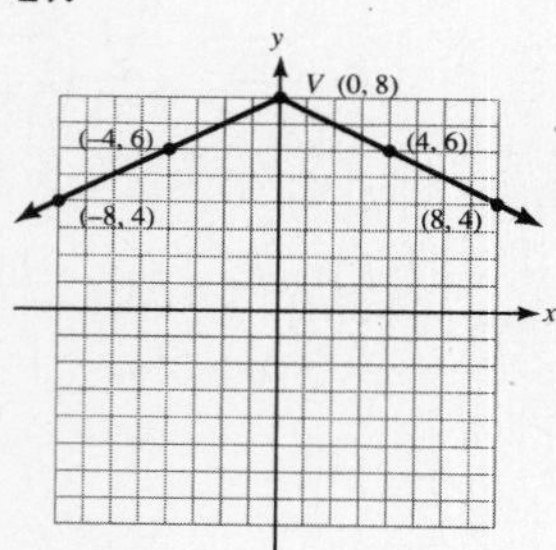

29.

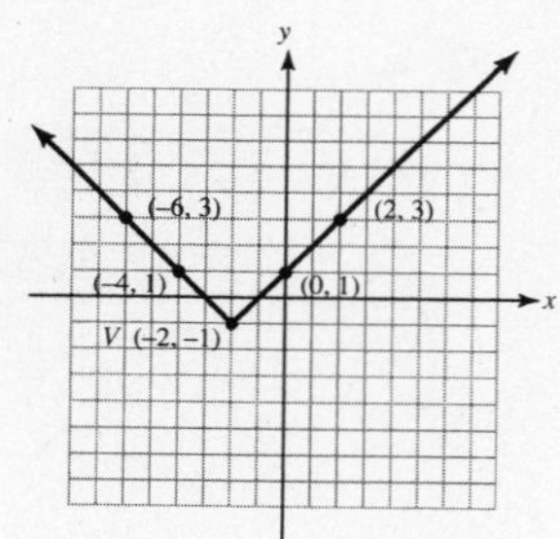

31.

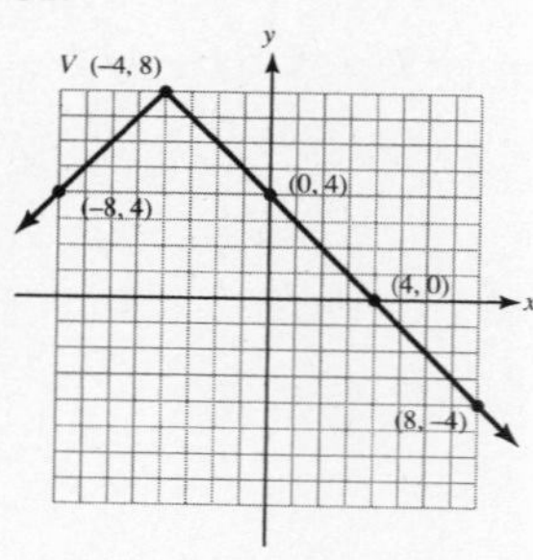

33.

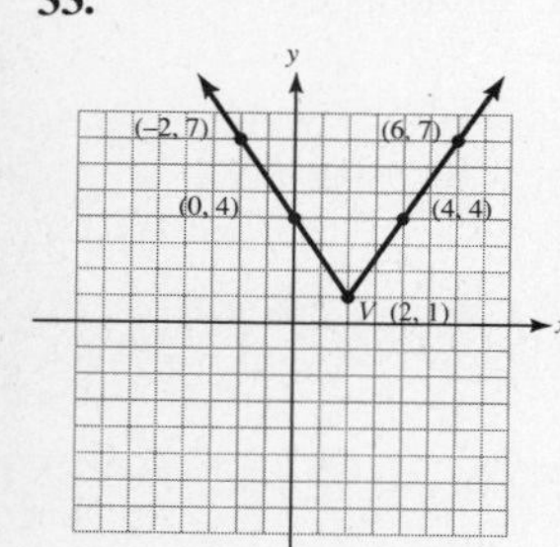

35.

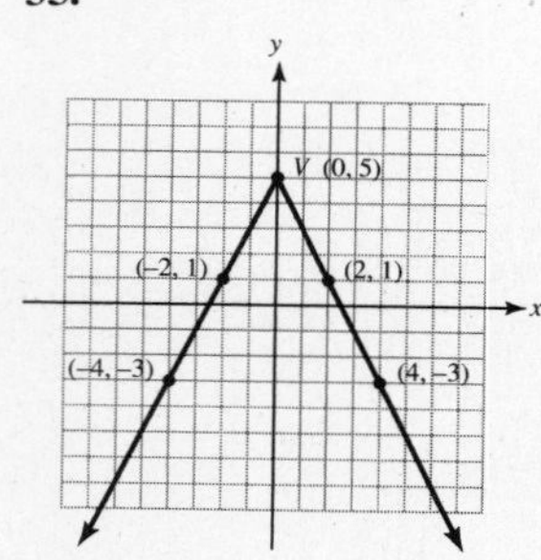

37.

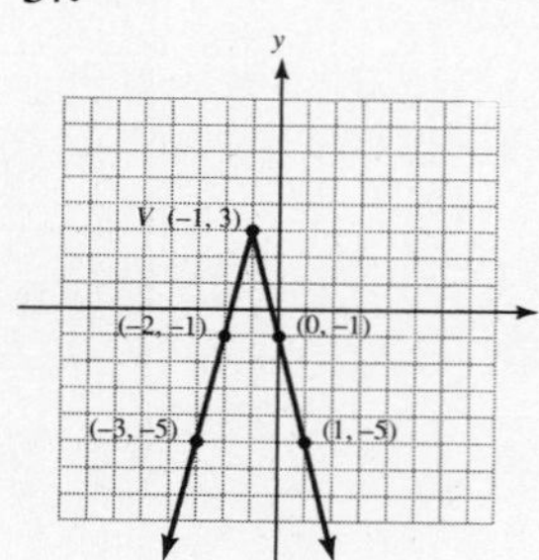

39.

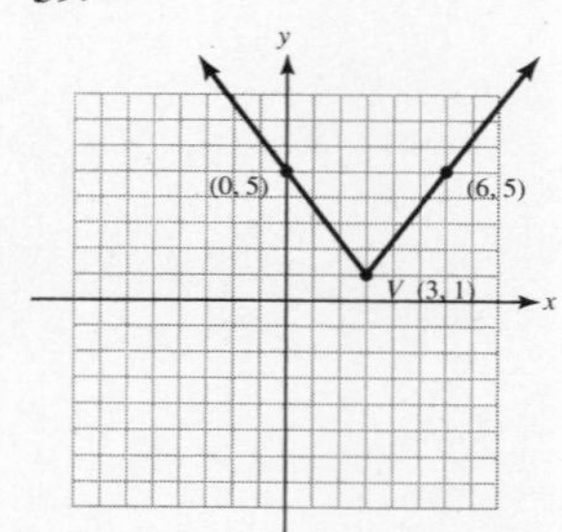

41.

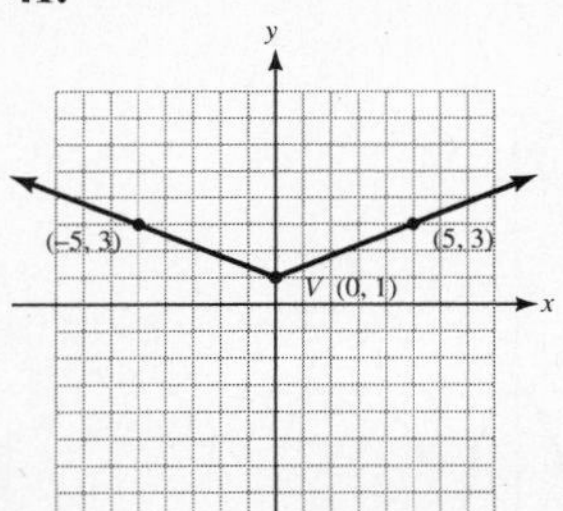

43.

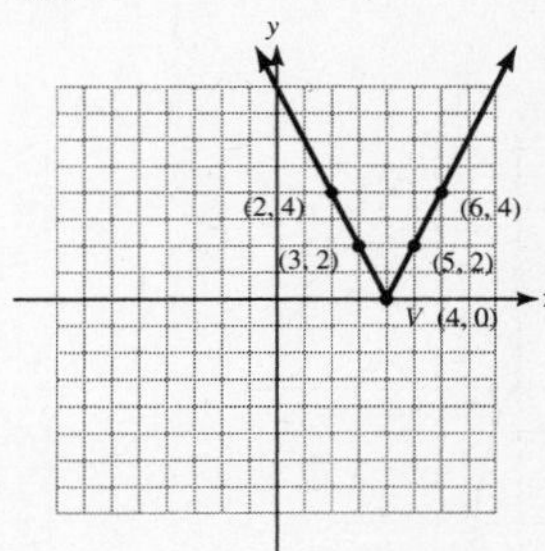

45.

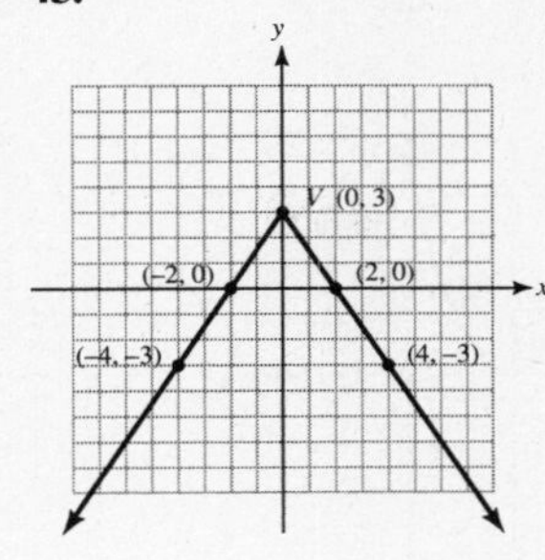

47.

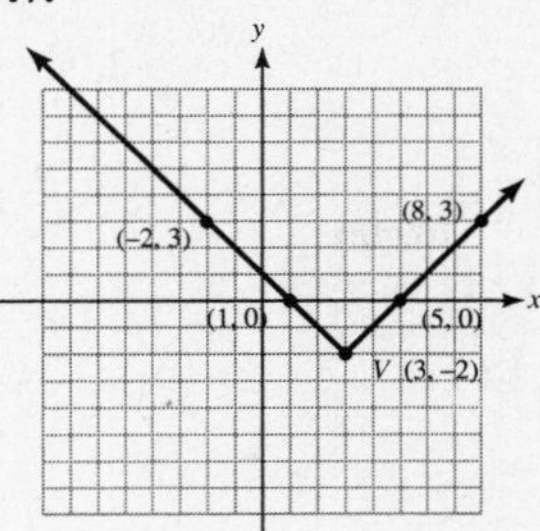

49.

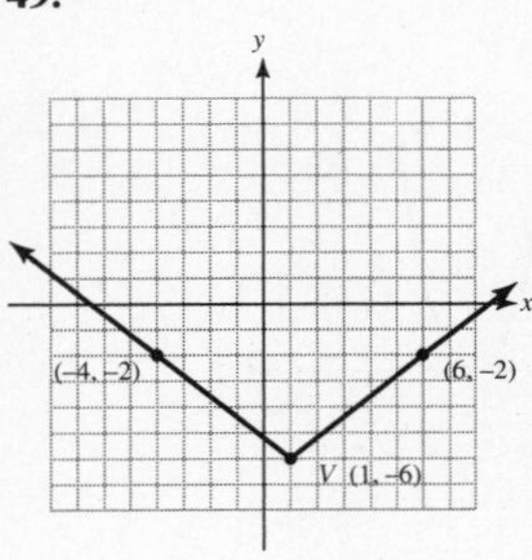

Section 8-1. Ten Review Exercises

1. $\dfrac{\sqrt{6}}{2}$ **2.** $\sqrt[3]{5}$ **3.** $\dfrac{\sqrt[4]{21}}{3}$ **4.** $8 + 2\sqrt{7}$ **5.** $\sqrt{2t}$ **6.** $\dfrac{-v}{4u^2}$ **7.** $\dfrac{-972}{a^3}$ **8.** $4k + 12k^{1/2} + 9$ **9.** $25m - 9$
10. $\dfrac{-x^{10}}{32y^5}$

Section 8-1. Supplementary Exercises

1. $y = \dfrac{2}{3}|x + 6|$ **3.** $y = \dfrac{4}{3}|x| - 7$ **5.** $y = \dfrac{-4}{3}|x + 5|$ **7.** $y = |x + 5| - 4$

9.

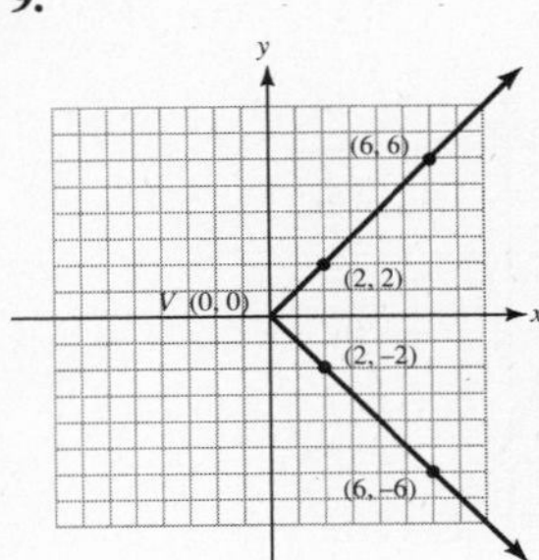

11.

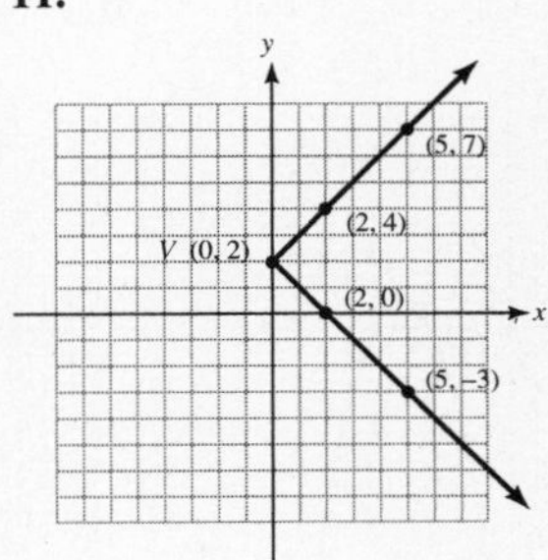

13.

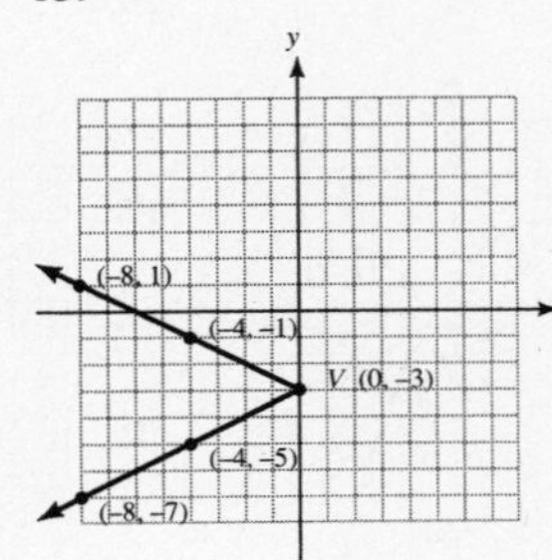

15.

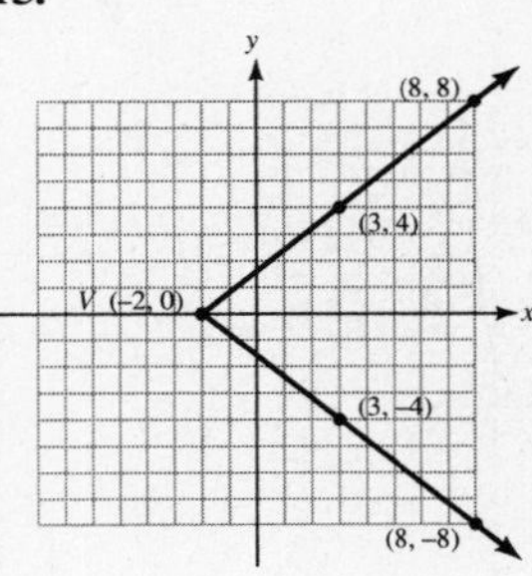

17.

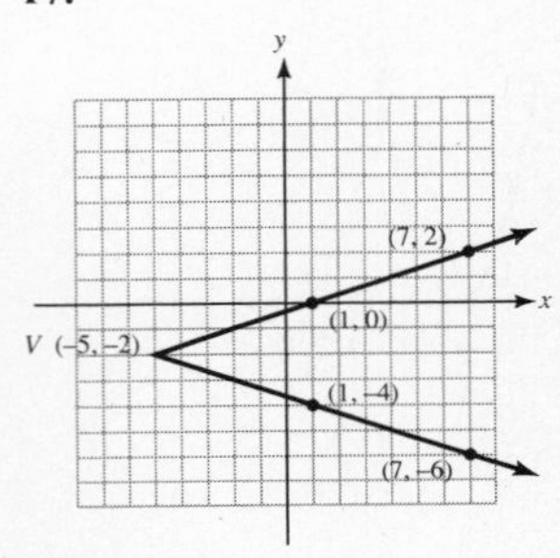

19.

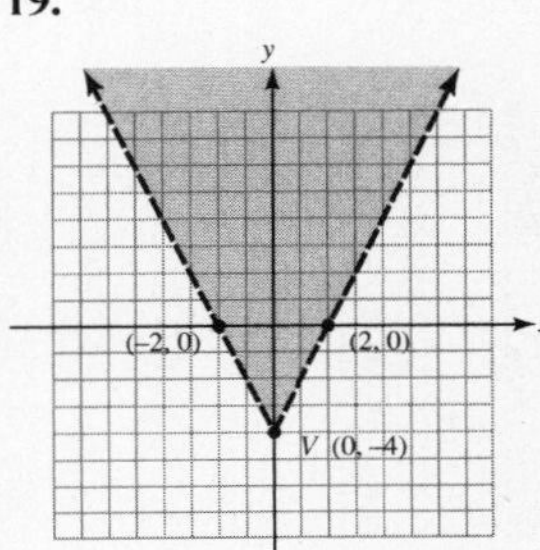

21.

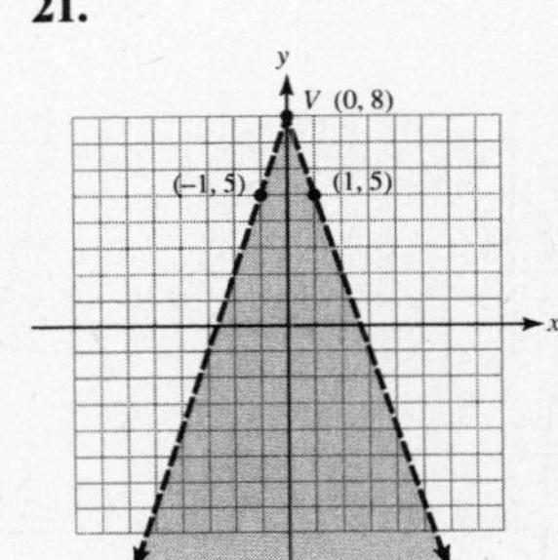

23.

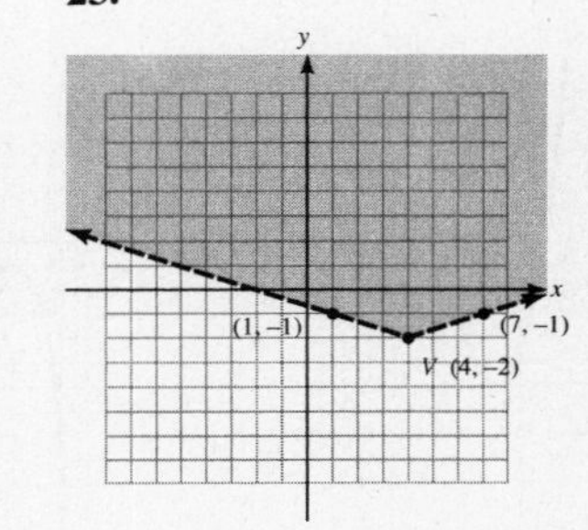

Section 8-2. Practice Exercises

1.

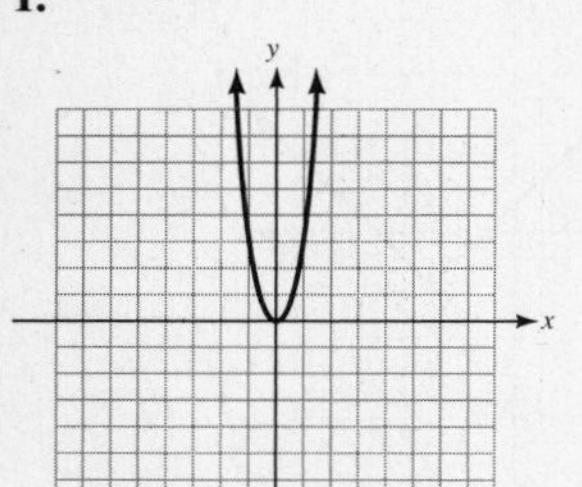

3.

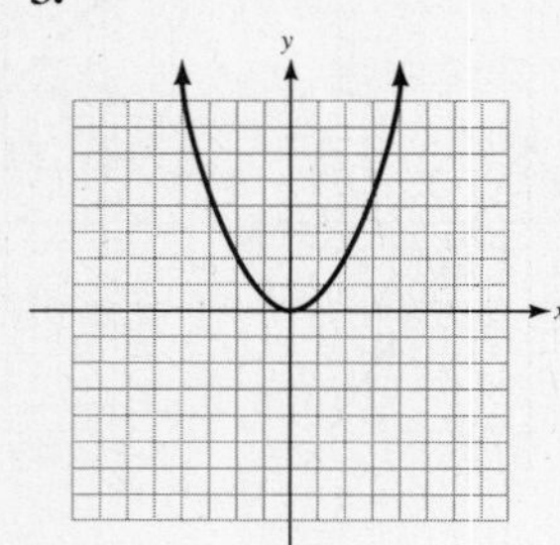

5.

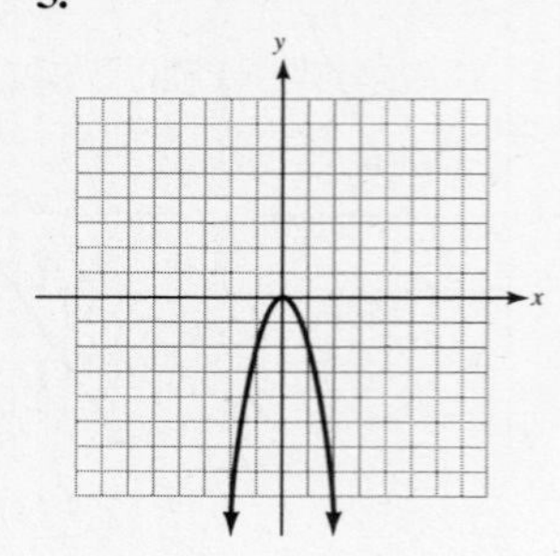

7.

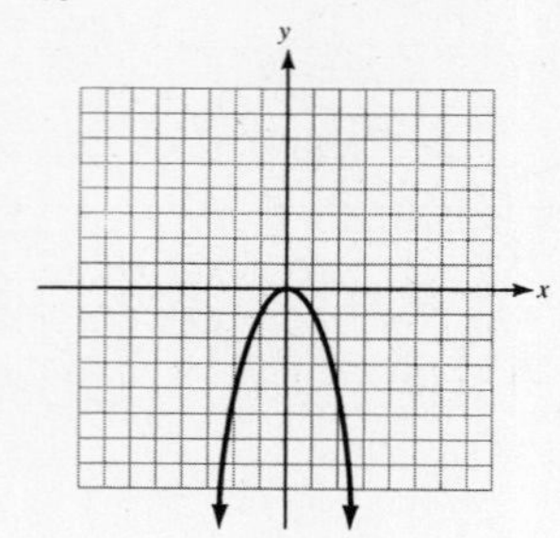

9.

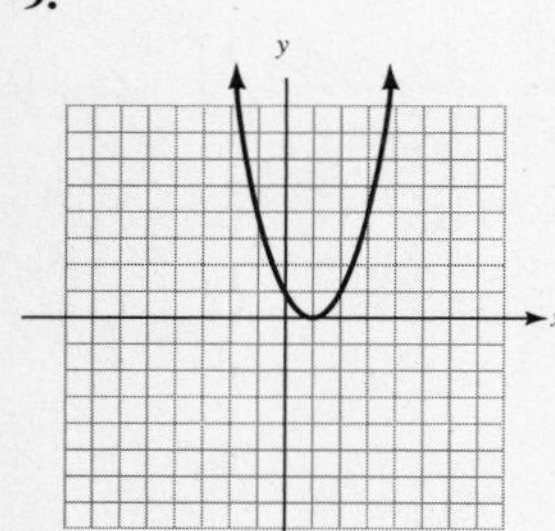

11.

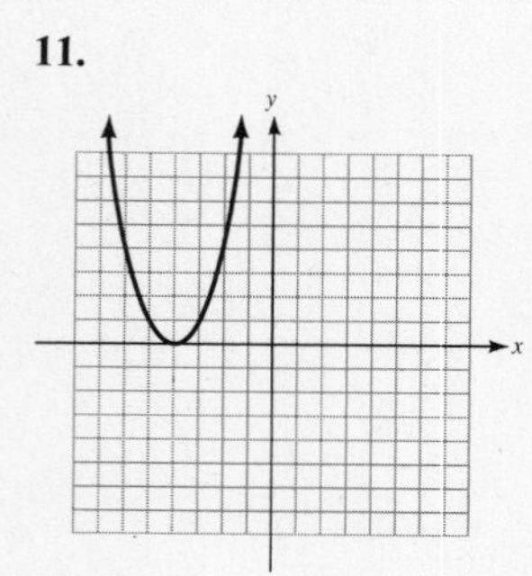

13.

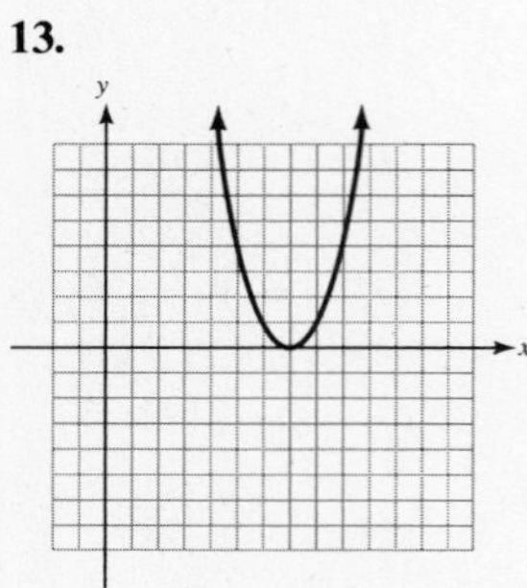

15.

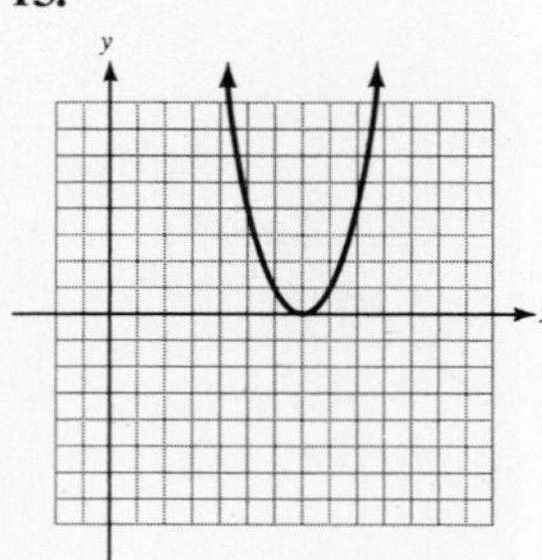

17.

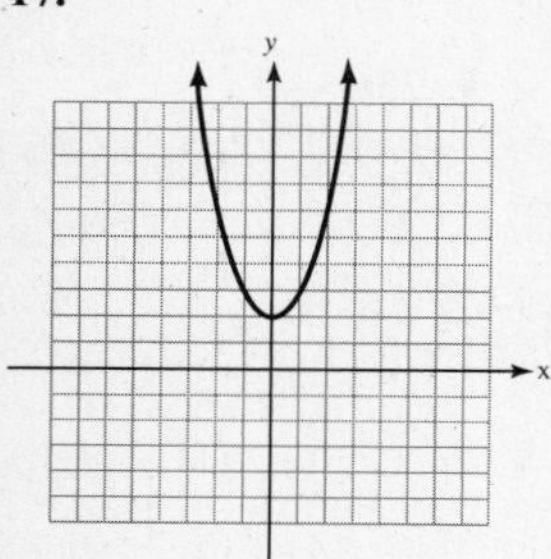

19.

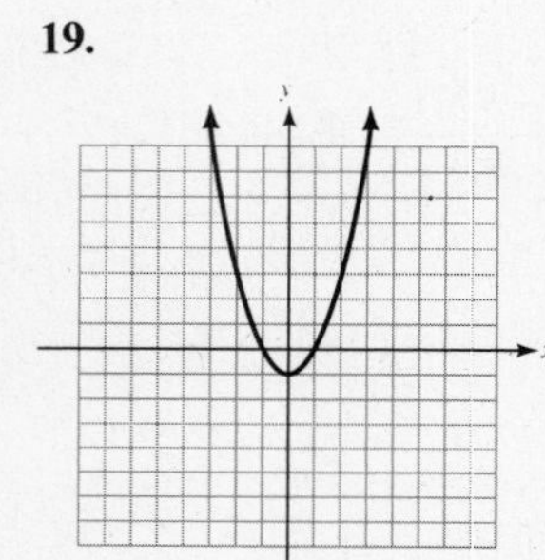

21.

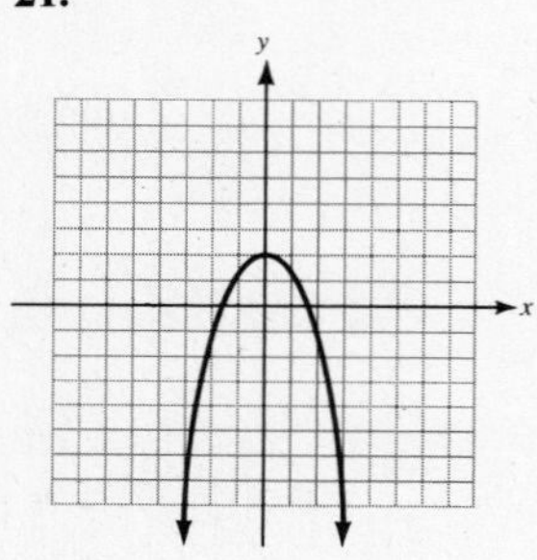

23.

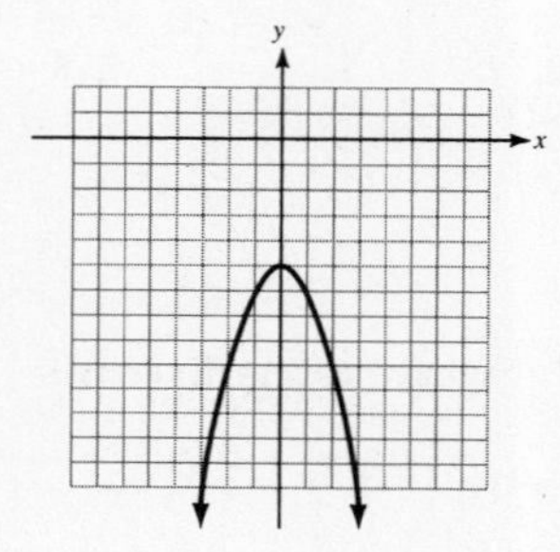

25.

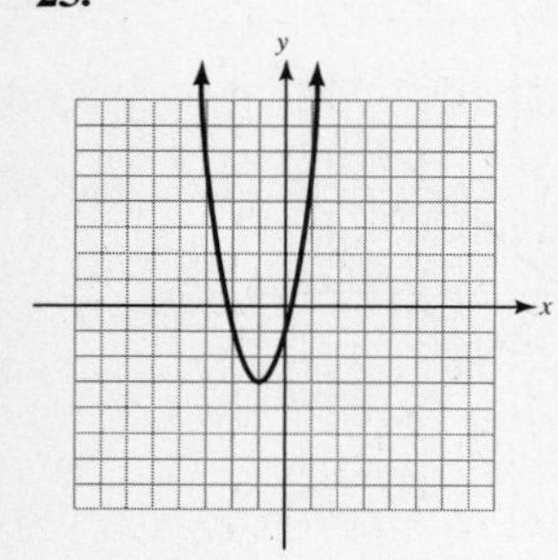

27.

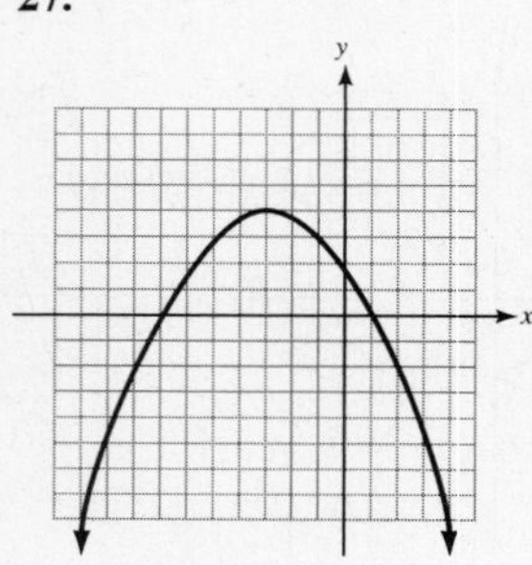

29.

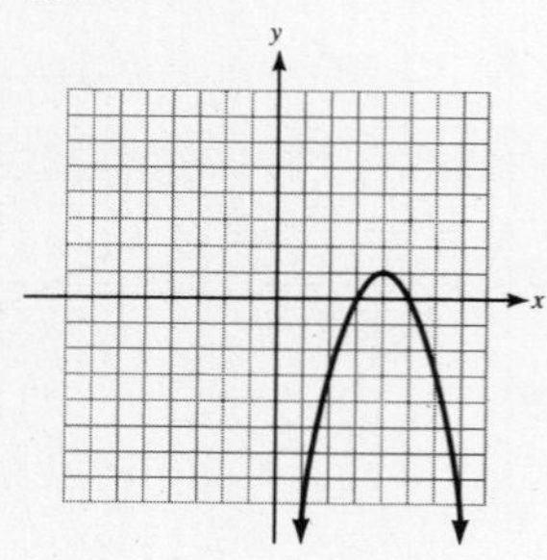

31.

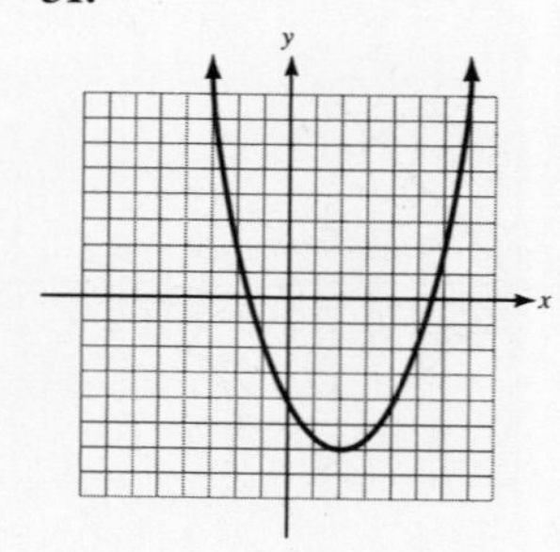

33.

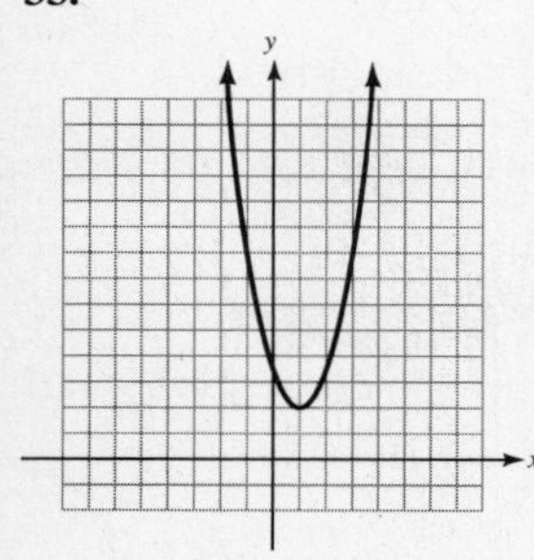

35.

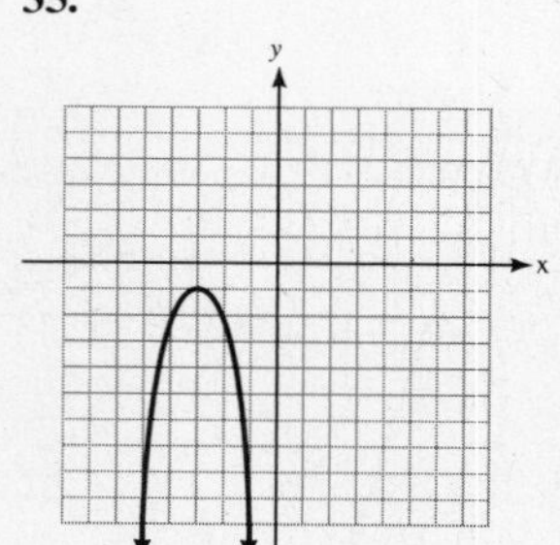

37.

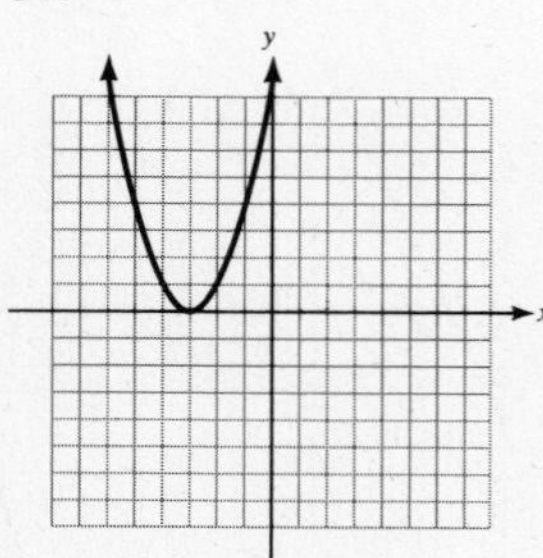

39.

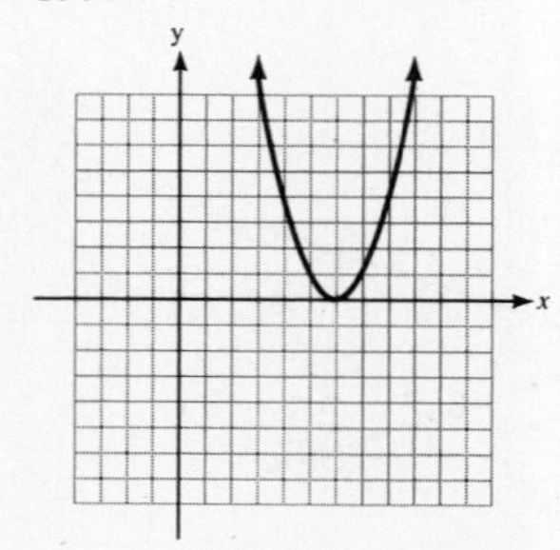

41.

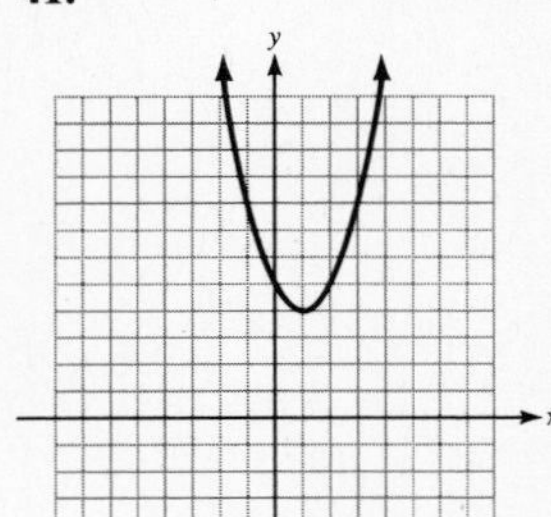

43.

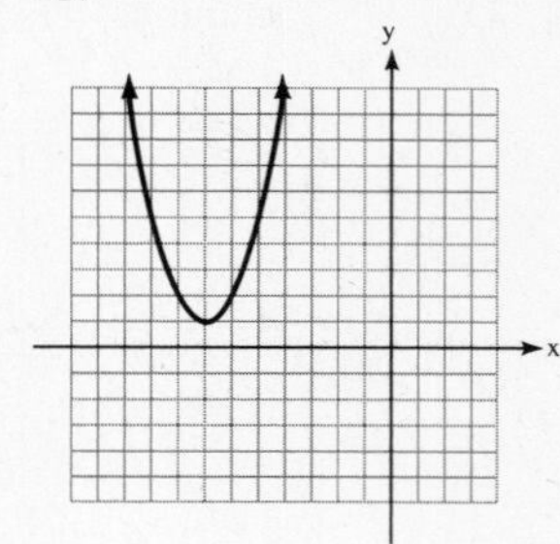

45.

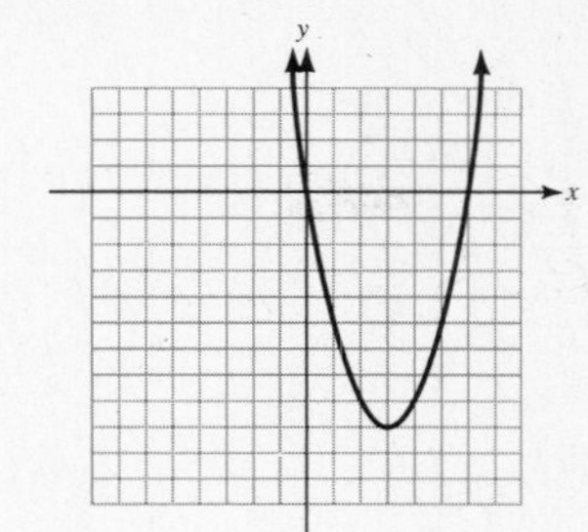

47.

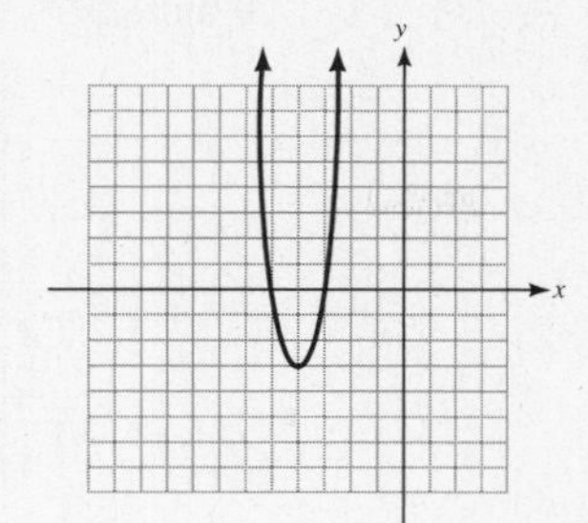

49.

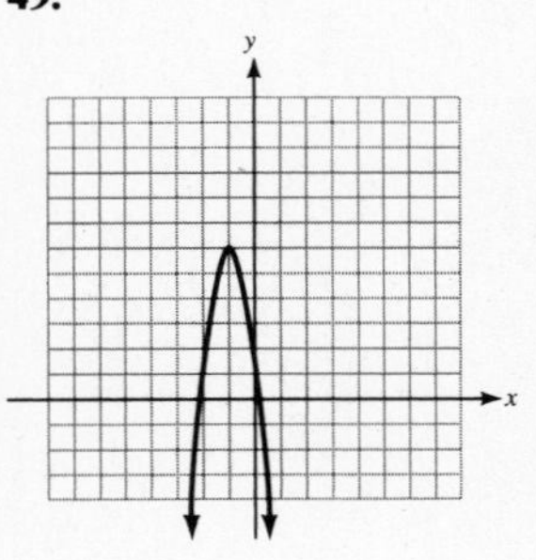

Section 8-2. Ten Review Exercises

1. 8 **2.** 8 **3.** 18 **4.** $\{5\}$ **5.** $\{-3, 5\}$ **6.** $\{-1, 3\}$

7.

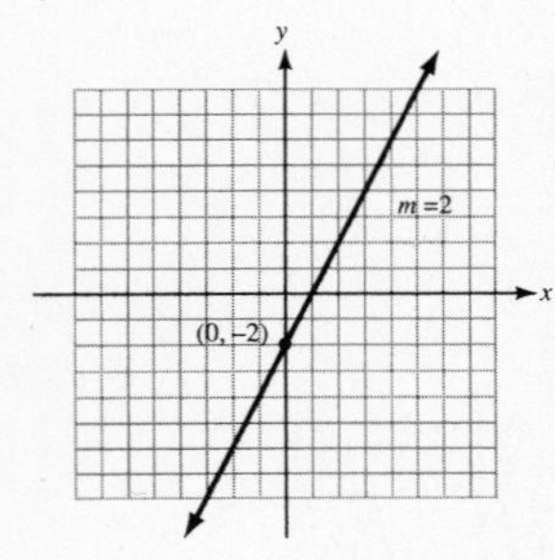

8.

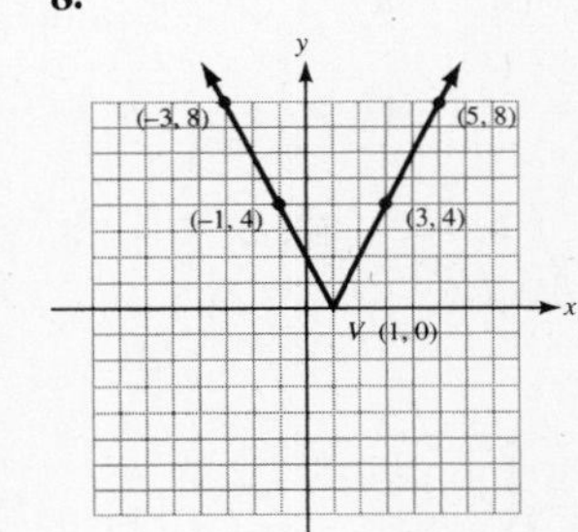

9.

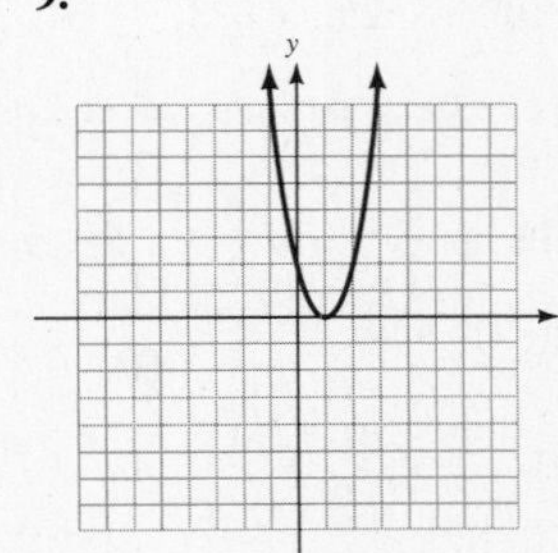

10.

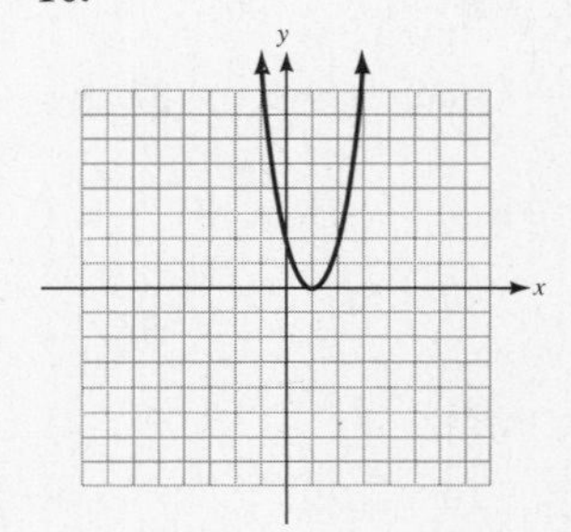

Section 8-2. Supplementary Exercises

1. $y = x^2 - 4x + 8$ **3.** $y = -2x^2 + 3$ **5.** $y = \frac{-1}{2}x^2 - 2x - 2$ **7.** $y = 3x^2 + 12x + 9$ **9.** $y = 8x - 16$

11. $y = -4x + 11$

13.

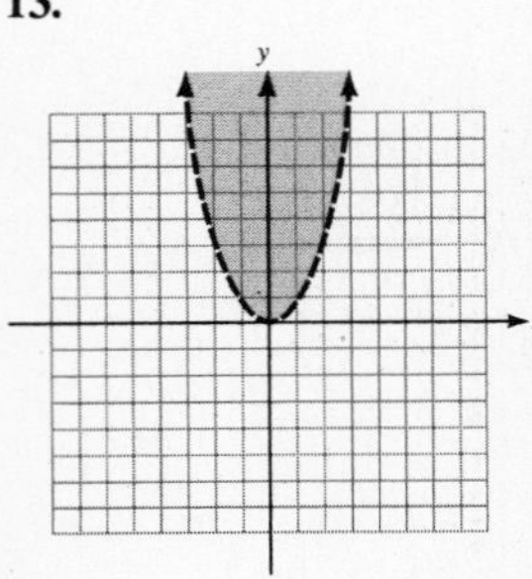

15.

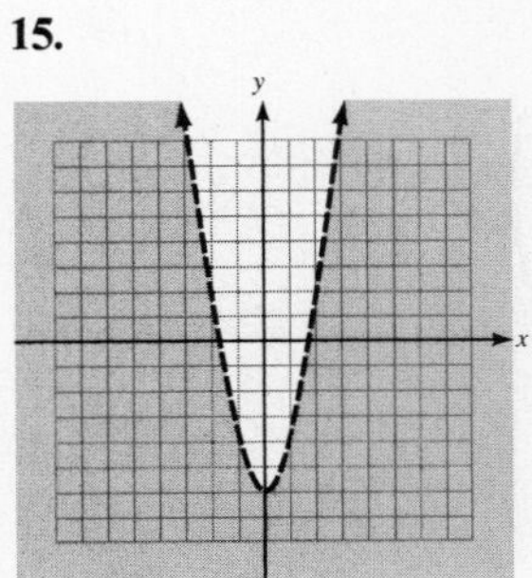

17.

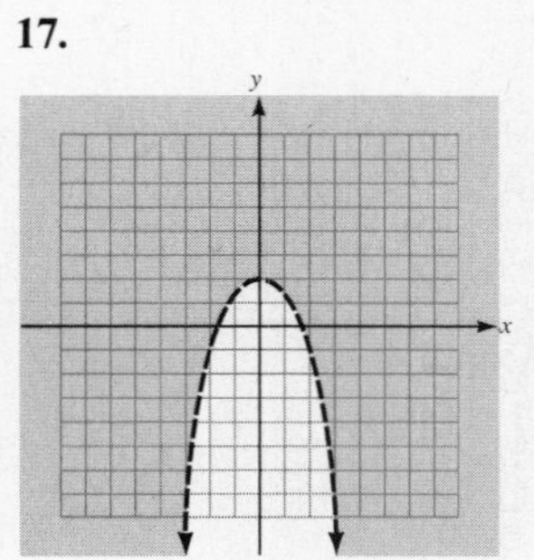

19.

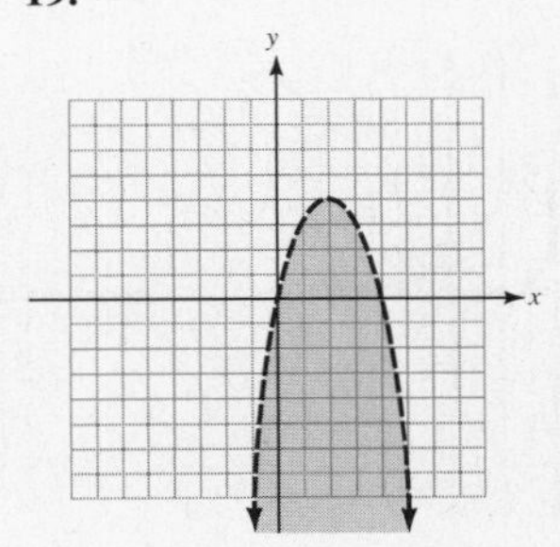

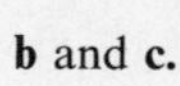

21. a. $-4, \frac{-1}{2}, 0, \frac{1}{2}, 4$ **b and c.**

23. a. $\frac{27}{4}, 2, 0, -2, \frac{-27}{4}$ **b and c.**

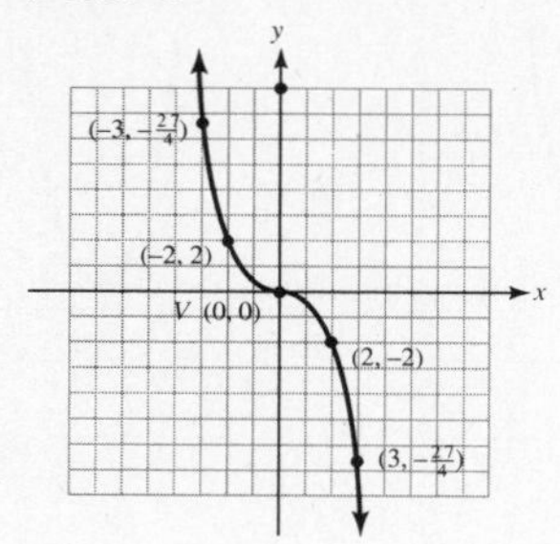

25.

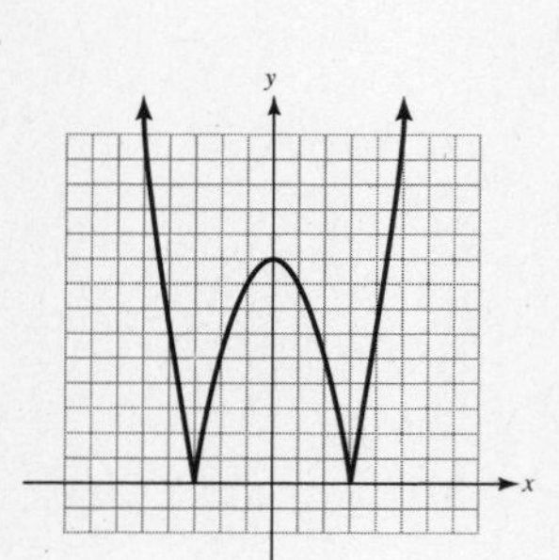

27.

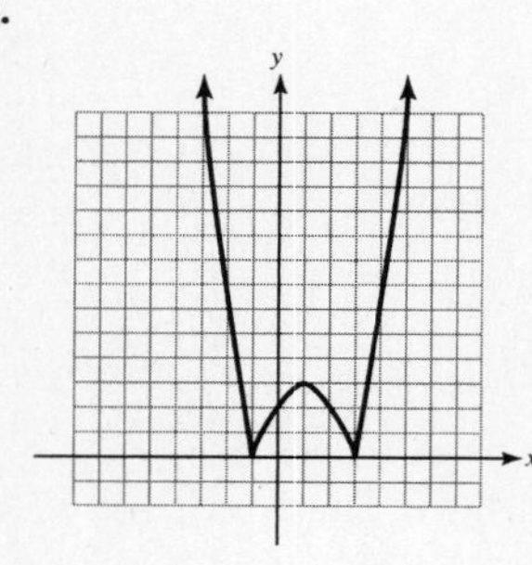

29.

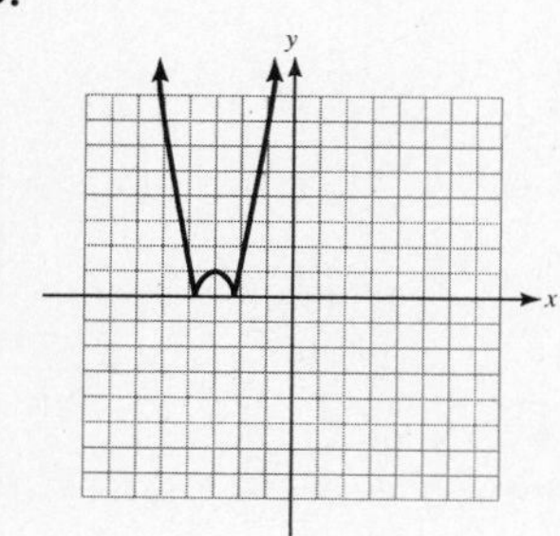

Section 8-3. Practice Exercises

1. All x; $x \neq -7$ **3.** All x; $x \neq 15$ **5.** All x; $x \neq 0, 3$ **7.** All x; $x \neq -5, 5$ **9.** All x **11.** All x; $x \neq 2, 5$ **13.** All x; $x \neq -2, 0, 8$ **15.** All x; $x \neq 0$ **17.** All x **19.** All x; $x \neq 2$ **21. a.** $-2, -5, -29, -299$ **b.** $4, 7, 31, 301$ **23. a.** $-1, -3, -19, -199$ **b.** $3, 5, 21, 201$ **25. a.** $-2, -5, -29, -299$ **b.** $4, 7, 31, 301$ **27. a.** $-1, -2, -10, -100$ **b.** $1, 2, 10, 100$ **29.** $x = 8$ **31.** $x = -6$ **33.** $x = -2$ and $x = 9$ **35.** $x = \frac{-3}{2}$ and $x = 5$ **37.** $x = -2$ **39.** $x = -3, x = 0$, and $x = 3$ **41. a.** 1 **b.** 1 **43. a.** 2 **b.** 2 **45. a.** 0 **b.** 0 **47. a.** 0 **b.** 0 **49. a.** $\frac{3}{2}$ **b.** $\frac{3}{2}$ **51. a.** $\frac{1}{4}$ **b.** $\frac{1}{4}$ **53. a.** 2 **b.** 2

55.

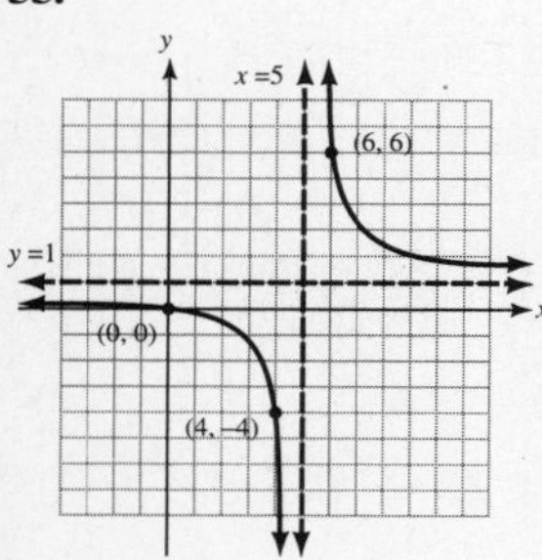

57.

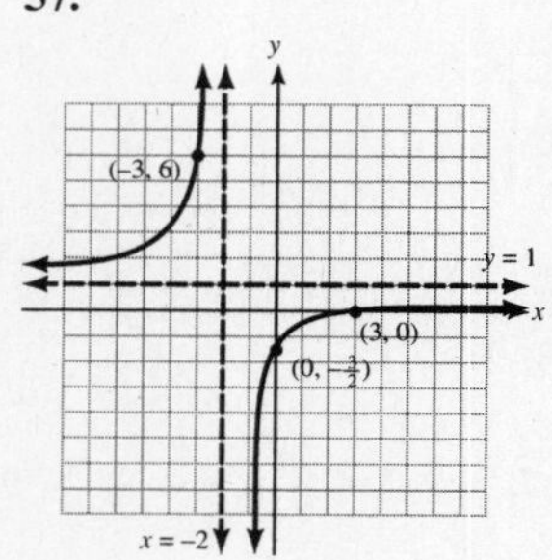

59.

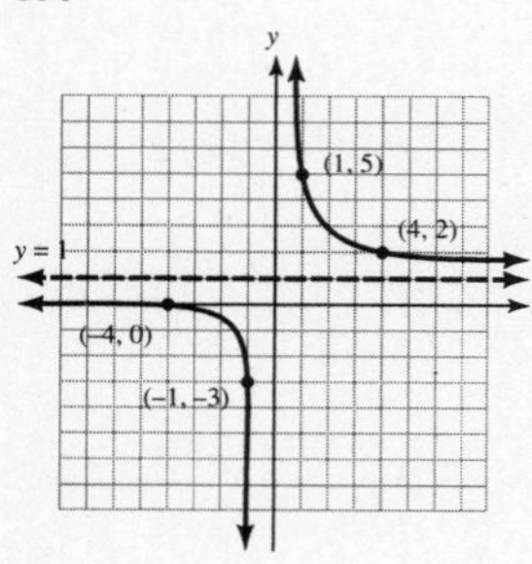

61.

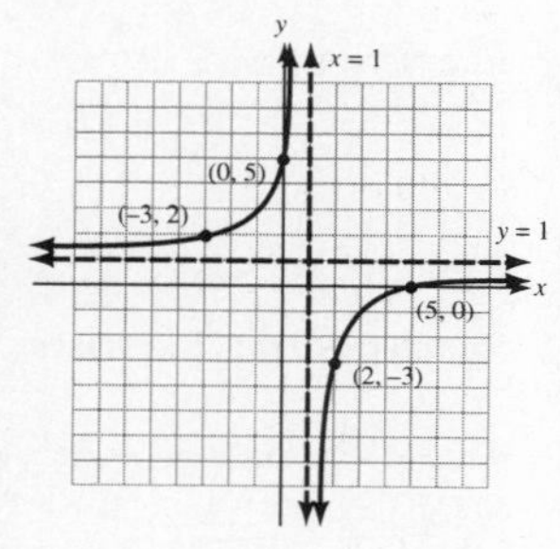

63.

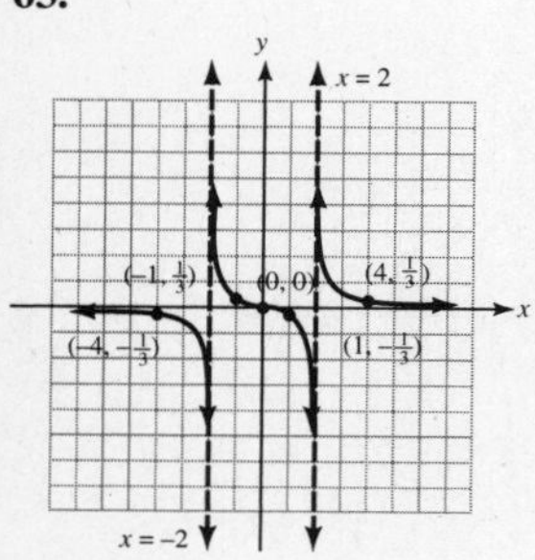

65.

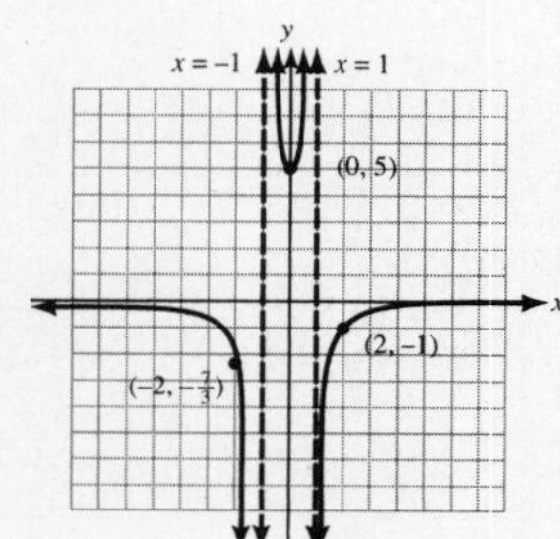

67.

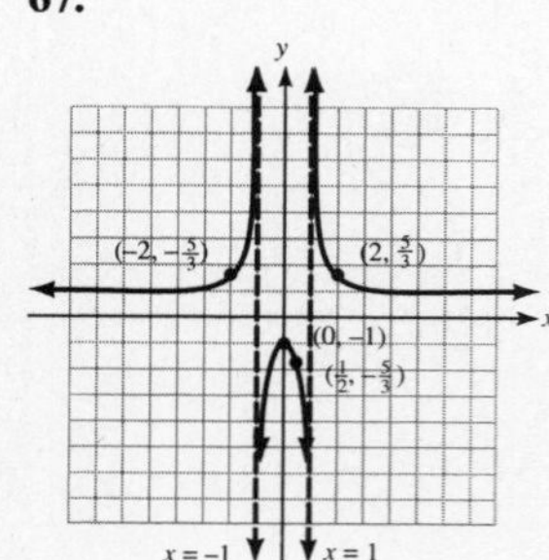

69.

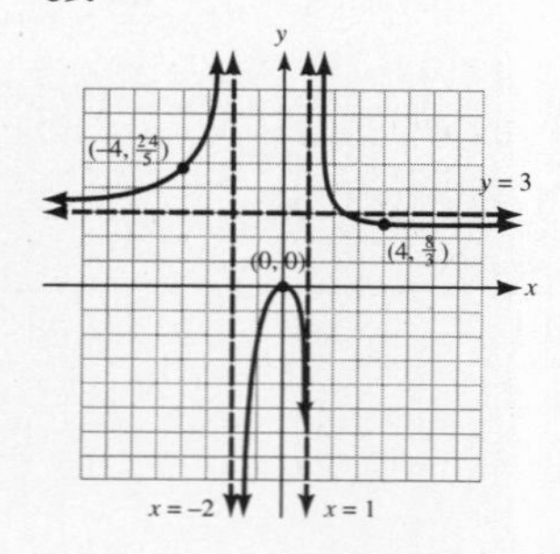

Section 8-3. Ten Review Exercises

1. $\frac{-3-2\sqrt{3}}{2}$ **2.** $\frac{i\sqrt{3}}{2}$ **3.** $-5+3\sqrt{2}$ **4.** $-5+3i\sqrt{2}$ **5.** $-12+7\sqrt{2}$ **6.** $-4+16i$ **7.** $\frac{8t+4}{t(t-4)(t+4)}$

8. $\frac{u+v}{u}$ **9.** $\frac{x+2}{x-2}$ **10.** $4b^2-7b+3+\frac{b+2}{2b^2-1}$

Section 8-3. Supplementary Exercises

1. $10 \neq 0$ **3.** For all x, $x^2+4 \neq 0$ **5.** $y = x-3$, where $x \neq 0$ **7.** $y = x-2$, where $x \neq 3$

9. $y = 2x-3$, where $x \neq \frac{-3}{2}$ **11.** $y = 2x+3$, where $x \neq \pm\sqrt{2}$ **13.** $y = x + \frac{-3x}{x+1}$ **15.** $y = 3x + \frac{6x}{3x-2}$

Section 8-4. Practice Exercises

1. a. -17 **b.** -5 **c.** 15 **d.** $4x+4a-5$ **3. a.** -22 **b.** 2 **c.** 122 **d.** $(x+a)^3-(x+a)+2$ **5. a.** 0 **b.** 9 **c.** -16 **d.** $9-(x+a)^2$ **7. a.** $3\sqrt{3}$ **b.** 6 **c.** $\sqrt{11}$ **d.** $\sqrt{36-(x+a)^2}$ **9. a.** -3 **b.** 0 **c.** $\frac{5}{9}$

d. $\frac{x+a}{x+a+4}$ **11. a.** 11 **b.** 10 **c.** $\frac{25}{3}$ **d.** $\frac{-1}{3}(x+a)+10$ **13.** $f^{-1} = \{(2, 1), (-1, 2), (-4, 3), (-7, 4), (-10, 5)\}$

15. $f^{-1} = \{(8, -2), (5, -1), (4, 0), (-5, 1), (-8, 2)\}$ **17.** $f^{-1} = \{(0, -3), (1, -2), (2, 1), (3, 6), (4, 13)\}$

19. $f^{-1} = \left\{\left(\frac{1}{9}, -3\right), \left(\frac{1}{4}, -2\right), (1, -1), \left(4, \frac{-1}{2}\right), \left(9, \frac{-1}{3}\right)\right\}$ **21. a.** $x \geq 0$ **b.** $y \geq -2$ **23. a.** $x \geq 0$ **b.** All y

25. a. All x, where $x \neq 0$ **b.** All y, where $y \neq -1$ **27.** $f^{-1}(x) = \frac{x+10}{3}$ **29.** $f^{-1}(x) = \frac{-5(x-4)}{2}$ **31.** $f^{-1}(x) = \frac{3x+8}{2}$

33. a. $f^{-1}(x) = \frac{8x}{x-1}$ **b.** 4 **35. a.** $f^{-1}(x) = \frac{x+3}{x-1}$ **b.** 5 **37. a.** $f^{-1}(x) = \frac{7}{x-1}$ **b.** 7 **39. a.** $f^{-1}(x) = \frac{2x-6}{x+1}$

b. 3 **41. a.** 1 **b.** $f^{-1}(x) = \sqrt[3]{x-2}$ **c.** -1 **43. a.** 27 **b.** $f^{-1}(x) = \sqrt[3]{x}-1$ **c.** 2 **45. a.** 2 **b.** $f^{-1}(x) = x^3-1$ **c.** -9 **47. a.** 33 **b.** $f^{-1}(x) = \sqrt[5]{1-x}$ **c.** 2 **49.** One-to-one **51.** Not one-to-one **53.** Not one-to-one **55.** One-to-one **57. a.** Function **b.** One-to-one **59. a.** Function **b.** Not one-to-one **61. a.** Function **b.** One-to-one **63. a.** Function **b.** Not one-to-one

Section 8-4. Ten Review Exercises

1. $(x-5)(x+1)$ **2.** $\{-1, 5\}$ **3.** $x = -1$ and $x = 5$ **4.** $x = -1$ and $x = 5$ **5.** 17 **6.** $m\left(\frac{1}{2}, 0\right)$ **7.** $\frac{-8}{15}$

8. $\frac{15}{8}$ **9.** $8x+15y = 4$ **10.** No

Section 8-4. Supplementary Exercises

1. a. $f^{-1}(x) = \frac{1}{4}(x+3)$ **b.** 1 **c.** 0 **3. a.** $f^{-1}(x) = \sqrt[3]{x+2}$ **b.** 1 **c.** 2 **5. a.** $f^{-1}(x) = \frac{5x}{2-x}$ **b.** -15

c. -7 **7. a.** $f^{-1}(x) = x^2$ **b.** 25 **c.** 16 **9. a.** $f^{-1}(x) = \frac{1}{2}(x^2+1)$ **b.** 313 **c.** 1 **11. a.** $f^{-1}(x) = \frac{1}{3}(6-x^2)$

b. -1 **c.** 2 **13. a.** $f^{-1}(x) = \sqrt{x-1}$ **b.** 10 **c.** 7 **15. a.** $f^{-1}(x) = -\sqrt[4]{6-x}$ **b.** 1 **c.** -3 **17. a.** Function **b.** One-to-one **19. a.** Function **b.** Not one-to-one **21. a.** Not a function **23. a.** Function **b.** Not one-to-one **25. a.** Function **b.** One-to-one **27.** $x \geq 0$ **29.** $0 \leq x \leq 3$

Chapter 8 Review Exercises

1.

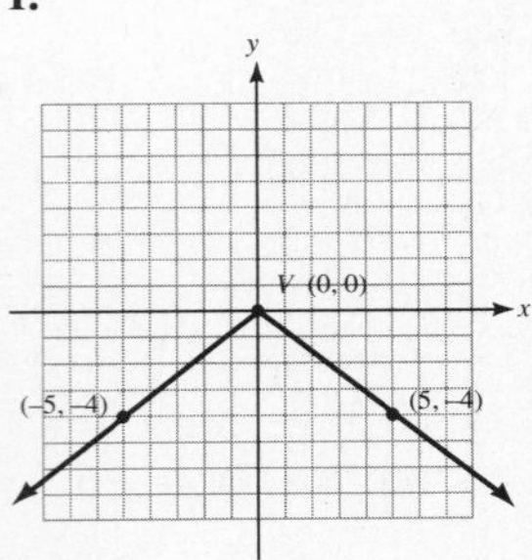

3.

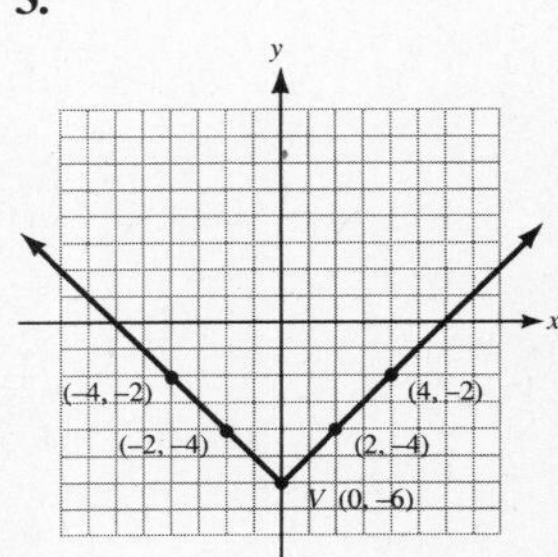

5.

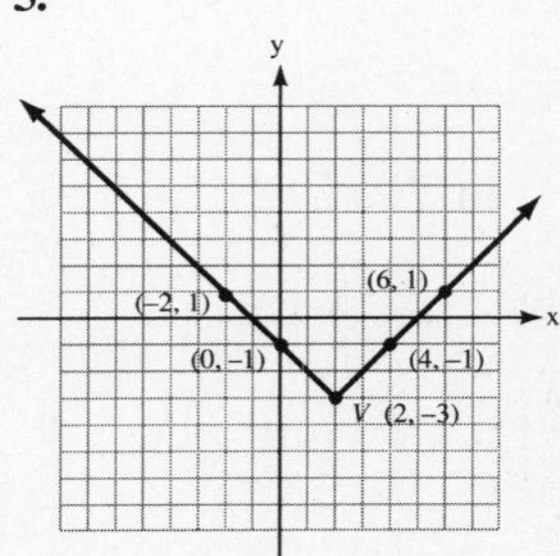

7.

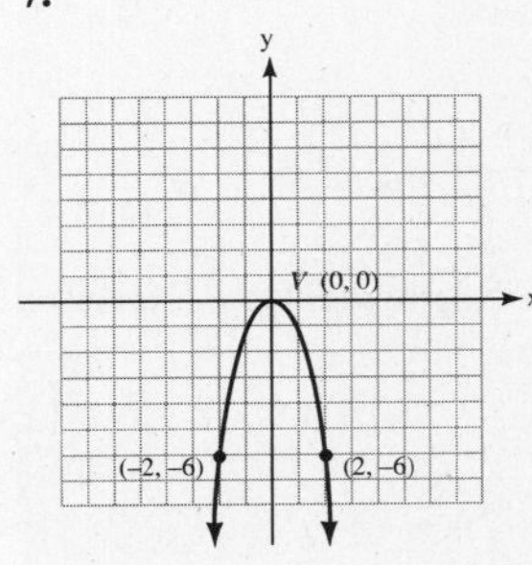

9.

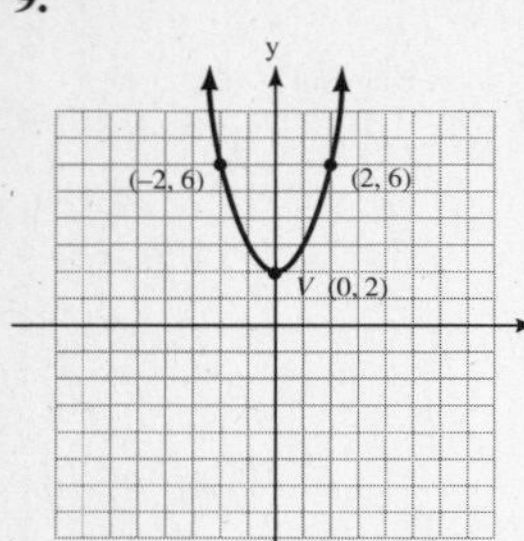

11.

13. a. All x; $x \neq -10$ **b.** $x = -10$ **c.** 2, 2

15. a. All x; $x \neq \dfrac{-5}{2}, \dfrac{5}{2}$ **b.** $x = \dfrac{-5}{2}$ and $x = \dfrac{5}{2}$ **c.** $\dfrac{1}{4}, \dfrac{1}{4}$

17.

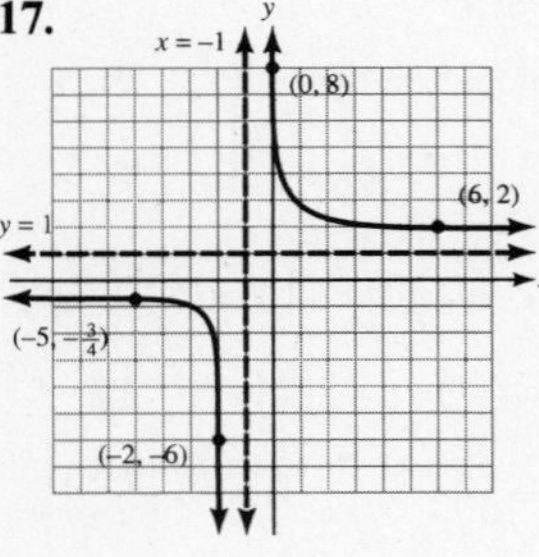

19. a. 0 **b.** -5 **c.** $\dfrac{x+a+2}{x+a-4}$

21. $f^{-1} = \{(-6, -5), (-4, 0), (-2, 5), (0, 10)\}$ **23.** Domain: $x \geq -5$; Range: All y **25.** $f^{-1}(x) = \dfrac{x}{x-2}$

27. $f^{-1}(x) = 1 - x^2$ **29.** $f^{-1}(x) = \sqrt[3]{\dfrac{1-x}{x}}$

Chapter 9 Systems of Linear Equations and Inequalities

Section 9-1. Practice Exercises

1. Yes **3.** Yes **5.** No **7.** Yes **9.** Yes **11.** $\{(2, 1)\}$ **13.** $\{(4, -1)\}$ **15.** $\{(4, 4)\}$ **17.** $\{(5, -3)\}$
19. $\{(8, -7)\}$ **21.** Inconsistent **23.** Dependent **25.** Independent and consistent **27.** Dependent

29. Inconsistent **31.** $\{(3, -1)\}$ **33.** $\{(-4, 2)\}$ **35.** $\left\{\left(5, \dfrac{1}{3}\right)\right\}$ **37.** $\left\{\left(\dfrac{-1}{2}, -2\right)\right\}$ **39.** $\{(0, 10)\}$ **41.** Inconsistent

43. $\{(-6, 10)\}$ **45.** Dependent **47.** $\{(-1, 9)\}$ **49.** $\left\{\left(\dfrac{2}{3}, \dfrac{-2}{3}\right)\right\}$ **51.** $\left\{\left(\dfrac{-2}{3}, \dfrac{-1}{2}\right)\right\}$ **53.** Inconsistent

55. $\left\{\left(-3, \dfrac{-3}{4}\right)\right\}$ **57.** $\left\{\left(\dfrac{-1}{3}, 6\right)\right\}$ **59.** $\left\{\left(\dfrac{-4}{3}, \dfrac{3}{4}\right)\right\}$

Section 9-1. Ten Review Exercises

1. $(x-4)(x+2)$ **2.** $\{4, -2\}$ **3.** $y = (x-1)^2 - 9$ **4.** $V(1, -9)$ **5.** $x = 1$ **6.**

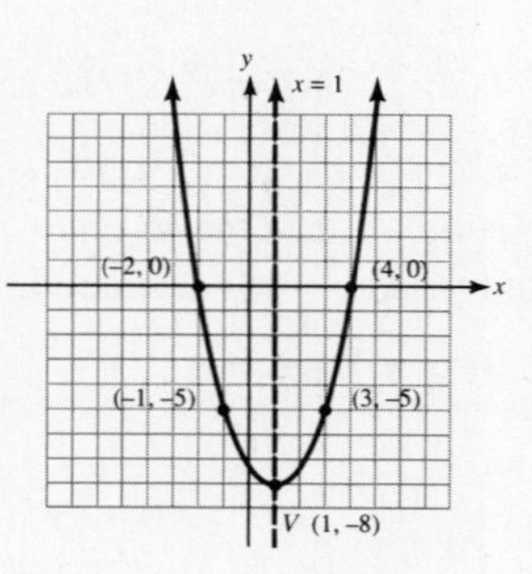

7. $-32t$ **8.** $10u + 5$ **9.** $\dfrac{9}{w(w-4)}$ **10.** $\dfrac{y+2}{y-2}$

Section 9-1. Supplementary Exercises

1. $\{(-2, 3)\}$ **3.** $\{(2, 4)\}$ **5.** $\left\{\left(\dfrac{1}{2}, \dfrac{-7}{2}\right)\right\}$ **7.** $\left\{\left(\dfrac{1}{2}, \dfrac{1}{3}\right)\right\}$ **9.** $\left\{\left(\dfrac{-1}{9}, \dfrac{-1}{4}\right)\right\}$ **11.** $\{(3, -4)\}$ **13.** (1) $3x - y = 2$
(2) $x - 2y = -6$

15. (1) $3x - 2y = -6$
(2) $2x + 3y = -17$

17. (1) $2x - 3y = -12$
(2) $2x + 3y = 12$

19. a. Yes **b.** Yes **21. a.** No **b.** Yes

Section 9-2. Practice Exercises

1. Yes **3.** No **5.** Yes **7.** $\{(2, 3, -1)\}$ **9.** $\{(-2, 5, 4)\}$ **11.** $\{(5, 0, -3)\}$ **13.** Inconsistent **15.** $\{(3, 10, -6)\}$
17. $\{(-8, 6, 10)\}$ **19.** Dependent **21.** $\left\{\left(\frac{1}{2}, \frac{1}{5}, \frac{1}{3}\right)\right\}$ **23.** $\left\{\left(\frac{-2}{3}, \frac{1}{2}, \frac{3}{5}\right)\right\}$ **25.** $\left\{\left(\frac{-1}{2}, \frac{1}{2}, \frac{2}{3}\right)\right\}$

Section 9-2. Ten Review Exercises

1. All x **2.** $\frac{-5}{4}$ **3.** $f(x) = \frac{-5}{4}x - 3$ **4.** $\frac{-12}{5}$ **5.** $g(x) = \frac{4}{5}x + \frac{11}{5}$ **6.** $f^{-1}(x) = \frac{-4}{5}x - \frac{12}{5}$ **7.** $g^{-1}(x) = \frac{5}{4}x - \frac{11}{4}$
8. $\left(\frac{-4}{3}, \frac{-4}{3}\right)$ **9.** $2\sqrt{41}$ **10.** $(0, -3)$

Section 9-2. Supplementary Exercises

1. $\{(x, 2 - x, 4 - 2x)\}$ **3.** $\left\{\left(x, \frac{7x - 7}{2}, \frac{4x - 7}{2}\right)\right\}$ **5.** $\left\{\left(\frac{-5z - 4}{7}, \frac{9z + 24}{7}, z\right)\right\}$ **7.** $\left\{\left(\frac{16 - 9z}{5}, \frac{11z - 34}{5}, z\right)\right\}$
9. Burgers cost \$2.50; fries cost \$0.50; drinks cost \$1.00.

Section 9-3. Practice Exercises

1. 8 and 12 **3.** -5 and 6 **5.** $\frac{1}{2}$ and $\frac{5}{3}$ **7.** -4, 1, and 6
9. Apples cost \$0.45 a pound; oranges cost \$0.23 a pound. **11.** 8% on the mutual fund; 9.5% on the time certificate
13. 20% caustic chemical in A; 35% caustic chemical in B
15. X costs \$0.75 per pound; Y costs \$1.20 per pound; Z costs \$1.75 per pound. **17.** $y = -x^2 - 2x + 4$ **19.** $y = x^2 - 4x + 3$
21. $x^2 + y^2 - 2x + 2y = 2$ **23.** $x^2 + y^2 + 2x - 2y = 23$ **25.** Plane 190 mph; wind 35 mph
27. Bus 45 mph; plane 420 mph **29.** 2 hr by car; 4 hr by plane **31.** Desiree is 17; David is 24.
33. Henry is 20; George is 32. **35.** Par Bleu is 5; Zanetto is 11. **37.** The locket is 75; the pin is 110; the ring is 125.

Section 9-3. Ten Review Exercises

1. $\frac{x - 3}{x - 2}$ **2.** -2 and 2 **3.** $\frac{x - 3}{x - 2}$ **4.** $\{3\}$ **5.** $x = 2$ **6.** All x; $x \neq 2$ **7.** $y = 1$ **8.** All y; $y \neq 1$

9.

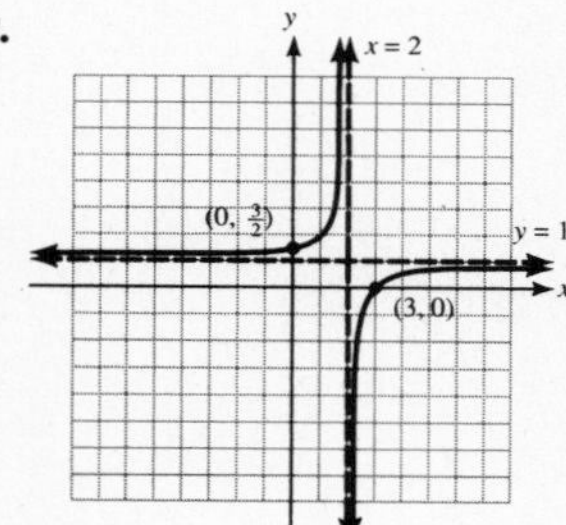

10. $y = \frac{2x - 3}{x - 1}$

Section 9-3. Supplementary Exercises

1. a. The sum of the integers is 42. **b.** The difference between 2 times the first and 3 times the second is 9.
c. The integers are 27 and 15. **3. a.** The difference between the first and 2 times the second is 9.
b. The sum of 3 times the first and 4 times the second is 7. **c.** The integers are 5 and -2. **5.** $x^2 + 4y^2 = 4$
7. $9x^2 + 4y^2 = 36$

Section 9-4. Practice Exercises

1. a. Yes **b.** Yes **3. a.** No **b.** Yes **5. a.** No **b.** No **7. a.** Yes **b.** Yes **9. a.** Yes **b.** Yes

11.

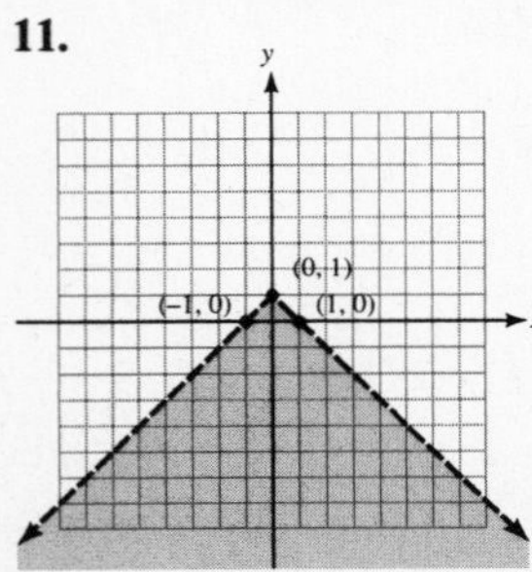

13.

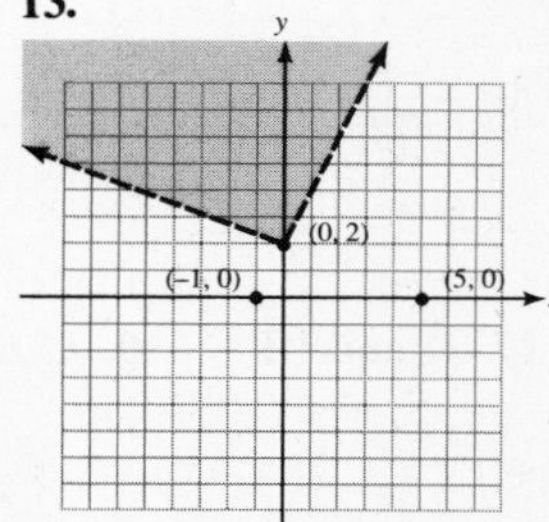

15.

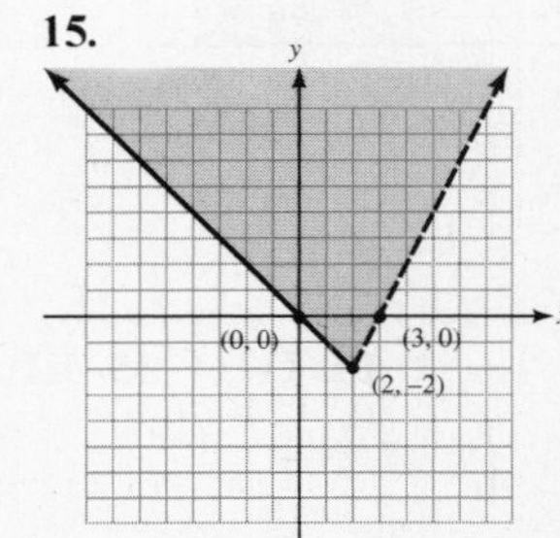

17.

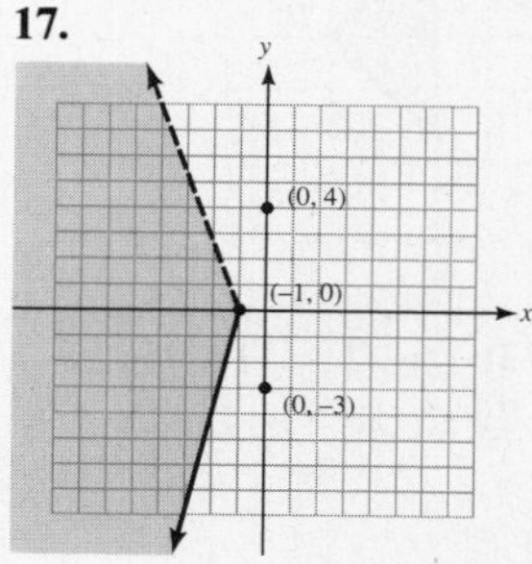

19.

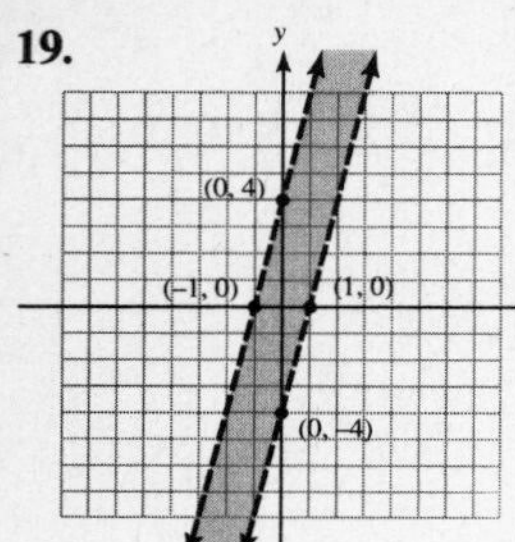

21.

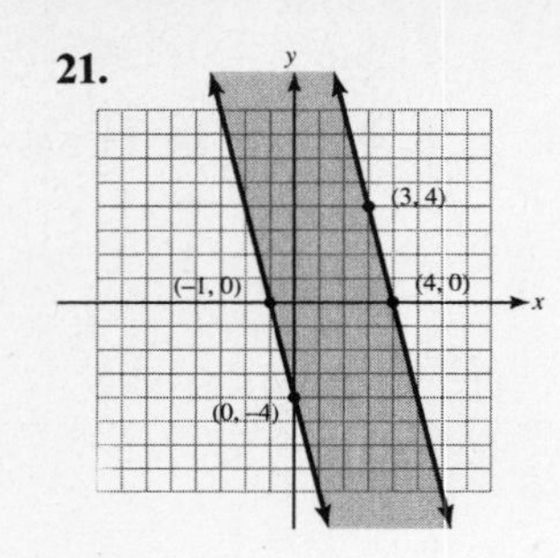

23. $\varnothing$

25.

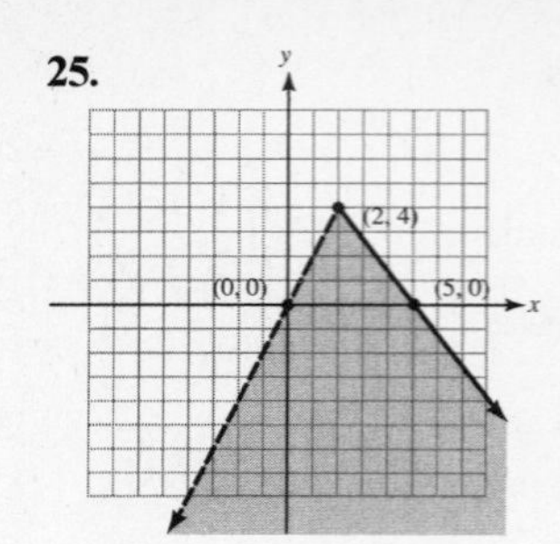

27.

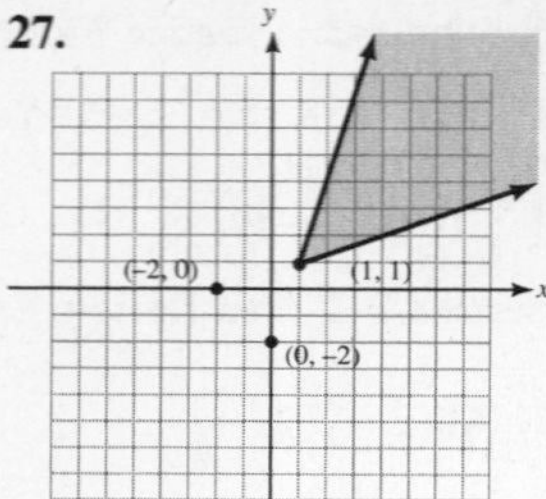

29.

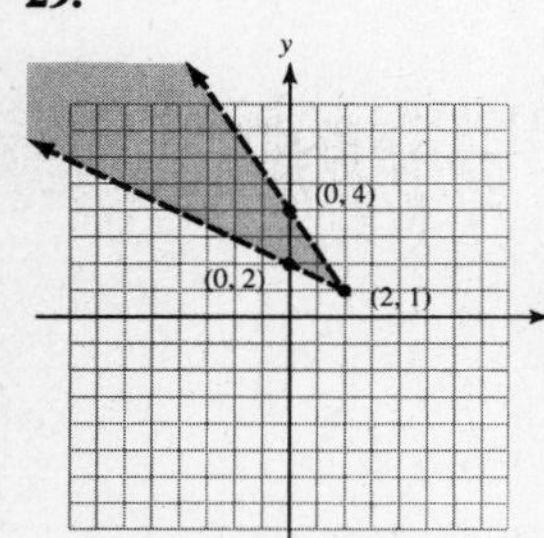

31.

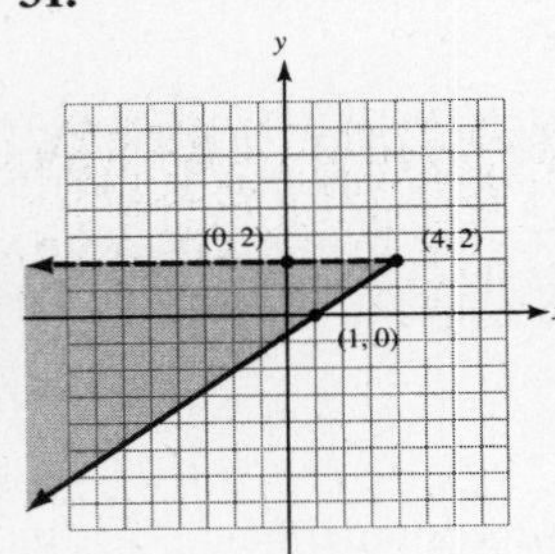

33.

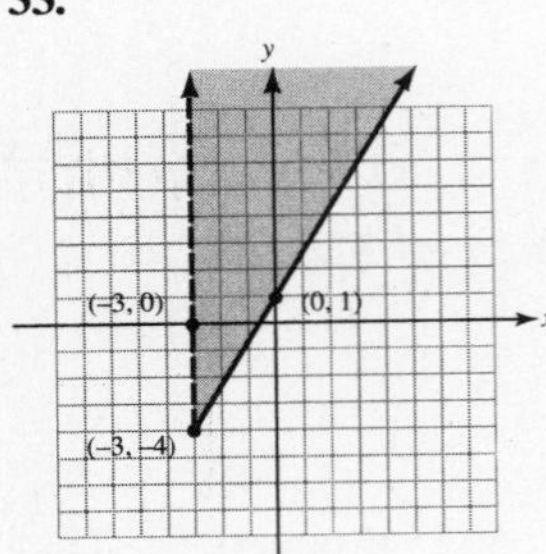

35.

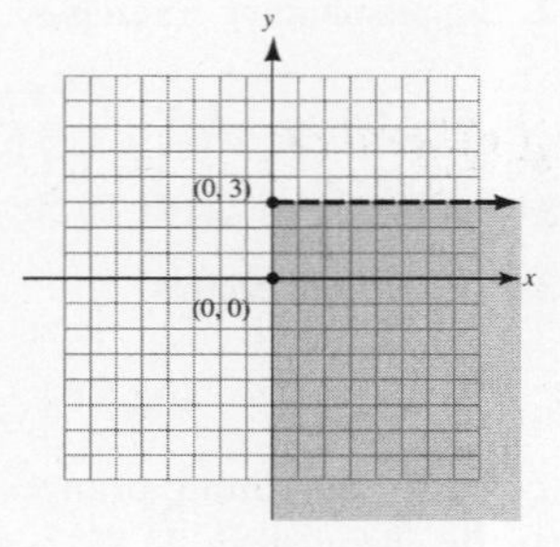

37.

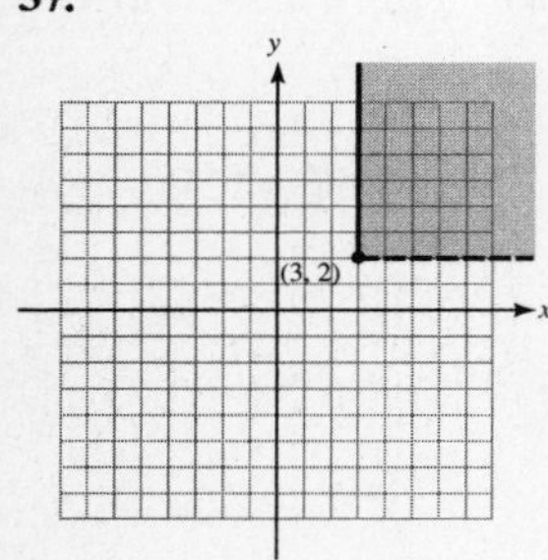

39.

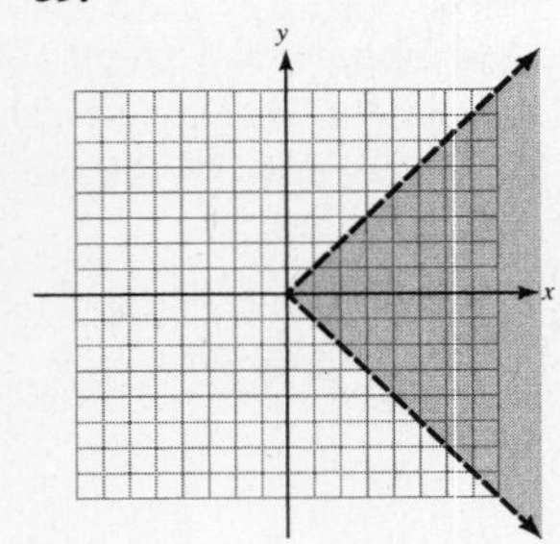

Section 9-4. Ten Review Exercises

1. $\left\{\frac{3}{8}\right\}$ **2.** $\left\{-4, \frac{16}{3}\right\}$ **3.** $\left\{\frac{2}{5}, \frac{3}{4}\right\}$ **4.** $\{2\}$ **5.** $\{7\}$ **6.** $1 \leq t \leq 5$

7. $-2 < v < \frac{2}{3}$ **8.** $x \leq -13$ or $x \geq 3$

9. $-1 < y < \frac{21}{5}$ **10.** $-5 \leq z \leq 5$

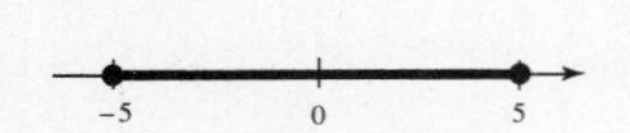

Section 9-4. Supplementary Exercises

1.

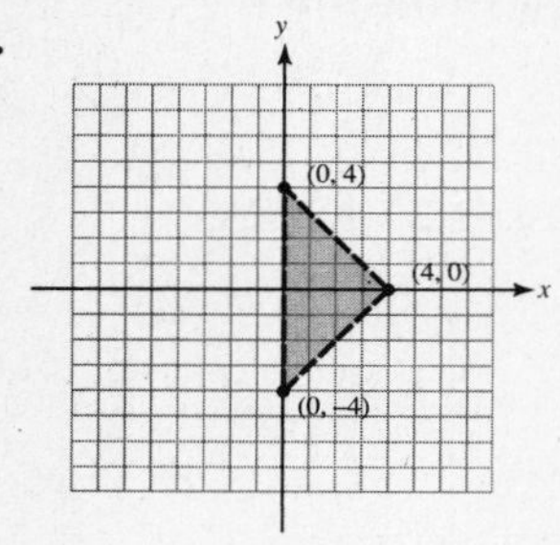

3.

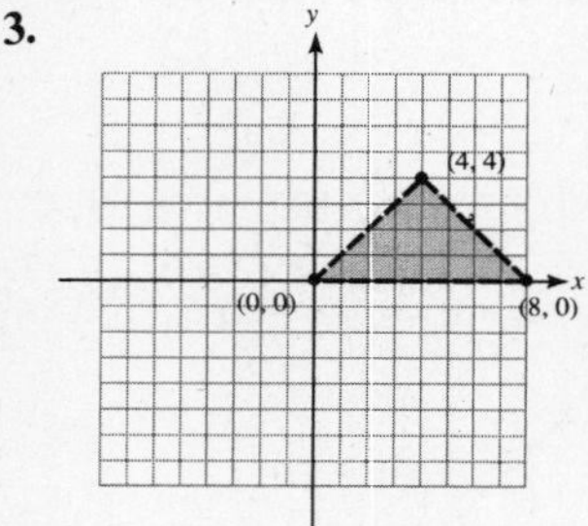

5.

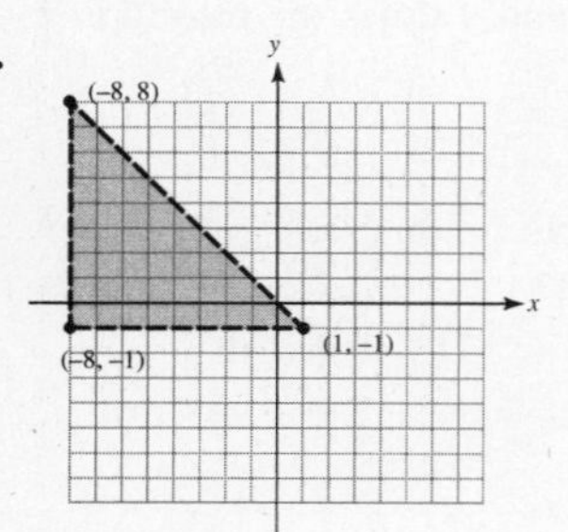

7. (1) $x + y > -4$
(2) $2x - y < 4$

9. (1) $2x + 3y \leq -2$
(2) $2x - 7y \leq 18$

11. (1) $2x - y > -6$
(2) $x + y < 6$
(3) $x - 5y < 6$

13. IV **15.** III **17.** II and III **19.** I and II

Section 9-5. Practice Exercises

1. 14 **3.** 0 **5.** -3 **7.** 1.02 **9.** $\frac{-1}{6}$ **11.** $\sqrt{6}$ **13.** 20 **15.** 13 **17.** 3 **19.** 5 **21.** $\{(3, -5)\}$

23. $\{(-6, 7)\}$ **25.** $\{(10, 0)\}$ **27.** $\left\{\left(\frac{2}{3}, \frac{1}{4}\right)\right\}$ **29.** $\left\{\left(\frac{3}{2}, \frac{-1}{3}\right)\right\}$ **31.** $\{(2, -1, 3)\}$ **33.** $\{(-3, 5, 0)\}$ **35.** $\{(0, -4, -2)\}$

37. $\{(3, 2, -1)\}$ **39.** $\{(-1, -2, 2)\}$ **41.** Inconsistent **43.** Dependent **45.** $\{(1, 5)\}$ **47.** Inconsistent
49. Dependent

Section 9-5. Ten Review Exercises

1. 0 **2.** $\frac{2a^5}{3}$ **3.** $\frac{-u^8v^5}{5}$ **4.** $4x^2 - 3x$ **5.** $3a^2 + ab + 40b^2$ **6.** $\frac{5mn}{5m + n}$ **7.** $\frac{5}{2t + 3}$ **8.** $\frac{2u + 3}{3u + 2}$ **9.** $\frac{4}{9}$ **10.** $\frac{1}{8}$

Section 9-5. Supplementary Exercises

1. 85 **3.** -89 **5.** -80 **7.** 0 **9.** 249 **11.** -18 **13.** -9 **15.** -14 **17.** -11 **19.** 18

Section 9-6. Practice Exercises

1. $\left[\begin{array}{rr|r} -4 & 5 & 30 \\ 7 & -9 & -54 \end{array}\right]$ **3.** $\left[\begin{array}{rrr|r} 1 & 3 & 2 & 2 \\ 0 & 2 & -4 & -12 \\ 2 & 0 & 3 & 1 \end{array}\right]$ **5.** (1) $x + 3y = 5$
(2) $2x - 3y = 7$
7. (1) $3x + 2y = -5$
(2) $-3y + 4z = 6$
(3) $x - 8z = 10$
9. Yes

11. $\{(3, -2)\}$ **13.** $\{(-5, -8)\}$ **15.** $\left\{\left(\frac{1}{2}, \frac{4}{3}\right)\right\}$ **17.** $\{(3, 5, 2)\}$ **19.** $\{(4, 2, -2)\}$ **21.** $\{(-1, 23, 16)\}$ **23.** $\{(3, 6, 4)\}$

25. Inconsistent **27.** Dependent **29.** Inconsistent

Section 9-6. Ten Review Exercises

1. $\frac{\sqrt{2x}}{2x}$ **2.** $\frac{\sqrt{x + 2}}{x + 2}$ **3.** $\frac{\sqrt{x}}{2x}$ **4.** $\frac{\sqrt{2}}{2x}$ **5.** $\frac{x + \sqrt{2}}{x^2 - 2}$ **6.** $\frac{\sqrt{x} - \sqrt{2}}{x - 2}$ **7.** 13 **8.** $\frac{-5}{12}$ **9.** $y = \frac{-5}{12}x + \frac{23}{12}$ **10.** $\frac{23}{5}$

Section 9-6. Supplementary Exercises

1. Multiply row 1 by $\frac{1}{3}$. **3.** Multiply row 2 by -1 and add to row 1. **5.** Multiply row 1 by -2 and add to row 2.

7. Multiply row 2 by 2. **9.** Interchange rows 1 and 3. **11.** Multiply row 3 by -3 and add to row 1. **13.** $\{(2a, -a, a)\}$
15. $\{(b, 2b, 3b)\}$ **17.** $\{(2a, 3b)\}$ **19.** $\{(a + b, a - b)\}$

Chapter 9 Review Exercises

1. No **3.** $\{(-3, 6)\}$ **5.** Inconsistent **7.** $\left\{\left(\frac{-2}{3}, -4\right)\right\}$ **9.** No **11.** Yes **13.** $\left\{\left(\frac{-1}{2}, 0, \frac{2}{3}\right)\right\}$ **15.** $\frac{7}{4}$ and $\frac{2}{3}$

17. First game, 185; second game, 168; third game, 222 **19.** $y = -2x^2 - 12x - 13$ **21. a.** Yes **b.** No

23.

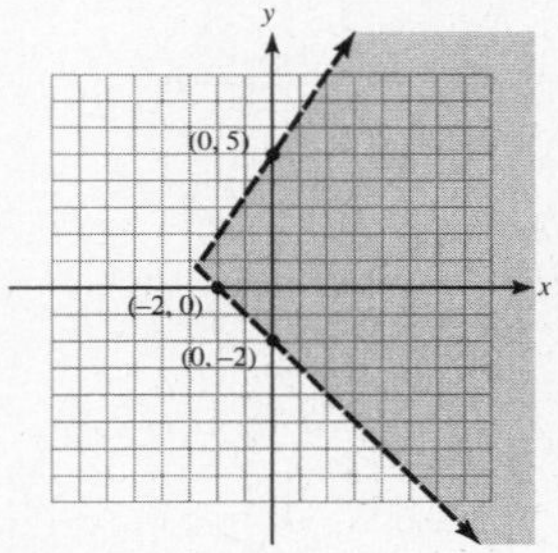

25.

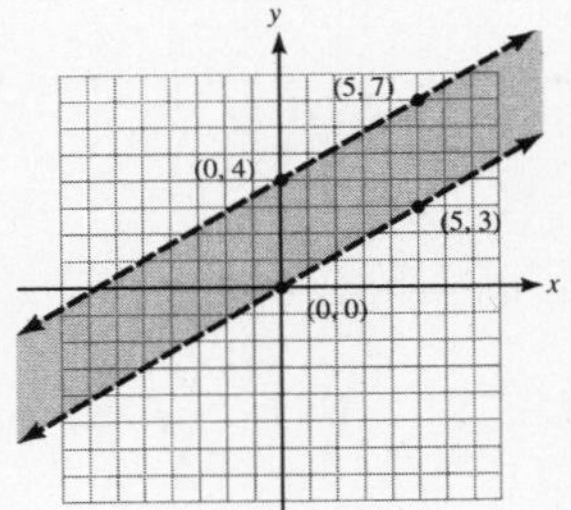

27. 48 **29.** 53 **31.** $\left\{\left(4, \frac{-2}{3}\right)\right\}$ **33.** $\{(-2, 1, 4)\}$

35. $\left[\begin{array}{rr|r} -10 & 3 & 5 \\ 7 & -6 & 16 \end{array}\right]$ **37.** (1) $x - 6y + 4z = -8$
(2) $2x - 5z = 3$
(3) $3x + 4y - z = 7$
39. Yes **41.** $\{(-4, 7)\}$ **43.** $\{(2, -1, 4)\}$

Section 9-7. Practice Exercises

1. a. $D(14, 0)$ **b.** 28 **3. a.** $D(10, 6)$ **b.** 52 **5.** 25 **7.** 12 **9.** 18 **11.** 11

Section 9-7. Ten Review Exercises

1. $2t - 6$ **2.** $\left\{\frac{9}{2}\right\}$ **3.** $t < 5$ **4.** $\left\{\frac{1}{2}, \frac{11}{2}\right\}$ **5.** $t < 0$ or $t > 6$ **6.** Linear function in t and u **7. a.** 2 **b.** -6

8. $(3x - 4)(x + 2)$ **9.** $\left\{\frac{4}{3}, -2\right\}$ **10.** Quadratic function in x and y; parabola

Section 9-7. Supplementary Exercises

1. \$3,000 in Bank A; \$2,000 in Bank B **3.** 180 at \$0.50; 210 at \$0.90 **5.** 0 **7.** 9

Chapter 10 Exponential and Logarithmic Functions

Section 10-1. Practice Exercises

1. a. 125 **b.** $\frac{1}{5}$ **c.** $\frac{1}{25}$ **d.** 1 **3. a.** $\frac{1}{4}$ **b.** 64 **c.** 2 **d.** $\frac{1}{2}$ **5. a.** $\frac{1}{9}$ **b.** 3 **c.** 1 **d.** 9 **7. a.** 1 **b.** 256 **c.** $\frac{1}{2}$ **d.** 8

9.

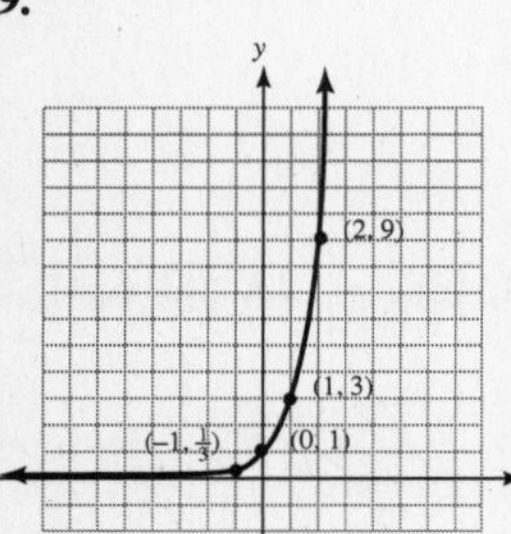

11.

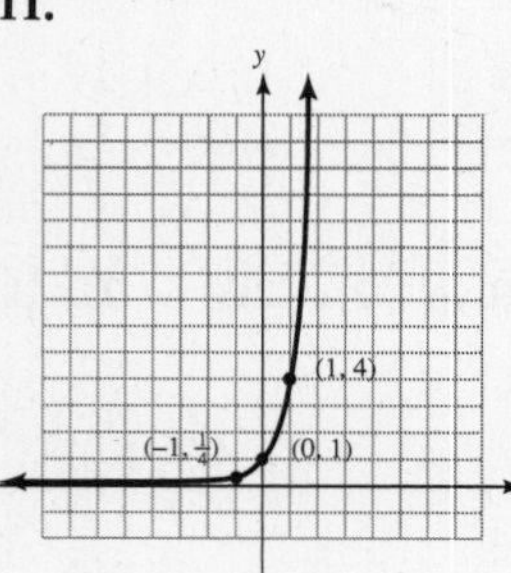

13.

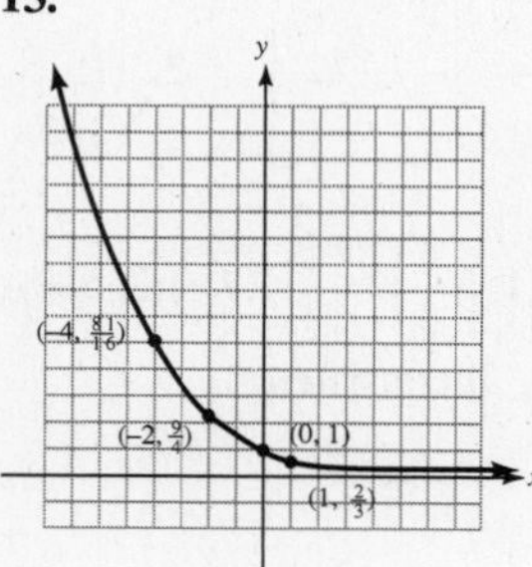

15.

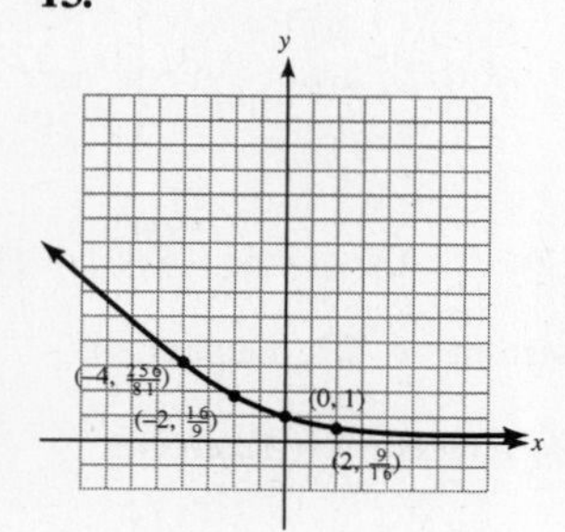

17. 10^{5x} **19.** 5^{4x} **21.** 7^{8-x} **23.** 1 **25.** $\{3\}$ **27.** $\{5\}$ **29.** $\{-2\}$ **31.** $\left\{\frac{3}{2}\right\}$ **33.** $\{-2\}$ **35.** $\{3\}$

37. $\{-3, 3\}$ **39.** $\left\{\frac{2}{3}\right\}$ **41.** $\left\{\frac{1}{2}\right\}$ **43.** $\{-2, 2\}$ **45.** $\{4\}$ **47.** $\{1\}$ **49.** $\{2\}$ **51.** $\{3\}$ **53.** $\{2\}$

Section 10-1. Ten Review Exercises

1. t^5 **2.** 5^{5t} **3.** u^3 **4.** 10^{3u} **5.** v^6 **6.** 7^{6v} **7.** $3\sqrt{5}$ **8.** $3k^{1/2}$ **9.** $43 + 30\sqrt{2}$ **10.** $9z + 30z^{1/2} + 25$

Section 10-1. Supplementary Exercises

1.

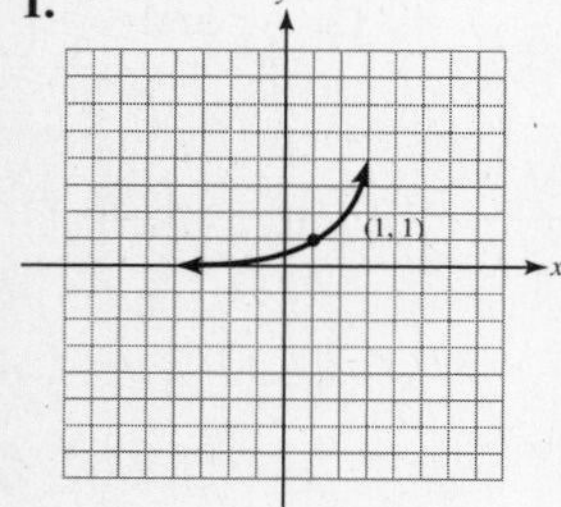

3.

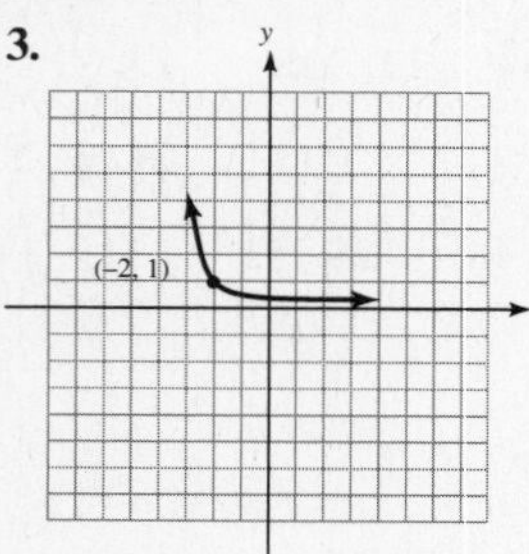

5.

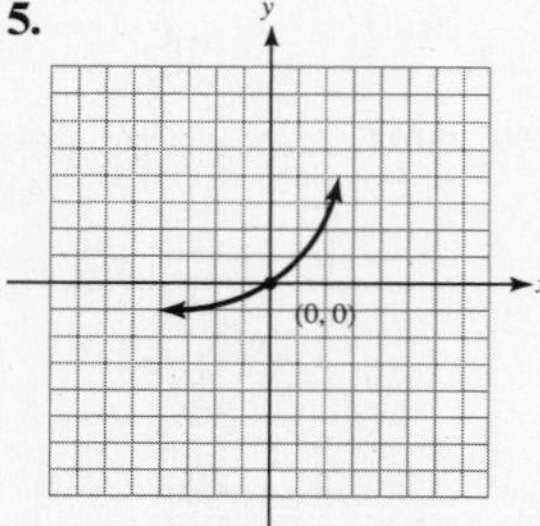

7.

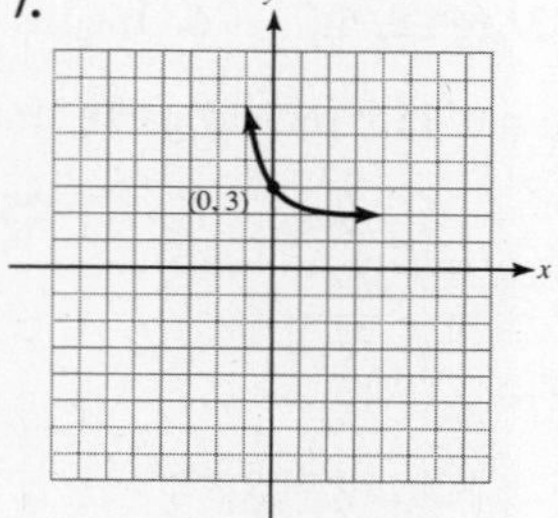

9.

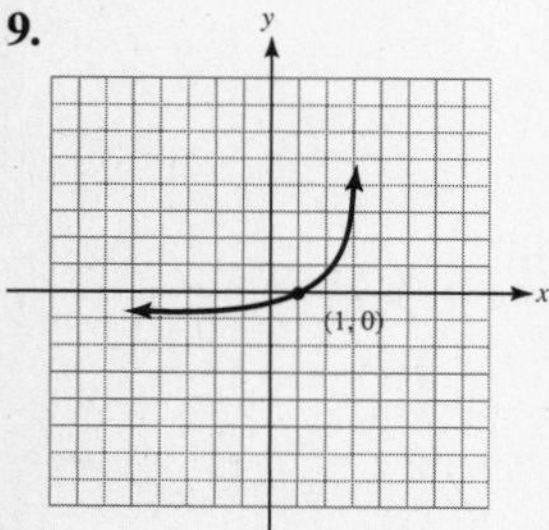

11.

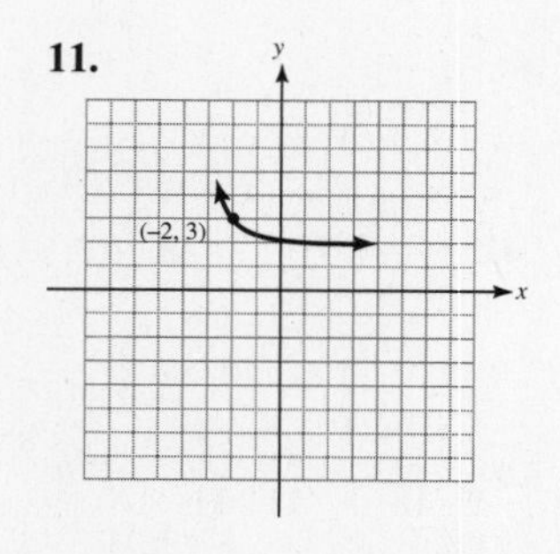

13. a. 8 **b.** 7 **c.** No **15. a.** 2^x **b.** $2x + 1$

Section 10-2. Practice Exercises

1. $\log_{10}100 = 2$ **3.** $\log_{32}2 = \frac{1}{5}$ **5.** $\log_7\frac{1}{7} = -1$ **7.** $\log_{10}10 = 1$ **9.** $\log_{16}8 = \frac{3}{4}$ **11.** $\log_{20}1 = 0$

13. $\log_8\frac{1}{4} = \frac{-2}{3}$ **15.** $\log_4 32 = 2.5$ **17.** $\log_{10}3 = 0.4771$ **19.** $\log_{10}5{,}000 = 3.6990$ **21.** $\log_{10}0.003 = -2.5229$

23. $10^3 = 1{,}000$ **25.** $5^{-2} = \frac{1}{25}$ **27.** $8^1 = 8$ **29.** $12^0 = 1$ **31.** $\left(\frac{1}{2}\right)^{-6} = 64$ **33.** $216^{4/3} = 1{,}296$ **35.** $8^{2/3} = 4$

37. $9^{-3/2} = \frac{1}{27}$ **39.** $10^{0.7782} = 6$ **41.** $10^{2.6021} = 400$ **43.** 6 **45.** 625 **47.** 10 **49.** -3 **51.** 16 **53.** 25

55. 1 **57.** $\frac{1}{20}$ **59.** 10 **61.** -6 **63.** $\frac{1}{8}$ **65.** $\frac{1}{3}$ **67.** 0 **69.** $\frac{5}{2}$ **71.** $\frac{1}{125}$ **73.** 2 **75.** 8 **77.** 2.1505

79. 10 **81.** 7.25

83.

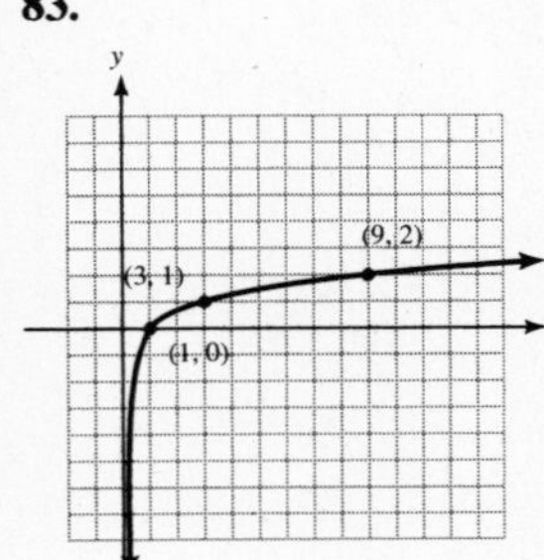

85.

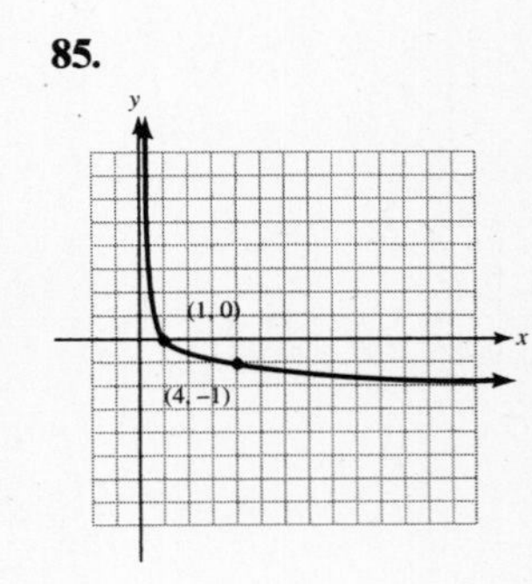

87.

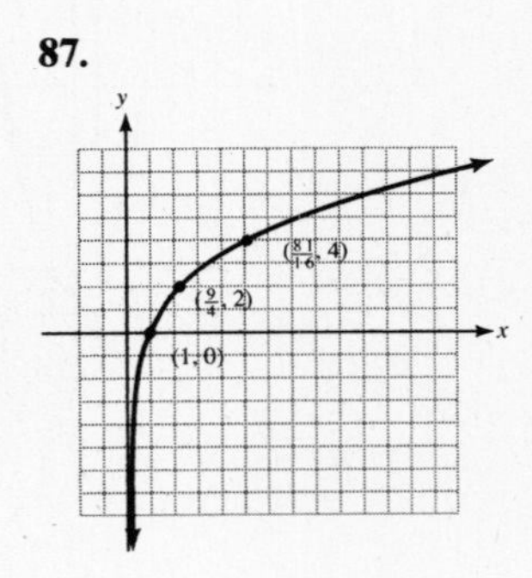

89.

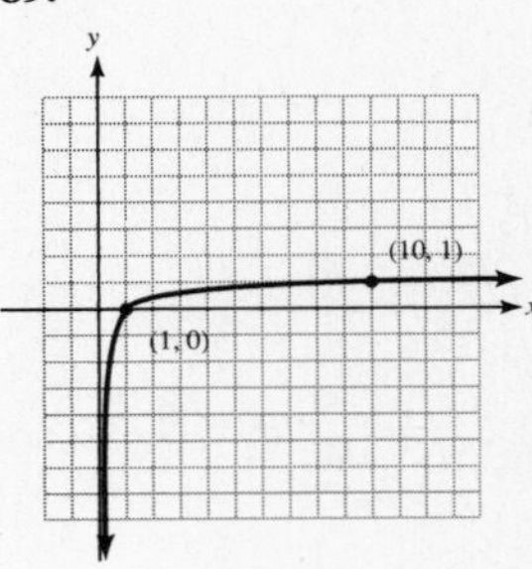

Section 10-2. Ten Review Exercises

1. ± 8 **2.** 6 **3.** 3 **4.** 3 **5.** ± 5 **6.** 4 **7.** $(t-3)(t^2+3t+9)$ **8.** $(2t-1)(t^2+3)$ **9.** $(u-2)(u+2)(u^2+4)$
10. $(10-x+y)(10+x-y)$

Section 10-2. Supplementary Exercises

1.

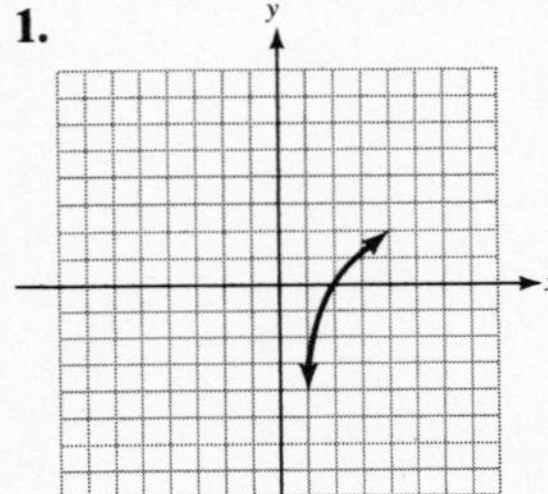

3.

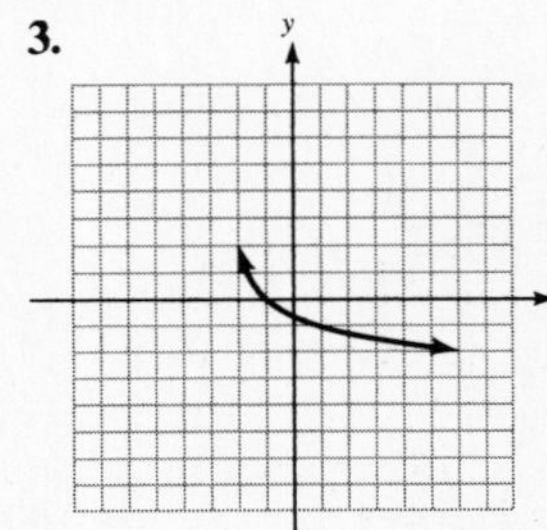

5.

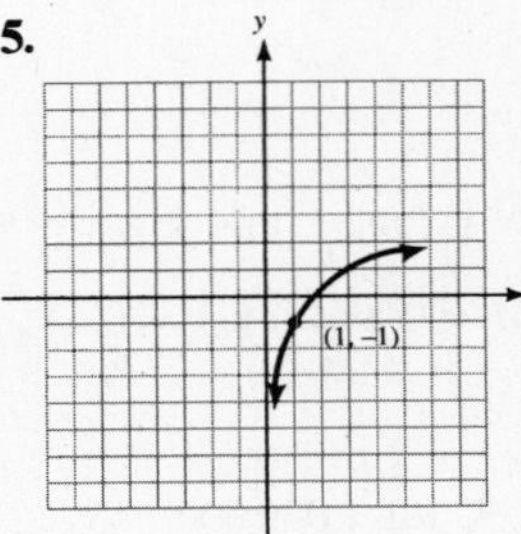

7.

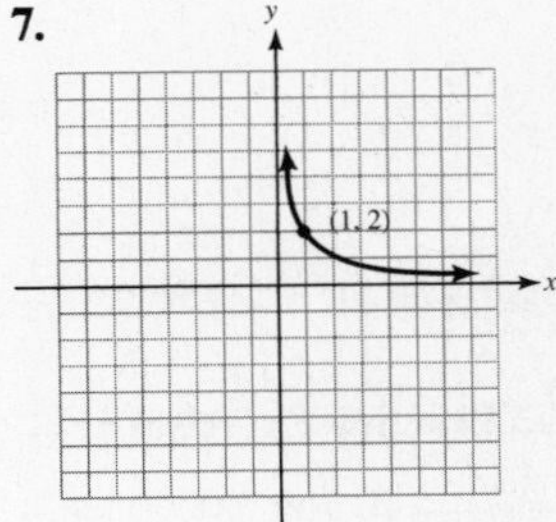

9.

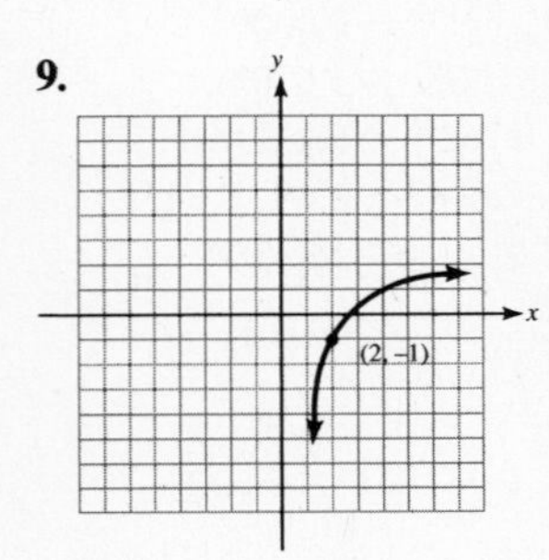

11.

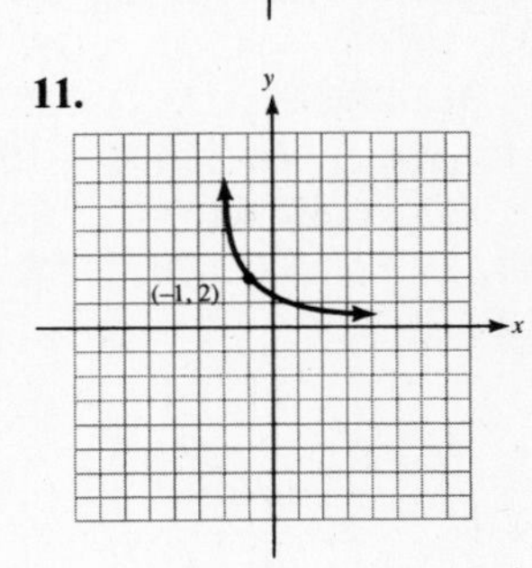

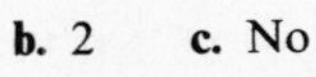
13. a. 4 **b.** 2 **c.** No

Section 10-3. Practice Exercises

1. $\log_2 2 + \log_2 3$ **3.** $\log_2 5 - \log_2 8$ **5.** $2\log_3 7$ **7.** $-\log_{10}9$ **9.** 1 **11.** 0 **13.** $\log_3 7 + \log_3 10$ **15.** $7\log_3 10$

17. $\log_{10}3 - \log_{10}7$ **19.** $\frac{1}{3}\log_{10}7$ **21.** $\log_2 6 + \log_2 5 - \log_2 11$ **23.** $\frac{1}{2}(\log_3 5 + \log_3 x)$ **25.** $\log_5 10 + 2\log_5 x$

27. $\frac{1}{3}\log_{10}2 - 2\log_{10}x$ **29.** $-(\log_2 7 + \log_2 x)$ **31.** $\log_{10}6$ **33.** $\log_3 5t$ **35.** $\log_2 0.7$ **37.** $\log_5 \frac{6t}{7}$ **39.** $\log_3 25$

41. $\log_{10}\sqrt{6}$ **43.** $\log_5(3x)^2$ **45.** $\log_2\left(\frac{y}{5}\right)^3$ **47.** $\log_3\sqrt{10x}$ **49.** $\log_{10}\frac{\sqrt[3]{5}}{9}$ **51.** $\log_2\frac{\sqrt{3x}}{10}$ **53.** $\log_3\frac{2x}{11y}$

55. $\log_5\frac{\sqrt{10y}}{3x}$ **57.** $\log_2\frac{10t^3}{7u^2}$ **59.** $(\log_3 4k)(\log_3 10k)$ **61. a.** $\frac{\log_{10}5}{\log_{10}9}$ **b.** $\frac{\log_e 5}{\log_e 9}$ **63. a.** $\frac{\log_{10}23}{\log_{10}2}$ **b.** $\frac{\log_e 23}{\log_e 2}$

65. a. $\frac{-\log_{10}2}{\log_{10}5}$ **b.** $\frac{-\log_e 2}{\log_e 5}$ **67. a.** $\frac{\log_{10}17}{-\log_{10}3}$ **b.** $\frac{\log_e 17}{-\log_e 3}$

Section 10-3. Ten Review Exercises

1. 0 **2.** 0 **3.** $\{-2, 4\}$ **4.** $(t-1)^2 - 9$ **5.**

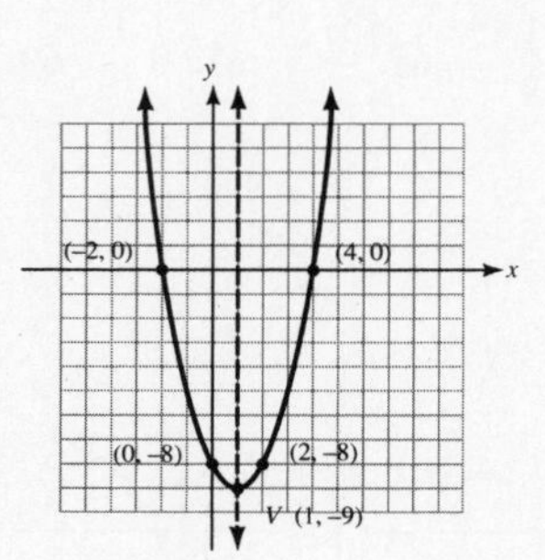

6. $6 + y - 2y^2$ **7.** $-4 + \sqrt{5}$ **8.** $8 + i$

9. $-16 - 7\sqrt{5}$ **10.** $\frac{4 + 7i}{5}$

Section 10-3. Supplementary Exercises

1. 0.78 **3.** −0.4 **5.** 0.60 **7.** 2.10 **9.** 0.24 **11.** 1.48 **13.** −0.30 **15.** 1.0 **17.** 0.45 **19.** 0.79

21. $6 \neq 5$ **23.** $3 \neq 2$ **25.** $5 \neq 6$ **27.** $0 \neq 7$ **29.** $\frac{\log_2 15}{\log_2 60}$ **31.** $(\log_{10}2x)(\log_{10}3y)$ **33.** $\frac{4\log_2 x}{\log_2 10}$

35. $\left(\log_3\frac{36x}{5}\right)\left(\log_3\frac{\sqrt{10y}}{7}\right)$

37. Let $M = \log_b x$; then $b^M = x$
Let $N = \log_b y$; then $b^N = y$
$\frac{b^M}{b^N} = \frac{x}{y}$, and $b^{M-N} = \frac{x}{y}$

$M - N = \log_b\frac{x}{y}$, and $\log_b x - \log_b y = \log_b\frac{x}{y}$

39. $\log_b\frac{1}{x} = \log_b x^{-1} = -1 \cdot \log_b x = -\log_b x$

41. If $\log_b 1 = M$, then $1 = b^M$. Therefore, $M = 0$ and $\log_b 1 = 0$.

Section 10-4. Practice Exercises

1. $\log 100 = 2$ **3.** $\log 0.00001 = -5$ **5.** $\log 1{,}000{,}000 = 6$ **7.** $\log 31.6 = 1.5$ **9.** $\log 0.0079 = -2.1$
11. $10^{1.1761} = 15$ **13.** $10^{0.1703} = 1.48$ **15.** $10^{-0.1580} = 0.695$ **17.** $10^{3.6532} = 4{,}500$ **19.** $10^{-3.0458} = 0.0009$
21. 0.6946 **23.** 0.2833 **25.** 1.8751 **27.** 2.1673 **29.** 3.7993 **31.** 4.5694 **33.** −0.2518 **35.** −1.6308
37. −2.0809 **39.** −4.0362 **41.** 3.75 **43.** 9.17 **45.** 79.0 **47.** 514 **49.** 18,900 **51.** 0.762 **53.** 0.00415
55. 0.0000955 **57.** 0.0306 **59.** 0.000692 **61.** 2.8904 **63.** −1.3863 **65.** 6.1570 **67.** −2.6187 **69.** 8.8451
71. −7.6009 **73.** 1 **75.** 5.6664 **77.** 0.9163 **79.** 4.7918 **81.** 3.49 **83.** 1.20 **85.** 11.9 **87.** 98,700
89. 0.705 **91.** 0.239 **93.** 0.130 **95.** 0.0235 **97.** 1 **99.** 15.2

Section 10-4. Ten Review Exercises

1. $\frac{1}{xy^4}$ **2.** $\frac{4}{x^6}$ **3.** $-3\sqrt{3}$ **4.** 30 **5.** $\frac{5}{2}$ **6.** $60\sqrt{3}$ **7.** $5u + 11$ **8.** $\frac{2u+3}{2u-3}$ **9.** $-11 + 5i$ **10.** $\frac{8-i}{5}$

Section 10-4. Supplementary Exercises

1. 5.43 **3.** −0.325 **5.** 4.16 **7.** −1.34 **9.** 0.802 **11.** 4.11 **13.** 5.93 **15.** −3.70 **17.** 1.46 **19.** −2.58
21. 4.692 **23.** 2.179 **25.** −2.226 **27.** 24.3 **29.** 1.26 **31.** 0.0646 **33.** 17.2 **35.** 17.6 **37.** 1.95
39. 46.1 **41.** 13.1 **43.** 3,120

Section 10-5. Practice Exercises

1. a. $\left\{\frac{\log 12}{\log 2}\right\}$ **b.** 3.58 **3. a.** $\left\{\frac{\log 15}{\log 36}\right\}$ **b.** 0.756 **5. a.** $\left\{\frac{\log 4}{\log 8}\right\}$ **b.** 0.667 **7. a.** $\left\{\frac{\log 133}{\log 7}\right\}$ **b.** 2.51

9. a. $\left\{\frac{\log 17}{\log 4}\right\}$ **b.** 2.04 **11. a.** $\left\{\frac{\log 2{,}400}{\log 125}\right\}$ **b.** 1.61 **13. a.** $\left\{\frac{\log 24}{\log 3}\right\}$ **b.** 2.89 **15. a.** $\left\{\frac{\log 3}{\log 6}\right\}$ **b.** 0.613

17. a. $\left\{\frac{\log 96}{\log 1.5}\right\}$ **b.** 11.3 **19. a.** $\left\{\frac{\log 0.02}{\log 4}\right\}$ **b.** -2.82 **21. a.** $\left\{\frac{\log 0.05}{\log 2.5}\right\}$ **b.** -3.27 **23. a.** $\left\{\pm\sqrt{\frac{\log 48}{\log 2}}\right\}$

b. ± 2.36 **25. a.** $\left\{\pm\sqrt{\frac{\log 15}{\log 6}}\right\}$ **b.** ± 1.23 **27. a.** $\left\{\left(\frac{\log 75}{\log 4}\right)^2\right\}$ **b.** 9.70 **29. a.** $\left\{\left(\frac{\log 380}{\log 5}\right)^2 + 1\right\}$ **b.** 14.6

31. {3} **33.** {1} **35.** {2} **37.** {4} **39.** {11} **41.** {3} **43.** {8} **45.** $\left\{\frac{1}{2}, 4\right\}$ **47.** $\left\{\frac{3}{2}, \frac{4}{3}\right\}$ **49.** {2}

51. {1} **53.** {2, 8} **55.** {3} **57.** {3, 7} **59.** $\{10, \sqrt{10}\}$

Section 10-5. Ten Review Exercises

1. $2m - 10$ **2.** {5} **3.** $(x + 5b)(x - 3a)$ **4.** $\{-5b, 3a\}$ **5.** $\frac{2 - 5t}{t(t - 1)(t + 1)}$ **6.** $\left\{\frac{2}{5}\right\}$ **7.** $5 < x < 11$

8. $x \leq -4$ or $x \geq -1$

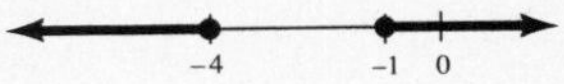

9. $x < -3$ or $x > 10$

10. $\frac{-3}{2} < x < 2$

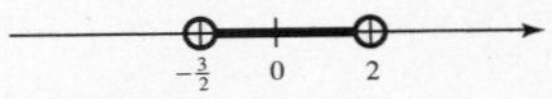

Section 10-5. Supplementary Exercises

1. $y = 2x^3$ **3.** $y = 0.5x^2$ **5.** $y = 1.2x^{2.5}$ **7.** $y = x^{0.3}$ **9. a.** D **b.** D **c.** F **d.** I **e.** G

Section 10-6. Practice Exercises

1. $r = \frac{\log S - \log P}{t \log a}$ **3.** $t = \frac{\log A - \log P}{r \log e}$ **5.** $k = \frac{\log B - \log A}{t \log(1 + i)}$ **7.** $h = \frac{t \log 2}{\log k - \log B}$ **9.** $r = \frac{t \log e}{\log X + \log Y - \log k}$

11. 5.6 **13.** 8.9 **15.** 2.1 **17.** 9.3 **19.** 1.6×10^{-3} **21.** 3.2×10^{-5} **23.** 6.3×10^{-8} **25.** 7.9×10^{-10}

27. \$13,400.96 **29.** \$3,869.68 **31.** \$6,756.00 **33.** 8 yrs **35.** 609 **37.** 5,742 **39.** 626,591 **41.** 8.66 yrs

43. 121 mg **45.** -0.006

Section 10-6. Ten Review Exercises

1. Parabola **2.** $y = (x - 3)^2 - 4$ **3.** $V(3, -4)$ **4.** $x = 3$ **5.** 1 and 5 **6.** 5 **7.** All x **8.** All y; $y \geq -4$

9. $\frac{-4}{3}$ **10.** $f(x) = \frac{-4}{3}x + \frac{1}{3}$

Section 10-6. Supplementary Exercises

1. $y = 1.5e^x$ **3.** $y = 2e^{0.5x}$ **5.** $y = 5e^{-x}$ **7.** $y = 4.2e^{0.3x}$

Chapter 10 Review Exercises

1. a. $\frac{1}{100}$ **b.** 1 **3.**

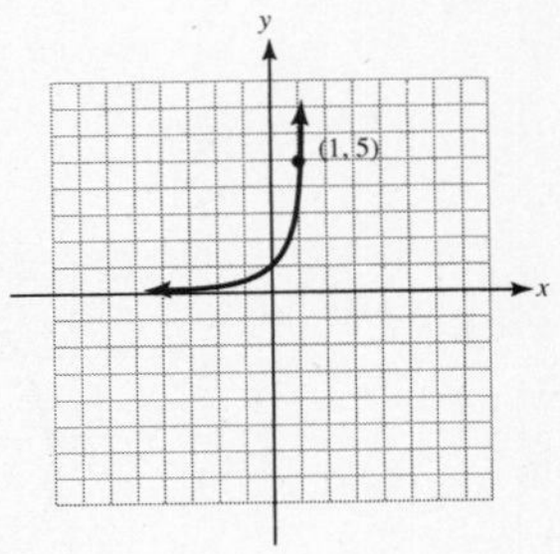

5. 5^{6x} **7.** $\left\{\frac{3}{4}\right\}$ **9.** {3} **11.** $\left\{\frac{2}{5}\right\}$ **13.** $\log_5 625 = 4$

15. $\log_{10} 1 = 0$ **17.** $3^5 = 243$ **19.** $10^{2.6928} = 493$ **21.** 4 **23.** $\frac{1}{16}$ **25.** 10

27.

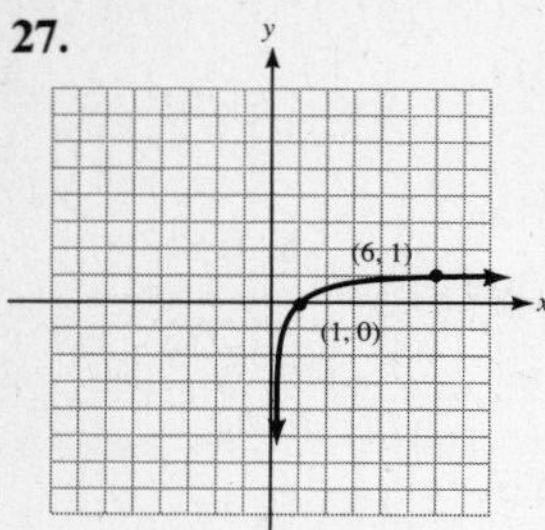

29. $\log_{10}3 + \log_{10}x - \log_{10}5$ **31.** $\frac{1}{2}(\log_3 2 + \log_3 x + \log_3 y)$ **33.** $\log_2\sqrt{15}$ **35.** $\log_{10}8y^2$

37. a. $\frac{\log 93}{\log 8}$ **b.** $\frac{\ln 93}{\ln 8}$ **39.** 5.6 **41.** 1.1 **43.** $\log 631 = 2.8$ **45.** $10^{3.8943} = 7{,}840$ **47.** 1.4472 **49.** -1.9066

51. 736 **53.** 0.0329 **55.** 4.96 **57.** 4.08 **59.** 2.05 **61.** 0.771 **63. a.** $\left\{\frac{\log 1{,}500}{\log 12}\right\}$ **b.** 2.94 **65. a.** $\left\{\frac{\log 6}{\log 24}\right\}$

b. 0.564 **67.** $\{14\}$ **69.** $\{5\}$ **71.** $t = \frac{\log I - \log A}{k \log e}$ **73.** 1.4 **75.** 5.0×10^{-4} **77.** $33,120.60 **79.** 173 yrs

Chapter 11 Quadratic Relations

Section 11-1. Practice Exercises

1. a. $y = \frac{1}{4}x^2$ **b.**

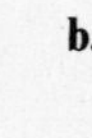

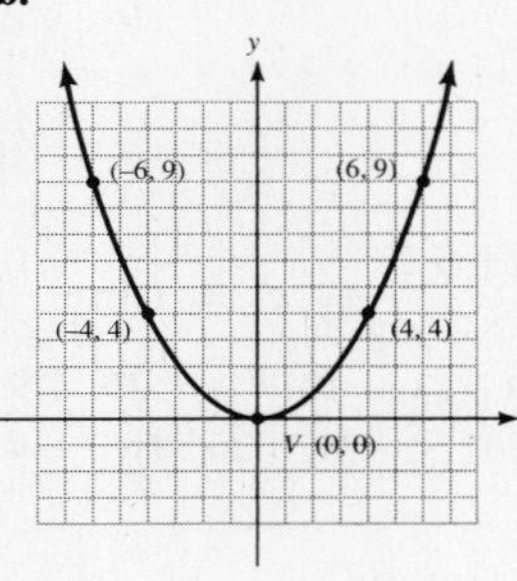

3. a. $y = \frac{-1}{8}x^2$ **b.**

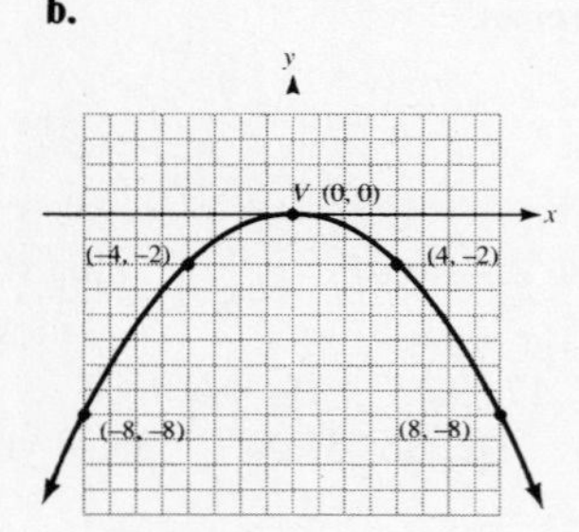

5. a. $y = \frac{-1}{3}x^2$

b.

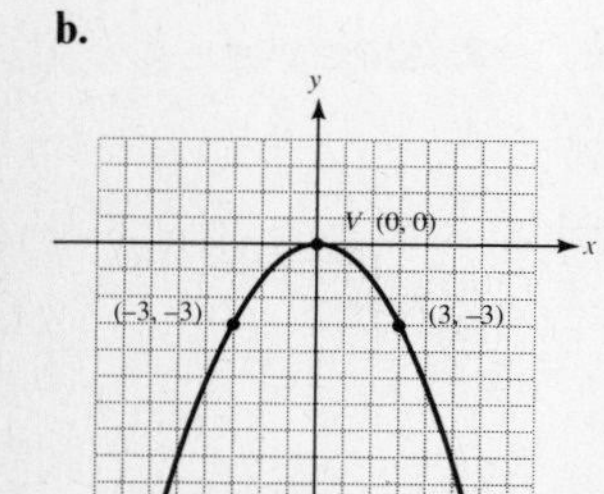

7.

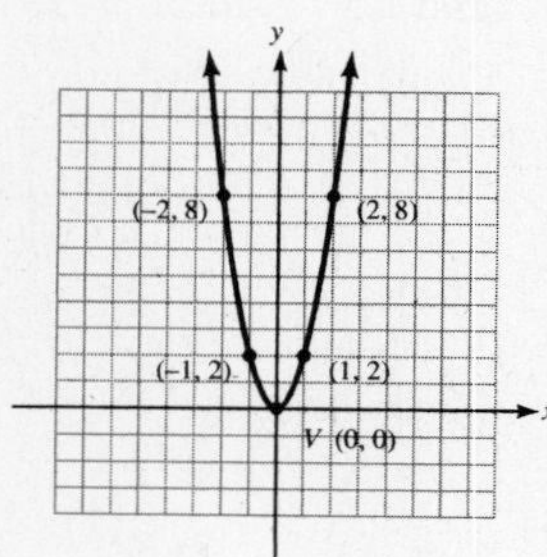

9.

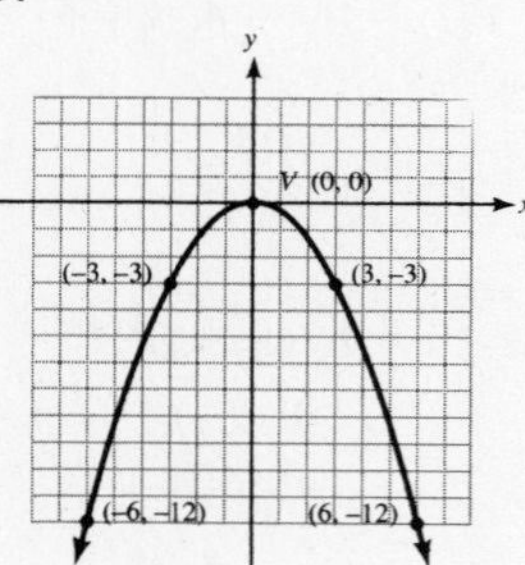

11.

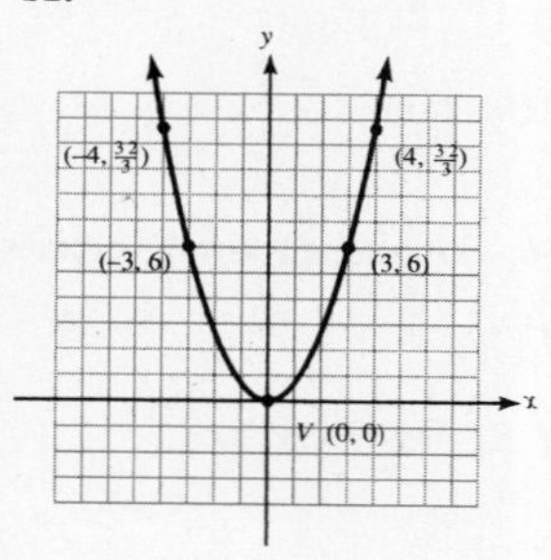

13.

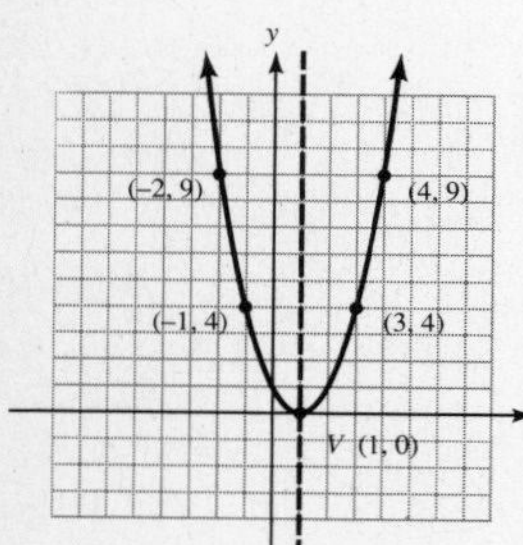

15.

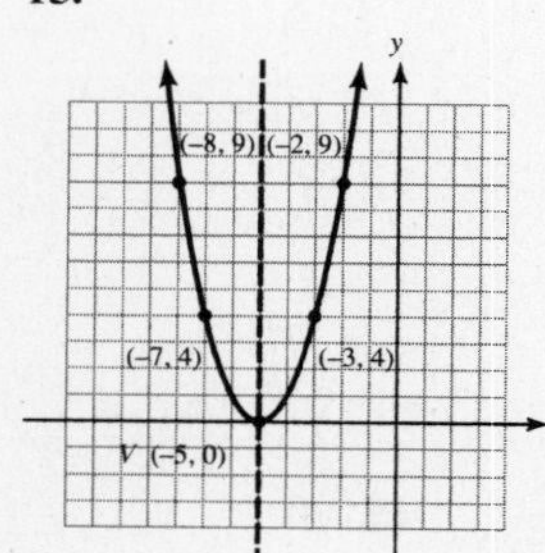

17.

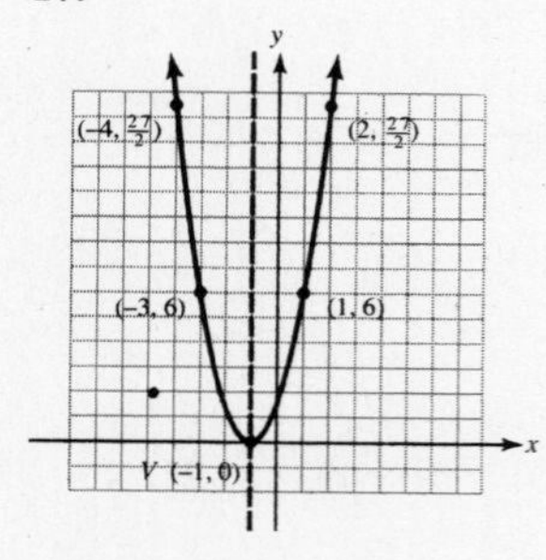

19.

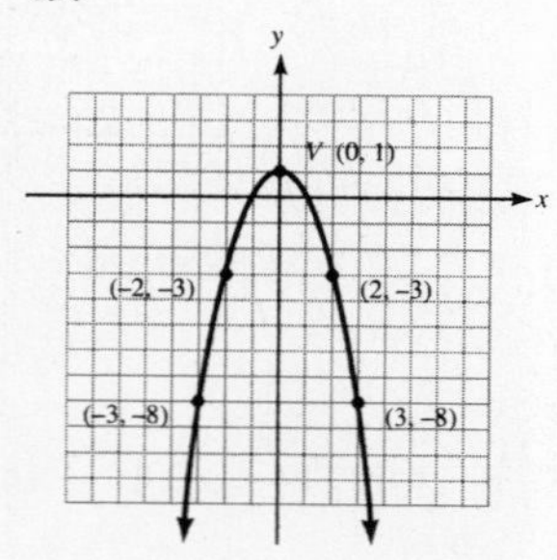

21.

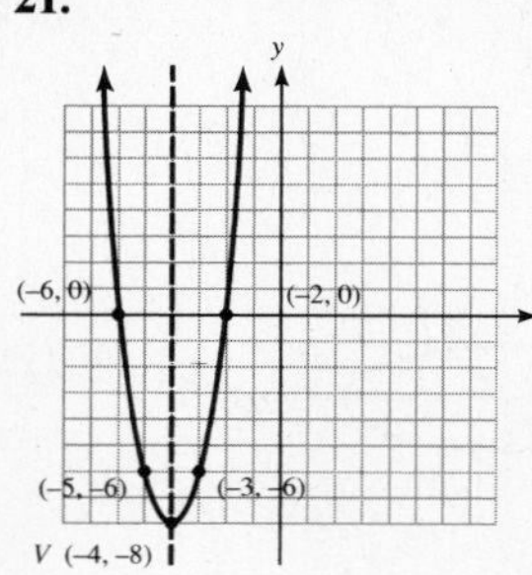

23.

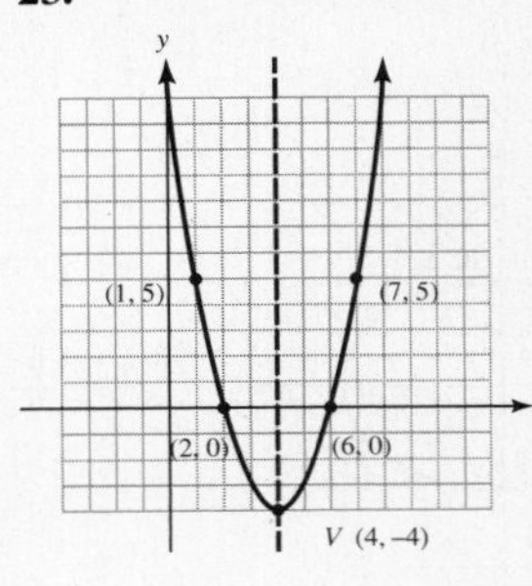

25.

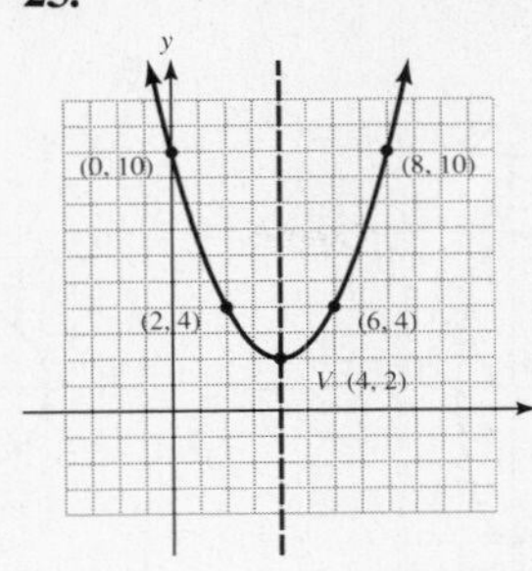

27.

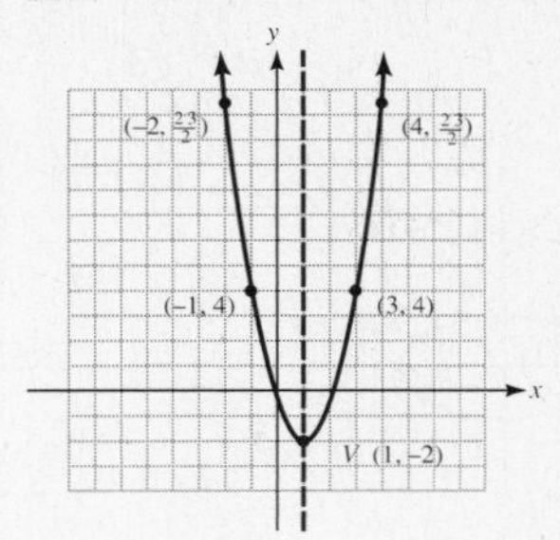

29.

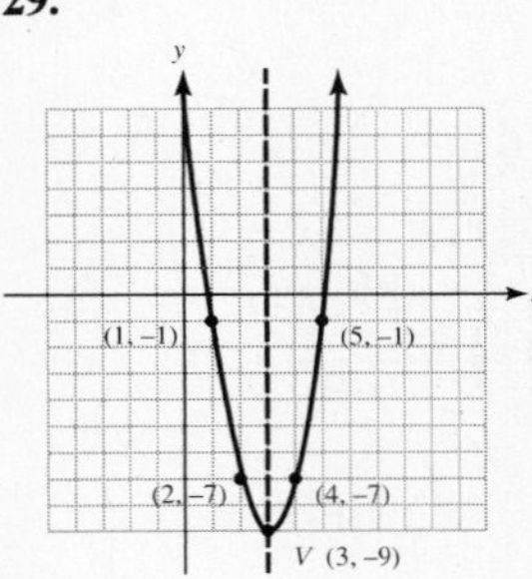

31.

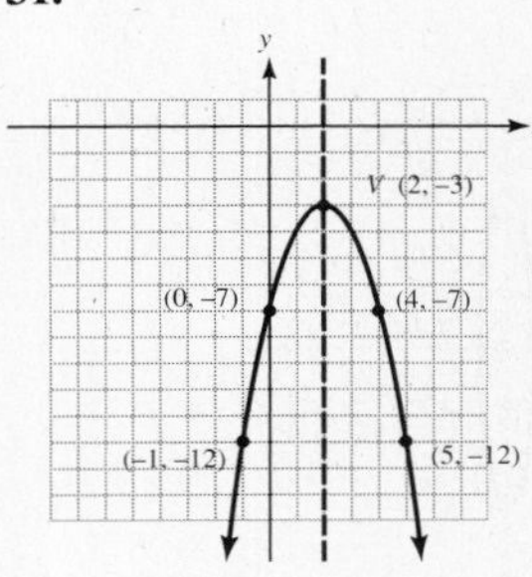

33.

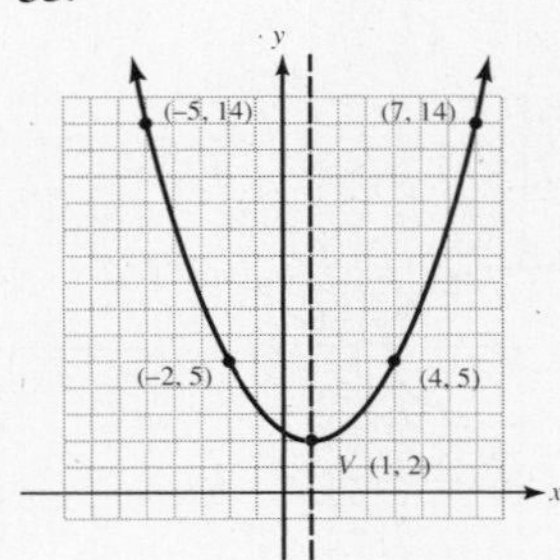

Section 11-1. Ten Review Exercises

1. $\{5\}$ **2.** $\{5, 11\}$ **3.** $\left\{\frac{2}{5}\right\}$ **4.** $\left\{\frac{-2}{3}, \frac{-3}{2}\right\}$ **5.** $\{16\}$ **6.** $\left\{\frac{2}{3}\right\}$ **7.** $\left\{\frac{\log 96}{\log 125}\right\}$ **8.** $\{2\}$

9. $\frac{-3}{2} < x < 2$ **10.** $x < -8$ or $x > -2$

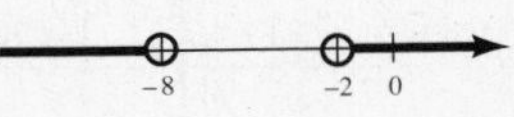

Section 11-1. Supplementary Exercises

1.

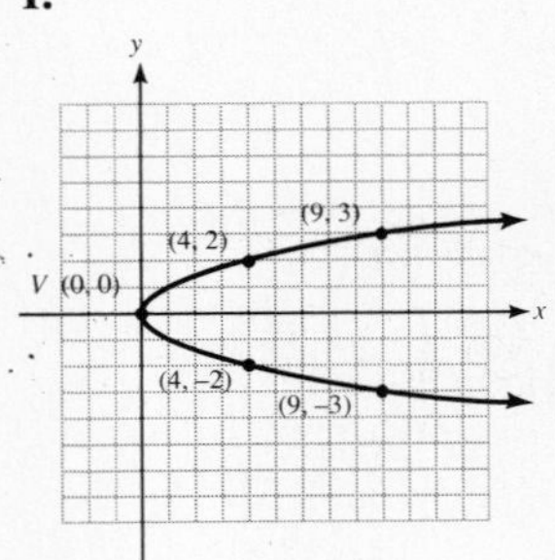

3.

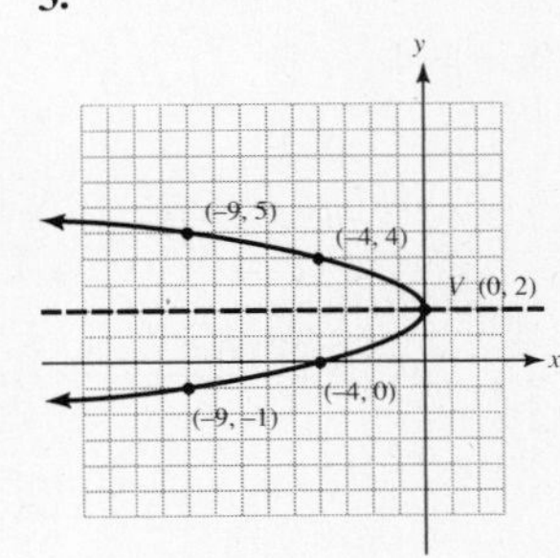

5.

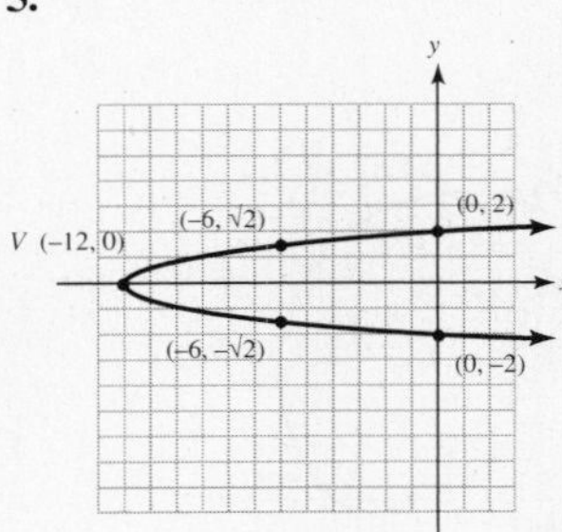

7.

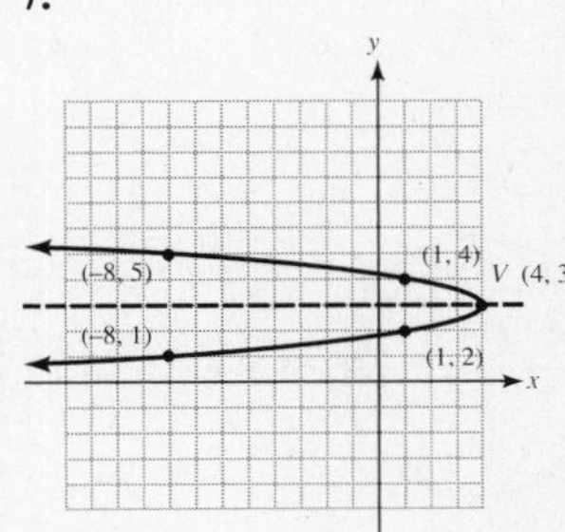

9.

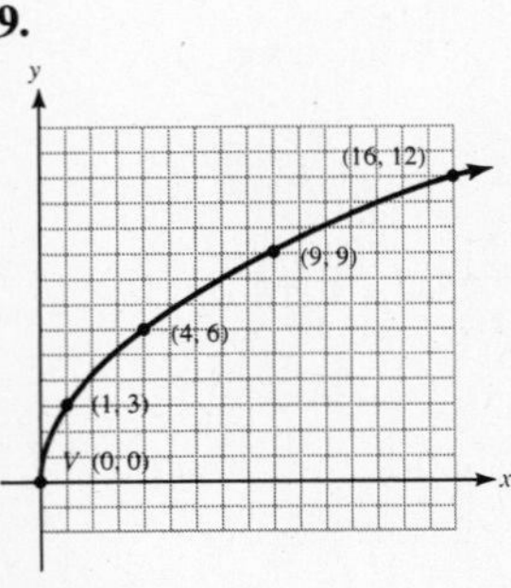

11.

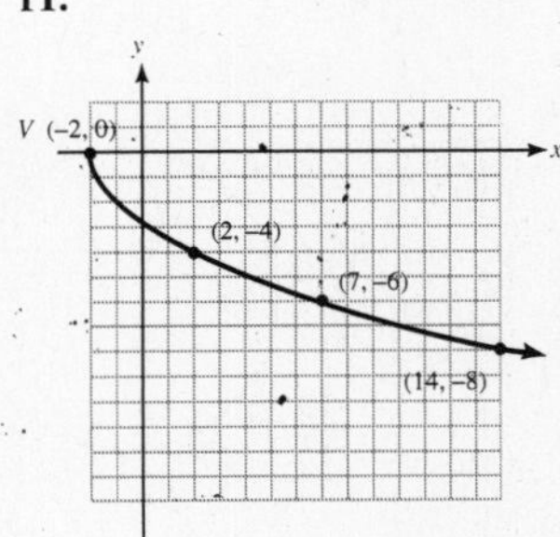

13. $y = \frac{47}{441{,}000}x^2$ **15.** $y = \frac{47}{441{,}000}(x + 2{,}100)^2$

Section 11-2. Practice Exercises

1.

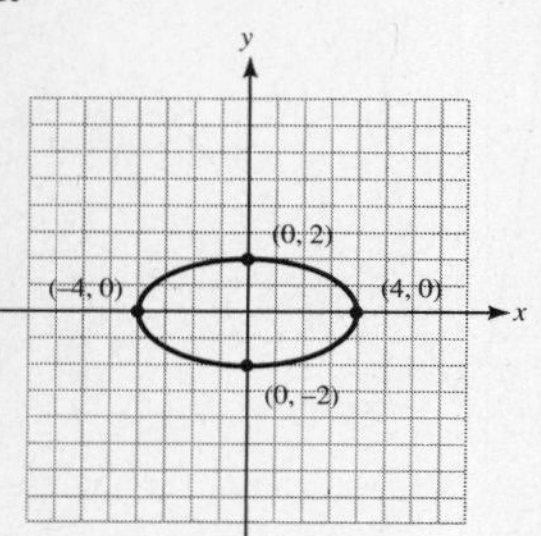

3.

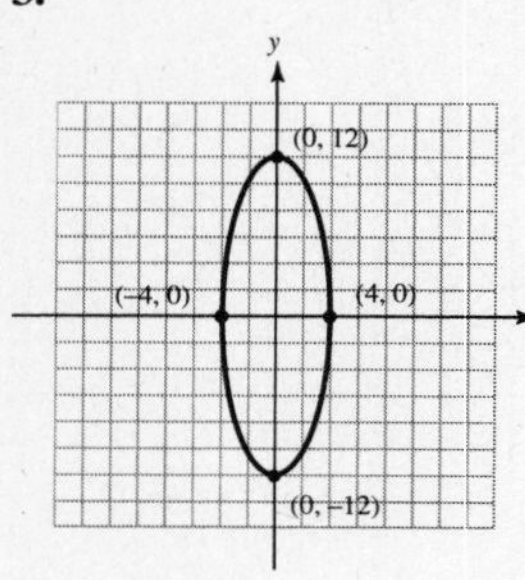

5.

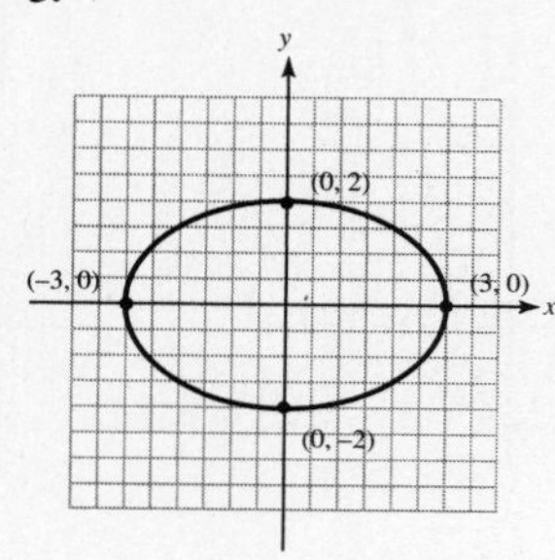

7.

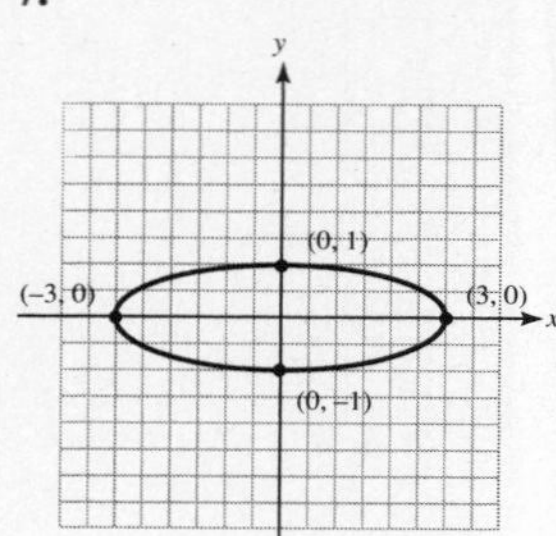

9.

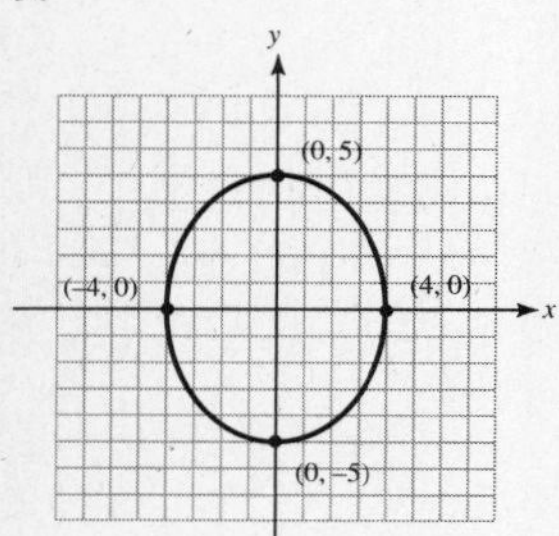

11.

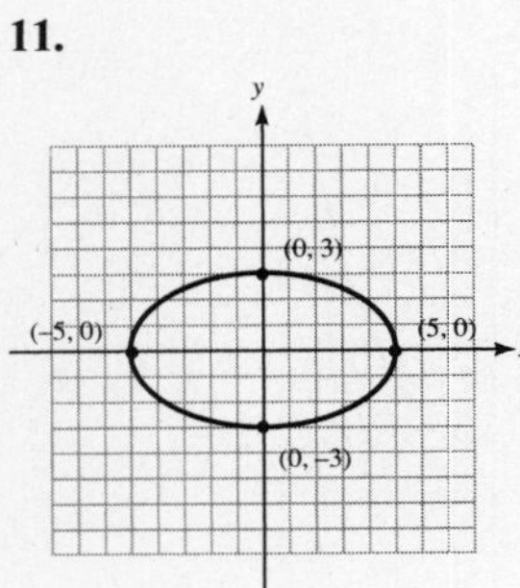

13.

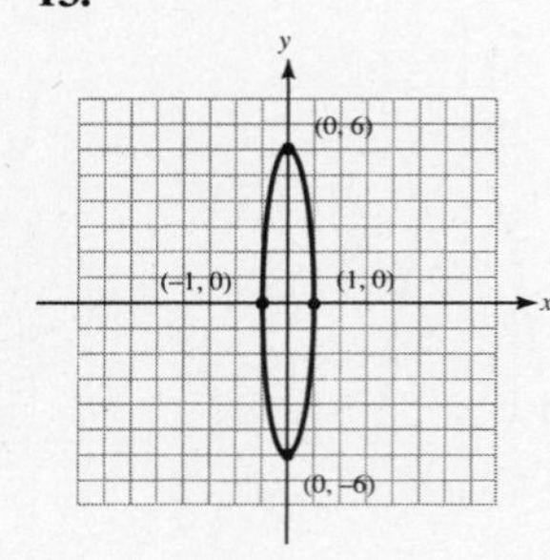

15.

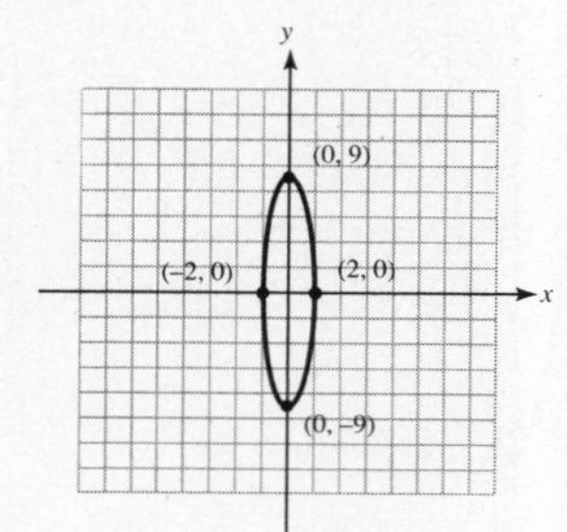

17.

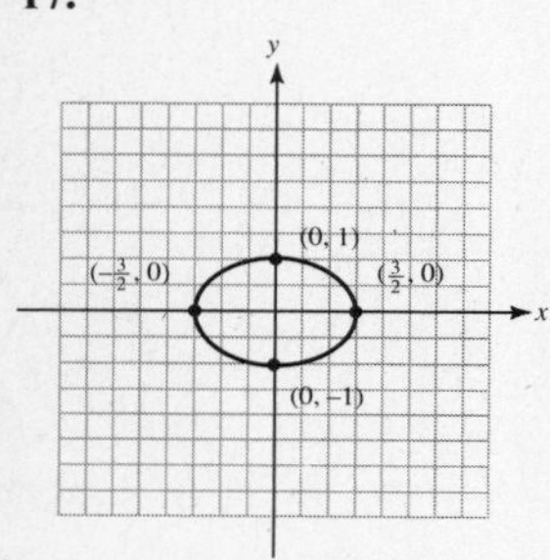

19.

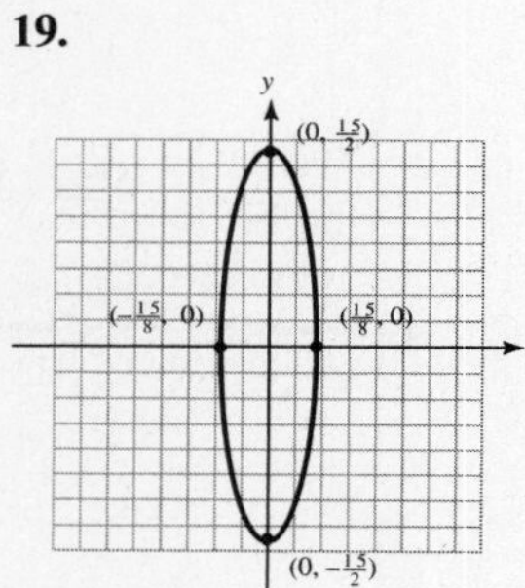

21.

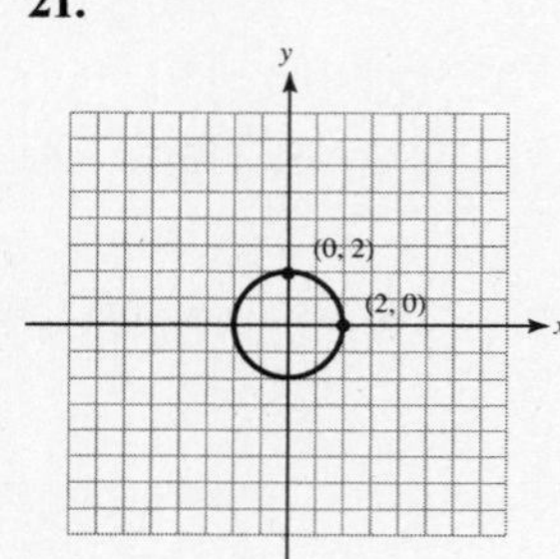

23.

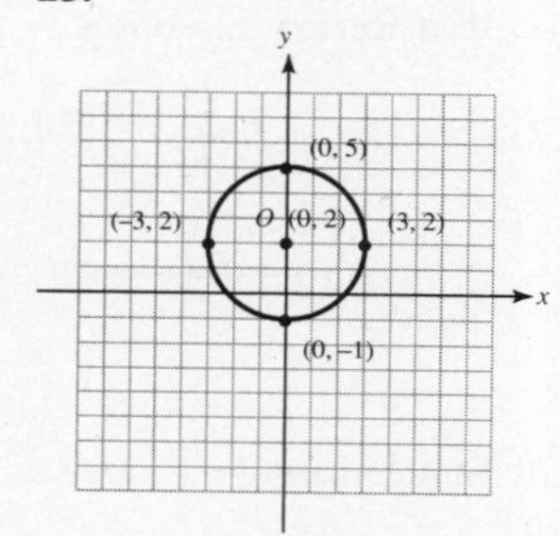

25.

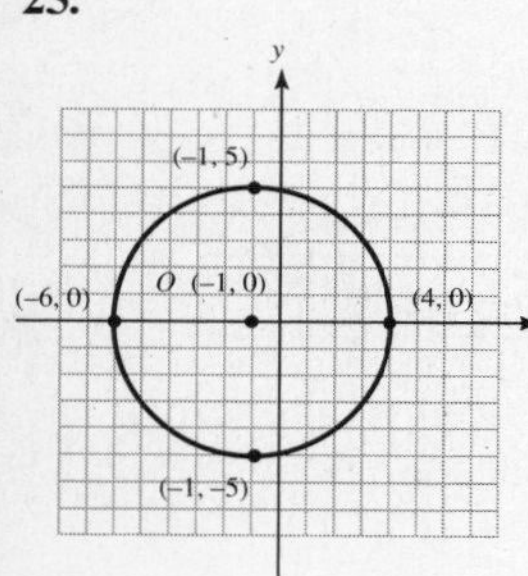

27.

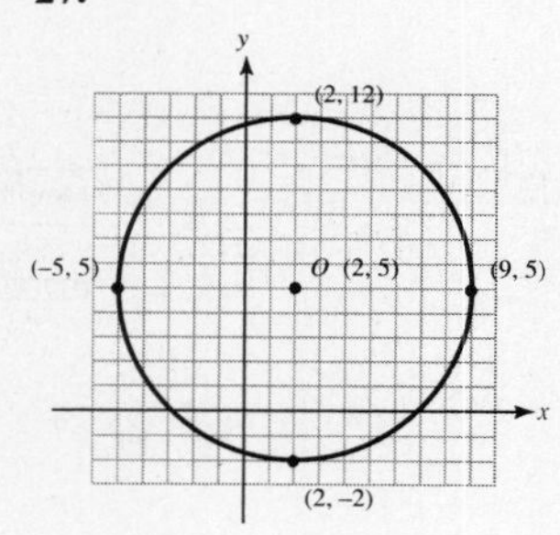

29. a. (5, 0); $r = 4$ **b.**

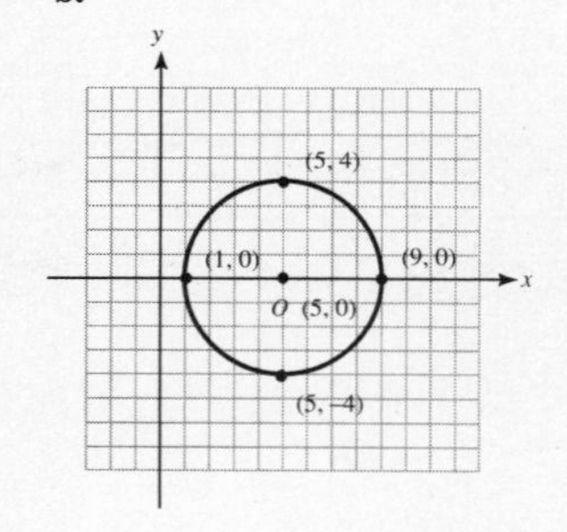

31. a. $(0, -3)$; $r = 5$ **b.**

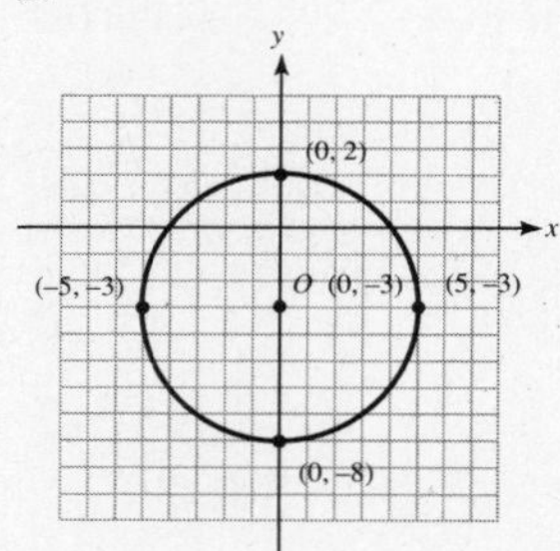

33. a. $(1, -2)$; $r = 6$ **b.**

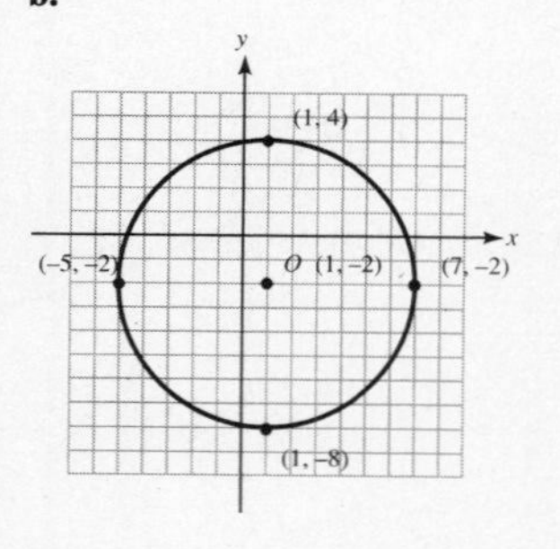

35. a. $(-2, 1)$; $r = 10$ **b.** 

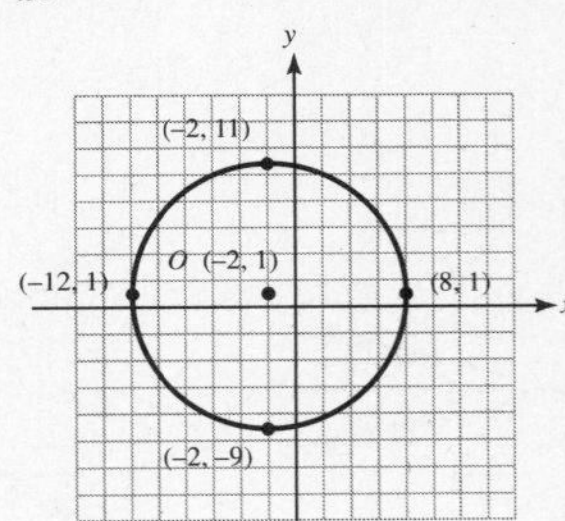

37. a. $(-4, -5)$; $r = 2$ **b.**

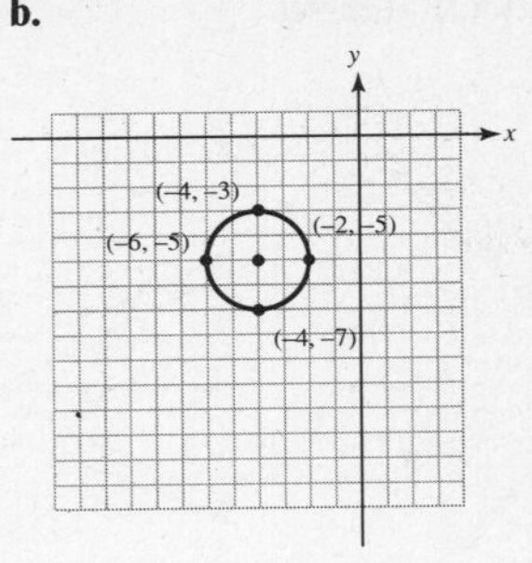

39. a. $\left(\frac{3}{2}, \frac{-5}{2}\right)$; $r = 3$ **b.**

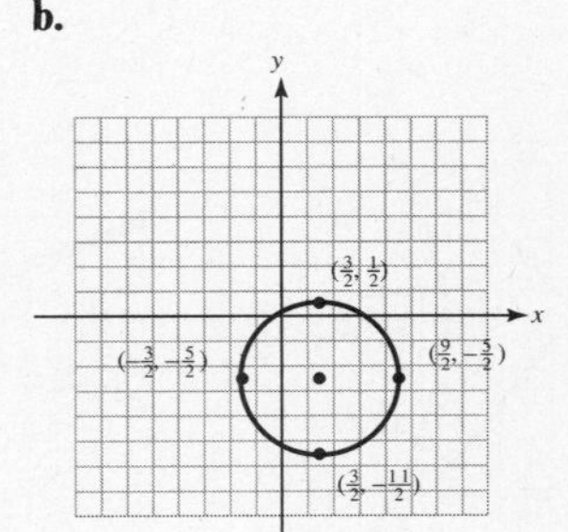

Section 11-2. Ten Review Exercises

1. $\frac{2}{9t}$ **2.** $16u^2$ **3.** All x; $x \neq \frac{2}{3}$ **4.** $f^{-1}(x) = \frac{2x + 3}{3x - 1}$ **5.** All x; $x \neq \frac{1}{3}$ **6.** All y; $y \neq \frac{2}{3}$ **7.** $f(x) = \frac{4}{3}x - 1$ **8.** $\frac{4}{3}$ **9.** -1 **10.** $\frac{3}{4}$

Section 11-2. Supplementary Exercises

1.

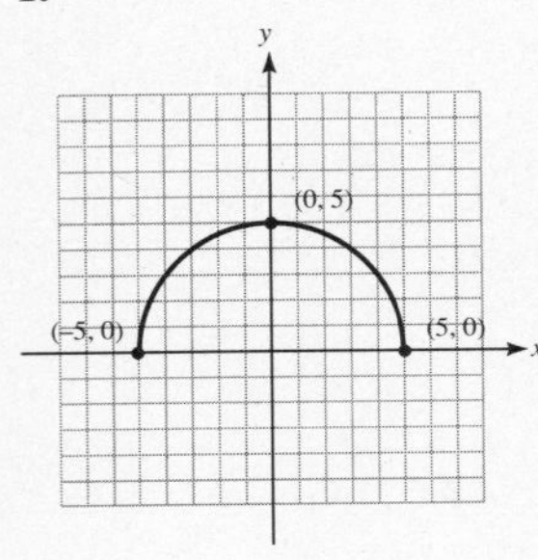

3.

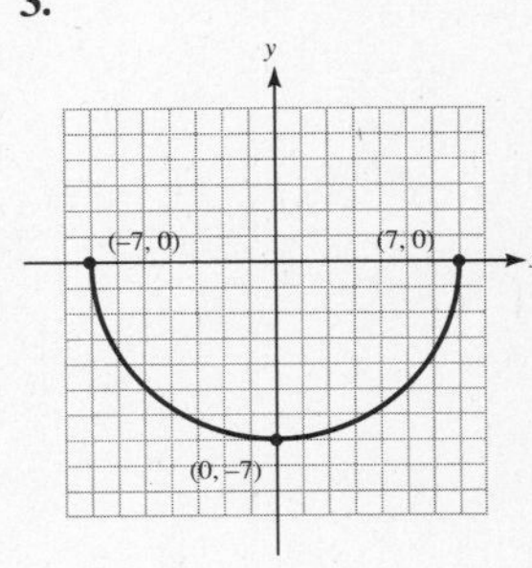

5.

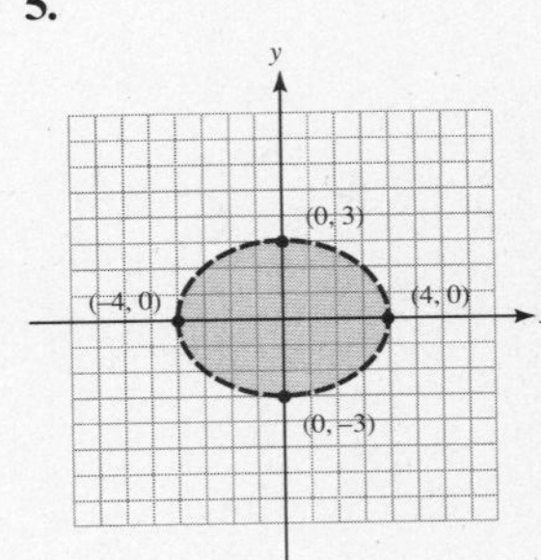

7.

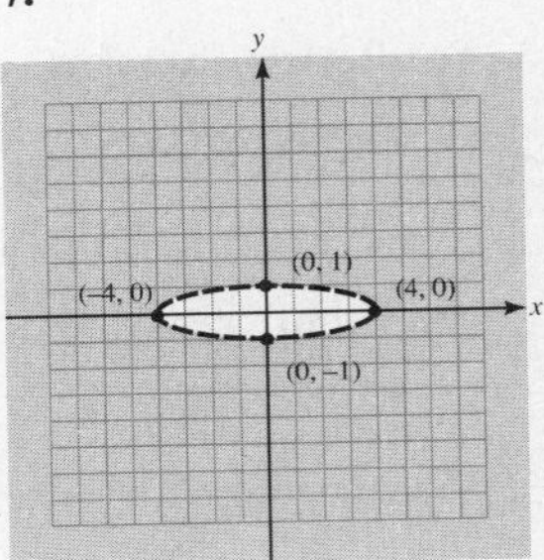

9.

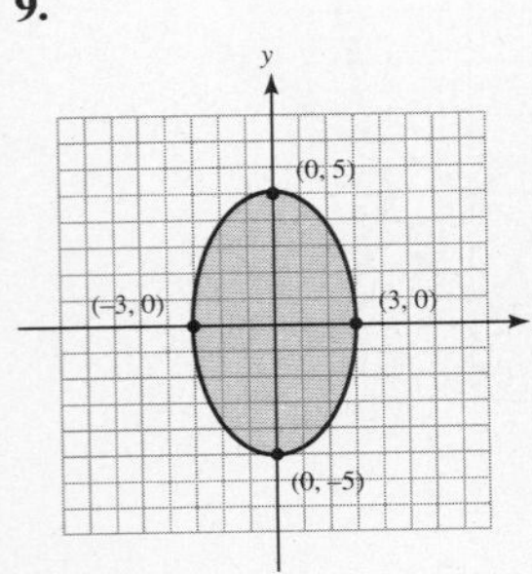

11.

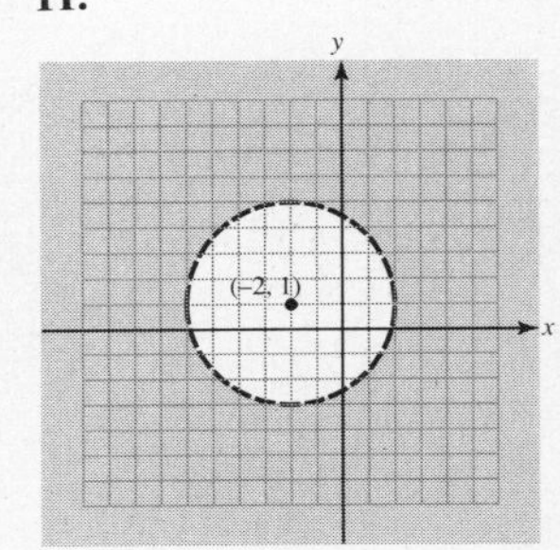

13.

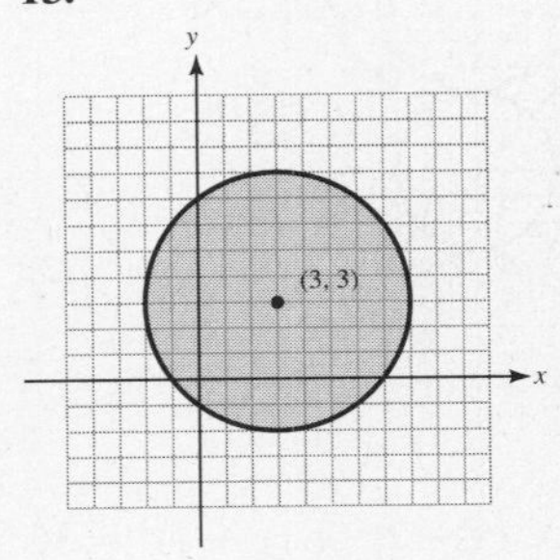

Section 11-3. Practice Exercises

1.

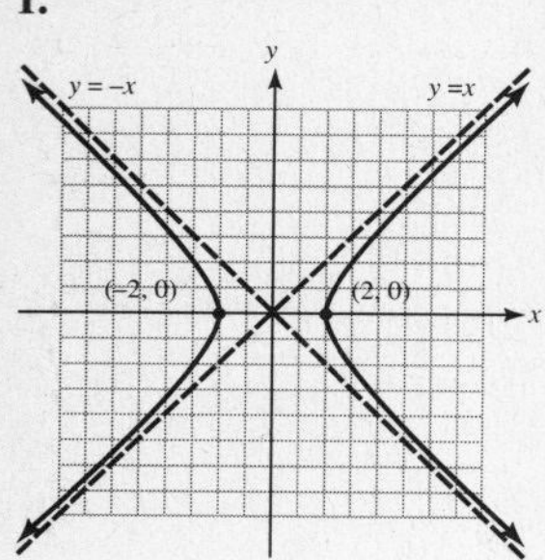

3.

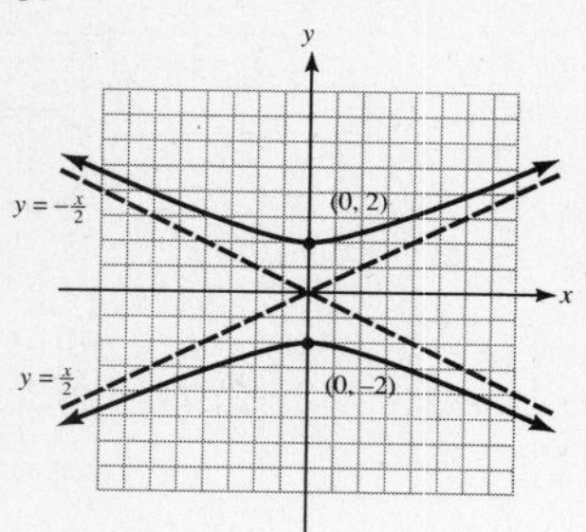

5.

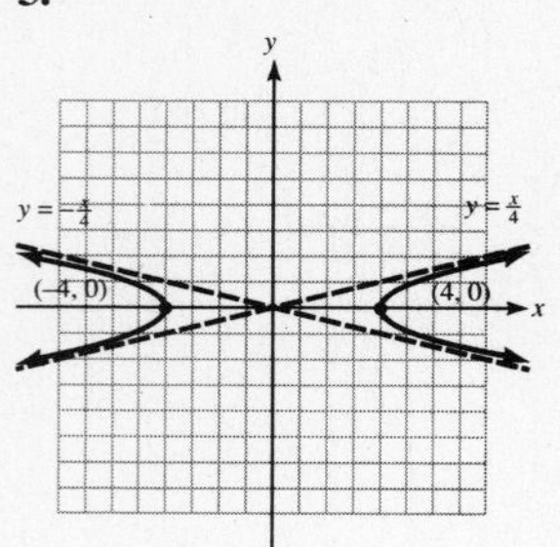

7.

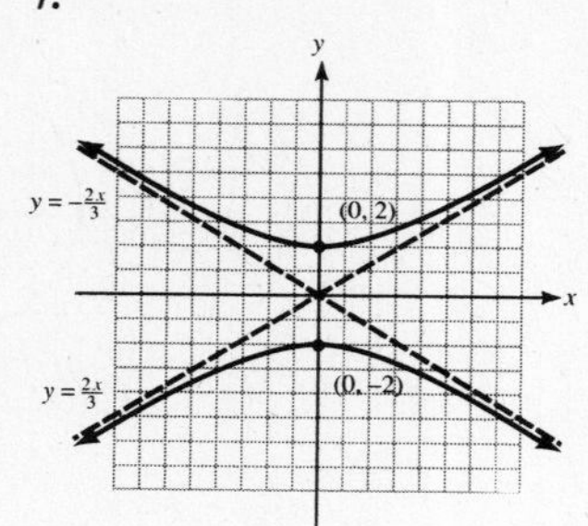

9.

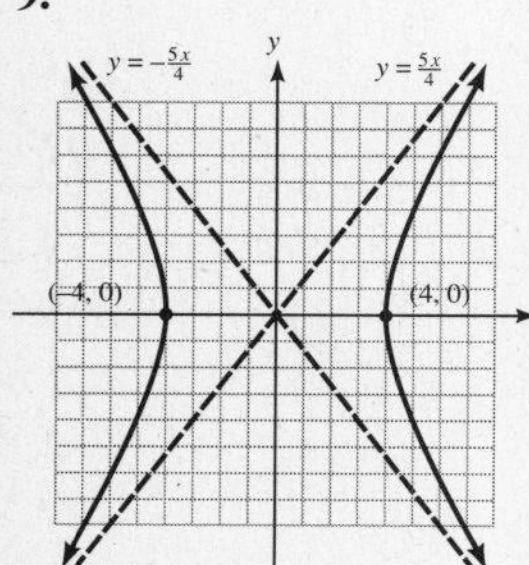

11.

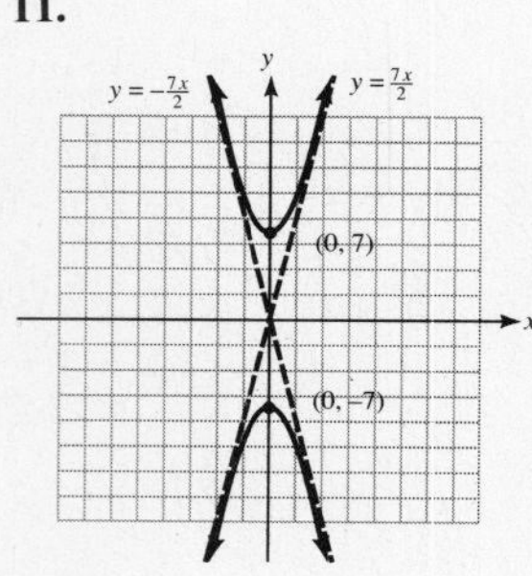

13.

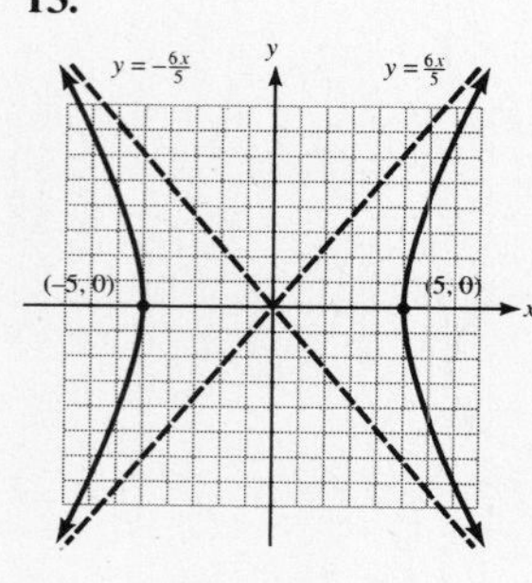

15.

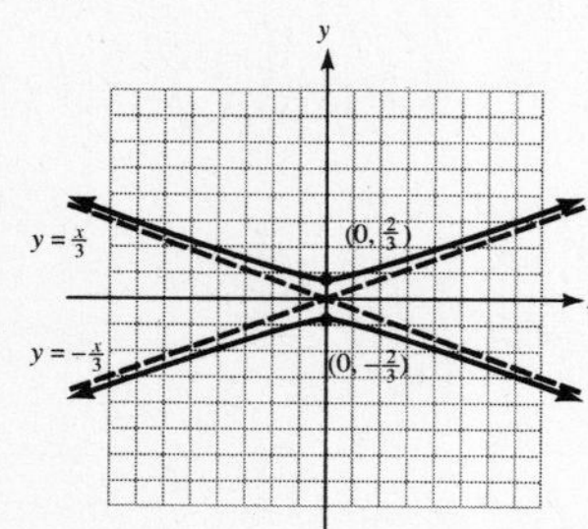

17.

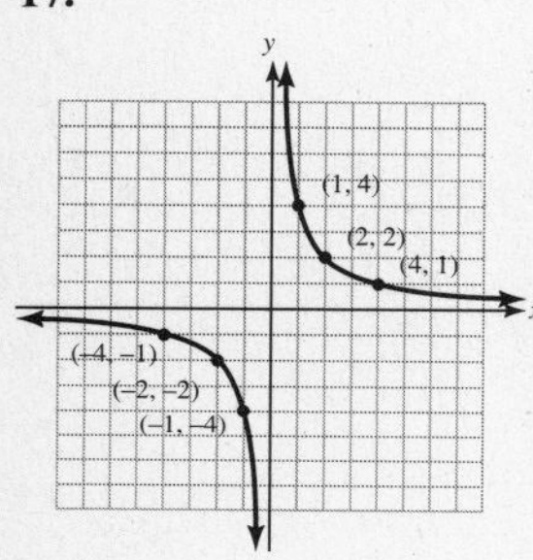

19.

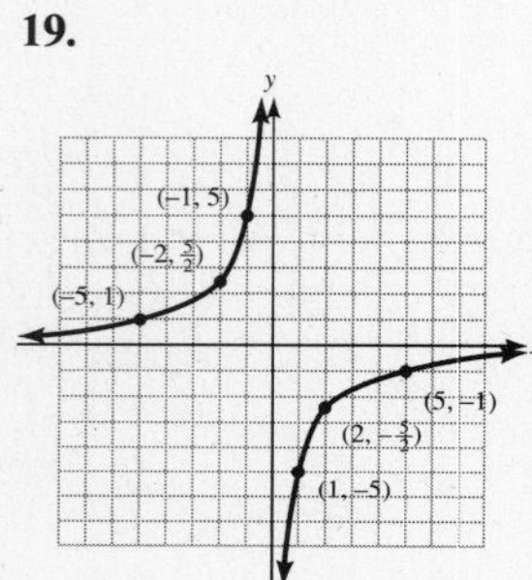

21.

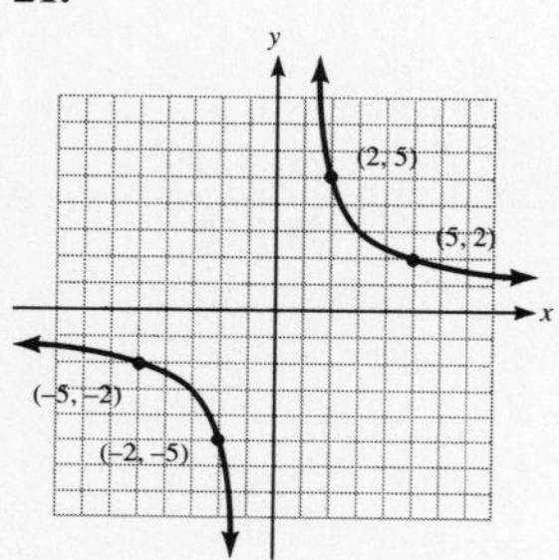

23.

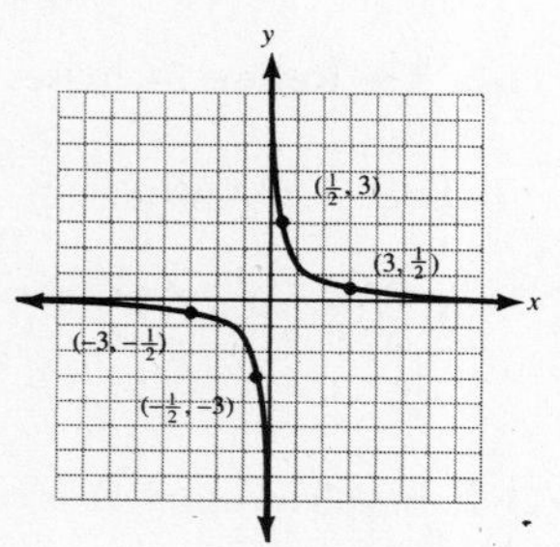

Section 11-3. Ten Review Exercises

1. -6 **2.** 20 **3.** $2t^3 + 5t^2 - 3 + \dfrac{7}{3t - 1}$ **4.** $\dfrac{6u}{u - 6}$ **5.** $\{-3, 9\}$ **6.** $\{0\}$ **7.** $\{-2\}$ **8.** $\{2 \pm 7i\}$

9. All x; $x \neq -2$ **10.** $f^{-1}(x) = \dfrac{2x}{1 - x}$

Section 11-3. Supplementary Exercises

1.

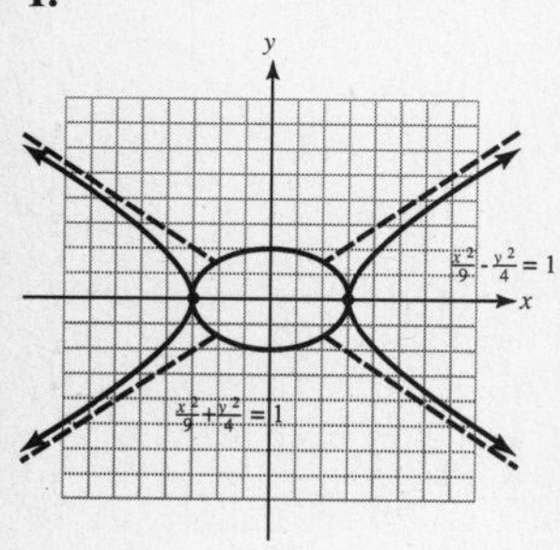

3.

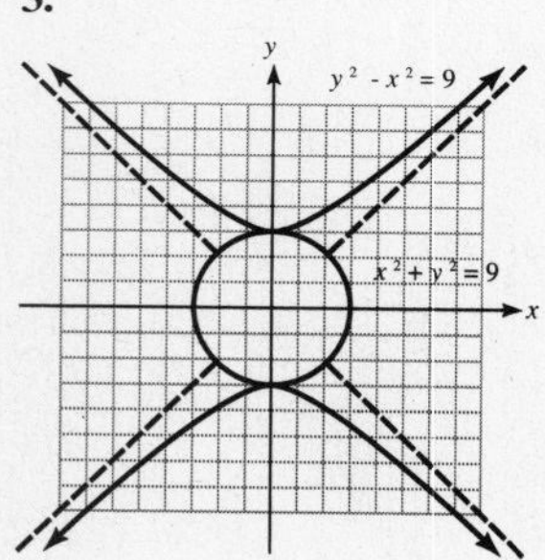

5.

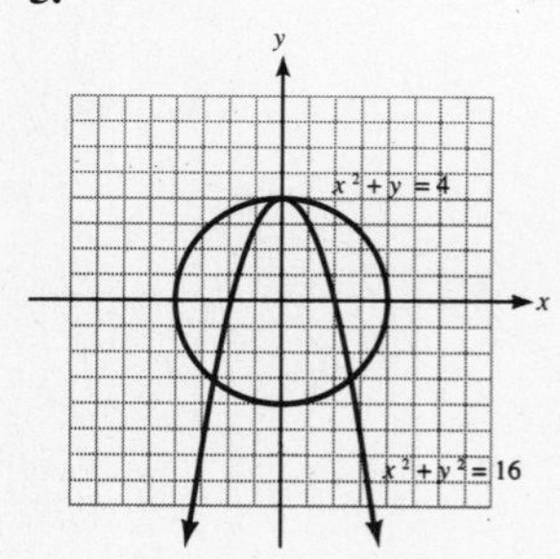

7.

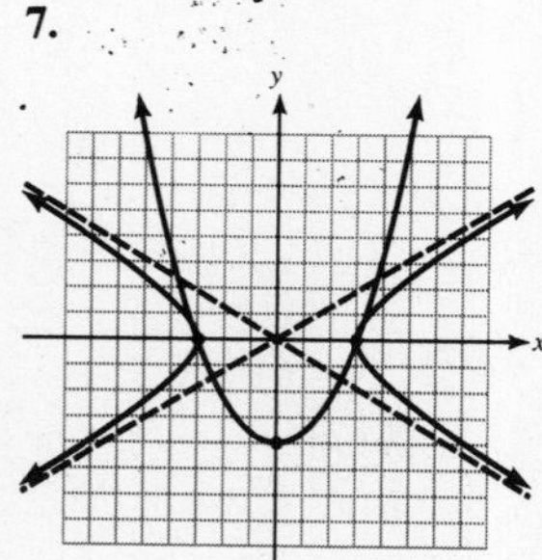

9. $\dfrac{x^2}{16} + \dfrac{y^2}{36} = 1$ **11.** $\dfrac{x^2}{9} - \dfrac{y^2}{16} = 1$ **13.** $(x - 2)^2 + (y + 1)^2 = 36$ **15.** $\dfrac{x^2}{4} + \dfrac{y^2}{16} = 1$

Section 11-4. Practice Exercises

1. $\{(3, -2)\}, (-2, 3)\}$ **3.** $\left\{\left(\frac{-1}{2}, \frac{5}{2}\right), (-1, 2)\right\}$ **5.** $\{(3, 2), (1, 6)\}$ **7.** $\left\{\left(\frac{-1}{2}, \frac{-9}{2}\right), (4, 6)\right\}$ **9.** $\left\{\left(\frac{-18}{5}, \frac{-1}{5}\right), (3, 2)\right\}$
11. $\left\{\left(\frac{-4}{3}, 9\right), (3, -4)\right\}$ **13.** $\{(3, -2), (-1, 6)\}$ **15.** $\left\{\left(\frac{2i\sqrt{5}}{5}, -i\sqrt{5}\right), \left(\frac{-2i\sqrt{5}}{5}, i\sqrt{5}\right), (1, 2), (-1, -2)\right\}$
17. $\{(1 + 2\sqrt{2}, 1 - 3\sqrt{2}), (1 - 2\sqrt{2}, 1 + 3\sqrt{2})\}$ **19.** $\{(3, 2i), (-3, 2i), (3, -2i), (-3, -2i)\}$
21. $\{(2\sqrt{7}, 3\sqrt{5}), (2\sqrt{7}, -3\sqrt{5}), (-2\sqrt{7}, 3\sqrt{5}), (-2\sqrt{7}, -3\sqrt{5})\}$ **23.** $\{(5i, i\sqrt{2}), (5i, -i\sqrt{2}), (-5i, i\sqrt{2}), (-5i, -i\sqrt{2})\}$
25. $\left\{\left(3, \frac{-5}{2}\right), \left(-3, \frac{-5}{2}\right), (0, 2)\right\}$ **27.** $\{(3, -6), (3, 2)\}$ **29.** $\{(2, 3), (-2, -3), (1, 2), (-1, -2)\}$
31. (5 and 7) or $\left(\frac{5}{3} \text{ and } 21\right)$ **33.** 10 ft by 14 ft **35.** $AC = 6$ m; $BC = 9$ m **37.** 25 d; \$30 **39.** 2 o; 8 o

Section 11-4. Ten Review Exercises

1. $\sqrt{5t}$ **2.** $3(\sqrt{10} - \sqrt{3})$ **3.** $2\sqrt[3]{7}$ **4.** $7\sqrt{3}$ **5.** $16y^2 - 19$ **6.** $\frac{9u^2 + 3u + 1}{2u + 1}$ **7.** $\frac{k}{k - 1}$ **8.** $\frac{2 - \sqrt{5}}{3}$
9. $2 + 3i$ **10.** $\frac{-\sqrt{10}}{5}$

Section 11-4. Supplementary Exercises

1. a. (1) $x - 2y = 0$ (2) $\frac{x^2}{9} + \frac{y^2}{4} = 1$ **b.** $\{(1.92, 1.46), (-1.92, -2.46)\}$ **3. a.** (1) $x^2 + y^2 = 16$ (2) $x^2 - y = 0$ **b.** $\{(-1.88, 3.53), (1.88, 3.53)\}$
5. a. (1) $\frac{y^2}{36} + \frac{x^2}{9} = 1$ (2) $\frac{x^2}{36} + \frac{y^2}{9} = 1$ **b.** $\{(-2.68, -2.68), (-2.68, 2.68), (2.68, -2.68), (2.68, 2.68)\}$
7. No **9.** Yes **11.** Yes

Chapter 11 Review Exercises

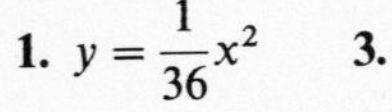
1. $y = \frac{1}{36}x^2$

3.

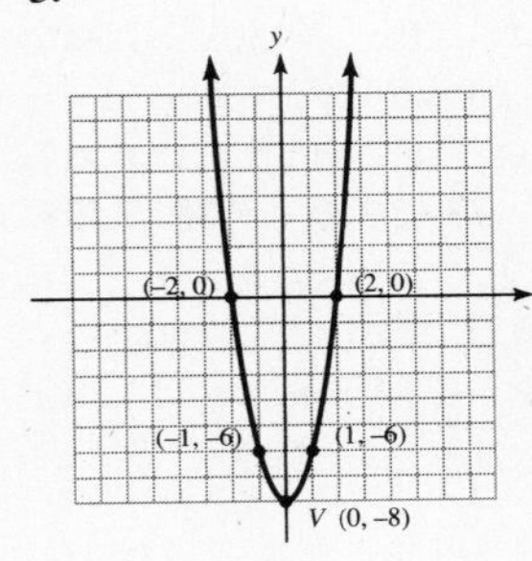

5.

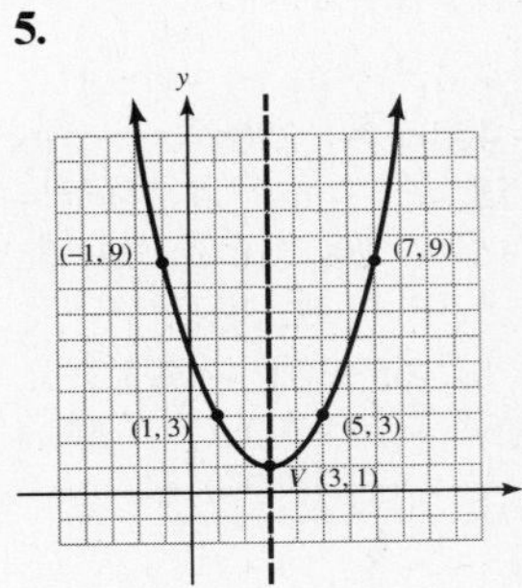

7.

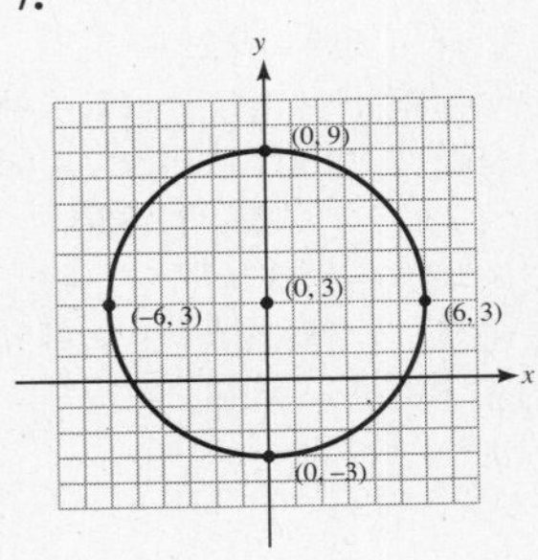

9. a. $(5, -2)$, $r = 4$ **b.**

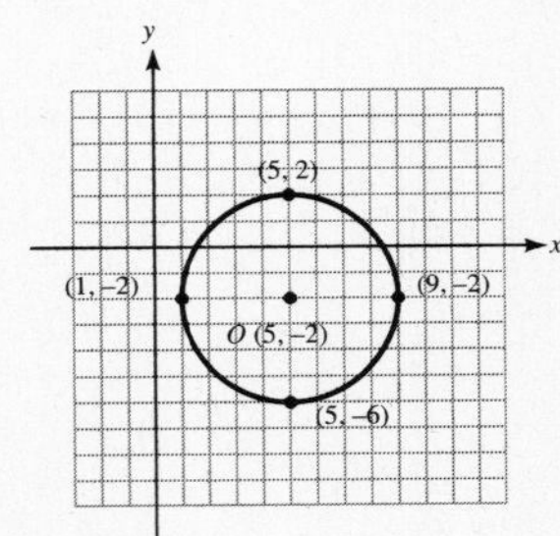

11.

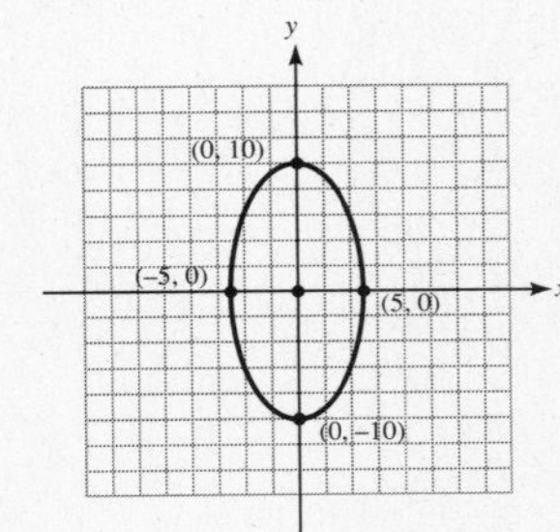

13.

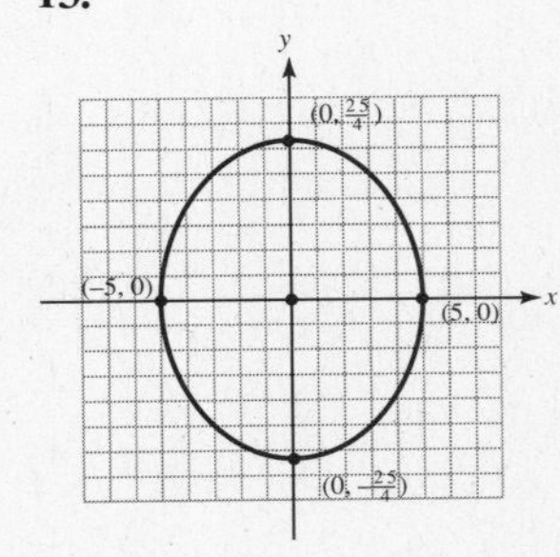

15.

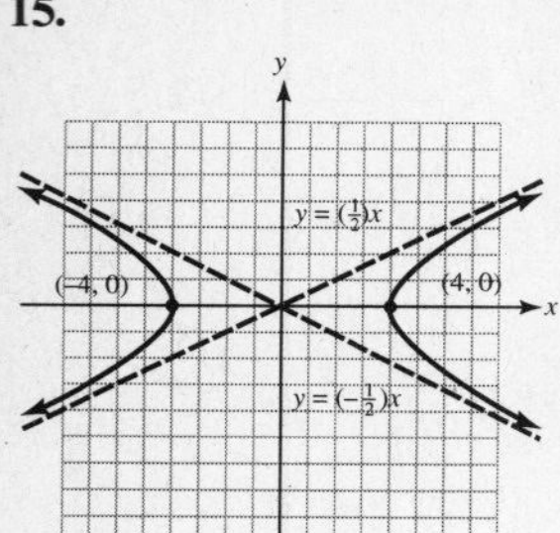

17.

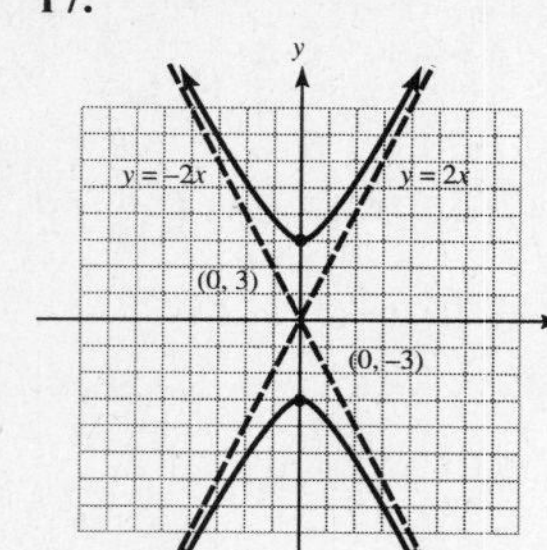

19.

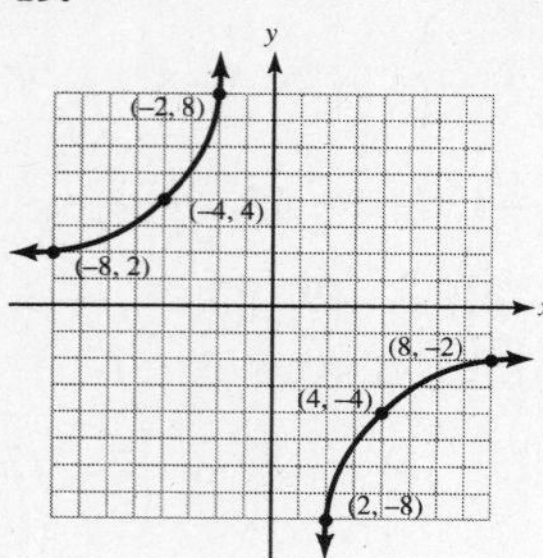

21. $\left\{(-1, -2), \left(\frac{11}{5}, \frac{-2}{5}\right)\right\}$

23. $\left\{\left(\frac{3}{2}, 4\right), \left(\frac{-3}{2}, -4\right), (2i, -3i), (-2i, 3i)\right\}$ **25.** $\left(\frac{4}{3} \text{ and } \frac{3}{2}\right)$ or (1 and 2)

Appendix A Sequences and Series

Exercises

1. 5, 7, 9, 11, 13 **3.** 9, 16, 35, 72, 133 **5.** 3, 5, 9, 17, 33 **7.** $\frac{-1}{2}, \frac{2}{3}, \frac{-3}{4}, \ldots, \frac{10}{11}$ **9.** $\sqrt{2}, \sqrt{3}, 2, \ldots, \sqrt{11}$
11. $\frac{-1}{3}, \frac{-3}{4}, -1, \frac{-7}{6}, \ldots, \frac{1-2n}{2+n}, \ldots$ **13.** $2, 6, 12, 20, \ldots, n^2 + n, \ldots$ **15.** $-2, 1, \frac{-8}{9}, 1, \ldots, \frac{(-2)^n}{n^2}, \ldots$
17. $0, \sqrt{3}, 2\sqrt{2}, \sqrt{15}, \ldots, \sqrt{n^2-1}, \ldots$ **19.** $\frac{1}{2}, \frac{\sqrt{2}}{3}, \frac{\sqrt{3}}{4}, \frac{2}{5}, \ldots, \frac{\sqrt{n}}{n+1}, \ldots$ **21.** 0, log 2, log 3, log 4, . . . , log n, . . .
23. 3 **25.** $\frac{-3}{2}$ **27.** 76 **29.** 9.9 **31.** $-22\sqrt{3}$ **33.** 21 **35.** 30 **37.** 35 **39.** 3 **41.** -5 **43.** 243
45. 0.0243 **47.** 32 **49.** 60 **51.** 60 **53.** 170 **55.** 35 **57.** 110 **59.** log 720 **61.** $\sum_{n=1}^{8} 2^n$
63. $\sum_{n=1}^{10} (-1)^n(3n-2)$ **65.** 790 **67.** 82 **69.** -440 **71.** 225 **73.** 35.4 **75.** $231\sqrt{5}$ **77.** 1365 **79.** 547
81. 126 **83.** 546 **85.** $\frac{3,165}{16}$ **87.** 3 **89.** $\frac{-2}{3}$ **91.** No sum **93.** $\frac{-50}{3}$ **95.** No sum **97.** $32 + 16\sqrt{2}$
99. 12 **101.** 12 **103.** 20 **105.** No sum **107.** 700 **109.** -63

Appendix B The Binomial Theorem

Exercises

1. $k^3 + 3k^2 + 3k + 1$ **3.** $z^3 - 6z^2 + 12z - 8$ **5.** $y^4 + 4y^3 + 6y^2 + 4y + 1$ **7.** $x^4 - 20x^3 + 150x^2 - 500x - 625$
9. $16m^4 + 32m^3n + 24m^2n^2 + 8mn^3 + n^4$ **11.** $243x^5 - 405x^4y + 270x^3y^2 - 90x^2y^3 + 15xy^4 - y^5$
13. $t^{12} + 12t^{10} + 60t^8 + 160t^6 + 240t^4 + 192t^2 + 64$ **15.** $u^{18} - 6u^{15} + 15u^{12} - 20u^9 + 15u^6 - 6u^3 + 1$ **17.** 24
19. 5,040 **21.** 30,240 **23.** 2 **25.** 11,880 **27.** 15 **29.** 210 **31.** 3,003 **33.** 2,300 **35.** 2,598,960
37. $12x^5$ **39.** $5{,}103t^5$ **41.** $-1{,}512k^{10}$ **43.** $5{,}376x^3y^6$ **45.** $30x^{18}y^2$
47. $11{,}520x^8y^2$ **49.** $15{,}625k^6 + 18{,}750k^5 + 9{,}375k^4$ **51.** $65{,}536z^8 - 393{,}216z^7 + 1{,}032{,}192z^6$
53. $2187u^7 + 25{,}515u^6 + 127{,}575u^5$ **55.** $512v^9 - 16{,}128v^8 + 225{,}792v^7$ **57.** $1{,}024m^{10} + 5{,}120m^{9n} + 11{,}520m^8n^2$
59. $64k^{12} - 576k^{10} + 2160k^8$

Index

A 0 B 1 C 2 D 3 E 4 F 5 G 6 H 7 I 8 J 9